ship

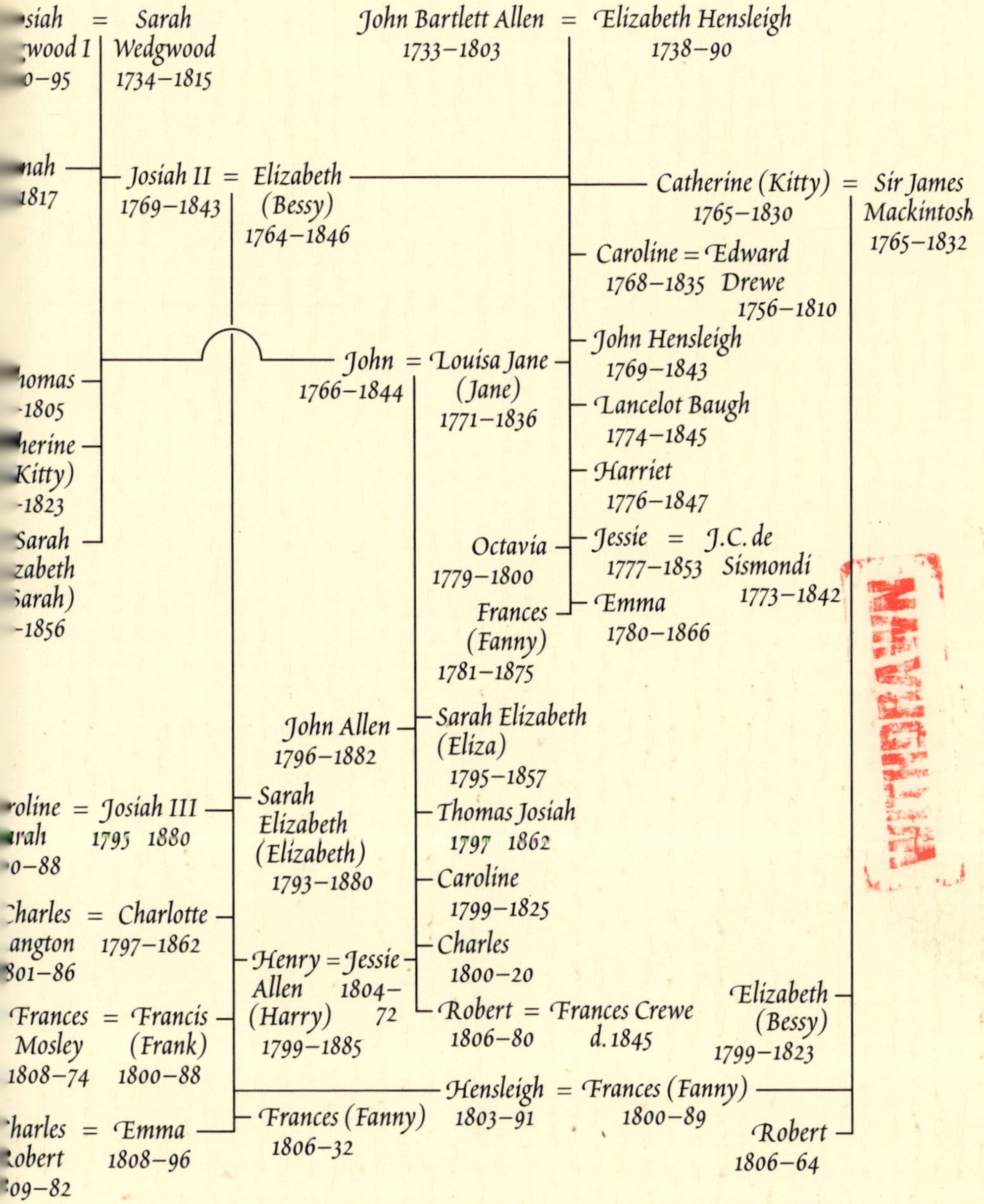

…siah = Sarah
…wood I Wedgwood
…0–95 1734–1815
John Bartlett Allen = Elizabeth Hensleigh
1733–1803 1738–90
…nah
…1817
Josiah II = Elizabeth
1769–1843 (Bessy)
1764–1846
Catherine (Kitty) = Sir James
1765–1830 Mackintosh
1765–1832
Caroline = Edward
1768–1835 Drewe
1756–1810
John Hensleigh
1769–1843
…homas
…1805
John = Louisa Jane
1766–1844 (Jane)
1771–1836
Lancelot Baugh
1774–1845
…herine
…Kitty)
…1823
Harriet
1776–1847
…Sarah
…zabeth
…arah)
…1856
Octavia
1779–1800
Jessie = J.C. de
1777–1853 Sismondi
1773–1842
Frances
(Fanny)
1781–1875
Emma
1780–1866
John Allen
1796–1882
Sarah Elizabeth
(Eliza)
1795–1857
…roline = Josiah III
…rah 1795–1880
…0–88
Sarah
Elizabeth
(Elizabeth)
1793–1880
Thomas Josiah
1797–1862
Caroline
1799–1825
…harles = Charlotte
…angton 1797–1862
…801–86
Henry = Jessie
Allen 1804–
(Harry) 72
1799–1885
Charles
1800–20
Frances = Francis
Mosley (Frank)
1808–74 1800–88
Robert = Frances Crewe
1806–80 d.1845
Elizabeth
(Bessy)
1799–1823
Hensleigh = Frances (Fanny)
1803–91 1800–89
…harles = Emma
…obert 1808–96
…09–82
Frances (Fanny)
1806–32
Robert
1806–64

THE CORRESPONDENCE OF
CHARLES DARWIN

This edition of the Correspondence of Charles Darwin is sponsored by the American Council of Learned Societies. Its preparation is made possible by the co-operation of the Cambridge University Library and the American Philosophical Society.

Advisory Committees for the edition, appointed by the Council, have the following members:

United States Committee

Whitfield J. Bell Jr
Frederick B. Churchill
John C. Greene
Ernst Mayr
Frank H. T. Rhodes
Marsha Richmond

British Committee

Gillian Beer
W. F. Bynum
Owen Chadwick
Peter J. Gautrey
Richard Darwin Keynes
Desmond King-Hele
G. E. R. Lloyd

Support for editing has been received from the Alfred P. Sloan Foundation, the Andrew W. Mellon Foundation, the National Endowment for the Humanities, the National Science Foundation, the Pew Charitable Trusts, the British Academy, the Isaac Newton Trust, the Natural Environment Research Council, the Royal Society of London, the Stifterverband für die Deutsche Wissenschaft, and the Wellcome Institute for the History of Science. The National Endowment for the Humanities's grants (Nos. RE-23166-75-513, RE-27067-77-1359, RE-00082-80-1628, RE-20166-82, RE-20480-85, RE-20764-89, RE-20913-91, and RE-21097-93) were from its Program for Editions; the National Science Foundation's funding of the work was under grants Nos. SOC-75-15840, SES-7912492, SES-8517189, and SBR-9020874. Any opinions, findings, conclusions, or recommendations expressed in this publication are those of the editors and do not necessarily reflect the views of the grantors.

Down House hothouse. Engraving.
From *Century Illustrated Monthly Magazine*, January 1883.
(By permission of the Syndics of Cambridge University Library.)

THE CORRESPONDENCE OF
CHARLES DARWIN

VOLUME 11 1863

CAMBRIDGE
UNIVERSITY PRESS

PUBLISHED BY THE PRESS SYNDICATE OF THE UNIVERSITY OF CAMBRIDGE
The Pitt Building, Trumpington Street, Cambridge, United Kingdom

CAMBRIDGE UNIVERSITY PRESS
The Edinburgh Building, Cambridge CB2 2RU, UK http://www.cup.cam.ac.uk
40 West 20th Street, New York, NY 10011-4211, USA http://www.cup.org
10 Stamford Road, Oakleigh, Melbourne 3166, Australia

First published 1999

Printed in the United Kingdom at the University Press, Cambridge

Typeset in Monotype Baskerville 10/12 pt

A catalogue record for this book is available from the British Library

ISBN 0 521 59033 7 hardback

CONTENTS

ILLUSTRATIONS

CALENDAR LIST OF LETTERS

The following list is in the order of the entries in the *Calendar of the correspondence of Charles Darwin*. It includes all those letters that are listed in the *Calendar* within the date range covered by this volume of the *Correspondence*. Some letters have been redated since the publication of the *Calendar*, so this list is necessary to enable users of the *Calendar* to locate such letters in the *Correspondence*. Alongside the *Calendar* numbers are the corrected dates of each letter. A date printed in italic type indicates that the item has been omitted from this volume.

Letters acquired after the publication of the *Calendar*, in 1985, have been given numbers corresponding to the chronological ordering of the original *Calendar* listing with the addition of an alphabetical marker. These letters are summarised in a 'Supplement' to a new edition of the *Calendar* (Cambridge University Press, 1994). The marker 'f' denotes letters acquired after the *Calendar* went to press in 1994.

2027. [24 March 1863]
2662. 23 January [1863 or 1864]
3367. 29 [March–June 1863 or 1864]
3399. 21 [January 1863]
3772. [after 19 January 1863]
3776. 20 [January 1863]
3852. 10 [January 1863]
3858. *Part of 3991.*
3884. *[1 March – 15 May 1862]*
3885. Part of 4248
3886. [1 April – 16 June 1863?]
3887. [before 11 May 1863]
3888. *Part of 4495.*
3889. *Same as 4738*
3889a. *27 [February or March 1861]. To appear in Supplement to vol. 13.*
3890. *[31 December 1862]*
3891. *[27 or 28 December 1862]*
3892. [12 January 1863]
3893. [16? February 1863]
3894. [26 March 1863]
3895. *[before 10 October 1862]*
3896. [before 25 February 1863]
3897. 2 January [1863]
3898. 3 January [1863]
3899. 3 January [1863]
3900. 4 January [1863]
3901. 5 [and 6] January [1863]
3902. 6 January 1863
3903. 6 January [1863]
3904. 6 January 1863
3905. 7 January [1863]
3906. 7 January [1863]
3907. Part of 3950
3908. 8 January [1863]
3909. 9 January 1863
3910. 11 January [1863]
3911. 12 January [1863]
3912. 12 January 1863
3913. 13 January [1863]
3914. 14 January [1863]
3915. 14 January [1863]
3916. 14 January 1863
3917. 14 January [1863]
3918. 15 January [1863]
3919. [15 January 1863]
3920. 16 January 1863
3921. 16 January 1863
3922. 17 [January 1863]
3923. 17 January 1863
3924. 17 January 1863
3925. 17 January [1863]
3926. 18 January [1863]
3927. 19 January [1863]
3928. 20 [January 1863]
3929. 20 January [1863]
3930. Part of 3934
3931. *Part of 3868*

INTRODUCTION

At the start of 1863, Charles Darwin was actively working on the manuscript of *The variation of animals and plants under domestication*, anticipating with excitement the construction of a hothouse to accommodate his increasingly varied botanical experiments, and continuing a massive scientific correspondence. Six months later the volume of his correspondence dropped markedly, reflecting a decline in his already weak health. Darwin then began punctuating letters with explanations that he was 'unwell & must write briefly' (letter to John Scott, 31 May [1863]), and in a letter of 23 [June 1863] he wrote to his close friend Joseph Dalton Hooker: 'I am languid & bedeviled . . . & hate everybody'. Although Darwin did continue his botanical pursuits over the summer, and persevered with his work on *Variation* until 20 July, his letter-writing dwindled considerably. The correspondence and Darwin's scientific work diminished even further when he and his family departed on 2 September for more than a month at a hydropathic establishment in Malvern Wells, Worcestershire, where he underwent a course of the water-cure. The treatment was not effective and Darwin remained ill for the rest of the year.

The first five months of 1863 contain the bulk of the correspondence in this volume. In the first part of the year, the letters illustrate Darwin's preoccupation with the challenges presented by the publication in February of books by his friends Charles Lyell, the respected geologist, and Thomas Henry Huxley, the zoologist and anatomist. Lyell's *Antiquity of man* and Huxley's *Evidence as to man's place in nature* both had a direct bearing on Darwin's species theory and on the problem of human origins. Specifically, Darwin wondered how far the public had been led by his theory to accept the notion of human descent from 'some Quadrumanum animal', as he put it in a letter to J. D. Hooker of 24[–5] February [1863]. When Huxley's book described the detailed anatomical similarities between humans and apes, Darwin was full of praise. He especially admired its discussion of the implications of ape ancestry for prevailing views of human dignity and intelligence, exclaiming to Huxley: 'I declare I never in my life read anything grander' (letter to T. H. Huxley, 26 [February 1863]). In the same letter, he gave his reaction to the *Antiquity of man*, complaining that he was 'fearfully disappointed at Lyells excessive caution in expressing any judgment on Species or origin of man'. Darwin's concern about the popular reception of his transmutation theory led him, after some consideration, to briefly play a public role in the controversies that embroiled Britain's scientific circles following the publication of Lyell's and Huxley's books.

Three years earlier Darwin had predicted that Lyell's forthcoming discussion of human antiquity would 'horrify the world' far more than *Origin* had (see *Correspondence* vol. 8, letter to Charles Lyell, 10 January [1860]). In the same letter he reminded Lyell of his belief that human beings were 'in the same predicament with other animals' and that he had made this suggestion in *Origin* with the famous line: 'Light will be thrown on the origin of man and his history' (p. 488). Since the publication of *Origin* in 1859, new evidence that early humans had coexisted with animals now extinct had been rapidly accumulating. Lyell's argument for a greater human antiquity than was commonly accepted was based on recent discoveries in England and France of primitive stone tools in association with the bones of extinct mammals, as well as on evidence collected earlier in the century. Lyell's *Antiquity of man* and Huxley's *Evidence as to man's place in nature* directly confronted experts and non-experts alike with the possibility of human descent from an ape-like animal, while dating human origins to a time far earlier than that decreed by scripture. Victorian anxiety regarding human origins was further increased by the discovery in March 1863 of the Moulin-Quignon jaw, the first human fossil found in association with bones from animals like the woolly mammoth and cave bear (see letter from Jacques Boucher de Perthes, 23 June 1863). Although English experts subsequently decided the jaw was a forgery, publications in learned journals and the press during the first half of 1863 focused attention even more closely on Darwin's arguments for species change.

In this context, Lyell's discussion of the origin of species particularly dismayed Darwin. In the first and most detailed of several letters to Lyell discussing *Antiquity*, Darwin made a list of criticisms, including the objection that one statement led the reader to think that the author was 'far from believing that man was descended from any animal'. Another sentence suggesting that human intelligence appeared in a sudden leap from that of inferior animals made him 'groan' (letter to Charles Lyell, 6 March [1863]). Darwin reiterated in a later letter that it would not have mattered if Lyell had thrown doubt on the significance of variation and natural selection, if only he could have permitted himself to say 'boldly & distinctly out that species were not separately created' (letter to Charles Lyell, 17 March [1863]). Public perceptions of creation, and the origin of species particularly, worried Darwin; he told Hooker that he had once thought Lyell 'could do more to convert the Public than all of us', but that after reading the book he wished his one-time mentor had not said a word (letter to J. D. Hooker, 24[–5] February [1863]).

Darwin did not relish telling Lyell of his disappointment; when he did write, he tempered his criticisms by acknowledging: 'I know you will forgive me for writing with perfect freedom; for you must know how deeply I respect you, as my old honoured guide & master' (letter to Charles Lyell, 6 March [1863]). Nevertheless, Darwin's regret was profound that the colleague and friend who had first advised him in 1856 to write his essay on species could not now support his theory more definitively.

Lyell, for his part, admitted that his feelings prevented him from discussing human descent from the 'brutes', but added that he would bring many towards Darwin who would have rebelled against stronger statements regarding species change (letter from Charles Lyell, 11 March 1863). The botanist Asa Gray, Darwin's friend in the United States, agreed that Lyell's approach would sway many towards a new way of thinking, while Huxley's book would scare them off (see letter from Asa Gray, 20 April 1863). In May, Darwin responded to Gray that Lyell's and Gray's indecision regarding change of species by descent put him 'into despair' (letter to Asa Gray, 11 May [1863]). In the same letter, he assured Gray that the essential question was not natural selection, but '*Creation* **or** *Modification*'.

Darwin was not alone in feeling disaffected towards Lyell and his book. In a February letter to the *Athenæum*, a weekly review of science, literature, music, and the arts, the prominent anatomist Richard Owen denounced the account provided in *Antiquity of man* of the ongoing debate between Owen, Huxley, and others concerning the comparative anatomy of the human and simian brain. He charged Lyell with failing to put across Owen's position, a failure which he attributed to Lyell's lack of expertise in the subject. 'The worst of it is', Hooker wrote to Darwin, 'I suppose it is virtually Huxley's writing, & that L. will find great difficulty in answering Owen *unaided*' (letter from J. D. Hooker, [23 February 1863]). Hugh Falconer was also preparing a denunciation of Lyell's book on the grounds that it gave insufficient credit to his own research and that of Joseph Prestwich. Hooker wrote: 'I fear L. will get scant pity even from his own side, for F spoke to me the other night in the most strongly slighting terms of so much of Lyell's book being written by others' (letter from J. D. Hooker, [23 February 1863]). Falconer published his criticisms in a letter in the *Athenæum* on 4 April, concluding with an invitation to Lyell to present a single fact or original observation that he had contributed to the proofs of human antiquity. Darwin and Hooker repeatedly exchanged regrets about Falconer's action towards Lyell, with Darwin commenting: 'It is wretched to see men fighting so for a little fame' (letter to J. D. Hooker, 17 March [1863]).

Falconer and Owen were already engaged in a dispute over whose name for a recently discovered fossil elephant should have priority. When Falconer's account of the elephant appeared in the *Natural History Review* in January, Darwin, who was already ill-disposed towards Owen following his 1860 review of *Origin*, wrote to Falconer: 'You would laugh, if you could have seen how indignant all Owen's lies and mean conduct about E. Columbi made me. . . . The case is come to such a pass, that I think every man of science is bound to show his feelings by some overt act, and I shall watch for a fitting opportunity' (letter to Hugh Falconer, 5 [and 6] January [1863]).

Falconer, Darwin, and others found an additional reason to be annoyed with Owen when it became clear that Owen's November 1862 description of the recently discovered *Archaeopteryx* had missed some essential elements of the Jurassic fossil. When informed by Falconer of Owen's 'slip-shod and hasty account' of the find, Darwin asked, 'Has God demented Owen, as a punishment for his crimes . . . ?' (letter from Hugh Falconer, 3 January [1863], and letter to Hugh Falconer, 20 [January 1863]). Aside from Owen's apparent oversights, Falconer enthusiastically observed that the fossil was 'a strange being à la Darwin', a transitional form between reptiles and birds (letter from Hugh Falconer, 3 January [1863]). Darwin was delighted by Falconer's description of the 'wondrous Bird', commenting that: 'no group is so isolated as Birds'; Darwin thought that now perhaps a significant gap had been filled in the fossil record (letter to Hugh Falconer, 5 [and 6] January [1863]). Only until March did Darwin take much notice of the now famous fossil in his correspondence and publications (see letter to J. D. Dana, 20 February [1863], and letter to Charles Lyell, 6 March [1863]); Owen's perceived failings, however, continued to capture his and others' attention.

In March, after hearing from Lyell that Owen in his paper on the aye-aye claimed the 'whole credit of making out the derivation or origin of species', Darwin considered writing a letter to the *Athenæum* in response (letter to J. D. Hooker, 5 March [1863]). He later expressed disappointment that the paper gave him no scope to attack Owen even though he 'had partly composed such *a good letter* (!)' (letter to J. D. Hooker, 17 March [1863]). At the same time Darwin admitted his mixed feelings: 'I long to be in the same boat with all . . . my friends ie at open war; but at same time I rejoice not to be annoyed at public quarrel, & it would annoy me much.'

Darwin did find an opportunity to enter his own protest against Owen with the appearance of an anonymous review in the *Athenæum* of William Benjamin Carpenter's book on Foraminifera (see letter to J. D. Hooker, [29 March 1863], and Appendix VII). The reviewer, soon identified as Owen, accused Carpenter, a physician and naturalist, of slavishly following Darwin's ideas by concluding that the varied Foraminifera probably derived from a few ancestral types. Owen also censured Darwin for subscribing to an 'occult cause' when he wrote that all organic forms had probably 'descended from some one primordial form, into which life was first breathed' (*Origin*, p. 484). Owen preferred Jean Baptiste de Lamarck's explanation of the origin of life: heterogeny, or the spontaneous generation of species from non-living matter; and he referred to the beginning of life in the 'vitally acting slime' settled at the bottom of seas, lakes, and rivers (Appendix VII).

Darwin told Hooker of his regret that he had 'truckled to public opinion & used Pentateuchal term of creation, by which I really meant "appeared" by some wholly unknown process.— It is mere rubbish thinking, at present, of origin of life; one might as well think of origin of matter.—' (letter to J. D. Hooker, [29 March 1863]). Owen's endorsement of Lamarck rankled all the more because Darwin was already disturbed by Lyell's presentation of his theory in *Antiquity of man* as a modification of Lamarck's. In complaining of this to Lyell, he described Lamarck's 1809 publication as a 'wretched book' from which he had 'gained nothing' (letter to Charles Lyell, 12–13 March [1863]).

When Carpenter's answer to Owen's review was judged too weak a response to the man whom Hooker referred to as a 'poor miserable devil of a scotched viper', Darwin wrote a letter to the *Athenæum* in opposition to Owen's review, in which he sought to advance his theory by making clear the difference between his view of transmutation and Lamarck's and by softening the effect of Lyell's failure to fully support him in the first edition of *Antiquity of man* (letter from J. D. Hooker,

[23 February 1863], and letter to *Athenæum*, 18 April [1863]). He told Gray: 'Under the cloak of a fling at Heterogeny I have sent a letter to Athenæum in defence of myself, & I take sly advantage to quote Lyells *amended* verdict on the Origin' (letter to Asa Gray, 20 April [1863]). Darwin quoted a sentence from the second edition of *Antiquity of man* (C. Lyell 1863b, p. 469), published in April, in which Lyell had added that he fully expected it would become the 'generally received opinion of men of science' that organic change had 'been brought about by the subordinate agency of such causes as Variation and Natural Selection'. Darwin explained his use of the quote to Lyell: 'you see that I am determined, as far as I can, that the Public shall see how far you go' (letter to Charles Lyell, 18 April [1863]).

Hooker implored him not to write any more letters to the *Athenæum*, stressing how much he disliked bringing science before the public in this way (see letter from J. D. Hooker, [7 May 1863], and Appendix VII). He also suspected that Owen's reply had won favour with the public. Darwin agreed, and regretted that he had written to the *Athenæum* in response to Owen: 'I have been gnashing my teeth at my own folly' (letter to J. D. Hooker, [9 May 1863]). After his venture into the popular press, he admired a satiric rendition in *Public Opinion* of the recent quarrels that he thought showed that 'a scientific man had better be trampled in dirt than squabble' (see Appendix VIII).

While Darwin worried about the effect of the quarrels on public perceptions, his theory was gathering support in influential scientific circles. George Bentham devoted the first part of his presidential address at the Linnean Society of London to British and foreign evaluations of Darwin's theory. While preparing the lecture, Bentham had warned Darwin that though he appreciated the force of Darwin's arguments, he still could not satisfy himself on all points (see letter from George Bentham, 21 April 1863). However, when he read the address, Darwin was greatly pleased with its positive approach to both transmutation and natural selection: 'I verily believe that your address, written as it is, will do more to shake the unshaken & bring on those leaning to our side, than anything written directly in favour of transmutation. I can hardly tell why it is, but your address has pleased me as much as Lyell's book disappointed me,—that is the part on species, though so cleverly written' (letter to George Bentham, 19 June [1863]).

Darwin was also pleased by the publication of *The naturalist on the river Amazons*, a book that he had encouraged Henry Walter Bates to write. When the book appeared in April 1863, Darwin gave it fulsome praise: 'My criticisms may be condensed into a single sentence, namely that it is the best book of Natural History Travels ever published in England.' He added that Bates had 'spoken out boldly on Species; & boldness on this subject seems to get rarer & rarer' (letter to H. W. Bates, 18 April [1863]), dismissing a reviewer in 'that d——d Athenæum' who accused Bates of bending his facts to support Darwin's theory (letter to J. D. Hooker, [17 April 1863]). Darwin felt equally strongly about the quality of Bates's work on mimicry in butterflies, which had been published in 1862 (see *Correspondence* vol. 10). He sent a copy to Asa Gray to review in an American journal, and also wrote a long unsigned review of it himself ('Review of Bates on mimetic butterflies') for the *Natural History Review* (see letter to H. W. Bates, 12 January [1863]). Darwin added Bates's name to his list of the 'half-a-dozen real downright believers in modification of Species in all England . . . who dare speak out' (letter to J. D. Hooker, [9 May 1863]). The others listed were himself, Hooker, Huxley, Alfred Russel Wallace, and John Lubbock.

Darwin had also found a supporter in New Zealand. Julius von Haast, a German working as a provincial geologist in the British colony, paid tribute to Darwin's work when he spoke to the Philosophical Institute of Canterbury in September 1862 (see letter to Julius von Haast, 22 January 1863); he discussed the importance of applying Darwin's theory to geological and palaeontological discoveries made in New Zealand. Haast's arduous explorations and geological, zoological, and botanical observations greatly interested Darwin, who applauded him as a 'glorious species man', while Haast extolled Darwin as the 'noble champion of true philosophic enquiry' (letter to J. D. Hooker, 23 April [1863], and letter from Julius von Haast, 21 July [–7? August] 1863). Darwin was subsequently elected an Honorary Member of the Philosophical Institute of Canterbury, New Zealand, of which Haast was a founding member (see letter from the secretary of the Philosophical Institute of Canterbury, New Zealand, 14 September 1863).

Additional notices of honours arrived at Down House in 1863. From Shropshire, where Darwin first began observing nature, he was invited to become an honorary member of the Severn Valley Naturalists Field Club (see letter from George Maw, 19 February 1863). Other honours included his election as a corresponding member of the Royal Prussian Academy of Sciences, Berlin (see Appendix III), and of the Société des Sciences Naturelles de Neuchâtel (see letter from La Société des Sciences Naturelles de Neuchâtel, 14 May 1863).

The foreign reception of *Origin* was reflected not only in awards, but in Darwin's correspondence. When Darwin felt that Lyell failed to uphold his views on transmutation in *Antiquity of man*, he did not hesitate to tell him of recent reports from other countries suggesting growing international interest in his theory (see letter to Charles Lyell, 17 March [1863]). In January 1863, the geologist Friedrich Rolle, who had recently published a popular exposition of Darwin's views, wrote with positive news of reactions to his theory in Germany (see letter from Friedrich Rolle, 26 January 1863). Darwin was not surprised to hear from Hooker that the French botanists Joseph Decaisne and Charles Naudin thought little of his theory (see letter to J. D. Hooker, 30 January [1863]), but he was happy that the respected Swiss botanist Alphonse de Candolle sent information on the 'progress of species-question', citing the French palaeontologist Gaston de Saporta's prediction that the role of natural selection would ultimately be accepted in France. Candolle had sent his monograph on oaks (A. de Candolle 1862b), which included a generally favourable discussion of *Origin* with what Darwin called 'prudent reservations' (letters to Alphonse de Candolle, 14 January [1863] and 31 January [1863], and letter to Asa Gray, 31 May [1863]). Asa Gray reviewed Candolle's memoir in the

American Journal of Science and Arts (A. Gray 1863e), continuing his practice of making discussions of Darwin's views known to American readers.

Less encouraging news arrived in November when Darwin heard that his nomination for the Royal Society's Copley Medal had been unsuccessful (see letter from E. A. Darwin to Emma Darwin, 11 November [1863]). The council of the Royal Society voted instead for the geologist Adam Sedgwick; Darwin suspected that Owen, a member of the council, had placed Sedgwick in opposition to him (see letter to J. D. Hooker, [22–3 November 1863]). However, it is certain that one influential figure in Darwin's failure to win the award was Edward Sabine, President of the Royal Society (see letter from Edward Sabine to John Phillips, 12 November 1863). Characteristically, Darwin claimed to care more about the resulting sympathy and praise from close friends than about honours like the Copley Medal (see letter to J. D. Hooker, 5 [December 1863]).

Darwin continued to pay particularly close attention to specific criticisms from supporters or near-supporters of his theory. Hugh Falconer suggested that a 'law of Phyllotaxis' governed the arrangement of leaves around a stem in a series of fixed angles for every plant, and stated that there must then be 'in nature, a deeper seated and innate principle, to the operation of which "Natural Selection" is merely an adjunct' (Falconer 1863a, p. 80). Darwin, challenged by this statement, began puzzling over the angles of leaves, asking the professional botanists Gray, Hooker, and Daniel Oliver for references on phyllotaxy, and setting his son George, the mathematician in the family, to explaining the phenomenon (see letter from Daniel Oliver, 17 February 1863, letter to Asa Gray, 20 April [1863], letter to J. D. Hooker, [9 May 1863], and memorandum from G. H. Darwin, [before 11 May 1863]). As he struggled with leaf angles, fractions, diagrams, and shoot dissections, his letters grew full of complaints that the problem was 'enough to drive the quietest man mad' (letter to Asa Gray, 11 May [1863]). Hooker and Gray agreed wholeheartedly, but Hooker sent another reference, acknowledging that Darwin was a 'hard headed man' (letter from J. D. Hooker, [24 May 1863]). Darwin finally decided that Falconer's remark was much exaggerated; he conceded to his critic that it was a 'wonderful problem' but he himself had 'made out . . . nothing' and wished someone could give a rational explanation of phyllotaxy (letter to J. D. Hooker, 29 May [1863], and letter to Hugh Falconer, [25–6 August 1863]).

Another criticism that continued to exercise Darwin was Huxley's assertion, first made in his 1860 review of *Origin*, that in order to prove the emergence of new species by natural selection, it was necessary to provide a case in which two forms of the same species of plant or animal, produced by selective breeding, were either unable to cross or else formed sterile hybrids. Huxley made this point again in his six lectures to working men, given at the Museum of Practical Geology at the end of 1862, and published as a book in early 1863 (T. H. Huxley 1863a). Though Darwin was otherwise extremely pleased with the book, he continued to disagree with Huxley about sterility as a fair test for species, pointing out that in his work on domesticated animals he had come to the conclusion that 'there are almost certainly several cases of 2 or 3 or more species blended together & now perfectly fertile together' which led him to believe 'that there must be something in domestication ... which eliminates the natural sterility of species, when crossed' (letter to T. H. Huxley, 10 [January 1863]). He reminded Huxley again of the German botanist Karl Friedrich von Gärtner's experiments, which had produced infertile hybrids in *Verbascum* and *Zea* (see *Correspondence* vol. 10, Appendix VI). However, when *Evidence as to man's place in nature* was published in February 1863, Huxley again argued that natural selection required experimental proof that physiological species could be produced by selective breeding.

Huxley's criticism provided additional impetus for Darwin to continue the work on dimorphic plants, and on cross and hybrid sterility, that had already occupied much of his time in 1861 and 1862. With the publication in 1862 of his paper on *Primula* ('Dimorphic condition in *Primula*'), he had demonstrated that two different flower forms, one with a long style and short stamens, the other with a short style and long stamens, were only fully fertile when crossed with each other. In February 1863, his paper on *Linum* ('Two forms in species of *Linum*') was read before the Linnean Society. In the paper, Darwin presented experimental results that suggested a high degree of sterility when the same flower forms were crossed (see Appendix IV). Darwin continued to investigate the true nature of sterility, a question he had been struggling with in 1861 and 1862; he wanted to determine experimentally whether it was selected for or whether, as he had argued in *Origin* (p. 245), it was 'incidental on other acquired differences' (see *Correspondence* vol. 10, Appendix VI). In addition to crossing varieties of *Primula* in 1863, he continued working with a plant in another genus, the purple loosestrife, *Lythrum salicaria.* This species provided the 'oddest case of reproduction ever noticed.—a triple marriage' between three hermaphrodite flower forms; during the summer of 1863, Darwin worked 'like a Trojan' to finish the large series of crosses between the three forms that he had started the previous year (letter to Asa Gray, 4 August [1863]). The results were published in his 1865 paper 'Three forms of *Lythrum salicaria*'.

In the process of acquiring botanical specimens, as well as sharing observations and experimental results, Darwin continued corresponding with a wide range of gardeners, horticulturists, and hobbyists, in addition to professional botanists. Lydia Ernestine Becker, who was soon to be involved in the women's suffrage movement, began writing to him in 1863 with field observations and specimens of the dimorphic *Lychnis.* A correspondence also began with the Scottish hybridiser Isaac Anderson-Henry, and Darwin continued the exchange of letters that had started the previous year with the Hertfordshire nurseryman Thomas Rivers.

Darwin had found a protégé and collaborator in the examination of hybridity and sterility at the end of the previous year. John Scott, a gardener at the Royal Botanic Garden, Edinburgh, had initiated the correspondence in November 1862 with a letter correcting Darwin's description of reproduction in the orchid genus *Acropera* (see *Correspondence* vol. 10). Their 1863 letters reveal Darwin's excitement at finding another person in Britain who shared his interests in hybridity, and who was more than willing to carry out breeding experiments. Though the two men discussed a multitude of botanical subjects, the crossing experiments that Scott had begun on the primrose family after reading Darwin's paper, 'Dimorphic condition in *Primula*', as well as investigations of orchid pollination, occupied much of their detailed and lengthy botanical correspondence in 1863. Darwin eventually communicated Scott's *Primula* work to the Linnean Society in a paper that was read in February 1864. He had already promised Scott that he would delay publication of his own continuing work with *Primula* crosses, the results of which were published in 1868 (see letter to John Scott, 25 and 28 May [1863]). Scott's experiments confirmed the results of Darwin's earlier paper on *Primula*; his additional crosses between differently coloured varieties of *P. vulgaris*, in which he found 'the *very zero of fertility between varieties of a species*', were the sort of result that both he and Darwin hoped would counteract Huxley's criticism (letter from John Scott, 23 July [1863]). Darwin reviewed the draft of Scott's paper with satisfaction (see letter to John Scott, 7 November [1863]).

Scott had referred directly to Huxley's reservations about natural selection in the introduction to his paper on orchids (Scott 1863a), completed earlier in 1863, which Darwin had also encouraged him to write (see letter to John Scott, 12 April [1863]). In this paper, Scott provided evidence of self-sterility in *Maxillaria* species that were fertile when crossed with another species in the same genus. Darwin's reaction was enthusiastic: 'The facts will be of *highest* use to me: I feel convinced that your paper will have permanent value' (letter to John Scott, 31 May [1863]). Scott received a different response in Scotland; he warned Darwin that at the Edinburgh Botanical Society, where he read his orchid paper, anything that 'savours of the "*Origin*" is not at all palatable!' (letter from John Scott, [3 June 1863]).

Darwin's early confidence in Scott rested not only on their common theoretical interests but on Scott's powers of experimental observation. When Scott expressed growing frustration with his position as gardener, hoping for a foreign appointment that would offer more independence and opportunity for his experimental work, Darwin was sympathetic. He consulted Hooker on the advisability of this 'truly remarkable man' taking a position offered in Darjeeling, India (see letter from John Scott, 22 May 1863, and letter to J. D. Hooker, 23 May [1863]). Hooker advised against the appointment; Darwin replied that if Scott 'had leisure he would make a wonderful observer; to my judgment I have come across no one like him' (letter to J. D. Hooker, 29 May [1863]).

Darwin continued his own study of orchid pollination and was aided by the completion of his new hothouse in February; this enabled him to grow more tropical plants than before (see Appendix VI). He was fascinated with the adaptations in orchids for cross-pollination by insects because of his conviction that natural selection had brought about modifications leading to an apparent perfection of structure and coadaptation. The cross-pollination of orchids, as well as dimorphism in other flowers, provided evidence for his assertion that nature 'abhors perpetual self-fertilisation' (*Orchids*, p. 359); he had discussed the indispensibility of intercrossing to natural selection in *Origin*, pp. 96–109, and he continued to observe individuals of the same species crossing with one another in a variety of plants. With additional study of *Cypripedium*, *Acropera*, and the bee orchid, *Ophrys apifera*, he gathered information that would ultimately be published in his paper 'Fertilization of orchids' and in the second edition of *Orchids*. Darwin also exchanged information on orchids, melastomas, and other tropical and sub-tropical plants with observers from warmer parts of the world. Correspondence was established in 1863 with Roland Trimen, a civil servant in Cape Town, whose work on *Disa* and *Herschelea* Darwin communicated to the Linnean Society (see letter to Roland Trimen, 23 May [1863]). Darwin began to exchange letters with Hermann Crüger, a German botanist in Trinidad, and continued writing to George Henry Kendrick Thwaites, the director of the Peradeniya botanic gardens in Ceylon. Closer to home, Darwin arranged for a paper on orchids written by a German correspondent, the botanist Friedrich Hildebrand, to be published in the English journal *Annals and Magazine of Natural History.* In addition to following Darwin's orchid work, Hildebrand told him he had repeated all Darwin's *Linum* experiments with the same results (see letter from Friedrich Hildebrand, 16 July 1863). In England, Darwin asked Philip Henry Gosse, a writer and naturalist in Torquay, about reproduction in *Ophrys apifera*, his 'greatest puzzle', an orchid that seemed perfectly adapted for both self-pollination and cross-pollination (letter to P. H. Gosse, 2 June [1863]). The observation of insects entering and leaving flowers filled Darwin with pleasure. After describing a bee's movements in relation to the pollinia, movements that confirmed Gray's theory that *Cypripedium* was pollinated by insects crawling through the flower, Darwin concluded: 'It was beautiful' (letter to Asa Gray, 20 April [1863]).

Darwin was also investigating different mechanisms for cross-pollination by searching for and studying dichogamous flowers, those in which the stamens and pistils mature at different times (see letter to Asa Gray, 11 May [1863]). The fertility of unopening, self-pollinated flowers (later called cleistogamic flowers) relative to that of the opening, cross-pollinated flowers on the same plant, as in some *Oxalis* and *Viola* species, had interested Darwin since 1860; it continued to capture his attention (see letter to John Scott, 12 April [1863]). Additionally, Darwin investigated the fertility of peloric flowers to determine whether the change in flower structure from irregular to regular was associated with a change in fertility when crossed with other plants in the same species (see letter to Isaac Anderson-Henry, 20 January [1863]). These botanical observations contributed to his understanding of intercrossing, the subject of several papers published later, and of *Forms of flowers* and *Cross and self fertilisation*, both published in the 1870s.

Darwin regularly discussed a wide range of botanical topics with Hooker, Gray, and Oliver. These correspondents were the first to hear about Darwin's 'chief Hobby-horse' of the summer, one that he said he owed to Gray's 1858 paper on the coiling of tendrils in *Echinocystis*. Darwin had placed the gourd *Echinocystis lobata* in his study to watch the spontaneous gyratory movements of its tendrils (see letter to J. D. Hooker, 25 [June 1863]). Darwin observed: 'their irritability is beautiful, as beautiful in all its modifications as anything in orchids' (letter to Asa Gray, 4 August [1863]). He acquired tropical 'tendrilliferous' plants that could grow in his hothouse from Hooker, and set about studying the various types of movement of tendrils, stems, and leaves in different plants (see letter from J. D. Hooker, [31 July 1863]). When he tried to determine the originality of his observations, Hooker and Oliver responded that they knew little; Hooker suggested that Darwin's discoveries were 'most curious & novel' (letter from J. D. Hooker, [21 July 1863]). Gray had noticed more tendril movements and supplied additional references, expressing astonishment at his English colleagues' lack of knowledge about these types of plant movements: 'What are Hooker & Oliver (the latter a Professor too) about, and where have they lived not to know anything of them?' (letter from Asa Gray, 1 September 1863). Gray encouraged Darwin's work, assuring Darwin: 'it will be fruitful in your hands'.

Darwin was considering the movements as adaptations for climbing, and was beginning to puzzle over the various types of sensitivity in different plant parts. The detail of his note-taking, the diagrams, and the careful timing of the oscillations, suggest how important the observations would be to the later publications 'Climbing plants' and *Movement in plants*. At the same time, he explained to Hooker: 'it is just the sort of niggling work which suits me & takes up no time & rather rests me whilst writing' (letter to J. D. Hooker, 14 July [1863]). Darwin's pleasure in watching these 'wonderfully crafty & sagacious' plants was cut short in September by his departure with his family for Malvern Wells, but even when ill at home in November, he watched them in his study as they twined past volumes on the bookcase and around the head of the sofa (letter to W. E. Darwin, [25 July 1863], and letter to J. D. Hooker, [22–3 November 1863]).

Many of Darwin's botanical observations and experiments of 1863 naturally overlapped with his writing of *Variation*, particularly when he was working on the chapter he called 'Crossing & Sterility' (see Appendix II). When Darwin finished, by 23 January 1863, a chapter on bud-variation, or sports, he had completed the first half of the book, which was to amass 'a large body of facts' on variation in domestic animals and plants (*Variation* 1: 1). He then turned to general discussions, completing three sections, on inheritance, crossing and sterility, and selection, by the end of July 1863. While it was not unusual for Darwin to complain to his correspondents about how slowly the writing was going, and how he feared he would never finish, he struck a more optimistic note in a letter to Julius von Haast of 18 July [1863], in which he said he was working very hard at *Variation* and hoped to begin printing in six months. However, the two-volume work was not published until 1868.

Darwin relied on members of his family for help in his research throughout the year. His eldest son, William Erasmus Darwin, a banker in Southampton, had been pressed into service to make observations in the field on dimorphic plant species. William was a conscientious observer, as is indicated by his botanical notebook and sketchbook (DAR 117 and DAR 186: 43). Many of his observations were later cited in *Forms of flowers*, and his help with the examination of the trimorphic genus *Lythrum* was noted in 'Three forms of *Lythrum salicaria*'. George contributed his mathematical expertise to his father's investigation of phyllotaxy, while Henrietta was often at her father's side in the hothouse and garden. Henrietta took an interest in science, reading and commenting on Huxley's books and Lyell's *Antiquity of man* (see letter from T. H. Huxley, 25 February 1863, and letter to Charles Lyell, 12–13 March [1863]). Emma was a steady help to Darwin, writing many letters that her husband dictated when he was too ill to hold a pen. During the autumn of 1863, they collaborated in a new endeavour, producing an appeal against the use of steel vermin traps and the animal suffering caused by them (see Appendix IX). Francis Darwin later wrote of his father's strong feelings regarding torment of both animals and humans: 'It was indeed one of the strongest feelings in his nature, and was exemplified in matters small and great, in his sympathy with the educational miseries of dancing dogs, or in his horror at the suffering of slaves' (*LL* 3: 199).

Darwin's aversion to slavery had led him to side with the Union cause in the American Civil War, discussion of which figured prominently in his correspondence with Asa Gray in Cambridge, Massachusetts. At the start of the year, Darwin displayed an interest in war-related issues, particularly after reading a tract written on relations between Britain and the United States by Gray's father-in-law, the lawyer Charles Greely Loring (see letter to Asa Gray, 23 February [1863], and Loring 1862). However, his tolerance of Gray's fervent approach to the war declined in 1863. In April, he complained that a note of Gray's sounded 'very savage against England' (see letter to Asa Gray, 20 April [1863]). Darwin evidently refrained from discussing the subject again until he read Gray's statement that he wished he had a son so that he could send him to the war (see letter from Asa Gray, 7 July 1863). Darwin shared this letter with Hooker, who had long since ceased to discuss politics or the American Civil War in his correspondence with Gray: 'What pleasant letters Asa Gray writes. One might as well write to a madman as to him about the war. It is a holy war & everything sh^d. be sacrificed to it' (letter to J. D. Hooker, 26 [July 1863]). Darwin then confessed to Gray that he could not eagerly support the north owing to the boasting and the abuse of England, and added that 'the treatment of the free coloured population, and the not freeing Maryland slaves stops all my enthusiasm' (letter to Asa Gray, 4 August [1863]). He urged Gray not to hate England too much, claiming: 'she is the mother of fine children all over the world'.

Despite his declining health, Darwin continued to be occupied with his various botanical subjects throughout the summer, until he told Hooker in a letter of 25 [August 1863] that after having had 'a deal of sickness of late; every morning for a fortnight', he had resolved to take his household to Malvern the following week. Three letters in August from John Goodsir, professor of anatomy at the University of Edinburgh, indicate that Darwin had sent him a slide and a phial of vomit for analysis. Goodsir found no micro-organisms other than normal inhabitants of the mouth and stomach. On the way to Malvern Wells, Darwin stopped in London overnight to consult George Busk, former Hunterian Professor at the Royal College of Surgeons, whom Goodsir had mentioned, and whom Hooker had recommended as having 'the most fertile brain of any man I know in regard of all such matters as your stomach' (see letter from John Goodsir, 21 August [1863]; letter from J. D. Hooker, [27 August 1863]). Unfortunately Darwin took a turn for the worse once he arrived in Worcestershire. Only by the end of September was he able to write a letter in his own hand and recount his condition over the summer and early autumn: 'I got worse & worse at home & was sick every day for two months; so came here, where I suddenly broke down & could do nothing; but I hope I am now very slowly recovering, but am very weak' (letter to A. R. Wallace, [29? September 1863]).

Darwin wrote one other letter in his own hand before the end of October; this was a heart-felt note to Hooker, who had recently suffered the death of his six-year-old daughter (see letter to J. D. Hooker, [4 October 1863]). The Darwins' daughter, Annie, had died at Malvern in 1851, and Hooker's news was a powerful reminder of their loss (see *Correspondence* vol. 5). Unable to find Annie's gravestone in 1863, they wrote to Darwin's cousin, William Darwin Fox, who had visited the grave seven years earlier. Fox sent a helpful letter that guided them to where it was hidden by overgrown trees and shrubs (see letter from W. D. Fox, 7 September [1863]). Emma wrote back: 'This has been a great relief' (letter from Emma Darwin to W. D. Fox, [29 September 1863]).

When the Darwins returned home, Charles fared little better, and most letters were dictated to Emma. Darwin only managed one of his typically long letters to Hooker by writing while lying down (see letter to J. D. Hooker, [22–3 November 1863]). He told Roland Trimen in a letter of 25 November [1863] that he had been 'ordered to do nothing for 6 months' by his doctors, mentioning that even writing the letter was 'against rules'. George Busk had diagnosed Darwin as having 'Waterbrash', a 'diseased secretion of the stomach', which accompanied 'various kinds of general derangement'. Busk prescribed medication and also advised him to consult the physician Dr William Brinton, a stomach specialist at St Thomas's Hospital, London (letter from George Busk, [*c.* 27 August 1863]). Brinton, who visited Down House in early November, evidently advised Darwin to curtail his activities.

Dr Brinton visited Down House again in early December, when Darwin had been 'vomiting every day for eleven days & some days many times after every meal' (letter to J. D. Hooker, 5 [December 1863]). Nevertheless, Brinton, he said, had told him that there was no reason why he should not recover his previous state of health. Brinton's words must have encouraged Darwin; he wrote Hooker a somewhat more lively letter than usual: 'I sh^{d}. like to live to do a little more work & often I feel sure I shall & then again I feel that my tether is run out' (letter to J. D. Hooker, 5 [December 1863]). He also triumphantly informed Hooker that an Italian edition of *Origin* was being prepared, which would be the fifth foreign edition. Recalling the controversies of the first five months of the year, Darwin added: 'Owen will not be right in telling Longmans that Book w^{d} be utterly forgotten in ten years'.

However, Darwin would not return at once to gathering more support for *Origin*, or to the observations and experimental work that he hoped would further bolster his theory. The few remaining letters from him in December were short, and dictated to Emma. By the end of the year, Emma admitted to Hooker (letter from Emma Darwin to J. D. Hooker, 26 December [1863]) that Darwin's health fluctuated and showed little improvement; he usually stayed in his bedroom, tolerating only short visits even from his sons. She was pessimistic about his prospects of health over the next few months: 'When not uncomfortable his spirits are wonderfully good, but I am afraid he may remain just as he is very long before there is a struggle in his constitution & that the sickness is conquered.'

ACKNOWLEDGMENTS

The editors are grateful to Mr George Pember Darwin for permission to publish the Darwin letters and manuscripts. They also thank the Syndics of the Cambridge University Library and other owners of the manuscript letters who have generously made them available.

The work for this edition has been supported by grants from the National Endowment for the Humanities (NEH), the National Science Foundation (NSF), and the Wellcome Trust. The Alfred P. Sloan Foundation, the Pew Charitable Trusts, and the Andrew W. Mellon Foundation provided grants to match NEH funding, and the Mellon Foundation awarded grants to Cambridge University that made it possible to put the entire Darwin correspondence into machine-readable form. Research and editorial work have also been supported by grants from the Royal Society of London, the British Academy, the Isaac Newton Trust, and the Natural Environment Research Council. The Royal Society has helped meet the costs of publication. The Stifterverband für die Deutsche Wissenschaft has provided funds to edit Darwin's correspondence with German naturalists.

The Cambridge University Library and the American Philosophical Society (APS) have generously made working space and many services available to the editorial staff.

Since the project began in 1975, we have been fortunate in benefiting from the interest, experience, and practical help of many people in many places, and the editors hope that they have adequately expressed their thanks to them individually as the work proceeded.

Without the expert help of John L. Dawson of the Literary and Linguistic Computing Centre of Cambridge University, the computerisation of the correspondence would not have been possible, for the work on both the *Calendar* and the *Correspondence* required the solution of many novel technical problems. Matthew Bernstein, Peter Dunn, Patricia Killiard, Adrian Miller, Harry Powell, and Chris Sendall have helped maintain and extend our access to the Library's electronic resources, and provided essential technical support. Simon Buck has offered invaluable assistance in preparing automatic typesetting procedures and has acted as typesetting advisor to the project.

English Heritage has responded most generously to requests for information and for material from the collections at Down House, Downe. Richard Darwin Keynes kindly made available Darwin family material in his possession. Ursula Mommens has also provided letters and other materials that belonged to her grandfather, Francis Darwin.

Libraries all over the world have given help that was literally indispensable by making available photocopies of Darwin correspondence and other manuscripts in their collections. The institutions and individuals that furnished copies of letters for this volume can be found in the List of provenances. The editors are extremely indebted to them all. We are also grateful to the many people who have transmitted information regarding the whereabouts of particular letters and, in many cases, have generously provided copies of such letters for the project's use.

The editors make daily use of the incomparable facilities of the Cambridge University Library and have benefited greatly from its services and from the help and expertise of its staff, particularly the staff of the Manuscripts Department. We are especially grateful to the former University Librarian, Frederick W. Ratcliffe, to his successor, Peter K. Fox, and to the Under-librarian of Manuscripts, Patrick Zutshi, for their generous support. Other members of the library's staff who have frequently responded to the editors' requests are: Marjolein Allen, Teri Barringer, Gerry Bye, Kathleen Cann, Colin Clarkson, Barry Eaden, Roger Fairclough, David J. Hall, John Hall, Tony Harper, Brian Jenkins, Morag Law, Elisabeth Leedham-Green, David Lowe, Peter Meadows, Mark Nicholls, William Noblett, Adam Perkins, Jayne Ringrose, Clive Simmonds, Anne Taylor, Nicola Thwaite, Godfrey Waller, Cynthia Webster, and Jeremy Wong. The fetchers in the Manuscripts reading room have also patiently dealt with the editors' often complex requirements.

At the American Philosophical Society Library, a splendid collection of Darwiniana and works in the history of science has been continuously available to the editorial staff since the inception of the project. Whitfield J. Bell Jr, secretary of the society until 1983, serves on the United States Advisory Committee for the project and has done his utmost to further its work. The editors have also benefited from the co-operation of the APS librarian Edward Carter II, Roy C. Goodman, Carl F. Miller, Elizabeth Carroll-Horrocks, and Bertram Dodelin, all of the APS Library.

The editors would like to acknowledge the assistance of Rodney Dennis, Jennie Rathbun, and Susan Halpert of the Houghton Library, Constance Carter of the Science Division of the Library of Congress, and Judith Warnement and Jean Cargill of the Gray Herbarium of Harvard University, who have all been exceptionally helpful in providing material from the collections in their charge.

In Britain, the editors have often received assistance from Mary Sampson, archivist of the Royal Society; Rex E. R. Banks and John Thackray of the Natural History Museum Library; Gina Douglas, librarian of the Linnean Society of London; and Sylvia FitzGerald, Cheryl Piggott, Lesley Price, and Leonora Thompson of the Library of the Royal Botanic Gardens, Kew. We would also like to thank Richard G. Williams, Mrs Felton, and Anne Barrett (college archivist) of Imperial College of Science, Technology and Medicine; Christine Fyfe, archivist of Keele University; and Dr and Mrs Courtney of Christ's College, Cambridge. Invaluable assistance has been received from local studies librarians and county archivists, too numerous to name individually, although Elizabeth Silverthorne of Bromley Central Library and Donald Gibson of the Centre for Kentish Studies deserve special mention for their repeated assistance. Virginia Murray of John Murray (publishers) Ltd has graciously provided the editors with copies of the Darwin correspondence in the Murray archive and with related information. Leonard G. Wilson kindly provided the editors with photocopies of manuscript material in the possession of the family of Sir Charles Lyell. Quentin Keynes and Joseph Sakmyster have generously provided photocopies of newly found letters.

Among the others who have advised and assisted the editors in their work are Janet Browne, Mary Dennison, Nick Gill, Ruth Goldstone, Julius Held, Sandra Herbert, Martin Rudwick, R. J. G. Savage, Jim and Anne Secord, Garry J. Tee, Hugh Torrens, and Morton N. Cohen. The editors are also pleased to acknowledge the invaluable support of the members of the British and United States Advisory Committees. Thanks are due to Joy Harvey, David Kohn, William Montgomery, and Marsha Richmond, all former associate editors with the Darwin Correspondence Project, for their continuing interest and help. Gratitude is expressed to C. A. Tripp for providing the editors with an invaluable research tool by enabling them to access texts through search and retrieval software. The editors also gratefully acknowledge the valuable assistance of Kate Fletcher, Mario Di Gregorio, Philip Hardie, Arne Hessenbruch, Thomas Junker, Anna-K. Mayer, Paul White, and Béatrice Willis and extend thanks to Jean Macqueen for providing the index to the volume.

Finally, we would like to acknowledge the contribution of colleagues who left the staff during the preparation of this volume. Hedy Franks has been with us since 1981, transcribing the letters collected, and, recently, as Office Manager. We wish her well for her retirement. During 1997 we also lost the valuable services of Sarah Lavelle and Jonathan Topham, both of whom we wish well in all they undertake in the future.

LIST OF PROVENANCES

The following list gives the locations of the original versions of the letters printed in this volume. The editors are grateful to all the institutions and individuals listed for allowing access to the letters in their care. Access to material in DAR 261 and DAR 263, formerly at Down House, Downe, Kent, England, is courtesy of English Heritage.

19th Century Shop (dealers), Baltimore, MD, USA
Alexander Turnbull Library, National Library of New Zealand / Te Puna Mātauranga o Aotearoa, Wellington, New Zealand
B. Altman & Co. (dealers), New York, NY, USA
American Philosophical Society, Philadelphia, PA, USA
Archiv der Berlin-Brandenburgischen Akademie der Wissenschaften, Berlin, Germany
Athenæum (publication)
Beinecke Rare Book and Manuscript Library, Yale University
British Library, The Manuscript Collections, Great Russell Street, London, England
Brotherton Collection, Leeds University Library, Leeds, England
Bulletins de la Société d'Anthropologie de Paris (publication)
Cambridge University Library, Manuscripts department, Cambridge, England
Canterbury Museum, Canterbury, New Zealand
Charterhouse Archives, Charterhouse, Godalming, Surrey, England
Christie's (dealers)
Christ's College Library, Cambridge, England
Cleveland Health Sciences Library, Cleveland, OH, USA
Mrs Carole Cockett (private collection)
Mrs H. Codd (private collection)
Columbia University, Rare Books and Manuscript Library, New York, NY, USA
CUL. *See* Cambridge University Library
Dana Family Papers, Manuscripts and Archives, Yale University Library, New Haven, CT, USA
DAR. *See* Cambridge University Library
Ernst-Haeckel-Haus, Institut für Geschichte der Medizin, Naturwissenschaft und Technik, Jena, Germany
Eva Crane IBRA Library, International Bee Research Association, Cardiff, Wales
Forschungsinstitut und Natur-Museum Senckenberg, Frankfurt, Germany
Gardeners' Chronicle and Agricultural Gazette (publication)
Gray Herbarium of Harvard University, Cambridge, MA, USA

Harold B. Lee Library, Brigham Young University, Provo, UT, USA
Eilo Hildebrand (private collection)
Imperial College of Science, Technology and Medicine, South Kensington, London, England
Col. James Innes (private collection)
John Innes Institute, Norwich, England
John Murray Archive, Albemarle Street, London, England
Journal of Horticulture and Cottage Gardener (publication)
Quentin Keynes (private collection)
La Fondation Augustin de Candolle
Legends (dealers), Sedona, AZ, USA
Liverpool Libraries and Information Services, Central Library, Liverpool, England
LuEsther T. Mertz Library at The New York Botanical Garden, Bronx, New York, NY, USA
K. M. Lyell ed. 1881 (publication)
Maggs Brothers (dealers), Berkeley Square, London, England
National Library of Scotland, Edinburgh, Scotland
Progressus rei botanicæ (publication)
Sybil Rampen (private collection)
Paul C. Richards (dealers), Templeton, MA, USA
Royal Botanic Gardens, Kew, Richmond, Surrey, England
Royal Entomological Society of London, Queen's Gate, London, England
Royal Horticultural Society, Vincent Square, London, England
Shrewsbury School, Shrewsbury, Shropshire, England
Barton L. Smith (private collection)
Smithsonian Institution Libraries, Washington DC, USA
Sotheby's (dealers)
J. A. Stargardt (dealers), Marburg, Germany
John Wilson (Autographs) (dealers), Oxford, England

A NOTE ON EDITORIAL POLICY

The first and chief objective of this edition is to provide complete and authoritative texts of Darwin's correspondence. In so far as it is possible, the letters have been dated, arranged in chronological order, and the recipients or senders identified. Darwin seldom wrote the full date on his letters and, unless the addressee was well known to him, usually wrote only 'Dear Sir' or 'Dear Madam'. After the adoption of adhesive postage stamps in the 1840s, the separate covers that came into use with them were usually not preserved, and thus the dates and the names of many recipients of Darwin's letters have had to be derived from other evidence. The notes made by Francis Darwin on letters sent to him for his editions of his father's correspondence have been helpful, as have matching letters in the correspondence, but many dates and recipients have had to be deduced from the subject-matter or references in the letters themselves. These tasks, together with deciphering the handwriting of Darwin and others, have presented the most troublesome editorial problems.

Whenever possible, transcriptions have been made from manuscripts. If the manuscript was inaccessible but a photocopy or other facsimile version was available, that version has been used as the source. In many cases, the editors have had recourse to Francis Darwin's large collection of copies of letters, compiled in the 1880s. Other copies, published letters, or drafts have been transcribed when they provided texts that were otherwise unavailable.

The method of transcription employed in this edition is adapted from that described by Fredson Bowers in 'Transcription of manuscripts: the record of variants', *Studies in Bibliography* 29 (1976): 212–64. This system is based on accepted principles of modern textual editing and has been widely adopted in literary editions.

The case for using the principles and techniques of this form of textual editing for historical and non-literary documents, both in manuscript and print, has been forcefully argued by G. Thomas Tanselle in 'The editing of historical documents', *Studies in Bibliography* 31 (1978): 1–56. The editors of the *Correspondence* followed Dr Tanselle in his conclusion that a 'scholarly edition of letters or journals should not contain a text which has editorially been corrected, made consistent, or otherwise smoothed out' (p. 48), but they have not wholly subscribed to the statement made earlier in the article that: 'In the case of notebooks, diaries, letters and the like, whatever state they are in constitutes their finished form, and the question of whether the writer "intended" something else is irrelevant' (p. 47). The editors have preserved the spelling, punctuation, and grammar of the original, but they have found it impossible to set aside entirely the question of authorial intent. One obvious reason is that in reading Darwin's writing, there must necessarily be reliance upon both context and intent. Even when Darwin's general intent is clear, there are cases in which alternative readings are, or may be, possible, and therefore the transcription decided upon must to some extent be conjectural. Where the editors are uncertain of their transcription, the doubtful text has been enclosed in italic square brackets.

A major editorial decision was to adopt the so-called 'clear-text' method of transcription, which so far as possible keeps the text free of brackets recording deletions, insertions, and other alterations in the places at which they occur. Darwin's changes are, however, recorded in the back matter of the volume, under 'Manuscript alterations and comments', in notes keyed to the printed text by paragraph and line number. All lines above the first paragraph of the letter (that is, date, address, or salutation) are referred to as paragraph '0'. Separate paragraph numbers are used for subscriptions and postscripts. This practice enables the reader who wishes to do so to reconstruct the manuscript versions of Darwin's letters, while furnishing printed versions that are uninterrupted by editorial interpolations. The Manuscript alterations and comments record all alterations made by Darwin in his letters and any editorial amendments made in transcription. No attempt has been made to record systematically all alterations to the text of copies of Darwin letters included in the correspondence, but ambiguous passages in copies are noted. The editors believe it would be impracticable to attempt to go further without reliable information about the texts of the original versions of the letters concerned. Letters to Darwin have been transcribed without recording any of the writers' alterations unless they reflect significant changes in substance or impede the sense; in such cases footnotes bring them to the reader's attention.

Misspellings have been preserved, even when it is clear that they were unintentional: for instance, 'lawer' for 'lawyer'. Such errors often indicate excitement or haste and may exhibit, over a series of letters, a habit of carelessness in writing to a particular correspondent or about a particular subject.

Capital letters have also been transcribed as they occur except in certain cases, such as 'm', 'k', and 'c', which are frequently written somewhat larger than others as initial letters of words. In these cases an attempt has been made to follow the normal practice of the writers.

In some instances that are not misspellings in a strict sense, editorial corrections have been made. In his early manuscripts and letters Darwin consistently wrote 'bl' so that it looks like 'lb' as in 'albe' for 'able', 'talbe' for 'table'. Because the form of the letters is so consistent in different words, the editors consider that this is most unlikely to be a misspelling but must be explained simply as a peculiarity of Darwin's handwriting. Consequently, the affected words have been transcribed as normally spelled and no record of any alteration is given in the textual apparatus. Elsewhere, though, there are misformed letters that the editors have recorded because they do, or could, affect the meaning of the word in which they appear. The main example is the occasional inadvertent crossing of 'l'. When the editors are satisfied that the intended letter was 'l' and not 't', as, for example, in 'stippers' or 'istand', then 'l' has been transcribed, but the actual form of the word in the manuscript has been given in the Manuscript alterations and comments.

If the only source for a letter is a copy, the editors have frequently retained corrections made to the text when it is clear that they were based upon comparison with the original. Francis Darwin's corrections of misreadings by copyists have usually been followed; corrections to the text that appear to be editorial alterations have not been retained.

Editorial interpolations in the text are in square brackets. Italic square brackets enclose conjectured readings and descriptions of illegible passages. To avoid confusion, in the few instances in which Darwin himself used square brackets, they have been altered by the editors to parentheses with the change recorded in the Manuscript alterations and comments. In letters to Darwin, square brackets have been changed to parentheses silently.

Material that is irrecoverable because the manuscript has been torn or damaged is indicated by angle brackets; any text supplied within them is obviously the responsibility of the editors. Occasionally, the editors are able to supply missing sections of text by using ultraviolet light (where text has been lost owing to damp) or by reference to transcripts or photocopies of manuscript material made before the damage occurred.

Words and passages that have been underlined for emphasis are printed in italics in accordance with conventional practice. Where the author of a letter has indicated greater emphasis by underlining a word or passage two or more times, the text is printed in bold type.

Paragraphs are often not clearly indicated in the letters. Darwin and others sometimes marked a change of subject by leaving a somewhat larger space than usual between sentences; sometimes Darwin employed a longer dash. In these cases, and when the subject is clearly changed in very long stretches of text, a new paragraph has been started by the editors without comment. The beginnings of letters, valedictions, and postscripts are also treated as new paragraphs regardless of whether they appear as new paragraphs in the manuscript. Special manuscript devices delimiting sections or paragraphs, for example, blank spaces left between sections of text and lines drawn across the page, are treated as normal paragraph indicators and are not specially marked or recorded unless their omission leaves the text unclear.

Additions to a letter that run over into the margins, or are continued at its head or foot, are transcribed at the point in the text at which the editors believe they were intended to be read. The placement of such an addition is only recorded in a footnote if it seems to the editors to have some significance or if the position at which it should be transcribed is unclear. Enclosures are transcribed following the letter.

The hand-drawn illustrations and diagrams that occur in some letters are reproduced as faithfully as possible and are usually positioned as they were in the original text. In some cases, however, it has been necessary to reduce the size of a diagram or enhance an outline for clarity; any such alterations are recorded in footnotes. The location of diagrams within a letter is sometimes changed for typesetting reasons. Tables have been reproduced as close to the original format as possible, given typesetting constraints.

Occasionally punctuation marking the end of a clause or sentence is not present in the manuscript, but the author has made his or her intention clear by allowing, for example, extra space or a line break to function as punctuation. In such cases, the editors have inserted an extra space following the sentence or clause to set it off from the following text.

Darwin consistently used the term 'fertilisation' for both fertilisation (the fusion of sperm with egg) and pollination (the transfer of pollen from anther to stigma); the first recorded usage of the term 'pollination' was in 1875 (*OED*). Darwin's use of 'fertilisation' in his letters and publications often, but not always, actually describes pollination. In the footnotes, the editors have attempted to use the modern terms in the context of Darwin's descriptions of experiments. When he is quoted directly, his original word usage is never altered.

Some Darwin letters and a few letters to Darwin are known only from entries in the catalogues of book and manuscript dealers or mentions in other published sources. Whatever information these sources provide about the content of such letters has been reproduced without change.

For every Darwin letter, the text that is available to the editors is always given in full. The editors have occasionally included letters that are not to or from Darwin if they are relevant to the published correspondence. Letters that can only be given a wide date range, in some instances spanning several decades, will be printed together in one of the concluding volumes.

The format in which the transcriptions are printed in the *Correspondence* is as follows:

1. *Order of letters*. The letters are arranged in chronological sequence. A letter that can be dated only approximately is placed at the earliest date on which the editors believe it could have been written. The basis of a date supplied by the editors is given in a footnote unless it is derived from a postmark, watermark, or endorsement that is recorded in the physical description of the letter (see section 4, below). Letters with the same date, or with a range of dates commencing with that date, are printed in the alphabetical order of their senders or recipients unless their contents dictate a clear alternative order. Letters dated only to a year or a range of years precede letters that are dated to a particular month or range of months, and these, in turn, precede those that are dated to a particular day or range of dates commencing with a particular day.

2. *Headline*. This gives the name of the sender or recipient of the letter and its date. The date is given in a standard form, but those elements not taken directly from the letter text are supplied in square brackets.

3. *The letter text*. The transcribed text follows as closely as possible the layout of the source, although no attempt is made to produce a type-facsimile of the manuscript: word-spacing and line-division in the running text are not adhered to. Similarly,

the typography of printed sources is not replicated. Dates and addresses given by authors are transcribed as they appear, except that if both the date and the address are at the head of the letter they are always printed on separate lines with the address first, regardless of the manuscript order. If no address is given on a letter by Darwin, the editors have supplied one, when able to do so, in square brackets at the head of the letter. Similarly, if Darwin was writing from an address different from the one given on the letter, his actual location is given in square brackets. Addresses on printed stationery are transcribed in italics. Addresses, dates, and valedictions have been run into single lines to save space, but the positions of line-breaks in the original are marked by vertical bars.

4. *Physical description*. All letters are complete and in the hand of the sender unless otherwise indicated. If a letter was written by an amanuensis, or exists only as a draft or a copy, or is incomplete, or is in some other way unusual, then the editors provide the information needed to complete the description. Postmarks, endorsements, and watermarks are recorded only when they are evidence for the date or address of the letter.

5. *Source*. The final line provides the provenance of the text. Some sources are given in abbreviated form (for example, DAR 140: 18) but are listed in full in the List of provenances unless the source is a published work. Letters in private collections are also indicated. References to published works are given in author–date or short-title form, with full titles and publication details supplied in the Bibliography at the end of the volume.

6. *Darwin's annotations*. Darwin frequently made notes in the margins of the letters he received, scored significant passages, and crossed through details that were of no further interest to him. These annotations are transcribed or described following the letter text. They are keyed to the letter text by paragraph and line numbers. Most notes are short, but occasionally they run from a paragraph to several pages, and sometimes they are written on separate sheets appended to the letter. Extended notes relating to a letter are transcribed whenever practicable following the annotations as 'CD notes'.

Quotations from Darwin manuscripts in footnotes and elsewhere, and the text of his annotations and notes on letters, are transcribed in 'descriptive' style. In this method the alterations in the text are recorded in brackets at the places where they occur. For example:

'See Daubeny ['vol. 1' *del*] for *descriptions of volcanoes in [*interl*] S.A.' *ink*

means that Darwin originally wrote in ink 'See Daubeny vol. 1 for S.A.' and then deleted 'vol. 1' and inserted 'descriptions of volcanoes in' after 'for'. The asterisk before 'descriptions' marks the beginning of the interlined phrase, which ends at the bracket. The asterisk is used when the alteration applies to more than the immediately preceding word. The final text can be read simply by skipping the material in brackets. Descriptive style is also used in the Manuscript alterations and comments.

Volumes of the *Correspondence* are published in chronological order. Each volume is self-contained, having its own index, bibliography, and biographical register. A chronology covering the period of each volume is supplied, and appendixes give additional material as appropriate to assist the understanding of the correspondence. A cumulative index is planned once the edition is complete. References are supplied for all persons, publications, and subjects mentioned, even though some repetition of material in earlier volumes is involved.

If the name of a person mentioned in a letter is incomplete or incorrectly spelled, the full, correct form is given in a footnote. Brief biographies of persons mentioned in the letters, and dates of each correspondent's letters to and from Darwin, are given in the Biographical register and index to correspondents. For correspondence with companies or institutions, both the company or institution and the individual correspondent are listed in the Biographical register. Where a personal name serves as a company name, it is listed according to the family name but retains its original order: for example, 'E. Schweizerbart'sche Verlagsbuchhandlung' is listed under 'S', not 'E'.

Short titles are used for references to Darwin's books and articles and to collections of his letters (e.g., *Descent*, 'Parallel roads of Glen Roy', *LL*). They are also used for some standard reference works and for works with no identifiable author (e.g., *Alum. Cantab.*, *Wellesley index*, *DNB*). For all other works, author–date references are used. References to the Bible are to the authorised King James version unless otherwise stated. The full titles of all the books referred to are given in the Bibliography.

Occasional supplements will be published containing letters that have come to light or have been redated since the relevant volumes of the *Correspondence* appeared. The first such supplement was included in Volume 7 and covered the years from 1821 to 1857.

The editors use the abbreviation 'CD' for Charles Darwin throughout the footnotes. A list of all abbreviations used by the editors in this volume is given on p. xlii.

The Wedgwood

- Robert Waring Darwin 1766–1848 = Susannah Wedgwood 1765–1817
  - Marianne 1798–1858 = Henry Parker 1788–1856
    - Robert b. 1825
    - Henry 1827–92
    - Francis 1829–71
    - Charles b. 1831
    - Mary Susan 1836–93
  - Caroline Sarah 1800–88 = Josiah III (Jos) 1795–1880
    - Sophy Marianne 1838–9
    - Katherine Elizabeth Sophy (Sophy) 1842–1911
    - Margaret Susan 1843–1937
    - Lucy Caroline 1846–1919
  - Susan Elizabeth 1803–66
  - Erasmus Alvey 1804–81
  - Charles Robert 1809–82 = Emma 1808–96
    - William Erasmus 1839–1914
    - Anne Elizabeth 1841–51
    - Mary Eleanor Sept.-Oct. 1842
    - Henrietta Emma (Etty) 1843–1927
    - George Howard 1845–1912
    - Elizabeth 1847–1926
    - Francis (Frank) 1848–1925
    - Leonard 1850–1943
    - Horace 1851–1928
    - Charles Waring 1856–8
  - Emily Catherine (Catherine) 1810–66

rwin Families up to 1863

- *Josiah Wedgwood II* 1769–1843 = *Elizabeth (Bessy) Allen* 1764–1846
  - *Sarah Elizabeth (Elizabeth)* 1793–1880
  - *Mary Anne* 1796–8
  - *Charlotte* 1797–1862 = *Charles Langton* 1801–86
    - *Edmund* 1841–75
  - *Henry Allen (Harry)* 1799–1885 = *Jessie Wedgwood* 1804–72
    - *Louisa Frances* 1834–1903
    - *Caroline* b. 1836
    - *John Darwin* 1840–70
    - *Anne Jane* 1841–77
    - *Arthur* 1843–1900
    - *Rowland* 1847–1921
  - *Frances Mosley* 1808–74 = *Francis (Frank)* 1800–88
    - *Godfrey* 1833–1905
    - *Amy* 1835–1910
    - *Cicely Mary* 1837–1917
    - *Clement* 1840–89
    - *Lawrence* 1844–1913
    - *Constance Rose* 1846–1903
    - *Mabel Frances* 1852–1930
  - *Hensleigh* 1803–91 = *Frances (Fanny) Mackintosh* 1800–89
    - *Frances Julia (Snow)* 1833–1913
    - *James Mackintosh (Bro)* 1834–64
    - *Ernest Hensleigh* 1838–98
    - *Katherine Euphemia (Effie)* 1839–1931
    - *Alfred Allen* 1842–92
    - *Hope Elizabeth* 1844–1934
  - *Frances (Fanny)* 1806–32

ABBREVIATIONS

AL	autograph letter
ALS	autograph letter signed
LS	letter in hand of amanuensis, signed by sender
LS(A)	letter in hand of amanuensis with additions by sender
Mem	memorandum
(S)	signed with sender's name by amanuensis
CD	Charles Darwin
CUL	Cambridge University Library
DAR	Darwin Archive, Cambridge University Library
del	deleted
illeg	illegible
interl	interlined
underl	underlined

TRANSCRIPTION CONVENTIONS

[some text]	'some text' is an editorial insertion
*[*some text*]*	'some text' is the conjectured reading of an ambiguous word or passage
[some text]	'some text' is a description of a word or passage that cannot be transcribed, e.g., '*3 words illeg*'
⟨ ⟩	word(s) destroyed
⟨some text⟩	'some text' is a suggested reading for a destroyed word or passage
⟨*some text*⟩	'some text' is a description of a destroyed word or passage, e.g., '*3 lines excised*'

THE CORRESPONDENCE OF CHARLES DARWIN 1863

To Asa Gray 2 January [1863][1]

Down Bromley Kent
Jan 2^{d}.

My dear Gray

I have been rather extra unwell of late[2] & overburdened with letters, but I cannot rest without thanking you for your two notes of Nov. 24th & Dec. 9th (the former with answers to several queries)[3] & especially I must thank you for Box with plants (I wrote & thanked Capt. Anderson)[4] which arrived in first-rate order & are planted. Positively the Mitchella looked as fresh, as if dug up the day before! What a pretty little creeper it is with its scarlet berries.[5] Miss Cooper, I remember, mentions it.—[6] I hope the Cypripediums will flower & I have this evening been thinking how I will try them, viz by putting live insects in & stopping up end of slipper & catching them as they come out of lateral orifices, & then, if they are smeared with pollen, I will put them in again & so make them go the round & then examine stigma.—[7]

I have just finished paper for Linn. Soc. on dimorphism of Linum,—much better case than Primula:[8] I see Planchon says that L. Lewisii (var. of perenne as he calls it, I doubt not falsely) bears on *same* plant flowers with long, & short & equal (to anthers) pistils;[9] I wish I could get seed of this arctic plant; I sh^{d} like to see this new case.—[10] I have Amsinckia growing well in my greenhouse & your Mitchellas; so I shall have as much as I want in these two Families.[11] I hear Cinchona is dimorphic & have written to Thwaites in Ceylon to try the pollen.—[12] I will send my Linum paper whenever published.—[13]

I was heartily glad to receive a note a week or so ago from Dana, giving a moderately good account of himself.—[14]

I thank you most heartily for all your extraordinary kindness in helping Leonard so much in his passion for stamps.[15] He has just exchanged one of Blood's for, I believe 9 rare stamps![16] I, also, of late troubled you with an extra number of questions &c.— It seems quite strange that I have only one trifle to ask you tonight, if you can remember it, viz to weigh in grains one of your wild Fragaria virginiana.—[17]

Have you ever attended much to garden plants; if you have ever noticed any what some gardeners call sports & what I shall call "**bud**-*variations*", I sh^{d} be glad of case to add to my large collection of facts, which seem to me of value in regard to theory of variability.[18] Hooker is in great spirits at having finished Welwitschia[19]

& is going to Paris.—[20] Good night my good & very kind friend, I am tired. I fear the last has been a dreadful battle & defeat.[21] When will peace come! But then Slavery, I know not what to wish. I wish to Heaven the north did not hate us so, I, for one, could wish more heartily for you then; even though I doubt the war being now justifiable. But thank Heaven wishes make no difference. Dr. Boott seems in despair & hates to hear of all the bloodshed.[22] We in north England seem tiding over our difficulty far better than anyone ever ventured to hope. The subscriptions have been gigantic.[23] Good night again. I cannot help still wondering that you or anyone in U. States can care for science at present.—

I am building a small Hot-House, so that if I have strength I shall have better means for my little experiments.[24]

Ever yours very truly | C. Darwin

If flowers of an Oak or Beech tree had fine grand well-colored corolla & calyx, would they be still classed as low in Vegetable Kingdom? This query, I daresay, shows my profound ignorance.—

Gray Herbarium of Harvard University (56)

1 The year is established by the relationship of this letter to the letters from Asa Gray, 24 November 1862 and 9 December 1862 (*Correspondence* vol. 10).

2 CD reported feeling unwell in late December 1862 (see *Correspondence* vol. 10, letter from J. D. Hooker, [21 December 1862], and letter to W. B. Tegetmeier, 27 [December 1862]). In a letter to William Erasmus Darwin of [13 December 1862] (DAR 219.1: 69), Emma Darwin commented: 'Your father is constantly plagued with slight attacks of exema which make him very uncomf—'.

3 See *Correspondence* vol. 10, letters from Asa Gray, 24 November 1862 and 9 December 1862.

4 Gray had sent CD a box containing specimens of *Mitchella* and *Cypripedium* via James Anderson, captain of the transatlantic steamer *Africa* (see *Correspondence* vol. 10, letter from Asa Gray, 9 December 1862, and letter to James Anderson, 23 December [1862]).

5 In his letter of 11 October 1861 (*Correspondence* vol. 9), Gray suggested that *Mitchella repens* would be a good case of flower dimorphism for CD to experiment upon; he sent CD specimens of the plant with his letter of 15 July [1862] (*Correspondence* vol. 10). CD's experimental notes on *M. repens*, made in 1864 and 1865, are in DAR 110: B84–93; his results are given in *Forms of flowers*, pp. 125–7.

6 CD had read Susan Augusta Fenimore Cooper's *Journal of a naturalist in the United States* (Cooper 1855) the previous year (see *Correspondence* vol. 10, letter to Asa Gray, 6 November [1862]). Cooper referred to *Mitchella* as the 'partridge-plant' (Cooper 1855, 2: 9).

7 CD intended to test Gray's hypothesis that pollination in *Cypripedium* (the lady's slipper orchid) was effected by small flies and bees crawling through the flower (A. Gray 1862a, pp. 427–8), rather than by a larger insect inserting its proboscis as CD had suggested in *Orchids*, pp. 274–5. See *Correspondence* vol. 10, letter from Asa Gray, 17 November 1862, letter to Asa Gray, 26[–7] November [1862], and CD's experimental note dated 19 April [1863] in DAR 70: 112–13. See also letter to Asa Gray, 20 April [1863], 'Fertilization of orchids', pp. 155–7 (*Collected papers* 2: 152–3), and *Orchids* 2d ed., pp. 229–31.

8 CD's paper, 'Two forms in species of *Linum*', was read before the Linnean Society of London on 5 February 1863, and was subsequently published in the *Journal of the Proceedings of the Linnean Society (Botany)* 7 (1863): 69–83 (*Collected papers* 2: 93–105). CD also refers to his paper 'Dimorphic condition in *Primula*', which was read in 1861 and published in the same journal in 1862.

9 Jules Emile Planchon described the different forms of flowers in *Linum lewisii*, a flax from western North America, in Planchon 1847–8, p. 175; CD discussed Planchon's observations in 'Two forms in species of *Linum*', pp. 82–3 (*Collected papers* 2: 104–5). CD had been studying dimorphism in different

plants of the same species, rather than two different forms found on the same plant (see 'Dimorphic condition in *Primula*', and n. 8, above). Gray thought that *L. lewisii* was the same species as *L. perenne* (see letter from Asa Gray, 27 January 1863). See also Baker 1965, pp. 147–50.

10 Although CD did not acquire seed of *Linum lewisii*, John Scott, who was head of the propagating department at the Royal Botanic Garden, Edinburgh, offered to conduct experiments on the species on CD's behalf (see letter from John Scott, [3 June 1863], and letter to John Scott, 6 June [1863]). See also *Forms of flowers*, p. 101.

11 Following Gray's suggestion that *Amsinckia spectabilis* might be dimorphic, CD had been experimenting with the plant since March 1862 (see *Correspondence* vol. 9, letter from Asa Gray, 9 November 1861, and letter to J. D. Hooker, 18 [December 1861], and *Correspondence* vol. 10, letter to Asa Gray, 15 March [1862]). His results are given in *Forms of flowers*, pp. 110–11. CD's references to 'families' were either to groups that were then more often called 'orders' (*EB*), or to a sub-family or tribe. Elsewhere, he referred to *A. spectabilis* as a member of the Boraginaceae (DAR 110: B2), and to *Mitchella* as a member of the Rubiaceae (madders) (*Forms of flowers*, pp. 284–5).

12 CD was investigating the possible occurrence of dimorphism in a number of plants, including *Cinchona*, a member of the Rubiaceae (see *Correspondence* vol. 10, letters to G. H. K. Thwaites, 15 June [1862] and 29 December [1862]). See also *Correspondence* vol. 9, letter from Asa Gray, [27 and 29 August] and 2 September [1861]. George Henry Kendrick Thwaites was superintendent of the Peradeniya Botanic Gardens, Ceylon; *Cinchona* trees were cultivated for the medicinal qualities of quinine produced by the bark (*EB*).

13 'Two forms in species of *Linum*' was published in the *Journal of the Proceedings of the Linnean Society (Botany)* issued on 13 May 1863 (*General index to the Journal of the Linnean Society*, p. vi). However, CD obtained a number of offprints of the paper in mid-April for distribution; Gray's name appears on CD's presentation list (see Appendix IV).

14 Letter from J. D. Dana, 4 December 1862 (*Correspondence* vol. 10). In 1859, James Dwight Dana had suffered a nervous breakdown from which he never fully recovered (see *Correspondence* vol. 7, letter to Charles Lyell, 29 [December 1859], and letter to J. D. Dana, 30 December [1859]).

15 CD refers to his twelve-year-old son, Leonard Darwin. At CD's request, Gray had, since the summer of 1862, sent United States postage stamps for Leonard's collection (see, for example, *Correspondence* vol. 10, letter to Asa Gray, 10–20 June [1862], and letter from Asa Gray, 2–3 July 1862). Gray's letter of 24 November 1862 (*Correspondence* vol. 10) had been sent using a 'peculiar stamp . . . for Leonard's collection'.

16 CD refers to stamps issued by the Philadelphia carriers D. O. Blood & Co. (see Sutton 1966, p. 41).

17 CD was seeking information about cultivated plants for a chapter of *Variation* (*Variation* 1: 332–72). Since beginning work on this chapter in October 1862 (see *Correspondence* vol. 10, Appendix II), he had twice asked Gray about the size difference between *Fragaria vesca*, a wild European and North American strawberry, and *F. virginiana*, a wild North American strawberry from which considerably larger fruits had been cultivated (see *Correspondence* vol. 10, letters to Asa Gray, 23 November [1862] and 26[–7] November [1862]). Gray replied to CD's queries in the missing section of his letter of 29 December 1862 (*Correspondence* vol. 10), which CD received later in January. CD mentioned Gray's comments in *Variation* 1: 351 n. 100. See also letter to *Journal of Horticulture*, [before 25 November 1862] (*Correspondence* vol. 10), and the annotated articles on strawberries in CD's issues of the *Gardeners' Chronicle* (8 November 1862, p. 1053, and 22 November 1862, p. 1101), which are in the Cory Library, Cambridge Botanic Garden.

18 CD defined bud-variation as 'all those sudden changes in structure or appearance which occasionally occur in full-grown plants in their flower-buds or leaf-buds' (*Variation* 1: 373). He began writing a draft of the chapter of *Variation* dealing with 'bud-variation' on 21 December 1862 (see *Correspondence* vol. 10, Appendix II), soliciting examples from several of his correspondents, including Thwaites, George Maw, John Scott, Thomas Rivers, and Hugh Falconer (see *Correspondence* vols. 9 and 10).

19 Joseph Dalton Hooker had been working on *Welwitschia* since as early as October 1861 (see, for instance, *Correspondence* vol. 9, letter to J. D. Hooker, 23 October [1861] and n. 4, and *Correspondence*

vol. 10, letter from J. D. Hooker, 20 August 1862). He read his account of this unusual plant from the south-west coast of Africa before the Linnean Society over the course of two meetings on 18 December 1862 and 16 January 1863 (J. D. Hooker 1863a). See *Correspondence* vol. 10, letter from J. D. Hooker, [21 December 1862].

20 On 17 January 1863, Hooker travelled to Paris with George Bentham for ten days (Jackson 1906, p. 193).

21 CD refers to the battle of Fredericksburg, fought on 13 December 1862; the Union army suffered one of its worst defeats of the American Civil War, with nearly 13,000 casualties (see McPherson 1988, pp. 571–4).

22 Gray had been sending copies of American newspapers to CD, and had requested that they afterwards be sent to the American-born physician and botanist, Francis Boott (see *Correspondence* vol. 10, letter from Asa Gray, 24 November 1862). However, in his letter of 26 December 1862 (*ibid.*), Boott asked that CD stop sending them since he could not endure the 'details of hopeless bloodshed'.

23 The scarcity of raw cotton from the Confederate States of America caused widespread unemployment in the textile industry of Lancashire, in northern England, and several charities had successfully been raising money for relief of distress in affected areas; between June and December 1862, the Cotton Districts Relief Fund had raised £593,000 (Longmate 1978, pp. 79–80 and 146–8). According to his Classed account book (Down House MS), during the height of the 'cotton famine' CD gave £32 to relief funds for the Lancashire cotton workers, making contributions on 14 July, 19 September, and 23 November 1862.

24 At the end of 1862, CD planned to build a hothouse in the garden at Down House (see *Correspondence* vol. 10, letter to J. D. Hooker, 24 December [1862]). The construction was undertaken in 1863, supervised by John Horwood, the gardener to CD's neighbour, George Henry Turnbull (see letter to G. H. Turnbull, [16? February 1863], and Appendix VI). The hothouse was built behind Down House, in the kitchen garden beyond the lawn (*LL* 1: 321).

From Hugh Falconer 3 January [1863][1]

21 Park Crescent N.W.
3d. Jany.

My Dear Darwin

A happy new year to you—and many happy returns of the season to you and yours!

I ought long ago to have replied to your query about cases of *dimorphism*.[2] I wanted to overhaul my notes—and see if they contained any thing worth sending to you. But I have found nothing sufficiently precise to bear upon the case. There is a large showy species of Linum, *L. trigynum*, common in India—with large yellow flowers—which is very variable in the number of the styles—and if my recollection is right—in their development also.[3] You could easily get the plant through Hooker[4]—and it would not trouble you much to rear it, as it flowers with Green House heat. I have a notion that it might yield you something in your line worth observing.

Nor can I give you a more satisfactory reply, to your query, about 'sporting buds'[5] I have no case, that I can bring to mind.

I was sorry to hear from your brother, of the efflorescence which has been troubling you—and which he tells me is one of the reasons, that has prevented you from coming to town.[6] You were never more missed—at any rate by me—for there has been this grand *Darwinian* case of the *Archæopteryx* for you and me to have a long jaw about. Had the Solenhofen quarries been commissioned—by august command—to turn out a strange being à la Darwin—it could not have executed the behest more handsomely—than in the *Archæopteryx*.[7] This is sober earnest—and that you should not have been in to town—and see it and talk over it with me, is a criminal proceeding. You are not to put your faith in the slip-shod and hasty account of it given to the Royal Society.[8] It is a much more astounding creature—than has entered into the the conception of the describer—who compares it with the Raptores & Passeres. & Gallinaceæ, as a round winged (like the last) 'Bird of flight.'[9] It actually had at least two long free digits to the fore limb—and those digits bearing claws as long and strong as those on the hind leg. Couple this with the long tail—and other odd things,—which I reserve for a jaw—and you will have the sort of misbegotten-bird-creature—the dawn of an oncoming conception à la Darwin.[10] But I will not say more about it till you show yourself in town. A ludicrous event has turned up. John Evans appears to have hit upon the very obvious cast of the interior of the skull—undetected by the describer, and before Owen's paper is out, we have M^r. Mackie describing the hemispheres—and optic lobes of Archæopteryx! Look at the Geologist.[11]

The Germans generally have a spite at Darwinianism. It killed poor Wagner. but on his death-bed, he took consolation in denouncing it as a *phantasia*.[12] Even Von Martius[13] has a shot at you. I send the paper containing it—being a sort of éloge upon Wagner.[14] Kindly return it when done with.

You will have seen my elephant paper out in the Nat Hist. Review[15]—with some of the evidence given which you thought to be wanting in what I sent you. I have not altered the latter, but I have gone into the value of the American measure of variety—and some other points.[16]

Lyell is hard at work in finishing his book.[17] I fancy he has had much to alter and adjust—and I have got a quagmire-*ish* kind of feeling, that he—and the Subject Homo—will be bogged in it—alike.

My Dear Darwin | Yours very Sin^ly | H Falconer

P.S. The last parag. is entre nous—so please do not repeat it.—

DAR 164: 10

[1] The year is established by the reference to Falconer 1863a (see n. 16, below).

[2] In his letter of 14 November [1862] (*Correspondence* vol. 10), CD asked Falconer whether he knew of any cases of dimorphism in plants analogous to those described in his paper, 'Dimorphic condition in *Primula*'.

[3] CD told Falconer that he was preparing a paper on dimorphism in species of *Linum* in his letter of 14 November [1862] (*Correspondence* vol. 10). However, he made no reference to *Linum trigynum* in his paper, 'Two forms in species of *Linum*' (see preceding letter and n. 8).

[4] Joseph Dalton Hooker was assistant director of the Royal Botanic Gardens, Kew (R. Desmond 1994).

[5] See *Correspondence* vol. 10, letter to Hugh Falconer, 29 December [1862]. See also preceding letter and n. 18.

6 Erasmus Alvey Darwin had apparently told Falconer about CD's eczema (see letter to Hugh Falconer, 5 [and 6] January [1863]). See also *Correspondence* vol. 10, letter to J. D. Hooker, 30 [June 1862].

7 Falconer refers to the fossil, recently described by Richard Owen (Owen 1862a), that displayed both reptilian and bird-like features, including feathers connected to a long, jointed tail. The specimen was found in 1861 in the Jurassic lithographic limestone near Solenhafen, Bavaria, and was named *Archæopteryx macrura* by Owen (*ibid.*, p. 33). The fossil had already achieved considerable notoriety as a 'feathered reptile' (Mackie 1863, p. 2); it was purchased by Owen on behalf of the British Museum for the sum of £400, and was brought to London in October 1862 (Rupke 1994, pp. 71–4). See also A. Desmond 1982, pp. 124–31. There is an annotated copy of Owen 1862a, which was not published until later in 1863 (Royal Society, Register of papers), in the Darwin Library–CUL. An abstract was published in the *Proceedings of the Royal Society of London* 12 (1863): 272–3.

8 Owen's paper on *Archaeopteryx*, read before the Royal Society of London on 20 November 1862 (Owen 1862a), was the earliest account of the fossil based on first-hand observation. Owen compared *Archaeopteryx* with pterodactyls and birds, concluding that: 'The best-determinable parts of its preserved structure declare it unequivocally to be a Bird, with rare peculiarities indicative of a distinct order in that class' (*ibid.*, p. 46). Falconer's objections to Owen's conclusions emerged in his role as one of the paper's referees; he protested that the author overemphasised the ornithological evidence (see Rupke 1994, pp. 73–4).

9 In his paper (Owen 1862a), Owen frequently compared features of *Archaeopteryx* with those of familiar and extant birds; for example, he noted that it was similar in size to the peregrine falcon and rook (p. 34), representing Falconer's 'Raptores & Passeres', and he believed that the wings corresponded in form and proportion to those of gallinaceous birds, such as the quail or grouse (p. 36).

10 Unlike Owen, Falconer was impressed by indications of the fossil's intermediate position between reptiles and birds. After hearing Owen's paper (Owen 1862a), Hooker reported that the 'general opinion was that Owen demonstrated its ornithic affinity and proved it to be a bird with the tail-feathers set on a jointed tail instead of the truculent hump that most birds have, but some say that there are peculiar bones or organs amongst the bones that may yet prove it to be Reptilian' (L. Huxley ed. 1918, 2: 32).

11 Owen reported in his paper that the head of *Archaeopteryx* was absent from the specimen (Owen 1862a, p. 34). However, John Evans, an archaeologist and geologist who examined the fossil shortly after hearing Owen's account, claimed to have discovered the cast of the specimen's skull. He did not publish an account of his findings for two years (John Evans 1865), but the geologist and palaeontologist Samuel Joseph Mackie reported in the January 1863 issue of *The Geologist* that Evans had found portions of the skull in the fossil slab (Mackie 1863, pp. 7–8; see also Joan Evans 1943, pp. 115–16). Owen made reference to Evans's findings at the end of the published version of his paper (Owen 1862a, p. 46), but did not express an opinion as to their validity. He also added engravings of the skull to plate 1, fig. 1, n and n′ (Owen 1862a). See also De Beer 1954.

12 The German zoologist and palaeontologist Johann Andreas Wagner was known for his opposition to the theory of the transmutation of species, and to *Origin* (see J. A. Wagner 1861a, and Gregory 1977, pp. 35, 38–9). Wagner, who had for many years studied fossil remains from the Solenhofen stone, published a description of *Archaeopteryx* based on the verbal report of a friend who examined the specimen (J. A. Wagner 1861b). The paper was later translated and published in the *Annals and Magazine of Natural History* with the title, 'On a new fossil reptile supposed to be furnished with feathers' (J. A. Wagner 1862). He named the creature *Griphosaurus*, meaning a saurian that is an 'enigma'. Wagner concluded (J. A. Wagner 1862, pp. 266–7):

> I must add a few words to ward off Darwinian misinterpretations of our new Saurian. At the first glance of the *Griphosaurus* we might certainly form a notion that we had before us an intermediate creature, engaged in the transition from the Saurian to the bird. Darwin and his adherents will probably employ the new discovery as an exceedingly welcome occurrence for the justification of their strange views upon the transformations of animals. But in this they will be wrong. . . . I am entitled to ask of the Darwinians . . . to show me, first of all, the intermediate steps by which

> the transition of some one living or extinct animal from one class into another was effected. If they cannot do this (as they certainly cannot), their views must be at once rejected as fantastic dreams, with which the exact investigation of nature has nothing to do.

See also n. 14, below.

13 Karl Friedrich Philipp von Martius.

14 Wagner died on 19 December 1861 (*ADB*), only weeks after the reading of his *Archaeopteryx* paper on 9 November (J. A. Wagner 1861a; see n. 12, above). Martius was responsible for delivering eulogies following the death of distinguished associates of the Royal Bavarian Academy (*DSB*); his eulogy upon Wagner discussed the palaeontologist's insistence on the immutability of species, his diluvialism, and his use of the Bible in arguing against CD's views (Martius 1862, pp. 8–10).

15 Falconer's paper on American fossil elephants (Falconer 1863a) was published in the January 1863 number of the *Natural History Review*, CD's annotated copy of which is in the Darwin Library–CUL.

16 Falconer had sent CD a manuscript portion of his paper on the American fossil elephants in September 1862 (see *Correspondence* vol. 10, letter from Hugh Falconer, 24–7 September [1862]). In the published paper (Falconer 1863a), Falconer emphasised the lack of variability in the European fossil elephants. Moreover, while praising CD and avowing a belief in the development of species by modified descent, he questioned the adequacy of natural selection as the sole mechanism for species change (Falconer 1863a, pp. 77–81). In response to the manuscript copy sent by Falconer, CD had urged him to consider the importance of variability to natural selection: 'when you speak of "moderate range of variation", I cannot but think that you ought to remind your readers . . . what the amount is, *including case of the* American Bog Mammoth' (see *Correspondence* vol. 10, letter to Hugh Falconer, 1 October [1862]; see also letter to Charles Lyell, 1 October [1862]). The published paper included a section in which Falconer discussed possible varieties of American fossil elephants in addition to the two generally recognised species (Falconer 1863a, pp. 63–7).

17 Charles Lyell's *Antiquity of man* (C. Lyell 1863a) was published on 9 February 1863 (*Athenæum*, 7 February 1863, p. 176).

To Joseph Dalton Hooker 3 January [1863]

Down
Jan 3^d.

My dear Hooker

I am burning with indignation & must exhale. If you have not already read, do read the first part of Falconer's paper on Elephants in N. H. Review & mark Owen's whole conduct.—[1] I could not get to sleep till past 3 last night from indignation. Thinking over his conduct in this case, in the Brain-case[2] & towards Mantell[3] Nasmyth,[4] Huxley,[5] you & self,[6] & review on Lyell,[7] the Terlepeton case[8] &c &c, I declare I think every honest man of science is almost bound to show his sense of Owen's character. I have made up my mind, as far as I can at this distance of time, to attend when next Council of Royal is elected, & if no else does, vote for some other man;[9] & if Owen were to come & speak to me I would tell him for what I came. But *possibly* he may answer Falconer, & explain. I have read only about $\frac{1}{4}$ of Falconer's paper.—[10] The Reviews seem good in this number.—[11]

Now for pleasanter subjects; we were all amused at your defence of stamp collecting & collecting generally.[12] Henrietta[13] had audacity to say "well I think D^r. Hooker shows that it leads all sorts of vice; yet I shall go on collecting plants, for

I love to look at them." I ought to say nothing against collecting for $\frac{6}{7}$th of my children collect, & I collected seals, franks, coins minerals, shells insects & God knows what else. But by Jove I can hardly stomach a grown man collecting stamps. Who would ever have thought of your collecting Wedgwood ware![14] but that is wholly different like engravings or pictures. We are degenerate descendants of old Josiah W. for we have not a bit of pretty ware in the house.—[15]

When you see Mr Oldfield pray thank him; my questions were foolish; but not rarely foolish questions, lead, I find, to good results.—[16]

Notwithstanding the very pleasant reason you give for our not enjoying a holiday, namely that we have no vices, it is a horrid bore.—[17] I have been trying for health sake to be idle with no success. What I shall soon have to do, will be to erect a tablet in Down church "sacred to the memory &c" & officially die, & then publish books "by the late Charles Darwin"; for I cannot think what has come over me of late; I always suffered from the excitement of talking, but now it has become ludicrous. I talked lately for 1½ hours (broken by tea by myself) with my nephew[18] & I was shaking & vomiting half the night— It is a fearful evil for self & family.

Goodnight | Ever yours | C. Darwin

My children's dried flowers get a little mouldy; is it not good to wash them with corrosive Sublimate[19] (how much?) in spirits? or in water? Sometime tell me.—

Endorsement: '/63.'
DAR 115: 178

[1] In his paper on fossil elephants, which was published in the January 1863 number of the *Natural History Review*, Hugh Falconer claimed (Falconer 1863a, pp. 43–9) that Richard Owen and his protégé, Charles Carter Blake, had abused the law of priority in zoological nomenclature. He explained that Owen and Blake had supplanted the species name Falconer had first given to the fossil elephant, *Elephas columbi* (Falconer 1857a, p. 319), with the name *E. texianus* (Owen 1858 and 1861b, and Blake 1861). In 1862, Blake argued that, according to the rules of zoological nomenclature as defined by the British Association for the Advancement of Science, the name *E. columbi* could be changed even without 'published priority', on the grounds that it was not clearly defined and was likely to propagate errors (Blake 1862, p. 58). Falconer refuted the claims of both Owen and Blake for the name *E. texianus*, dismissing their reasons as 'light and trivial' (Falconer 1863a, p. 49).

[2] CD refers to Owen's long-running dispute with Thomas Henry Huxley, George Rolleston, and William Henry Flower on the comparative anatomy of human and simian brains, now often referred to as the hippocampus controversy (see *Correspondence* vols. 8–10, and Rupke 1994, pp. 270–86).

[3] CD refers to the geologist and palaeontologist Gideon Algernon Mantell. In addition to disputes between the two men regarding the nature of particular fossils, in 1850 Owen attempted to reproduce without permission illustrations of fossil reptiles from a publication by Mantell. Owen was also thought to be the author of an obituary of Mantell that appeared in the *Literary Gazette*, 13 November 1852, p. 842; the article discussed Mantell's 'weaknesses', dismissing him as an enthusiast with an 'overweaning estimate' of the value of his own work (see A. Desmond 1982, p. 208 n. 13). On Owen's relationship with Mantell, see Spokes 1927, pp. 204–8, 221–7; Curwen ed. 1940, pp. 245–7, 260–2; Benton 1982; A. Desmond 1982, pp. 24, 208 n. 13; and Rupke 1994, pp. 6–8, 125–6. See also n. 8, below.

4 In 1839, a priority dispute developed between Owen and the dentist-surgeon Alexander Nasmyth; both men claimed to have been the first to outline a new theory of the ossific transformation of the cells of the pulp into dental ivory, that is, the development of teeth (see Owen 1839 and Nasmyth 1841). The debate was conducted in the *Lancet* between 6 June and 4 July 1840, and in the *London Medical Gazette* between 5 June and 17 July 1840. These exchanges culminated with Owen ridiculing Nasmyth's efforts in the scientific arena (*London Medical Gazette*, 17 July 1840, pp. 657*–673*); Nasmyth subsequently claimed that Owen had prevented the publication of his research in the official report of the British Association's 1839 meeting (see Nasmyth 1841, pp. iii–xvi). See also letter from J. D. Hooker, [6 March 1863] and n. 7.

5 T. H. Huxley and Owen had maintained a deep-seated antipathy since the early 1850s (see A. Desmond 1982, pp. 19–55, A. Desmond 1994–7, and Rupke 1994; see also n. 2, above).

6 CD refers in part to Owen's anonymous review of *Origin* ([Owen] 1860a). CD thought the review 'malignant', and that it misrepresented and misquoted his work; in his letter to Charles Lyell of 10 April [1860] (*Correspondence* vol. 8), CD referred to the review: 'It is painful to be hated in the intense degree with which Owen hates me'. For accounts of the additional differences between CD and Owen, see Hull 1973, pp. 171–215, and Rupke 1994. CD also strongly objected to the 'slighting way' Owen alluded to J. D. Hooker 1859, in which Hooker announced his support for the theory of natural selection (see *Correspondence* vol. 8, letter to Asa Gray, 18 May [1860]). See also *Correspondence* vol. 8, letter to T. H. Huxley, 9 April [1860]. There is an annotated copy of [Owen] 1860a in the Darwin Pamphlet Collection–CUL.

7 In an anonymous review ([Owen] 1851) of the eighth edition of Charles Lyell's *Principles of geology* (C. Lyell 1850), and of C. Lyell 1851a and 1851b, Owen attacked Lyell's anti-progressionism and uniformitarianism.

8 *Telerpeton* (*Leptopleuron*) was a fossil reptile discovered in Scotland in 1851, which Owen and Mantell may both have been asked to describe; a priority dispute followed (Benton 1982).

9 Hooker and CD had discussed the possibility of 'organising an opposition' to Owen's election to the council of the Royal Society of London in November 1862 (see *Correspondence* vol. 10, letter from J. D. Hooker, [15 and] 20 November [1862], and letter to J. D. Hooker, 24 [November 1862] and nn. 9 and 10). Owen was not re-elected at the 30 November 1863 anniversary meeting (see *Proceedings of the Royal Society* 13: 39), although CD did not attend the meeting (see 'Journal' (Appendix II)).

10 Falconer 1863a.

11 CD refers to the January 1863 number of the *Natural History Review*, which included a review of the first part of volume 1 of Bentham and Hooker 1862–83. CD's annotated copy of this number of the journal is in the Darwin Library–CUL. See also following letter and n. 5.

12 See *Correspondence* vol. 10, letter from J. D. Hooker, [27 or 28 December 1862].

13 CD's daughter, Henrietta Emma Darwin, was 19 years old.

14 Hooker had written that he was collecting Wedgwood ware 'solely because they are pretty & I love them' (*Correspondence* vol. 10, letter from J. D. Hooker, [27 or 28 December 1862]).

15 Both CD and Emma Darwin were grandchildren of the master-potter, Josiah Wedgwood I.

16 The botanist Augustus Frederick Oldfield, who had travelled widely in Australia and Tasmania, was a frequent visitor to the Royal Botanic Gardens at Kew; Hooker offered to convey any questions that CD might have for him (see *Correspondence* vol. 10, letter from J. D. Hooker, [21 December 1862] and n. 5). In his letter to Hooker of 24 December [1862] (*ibid.*), CD asked several questions about the diet and behaviour of Australian aborigines. Hooker sent Oldfield's replies to CD's queries with his letter of [27 or 28 December 1862] (*ibid.*).

17 See *Correspondence* vol. 10, letter from J. D. Hooker, [31 December 1862].

18 Henry Parker, the son of CD's sister, Marianne Parker, visited Down House on 29 December 1862 (see *Correspondence* vol. 10, letter to J. D. Hooker, 29 [December 1862]).

19 Corrosive sublimate (mercuric chloride) was used in herbaria to provide protection for dried specimens against fungal and insect attack (*EB*).

To John Lubbock 4 January [1863][1]

Down. | Bromley. | Kent. S.E.
Jan 4th

Dear Lubbock

I have got an uneasy feeling that I wrote surlily to you the other day.—[2] I really forget what I said except begging you not to return my Review.[3] If I did write surlily, I wrote with a most false spirit towards you; but I was miserably uncomfortable: I **know** that you will forgive me, if I did write so; so do not answer this; but if you wish me to alter anything tell me & mark passages in the Review.—& I will do my best.—

What a capital number of N. H. Review![4] I like your articles much;[5] they tell just what a general reader would like to hear; you touch up capitally, in first-rate style, Mr Wilson's reasoning powers.[6] I have got only as far as $\frac{2}{3}$ through Falconer'.[7]

I am fairly burning with indignation at Owen about Elephas Columbi;[8] if he does not answer & explain, I think every man of science ought somehow to show his disapproval of his whole line of conduct.— I have pretty nearly made up my mind what I will do.—[9] Good night

Yours affecty | C. Darwin

How strong Falconer is coming out[10] & how very well he writes.—

DAR 263: 58

[1] The year is established by the references to [Lubbock] 1863a and [Lubbock] 1863b (see n. 5, below).

[2] The letter to Lubbock has not been found.

[3] CD refers to the review of Bates 1861 that Lubbock had asked him to write for the *Natural History Review* (see *Correspondence* vol. 10, letter from John Lubbock, 15 December 1862). In his letter to Lubbock of 16 [December 1862] (*Correspondence* vol. 10), CD agreed to write the article; 'Review of Bates on mimetic butterflies' appeared in the April 1863 number of the *Natural History Review.*

[4] CD refers to the January 1863 number of the *Natural History Review*, CD's annotated copy of which is in the Darwin Library–CUL.

[5] Lubbock had written two review articles for the January 1863 number of the *Natural History Review.* The first ([Lubbock] 1863a), was a survey of recent books on North American archaeology, and the second ([Lubbock] 1863b) was a review of D. Wilson 1862 (see n. 6, below).

[6] In his review of Daniel Wilson's *Prehistoric man: researches into the origin of civilisation in the old and new world* (D. Wilson 1862), Lubbock recommended the book as an introduction to more specialised works on archaeology. However, he criticised the author's tendency to make obscure or contradictory statements, noting that Wilson, while 'clear enough on purely scientific questions, becomes unintelligible as soon as he treads on sacred ground' ([Lubbock] 1863b, p. 27). Lubbock was also critical of the last chapter (entitled 'Guesses at the age of man'), in which Wilson argued that the supposition that the human species had evolved from other organic forms by a process of natural selection made it difficult to determine the boundary between humans and apes, and, consequently, to place chronological limits on the human species, whereas the belief that God made man in his own image placed moderate, if undefined bounds on human antiquity. Lubbock responded ([Lubbock] 1863b, p. 30):

> We do not perceive the force of the argument, that moderate limits 'must be assigned to the existence of the race,' because 'God made man in his own image;' nor can we too strongly reprobate the attempt to fix a stigma of irreligion on the theory of Natural Selection.

[7] Falconer 1863a.

[8] Richard Owen; see preceding letter and n. 1.

9 See preceding letter and n. 9.

10 CD refers to Falconer 1863a; for CD's reaction to Hugh Falconer's endorsement of evolution, see *Correspondence* vol. 10.

To Hugh Falconer 5 [and 6] January [1863][1]

Down
Jan. 5th

My dear Falconer

I finished your Elephant paper last night[2] and you must let me express my admiration at it. All the points strike me as admirably worked out, and very many most interesting. I was particularly struck with your remarks on the character of the ancient Mammalian Fauna of N. America; it agrees with all I fancied was the case, namely a temporary irruption of S. American forms into N. America, and conversely.[3] I chuckled a little over the specimen of M. Andium "hesitating" between the two groups.[4] I have been assured by Mr. Wallace that abundant Mastodon remains have been found at Timor, and that is rather close to Australia:[5] I rejoice that you have smashed that case.[6] It is indeed a grand paper. I will say nothing more about your allusions to me, except that they have pleased me quite as much in print as in M.S.[7] You must have worked very hard; the labour must have been extreme, but I do hope that you will have health and strength to go on. You would laugh, if you could have seen how indignant all Owen's lies and mean conduct about E. Columbi made me.[8] I did not get to sleep till past 3 o'clock. How well you lash him; firmly and severely with unruffled temper, as if you were performing a simple duty.[9] The case is come to such a pass, that I think every man of science is bound to show his feelings by some overt act, and I shall watch for a fitting opportunity.

Ever my dear Falconer | Yours most truly | Ch. Darwin

P.S. I have kept back for a day the enclosed owing to the arrival of your most interesting letter.[10] I knew it was a mere chance whether you could inform on points required; but no one other person has so often responded to my miscellaneous queries.[11] I believe I have now in my greenhouse L. trigynum, which came up from seed purchased as L. flavum, from which it is wholly different in foliage.[12] I have just sent in paper on Dimorphism of Linum to the Linn. Soc.[13] and so I do not doubt your memory is right about L. trigynum:[14] the functional difference in the two forms of Linum is really wonderful. I assure you I quite long to see you and a few others in London: it is not so much the eczema which has taken off the epidermis a dozen times clean off; but I have been knocked up of late with extraordinary facility and when I shall be able to come up I know not.[15] I particularly wish to hear about the wondrous Bird;[16] the case has delighted me, because no group is so isolated as Birds. I much wish to hear when we meet which digits are developed; when examining birds two or three years ago, I distinctly remember writing to Lyell that some day a fossil bird would be found with end of wing cloven, i.e. the bastard wing and other part both well developed.[17] Thanks for Von Martius returned by this post, which I was glad to see.[18] Poor old Wagner always attacked me in a proper spirit and sent me two or three little brochures, and I thanked him cordially.[19] The Germans seem much stirred up on the subject. I received by the same post almost a little volume on the *Origin*.[20]

I cannot work above a couple of hours daily and this plays the deuce with me.

Farewell my good old friend | C. D.

I put in another query, on *remotest* possibility of information; if I do *not* hear I shall understand you have none.[21]

P.S. 2nd. I have worked like a slave and been baffled like a slave in trying to make out the meaning of two very different sets of stamens in some Melastomaceae.[22] I must tell you one fact. I counted 9000 seeds, one by one, from my artificially fertilised pods. There is something very odd but I am as yet beaten. Plants from two pollens grow at different rates![23] Now what I want to know is, whether in individuals of the *same species*, growing together, you have ever noticed any difference in the position of the pistil or in the size and colour of the stamens?

Copy
DAR 144: 29

1 The year is established by the references to Falconer 1863a. The postscripts were added on 6 January, the day after the letter was begun.

2 Falconer 1863a. CD's annotated copy of the number of the *Natural History Review* in which this paper appeared is in the Darwin Library–CUL.

3 In his paper, Falconer described fossil mammals associated with *Elephas columbi* in North and Central America (Falconer 1863a, pp. 61–3). He also noted the discovery of an elephant molar in French Guiana, as well as mastodon fossils in Honduras thought to be the same species as *Mastodon giganteus* of North America, and suggested that *E. columbi*, 'the fossil Elephant of Georgia', might have ranged 'still further south than Mexico, into Guiana' (Falconer 1863a, p. 60). See CD's annotations to his copy of the paper; see also *Origin*, pp. 365–82, for CD's discussion of his theory of the trans-tropical migration of temperate forms during a global glacial epoch.

4 CD refers to the footnote (Falconer 1863a, p. 100 n.) in which Falconer stated that in 1857, he had called attention to the 'exceptional character' of certain specimens of *Mastodon andinum*, which appeared to be 'hesitating' between the two subgenera *Tetralophodon* and *Trilophodon*, which were distinguished by the number of ridges on the molar crowns (Falconer 1857a, p. 313).

5 Alfred Russel Wallace had given CD information about Carl Friedrich Adolph Schneider, who claimed to have found teeth of *Mastodon* in Timor in the Malay Archipelago (see *Correspondence* vol. 9, letter from A. R. Wallace, 30 November 1861, and Schneider 1863). See also letter from A. R. Wallace, 14 January [1863] and n. 6.

6 On the basis of a single fossil molar tooth purchased from a native by Paul Edmund de Strzelecki during a geological expedition in Australia, Richard Owen proposed the species *Mastodon australis* (Owen 1844). Falconer argued that because of the lack of direct testimony as to its discovery, the specimen did not constitute evidence of an Australian species. He asked, if the fossil was indeed Australian, how 'the *Mastodon* alone, of all the higher placental mammals', had overcome 'the barriers of marsupial isolation, characteristic of the great southern island' (Falconer 1863a, p. 101). Falconer maintained that Owen's *M. australis* was a specimen of *M. andinum*, a South American species (Falconer 1863a, pp. 96–101). See also *Correspondence* vol. 10, letter to Hugh Falconer, 14 November [1862] and n. 13.

7 Falconer 1863a, pp. 77–81. While disputing the adequacy of natural selection to account for the origin of species, Falconer at the same time stated that CD, 'beyond all his contemporaries', had

inspired 'the philosophical investigation of the most backward and obscure branch of the Biological Sciences of his day', and had laid 'the foundations of a great edifice' (*ibid.*, p. 80). Falconer sent CD a manuscript portion of his paper on American fossil elephants in September 1862 (see *Correspondence* vol. 10, letter from Hugh Falconer, 24–7 September [1862], and letter to Hugh Falconer, 1 October [1862], and this volume, letter from Hugh Falconer, 3 January [1863]).

8 For CD's comments on the charges made by Falconer against Owen, see the letter to J. D. Hooker, 3 January [1863] and n. 1.

9 Falconer 1863a, pp. 43–9.

10 Letter from Hugh Falconer, 3 January [1863]; the enclosure has not been found.

11 In his letter to Falconer of 14 November [1862] (*Correspondence* vol. 10), CD asked whether Falconer knew of any cases of sexual dimorphism analogous to those described by CD in his paper, 'Dimorphic condition in *Primula*'; CD also asked Falconer for information on 'sports' or 'bud-variations' (see *Correspondence* vol. 10, letter to Hugh Falconer, 29 December [1862]).

12 Falconer had mentioned to CD the variability in the number of styles, and possible dimorphism, in the flax species *Linum trigynum* (see letter from Hugh Falconer, 3 January [1863]).

13 CD's paper, 'Two forms in species of *Linum*', was read before the Linnean Society on 5 February 1863.

14 See n. 12, above. Although CD made no reference to *Linum trigynum* in 'Two forms in species of *Linum*', the dimorphic character of the flower was mentioned in *Forms of flowers*, p. 100.

15 See letter from Hugh Falconer, 3 January [1863]; CD visited London from 4 to 14 February 1863 (see 'Journal' (Appendix II)).

16 CD refers to *Archaeopteryx*, a recently discovered fossil that Falconer had described to him (see letter from Hugh Falconer, 3 January [1863] and n. 7).

17 In 1859, CD wrote to Charles Lyell: 'I believe that if ever fossil birds are found very low down in series, they will be seen to have a double or bifurcated wing' (see *Correspondence* vol. 7, letter to Charles Lyell, 11 October [1859]).

18 Martius 1862. See letter from Hugh Falconer, 3 January [1863] and n. 14.

19 See letter from Hugh Falconer, 3 January [1863] and n. 14. No correspondence between CD and Johann Andreas Wagner has been found, nor are there any papers by Wagner in the Darwin Pamphlet Collection–CUL.

20 The reference is probably to Rolle 1863 (see letter to [F. E. Suchsland], [after 19 January 1863], and letter from Friedrich Rolle, 26 January 1863).

21 The enclosure has not been found.

22 In October 1861, CD began to investigate the occurrence of two different sets of stamens in the flowers of members of the Melastomataceae, believing that the family might exhibit a novel form of dimorphism (see *Correspondence* vol. 9, letter to J. D. Hooker, 17 November [1861], and *Correspondence* vol. 10, letter to George Bentham, 9 February [1862]). Although he continued the research throughout 1862 and 1863, he reached no definite conclusion and did not publish on the subject (see *Cross and self fertilisation*, p. 298 n., and *ML* 2: 292–302).

23 CD's notes on the Melastomataceae are in DAR 205.8; one of these notes records that of the seeds gathered in April and May 1862, CD counted 8911 (DAR 205.8: 30). See also *Correspondence* vol. 10, letter to John Scott, 11 December [1862].

From J. D. Hooker 6 January 1863

Kew
Jany 6/63

My dear Darwin

I have read Falconer most interesting & most prosy paper,[1] I do wish he had cut it into 4—for a non-Zoologist like me it is an apalling thing to have 70 pages of such an article in one quarter.[2] The "Review" is much better than usual, but confound them they have printed the Index of New genera of *last volume on last page* of first number of this volume!. it is too bad of Huxley.[3]

As to Owen he certainly is in a degrading position, but really his conduct is so *small* as well as detestable that I cannot even get up indignation.[4] I hardly know how to speak to him when I see him, or of him when I am asked by some great swell to echo his praises. Falconer is evidently consumedly riled. I am most curious to know what you think of Falconers observations on you[5]—he seems to me to have just awakened to the fact that there is something in you, & he too thinks that you make Nat Selection work independently & do every thing without variation to work upon.[6] I have not seen F. for ages.

I have finished de Tocqueville Democ. in America[7] & cannot help thinking how differently he might have written had he read the Origin & applied it—all his fallacies are attributable to ignorance of its principles. Specially his want of perception that the versatility & variety of resources each Yankee to which he attributes all their excellencies, more or less, possesses is simply the result of want of competition, & that when the land is filled with people this superiority will vanish, each will be good at his speciality only & the evil effects of Republicanism will burst out all over the peoples & communities. I do not believe that any nation can last for ever, either under a Republic or Monarchy, (both being bad) but ⟨by⟩ my notions I think the ⟨ ⟩ will be the longest lived, ⟨ ⟩ always turn up first—⟨*2 words missing*⟩ **growing** community ⟨ ⟩ started as a *grown* commun⟨ity⟩ was too precocious all along.

Then too all de Tocquevilles ⟨com⟩parative raticinations are frustrated by the growth of Englands colonies, which he (frenchman like) utterly ignores. Then too he is utterly wrong in assuming that the greater proportion of foreign traders to home traders in the American ports, compared to the English ports is a good sign for America[8]—& forgets that our home traders are the result of the boundless resources of different parts of our own country.— He says that all Americans entertain the most "virulent hatred" against the ⟨Eng⟩lish nation!—[9] this was just ⟨30 years⟩ ago!

⟨I am⟩ extremely concerned to hear ⟨of your⟩ failing powers of enduring ⟨conve⟩rsation—that will however ⟨soon⟩ come right again as far as ⟨ev⟩er you can be right: I do not doubt—but cannot help laughing at your plan of the Epitaph,[10] how jolly it would be to hear your own works criticized after death, & yourself abused for what you never said or did & praised for what you never wrote or thought of.

I have not been long waiting for a place to scratch, for the New Zeald Govt. have asked me to write a manual of the N. Z. Flora like Bentham's Flora of Hong Kong.[11] I am not sorry to do this, as it will enable me to correct many errors, add many Southern Island species, & interest me further in New Zealand.[12] They have voted £500— th⟨at⟩ gives me £150 a volume ⟨for⟩ 2 vols of about 600 pages each & the other £200 goes to publ⟨isher⟩ as no one will publish with⟨out⟩ that *bonus* per volume—for which the publisher gives 100 copies.[13] This is equivalent to Govt subscribing for 100 copies at £1 per volume per copy. Of course writing so much is a frightful bore, but I may as well do that as worse, & please tell Henrietta that I have no scope for vice in the undertaking.[14]

I am quite aware of your insensibility to Wedgewood ware,[15] Were it otherwise I do not think I could have *gone into* this foible, for I should have bored you out of your life to beg buy borrow & steal for me (do not tell Henrietta.) A⟨s⟩ it is I do not go further ⟨than⟩ little Medallions & such ⟨m⟩atters—such gorgeous things ⟨as⟩ you had on slates; are not ⟨fo⟩r the like of me; & as to the chimney pots on your chimney piece in the dining room they are not worth carriage.

We have hatched some leaf-insects of Java— Can you tell us what to feed them on?

Ever yours affec | J D Hooker

Send to Bolton & Barnett in Holborn bars, for a Quart bottle of the poison he supplies to the Kew Herbarium & wash **all** plants with that as soon as dry, & they will neither mould nor be devoured by beasts.[16] Use it with a great **big** Camel hair brush—not a tin mounted one—& just wash the specimens lightly over.

DAR 101: 88–91

[1] Falconer 1863a.

[2] The *Natural History Review*, in which Falconer 1863a appeared, was published in quarterly numbers.

[3] The 'Index to new genera described in the works enumerated' was listed on page 176 of the January 1863 number of the *Natural History Review*; the index entries corresponded to new botanical publications described in the last number of 1862. Thomas Henry Huxley was editor-in-chief of the new series of the *Natural History Review*. The series had been beset by problems, for which Hooker considered Huxley responsible (see *Correspondence* vol. 10, letter from J. D. Hooker, [24 July 1862] and n. 10, and L. Huxley ed. 1900, 1: 209–10).

[4] CD had told Hooker that he was 'burning with indignation' at the conduct of Richard Owen towards Hugh Falconer (see letter to J. D. Hooker, 3 January [1863] and n. 1).

[5] Hooker refers to Falconer 1863a, pp. 77–81; see preceding letter and n. 7.

[6] In his paper, Falconer praised CD, but also argued (Falconer 1863a, p. 80):

> the means which have been adduced to explain the origin of species by 'Natural Selection,' or a process of variation from external influences, is inadequate … it is difficult to believe, that there is not in nature, a deeper seated and innate principle, to the operation of which 'Natural Selection' is merely an adjunct.

On Hooker's view of the relationship between natural selection and variation in CD's theory, see also *Correspondence* vol. 10, letter from J. D. Hooker, 26 November 1862. See also this volume, letter from Hugh Falconer, 3 January [1863] and n. 16.

[7] Alexis Henri Charles Maurice Clérel, comte de Tocqueville, was a French politician and writer who travelled in the United States between 1831 and 1833, and in 1835 published his influential *De la démocratie en Amérique* (*NBU*). An English edition of *Democracy in America*, translated by Henry Reeve, was published in 1862 (H. Reeve trans. 1862). See also *Correspondence* vol. 10, letters from J. D. Hooker, 26 November 1862 and [21 December 1862].

[8] Tocqueville viewed the rapid growth of shipping in America as a strong element in the country's commercial prosperity (H. Reeve trans. 1862, 1: 505–15), and included statistics on the greater proportion of foreign to home traders in United States ports in relation to English ports (*ibid.*, p. 507 n. 3),

concluding that the Anglo-Americans were 'born to rule the seas, as the Romans were to conquer the world' (*ibid.*, p. 515).

9 In noting the growing interdependence of English manufacturing and American prosperity, Tocqueville wrote that, nevertheless, 'nothing can be more virulent than the hatred which exists between the Americans of the United States and the English' (H. Reeve trans. 1862, 1: 514).

10 See letter to J. D. Hooker, 3 January [1863].

11 J. D. Hooker 1864–7. George Bentham, Hooker's colleague at the Royal Botanic Gardens, Kew, was the author of a flora of Hong Kong (Bentham 1861).

12 Hooker had been in New Zealand in 1841 during the voyages of discovery of HMS *Erebus* and *Terror*; *Flora Novæ Zelandiæ* formed the second part of Hooker's botanical account of the voyage (J. D. Hooker 1853–5).

13 *Handbook of the New Zealand flora* (Hooker 1864–7) was published in two parts by Lovell Reeve & Co. of Covent Garden, London.

14 Henrietta Emma Darwin, CD's nineteen-year-old daughter, had commented that Hooker's remarks on collecting showed that it led to 'all sorts of vice' (see letter to J. D. Hooker, 3 January [1863]).

15 Hooker had started to collect Wedgwood pottery (see *Correspondence* vol. 10, letter from J. D. Hooker, [27 or 28 December 1862]). In his letter to Hooker of 3 January [1863], CD described himself and Emma Darwin, grandchildren of the master-potter Josiah Wedgwood I, as 'degenerate descendants', noting that they had 'not a bit of pretty ware in the house'.

16 Bolton & Barnitt were importers and makers of chemicals and photographic apparatus; their premises were located at 146 Holborn Bars, London (*Post Office London directory* 1861). In his letter to Hooker of 3 January [1863], CD asked for advice about using corrosive sublimate to prevent mould from growing on his children's dried flowers.

From John Lubbock 6 January [1863][1]

15, Lombard Street. E.C.
6 Jan. /62

My dear Mr. Darwin

It would require a series of surly letters before I should think that you meant anything, but in this case I had not noticed it, but am glad that you should have thought so as it produced your kind letter received this morning.[2]

I am very glad that you approve both of my articles & of the number generally, but we have not it seems to me enough variety.[3] Falconer's article is certainly very interesting & capitally written.[4] I wonder what you are thinking of doing about Owen?[5]

Do get better[6] & believe me | Yours affec | John Lubbock

DAR 170: 24

1 The year is established by the relationship between this letter and the letter to John Lubbock, 4 January [1863]; Lubbock wrote '/62' in error.

2 In his letter to Lubbock of 4 January [1863], CD apologised for writing 'surlily'.

3 Lubbock was one of the editors of the *Natural History Review*. The number for January 1863 contained two review articles by Lubbock ([Lubbock] 1863a and [Lubbock] 1863b), which CD praised in his letter to Lubbock of 4 January [1863].

4 Falconer 1863a was also published in the January 1863 number of the *Natural History Review*.

[5] See letter to John Lubbock, 4 January [1863]. For CD's comments on the conduct of Richard Owen towards Hugh Falconer and others, and his intended action, see the letter to J. D. Hooker, 3 January [1863].

[6] In his letter to Lubbock of 4 January [1863], CD wrote that he had been 'miserably uncomfortable'. See also letter to J. D. Hooker, 3 January [1863], and letter to Hugh Falconer, 5 [and 6] January [1863].

From John Scott 6 January 1863

Botanic Gardens, Edinburgh
January 6th 1863.

Sir,

I send off by train to-night a small box with three plants of P. Scotica and three of P. farinosa.[1] I am sorry that I cannot get any more for you at present, but if I can possibly do so in the course of the ensuing spring; you may depend upon receiving them! On one of the plants of P. Scotica, you will observe a few capsules. Perhaps they may not contain many perfect seeds being the produce of late autumn flowers, in cold, damp weather. ⟨You⟩ will, however, see by these withered and attached ⟨f⟩lowers ⟨ ⟩ dimorphic condition.[2] I have not had an ⟨*two lines destroyed*⟩ seen an approach to the dimorphic ⟨ ⟩. I have ⟨not had⟩ ⟨an opp⟩ortunity of examining it!

I also send in the box with ⟨ ⟩ flowers of the *Gongora atropurpurea*, which I think ⟨will⟩ interest you, unless you have already examined it. I believe your remarks on *Acropera luteola* are equally applicable to it⟨.⟩ I find the *genitalia* of *Gongora atropurpurea* very similar to those of *A. luteola* according to your description.[3] But all my attempts to fertilise *it* have as *yet* been unsuccessful. I have never succeeded in inserting a pollen-mass into the stigmatic cavity. It is decidedly more contracted than the majority of those I have examined of *A. Loddigesii.*[4] In the latter there are usually a few flowers sufficiently open to permit a partial insertion of a pollen-mass; and so it appears to be with *A. luteola* though in a still less degree; as I observe you had to allow them to shrink a little, and even then rarely succeeded.[5] I am now applying viscous matter from other *orchids* to stigmatic cavity of *G. atropurpurea*, and thus attaching its pollen-mass, in the hope of thus effecting its fertilisation. This will also aid in preventing the desiccation of the pollen-masses, which takes place ve⟨ry r⟩apidly in the hot, dry atmosphere where the plant is;[6] and thus favours the independent production and protrusion of the pollen-tubes. I may be wrong in entertaining this hope, yet from the success attending the experiments on *Acropera*, in one of which the *pollen mass*, scarcely—if at all—penetrated stigmatic cavity,[7] I am anxious to to give it a fair test. This opinion, is furthermore supported by the following experiment which seems to show ⟨that⟩ the *pollen tubes* may penetrate the column independent of the *stigmatic* ⟨ ⟩. Having lately observed Dr Hooker's experiments on the fertilisation ⟨of⟩ Meconopsis after the stigma had been cut off,[8] I ⟨*three words destroyed*⟩

Amongst ⟨others⟩ I have tried ⟨*three words destroyed*⟩ Max⟨illaria⟩ ⟨ ⟩stii; with the latter I have ⟨*three words destroyed*⟩ of the former, at present gives fair indications, ⟨ ⟩ ⟨I⟩ believe some of my experiments at your suggestion ⟨ ⟩ then by applying *pollen-mass* to the *rostellum*, now promise to be successful. This will be somewhat strange, in considering how much longer the period is before indications of its being effected are given by this mode in comparison with the normal.[9]

I have been making various *random experiments* of late on the *fertilisation* of *distinct genera*, for the purpose of elucidating in part, your view, "of sterility not being a special endowment':[10] Among many failures, I really believe I have succeeded in fertilising the *Stanhopea oculata*, with *Lælia anceps*.[11] I will state my grounds for this hope. There were *three flowers* experiment⟨ed⟩ upon one self-impregnated, the other two with pollen from *Lælia*. the other flowers on the raceme, I did not operate upon. Now in the latter as soon as the *perianth* drooped, the *ovary* and *labellum* quickly followed, first exhibiting a general flaccidity, then becoming dry and shrivelled. Those impregnated, on the other hand, exhibited no such signs, but are still fresh. The *ovaries*, are swelling, certainly very slowly. Nevertheless the *two hybrid products*, are still equally as *large* as the self-impregnated; and they were all experimented upon on the same day, viz; the *13th Dec*. Whether they ever reach maturity or not, they are evidently affected by the pollen of *Lælia*, otherwise they would have withered long ere this, as is invariably the case with the unfertilised ovaries of Stanhopea. Should they not reach maturity, I will if you wish transmit the ovaries for examination; otherwise—and I have much faith in its being so—I will try if I can possibly get the seeds to germinate; as they certainly would be most anomalous products. By the way, in talking of hybridisation, I will have the so-called *Bryanthus erectus* in flower shortly, and I intend testing its capabilities for fertilisation. I am aware of Mr. Anderson's failures, but there is no harm in again attempting it.[12] I will also *re-cross* its *parents*, and see if I can succeed in raising anything from them. Perhaps I may thus out which was the *male* and which the *female*, of the *Bryanthus*, as I suppose it is not known, and the raiser is now long dead.[13] I will be glad if you suggest to me any experiments on these points, that you think would be of interest.

The seed capsule of *Acropera* is now pretty large—2¾ inches long, by 1½ inches in circumference.[14] I am sorry that those for comparison should be so far behind it. I expected to have a few capsules of a *Maxillaria* for comparison, but not one of them set; and so it was with the *Oncidium ornithorynchum*, although I impregnated many flowers of the latter, not a *single capsule* set. I have now *succeeded* in *fertilising* the latter, however, by applying pollen from *Oncidium sphacelatum*.[15] I do not lay much *stress* on this experiment *yet*, as my former failures may be owing to uncongenial conditions; but, as further opportunities present themselves, I will carefully attend to them. This far, however, I may safely say, that we have at least an instance, of *equal*—if not higher *susceptibility* to *others* than *own pollen*. I am determined to devote in future a little more attention to the Orchids. I think curious results may be derived from them.[16] I am only sorry that I have not a better collection to experiment upon; for we are indeed poor, and not only in them, but in the majority of the *florists* favourites. We are very differently situate here in this respect than Kew. Thus first to take the few things you have directed my attention to, I have only at present the Primulas.[17] I am now, however, promised from a friend, plants of peloric *Antirrhinums* and *Columbines*,[18] and likewise Verbascums—one species at least with its varieties, perhaps I may yet get others.[19] Do you know anything of the two varieties of *Maize*, Gartner experimented upon? If I asked you in my last this question about the *Verbascum*, it was the *Maize* I intended.[20]

Had I remembered the *Begonia frigida*, I should have asked Mr. McNab,[21] before he visited Kew last to have seen if they could have spared a plant for the garden here, but now it is too late for this season. I am sorry as it would be an interesting subject for experiment, and I really feel shy in my circumstances to ask Mr. McNab specially to see whether they could afford us a plant of it or not. And here I may take the opportunity of stating in reply to your kind enquiries into my personal circumstances,[22] that in consequence of certain loses in business sustained by my parents, we were thus brought down from easy circumstances, to a comparatively humble position.[23] My parents died while I was very young, leaving me in charge of a relative. After receiving an ordinary education, I became a gardener[24]—more for the purpose of gratifying a predilection for Natural History—than any love for this line of life. But I thus thought that I might have better opportunities for pursuing these branches of Science, than by any other I had it in my power to engage in. And, so far, I am happy to say—I have not been disappointed, and I now through the kindness of Prof. Balfour,[25] and Mr. McNab, enjoy great facilities for such pursuits. At present, I have the charge of the propagating department in the Botanic Gardens here, and I have thus excellent opportunities for performing experiments, having a sufficient range of temperature for all ordinary purposes. I may state that I was all but engaged a few months back by Prof. Balfour, to go to Madras, as Superintendent of the Horticultural Gardens there. I was ultimately disappointed of this situation, however, in consequence of the re-engagement of the then Superintendent.[26] It is therefore hard to say what part of the world I may next be sent to, but time will show;[27] And then, perhaps you will kindly favour me with a few suggestions hints, and advices, on observations that I may then have it in my power to make. And now by giving you excuse for this detailed account of personal affairs I will only ask you in conclusion to excuse my non-reference to one or two points in your last, which I will do at an early opportunity.

I remain | Sir, | Yours very respectfully | John Scott.

I am unfortunately perfectly ignorant of the *staminal* conditions of our Melastomads.[28] we have indeed very few of them in the Gardens here. But if the few that we have present any differences, I will attend to them when they again flower.

J. S.

The Primulas being all very weak it would not be advisable to plant them in the garden at present—they had better be kept in pots till spring. It is difficult to get strong plants of *Scotica*, *especially* it does not thrive long under cultivation.[29] I am sorry that I had no stronger to send you at present. Such as they are you will please to accept in slight acknowledgement of your kindness to me.

J. Scott

DAR 177: 81, 83

CD ANNOTATIONS
1.1 I send . . . examining it! 1.9] *crossed pencil*
2.4 But all . . . *A. Loddigesii*. 2.7] *double scored red crayon*
2.11 applying viscous . . . this hope. 4.5] *crossed blue crayon*
5.7 as my former . . . *own pollen*. 5.10] *scored pencil*
Top of letter: 'Primula | Gongora | *Pollen tubes | *[*Laelia*]* [*left brace added*] | Generic crosses | Bryanthus | Acropera | Oncidium | *Maize I know | nothing [*left brace added*] | Personal details | Melastomas | Sports [*circled*]' *pencil del pencil*

CD note:[30]
Gongora atropurpurea. Jan 8th 63. Bot. Garden Edinburgh. Labellum astonishing—elongated whenever arched over— viscid [*added*] disc much larger than in Acropera, to which evidently closely allied.— No movement of depression— *Mouth of stigma [*underl red crayon*] contracted (see J. Scott)—but viscid within in ½ ⟨ro⟩tten flower— In *a fresh [*underl red crayon*] a quantity of this *viscid fluid exuding [*underl red crayon*] from mouth ⟨ ⟩ stigma— Ovarium very *[*thin*]* ⟨ ⟩ put in spirits to examine ovarium & utriculi— Gongora [*red crayon, circled red crayon*] | *He cut open at *[*illeg*]* & *[*fertilised it*]* [*red crayon*]

1 In his letter of 17 December [1862] (*Correspondence* vol. 10), Scott promised to send CD specimens of *Primula farinosa* and *P. scotica* (bird's-eye primrose and Scottish primrose); CD had been trying to acquire specimens of *P. farinosa* since 1861, as part of his examination of dimorphism in *Primula* (see *Correspondence* vol. 9, letter to Daniel Oliver, 11 September [1861], and *Correspondence* vol. 10, letter from Daniel Oliver, 10 April 1862).

2 See *Correspondence* vol. 10, letter from John Scott, 6 December [1862]. *Primula scotica* is not dimorphic; in a paper detailing his work on *Primula*, communicated by CD to the Linnean Society on 5 February 1864, Scott described the native plant, *P. scotica*, as 'an indigenous illustration of the "non-dimorphic" structure' in which any variability of stamen length was matched by the style length (Scott 1864a, p. 82).

3 CD described *Acropera* as the 'opprobrium' of his work on orchids, as all the parts seemed 'determinately contrived' so that the plant should never be pollinated (*Orchids*, p. 203); he inferred that *A. luteola* was dioecious, and that the specimen he examined was male (*Orchids*, pp. 208–9). In his letter of 11 November 1862 (*Correspondence* vol. 10), Scott informed CD of his own experiments on a closely allied plant, *A. loddigesii*; after achieving pollination in two out of six flowers of a raceme, Scott suggested that these species were those in which some of the flowers were male, but others were 'truly hermaphrodite'. When CD thanked Scott for his letter, he wrote: 'Your fact has surprised me greatly, & has alarmed me not a little' (see *Correspondence* vol. 10, letter to John Scott, 12 November [1862]). Scott subsequently offered to send the ovary of *A. loddigesii* to CD and to continue experimenting on *Acropera* (see *Correspondence* vol. 10, letter from John Scott, 15 November [1862]). See also *Correspondence* vol. 10, letter to John Scott, 19 November [1862], letter from John Scott, [20 November – 2 December 1862], and n. 4, below. See CD note, above, in which he records his observations of the *Gongora* specimen sent by Scott.

4 Scott had successfully pollinated *Acropera* where CD had failed (see *Orchids*, pp. 205–6); by inserting only a portion of a stipe with the pollinia attached, Scott found that the pollinia then adhered to the mouth of the stigmatic chamber (see *Correspondence* vol. 10, letter from John Scott, 11 November 1862). Scott's attempts to pollinate *Gongora atropurpurea* were unsuccessful. See 'Fertilization of orchids', p. 153

(*Collected papers* 2: 150); see also *Correspondence* vol. 9, letter to J. D. Hooker, 18 [December 1861]. *Acropera loddigesii* is now synonymous with *G. galeata* (Bailey and Bailey 1976).

5 See *Orchids*, pp. 205–6.

6 *Gongora atropurpurea* is native to the tropical Americas (Bailey and Bailey 1976). Scott and CD had discussed the importance to successful pollination of viscid matter on the mouth of the stigmatic cavity (see *Correspondence* vol. 10, letter from John Scott, 11 November 1862, and letters to John Scott, 12 November [1862] and 19 November [1862]).

7 In his letter of 11 November 1862 (*Correspondence* vol. 10), Scott described his success in pollinating *Acropera loddigesii* (see also n. 4, above).

8 Scott refers to J. D. Hooker 1854a, in which Hooker described pollinating the ovules directly after cutting off the stigma in several plants of the Papaveraceae (poppies), including *Meconopsis*. See letter from John Scott, [1–11] April [1863].

9 In his letter to John Scott, 3 December [1862] (*Correspondence* vol. 10), CD suggested placing a pollen mass directly on the rostellum (the modified stigma of one of the three carpels); under natural conditions, the pollen enters the stigma of one of the two pistils. See also *ibid.*, letter to Asa Gray, 26[–7] November [1862].

10 Scott refers to chapter 8 of *Origin* (pp. 245–78), in which CD argued against the view that the function of sterility was to prevent the 'confusion of all organic forms'. In *Origin*, p. 245, CD expressed his hope to show that sterility was not a 'specially acquired or endowed quality', but was 'incidental on other acquired differences'. See also *Correspondence* vol. 10, Appendix VI.

11 See *Correspondence* vol. 10, letter to John Scott, 19 December [1862], for CD's interest in the 'relation of well-marked, but undoubted varieties in fertilising each other'. He encouraged Scott to continue this type of experiment, of which the cross of *Stanhopea* and *Laelia* was an example. See also *ibid.*, letters from John Scott, 6 December [1862] and 17 December [1862].

12 In his letter of 26–7 January 1863, Isaac Anderson-Henry described a successful cross that yielded the hybrid *Bryanthus erectus*. Scott later reported that, having crossed *Rhododendron chamaecistus* and *Menziesia empetrifolia*, both members of the Ericaceae (heaths), to produce *B. erectus*, he 'repeatedly failed' in setting seed with its own pollen, or the pollen of either parent (Scott 1864c, pp. 199–200 n.).

13 The Edinburgh nurseryman James Cunningham, who died in 1851, was the first to cross *Phyllodoce* (*Menziesia*) *caerulea* and *Rhodothamnus chamaecistus* to produce the hybrid *Bryanthus erectus* (*Gardeners' Chronicle*, 1 November 1851, p. 695).

14 Scott had promised to send CD the *Acropera* seed capsule when fully ripened for dissection, along with the seed capsule of another orchid species for comparison (see *Correspondence* vol. 10, letters from John Scott, 15 November [1862] and [20 November – 2 December 1862]). See also n. 3, above.

15 Scott described this and further experiments with the two *Oncidium* orchid species in Scott 1863a, pp. 547–8, and Scott 1864b.

16 In his letter to Scott of 19 December [1862] (*Correspondence* vol. 10), CD had suggested that Scott might undertake crossing experiments with a variety of species. In pollinating plants with their own pollen and with pollen from other varieties and species, Scott, like CD, hoped to answer Thomas Henry Huxley's objections to natural selection (T. H. Huxley 1863b, p. 107). See letter from John Scott, 21 May [1863] and nn. 10–11. See also Scott 1864b, and *Correspondence* vol.10, Appendix VI. There are copies of Scott 1863a and 1863b in the Darwin Pamphlet Collection–CUL.

17 With encouragement from CD (see *Correspondence* vol. 10, letter to John Scott, 3 December [1862]), Scott later published a paper on dimorphism in *Primula* (Scott 1864a). There is a copy of Scott 1864a in the Darwin Pamphlet Collection–CUL.

18 In his letter of 19 December [1862] (see *Correspondence* vol. 10), CD suggested that Scott experiment with peloric flowers.

19 In his letters to Scott of 11 December [1862] and 19 December [1862] (*Correspondence* vol. 10), CD suggested that Scott experiment with maize and *Verbascum* (see n. 20, below). Scott later published a paper on several species and varieties of *Verbascum* (Scott 1867), as well as a paper on *Zea mays* (Scott 1863b).

[20] See *Correspondence* vol. 10, letter from John Scott, 17 December [1862]. In his letters to Scott of 11 December [1862] and 19 December [1862] (*ibid.*), CD suggested that Scott repeat the experiments conducted by Karl Friedrich von Gärtner on the fertility of varieties of *Verbascum* and maize (Gärtner 1844 and 1849), and referred him to his abstract of Gärtner's work on these plants in *Origin*, pp. 269–71. The 'two varieties of maize' referred to by Scott were a dwarf variety with yellow seeds and a tall variety with red seeds. When these varieties were artificially crossed by Gärtner, only a few seeds were produced. However, the hybrid plants raised from this seed were '*perfectly* fertile'; therefore, as CD pointed out, these two varieties could not be considered 'specifically distinct' (*Origin*, p. 270).

[21] James McNab was curator of the Royal Botanic Garden, Edinburgh (R. Desmond 1994). CD was interested in the fertility of abnormal flowers like *Begonia frigida*, which produced male and female flowers on the same inflorescence. He had acquired a specimen from the Royal Botanic Gardens, Kew (see *Correspondence* vol. 10, letter to J. D. Hooker, 12 [December 1862], and letter from J. D. Hooker, [14 December 1862]). In his letter to Scott of 19 December [1862] (*Correspondence* vol. 10), CD mentioned that his *Begonia* plant was unlikely to set its seeds, and so suggested that Scott investigate the fertility of 'any monster flower', particularly *Celosia.*

[22] In his letter to Scott of 19 December [1862] (*Correspondence* vol. 10), CD offered 'pecuniary assistance' with purchases for any experiments Scott conducted at CD's request.

[23] An obituary in the *Transactions of the Botanical Society* [*of Edinburgh*] 14 (1883): 160–1, states that Scott's father was a tenant farmer, his family occupying the same farm for a century.

[24] Scott became an orphan at the age of 4. He was brought up by an aunt, and educated at a parish school until he became an apprentice gardener at the age of 14 (*Transactions of the Botanical Society* [*of Edinburgh*] 14 (1883): 160).

[25] John Hutton Balfour was regius keeper of the Royal Botanic Garden, Edinburgh (*DNB*).

[26] At a meeting of the committee of the Agri-horticultural Society of Madras on 12 March 1862, it was decided to re-engage Robert N. Brown as superintendent of its gardens for a period of three years (Shaw 1877, p. 46).

[27] In 1864, Scott became head of the herbarium at the botanic garden in Calcutta (*Transactions of the Botanical Society* [*of Edinburgh*] 14 (1883): 160–1).

[28] CD was interested in dimorphism in Melastomataceae (see letter to Hugh Falconer, 5 [and 6] January [1863] and n. 22). In his letter to John Scott, 11 December [1862] (*Correspondence* vol. 10), CD wrote that despite his hard work at seeking the meaning of the 'two sets of very different stamens', he was 'shamefully beaten'.

[29] See n. 1, above; the specimens sent to CD by Scott did not thrive (see letter to Daniel Oliver, 20 [February 1863], and letter to John Scott, 24 March [1863]).

[30] The note is with Scott's letter in DAR 177: 81. See nn. 3 and 4, above.

To James Dwight Dana 7 January [1863][1]

Down Bromley Kent
Jan 7th

My dear Prof. Dana

I was most truly rejoiced to hear by your letter of Dec. 4th. that your health is considerably reestablished & that you are at work on Science again.—[2] From one to three hours per day must be a great change to you; but for me during many years three hours has been a most unusually hard days work. I hope to God that your health will steadily, though slowly must be expected, improve.— I have received the printed Corrigenda; but I am sorry to say that your Manual has not arrived.[3] I wrote to Geolog. Soc & it has not there arrived, for the Society as I heard this morning.[4] I enclose a Photograph at your request; it was made by my eldest son & is the only one which I have.[5] One almost too large for Post has been made in London.—[6]

My health of late has been very indifferent, & I have not seen one man of Science for months; so I really have no news. Man is our great subject at present; & Lyell has been working very hard & I cannot conceive why his Book has not appeared.[7] Murray on day of sale disposed of 4000 Copies!![8] The fossil Bird with the long tail & fingers to its wings (I hear from Falconer that Owen has not done the work well) is by far the greatest prodigy of recent times.[9] It is a grand case for me; as no group was so isolated as Birds; & it shows how little we know what lived during former times. Oh how I wish a skeleton could be found in your so-called Red Sandstone footstep-beds.—[10]

I am not at all surprised that you have not read the "Origin";[11] all my friends say it takes much thought (which rather surprises me) & most have had to read it two or three times.— I am at present at work on dry facts & dry Bones, preparing a work to be entitled "Variation under Domestication".—[12] Mr Esquereux has sent me some Books; I do not know his address; will you be so kind as to address & stamp the enclosed letter.—[13]

With cordial good wishes, believe me, Your's very sincerely | Ch. Darwin

Dana Family Papers, Manuscripts and Archives, Yale University Library

1 The year is established by the relationship between this letter and the letter from J. D. Dana, 4 December 1862 (*Correspondence* vol. 10); see n. 2, below.

2 See letter from J. D. Dana, 4 December 1862 (*Correspondence* vol. 10). In 1851, Dana, who was professor of natural history at Yale University, suffered a nervous breakdown from which he never fully recovered (*DAB*).

3 In his letter of 4 December 1862 (*Correspondence* vol. 10), Dana expressed the hope that CD had already received a copy of his *Manual of geology* (Dana 1863a). However, the book did not arrive until February (see letter to J. D. Dana, 20 February [1863]). There is a copy of Dana 1863a in the Darwin Library–Down House.

4 The letters have not been found.

5 William Erasmus Darwin photographed CD in April 1861 (see *Correspondence* vol. 9, letter to Asa Gray, 11 April [1861] and n. 19); the photograph is reproduced as the frontispiece to *Correspondence* vol. 9.

6 CD may refer to the photograph produced by Maull & Polyblank *circa* 1857; the photograph is reproduced from a later print by Maull and Fox as the frontispiece to *Correspondence* vol. 8. See also *Correspondence* vol. 10, letter from E. A. Darwin, [April–May? 1862].

7 Charles Lyell's *Antiquity of man* (C. Lyell 1863a) was published on 9 February 1863 (*Athenæum*, 7 February 1863, p. 176). CD may also refer to Thomas Henry Huxley's book, *Evidence as to man's place in nature* (T. H. Huxley 1863b), which was also published in February 1863 (*Publishers' Circular* 26 (1863): 112).

8 CD refers to the trade sale held on 4 November 1862 by the publisher John Murray (*Athenæum*, 8 November 1862, p. 595).

9 CD refers to the discovery of the fossil bird *Archaeopteryx*; Hugh Falconer dismissed Richard Owen's account of the specimen before the Royal Society of London (Owen 1862a) as 'slip-shod and hasty' (see letter from Hugh Falconer, 3 January [1863] and nn. 7–14).

10 The Triassic sandstones of the Connecticut River Valley were renowned for the large bird-like footprints imprinted in the rock (Hitchcock 1836).

[11] See letter from J. D. Dana, 4 December 1862 and n. 2. CD sent Dana a presentation copy of *Origin* in 1859 (see *Correspondence* vol. 7, letter to J. D. Dana, 11 November [1859], and *Correspondence* vol. 8, Appendix III).

[12] In January 1860, CD had begun work on *Variation*, the first part of a projected three-part work in which he intended to give the facts and authorities on which he had based the theory of natural selection (*Correspondence* vol. 8). It was published in 1868, with the full title *The variation of animals and plants under domestication.*

[13] The reference is to Leo Lesquereux, the Swiss-born palaeobotanist and bryologist, who settled in the United States (*DSB*). The letter to Lesquereux has not been found and the books referred to have not been identified. However, in his letter of 14 December 1864 (*Correspondence* vol. 12), Lesquereux thanked CD for a letter written 'about one year ago (7th Jany)', in which CD had acknowledged 'the communication of some Geological papers'. There are a number of Lesquereux's papers in the Darwin Pamphlet Collection–CUL; the reference may be to the part of Lesquereux 1859–63 preserved in the collection.

To Thomas Rivers 7 January [1863][1]

Down. | Bromley. | Kent. S.E.
Jan 7th

My dear Sir

I thank you much for your letter & the parcel of shoots.[2] The case of the Yellow Plum is a treasure & is now safely recorded on your authority in its proper place;[3] in contrast with A. Knight's case of yellow magnum bonum sporting into red.[4] I could see no difference in the shoots, except that those of the Yellow were thicker, & I presume that this is merely accidental: as you do not mention it, I further presume that there are no differences in leaves or flowers of the two Plums.—

I am very glad to hear about the Yellow ash,[5] & that you yourself have seen the Jessamine case;[6] I must confess that I hardly fully believed in it; but now I do; & very surprising it is.

In an old French Book, published in Amsterdam in 1786 (I think) there is an account apparently authentic & attested by the writer as an eye-witness, of Hyacinth bulbs of two colours being cut in two & grafted & they sent up single stalk with differently coloured flowers on the two sides, & some flowers parti-coloured.[7] I once thought of offering 5£ reward in Cottage Gardener for such a plant;[8] but perhaps it would seem too foolish: no instructions are given when to perform the operation; I have tried 2 or 3 times & utterly failed.— I find that I have a grand list of "bud-variations", & tomorrow shall work up such cases as I have about Roses—sports which seem very numerous, & which I see you state to occur comparatively frequently.[9]

When a person is very good-natured, he gets much pestered,—a discovery which I daresay you have made, or anyhow will soon make; for I do want very much to know, whether you have sown seed of any Moss Roses, & whether the seedlings were moss-roses.— Has a common Rose produced by *seed* a moss-rose?[10]

What can be the origin of the Austrian Bramble, which seems always to have imperfect pollen (at least I have found it so) & which sport into a yellow rose:[11] may not this be case like Laburnum?—[12]

If any light comes to you about very slight changes in the buds, pray have kindness to illuminate me: I have cases of 7 or 8 varieties of the Peach which have produced by "bud-variation" Nectarines;[13] & yet only one single case (in France) of a Peach producing another closely similar peach (but later in ripening).[14] How strange it is that a great change in the Peach should occur not rarely & *slighter* changes apparently very rarely! How strange that no case seems recorded of new Apples or Pears or Apricots by "bud-variation"! How ignorant we are! But with the many good observers now living our children's children will be less ignorant, & that is a comfort. I am ashamed of myself to be so troublesome, & am grateful. (*not* for favour to come)—

My dear Sir | Yours sincerely | Ch. Darwin

DAR 185: 81

[1] The year is established by the relationship between this letter and the letter to Thomas Rivers, 11 January [1863].

[2] The letter from Rivers, presumably a reply to CD's letter to Rivers of 28 December [1862] (*Correspondence* vol. 10), has not been found. The shoots Rivers sent were apparently from the Early Prolific plum tree that produced fruits of two different colours (see nn. 3 and 4, below).

[3] CD recorded Rivers's information in the chapter of *Variation* on bud-variation, or 'all those sudden changes in structure or appearance which occasionally occur in full-grown plants in their flower-buds or leaf-buds' (*Variation* 1: 373). CD wrote (p. 375) that in January 1863, Rivers had informed him that:

> a single tree out of 400 or 500 trees of the Early Prolific plum, which is a purple kind, descended from an old French variety bearing purple fruit, produced when about ten years old bright yellow plums; these differed in no respect except colour from those on the other trees, but were unlike any other known kind of yellow plum.

See also *Gardeners' Chronicle*, 10 January 1863, p. 27, where Rivers reiterated his observation; CD's annotated copy of this issue is in the Cory Library, Cambridge Botanic Garden.

[4] In *Variation*, CD compared Rivers's account with Thomas Andrew Knight's observation that 'a tree of the yellow magnum bonum plum, forty years old, which had always borne ordinary fruit, produced a branch which yielded red magnum bonums' (*Variation* 1: 375). See also Knight 1815.

[5] In his letter to Rivers of 28 December [1862] (*Correspondence* vol. 10), CD inquired about cases of a tree stock being affected by grafts. The letter from Rivers that included this information has not been found. However, the yellow ash case is described in the chapter on bud-variation in *Variation* 1: 394:

> Mr. Rivers, on the authority of a trustworthy friend, states that some buds of a golden-variegated ash, which were inserted into common ashes, all died except one; but the ash-stocks were affected, and produced, both above and below the points of insertion of the plates of bark bearing the dead buds, shoots which bore variegated leaves.

[6] In *Variation* 1: 394, CD wrote: 'It is notorious that when the variegated Jessamine is budded on the common kind, the stock sometimes produces buds bearing variegated leaves; Mr. Rivers, as he informs me, has seen instances of this'. CD asked about the variegated Jessamine in his letter to Rivers of 28 December [1862] (*Correspondence* vol. 10).

[7] Saint-Simon 1768, p. 124; CD cited this work in *Variation* 1: 395.

[8] CD refers to the *Journal of Horticulture, Cottage Gardener, and Country Gentleman.*

[9] No letter from Rivers including this information has been found. CD may be referring to Rivers 1837; Rivers is mentioned in CD's discussion of bud-variation in roses as having 'informed' CD about particular cases (*Variation* 1: 379–81). CD was interested in bud-variations and 'certain anomalous modes of reproduction and variation'; he noted that gardeners 'call such changes "Sports"' (*Variation* 1: 373).

[10] No reply by Rivers to this query has been found. However, in *Variation* 1: 380, CD discussed the relationship between various moss-roses and the Provence rose, twice citing Rivers. See the letter to Thomas Rivers, 11 January [1863] and n. 4.

[11] The Austrian bramble (*Rosa lutea*) is mentioned in *Variation* 1: 381. CD's 'imperfect pollen' was pollen that did not germinate.

[12] The flowers of the Adam's laburnum (*Cytisus adami*), a sterile hybrid, were known to revert spontaneously to those of its fertile parent species, *C. laburnum* and *C. purpureus* (see *Variation* 1: 387–91); the originator of *C. adami* claimed that it had been raised not by normal hybridisation, but by grafting a bud of *C. purpureus* onto a stock of *C. laburnum*. If this was the case, CD wrote in *Variation* 1: 390, 'we must admit the extraordinary fact that two distinct species can unite by their cellular tissue, and subsequently produce a plant bearing leaves and sterile flowers intermediate in character between the scion and stock, and producing buds liable to reversion'. CD discussed the case in his letter to Rivers of 28 December [1862] (*Correspondence* vol. 10).

[13] See *Variation* 1: 340–1 and 374–5. In his letter to Rivers of 23 December [1862] (*Correspondence* vol. 10), CD cited the example of a nectarine appearing on a peach tree to illustrate what he meant by a 'bud-variation'.

[14] See *Variation* 1: 375.

From Hugh Falconer 8 January [1863][1]

21 Park Crescent N.W.
8th. Jany.

My Dear Darwin,

I am charmed with the heartiness of your indignation with 'Dirty Dick'—and your approval of the dose which I have administered to him.[2] It is after all but part of the indictment—and I have let him off easy this time. But it is really apalling to see such disregard of fact—and contempt of the consequences of being shown up—in a man of Science.[3] He has now got hold of a silly empty-headed young man—Carter Blake[4]—to work upon—for his dirty work—and he will be worse than ever. But we must drop him—or my note will be spoilt.

Many thanks for all the kind things you have been pleased to say about the Elephant paper.[5] Is it not an odd thing—that there should not have been found a single Edentate—big or little either in this Pliocene or miocene Fauna of Niobrara and Nebraska;[6] and fancy a *Rhinoceros*, said to be indistinguishable nearly from the Indian Unicorn being found, with an *Indianoid* Elephant & Mastodon, in the Niobrara Pliocene! the Mastodon is described by Leidy as being exactly like *M. Sivalensis*![7]

I am very sorry to hear of the aggravated form of this *Eczema* and of the facility with which you are knocked up.[8] But the change of a day or two to London might be useful, in interrupting the established routine of your sufferings. If you do not come up, I must run down for a day to see you.[9]

Your 'dimorphic' observations are of surpassing interest—and great importance.[10] You will really inaugurate a new era of observation, in Botany. Instead of fiddling about the varieties or species of *Rubus* & Rosa &c—the Country Botanists—an honest & intelligent class—will find, that there is a new avenue of observation opened to them, which will yield them a real reward—and at the same time advance Science. What you mention to me, of your present puzzle, is of great interest—but I am sorry to say that I cannot contribute a fact to you, in reference to what you tell me of your observation on the *Melastomas*.[11] When the mind has been unstrung for 8 or 10 years, from Botanical observations many facts—pass out of recollection[12]

With kindest regards | My Dear Darwin | Yours very Sin[ly] | H Falconer

DAR 164: 11

[1] The year is established by the relationship between this letter and the letter to Hugh Falconer, 5 [and 6] January [1863].

[2] In the letter to Hugh Falconer, 5 [and 6] January [1863], CD condemned Richard Owen for his 'lies and mean conduct' over the priority dispute regarding *Elephas columbi* (see letter to J. D. Hooker, 3 January [1863] and n. 1). He also commended Falconer's 'unruffled' response in Falconer 1863a, pp. 46–9.

[3] Falconer had named the specimen *Elephas columbi*, and argued that Owen's attribution of the name *E. texianus* disregarded the evidence that the fossil species had ranged from Georgia, in the southern United States, to Mexico. Falconer also felt that Owen and his protégé, Charles Carter Blake, had inadequately and imperfectly described the dental characteristics of the specimen (see Falconer 1863a, p. 49). See also letter to J. D. Hooker, 3 January [1863] and nn. 1–10.

[4] Blake, a young colleague of Owen's, was involved in the dispute between Owen and Falconer regarding *Elephas columbi* (see letter to J. D. Hooker, 3 January [1863] and n. 1). See also Blake 1862.

[5] See letter to Hugh Falconer, 5 [and 6] January [1863]; the reference is to Falconer 1863a.

[6] Falconer contrasts the absence of examples of the order Edentata (sloths, anteaters, armadillos) in the older Pliocene and the Miocene of the Niobrara and Nebraska deposits, with the abundance of large Edentata in younger Pliocene fauna of both North and South America. In his letter to Hugh Falconer of 5 [and 6] January [1863], CD had been 'particularly struck' by Falconer's remarks in his elephant paper (Falconer 1863a, pp. 60–3) on the ancient mammalian fauna of North America, and their implications for CD's theory of trans-tropical migration. See n. 7, below.

[7] Falconer refers to Leidy 1858, pp. 28–9. Joseph Leidy, a North American palaeontologist, described two lower jaws from the Niobrara deposits that resembled the existing Indian rhinoceros; the one-horned rhinoceros of Asia had previously been described as a 'unicorn' (see, for example, W. Haughton 1862, p. 365). Leidy wrote that the tooth of the new species of mastodon from the Niobrara resembled 'the corresponding one of *M. angustidens*, of Europe, or of *M. sivalensis*, of the Sivalic Hills, of India' (Leidy 1858, p. 28). See also *Proceedings of the Academy of Natural Sciences of Philadelphia* (1858): 10. Leidy 1858 was summarised in the January 1863 number of *Annals and Magazine of Natural History* (1863): 148–50; CD's note on this summary is in DAR 205.9: 367.

[8] See letter to Hugh Falconer, 5 [and 6] January [1863].

[9] On 4 February 1863, CD went to London for ten days (see 'Journal' (Appendix II)).

[10] In his letter to Falconer of 5 [and 6] January [1863], CD mentioned that he had sent his paper on dimorphism in *Linum* ('Two forms in species of *Linum*') to the Linnean Society; the paper was read before the society on 5 February 1863 and published in the *Journal of the Proceedings of the Linnean Society (Botany)* 7 (1863): 69–83. Falconer also refers to CD's paper 'Dimorphic condition in *Primula*', which was published in March 1862.

[11] CD was investigating the occurrence of two different sets of stamens in flowers of Melastomataceae, and suspected that members of this family might exhibit a novel form of dimorphism (see letter to Hugh Falconer, 5 [and 6] January [1863] and n. 22).

[12] Falconer had been superintendent of the Calcutta Botanic Garden and professor of botany at the Calcutta Medical College from 1848 to 1855 (*DNB*).

From Hugh Falconer 9 January 1863

21 Park Crescent.
9th. Jany/63

My Dear Darwin.

In my reply to your last note, I think I omitted to reply to your query about the free digits of Solenhofen Bird Creature.[1] I have not yet examined it with the close scrutiny the case requires—nor can one well expect to do that until Owens paper is out: but the free digits are the two next the *Axis* i.e. the Radial digits: the ulnar one and outer having borne the big quill feathers.[2] But mind this is only approximative.

Should you be writing at any time to Wallace I wish you would keep me in mind—for a further enquiry of him—regarding the Timor Mastodon remains[3] Has he ever seen them? Where are they? or where can they be seen or hear about—or described?

Yours Ever Sinly | H Falconer

DAR 164: 12

[1] Falconer refers to *Archaeopteryx*, the Jurassic fossil recently discovered in the Solenhofen quarries of Bavaria (see letters from Hugh Falconer, 3 January [1863] and 8 January [1863]). In his letter to Falconer of 5 [and 6] January [1863], CD asked which digits on the forelimb had split from the wing and developed as claws.

[2] Richard Owen described *Archaeopteryx* in a paper read before the Royal Society of London on 20 November 1862; the paper was published in part 1 (1863) of the society's *Philosophical Transactions* (Owen 1862a). For Owen's description and illustrations of the two free digits, see pp. 38–9, plate 1 (fig. 1, 1 and 2) and plate 2 (fig. 1, 1 and 2) (Owen 1862a). There is an annotated copy of the number of the *Philosophical Transactions* containing Owen 1862a in the Darwin Library–CUL; CD would not have received this issue until mid-1863 (see letter from Hugh Falconer, 3 January [1863], n. 7).

[3] In his letter to Falconer of 5 [and 6] January [1863], CD reported Alfred Russel Wallace's information that 'abundant Mastodon remains' had been found at Timor in the Malay Archipelago (see *Correspondence* vol. 9, letter from A. R. Wallace, 30 November 1861). Falconer was interested in information about potential Timor specimens in connection with his argument against the existence of a fossil elephant in nearby Australia (see *Correspondence* vol. 10, letter to Hugh Falconer, 14 November [1862], and this volume, letter to Hugh Falconer, 5 [and 6] January [1863] and n. 5). See also Falconer 1863a, pp. 96–101.

To Thomas Henry Huxley 10 [January 1863][1]

Down Bromley Kent
Dec. 10th

My dear Huxley

You will be weary of notes from me about the little book of yours.[2] It is lucky for me that I expressed, before reading no VI, my opinion of its absolute excellence, & of its being well worth wide distribution & worth correction (not that I see where you could improve), if you thought it worth your valuable time. Had I read no VI, even a rudiment of modesty would, or ought to, have stopped me saying so much.[3] Though I have been well abused, yet I have had so much praise, that I have become a gourmand, both as to capacity & taste; & I really did not think that mortal man could have tickled my palate in the exquisite manner with which you have done the job.[4] So I am an old ass, & nothing more need be said about this.— I agree entirely with all your reservations about accepting the doctrine,[5] & you might have gone further with further safety & truth. Of course I do not **wholly** agree about sterility.[6] I hate beyong all things finding myself in disagreement with any capable judge, when the premises are the Same; & yet this will occasionally happen.

Thinking over my former letter to you,[7] I fancied (but I now doubt) that I had partly found out cause of our disagreement, & I attributed it to your naturally thinking most about animals, with which the sterility of the hybrids is much more conspicuous than the lessened fertility of the first cross.[8] Indeed this could hardly be ascertained with mammals, except by comparing the product of whole life; & as far as I know this has only been ascertained in case of Horse & ass, which do produce fewer offspring in lifetime than in pure breeding. In plants the test of first cross seems as fair, as test of sterility of hybrids. And this latter test applies, I will maintain to the death, to the crossing of vars. of Verbascum & vars, selected vars, of Zea.—[9] You will say go to the Devil & hold your tongue.— No I will not hold my tongue; for I must add that after going for my present book all through domestic animals;[10] I have come to conclusion that there are almost certainly several cases of 2 or 3 or more species blended together & now perfectly fertile together. Hence I conclude that there must be something in domestication,—perhaps the less stable conditions,—the very cause which induces so much variability,—, which eliminates the natural sterility of species, when crossed.[11] If so, we can see how unlikely that sterility should arise between domestic races. Now I will hold my tongue.—

p 143. ought not "Sanscrit" to be "Aryan"?[12]

What a capital number the last N.H. Review is.[13] That is a grand paper by Falconer.[14] I cannot say how indignant Owen's conduct about E. Columbi has made me.[15] I believe I hate him more than you do,[16] even perhaps more than good old Falconer does. But I have bubbled over to one or two correspondents on this head, & will say no more.—[17] I have sent Lubbock a little Review of Bates' paper in Linn. Transact. which L. seems to think will do for your Review.—[18] Do inaugurate a great improvement, & have pages cut, like the Yankees do;[19] I will heap blessings on your head. Do not waste your time in answering this.—

Ever yours | C. Darwin

Imperial College of Science, Technology and Medicine Archives (Huxley 5: 183)

[1] The date is established by the reference to T. H. Huxley 1863a (see n. 2, below), and by the relationship between this letter and the letter to T. H. Huxley, 28 December [1862] (*Correspondence* vol. 10; see nn. 3 and 7, below). CD wrote 'Dec.' in error.

[2] Between 10 November and 15 December 1862, Huxley delivered six lectures at the Museum of Practical Geology in London. The lectures were transcribed by the shorthand writer J. Aldous Mays and published in six parts by Robert Hardwicke in 1862; they were subsequently bound together and sold as a separate volume (T. H. Huxley 1863a). Huxley sent the first three parts with his letter to CD

of 2 December 1862 (*Correspondence* vol. 10); for CD's comments, see *Correspondence* vol. 10, letters to T. H. Huxley, 7 December [1862], 18 December [1862], and 28 December [1862]. CD's annotated copies of the six parts of T. H. Huxley 1863a are in the Darwin Library–CUL (see *Marginalia* 1: 425).

3 See *Correspondence* vol. 10, letter to T. H. Huxley, 28 December [1862]. Huxley's sixth lecture was entitled: 'A critical examination of the position of Mr. Darwin's work, "On the Origin of Species," in relation to the complete theory of the causes of the phenomena of organic nature' (T. H. Huxley 1863a, pp. 133–57).

4 In the last lecture of the series (see n. 3, above), Huxley examined CD's method of inquiry in *Origin*, and concluded (T. H. Huxley 1863a, pp. 156–7) by comparing CD's work to that of Georges Cuvier (Cuvier 1817) and Karl Ernst von Baer (Baer 1828–37):

> Mr. Darwin's work is the greatest contribution which has been made to biological science since the publication of the 'Règne Animal' of Cuvier, and since that of the 'History of Development,' of Von Baer. I believe that if you strip it of its theoretical part it still remains one of the greatest encyclopædias of biological doctrine that any one man ever brought forth; . . . it is destined to be the guide of biological and psychological speculation for the next three or four generations.

5 Huxley argued that CD had yet to demonstrate sterility between the offspring of hybrids to prove the role of natural selection in speciation (T. H. Huxley 1863a, pp. 146–9, and n. 6, below); Huxley added that he 'provisionally' accepted CD's general hypothesis (*ibid.*, pp. 151–2).

6 For Huxley, CD's views would not be 'beyond the reach of all possible assault' (T. H. Huxley 1863a, p. 147), until CD could demonstrate:

> the possibility of developing from a particular stock, by selective breeding, two forms, which should either be unable to cross one with another, or whose cross-bred offspring should be infertile with one another.

Only then, Huxley argued, would artificial selection have proved the role of natural selection in the origin of new species (see T. H. Huxley 1863a, pp. 146–50, and *Correspondence* vol. 10, Appendix VI).

7 Letter to T. H. Huxley, 28 December [1862] (*Correspondence* vol. 10). CD and Huxley had been debating the significance to natural selection of hybrid sterility since January 1860 (see *Correspondence* vol. 10, Appendix VI). See also n. 6, above.

8 In his letter to Huxley of 28 December [1862] (*Correspondence* vol. 10), CD wrote: 'To get the degree of sterility you expect in recently formed varieties seems to me simply hopeless'.

9 CD refers to Karl Friedrich von Gärtner's experiments with two varieties of maize (*Zea mays*) that were found to be almost completely infertile when crossed (see *Origin*, pp. 269–70). CD also described in *Origin* Gärtner's experiments with *Verbascum* (pp. 270–1):

> yellow and white varieties of the same species of Verbascum when intercrossed produce less seed, than do either coloured varieties when fertilised with pollen from their own coloured flowers. Moreover, . . . when yellow and white varieties of one species are crossed with yellow and white varieties of a *distinct* species, more seed is produced by the crosses between the same coloured flowers, than between those which are differently coloured.

See also *Variation* 2: 105–7.

10 CD had been intermittently preparing the first volume of *Variation* (published in 1868) since January 1860 (see *Correspondence* vols. 8–10, Appendix II). The domestic breeds that CD discussed included dogs and cats (chapter 1), horses and asses (chapter 2), pigs, cattle, sheep, and goats (chapter 3), domestic rabbits (chapter 4), domestic pigeons (chapters 5 and 6), fowls (chapter 7), and other birds and fish (chapter 8).

11 In *Variation* 1: 2–3, CD discussed the conditions imposed on cultivated plants and domesticated animals in contrast to organic beings living under natural conditions. He developed this idea further in a later chapter of *Variation* entitled: 'On the advantages and disadvantages of changed conditions of life: sterility from various causes' (*Variation* 2: 145–72).

12 T. H. Huxley 1863a, pp. 142–3. Huxley contrasted the doctrines of special creation and descent by modification, illustrating his point by references to the fossil record, homologies in the parts

of animals, and the derivation of languages from pre-existing languages. In the passage to which CD refers, Huxley had written that the English and Greek languages were derived from Sanskrit. There is an annotated copy of T. H. Huxley 1863a in the Darwin Library–CUL (see *Marginalia* 1: 425).

13 CD refers to the number of the *Natural History Review* for January 1863. Huxley became editor-in-chief of the new series of the *Natural History Review* in 1861 (see *Correspondence* vol. 9).

14 CD refers to Hugh Falconer's paper on American fossil elephants (Falconer 1863a).

15 CD refers to a priority dispute between Richard Owen and Falconer regarding the naming of the fossil elephant *Elephas columbi* (see letter to J. D. Hooker, 3 January [1863] and n. 1, letter to John Lubbock, 4 January [1863], and letter to Hugh Falconer, 5 [and 6] January [1863]).

16 For details of the disputatious relationship between Owen and Huxley, see A. Desmond 1982 and 1994–7, and Rupke 1994.

17 See n. 15, above.

18 Henry Walter Bates's account of mimetic butterflies, which had appeared in the *Transactions of the Linnean Society of London* (Bates 1861), was reviewed by CD for the *Natural History Review* ('Review of Bates on mimetic butterflies'); it appeared in the number of the journal for April 1863. See *Correspondence* vol. 10, letter from John Lubbock, 15 December 1862, and letter to John Lubbock, 16 [December 1862], and this volume, letter to John Lubbock, 4 January [1863].

19 Like many periodicals, the *Natural History Review* was printed in octavo (that is, with eight leaves to a sheet); subsequent folding made it necessary for the reader to cut some of the page edges. From the 1850s, North American publications had the pages trimmed after binding (see Tebbel 1972, pp. 260–1).

To Thomas Rivers 11 January [1863][1]

Down. | Bromley. | Kent. S.E.
Jany. 11th

My dear Sir

How rich & valuable a letter you have most kindly sent me.[2] The case of Baronne Prevost with its different shoots, foliage, spines & flowers will be grand to quote.[3] I am extremely glad to hear about the seedling moss-roses.[4] That case of seedling like Scotch Rose, unless you are sure that no Scotch rose grew near (& it is unlikely that you can remember) must, one would think, have been a cross.—

I have little compunction for being so troublesome,—not more than a grand Inquisitor has in torturing a Heretic—for am I not doing a real good public service in screwing crumbs of knowledge out of your wealth of information?

Believe me | Yours cordially obliged | Ch. Darwin

P.S. Since the above was written I have read your paper in G. Chron:[5] it is admirable & will, I know, be a treasure to me: I did not at all know how strictly the character of so many plums is inherited.[6]

On my honour when I began this note I had no thought of troubling you with a question, but you mention one point so interesting & which I have had occasion to notice that I must supplicate for a few more facts to quote on your authority. You say that you have one or two seedling peaches approaching very nearly to thick-fleshed Almonds.[7] (I know about A. Knight & the Italian Hybrid cases.)[8] Now did any Almond grow near your mother Peach? But especially I want to know whether you remember what shape the *stone* was, whether *flattened* like that of an almond;

this Botanically seems the most important distinction. I earnestly wish to quote this.—[9] Was the flesh at all sweet? Forgive if you can.—

Have you kept these seedling Peaches; if you would give me next summer a fruit, I would have it engraved.[10]

P.S. 2d. | I know that I am quite unreasonable; but I cannot resist asking one other question, for the chance of information as a guide for experiment. The varieties of most plants, if grown close to other varieties, yield seed which does not come all true; but shows the influence of a cross.— Have you ever observed any varieties of any plants (except papilionaceous plants) which can safely be grown close to other varieties for several generations, & yet are not affected by crossing?—[11]

DAR 185: 82

[1] The year is established by the reference to Rivers's letter to the *Gardeners' Chronicle* (see n. 5, below).

[2] Rivers's reply to CD's letter of 7 January [1863] has not been found.

[3] In his letter to Rivers of 7 January [1863], CD asked specifically about bud-variation in roses. The Baronne Prevost case was described in *Variation* 1: 381.

[4] In his letter to Rivers of 7 January [1863], CD asked whether Rivers had 'sown seed of any Moss Roses, & whether the seedlings were moss roses'. He also wondered whether 'a common Rose produced by *seed* a moss-rose'. In *Variation* 1: 380, CD reported Rivers's information that 'his seedlings from the old single moss-rose almost always produced moss-roses', and also that he raised two or three roses of the Provence class from the seed of the old single moss-rose.

[5] Rivers's article on seedling plums appeared in the *Gardeners' Chronicle*, 10 January 1863, p. 27.

[6] CD scored portions of Rivers's article in his copy of the *Gardeners' Chronicle*, 10 January 1863, p. 27. Rivers described the Common Damson plum from Kent, the Quetsche Plum, and the Petite Mirabelle; the seedlings of these varieties deviated only slightly from the parents. His note that some Damson seedlings reverted to the Sloe Plum, from which he thought the Damson originated, is also scored in CD's copy (see also *Variation* 1: 345); CD was also interested in Rivers's comment that the Saint Catherine Plum 'reproduces itself from seed without the slightest variation in habit, so that one would think it a species.' CD's annotated copy of this number of the *Gardeners' Chronicle* is in the Cory Library, Cambridge Botanic Garden. This letter is cited in *Variation* 1: 347.

[7] In his article, Rivers had cited the example of seedling peaches retaining the appearance of 'thick-fleshed almonds' as evidence in support of his hypothesis that 'if left to a state of nature', domesticated fruits would return to their 'normal [i.e., original] state' (*Gardeners' Chronicle*, 10 January 1863, p. 27). See also *Variation* 1: 338 n. 24.

[8] Thomas Andrew Knight had raised, from a sweet almond pollinated by a peach, a tree that produced fruit like that of a peach tree; he suggested that the peach tree was a modified almond (see Knight 1817 and 1821, and *Variation* 1: 338).

[9] In *Variation* 1: 338–9, CD acknowledged Rivers's information, and described 'several varieties which connect the almond and the peach'.

[10] Rivers sent CD two peach seedlings (see letter from Thomas Rivers, 21 January 1863); prior to the publication of *Variation*, he sent peach fruits. Drawings of a series of peach and almond stones were reproduced in *Variation* 1: 337; CD acknowledged Rivers's help in providing some of the specimens illustrated (p. 338).

[11] CD had already determined that different varieties of peas rarely crossed (see *Natural selection*, pp. 67–71, and *Variation* 1: 329–30). He continued to question whether varieties of other plants were restricted to self-pollination (see, for instance, *Correspondence* vol. 9, letters to *Journal of Horticulture*, [before 14 May 1861] and [17 May 1861]). See also *Variation* 2: 91, 104–9, and *Cross and self fertilisation*, pp. 8, 382.

To Henry Walter Bates 12 January [1863][1]

Down. | Bromley. | Kent. S.E.
Jan 12th

Dear Bates

One line to say that I have just heard from Asa Gray, that if he had a copy of your paper he would *endeavour* to get Prof. Haldeman to review in Sillimans Journal of Science.[2] **If** you have a spare copy & think it worth **chance**, post by Book Post it to

Prof. Asa Gray
Cambridge
Massachusetts
U. States

I may as well tell you, as you will be sure to find out from resemblance to my letter that I have sent little Review of your Paper to N. Hist. Review.—[3] But I have reason to believe Editors will modify some part.—

If you send copy to Asa Gray, tell me; otherwise do not trouble yourself to write.—

By the way here is question sometime for you to answer, *if you can*, do Bees or Lepidoptera visit flowers of Melastomatads; if you shd. remember what genera of plants, please state;[4] all a mere chance. whether you can answer, I know, I hope Book progresses[5]

In Haste | Yours | C. Darwin

Cleveland Health Sciences Library (Robert M. Stecher collection)

[1] The year is established by the reference to the reviews of Bates 1861 (see nn. 2 and 3, below).

[2] The letter from Asa Gray has not been found; the information referred to by CD was probably contained in the missing section of the letter from Asa Gray, 29 December 1862 (*Correspondence* vol. 10). Bates's paper appeared in the *Transactions of the Linnean Society of London* in November 1862 (Bates 1861). CD had asked Gray to find a colleague to review Bates 1861, an account of mimetic resemblances in species of Amazonian *Lepidoptera*, which CD thought 'well worth some labour in studying' (see *Correspondence* vol. 10, letters to Asa Gray, 23 November [1862] and 26[–7] November [1862]). The review of Bates 1861 published in the September 1863 number of the *American Journal of Science and Arts* (commonly referred to as 'Silliman's journal', after its founder, Benjamin Silliman) was written by Gray himself (A. Gray 1863a), and not by Samuel Steman Haldeman, professor of natural sciences at Delaware College (*DAB*).

[3] CD refers to the letter to H. W. Bates, 20 November [1862] (*Correspondence* vol. 10). CD's review of Bates 1861 ('Review of Bates on mimetic butterflies') appeared in the April 1863 number of the *Natural History Review.*

[4] CD suspected that some species of Melastomataceae might exhibit a novel form of flower dimorphism; he had been experimenting with the family since 1861 (see letter to Hugh Falconer, 5 [and 6] January [1863] and n. 22). Melastomataceae are very common in South America, where Bates had spent eleven years as a collector.

[5] Bates's account of his eleven years in the Amazon region of South America (Bates 1863) was published in April 1863 (*Publishers' Circular* 26: 193). See also letter from H. W. Bates, 17 January [1863].

From J. D. Hooker [12 January 1863][1]

[Royal Gardens Kew]

Dr. Darwin

Can you answer this[2] I have just finished Huxleys Lectures—[3] the 3 last are tremendously good,[4] how lucidly & vigorously he writes— if he took more pains he would be a Scientific Buckle:[5] & Oh My what praise of you—& all merited too, richly.—[6] I hope you are better.[7] I am off to freeze in Paris on Saturday with Bentham for 10 days—[8]

Ever yours affec | J D Hooker

What is the sum of our knowledge regarding qualities *induced in the individual* being in any degree hereditary?

Monday.

DAR 101: 98

CD ANNOTATION
End of letter: '(Medallions)[9] | Naudin'[10]

[1] The date is established by the reference to Hooker's planned trip to Paris (see n. 8, below); 12 January 1863 fell on a Monday.

[2] No enclosure containing Hooker's query has been found; however, see the letter to J. D. Hooker, 13 January [1863].

[3] T. H. Huxley 1863a. See letter to T. H. Huxley, 10 [January 1863] and n. 2.

[4] T. H. Huxley 1863a, pp. 83–157.

[5] Hooker refers to the historian Henry Thomas Buckle, the author of *History of civilisation in England* (Buckle 1857–61), a work that CD greatly admired (see, for example, *Correspondence* vol. 7, letters to J. D. Hooker, 23 February [1858] and 31 March [1858], and *Correspondence* vol. 10, letter to J. D. Hooker, 7 March [1862]).

[6] See letter to T. H. Huxley, 10 [January 1863] and n. 4.

[7] In his letter to Hooker of 3 January [1863], CD wrote that the poor state of his health was 'ludicrous', and the slightest excitement caused 'shaking & vomiting'.

[8] Hooker and George Bentham departed for Paris on 17 January 1863 (Jackson 1906, p. 193).

[9] See letter to J. D. Hooker, 13 January [1863] and n. 34.

[10] Charles Victor Naudin. See letter to J. D. Hooker, 13 January [1863] and n. 36.

From George Varenne Reed 12 January 1863

Hayes Rectory | SE
Jan 12. 1863

My dear Mr Darwin

I am much obliged to you for your cheque which reached me safely.[1] I am very sorry to hear you consider Horace unfit for School for years![2] & for his own sake, but I quite miss them when they do not come and shall therefore be delighted to have him as often as he is able.[3]

With our united best wishes | I remain | yours very sincerely | G V Reed

G V R | Jan 12./63 | £13″2″6

DAR 176: 78

[1] Reed, rector of the nearby parish of Hayes, Kent, tutored the younger Darwin boys before they attended Clapham Grammar School. A payment of £13 2*s.* 6*d.* for 'Two Boys' is recorded in CD's Classed account book (Down House MS), dated 9 January 1863; it represents the fees for Leonard and Horace Darwin from 6 October 1862 to 6 January 1863. Although Leonard had started at Clapham Grammar School in January 1862, he was sent home in June suffering from scarlet fever, a condition from which he was convalescent throughout the summer and autumn. Consequently he returned to Reed for the remainder of the year (see *Correspondence* vol. 10, letter from Charles Pritchard, 17 June [1862]). See also Moore 1977, and the letter from Emma Darwin to William Erasmus Darwin, [3 February 1862], in DAR 219.1: 70.

[2] CD's letter to Reed has not been found. The Darwins' youngest son, Horace, had been ill for much of the previous year (see Emma Darwin's diary (DAR 242); *Correspondence* vol. 10, letter from Henry Holland, 26 March [1862]; and the letter from Emma Darwin to William Erasmus Darwin, [7? April 1862], in DAR 219.1: 51). See also Leonard Darwin's 'Notes on Horace Darwin', Cambridge University Library, Cambridge Scientific Instrument Company Archives, box 3. In January 1863, Horace was 11 years old.

[3] Horace accompanied Leonard to be tutored by Reed between October 1862 and January 1863 (see n. 1, above). However, according to CD's Classed account book (Down House MS), he was not tutored by Reed again until after Easter 1864. The next payment to Reed was made on 16 August 1864 when he was paid £12 12*s.* for the period 19 April to 19 July 1864.

To J. D. Hooker 13 January [1863]

Down
Jan 13th

My dear Hooker

I send very imperfect answer to question, & which I have written on foreign paper, to save you copying & you can send when you write to Thomson in Calcutta.—[1] Hereafter I shall be able to answer better your question about qualities induced in individual being inherited:[2] gout in man,—loss of wool in sheep (which begins in 1st generation & takes 2 or 3 to complete) probably obesity (for it is rare with poor); probably obesity & early maturity in Short-horn Cattle, &c.—

I am very glad you like Huxley's Lectures;[3] I have been very much struck with them; especially with the philosophy of induction.—[4] I have quarrelled with him with overdoing sterility & ignoring cases from Gärtner & Kölreuter about sterile varieties.[5] His geology is obscure;[6] & I rather doubt about man's mind & language.—[7] But it seems to me *admirably* done, & as you say "oh my" about the praise of the Origin:[8] I can't help liking it, which makes me rather ashamed of myself.—

I enclose Asa Gray;[9] only last page & ½ will interest you; but look at *red* (?) & rewrite names.[10] *Do not allude to Gray* that you have seen this letter, as he might not like it, as he speaks of your being wrong (& converted, alas not so!) about Crossing.[11] The sentence about Strawberries made me look at Bentham, & I have enclosed remark for him;[12] I can assure him his remark would make any good horticulturist's hair stand on end.[13] It is marvellous to see Asa Gray so cock-sure about the doom of Slavery.—[14]

You wrote me a famous long letter a few days ago: Emma is going to read De TocVille & so was glad to hear your remarks.—[15] I am glad to hear that you are going to do some work which will bring a little grist to the mill; but good Heavens how do you find time with Genera Plantarum, official work, friends, & Heaven knows what![16]

Many thanks about Poison for Plants.—[17] I know nothing about leaf-insects, except that they are carnivorous.— Andrew Murray knows.—[18]

You ask what I think about Falconer;[19] of course I am much pleased at the very kind way he refers to me;[20] but, as I look at it, the great gain is for any good man to give up immutability of species: the road is then open for progress; it is comparatively immaterial whether he believes in N. Selection; but how any man can persuade himself that species change unless he sees how they become adapted to their conditions is to me incomprehensible.—[21] I do not see force of Falconer's remarks about spire of shells, Phyllotaxis, &c:[22] I suppose he did not look at my chapter on what I call laws of variation.—[23]

How very well Falconer writes: by the way in one of your letters you insisted on importance of style;[24] I have just been struck with excellent instance in Alex. Braun on Rejuvenescence in Ray Soc 1853; I have tried & literally I *cannot* read it.[25] Have you read it?

I have just received long pamphet by Alph. De Candolle on Oaks & allies,[26] in which he has worked out in very complete & curious manner individual variability of species, & has wildish speculations on their migrations & duration &c.[27] It is really curious to see how blind he is to the conditions or struggle for life; he attributes the presence of all species of all genera of trees to dryness or dampness! At end he has discussion on "Origin";[28] I have not yet come to this, but suppose it will be dead against it. Should you like to see this pamphlet?

My hot-house will begin building in a week or so,[29] & I am looking with much pleasure at catalogues to see what plants to get: I shall keep to curious & experimental plants. I see I can buy Pitcher plants for only 10^s^·6![30] But the job is whether we shall be able to manage them. I shall get Sarracenia Dichœa your Hedysarum, Mimosa & all such funny things,, as far as I can without great expence.[31] I daresay I shall beg for *loan* of some few orchids; especially for Acropera Loddigesii.[32] I fancy orchids cost awful sums; but I must get priced catalogue. I can see hardly any Melastomas in catalogues.—[33]

I had a whole Box of small Wedgwood medallions; but drat the children everything in this house gets lost & wasted; I can find only about a dozen little things as big as shillings, & I presume worth nothing; but you shall look at them when here & take them if worth pocketing.[34]

You sent us a gratuitous insult about the "chimney-pots" in dining room, for you *shan't* have them; nor are they Wedgwood ware.—[35]

Remember Naudin[36]

When you return you must remember my list of experimental seeds.—[37] I hope you will enjoy yourself[38]

Goodnight my dear old friend | C. Darwin

You have not lately mentioned M^rs^. Hooker: remember us most kindly to her.—[39]

Endorsement: '/63.'
DAR 115: 179

[1] The reference is to the surgeon and botanist Thomas Thomson; Thomson lived in Calcutta only until 1860 or 1861 (*DNB*). The enclosure has not been found. See also letter from J. D. Hooker, [12 January 1863] and n. 2.

[2] See letter from J. D. Hooker, [12 January 1863]. On 23 January 1863, CD began writing up his 'Chapter on Inheritance' for *Variation*, eventually published as chapters 12–14 (*Variation* 2: 1–84; see 'Journal' (Appendix II)).

[3] Thomas Henry Huxley presented an evening lecture series for working men at the Museum of Practical Geology in London during November and December 1862; the lectures were published as T. H. Huxley 1863a. See letter to T. H. Huxley, 10 [January 1863], and letter from J. D. Hooker, [12 January 1862].

[4] T. H. Huxley 1863a, pp. 55–67. Huxley's discussion of induction formed part of the third lecture, delivered on 24 November 1862 ('The method by which the causes of the present and past conditions of organic nature are to be discovered.— The origination of living beings'). There is a lightly annotated copy of T. H. Huxley 1863a in the Darwin Library–CUL (see *Marginalia* 1: 425).

[5] Huxley argued that the origin of species through natural selection could not be proven until artificial selection produced from a common stock varieties that were sterile with one another (T. H. Huxley 1863a, pp. 146–50). CD, by contrast, was impressed by the plant hybridisation experiments conducted by Karl Friedrich von Gärtner and Joseph Gottlieb Kölreuter (*Origin*, pp. 246–9, 257–9, 270–4; Gärtner 1844 and 1849; Kölreuter 1761–6). See letter to T. H. Huxley, 10 [January 1863], and *Correspondence* vol. 10, Appendix VI.

[6] T. H. Huxley 1863a, pp. 29–52. CD refers particularly to pages 39–41, and to figure 5 on page 40, which he thought would be confusing to a non-geologist. See letters to T. H. Huxley, 7 December [1862] and n. 7, and 18 December [1862] (*Correspondence* vol. 10).

[7] T. H. Huxley 1863a, pp. 153–6. While arguing that 'man differs to no greater extent from the animals which are immediately below him than these do from other members of the same order', Huxley wrote that it was largely the power of language that distinguished man 'from the whole of the brute world' (*ibid.*, pp. 154–5).

[8] See letter from J. D. Hooker, [12 January 1863], and letter to T. H. Huxley, 10 [January 1863] and n. 4.

[9] Letter from Asa Gray, 29 December 1862 (*Correspondence* vol. 10).

[10] In Asa Gray's letter, CD marked some of the plant names with marginal crosses in red crayon, and Hooker clearly printed the names 'Abronia', 'Nyctaginia', 'Pavonia' for *Pavonia hastata*, and '*Ruellia*'. These were plants in which the plants flowering earlier in the season were pollinated in the bud (see *Correspondence* vol. 10, letter from Asa Gray, 29 December 1862).

[11] In November and December 1862, CD and Hooker debated the effects of crossing on variation, with Hooker maintaining that self-fertilisation did not favour variation, 'whereas crossing tends to variation by *adding* differences' (see *Correspondence* vol. 10, letter from J. D. Hooker, 26 November 1862). CD agreed with Gray (A. Gray 1862d, p. 420) that:

> free cross-breeding of incipient varieties *inter se* and with their original types is just the way to blend all together, to repress all salient characteristics as fast as the mysterious process of variation originates them, and fuse the whole into a homogeneous form.

See *Correspondence* vol. 10, letter to Asa Gray, 26[–7] November [1862].

[12] The letter from Asa Gray, 29 December 1862 (*Correspondence* vol. 10), is incomplete; Gray's statement concerning strawberries was made in a postscript that has not been located. However, in his account of strawberries in *Variation* 1: 351–4, CD considered it unlikely that hybrids of European and American strawberries were fertile enough to be worth cultivation. This fact was surprising to him 'as these forms structurally are not widely distinct, and are sometimes connected in the districts where they grow wild, as I hear from Professor Asa Gray, by puzzling intermediate forms' (*Variation* 1: 352). CD probably consulted George Bentham's *Handbook of the British flora* (Bentham 1858; see n. 13, below).

The enclosure for Bentham has not been found. See also letter to Asa Gray, 2 January [1863] and n. 17.

13 In his *Handbook of the British flora*, Bentham wrote that while several wild and cultivated strawberries had been proposed as species, 'the great facility with which fertile hybrids are produced, gives reason to suspect that the whole genus . . . may prove to consist but of one species' (Bentham 1858, pp. 191–2). CD's annotated copy of Bentham 1858 is in the Rare Books Room–CUL (see *Marginalia* 1: 51).

14 The letter from Asa Gray, 29 December 1862 (*Correspondence* vol. 10), is incomplete; the portion containing Gray's statement regarding events in the United States has not been found. Gray may have commented on Abraham Lincoln's emancipation proclamation, which was to come into effect on 1 January 1863; from that time all slaves in territories still in rebellion were to be freed (see Denney 1992, pp. 248, 251).

15 Emma Darwin. CD refers to Alexis Henri Charles Maurice Clérel, comte de Tocqueville's *Democracy in America* (H. Reeve trans. 1862). See *Correspondence* vol. 10, letter from J. D. Hooker, [21 December 1862], and this volume, letter from J. D. Hooker, 6 January 1863. CD had read Tocqueville's *De la démocratie en Amérique* (Tocqueville 1836) in February 1849 (see *Correspondence* vol. 10, letter to J. D. Hooker, 24 December [1862], and *Correspondence* vol. 4, Appendix IV, 119: 22b).

16 Hooker had been commissioned to write a flora of New Zealand (J. D. Hooker 1864–7; see letter from J. D. Hooker, 6 January 1863). At the same time, Hooker was at work on *Genera plantarum* (Bentham and Hooker 1862–83), and also had official duties in his capacity as assistant director of the Royal Botanic Gardens, Kew.

17 In his letter to Hooker of 3 January [1863], CD asked for advice about how to prevent mould from growing on his children's dried flower collections; for Hooker's reply, see his letter of 6 January 1863. In his Account book–cash accounts (Down House MS), on 16 January 1863, CD recorded a payment of 9*s.* for 'Poison for plants' to the London importers and makers of chemical and photographic apparatus, Bolton & Barnitt of Holborn Bars, London.

18 Hooker had asked CD what he should feed newly hatched leaf insects from Java (see letter from J. D. Hooker, 6 January 1863). Andrew Murray was a botanist and entomologist with expertise in Coleoptera and insects harmful to crops (*DNB*).

19 Hooker had asked CD's opinion of Falconer 1863a (see letter from J. D. Hooker, 6 January 1863).

20 Falconer 1863a, pp. 77–81 (see n. 21, below). See also letter to Hugh Falconer, 5 [and 6] January [1863] and n. 7.

21 In his article on fossil and recent elephants, Hugh Falconer praised CD and his theory of modified descent (Falconer 1863a, pp. 77, 80). At the same time, he argued that natural selection was an inadequate explanation for the origin of species since some species subject to variable conditions over time, such as the mammoths, had remained unchanged (Falconer 1863a, p. 80).

22 While Falconer conceded that forms like the mammoth and other extinct elephants were 'modified descendants of earlier progenitors' (Falconer 1863a, p. 80), he continued to argue against the adequacy of natural selection to explain this modification:

> The law of Phyllotaxis, which governs the evolution of leaves around the axis of a plant, is nearly as constant in its manifestation, as any of the physical laws connected with the material world. Each instance, however different from another, can be shown to be a term of some series of continued fractions. When this is coupled with the geometrical law governing the evolution of form, so manifest in some departments of the animal kingdom, *e. g.* the spiral shells of the Mollusca, it is difficult to believe, that there is not in nature, a deeper seated and innate principle, to the operation of which 'Natural Selection' is merely an adjunct.

23 *Origin*, pp. 131–70.

24 The reference has not been identified.

25 Braun 1851. The English title of the article was 'Reflections on the phenomena of rejuvenescence in nature, especially in the life and development of plants' (Henfrey trans. 1853). There is an annotated copy of Arthur Henfrey's translation of Braun 1851 in the Darwin Library–CUL (see *Marginalia* 1: 366–7).

26 Alphonse de Candolle sent CD copies of A. de Candolle 1862a and 1862b. See *Correspondence* vol. 10, letter from Alphonse de Candolle, 18 September 1862; see also following letter. CD's annotated copies of A. de Candolle 1862a and 1862b are in the Darwin Pamphlet Collection–CUL.

27 A. de Candolle 1862a, pp. 326–53. See following letter and n. 6.

28 A. de Candolle 1862a, pp. 354–61, 363. See *Intellectual Observer* 3 (1863): 81–6, for a translation of the last portion of A. de Candolle 1862b. See also following letter and n. 7.

29 See *Correspondence* vol. 10, letter to J. D. Hooker, 24 December [1862]), and this volume, letter to Asa Gray, 2 January [1863] and n. 24, and Appendix VI.

30 See DAR 157.1: 129 and 130 for CD's botanical notes on experiments with *Nepenthes* (pitcher plants).

31 CD had experimented on the power of movement in *Hedysarum* and *Mimosa* in 1862 (see *Correspondence* vol. 10).

32 CD was keen to obtain fresh flowers of *Acropera*; for CD's continuing investigation of this orchid genus, see *Correspondence* vol. 10, letter from John Scott, 11 November 1862, and letter to John Scott, 12 November [1862], and this volume, letter from John Scott, 6 January 1863 and nn. 3 and 4.

33 See letter to Hugh Falconer, 5 [and 6] January [1863] and n. 22.

34 Hooker had started to collect Wedgwood ware and was particularly interested in medallions. See *Correspondence* vol. 10, letter from J. D. Hooker, [27 or 28 December 1862], and this volume, letter to J. D. Hooker, 3 January [1863], and letter from J. D. Hooker, 6 January 1863.

35 See letter from J. D. Hooker, 6 January 1863.

36 With his letter to Hooker of 24 December [1862] (*Correspondence* vol. 10), CD enclosed a 'memorandum of enquiry' for Charles Victor Naudin, whom Hooker hoped to meet during his forthcoming visit to Paris (see n. 38, below).

37 In his letter to Hooker of 3 November [1862] (*Correspondence* vol. 10), CD enclosed a list of the seeds he wanted for experiments on sensitivity in plants. See also *ibid.*, letter to J. D. Hooker, [10–]12 November [1862].

38 Hooker and Bentham departed for Paris on 17 January 1863 (Jackson 1906, p. 193).

39 Since the death of her father, John Stevens Henslow, in May 1861, Frances Harriet Hooker had been suffering from depression and ill-health (see *Correspondence* vols. 9 and 10).

To Alphonse de Candolle 14 January [1863][1]

Down. | Bromley. | Kent. S.E.
Jan. 14th

My dear Sir

I thank you most sincerely for sending me your Memoir.—[2] I have read it with the liveliest interest, as is natural for me; but you have the art of making subjects, which might be dry, run easily. I have been fairly astonished at the amount of individual variability in the oaks.[3] I never saw before the subject in any department of nature worked out so carefully. I noted with delight case of achenia &c &c.—[4] What labour it must have cost you! You spoke in one letter of advancing years; but I am very sure that no one would have suspected that you felt this.—[5] I have been interested with every part; though I am so unfortunate as to differ from most of my contemporaries in thinking the the vast continental extensions of Forbes, Heer & others are not only advanced without sufficient evidence, but are opposed to much weighty evidence.—[6]

You refer to my work in the kindest & most generous spirit.— I am fully satisfied at the length in belief to which you go, & not at all surprised at the prudent reservations which you make.[7] I remember well how many years it cost me to go round from old beliefs. It is encouraging to me to observe that everyone who has gone an inch with me; after a period goes a few more inches or even feet.— But the great point, as it seems to me, is to give up the immutability of specific forms; as long as they are thought immutable, there can be no real progress in 'epiontology".[8] It matters very little to anyone except myself, whether I am a little more or less wrong on this or that point; in fact I am sure to be proved wrong in many points. But the subject will have, I am convinced, a grand future.—

Considering that Birds are the most isolated group in the animal kingdom, what a splendid case is this Solenhofen bird-creature with its long tail & fingers to its wings![9] I have lately been daily & hourly using & quoting your Geograph. Bot., in my book on "Variation under Domestication".[10]

with cordial thanks & sincere respect | I remain Dear Sir | Yours very faithfully Ch. Darwin

La Fondation Augustin de Candolle

[1] The year is established by the reference to A. de Candolle 1862a (see n. 2, below).

[2] A. de Candolle 1862a. At some point, CD also received A. de Candolle 1862b, a shorter notice on oaks. Candolle originally undertook the review of *Quercus*, the oak genus, and its relatives, as part of his work for *Prodromus systematis naturalis regni vegetabilis* (Candolle and Candolle 1824–73). In his letter of 18 September 1862 (*Correspondence* vol. 10), Candolle promised CD copies of the publications on *Quercus*. CD's annotated copies of Candolle's accounts of the oak family are in the Darwin Pamphlet Collection–CUL.

[3] CD cited A. de Candolle 1862a in *Variation* 1: 387, in regard to changes in oaks with age.

[4] CD refers to the discussion of the acorn in A. de Candolle 1862a, pp. 222–5; this section is heavily annotated in CD's copy of the work, which is in the Darwin Pamphlet Collection–CUL.

[5] See *Correspondence* vol. 10, letter from Alphonse de Candolle, 13 June 1862.

[6] For Candolle's discussion of the distribution of the Cupuliferae, see A. de Candolle 1862a, pp. 326–53. CD refers to Edward Forbes's account of European flora and fauna distribution patterns, in which he proposed that in the Miocene era there existed a land-bridge connecting the Iberian peninsula with the Azores and with Ireland (E. Forbes 1846). Similarly, working on the biogeography of Madeira, Oswald Heer invoked the theory of a former land-extension linking that island with the European mainland (Heer 1855). Charles Lyell was also a supporter of a land-bridge theory, and Joseph Dalton Hooker used it to explain distribution patterns in the southern Pacific (J. D. Hooker 1844–7 and 1853–5). CD did not completely oppose the concept of land-bridges and accepted, for example, the probability that such a connection formerly linked North America and Asia. However, he vigorously criticised what he thought was the *ad hoc* invocation of former land-extensions, even former continents, without fully considering other means that might account for observed plant and animal distributions. CD had been debating the land-bridge theory with Hooker since 1846 and with Lyell since 1856 (see *Correspondence* vols. 3 and 6). See especially *Correspondence* vol. 6, letters to Charles Lyell, 16 [June 1856] and 25 June [1856], and *Origin*, pp. 352–6. See also *Correspondence* vol. 10, letter to J. D. Hooker, 4 November [1862] and n. 6; for CD's comments to Hooker on A. de Candolle 1862a, see the preceding letter.

[7] A. de Candolle 1862a, pp. 354–61, 363. Candolle described CD's account of the origin of species as the most modern, and, at the same time, the most ingenious and complete of the systems founded on what he termed an 'évolution' of organised beings over the course of time (A. de Candolle 1862a, p. 354). His reservations concerned what he saw as a lack of direct evidence that natural selection was the only means by which modified forms were slowly changed through what seemed to be confoundingly vast periods of time (A. de Candolle 1862a, pp. 358–61, 363). See *Intellectual Observer* 3 (1863): 81–6, for a translation of the last portion of A. de Candolle 1862b. See also *Correspondence* vol. 10, letter from Alphonse de Candolle, 13 June 1862.

8 Attempting to unify biogeography and the 'geography of organic beings', Candolle defined 'épi-ontologie' as the study of the distribution and succession of organised beings from their origin up to the present time (A. de Candolle 1862a, pp. 363–5).

9 CD refers to the reptile-like fossil bird *Archaeopteryx*, found in the Jurassic rocks of the Solenhofen quarries, Bavaria, in 1861. See letter from Hugh Falconer, 3 January [1863].

10 On 21 December 1862, CD began the chapter of *Variation* discussing bud-variation (see *Correspondence* vol. 10, Appendix II). However, his frequent references to A. de Candolle 1855 appear in the two previous chapters on cultivated plants (*Variation* 1: 305–72); he had started working on those sections in October 1862 (see *Correspondence* vol. 10, Appendix II). CD's annotated copy of A. de Candolle 1855 is in the Darwin Library–CUL (see *Marginalia* 1: 106–53).

To Smith, Elder & Company 14 January [1863]

Down. | Bromley. | Kent. S.E.
Jan. 14th

Dear Sir

I have had on my file a little unpaid account ever since 1858.[1] When I received it I asked you to be so good as to let me have account of any sales of my three Geological Books.—[2] Will you be so kind as to let me now hear on this subject.—[3] I presume a few copies will have been sold during the several years since I had an account.— When you see what sale there has been, you can, if you please, make me any offer for the stock in hand, & so finally settle accounts, as we did many years ago for the Zoology of the Voyage of the Beagle.—[4]

Pray believe me | Dear Sir | Yours very faithfully | Charles Darwin

Postmark: JA 15 63
National Library of Scotland (MS 23181: 3)

1 The statement of account has not been found; however, see *Correspondence* vol. 4, letter to Smith, Elder & Company, [16 February 1849] and n. 3. Smith, Elder & Company published CD's geology and zoology of the *Beagle* voyage (see nn. 2 and 4, below).

2 CD's letter has not been found. Smith, Elder & Company published the three parts of CD's account of the geology of the voyage of the *Beagle*: *Coral reefs* in 1842, *Volcanic islands* in 1844, and *South America* in 1846. In 1851, the three parts were combined and reissued, retaining the old pagination, although the title and half-title leaves of the parts were discarded and replaced by a new general title leaf, as *Geological observations on coral reefs, volcanic islands, and on South America* (Freeman 1977). See *Correspondence* vols. 2–4 for some of the transactions between CD and Smith, Elder & Company.

3 See letter from Smith, Elder & Company, 3 March 1863.

4 Smith, Elder & Company also published the five parts of *Zoology* (Freeman 1977; see *Correspondence* vols. 2 and 3). However, the correspondence referred to has not been found, and, in CD's account books, there is no explicit reference to a final settlement or sale of the stock of *Zoology*.

From J. P. Thom 14 January 1863

Home News Office | 124 Bishopgate st | London, E.C.
14th Jan. '63.

My dear Mr Darwin,

Dr Lane told me last night that you had sent to him a few days ago for me a cheque on your banker for £20.[1] This I consider to be an act of such true and genuine kindness on your part that I do not expect to be able to find phrases which will express forcibly enough my sincere appreciation of it. I will therefore only say that I thank you with all my heart. This generous gift is one which, coming from you, I will freely tell you I regard as as one of the most flattering testimonials I have ever received in my life. I will only add that it will also be of the greatest possible service to me at present.

Before I leave England (which I expect to do early in March) I hope to write to you again and tell you something in reference to my plans which will, I believe, tend to show that they are feasible.[2]

Meanwhile believe me, my dear Mr Darwin, | Ever yours truly & gratefully | J. P. Thom

Charles Darwin, Esq

DAR 178: 107

[1] The physician Edward Wickstead Lane was the proprietor of a hydropathic establishment at Sudbrook Park, Petersham, Surrey. CD and Thom became acquainted in 1857 or 1858 at Lane's previous establishment at Moor Park, near Farnham, Surrey (see *Correspondence* vol. 6, Appendix II, and *Correspondence* vol. 7, letter to W. D. Fox, 24 June [1858]). There is an entry in CD's Account book–banking accounts (Down House MS) for 1 January 1863, recording the payment of £20 to Thom.

[2] Thom was planning to emigrate to Queensland, Australia; a fellow patient at Sudbrook Park, Mary Butler, had asked CD if he could recommend an appointment for Thom (see *Correspondence* vol. 10, letter from Mary Butler, [before 25 December 1862]). No further correspondence on this matter has been found.

From Alfred Russel Wallace 14 January [1863][1]

5, Westbourne Grove Terrace, W.
January 14th.

My dear Mr. Darwin

I am very sorry indeed to hear you are still in weak health.[2] Have you ever tried mountain air. A residence at 2000 or 3000 ft. elevation is very invigorating.

I trust your family are now all in good health, & that you may be spared any anxiety on that score for some time—[3] If you come to Town I shall hope to have the pleasure of seeing you—

I am now in much better health but find sudden changes of weather affect me very much bringing on ague & fever fits—[4] I am now working a little but having fresh collections still arriving from Correspondents in the E. it is principally the drudgery of cleaning packing & arrangement.[5]

On the opposite page I give all the information I can about the Timor fossils so that you can send it entire to Dr. Falconer.[6]

With best wishes for the speedy recovery of your health | I remain | My dear Mr. Darwin | Yours very faithfully | Alfred R. Wallace

Charles Darwin Esq.

DAR 106: B7

[1] The year is established by the relationship between this letter and the letter from Hugh Falconer, 9 January 1863 (see n. 6, below).

[2] The letter from CD has not been found. In his letter to Hugh Falconer of 5 [and 6] January [1863], CD described himself as having been 'knocked up of late with extraordinary facility'. See also letter from Hugh Falconer, 3 January [1863], letter to J. D. Hooker, 3 January [1863] and letter to J. D. Dana, 7 January [1863].

[3] Other members of the Darwin family had been ill in 1862, particularly Emma and Leonard, who contracted scarlet fever (see *Correspondence* vol. 10).

[4] After eight years in the Malay Archipelago, Wallace returned to London in the spring of 1862 in a weak state of health (see Wallace 1905, 1: 386, and *Correspondence* vol. 10, letters from A. R. Wallace, 7 April 1862 and [after 20 August 1862]).

[5] Upon his return to London in 1862, Wallace unpacked and sorted his collections while also writing papers (Wallace 1905, 1: 385–6).

[6] The enclosure has not been found. In his letter to Hugh Falconer, 5 [and 6] January [1863], CD mentioned that Wallace had informed him of Mastodon remains in Timor. In his letter to CD of 30 November 1861 (*Correspondence* vol. 9), Wallace had reported an account by Carl Friedrich Adolph Schneider, who claimed to have found teeth of *Mastodon* on this island of the Malay archipelago (see Schneider 1863). Falconer asked CD to keep him 'in mind' for more information the next time he wrote to Wallace (see letter from Hugh Falconer, 9 January 1863 and n. 3).

From J. D. Hooker [15 January 1863][1]

Royal Gardens Kew
Thursday.

Dear Darwin

I return A Gray with the names, as far as I can make them out.—[2] What a fertile man he is, & what a sanguine one about the war & Slavery![3]

I should like vastly to have a talk with you one day about, variation.

I quite agree that Huxley is still *uninstructed* on the subject, & shall tell him so.[4] Carpenter is better fitted than he to deal fully with a subject he has no practical acquaintance with:[5] but then what spirit what force Huxley commands & compells his audience with.

I liked the bits about Man's mind & language[6]

Strawberries are awful cases,—& suggest to me the desirability of crossing *native* American with native English specimens. Write to A Gray for seeds of *native* specimens Europ: plants & sow them & cross them with English-grown ones.[7]

Murray came & saw leaf Insect it is a Phyllium that eats **leaves**.![8]

Falconer is a Scotchman.[9] A thousand thanks for answer for Thomson, he is at Hastings—has been home for a year & very ill off & on.[10]

You will give me *deadly* offence if you do not send me your Catalogue of the plants you want before going to Nurserymen.[11] If by pitcher plants you mean Nepenthes, I can give you a lot of excellent seedling & 2 year old plants half a dozen I dare say.[12]

My wife has been wonderfully well of late—though rather neuralgic.—[13] Willy is improving rapidly I find.[14] I want a good semi ladies school not too far from

London for Charlie[15] at $7\frac{3}{4}$ a splendid boy in all ways— Can you or Mrs Darwin help me.

I have ADC on Oaks but not read him yet.—[16]

I shall not forget Naudin[17]

Ever affec | J D Hooker

∞ thanks for little Medallions—the last things I thought you would have![18]

DAR 101: 101–2

1 The date is established by the relationship between this letter and the letter to J. D. Hooker, 13 January [1863], and by the reference to Hooker's forthcoming visit to Paris, which began on 17 January 1863 (see n. 17, below); the intervening Thursday was 15 January.

2 With his letter to Hooker of 13 January [1863], CD enclosed Asa Gray's letter of 29 December 1862 (*Correspondence* vol. 10), in which Gray had provided the names of some plants pollinated in the bud; CD asked Hooker to clarify the names.

3 Hooker refers to the letter from Asa Gray, 29 December 1862 (*Correspondence* vol. 10; see n. 2, above). Gray's letter is incomplete; the portion containing Gray's statement about events in the United States has not been found. However, it may have concerned the emancipation proclamation that took effect from 1 January 1863 (see letter to J. D. Hooker, 13 January [1863] and n. 14).

4 The reference is to Thomas Henry Huxley's six lectures to working men (T. H. Huxley 1863a) delivered at the Museum of Practical Geology in London between 10 November and 15 December 1862. The lectures were entitled 'On our knowledge of the causes of organic phenomena'. In his letter to Hooker of 13 January [1863], CD praised the lectures but admitted he had 'quarrelled' with Huxley about 'overdoing sterility' (see T. H. Huxley 1863a, pp. 146–50). See letter to T. H. Huxley, 10 [January 1863] and nn. 5–9, and *Correspondence* vol. 10, letters to T. H. Huxley, 18 December [1862] and 28 December [1862], and Appendix VI. There is an annotated copy of T. H. Huxley 1863a in the Darwin Library–CUL (see *Marginalia* 1: 425).

5 Hooker refers to the physician and zoologist William Benjamin Carpenter. In his microscopic study of Foraminifera (Carpenter 1862), Carpenter described and classified the group, fossil and recent, and evaluated the immense range of variation displayed. There is a copy of Carpenter 1862 in the Darwin Library–Down.

6 T. H. Huxley 1863a, pp. 153–6. In his letter to Hooker of 13 January [1863], CD expressed doubts about Huxley's statements on the human mind and language.

7 In a postscript to his letter of 29 December 1862 (*Correspondence* vol. 10), Gray replied to CD's query regarding strawberries; CD had asked whether *Fragaria vesca* and *F. virginiana* differed '*Botanically*', and whether anyone had succeeded in crossing them (*ibid.*, letter to Asa Gray, 26[–7] November [1862]). Gray's postscript has not been found. For CD's conclusions on the fertility of hybrids of American and European strawberries, see *Variation* 1: 351–4. See also letter to Asa Gray, 2 January [1863] and n. 17, and letter to J. D. Hooker, 13 January [1863] and n. 12.

8 *Phyllium* is a genus of leaf-insect from south-east Asia and New Guinea. In his letter of 6 January 1863, Hooker reported that they had hatched some leaf-insects from Java at the Royal Botanic Gardens, Kew, and asked whether CD knew what they should be fed. CD thought they were carnivorous, but suggested he ask the entomologist Andrew Murray (see letter to J. D. Hooker, 13 January [1863]).

9 Hugh Falconer. See letter to J. D. Hooker, 13 January 1863.

10 Thomas Thomson had been superintendent of the Calcutta botanic garden and professor of botany at the Calcutta Medical College until 1860 or 1861, when he returned to Britain (*DNB*, and *Kew Bulletin* (1895): 236). Thomson had asked for some information and Hooker forwarded the request to CD (see letter from J. D. Hooker, [12 January 1863], and letter to J. D. Hooker, 13 January [1863]).

11 In his letter to Hooker of 13 January [1863], CD told Hooker of the imminent construction of a hothouse at Down House, and spoke of his plans for purchasing experimental plants. See also letter to Asa Gray, 2 January [1863] and n. 24, and Appendix VI.

12 In his letter to Hooker of 13 January [1863], CD reported that he could buy pitcher plants for only 10*s.* 6*d.*

13 Since the death of her father, John Stevens Henslow, in May 1861, Frances Harriet Hooker had been suffering from depression and ill-health (see *Correspondence* vols. 9 and 10).

14 Hooker had been concerned about the development of his nine-year-old son, William Henslow Hooker, whom he described as a 'standing protest against the "Origin of Species"', appearing to care 'for no one thing in life' (see *Correspondence* vol. 10, letter from J. D. Hooker, [21 December 1862]).

15 Hooker refers to his son, Charles Paget Hooker.

16 Hooker refers to Alphonse de Candolle's review of the oak genus and its relatives (A. de Candolle 1862a). For CD's comments on A. de Candolle 1862a, see the letter to J. D. Hooker, 13 January [1863], and the letter to Alphonse de Candolle, 14 January [1863].

17 Hooker and George Bentham were preparing to leave for a ten-day trip to Paris where they planned to visit Charles Victor Naudin. In his letter to Hooker of 13 January [1863], CD reminded him of the 'memorandum of enquiry' for Naudin, which he enclosed with his letter to Hooker of 24 December [1862] (*Correspondence* vol. 10).

18 Hooker had started to collect Wedgwood ware (see *Correspondence* vol. 10, letter from J. D. Hooker, [27 or 28 December 1862], and this volume, letter from J. D. Hooker, 6 January 1863). In his letter to Hooker of 3 January [1863], CD described Emma Darwin and himself as 'degenerate descendants of old Josiah W.', because of their insensibility to the pleasure of Wedgwood ware. In his letter to Hooker of 13 January [1863], CD offered 'about a dozen little things as big as shillings' that had survived at Down House.

To Thomas Rivers 15 January [1863][1]

Down. | Bromley. | Kent. S.E.
Jan. 15th

My dear Sir

You not only give me information of much value, but you give it in the kindest manner possible.[2] All that you say about peaches is particularly interesting, as the case struck me much in many respects. I have alluded in my M.S. from Gard. Chronicle to the Double-flowering peaches of China, though I have never seen them.[3] The case struck me as good in showing what man can do by continued selection in two different lines on the same species, viz flowers & fruits—[4]

You are so kind as to offer me two trees of the Double peach; but would they flower & *fruit* if transplanted this time of year? if so & you would give me them, I shd. be delighted: otherwise I would not rob you & would remind you to give me one or two fruit for engraving stones.[5] I am so ignorant that I do not know whether the almond fruits in England; if you keep any almond trees & would send me one, instead of one of the two peaches it would be even a more valuable present.— If you do send either please address them, thus

C. Darwin Es
care of Down Postman
Per Rail. Bromley
Kent

I have a greenhouse (& am going immediately to build small hot-house for experimental purposes)[6] please have label tied to tree, if sent, saying whether I had better put trees in large pots & keep in greenhouse for fruiting.—

I can only thank you for all your kindness.— It seems almost childish to ask you whether you would like to have any book published by me, my Journal, or the Origin of Species, or Book on Orchids.—[7] I do not offer my geological or pure Zoological books, as you would not care for them.—[8]

Believe me | Yours gratefully | Ch. Darwin

There is a very curious account in Gard. Chronicle 1860 p. 672 of a rose-shoot intermediate in character between white Banksian & R. Devoniensis, springing from junction when the latter had been budded on former. It is there said that Banksian often affects roses budded on it.—[9]

DAR 185: 83

[1] The year is established by the relationship between this letter and the letter to Thomas Rivers, 11 January [1863].

[2] The letter from Rivers, a reply to CD's letter to Rivers of 11 January [1863], has not been found.

[3] See letter to Thomas Rivers, 11 January [1863]. A discussion of the double-flowering Chinese peach in *Variation* 1: 338–9, mentions Rivers; CD also referred to a description of these ornamental trees on page 216 of the 28 March 1857 issue of the *Gardeners' Chronicle* (see *Variation* 1: 343–4).

[4] In the chapter on bud-variation in *Variation*, which CD started writing in December 1862 (see *Correspondence* vol. 10, Appendix II), he described the double-flowered Chinese peaches that had been propagated for ornamental value (*Variation* 1: 343–4). In a later chapter (*Variation* 2: 217–18), CD noted:

> The power of long-continued selection, whether methodical or unconscious, or both combined, is well shown in a general way, namely, by the comparison of the differences between the varieties of distinct species, which are valued for different parts. . . . The Chinese double-flowering peaches . . . show that varieties of this tree have been formed, which differ more in the flower than in fruit. If, as is highly probable, the peach is the modified descendant of the almond, a surprising amount of change has been effected in the same species, in the fleshy covering of the former and in the kernels of the latter.

[5] In his letter to Rivers of 11 January [1863], CD asked for the fruit of some of the peaches Rivers had raised, which CD wanted to use for illustrations in *Variation*. Engravings of peach and almond stones appeared in *Variation* 1: 337. CD noted that he was indebted to Rivers for some of the specimens figured (*Variation* 1: 338).

[6] CD had resolved to build a hothouse at the end of 1862 (see letter to Asa Gray, 2 January [1863] and n. 24).

[7] CD refers to *Journal of researches*, the third edition of *Origin*, and *Orchids*.

[8] The references are to *Zoology*, *Geology of the 'Beagle'*, *Fossil Cirripedia* (1851 and 1854), and *Living Cirripedia* (1851 and 1854).

[9] CD refers to a letter in the *Gardeners' Chronicle*, 21 July 1860, p. 672. In *Variation* 1: 396, CD described this case as the 'most reliable instance' that he knew of the 'formation of a graft-hybrid'; he likened it to reports of *Cytisus adami* (Adam's laburnum), a graft-hybid resulting from a *C. purpureus* scion being budded to a stock of *C. laburnum* (*Variation* 1: 387–91, 396).

From George Bentham 16 January 1863

25. Wilton Place, S.W.
Jan[y] 16/63

My dear Darwin

I was much pleased to hear that you had prepared a paper for us which M[r] Kippist will have announced for reading—on the 5th Feb[y] which I trust may suit you—[1] Should anything occur to prevent it would you kindly let him know. I am going with Jos. Hooker over to Paris for a few days but I shall be sure to be at my post on the 5th[2]

Yours very sincerely | George Bentham

DAR 160: 154

[1] Richard Kippist was librarian of the Linnean Society (R. Desmond 1994). CD's paper, 'Two forms in species of *Linum*', was read at a meeting of the society on 5 February 1863.

[2] Bentham was president of the Linnean Society. Bentham and Joseph Dalton Hooker began a ten-day trip to Paris on 17 January 1863 (Jackson 1906, p. 193).

From John Scott 16 January 1863

Edinburgh | Botanic Gardens
Jany. 16th. 1863.

Sir,

The flowers of Lælia were all experimented upon ere I received your last, and there is no prospect of us having more of its section in flower for some time. I consequently cannot try the experiment you have lately suggested.[1] I have, however, cut a portion of the column from two flowers of Lælia fertilised—as I had thought—by application of pollen-mass to rostellum.[2] But I find I have been greatly deceived on an examination of these—by the mode at least in which this has been effected, though the pollen-mass was carefully applied to the latter organ. The difficulty which I anticipated in a former letter to you has presented itself in such a manner that I fear all our attempts at rostellic fertilisation will be baffled.[3] A latent provisional force, an almost conscious sympathy—if I may so express myself—seems to exist between pollen and stigma, and is strikingly evoked, when these are not directly and normally applied to each other. There have we an interesting exhibition of that latent instinctive power, which nature has more or less lavishly bestowed upon vegetal Life, in anticipation of certain contingencies, by which she overcomes these, and enables the plant to perform its functional requirements. In the case under consideration, I find that in one of the columns a process extends itself from each of the stigmas—i.e. each side of stigmatic cavity—to the rostellum, and thus reaches the pollen-mass, into which the pollen-tubes insinuate themselves and are thus conducted into their normal route. In the other no such process is developed, nevertheless, the pollen-tubes are abundant in the conducting tissue of the style. Whether they extend themselves directly to the stigma; or pass along the anterior surface of the modified pistil—or rostellum—to it I cannot discover. I am inclined, to favour the former view, as I cannot find after a careful examination any trace of a pollen-tube passing down the anterior surface of rostellum to stigma; and I observed a single pollen-tube protruding quite freely from a grain upon rostellum towards the stigma. I am sorry that I have not examined some of these flowers earlier, as I might then have been able to speak more definitely. The latter case is not a little perplexing, as with the solitary exception I have mentioned, I can discover no trace now of pollen-tubes above the stigmatic-tissue where they most abruptly end abundantly.[4] I will now attempt to describe a little more particularly one or two sections of them and you will perhaps suggest some means of preventing the recurrence of this deception, when I have an opportunity of experimenting upon Cattleya, and thus not let us again be so teasingly baffled, but force the tubes either to penetrate the rostellum, or remain inert, and thus satisfactorily show its real functional condition.[5] I have an idea of applying a thickish solution of gum to stigma and allowing it to harden, before applying pollen-mass to rostellum.

In the column first referred to then, I find the stigmas and pollen-mass—the latter of course being applied to rostellum—connected by a firm yellowish process extended from the former part. This contains an abundance of elliptical and fusiform cells, similar to those found in the upper portion of the conducting tissue of the styles. It is surprising to see how these cells become elongated at times, when we follow them down the style, and how much they then resemble in certain respects the pollen tubes. Indeed, a hasty observer might almost pass them over for the latter were it not their closed extremities, and pale yellow nuclei. Their function is puzzling! Through this interpolated process then the pollen-tubes insinuate themselves and pass down into the style. *Not a single tube is to be discovered in the tissue of rostellum.* In the other column fertilisation is effected by a somewhat different process. There, I can discover no connection between *pollen-grains* and *tubes*, which nevertheless appear in great abundance in the stigmatic tissue. I made a number of sections of this column, to see if I could discover any trace of them along the anterior surface of rostellum, which I thought they might have passed along and then reached the stigma, but in vain. I therefore, know not whether of the conjectures I have already offered is the more probable, but this much I vouch, that there is an *abundance* of *tubes* in the *style, and not one to be observed in the tissue* of *rostellum*

I see in turning to your observations on the structure of the rostellum, that you remark on the absence of the coherent spindle-formed utriculi in that organ, and that this may probably account for its infertility.[6] My experiments so far favour this view. I was perfectly ignorant of the opinion you there state, and was much puzzled with the occurrence of these peculiar cells, as none of the few works I have had an opportunity to consult mentioned even the existence of such. You have, however, afforded us a rich store in your treatise upon these plants; but I have not as yet had much time for digesting the structural portion;[7] I have engaged my spare time principally with that explicating their varied means for fertilisation

I will be glad to afford you all the information I have on the subject of variation in plants,[8] which I intended mentioning in my last[9] had time permitted. And now I ask you to favour me with the nature of the information you require Is it simply lists of plants now presenting variations with us, or must I furnish you the history of *each*? If the former, I think I could furnish you with a number. That is supposing you to accept such changes as variegation of leaves, changes in colour of flower and the like. Have you done much amongst the *Ferns* in regard to variation, or do you intend treating upon them at all, as I see no allusion to them in "*Origin*"[10] The facilities they afford for reproducing variations is wonderful and affords a strong argument in favour of the view I have proposed—'on influence of sexual relations in transmitting variations"—[11] May I take the liberty of asking you if any more probable explanation has occurred to you for their peculiar facility, than that I have proposed? For a time they almost exclusively engaged my attention. I will be glad to hear how you intend treating them.

I thank you for your promise, to remember with a copy of your paper on Linum.[12] I am afraid my late paper on Drosera & Dionæa, will not be of much interest to you, such as it is however, I will send to you when printed, but they are very tardy in printing their 'Proceedings' here. As I see your name occasionally in the *Gard. Chron.* you may have observed a short notice of it in *No. 2. 1863* under *Bot. Soc. of Edin.*[13]

I am, Sir. J. Scott.

DAR 177: 82

CD ANNOTATIONS
1.9 A latent ... and stigma, 1.11] *double scored pencil*
1.24 and I observed ... stigma. 1.26] *scored pencil*
1.33 but force ... condition. 1.35] *scored pencil*; 'Plaster of Paris'[14] *pencil*
1.35 I have ... rostellum. 1.36] *double scored pencil*
2.1 In the column ... part. 2.3] 'Read Asa Gray'[15] *pencil*
2.12 There, ... *tubes*, 2.13] *cross in margin, pencil*
3.1 you remark on 3.2] '*[*from this*]*' *brown crayon*
3.1 you remark ... infertility. 3.3] *scored brown crayon*; 'Linum—' *brown crayon*
4.1 I will be ... plants, 4.2] *double scored brown crayon*
4.1 I will be ... the like. 4.7] 'in tubers—Bulbs—Suckers'[16] *pencil*
4.6 variegation of leaves] *underl brown crayon*
4.7 Have you ... in "*Origin*" 4.8] *double scored brown crayon*
5.2 I am afraid ... 'Proceedings' here. 5.4] *double scored brown crayon*

1 See *Correspondence* vol. 10, letter to John Scott, 19 December [1862], for CD's advice to Scott regarding the examination of pollination in *Laelia*. See also letter from John Scott, 17 December [1862] (*ibid.*). In his letter to CD of 6 January 1863, Scott detailed his work with *Laelia*, in particular his attempt on 13 December 1862 to cross *Stanhopea oculata* with *Laelia anceps*. CD's response to Scott's letter of 6 January 1863 has not been found; see CD's annotations to that letter for an indication of his reply.

2 In his letter to Scott of 3 December [1862] (*Correspondence* vol. 10), CD asked Scott to try to split the labellum of a *Cattleya*, or an allied orchid, and place a single pollen-mass on the 'large tongue-like Rostellum', which CD thought was a modified stigma. CD wondered whether the rostellum retained 'some of *its primordial function* of being penetrated by pollen-tubes'. With the closely related *Laelia anceps* soon flowering, Scott chose to work with this species instead (see *ibid.*, letters from John Scott, 6 December [1862] and 17 December [1862]).

3 In his letter to CD of 17 December [1862] (*Correspondence* vol. 10), Scott reported that he had pollinated two flowers of *Laelia* in the manner suggested by CD (see n. 2, above). Scott expressed concern that the pollen-tubes, instead of penetrating the rostellum, would 'pass along its surface, down sides of stigmatic cavity, and penetrate stigmas proper'. He asked CD whether a 'preventive for such an

occurrence' could be engineered; CD replied that if Scott examined the column of a specimen in which seed capsules had set, he might be able to determine whether pollen-tubes had penetrated the rostellum (see *ibid.*, letter to John Scott, 19 December [1862]).

4 In *Cattleya* and the related *Laelia*, the stigmatic tissue lies below the rostellum (see *Orchids*, p. 161 fig. 29); Scott was investigating how the pollen-tubes travelled from the rostellum, through or past the stigma, to the conducting tissue of the style.

5 In the final chapter of *Orchids*, chapter 7, CD's account of the rostellum included his argument that the structure was a modified stigma (pp. 307–23).

6 CD wrote: 'The utriculi are believed to be connected with the penetration of the pollen-tubes; and their absence in the rostellum probably accounts for its infertility' (*Orchids*, p. 311). He also noted that the layer of cells secreting viscid matter was thicker on the rostellum than on the stigma.

7 Part of chapter 7 of *Orchids* (pp. 307–30) was devoted to the examination of orchid structure, particularly the structure of the rostellum and pollen-masses.

8 CD was writing chapter 11 of *Variation* on 'bud-variation', which he began on 21 December 1862 (see *Correspondence* vol. 10, Appendix II); in his letter to Scott of 19 December [1862], CD asked him to provide any cases of 'what Gardeners sometimes call sports, & which I shall call "bud-variation"'.

9 Letter from John Scott, 6 January 1863.

10 CD mentioned ferns in *Variation* 1: 383, citing Scott 1862a.

11 In Scott 1862a, pp. 357–60, Scott sought to explain the facility with which ferns, when propagated by spores, inherited the morphological peculiarities of the part of the plant on which the spores were produced. Scott attributed the more frequent inheritance of such peculiarities in ferns, as compared to seed plants, to the nature of the reproductive structures. He argued that the separation of the reproductive structures in seed plants, and their necessary union, meant that any individual peculiarities would tend to be lost by blending inheritance. In ferns, by contrast, he argued that because the spore was the product of a single lateral organ, individual peculiarities would tend to be inherited unmodified. Scott sent a copy of his paper to CD in December 1862, noting the impact on the theory of natural selection of his views regarding blending inheritance in seed plants (see *Correspondence* vol. 10, letter from John Scott, 6 December [1862]).

12 The letter in which this offer was made has not been found (see n. 1, above). However, Scott's name appears on CD's presentation list for the *Linum* paper (see Appendix IV). See also Scott 1863a, p. 547 and n.

13 Scott's paper, 'On the propagation and irritability of *Drosera* and *Dionæa*', was read at a meeting of the Botanical Society of Edinburgh on 11 December 1862, and was published as an abstract (Scott 1862b). An abstract of Scott's paper appeared in the *Gardeners' Chronicle*, 10 January 1863, p. 30.

14 See n. 3, above. See also letter to John Scott, 21 January [1863].

15 A. Gray 1862a. See letter to John Scott, 21 January [1863] and n. 11.

16 See n. 8, above. See also letter to John Scott, 21 January [1863].

From Isaac Anderson-Henry 17 January 1863

Hay Lodge, Trinity, Edinburgh.
Jany 17/63

Dear Sir

I have much pleasure in acknowledging receipt of your very polite note to me of the 14th Inst.[1] If the Strawberries are of no use, and if I have anything else that is;—or if there is any information I can give you,—I will feel honoured and gratified in communicating it[2]

Your treatise on the *Primulaceæ* of which you so kindly sent me a copy was to me a communication of no ordinary interest.[3] I had always intended writing my views to you on the subject. Many years ago I tried my hand on that family, but made nothing of them. The *Linum* tribe to which you alluded near the close[4] had, the Summer before last occupied me a good deal. I was most anxious to incorporate a dash of the fine colour of *L. rubrum* on to L. album a perennial—L *Corymbiflorum* a yellow sps—or L. *trygynum* a brighter yellow still— But a more difficult intractable race I seldom before grappled with. Your treatise threw in upon me a gleam of light—and I suspect I had not used the *sympathetic* anthers—the mystery of the long and short ones being unknown to and unsuspected by me.[5] Last Summer I had determined to test the matter as regards the *Linum* tribe. But having gone on a pretty extensive tour on the Continent my operations in that way were prevented from being carried out. I may try my hand anew this Summer. I may observe, however, that of one Cross I did save seeds viz[t] between the *L. Album* (seed bearer) and L. *Rubrum*. But as the former was prone to to self fertilisation, I am not very confident of the Cross being effected. The plants of this cross are up—four in number & the foliage offers a difference from the female parent. So I hope it is true. I may have hit the right anthers. But if I remember rightly—the Anthers are far from being uniformly short & long as in the *Primulæ*. It calls for and deserves a close investigation which I mean to give it this season

I have got your admirable Book on the hybridisation of Orchids.[6] I never felt myself so small, as I did on reading it, in all my life. You have achieved immortality by such a work. A tithe of its discoveries would be an enduring monument to any other man. I rejoice to see you still at work—and what next it is you are to astound & astonish us with I, & all the world stand on tiptoe of expectation to learn.

If you have occasion ever to come this way do honour me with a visit— Save D[r] Hooker & D[r] Lindley none in all England will be more welcome to me.[7]

Believe me | most faithfully yours | I Anderson Henry.

I have for many years tried to incorporate the Raspberry with the Strawberry. I have one brood of plants with wiry foliage some having 4 divisions in the leaf. Myatts Pine was here also the seed bearer—a soft foliaged of variety. This coming Season I am to have an alliance if possible between the Bramble & raspberry. When I was at Woodend (my place in Perthshire) I found an upright growing variety of the Bramble which I had removed into the Garden for my intended manipulations. I will be glad to let you know the result.

If you would wish me to try anything to serve you in any way I will very gladly undertake it

Charles Darwin Esqre F.R.S. | &c &c &c

DAR 159: 60

[1] CD's letter has not been found.

[2] CD was working on *Variation* and completed a draft of his 'long Chapter' on plant variation, eventually published as chapters 9 and 10, in December 1862 (*Correspondence* vol. 10, Appendix II); strawberries

were discussed in *Variation* 1: 351–4. CD may have written to Anderson-Henry seeking the same information that he sought from readers of the *Journal of Horticulture* (see *Correspondence* vol. 10, letter to the *Journal of Horticulture*, [before 25 November 1862]). Anderson-Henry evidently offered to send CD a strawberry hybrid in response (see letter to Isaac Anderson-Henry, 20 January [1863]). See also letter to Asa Gray, 2 January [1863] and n. 17.

3 'Dimorphic condition in *Primula*'; Anderson-Henry's name is included on CD's presentation list for this paper (*Correspondence* vol. 10, Appendix III).

4 See 'Dimorphic condition in *Primula*', p. 96 (*Collected papers* 2: 62–3).

5 In 'Two forms in species of *Linum*', CD examined the fertility of dimorphic forms, the 'short-styled' and 'long-styled', and discussed the length of the stamens and styles. CD was due to read his paper, 'Two forms in species of *Linum*', before the Linnean Society in February 1863 (see letter from George Bentham, 16 January 1863).

6 *Orchids*.

7 Joseph Dalton Hooker and John Lindley.

From H. W. Bates 17 January [1863][1]

10 Hollis place | Prince of Wales' road | N.W.
Jany 17th

My Dear Mr Darwin

Your last has been sent to me here where I have fixed myself for a few weeks to see my book through the press.[2] I hope you will put yourself to no further trouble about the article in Linnean trans. your writing a notice for Nat. Hist. rev. consuming time which is required for your great work, is an act of kindness which I feel most keenly.[3] I send today a copy to Prof. Asa Gray.[4]

With regard to insects frequenting flowers of Melastomads: this order of plants is certainly less visited by bees & Lepidoptera (& all orders) than any other order having flowers.[5] With regard to the general fact of neglect of Melastomad flowers there cannot be any doubt as the bushes are excessively numerous in the woods where I collected daily.

Butterflies beetles & bees prefer flowers of Myrtaceæ— there is a sweet-smelling blossom of a tree I supposed to be a myrtle which always swarms with insects when the flowers of adjoining Melastoma bushes have not a single visitor.

Bees *Euglossa, Melipona, Cenchris, Megachile, Augochlora* in the real virgin forests of the Amazons are more frequently seen at sap exuding from trees,[6] at excrement of birds on leaves & on moist sand at edge of water than at flowers. The larger butterflies accompany them; some of these latter crowd on fine flowers of some creeping plant (I think a kind of Combretaceæ). Blossoms of Inga are the favourite resort of some floral beetles (antichira).[7]

Not having my note books here I cannot give you more details on this subject. —Mr Murray has ordered so many fine illustrations for my book & these have been (by my residing away from London) so slow in executing that the printing of my book has been much hindered.[8]

Yours sincerely | H W Bates

DAR 205.8: 67 (Letters)

CD ANNOTATIONS
1.1 Your ... Gray. 1.5] *crossed pencil*
5.1 Not having ... hindered. 6.3] *crossed pencil*

[1] The year is established by the reference to Bates 1863 (see n. 2, below).
[2] Letter to H. W. Bates, 12 January [1863]. Bates refers to his book *The naturalist on the river Amazons* (Bates 1863).
[3] In his letter to Bates of 12 January [1863], CD informed him that he had written the review of Bates 1861 that was to appear in the next number of the *Natural History Review* ('Review of Bates on mimetic butterflies').
[4] See letter to H. W. Bates, 12 January [1863] and n. 2. CD had asked Asa Gray if he could find a colleague in the United States to review Bates's account of mimetic resemblances in Amazonian *Lepidoptera* (Bates 1861; see *Correspondence* vol. 10, letters to Asa Gray, 23 November [1862] and 26 November [1862]). Gray himself wrote the review of Bates 1861 that was published in the September 1863 number of the *American Journal of Science and Arts* (A. Gray 1863a).
[5] See letter to H. W. Bates, 12 January [1863] and n. 4; Bates observed these tropical and sub-tropical plants during his travels in the Amazon region.
[6] Bates lists New World bee genera. 'Cenchris' is probably a mistake for *Centris*.
[7] *Antichira* was a genus of Coleoptera well represented in Central America. Bates later described the beetles as members of the Lamellicornia tribe and the family Rutelidae (see Bates 1886–90, 2 (pt 2): 265).
[8] John Murray published Bates 1863 in April 1863 (*Publishers' Circular* 26: 193). The book included forty-two illustrations.

To Thomas Rivers 17 [January 1863][1]

Down
17th

My dear Sir

Will you permit me to ask one more question;[2] Can you distinguish generally, always, or never a nectarine tree from a Peach before it *flowers* or before it *fruits*. I sh^d. like to quote your answer.—[3]

Yours sincerely | C. Darwin

John Wilson (Autographs) (catalogue 61, 21 July 1989, item 50)

[1] The date is established by the relationship between this letter and the letters to Thomas Rivers, 11 January [1863] and 15 January [1863].
[2] In December 1862, CD began writing chapter 11 of *Variation*, dealing with bud-variation, and initiated a correspondence with Rivers, a renowned Hertfordshire nurseryman (see *Correspondence* vol. 10, letters to Thomas Rivers, 23 December [1862] and 28 December [1862]). See also letters to Thomas Rivers, 7 January [1863], 11 January [1863], and 15 January [1863].
[3] In *Variation* 1: 340, CD cited Rivers as the authority for his statement that the fruit of the peach and nectarine differed in appearance and flavour, but the trees differed 'in no other respects', and could not even be distinguished 'whilst young'.

From Henri Louis Frédéric de Saussure[1] 17 January 1863

Genève.
17 janv 63.

Monsieur

J'ai bien reçu votre lettre du 8e Janv. et vous en suis bien reconnaissant.[2]

Le travail sur le *mexique* contiendra qque géologie et pourra vous intéresser par la suite.[3] Quant à celui des oiseaux il y en a qques exemplaires chez Rothschild à Paris (14 rue de Buci)[4] Il est presque épuisé.

Dans ce moment j'ai divers travaux sur le chantier, mais il en est un qui m'occupe depuis longtemps et qui vous interessera. C'est la *filiation* des genres, des espèces et des moeurs dans les guêpes.[5] Il y a dans ce groupe des transitions singulières et des transformations spécifiques et morales évidentes.

J'avais pensé que ma carte du Mexique devrait intéresser dans ce moment et j'en ai fait un depôt chez Stanford, *6 Charing Cross*.[6] Mais il parait qu'en Angleterre on se soucie peu de ce pays car cette carte a bien de la peine à se vendre quoique fort belle et d'une exécution soignée, et ajoutant bien des choses a la géographie de ces contrées.

Je reste Monsieur votre très dévoué | H. de Saussure

Oserais-je vous prier de faire parvenir la ci-jointe lettre au British Museum.[7] Elle n'est pas pressée et vos courses vous conduisent sans doute à cet etablissement.

DAR 177: 40

[1] For a translation of this letter, see Appendix I.
[2] CD's letter has not been found.
[3] The reference is to the first volume of Saussure 1858–71, a memoir on the natural history of Mexico, the Antilles, and the United States.
[4] Saussure 1858. CD's lightly annotated copy of Saussure 1858 is in the Darwin Pamphlet Collection–CUL. Saussure refers to the Paris publisher, Jules Rothschild.
[5] Saussure had already published a major study on wasps (Saussure 1852–8); his synopsis of American wasps formed the fourth part of Saussure 1858–71.
[6] Saussure 1862. The London bookseller, stationer, and map engraver, Edward Stanford, was an agent for ordnance maps, geological survey maps, and Admiralty charts (*Post Office London directory* 1861).
[7] The enclosed letter has not been found.

From Hugh Falconer 18 January [1863][1]

The Athenæum
18th. Jan^y^

My Dear Darwin

Many thanks for your note with the enclosure from Wallace which I shall try and hunt down—I mean the *Mastodon*.[2] But I do not fear him quoad the Australian Case.[3] That was an imposter from the outset—probably picked up in the valley of Tarija, when a certain enterprising traveller—*not* Charles Darwin—was there.[4]

Wallace—whom I called up at Cambridge—was disposed to stick up for the Sumatran Elephant—but I do not think the case will hold—beyond a moderate measure of variation, not of specific value.[5]

John Evans has found *a jaw with teeth* in the line between the angle of the perfect leg and wing of *Archæopteryx*!![6]

Waterhouse I am told pronounces it to be a *fish's* jaw.[7] Fancy only a feathered Fish! But joking apart it is odd, that this jaw should present itself alongside of the fossil.— I enclose John Evans' sketch.[8] Kindly return it to me. I am sure you will be amused—at so many signs of the shallow examination given to the precious object in the first instance.[9]

I do hope and trust that you will seize an occasion, to do what you hint at doing in one of your notes,[10]—i.e. to strike in Charles 'with the Strong Arm' against the Charlatan pretensions of the common enemy of British Naturalists.[11]

I have been collating your charming lines, forming the commencement of your little paper on the 'sac' in the cirripedes N.H.R.—with the avowal of the blunder about British Foss. monkeys.[12] What a contrast in the Candour of the one—and the double eyed *Fouchè-ism* of the other![13] Had Satan himself been compelled to a confession, that is the style he would have adopted. But we are philosophes—and must not use naughty words—for fear of the example.

My Dear Darwin | Yours Ever Sin^{ly} | H Falconer

PS. When you write thank Wallace very much.

DAR 164: 13

1 The year is established by the relationship between this letter, the letter from Hugh Falconer, 9 January 1863, and the letter to A. R. Wallace, 14 January [1863] (see n. 2, below).

2 CD's note has not been found. The reference is probably to Schneider 1863; in his letter to CD of 14 January [1863], Alfred Russel Wallace enclosed information for Falconer on Carl Friedrich Adolf Schneider's reported find of mastodon fossil teeth in Timor, in the Malay Archipelago. See also *Correspondence* vol. 9, letter from A. R. Wallace, 30 November 1861. See letter to Hugh Falconer, 5 [and 6] January [1863], and letter from Hugh Falconer, 9 January 1863.

3 In 1844, on the basis of Paul Edmund de Strzelecki's report of a single fossil tooth found in Australia, Richard Owen proposed the species *Mastodon australis* (Owen 1844). He thereafter cited the supposed species as evidence of the remarkable distribution of the Proboscidean order. Falconer questioned the provenance of the fossil, arguing that the tooth was from the South American species *M. andium* (Falconer 1863, pp. 96–101). See also letter to Hugh Falconer, 5 [and 6] January [1863] and n. 6, and letter from Hugh Falconer, 9 January 1863 and n. 3.

4 Tarija is in Bolivia. Falconer refers to Strzelecki (see n. 3, above), who travelled up the west coast of the Americas, from Chile to California, in 1836 (*ADB*).

5 In October 1862, Wallace stayed at Magdalene College, Cambridge, as a guest of Alfred Newton, and attended his first meeting of the British Association for the Advancement of Science (Wallace 1905, 2: 45–6). Sumatra, like Timor (see n. 2 above), is part of the Malay Archipelago.

6 See letter from Hugh Falconer, 3 January [1863] and n. 11. Falconer refers to *Archaeopteryx*, the fossil that was both reptilian and bird-like; John Evans discovered the animal's cranium, which Owen had overlooked (Owen 1862a), after Samuel Joseph Mackie reported Evans's discovery of the skull in the *Geologist* (Mackie 1863). Evans also identified the creature's jaw and teeth (see the letter from Hugh Falconer to John Evans, 15 January 1863, in the Falconer Museum, Forres, Moray, Scotland (H.363c), and John Evans 1865, pp. 418–21). Between Owen's description of *Archaeopteryx*, presented to the Royal Society of London on 20 December 1862, and his publication of this description in the second half of 1863, an engraving of the jaw was added to plate 1 (Owen 1862a, fig. 1, p, and fig. 3, p′). Owen also mentioned Evans's discovery of the jaw in the text (Owen 1862a, p. 46), but did

not offer his own opinion regarding its nature. There is an annotated copy of Owen 1862a in the Darwin Library–CUL.

7 George Robert Waterhouse was keeper of the department of geology at the British Museum. In an article published in the January number of the *Geologist*, prior to Evans's discovery of the jaw, Mackie wrote that both Owen and Waterhouse had inferred that the fossil was a 'true bird' (Mackie 1863, p. 4). Evans discussed whether the jaw belonged to the rest of the fossil (Evans 1865, pp. 419–21), and argued that if the jaw was part of *Archaeopteryx*, the evidence of its ornithological nature was diminished since birds lacked teeth.

8 The sketch has not been found.

9 See n. 6, above.

10 CD had mentioned to Joseph Dalton Hooker that he was thinking of voting against Owen when the council of the Royal Society was next elected (see letter to J. D. Hooker, 3 January [1863] and n. 9); in his letter to Falconer of 5 [and 6] January [1863], CD expressed a less specific plan of slighting Owen 'by some overt act'.

11 The allusion may be to William Shakespeare's *Julius Caesar*, 4.2.121–4: 'There is no terror, Cassius, in your threats,/For I am armed so strong in honesty/That they pass by me as the idle wind,/Which I respect not' (Wells and Taylor eds. 1988).

12 In his letter to the *Natural History Review*, [before 10 October 1862] (*Correspondence* vol. 10; *Collected papers* 2: 85–7), CD noted how a sac on a particular barnacle, or cirripede, that he described as auditory, had recently been described by the Russian zoologist August David Krohn as ovarian (Krohn 1859). CD wrote: 'As Dr. Krohn is no doubt a much better dissector than I am, I fully admitted my error and still suppose that he is right', but he also urged other scientists to continue to investigate the problem. Falconer contrasted CD's style with Owen's in his letter in the 'Miscellaneous' section of the *Annals and Magazine of Natural History* (Owen 1862b); Owen announced that the teeth he had described as belonging to a fossil monkey from the Eocene sands of Kyson, Suffolk (Owen 1840), were 'most probably' the lower molars of a species of *Hyracotherium*, an Eocene horse, but that the molars were easily confused.

13 Joseph Fouché was a French statesman noted for his adroit and unprincipled manoeuvering and duplicity (*EB*).

To Asa Gray 19 January [1863][1]

Down Bromley Kent
Jan 19th.

My dear Gray

When I look over your letter of Dec 29th,[2] & see all the things you tell me & all the trouble which I have caused you, overworked as you are, upon my life I am ashamed of myself.— I will be less unreasonable for the future; but with lots of gold close beneath the surface it is hard not to dig for it. I was glad to hear of Platanthera & the Butterfly, & hearty thanks for the Indian corn; what little grains! I knew nothing about "glucose partly replacing starch".[3] I have several odds & ends to say: Bates has forwarded copy of his paper to you for Haldeman.[4] I have reviewed for next Nat. Hist Review, & five pages will therein, if you care, give the cream of his case.—[5] Do not trouble yourself to weigh, as I asked, fruit of your wild Strawberry; what you say suffices for my purpose.—[6] I was muddled about 1st part of your Review of Orchids[7]—it is not worth explaining how I came to be so; but my memory returned the day after my letter went; I confounded one Review with another.—[8]

You ask about "Dichogamy", term & facts are given by C. K. Sprengel in his Entdeckte Geheimniss der Natur,[9]—a curious old book full of truth with some little nonsense.— I have been thinking how interesting it would be to experiment on the 3 kinds of flowers of Linum Lewisii, but I fear it would be impossible to get seed.—[10] I have been at those confounded Melastomas again; throwing good money (ie time) after bad.[11] Do you remember telling me you could see no nectar in your Rhexia;[12] well I can find none in Monochætum, & Bates tells me that the flowers are in the *most marked* manner neglected by Bees & Lepidoptera in Amazonia.[13] Now the curious projection or horns to the stamens of Monochætum are full of fluid, & the suspicion occurs to me that Diptera or small hymenoptera may puncture these horns like they puncture (proved since my orchid book was published) the dry nectaries of true Orchis.—[14] I forget whether Rhexia is common; but I very much wish you would next summer watch on a warm day a group of flowers & see whether they are visited by small insects & what they do.— I sh$^{d.}$ die a happier man, if I could understand these Melastomas. Oh dear, I wish poor wounded Mr Rothwick would not care for such trifles as the welfare of his country, & would stick to flowers![15]

Well, your President has issued his fiat against Slavery—[16] God grant it may have some effect.— I fear it is true that very many English do not now *really* care about Slavery; I heard some old sensible people saying here the same thing; & they accounted for it (& such a contrast it is to what I remember in my Boy-hood) by the present generation never having seen or heard much about Slavery.—[17] I sometimes cannot help taking most gloomy view about your future. I look to your money depreciating so much that there will be mutiny with your soldiers & quarrels between the different states which are to pay.[18] In short anarchy & then the South & Slavery will be triumphant. But I hope my dismal prophecies will be as utterly wrong as most of my other prophecies have been. But everyone's prophecies have been wrong; those of your Government as wrong as any.— It is a cruel evil to the whole world; I hope that you may prove right & good come out of it.— Do not trouble or tire yourself to write to me,—though I never receive a letter from you without real pleasure & kind instruction.—

Farewell | Yours most sincerely | C. Darwin

Gray Herbarium of Harvard University (57)

1 The year is established by the relationship of this letter to the letter from Asa Gray, 29 December 1862 (*Correspondence* vol. 10).

2 Letter from Asa Gray, 29 December 1862 (*Correspondence* vol. 10).

3 See *Correspondence* vol. 10, letter from Asa Gray, 29 December 1862, and letter to Asa Gray, 6 November 1862. Gray's reference to glucose and starch has not been found, but see the letter to John Scott, 16 February [1863].

4 Gray was one of the editors of the *American Journal of Science and Arts*, and had offered to ask Samuel Steman Haldeman to review Bates 1861 for the journal (see letter to H. W. Bates, 12 January [1863] and n. 2, and letter from H. W. Bates, 17 January [1863]).

5 'Review of Bates on mimetic butterflies' was published in the April 1863 number of the *Natural History Review*.

[6] See letter to Asa Gray, 2 January [1863] and n. 17. In the section on strawberries in *Variation* 1: 351–4, CD stated that he was informed by Gray that the fruit of *F. virginiana* was 'only a little larger' than that of the common wood strawberry, *F. vesca* (*Variation* 1: 351 n. 100).

[7] A. Gray 1862b.

[8] A. Gray 1862a and 1862b. See *Correspondence* vol. 10, letter from Asa Gray, 10 November 1862, and letter to Asa Gray, 26[–7] November [1862].

[9] Sprengel 1793. CD refers to the letter from Asa Gray, 29 December 1862 (*Correspondence* vol. 10).

[10] CD completed his paper, 'Two forms in species of *Linum*', in December 1862. The paper was to be read before the Linnean Society on 5 February 1863 (see letter from George Bentham, 16 January 1863). Citing Planchon 1847–8, CD described *L. lewisii* as bearing, on the same plant, some flowers with anthers and stigmas of the same height, and others with styles either longer or shorter than the stamens; this he called 'a unique case' (p. 82; *Collected papers* 2: 104–5). CD does not appear to have conducted the trials to which he refers; instead, John Scott, who was head of the propagating department at the Royal Botanic Garden, Edinburgh, offered to conduct experiments on the species on CD's behalf (see letter from John Scott, [3 June 1863], and letter to John Scott, 6 June [1863]). See also letter to Asa Gray, 2 January [1863] and nn. 9 and 10.

[11] CD was interested in what he considered might be a novel form of flower dimorphism in Melastomataceae (see letter to Hugh Falconer, 5 [and 6] January [1863] and n. 22).

[12] See *Correspondence* vol. 10, letter from Asa Gray, 4 August 1862 and n. 3.

[13] See letter from H. W. Bates, 17 January [1863].

[14] Treviranus 1863a. CD's heavily annotated copy of Treviranus 1863a is in the Darwin Pamphlet Collection–CUL.

[15] Joseph Trimble Rothrock, a student and assistant of Gray's, had made some observations for CD in the past, but in July 1862 he enlisted in the 131st Pennsylvania Infantry (*DAB*; see *Correspondence* vol. 10, letters from Asa Gray, 5 September 1862 and 22 September 1862).

[16] In September 1862, Abraham Lincoln announced his emancipation proclamation, in which he decreed that from 1 January 1863, all slaves in territories still in rebellion would be free (Denney 1992, pp. 248, 251).

[17] Many people in Britain were hostile to the Union for economic reasons; trade with the Confederate States had been prosperous prior to the outbreak of war (see McPherson 1988, pp. 548–53).

[18] In addition to issuing bonds and raising taxes to fund the war effort, the gold standard was abandoned by the United States government in 1861, and, in February 1862, the Legal Tender Act was passed. By the end of the war, the Union authorities had sanctioned the issue of a total of $3.2 billion in legal tender notes, or 'greenbacks', to pay for military spending. As soon as they were printed, the value of greenbacks depreciated; by the end of the war a greenback dollar was worth only 35 cents in gold (Poulson 1981, pp. 352–7, and McPherson 1988, pp. 442–50).

To [Friedrich Emil Suchsland][1] [after 19 January 1863][2]

[Down]

Dear Sir

you have sent me, apparently by mistake Dr. Rolle's Book, which I return by this post.— The Author has sent me copies . . .[3]

Incomplete[4]
J. A. Stargardt (catalogue 618, item 441)

[1] Suchsland was head of the publishing firm Johann Christian Hermann'sche Verlagsbuchhandlung, based in Frankfurt-am-Main, which published Rolle 1863 and 1866 (see n. 3, below).

[2] The date is established by reference to the publication date of the last two parts of Rolle 1863; the earlier instalments were published in 1862 (see n. 3, below).

[3] Suchsland sent CD copies of two books by Friedrich Rolle (Rolle 1863 and 1866) before 16 March 1866 (see letter from Rudolf Suchsland, 16 March 1866, *Calendar* no. 5035). Rolle himself had sent CD a copy of Rolle 1863, but only the first part of Rolle 1866; Suchsland sent the remaining parts (see letters from Friedrich Rolle, 26 January 1863 and 28 January 1866 (*Calendar* no. 4986)). The publication of the parts of Rolle 1863 was announced on 8 September 1862, 27 October 1862, and 19 January 1863 (*Börsenblatt für den Deutschen Buchhandel* 29 (1862): 1862, 2258; 30 (1863): 113), and CD mentioned them in his letters to Rolle of 17 October [1862] (*Correspondence* vol. 10) and 30 January [1863]. There is a lightly annotated copy of Rolle 1863 in the Darwin Pamphlet Collection–CUL.

[4] According to the catalogue description, the original autograph letter consists of one page.

To Isaac Anderson-Henry 20 January [1863][1]

Down, Beckenham, Kent. S.E.
Jan. 20th. 1863.

Dear Sir,

I thank you cordially for your singularly kind letter.[2] But your kindness leads me to overrate what I have done in Natural history.— I was much surprised at the appearance of the leaf of the hybrid strawberry;[3] and was much tempted to accept your kind offer and declined only on reflection, that as my health is weak and I have many other points to attend to, I feared I should have neglected your specimen. Perhaps you will be so good as to inform me whether this hybrid produces next spring *fertile* seeds. If I understand right the supposed progeny from the strawberry and raspberry have not yet fruited: but if this be really a cross you will indeed have effected a prodigy, and it will be very curious if you cross *Blackberry and Raspberry*;[4] I shall much like to hear the result. I may mention that this past spring I tried again the crosses on *Primula* with the same result rather more strongly marked; and I have gone on, now for three generations breeding them what I call homomorphically, with some curious results, which I shall publish whenever I have time.[5] I have sent a paper on *Linum* to the Linnæan Society;[6] when it is published I will do myself the pleasure of *sending to you a copy*; and it will I should think be in good time for your experiments.[7] I cannot say how glad I am that you will make some experiments on the subject.

It does not absolutely follow in making a cross between distinct species, that the same rules would follow in the *fertility of the pollen*. I hope that you will try and mark separately (excluding insects, as you know better than I do, the necessity) the *two* kinds of pollen of one species on the stigma of the other; and see in making hybrids what the difference is in fertility, and in the character of the hybrid seedlings. This would be an entirely new field for observation and discovery. You will see in my paper that some species of *Linum* are *not* dimorphic, and are self-fertile; and so it is in some other genera.—[8] You refer to *L. rubrum*; I am not a Botanist and have called one of the species on which I have experimented *L. grandiflorum*, which is crimson and not uncommon in flower gardens; I hope I have not made a mistake in name.—[9]

You *most* kindly permit me to mention any point on which I want information. If you are so inclined, I am curious to know from systematic experiments, whether Mr.

D. Beaton's statement that the pollen of two shortest anthers of scarlet *Pelargonium* produce dwarf plants, in comparison with plants produced from *same* mother-plants by the pollen of longer stamens from same flower.[10] It would aid me much in some laborious experiments on *Melastomas*.[11] I confess I feel a little doubtful: at least I feel pretty nearly sure that I know the meaning of short stamens in most plants. This summer (for another object) I crossed Queen of Scarlet Pelargonium with pollen of *long* and *short* stamens of Multiflora alba, and it so turns out that plants from *short*-stamens are the tallest; but I believe this to have been mere chance.—[12] My few crosses in Pelargonium were made to get seed from the central *peloric* or regular flower (I have got one from peloric flower by pollen of peloric); and this leads me to suggest that it would be very interesting to test fertility of *peloric* flowers in three ways, own peloric *pollen on peloric stigma—common pollen on peloric stigma*—peloric pollen on common stigma of same species. *My object is* to discover whether with change of structure of flower there is any change in fertility of pollen or of female organs. This might, also, be tested by *trying peloric and common pollen* on stigma of a distinct species, and conversely.[13] I believe there is a *peloric* and common variety of *Tropæolum*; and a *peloric* or upright and common variation of some species of *Gloxinia*.—[14] And the *medial peloric* flowers of Pelargonium; and probably others unknown to me.—

To recur to *Linum*; if you cross distinct species, it would, I think be advisable to *take* **two** *dimorphic* species; and *not* **one** *dimorphic* and the other self-fertile. I have reason to suspect L. trigynum is dimorphic but it has not yet flowered with me.—[15]

Your kindness has led me to trespass at unreasonable length on your time, and, | I remain dear Sir, | yours truly obliged. | Ch. Darwin.

P.S. I thank you for your most kind invitation, which I fear I shall never profit by.—[16]

P.S. Have you any *quite* sterile Hybrid plant with a rather large stigma, (and of which I could procure either one or both pure parent flowers) as I very much wish to compare by dissection certain minute parts of stigma: if so, and you would have great kindness to send me by Post in a little *tin* box *some flowers, it would be a very great favour*.[17]

Copy
DAR 145: 1

[1] The text of the letter is taken from a copy commissioned by Francis Darwin for his editions of CD's letters; part of the letter was published in *ML* 2: 297–8. The copyist dated the letter 20 January 1863; however, when Francis edited the transcription he added pencil brackets to indicate that the year had been added by the copyist and was not present in the original letter. The year is confirmed by the relationship between this letter and the letter from Isaac Anderson-Henry, 17 January 1863.

[2] Letter from Isaac Anderson-Henry, 17 January 1863.

[3] In his letter of 17 January 1863, Anderson-Henry informed CD that he had hybridised the raspberry and strawberry, resulting in 'one brood of plants with wiry foliage some having 4 divisions in the leaf'.

[4] In his letter of 17 January 1863, Anderson-Henry informed CD that this 'coming Season' he planned to attempt a cross between 'the Bramble & raspberry'.

5 CD first published an account of *Primula* in 1862 ('Dimorphic condition in *Primula*'). CD's subsequent experimental results on homomorphic crosses in *Primula* (that is, crosses between plants of the same species with the same flower forms) were discussed in 'Illegitimate offspring of dimorphic and trimorphic plants'. His notes on these experiments are preserved in DAR 108, 110, and 111. CD also suggested additional *Primula* experiments to John Scott, encouraging him to publish the results (see, for instance, *Correspondence* vol. 10, letter to John Scott, 3 December [1862], and Scott 1864a).

6 'Two forms in species of *Linum*' was read before the Linnean Society on 5 February 1863 (see letter from George Bentham, 16 January 1863).

7 See letter from Isaac Anderson-Henry, 17 January 1863. Anderson-Henry's name appears on CD's presentation list for 'Two forms in species of *Linum*' (see Appendix IV).

8 'Two forms in species of *Linum*', p. 82 (*Collected papers* 2: 104).

9 *Linum grandiflorum* is the name CD used for this plant both in 'Two forms in species of *Linum*' and in *Forms of flowers*.

10 Donald Beaton was a regular contributor to the *Journal of Horticulture*; the reference is to Beaton 1861, pp. 312–13. See also Beaton 1860, pp. 254–5, and *Correspondence* vol. 9, letters to *Journal of Horticulture*, [17 May 1861] and [before 9 July 1861], and letters to J. D. Hooker, 14 May [1861] and 18 [May 1861]. See Appendix V for Beaton's responses to CD in the *Journal of Horticulture* in 1863.

11 CD continued his study of the Melastomataceae, begun in 1861, in January 1863 (see letter to Hugh Falconer, 5 [and 6] January [1863] and n. 22).

12 CD conducted a series of crosses with different varieties of *Pelargonium* in May 1862 (see *Correspondence* vol. 10, letter to Daniel Oliver, 8 June [1862], and letter to Asa Gray, 1 July [1862]). CD's notes on these experimental crosses are preserved in DAR 51: B4–13. See also the annotated article on pelargoniums in CD's copy of the *Gardeners' Chronicle*, 15 February 1862, pp. 139–40, which is in the Cory Library, Cambridge Botanic Garden.

13 These experiments on regular (peloric) flowers found on the same plant with primarily irregular flowers, were discussed in *Variation* under the subheading 'Monstrosities as a cause of sterility' (*Variation* 2: 167).

14 CD also discussed the peloric *Tropaeolum* with Charles William Crocker, a former gardener at the Royal Botanic Gardens, Kew (see *Correspondence* vol. 10, letter from C. W. Crocker, 24 November 1862 and n. 4). The *Gloxinia* referred to was *G. speciosa* (*Variation* 1: 365 and 2: 167). See also *Correspondence* vol. 10, letter to John Scott, 19 December [1862].

15 CD discussed the crossing of dimorphic *Linum* species in *Forms of flowers*, pp. 81–101. See *ibid.*, p. 100, for a discussion of *L. trigynum*.

16 Anderson-Henry invited CD to pay him a visit if he ever came to Scotland (see letter from Isaac Anderson-Henry, 17 January 1863).

17 In his letter to John Scott of 21 January [1863], CD stated that he wished to observe the state of the tissues of the stigma of a sterile hybrid for comparison with 'the same in fertile parent-species', in order to show the 'difference in female organs of hybrid and pure species'.

To Hugh Falconer 20 [January 1863][1]

Down.
20th

My dear Falconer

It was really very good of you to tell me of the jaw.[2] What a strange accident if a fish's jaw has got there; but if it belongs to the Archæopteryx, it will surely show some great peculiarity.[3] I hope it may belong to it. Already a German author has advanced the case as a grand argument in favour of the "Origin".[4] Has God demented Owen, as a punishment for his crimes, that he should overlook such a point?[5] I hope to be in London on the first week in February and to nothing do I look forward with more pleasure than seeing you.[6]

I have a pile of letters to answer so farewell my good old friend. | C. Darwin.

Copy
DAR 144: 30

[1] The date is established by the relationship between this letter and the letter from Hugh Falconer, 18 January [1863].
[2] See letter from Hugh Falconer, 18 January [1863] and n. 6.
[3] See letter from Hugh Falconer, 18 January [1863] and n. 7.
[4] Friedrich Rolle discussed the relevance of *Archaeopteryx* to *Origin* in the postscript to Rolle 1863, p. 263. There is an annotated copy of Rolle 1863 in the Darwin Pamphlet Collection–CUL.
[5] Richard Owen. See n. 2, above.
[6] According to Emma Darwin's diary (DAR 242), CD, Emma, Henrietta, and Horace Darwin stayed in London from 4 to 14 February 1863. See 'Journal' (Appendix II). See also letter from E. A. Darwin, 21 [January 1863].

To Daniel Oliver 20 [January 1863][1]

Down. | *Bromley.* | *Kent. S.E.*
20th

Dear Oliver

Many thanks. I have some analogous cases, but I am very glad to have what you have sent.—[2] Have I not been very good, & not given you any trouble for an age?[3]

Ever yours sincerely | C. Darwin

I have been this morning copying out References on subjects which concern me from N. Hist. R.—[4] What an enormous benifit you have conferred on everyone by your gigantic labours. No wonder that you are "omniscient".—[5]

Do you remember my suggesting that if ever you had *spare* time (so likely this is!!) that you might do good work by discussing F. Water plants.—[6] Now it occurred to me the other day (& I then much wished to know,) whether there is not an unusual proportion of such plants with separated sexes. I merely mention this, as showing one *little* point which would turn up in such an examination.

Cleveland Health Sciences Library (Robert M. Stecher collection)

[1] The date is established by the relationship between this letter and the letter from Daniel Oliver, 22 January 1863.
[2] The reference has not been identified.
[3] There are no extant letters from CD to Oliver from the period between the letter to Daniel Oliver, 23 [November 1862] (*Correspondence* vol. 10), and this letter.
[4] CD refers to Oliver's 'Botanical bibliography' in the *Natural History Review* ([Oliver] 1862a; see letter from Daniel Oliver, 22 January 1863).
[5] See *Correspondence* vol. 10, letter to J. D. Hooker, 11 June [1862].
[6] See *Correspondence* vol. 10, letter to Daniel Oliver, 12 [April 1862].

From Erasmus Alvey Darwin 21 [January 1863][1]

21st.

Dear Charles.

I shall be most glad to have you and hope much the experiment will succeed tho' you must give it fair play & not work yourself to death—[2] We were rather meditating a dinner with Dr Falconer as basis but it might just as well be the week after as we have settled nothing.[3]

The juvenile world is very happy in the thoughts of the ball tomorrow & I hope Emma will not break down.[4]

When you come order your newspaper here as mine goes off to Algiers—[5]

Yours affec | E D

DAR 105: B15–16

[1] The date is established by the reference to CD's intention to stay at Erasmus's house in London (see n. 2, below) and by the reference to a ball taking place 'tomorrow' (see n. 4, below).

[2] According to Emma Darwin's diary (DAR 242), CD, Emma, Henrietta, and Horace Darwin stayed at Erasmus's house at 6 Queen Anne Street, London, from 4 to 14 February 1863.

[3] See letter to Hugh Falconer, 20 [January 1863].

[4] On 22 January 1863, Emma Darwin recorded in her diary (DAR 242) a visit to London 'for ball at C.P.' Cumberland Place was the home of Hensleigh and Frances Wedgwood.

[5] Erasmus may have been sending his newspapers to his invalid cousin, James Mackintosh Wedgwood, who had left in November 1862 to spend the winter in Algiers (see the letter from Emma Darwin to William Erasmus Darwin, [13 November 1862], in DAR 219.1: 65). The winter of 1862 to 1863 saw an influx of British into Algeria, many of whom were 'seeking health in more genial climes' (see Rogers 1865, p. viii).

From Thomas Rivers 21 January 1863

Nurseries, Sawbridgeworth, Herts, | Great Eastern Railway.[1]

Jany 21/63

My dear Sir/

I have given orders to have sent to you

1 double rose flowered Chinese peach
1 Amygdalus communis dulcis

The trees are from pots & if potted & placed under glass they will bloom well.[2]

I shall be delighted to receive from you the "Origin of Species" with your name attached to it.[3] I had it from Mudies when first published & was on the eve of ordering it from Longmans when yr. last arrived[4] a bought copy I should of course have valued but a copy as your gift *will* be to me *invaluab⟨le.⟩*

The little package of trees will be paid to London from Bromley pr. Down Postman. Pray communicate me if you have the least wish to have a tree or shrub I have so many thousands that it is always a *pleasure* to give—a *duty* to sell

I am My dr Sir | Yrs. very truly | Thos. Rivers

The strawberry hybrids noticed in Journal of Horticulture (I mean a week or two since) are curious but the hybridisms are not accurate enough[5] one talks of the "Russian Alpine" & the "High Wood" strawberries what are they?

Postmark: JA 21 63
DAR 176: 160

CD ANNOTATIONS
4.1 I am . . . Rivers] 'Shoots | Origin—' *pencil*
Top of letter: '(Railway)' *ink*

[1] In this letter and later ones, the location of Rivers's nursery is followed on the letterhead by: '*Harlow Station is the most convenient for passengers*'.
[2] The letter in which Rivers offered to send CD peach trees, a reply to CD's letter to Rivers of 11 January [1863], has not been found. However, see the letter to Thomas Rivers, 15 January [1863], in which CD expressed an interest in peach trees, as well as *Amygdalus*, an almond tree.
[3] In his letter to Rivers of 15 January [1863], CD offered Rivers a copy of one of his works.
[4] Rivers refers to Mudie's Select Library, New Oxford Street, London, and the publishers Longman, Green, Longman, Roberts & Green, of Paternoster Row, London (*Post Office London directory* 1861).
[5] This notice, from William Smith of York (W. Smith 1862), was in response to CD's appeal for information on strawberry crosses in his letter to the *Journal of Horticulture*, [before 25 November 1862] (*Correspondence* vol. 10). When Smith replied, he also sent three runners from one of his crosses, suggesting that the editor send them to CD. His notice was followed by an editorial note: 'Mr. Darwin, to whom we forwarded all the runners, is very much obliged, and has planted them' (*Journal of Horticulture*, 30 December 1862, p. 779). During November 1862, CD was preparing a draft of chapter 10 of *Variation*, in which he included a section on cultivated strawberries (*Variation* 1: 351–4).

To John Scott 21 January [1863][1]

Down Bromley Kent
Jan. 21st

Dear Sir

I thank you for your *very interesting* letter;[2] I must answer as briefly as I can, for I have a heap of other letters to answer. I strongly advise you to follow up & publish your observations on the pollen-tubes of orchids; they promise to be very interesting.[3] If you could prove what I only conjectured (from state of utriculi in Rostellum & in stigma of Catasetum & Acropera) that the utriculi somehow induce or are connected with penetration of pollen-tubes you will make an important physiological discovery.[4] I will mention, as worth your attention (& what I have anxiously wished to observe, if time had permitted & still hope to do) viz the state of tissues or cells of stigma in an utterly sterile hybrid, in comparison with the same in fertile parent-species: to test these cells, immerse stigmas for 48 hours in spirits of wine; I shd. expect in hybrids that the cells would not show coagulated contents. It would be interesting discovery to show difference in female organs of hybrids & pure species. Anyhow it is worth trial & I recommend you to make it & publish if you do.—[5]

The pollen-tubes directing themselves to stigma is also very curious; though not quite so new; but well worth investigation when you get Cattleya &c in flower.—[6] I say not so new; for remember small flower of Viola & Oxalis; or better, see Bibliography of Nat. Hist. Review Part VIII. p. 419—(Oct 1862) for quotation from M. Baillon, on pollen tubes finding way from anthers to stigma in Helianthemum.[7] I sh^{d}. doubt gum getting solid from continued secretion. Why not sprinkle fresh Plaister of Paris & make impenetrable crust?[8] You might modify experiment by making little hole on one lower corner, & see if tubes find it out.— See in my future paper on Linum pollen & stigma recognising each other.—[9] If *you* will tell me that pollen smells the stigma I will try & believe you; but I will not believe the Frenchman (I forget who) who says that Stigma of Vanilla actually attracts mechanically by some unknown force the solid pollen-masses to it![10]

Read Asa Gray in 2^{d} Review of my Orchis Book on pollen of Gymnadenia penetrating rostellum;[11] I can, if you like, lend you these Reviews; but they must be returned.— R. Brown, I remember, says pollen-tubes separate from grains before the lower ends of tubes reach ovules.—[12] I saw, & was interested by, abstract of your Drosera paper:[13] we have been at very much same work.—

With respect to **bud**-variation: perhaps it would give you the least trouble first to send me mere list. I have devoted whole Chapter to subject.[14] Perhaps it will be best to specify cases which interest me most—variation by modified buds as bulbs or tubers—or underground as suckers—anything on inheritance from seed of the varying buds—, whether parents are crossed plants, & whether the variation is case of Reversion. Of course the more marked the variation is so much better. I do not care for mere zoned leaves, unless something unusual about them.

I have hitherto just alluded to every case of change in colour in flower, especially if accompanied by any other change. Mr Salter & Mr Rivers are aiding me.—[15]

With respect to Ferns I am so ignorant that I hardly know what to do. Am I right in supposing (probably I am wrong) that a spore (whether spore be unfertilised ovule or bud) from a variagted branch produces a thallus, & this produces the two sexual individuals & from their union a variegated fern is produced; if so, the case would *not* come under variation independent of sexual union. Please briefly illuminate my ignorance.—[16]

M^{r}. Bridgman sent me his paper.—[17]

Asa Gray has sent me a **few** white & red seed of N. England Popping very small seeded maize;[18] shall you experiment on this; if so these would be good to cross with some large kind of different colour.— Shall I send??

Yesterday I had very kind letter from Mr I. Anderson Henry, of Hay Lodge, Trinity offering to try any experiment for me.[19] I have suggested peloric flowers & pollen of short stamens of Pelargonium.—[20] Please tell me in *strict confidence* whether you think him a good observer; from his short papers, I say to you *in confidence* that I doubt it.—[21] You will see that it is of paramount importance for me not to waste time in suggesting experiments &c & asking for information from an inaccurate observer.—

Excuse this hurried letter. | Dear Sir | Yours very faithfully | Chs. Darwin

I hardly know what to say on your view of male & female organs & variability.—[22] I must think more over it. But I was amused by finding the other day in my Portfolio devoted to Bud-variation, a slip of paper dated June 1860, with some such words as these "May not permanence of grafted buds be due to the two sexual elements, derived from different part not having come into play?"[23] I had utterly forgotten, when I read your paper, that any analogous notion had ever passed through my mind—nor can I now remember, but the slip shows me that it had.—

DAR 93: B56–7, B75–6

1 The year is established by the relationship between this letter and the letter from John Scott, 16 January 1863.

2 Letter from John Scott, 16 January 1863.

3 In his letter of 16 January 1863, Scott provided CD with an account of some of his observations on the pollination of orchids and the paths of pollen tubes (see n. 4, below). These observations were undertaken at CD's suggestion to discover whether the rostellum retained any of its rudimentary stigmatic functions. In his letter to Scott of 3 December [1862] (*Correspondence* vol. 10), CD asked Scott to try to split the labellum of a *Cattleya*, or an allied orchid, and place a single pollen-mass 'carefully into the large tongue-like Rostellum, & see if pollen-tubes will penetrate'. Scott published accounts of his work on orchid pollination in Scott 1863a and 1864b, copies of which are in the Darwin Pamphlet Collection–CUL.

4 For CD's discussion of the stigma and rostellum in *Catasetum* and *Acropera*, see *Orchids*, pp. 204–7 and 237–8. He discussed the possible derivation of the rostellum from the stigma on pp. 307–23. In his letter of 16 January 1863, Scott told CD that his observations tended to confirm the view that the absence of the coherent spindle-formed utriculi from the rostellum was related to the infertility of the organ, that is, that pollen tubes did not penetrate it. See letter from John Scott, 16 January 1863 and n. 6.

5 CD had previously used this method to examine stigmas, though not to compare those of a sterile hybrid and a fertile parent (see *Orchids*, pp. 206–7). See also *Correspondence* vol. 10, letter to Daniel Oliver, 8 June [1862], and this volume, letter to Isaac Anderson-Henry, 20 January [1863].

6 See letter from John Scott, 16 January 1863.

7 The references are to Daniel Oliver's bibliographic review of botanical literature in the *Natural History Review* ([Oliver] 1862a, p. 419), and to Baillon 1861, p. 56.

8 In his letter to CD of 16 January 1863, Scott suggested putting a 'thickish solution of gum to stigma and allowing it to harden, before applying pollen-mass to rostellum', in order to test whether pollen tubes would penetrate the rostellum.

9 'Two forms in species of *Linum*', p. 75 (*Collected papers* 2: 98). CD's paper on *Linum* was read before the Linnean Society on 5 February 1863 (see letter from George Bentham, 16 January 1863).

10 The reference has not been identified. In his letter of 16 January 1863, Scott wrote: 'A latent provisional force, an almost conscious sympathy . . . seems to exist between pollen and stigma, and is strikingly evoked, when these are not directly and normally applied to each other'.

11 A. Gray 1862a, p. 426.

12 Robert Brown 1831b, p. 705.

13 An abstract of Scott's paper, 'On the propagation and irritability of *Drosera* and *Dionæa*', read at a meeting of the Botanical Society of Edinburgh on 11 December 1862, appeared in the *Gardeners' Chronicle*, 10 January 1863, p. 30. An abstract was also published in the *Transactions of the Botanical Society* [*of Edinburgh*] 7: 429–30 (Scott 1862b).

14 CD began writing chapter 11 of *Variation*, dealing with 'bud-variation', on 21 December 1862 (*Correspondence* vol. 10, Appendix II); in his letter to Scott of 19 December [1862], CD asked Scott to provide him with any cases of 'what Gardeners sometimes call sports, & which I shall call "bud-variation"'.

In his letter of 16 January 1863, Scott asked whether CD merely wanted lists of plants presenting variations or 'the history of *each*'.

15 Scott had asked whether CD was interested in changes in colours of flowers (see letter from John Scott, 16 January 1863). CD refers to John Salter, proprietor of the Williams Street nursery in Hammersmith, London (R. Desmond 1994). Salter was cited several times in chapter 11 of *Variation* on bud-variations in pelargoniums, chrysanthemums, and *Phlox* (see *Variation* 1: 378–9); however, no correspondence on these points has been found. CD also refers to the Hertfordshire nurseryman, Thomas Rivers, whom he cited on bud-variation in roses (see *Variation* 1: 379–81). For their correspondence on bud-variation, see the letters to Thomas Rivers, 7 January [1863], 11 January [1863], and 15 January [1863].

16 In his letter of 16 January 1863, Scott asked CD whether he intended to discuss ferns. See also letter to John Scott, 16 February [1863] and nn. 7–8. Ferns were briefly discussed in *Variation*; the reader was referred to Bridgman 1861 and Scott 1862a (*Variation* 1: 383).

17 William Kencely Bridgman; see n. 16, above.

18 See *Correspondence* vol. 10, letter from Asa Gray, 29 December 1862. See also letter to Asa Gray, 19 January [1863] and n. 3.

19 Letter from Isaac Anderson-Henry, 17 January 1863.

20 See letter to Isaac Anderson-Henry, 20 January [1863].

21 CD may refer to Isaac Anderson-Henry's contributions to the discussions regarding variegation in plants in the *Gardeners' Chronicle* and the *Journal of Horticulture* in 1861 (*Gardeners' Chronicle*, 11 May 1861, pp. 432–3; *Journal of Horticulture* n.s. 2 (1861): 41–3). Scott's reply has not been found (see letter to John Scott, 16 February [1863]).

22 Scott enclosed a copy of his paper on fern spores (Scott 1862a) with his letter to CD of 6 December [1862] (*Correspondence* vol. 10). He argued that ferns, the product of a single organ, presented the most favourable opportunity for reproducing individual variations (see letter from John Scott, 16 January 1863 and n. 11). There is an annotated copy of Scott 1862a in the Darwin Pamphlet Collection–CUL.

23 CD's note has not been found. CD discussed bud-variation in chapter 11 of *Variation* (*Variation* 1: 373–411).

To Julius von Haast 22 January 1863

Down. | Bromley. | Kent. S.E.
Jan 22.—/63

Dear Sir

I thank you most sincerely for sending me your address & the Geological Report.[1] I have seldom in my life read anything more spirited & interesting than your address. The progress of your colony makes one proud, & it is really admirable to see a Scientific Institution founded in so young a nation.—[2] I thank you for the very honourable notice of my "Origin of Species".—[3] You will easily believe how much I have been interested by your striking facts on the old Glacial period.[4] And I should suppose the world might be searched in vain for so grand a display of Terraces.—[5] You have indeed a noble field for Scientific research & discovery.— I have been extremely much interested by what you say about the tracks of supposed mammalia.—[6] Might I ask if you succeed in discovering what the creatures are, you would have the great kindness to inform me.— Perhaps they may turn out something like the Solen-hofen bird-creature with its long tail & fingers with claws to its wings![7] I may mention that in S. America in completely uninhabited regions I found spring Rat-traps, baited with *cheese* were very successful in catching the smaller mammals.[8]

I would venture to suggest to you to urge on some of the capable members of your Institution to observe & annually record the rate & manner of spreading of European weeds & insects, & especially to observe *what native plants most fail*: this latter point has never been attended to. Do the introduced Hive-bees replace any other insect? &c &c.— All such points are, in my opinion, great desiderata in science.—[9]

What an interesting discovery that of the remains of Prehistoric man![10]

Believe me, Dear Sir, with the most cordial respect & thanks | Yours very faithfully | Charles Darwin

Alexander Turnbull Library, National Library of New Zealand (Haast family papers, MS-Papers-0037-051)

[1] Haast sent CD the text of an address delivered before the Philosophical Institute of Canterbury in September 1862 (J. F. J. von Haast 1862a; see n. 2, below). CD also refers to Haast's report on his explorations during his first nineteen months as provincial geologist in Canterbury, New Zealand (J. F. J. von Haast 1862b; see n. 4, below). Haast sent CD copies of these publications in November 1862, a month before he sent a letter mentioning them; CD did not receive that letter until June 1863 (see *Correspondence* vol. 10, letter from Julius von Haast, 9 December 1862, and this volume, enclosure to letter from Julius von Haast, 5 March 1863, and letter from J. D. Hooker, 19 June 1863). CD's annotated copies of J. F. J. von Haast 1862a and 1862b are in the Darwin Pamphlet Collection–CUL.

[2] Haast was one of the founding members of the Philosophical Institute of Canterbury (*DSB*; H. F. von Haast 1948, pp. 220–30). The society held its first meeting at the offices of the Geological Survey of Canterbury in August 1862 and elected Haast as its first president. At its inaugural dinner in September 1862, Haast delivered the address, spelling out the institute's goals, which included the formation of a museum of economic geology and natural history, a library, a chemistry laboratory, and an astronomical observatory. He also presented a review of the current state of knowledge respecting New Zealand's natural history (J. F. J. von Haast 1862a).

[3] In his inaugural address, Haast described *Origin* 'as the great work of the age', noting that prior to its publication, natural history 'as a science' was 'as little advanced as astronomy was before the labours of Galileo, Kepler, and Newton' (J. F. J. von Haast 1862a, p. 7).

[4] Between January and May 1862, Haast conducted an exploration of New Zealand's Southern Alps. The mountain range offered considerable evidence of extensive Pleistocene glaciation (see J. F. J. von Haast 1862b, pp. 127–31, and H. F. von Haast 1948, pp. 199–219). See also *Correspondence* vol. 9, letter from J. D. Hooker, [29 December 1861], and *Correspondence* vol. 10, letter from J. D. Hooker, 20 September 1862, and letter from Julius von Haast, 9 December 1862. In *Origin*, p. 373, CD referred to indications of former glacial action in New Zealand, and in the fourth edition (pp. 442–3), he cited the 'excellent researches' of Haast and James Hector, stating that 'we now know . . . that enormous glaciers formerly descended to a low level in New Zealand'.

[5] Haast observed littoral, alluvial, and glacial terraces in the province of Canterbury (J. F. J. von Haast 1862b, pp. 128–30).

[6] In his address (J. F. J. von Haast 1862a, p. 6), Haast referred to the 'indigenous quadruped' called 'Kaureke' by the Maori, which he thought might be a badger or otter, though he had only seen its tracks, and described his own observation of a smaller mammal's tracks. CD expressed his interest in endemic island organisms in *Origin*, pp. 201–2.

[7] CD refers to *Archaeopteryx*. See letter from Hugh Falconer, 3 January [1863].

[8] See, for example, *Journal of researches*, p. 351.

[9] Haast was interested in importing European plants and animals that would benefit New Zealand; he suggested the establishment of acclimatisation societies for the introduction of 'useful animals' (J. F. J. von Haast 1862a, p. 3). In *Origin*, CD discussed the replacement of native New Zealand

species (pp. 201–2), as well as the geographical distribution of animals and plants on oceanic islands (pp. 388–406). After discussing the naturalisation of plants on New Zealand and St Helena, he wrote (p. 390):

> He who admits the doctrine of the creation of each separate species, will have to admit, that a sufficient number of the best adapted plants and animals have not been created on oceanic islands; for man has unintentionally stocked them from various sources far more fully and perfectly than has nature.

10 In his address to the Philosophical Institute of Canterbury, Haast referred to reports of the discovery of pre-Maori stone implements in Wellington Province, and proposed the possibility of inhabitants or visitors before the Maori as a subject worthy of inquiry (J. F. J. von Haast 1862a, p. 7).

To John Murray 22 January [1863][1]

Down Bromley | Kent
Jan 22[d]

My dear Sir

Will you be so good as to send the "Origin", *pasting in enclosed paper*, & addressed to

Th. Rivers Esq[re][2]
Nursery Grounds
Sawbridgeworth
Herts.

I sh[d]. be curious to hear some time how the Orchis-Book has gone off,—whether any more copies have sold.[3]

It is too stiff reading for the Public; if praise from Botanists would sell, it would go off well.—

My dear Sir | Yours very sincerely | Ch. Darwin

John Murray Archive (Darwin 127)

1 The year is established by the relationship between this letter and the letter from Thomas Rivers, 21 January 1863 (see n. 2, below).

2 In his letter to Thomas Rivers of 15 January [1863], CD offered a copy of one of his works in appreciation of Rivers's information regarding domestic plant variation. Rivers replied on 21 January 1863, saying he would be 'delighted' to receive an autographed copy of *Origin*.

3 *Orchids* was published on 15 May 1862 (*Publishers' Circular* 25: 247); 1500 copies were offered for sale (see *Correspondence* vol. 10, letter to John Murray, 9 April [1862], and Appendix IV). There is a statement of Murray's account for *Orchids* in DAR 171.3: 524–5. On 30 June 1862, Murray had 685 copies in stock; by 27 February 1866, there were still 600 copies remaining.

From Daniel Oliver 22 January 1863

Richmond, S.W.
22. 1. 1863

My dear Sir

You *very greatly* over-rate the labour of my Bibliography & yet more greatly extravagant is your notion as to my knowing many things.—[1] Alas! I am *dreadfully* ignorant & my memory *sadly* imperfect.— Your suggestion about the *sexes* of water plants is interesting.[2] In this relation it would be important to distinguish species the flowers of wh. expand **under** the surface,—& this w^d. reduce the no. of "aquatics" very greatly. Crowfoot, Water-Lily, *Utricularia*, Vallisneria, Elodea & the like on the one hand—*Naias*, *Zannichellia*, *Subularia* (Cruciferae), *Zostera* &c. on the other.[3] I should have liked to examine *Subularia* & thought about it before going down to the Lakes but saw none when there last time.

I am not *aware* that any sexual difference has been noticed in its flowers.[4] Acct. w^d. have to be taken of the *Algae* in such great preponderance submerged altogether,—fertilising of course by antherozoids. **Some** (as some *Fuci*) are monoecious & I fancy have both sexes in same *conceptacle*.[5]

My prospect of spare time keeps far off. Tomorrow even^g. (4 Fridays running) I have to lecture at Norwich (a thing you need not fear I shall get in to the habit of)[6] On Saturd^y. even^g. I have lectures to my class at U. College nearly done.[7] I have undertaken to work up the botany of Amomums with Hanbury.[8] the relations of *Viscums* &c to *Gnetaceae* I have to make up my mind about, &c. &c. &c.[9]

Very sincerely yours | Danl. Oliver.

M^r. Watson (of Ditton)[10] sends me today some 2 score baby *Brambles* (seedling) to shew they are not the surpassingly rare things *he* thinks M^r. Bentham makes out (Linn. Soc^y. address./62)[11]

An obs. of mine in a Review of British Floras suggested his looking for them.—[12]

DAR 173: 19

[1] In his letter to Oliver of 20 [January 1863], CD praised Oliver's bibliographic review of botanical literature ([Oliver] 1862a).

[2] In the letter to Daniel Oliver, 20 [January 1863], CD asked whether there was an unusual proportion of freshwater plants with separated sexes.

[3] The flowers of the first group open on the surface of the water, while those of the second group expand while submerged; Oliver considered only the latter to be truly aquatic. There is a note on aquatic plants in DAR 111: A69. See also letter to Daniel Oliver, 20 [January 1863].

[4] See *Forms of flowers*, p. 311 n.

[5] Oliver refers to the motile male gametes (antherozoids) of some algae, and to some seaweeds containing male and female sex organs in the same conceptacle, a cavity in the thallus.

[6] At the Assembly Hall in Norwich, Oliver lectured on the topic of dimorphism in flowers (*Norfolk Chronicle*, 14 February 1863, p. 2).

[7] Oliver was professor of botany at University College London. The course of ten lectures for advanced students began on 8 November 1862 (*Athenæum*, 8 November 1862, p. 577).

[8] Oliver and Hanbury 1863. The reference is to the pharmacist Daniel Hanbury.

[9] Oliver read a paper on the Loranthaceae (which then included *Viscum*) and its relationship to the Gnetaceae before the Linnean Society on 15 January 1863 (Oliver 1863a).

[10] Hewett Cottrell Watson lived in Thames Ditton.

[11] In his anniversary address to the Linnean Society, George Bentham discussed the subject of biological classification; he cautioned against remodelling the taxonomic system by creating new species and new names (Bentham 1862). Bentham was referring to the recent contribution on *Rubus* by Philipp Jakob Müller (Müller 1859), a 225-page account of 239 species of the genus, and to Müller 1861, a 40-page account describing 31 new bramble species. In the conclusion to *Origin*, CD predicted

a 'considerable revolution in natural history' once his views were acknowledged, resulting in the cessation of the 'endless disputes' about 'whether or not some fifty species of British brambles are true species' (*Origin*, p. 484).

12 In the January 1863 number of the *Natural History Review*, Oliver reviewed new editions of a number of standard British floras, including the fifth edition of Charles Cardale Babington's *Manual of British botany* (Babington 1862) and the eighth edition of *The British flora* by William Jackson Hooker and George Arnott Walker Arnott (W. J. Hooker and Arnott 1860). Here ([Oliver] 1863b, p. 38), Oliver criticised the 'wholesale manufacture of species', which caused such 'mischief' when it came to philosophical generalisation or comparison.

From Francis Boott 23 January 1863

24 Gower St.
Jany 23. 1863.

My Dear Darwin

My Son[1] is troubled in his mind about the Cub, supposed to be the progeny of a Lioness & a mastif, & I have promised him to appeal to you for his relief.[2] The Letter of Frank Buckland in the Field paper, I have sent to Lyell, with a request that he will send it on to you.[3] You may have seen it, but I found yesterday that Lyell had not heard of it.

You will see a mention of the Cub in the note enclosed from my Son.[4] In it he alludes to a photograph of the cub & his owner,[5] which I have seen. If this strange birth is what Buckland considers it,[6] is it not strange that cats & Dogs have not paired?

Do not return the paper, which Lyell will send, but pray say something to comfort my Son, if you know more than Buckland imparts.

I found Lyell with one of his last *proofs*, & full of scorn of Owen, for the proof had wood cuts of Brains, which brought him before us—[7]

Yrs sincerely | F. Boott

Charles Darwin Esq

DAR 160: 254

1 Boott's son has not been identified.

2 Boott refers to a report of two cubs, the result of an alleged cross between an African lion transported to England, and a large dog, itself the product of a true English Mastiff and a St Bernard (*Bell's Life in London*, 16 November 1862, p. 5; *Field*, 6 December 1862, p. 515). No letter from CD on this subject has been found.

3 Boott refers to Charles Lyell and to the article by Francis Trevelyan Buckland entitled 'Supposed hybrid between a lion and mastiff', which appeared in the *Field*, 6 December 1862, p. 515 (see n. 2, above).

4 The enclosure has not been found.

5 In 1859, William Henry Patten-Saunders brought the lioness from Africa as a present for his mother (*Bell's Life in London*, 16 November 1862, p. 5). A photograph of the cub has not been found.

6 Buckland had been invited to inspect the creatures, and although the owner emphasised the lion-like qualities of the cub, Buckland confessed that in his own 'humble judgement', the animals were 'in

external appearance, teeth, &c., amazingly like ordinary dogs, with much of the mastiff-mongrel about them' (*Field*, 16 December 1862, p. 515).

7 *Antiquity of man* (C. Lyell 1863a) was published on 9 February 1863 (*Athenæum*, 7 February 1863, p. 176). The last chapter considered the transmutation doctrine in relation to humans (pp. 471–506, with woodcuts of brain sections on pp. 482–3). In it, Lyell reviewed the public controversy between Richard Owen and Thomas Henry Huxley concerning the similarities and differences between human and simian brains, now often referred to as the 'hippocampus controversy' (C. Lyell 1863a, pp. 480–93). On the hippocampus controversy, see *Correspondence* vols. 8–10, A. Desmond 1982, and Rupke 1994.

To T. H. Huxley 23 January [1863 or 1864][1]

Down Bromley Kent
Jan. 23d

My dear Huxley

I have signed the enclosed with much pleasure.[2]

I cordially thank you for your kind, most kind, expressions about the medal.[3] Sympathy & not fame is the true & good reward of labour.—

Yours most truly | Ch. Darwin

Imperial College of Science, Technology and Medicine Archives (Huxley 5: 254)

1 The date range is conjectured by CD's mention of the Copley Medal (see n. 3, below).

2 The enclosure has not been found.

3 The letter from Huxley has not been found. CD was nominated for the Royal Society of London's Copley Medal in 1862 and 1863 but his candidature was not successful (Royal Society, Council minutes, vol. 3 (1870), pp. 124, 160). CD was nominated again in 1864 and awarded the medal in November of that year (see *Correspondence* vol. 12, letter from Edward Sabine, 3 November 1864).

From A. R. Wallace [23 January 1863?][1]

5. Westbourne Grove Terrace, W.
Friday evening

My dear Mr Darwin

It has just occurred to me, that I *have seen* insects on *Melastomas*.[2] A small shrubby species is abundant in all the cleared waste grounds in the Malay Islands, & I now distinctly recollect having frequently taken small *Cetoniadæ* of the genus *Valgus* from the flowers. The large wood boring bees (*Xylocopa*) also visit the flowers, & the whole plants often swarm with small black ants.

I remember also distinctly having seen the flowers much eaten by insects.

The Plant however is so common close to Singapore that I will write to one of my friends there to observe & make a list of all the insects, that visit it.[3]

Hoping these few facts may be useful | I remain | My dear Mr Darwin | Yours very faithfully | Alfred R. Wallace

Charles Darwin Esq

DAR 205.8: 70 (Letters)

[1] The date is conjectured from the relationship between this letter and the following letter (see n. 2, below). In 1863, 23 January was a Friday.
[2] CD had recently sent inquiries to a number of correspondents regarding dimorphism and insect pollination of Melastomataceae (see, for example, letter to Hugh Falconer, 5 [and 6] January [1863], and letter to H. W. Bates, 12 January [1863]). He may have asked Wallace for information on this subject in a letter written between 9 and 13 January 1863 that has not been found. See also letter from A. R. Wallace, 14 January [1863], the following letter and n. 1, and letter to H. W. Bates, 26 January [1863].
[3] See letter from A. R. Wallace, 26 September 1863.

From H. W. Bates 24 January 1863

10 Hollis place | Prince of Wales road | N.W.
Jan 24 1863

My Dear Mr Darwin

Although I cannot supply you with any information on the subject you mention I think it well to write saying that the day before receiving your last on repeating your enquiry to Mr Wallace he said he had certainly noticed Melastomæ in the Malay Archipelago to be frequented by small Hymenoptera.[1] Neither he nor I could think of watching where & how bees found the fluid.

I cannot think of any one on the Amazons likely to know what is a Melastoma: there is however a Pará gentleman (now in England) who lives in the forest when at home & could pursue the enquiry if he knew the plants; him I have written to so probably you will hear from him.[2] He knows the reward is purely honorary.

We had a pretty little Darwinian discussion at last meeting of Zoological: an impartial auditor must have thought the Darwinians had the best of it[3]

Yours sincerely | H W Bates

P.S. I have committed the folly of getting married & as lady & self have both a strong desire to live in or near London,[4] I am going to try hard to get employment to add to my small independent income sufficient to live here & devote all time to science

First vol. of book nearly finished printing.[5] All illustrations done but 3 or 4 & the map which require 4 or 5 days more. M.S. finished except a few lines to the last & toughest discussion & preface.[6] I am told 1250 copies to be struck off but do not know if this be correct. The illustrations are extremely beautiful I have watched over nearly all & some of them are quite in an original style for books of travel[7]

DAR 160: 73

CD ANNOTATION
2.1 Amazons ... Bates 4.1] *crossed ink*

[1] In his letter to Bates of 12 January [1863], CD asked which insects he had observed visiting flowers of species of Melastomataceae during his years on the Amazon. For Bates's response, see the letter from H. W. Bates, 17 January [1863]. Evidently, CD wrote to Bates again, but the letter has not been found. Bates refers to his friend Alfred Russel Wallace, who returned to Britain in 1862 after seven years of

collecting in the Malay Archipelago. See preceding letter. For CD's interest in Melastomataceae, see the letter to Hugh Falconer, 5 [and 6] January [1863] and n. 22.

2 Bates wrote about the city and province of Pará (an 800,000 square mile region of northern Brazil) in the first three chapters of Bates 1863 (pp. 1–111). The 'Pará gentleman' has not been identified.

3 At a meeting of the Zoological Society of London on 13 January 1863, Wallace read a paper on the birds of Bouru (Wallace 1863a; present day Buru, one of the Moluccan islands, is in the Indian Ocean). A report of the meeting appeared in the *Intellectual Observer* 3: 62:

> In a paper descriptive of some new species of birds collected in Bourou, Mr. Wallace called attention to several new cases of mimetic analogy, or the imitation of one species of the plumage of another. In describing these specimens Mr. Wallace maintained, that the numerous cases of imitation that exist were incapable of being accounted for on any supposition except by that which accounts for the variation of species by natural selection; also that this imitation was a protection against the natural enemies of the imitating animal; that those animals most closely resembling the aggressive species were the most certain to escape destruction and perpetuate their variety; and that this method of natural selection acting through long periods produced the close imitation between species in totally distinct genera or orders, which appears capable of explanation on no other hypothesis.

4 Bates married Sarah Ann Mason on 19 January 1863 (Woodcock 1969, p. 254).

5 *The naturalist on the river Amazons* (Bates 1863) was published between 1 and 14 April 1863 (*Publishers' Circular* 26: 193).

6 The preface to Bates 1863, 1: iii–vi, dated January 1863, mentions Wallace's proposal of the expedition and their choice of divergent routes after two years in South America. Bates also describes the nature of his collections and CD's encouragement to write the volume. The last discussion in Bates 1863 (2: 415–17) is an account of his ambivalence about returning to Britain.

7 Bates's two volumes include forty-two illustrations; he acknowledged the artists in Bates 1863, 1: vi. See also letter from H. W. Bates, 17 January [1863], and letter to H. W. Bates, 9 April [1863].

From J. D. Hooker 24 January 1863

Paris
Jany 24/63

Dr Darwin

I gave your letter to Naudin,[1] who toute suite brought it back, 1 to be deciphered, 2 to be put into English 3d to be translated. however thanks to Bentham that did not take long,[2] & so he took it home for his private eating, & will I hope give me an answer to take back to you—[3] N. being stone deaf, I cannot do much business with him— I have had a long talk on tablets with him however,[4] & with Decaisne too,[5] both have much curious matter, but neither appreciate your book as they should, & will when they read it in its french garb I hope.[6]

Decaisne is writing a paper for the Institute on fruit trees, which will I doubt not contain much curious matter.[7] Naudin *says* that he has discovered the physiological cause of change becoming specific; ie of vars. no longer breeding together.—[8] good if true.— Decaisnes observations upon the absolute hereditary transmission of minute characters in some garden varieties of Lettuces &c are most curious— he has also good matter about permanence of types of fruits.[9]

Every thing here is quiet—gay & beautiful to the eye, industry activity & propriety meet the eye every where to an extraordinary degree where the vice, misery,

& poverty of every large town are no where appears— you see more in half an hour of the best part of London than in the back slums of all Paris.

How dreadful the New York papers are, we see them here, & I read & moralize over them by the hour— I believe that a Republican is the worst form of govt. that can be *given* to a people, but perhaps the best they can *make for themselves* the mistake is to suppose that the Americans made it for themselves—they never did so; they accepted it from the hands of the few great men of that day, & so long as there was no struggle for existence it was never put to the test— when the struggle came they found out that what they accepted for a working theory, had not taken root enough in the hearts of the people to be upheld at any price.[10] Lincoln's Emancipation proclamation is the most damnable thing ever done.[11] Really there is no bright spot in this sad sad world but in shops that sell Wedgewood ware, which I have been haunting with some success.[12] As I know that you will listen to nothing from me after this I will shut up

Ever dear Darwin | Yrs affec | J D Hooker

I shall return in a day or two.

DAR 101: 99–100

CD ANNOTATIONS
End of letter: '['Na' *del*] Decaisne on Lettuce & Fruit trees' *ink*
Verso of last page: 'Victoria Lily'[13] *brown crayon*; 'Cape orchids'[14] *ink*; 'Bates'[15] *pencil*

1 In his letter to Hooker of 24 December [1862] (*Correspondence* vol. 10), CD enclosed a 'memorandum of enquiry' for Charles Victor Naudin, whom Hooker planned to meet during a visit to Paris. The memorandum has not been found.

2 Hooker and George Bentham left together for a ten-day trip to Paris on 17 January 1863 to work on the second volume of *Genera plantarum* (Bentham and Hooker 1862–83; Jackson 1906, p. 193). Bentham read botanical literature in fourteen modern European languages; he was fluent in French, having lived in France as a young man from 1814 to 1827 (*DNB*).

3 Naudin's response to CD's inquiries has not been found. However, see the letter to Isaac Anderson-Henry, 2 February [1863], and the letter to C. V. Naudin, 7 February 1863.

4 Tablets: 'sometimes . . . applied vaguely to a note-book' (*OED*).

5 Joseph Decaisne was professor of plant cultivation at the Jardin des Plantes, Muséum d'Histoire Naturelle, Paris (*DBF*).

6 A French edition of *Origin*, translated by Clémence Auguste Royer, was published in May 1862 (Royer trans. 1862; J. Harvey 1997). For CD's earlier discussions of Naudin's theory of transmutation, see *Correspondence* vol. 7, letter to Charles Lyell, 22 [December 1859] and nn. 6 and 7, and letter to J. D. Hooker, 23 [December 1859], and *Correspondence* vol. 8, letter to J. D. Hooker, 31 [January 1860] and n. 2, and Appendix IV.

7 Hooker may be referring to Decaisne 1863, a paper on the pear trees of the Muséum d'Histoire Naturelle, or to portions of Decaisne 1858–75, which was published in 129 parts before being collected in nine volumes.

8 Naudin discussed his research on hybridity in Naudin 1862 and 1863. There are annotated copies of these works in the Darwin Library–CUL. Naudin 1862 was also published later as part of Naudin 1865, an annotated copy of which is in the Darwin Pamphlet Collection–CUL (see *Marginalia* 1: 638–9). CD cited Naudin extensively on hybridity in *Variation.* See also letter from C. V. Naudin, 26 June 1862 (*Correspondence* vol. 10).

9 See n. 7, above. CD cited Decaisne's work on fruit (Decaisne 1863) in *Variation* 1: 350.

10 See also letter from J. D. Hooker, 6 January 1863.
11 Abraham Lincoln's emancipation proclamation, which freed slaves in all territories still in rebellion against the United States government, took effect from 1 January 1863 (Denney 1992, pp. 248, 251).
12 Hooker had started to collect Wedgwood ware (see *Correspondence* vol. 10, letter from J. D. Hooker, [27 or 28 December 1862], and this volume, letter from J. D. Hooker, 6 January 1863).
13 *Victoria regia*. See letter to J. D. Hooker, 30 January [1863].
14 See letter to J. D. Hooker, 30 January [1863].
15 Henry Walter Bates; see letter to J. D. Hooker, 30 January [1863].

To Hermann Crüger 25 January [1863][1]

Down | Bromley Kent.
Jany 25

Dear Sir.

I hope that you will excuse the liberty I take in begging a favour of you— I have been investigating the manner of fertilization and nature of the pollen in Melastomads, especially in Heterocentron & Monochætum.[2] I can find no nectar in these flowers and Prof Asa Gray writes to me that he can find none in the Northern U.S. sps of Rhexia.[3] It would take up too much space to explain why I am very anxious to know whether insects visit the flowers of this order.— Mr Bates informs me that in the region of the Amazon's he hardly ever saw bees or Butterflies sucking the many flowers of the many plants of this order.—[4] I dare say some grow in your garden under your superintendence.[5] Would you have the kindness to watch them when the weather is fine, occasionally for 5 or 10 minutes & inform me of the result? The species which interest me most are those which have the singular horns or projections to the anther; & four stamens of one colour & shape & four stamens of another colour and shape— The suspicion has crossed me that small insects such as flies or minute sand wasps may penetrate with their proboces or ⟨ ⟩ the horns of the anthers which are filled with fluid [6] If this should prove to be the case, I should very much wish to hear in what position the insect stands on the flower. I do not know whether you have any native species of this order; as such species, or species of allied genera would be best to watch.

Hoping that you will excuse the liberty which I have taken in writing to you & that you will grant me this favour if in your power I Remain | yours ⟨ ⟩

P.S. | If you raise any melastomads from seed in considerable numbers would you be so good as to observe whether on the different individual plants, there is any difference in the position of ⟨ ⟩ pis⟨ti⟩l or stamens; all the ⟨flo⟩wers on one plant having the pistil in one position & *al⟨l⟩* the flowers ⟨o⟩n another plant having the pistil in different position?[7]

Charles Darwin

Endorsement: '25 Jany 63 | T 22 Feby | U 23^d^—'
Copy[8]
DAR 143: 358

1 The year is established by the relationship between this letter and the letter from Hermann Crüger, 23 February 1863.
2 For CD's interest in Melastomataceae, see letter to Hugh Falconer, 5 [and 6] January [1863] and n. 22, and letter to H. W. Bates, 12 January [1863]. For his experimental notes on these two genera, see DAR 205.8.
3 See letter from Asa Gray, 4 August 1862 (*Correspondence* vol. 10). Although most members of the Melastomataceae were tropical, *Rhexia* was a native of North America.
4 See letter from H. W. Bates, 17 January [1863].
5 Crüger was director of the botanic garden and government botanist in Trinidad (R. Desmond 1994).
6 CD was wondering whether insects pollinated the flower when seeking nectar in the horns of the anthers (see letter to Asa Gray, 19 January [1863]). See also letter to Roland Trimen, 16 February [1863] and n. 4, and *Orchids*, pp. 50–3 and 281–2.
7 CD began investigating the position of the pistil in the flowers of melastomataceous plants in 1862 (see, for example, *Correspondence* vol. 10, letter to George Bentham, 3 February [1862], and letter to Asa Gray, 16 February [1862]).
8 The letter is a copy made from the original in November 1887 by John Hinchley Hart of the Royal Botanic Garden, Trinidad. On the copy he recorded that the original was 'worm eaten'.

To Thomas Rivers 25 January [1863][1]

Down. | *Bromley.* | *Kent. S.E.*
Jan. 25th

My dear Sir

I received safely the two fine trees yesterday evening & heartily obliged I am to you for them.[2] They are now in my green-house & I hope will fruit, for I am very curious to see that of the Double-peach.— I am, also, very much obliged for a former note with sketch from memory of the fruit.—[3] I hope you will soon receive the "Origin".—[4]

My dear Sir | Yours very sincerely obliged | Ch. Darwin

Maggs Brothers (catalogue 1086)

1 The year is established by the relationship between this letter and the letter from Thomas Rivers, 21 January 1863.
2 See letter from Thomas Rivers, 21 January 1863.
3 The note and sketch have not been found.
4 In his letter to Rivers of 15 January [1863], CD offered him one of his publications. See also letter from Thomas Rivers, 21 January 1863, and letter to John Murray, 22 January [1863].

From Isaac Anderson-Henry 26–7 January 1863

Hay Lodge, | *Trinity,* | *Edinburgh.*
Jany 26/63

My dear Sir

I had much pleasure in receiving your Letter of the 20th.[1] As for the *Strawberry* I do not set much by it. I have little doubt of its seeds being fertile—[2] "Myatts Pine" was *not* an *alpine*— And, for aught I know the Chilian sps crossed upon it might have been some remote ally. I have a plant in my warm Pits being forced—a very singular dwarf form of the Cross; and I will send you a berry when ripe, if you wish

The *Strawberry crossed with the Raspberry* (of which cross I have large plants above 5 years old, has not yet flowered—[3] It is about coeval with the above cross which has been fruiting for the bygone 3 or 4 years. Its tardiness & its wiry stalked foliage encourages the hope that it is not quite a *miss*. If it fruit this year I will report to you.

In this family I have made numerous attempts besides the above.— I tried to X the *Potentilla* with the Strawberry. I crossed, as I believed, the latter with *Rubus glabratus*—an *Andean* species, which I believe no one else has, as I raised it from seeds transmitted by Dr Jameson of Quito.[4] Of this cross I have several plants; but tho' different in foliage from its female parent, I can see little of the male parent, the *rubus glabratus*, which is a very dwarf bramble with deep dark shining foliage. Here I was at much pains to prevent self fertilization, by carefully opening and cutting off the male organs of the flower ere the pollen passes from its green,—unripe condition. And, as this tube is much infested by an insect of the *bee* family a rapacious eater of pollen I invariably pot my plants under experiment, remove them to the Greenhouse,—& farther admit no other flower on the same plant, from which pollen might be transported.— Yet with all my care I may have been thwarted by some of these pests, whom the better to beguile, I invariably dismantle the bloom of the flower operated on by cutting away the corolla

The *Blackberry* & Raspberry I have strong hopes of.[5] I tried it early last year at Woodend, my place in Perthshire, putting gauze over the blooms, the Raspy plant acted on being in the open garden, but birds interfered with some, and when I returned after a 7 or 8 Weeks absence on the Continent, I found only one fruit & but with one seed partially ripe, which I sowed but which has never yet sprung I do not know if you have found, or whether you have tried these severe crosses & found, like me, that tho you get *ripe* seeds you can't get them to vegetate. The instance I am now to notice bears upon a question you put to me about the *Short anthers*[6]

About the month of March 1851 I crossed—I should rather say *hybridised*, if not *muled* the *Rhodothamnus Chamæcistus* with pollen of a *large species of Rhododendron* I had from M. Van Houtte,[7] which, as it resembled, I named "R. Nilaghericum" which, however, it was not.[8] I find from my Garden book for that year, the following entry under date "May 10 1851— Rhod. Chamæcistus. Counted 7 pods all finely set with seed—(pods size of sweet peas—all (save one) marked *Castrate*, and all crossed with *my R Nilaghericum*— **Short stamens**" I had also crossed the same R. *Chamæcistus* with *R. glaucum* (a fine dwarf Rhodn. one of Dr Hookers Sikkim sps)[9] at the same time with the above; and the memorandum goes on. Or again looking at this *mem.* I find I have written **R** *glau***cum** for "**K.** *glauc***a**"—ie *Kalmia glauca*, with which the cross was attempted & *not* with *Rhod glaucum* as I had erroneously read & have here written[10] "Those *cast* and done with K glauca (blue silk) appear to have failed, a certain proof of the *others* being true"— The seeds of *Rhodothamnus* Chamæcistus from the cross with the large Rhod^{n}., (called by me R Nilaghericum) a white flowered Indian species came to *full maturity*. But tho I sowed them & watched them for *years* they never vegetated—not one of them. Now there was a very improbable looking cross effected by the *short anthers* of a large *Rhododendron*, having large white flowers, fully as large as those of R. cinnamomeum—on this heath like Rhodothamnus—effected at least to the point of producing seed beautifully ripened, but which proved abortive The seeds I remember well were so well ripened that, as is the case with seeds of R. Chamæcistus, they were quite round and ran along paper like small shot infinitely less of course

I had attempted the same cross in the month of March in the immediately previous year 1850—and I have noticed this in an article I wrote for "The Book of the Garden vol. II, p 319[11] and which Dr Lindley has done me the honour to cite, and to give at considerable length in "The Theory & Practice of Horticulture" (vid 2^{d} Ed^{n}. p 490)[12] These seeds too failed to vegetate. I find I have first noticed the capability of working with these *short stamens* by the following *Mem* under date "27 April 1850 & May 9th" Discovered that the *Short stamens* of *Rhod^{n}. Cinnamomeum* (allied to R arboreum) & and particularly of *R Catawbiense*—crossed the small Rhodoth. Chamæcistus—which appear now (11 May) to be fully fertilized & pods swelling—the operation having been performed about the 27 April and cast^{n}. performed. See also operation on *Azalea Phœnicea* by the *Short Stamens* of R *Cinnamomeum*. Tried also on *Menziesia* but failed". I find another Mem of May 1850 which refers to the cross in the article given by Dr Lindley.[13] "The *first* (cross) on my own old (separate Plant) of R Chamæcistus by R. Nilaghericum (so called by me) now well on to ripening performed about 5 March, supposed to have been done with the *Short stamens*" Again, of same date, I notice a cross on the same *R. Chamæcistus* by *Azalea* procumbens which "appears to hold". But I forget what came of this—a failure probably. I find that I tried the same R Chamæcistus with *Erica Odororosea* which I note "appears also to hold".

These operations are somewhat akin to the one effected by Mr Cunningham of Comely bank Nursery here, by which he produced between the Menziesea (some say M *Cærulea* some M *Empetriformis*) and the same *R Chamæcistus*,—the *Bryanthus erectus*.[14] Mr C. rather imposed on Dr Graham by this hybrid and was not fond of admitting it to be a hybrid *at all while* Dr G. lived—[15] He did so to me however, for happening to be in his nursery he showed me the plant under a hand glass—and asked me, as he had done D^{r} G, to say what it was— And as I had been myself at work before on a Cross between the same parents—I at once guessed it to be a hybrid between the M. Cærulea & R Chamæcistus when he owned I was not far off it. I found out afterwards that he had made the *Menziesia* the seed bearer, while I had failed, from making it the *male* parent. I returned to the experiment and made it on the *Menziesea Cærulea* (M^{r} C's experiment must have been on M *empetriformis*) I succeeded fully & sowed the ripened seeds on 18 June 1850 and had 4 young plants through on 10 Septr. same year. These were afterwards destroyed most ignobly by a marauding snail— So, if we make crosses we must learn to bear them—a lesson I am often taught and am now *smarting* under from this morning discovering that perhaps the *only* plant of *Eccremocarpus longifolus*[16] in Europe, which after some 6 months waiting for had sprung with me, is devoured root & branch if I may so speak of a thing in its seed leaves, by some wretched grub, even tho I had it covered up with glass. I saw it yesterday & today it is *non est*.

But of these *short stamens*:—from the above results I took a fancy to them, believing their pollen to have more activity in it—or at least to be better adapted, from its supposed smaller granules, for fertilising *Congeners* of a lesser growth. Please observe my whole working with them was among the **Rhod^n^. family** whose flowers are alike in *form* & similar in having almost invariably 2 short stamens,—with the Pelargonia, I entertained the firm belief also that as they were themselves of miniature dimensions, they would be likely to produce a dwarfish progeny[17]

The magnificent scented *Rhod^n^. Edgworthii*, (of Hooker fil.) with foliage of a distinct cast from all other species fascinated everyone.[18] But it had one great drawback being "leggy" to an undue degree. I set to work upon it, making it the male parent and effected various crosses with its pollen on *R. Ciliatum* a Dwarf sikkim sp^s^., using the short stamens. Now as noticed in article in the Cottage Gardener,[19] I got a brood of dwarfs some of which I yet have not 2 inches high. I *believe* these to be the progeny of the Short anthers—and had determined on watching it better ere saying aught about it, when Donald Beaton drew me out by some remarks I sent the same publication as confirmation of a like observation made by him on the Pelargoniæ tribes.[20] But it is right to observe that I at first wrought with long & short anthers from getting pollen from friends who were before me in blooming the R Edgworthii. But I know when I could get them I wrought with *Short* & *Shorter* stamens as I could get them in preference to the longest ones. Your expressed doubt however of our friend Donalds accuracy demands a precise testing of the thing afresh.[21] And I will try it: But the sad sad thing is, "life is too short for the Experiment among my favourite Rhod^n^. tribes—but the Pelargonium family may be tried.

I fear I am trying your patience sadly in giving you a bundle of chaff to look over for the one or two grains you may be in quest of in this absurdly too long Letter— But I must digress again by mentioning a fact in regard to crossing which I found invariable with all I tried upon *R. Edgworthii* that while its pollen would readily fertilise several (not *all* of its congeners) I could never effect *one* cross on it by foreign pollen. I have found this in *many* plants. I noticed above the same result where *Rhodothamnus Chamæcistus* was made the female to *Menziesia*—(now *Phyllodoce*) *Cærulea* while inverted the cross held & produced seeds which vegetated.[22] But I find I must refer you to the Book of the Garden Vol II p 320 for this Experiment,[23] for D^r^. Lindley does not quote my *entire* article—.[24] I find my article there, is given at length in the 26th Vol (for 1858) of Harrisons Floricultural Cabinet p 60.[25] My ideas *then* ran after Lamarck.[26] But I have been so often baffled in attempts, where I expected success as inevitable, that I have been forced to give up a great part of that bewitching theory. Yet there are *times* & *seasons* & *influences* when I have effected crosses I would have attempted in vain at other times & diverse circumstances. Have you considered this?— But here I am repeating myself in the Article referred to—& must now forbear to tire you longer

To return. I will most gladly try my hands on the *pelargoniums* both for the *above* experiment of the *Short anthers*—and to test the fertility of the *Central* flower, *peloria*, of that tribe.[27] In all my crosses I invariably avoided *that central* flower, just from the impression I had of its infertility

I shall be delighted to receive your article on the *Linums*.[28] They are a most anomalous tribe; but I have no doubt you will throw new light upon it. I will cheerfully try the experiment you point at and report the result.[29] What I meant by *L rubrum* is the same which you with equal correctness name L grandiflorum. I believe the true name is L *rubrum grandiflorum*—an annual, and the only crimson sp.s I know of in the family

You ask me if I have any quite sterile hybrid with a rather large *stigma*. I know of none nearer your mark than Mr Cunninghams *mule* the above Bryanthus erectus, which I am sorry I have lost; but it will be found in almost every Nursery. It is a thing of high interest. I *have* a singular hybrid—which I have kept for nearly 10 years—it is between *Phlox subulata* (if that be a true Phlox)—a thing with awl shaped or grass like foliage, and P. *verna* a broad leaved dwarf species, with rose coloured flowers. It flowered one year & bore no seed, the bloom had all the appearance of sterility—& so far as I remember had its sexual organs quite imperfect. I have various things coming on of which, if true, will put fertility to its severest test of this again.

27th. I have this morning the pleasure of your Letter of yesterday[30] Dr Jamesons address—as I address him is;

'William Jameson Esq MD.[31]
Director of The Mint
Quito
Ecuador via Panama

And the mail leaves *here* on the 1st & 16th. of every month—being the West India mail. By all means; use my name. I believe Dr Jameson must be on the wintry side of 60 but the "sacred fire" burns in him as intensely as ever. He has been and I believe is still Professor of Botany and of Chemistry in the University of Quito. It is years since he told me that he had ceased to speak English for 30 years having all that time lectured in Spanish; but he writes it well. He is a good naturalist. I sent an article of his on the River system of the Napo a new territory to Dr Shaw who inserted it in the Transactions or Journal of the Royal Geographical society.[32] But the Dr complains bitterly of his new avocations with Mint as affording no time for his Explorations either up or down the glorious mountains amid which he is perched halfway up. How it will delight him to hear from *you*. Depend upon him if it be in his power at all to serve you. He has sent me several species of those Melastomaceous things[33]—of which I have 3 or 4 sp.s of the tribe *Rhexia*. I have not flowered one of them—but will if you desire it send you blooms when they flower

Meantime, & asking your kind indulgence for so severe an infliction on your valuable time, I remain | Yours most faithfully | I. Anderson Henry

If I can find any specimen to answer of the male tribes I will either send you a *bit* of the plant or the flowers as you may desire

Charles Darwin Esq FR.Sc. | &c &c &c

DAR 159: 61

CD ANNOTATIONS

1.1 I had ... sprung 4.5] *crossed ink*
3.12 Yet with ... corolla 3.14] *double scored pencil*
4.6 I do not ... *anthers* 4.9] 'Bigeneric crosses' *added ink*
7.1 These ... far off it. 7.10] 'Menziesia cærulea & Rhodothamnus chæmæcistus | Bigeneric forms' *added ink*
10.3 But I ... *Cærulea* 10.8] *scored ink*; 'Reciprocal infertile' *added ink*
10.8 But I find ... Experiment, 10.9] *double scored brown crayon*
10.10 I find ... p 60. 10.11] *double scored brown crayon*; 'I. Anderson Henry Esqre | Books on Hybrid by do' *added at top of page, ink*
11.1 To return. ... tribe. 11.3] *crossed ink*
12.1 I shall ... Phlox) 13.5] *crossed ink*
13.1 I know ... Nursery. 13.3] *scored brown crayon*
15.13 I have 3 ... flower 15.14] *double scored brown crayon*

1 Letter to Isaac Anderson-Henry, 20 January [1863].
2 In a letter written on 14 January 1863, which is now missing (see letter from Isaac Anderson-Henry, 17 January 1863), CD may have sought information from Anderson-Henry on strawberry crosses (see also *Correspondence* vol. 10, letter to *Journal of Horticulture*, [before 25 November 1862]). Anderson-Henry discussed strawberry hybrids in his letter to CD of 17 January 1863. See letter to Isaac Anderson-Henry, 20 January [1863] and n. 3.
3 CD was curious as to whether Anderson-Henry's strawberry–raspberry cross would produce fertile seeds in the spring (see letter to Isaac Anderson-Henry, 20 January [1863]).
4 William Jameson.
5 In his letter of 20 January [1863], CD expressed curiosity about Anderson-Henry's proposed attempt to hybridise a raspberry and a blackberry.
6 See letter to Isaac Anderson-Henry, 20 January [1863].
7 Anderson-Henry refers to the Belgian horticulturist Louis Benoit van Houtte.
8 Anderson-Henry misspelled *Rhododendron nilagiricum.*
9 Joseph Dalton Hooker is credited with introducing the rhododendrons of Sikkim to Britain; he collected many species during his Himalayan expedition from 1848 to 1851, publishing an account of these Himalayan shrubs in J. D. Hooker 1849 (*DNB*).
10 This sentence was written as an addition in the margin. In the following sentence, Anderson-Henry crossed out 'R. glaucum', and substituted 'K. glauca'.
11 Anderson[-Henry] 1853.
12 John Lindley cited Anderson-Henry's observations on hybridising (Anderson[-Henry] 1853) in Lindley 1855, pp. 490–4.
13 See n. 12, above.
14 James Cunningham of the Cunningham & Fraser nursery, Comley Bank, Edinburgh, successfully propagated the hybrid *Bryanthus erectus* (*Gardeners' Chronicle*, 1 November 1851, p. 695). *Bryanthus erectus* is now known as *Phyllothamnus erectus*, an intergeneric hybrid. This dwarf evergreen shrub reportedly

resulted from a cross between *Rhodothamnus chamaecistus* and *Phyllodoce caerulea* (also known as *Menziesia caerulea*); however, it was generally believed that *P. empetriformis* (also known as *M. empetriformis*) was the second parent (see Bean 1970–88, 3: 173).

15 Robert Graham was professor of botany at the University of Edinburgh from 1820 until his death in 1845 (R. Desmond 1994).

16 Anderson-Henry refers to *Eccremocarpus longiflorus*, a Peruvian evergreen climber of the Bignoniaceae (*Index Kewensis*).

17 CD's question about short stamens concerned pollination in *Pelargonium* (see letter to Isaac Anderson-Henry, 20 January [1863]), and whether short stamens produced dwarf plants. See n. 20, below.

18 *Rhododendron edgworthii* was described in J. D. Hooker 1849.

19 Anderson-Henry refers to his article, 'Variegation, cross-breeding, and muling of plants', in the *Journal of Horticulture and Cottage Gardener* n.s. 2 (1861): 41–3.

20 Donald Beaton was one of the gardening editors of the *Journal of Horticulture*. For Beaton's observations on *Pelargonium*, see the letter to Isaac Anderson-Henry, 20 January [1863] and n. 11.

21 See letter to Isaac Anderson-Henry, 20 January [1863] and n. 13.

22 See n. 14, above.

23 Anderson[-Henry] 1853.

24 Lindley 1855.

25 Anderson[-Henry] 1853 was republished in 1858 in the *Horticultural Cabinet and Florists' Magazine*, which was edited by Joseph Harrison.

26 Anderson-Henry refers to Jean Baptiste de Lamarck's theory of transmutation (Lamarck 1809).

27 For CD's interest in peloric flowers, see the letter to Isaac Anderson-Henry, 20 January [1863].

28 In his letter of 20 January [1863], CD promised to send Anderson-Henry a copy of his paper on dimorphism in *Linum* when it was printed; 'Two forms in species of *Linum*' was read at a meeting of the Linnean Society on 5 February 1863 (*Collected papers* 2: 93–105). For CD's presentation list for the *Linum* paper, see Appendix IV.

29 See letter to Isaac Anderson-Henry, 20 January [1863].

30 The letter has not been found.

31 Although CD did write to Jameson (see letter from Isaac Anderson-Henry, 17 April 1863), the correspondence has not been found.

32 Henry Norton Shaw was secretary of the Royal Geographical Society and editor of its journal, in which Jameson 1858 was published.

33 CD was interested in botanical collectors to whom he could apply for seeds '*in the native land*'; in the early 1860s, he was keen to obtain Melastomataceae seeds for experiments on flower dimorphism (see *Correspondence* vol. 10, letter to J. D. Hooker, 14 [October 1862]).

To H. W. Bates 26 January [1863][1]

Down Bromley Kent
Jan 26th

Dear Bates

I congratulate you very sincerely on your marriage.[2] Judging from my own experience it is the best & almost only chance for what share of happiness this world affords. I hope you may succeed, for the sake of science, in getting fixed near London.—[3]

I am heartily glad to hear that your great labours over your Book are drawing to a close:[4] I know that I for one shall read it with real interest.— Pray thank Wallace when you see him about Melastomas;[5] I have in truth given more trouble on this, than case deserves, & am truly obliged to you. The fact is I cannot endure being beaten by a beggarly flower to the degree, which these confounded Melastomas have beaten me.— I am pleased to hear of discussion on species at Zoolog. Soc.—[6]

With my renewed congratulations & hopes for your happiness, believe me | Yours sincerely | Ch. Darwin

Cleveland Health Sciences Library (Robert M. Stecher collection)

[1] The year is established by the relationship between this letter and the letter from H. W. Bates, 24 January 1863.

[2] Bates married Sarah Ann Mason on 19 January 1863 (see letter from H. W. Bates, 24 January 1863 and n. 4).

[3] See letter from H. W. Bates, 24 January 1863

[4] Bates 1863. See letter from H. W. Bates, 24 January 1863 and n. 5.

[5] See letter to H. W. Bates, 12 January [1863], letters from H. W. Bates, 17 January [1863] and 24 January 1863, and letter from A. R. Wallace, [23 January 1863?].

[6] See letter from H. W. Bates, 24 January 1863 and n. 3.

To Francis Trevelyan Buckland 26 January [1863][1]

[letter . . . to Francis Buckland, *Down, 26 January n.y.*, recalling that they were introduced 'by your late honoured Father' many years ago,[2] writing as a fellow-naturalist to ask Buckland's help in identifying an article in *The Field* about the fins of fishes growing again after being cut off,[3] and enquiring whether he has heard of the re-growth of organs 'such as tail or finger or toe' in the mammalia or birds, 'I am privately informed that regrowth occurs with *monstrous* additional fingers with men'[4]]

I hope that you will excuse my venturing thus to trouble you, & I beg leave to remain, Dear Sir | Yours faithfully, | Charles Darwin

A friend sent me a few days ago the Field with your[5]

[Darwin in a postscript refers also to Buckland's letter in *The Field* about a 'Dog-Lion' and a reported sighting in Russia.][6]

Incomplete[7]
Christie's, London (23 June 1993, lot 146)

[1] The year is established by the relationship between this letter and the letter to F. T. Buckland, 1 February [1863]. See also n. 3, below.

[2] Buckland's father, the geologist William Buckland, died in 1856 (*DNB*).

[3] The reference has not been identified. See letter to J. J. Briggs, 2 February [1863]. In the chapter on inheritance in *Variation*, CD stated (*Variation* 2: 15–16):

> Lastly, as I have been informed by Mr. J. J. Briggs and Mr. F. Buckland, when portions of the pectoral and tail fins of various freshwater fish are cut off, they are perfectly reproduced in about six weeks' time.

[4] See *Variation* 2: 16, 57.

[5] The letter text from 'I hope' to 'with your' is transcribed from a partial facsimile which accompanies the catalogue description.

[6] See letter from Francis Boott, 23 January 1863, and nn. 2, 3, 5 and 6.
[7] The catalogue description notes that the original letter is four pages long.

From Thomas Rivers 26 January 1863

Nurseries, Sawbridgeworth, Herts, | Great Eastern Railway.
Jan^ry^. 26/63

My dear Sir/

Pray accept my best thanks for your book which I shall now read with fresh interest[1]

I remember when I read it on its first being published there were several subjects supporting y^r^. views on which I was nearly tempted to address you. If on again reading it I go into the same train of thought I shall venture to do so.

I have always thought that our naturalists & botanists have never made sufficient allowance for the changes in animals & plants brought on by site & soil & climate hence has arisen the enormous number of so-called species. Some thirty years since I used to make an annual journey through nearly all the counties in England & I then used to amuse myself by noticing the effect of site on men this I did by going into the markets & observing the country people come in, selecting a town near the centre of any particular county. After leaving Essex a county of mixed people I found that Suffolk had a race which I used to distinguish by features which I always carried in my memory. Then came Norfolk & at Norwich Market I saw the Norfolk race very distinct from the Suffolk Then came Lincolnshire & at Boston the centre of the flat rich land of the county I used to see the large boned large featured men peculiar to the soil & so I used to observe the race-like character of the Nottinghamshire, Yorkshire & Leicestershire the sallow heavy browed men of the latter county always struck me & then Lancashire & Cumberland & some other counties my observations always taken in the markets & of the country people—made me such an adept in judging of a cast of features appertaining to particular counties that I have often surprised men who had long lived in London by telling them what county they came from. I then used to reason in my uncultured natural way if the human-race can be so affected by site in this small island how vast the effect must be in climates & sites differing as they do in the world & carried on for thousands of years All this I well know must be trifles to you but my being able to distinguish people in counties has always appeared to me odd but it led me to look into the changes brought on in plants & trees by site & soil & climate & led me into a correspondence with Morren[2] years ago.

so pray pardon me I am My dear Sir Y^rs^. very truly Th^os^. Rivers

Postmark: JA 26 63
DAR 176: 161

[1] CD had arranged for a copy of *Origin* to be sent to Rivers (see letter to Thomas Rivers, 25 January [1863], and letter to John Murray, 22 January [1863]).
[2] Charles François Antoine Morren.

From Friedrich Rolle[1] 26 January 1863

Bad Homburg bei Frankfurt am M⟨ain⟩
den 26ten. Januar 1863.

Geehrtester Herr!

Entschuldigen Sie, wenn ich nicht eher Ihr gütiges Schreiben vom vorigen Herbst beantworte.[2] Ich wollte erst den Schluss meines kleinen Werkes "Darwin's Lehre" abwarten.[3]

Ich war sehr erfreut zu sehen, dass meine Arbeit Ihren Beifall gefunden hat und ich hoffe dass Sie im Schluss-Hefte noch weitre Gegenstaende antreffen, welche Ihre Aufmerksamkeit auf sich ziehen. Sie werden manche Versuche finden der Palaeontologie die physiologische Seite abzugewinnen.

Ich bin eigentlich Geologe und Palaeontologe. Meine Hauptaufgabe war in den letzten Jahren die Ermittelung der Genealogie der europaeischen Meeres-Fauna und ihre Ableitung von den Tertiaeren Fossilien. Ich habe darüber nur wenig veroeffentlicht, denn ich verliess im letzten Frühjahre vor Beendung dieser Arbeiten das Kaiserl. Koenigl. Hofmineralien-Cabinet zu Wien, an dem ich fünf Jahre lang als Assistent, spaeter als Adjunkt stand.[4] Ich habe mich seither in meiner Vaterstadt Homburg als Naturalienhaendler niedergelassen und hoffe mir auf diesem Wege ein unabhaengiges Fortkommen gewinnen zu koennen.

Ich wandte mich damals an den Buchhaendler H[rn]. Suchsland in Frankfurt um den Auftrag zu einem wissenschaftlichen Werke.[5] H[r]. Suchsland hatte in Zeitschriften von Ihrer Lehre Kenntniss erhalten und äusserte viel Antheil daran. Er übertrug mir daher eine Erläuterung derselben.

So fiel mir die Aufgabe zu, für die Erweiterung und Ausbreitung Ihrer Theorie zu wirken. Eigentlich ist der thaetigste Vertreter Ihrer Ansichten mein Freund Doctor Gustav Jaeger in Wien, ein geborner Schwabe, der Gründer und jetzt Director des zoologischen Gartens in Wien ist.[6]

Gustav Jäger und ich waren schon Ihrer Ansicht, ehe wir noch Ihr Werk erhielten. Wir gedachten selbst ein gemeinsames Werk über die Genealogie der Thiere zu schreiben, Jaeger sollte den zoologischen und physiologischen Theil, ich wollte den geologischen und palaeontologischen Theil bearbeiten. Aber unsre Absicht gelangte nicht zu vollstaendiger Ausführung.[7]

Als Ihr Werk erschien, übernahm es Jäger die Theorie in Wien durch oeffentliche Vortraege und in Zeitschriften zu erläutern. Ein Herr Pelzeln in Wien trat als sein Gegner auf.[8] Aber die oeffentliche Meinung sprach für Jäger. H[r]. Pelzeln fand nicht einmal einen Raum zur Veroeffentlichung seiner Einwaende. Er liess seine Erwiederung daher als eigene Flugschrift erscheinen. (August von Pelzeln, Bemerkungen gegen Darwin's Theorie vom Ursprung der Spezies. Wien bei Pichler 1861. 8[o]. 17 pag.)[9]

Diese Flugschrift ist aber so arm, so albern und zugleich so fanatisch geschrieben, dass man sie am besten gar nicht aufführt und Jaeger hat auch darauf nicht geantwortet.

Jaeger hielt damals auch auf der Versammlung der deutschen Ornithologen zu Stuttgart einen Vortrag; als seine Gegner traten Professor Blasius (von Braunschweig) und Professor Altum (von Münster) auf.[10]

Seit dieser Zeit hat Jaeger nichts mehr darin gethan, weil er zusehr mit der Begründung des Wiener Zoologischen Gartens beschaeftigt war. Doch beabsichtigt er demnaechst wieder hervorzutreten und ich bin bemüht ihn anzutreiben.

Unsre Professoren in Deutschland sind theils Ihre Gegner, theils schweigen sie.

Aehnliche Ansichten hat von jeher Professor Quenstedt in Tübingen gehabt. Aber er arbeitet ohne Methode und macht viele Fehler, daher er auch nie allgemeine Geltung in Deutschland gefunden hat.[11]

Für Ihre Ansichten haben sich namentlich noch D^r. Weinland in Frankfurt (ehemals Assistent von Agassiz) und D^r. Hensel ⟨in⟩ Berlin ausgesprochen.[12] Weinland hat ⟨mich⟩ bei meiner Arbeit sehr unterstützt⟨. Hensel⟩ kenne ich nicht von Person; Sie f⟨inden⟩ einige wichtige Bemerkungen in Hensel's Schrift, über Hipparion in den Abhandlungen der Berliner Akademie.[13] Er bemerkt, dass das Pferd von den Pachydermen entfernt steht. Seine abnorme Fuss-Bildung zeigt das Pferd schon als Embryo. Es gibt aber monstroese Bildungen am Pferde, bei der die normale Pachydermen-Bildung des Fusses sichtbar wird. Dies ist ein Nachklang der Abstammung des Pferdes von Pachydermen. Die fossile Gattung Hipparion bestaetigt dies und liefert einen deutlichen Uebergang vom Pferd zu Rhinoceros, Hippopotamen und andren Pachydermen. Hensel sagt auch, dass rudimentaere Organe für jeden Naturforscher, der auf dem Boden der Realitaet steht, auf keine andre Weise naturgemaess erklaert werden koennen als durch Abstammung.[14]

Sollte ich Ihnen irgendwie gefaellig sein koennen, so soll es mit Vergnügen geschehen. Besonders über tertiaere Mollusken oder andre palaeontologische Gegenstaende bin ich bereit.

Entschuldigen Sie, dass Ich Ihnen deutsch antworte, das Englische ist mir sehr wenig gelaüfig, Sie würden mein Englisch vielleicht nicht verstehen.

Es empfiehlt sich Ihnen, geehrtester Herr | Ihr hochachtungsvoller Diener | D^r. Friedrich Rolle.

DAR 176: 201

1 For a translation of this letter, see Appendix I.

2 Letter to Friedrich Rolle, 17 October [1862] (*Correspondence* vol. 10).

3 Rolle 1863 was published in instalments, the first and second of which had been sent to CD before 17 October 1862. The publication of the third and fourth parts was announced on 19 January 1863 (*Börsenblatt für den Deutschen Buchhandel* 30 (1863): 113). In translation, its title read: 'Charles Darwin's theory of the origin of species in the plant and animal kingdoms and its application to the history of creation'. In the foreword to the book (Rolle 1863, p. [ii]), Rolle referred to CD's approval of his efforts (see *Correspondence* vol. 10, letter to Friedrich Rolle, 17 October [1862] and n. 3). CD's annotated copies of all four parts of Rolle 1863 are in the Darwin Pamphlet Collection–CUL.

4 For a description of Rolle's life in Vienna at the Kaiserlich-königliche Hofmineraliencabinet (Royal Imperial Cabinet of Mineralogy), and for a discussion of his reasons for leaving academic life and moving to Homburg in March 1862, see Martin and Uschmann 1969, pp. 18–26.

[5] The publisher and bookseller Friedrich Emil Suchsland was head of the Frankfurt firm Johann Christian Hermann'sche Verlagsbuchhandlung (Martin and Uschmann 1969).

[6] Gustav Jäger announced his support for transmutation and CD's theory in lectures given at the annual meeting of the Deutsche Ornithologen-Gesellschaft in Stuttgart on 18 September 1860 (see n. 10, below), and at the Verein zur Verbreitung naturwissenschaftlicher Kenntnisse in Vienna on 10 and 15 December 1860 (Jäger 1862; see n. 8, below). Jäger was one of the directors of the zoological garden in Vienna, which opened on 25 May 1863 (see Weinreich 1993, pp. 48–64).

[7] *Zoologische Briefe* (Jäger 1864) included a chapter in which Jäger argued in support of the transmutation of species theory; the author claimed that this was written before the publication of *Origin* in 1859 (Jäger 1864, pp. 36–56). There is a copy of Jäger 1864 in the Darwin Pamphlet Collection–CUL. For a discussion of Jäger's reaction to *Origin*, see Weinreich 1993, pp. 36–47.

[8] Rolle refers to the confrontation between August Pelzel von Pelzeln and Jäger at the meeting of the Zoologisch-botanische Gesellschaft in Vienna on 5 December 1860. Pelzeln had given a lecture in which he criticised CD's theory, and which was later published as Pelzeln 1861; CD's annotated copy of this work is in the Darwin Pamphlet Collection–CUL. According to the report in the *Verhandlungen der kaiserlich-königlichen zoologisch-botanischen Gesellschaft in Wien* 10 (1860): 98, Jäger strongly opposed Pelzeln's view and offered to speak in support of CD's theory (see Jäger 1862; see also n. 7, above).

[9] Pelzeln 1861. See n. 8, above.

[10] Jäger's lecture in support of CD's theory was given at the annual meeting of the Deutsche Ornithologen-Gesellschaft on 18 September 1860. In the subsequent discussion, Bernhard Altum and Johann Heinrich Blasius maintained that CD's theory was a purely speculative and unsubstantiated hypothesis (see *Bericht über die XIII. Versammlung der Deutschen Ornithologen-Gesellschaft zu Stuttgart vom 17. bis 20. September 1860*, pp. 6, 40–6). In the late 1860s, Altum, a clergyman and lecturer in zoology, became known as one of the most uncompromising opponents of the theory of evolution in Germany (see Stresemann 1975, pp. 273–4, 328–32).

[11] Friedrich August Quenstedt had been Rolle's tutor at the University of Tübingen. In 1856, he published a popular book in which he made a number of remarks in favour of biological evolution and common descent (see Quenstedt 1856, pp. 227–9).

[12] The references are to David Friedrich Weinland and Reinhold Friedrich Hensel. Weinland, a lecturer in zoology at the Senckenberg Museum, Frankfurt, had been an assistant to Louis Agassiz in Cambridge, Massachusetts, between 1855 and 1858 (*Zoologischer Beobachter* 57 (1916): 160). In February 1861, he published a short but favourable review of *Origin* (Weinland 1861). Hensel had been a tutor in natural history at the University of Berlin until 1860; three years later, he was commissioned to conduct zoological research in southern Brazil (*Leopoldina* 18 (1882): 19–21).

[13] Hensel 1860.

[14] Hensel mentioned CD's theory briefly when he considered the importance of rudimentary organs and maintained that biological traits were gradually changing over time (see Hensel 1860, pp. 69–70).

To William Bernhard Tegetmeier 26 [January 1863][1]

Down Bromley Kent
26th

My dear Sir

I have got the vaguest idea that I have read of a cut-off *extra* or monstrous toe on a fowl growing again.—[2] Have you heard of such a case? If you have, kindly inform me; if not, please do not write, as I shall understand silence

My dear Sir | Yours very sincerely | C. Darwin

Shall you do anything about Spanish Cock & white Silk-Hen?[3]

I saw in Report of Philoperisteron a notice of curious pigeon with long wings.—[4] Do you know owner, could you get me a longer description, with any measurement of wing from tip to tip & measurement of body from base of beak to base of tail. But it would hardly worth trouble without the Bird was dead.—

The LuEsther T. Mertz Library at The New York Botanical Garden (Charles Finney Cox collection, Tegetmeier ser. 2: 24)

[1] The date is established by the reference to the report of the Philoperisteron Society's annual show (see n. 4, below).

[2] For CD's interest in polydactylism, see, for instance, the letter to T. H. Huxley, [8 February 1863] and nn. 2 and 3.

[3] In his letter to Tegetmeier of 27 [December 1862] (*Correspondence* vol. 10), CD offered £5 to 'cover expences of experiment'; CD encouraged Tegetmeier to cross a 'Spanish Cock & a couple of *white* Silk hens', adding that he would be 'most grateful to hear whether the offspring breed well'. See also letter to W. B. Tegetmeier, 19 February [1863] and n. 2.

[4] CD refers to the report in the *Journal of Horticulture*, 20 January 1863, pp. 58–9, of the Philoperisteron Society's annual show, held at the Freemason's Hall, Lincoln's Inn Fields, London, on 15 January 1863. The report provided details of a Belgian pigeon combining the colour of the archangel variety with the head of the owl or barb pigeon; its most 'striking pecularity' was 'the extraordinary length of the tail and flight feathers, the latter crossed beyond the tail giving the animal the appearance of a gigantic Swift (Cypselus), or Longwinged Hawk'.

From W. B. Tegetmeier [after 26 January 1863][1]

Muswell Hill | London N

My dear Sir

I never heard of such a case nor can I believe it possible,[2] but I think the idea very likely to have arisen from many dorkings having *two* or *three* extra toes partially developed, one of these (the longest) may have been removed accidentally and the others having been noticed afterwards may have been mistaken for new growths—[3] Some old Dorking breeders always used to cut off the extra toes and they never grew again.

I have been to Ireland or should have replied to your former letter before— I should be very glad to undertake the Spanish and Silk hen experiment, and will at once fit up a place for their reception—.[4]

I cannot see your motive for suggesting turbits and carriers for the pigeon experiment as I do not think there would be any probability of a sterile cross— doubtless however you have some good reason[5]

The Pigeon at the Philoperisteron was *very* peculiar— I wrote the account in the Cottage Gardener.[6] And will get you a much more detailed account of measurement—. Mr Wallace[7] was there and examined it with some interest.

I shall not see owner for fortnight but will ask him to bring bird to town and supply you with full details

Believe me | Yours very truly | W B Tegetmeier

DAR 178: 56

[1] The date is established by the relationship between this letter and the preceding letter.
[2] See preceding letter.
[3] Dorkings were a square-bodied breed of poultry, typically with five claws on each foot (*Variation* 1: 227).
[4] See preceding letter and n. 3.
[5] See *Correspondence* vol. 10, letter to W. B. Tegetmeier, 27 [December 1862] and Appendix VI. See also letter to W. B. Tegetmeier, 19 February [1863].
[6] See preceding letter and n. 4.
[7] Alfred Russel Wallace.

To *Journal of Horticulture and Cottage Gardener* [before 27 January 1863][1]

Few facts in vegetable physiology are more remarkable than the well-ascertained influence of the pollen of one species or variety on the seed and fruit of another species or variety whilst still attached to the female plant. There are several old accounts, and the case has been well proved by Gärtner of the colour of the pea in one variety of the Garden Pea, being changed by the direct action of the pollen of another differently-coloured variety.[2] So, again, the famous St. Valery Apple tree produces many different kinds of fruit, according to the nature of the pollen used; for the singularly-constructed flowers yield no pollen, and they are annually fertilised by a party of French girls, who bring pollen from other trees, and mark with ribbons the flowers thus fertilised.[3] About a year ago Mr. Beaton gave an analogous case, far more remarkable than any hitherto recorded, for he showed (if my memory does not deceive me) that the pollen of one species acted on the footstalk of the seed-capsule of another species, and caused it slowly to assume a position which it would not otherwise have acquired.[4] I forget the name of the plant, and have vainly spent an hour in trying to find the passage, though I am sure I marked it. Will Mr. Beaton have the kindness to repeat the statement? and I am sure it is worth repetition.[5] If he grant this favour, will he inform us whether his observations were made on several flowers, and during one or more years? I remember some difficulty in finding the name of the plant in such catalogues as I happened to have at hand, which led me to suppose that it had, like too many plants, more names than one.—

Charles Darwin.

Journal of Horticulture and Cottage Gardener n.s. 4 (1863): 70

[1] CD's letter appeared in the *Journal of Horticulture* on 27 January 1863, under the title 'Influence of pollen on the appearance of seed'.
[2] Gärtner 1849, pp. 80–7. Karl Friedrich von Gärtner's experiments with peas are discussed in Appendix V (see, especially, 'Beaton's first response to the letter from Charles Darwin to the *Journal of Horticulture and Cottage Gardener*, [before 27 January 1863]' and n. 7). In *Variation* 1: 397–403, CD included a section entitled: 'On the direct or immediate action of the Male element on the Mother Form', in which he cited Gärtner's work (*Variation* 1: 397).

[3] The case of the artificially pollinated St Valery apple is mentioned in *Variation* 1: 350 and 401.

[4] In an article on crossing flowers, Beaton reported an observation that he termed 'the most curious thing I know of among plants' (Beaton 1860, p. 254):

> The pods of *Imatophyllum miniatum* stand erect as the umbels of flowers, and the pods of *I. cyrtanthiflorum* hang down as the flowers do. By crossing the two, the pods of the former become as pendent as those of the latter

CD double scored this passage in the margin of his unbound copy of the journal, which is in the Darwin Library–CUL.

[5] For Beaton's replies, see Appendix V.

From Asa Gray 27 January 1863

Cambridge [Massachusetts]
Jany. 27, 1863

My Dear Darwin.

I have been far too busy to write letters—have been interrupted too by visitors, etc—

Your last is of Jany. 2^d [1]

I am glad the plants reached you safely.[2]

Linum Lewisii, I suppose to be only *L. perenne*. Why do you think not?[3]

I do not recal any *bud-variations*.[4]

You "wish to Heaven the North did not hate us so".[5] We equally wish the English did not hate *us* so. Perhaps we exaggerate the ill will in England against us. You certainly over-estimate that of U.S. against England,—which an influential part of your press exaggerates and incites for the worst purposes. But, after all after the first flurry,[6] we think and say very little about you, and shall live in peace with you, if you will let us. There should have been and might have been the most thorough good-will between us. I do not think it is all *our* fault that it is not so.

In reply to your question—[7]

If Oak and Beech had large-colored corolla, &c—I know of no reason why it would be reckoned a low form, but the contrary, quite. But we have no basis for high & low in any class—say Dicotyledons, except perfection of development or the contrary in the floral organs,—and even the envelopes,—and as we know these may be reduced to any degree in any order or group, we have really that I know of, no philosophical basis for high & low. Moreover, the vegetable kingdom does not *culminate*, as the *animal* kingdom does. It is not a *kingdom*, but a commonwealth—a *democracy*, and therefore puzzling and unaccountable from the former point of view.

I have just read De Candolle's paper on Oaks & Species, & origin.—[8] Well, he has got on about as far towards you as I have.[9] It is clear enough that (as I thought at first) *Derivation* of *species* is to be the word. And *Nat. Sel.*, admitted. The only question, whether this is enough.[10]

Ever Your attached friend | A. Gray

[Enclosure 1]

No 3 Myrtle St, Boston
Jan. 7th., 1863

Prof. Asa Gray.
Dear Sir,

The butterfly which I showed you the other day with the pollinia of Platanthera Hookeri attached, I discover to be ⟨a n⟩ew species of *Nisoniades*, which I shall pu⟨blish⟩ very shortly under the name of *Persius*[11]

I forgot to remark when I saw you that if these pollinia had been atta⟨ched to⟩ any part of the body covered by scales, their weight would doubtless have dragged away the scales ⟨from⟩ their attachment, wasting the pollinia ⟨ ⟩ they were attached to the ⟨*exter*⟩*na*⟨*l*⟩ ⟨ ⟩ body, except the tongue.

Recalling our conversation,[12] I should like to copy a remark from my notes on ⟨Platan⟩thera orbiculata, made at the ⟨ ⟩ "the pollinia on being removed ⟨see⟩med to have their angle of verticality changed, not by a rotation of th⟨e⟩ ⟨ ⟩ pedicel, but by the weight of the ponderous masses of pollen at the tip, which passed down to a nearly horizontal position, or as far as the elasticity of the thread would allow. ⟨A⟩ll the other movements were the same (as described by Dar⟨win⟩ but the disks of the pollinia were so widely seperated that I should hardly think they could be often detached (because not stuck) by ⟨insect⟩ smaller than a bee, yet in a spike I examined containing 23 flowers, all open, none had ⟨b⟩oth pollinia remaining and only 4 or 5 had ⟨a⟩ny— upo⟨n⟩ the stigmatic surface of one were plainly seen the hairs and scales of a Lepidopterous insect."[13]

Very truly and respectfully yours | Sam. Scudder—

This note gives the name of the Lepidopt. which I wrote of as having pollinia of Platanthera on his eyes.[14]

[Enclosure 2][15]

On your principle of Nat. Selection being *not subordinated to intention*, are we to suppose, in accounting for the naked eyes,—that the ancestors of this butterfly had all its scales pulled off their eyes by the pollinia of orchids, and the race was *naturally-selected* through the advantage of seeing their way to the flowers?[16]

DAR 165: 129, 130

CD ANNOTATIONS

Enclosure 1

2.2 their weight . . . tongue. 2.4] *scored pencil*; *cross in margin, blue crayon*

Top of letter: '⟨*four words destroyed*⟩ except on eyes and proboscis' *ink*

Enclosure 2

1.1 'On your principle'] *cross in margin, blue crayon*

[1] Letter to Asa Gray, 2 January [1863].

[2] Gray had sent CD a box of living plants for him to grow and examine (see letter to Asa Gray, 2 January [1863] and nn. 4–7).

[3] See letter to Asa Gray, 2 January [1863] and n. 9 .

[4] See letter to Asa Gray, 2 January [1863] and n. 18.

5 See letter to Asa Gray, 2 January [1863].
6 Gray refers in part to the *Trent* affair of November 1861, which created a diplomatic crisis between Britain and the United States (see McPherson 1988, pp. 389–91, and *Correspondence* vol. 9).
7 CD had asked Gray whether, if oak or beech trees had colourful flowers, they would still be classified as 'low' (see letter to Asa Gray, 2 January [1863]). See also letter to Asa Gray, 23 February [1863] and n. 7.
8 A. de Candolle 1862a; there is an annotated copy of this and A. de Candolle 1862b in the Darwin Pamphlet Collection–CUL. There is also an annotated copy of Gray's review of A. de Candolle 1862a and 1862b (A. Gray 1863d) in the Darwin Pamphlet Collection–CUL.
9 See n. 8, above. For CD's view of Candolle on natural selection, see the letter to J. D. Hooker, 13 January [1863], and the letter to Alphonse de Candolle, 14 January [1863].
10 CD and Gray had long debated teleology and the possibility of an intentional design in nature (see *Correspondence* vols. 8 and 9).
11 The author of the enclosure, Samuel Hubbard Scudder, showed Gray the butterfly with orchid pollinia attached, on 29 December 1862 (see *Correspondence* vol. 10, letter from Asa Gray, 29 December 1862). Scudder published his description of *Nisoniades persius*, a member of the Hesperidae, in the *Proceedings of the Essex Institute* 3 (1863): 170 (see also Scudder 1889, pp. 1468–76).
12 Gray had speculated that the pollinia belonged either to *Platanthera hookeri* or to *P. orbiculata*, but probably to the former owing to its size (see *Correspondence* vol. 10, letter from Asa Gray, 29 December 1862).
13 CD discussed movements of the pollinia of various genera in *Orchids*, pp. 80, 90–1, and 335–9, including the genus *Habenaria*, synonymous with *Platanthera* (A. Gray 1862c, pp. 259–61). In the second edition of *Orchids* (*Orchids* 2d ed., pp. 75–7), CD added information from A. Gray 1862a, 1862b, and 1863b on pollination in American species of *Platanthera*, and also mentioned the *Nisoniades* that Gray had seen with the pollinia attached to the eyes. See also 'Fertilization of orchids', pp. 147–8 (*Collected papers*, pp. 144–5).
14 This sentence was appended by Gray to the bottom of Scudder's letter. Gray refers to his letter to CD of 29 December 1862 (*Correspondence* vol. 10).
15 Gray enclosed an additional note.
16 See n. 10, above.

From Thomas Francis Jamieson 28 January 1863

Ellon
28. Jan 63

My Dear Sir,

I am much gratified by your Commendation of my Glen Roy paper, and am still more glad to think that you found nothing in it to give you offence, which would have vexed me much.[1]

—If it is printed I hope it may induce some others to visit the district which is remarkably rich in geological points of interest & is not nearly exhausted yet—[2] I wish it was nearer to me

I may mention that I have got the lectureship for which I was applying.[3]

—In reading over some of Liebig's writings lately the following idea occurred to me as a *possible* mode of influencing to *some* degree change in species—[4]

Suppose a country to be covered with a certain kind of vegetation—say forests of certain species of trees—&c. and let these flourish for so many ages— It seems not improbable that they will exhaust in course of time certain ingredients of the soil that are necessary for their growth, & will then give away to others that are differently constituted & can still find plenty of the sort of mineral food they require.[5]

—In this way the features of the vegetation may change—

Then take the animals—say *sloths* for example feeding upon the leaves of these forests—having to live upon trees of a different nature & containing different proportions or kinds of ingredients, & having different fashions of growth— they will be influenced in turn—& thus the herbivora will change— And then the Carnivora that prey upon them will have to suit themselves to their victims, & alter their habits, & their digestive organs accordingly— The Omnivora will change most slowly of all, having a wider range of food.

—In this way first the vegetation changes, then the herbivora, next the Carnivora & last of all the Omnivora.

I don't suppose the speculation is of any value, only it may amuse you for an idle moment I think there is no doubt but that species have *as a rule* arisen in some way out of previous forms.

I am | My Dear Sir | Yours very truly | Thos. F. Jamieson

Charles Darwin Esq

DAR 168: 45

1 Jamieson's paper on the so-called 'parallel roads' of Glen Roy, a series of terraces running parallel to one another along the sides of the glen in Lochaber, Scotland (Jamieson 1863), was read at the Geological Society of London on 21 January 1863. Since CD had published a major study on the subject in 1839 ('Parallel roads of Glen Roy'), it was referred to him before being published (see *Correspondence* vol. 10, letter from A. C. Ramsay, 13 December 1862). CD's referee report is no longer extant.

2 Jamieson 1863 appeared in the *Quarterly Journal of the Geological Society of London* later in the year.

3 In 1862, CD wrote a testimonial in support of Jamieson's application to become the Fordyce lecturer in agriculture at the University of Aberdeen (see *Correspondence* vol. 10, letter to T. F. Jamieson, 21 November 1862).

4 Justus von Liebig, professor of chemistry at the University of Munich, was then the leading experimental chemist in Europe. His treatise on chemistry, agriculture, and physiology, published in Britain as *Organic chemistry in its applications to agriculture and physiology* (Playfair trans. 1840), brought together a vast amount of knowledge and practice in a coherent form, making it 'one of the most important books in the history of agriculture' (*DSB*). Jamieson may refer to a seventh, expanded edition that appeared in 1862 (Liebig 1862). See also Liebig 1840–4.

5 For a discussion of Liebig's ideas on soil exhaustion and the reception of Liebig 1862, see Finlay 1991.

To J. D. Hooker 30 January [1863]

Down
Jan. 30th

My dear Hooker

Please address enclosed & post.—[1]

I have to thank you for two very pleasant notes:[2] it was very good of you to write from Paris.—[3] I fear Naudin has not responded;[4] but I hope he may experiment, for I clearly see that he trusts too much to resemblance to parents & does not think enough of actual fertility or number of seeds. I hope to Heaven, that he may explain sterility, but do not expect it.—[5] I am not surprised at all at Decaisne &

Naudin thinking little of "Origin";[6] there has always appeared to me something antagonistic to a Frenchman in the way in which Englishman writes. On other hand I have just had another letter from A. Decandolle;[7] he speaks of "us" as believing in mutability in glorious way, & reports that a Count Saperda, who is writing on Fossil plants goes the whole hog.—[8]

What do you think, Bates writes *in a P.S.* that he is married![9]

I have just had a letter or rather M.S. with capital drawings by a young civilian at the Cape of Good Hope about the orchids there; with most curious account of their structure;[10] they seem to me more singular creatures than any that I have seen, viz Satyrium, Disperis &c.— My hot-house will soon be ready, & the thought gives me not a little pleasure.[11] I shall be *most grateful* for Nepenthes, or anything experimental:[12] anything ornamental which, however, I shall avoid of course I must not have from Kew— I wish you would remind H. *Gower*? (your superintendent) that he really ought to try again on several flowers of Victoria Lily own pollen & pollen from distinct plant or distinct flower,—his result was so curious.—[13]

We are all going to London on Tuesday evening for a week, chiefly to see if change will do me good;[14] I hope to read my paper on Linum on the 5th at Linn. Soc.; but Heaven knows whether I shall be able.—[15] If I do get a bit stronger, I **must** come & pay you a visit of an hour & see Hot-Houses.—[16]

I see it announced that Wellwitschia is actually published![17]

I suppose you saw Haast's address it seemed to me very good.— Splendid Glacial action & how wonderful the old flint tools.[18]

I am tired—goodnight.— Anhow I hope I shall see you at Linn Soc. for come I will if I can, however bad sickness may come on— goodnight | C. Darwin

Jameson uses well your Himalayan glacial work in his Glen Roy paper.—[19]

DAR 115: 180
Endorsement: '/63.'

[1] The enclosure has not been identified.

[2] Letters from J. D. Hooker, [15 January 1863] and 24 January 1863.

[3] See letter from J. D. Hooker, 24 January 1863 and n. 2.

[4] In his letter to Hooker of 24 December [1862] (*Correspondence* vol. 10), CD enclosed a 'memorandum of enquiry' for Charles Victor Naudin, whom Hooker planned to meet during a ten-day trip to Paris. See also letter from J. D. Hooker, 24 January 1863 and n. 3.

[5] See letter from J. D. Hooker, 24 January 1863 and n. 8.

[6] Joseph Decaisne. See letter from J. D. Hooker, 24 January 1863 and n. 6.

[7] Alphonse de Candolle's letter has not been found; however, see the letters to Alphonse de Candolle, 14 January [1863] and 31 January [1863].

[8] Louis Charles Joseph Gaston de Saporta. Saporta had recently published two papers on Tertiary flora (Saporta 1862a and 1862b); he expressed his general views on species and species change in Saporta 1862b, pp. 16–17, without reference to CD's theory. There is a presentation copy of Saporta 1862b in the Darwin Pamphlet Collection–CUL, bearing the inscription: 'M. Darwin | hommage de l'auteur qui partage et propage ses idèes relative á la notion de l'espéce' (compliments of the author, who shares and spreads his ideas relative to the notion of species). Saporta and CD did not correspond until 1868 (see *Calendar*, and Conry 1972).

[9] See letter from H. W. Bates, 24 January 1863.

10 Roland Trimen's letter and manuscript have not been found. However, see the letter to Roland Trimen, 31 January [1863]; part of the letter from Trimen was published as Trimen 1863. See also letter from W. H. Harvey, 3 February 1863.

11 At the end of 1862, CD resolved to build a hothouse (see letter to Asa Gray, 2 January [1863] and n. 24).

12 See letter from J. D. Hooker, [15 January 1863] and n. 12.

13 *Victoria regia*. William Hugh Gower was a foreman at the Royal Botanic Gardens, Kew. See letter from W. H. Gower, 23 November 1861 (*Correspondence* vol. 9).

14 According to Emma Darwin's diary (DAR 242), CD, Emma, Henrietta, and Horace Darwin stayed at Erasmus Alvey Darwin's house at 6 Queen Anne Street, London, from 4 to 14 February 1863.

15 CD's paper on dimorphic flowers in *Linum*, 'Two forms in species of *Linum*', was read at a meeting of the Linnean Society on 5 February 1863, which CD was too ill to attend (see letter to W. B. Tegetmeier, 19 February [1863] and n. 6).

16 According to Emma Darwin's diary (DAR 242), the Darwins visited the Royal Botanic Gardens, Kew, on 11 February 1863.

17 J. D. Hooker 1863a. The first part of volume 24 of the *Transactions of the Linnean Society of London* was published on 30 January 1863 (Raphael 1970, p. 76).

18 J. F. J. von Haast 1862a. See letter to Julius von Haast, 22 January 1863.

19 Jamieson 1863. Thomas Francis Jamieson's paper on the parallel roads of Glen Roy, Scotland, was referred to CD before being published (see preceding letter and n. 1). During his expedition to the Himalayas from 1848 to 1850, Hooker made a number of geological observations in Sikkim, particularly on the phenomena of terraced hillsides, which he explained as former glacial lake shorelines (see J. D. Hooker 1854b, 1: 242–4; 2: 116–21).

From Thomas Rivers 30 January 1863

Nurseries, Sawbridgeworth, Herts, | *Great Eastern Railway.*
Jany 30/63

My dear Sir/

I am now going to pester you with a question.[1] In the cultivation of peach trees in enclosed houses I have found that they cannot be kept in health unless the fresh air is admitted so that it enters at a lower level & makes its way *under* the leaves as at *a*

Mem the lines | are to be supposed the two walls of a house

rising as it becomes rarefied by the heat of the house, & that air admitted at the upper part of the wall of a glass roofed house as at *b* will not keep peach trees in health this has induce⟨d⟩ me to think that the lungs ⟨of⟩ the leaves (so to speak) are on the under surface— there is as you know much difference in the appearance of the two surfaces. Now can you by employing a powerful microscope & your power of mind do anything for me in this matter?

Referring to my last[2] I was brought to my, I fear, idle reasoning on the effects of soil & climate on our race by observing for several years past English gardeners & nurserymen visiting here who had lived from seven to ten years in the United States of America some from the far west others in the sea-board states with only one exception & he was a Scotchman, I remarked that in their features & figures they had assumed the genuine American type the cheekbones in most cases high the nose thin & if in the least inclining to the aquiline this peculiarity much agravated. the cheeks hollow the eyes sunken everything belonging to them reminding one of the normal Indian race. I have often expressed my surprise to those men that soil & climate would have so changed them & they have admitted that they felt a change not only in their persons but in a rush of fierce energy unknown to us at home.

These facts are all straws but they may give a hint however trifling to your reasoning powers

I am My dear Sir | Y^{rs}. very truly | Th^{os}. Rivers

Postmark: JA 30 63
DAR 176: 162

CD ANNOTATION
On cover: 'Weeping Trees' *ink*

[1] CD began a correspondence with Rivers at the end of 1862, first inquiring about bud-variations (see *Correspondence* vol. 10, letter to Thomas Rivers, 23 December [1862]).
[2] See letter from Thomas Rivers, 26 January 1863.

To Friedrich Rolle[1] 30 January [1863][2]

Down. Bromley. Kent. S.E.
Jan. 30th.

Dear Sir

You must excuse me writing briefly, as I am much engaged and far from well. I thank you sincerely for your sketch of the progress of opinion on the change of species in Germany, which has interested me much.[3] If you have any communication with D^{r}. Jaeger, be so kind as to present my compliments to him and say how glad I am to hear that he is on my side.[4]

Pray also remember me kindly to D^{r}. Weinland.[5] I have read and been much pleased with the later parts of your work which seem to me well done.— You have used excellently D^{r}. Rütimeyer's admirable memoir[6]

With my best wishes for your success, I remain, dear Sir, yours very faithfully | Charles Darwin

Copy
Forschungsinstitut und Natur-Museum Senckenberg

[1] The original letter to Rolle has not been found. The text reproduced here is based on a copy made by Rolle.

[2] The year is established by the relationship between this letter and the letter from Friedrich Rolle, 26 January 1863. On the transcription of the letter, there is a note by Rolle that states: 'Von Herrn Ch. Darwin d. 2 Februar 1863'.

[3] See letter from Friedrich Rolle, 26 January 1863.

[4] In his letter of 26 January 1863, Rolle discussed the reception of CD's theory in Germany and Austria, and emphasised Gustav Jäger's efforts to promote CD's theory.

[5] Rolle had mentioned David Friedrich Weinland as one of CD's supporters in Germany (see letter from Friedrich Rolle, 26 January 1863).

[6] CD refers to the third and fourth parts of Rolle 1863 (see letter from Friedrich Rolle, 26 January 1863 and n. 3). Rolle cited Ludwig Rütimeyer's work (Rütimeyer 1861) in corroboration of CD's theory (see Rolle 1863, pp. 126–32). There is an annotated presentation copy of Rütimeyer 1861 in the Darwin Library–CUL (see *Marginalia* 1: 718–26). See also *Correspondence* vol. 9, letter to Ludwig Rütimeyer, 5 December [1861].

From Isaac Anderson-Henry 31 January 1863

Hay Lodge, | Trinity, | Edinburgh.
Jany 31/63

My dear Sir

I cannot deny myself the pleasure of saying however briefly at present, how much gratified I am by the receipt, this morning, of yours of the 29th.[1] Every line of it is suggestive of things to do, as well, alas, as of opportunities neglected. That is a new idea to me of *counting* the seeds.[2] Many, many a time have I had my attention called to the paucity of seeds on crossing alien things—and even I observed where fewest they were the most obstinate in vegetating. And yet in that very *severe* cross of Rhodothamnus Chamæcistus with the large species of Rhodn. the seeds were abundant & ripe—[3] the seeds of this cross were not long & elliptical in form as in the *male* parent (Rhodn.) but rounded as is the characteristic of the female parent

I yesterday got the same *Bryanthus* from a Nursery with *Menziesea* Cærulea—The *M empetriformis* I have ordered from Veitch London. So I shall again test the experiment.[4] I do believe, and all here do,—that the *empetriformis* was what Mr Cunningham wrought upon.[5] His nephew, I think I mentioned,[6] said he really believed his Uncle was unable to say which,—whether that or the *M. Cærulea* was the female parent.

Having since I wrote you, answered some 20 questions submitted to me a M Neumann, describing himself as "Chef de Culture Jardins des Plantes", Paris,[7] an intelligent and most earnest Enquirer into these mysteries, I am perplexed whether certain explanations I gave were to you or to him. For this was a second Letter from the same Gentn. who wished what light I could give him for some Conference to be held at the end of this month. He besought of me for leave to publish my Letters,[8]—very hasty things they were, but I hope their imperfections will be concealed under a French petticoat

This is the day I send off my despatches to Dr Jameson, and I shall mention your desire to correspond with him.[9] These *Rhexias* relaise Humboldts fear (I think it is Humboldts)[10] that certain plants from these elevated Plateaus are difficult to bring into bloom in Europe—& *vice versa*, D^r^ J mentions the same difficulty there with some European fruits. I have one thing more than 10 years old not l foot high, a *Rhexia*,[11] which shows no symptom of blooming yet.

But I must defer till next Letter further remarks on your Letter, merely observing now, that I will most gladly cooperate not only in the Experiments you have suggested but in any others you may point out. Certainly D. B.[12] is loose & as fast as he is loose—often wrong & incorrigible when he is. Yet there is matter in him, tho not the *root* of it. Long before Donald stumbled on about these "short anthers" I had not only come to regard them as the parents of my puny things—[13] But I took up the idea that each separate pair had their functions separate from others—some conducing to to precocious fertility—some to more robust & others to a more tender constitution;—some to finer flowers & fruit & some to hardier & some to more tender constitutions. Hence it may be that Cultivation, in due time tends so much to improve our cultivated fruits & flowers. Again (do forgive my trespassing on your valuable time) I have ever held that there is much in *times* & *seasons* for accomplishing crosses which at other times I w^d^ never endeavour to essay. Is it *Ozone*, which, at times they say (and I think in Spring most) intermixes with the air in greater degree, that may produce these general influences. Some days I have found several hopeless seeds, long sown, to spring & crosses to take which, at other times it were in vain to try. Do turn your powers of mind and observation to atmospheric conditions and influences

But now I must come to the single line I intended to write. Did I tell you of an Association formed here to send a Collector to Vancouver Island & B. Columbia.[14] I intended to do so and to send a Print of our proceedings—but I forget if I did either. I now enclose the Print.[15] It occurs to me that you might kindly throw out some valuable suggestions for what our Collector,—an intelligent Botanist, might look after— And he at same time might pick up seeds of some things for crossing—those strawberries e.g.[16] He sails immediately ie in some 10 or 14 days going by Cape Horn direct then for Vancouver Island

It is odd that if these American species won't cross with ours, I certainly failed with some cross I attempted on F Indica a beautiful but insipid fruited thing.[17] One cross I have fought & wrought with for years,—to cross Australian Veronicas with our native suffruticose kinds e.g. V *fruticulosa* to V. saxatalis. Yet I have plants much dwarfer than the female (V Speciosa. N Zealand) but the foliage tho' smaller is that of the Mother, and I fear it is *her* repeated— we will see

But I must resist all temptation to go into the other points in your letter till another time. Oh that I had sooner had the incentives applied which your intelligence & ardour now so late inspire me with. But now I have the time I never had before, and with the electric light which you hold up & the spark communicated my time may not be wasted on merely producing a beautiful thing, the admiration of a day then to be forgotten forever.

Believe me | Yours very Sincerely & most obliged | Is Anderson Henry

Chas Darwin Esq | F.R.S.

DAR 159: 62

[1] CD's letter has not been found, but see the letter from Isaac Anderson-Henry, 26–7 January 1863.

[2] CD often counted seeds as a measure of the fertility of a cross and advised others to do the same (see 'Dimorphic condition in *Primula*', and *Correspondence* vol. 10, letter to John Scott, 3 December [1862]).

[3] In his letter of 26–7 January 1863, Anderson-Henry described a cross he attempted in 1851 between *Rhodothamnus chamaecistus* and a large species of *Rhododendron*.

[4] The reference is to the Chelsea nurserymen, James Veitch and James Veitch Jr. Anderson-Henry earlier described his attempts to repeat the cross made by the Edinburgh nurseryman, James Cunningham, which resulted in the hybrid *Bryanthus erectus* (see letter from Isaac Anderson-Henry, 26–7 January 1863 and n. 14).

[5] See n. 4, above.

[6] No previous mention of Cunningham's nephew has been found. A later letter mentions that the nephew, who has not been identified further, had started working at a nearby nursery, Peter Lawson & Sons (see letter from Isaac Anderson-Henry, 24 April 1863, and *Post Office Edinburgh directory* 1867–8).

[7] Louis Neumann was a gardener at the Jardin des Plantes, the botanical department of the Muséum d'Histoire Naturelle. Neumann's inquiries to Anderson-Henry were probably made under the direction of Bernard Verlot, who was the chef de culture at the Jardin des Plantes (see n. 8, below).

[8] Anderson-Henry's replies to Neumann's inquiries were published in Verlot 1864, pp. 10–11; Verlot's essay was prepared as an entry for an open competition of the Société Impériale et Centrale d'Horticulture in Paris (see letter from John Scott, 3 March 1863, n. 7). The deadline for the essay was 1 February 1863, with the prize a gold medal worth 300 francs.

[9] Anderson-Henry refers to the Scottish botanist William Jameson, who was living in Ecuador. See letter from Isaac Anderson-Henry, 26–7 January 1863.

[10] Alexander von Humboldt.

[11] CD also experimented with *Rhexia* (see letter to Asa Gray, 19 January [1863], and letter from Isaac Anderson-Henry, 26–7 January 1863 and n. 33).

[12] Donald Beaton.

[13] In his letter to Anderson-Henry of 20 January [1863], CD stated that he was 'a little doubtful' about Beaton's claim that the pollen of the two shortest anthers of scarlet Pelargoniums produced dwarf plants 'in comparison with plants produced from *same* mother-plants by the pollen of longer stamens from same flower'. CD expressed a wish to have this point tested by 'systematic experiments'. Anderson-Henry responded at length to CD's comments regarding short anthers in his letter of 26–7 January 1863. See also Appendix V.

[14] Anderson-Henry refers to the British Columbia Botanical Association of Edinburgh, a shareholding venture launched in 1849 for the further exploration of Vancouver Island and British Columbia. Anderson-Henry was secretary to the association (*Journal of Horticulture* n.s. 3: 451). The collector chosen was Robert Brown; an account of Brown's expedition appeared in the *Transactions of the Botanical Society* [*of Edinburgh*] (McNab 1872).

[15] The enclosure has not been found.

[16] See letter to Isaac Anderson-Henry, 20 January [1863].

[17] *Fragaria indica* was the Indian, or yellow, strawberry. For CD's and Anderson-Henry's interest in strawberry hybrids, see the letter to Asa Gray, 2 January [1863] and n. 17, and the letter from Isaac Anderson-Henry, 26–7 January 1863 and nn. 2 and 3.

To Alphonse de Candolle 31 January [1863][1]

Down Bromley Kent
Jan. 31st

My dear Sir

I thank you for so kindly writing to me; & many parts of your letter have interested me much.—[2] I take a keen interest in progress of species-question, & am delighted to hear of the Count Saporta's views.—[3] I really think that I believe in as much migration as even you believe in, & as shown in your admirable great work;[4] only I do not believe nearly *so much* in continental extensions & I believe more (not very much more, I begin to suspect, & it pleases me greatly) than you do in modification in form.—[5] That is good remark on melons, & I shall probably use it, in conjunction with improvement of the pear.[6] Also many thanks about oranges: I confess that I thought you rather too sceptical about Macfayden's statement in W. Indies; I must change my tone.—[7]

what you say about antiquity of man, has occurred to me: it must make considerable difference in our views on domesticated animals & plants.— How very interesting the Swiss discoveries of the old Lake habitations have been; but I know only Rütimeyers admirable essay,[8] in which there is some little on Plants by Dr Christ:[9] I am very much obliged for your hint & must enquire; but I find German very slow reading & it wastes much time.[10] I received a few days ago an Address from New Zealand, & traces of Prehistoric man have been found there![11]

I thank you for your Photograph, which I am glad to add to some others.[12] I have never had a proper "carte photographique" taken of myself; but I enclose one done 2 or 3 years ago by my son, if worth your acceptance.[13]

I am almost tired of my present big book (though a mere trifle to yours) on Variation under Domestication; for I have not sufficient knowledge to treat the plant-part well, but I have done so much that I will finish it.—[14]

Pray believe me, my dear Sir, with my best thanks & sincere respect.— | Yours very faithfully | Ch. Darwin

La Fondation Augustin de Candolle

1 The year is established by the reference to J. F. J. von Haast 1862a (see letter to Julius von Haast, 22 January 1863). See also n. 2, below.

2 Candolle's letter has not been found. However, see the letter to J. D. Hooker, 30 January [1863].

3 Gaston de Saporta. In the letter to J. D. Hooker, 30 January [1863], CD stated that he had just received 'another letter from A. Decandolle' in which Candolle reported that on the question of mutability, Saporta 'goes the whole hog'.

4 A. de Candolle 1855. CD's heavily annotated copy of this work is in the Darwin Library–CUL (see *Marginalia* 1: 106–53).

5 See letter to Alphonse de Candolle, 14 January [1863].

6 In *Variation*, CD discussed the improvement of cultivated foods over hundreds of years through 'long-continued selection', reporting (*Variation* 2: 215–16):

> Alphonse De Candolle informs me that he has lately seen on an ancient mosaic at Rome a representation of the melon; and as the Romans, who were such gourmands, are silent on this fruit, he infers that the melon has been greatly ameliorated since the classical period.

7 CD refers to James Macfadyen's assertion that oranges in Jamaica were either sweet or bitter, depending on the soil in which they were grown (Macfadyen 1830, p. 302, and Macfadyen 1831, p. 111). In *Variation* 1: 335, CD thought Macfadyen's claim an error because Candolle had since 'received accounts from Guiana, the Antilles, and Mauritius, that in these countries sweet oranges faithfully transmit their character'.

8 Rütimeyer 1861. There is a heavily annotated copy of Rütimeyer 1861 in the Darwin Library–CUL; CD cited this work extensively in *Variation* (see *Marginalia* 1: 718–26).

9 CD refers to Konrad Hermann Heinrich Christ, the plant geographer who wrote a section in Rütimeyer 1861 on plant remains found in the Swiss lake-dwellings (pp. 224–9). The dwellings had been villages built on piles above lakes and were thought to date from the Stone Age (see also C. Lyell 1863a, pp. 17–26). CD quoted a passage of Christ's in *Variation* 1: 318.

10 The reference has not been identified.

11 J. F. J. von Haast 1862a. See letter to Julius von Haast, 22 January 1863.

12 CD's photograph of Candolle has not been found.

13 William Erasmus Darwin's photographic portrait of his father, made on 11 April 1861, is reproduced as the frontispiece to volume 9 of the *Correspondence*.

14 CD refers in part to the chapters in *Variation* on cultivated plants and bud-variation (*Variation* 1: 305–411), which he completed by 23 January 1863 (see *Correspondence* vol. 10, Appendix II, and this volume, Appendix II). *Variation* was published in two volumes in 1868.

To Roland Trimen 31 January [1863][1]

Down. | *Bromley.* | *Kent. S.E.*
Jan. 31st

My dear Sir

I thank you most sincerely for your pleasant letter & M.S. on Orchids. Your sketches seem to me very good, & wonderful under circumstances of their execution.[2] I cannot say how much interested I have been in studying your descriptions. I think I understand all; but these orchids (except Eulophia) are so *surprisingly* different from anything that I have seen that I could hardly make them out for some time & even fancied in some cases that you had miscalled upper sepal & Labellum. But at last I see my way. I am no more a Botanist than you say you are, and I know nothing of any orchids except those seen by me. Therefore I was astonished at the upper sepal being produced into a nectary;[3] even more astonished at stigma standing high above the pollinia &c &c.— How curious the pollinium of *Disperis*!—[4] What beautiful and new contrivances you show, & how well you have studied them! Upon the whole I think No V. & VI unnamed (I have sent your drawings to Prof. Harvey to name for me) have interested me most:[5] everything seems to occur in a reversed direction compared with our true Orchis.— You do not mention any movement of the pollinia, when attached to an object; & as you are so acute an observer, I infer that there are no such movements;[6] & indeed in those you describe such movements would be superfluous. If you have time to wander about do watch some kinds & see insects do the work. Those with long nectaries would be probably hopeless to watch as probably fertilised by Moths.—[7] But since my publication I have ascertained that with Orchis, Diptera are chief workmen.—[8] They certainly do puncture the walls of nectary, & so get juice. *Disperis* would be grand to watch.

& discover what attracts insects— You draw so well, & have so seized on the subject that you ought really to take up 2 or 3 of the most distinct genera, & watch them, experiment on them by mutilation of parts, & describe them & send over an excellent paper to Linnean Soc^y or some other Soc^y.—[9] I have so much other work, that I hardly know whether I shall ever publish again,—not but what I have already collected some curious new matter; for the subject delights me—& I cannot resist observing.[10]

I am very glad to hear that you do not now think me so dangerous a person![11] You will gradually, I can see, become as depraved, as I am.— I believe, or am inclined to believe, in one or very few primordial forms, from community of structure & early embryonic resemblances in each great class..—

With most cordial thanks I remain my dear Sir | Yours sincerely | Ch. Darwin

P.S. Would it be asking too great a favour to beg you to put 2 or 3 flowers of *Satyrium* or your No V. or VI in bottle with spirits & water, & send home by any opportunity. I would then compare your drawings & add some remarks on your authority, if I ever publish again— But I hope, what will be much better, to see a paper by yourself.—

If you come across Bonatea pray study it— it seems most extraordinary in description.—[12]

Royal Entomological Society of London (Trimen papers, box 21: 54)

1 The year is established by the reference to Trimen 1863 (see n. 2, below).

2 Trimen's letter, the manuscript of his observations on South African orchids, and the accompanying sketches have not been found. CD later reworked some of Trimen's observations as a paper for the Linnean Society, noting that it was 'drawn up from Notes and Drawings sent to C. Darwin' (see Trimen 1863, p. 144); however, the published illustrations were some of those sent with Trimen's letter to CD of 16 March 1863. See also letter to Roland Trimen, 23 May [1863].

3 See n. 6, below.

4 See *Orchids* 2d ed., p. 265.

5 CD's letter to William Henry Harvey and Trimen's sketches have not been found. However, see the letter from W. H. Harvey, 3 February 1863. Harvey was professor of botany at Trinity College, Dublin; he had collected extensively in the Cape colony between 1836 and 1842 when he was colonial treasurer (*DNB*).

6 In Trimen 1863, pp. 146–7, Trimen described pollinia movements in *Disa*; CD discussed these observations in *Orchids* 2d ed., p. 77:

> The superb flowers of *Disa grandiflora* have been described and figured by Mr. Trimen. The posterior sepal, instead of the labellum, is developed into a large nectary. In order that insects may reach the copiously stored nectar, they must insert their proboscides on either side of the column; and in accordance with this fact the viscid discs are turned outwards in an extraordinary manner. The pollinia are crooked, and when removed bend downwards from their own weight, so that no movement is necessary for placing themselves in a proper position.

See also 'Fertilization of orchids', p. 144 (*Collected papers* 2: 141).

7 In Trimen 1863, p. 147, Trimen suggested that the orchids with long nectaries might also be attractive to 'some day-flying Hymenopterous or Lepidopterous insect'. CD later published a comment on Trimen's observation of a 'Dipterous insect, allied to *Bombylius*, frequenting the flowers' ('Fertilization of orchids', p. 144; *Collected papers* 2: 141).

[8] CD refers to *Orchids*, published in May 1862; he included this point in 'Fertilization of orchids', p. 142 (*Collected papers* 2: 139–40). See also *Correspondence* vol. 10, letter to Asa Gray, 1 July [1862].
[9] See n. 2, above.
[10] In 1869, CD published additional material on orchids as 'Fertilization of orchids' (*Collected papers* 2: 138–56); this was originally prepared for the French edition of *Orchids* (Rérolle trans. 1870). The second edition of *Orchids*, incorporating this and other new material, was published in 1877 (Freeman 1977).
[11] Edward Bagnall Poulton's anniversary addresses (Poulton 1909) included an account by Trimen describing the first time he met CD; in his letter (now missing, see n. 2, above), Trimen appears to have told CD of a friend's admonition that CD was 'the most dangerous man in England' (see *ibid.*, pp. 213–15).
[12] Trimen published an account of the structure of the flower *Bonatea* in Trimen 1864; there is an annotated copy of this paper in the Darwin Pamphlet Collection–CUL.

From F. T. Buckland [before 1 February 1863][1]

and will get you to see my Salmon hatching at The Field window.[2] By the way the Editor of The Field[3] requests me to inform you that he would feel much pleased if you would send any natural history Quæries which would be likely to elicit discussion as there are so many correspondents who have every opportunity of practical observation

Yours ever | most truly | F Buckland

I have been asked to examine a cross between a duck & a fowl—*it wont do* my experience show me that people coin a story & tell it so many times that at last *they believe it* themselves. This is case with

1 Viper swallowing her young.[4]
2 Dog & Fox Cross
3 Toad in a hole question[5]

cum multis aliis[6] &c &c &c &c

Incomplete
DAR 160: 356

CD ANNOTATION
4.2 2 Dog . . . Cross] *cross in margin, pencil*

[1] The date is established by the relationship between this letter and the following letter, and by reference to the display in the window of the office of the *Field* (see n. 2, below). This letter probably formed part of the reply to the letter to F. T. Buckland, 26 January [1863].
[2] In the window of the office of the *Field*, at 346 the Strand, London, where he worked as a staff writer, Buckland exhibited 'a rude, but yet effective, apparatus, whereby the curious yet simple process of hatching salmon and trout from the egg, might be witnessed' (*Field* 21 (1863): 53). CD may have informed Buckland that he was planning to visit London in early February in the missing portion of his letter to F. T. Buckland, 26 January [1863] (see following letter).
[3] John Henry Walsh was editor of the *Field* (Rose 1953).

[4] An account of a viper swallowing its young was published in the *Natural History Review* in January 1862 ([Lubbock] 1862b), and reported in the *Field* 19 (1862): 300, the editor remarking, 'We dread a renewal of this apparently interminable controversy'.

[5] Buckland refers to the myth, prevalent in popular natural history books of this period, that toads or frogs contemporaneous with coal or rock formations had been exhumed alive in modern times (see Barber 1980, p. 18). This story had enjoyed a recent revival following the exhibition of a piece of coal supposed to have contained a living frog, at the International Exhibition of 1862. *The Times* carried several letters in response to the exhibit in September 1862, including one from Buckland attempting to dispel the myth (*The Times*, 16 September 1862, p. 7).

[6] *Cum multis aliis*: 'with many others'.

To F. T. Buckland 1 February [1863][1]

Down. | *Bromley.* | *Kent. S.E.*
Feb. 1st

My dear Mr Buckland

I thank you sincerely for your very kind note & information.[2] I will write about the fins.—[3] Pray present my thanks to the Editor of the Field;[4] & will, if I require it, avail myself of his kind permission. I formerly did occasionally receive some good facts through that channel.[5] But my health is very weak & I find myself overdone & almost smothered with facts & necessary enquiries, so that I am trying to restrict, as much as I can, the scope of my present work, which is on a large subject "Variation under Domestication".—[6]

As you truly say the power of lying is great; I thought that dog & fox cross did not come under that grand category; but I forget my facts, except that one good observer Hon. & Revd. Dr Herbert assured me that he had seen an animal which he felt assured was such a cross.—[7]

I should much enjoy having a little talk with you, when in London.— I am coming up this week & will endeavour to see you at Athenæum or elsewhere; but I have not been well enough for nearly a year to stay in London & in consequence have got far more to do than I shall be able to get through.—[8]

With sincere thanks, pray believe me | Yours sincerely | Charles Darwin

American Philosophical Society (Getz 3961)

[1] The year is established by the reference to the preceding letter, and by the reference to CD's planned visit to London (see n. 8, below).

[2] See preceding letter, and letter to F. T. Buckland, 26 January [1863].

[3] See letter to J. J. Briggs, 2 February [1863]. Buckland had evidently suggested that CD write to John Joseph Briggs on the subject of fish fins in the missing portion of the preceding letter.

[4] The editor of the *Field*, John Henry Walsh, had invited CD to use the paper's columns for his natural history queries (see preceding letter).

[5] The *Field* published several of CD's letters in 1861 (see *Correspondence* vol. 9, letters to the *Field*, [before 27 April 1861], [before 4 May 1861], [before 25 May 1861], and [before 15 June 1861]). CD cited the information he received from the *Field* in *Variation* 1: 58, 253.

[6] *The variation of animals and plants under domestication*, eventually published in two volumes in 1868, was intended to be the first part of CD's 'big book' on species (see *Variation* 1: 3–10). CD began writing

drafts of the chapters on inheritance (chapters 12–14) in January 1863, completing the work on 1 April (see 'Journal' (Appendix II)).

7 See preceding letter. CD had received information from Brian Houghton Hodgson that dog and fox crosses had occurred in India (see *Correspondence* vol. 4, letter from J. D. Hooker, 13 October 1848; see also *Correspondence* vol. 2, Questions for Mr Wynne, [February–July 1838]). In reviewing the evidence for the intercrossing of dogs with other species in *Variation*, CD repeated two further anecdotes indicating that some breeds of dog might interbreed with foxes (see *Variation* 1: 31–3). William Herbert's observation of dog and fox crosses (Herbert 1837, pp. 339–40) was not cited, and no letter from Herbert on this subject has been found.

8 CD was in London between 4 and 14 February 1863 (see 'Journal' (Appendix II)).

To Thomas Rivers 1 February [1863][1]

Down. | Bromley. | Kent. S.E.
Feb. 1st

My dear Sir

I can answer your query on much higher authority than any observations which I could make in months of time.[2] The "stomata" or mouths, which by their lips have power of opening & closing, & which when opened put the spaces within the leaf into free communication with the open air, are far more abundant on the *lower* than on the *upper* surface of leaf.— "More commonly there are few or none on the upper side."[3] In the white lily it has been calculated that there are in square inch of surface, 60,000 stomata on lower surface, & only 3,000 on upper surface; in the apple there are 24,000 to the inch: in some plants 170,000 to the square inch.![4]

I have often marvelled over the American type of features.—[5]

I was heartily glad to see your handwriting, for I was meditating a query: I am taking "Weeping trees", as an example how inexplicable the laws of inheritance are; some weeping trees reproducing themselves almost truly by seed, & some quite failing to do so.—

Can you give me any certain facts on character of seedlings from a weeping trees?[6]

Also have you ever sowed from a sporting branch or bud (i.e. case of bud-variation) & if so what was result:—[7] I know case of Boston Nectarine.—[8]

My dear Sir | Yours sincerely | Ch. Darwin

If no answer, I will understand no facts.—

Sotheby's, London (23–4 July 1987)

1 The year is established by the relationship between this letter and the letter from Thomas Rivers, 30 January 1863.

2 In his letter of 30 January 1863, Rivers asked CD to confirm his observation that the 'lungs' of plants were situated on the under-surface of the leaves. CD refers to Asa Gray (see nn. 3 and 4, below).

3 CD's information on stomata was taken from Gray's *First lessons in botany and vegetable physiology*, an annotated copy of which is in the Darwin Library–CUL (A. Gray 1857; see *Marginalia* 1: 347). Gray's observation, 'More commonly there are few or none on the upper side' continued, 'direct sunshine evidently being unfavorable to their operation' (A. Gray 1857, p. 157).

4 CD quoted the figures from A. Gray 1857, pp. 156–7.

5 In his letter to CD of 30 January 1863, Rivers discussed the effects of soil and climate on the North American population.

6 In chapter 12 of *Variation*, CD cited the weeping habit of trees as an example of 'how feeble, capricious, or deficient the power of inheritance sometimes is' (*Variation* 2: 17), the weeping habit being transmitted to the seedlings sometimes strongly and sometimes feebly. Rivers provided CD with the 'crowning case' of the vagaries of inheritance in trees with his information on varieties of weeping ash (*Variation* 2: 19).

7 By the term 'bud-variation', CD meant 'all those sudden changes in structure or appearance which occasionally occur in full-grown plants in their flower-buds or leaf-buds' (*Variation* 1: 373). CD made several inquiries on the subject of bud-variation in December 1862 (see *Correspondence* vol. 10, letter to John Scott, 19 December [1862], letter to Thomas Rivers, 23 December [1862], and letter to Hugh Falconer, 29 December [1862]). Rivers was able to provide CD with several examples of bud-variations, including cases where fully grown peach trees had suddenly produced nectarines and vice versa (*Variation* 1: 340), purple plums had produced yellow plums (*Variation* 1: 375), and various rose varieties produced other varieties (*Variation* 1: 380–1, 409–10).

8 The Boston nectarine was reported to have been produced from a peach stone, and to have produced by seed a closely allied nectarine. CD cites Knight 1826, p. 394, and Downing 1845, p. 502, on this point in *Variation* 1: 340.

To Isaac Anderson-Henry 2 February [1863][1]

Down, Beckenham, Kent S.E.[2]
Feb. 2nd. 1863.

My dear Sir,

I am not well today and leave home tomorrow for a week or ten days to try and get some rest;[3] so I must write briefly; and thank you for your very kind letter briefly.[4]

I have really not knowledge sufficient about Columbia to make any suggestions.[5] If the Latitude had been given, I should have strongly urged (but this will have occurred to you) to have collected seeds from individual trees growing at great heights. I allude, of course to Hooker's observation that seeds collected at different heights of some species produced seedlings with different *constitution*.[6] Mr. Thwaites of Botanical gardens of Ceylon tells me that this is the case in that island.[7] It might *even* be worth attending to in your present expedition as a point of science; i.e. to compare character of seedlings from great height and *low places*.— Many thanks for corrections about the Menziesia, and about the hybrid sent to Hooker.—[8]

Here is odd chance! I have made two or three trials to see whether seeds from *short anthers* germinated at different rates, as yet with no result; and I have tried ozonised water, with no result.—[9] *Not* that I have tried nearly enough to come to any conclusion *worth* anything. Pray forgive brevity, and, with cordial thanks.

Believe me, | Yours Sincerely. | C. Darwin.

Mr. *Neumann* I believe to be the man who has worked at fertilization of stove orchids and in good gardens.[10] I suspect he is working under M. Naudin (who is Decaisne right hand man and capital Botanist)[11] M. Naudin writes to me, that he is going to publish on Hybridity and he believes he has discovered the physiological cause of sterility of Hybrids!!![12] I doubt.

If by any chance you have raised seedlings from any *"weeping" tree*, I should be grateful for information to quote on degrees of inheritance of "weeping" quality.—[13]

Pray forgive this wretchedly untidy note; but I am good for nothing.

Copy
DAR 145: 2

[1] The text of the letter is taken from a copy made for Francis Darwin when he was preparing editions of his father's letters (*LL* and *ML*). The copyist included the year in the date line, although this was not CD's usual practice. The year is confirmed by the relationship between this letter and the letter from Isaac Anderson-Henry, 31 January 1863.

[2] The copyist wrote 'Beckenham' in error; the embossed address on CD's writing paper at this time read 'Bromley'.

[3] On 2 February 1863, Emma Darwin recorded in her diary (DAR 242) that CD was 'very languid in mg' and, on 3 February 1863, 'Ch. ditto'. CD was in London from 4 to 14 February 1863 (see 'Journal' (Appendix II)).

[4] Letter from Isaac Anderson-Henry, 31 January 1863.

[5] In his letter of 31 January 1863, Anderson-Henry informed CD that a collector (Robert Brown) had been sponsored by the British Columbia Botanical Association of Edinburgh to gather plants and seeds in Vancouver Island and British Columbia. Anderson-Henry asked CD whether he had any suggestions regarding what Brown 'might look after'.

[6] Joseph Dalton Hooker's observations that altitude materially altered the habits of plants of the same species, and that seeds gathered at high altitudes produced hardier plants than seeds gathered at lower altitudes, were made in J. D. Hooker 1852, pp. 69–71, and J. D. Hooker 1853, pp. xi–xii. There is a copy of J. D. Hooker 1853 in the Darwin Library–CUL.

[7] George Henry Kendrick Thwaites was superintendent of the Peradeniya botanic gardens in Ceylon (*DNB*). Thwaites communicated his observations on the acclimatisation of plants at different altitudes in a letter to CD of 28 December 1857, but the letter has not been found (see *Correspondence* vol. 7, letter to G. H. K. Thwaites, 7 February [1858]). Thwaites's observations were cited in *Origin*, p. 140, and in the manuscript of CD's 'big book' on species (see *Natural selection*, p. 286).

[8] For Anderson-Henry's comments on *Menziesia*, see the letter from Isaac Anderson-Henry, 26–7 January 1863. The hybrid sent to Hooker was probably sent in response to CD's request for 'any *quite* sterile Hybrid plant with a rather large stigma' (see the letter to Isaac Anderson-Henry, 20 January [1863]). CD wanted to compare the stigmas of a hybrid plant with the stigmas of its parent-species by dissection (see letter to John Scott, 21 January [1863]). In a letter to Anderson-Henry, now missing, but written before 27 January 1863 (see letter from Isaac Anderson-Henry, 26–7 January 1863, n. 30), CD may have asked that the plant be sent to Hooker at the Royal Botanic Gardens, Kew, where the parents of the hybrid could be located and where the dissection could be conducted conveniently. According to Emma Darwin's diary (DAR 242), CD visited the Royal Botanic Gardens, Kew, on 11 February 1863.

[9] In his letter to CD of 31 January 1863, Anderson-Henry remarked on his hypothesis that pollen from the short anthers of plants might have the property not only of producing dwarf offspring, but also of effecting 'precocious fertility' and differences in the hardiness of offspring. Anderson-Henry also suggested that increases in atmospheric ozone might promote the germination of seeds.

[10] In his letter of 31 January 1863, Anderson-Henry informed CD that he had been corresponding with Louis Neumann, a gardener at the Jardin des Plantes, the botanical department of the Muséum d'Histoire Naturelle in Paris. CD knew of Neumann's work with orchids through the publications of Adolphe Théodore Brongniart, the professor of botany at the Muséum d'Histoire Naturelle (see *Collected papers* 2: 70 n. 6).

[11] The references are to Charles Victor Naudin and Joseph Decaisne. Naudin was aide-naturaliste and Decaisne was professor of plant cultivation at the Muséum d'Histoire Naturelle (*DBF*, *Taxonomic*

literature). However, Neumann's inquiries to Anderson-Henry were probably made under the direction of Bernard Verlot, chef de culture at the Jardin des Plantes (see Verlot 1864). See also letter from Isaac Anderson-Henry, 31 January 1863 and n. 8.

12 Naudin's letter has not been found; however, see the letter to C. V. Naudin, 7 February 1863. See also letter from J. D. Hooker, 24 January 1863.

13 CD wanted information on weeping trees for chapter 12 of *Variation*, which he had begun writing on 23 January 1863 (see 'Journal' (Appendix II), and *Variation* 2: 18–19). See also preceding letter and n. 6.

To John Joseph Briggs 2 February [1863][1]

Down. | *Bromley.* | *Kent. S.E.*
Feb. 2d

Dear Sir

I hope that you will forgive the liberty which I take in addressing you.— Mr F. Buckland tells me that he thinks that you would give me one piece of information.[2] I remember seeing in The Field an excellent article, written I believe by you, on the regrowth of the fin's of fishes when cut off.— Could you inform me of the year & page in The Field in which the article appeared.[3] Unfortunately I have not kept the copies, & if it be not asking too great a favour, I should be very much obliged if you would inform me (1) what kinds of fish were tried (2) what fins whether pectoral, dorsal &c. were cut (3) whether whole or half or quarter of fin was cut off (4) whether the bony rays were again formed, & whether the fin ultimately appeared perfect. I wish to quote the fact on your authority in a work which I am preparing for publication, & should be grateful for any information.

Hoping that you will forgive the liberty, which I have taken, I beg leave to remain, Dear Sir | Yours faithfully & obliged | Charles Darwin

American Philosophical Society (286)

1 The year is established by reference to CD's inquiry regarding the regrowth of fish fins, the results of which were incorporated into chapter 12 of *Variation* (*Variation* 2: 15–16). CD began writing the chapter on 23 January 1863 (see 'Journal' (Appendix II)). See also letter to F. T. Buckland, 1 February [1863].

2 Francis Trevelyan Buckland presumably included this information in the missing portion of his letter to CD of [before 1 February 1863].

3 The reference may be to a letter published by Briggs in the *Field* 20 (1862): 412, in which Briggs reported the results of experiments with salmon fins. CD did not cite the letter in *Variation*, but stated that he had been informed by Briggs and Buckland that 'when portions of the pectoral and tail fins of various freshwater fish are cut off, they are perfectly reproduced in about six weeks' time' (*Variation* 2: 15–16). CD's interest in fish fins formed part of an investigation of polydactylism in humans. CD argued that 'supernumary digits' in humans retained an 'embryonic condition' that enabled such digits to regrow following amputation, a property analogous with the regrowth of fins in fish.

To *Journal of Horticulture and Cottage Gardener* [before 3 February 1863][1]

In my last communication I said that Gärtner had proved that the colour of the Pea in one variety of the garden Pea may be changed by the direct action of the pollen of another differently-coloured variety.[2] Mr. Beaton[3] authoritatively remarks on this: “Gärtner never found that—he only asserted it; and when he was pushed to the proof he lowered his sails, made a second edition of his great work, and confessed many of his errors.”[4] He adds, “No cross-breeder of any practice in England at the present day would like to have his name associated with that of Gärtner for or against any exploit in crossing.”

I should have taken no notice of this, although I should be sorry to lie under the imputation of having made an entirely incorrect statement, and although it is not pleasant to be flatly contradicted; but I wish much to be allowed to endeavour to vindicate the memory of one of the most laborious lovers of truth who ever lived. It is painful to see a long life of honest labour repaid by contumely from a fellow-experimentalist, who, I suppose—anyhow I hope—never read one page of the great original work—namely, the “Bastarderzeugung,” published in 1849, a mine of wealth to all who will explore it.[5]

Gärtner, when young, and at the very commencement of his long work, committed a very foolish action; he crossed a number of plants belonging to distinct genera without having taken due precaution to exclude insects, and when he found their capsules full of seed, he thought that he had succeeded in crossing them. With the enthusiasm of a beginner he most unwisely published the result, and to this first paper Dr. Herbert has alluded with proper blame.[6] When Gärtner found his seedlings came up pure, he, like an honest and excellent man (as all who knew anything of his life will admit that he was), publicly confessed his error.[7]

Gärtner’s great and last work, entitled “Versuche über die Bastarderzeugung,”[8] contains in 790 closely-printed pages the detailed results of nine thousand distinct experiments in crossing, together with admirable observations on the whole subject of hybridisation. This is a greater number of experiments than, as I believe, have ever been published by any other man, even by Kölreuter,[9] and a far greater number than those published by Dr. Herbert.[10] One great superiority in Gärtner’s work over those of Kölreuter, Herbert, and others consists in his having actually taken the trouble to count the seeds in the capsules of every cross and hybrid which he made. He kept an exact record at the time of making each experiment; and this I have reason to believe was not done by Herbert, and certainly has been very far from the case with other English experimentalists.

I cannot resist here mentioning—as some who honour, as I do, the memory of Dr. Herbert, might like to hear the fact—that I have reason to believe that the last words ever uttered by Herbert were on his favourite subject of crossing. I called on him in London, and saw that he was very feeble. I wished to leave him, but he stopped me, and talked with much interest on this subject. An hour or two afterwards, as far as I could judge by the published account, he was found dead in the chair in which I left him.[11]

But to return to the Pea-question. An account of the various crosses made by Gärtner (he selected the most constant varieties) between differently coloured Peas, with the results given in detail, will be found at page 81 to 85 in his “Bastarderzeugung.” Gärtner was led to try these experiments from doubting the accuracy of

Wiegmann's statements,[12] and he found many of them incorrect; but he was compelled to believe in the Pea case;[13] not that Peas can be crossed with Vetches, to which other statement of Wiegmann Mr. Beaton alludes. I may add that Gärtner knew of the account, published in vol. v., pages 234, 237 of the "Transactions of the Horticultural Society of London," on the influence of pollen on Peas.[14] In an old volume of the "Philosophical Transactions," vol. xliii., page 525, there is a full account, with every appearance of truth, of Peas in adjoining rows affecting each other.[15] The Rev. M. J. Berkeley has, as I have been informed, subsequently to the publication of Gärtner's book, tried again the Pea-experiment with the same result.[16]

—Charles Darwin, Down, Bromley, Kent.

Journal of Horticulture and Cottage Gardener n.s. 4 (1863): 93

1 The letter was published in the issue of 3 February 1863, under the heading 'Vindication of Gärtner—effect of crossing peas'.

2 See letter to *Journal of Horticulture*, [before 27 January 1863]. The reference is to the German botanist Karl Friedrich von Gärtner and his experiments on the crossing of different varieties of peas in Gärtner 1849. See also n. 4, below.

3 Donald Beaton was a gardener with expertise in hybrid plants, and editor of 'The flower garden' section of the *Journal of Horticulture*; his replies to CD's letter to the *Journal of Horticulture*, [before 27 January 1863], are reproduced in Appendix V.

4 It is not clear to which of Gärtner's publications Beaton refers. Gärtner's experiments on the direct action of pollen in varieties of peas were published in Gärtner 1826 and 1849; he had originally argued that foreign pollen never affected the appearance of the seed in the female parent (Gärtner 1826, pp. 36, 61), but in Gärtner 1849, he revised this opinion, and conceded that in one variety of garden pea the direct action of foreign pollen on the female parent had been observed (Gärtner 1849, p. 84). Beaton's assessment of the value of Gärtner's work appears to have been derived largely from William Herbert's discussion of Gärtner's experiments in Herbert 1837, pp. 348–52. See also Appendix V, 'Beaton's first response to the letter from Charles Darwin to the *Journal of Horticulture and Cottage Gardener*, [before 27 January 1863]', nn. 3 and 7).

5 There is a heavily annotated copy of Gärtner 1849 in the Darwin Library–CUL (see *Marginalia* 1: 256–98). CD had earlier referred to Beaton's repeated criticisms of Gärtner in the *Journal of Horticulture*, telling Joseph Dalton Hooker: 'He has made me indignant by the way he speaks of Gärtner, evidently knowing nothing of his work' (see *Correspondence* vol. 9, letter to J. D. Hooker, 14 May [1861]).

6 The references are to Gärtner 1826 and Herbert 1837, pp. 348–52. See also Appendix V, 'Beaton's first response to the letter from Charles Darwin to the *Journal of Horticulture and Cottage Gardener*, [before 27 January 1863]', n. 5.

7 Gärtner 1827, pp. 74–5.

8 Gärtner 1849.

9 Joseph Gottlieb Kölreuter's major work on plant sexuality and hybridisation, *Vorläufige Nachricht* (Kölreuter 1761–6), was frequently cited by CD in *Variation* and in the manuscript of his 'big book' on species (*Natural selection*); there is a heavily annotated copy of Kölreuter 1761–6 in the Darwin Library–CUL (see *Marginalia* 1: 458–71).

10 William Herbert published most of the results of his experimental work on plant hybridisation in the transactions and journal of the Horticultural Society (Herbert 1818, 1819, and 1846) and in Herbert 1837, pp. 335–80. Annotated copies of Herbert 1837 and part of Herbert 1846 are preserved in the Darwin Library–CUL (for Herbert 1837, see *Marginalia* 1: 372–6). CD's 1855 memorandum on the experimental work of Kölreuter, Herbert, and Gärtner is in DAR 116.

11 CD refers to his visit to Herbert's house in Hereford Street, Park Lane, London, on 28 May 1847 (see *Correspondence* vol. 4, letter to J. D. Hooker, [2 June 1847]).

12 Wiegmann 1828.

13 CD refers to Arend Friedrich Wiegmann's statement that pollen of a foreign variety exerted an immediate effect upon the colour and other characters of the fruit and seeds of the female parent in some varieties of legumes (Wiegmann 1828, pp. 23, 29). Gärtner later substantiated the claim regarding the effect of foreign pollen on varieties of peas, but not Wiegmann's claim to have crossed peas and vetches (see Appendix V, 'Beaton's first response to the letter from Charles Darwin to the *Journal of Horticulture and Cottage Gardener*, [before 27 January 1863]', n. 6).

14 Goss 1822.

15 CD refers to a note by Cromwell Mortimer reporting Thomas Henchman's observation to the effect that blue- and white-seeded peas grown separately but near one another produced pods containing the two pea forms. Mortimer's note appears as a footnote to Cooke 1745, p. 526 n., in which the effect of the pollen of one form of apple tree on a neighbouring different form is discussed. CD provided a summary of this and other evidence for the direct influence or immediate action of foreign pollen on the female parent in *Variation* 1: 397–401.

16 Miles Joseph Berkeley described the effect of pollen on the seed-coats of peas in *Gardeners' Chronicle and Agricultural Gazette*, 24 June 1854, p. 404. CD had corresponded with Berkeley on the topic (see *Correspondence* vol. 5, letter to M. J. Berkeley, 7 April [1855], and *Correspondence* vol. 6, letter to M. J. Berkeley, 29 February [1856]).

From William Henry Harvey 3 February 1863

Trin. Coll. Dublin
Feb. 3. 1863

Dear Darwin

I am *right glad* to find you have got so capital a worker on Cape Orchids,[1] which I have been longing to have investigated on the spot, from y^r^ point of view.[2] The Ophrydeæ there are almost *endless* in extraordinary modifications of parts & well worth study. The two now sent are *comparitively simple* in modifications. Both are of the large genus *Disa*, & I feel confidente in calling them (Pl. V.) *D. barbata* & (Pl. VI) *D. cornuta*, both common near Capetown.[3]

Tell M^r^. Trimen to *dry specimens* of everything he *draws*, and to send the specimens with No^s^ corresponding to the drawing. Then if he finds any novelty, we shall make sure of it,—& we shall also be able to *name* his sketches without **guess**. Nectariferous back sepals are quite frequent among Cape Orchids—and correspondently depauperated labella. The labellum is *often* a mere little tongue 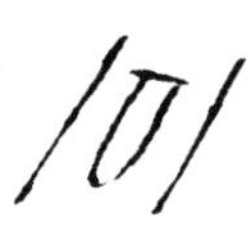

—sometimes a mere thread 

—and sometimes as as in *Brownlia*, nearly disappears altogether, & is adnate to the column.

In Satyrium the *two spurres affair* is a *true* labellum—the sepals & petals small & crowded together at the *front* of flower—the *opposite* to Disa.

Yours truly | W. H. Harvey

Copy[4]
Royal Entomological Society of London (Trimen papers, box 21: 78)

CD ANNOTATION
End of letter: '(Copy)' *added ink*; *square brackets in MS*

1 Harvey refers to Roland Trimen. Having read *Orchids*, Trimen sent CD a letter, probably early in January 1863, with which he enclosed a manuscript and sketches detailing observations he had made on a number of South African orchids. CD sent Trimen's sketches to Harvey, asking him to identify two unnamed species (see letter to Roland Trimen, 31 January [1863]). Neither Trimen's letter to CD nor CD's letter to Harvey has been found. Harvey was the author of several works on the taxonomy of South African plants (see W. H. Harvey 1838 and 1859–63, and W. H. Harvey and Sonder 1859–65).

2 Harvey's initial response to *Origin* had been hostile (see *Correspondence* vol. 8, letter from W. H. Harvey, 24 August 1860, and Appendix VII). Harvey was more favourable to CD's views after reading Asa Gray's interpretation of *Origin* (A. Gray 1860a), which argued that natural selection provided evidence of purposeful design in nature (see *Correspondence* vol. 8, letter to Asa Gray, 26 September [1860], n. 14). CD sought to strengthen this favourable impression by ensuring that Harvey received a copy of Gray's pamphlet *Natural selection not inconsistent with natural theology* (A. Gray 1861a; see *Correspondence* vol. 9, letter to Asa Gray, 23 [January 1861], and Appendix III). Harvey was also on CD's presentation list for *Orchids* (see *Correspondence* vol. 10, Appendix IV).

3 Harvey worked at the colonial treasury in Cape Town between 1836 and 1842 (*DNB*).

4 CD enclosed this copy, made by Henrietta Emma Darwin, with the letter to Roland Trimen, 16 February [1863].

From Thomas Rivers [3 February 1863]

You should live near a large nursery & your mind would find abundance of food.

when I first read the "Origin" I was amused at what I had observed with regard to "selection".[1] A patch of seedling trees if not transplanted seems to illustrate this (but perhaps I am taking a wrong view) the first year they are all equal in two or three years several will have pushed up—not confined to the outside of the patch which is easily accounted for by their finding more food— at the end of five or six years one or two or three will have smothered nearly all their brethren & then one alone will often be left. I have observed in what we call "pans" (flat pots of seedling) of pines which have stood a few years in one place forgotten the first year all equal the second & third a few more adventurous have made roots through the holes at the bottom & at last one gigantic fellow spreads over all & the underlings die.

Incomplete[2]
DAR 46.1: 95

CD ANNOTATION
Top of letter: 'Th. Rivers. Feb. 3^{d}— 1863' *ink*

1 CD sent Rivers a copy of *Origin* in January 1863 to thank him for his assistance with the chapter on bud-variation for *Variation* (see letter to Thomas Rivers, 15 January [1863]). Rivers had first read *Origin* several years earlier (see letter from Thomas Rivers, 21 January 1863).
2 Some indication of the contents of the missing portion of the letter is given by the letter to Rivers of [14 February 1863], in which CD thanked him for 'capital' information concerning inheritance in an ash tree and a species of thorn. CD made extensive use of this information in *Variation* 2: 18–19.

To Charles Lyell 4 [February 1863]

Down
4th

My dear Lyell

I have just received the great book.—[1] very sincere thanks for it. I had intended starting for London yesterday, but have been unable & doubt whether I can this evening.[2] But you will see me some morning very soon at your breakfast time, if you sh^{d}. be disengaged—

I have turned over pages on species & am very much pleased to see you to hit on many of the points which seem to me most important & not generally touched on by others. I have read last chapt. with very great interest.[3] By Jove how black owen will look.[4] You are quite civil to him: more civil that I could be. I am getting more savage against him, even than Huxley or Falconer.—[5] He ought to be ostracised by every naturalist in England.

You will, I feel sure, give the whole subject of change of species an enormous advance. Farewell | C. Darwin

Your book looks beautiful & I am impatient to begin reading it; but I must get a little more strength.—

Endorsement: '4 Feby 1863'
American Philosophical Society (287)

1 The reference is to Lyell's *Antiquity of man* (C. Lyell 1863a). Evidently CD was sent a pre-publication copy, since, according to an advertisement in the second edition, the publication date of the first edition was 6 February 1863 (C. Lyell 1863b, p. vii). However, John Murray placed a notice in the *Athenæum*, 7 February 1863, p. 176, stating that the work would appear on 9 February. CD's annotated copy of C. Lyell 1863a is in the Darwin Library–CUL (see *Marginalia* 1: 525–7).
2 CD went to London on 4 February, where he stayed at 6 Queen Anne Street, returning to Down on 14 February 1863 (see 'Journal' (Appendix II)). On 2 February 1863, Emma Darwin recorded in her diary (DAR 242) that CD was 'very languid in mg' and, on 3 February 1863, 'Ch. ditto'.
3 Chapter 24 of *Antiquity of man* is entitled, 'Bearing of the doctrine of transmutation on the origin of man, and his place in the creation' (C. Lyell 1863a, pp. 471–506).
4 Richard Owen had argued that humans should be placed in a distinct sub-class of the mammalia on the premise that the human brain was uniquely distinguished from the brains of apes by virtue of the 'hippocampus minor' and other features (Owen 1857). Lyell reviewed the controversy surrounding Owen's thesis in some detail in the last chapter of *Antiquity of man*, supporting Owen's chief opponent in the debate, Thomas Henry Huxley (C. Lyell 1863a, pp. 480–500). Owen responded with a long

and vitriolic letter to the *Athenæum*, 21 February 1863, pp. 262–3, which fuelled further controversy (see letter from J. D. Hooker, [23 February 1863] and n. 2; Bynum 1984, pp. 154–8; and L. G. Wilson 1996b, pp. 204–6).

5 Huxley and Hugh Falconer were both engaged in controversies with Owen. See *Correspondence* vol. 9, letter to T. H. Huxley, 1 April [1861], and letter to J. D. Hooker, 23 [April 1861]; *Correspondence* vol. 10, letter from T. H. Huxley, 9 October 1862 and n. 4; and this volume, letter to J. D. Hooker, 3 January [1863] and nn. 1 and 2, and letter to Hugh Falconer, 5 [and 6] January [1863]. See also Rupke 1994 and Wilson 1996b.

From Bartholomew James Sulivan 4 February [1863][1]

Board of Trade. S.W.
Feby 4

My dear Darwin

You will I have no doubt be surprised when I tell you that it is most probable I shall shortly leave England as *Commodore* to *command the Squadron* on *West Coast* of *S america*[2] The Station being from *Equator* to *Cape Horn*. The Admiral taking the North Pacific— Will it not be singular if should once more go over our old ground—[3] The fact is I have been for more than two months suffering from Bronchial complaint which though slight will not get better as every time it seems better I get a fresh cold. about six weeks since I was very unwell. I was saying at home that as it was the third winter I had been so frequently attacked this way I should have to avoid the winter in England—& I wished there was any appointment in a good climate that I could get as it would set me up again as twice I had been set up by going to S america But I knew of nothing likely and therefore we were seriously discussing my having to give up before next winter. The *very next day* I saw in the paper that a Commodore *was to be sent to the Chili Station*. It seemed of all things the very best as it would also enable me to serve my time for Active Flag, which next winter would be too late for.[4] after hesitating for several days I found no one was nominated, so I went in for it—and at first thought it was quite settled, for Senior Naval Lord & Private Secy—both nominated me for it to First Lord, which is generally conclusive,[5] but they first postponed the appt. to April on account of the financial years account preventing their doing it this year—and also to prepare a ship, and then they would not tell me if I were to go or not but I have heard some interest was rising for another— However as four admirals at admy have concurred in telling 1st Lord I have a stronger claim to it than any one, and as the Senior Naval Lord has told me what the limits of station are & other particulars it ought to be pretty certain. I have also heard from several that it is said at admiralty I am to go: and so it rests till March.

I have not spoken about it outside Admiralty—for if I do not get it I would rather it was not generally known, so I only tell you for your private information that you may turn over in your all mind all the *Geological instructions* you wish to give me. The ship will be "Curacoa" a Screw frigate, and if I go I shall probably go through Straits of M. and so see much of our old ground—[6]

If they will give me leave my wife[7] and girls will go with me as the Sea Voyage & the climate of Valpo[8] might do the girls much good.

I hope you are better now—& that our talking so much did not do you any more than a Scrap of injury. I hear that Wickham and his family have not been well in France[9]

with kind regards to all Your family believe me | Yours very sin^y | B J Sulivan

DAR 177: 280

[1] The year is established by the reference to Sulivan's visit to Down House in October 1862 (see n. 9, below).

[2] Sulivan was a naval officer in the marine department at the Board of Trade (Sulivan ed. 1896, pp. 375, 379, and *DNB*). A commodore is an officer intermediate in rank between a captain and an admiral.

[3] Sulivan was second lieutenant during the visit of HMS *Beagle* to South America and during its surveying voyage of 1831–6 (Freeman 1978).

[4] Sulivan was attempting to qualify for the active list of admirals (Sulivan ed. 1896, p. 377).

[5] The first lord of the admiralty was Edward Adolphus Seymour Seymour, his private secretary was John Moore, and the senior naval lord was William Hall Gage (*Navy list*, *Post Office London directory* 1863).

[6] Sulivan refers to the Straits of Magellan; the *Beagle* sailed through the Straits between 27 January and 13 February 1834, and between 25 May and 10 June 1834 (*Narrative*, Appendix, pp. 23–4, 29–30).

[7] Sulivan married Sophia Young in 1837 (*County families* 1871).

[8] The reference is to Valparaiso, Chile.

[9] On 21 October 1862, Sulivan, and two of the other officers who had served with CD on board HMS *Beagle*, Arthur Mellersh and John Clements Wickham, visited Down House. CD suffered ill health following the visit (see *Correspondence* vol. 10, letter from B. J. Sulivan, 18 October [1862] and n. 1, and letter to John Lubbock, 23 October [1862]). Wickham had retired to the south of France; he died of a stroke on 6 January 1864 (*Aust. dict. biog.*).

From J. D. Dana 5 February 1863

New Haven,
Feby 5. 1863.

My dear Mr Darwin—

The arrival of your photograph has given me great pleasure, and I thank you warmly for it.[1] I value it all the more that it was made by your son. He must be a proficient in the photographic art; for I have never seen a finer black tint on such a picture.—

I hope that ere this you have the copy of the Geology* (and without any charge of expenses, as was my intention)[2] I have still to report your book unread; for my head has all it can now do in my college duties.[3]

I have thought that I ought to state to you the ground for my assertion on page 602, that Geology has not afforded facts that sustain the view that the system of life has been evolved through a method of development from species to Species.—[4] There are three difficulties that weigh on my mind, and I will mention them.

1. The absence, in the great majority of cases, of those transitions by small differences required by such a theory.— As the life of America & Europe has

been with few exceptions independent, one of the other, it is right to look for the transitions on each Continent separately.— The reply to this difficulty is that the Science of Geology is comparatively new and facts are daily multiplying. But this admits the proposition that Geology does not yet afford the facts required.—[5]

2. The fact of the commencement of types in some cases by their higher groups of species instead of the lower.— As fishes began with the Selachians or Sharks (the highest order of fishes & the Ganoids, which are above the true level of the fish between fishes & Reptiles. In the introduction of land plants, there were Acrogens & Conifers and intermediate types, but not the lower grade of mosses—seemingly the natural stepping-stone from the Seaweeds. The fishes, Lepidodendron, Sigillarids, are examples of those intermediate or *comprehensive* types, with which great groups often began, and seem to explain the true relations of such types:—that they were *not transitional* forms in the system of life, but rather the *commencing* forms of a type.— If I advocated your theory, I think I should take the ground that there were certain original points of divergence from time to time introduced into the System, as indicated by the Comprehensive types.[6]

3. The fact that with the transitions in the strata & formations, the exterminations of species often cut the threads of genera, families & tribes, and sometimes, also, of the higher groups of Orders, classes and even Subkingdoms; and yet the threads have been started again in new species. The transition after the Carboniferous age was one apparently of complete extermination both in America & Europe, where all threads were cut; & yet life was reinstated and partly by renewing with species old genera in all the classes & subkingdoms, besides adding new types.[7]

You thus see that I have not spoken positively on page 602, without thinking I had some foundation for it. I speak merely of the geological facts that bear on the or any theory of development, not of facts from other Sources.—

You say in your letter that according to Mr Falconer, Prof. Owen has not done his work well with the Reptilian bird.[8] I should be very glad to know what are Mr Falconer's views.[9] I should like also to have his present opinions with respect to the Mesozoic mammals of England, or at least, to be informed whether he sustains the conclusions he first published on the subject. I have quoted from Owen in my book because his publications were more recent, not that I have greater confidence in his opinions or knowledge.—[10]

With earnest wishes for your health and happiness, I remain | Sincerely yours | James D. Dana

* It was sent through Trubner & Co. 60 Paternoster Row[11]

Dana Family Papers, Manuscripts and Archives, Yale University Library

CD ANNOTATION

8.1 You say . . . bird. 8.2] *scored*

[1] CD enclosed a photograph of himself, made by William Erasmus Darwin in April 1861, with his letter to Dana of 7 January [1863].

[2] The reference is to Dana's *Manual of geology* (Dana 1863a; see letter to J. D. Dana, 7 January [1863] and n. 3). CD did not receive the book until later in the month (see letter to J. D. Dana, 20 February [1863]). There is a copy of Dana 1863a in the Darwin Library–Down.

[3] CD had sent Dana a presentation copy of *Origin* in November 1859 (see *Correspondence* vol. 7, letter to J. D. Dana, 11 November [1859], and *Correspondence* vol. 8, Appendix III). Dana, who was professor of geology at Yale University, had suffered a nervous breakdown in 1859 (see *Correspondence* vol. 7, letter to Charles Lyell, 29 [December 1859] and n. 10).

[4] Dana had briefly stated his objections to 'a theory that derives species from others', concluding that geology led to 'no other solution of the great problem of creation, whether of kinds of matter or of species of life, than this:— DEUS FECIT [God made it]' (Dana 1863a, p. 602).

[5] CD had attempted to address the problem of the absence of transitional forms in the geological record in *Origin*, admitting that 'Geology assuredly does not reveal any such finely graduated organic chain; and this, perhaps, is the most obvious and gravest objection which can be urged against my theory' (*Origin*, p. 280; see also pp. 172–9, 279–82, 292–302, and letter to J. D. Dana, 20 February [1863]).

[6] Dana defined 'comprehensive types' as 'types comprehending, along with their own characteristics, some of those of other tribes which were yet uncreated, but which were to exist in the future unfolding of the system of life' (Dana 1863a, p. 203).

[7] In *Origin*, p. 316, CD argued that genera and families did not reappear once they had disappeared, adding: 'I am aware that there are some apparent exceptions to this rule, but the exceptions are surprisingly few . . . and the rule strictly accords with my theory.'

[8] See letter to J. D. Dana, 7 January [1863]. The references are to the paper on *Archaeopteryx* read by Richard Owen before the Royal Society of London on 20 November 1862 (later published as Owen 1862a), and to remarks made by Hugh Falconer in a letter to CD of 3 January [1863]. Falconer criticised Owen for understating the transitional character of the *Archaeopteryx* fossil, and for overlooking the cast of the interior of the skull.

[9] Following a résumé of the reports made by Johann Andreas Wagner and Hermann von Meyer on the *Archaeopteryx* fossil, published in the January 1863 issue of the *American Journal of Science and Arts*, Dana provided a commentary on the interpretation of the fossil's avian and reptilian characteristics ('Discovery of remains of vertebrated animals provided with feathers, in a deposit of Jurassic age', *American Journal of Science and Arts* 35 (1863): 129–33).

[10] Falconer had described two Mesozoic fossil mammal species of the genus *Plagiaulax*, concluding that they were herbivorous marsupials (Falconer 1857b). Owen, however, argued that the structure of the lower jaw and teeth of *Plagiaulax* indicated that it was a carnivorous marsupial (Owen 1860b, p. 321). Dana had followed Owen's interpretation in his *Manual of geology* (Dana 1863a, p. 463), and was apparently unaware of Falconer's subsequent refutation of Owen's deductions (Falconer 1862).

[11] The London publisher and bookseller Nicholas Trübner specialised in American literature (*DNB*).

From William Darwin Fox 6 February [1863][1]

Delamere Rectory | Northwich | Cheshire
Feby 6

My dear Darwin

It is now long since I have heard any thing of you.[2] Is there the least hope of your being in or near London from middle of next week this the following one?[3] I should so very much like to see you again. I am going to take my 3rd Boy to a school near Maidenhead, and must be a few days in or about London.[4] How our children grow up— My eldest has been now near two years at Oxford[5]

—My 2^d is in 6th Form at Repton, and promises to be clever—[6] My 3^d, as I said going to Maidenhead— My 2 eldest girls each with a fine boy—and the other 10 running up like willows.[7]

Yours I suppose are doing rapidly—while we (as one of my little ones said to her Mother a few days since) go down as they go up.

Are you a Believer in Free Martins—[8] I have always found them true in practise till now. A Lady near here, whom I know well, and her twin Brother also—has one from Baby & just had a miscarriage of another.

How are Caroline Susan & Catherine?[9] What ages it is since I have seen them—The former never since her Marriage. I sh^d much like to see her & her two girls.[10]

Susan I mean to offer to pay a visit to ere long, if she will have me.

That I may not tire you out I will conclude with our kindest regards to Mrs Darwin—from your old friend & cousin | W D Fox

Mr Woodds | Hillfield | Hampstead | NW | *after next Wednesday*[11]

DAR 164: 176

1 The year is established by the reference to Samuel William Darwin Fox's time at Oxford University (see n. 5, below).

2 See *Correspondence* vol. 10, letter to W. D. Fox, 20 [September 1862].

3 CD arranged to meet Fox in London on 13 February 1863 (see letter to W. D. Fox, [10 February 1863], and letter from W. D. Fox [11 February 1863]).

4 Fox refers to his thirteen-year-old son, Robert Gerard Fox (*Darwin pedigree*).

5 Samuel Fox entered Wadham College, Oxford, in 1861 (*Alum. Oxon.*).

6 The reference is to Repton School in Derbyshire, which Charles Woodd Fox had attended since 1859 (*Repton School register*).

7 Fox refers to Eliza Ann Sanders and her son Charles Henry Martyn Sanders, and to Harriet Emma Overton and her son Frederick Arnold Overton. By 1863, Fox had fifteen children from two marriages (*Darwin pedigree*).

8 Freemartin or free martin: a 'hermaphrodite or imperfect female' (*OED*).

9 Fox refers to CD's sisters, Caroline Sarah Wedgwood, Susan Elizabeth Darwin, and Emily Catherine Darwin.

10 Caroline Sarah Darwin married Josiah Wedgwood III in 1837. There were three surviving daughters of the marriage, Katherine Elizabeth Sophy, Margaret Susan, and Lucy Caroline Wedgwood (*Darwin pedigree*).

11 Charles Henry Lardner Woodd was a brother of Fox's second wife, Ellen Sophia Fox (*Burke's landed gentry* 1952).

To Charles Victor Naudin[1] 7 February 1863

le 7 février 1863

Cher Monsieur,

Une absence de chez moi m'a empéché de vous offrir mes remercimen(t)s sincèr(e)s pour la peine que vous avez pris(e) en m'écrivant votre lettre du 2 Février si remplie d'information(s) d'une si grande valeur.[2]

Il me fait grand plaisir d'avoir votre opinion (autant que vous êtes capable de l'exprimer) sur le croisement des variétés de Melons; et je m'en servirai pour l(a) citer.[3]

Récemment j'ai fait grand usage de vos Mémoires admirables sur les Cucurbitacae,[4] autant pour la citation que pour m'informer. Je vous remercie bien de votre intention de m'envoyer votre nouveau Mémoire.[5] J'anticipe avec grand intérêt votre ouvrage sur l'Hybridation.[6]

Agréez mes remerciements cordi(aux) pour votre extrême bonté et croyez moi, avec beaucoup de respect, votre très obéissant serviteur.

Charles Darwin.[7]

Progressus rei botanicæ 4 (1913): 94

[1] For a translation of this letter, see Appendix I.

[2] Naudin's letter has not been found. CD was in London from 4 to 14 February 1863, staying at Erasmus Alvey Darwin's house at 6 Queen Anne Street (see 'Journal' (Appendix II)).

[3] CD had sent a memorandum to Naudin requesting information on crossing varieties of melon (see *Correspondence* vol. 10, letter to J. D. Hooker, 24 December [1862], and this volume, letter from J. D. Hooker, 24 January 1863). CD cited Naudin's study of varieties of melon (Naudin 1859) in *Variation* 1: 359–60. CD's notes on Naudin 1859 are preserved in DAR 205.7 (2): 146. CD cited additional information supplied by Naudin in *Variation* 2: 108.

[4] CD refers to Naudin 1856 and 1862. In discussing the variations of cucurbitaceous plants in *Variation* 1: 357–60, CD quoted extensively from Naudin's experiments on the genus *Cucurbita* (Naudin 1856). CD's notes on Naudin's article are in DAR 205.7 (2): 141. CD wrote a draft of this chapter between October and December 1862 (see *Correspondence* vol. 10, Appendix II); it dealt with variation in fruits, ornamental trees, and flowers.

[5] Naudin may have referred to his forthcoming publication on hybridity (Naudin 1863). There is an annotated copy in the Darwin Pamphlet Collection–CUL.

[6] Naudin's researches on the hybridisation of plants were published in Naudin 1863 and 1865, of which there are annotated copies in the Darwin Library–CUL. For Naudin 1865, see *Marginalia* 1: 638–9. CD's opinion of Naudin's work is discussed in *Correspondence* vol. 10, letter to J. D. Hooker, 30 [June 1862], and letter to Asa Gray, 1 July [1862].

[7] A footnote in Blaringhem 1913, p. 94, notes that only the signature of this letter is in CD's hand.

From James Paget 7 February 1863

1 Harewood Place | Hanover Square | W.
Feby. 7. 1863

Dear Darwin

I send herewith the genealogical table,[1] and with it, a book by the same writer, D^{r}. Dobell,[2] which he begs me to present to you saying "I have from the first wished to have an opportunity of giving him a copy, having taken his name so much in vain:[3] and will you say that I much desire to know what he has to say to the part I have doubled-down"—[4]

The facts which I have ascertained in relation to the inheritance of cancer[5]

Incomplete[6]
DAR 174: 4

[1] The table showed the inheritance of fingers with thickened joints in a family through five generations; it was taken from Dobell 1862. CD incorporated the information into a section he was writing on polydactylism in chapter 12 of *Variation* (*Variation* 2: 12–16), which he began on 23 January 1863 (see 'Journal' (Appendix II)). Dobell 1862 is cited in *Variation* 2: 14 n. and 36 n. There is an annotated copy of Dobell 1862 in the Darwin Pamphlet Collection–CUL. See also letter to T. H. Huxley, [8 February 1863], and letter from H. B. Dobell, 12 May 1863.

[2] Dobell 1861. There is a copy of Horace Benge Dobell's *Lectures on the germs and vestiges of disease* in the Darwin Library–Down.

[3] Dobell, a medical lecturer and physician at the Royal Infirmary for Diseases of the Chest in London (*Medical directory* 1863), had attempted to link CD's theory of natural selection with his own theory of disease (see Dobell 1861, pp. 46–7, 169–70, and letter to H. B. Dobell, 16 February [1863]).

[4] An indication of the part referred to may be gained from the letter to H. B. Dobell, 16 February [1863] and nn. 3–4.

[5] Paget's findings on the hereditary transmission of cancer (Paget 1857) are cited in *Variation* 2: 79–80. CD used the evidence of hereditary diseases in humans to illustrate the point that disadvantageous characteristics are often inherited (*Variation* 2: 7–10).

[6] An indication of CD's original inquiry is given in the letter from James Paget, 16 March 1863.

From Camille Dareste[1] 8 February 1863

Lille;
ce 8 février 1863

Monsieur

Quoique je n'aie pas l'honneur d'être connu de vous, je tiens cependant à vous témoigner la satisfaction que j'ai éprouvée en lisant votre admirable livre sur l'origine des Espèces. Je m'occupe depuis vingt ans, de l'étude des sciences naturelles; et j'étais depuis bien longtemps arrivé par mes observations et par mes réflexions personelles, à considérer l'idée de la fixité de l'espèce, si généralement admise par les naturalistes, comme tout à fait en contradiction avec les faits. Mais si je ne pouvais me contenter des théories existantes, je ne voyais pas cependant comment on pouvait les remplacer; et je me voyais hésitant entre des doctrines qui ne sont plus admissibles, et la plus grande incertitude relativement aux doctrines qu'il fallait leur substituer. Votre livre a fait briller pour moi la lumière au sein des ténèbres; en me montrant où il fallait aller chercher le guide qui jusqu'alors m'avait fait défaut dans mes réflexions sur la nature des espèces.

Je prends, à ce sujet, la liberté de vous adresser un petit travail que je viens de publier.[2] C'est un mémoire que l'Académie des Sciences de Paris a couronné, il y a un mois.[3] Si vous voulez bien prendre la peine de consulter ces opuscules, vous verrez que j'ai cherché à résoudre, par la voie expérimentale, des questions tout à fait comparables à celles que vous traitez dans votre livre.[4] Je suis arrivé en effet, après de bien longues expériences, à produire, comme Geoffroy Saint-Hilaire l'avait fait il y a trente ans,[5] un grand nombre de monstruosités artificielles; et j'espère, en poursuivant ces études, pouvoir quelque jour, expliquer la formation d'un grand nombre d'anomalies Ces travaux, si je puis les poursuivre, me donneront quelque jour, je l'espère, la possibilité d'appliquer la méthode expérimentale à l'examen d'une partie de la science où vous vous êtes fait une si belle place.

Je regrette, Monsieur, d'être obligé de vous écrire en français: mais si je sais assez d'anglais pour avoir pu lire votre livre avant qu'il n'ait été traduit,[6] je ne connais pas assez votre langue pour la parler, et encore moins pour l'écrire

Je suis, avec un profond respect | Votre très humble et très obéissant serviteur | Camille Dareste.

Professeur à la faculté des sciences de Lille.

M. Charles Darwin Esqre.

DAR 162: 42

1 For a translation of this letter, see Appendix I.

2 Dareste 1863. There is a lightly annotated copy of this work, in which Dareste provides an account of monstrosities artificially induced in chicken embryos, in the Darwin Pamphlet Collection–CUL.

3 Dareste refers to the Alhumbert Prize (*Comptes rendus hebdomadaires des séances de l'Académie des Sciences* 55: 979–82).

4 Dareste's experiments on hen's eggs were published in two complementary pamphlets as Dareste 1862 and 1863. The wording of this letter, and the letter to Camille Dareste, 16 February [1863], suggests that Dareste only sent the second pamphlet; however, Dareste appears to assume that CD also had access to the earlier pamphlet. There are annotated copies of Dareste 1862 and 1863 in the Darwin Pamphlet Collection–CUL. CD discussed the implications of embryology for the theory of natural selection in *Origin*, raising the question of the relationship of embryos to external conditions (*Origin*, pp. 439–50). CD cited Dareste's experiments in *Variation* to illustrate the point that external conditions could be the direct cause of modifications in the structure of organisms (see *Variation* 2: 289).

5 Etienne Geoffroy Saint-Hilaire and his son, Isidore Geoffroy Saint-Hilaire, pioneered the scientific study of monsters (see E. Geoffroy Saint-Hilaire 1825a, 1825b, 1825c, 1826, 1827a, and 1827b, and I. Geoffroy Saint-Hilaire 1832–7). By artificially generating monstrosities in embryos, they hoped to throw light on the laws governing the production of monsters and new species (see E. Geoffroy Saint-Hilaire 1825a and I. Geoffroy Saint-Hilaire 1832–7, 3: 500–6).

6 A French translation of *Origin* (Royer trans. 1862) was published on 31 May 1862 by Guillaumin et Cie (*Journal Générale de l'Imprimerie et de la Librarie* 2d ser. 6 (pt 3): 341).

To T. H. Huxley [8 February 1863][1]

6. Queen Anne St
Sunday night

My dear Huxley

Reflecting over the plate of the Ray fins, I suspect that I have been blundering; & that in six-fingered men the increase is generally confined to metacarpals & digits. If so Fish would do??[2] But I have written to Paget to look at Vrolik's Dutch book on double monsters & on six-fingered cases & tell me how this is.—[3] I cannot remember whether I assumed or knew that carpals are ever or generally increased in number.—

I did so enjoy my visit to you this morning, & it was not by any means solely those sweet pats of soft butter.—[4]

Ever yours | C. Darwin

When I hear from Paget I will tell you.—

Imperial College of Science, Technology and Medicine Archives (Huxley 5: 19)

[1] The date is established by the relationship between this letter and the letter to H. B. Dobell, 16 February [1863], and by the reference to CD's visit to Huxley's London home. CD was in London from 4 to 14 February 1863 (see n. 4, below); the only Sunday in this period was 8 February.

[2] CD was preparing a section on polydactylism for chapter 12 of *Variation* (*Variation* 2: 12–18). See also n. 3, below, and the letter to J. J. Briggs, 2 February [1863]. CD noted that fish fins, unlike the normal digits of humans, mammals, birds, and reptiles, had the power of regrowth. From this observation CD inferred that supernumerary digits in humans retained an 'embryonic condition', and resembled the normal digits of lower vertebrates (see *Variation* 2: 16, and letter to T. H. Huxley, 16 February [1863]).

[3] CD's letter to James Paget has not been found; however, see the letter to H. B. Dobell, 16 February [1863]. In *Variation* 2: 12 n. 26, CD stated: 'Vrolik has discussed this point [gradations of polydactylous features] at full length in a work published in Dutch, from which Mr. Paget has kindly translated for me passages.' Paget's letter of 7 February is incomplete; however, the reference is probably to Vrolik 1840–2. Willem Vrolik was professor of anatomy at the University of Amsterdam (*NNBW*).

[4] CD was staying at 6 Queen Anne Street, London, the home of his brother Erasmus Alvey Darwin, between 4 and 14 February 1863 (see 'Journal' (Appendix II)).

To W. D. Fox [10 February 1863]

6. Queen Anne St | W.
Tuesday Eveng

My dear Fox

Your note has been forwarded here, where we have all been staying for a week.[1] I have been bad enough of late & came here to see whether a change w^{d}. do me some good & it has succeeded.— I suppose from your note that you arrive at Mr Woodd's on Wednesday Evening.— [2] We return home on Friday (*possibly*, but improbably on Saturday—) & leave this house at 2°30$'$ for the Train.—

Now would it suit you to come here to lunch at 1° or 1°15$'$ on Friday; or on Thursday, but on that day Emma will probably be out.— I do hope you will be able to come for I shall be so sincerely glad to see you.— Can you come to Down; we sh^{d}. be very glad; but my stomach has got to such a pitch that I can seldom stay, not even with nearest relations, for more than half-hour at a time. Let me have a line, that I may be sure to be at home, whenever you can come.—

My dear old friend | Ever yours | C. Darwin

Postmark: FE 11 63
Christ's College Library, Cambridge (Fox 136)

[1] Letter from W. D. Fox, 6 February [1863]. CD stayed at 6 Queen Anne Street, London, the home of his brother Erasmus Alvey Darwin, from 4 to 14 February 1863 (see 'Journal' (Appendix II)).

[2] The reference is to Fox's brother-in-law, Charles Henry Lardner Woodd, with whom Fox was staying while in London (see letter from W. D. Fox, 6 February [1863]).

From Henry Holland [10 February 1863][1]

Q Anne Street[2]
Tuesday

My dear Charles,

After you left me yesterday I found the note, which I had sought for unsuccessfully, touching the particular form of Atavism, in *alternate generations*.[3]

It was *Burdach* (whose name you well know) who described this result from averages largely taken.[4] My note does not go beyond this mere fact; but having the name of the observer, you will be able easily to find further information, if the point is of any value to you.

When you happen to be in Town again, let me know it, that I may have the satisfaction of talking over with you many matters of *fact* & *thought*, on this most interesting & obscure of all subjects—the hereditary transmission of physical characters, of instincts, & of mental faculties.

Ever yours aff[y] | H Holland

DAR 166: 243

[1] The date is established by the reference to CD's stay in London (see n. 3, below); the only Tuesday during this time was 10 February.

[2] Holland had apparently called at Erasmus Alvey Darwin's house, 6 Queen Anne Street, where CD was staying at that time (see n. 3, below), and left this note.

[3] CD was in London from 4 to 14 February 1863. Between January and April 1863, CD wrote a draft of the chapters on inheritance for *Variation* (see 'Journal' (Appendix II)).

[4] Reversion, or atavism, is discussed in *Variation* 2: 28–61; however, the work of Karl Friedrich Burdach is not cited in this discussion, and the reference has not been identified.

From B. J. Sulivan 10 February [1863][1]

Board of Trade. S.W.
Feby 10

My dear Darwin

I send you a ticket—and if it would be at a time when I could go with you I should be glad to show your boy the model[2] I could perhaps explain it to him better than you can.

I could go any day *but Friday*—or I would take him instead of sending the ticket If he could fix a day and hour & call here I will go with him if I can any time between *11* & *12* tomorrow or Thursday would suit me best.

I am sorry to hear such an account of you. have you ever tried *turkish bath* for I hear wonderful stories about it.

Nothing new as to my going from Adm[y].—but report says it is decided I am to go—[3]

Believe me | Yours very sincerely | B. J. Sulivan

I send a few extra in case you have at any time friends who would *like to go*

DAR 177: 281

1 The year is established by the relationship between this letter and the letter from B. J. Sulivan, 4 February [1863].
2 The reference is to Horace Darwin, then aged 11, who had accompanied his parents to London (Emma Darwin's diary (DAR 242)). The model referred to has not been identified, but Horace had developed an interest in engineering (Notes on the early life of Horace Darwin by Leonard Darwin, Cambridge Scientific Instrument Company Archive–CUL, Box 3).
3 Sulivan was waiting to hear whether he was to be appointed as commodore to command the British naval squadron on the west coast of South America (see letter from B. J. Sulivan, 4 February [1863]); he was unsuccessful in his application (see Sulivan 1896 ed., p. 377).

From W. D. Fox [11 February 1863][1]

Hillfield | Hampstead | NW

My dear Darwin

Many thanks for your note. I am truly sorry to hear you are so poorly.[2]

I scarcely hoped to see you, and am very glad to be able to do so. I will certainly come on Friday about 1—and shall rejoice to see Er⟨as⟩mus at same time.[3]

Many thanks for your ⟨invite⟩ to Down, but it is impossible for me to accept it—nor do I think it w^{d} be anything but a teaze to you if I did.

Had I not been able to get a glimpse of you in Town, I might have tried to run down for a few hours.

Ever my dear Darwin | Yours W D Fox

DAR 164: 177

1 The date is established by the relationship between this letter and the letter to W. D. Fox, [10 February 1863].
2 See letter to W. D. Fox, [10 February 1863].
3 CD stayed with his brother, Erasmus Alvey Darwin, at 6 Queen Anne Street, London, from 4 to 14 February 1863 (see 'Journal' (Appendix II)).

From Henrietta Grace Powell 11 February 1863

1, Hyde Park Gate South, | W.

My dear Sir

There are some points in your most interesting book, on which we were wishing much to know more, when my dearest Husband died, and I could not venture to ask you to take the trouble to come to see me alone.[1] I was sorry indeed to hear from M^{r}. Wedgewood[2] that your health is so delicate as to render it fruitless to ask you to any *meal*. May I however hope that you will give me the pleasure of seeing you here next Sunday afternoon? M^{r}. Huxley and the Bishop of Natal are coming to have a quiet talk with me between 2—and 6 oclock and you shall not be teazed with any *eating*![3]

I need scarcely add how pleased I should be to see M^rs. Darwin also should she be in town.

Truly your's | Henrietta G Powell.

February 11. 1863.
Address | M^rs. Baden Powell | 1. Hyde Park Gate South | Kensington Gore

DAR 160: 13

[1] The reference is to the former Savilian Professor of geometry at Oxford University, Baden Powell (*DNB*). CD and Powell had corresponded about *Origin* before the latter's death in June 1860 (see *Correspondence* vol. 8, letter to Baden Powell, 18 January [1860]). Powell's enthusiastic reception of *Origin* is discussed in Corsi 1988, pp. 283–4.
[2] Powell probably refers to Hensleigh Wedgwood.
[3] Thomas Henry Huxley and John William Colenso. CD did not accept this invitation, and returned to Down on Saturday 14 February (see 'Journal' (Appendix II)).

From Frederick Smith 11 February 1863

Brit Museum
11 Feby. 1863.—

My Dear Sir

Up to this time the work referred to has not been found—[1] M^r. Watts however wishes me to leave the matter in his hands until tomorrow Morn^g. when he hopes at least to be able to say where the work is to be got at[2]

Yours very Truly | Fred^k Smith

Chas. Darwin Es^re.

DAR 177: 196

[1] CD visited the British Museum while he was in London between 4 and 14 February (see letter from S. P. Woodward, 14 February 1863, letter to H. B. Dobell, 16 February [1863], and 'Journal' (Appendix II)). Smith, an entomologist, was assistant keeper of zoology in the department of natural history at the British Museum (*Post Office London directory* 1861).
[2] Thomas Watts was assistant keeper at the British Museum's department of printed books (*Post Office London directory* 1861). The work CD wished to consult has not been identified.

To George Gabriel Stokes [12 February 1863?][1]

6. Queen Anne St | London. W.
Thursday

My dear Prof. Stokes

Absence from home has prevented me from sooner thanking you most sincerely for the trouble which you have so kindly taken for me. I was rather crazy with curiosity to know what the chances were.—[2]

I believe your way of stating the problem is rather better for me.— I think I understand your two letters.[3] The second way of calculating the case is much the best for me.—

I have made a copy for myself of your M.S. sentence & have altered the few words & figures which are necessary. I cannot suppose that I have made any blunder; so if I do **not** receive your sentence back, I shall understand that it is right.

with my sincere thanks | pray believe me | Yours truly obliged | Ch. Darwin

CUL (Add 7656: D 76)

1 The date is conjectured from the address; CD stayed at 6 Queen Anne Street, London, between 4 and 14 February 1863 (see 'Journal' (Appendix II)). In 1863, 5 and 12 February were Thursdays. The wording of the letter suggests that Stokes's letter was forwarded from Down, making 12 February the most likely date.

2 Stokes was a mathematical physicist and Lucasian professor at Cambridge University (*DNB*). CD had apparently asked Stokes to determine the statistical chances of the same peculiarity recurring in a family. In *Variation* 2: 5, CD stated on Stokes's authority that, given a peculiarity occurring on average in one individual in a million, and a population of sixty million, consisting of ten million families of six members each, 'the odds will be no less than 8333 millions to 1 that in the ten million families there will not be even a single family in which one parent and two children will be affected by the peculiarity in question'. See also the manuscript of CD's 'big book' on species (*Natural selection*, pp. 480–1 n.), in which Stokes is cited with regard to the extreme improbability of the same rare peculiarity appearing in father and child without some genetic connection.

3 The letters from Stokes have not been found.

From Ludolph Christian Treviranus[1] 12 February 1863

Bonn (Rheinish Prussia)
Febr. 12. 1863.

Most honoured Sir

In due time I received Your kind letters, dated 18. and 24. Jun. p. Year.[2] In the latter of them You expressed a wish, to have any knowledge from what I should publish about the matter, treated in so excellent a manner by You in several publications.[3] In consequence I have availed myself to send You by care of Mrs. Williams & Norgate[4] two copies of N. 1. and 2. of Botanische Zeitung 1863.[5] one for You and the other for Dr. Jos. Hooker, to whom You will be so kind, to present it with my kindest regards.[6] In this paper You will find, that, besides making known to my countrymen Your most valuable labours about the aid of insects in fertilising flowers, I have ventured to detail, in what points I coincide with Your opinions and in which I differ. Yet in doing so, I have never uttered any differing view, without such reasons, as seemed sufficiently founded to my mind and without expressing me deep and unrefreined esteem for Your merits upon this widely extending field for observation[7]

Believe me, dear and honoured Sir, Your | very faithful servant | L. C. Treviranus | Prof: Bot. Bonn.

DAR 178: 182

1 Treviranus wrote 'ÿ' for 'y' throughout; these letters have been transcribed as 'y' in accordance with English practice.

[2] CD's letters to Treviranus have not been found; CD probably discussed the proposal that Treviranus prepare a translation of *Orchids* for the Stuttgart publishing firm, E. Schweizerbart'sche Verlagsbuchhandlung. However, Heinrich Georg Bronn, the translator of *Origin*, had already completed most of the translation when the proposal was made (see *Correspondence* vol. 10, letter from H. G. Bronn, 21 June 1862, and letter from E. Schweizerbart'sche Verlagsbuchhandlung, 11 July 1862).

[3] Treviranus was particularly interested in CD's work on the pollination of flowers. His review of *Origin* was the first review of the book to appear in a German botanical journal (Treviranus 1861). In Treviranus 1863a (see n. 5, below), he provided commentaries on several of CD's botanical publications, including letter to *Gardeners' Chronicle*, [before 13 November 1858] (*Correspondence* vol. 7), *Origin*, 'Dimorphic condition in *Primula*', and *Orchids*.

[4] Williams & Norgate was a firm of London booksellers and publishers, specialising in foreign scientific literature (*Modern English biography*).

[5] The reference is to Treviranus 1863a, which appeared in the *Botanische Zeitung*. There is a lightly annotated copy of the work in the Darwin Pamphlet Collection–CUL.

[6] Joseph Dalton Hooker.

[7] In his article, Treviranus discussed CD's doctrine that 'nature ... abhors perpetual self-fertilisation' (*Orchids*, p. 359; see also *Origin*, p. 97). His main objection was that CD had not explained why the hermaphrodite character should be so common in plants if self-fertilisation was significantly injurious (see Treviranus 1863a, p. 2).

From Isaac Anderson-Henry 14 February 1863

Woodend, Maderty, Crieff.
February 14/63

My dear Sir

I received your Letter dated the 2^{d} Inst on the eve of my starting for this place, and was truly sorry to observe that your health had imposed on you the necessity of going away for a change of air.[1] I hope you have derived the much to be desired benefit from that change, and are now again returned to the scene of your labours in fit frame to pursue them with that ardour & success which have hitherto so distinguished them. I find it necessary too to make a change occasionally, and here, in a solitude, I can recruit at leisure, tho' as Chairman of a Parochial Board here I am not without some smatter of Business to vary the quiet monotony of Country life[2]

But here I am away from my laboratory—and at a distance from it I find myself unfitted to resume the all engrossing subjects to which your Letters allude. But depend upon my aiding you in all I can in the field you are now upon where brighter laurels than have yet been gathered await you

Singularly, the frenchman M. Neumann,[3]—your information as to whom obliges & gratifies me much,—put that question to me you do—whether I knew an instance of seed from a weeping tree reproducing a weeper? I must make the same answer to you I did to him— I am unacquainted with such a result & never myself tried the experiment

I shall set to work on the experiments you alluded to—and my Gardener at Hay Lodge writes me of the arrival of a hamper of plants from London containing some of the subjects[4]

Meantime with fervent wishes for your better health | I am | yours most faithy | I. anderson Henry

I return to Hay Lodge next Wednesday

Charles Darwin Esqre. F.R.S. &c.

DAR 159: 63

[1] See letter to Isaac Anderson-Henry, 2 February [1863]. CD was in London from 4 to 14 February 1863 (see 'Journal' (Appendix II)).

[2] Following his marriage to the heiress of the landed estate of Woodend, Perthshire, Anderson-Henry divided his time between their country home and his house in Edinburgh (*Gardeners' Chronicle and Agricultural Gazette*, 22 March 1873, p. 399). Parochial boards, established in each parish by the Scottish Poor Law Act of 1845, were responsible for the administration of relief to the poor and the sick (see Clarke 1955).

[3] In his letter of 31 January 1863, Anderson-Henry informed CD that he had been corresponding with Louis Neumann, a gardener at the Jardin des Plantes, the botanical department of the Muséum d'Histoire Naturelle in Paris. CD had requested information on weeping trees to incorporate into chapter 12 of *Variation*, which he had begun writing on 23 January (see letter to Isaac Anderson-Henry, 2 February 1863, 'Journal' (Appendix II), and *Variation* 2: 18–19). See also letter to Thomas Rivers, 1 February [1863]. The results of Neumann's extensive inquiries on weeping trees were published in Verlot 1864. See also letter from John Scott, 3 March 1863.

[4] In his letter to Isaac Anderson-Henry, 20 January [1863], CD gave details of a series of experiments which he hoped Anderson-Henry would undertake. Alexander Donald was head gardener to Anderson-Henry at Hay Lodge, Edinburgh (Census returns 1871 (General Register Office for Scotland: 692/1/19/119)).

From Edwin Brown 14 February 1863

Burton on Trent
14 Feby 1863

Dear Sir

I send you by this same post a copy of my second paper on the Mutability of Race Forms[1] Perhaps one or two of the points mentioned may be of interest to you— You will be amused to see what a wasps nest I have got into— the remarks however are so wild & puerile that I have announced my intention of leaving them to their own refutation Mr Westwood will probably give me a rejoinder[2]

I believe it is the right way to go to work to carry the war into the enemies' camp & after showing a *prima facie* case against *im*mutability to defy the advocates to prove the affirmative of their case

The speakers mentioned several things as facts which I doubt very much One man (Mr Gregson) I know to be most reckless of the truth of his assertions[3]

It is no wish of mine to advance anything as a *New* theory— I merely wish to establish the fact of mutability from whatsoever cause ⟨it⟩ may arise You have shown most admirably how variation may arise in one way all of which I fully believe but I think that is only a portion of a wider law[4]

My friend H W Bates & myself had each curiously enough fixed upon the Carabidae & the Vanessae as groups particularly well suited for working out the laws of relationship— I have given up to him the Vanessae & he has I think left the Carabs to me[5]

I am Dr Sir | Yours very truly | Edw Brown

Chas Darwin Esq

DAR 160: 325

[1] The reference is to E. Brown 1862b; a lightly annotated copy is preserved in the Darwin Pamphlet Collection–CUL. Brown sent CD a copy of his first paper, 'On the mutability of specific or race forms', in October 1862 (E. Brown 1862a; see *Correspondence* vol. 10, letter to H. W. Bates, 15 October [1862]). In his second paper, Brown replied to his critics in the Northern Entomological Society, where his case for the mutability of species was pronounced 'not proven'. The discussions on the issue are reproduced in the *Proceedings of the Northern Entomological Society* of 28 July 1862, p. 18, 4 October 1862, pp. 8–13, and 22 December 1862, pp. 14–23.

[2] John Obadiah Westwood was the principal critic of Brown's first paper, arguing in a letter published in the *Proceedings of the Northern Entomological Society*, 4 October 1862, pp. 9–13, that certain species had remained essentially unchanged through thousands of years of existence, and demanding evidence of the mutability of species. Brown attempted to answer these objections (E. Brown 1862b, pp. 7–12), but there is no evidence that Westwood responded.

[3] The reference is to Charles Stuart Gregson. Gregson's remarks appeared in the *Proceedings of the Northern Entomological Society*, 4 October 1862, p. 9, and 22 December 1862, pp. 16–19.

[4] Brown took the view that in emphasising variation by natural selection, CD had neglected the influence of climate, food, and 'the accumulative effects of those apparently causeless individual variations that take place at every generation' (E. Brown 1862a, p. 9).

[5] Henry Walter Bates described Brown as 'my earliest Naturalist friend' (see *Correspondence* vol. 10, letter from H. W. Bates, 17 October 1862). Carabidae are a family of beetles and Vanessae are a family of butterflies, now known as Nymphalidae (Westwood 1838, p. 405; Grzimek ed. 1975, pp. 546, 553).

CD memorandum 14 February 1863

Feb 14th.— 1863

I have agreed to consider this Bond of Prof. Ansted dated April 19th 1855 cancelled;[1] Prof Ansted informing me that when the new Company for the Patent Stone is formed, he will endeavour to dispose of some of his shares & pay me what he can.[2]

Charles Darwin

DAR 210.10: 24

[1] David Thomas Ansted was formerly professor of geology at King's College, London (*DNB*). Ansted was co-owner, with Frederick Ransome, of the Patent Siliceous Stone Company, founded in 1848. CD purchased shares and provided several loans for the company, which had a troubled history, having to be reorganised and relaunched several times. The bond to which CD refers was returned to Ansted (see letter from D. T. Ansted, 23 April 1863). According to CD's Investment book (Down House MS), CD lent Ransome £250 in November 1854 and this loan was transferred to Ansted in April 1855. See also n. 2, below.

[2] The arrangements for cancelling the bond were not finally settled until April (see the letter to D. T. Ansted, 15 April 1863, and the letters from D. T. Ansted, 13 April 1863, 17 April 1863, and 23 April

1863). In CD's Investment book (Down House MS), the last entry concerning Ansted and the Patent Siliceous Stone Company is dated May 1863, and reiterates the settlement recorded in this document: 'I have cancelled Ansted Bond on understanding that if he can ever sell the shares, he will repay me.' Ransome's new company, the Patent Concrete Stone Company, went into production at new premises in East Greenwich in 1867 (*Engineering*, 28 June 1867, pp. 671–2).

To Thomas Rivers [14 February 1863][1]

Down. | Bromley. | Kent. S.E.
Saturday

My dear Sir

Absence from home, for a little rest, for the last nine days has prevented me from thanking you sooner for your last letter.[2]

You could not by any possibility have given me a more curious case of inheritance than that of the Ash, which produced weeping seedlings & itself lost the weeping peculiarity![3] It is capital for my purpose. I am also very glad to hear of the Thorn.—[4] I am nearly sure I have already in my M.S. index from "Loudon's Gard Mag." your first case of 20,000, or 30,000 seedlings from the common weeping ash.—[5] I wish I could get authentic information on the weeping Elm.—[6]

What you say of seedlings conquering each other well illustrates the "struggle for existence" & "natural selection".[7] I have often & often looked at a crowd of natural seedlings with just such feelings & reflexions as yours.—

With hearty thanks for your capital facts | My dear Sir | Yours sincerely | Ch. Darwin

19th Century Shop (catalogue 5, 1988)

1 The date is established by CD's reference to his stay in London (see n. 2, below); in 1863, 14 February was a Saturday.

2 Letter from Thomas Rivers, [3 February 1863]. CD was in London from 4 to 14 February 1863 (see 'Journal' (Appendix II)).

3 The letter from Rivers of [3 February 1863] is incomplete; however, CD quoted the information on the weeping ash contained in the missing portion of the letter in *Variation* 2: 19. CD cited the weeping habit of trees as an example of 'how feeble, capricious, or deficient the power of inheritance sometimes is', the weeping habit being transmitted to the seedlings sometimes strongly and sometimes weakly (see *ibid.*, pp. 17–19).

4 In the missing portion of his letter of [3 February 1863], Rivers evidently provided information on the young trees resulting from a cross between a species of weeping thorn (*Crataegus oxyacantha*) and a 'not-weeping variety' (see *Variation* 2: 18).

5 In 1834, Rivers reported that he had sown over 20,000 seedlings of the weeping ash (*Fraxinus excelsior*), none of which inherited the weeping characteristic. The letter reporting this information was published in the *Gardener's Magazine* 10 (1838): 408, and was cited in *Variation* 2: 19. CD's copy of the volume has a manuscript index pencilled at the back and is preserved in the Darwin Library–CUL. John Claudius Loudon founded the *Gardener's Magazine* in 1826 and edited it until his death in 1843 (*DNB*).

6 Rivers later raised a number of seedlings from different varieties of weeping elm at CD's request, none of which inherited the weeping habit (see *Variation* 2: 19).

7 See letter from Thomas Rivers, [3 February 1863].

From Samuel Pickworth Woodward 14 February 1863

British Museum.
Feb.y 14/63

Dear Sir,

I have been unable to find the Review of your 'Orchids' which I promised to send—or at least the *copy of it* just then given me by one of my colleagues— But if you will not mind the trouble of returning it I will send a rough proof—& also (by way of complement) another article *contra Scottos*—which it is to be hoped they enjoyed, altho' not nearly equal to their deserts.[1]

When I first read the 'Origin' I made a list of errata & *objections*—but found them not worth sending.[2] I have only seen the *first ed.* & you must have long since corrected such errata as p. 49 *Primula acuulis* or p. 29 where "Herefords" were spoken of as if they were "Short-horns"—[3] Did you notice when you came to see the Assyrian relics, the little bas-relief of the Tibetan mastiff?[4] Its character is well represented—in contrast with ⟨the⟩ Syrian hunting dogs—

There was an obscure expression at p. 429 where you speak of no "single insect of a new order being found in Australia"— If any new order (i.e unknown elsewhere) should be f⟨ound⟩ there, it would surely be represented by *many specie⟨s.⟩* The phrase appears to require inversion—[5]

Respecting that endless problem—the honeycomb! Mr Walter Mitchell objects that if the form of the cell results only from the necessity of the case (as I hold) then the chances are a million to one against the result we get—[6] The cell of the humble bee is elliptical—& the chances are that the bottom of the honey bees' cell will be elliptical—or more or less obtuse than it is now. But I cannot see the force of this objection as the chances are the same *against* **any** form you may please to imagine—& not more against the true sphere than the spheroid— The shape & proportions of the honey bee being what they are it must strike a circle—or excavate a segment of a sphere— If it were more acute or more obtuse it would make something different—but I can't see any reason why it shouldn't just as soon happen to be— what it is—

Mr Mitchell could understand the form of the bees' cell *if* wax had a *hexagonal* cleavage like *graphite* (which is **C**'.) But according to the accepted doctrine it *must* have a cleavage (whether apparent to us or not) at right angles to the incidence of pressure—& that (according to my notion) will be such as to give hexagonal cleavage—

⟨*six lines excised*⟩

If you come to a — edn. I hope you will erase all mention of Buckman's exp.ts at Cirencester—[7] it will be better not to mention his name than give it with the admission made in ed. 1. I went to see *my old* "B.G." at Cisseter under Buckman's managem.t, & took care to examine *it alone*—[8] "What I saw there I will not declare"—[9] But it will be sufficient to say that one of the students (who owed the Prof. a grudge) confessed to R. Tomes (*vespertilio*)[10] that **he** himself had mixed the seeds intended for experiment in the Botanic Garden—

Have you met Mr Lord? the traveller in Oregon—[11] He brought up the other day, (just before you came) two varieties of Musk rat— One builds houses out in the water, like the Beaver— the other burrows in the bank & only makes sub-aqueous *runs*— Yet Waterhouse would not admit any difference in the animals![12]

⟨*six lines excised*⟩ roots chiefly, are not nearly so formidable as described— Whereas the same species on the East flank of the Rocky mtns feeding on Bisons (left by the Indian Hunters) become large & heavy & very savage.[13]

Yours sincerely | S. P. Woodward

Cha[s] Darwin Esq[e]

Incomplete
DAR 181: 154

CD ANNOTATIONS
1.1 I have ... p. 49 2.3] *crossed pencil*
2.3 p. 29 ... "Short-horns"— 2.4] *double scored pencil*
3.1 There was ... *specie*⟨*s.*⟩ 3.3] *double scored pencil*
6.1 If you ... animals! 7.4] *scored pencil*
8.1 roots ... savage. 8.3] *scored pencil*
Top of first page: 'Musk Rat | *2 vars in Habits* | *Bees different Habits* | Buffaloes | W. of Rocky Mountains' *ink*

1 The review and article to which Woodward refers have not been identified. While he was in London between 4 and 14 February 1863, CD visited the British Museum (see n. 4, below), where Woodward worked as an assistant in the department of geology and mineralogy (*DNB*).

2 CD sent Woodward a presentation copy of *Origin* (see *Correspondence* vol. 8, Appendix III).

3 In the first edition of *Origin* (1859), CD referred in error to the primrose and cowslip, respectively, as 'Primula veris and elatior' (*Origin*, p. 49). This should have read 'Primula vulgaris and veris'. *P. acaulis* was a synonym for *P. vulgaris*. CD corrected his mistake in the second edition (1860). Concerning the Hereford breed of cattle, CD had written: 'Ask, as I have asked, a celebrated raiser of Hereford cattle, whether his cattle might not have descended from long-horns, and he will laugh you to scorn' (*Origin*, p. 29). The sentence remained unaltered in subsequent editions.

4 CD visited the British Museum while he was staying in London between 4 and 14 February (see 'Journal' (Appendix II), and letter from Frederick Smith, 11 February 1863). The reference is to an Assyrian bas-relief, dated *circa* 640 B.C., taken from the tomb of the son of Esar-haddon, king of Assyria. In *Variation*, CD stated that he had seen drawings and clay models of the antiquity in the British Museum, adding that there was some doubt as to whether the dog depicted was a 'Thibetan mastiff' (*Variation* 1: 17, n. 4). An illustration of the bas-relief is given in Nott and Gliddon 1854, p. 392, an annotated copy of which is in the Darwin Library–CUL (see *Marginalia* 1: 603–6). CD used the evidence of the representation of dogs in antiquities to argue that the different breeds existing in modern times were of extreme antiquity and multiple origin (*Variation* 1: 33).

5 The phrase referred to occurs in the following sentence (*Origin*, p. 429):

> As showing how few the higher groups are in number, and how widely spread they are throughout the world, the fact is striking, that the discovery of Australia has not added a single insect belonging to a new order; and that in the vegetable kingdom ... it has added only two or three orders of small size.

This sentence was modified in subsequent editions of *Origin*, but not in the form suggested by Woodward. CD changed 'order' to 'class' in the second edition, and in the third edition, 'only two or three orders' is replaced by 'only two or three families'. There were further slight changes to the sentence in the fourth, fifth, and sixth editions (see Peckham ed. 1959, p. 670).

[6] Walter Mitchell was a clergyman and natural philosopher with a special interest in crystallography (*Alum. Cantab.*). CD discussed bee-cell construction in *Origin*, pp. 224–35. In his discussion, CD sought to demonstrate that the hexagonal geometry of the honeycomb constructed by hive-bees could be explained by reference to simple instincts and natural selection. CD made several minor changes to the text in subsequent editions, but did not discuss the point raised by Mitchell (see Peckham ed. 1959, pp. 402–12).

[7] James Buckman was professor of geology and botany at the Royal Agricultural College, Cirencester (*DNB*). In *Origin*, p. 10, discussing the effect of external conditions as a cause of variation, CD described 'extremely valuable' research by Buckman that appeared to show that species of the same genus, differing widely in appearance in their natural habitat, became indistinguishable from one another when grown under the same conditions in experimental plots (Buckman 1857). CD had held the 'deepest & most lively interest' in these experiments (see *Correspondence* vol. 6, letter to James Buckman, 4 October [1857]). There is an annotated copy of Buckman's paper in the Darwin Pamphlet Collection–CUL. The reference to Buckman's work was removed from the fourth edition of *Origin* (Peckham ed. 1959, p. 82).

[8] Between 1845 and 1848, Woodward occupied the chair of geology and natural history at the newly established Royal Agricultural College, Cirencester (Woodward 1881, pp. 286–90). 'B.G.' is an abbreviation for botanic garden.

[9] The source of the quotation has not been identified.

[10] Robert Fisher Tomes had published many papers on the Vespertilionidae (*Royal Society catalogue of scientific papers*); *Vespertilio* is a genus of bats.

[11] In 1858, John Keast Lord was appointed naturalist to the British North American boundary commission sent to British Columbia and Washington Territory (*DNB*). Lord travelled through Oregon in May 1860 (Lord 1866).

[12] Lord displayed two specimens of muskrat at a meeting of the Zoological Society of London on 24 March 1863, claiming that although the structural differences were trivial, his new species *Fiber osoyooensis* differed from the well-known *F. zibethecus* in habit, size, colour, and distribution, and therefore was a separate species, not a well-marked variety of the same species (Lord 1863). George Robert Waterhouse was keeper of geology in the department of natural history at the British Museum and a vice-president of the Zoological Society (*Post Office London directory* 1863, *British imperial calendar* 1863).

[13] This may be a reference to the prairie wolf; in *Variation* 1: 22, CD states:

> North America is inhabited by a second kind of wolf, the prairie-wolf (*Canis latrans*), which is now looked at by all naturalists as specifically distinct from the common wolf; and is, according to Mr. J. K. Lord, in some respects intermediate in habits between a wolf and a fox.

To J. D. Hooker 15 February [1863][1]

Down Bromley Kent
Feb. 15th | Sunday

My dear Hooker

We got home last evening, all wonderfully improved by our London trip.—[2] I cannot tell you how thoroughily I enjoyed seeing you at Kew.—[3]

I enclose scrap from A. Gray: he is easing down:[4] I have thought you might like to see what he says on democracy of Plants.— Please return it. But I write now, because the new Hothouse is ready & I long to stock it, just like a school-boy.—[5] Could you tell me pretty soon what plants you can give me; & then I shall know what to order.[6] And do advise me how I had better get such plants as you can *spare*. Would it do to send my tax-cart early in morning,[7] on a day that was not frosty, lining the cart with mats; & arriving here before night.

I have no idea whether this degree of exposure & of course the cart w^d be cold, would injure stove plants: they would be about 5 hours (with bait) on journey home.— Will you have kindness to consult M^r. Gower?[8] It would give all you at Garden less trouble just to pack them in bottom of tax-cart, but I sh^d. be very sorry if they were injured & would much rather of course wait till Spring.—

Remember Gloxinias, 2 or 3 plants (of same species) of *erect* & of *drooping*.[9]

I thought I had overrated adaptation of orchids, but I believe I have underrated it; for A. Gray sends me letter of good entomologist, who has been observing & he most truly remarks that pollinia must be placed to adhere either to eye or proboscis, for rest of body covered with loose scales; & by Jove all the pollinia that I have seen were attached to these two parts.—[10]

I have 15 letters to write, so farewell. My dear old friend | C. Darwin

DAR 115: 181

[1] The year is established by the reference to CD's return from London (see n. 2, below).
[2] CD had been in London, staying with his brother Erasmus Alvey Darwin, from 4 to 14 February 1863 (see 'Journal' (Appendix II)).
[3] Emma Darwin recorded in her diary (DAR 242) that they visited the Royal Botanic Gardens, Kew, on 11 February 1863.
[4] See letter from Asa Gray, 27 January 1863. The reference is to Asa Gray's view of British attitudes to the American Civil War (see *Correspondence* vol. 10).
[5] The hothouse was constructed under the supervision of John Horwood, George Henry Turnbull's gardener (see *Correspondence* vol. 10, letter to J. D. Hooker, 24 December [1862], and this volume, letters to J. D. Hooker, 13 January [1863] and 30 January [1863], and letter to G. H. Turnbull, [16? February 1863]). See also Appendix VI.
[6] CD had apparently given Hooker a list of plants he wished to obtain during his visit to Kew (see letter from J. D. Hooker, [16 February 1863]).
[7] Tax-cart: 'a spring-cart paying a low rate of duty' (*OED*).
[8] William Hugh Gower was a gardener at the Royal Botanic Gardens, Kew (R. Desmond 1994).
[9] CD probably made this request when he visited the Royal Botanic Gardens, Kew (see nn. 2 and 3, above). For CD's interest in species of *Gloxinia*, see *Correspondence* vol. 10, letter to John Scott, 19 December [1862], and this volume, letter to Isaac Anderson-Henry, 20 January [1863], and letter to John Scott, 20 [June 1863].
[10] With his letter to CD of 27 January 1863, Gray enclosed a letter from Samuel Hubbard Scudder detailing this observation, illustrated with reference to a butterfly of the genus *Nisoniades* that had the pollinia of *Platanthera hookeri* attached (see the first enclosure with the letter from Asa Gray, 27 January 1863). CD included the information in *Orchids* 2d ed., p. 76.

To J. J. Briggs 16 February [1863]

Down. | Bromley. | Kent. S.E.
Feb. 16th

Dear Sir

Absence from home during the last ten days must plead my excuse for not having earlier thanked you most sincerely for your kind & to me very valuable letter, which gives me just the facts, which I wanted & which will be very useful to me to quote.[1]

with my renewed thanks | Pray believe me | Dear Sir | Yours faithfully & obliged | Charles Darwin

Postmark: Derby FE 17 63
Mrs H. Codd (private collection)

[1] The letter from Briggs has not been found. However, see the letter to J. J. Briggs, 2 February [1863], and *Variation* 2: 15–16.

To Camille Dareste 16 February [1863][1]

Down. | Bromley. | Kent. S.E.
Feb 16

Dear & respected Sir.

I thank you sincerely for your letter & your pamphlet.—[2] I had heard (I think in one of M. Quatrefage's books)[3] of your work & was most anxious to read it, but did not know where to find it. You could not have made me a more valuable present. I have only just returned home & have not yet read your work when I do if I wish to ask any questions I will venture to trouble you.— Your approbation of my book on Species has gratified me extremely—[4] Several naturalists in England, North America & Germany have declared that their opinions on the subject have in some degree been modified, but as far as I know my book has produced no effect whatever in France & this makes me the more gratified by your very kind expression of approbation.[5]

Pray believe me, Dear Sir. | With much respect. | Yours faithfully & obliged. | Ch Darwin

Copy
DAR 143: 368

[1] The year is established by the relationship between this letter and the letter from Camille Dareste, 8 February 1863.
[2] Letter from Camille Dareste, 8 February 1863. CD refers to Dareste 1863, an annotated copy of which is in the Darwin Pamphlet Collection–CUL.
[3] Dareste is briefly mentioned in Armand de Quatrefages's *Métamorphoses de l'homme et des animaux*, but Quatrefages does not refer to Dareste's embryological researches (Quatrefages 1862, p. 274 n.). There is an annotated copy of this work in the Darwin Library–CUL (see *Marginalia* 1: 691).
[4] See letter from Camille Dareste, 8 February 1863.
[5] On the reception of *Origin* in France, see Conry 1974, Farley 1974, Stebbins 1974, and Corsi and Weindling 1985.

To Horace Benge Dobell 16 February [1863][1]

Down. | Bromley. | Kent. S.E.
Feb. 16th

Dear Sir

Absence from home & consequent idleness are the causes that I have not sooner thanked you for your very kind present of your Lectures.[2] Your reasoning seems quite satisfactory (though the subject is rather beyond my limit of thought & knowledge) on the V.M.F. not being "a given quantity."[3] And I can see that the conditions of Life must play a most important part in allowing this quantity to increase as in the budding of a tree &c.— How far these conditions act on "the *forms* of organic life" (p. 46) I do not see clearly.—[4] In fact no part of my subject has so completely puzzled me as to determine what effect to attribute to (what I vaguely call) the direct action of the conditions of life. I shall before long come to this subject & must endeavour to come to some conclusion, when I have got the mass of collected facts in some sort of order in my mind.[5] My present impression is that I have underrated this action, in the "Origin".—[6] I have no doubt when I go through your Volume, I shall find other points of interest & value to me.—

I have already stumbled on one case, (about which I want to consult Mr Paget) namely on the Regrowth of supernumerary digits.[7] You refer to "White on Regeneration &c. 1785."[8] I have been to Libraries of Royal & Linn. Soc, & to British Museum; where the Librarian got out your volume & made a special hunt, & can discover no trace of such a book.— Will you grant me the favour of giving me any clue, where I could see this Book? Have you it; if so & *the case is given* **briefly**, would you have great kindness to copy it.— I much want to know all particulars.[9] One case has been given me, but with hardly minute enough details, of a supernumerary little finger which has already been *twice* cut off, & now the operation will soon have to be done for the third time.[10]

I am extremely much obliged for the genealogical table;[11] the fact of the two cousins not, as far as yet appears, transmitting the peculiarity is extraordinary & must be given by me.—[12]

With very sincere thanks for your kindness.— | Pray believe me | Dear Sir | Yours truly obliged | Ch. Darwin

Barton L. Smith

[1] The year is established by the relationship between this letter and the letter from H. B. Dobell, 5 March 1863.

[2] The reference is to Dobell 1861, which was sent with the letter from James Paget, 7 February 1863, and is in the Darwin Library–Down.

[3] CD refers to 'Vitalized mode of Force', a term coined by Dobell for '*the force manifested in the production, maintenance, growth, repair, and reproduction of the animal organism*' (Dobell 1861, p. 9). Dobell argued that 'the V.M.F. may be altered in its attributes of quantity and quality by numerous causes' (*ibid.* pp. 40–1).

[4] CD refers to the following passage in Dobell 1861, p. 46:

> If I am right in affirming that the constitution of the animal, both in material and force, is dependent on the conditions of the external world, from which that force and material are accumulated, and if I am right in concluding that some of this force passes into the germs, and thus on into the organisms of the next generation; then the law of the *conditions of existence*, insisted on by [Georges] Cuvier, becomes invested with a new dignity and importance; for not only will these conditions determine the existence of the organic being which they surround, but will carry their influence forward to the next generation, which will thus be assimilated to the conditions into which it is born. In this manner, there will be maintained a constant correlation between the forms of organic life brought into the world and the conditions on which they depend for

their existence. It does not appear that Mr. Darwin has recognised the influence which may thus be exercised by the 'conditions of life.' If my hypothesis is correct, it will lend new importance to his theory of 'natural selection,' and it will supply, in part at least, an element which some of his critics have thought wanting.

5 CD's assessment of the 'Direct and definite action of the external conditions of life' was given in chapter 23 of *Variation* (*Variation* 2: 271–92). CD wrote a draft of the chapter between September and November 1863 (see 'Journal' (Appendix II)).

6 See *Origin*, pp. 131–4, 206. In *Origin*, p. 132, CD stated: 'How much direct effect difference of climate, food, &c., produces on any being is extremely doubtful. My impression is, that the effect is extremely small in the case of animals, but perhaps rather more in that of plants.' CD revised subsequent editions of *Origin* to give more significance to the influence of external conditions (see Peckham ed. 1959, pp. 275–80). See also *Correspondence* vol. 10, letters to J. D. Hooker, 18 March [1862] and 26 [March 1862].

7 CD had written to James Paget for information on supernumerary digits (see letter from James Paget, 7 February 1863, and letter to T. H. Huxley, [8 February 1863], n. 3). Regeneration of amputated supernumerary digits in humans is discussed in *Variation* 2: 14–16.

8 The reference is to C. White 1782; see n. 9, below.

9 See letter from H. B. Dobell, 5 March 1863, nn. 1 and 2.

10 The case is given in *Variation* 2: 15.

11 See letter from James Paget, 7 February 1863, n. 1.

12 CD cites this case in *Variation* 2: 36 n. 19: 'in a large family, fingers with thickened joints were transmitted to several members during five generations; but when the blemish once disappeared it never reappeared'. CD used the case to illustrate the fact that 'in some few cases the power of reversion wholly fails'.

From J. D. Hooker [16 February 1863][1]

Royal Gardens Kew
Monday.

Dear Darwin.

Very glad indeed to hear excellent acct. of the London trip.[2] I return the list marking out what we cannot give—[3] by all means send the Tax cart as you propose, *& whenever you please*—with mats &c. if you chuse a fine day they cannot hurt.[4]

Thanks for Asa Grays letter[5]—it is impossible to put him right.— we cannot tell him the plain truth—that England does not *hate* America the least bit, only despises & thinks very slightingly of her— we think ourselves far too superior to be able to hate her, she hates us with the hatred of envy— we have not enough to envy to make hate of. No paper tries to incite *hatred*, that I have seen, but most strive to keep up feeling of contempt.[6] The notion of Democracy of plants is good, but I demur to a Commonwealth & Democracy being synonymous.[7]

Please return the list with the man who brings the tax cart

I have had some capital pickings out of the Wedgewood Medallions some of the seals are most beautiful things[8]

Ever yours affec | J D Hooker

I will come & see your Hot house one Saturday soon.[9]

DAR 101: 103–4

CD ANNOTATIONS
3.1 Please . . . cart] *double scored ink*
Top of letter: '*Seeds*' *pencil*; 'Slide Hothouse | Bell-glasses.' *ink*

[1] The date is established by the relationship between this letter and the letters to J. D. Hooker, 15 February [1863] and [21 February 1863]; the intervening Monday was 16 February.
[2] See letter to J. D. Hooker, 15 February [1863]; CD had been in London from 4 to 14 February 1863 (see 'Journal' (Appendix II)).
[3] The list, which CD had apparently given to Hooker during his visit to the Royal Botanic Gardens, Kew (see letter to J. D. Hooker, 15 February [1863] and n. 6), has not been found; however, a contemporary list of hothouse plants in CD's hand, including 'Gloxinia droopy and upright' (see letter to J. D. Hooker, 15 February [1863], n. 9) is in DAR 255. For lists of CD's hothouse plants, see Appendix VI.
[4] See letter to J. D. Hooker, 15 February [1863].
[5] In his letter to Hooker of 15 February [1863], CD enclosed the letter from Asa Gray, 27 January 1863.
[6] See letter from Asa Gray, 27 January 1863; Hooker's views on the American Civil War had prevented him from discussing politics in his own correspondence with Asa Gray (see *Correspondence* vol. 10, letter from J. D. Hooker, [19 January 1862]). For Gray's views on the British response to the war, see *Correspondence* vol. 10, letter from Asa Gray, 10 November 1862, and this volume, letter from Asa Gray, 1 September 1863.
[7] In his letter of 27 January 1863, Gray stated that 'the vegetable kingdom does not *culminate* as the *animal* kingdom does. It is not a *kingdom*, but a commonwealth—a *democracy*'.
[8] Hooker had started to collect Wedgwood ware (see *Correspondence* vol. 10, letter from J. D. Hooker, [27 or 28 December 1862]). In his letter of 13 January [1863], CD informed Hooker that somewhere in Down House there was 'a whole Box of small Wedgwood medallions', and, 'if worth pocketing', Hooker should have them; CD may have given them to Hooker when he visited the Royal Botanic Gardens, Kew, on 11 February 1863 (see letter to J. D. Hooker, 15 February [1863] and n. 3).
[9] See letter to J. D. Hooker, 15 February [1863]. The hothouse was built behind Down House, in the kitchen garden beyond the lawn (*LL* 1: 321). Hooker visited Down on Sunday 22 March 1863 (see letters from J. D. Hooker, [15 March 1863] and [24 March 1863]).

To T. H. Huxley 16 February [1863][1]

Down. | *Bromley.* | *Kent. S.E.*
Feb. 16th

My dear Huxley

I called twice in Jermyn St. but unfortunately found you once lecturing & secondly out.—[2] It is by no means the carpal or tarsal bones which are increased in number, but generally only the digits & metacarpals.[3] Indeed until the whole hand becomes doubled or bifurcated, it appears that the carpal bones are not increased. This same doubling puzzles me; but yet, I think, the wonderfully strong inheritance & **thrice** repeated growth shows that there is something in the case.[4] May I say that the digits, (divided by many joints in the rays) are indefinite in number & very generally more than five in the pectoral fins of Fishes? How are Sharks &c in this respect? These being one of oldest orders would be best.—[5]

Now there is one other point, for which I shd. be very glad of information & bears on a remark made by you: I am told that Roget (no good authority) says in

Bridgewater Treatise that in Frogs or Toads there is a rudiment of a sixth digit.[6] Have you any specimen in spirits? I think I remember a tubercle behind. If there is one, could you spare time to dissect this tubercle & see whether in bone or other respect there is any reason to suppose that this really is rudiment of 6th digit. I have *heard* vaguely of two cases of six-toed frogs on *hinder* limb. If it should turn out a rudiment, you will, I know, allow me to quote you.

Ever yours | C. Darwin

Imperial College of Science, Technology and Medicine Archives (Huxley 5: 200)

[1] The year is established by the relationship between this letter and the letter from T. H. Huxley, 20 February 1863.

[2] Huxley was professor of natural history at the Royal School of Mines, Jermyn Street, London, and Hunterian professor of comparative anatomy at the Royal College of Surgeons (*DSB*). CD had taken breakfast with Huxley on 8 February 1863, while staying in London (see letter to T. H. Huxley, [8 February 1863]).

[3] In his letter to Huxley of [8 February 1863], CD discussed the anatomical features of polydactylism. CD was collecting information on polydactylism for chapter 12 of *Variation* (see *Variation* 2: 12–14; see also letter to J. J. Briggs, 2 February [1863], and letter from James Paget, 7 February 1863 and n. 1). CD began writing a draft of the chapters on inheritance for *Variation* (*ibid.* 2: 1–84) on 23 January 1863 (see 'Journal' (Appendix II)).

[4] In his discussion of polydactylism in *Variation* (see n. 3, above), CD reported a privately communicated case of an additional digit in a child regrowing after two amputations (*Variation* 2: 14–15).

[5] In *Variation* 2: 16, CD wrote: 'fishes sometimes have in their pectoral fins as many as twenty metacarpal and phalangeal bones, which, together with the bony filaments, apparently represent our digits with their nails'. CD did not mention the case of sharks in *Variation*, but made the general point that the latent tendency to form additional digits in mammals, including humans, suggested reversion to 'an enormously remote, lowly-organised, and multidigitate progenitor' (*Variation* 2: 17).

[6] The reference is to the 'Bridgewater Treatise' written by Peter Mark Roget, entitled *Animal and vegetable physiology considered with reference to natural theology* (Roget 1834, 1: 544). The 'Bridgewater Treatises' comprised eight works on natural theology published between 1833 and 1836 under the terms of a bequest made to the Royal Society of London by Francis Henry Egerton, the eighth earl of Bridgewater. According to the terms of the bequest, each author was to write and publish a treatise 'On the Power, Wisdom and Goodness of God, as manifested in the Creation' (*EB*; see also Topham 1993).

To John Scott 16 February [1863][1]

Down Bromley Kent
Feb. 16

My dear Sir

Absence from home has prevented me answering you sooner.[2] I should think that the capsule of Acropera had better be left till it shews some signs of opening, as our object is to judge whether the seeds are good; but I should prefer trusting to your better judgment.[3] I am interested about the Gongora which I hope hereafter to try myself as I have just built a small hot-house.—[4]

Asa Gray's observations on the Rostellum of Gymnadenia are very imperfect, yet worth looking at.[5] Your case of Imatophyllum is most interesting; even if the sport does not flower it will be worth my giving.[6] I did not understand, or I had forgotten, that a single frond on a fern will vary; I now see that the case does come under bud-variation & must be given by me.[7] I had thought of it only as proof inheritance in Cryptogams; I am much obliged for your correction & will consult again your paper & Mr Bridgeman's.[8]

I enclose vars. of Maize from Asa Gray.[9] Pray do not thank me for trusting you: the thanks ought to go the other way: I felt a conviction after your first letter that you were a real lover of Natural History.[10] If you can advance good evidence shewing that bi-sexual plants are more variable than uni-sexual, it will be interesting.[11] I shall be very glad to read the discussion which you are preparing.[12] I admit as fully as any one can do that cross-impregnation is the great check to endless variability; but I am not sure that I understand your view. I do *not* believe that the structure of Primula has any necessary relation to a tendency to a Diœcious structure; but seeing the difference in the fertility of the two forms, I felt bound unwillingly to admit that they might be a step towards Diœciousness; I allude to this subject in my Linum paper.[13] Thanks for your answers to my other queries. I forgot to say that I was at Kew the other day, & I find that they can give me Capsules of several Vanda[14]

Will you present my compliments to Mr Macnab & ask him whether he has ever sown seeds of any weeping trees; & if so whether he would have the kindness to inform me how far this character was inherited.[15] I am much pressed with letters & have therefore written as briefly as I could—

Pray believe me | My dear Sir | Yours very faithfully | Ch Darwin

When Acropera-pod is sent please *register* it in an old match-box, & I enclose stamps; as I have no right to trouble you with my whim to make the capsule extra safe—

Small red or white w$^{d.}$ be good to cross with *large* other-coloured kinds; if you can *force* forward the latter.[16]

A. Gray unfortunately does not name any of the maize: but calls the small grain "precocious northern form good for popping"; & I imagine the transparent grains are the "sweet corn" with starch replaced by glucose.—[17]

You had better tally them by colour & size & say how all are N. U. States—

LS(A)[18]
DAR 93: B55, B81–2

[1] The year is established by CD's reference to having visited the Royal Botanic Gardens, Kew, and by his request for information on weeping trees from James McNab (see nn. 14 and 15, below, and the letter from John Scott, 3 March 1863).

[2] Scott's letter has not been found. CD had been in London for ten days (see n. 14, below).

[3] Scott's letter has not been found; however, in his letter of 18 February [1863], Scott stated that in his 'last' he had asked if he should send the capsule of *Acropera* as it might be 'sometime before it matured'. Scott had succeeded, where CD had failed, in effecting fertilisation in *Acropera loddigesii* (see *Orchids*, pp. 203–10; see also *Correspondence* vol. 10, letter from John Scott, 11 November 1862 and n. 3, and letter to John Scott, 12 November [1862], and this volume, letter from John Scott, 6 January

1863, nn. 3 and 4). Scott's experiments led CD to revise his former conclusion that the genus *Acropera* was dioecious ('Fertilization of orchids', p. 153 (*Collected papers* 2: 150)). See also letter from John Scott, 18 February [1863] and nn. 4 and 5.

4 Work commenced on the construction of the hothouse in the kitchen garden at Down House towards the end of January 1863 (see Appendix VI). Scott had also been attempting to effect pollination in species of *Gongora* (see letter from John Scott, 6 January 1863).

5 Asa Gray's observations on *Gymnadenia tridentata* appeared in A. Gray 1862a, p. 426, and A. Gray 1862c, p. 260. See also letter to John Scott, 21 January [1863].

6 In his letter to Scott of 19 December [1862] (*Correspondence* vol. 10), CD asked Scott to assist him by providing examples of bud-variation. Scott's letter, to which this is a reply, has not been found; however, in *Variation* 1: 385, CD stated:

> Mr. John Scott informs me that in 1862 *Imatophyllum miniatum*, in the Botanic Gardens of Edinburgh, threw up a sucker which differed from the normal form, in the leaves being two-ranked instead of four-ranked. The leaves were also smaller, with the upper surface raised instead of being channelled.

7 In his letter of 16 January 1863, Scott asked CD whether he intended including ferns in his work on the variations of plants and animals, pointing out that the 'facilities they afford for reproducing variations is wonderful'. In his letter to Scott of 21 January [1863], CD stated that he had supposed that ferns did not produce bud-variations; Scott evidently explained the case of bud-variation in ferns in his reply, which has not been found. In *Variation* 1: 383, CD stated:

> Cryptogamic plants are liable to bud-variation, for fronds on the same fern are often seen to display remarkable deviations of structure. Spores, which are of the nature of buds, taken from such abnormal fronds, reproduce, with remarkable fidelity, the same variety, after passing through the sexual stage.

8 The reference is to Scott 1862a and Bridgman 1861. Both papers, of which there are annotated copies in the Darwin Pamphlet Collection–CUL, are cited in *Variation* 1: 383.

9 Gray sent the New England maize seeds with his letter to CD of 29 December 1862 (*Correspondence* vol. 10). See also n. 16, below. CD began writing on variation in maize in November 1862, and urged Scott to repeat Karl Friedrich von Gärtner's experiments on crossing varieties of maize (see *Correspondence* vol. 10, letters to John Scott, 19 November [1862] and 11 December [1862], and this volume, letter from John Scott, 6 January 1863, n. 20). In his letter to Scott of 21 January [1863], CD suggested that Scott should cross Gray's maize with 'some large kind of different colour'.

10 Letter from John Scott, 11 November 1862 (*Correspondence* vol. 10).

11 Scott gave a more detailed account of his views on variability in hermaphrodite and diclinous plants in his letter of 3 March 1863. See also letters from John Scott, 16 January 1863, n. 11, and 3 March 1863, n. 8. See also n. 12, below.

12 In a missing letter, Scott evidently informed CD that he was preparing a paper on the relationship between the form of reproduction and the heritability of variation in plants (see also letter from John Scott, 16 January 1863 and n. 11).

13 In 'Dimorphic condition in *Primula*', p. 95 (*Collected papers* 2: 61–2), CD suggested that dimorphism was probably a stage towards dioeciousness in plants. In 'Two forms in species of *Linum*', p. 83, which was read before the Linnean Society on 5 February 1863, CD argued (*Collected papers* 2: 105):

> That in some cases this dimorphism may be a step towards a complete separation of the sexes, I will not dispute; but good reasons could be assigned to show that there is no necessary connexion between reciprocal dimorphism and a tendency to dioecious structure. Although good is gained by the inevitable crossing of the dimorphic flowers, yet numerous other analogous facts lead me to conclude that some other quite unknown law of nature is here dimly indicated to us.

See also letter from John Scott, 18 February [1863].

14 CD had been in London from 4 to 14 February 1863 (see 'Journal' (Appendix II)). Emma Darwin recorded in her diary (DAR 242) that they visited the Royal Botanic Gardens, Kew, on 11 February 1863. CD wanted *Vanda* capsules for comparison with the *Acropera* capsule (see n. 3, above, and *Correspondence* vol. 10, letter to J. D. Hooker, 18 [November 1862]). See also *Correspondence* vol. 10, letter from J. D. Hooker, 20 August 1862, letter to J. D. Hooker, 22 [August 1862], and letter from John Scott, [20 November – 2 December 1862].

15 James McNab was curator of the Royal Botanic Garden, Edinburgh, and Scott's immediate superior (R. Desmond 1994); Scott was foreman in the propagating department. CD was preparing a draft of chapter 12 of *Variation* (see 'Journal' (Appendix II)) and was collecting information on the weeping habit of trees to demonstrate the variability of inheritance (see *Variation* 2: 17–19). See also letters from Isaac Anderson-Henry, 31 January 1863 and nn. 7 and 8, and 14 February 1863 and n. 3.

16 CD refers to varieties of maize, and to Scott's proposed crossing experiments (see n. 9, above).

17 This sentence does not appear in the extant part of the letter with which Gray enclosed the maize seeds (*Correspondence* vol. 10, letter from Asa Gray, 29 December 1862). See also n. 9, above.

18 The letter, up to the valediction, is in Emma Darwin's hand; thereafter it is in CD's hand.

To Roland Trimen 16 February [1863][1]

Down. | *Bromley.* | *Kent. S.E.*
Feb. 16th

Dear Sir

I have thought you would like to see copy enclosed of letter by Prof. Harvey giving names of your two orchids Pl. V. & VI, which were unnamed.—[2] Now that I hear that in Satyrium the nectaries belong to the true Labellum; the relation of the parts is to me very puzzling: discs, pollen-masses & stigmatic surface seem all on the wrong side.—[3] If you pursue the subject, I hope you will observe whether there is any relation (as in English Orchids) between the rapidity of the setting of the viscid matter & nectar being stored ready for suction or confined in cellular tissue.—[4]

I was at Kew 2 or 3 days ago & was telling Dr. Hooker & Mr H. Gower of your work;[5] they expressed a strong wish to try whether they could not cultivate some of your wonderful forms; & tempted me by saying that if they could flower them, I shd have plants to examine. I said I would mention the subject to you; but that of course I doubted whether you had time & inclination to get them dug up.— They said the roots might be packed in almost dry peaty soil or charcoal in mass, & sent to "Royal Gardens Kew, London", marking what they were, i.e. terrestrial orchids from the Cape.— They ought to be dug up, when completely dormant after seeding over.— It certainly would be a treat to see a blooming Satyrium, or Disperis & that odd unnamed form![6] They said the safest way of all, but more troublesome, to send them, would be to plant them in pots in a box, with a little glazed windows on two sides under charge of some passenger. The heat starting them would be the great risk. But it is not at all likely you could spare time from your own pursuits.[7]

Pray believe me, my dear Sir | Yours sincerely & obliged | Ch. Darwin

If you come across Bonatea pray study it—it seems most extraordinary in description.—[8]

Royal Entomological Society of London (Trimen papers, box 21: 55)

[1] The year is established by the relationship between this letter and the letter from W. H. Harvey, 3 February 1863.

[2] CD refers to the letter from W. H. Harvey, 3 February 1863. William Henry Harvey was a specialist in the taxonomy of South African plants. See letter from W. H. Harvey, 3 February 1863, n. 1, and letters to Roland Trimen, 31 January [1863] and 23 May [1863].

[3] Trimen added a marginal annotation to CD's letter concerning this point: 'These remarks are very much to the point. It always seemed to me that what is termed the labellum in Satyrium was much more like an upper sepal.' In his letter of 3 February 1863 (see n. 2, above), Harvey stated: 'In Satyrium the *two spurres affair* is a *true* labellum—the sepals & petals small & crowded together at the *front* of flower—the *opposite* to Disa.'

[4] CD's 'bold hypothesis' (*Orchids* p. 50) was that in species in which it took a long time for the viscid matter on pollinia to set, the nectar was more likely to be stored in cellular tissue, relatively inaccessible to insects, to allow time for the pollinia to attach firmly. Conversely, in species where the viscid matter set quickly, the nectar would be stored ready for suction. CD commented that, if true, this was 'a singular case of adaptation' (*Orchids* p. 53). See *Orchids*, pp. 50–3, 281–2.

[5] According to Emma Darwin's diary (DAR 242), CD visited the Royal Botanic Gardens, Kew, on 11 February 1863. Joseph Dalton Hooker was assistant director of the gardens, and William Hugh Gower was a gardener with specialist knowledge of orchids (R. Desmond 1994).

[6] In his letter of 16 March 1863, Trimen promised to send CD specimens of Cape orchids the following spring.

[7] Trimen worked as a civil servant in Cape Town, South Africa (*DNB*). See also letter from Roland Trimen, 16 March 1863.

[8] Trimen published his observations on *Bonatea* in Trimen 1864.

To George Henry Turnbull [16? February 1863][1]

Down
Monday

My dear Sir

My little hot-house is finished & you must allow me once again to thank you sincerely for allowing Horwood to superintend its erection.[2] Without his aid I should never have had spirit to undertake it;—and if I had should probably have made a mess of it.— It will not only be an amusement to me, but will enable me to try many little experiments, which otherwise would have been impossible.—[3]

With sincere thanks | My dear Sir | Yours sincerely & obliged | Ch. Darwin

DAR 261 (DH/MS 11: 5)

[1] The date is established by the relationship between this letter and the letters to J. D. Hooker, 30 January [1863] and 15 February [1863], and by reference to the payment of £4 5*s*. to John Horwood, on 16 February 1863 (see n. 2, below); in 1863, 16 February was a Monday.

[2] Horwood was Turnbull's gardener at The Rookery, Down, Kent. Payments to Horwood, 'for Hothouse', are detailed in CD's Account book–cash accounts (Down House MS). CD spent a total of £85 11*s.* 1*d.* constructing and equipping the hothouse in 1863 (see also Appendix VI).
[3] An indication of the plants on which CD intended to experiment is given by two contemporary lists of hothouse plants in CD's hand, reproduced in Appendix VI.

To T. H. Huxley [after 16 February 1863][1]

Do not forget, whenever you can screw out time— rudiment of 6th toe on *[front]* or Hind foot of Batrachian.—[2]

Incomplete

Imperial College of Science, Technology and Medicine Archives (Huxley 5: 202)

[1] The date is established by the relationship between this letter and the letter to T. H. Huxley, 16 February [1863].
[2] See letter to T. H. Huxley, 16 February [1863].

To Charles Lyell 17 [February 1863]

My dear Lyell

The same post that brought the enclosed,[1] brought Dana's pamphlet on same subject.—[2] The whole seems to me utterly wild. If there had not been the foregone wish to separate man, I can never believe that Dana or anyone wd have relied on so small a distinction, as grown man not using forelimbs for locomotion, seeing that monkeys use these limbs in all other respect for same purpose as man.—[3] To carry on analogous principles (for they are not identical, for in Crustacea the cephalic limbs are brought close to mouth) from Crustacea to the classification of Mammals seems to me madness.—[4] Who would dream of making fundamental distinction in Birds, from fore-limbs not being used at all in Birds, or used as fins in Penguins & for flight in other Birds?—

I get on slowly with your grand work,[5] for I am overwhelmed with odds & ends & letters—so farewell | My dear Lyell | C. Darwin—

17th Down Bromley Kent

Endorsement: '17—Feby 1863'
American Philosophical Society (288)

[1] CD may have enclosed the letter from J. D. Dana, 5 February 1863.
[2] CD refers to Dana 1863c, p. 66; there is a presentation copy of this work, bearing a New Haven postmark, in the Darwin Pamphlet Collection–CUL.

[3] In Dana 1863c, p. 66, James Dwight Dana stated that:

> while all other Mammals have both the anterior and posterior limbs organs of locomotion, in Man the anterior are transferred from the *locomotive* to the *cephalic* series. They serve the purposes of the *head*, and are not for locomotion ... Man, in this, stands *alone* among Mammals.

[4] Dana argued that 'cephalization' (see n. 3, above) was a fundamental principle in the classification of zoological life, illustrating his point with reference to crustaceans (Dana 1863c, p. 66).

[5] CD was reading *Antiquity of man* (C. Lyell 1863a). See also letters to Charles Lyell, 4 [February 1863] and 6 March [1863].

From Daniel Oliver 17 February 1863

Royal Gardens Kew
17.ii.1863

My dear Sir

Next to save you throwing away valuable time, the connection with the phyllotaxy difficulty it is not worth y^r. while reading Braun's Essay (Verjüngung) excepting at p. 116.—[1] the Essays referred to in note same page may throw further light,[2] but I think you have enough now for Dr Falconer[3]

Ever Yours Sincerely | D. Oliver

DAR 173: 20

[1] The reference is to Henfrey trans. 1853. In the letter to J. D. Hooker, 13 January [1863], CD mentioned that he had been attempting to read this English translation of Braun 1851, continuing: 'literally I *cannot* read it'. Apparently, CD had sought Oliver's help in searching the book for information on phyllotaxy (see n. 3, below). An annotated copy of Henfrey trans. 1853 is preserved in the Darwin Library–CUL (see *Marginalia* 1: 366–7).

[2] Braun cited Bravais and Bravais 1838, Naumann 1845, Schleiden 1842, and Silfverstråhle 1839 (Henfrey trans. 1853, pp. 116–17).

[3] CD's interest in phyllotaxy was stimulated by Hugh Falconer's discussion of *Origin* in his article 'On the American fossil elephant of the regions bordering the Gulf of Mexico', published in the *Natural History Review* in January 1863 (Falconer 1863a). See letter to J. D. Hooker, 13 January [1863] and n. 22. CD may have discussed Falconer's statements with Oliver during his visit to the Royal Botanic Gardens, Kew, on 11 February 1863 (Emma Darwin's diary (DAR 242)). CD wanted to understand the force of Falconer's objections to natural selection (see letter to Daniel Oliver, 20 [February 1863]), and began his own research on phyllotaxy. CD prepared a rough draft of a paper on phyllotaxy on 9 June 1863, which is in DAR 51: A1–32, together with a series of undated diagrams; however, the work was never published.

From George Henry Kendrick Thwaites 17 February 1863

Peradenia, Ceylon
17th Feby 1863

My dear M^r. Darwin,

It gives me very sincere pleasure to be able to gratify the wishes expressed in your welcome and interesting letter of the 29th. Dec^r.[1] and let me, in the first place, thank you for the information respecting the dimorphism of *Cinchona*. I shall not fail to act upon it when our plants come into flower.[2]

Dimorphism would appear to be common in the *Rubiaceæ*, & there would seem to be every degree of it as regards development of stamens & pistil in the respective flowers: In the genus *Discospermum* the ♂ flowers are always barren, though the ovary contains about 2 abortive ovules in each loculis, & it must have been from the examination of one of these that the genus *Diplospora* (synonymous with Discospermum) was constituted; The ovary of the ♀ flower contains several ovules.[3]

I send you in weak spirits the two forms of flowers of *Limnanthemum Indicum*, Enum.—p. 205[4] (hand Grisebr)—L. Kleinianum & L. Wightianum, Grisebr.)[5] The pollen is alike in both forms  and I find the same number of ovules (70–80) in the ovaries of both. I have not had an

Incomplete[6]
DAR 109: A94

CD ANNOTATIONS

1.1 It gives . . . flower. 1.4] *crossed ink*; 'General Cinchona' *added blue crayon*; 'Keep gradation' *added red crayon*; 'On into Dioiceous' *added blue crayon*

2.3 In . . . constituted; 2.6] *scored brown crayon*

2.6 The ovary . . . ovules. 2.7] 'more in Cinchona, as it does not follow at all /the two forms that [*illeg*] render /same/ plants w^{d} [*illeg del*] diœcious/'[7] *added ink*

3.1 I send . . . had an 3.4] *crossed pencil*

Top of letter: 'The meaning of different positions of anther in Sethia, is to [*over* 'the'] rake short anther on proboscis; whereas body w^{d} be dusted by long-anthers.—' *ink, del pencil*; 'Rubiaceæ' *pencil*; 'Limnanthemum' *pencil, del pencil*

1 In his letter to Thwaites of 29 December [1862] (*Correspondence* vol. 10), CD asked for examples of bud-variation; CD was writing a draft of chapter 11 of *Variation*, 'On bud-variation, and on certain anomalous modes of reproduction and variation' (*Variation* 1: 373–411), which he began on 21 December 1862 (*Correspondence* vol. 10, Appendix II). CD was seeking information on the question of whether introduced flowers from different climates were particularly apt to produce bud-variations. CD also asked Thwaites to repeat the names of the two genera (*Sethia* and *Limnanthemum*) that he had mentioned in an earlier letter as being dimorphic (*Correspondence* vol. 10, letter from G. H. K. Thwaites, 15 May 1862), and requested him to send specimens of the different forms (see n. 5, below, and letter to G. H. K. Thwaites, 30 March [1863]).

2 In his letter to Thwaites of 15 June [1862] (*Correspondence* vol. 10), CD noted that some trees of the genus *Cinchona* had long pistils and others had short pistils. CD stressed the probable importance of reciprocal pollination between the two forms of *Cinchona*, in order to get good seed and seedlings. He asked Thwaites, who was superintendent of the Peradeniya botanic gardens, Ceylon, to experiment with crossing the two forms, adding: 'The growth of Cinchona is so important for Mankind, that I am sure you will excuse my making this suggestion.' *Cinchona* was a source of quinine, which was widely used to treat malaria. In his letter to Thwaites of 29 December [1862] (*ibid.*), CD confirmed what he had previously written regarding dimorphism in *Cinchona*, adding that he now had 'proof that some of these dimorphic plants are *absolutely* sterile with their own-form pollen'.

3 *Cinchona*, *Discospermum*, and *Diplospora* are all members of the family Rubiaceae. Thwaites's information on dimorphism in Rubiaceae is cited in *Forms of flowers*, p. 286.

[4] The reference is to *Enumeratio plantarum Zeylaniae: an enumeration of the Ceylon plants* (Thwaites 1858–64).

[5] CD had asked Thwaites to send him specimens of dimorphic plants (see n. 1, above). *Limnanthemum indicum*, *L. kleinianum*, and *L. wightianum* were named by the German botanist August Heinrich Rudolph Grisebach (Grisebach 1845, p. 139); *L. kleinianum* and *L. wightianum* are synonyms for *L. indicum* (*Index Kewensis*). CD's notes on these specimens are in DAR 110: B9–10; see also *Forms of flowers*, p. 116.

[6] For an indication of the contents of the missing portion of the letter, see CD's annotations, and the letter to G. H. K. Thwaites, 30 March [1863].

[7] CD noted that, as one of the two forms of *Discospermum* was always barren, the species appeared to be strictly dioecious, and did not come within his definition of heterostyly (see *Forms of flowers*, p. 286).

To T. H. Huxley 18 [February 1863][1]

Down
18th—

My dear Huxley

Hurrah the Monkey Book has come—very many for your present of it.—[2] I long to read it, but am determined to refrain till I have finished Lyell, & I have got only half through it.[3] The Pictures are splendid.—

Ever yours | C. Darwin

Imperial College of Science, Technology and Medicine Archives (Huxley 5: 173)

[1] The date is established by reference to the publication of T. H. Huxley 1863b on 20 February 1863 (*Publishers' Circular*, 16 February 1863, p. 85, and *Athenæum*, 21 February 1863, p. 261).

[2] Huxley's *Evidence as to man's place in nature* (T. H. Huxley 1863b) was published in February 1863 (see n. 1, above) by the London booksellers and publishers Williams & Norgate. CD's annotated copy is in the Darwin Library–CUL (see *Marginalia* 1: 423–4).

[3] CD refers to Charles Lyell's *Antiquity of man* (C. Lyell 1863a) which he received at the beginning of the month (see letter to Charles Lyell, 4 [February 1863] and n. 1).

From John Scott 18 February [1863][1]

Edinburgh | Botanic Gardens
18th. Feby

Sir.

In my last, I asked you if I might now send the capsule of *Acropera*, thinking it might be sometime before it matured. I am now, however, forced to do so, in consequence of its unexpected partial dehiscence.[2] A few days before I last wrote, it changed colour slightly around the apex, this has slowly increased, though still as you will observe confined to the upper part. I, thus, had not the least suspicion of its bursting for some weeks at least, and I can assure you I was not a little vexed yesterday on observing what had happened, as I have thus lost a number of seeds. It is, I think, unusual to see them burst, before being at least partially coloured over. Perhaps the abnormalities of placentæ, utriculii, &c. which you first drew our attention to in this orchid, may have something to do with this.[3] I will be anxious to hear the results of your dissection.

I have no doubt you will be somewhat astonished—though not entirely unprepared—on actually seeing the straight, plump capsule produced by Acropera, in view of your careful dissections; and certainly, fully justifiable deductions.[4]

I have just placed a few seeds, which dropped into my hand, when I sent it off: under the microscope, a slight majority of these look to be quite perfect. These—30 or so—I will sow, and see if I can succeed in raising a few plants, which would be interesting as affording a chance for the solution of your query, as to what the other sexual forms of Acropera may be.[5] This chance will be increased if—as I believe—it have never been raised from seed in this country, all its representatives being mere divisions of the originally imported plant. Now, believing, as I do, that merely vegetative multiplication will never render the *slightest degree* of *permanency* to a *variation* when *tested* by *seeds*. It will, I believe, from information I have been collecting on this point—after years of this kind of multiplication have as great a tendency to produce the form from which it originally sported, as it would if tried by seeds while still in organic connection with its parent, i.e. supposing a "bud-variation."[6]

If then, Acropera, be as you are inclined to believe in a fluctuating state, we might expect some of the progeny at least to present the other forms. An important question, however, here presents itself. If Acropera be the male incipient form,[7] and not yet perfectly sterile, will it *when self-impregnated produce both forms*? I am inclined to believe it will not. I think, I have indications of this in a variety of the Primrose in our gardens, all of which are long-styled. You, however, will be able to correct me on this point.[8] I have great hopes that Parthenogenisis in plants, if established satisfactorily, will show this. I am led to hope for this result from my observation on the reproduction of variations in Cryptogams.[9]

Excuse this hurried note. | I remain | Sir, | Yours very respectfully | John Scott.

I have just received your letter with varieties of Maize.[10] I am glad that you have been able to afford me so many.

DAR 177: 84

CD ANNOTATION

4.5 I think, . . . long-styled. 4.6] *scored brown crayon*

1 The year is established by the relationship between this letter and the letter from John Scott, 3 March 1863.

2 Scott's letter has not been found; however, see the letter to John Scott, 16 February [1863] and n. 3. Dehiscence is the botanical term for the opening of a seed capsule.

3 Scott refers to CD's description of *Acropera* in *Orchids*, pp. 207–8.

4 CD had repeatedly failed in his attempts to effect pollination in two species of *Acropera*, namely, *A. loddigesii* and *A. luteola*. From this fact, and from dissections of *A. loddigesii* and *A. luteola*, CD deduced that all the flowers he had examined were male and had come from male plants, and that a female or hermaphrodite form of the same species must exist either under the same or under a

different name (see *Correspondence* vol. 9, letter to Daniel Oliver, 30 November [1861], and *Orchids*, pp. 203–10). Additionally, CD's study of the related orchid *Catasetum tridentatum*, revealed that this species had three forms of flowers: male, female, and hermaphrodite (see 'Three sexual forms of *Catasetum tridentatum*', and *Orchids*, pp. 236–48), which gave further credence to this interpretation. However, in November 1862, Scott wrote to CD informing him that he had succeeded in effecting pollination in *A. loddigesii*, by cutting open the stigmatic chamber and inserting the pollen masses (see *Correspondence* vol. 10, letter from John Scott, 11 November 1862, and letter to John Scott, 12 November [1862]), forcing CD to revise his earlier conclusion (see n. 5, below).

5 In *Orchids*, p. 209, CD remarked:

> What the female or hermaphrodite form of the Acropera luteola may prove to be—whether resembling in most respects the male, or whether it be at present named and masked as some distinct genus—it is impossible to say.

After Scott had sent capsules of *Acropera loddigesii* containing viable seed, obtained by pollinating some flowers with pollen from the same plant, CD revised his earlier conclusion that the sexes in this genus were separate (see n. 4, above, letter to John Scott, 20 [February 1863], and 'Fertilization of orchids', p. 153 (*Collected papers* 2: 150)).

6 Scott hoped to demonstrate that *Acropera* plants raised by vegetative division for a number of generations were more likely to reproduce individual variations than to reflect the normal range of forms of *Acropera* when produced by sexual reproduction. Scott and CD discussed at length the subject of variation and reproduction (see letters to John Scott, 21 January [1863] and 16 February [1863], and letter from John Scott, 3 March 1863). See also n. 9, below.

7 In his letter to Scott of 20 [February 1863], CD stated: 'It is more likely that I have made some dreadful blunder about Acropera than that it sh[d] be male not yet a perfect male'.

8 Throughout 1860 and 1861, CD had carried out a study of flower dimorphism and reproduction in *Primula*; he read his paper on the subject, 'Dimorphic condition in *Primula*', before the Linnean Society on 21 November 1861. CD continued to work on the subject during 1862 (see *Correspondence* vol. 10, letter to John Scott, 19 November [1862]). See also letters to John Scott, 16 February [1863] and 20 [February 1863].

9 In a paper on fern spores read before the Botanical Society of Edinburgh on 12 June 1862 (Scott 1862a), Scott argued that there would be a greater tendency for individual variation in plants produced by asexual reproduction (including parthenogenesis) than in plants produced by sexual reproduction. Illustrating his point with reference to ferns, Scott reasoned that the fern spore, as 'an independent self-developing organism', reproduced the peculiarities of the organ on which it originated, hence explaining the 'peculiar power which the spore possesses for the reproduction of accidental variations' (Scott 1862a, pp. 216–18). See also *Correspondence* vol. 10, letters from John Scott, 15 November [1862] and 17 December [1862], and this volume, letter from John Scott, 16 January 1863.

10 In his letter to Scott of 16 February [1863], CD sent specimens of North American maize seeds (see letter to John Scott, 16 February [1863] and n. 9).

From W. B. Tegetmeier 18 February 1863

Muswell Hill | N.
Feby 18/63

My dear Sir

I hoped to have the pleasure of seeing you at the Linnean society, and much regret the cause of your absence[1]

There were one or two points connected with my little experiment that I should have been very glad to have asked you respecting, such as your suggestion to employ Carriers and Turbits, instead of Barbs and Fantails.—[2]

In accordance with your suggestion I have procured some Silk fowls (two very good black skinned hens) and a Spanish Cock.[3] I am devoting a small paddock to them and will hatch several clutches of chickens— Do you think that the two hens will be sufficient or can you suggest any modification of the experiment.— Would it be desirable or necessary to have a second set to cross their offspring with those of the first—[4] If so I would make arrangements to procure them— —

Trusting to hear of improvement in your health | Believe me | Very truly yours | W B Tegetmeier

C Darwin Esq

DAR 178: 57

[1] Although CD was in London from 4 to 14 February 1863 (see 'Journal' (Appendix II)), illness prevented his attendance at the meeting of the Linnean Society on 5 February, at which his paper 'Two forms in species of *Linum*' was read (see letter to W. B. Tegetmeier, 19 February [1863]).

[2] On 1 December 1862, the council of the Royal Society of London resolved to grant £10 to Tegetmeier for 'experiments on the cross-breeding of pigeons' (Royal Society, Council minutes, 1 December 1862). CD sent Tegetmeier a manuscript list of the crosses he had made between different pigeon varieties, adding that: 'If I were going to try again, I would prefer Turbit with Carrier or Dragon' (*Correspondence* vol. 10, letter to W. B. Tegetmeier, 27 [December 1862] and n. 4); CD explained this preference in his letter to W. B. Tegetmeier of 19 February [1863]. CD's changing views on the causes of cross and hybrid sterility had prompted him to seek further experimental evidence on the question, especially with regard to animals (see *Correspondence* vol. 10, Appendix VI). Between 1863 and 1865, Tegetmeier carried out a series of crosses between varieties of pigeon designed to test the fertility of their hybrids; he tested the progeny of three generations without finding 'any sign of want of fertility' (see letter from W. B. Tegetmeier, 29 June – 7 July 1863, and letter from W. B. Tegetmeier, 13 March 1865, *Calendar* no. 4785). CD provided a brief summary of the results of his own crossing experiments with pigeons in *Variation* 1: 192 n. 19.

[3] In his letter of 27 [December 1862] (*Correspondence* vol. 10), CD asked Tegetmeier to obtain and cross 'a Spanish Cock & a couple of *white* Silk hens' in order to ascertain whether the offspring bred well and to test for prepotency in the transmission of silky plumage. CD had conducted similar trials in 1859 and 1860 (see CD's Experimental notebook (DAR 157a), pp. 41–2, 49–50). See also *Correspondence* vol. 7, letter to W. B. Tegetmeier, 13 September [1859], and *Correspondence* vol. 8, letter to W. B. Tegetmeier, 24 [February? 1860]. CD concluded that 'Silk-fowl' seldom transmitted their silky plumage when crossed with any other variety of chicken (*Variation* 2: 67). Tegetmeier's experiments confirmed CD's observations (Tegetmeier 1867, p. 224). There is an annotated copy of Tegetmeier 1867 in the Darwin Library–CUL (see *Marginalia* 1: 800–3).

[4] See letter to W. B. Tegetmeier, 19 February [1863].

From George Maw 19 February 1863

Benthall Hall, | n.r Broseley.
19th. Feby 63

Dear Sir

I have been requested by the promoters of the The "Severn Valley Naturalists field Club" which has just been established in this neighbourhood for the purpose of collecting facts relating to & reading papers upon the natural history of the southern half of Shropshire to write and ask you as a native of our county, whether you would do us the honor of allowing us to place your name on our list of honorary members.[1] The club includes several good naturalists & I trust that our operations will be really practical. I need not say what great pleasure it will give me to be able to communicate to the committee your assent to their wishes—

I am noting down a few particulars on a rather interesting fact I think I have noticed respecting the relative characters in different localities of the different strata of the Shropshire coal field. For instance that in two seams of coal where one is of unusual thickness it is to the disparity of another & distinct seam separated from it by several yards of intermediate strata

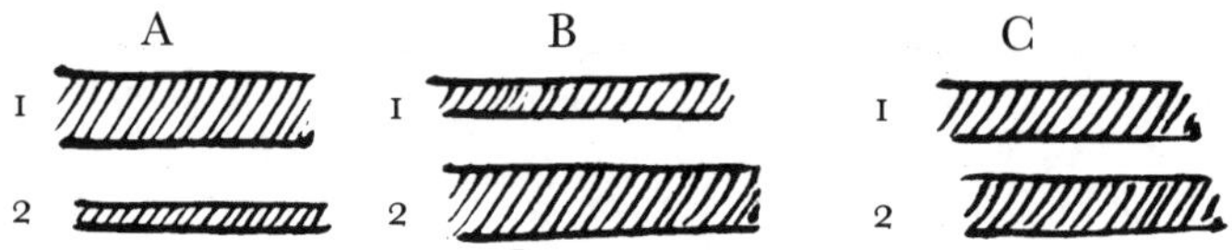

In a locality where as at A No 1 is thick No 2 is thin & where as at B No 1 is thin No 2 will be thick & at another locality C Nos 1 & 2 may be nearly equal—again Colliers in forming a pit can tell through many yards of strata the probable character of the seam of coal they are sinking to—these facts seem to imply that the several strata must have been simultaneously deposited & I fancy that this relationship of constituents will be found to exist thro many hundred feet of strata— I am collecting evidence on this point & fully expect to find that the sum of the thickness of the ten or twelve seams occurring in the fully developed parts of our coal field will vary but very slightly however much the proportionments into seams may differ in different localities— If the strata of our coal field had been formed by what we imagine to be an ordinary succession of deposits caused by an interrupted agency I do not see how this relationship amongst the constituent Strata can be accounted for

I am impressed with a vague idea of great simultaneous deposits & subsequent aggregation (whilst in suspension) into bands of coal clay Ironstone & Sandstone If every separate alternating band of Sandstone Limestone or Clay (as in the wealden for example)[2] has been the result of a distinct geological epoch—such a change in the contours of the earths surface as would be necessary to entirely shut out the source of supply of one material & bring another under disintegrating action appears inconsistent with the repose necessary to evenly deposit the thin & closely alternating bands of different materials occurring in such formations as the wealden or Lias—[3] In the formation of bands of Flints & Ironstones there must have been aggregation subsequent to deposition & yet these bands appear closely related in structure to the thin bands of Limestone & other stones occurring in the Lias &

several formations & these again insensibly merge into the structure of all stratified rocks.

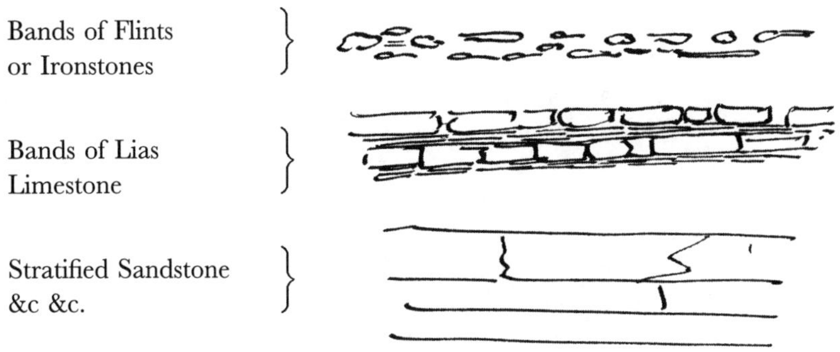

If stratified rocks are merely the result of the successive degradation of older rocks we should expect to find in the recent formations a uniform mixture equivalent to the average compositions of all the antecedent formations unless some power of re-aggregation once existed with which we now see no parallel

If such powers did exist a mere separation of strata or alternation of material does not imply a distinct act of deposit & it becomes a most interesting question (in addition to the evidences of organic remains) as to the range of strata in "a formation" included in each separate act of deposition whether the whole strata of a coal basin has been in simultaneous suspension whether deposited at two or three times or whether each stratum in which a different material occurs must be assumed to correspond with a distinct act of deposition I must ask your pardon for troubling you with these remarks but the subject is one which has deeply impressed me

Believe me Dr Sir | very truly yours. | George Maw.

Charles Darwin Esq | Down Bromley | Kent.

DAR 171: 97

[1] CD's reply, written on 23 February 1863, has not been found; however, he evidently accepted the offer of honorary membership of the Severn Valley Naturalists' Field Club (see letter from George Maw, 25 February 1863). The club was formed 'for the practical study of Natural History, Geology, and Archaeology' and held five meetings during 1863 (*Transactions of the Severn Valley Naturalists' Field Club* 1873).

[2] The Wealden is a thick series of Lower Cretaceous estuarine and freshwater deposits, named for its great development in the Weald of Kent and Sussex (*EB*).

[3] The Lias is the lowermost group of Jurassic strata, occurring across England from Dorset to Yorkshire (*EB*).

To W. B. Tegetmeier 19 February [1863]

Down Bromley Kent
Feb. 19th.

My dear Sir

I am delighted to hear that you have the Fowls:[1] as soon as you have chickens you could kill off the old Birds. I shd. think the 3 ample.— It would be better to cross some cocks & Hens of the half-breds from the two nests; so as not to cross *full* brother & sister. I have not much hope that they will be partly or wholly sterile, yet after what happened to me, I shd. never have been easy without a trial.—[2]

I suggested Turbits, because statements have been published that they are sometimes sterile with other breeds, & I mentioned Carriers, merely as a very distinct breed.[3] I thought Barbs & Fantails bad solely because I had made several crosses & found the $\frac{1}{2}$ breds perfectly fertile,—even brother & sister together. Did I send you (*I cannot remember*) a M.S. list of crosses; if so for Heaven sake return it.—[4] I get *slowly* on with my work; but am never idle.—[5]

I much wish I could have seen you at Linn. Soc; but I was that day very unwell.—[6] Pray do not forget to ask Poultry & Pigeon men (especially latter) whether they have ever matched two birds (for instance two almonds, Tumblers) & could not get them to breed, but afterwards found that both birds would breed when otherwise matched.—[7]

I hope the world goes pretty well with you.—

My dear Sir | Yours sincerely | C. Darwin

Endorsement: '*1863*'

The LuEsther T. Mertz Library at The New York Botanical Garden (Charles Finney Cox collection, Tegetmeier ser. 1: 27)

[1] Tegetmeier was about to conduct crossing experiments for CD and had informed him that Silk hens and a Spanish cock had been acquired for the trials (see letter from W. B. Tegetmeier, 18 February 1863 and n. 3, and *Correspondence* vol. 10, letter to W. B. Tegetmeier, 27 [December 1862]).

[2] In his letter of 18 February 1863, Tegetmeier asked CD for details of how to proceed with his crossing experiments (see n. 1, above). CD had previously conducted the cross that Tegetmeier was about to undertake, between a Spanish cock and a White Silk hen (see CD's Experimental notebook (DAR 157a), pp. 41–2, 49–50). The result of the experiment, carried out in 1859 and 1860, was the production of 'plenty of eggs & chickens; but two of these *seemed* to be quite sterile' (see *Correspondence* vol. 10, letter to W. B. Tegetmeier, 27 [December 1862]). CD added that at the time he was 'sadly overdone' with work, continuing: 'but [I] have ever since much reproached myself, that I did not preserve & carefully test the procre[a]tive power of these hens'.

[3] In his letter of 18 February 1863, Tegetmeier asked CD if he would explain his suggestion that it would be better 'to employ Carriers and Turbits, instead of Barbs and Fantails'. The supposed sterility of turbits when crossed with other breeds of pigeon is discussed in *Variation* 1: 192 n. 19. CD cited Temminck 1813–15 and Riedel 1824 as sources of this information. An annotated copy of Riedel 1824 is preserved in the Darwin Library–CUL (see *Marginalia* 1: 708–10).

[4] In his letter of 27 [December 1862] (*Correspondence* vol. 10), CD sent Tegetmeier a manuscript list of the crosses he and others had made with different pigeon varieties. The manuscript has not been found; however, see *Variation* 1: 192 n. 19.

5 CD refers to *Variation*, which was published in two volumes in 1868. CD had been engaged in writing a draft of the three chapters on inheritance since 23 January 1863 (*Variation* 2: 1–84); he completed the work on 1 April (see 'Journal' (Appendix II)).

6 See letter from W. B. Tegetmeier, 18 February 1863; CD refers to the meeting of the Linnean Society on 5 February at which his paper 'Two forms in species of *Linum*' was read. CD was in London from 4 to 14 February 1863 (see 'Journal' (Appendix II)).

7 See *Correspondence* vol. 10, letter to W. B. Tegetmeier, 27 [December 1862]. CD was anxious to find a case of 'two birds which when paired were unproductive, yet neither impotent', in an attempt to 'make two strains, both fertile, & yet sterile when one of one strain is crossed with one of the other strain' (*Correspondence* vol. 10, letter to T. H. Huxley, 28 December [1862]). This formed part of CD's search for experimental evidence to answer Thomas Henry Huxley's criticism of the theory of natural selection. Huxley had argued that natural selection could never be considered a *vera causa* for the origin of species until artificial selection had been shown to be capable of producing varieties of a species that were cross-sterile (see [T. H. Huxley] 1860, and *Correspondence* vol. 10, Appendix VI).

To J. D. Dana 20 February [1863][1]

Down. | *Bromley.* | *Kent. S.E.*
Feb. 20th

My dear Mr Dana

I received a few days ago your Book & this morning your pamphlet on Man & your kind letter.[2] I am heartily sorry that your head is not yet strong, & whatever you do, do not again overwork yourself.[3] Your book is a monument of labour, though I have as yet only just turned over the pages.—[4] It evidently contains a mass of valuable matter.—

With respect to the change of species I fully admit your objections are perfectly valid.[5] I have noticed them[6], excepting one of separation of countries, on which perhaps we differ a *little*.[7] I admit, that if we really now know the beginning of life on this planet, it is absolutely fatal to my views; I admit the same, if the geological record is not *excessively* imperfect; & I further admit that the a priori probability is that no being lived below our Cambrian era. Nevertheless I grow yearly more convinced of the general (with much incidental error) truth of my views: I believe in this from finding that my views embrace so many phenomena & explain them to a large extent. I am continually pleased by hearing of naturalists (within the last month I have heard of four) who have come round to a large extent to the belief of the modification—of Species.—[8] As my book has been lately somewhat attended to, perhaps it would have been better, if when you condemn all such views, you had stated that you had not been able yet to read it.[9] But pray do not suppose that I think for one instant that with your strong & slowly-acquired convictions, & immense knowledge, you would have been converted. The utmost that I could have hoped would have been that you might possibly have been here or there staggered. Indeed I should not much value any sudden conversion; for I remember well how many long years I fought against my present belief.—

With respect to Dr. Falconer, I fear I ought not to have said anything, as he lately told me that he shd. not interfere till Prof. Owen had published; so please do not repeat what I said.[10] I daresay Owen will work out everything carefully before he publishes in detail; but there is little doubt he was at *first* very careless. He overlooked a jaw with teeth which however may possibly not have belonged to this marvellous Bird, with its long tail & fingers to its wings.[11] As Birds are so isolated this case, as you may suppose, has pleased me.— You will see in Lyell's book that Owen has made a fearful mistake (not discovered by himself) about the British Eocene monkey.—[12] He has made such mistakes about the Elephants & Rhinoceroses, that I declare I am getting fearful of trusting him.[13] He has done the work of a giant; but I fear he has been too ambitious & not given time enough to most of his work.— I have not yet read Huxley's book;[14] but I hear it is very striking; but you will highly disapprove of it.—

With every good wish | pray believe me | Yours very sincerely | Ch. Darwin

I do not believe D[r] Falconer has changed his views about the Mesozoic mammals; but he has done more; for the specimens were presented to British Museum & are kept by Owen in his private room; & why he does not publish I cannot conceive.— In the last, or last but one, no[r] of the Geolog. Journal there is paper by Falconer on Plagiaulax, maintaining strenuously his former views.—[15]

Dana Family Papers, Manuscripts and Archives, Yale University Library

1 The year is established by the relationship between this letter and the letter from J. D. Dana, 5 February 1863.

2 The references are to Dana's *Manual of geology* (Dana 1863a), Dana 1863c, and the letter from J. D. Dana, 5 February 1863. CD had apparently received Dana 1863c by 17 February 1863 (see letter to Charles Lyell, 17 [February 1863]).

3 Dana had suffered a nervous breakdown in 1859, from which he was just beginning to recover (see *Correspondence* vol. 7, letter to J. D. Dana, 30 December [1859], *Correspondence* vol. 9, letter from Asa Gray, 31 December 1861, and this volume, letter from J. D. Dana, 5 February 1863).

4 CD refers to Dana's *Manual of geology* (Dana 1863a). There is an unannotated copy of Dana 1863a in the Darwin Library–Down.

5 In his letter of 5 February 1863, though, owing to ill health (see n. 3, above), he had not read *Origin*, Dana provided CD with an outline of his principal objections to the doctrine of the evolution of organic life.

6 Chapter 9 of *Origin* (pp. 279–311) dealt with the imperfection of the geological record. CD specifically addressed the problem posed to his theory by the absence of transitional forms in the geological record on pp. 172–9, 279–82, and 292–302.

7 Dana argued that the life forms of North America and Europe had generally developed independently (see letter from J. D. Dana, 5 February 1863); in *Origin*, CD emphasised 'the relationship, with very little identity, between the productions of North America and Europe'. This, he argued, was a reflection of past climatic changes affecting both continents, and the inter-migration of their flora and fauna during the warmer climate of the 'newer Pliocene period' (*Origin*, pp. 370–1).

8 Since the middle of January 1863, CD had received news of several naturalists expressing a belief in the modification of species, including Alphonse de Candolle, Gaston de Saporta, and Camille Dareste. In addition, Friedrich Rolle had written with details of CD's support in Germany (see letters to Alphonse de Candolle, 14 January [1863] and 31 January [1863], letter to J. D. Hooker, 30 January [1863], letter from Camille Dareste, 8 February 1863, and letter from Friedrich Rolle, 26 January 1863).

9 In Dana 1863a, pp. 601–2, Dana stated that:

> With regard to *the Origination of Species*, Geology suggests no theory of natural forces. It is right for science to search out Nature's methods, and strive to employ her forces—organic or inorganic—in

> the effort, vain though it prove, to derive thence new living species. The study of fossils has given no aid in this direction. It has brought to light no facts sustaining a theory that derives species from others, either by a system of evolution, or by a system of variations of living individuals, and bears strongly against both hypotheses.

10 In his letter to Dana of 7 January [1863], CD informed Dana that Hugh Falconer had dismissed Richard Owen's description of the recently discovered fossil bird *Archaeopteryx* as 'not done . . . well'. See also the letter from Hugh Falconer, 3 January [1863]. Richard Owen's description, which was read before the Royal Society of London on 20 November 1862, was not published until the second half of 1863 (Owen 1862a; Royal Society, Register of papers).

11 See letters from Hugh Falconer, 3 January [1863] and 18 January [1863].

12 In *Antiquity of man*, Lyell had written (C. Lyell 1863a, p. 500):

> The only reputed fossil monkey of eocene date, namely, that found in 1840 at Kyson, in Suffolk, and so determined by Professor Owen, has recently been pronounced by the same anatomist, after reexamination, and when he had ampler materials at his command, to be a pachyderm.

Lyell refers to Owen's description of the Eocene fossil in Owen 1840 and 1862b. See also the letter from Hugh Falconer, 18 January [1863] and n. 12.

13 CD probably refers to Owen's palaeontological work on the fossil elephant *Elephas columbi*, and the fossil rhinoceros *Rhinoceros leptorhinus*. With respect to *E. columbi*, Owen had overlooked Falconer's description of the fossil elephant and had renamed it *E. texianus*. Falconer interpreted this move as an attempt by Owen to usurp his priority in the description of the fossil, by substituting another, and in his view inferior, name (see letter to J. D. Hooker, 3 January [1863], n. 1). Falconer's critique of Owen's *E. texianus* was published in Falconer 1863a, pp. 45–9 (see also letter to Hugh Falconer, 5 [and 6] January [1863], and letter from Hugh Falconer, 8 January [1863]). Falconer may also have told CD of his doubts regarding Owen's identification of Clacton, Tuscan, and Rhenish specimens of fossil rhinoceros as *R. leptorhinus* (see Owen 1846b and Falconer 1868, 2: 317–20).

14 The reference is to Thomas Henry Huxley's *Evidence as to man's place in nature* (T. H. Huxley 1863b). See also letter to T. H. Huxley, 18 [February 1863].

15 Falconer had described two Mesozoic fossil mammal species of the genus *Plagiaulax*, concluding that they were herbivorous marsupials (Falconer 1857b). Owen, however, argued that the structure of the lower jaw and teeth of *Plagiaulax* indicated that it was a carnivorous marsupial (Owen 1860b, p. 321). Dana had followed Owen's interpretation in his *Manual of geology* (Dana 1863a, p. 463), and was apparently unaware of Falconer's subsequent refutation of Owen's conclusions in Falconer 1862. See also letter from J. D. Dana, 5 February 1863.

From T. H. Huxley 20 February 1863

The Athenæum

My dear Darwin

I am glad you have received the monkey book and I shall be right glad of criticisms by & bye if you have any time to spare for them—[1]

I am ashamed not to have answered your former letter but I fell ill on the Tuesday after you were with us—spent a day or so in bed and have not got right yet[2]

As soon as I am square again I will look up the fish business[3]

Ever | Yours faithfully | T H Huxley

Feb. 20th 1863

DAR 166: 297

[1] See letter to T. H. Huxley, 18 [February 1863]. The reference is to Huxley's *Evidence as to man's place in nature* (T. H. Huxley 1863b). CD stated his opinion on T. H. Huxley 1863b in his letter to Huxley of 26 [February 1863].
[2] Huxley refers to CD's letter of 16 February [1863]. CD visited Huxley on 8 February, during his stay in London (see letter to T. H. Huxley, [8 February 1863]).
[3] CD had consulted Huxley on the anatomy of fish fins, a subject which formed part of his researches on polydactylism (see letters to T. H. Huxley, [8 February 1863] and 16 February [1863], and letter to J. J. Briggs, 2 February [1863] and nn. 1 and 3).

From John Lubbock 20 February 1863

15, Lombard Street. E.C.
20 Feb/63

My dear Mr. Darwin

I could not answer your kind note at once, my plans being very undecided.[1] If however it would suit you I should much like to dine with you on Sunday.[2] Will you kindly send me a line to Chiselhurst.[3]

Yours affec— | John Lubbock

C Darwin Esq

DAR 170: 36

[1] The letter to Lubbock has not been found.
[2] Emma Darwin recorded in her diary (DAR 242) that John Lubbock dined at Down House on Sunday 22 February 1863.
[3] Lubbock resided in Chislehurst, Kent, a village about five miles north of Down.

To Daniel Oliver 20 [February 1863][1]

Down Bromley Kent
20th Friday Night.

Dear Oliver

Many thanks about Phyllotaxy. Your cases seem sufficient & when next in London, I shall hear what Falconer has to say.[2] Do not look for any more cases; but if you shd. *stumble* on them, please let me hear. I find the subject very difficult to understand; indeed I cannot understand several most simple points, such as whether there is ever more than one spire; but when Hooker comes he may be able to enlighten me.[3] I must learn the elements to understand force of Falconer's objections: he considered the laws as fixed as that of the attraction of gravity!—

I see what Treviranus says about Primula longiflora;[4] I shd. like to know (if you are up in Primula) whether this species is *closely* allied to P. Scotica; because Mr J. Scott of Bot. Garden of Edinburgh, has been carefully observing Primulas (& I feel a conviction that he is trust-worthy) & he says P. Scotica is never dimorphic, & is much surprised, as he says it is so like P. farinosa: he has sent me plants of both, but they look very sickly.[5]

By the way I see Mr Bentham makes P. Scotica *var*. of P. farinosa;[6] would it not be worth while to tell him of Mr Scotts observation;[7] for there can be no doubt that this difference indicates an important functional difference. Unless indeed P. farinosa presents 3 sexual forms; but then they all three would grow together.

Treviranus in his Review of the Orchids[8] does not seem to appreciate at all the prettiness of the adaptations, which seems to me the cream of the case.

Yours very sincerely | Ch. Darwin

DAR 261 (DH/MS 10: 41)

[1] The date is established by the relationship between this letter and the letters from Daniel Oliver, 17 February 1863 and 27 February 1863.

[2] Oliver had sent CD bibliographic references relating to phyllotaxy in his letter of 17 February 1863. The reference is to Hugh Falconer. See letter from Daniel Oliver, 17 February 1863 and n. 3.

[3] Joseph Dalton Hooker visited CD at Down House on 22 March 1863 (see letter from J. D. Hooker, [24 March 1863]).

[4] In his review of 'Dimorphic condition in *Primula*' (Treviranus 1863a, p. 4), Ludolph Christian Treviranus stated that *Primula longiflora* was homostyled. Treviranus had arranged for his review to be sent to CD earlier in the month (see letter from L. C. Treviranus, 12 February 1863). CD's annotated copy of Treviranus 1863a is in the Darwin Pamphlet Collection–CUL; at the head of the fourth page, CD noted that according to Treviranus *P. longiflora* was found 'alone' and was 'nondimorphic' and 'shortstyled'.

[5] John Scott sent CD three specimens each of *P. scotica* and *P. farinosa* on 6 January 1863 (see letter from John Scott, 6 January 1863). See also letter from John Scott, 17 December [1862] (*Correspondence* vol. 10). CD communicated Scott's observations on *P. scotica* to the Linnean Society on 4 February 1864 (Scott 1864a, p. 82).

[6] Bentham 1858, p. 354. CD's annotated copy of Bentham 1858 is in the Rare Books Room–CUL (see *Marginalia* 1: 51).

[7] George Bentham's classification of *Primula scotica* as a variety of *P. farinosa* remained unchanged in subsequent editions of the *Handbook of British flora*; however, in the fourth edition (Bentham 1878, p. 305), the form of statement was altered to read: 'Specimens from northern Scotland, with broader leaves, and shorter and broader lobes to the corolla, have been distinguished under the name of *P. scotica*, Hook.' Bentham worked with Oliver at the Royal Botanic Gardens, Kew.

[8] Treviranus 1863c; CD's annotated copy of this work is in the Darwin Pamphlet Collection–CUL, and a manuscript translation of the review is preserved in DAR 70: 38–52.

To John Scott 20 [February 1863][1]

Down Bromley Kent
20th

Dear Sir

What a magnificent capsule & good Heavens what a number of seed![2] I never before opened pod of larger orchids. It did not signify a few seed being lost, as it would be hopeless to estimate number in comparison with other species.—[3] If you sow any, had you not better sow a good many?[4] so I enclose small packet.—

I have looked at seeds: I never saw in British orchid nearly so many empty testæ; but this goes for nothing as unnatural conditions w^d^. account for it. I suspect, however from variable size & transparency that a good many of the seeds when dry (& I have put capsule on my chimney piece) will shrivel up. So I will wait for a month or two till I get capsule of some large Vandeæ for comparison.—[5]

It is more likely that I have made some dreadful blunder about Acropera than that it sh^d be male not yet a perfect male.[6] May there be some sexual relation between A. Loddigesii & luteola; they seem very close? I sh^d. very much like to examine capsule of unimpregnated flower of A. Loddigesii.— I have got both species from Kew; but whether we shall have skill to flower them I know not— One conjecture that that it is imperfect male, I still sh^d. incline to think it would produce by seed both sexes.—[7] But you are right about Primula (& a very acute thought it was) the long-styled P. Sinensis homorphically fertilised with own-form pollen, has produced during two successive homomorphic generations *only* long styled plants.—[8] The short-styled the same ie produced short-styled for 2 generations with exception of a single plant.— [9]

I cannot say about cowslip yet.—[10]

I sh^d. like to hear your case of the Primula: is it certainly propagated by seed?[11]

My dear Sir | Yours very faithfully | C. Darwin

DAR 93: B20–1

[1] The date is established by the relationship between this letter and the letters from John Scott, 18 February [1863] and 3 March 1863.

[2] Scott sent CD a seed capsule of the orchid *Acropera loddigesii* with his letter of 18 February [1863]. Scott had succeeded, where CD had failed, in effecting pollination in this genus (see *Correspondence* vol. 10, letter from John Scott, 11 November 1862, and this volume, letter to John Scott, 16 February [1863], n. 3).

[3] In his letter of 18 February [1863], Scott explained that the seed-capsule of *Acropera loddigesii* (see n. 2, above), had matured earlier than he expected and burst open, and that he had lost a number of seeds in the process.

[4] As Scott was packing the seed-capsule of the premature *Acropera* specimen (see nn. 2 and 3, above), '30 or so' of the seeds spilled into his hand; Scott planned to sow the seeds to see if he could raise the plants (see letter from John Scott, 18 February [1863]).

[5] CD had been promised capsules of several *Vanda* species on his visit to the Royal Botanic Gardens, Kew, on 11 February 1863 (see letter to John Scott, 16 February [1863] and n. 14).

[6] CD had repeatedly failed in his attempts to effect pollination in two species of *Acropera*, namely, *A. loddigesii* and *A. luteola*. From this fact, and from dissections of *A. loddigesii* and *A. luteola*, CD concluded that although no instance of the separation of the two sexes was known in orchids, these supposed species might actually be the male form of a dioecious species, of which the female form had not yet been identified (*Orchids*, pp. 206–9). After Scott had sent capsules of *A. loddigesii* containing viable seed, obtained by fertilising some flowers with pollen from the same plant, CD concluded that his former opinion was wrong, and that *Acropera* was a hermaphrodite (see letter from John Scott, 18 February [1863] and nn. 4 and 5, and 'Fertilization of orchids').

[7] In his letter of 18 February [1862], Scott posed the question: 'If Acropera be the male incipient form, and not yet perfectly sterile, will it *when self-impregnated produce both forms*?' Reasoning from the example of the transmission of forms in homomorphically fertilised primroses (see n. 8, below), Scott was inclined to believe that the self-pollinated *Acropera* would not produce both sexes from seed.

[8] Scott had observed that a self-pollinated variety of primrose in the Royal Botanic Garden, Edinburgh, produced only the long-styled form (see letter from John Scott, 18 February [1863]). Scott was aware

that CD had made homomorphic crosses with generations of *Primula sinensis* (see *Correspondence* vol. 10, letter to John Scott, 3 December [1862] and n. 6). CD's observations on the transmission of the long-styled form in *P. sinensis*, based on trials conducted in 1862, were published in 'Illegitimate offspring of dimorphic and trimorphic plants', pp. 410–11:

> I raised during February 1862, from some long-styled plants illegitimately fertilized by pollen from the same form, twenty-seven seedlings. These were all long-styled. They proved fully fertile or even fertile in excess. . . . From the foregoing twenty-seven plants, fertilized by their own-form pollen, I raised twenty-five seedling grandchildren; and these were all long-styled; so that from the two illegitimate generations fifty-two plants were raised, and all without exception proved long-styled.

9 CD's experimental crosses with the short-styled *Primula sinensis* are detailed in 'Illegitimate offspring of dimorphic and trimorphic plants', pp. 412–13. See also the experimental notes in DAR 108: 18, 27–8, 34–5.

10 CD had begun making homomorphic crosses with cowslips (*Primula veris*) in the spring of 1862 (see *Correspondence* vol. 10). CD's notes on these experiments are preserved in DAR 157a: 77 and DAR 108: 70. From his experiments CD concluded that 'the short-styled form of the cowslip, when self-fertilized, does not transmit the same form nearly so truly as does that of *P. sinensis*'. The long-styled form, when self-fertilised, was transmitted in 148 of the 152 plants raised, consisting of four generations ('Illegitimate offspring of dimorphic and trimorphic plants', p. 423). See also the letter to John Scott, 24 March [1863].

11 See letters from John Scott, 18 February [1863] and 3 March 1863.

To J. D. Hooker [21 February 1863][1]

Down Bromley Kent
Saturday

My dear Hooker

The Plants arrived **quite** safe last night. Pray thank Mr Gower for the trouble which he must have taken.—[2] I am fairly astounded at their number! why my hot-house is almost full!—[3] I have not yet even looked out their names; but I can see several things which I wished for, but which I did not like to ask for.

You have indeed set me up. I got my neighbour's gardener to come & do the Orchids.—[4]

Thanks, also, for very valuable seeds.— How very good of you to write to Paris about seeds. All that were on list were for experiments, which seem to me really worth trial.—[5]

Let me hear about Glass-man, as I must order some Bell-glasses soon on a venture; & for ventilation—iron-plate man.—[6]

I stewed so long admiring the plants, that I have got a bit of a headach.— so farewell | Ever yours | Ch. Darwin

P.S. We are in a puzzle; in neither of my Orchid-culture book is Acropera mentioned: is it grown in Basket, or pot or Block of wood? These Acroperas are so valuable to me, pray tell me.—[7]

Endorsement: 'Feby 21/63.'
DAR 115: 182

[1] The letter is dated by the relationship between this letter, the letter to J. D. Hooker, 15 February [1863], and the letter from J. D. Hooker, [23 February 1863]; the intervening Saturday was 21 February.

[2] CD had sent his tax-cart to the Royal Botanic Gardens, Kew, to collect plants for the new hothouse at Down House (see n. 3, below); William Hugh Gower, a foreman at the gardens, assisted Hooker in selecting and packing the plants for CD (see letters to J. D. Hooker, 15 February [1863] and 5 March [1863]).

[3] The building of CD's hothouse, in the kitchen garden at Down House, had recently been completed (see *LL* 1: 321, and letter to G. H. Turnbull, [16? February 1863]).

[4] The reference is to John Horwood, gardener to CD's neighbour, George Henry Turnbull. In a letter from Henrietta Emma Darwin to William Erasmus Darwin of [22 February 1863], which is in DAR 210.6: 109, Henrietta explained that the plants sent from the Royal Botanic Gardens, Kew (see n. 2, above), had been sent 'out of pots' and the gardener at Down House, Henry Lettington:

> had potted them all in common earth, the orchids I mean, & we had to send off instanter for Mr. Horwood to undo them & pot them as they like with their particular kind of moss & peat & the exact sized bits of charcoal.

In addition to supervising the construction of the hothouse (see letter to G. H. Turnbull, [16? February 1863]), Horwood had previously aided CD with some of his observations on orchids (see *Correspondence* vols. 9 and 10, and *Orchids*, p. 158 n.).

[5] CD may have given Hooker a list of the plants he required when he visited the Royal Botanic Gardens, Kew, on 11 February 1863. The list referred to has not been found; however, a contemporary list of hothouse plants in CD's hand is in DAR 255: 8 (see Appendix VI). See also letter from J. D. Hooker, [16 February 1863].

[6] CD probably discussed the requirements for his hothouse with the staff at the Royal Botanic Gardens, Kew, when he visited on 11 February 1863 (see 'Journal' (Appendix II)). In his letter of [23 February 1863], Hooker recommended the ventilation services of John Weeks & Co. of Sussex Terrace, Chelsea, London, and the glaziers James Powell & Sons of Fleet Street, Whitefriars, London.

[7] The references have not been identified. For Hooker's response to CD's question, see the letter from J. D. Hooker, [23 February 1863]. For CD's interest in *Acropera*, see, for example, the preceding letter, and letter to John Scott, 16 February [1863] and n. 3.

To Richard Frean 22 February 1863

Down
Feb 22/63.

Dear Sir

I am gratified by your approval of my book on species.[1] No doubt in many cases it is impossible to conjecture by what steps certain structures have been acquired. When we reflect on such cases as the stomach of the Pigeon secreting a nutritive fluid—the back of certain Batrachians secreting mucus on which the young feed, we can see that there may have been gradation in the formation of Mammary glands.—[2]

I have often thought over the case of the degeneration of civilized man; what you suggest, & especially for more care in marriage with more skill in detecting weak constitutions, and the germs of disease, seems to to be our only hope[3]

I have the &c &c | Yours faithfully | C. Darwin

Copy
DAR 144: 298

[1] Richard Frean's letter commenting on *Origin* has not been found. The text of this letter is taken from a copy made by Frean and sent to Leonard Darwin in 1882 for possible inclusion in Francis Darwin's edition of CD's letters (see the letter from Richard Frean to Leonard Darwin, November 1882, in DAR 144: 298).

[2] CD's interest in the development of mammary glands was stimulated by George Maw's review of *Origin*, which, in criticising the theory of natural selection, raised the question: 'How, for example, could mammiferous nutrition have been perfected by short stages?' ([Maw] 1861, p. 7596). In attempting to answer this point, CD referred Maw to Jeffries Wyman's researches on gestation in Batrachians (Wyman 1859), which showed the '*possibility* of such a transition' (*Correspondence* vol. 9, letter to George Maw, 19 July [1861]). See also *Correspondence* vol. 10, letter to George Maw, 3 July [1862].

[3] In his letter to Leonard Darwin of November 1882 (DAR 144: 298), Frean discussed the letter that elicited this response from CD:

> In writing I pointed out some difficulties which had occurred to me in the theory of Natural Selection, & then went on to show that if the theory be correct the degeneration of Civilized man was a necessary consequence.

From Hermann Crüger 23 February 1863

Trinidad
23 Feby 1863.

Dear Sir

In answer to your letter of 25th Jan, which reached me yesterday, I beg to say that I shall be delighted to undertake the examination you wish.[1] At this moment little can be done it being the beginning of the dry season, & scarcely any Melastomaceae are in flower, but as soon as I find an opportunity I shall attend to your request. We have a large number of Melastomaceae here some 50–60 Species. As far as I can see they are visited by insects of various classes, & only a few minutes ago I found that a sort of bee or humble-bee was very busy in the flowers of Heteronoma diversifolium De C.[2] which we cultivate here for ornament but which is a native & common in certain localities. It appeared to me that the insect was addressing itself principally to the basal appendices of the anthers, but I shall report later on this & other points when I shall have collected a certain number of facts.

I take this opportunity to send you a few remarks which I have had occasion to make in reading your two latest works, particularly the "British & foreign Orchideae".[3] Perhaps you will not find them altogether out of place from an individual who has made vegetable physiology his principal study for some fifteen years or more. That plants are fertilised by insects, is as evident to every body here as well as in Europe, but it is little imagined in Europe on what scale & to what extent it takes place in these latitudes. Besides the agency of butterflies & moths & bees & wasps we have the immense class of ants which are found everywhere visiting the flowers for some purpose or other. Nearly all fig-trees have their Cynips, which is apparently necessary for the fertilisation of these.[4] I have never found an insect nor good seeds in Ficus elastica, a native of the old world. Even the higher classes of animals contribute their share in this act, as

Humming birds f[or] instance. I am nearly certain that such Plants as Anguria, which have a very viscid & heavy pollen, are principally fertilized by Humming birds.

With regard to Orchids a few facts which I have noted will not be uninteresting to you. We have a few Epidendreae, a Schomburgkia, a Cattleya, & an Epidendron which are hardly ever known to open their flowers, but which nearly always set fruit, which however may drop afterwards. I find on examination, that these are fertilized by ants which creep into the flowers at the base of the sepals. In these cases, as in other Epidendreae the pollinia perform their function *in situ*, & I believe that it can only explained by the ants carrying the viscosity of the stigma to the pollinia, at least this is the only explanation *I* can find of the phenomenon.[5] The pollinia get pulpy without being removed from their bed, & the pollen tubes grow down over the anterior face of the style into the ovarium. I should like to know if there are any Orchids cultivated at Home which do not open. Here it is entirely confined to native species. In other Orchids Ants are also very active some at the bracts, some at the base at the sepals & petals & in the flowers.

The Catasetum I have examined repeatedly before I got your book, since then I have not succeeded getting good observations, the last season having been very unfavorable. The conclusions I have arrived at are rather different from yours, but I shall examine the question again in reference to your opinions.[6] In my opinion Myanthus barbatus & Catasetum tridentatum are distinct species, which assume both three forms, the female & intermediate form being much alike in the two species, but much smaller in My. barbatus. The so called female form of Catasetum tridentatum is in my opinion a fertile hermaphrodite form, the other the male. I believe that the flower of the former is nearly always fertilized by its own pollinia, which I have repeatedly found in a pulpy state & attached to the opening of the stigma which is hardly large enough to receive the pollinia of the male flower. Fecundation takes place in the female long before it reaches that aromatic state of maturity which attracts the noisy crowd of humble bees which surround it later for days, fighting for some substance or other in the flower. The female flower opens very early, when it is still very small & green, & immediately after the ovarium takes a rapid development. The anther although small is fully developed, the pollen apparently sound, there is a caudicle & gland. The pollinia are found soon after at the entrance of the stigmatic cavity as mentioned already, & in that pulpy state which precedes the emission of pollen tubes in these tribes. I believe that here also fecundation is brought on by ants, which are very active about these flowers.

On the other hand I have repeatedly seen that the male & hermaphrodite flowers are visited by the same species (2 or 3) of Humble bees, & I have also seen the gland & pollinia of the male flowers sticking to the back of them. There are in my opinion mechanical obstacles to their fecundating the hermaphrodite flowers.

These are my results before reading your book, since then I have found nothing contradicting them, but I shall cautiously reexamine the matter, & communicate my results if you wish it. I may mention that the intermediate form between the male & hermaphrodite is (here) rare, & that I have no good observations on it.

In conclusion I beg you will command my services in these matters, if f[or] instance you should like to have some materials in Spirit of wine, I shall be most happy to forward whatever I can get together.

I remain dear Sir | sincerely yours | H Crüger

Ch. Darwin Esq

DAR 161: 275

CD ANNOTATIONS
1.1 In answer . . . facts. 1.12] *scored blue crayon*
1.11 I shall report . . . facts. 1.12] 'Look for punctures.'[7] *added pencil*
2.1 I take . . . native species. 3.12] *crossed blue crayon*
2.12 Even . . . birds. 2.15] *double scored ink*
3.2 We have . . . afterwards. 3.4] *double scored pencil*
3.5 In these . . . pollinia, 3.8] *double scored pencil*
3.9 & the pollen . . . species. 3.12] *double scored pencil*
3.11 it is entirely confined 3.12] *underl pencil*
4.1 The Catasetum . . . book, 4.1] *scored blue crayon*
Top of letter: 'Melastoma' *added blue crayon*; 'O' *added red crayon*; 'Melast' *added & del, blue crayon*; '*Catasetum*—' *added blue crayon*
End of letter: 'D^r^. Hermann Cruger' *added ink*

1 In his letter to Crüger of 25 January [1863], CD asked for information regarding pollination in Melastomataceae. In particular, CD wanted Crüger to observe native species of Melastomataceae to see whether insects punctured the horns of the anthers with their probosces in order to obtain nectar (see also letter to Asa Gray, 19 January [1863]). Crüger was the government botanist and director of the botanic garden in Trinidad in the West Indies (R. Desmond 1994).

2 The species *Heteronoma diversifolium* (*Arthrostema ciliatum*) was named in Candolle and Candolle 1824–73, 3: 122.

3 Crüger refers to *Orchids*, which was published in May 1862, and probably also to *Origin*.

4 *Cynips* is a genus of gall-wasps (Grzimek ed. 1975, p. 413).

5 Crüger published his observations on pollination in *Schomburgkia*, *Cattleya*, and *Epidendrum*, members of the Epidendreae sub-family of orchids, in a paper communicated by CD to the Linnean Society, which was read on 3 March 1864 (Crüger 1864, p. 131). In the paper, Crüger revised his opinion that ants were responsible for pollination in these orchids, and concluded that self-pollination was common in this family. Crüger argued that the case of the Epidendreae undermined CD's doctrine that nature 'abhors perpetual self-fertilisation' (*Orchids*, p. 359). CD, however, had already reached the conclusion that Crüger's observations suggested he had underrated the power of tropical orchids to produce seed without insect pollination (see letter to *Journal of Horticulture*, [17–24 March 1863]). See also letter to John Scott, 24 March [1863]. CD cited Crüger 1864 in 'Fertilization of orchids', p. 152 (*Collected papers* 2: 149).

6 In *Orchids*, pp. 236–48, CD argued that *Catasetum tridentatum* was the male, *Monachanthus viridis* the female, and *Myanthus barbatus* the hermaphrodite form of a single species. Crüger's subsequent observations on *Catasetum* confirmed CD's conclusion, and demonstrated that pollination was effected by bees (see *Correspondence* vol. 12, letter from Hermann Crüger, 21 January 1864; see also Crüger 1864, pp. 127–9, and *Orchids*, 2d ed., pp. 205–6).

7 See n. 1, above.

To Asa Gray 23 February [1863][1]

Down Bromley Kent
Feb. 23^{d}

My dear Gray

Many thanks for your note of Jan 27th.—[2] The enclosures were forwarded.[3] The maize seed has proved a treasure; for besides seeing the kinds, a young man at Edinburgh will experiment on the mutual fertility of some of the varieties.—[4]

Pray thank, when you see, D^{r}. Scudder about pollinia (from whom I have since received a pamphlet):[5] that was very good remark about attachment possible only to eye or proboscis; & these are only two parts where I have seen attachment.[6] Thanks, also, about highness & lowness of oak-tree:[7] Hooker was pleased (to whom I mentioned your remark) about the "commonwealth" of Plants.—[8] When I send my Linum paper you will see about L. Lewisii.—[9]

If you have time to read you will be interested by parts of Lyell's Book on Man:[10] but I fear that the best part, about Glacial period, may be too geological for anyone except a regular geologist.[11] He quotes you at end with gusto.[12] By the way he told me the other day how pleased some had been by hearing that they could purchase your pamphlet.[13] The "Parthenon" also speaks of it as the ablest contribution to the literature of the subject.[14] It delights me when I see your work appreciated. The Lyells come here this day week,[15] & I shall grumble at his excessive caution: I feel sure that he admits almost fully the modification of species by variation & selection; & yet, though writing at length on subject, is afraid to say so; & he will not serve as guide to anyone.[16] The public may well say, if such a man dare not or will not speak out his mind, how can we who are ignorant, form even a guess on subject.— Lyell was pleased, when I told him lately that you thought that language might be used as excellent illustration of derivation of species; you will see that he has *admirable* chapter on this—[17]

I received a little while ago the correspondence between Loring & Field:[18] I cannot tell you how it has interested us all: it is so real; & it so curious to see two able & honest men differing so enormously. of course I side chiefly with the Englishman; but I never so well understood your horror of Disunion. It is very natural that you sh^{d}. dread becoming split up like Germany; but to us it does not seem quite so horrible. I think both correspondents underrate the very general belief entertained for many years in England, that your Government delighted in making us eat dirt, & that we had eat dirt about Boundary Line, Right of Search, Vancouver Is^{d} &c.[19] I believe that this has greatly checked all sympathy with you; & made the whole country fire up, when, as we thought, you had passed our swallowing powers in the Trent affair.—[20] after finishing the above Correspondence; I read Cairns excellent Lecture, which shows so well how your quarrell arose from Slavery.[21]

It made me for a time wish honestly for north; but I could never help, though I tried, all the time thinking how we sh^{d} be bullied & forced into a war by you, when you were triumphant. But I do, most truly think it dreadful that the South, with its accursed Slavery, shd. triumph, & spread the evil. I think if I had power, which thank God I have not, I would let you conquer the border states, & all west of Mississippi & then force you to acknowledge the Cotton States. For Do you not now begin to doubt whether you can conquer & hold them? I have inflicted a long tirade on you.—

The Times is getting more detestable,—but that is too weak a word,—than ever.[22] My good wife wishes to give it up; but I tell her that is a pitch of heroism, to which only a woman is equal to.. To give up the "Bloody Old Times" as Cobbett used to call it, would be to give up meat drink & air.—[23]

Farewell my dear Gray | Yours most truly | C. Darwin

Gray Herbarium of Harvard University (55)

1 The year is established by the relationship between this letter and the letter from Asa Gray, 27 January 1863.

2 Letter from Asa Gray, 27 January 1863.

3 The enclosures referred to have not been identified. Gray's letter of 27 January 1863 also included a letter from Samuel Hubbard Scudder to Gray (see n. 5, below), and a note from Gray commenting on Scudder's letter.

4 Gray sent CD maize seeds from New England with his letter of 29 December 1862 (*Correspondence* vol. 10). CD refers to John Scott at the Royal Botanic Garden, Edinburgh, to whom he had sent the seeds for crossing experiments (see letter to John Scott, 16 February [1863] and n. 9).

5 With his letter of 27 January 1863, Gray enclosed a letter he had received from Scudder giving detailed observations of insect pollination in two species of orchid: *Platanthera hookeri* and *P. orbiculata*. The pamphlet referred to by CD may be Scudder 1862a or 1862b. There are lightly annotated presentation copies of both papers in the Darwin Pamphlet Collection–CUL.

6 See enclosures to the letter from Asa Gray, 27 January 1863.

7 See letter to Asa Gray, 2 January [1863], and letter from Asa Gray, 27 January 1863. See also n. 8, below.

8 See letter from Asa Gray, 27 January 1863. CD enclosed Gray's letter with his letter to J. D. Hooker, 15 February [1863], and Joseph Dalton Hooker responded in his letter to CD of [16 February 1863].

9 See letter from Asa Gray, 27 January 1863. In his letter to Gray of 2 January [1863], CD expressed doubt about the view of the French botanist Jules Emile Planchon (Planchon 1847–8, p. 175) that *Linum lewisii* was only a variety of *L. perenne*, and noted Planchon's observation that *L. lewisii* was trimorphic, bearing 'on the *same* plant flowers with long, & short & equal (to anthers) pistils'. In 'Two forms in species of *Linum*', pp. 82–3 (*Collected papers* 2: 104–5), CD argued:

> now that we know the meaning of reciprocal dimorphism, [*L. lewisii*] surely deserves specific honours. . . . According to Planchon, the same plant bears some flowers with anthers and stigmas of the same height, and others with styles either longer or shorter than the stamens; so that the same individual plant is trimorphic. This, as far as I know, is a unique case. From analogy we may pretty safely predict the function of the three kinds of flowers: those with stigmas and anthers of the same height will be self-fertile; those with these organs of unequal height will require reciprocal fertilization.

CD sent Gray two copies of the paper, which was read before the Linnean Society on 5 February 1863, on 18 April 1863 (see letter to Asa Gray, 20 April [1863]).

10 C. Lyell 1863a.

11 Chapters 12 to 18 of Charles Lyell's *Antiquity of man* (C. Lyell 1863a, pp. 206–368) were concerned with the geological evidence of the Pleistocene glacial period, particularly the most recent glacial epoch, and examined the chronological relations between human and glacial history.

12 Chapter 24 of *Antiquity of man* (C. Lyell 1863a, pp. 471–506), 'Bearing of the doctrine of transmutation on the origin of man, and his place in the creation', concludes with a discussion of Gray's 'philosophical essay of great merit' (A. Gray 1861a). Quoting from the pamphlet, Lyell enthusiastically embraced Gray's argument that there was no inconsistency between natural selection and natural theology (C. Lyell 1863a, p. 506):

> The whole course of nature may be the material embodiment of a preconcerted arrangement; and if the succession of events be explained by transmutation, the pereptual adaptation of the organic world to new conditions leaves the argument in favour of design, and therefore of a designer, as valid as ever; 'for to do any work by an instrument must require, and therefore presuppose, the exertion rather of more than of less power, than to do it directly.'

13 CD probably met Lyell during his stay in London between 4 and 14 February 1863 (see letter to Charles Lyell, 4 [February 1863]). CD had supervised the arrangements for the publication of Gray's pamphlet (A. Gray 1861a), which was originally published as a series of articles in the American periodical *Atlantic Monthly*. CD and Gray shared the cost of having 500 copies printed, and 250 copies were put on sale at the London publishing firm Trübner & Co. (see *Correspondence* vol. 8, letter to Asa Gray, 11 December [1860], and *Correspondence* vol. 9, letter to Asa Gray, 17 February [1861], and Appendix III). See also *Correspondence* vol. 10, letter to Asa Gray, 15 March [1862].

14 In a review of Lyell's *Antiquity of man* in the *Parthenon*, 21 February 1863, pp. 233–5, the author stated (p. 235):

> Dr. Asa Gray, who has written what is probably the ablest contribution to the literature of Darwinism, seeks to calm the minds of those who are in alarm lest the doctrines of variation and natural selection should weaken the foundations of natural theology, for 'consistently with the derivative hypothesis of species, we may hold any of the popular views respecting the manner in which the changes in the natural world are brought about;' and still less can we fear that our derived spiritual life can be affected.

15 Charles and Mary Elizabeth Lyell had been invited to Down House for a few days, from 1 to 4 March 1863, but CD became ill with 'much sickness & weakness', and was obliged to retract the invitation (see letters to J. D. Hooker, 24[–5] February [1863] and 5 March [1863]).

16 Chapters 20 to 22 of *Antiquity of man* dealt with theories of transmutation, including the theory of the origin of species by variation and natural selection (C. Lyell 1863a, pp. 385–453); CD's annotated copy of this work is in the Darwin Library–CUL (see *Marginalia* 1: 525–7). In a letter from Henrietta Emma Darwin to William Erasmus Darwin of [22 February 1863], which is in DAR 210.6: 109, Henrietta recorded that CD was 'reading Sir Charles Lyell & is rather indignant that he does not say clearer that he believes in Papa.' CD confided to Hooker that he was 'deeply disappointed' that Lyell's 'timidity' had prevented him giving any judgment on the subject (see letter to J. D. Hooker, 24[–5] February [1863]).

17 CD refers to chapter 23 of *Antiquity of man*, 'Origin and development of languages and species compared' (C. Lyell 1863a, pp. 454–70). See *Correspondence* vol. 10, letters from Asa Gray, 4 and 13 October 1862 and 24 November 1862, and letter to Asa Gray, 6 November [1862].

18 CD refers to the published correspondence between the American lawyer Charles Greely Loring and the English lawyer Edwin Wilkins Field concerning relations between Britain and the United States following the outbreak of the American Civil War (Loring 1862). Loring was Gray's father-in-law (Dupree 1959, p. 180).

19 The United States and Britain had been involved in several border disputes, focusing particularly on Canada's borders with the United States, and including a dispute over the position of the boundary in the San Juan de Fuca Strait between Vancouver Island and United States territory to the south, a dispute which was not settled until 1871 (see H. C. Allen 1954, pp. 391–415). The 'Right of search' was the granting of reciprocal rights between nations to search ships for slaves, following agreements to bring an end to the international slave trade. Although the United States government subscribed to the ban on the slave trade, it consistently maintained the full freedom of the seas and refused

to concede the right of search, even in the face of British pressure. Following the outbreak of the Civil War, in an effort by the Union to influence British opinion in its favour, and because the representatives from the southern states no longer occupied positions in the legislature, a right of search was finally ratified in Washington in April 1862 (see Soulsby 1933).

20 In November 1861, the United States navy seized two Confederate envoys from the *Trent*, a British mail packet. The British demanded their release, and Britain and the United States stood on the brink of war with one another until news that the United States government had acquiesced to British demands put an end to the affair in January 1862. For an account of the *Trent* affair, see Ferris 1977; see also *Correspondence* vols. 9 and 10.

21 CD refers to John Elliot Cairnes's book, *The slave power*, the text of which had originally formed a course of lectures at the University of Dublin in 1861 (Cairnes 1862, p. vii). The book, which defended the Union cause, 'made a great impression both in England and America' (*DNB*).

22 Despite the editors' claims of neutrality, in its coverage of the American Civil War during February 1863, *The Times* adopted a hostile tone towards the Union cause (see *The Times*, 7 February 1863, p. 8, and 20 February 1863, p. 9). CD's complaint probably refers to the continued discussion of the war in *The Times* without 'a shade of feeling against slavery' (see *Correspondence* vol. 10, letter to Asa Gray, 16 October [1862]).

23 The journalist and reform campaigner, William Cobbett, maintained a written campaign against *The Times* newspaper, which he considered a 'vile ... atrocious paper'. He employed many derisory names to refer to it in the pages of *Cobbett's Political Register*: 'The Bloody Old Times' dates from about 1830. Cobbett gave it this title because of 'its having uniformly advocated punishment, cruelty, proscription and blood against all those, *in whatever country*, who were striving for freedom' (*Cobbett's Political Register*, 16 October 1830, p. 507; Spater 1982, 2: 542–3).

From J. D. Hooker [23 February 1863][1]

Royal Gardens Kew | Kew
Monday

Dr Darwin

I am deeply grieved to see Owen's letter in the Athenæum, because, be he right or be he wrong, it will do Lyell awful injury, & I am sure cut him & his family to the soul.[2]

I have not read that Chapter of Lyells, only glanced at it, but feared the result, from Falconer & Huxley both being *perfectly satisfied* that L. had gone quite far enough![3] The worst of it is that; I suppose it is virtually Huxley's writing, & that L. will find great difficulty in answering Owen *unaided*, & this is a dreadful position to be in.— he fought under Achilles shield![4] I fear L. will get scant pity even from his own side, for F spoke to me the other night in most strongly slighting terms of so much of Lyell's book being written by others.[5] I am most anxious to hear what you think if your head will allow you to send me one line I shall be thankful,[6] for I do feel quite heartsore about poor L.—assuming as I must that he will feel it deeply— What horrid accusations & how spitefully put, & God knows that there is some ground for Owen's putting it down to personal feeling.— Do you know I met Owen at Sabines on *Friday* night,[7]—he was *most* gracious, "Hookering" me across the table. Detesting the man's mind & conduct as I do, I cannot say I have the smallest ill-feeling towards him.— I can hate & respect; I cannot hate & despise—& I do on my conscience think that I despise Owen's mind and conduct too single mindedly to care one atom for his individuality— I look on him now as a poor miserable devil of a scotched viper, turning & poisoning with a bite what he can neither strangle nor gorge. But by the Lord I am shut up— poor dear Lyell—it is an awful accusation to have hove at one—hit or miss—in the way this is hove. How devilish ingenious the introduction is.

I am smothered with Examinations this week 53 men fr the Army & two 6 hours viva voce, besides 53 3 hours papers!—[8] I wish I were rich enough to throw these examinations overboard—& I dare say I shall spend them on Wedgewoods![9] after all—& that's something!

We grow Acropera in pots with moss, like Catasetum.[10]

As to the ventilators, I do not think them worth the expense, & intended telling you that removing $1\frac{1}{2}$ brick opposite the pipes, & having a removable plug of wood is all the same. I will get the name however[11]

P.S. No one can recollect the name of the ventilator maker, but the place is by Stanley bridge, King's Road Chelsea[12]

Glass men—Jas Powell & sons, Whitefriars City.[13]

Ever yours affec | J D Hooker

DAR 101: 105–7

[1] The date is established by the relationship between this letter and the letter to J. D. Hooker, 24[–5] February [1863].

[2] The reference is to a letter by Richard Owen, published in the *Athenæum* on 21 February 1863, pp. 262–3, in which Owen objected to remarks made by Charles Lyell in a section of *Antiquity of man* entitled 'Whether the structure of the human brain entitles man to form a distinct sub-class of the Mammalia' (C. Lyell 1863a, pp. 480–93). Lyell argued that Owen had based his case for a distinct difference between the anatomy of human and simian brains in Owen 1857, 1859a, 1860b, and 1861b on diagrams of a chimpanzee brain published by the Dutch anatomists Jacob Lodewijk Coenraad Schroeder van der Kolk and Willem Vrolik, long after these diagrams were proved to have been based on a distorted specimen. Owen argued that Lyell had been responsible for misstatements 'in order to impress his readers with the notion that I had a desire to promote and persist in promoting an error, and to mislead the public', and described Lyell's account as 'inferential calumny'. For an account of the debate surrounding Owen's work on simian brain anatomy, see Rupke 1994, pp. 266–86.

[3] Hooker refers to chapter 24 of Lyell's *Antiquity of man*, entitled 'Bearing of the doctrine of transmutation on the origin of man, and his place in the creation' (C. Lyell 1863a, pp. 471–506), in which Lyell provided a review of the controversy between Owen and Thomas Henry Huxley regarding the comparative anatomy of human and simian brains (see n. 2 above). Hooker also refers to the palaeontologist Hugh Falconer, who was also engaged in a dispute with Owen (see letter from Hugh Falconer, 3 January [1863], and letter to Hugh Falconer, 5 [and 6] January [1863]).

[4] The reference is to the Greek hero Achilles, and to the episode in Homer's *Iliad*, Book 16, where Patroclus takes Achilles' armour in order to assist the Greeks in battle against the Trojans, but is killed by Hector.

[5] Falconer publicly criticised *Antiquity of man* (C. Lyell 1863a) in a letter published in the *Athenæum* on 4 April 1863, pp. 459–60. Falconer complained that Lyell had given insufficient credit to the work of himself, Joseph Prestwich, John Evans and others in establishing human antiquity, and had not made any original contribution to the debate. For a discussion of Falconer's attack on Lyell, see Bynum 1984 and L. G. Wilson 1996a.

[6] For CD's response, see the letter to J. D. Hooker, 24[–5] February [1863].

7 Hooker refers to the president of the Royal Society, Edward Sabine, and to his meeting Owen on Friday 20 February 1863, the evening before Owen's letter appeared in the *Athenæum* (see n. 2, above).

8 Hooker served for many years as a scientific examiner for medical officers in the armed services (L. Huxley ed. 1918, 1: 387). Examinations for admission to the Army Medical Service were held at Chelsea Hospital in February 1863 (*Statistical, Sanitary, and Medical Reports* 5 (1865): 582–3). Hooker also held examinerships with the East India Company, the Apothecaries Company, and London University (L. Huxley ed. 1918, 1: 385, 537).

9 In 1862, Hooker began collecting Wedgwood ware (see *Correspondence* vol. 10, letter from J. D. Hooker, [27 or 28 December 1862]; see also this volume, letter to J. D. Hooker, 3 January [1863], and letter from J. D. Hooker, 6 January 1863).

10 In his letter to Hooker of [21 February 1863], CD asked whether *Acropera* orchids should be 'grown in Basket, or pot or Block of wood?' The *Acropera* orchids were included with the hothouse plants that CD received from the Royal Botanic Gardens, Kew, on 20 February 1863.

11 In his letter of [21 February 1863], CD asked Hooker to recommend a firm to assist with the ventilation of the new hothouse at Down House. See also n. 12, below.

12 The reference is to the firm of John Weeks & Co., horticultural engineers, of Sussex Terrace, Chelsea, London (*Post Office London directory* 1863).

13 Hooker refers to the glass manufacturers James Powell & Sons of Fleet Street, Whitefriars, London (*Post Office London directory* 1863). CD wished to purchase some 'bell-glasses' for experiments and had asked Hooker to recommend a supplier (see letter to J. D. Hooker, [21 February 1863]).

To John Lubbock 23 [February 1863][1]

Down
23^d

My dear Lubbock

I have no criticism, except one sentence not perfectly smooth.—[2] I think your introductory remarks very striking, interesting & novel. They interest me the more, because the vaguest thoughts of same kind had passed through my head; but I had no idea that they could be so well developed; nor did I know of exceptions.[3] Sitaris & meloe seem very good.[4] You have put whole case of metamorphosis in new light.—[5] I dare say what you remark about poverty of F. Water is very true.[6] I think you might write memoir on F. W. productions.[7] I suspect that the key-note is that land-productions are higher & have advantage in general over marine; & consequently land-productions have generally been modified with F. W. productions, instead of marine productions being directly changed into F. W. productions, as at first seems more probable, as the channel of immigration is always open from sea to rivers & ponds.—

Ever dear Lubbock | Yours most truly | C. Darwin

My talk with you did me a deal of good & I enjoyed it much.—[8]

DAR 263: 59

1 The date is established by the reference to the first part of Lubbock 1863–5 (see n. 2, below).

2 CD refers to the first part of Lubbock's paper 'On the development of Chloeon (Ephemera) dimidiatum' (Lubbock 1863–5), in which he identified over twenty stages through which the larvae of the pond insect *Chloëon dimidiatum*, one of the Ephemeridae, pass before reaching maturity. Lubbock read the first part of the paper before the Linnean Society on 15 January 1863, and may have requested

CD's comments on it when he dined at Down House on 22 February 1863 (see n. 8, below). There are two lightly annotated copies of Lubbock 1863–5 in the Darwin Pamphlet Collection–CUL.

3 In his introductory remarks (Lubbock 1863–5, pp. 61–6), Lubbock noted that entomologists believed it a general rule that every insect had four distinct periods of existence, each marked by an alteration of form, namely, the egg, the larva, the pupa, and the imago. Lubbock, however, drew attention to the fact that in several insects there was no such well-marked four-stage metamorphosis, and that not all insects left the egg in the same stage of 'embryonal development'. Lubbock argued that the condition of an insect at birth depended 'partly on the group to which it belongs, but perhaps still more on the manner in which it is to live' (*ibid.*, p. 63). Lubbock pointed out that several exceptions to the entomologists' simple model of four stages of development had already been established, including *Meloë* and *Sitaris* among the Coleoptera, *Lonchoptera* and 'Pupipara' (formerly a sub-order comprising the families Hippoboscidae and Nycteribiidae) among the Diptera, *Typhlocyba* and *Aphis* among the Hemiptera, *Psocus* among the Neuroptera, *Thrips* among the Thysanoptera, and *Gryllus campestris* among the Orthoptera.

4 CD refers to Lubbock's point regarding the genera *Meloë* and *Sitaris* (Lubbock 1863–5, pp. 63–4), which begin life as active hexapod larvae,

> but having introduced themselves into the cells of certain species of Hymenoptera, they undergo a retrograde metamorphosis, lose their legs, and emerge as grubs, not altogether unlike those whose places they have usurped.

5 In his copies of the first part of Lubbock 1863–5, CD annotated passages dealing with the connections between insect form during different stages of metamorphosis, and the insect's food supply and habits. Lubbock argued that both the speed at which different organs develop, and the form the metamorphosis took, depended upon adaptation to external conditions (*ibid.*, pp. 63–4); he developed this argument further in Lubbock 1866. In the fourth edition of *Origin*, CD cited Lubbock's studies of insect metamorphosis as examples of gradual developmental transformation, and of how embryos could be related to their 'conditions of existence' (*Origin* 4th ed., pp. 517, 520).

6 CD refers to Lubbock's remarks on the relative lack of freshwater fauna, compared with the variety of marine fauna (Lubbock 1863–5, pp. 64–5):

> The Mollusca are far less numerous and less varied; of the Fish the same may be said; compared with those of the sea, our freshwater Bryozoa are quite insignificant in numbers; the Hydrozoa are represented by only two genera. Among Crustacea, the Podophthalms have in this country but one freshwater species, the Isopods one, the Amphipods very few; Entomostraca, indeed, are well represented, but Cirrhipedes are altogether absent; neither the Actinozoa nor the Echinodermata have a single freshwater representative.

7 CD had previously encouraged Daniel Oliver to undertake a study of freshwater plants (see letter to Daniel Oliver, 20 [January 1863], and *Correspondence* vol. 10, letter to Daniel Oliver, 12 [April 1862]).

8 Emma Darwin recorded in her diary (DAR 242) that John Lubbock dined at Down House on 22 February 1863. See also letter from John Lubbock, 20 February 1863.

To J. D. Hooker 24[–5] February [1863]

Down Bromley Kent
Feb. 24th

My dear Hooker

I am astounded at your note.[1] I have not seen the Athenæum, but I have sent for it, & may get it tomorrow; & will then say what I think.[2] I have read Lyell's book.[3] The whole certainly struck me as a compilation; but of the highest class, for where possible the facts have been verified on the spot, making it almost an original work.[4] The Glacial chapters seem to me the best, & in parts magnificent.[5]

I could hardly judge about Man, as all the gloss of novelty was completely worn off.[6] But certainly the aggregation of the evidence produced a very striking effect on my mind. The Chapter comparing language & changes of species seems most ingenious & interesting.[7] He has showed great skill in picking out salient points in the argument for change of species; but I am deeply disappointed (I do not mean personally) to find that his timidity prevents him giving any judgment.—[8]

The whole discussion I look at as of no more value than a very good Review.— From all my communications with him, I must even think that he has really entirely lost faith in the immutability of species; & yet one of his strongest sentences is nearly as follows "if it should **ever** be rendered highly probable that species change by variation & natural selection" &c &c".—[9] I had hoped he would have guided the public as far as his own belief went. On the contrary, as the Parthenon says, he leaves the Public in a fog.[10] One thing does please me on this subject, is that he seems to appreciate your Work.—[11] No doubt the public, or a part, may be induced to think that as he gives to us a larger space than to Lamarck he must think there is something in our views.—[12] When reading the Brain chapter, it struck me forcibly that if he had said openly that he believed in change of species & as a consequence that man was derived from some Quadrumanum animal, it would have been very proper to have discussed by compilation the differences in the most important organ, viz the Brain. As it is, the chapter seems to me to come in rather by the head & shoulders.—[13] I do not think (but then I am as prejudiced as Falconer & Huxley or more so) that it is too severe; it struck me as given with judicial force.—[14] It might perhaps be said with truth that he had no business to judge on a subject, on which he knows nothing; but compilers must do this to a certain extent. (You know I value & rank high Compilers being one myself!)

I have taken you at your word, & scribbled at great length.[15] If I get Athenæum tomorrow, I will add my impression of Owen's letter.—

Thanks for answer about Glass-Man & Acropera &c.—[16] You cannot imagine what pleasure your plants give me (far more than your dead Wedgwood ware *can* give you):[17] Henrietta[18] & I go & gloat over them; but we privately confessed to each other, that if they were not our own, perhaps we sh^d. not see such transcendent beauty in each leaf.— Have you *Edwardsia tetraptera* from what Trevlranus says I sh^d. much like to observe it.—[19]

If you ever observe any plant with *glandular* hairs which secrete by their tips much viscid matter, will you remember that I sh^d. like a specimen.—[20]

I wonder that you can endure to speak in friendly manner to Owen;[21] but I daresay you are right; indeed, after more than twenty years experience, & permit me to add, admiration of the transparent honesty of your character, I have no doubt that you are right: I would rather trust to your instinctive conscience on such a point than to my own. I despise myself for hating him so much.—

The Lyells are coming here on Sunday Evening to stay till Wednesday.[22] I dread it, but I must say how much disappointed I am that he has not spoken out on Species, still less on Man. And the best of the joke is that he thinks he has acted with the courage of a martyr of old.— I hope I may have taken an exaggerated view of his timidity, & shall *particularly* be glad of your opinion on this head.— When I got his book, I turned over pages & saw he had discussed subject of Species, & said that I thought he could do more to convert the Public than all of us;[23] & now (which makes the case worse for me) I must in common honesty retract. I wish to Heaven he had said not a word on subject.—

Wednesday Morning. I have read the Athenæum. I do not think Lyell will be nearly so much annoyed as you expect. The concluding sentence is no doubt very stinging.[24] No one, but a good anatomist, could unravel Owen's long-winded & confused (as it seems to me) letter; at least it is quite beyond me. Nor do I trust a word he says without careful comparison with the papers he quotes. My impression, however, is clear that he has tacitly wholly changed his ground, & wishes every one falsely to believe that mere size of Brain was his object in the whole controversy.[25] Certainly Owen has not shaken my conviction, after reading what all the many anatomists have written, that he has been grossly wrong; & that he knows it & speaks falsely. Lyell's memory plays him false when he says all anatomists were astonished at Owen's paper;[26] it was often quoted with approbation. I *well* remember Lyell's admiration at this new Classification!! (do not repeat this) I remember it, because though I knew nothing whatever about the Brain,, I felt a conviction that a classification thus founded on a single character would break down; & it seemed to me a great error not to separate more completely the Marsupialia.—[27]

Owen is certainly a very clever fellow & whether right or wrong will completely bewilder the public; as I hear he has already bewildered Sir J. Lubbock.—[28] What an accursed evil it is that there sh[d]. be all this quarrelling within what ought to be the peaceful realms of Science.—

I will go to my own present subject of Inheritance & forget it all for a time.[29]

Farewell | My dear old friend | C. Darwin

Endorsement: '/63'
DAR 115: 183

[1] Letter from J. D. Hooker, [23 February 1863].

[2] In his letter of [23 February 1863], Hooker asked CD for his reaction to the letter published by Richard Owen in the *Athenæum*, 21 February 1863, pp. 262–3, objecting to remarks made by Charles Lyell in a section of *Antiquity of man* entitled 'Whether the structure of the human brain entitles man to form a distinct sub-class of the Mammalia' (C. Lyell 1863a, pp. 480–93).

[3] C. Lyell 1863a.

[4] Hugh Falconer criticised *Antiquity of man* for its reliance on work done by other scientists (see letter from J. D. Hooker, [23 February 1863] and n. 5).

[5] Chapters 12 to 18 of *Antiquity of man* (C. Lyell 1863a, pp. 206–368) discussed the geological evidence of the Pleistocene glacial period, and examined the chronological relations between human and glacial history.

[6] CD refers to chapter 24 of *Antiquity of man* (C. Lyell 1863a, pp. 471–506), 'Bearing of the doctrine of transmutation on the origin of man, and his place in the creation'.

[7] CD refers to chapter 23 of *Antiquity of man* (C. Lyell 1863a, pp. 454–70), 'Origin and development of languages and species compared'. See also letter to Asa Gray, 23 February [1863].

[8] CD expressed his disappointment at what he considered to be Lyell's 'excessive caution' in his letter to Asa Gray of 23 February [1863]. See also n. 9, below.

[9] Lyell concluded chapter 23 of *Antiquity of man* (C. Lyell 1863a, p. 469) with the statement that 'by no means' should one

> undervalue the importance of the step which will have been made, should it ever become highly probable that the past changes of the organic world have been brought about by the subordinate agency of such causes as 'Variation' and 'Natural Selection.'

CD underlined this passage in his copy of the book, which is in the Darwin Library–CUL (see *Marginalia* 1: 526–7).

[10] In a review of C. Lyell 1863a published in the *Parthenon*, 21 February 1863, pp. 233–5, the anonymous author commented (p. 234) that readers would be in a 'fog of bewilderment' as to the value of the evidence in support of the 'development theory', because Lyell had been so 'cautious and sparing of conclusion'.

[11] Lyell cited Hooker extensively in *Antiquity of man*, particularly as an authority on variation in plants (see, for example, C. Lyell 1863a, p. 418).

[12] In chapter 20 of *Antiquity of man*, Lyell discussed older theories of organic transmutation, and particularly the theory propounded by Jean Baptiste de Lamarck (C. Lyell 1863a, pp. 389–406). This chapter formed a preface to an extended discussion of CD's theory of the origin of species by natural selection and Hooker's supporting evidence (*ibid.*, pp. 407–23).

[13] In chapter 24 of *Antiquity of man* (C. Lyell 1863a, pp. 471–506), Lyell discussed the dispute between Owen and Thomas Henry Huxley concerning the comparative anatomy of human and simian brains, an account Hooker suggested was 'virtually Huxley's writing' (see letter from J. D. Hooker, [23 February 1863]).

[14] See letter from J. D. Hooker, [23 February 1863] and n. 3. At this time Huxley and Falconer were both engaged in disputes with Owen (see n. 13, above, letter from Hugh Falconer, 3 January [1863], and letter to Hugh Falconer, 5 [and 6] January [1863]).

[15] See letter from J. D. Hooker, [23 February 1863].

[16] See letter from J. D. Hooker, [23 February 1863] and nn. 10–13. The reference is to the glass manufacturers James Powell & Sons of Fleet Street, Whitefriars, London.

[17] On 20 February 1863, by arrangement with Hooker, CD received a cart-load of plants for his hothouse from the Royal Botanic Gardens, Kew (see letters to J. D. Hooker, 15 February [1863] and [21 February 1863]). CD also refers to Hooker's interest in collecting Wedgwood ware (see letter from J. D. Hooker, [16 February 1863] and n. 8).

[18] Henrietta Emma Darwin.

[19] In a section discussing *Orchids*, the German botanist Ludolph Christian Treviranus provided a description of how the flowers of *Edwardsia tetraptera* secreted nectar after their stamens fell (Treviranus 1863a, p. 10). CD's annotated copy of Treviranus 1863a is in the Darwin Pamphlet Collection–CUL. CD was interested in using the case to confirm his hypothesis that nectar could be present in cellular tissue, rather than contained in a nectary (see *Orchids* 2d ed., p. 41 n.). This was significant in explaining why many common orchids requiring insects to effect pollination nevertheless had nectaries which did not contain nectar. See also letter to Roland Trimen, 16 February [1863] and n. 4.

[20] CD had been working intermittently on insectivorous plants, especially *Drosera rotundifolia*, since the summer of 1860 (see *Correspondence*, vols. 8–10); however, he did not work extensively on the subject until 1872 (*LL* 3: 322). See also *Correspondence* vol. 10, letter to J. D. Hooker, 9 [April 1862]

[21] See letter from J. D. Hooker, [23 February 1863] and n. 7.

[22] Charles and Mary Elizabeth Lyell had been invited to Down House for a few days, from 1 to 4 March 1863, but CD became ill with 'much sickness & weakness', and was obliged to retract the invitation (see letter to J. D. Hooker, 5 March [1863]).

[23] See letter to Charles Lyell, 4 [February 1863]. CD gave his full reaction to C. Lyell 1863a in his letter to Charles Lyell, 6 March [1863].

[24] Owen concluded his letter to the *Athenæum* (see n. 2, above), with the statement:

> The tree is known by its fruit; and if such be the produce of the fair tree of knowledge which Lyell has so skilfully decked and set off to the world, it cannot be a sound and wholesome one.

[25] Owen's original claim was that on the basis of structural differences in the brain, the genus *Homo* should be classified as a distinct sub-class of the mammalia (see n. 27, below).

[26] In his letter to the *Athenæum* (see n. 2, above), Owen accused Lyell of implying that the anatomical world had been shocked by the inclusion of erroneous data in his reclassification of the mammalia according to cerebral characteristics (Owen 1857). In fact, Lyell did not comment on the initial reception of Owen's paper in *Antiquity of man* (C. Lyell 1863a, pp. 480–5).

[27] In Owen's reclassification of the mammalia (Owen 1857, p. 37), he divided the class into four sub-classes according to the structure of the brain: the 'Lyencephala', the 'Lissencephala', the 'Gyrencephela', and the 'Archencephala'. Humans were placed in the distinct sub-class of 'Archencephala', based principally on three cerebral characteristics that Owen argued belonged exclusively to humans (*ibid.*, pp. 19–20). See also letter to J. D. Dana, 5 April [1857], and letter to J. D. Hooker, 5 July [1857] (*Correspondence* vol. 6).

[28] John William Lubbock, CD's neighbour in Down; his son, John Lubbock, may have discussed his father's views with CD when he dined at Down House on 22 February 1863 (see preceding letter).

[29] CD was writing a draft of the chapters on inheritance for *Variation* (see 'Journal' (Appendix II)).

To T. H. Huxley [before 25 February 1863][1]

Lecture IV P. 89 Atavism[2]
Lecture VI P. 151 line 7 from top wetting **feet**: bodies?[3]

Miss Henrietta Darwin's criticisms.—[4]

You here & there use Atavism=Inheritance.—[5] Duchesne, who, I believe invented word in his Strawberry Book,[6] confined it, as everyone else has since done, to resemblance to grandfather or more remote ancestor, in *contradistinction*, to resemblance to parents.—

C. Darwin.

Incomplete[7]
Imperial College of Science, Technology and Medicine Archives (Huxley 5: 181)

[1] The date is established by the relationship between this letter and the following letter.

[2] The reference is to the published version of Huxley's 1862 lectures to working men at the Museum of Practical Geology (T. H. Huxley 1863a, p. 89), where he wrote:

> We notice the same thing occurring in the cases of the domestic animals—dogs, for instance, and their offspring. In all these cases of propagation and perpetuation, there seems to be a tendency in the offspring to take the characters of the parental organisms. To that tendency a special name is given—and as I may very often use it, I will write it up here on this black-board that you may remember it—it is called *Atavism*; it expresses this tendency to revert to the ancestral type, and comes from the Latin word *atavus*, ancestor.

The passage remained unchanged when the lectures were republished (T. H. Huxley 1893, pp. 397–8).

[3] In T. H. Huxley 1863a, p. 151, it was stated that:

> Lamarck thought that by a very simple supposition . . . he could explain the origin of the various animal species: he said, for example, that the short-legged birds which live on fish, had been

converted into the long-legged waders by desiring to get the fish without wetting their feet, and so stretching their legs more and more through successive generations.

The reference to 'feet' was changed to 'feathers' when the lectures were republished (T. H. Huxley 1893, p. 468).

4 Henrietta Emma Darwin. Huxley sent CD copies of his lectures, as they became available, in six separate parts (T. H. Huxley 1863a). CD commented on the first five lectures in his letters to Huxley of 7 December [1862], 18 December [1862], and 28 December [1862] (*Correspondence* vol. 10). CD's annotated copies of the six parts of T. H. Huxley 1863a are in the Darwin Library–CUL (see *Marginalia* 1: 423–4).

5 See n. 2, above. CD had been working on atavism in the course of preparing a draft of the chapters on inheritance for *Variation* (see *Variation* 2: 28–61, and letter from Henry Holland [10 February 1863]).

6 Duchesne 1766.

7 The first two lines are in Henrietta Emma Darwin's hand.

From T. H. Huxley 25 February 1863

Jermyn St
Feby 25th 1863

My dear Darwin

Please to say to Miss Henrietta Minor Rhadamanthus Darwin that I plead guilty to the justice of both criticisms & throw myself on the mercy of the Court[1]

As extenuating circumstances with respect to indictment N°. 1. see prefatory notice.[2]

Extenuating circumstance N° 2: that I picked up 'Atavism' in Pritchard years ago—and as it is a much more convenient word than 'Hereditary transmission of variations' it slipped into equivalence in my mind—and I forgot all about the original limitation[3]

But if these excuses should in your judgment tend to aggravate my offences suppress 'em, like a friend.

One may always hope more from a lady's tenderheartedness than from her sense of justice

Publisher has just sent to say that I must give him any corrections for second thousand of my booklet immediately—[4]

Why did not Miss Etty send any critical remarks on that subject by the same post? I should be most immensely obliged for them

Ever | Yours faithy | T. H. Huxley

DAR 166: 299

1 See preceding letter. The reference is to Henrietta (Etty) Emma Darwin. 'Rhadamanthus' is a character from Greek mythology: a son of Zeus and Europa and one of the judges in the lower world. The term is used allusively to denote an 'inflexible judge; a rigorous or severe master' (*OED*).

2 At Huxley's request, Robert Hardwicke, the publisher of Huxley's 1862 lectures to working men (T. H. Huxley 1863a), included a prefatory notice by Huxley explaining that while he had given Hardwicke's shorthand writer, J. Aldous Mays, permission to take lecture-notes with a view to publishing them, he had had 'no leisure to revise the Lectures, or to make alterations in them, beyond the correction of any important error in a matter of fact'.

[3] The reference has not been identified.

[4] Huxley's *Evidence as to man's place in nature* (T. H. Huxley 1863b), was first published on 20 February 1863 by the London publishers Williams & Norgate; it sold quickly and a second printing was called for almost immediately. The publishers launched the second printing on 21 March 1863 (*Publishers' Circular*, 16 February 1863, p. 85, and 1 April 1863, p. 181; see also L. Huxley ed. 1900, 1: 201–2). CD sent his comments on T. H. Huxley 1863b in his letter to Huxley of 26 [February 1863].

From George Maw 25 February 1863

Benthall Hall, | nr. Broseley.
25th. Feby 63.

Dear Sir.

I beg to thank you for your note of the 23rd. the contents of which will I know gratify the committee of our field club[1]

I have much pleasure in sending you a curious form of Lilium candidum in which the buds instead of being developed as perfect flowers have expanded into a spike of leaves in *color* & *texture* resembling the perianth but in arrangement & form more like the *stem* leaves[2]

I fear I have nothing more either of facts or specimens in this way that would be new to you but will in future be on the look out. With regard to the direct deposition of coal it has often occurred to me as being rather singular that the fire clays & shales intervening between the seams of coal are as full of organic *impressions* as the coal itself & yet unaccompanied by carbonaceous matter[3] I cannot help fancying that there must have been aggregation into seams after deposition similar to the aggregation of carbonate of Iron into thin strata & nodules—

yesterday evening I had the pleasure of being introduced to your friend Mr. Crotch who lectured at Bridgnorth on the "mutual relation of species" as a sort of popular exposition of the heads of the subjects treated of in your origin of species—[4]

Believe me to remain Dr Sir | yours very truly | George Maw

Charles Darwin Esq

DAR 171: 98

[1] CD's letter has not been found. Apparently, CD responded favourably to the invitation to become a honorary member of the Severn Valley Naturalists' Field Club (see letter from George Maw, 19 February 1863).

[2] CD had evidently asked Maw if he could supply examples of bud-variation (see letter to George Maw, 28 February [1863] and n. 2.) CD had solicited examples of bud-variations from several of his correspondents, including John Scott, Thomas Rivers, Hugh Falconer, and George Henry Kendrick Thwaites (see *Correspondence* vols. 9 and 10, and this volume, letter to Asa Gray, 2 January [1863], letter to Thomas Rivers, 1 February [1863], and letter to John Scott, 16 February [1863]). CD had recently completed a draft of chapter 11 for *Variation*, 'On bud-variation, and on certain anomalous modes of reproduction and variation' (see 'Journal' (Appendix II)).

[3] See letter from George Maw, 19 February 1863.

[4] Maw refers to the entomologist William Duppa Crotch. Bridgnorth is a town in Shropshire, south of Shrewsbury. See also letter to George Maw, 28 February [1863].

From J. D. Hooker [26 February 1863][1]

[Royal Gardens Kew]
Thursday

D[r] Darwin

A million thanks for your long & capital letter.[2] With all my attempts & wishes I have not been able to get through $\frac{1}{3}$ of Lyells book.[3] What I have read I like extremely, *especially* as you do the Glacial Chapter—[4] The fault is the unintentional impression conveyed, that untill he Lyell went & confirmed all & everybodies observations &c &c—they were little worth![5] Also that the book is far too long, or far too short—neither a summary nor a treatise

I am disappointed beyond measure at what you tell me of his withholding his own opinions on the origin & man questions—& am justly? wroth; for I have been holding Lyell up as a very godlike philospher for changing his views (under full conviction) after the 5th. decade of his life—for hoisting *his self* with his own petard"[6] & laying the gunpowder of variability of species under the fortress of the old "Principles" on which his horn was so justly exalted— to me what you tell me is a very *great disappointment*.[7]

I heard last night (from Murray) that he will not answer the Athenæum—[8] I am glad of this, for though I must confess it will be against Lyell & a terrible punishment to Owen!—it is his **only** course. The impression raised on all hands is already I can quite see unfavorable to him in various ways.— "he had no business to go into the subject if he could not defend his own position in it".— "he ought to acknowledge the ill-feeling & vindicate the holding it" &c &c &c— Then the not answering will be attributed to Lyell's known timidity. &c &c.

I am most comforted by what you think of Lyells not feeling it, as much as I suppose—[9] that is to me the great matter now, that "what's done can't be helped".

I feared its deeply hurting him—& preying on his mind; it would me, & I should make a clean breast of it in Athenæum As it is it will do Lyell an immense deal of injury I think.

As to the falseness futility & Jesuitry of Owens whole letter, it cannot be exaggerated—as you say, he has thrown a fog over the whole subject— I cannot understand his letter at all—but this was part of his plan—the first & last paragraphs are intelligeable enough![10]

I need not say that everyone says that Lyell's Brain-chapter was Huxley's writing, though not his words.[11] I have just received H's coarse-looking little book.—[12] not fit as somebody said to me, for a gentlemans table— I am sorry for this— Falconer, who has the most delicate & refined sense in such matters of any man I ever met,—is disgusted with the wood-cut of the shambles, & would let no young Lady look at it.[13]

But oh Lord I shall never stop at this rate.

Ever yours affec | J D Hooker

DAR 101: 108–10

[1] The date is established by the relationship between this letter and the letters to J. D. Hooker, 24[–5] February [1863] and from J. D. Hooker, [1 March 1863]; the intervening Thursday was 26 February.
[2] Letter to J. D. Hooker, 24[–5] February [1863].
[3] Hooker refers to Charles Lyell's *Antiquity of man* (C. Lyell 1863a).
[4] Chapters 12 to 18 of Lyell's *Antiquity of man* (C. Lyell 1863a, pp. 206–368) were concerned with the geological evidence of the Pleistocene glacial period, and examined the chronological relations between human and glacial history. CD expressed his admiration of these chapters in his letter to Hooker of 24[–5] February [1863].
[5] Hugh Falconer also criticised Lyell for giving insufficient credit to the work of others in C. Lyell 1863a (see letter from J. D. Hooker, [23 February 1863] and n. 5).
[6] Hooker alludes to Shakespeare's *Hamlet*, 3.4 (see Wells and Taylor eds. 1988, p. 689).
[7] See letter to J. D. Hooker, 24[–5] February [1863]. In *Principles of geology* (C. Lyell 1830–3), Lyell had published an important critique of the then current theories of transmutation.
[8] Hooker refers to John Murray, publisher of C. Lyell 1863a, and to Richard Owen's letter, published in the *Athenæum* on 21 February 1863, pp. 262–3, criticising Lyell's summary of Owen's work on the comparative anatomy of human and simian brains (see letter from J. D. Hooker, [23 February 1863] and n. 2).
[9] In his letter of [23 February 1863], Hooker expressed concern that Owen's letter to the *Athenæum* (see n. 8, above), would do Lyell 'awful injury'; in reply, CD wrote that he thought Lyell would not be 'nearly so much annoyed' as Hooker feared (see letter to J. D. Hooker, 24[–5] February [1863]).
[10] The opening and closing paragraphs of Owen's letter to the *Athenæum* described the means by which Owen had first learned of Lyell's attack on his work (see n. 8, above), and his subsequent feelings of injustice.
[11] Hooker refers to Thomas Henry Huxley and to C. Lyell 1863a, pp. 471–506. See also letter to J. D. Hooker, 24[–5] February [1863] and n. 13.
[12] The reference is to Huxley's *Evidence as to man's place in nature* (T. H. Huxley 1863b).
[13] The reference is to Hugh Falconer, and to a woodcut depicting a 'Butcher's Shop of the Anziques, Anno 1598'. In the illustration, a cannibal butcher is shown dismembering a human, while his shop displays a human head and body parts for sale (T. H. Huxley, 1863b, p. 55).

To T. H. Huxley 26 [February 1863][1]

Down
26th

My dear Huxley

I have just finished with very great interest "Man's Place".[2] I never fail to admire the clearness & condensed vigour of your style,—as one calls it, but really of your thought.— I have no criticisms; nor is it likely that I could have. But I think you could have added some interesting matter on the character or disposition of the young Ourangs which have been kept in France & England. I shd. have thought you might have enlarged a little on the later embryological changes in man & on his rudimentary structure—tail as compared with tail of higher monkeys intermaxillary bone, false ribs & I daresay other points—such as muscles of ears &c &c.—

I was very much struck with admiration at the opening pages of Part II (& oh what a delicious sneer, as good as a dessert, at p. 106.):[3] but my admiration is unbounded at p. 109 to 112.[4] I declare I never in my life read anything grander.

Bacon himself could not have charged a few paragraphs with more condensed & cutting sense than you have done.[5] It is truly grand. I regret extremely that you could not, or did not, end your book (not that I mean to say a word against the Geolog. Hist.) with those pages. With a Book, as with a fine day, one likes it to end with a glorious sunset.—

I congratulate you on its publication; but do not be disappointed if it does not sell largely; parts are highly scientific, & I have often remarked that the best book often do not get soon appreciated; certainly large sale is no proof of the highest merit.— But I hope it may be widely distributed & I am rejoiced to see in your note to Miss Rhadamanthus that a second thousand is called for of the little Book.—[6]

What a letter that is of Owen's in the Athenæum;[7] how cleverly he will utterly muddle & confound the public.— Indeed he quite muddled me, till I read again your "concise statement"[8] (which is capitally clear) & then I saw that my suspicion was true that he has entirely changed his ground to size of Brain. How candid he shows himself to have taken the slipped brain![9] I am intensely curious to see whether Lyell will answer. Lyell has been, I fear, rather rash to enter on subject on which he of course knows nothing by himself.

By Heavens Owen will shake himself, when he sees what an antagonist he has made for himself in you.—

with hearty admiration Farewell | C. Darwin

I am fearfully disappointed at Lyells excessive caution in expressing any judgment on Species or origin of man.—

Imperial College of Science, Technology and Medicine Archives (Huxley 5: 191)

1 The date is established by the relationship between this letter and the letter from T. H. Huxley, 25 February 1863.

2 CD refers to Huxley's *Evidence as to man's place in nature* (T. H. Huxley 1863b); CD's annotated copy is in the Darwin Library–CUL (see *Marginalia* 1: 423–4).

3 Part 2 of T. H. Huxley 1863b, pp. 57–118, examined 'the relations of man to the lower animals'. CD also refers to the paragraph in which Huxley ridiculed Richard Owen's axiom concerning the 'continuous operation of the ordained becoming of living things'. Owen first used the phrase in his presidential address to the British Association for the Advancement of Science in 1858 (Owen 1858, p. li); he also used the phrase in his unsigned review of *Origin* ([Owen] 1860a, p. 500). Huxley wrote (T. H. Huxley 1863b, p. 106):

> At the present moment, but one such process of physical causation has any evidence in its favour; or, in other words, there is but one hypothesis regarding the origin of species of animals in general which has any scientific existence—that propounded by Mr. Darwin. . . . and though I have heard of the announcement of a formula touching 'the ordained continuous becoming of organic forms,' it is obvious that it is the first duty of a hypothesis to be intelligible, and that a qua-quâ-versal proposition of this kind, which may be read backwards, or forwards, or sideways, with exactly the same amount of signification, does not really exist, though it may seem to do so.

4 CD refers to the conclusion of part 2 of *Evidence as to man's place in nature* (see n. 3, above), where Huxley addressed the 'repugnance' he believed the majority of his readers would feel in response to his conclusions regarding human origins. He dismissed the cry, 'We are men and women, not a mere better

sort of apes', as irrelevant, repeating that he saw no basis for a structural or 'psychical' line of demarcation between the animal world and humankind. Yet, he maintained, this should not detract from a sense of grandeur when contemplating humanity's place in nature in relation to that of the 'brutes', for 'whether *from* them or not, [humanity] is assuredly not *of* them' (T. H. Huxley 1863b, pp. 109–12).

5 CD refers to the philosopher and essayist, Francis Bacon.

6 Huxley's *Evidence as to man's place in nature* sold quickly, notwithstanding its 'highly scientific' nature (see letter from T. H. Huxley, 25 February 1863, n. 4). In his letter to CD of 25 February 1863, Huxley nicknamed Henrietta Emma Darwin 'Miss Henrietta Minor Rhadamanthus Darwin', after receiving her criticisms of T. H Huxley 1863a.

7 CD refers to Owen's letter published in the *Athenæum* on 21 February 1863, pp. 262–3, under the heading 'Ape-origin of man as tested by the brain'. In the letter Owen defended himself against the criticisms implicit in Charles Lyell's assessment (C. Lyell 1863a) of Owen's contribution to the debate on human and simian brain anatomy.

8 CD refers to the section in Huxley's *Evidence as to man's place in nature* entitled 'A succinct history of the controversy respecting the cerebral structure of man and the apes' (T. H. Huxley 1863b, pp. 113–18).

9 In his letter in the *Athenæum*, 21 February 1863, pp. 262–3, Owen defended his use of the chimpanzee brain drawn by Jacob Lodewijk Coenraad Schroeder van der Kolk and Willem Vrolik in 1849 (Schroeder van der Kolk and Vrolik 1849). Owen had originally used their diagrams to illustrate the degree to which the cerebrum covers the cerebellum in the highest apes (Owen 1861a). Lyell had pointed out Owen's error in drawing this conclusion from the dried and shrunken specimen represented by these diagrams (C. Lyell 1863a, pp. 481–6). In his letter to the *Athenæum*, Owen rejected Lyell's criticisms and claimed that he had used the diagrams solely to give an indication of the relative size of chimpanzee and human brains. See also letter to J. D. Hooker, 24[–5] February [1863] and n. 25.

From Francis Walker 26 February 1863

Church End | Finchley
Feb 26th. 63

My dear Sir,

I am sorry that I have so long delayed sending you the names of the flies that I received from you[1]

They are as follows,

1 Eulophus Westwoodii 2 Tetrastichus Diaphantus 3 di caudatus	} Hymenoptera chalcidites.
4 Scatopse brevicornis	} Diptera

They frequent flowers, but the three first are parasitic, or lay their eggs in the bodies of larvæ.—

Believe me | Yours truly | F. Walker

DAR 181: 3

CD ANNOTATION

Top of letter: 'Insects with pollinia of Musk. Orchis—'[2] *ink*

[1] CD visited the British Museum when he was in London at the beginning of February (see letter from Frederick Smith, 11 February 1863, and letter from S. P. Woodward, 14 February 1863), and may have asked Walker to identify the insects on that occasion. Walker had worked particularly on the Hymenoptera, Diptera, and the Chalcids, and compiled numerous catalogues of the British Museum's insect collections (*Entomologist's Monthly Magazine* 11 (1874): 140–1). CD acknowledged Walker's assistance in the identification of Diptera visiting British species of orchids in 'Fertilization of orchids', p. 142 (*Collected papers* 2: 139).

[2] CD was evidently collecting information on the insect pollinators of the musk orchis *Herminium monorchis* (see also *Correspondence* vol. 10, letter from Frederick Smith, 28 June 1862). His conclusions were incorporated into 'Fertilization of orchids', p. 145 (*Collected papers* 2: 142) and *Orchids* 2d ed., p. 61, where CD stated:

> When the first edition of this book appeared I did not know how the flowers were fertilised, but my son George has made out the whole process, which is extremely curious and differs from that in any other Orchid known to me. He saw various minute insects entering the flowers, and brought home no less than twenty-seven specimens with pollinia (generally with only one, but sometimes with two) attached to them. These insects consisted of minute Hymenoptera (of which *Tetrastichus diaphantus* was the commonest), of Diptera and Coleoptera, the latter being *Malthodes brevicollis*.

CD's notes on George Howard Darwin's observations, dated 22–7 June 1862, are in DAR 70: 32–6.

From Daniel Oliver 27 February 1863

Royal Gardens Kew
27.ii.1863

My Dear Sir.

You ask about *Primula longiflora*.[1] It is *not* near *P. scotica*. It is an Alpine plant with Corollas $\frac{1}{4}$ in. long or more The style consider^ly. exserted in most or all(?) our spec^ms.— *P. scotica* I have gathered in Sutherlandsh.[2] but never noticed dimorphism. From a few plants bro't home & cultivated it w^d. hardly be safe to settle it. I put in a few flowers wh. indicate varying length of style (I may be mistaken);—these are from M^r. Brown's Herbarium.[3]

I sh^d. much like a 2 page Abstract of your *Linum* paper for next No. of N.H.R. but doubt if I have chance of seeing it in type in time.[4]

Very Sincerely yours | D^l. Oliver

DAR 108: 178

[1] CD had asked Oliver whether *Primula longiflora* was '*closely* allied' to *P. scotica* after being informed that both species were non-dimorphic (see letter to Daniel Oliver, 20 [February 1863] and n. 4).

[2] The reference is to Sutherlandshire, Scotland.

[3] Oliver refers to the herbarium of Robert Brown, keeper of the botanical department at the British Museum until his death in 1858. Brown left his herbarium to his assistant and successor John Joseph Bennett. The collection was kept in a room in the basement of the British Museum, and 'working botanists' were allowed access (*Journal of Botany, British and Foreign* n.s. 5 (1876): 192; Mabberley 1985, p. 387).

[4] Oliver was one of the editors of the quarterly journal *Natural History Review*, and was preparing the April issue. 'Two forms in species of *Linum*', read before the Linnean Society on 5 February 1863, was not available in print until 13 May 1863 (*General index to the Journal of the Linnean Society* p. vi). CD,

however, arranged for a number of offprints of the paper to be distributed in mid-April; Oliver's name appears on the presentation list (see Appendix IV). The paper was reviewed in the July issue of the *Natural History Review* (*Natural History Review* 3: 476–8).

From Thomas Wright 27 February 1863

St. Margarets Terrace | Cheltenham
Feb 27/63

Dear Sir

The statement referred to in your favor of the 25th. inst. was not made by me, perhaps Dr. Perceval Wright F.L.S. was the observer, as I was not present at the Meeting for 1860.[1] If I had any observations or experiments upon the subject I should be delighted to place them at your disposal as I am sure no Naturalist makes a better use of his facts than yourself.

I am grateful to have it in my power to tell you that your valuable ⟨work⟩ on the Origin of Species has done excellent service to Natural Histo⟨ry;⟩ and even to those who are not disposed to agree with you in all your Conclusions relative to transmutation. It has I am happy to say given a great check to species manufacture and induced persons who formerly could see nothing but a new form in every slight variation of structure to reconsider their facts before proposing new sp to our already crowded lists this I think is a matter of great practical value for at the rate at which we were proceeding in former years nothg but confusion could have been the result. now I find that the good old doctrine I have been endeaving to inculcate for 30 years begins to get a fair hearing, namely that the forms ⟨c⟩alled species appear in all time to have had ⟨cer⟩tain limits of variations which they have ⟨exhi⟩bited under the influence of physical condit⟨ion⟩s and which limits are not the same for every sp. some forms presenting wide deviation from what may be called their typical characters, whilst others move in very limited orbits. I have endeavoured to inculcate this in my Monograph on the Echinodermata and have illustrated it very many examples.[2]

For some years past I have been studying the Ammonites with the view to publish a Monogr. on them but among them this law of variation is very great indeed so much so that I have not been able to collect materials for more than a very small number of sp. for unless I can obtain an Ammonite in its young, growing, adult and ancient condition I hold that no one is competent to speak of it as a species this I have been able to do with a few only out of the so called 100ds. of sp⟨.⟩ contained in the lists[3]

in much haste believe me always Yours Most truly | Thomas Wright

I expect if I am fortunate in obtaining a sufficient amount of material & make a beginning we shall have fine weeping and waling over "good species" which alas only turn out to be variations of other forms.— I expect my Monogr. on the Oolitic Asteridæ which has been done 8 Mos. and ought to have been out long ago will shew likewise that with a disposition to change there is likewise a law of constancy for you will find in some of the skeletons of the old Starfishes the persistent types

of existing structures which have not changed during all that vast period of time which has elapsed between the Liassic and the present period[4]
T. W.

DAR 181: 178

[1] CD's letter has not been found. Wright refers to the meeting of the British Association for the Advancement of Science held at Oxford between 26 June and 3 July 1860, at which Edward Perceval Wright served as one of the secretaries of Section D, 'Zoology and botany, including physiology'. On Tuesday 2 July 1860 in Section D, in the debate following Philip Lutley Sclater's paper on the geographical distribution of animals (Sclater 1860), the origin of species was discussed. A report of the exchanges in the *Athenæum* noted that Thomas Henry Huxley had challenged the assertion that domesticated forms, when allowed to run wild, returned to their primitive type; it was reported that Dr Wright replied: 'that he had tried experiments on the cultivated cabbage, and, although it degenerated, it never assumed the form of the genuine wild plant, *Brassica oleracea*' (*Athenæum*, 14 July 1860, p. 65). CD had probably written to Thomas Wright for clarification of the statement, unaware that he had mistaken the identity of the author.

[2] Wright refers to his monograph on oolitic Echinodermata, the first volume of which had been published in parts by the Palaeontographical Society (T. Wright 1857–80). See also n. 4, below.

[3] T. Wright 1878–86.

[4] The first part of the second volume of Wright's monograph on oolitic Echinodermata (see n. 2, above), was devoted to the 'Asteroidea' (starfishes), and was published in 1863 (T. Wright 1857–80). The Liassic denotes a period synonymous with the early Jurassic epoch, a period of geological time now defined as beginning 210 million years ago (*Collins dictionary of geology*).

From John Lubbock 28 February 1863

15, Lombard Street. E.C.
28 Feb/63

My dear Mr. Darwin

Unless I hear from you to the contrary I will come up to dinner on Monday.[1]

Yours affec | John Lubbock

DAR 170: 37

[1] Lubbock probably did not dine with the Darwins on 2 March 1863, as CD was ill; Emma Darwin recorded his condition in her diary (DAR 242): he was 'sick several times' during the course of the week, 'faint in night', and 'languid & heavy' every morning. Charles and Mary Elizabeth Lyell had been invited to stay at Down House for a few days at this time, from 1 to 4 March 1863, but their visit was cancelled (see letter to J. D. Hooker, 5 March [1863]).

To George Maw 28 February [1863][1]

Down. | Bromley. | Kent. S.E.
Feb. 28th

My dear Sir

I am much obliged for the curious Lily: I do not think it comes under my case, for I suppose it is not capable of propagation by bulbs.—[2]

When Dr Hooker comes here next he shall see it, so it shall be made best use of.—[3] I am glad you stick to your view, though I think ultimately you will have to give it up.[4] It is an old theory of mine, that no one can be a good & *original* observer unless he forms a multitude of theories & destroys a multitude of theories. Publication of theories is another question.

Almost the best papers I have ever read on Coal are some lately published in late numbers of Sillimans American Journal by Lesquereux.—[5] They would be worth your reading & you will like them all the better, as they give the "Origin of Species" a few little unpleasant kicks.—[6]

I have the pleasure of knowing Mr Crotch only by a little correspondence.[7] I am told he is a bold advocate of mutability of Species.—

My dear Sir | Yours sincerely | C. Darwin

Royal Horticultural Society

[1] The year is established by the relationship between this letter and the letter from George Maw, 25 February 1863.

[2] With his letter of 25 February 1863, George Maw enclosed a specimen of *Lilium candidum*. CD did not consider it a case of bud-variation because 'modifications which arise through bud-variation can generally be propagated . . . by grafting, budding, cuttings, bulbs, &c., and occasionally even by seed' (*Variation*, 1: 373), and this was not the case with the aberrant flowers of *L. candidum*.

[3] Joseph Dalton Hooker visited CD at Down House on 22 March 1863 (see letter from J. D. Hooker, [24 March 1863]).

[4] CD refers to Maw's theory of the deposition of coal (see letters from George Maw, 19 February 1863 and 25 February 1863).

[5] Lesquereux 1859–63. The Swiss bryologist and palaeontologist Leo Lesquereux sent CD the second and third parts of his seven-part series of articles on the coal formations of the United States, published in the *American Journal of Science and Arts*, founded and edited by Benjamin Silliman. The annotated copies are preserved in the Darwin Pamphlet Collection–CUL.

[6] Maw had published a review of *Origin* ([Maw] 1861) in which he made a number of criticisms that CD thought were 'capitally & very originally' put (see *Correspondence* vol. 9, letter to George Maw, 19 July [1861]). Lesquereux had argued (Lesquereux 1859–63, pt 3, pp. 380–1) that the coal formations of North America did not provide evidence of evolutionary change in plant species.

[7] See letter from William Duppa Crotch, 25 January 1861 (*Correspondence* vol. 9).

From J. D. Hooker [1 March 1863][1]

Kew
Sunday.

Dear Darwin

You will be glad to know that Lubbocks lecture was a complete success, & quite admirable I thought—[2] I never before recognized any satisfactory evidence of a probably distinct Stone, Iron & Bronze age. Lubbock put it all in the most satisfactory light that the subject admitted of, with excellent skill and judgement.—[3] (I wish he had written the chapters thereon in Lyell's book.).[4]

I have pretty nearly got through my Examination week's work & shall resume Lyell at once—[5]

Many thanks for A Grays letter—[6] I am always glad to see them, he never alludes to politics or "Darwinism" in his letters to me.[7] the War has made him very bumptious in many ways I fear.— & I often think that one of the worst effects of a war is the acerbation of feelings that it brings about. It certainly seems wonderfully to have dwarfed Gray's intellect— how he can so utterly misconceive the respective positions and attitudes of England & America is inconceivable to me, but it is impossible to undeceive him,—to attempt it would madden him.— His account of his "young Rothricks propensities is *sickening*,—"w^d^ make wounds rather than tend them"![8]

Rolleston's letter will please Lyell & I was glad to see it.[9] R. is a great favorite amongst the orthodox in Oxford, & is perhaps Owen's most formidable enemy, *socially* certainly he is so

P.S. I have progressed through Lyells Glacial Chapters, & can quite see they are far the best in the book & quite excellent.—[10] the subject is freely & boldly handled, & bears the stamp of his intellect his Experience & his **opinion** he throws himself into the subject—& carries his head erect through it like the Master he is— the wind up about Ramsay is excellent, & struck me as very original & able, but I am not convinced by it. it requires several readings.[11]

I do wish he could have begun or ended his book with the Glacial discussion & worked down to the recent times— The perplexity of post-pliocenes & post-tertiary is very irritating at the beginning for I must tell him how feeble the Cave, Stone, &c. discussion is, *comparatively*,[12] & how terribly disappointing the termination of the chapters on origin & on Man.—[13]which I have again read over, though still not with sufficient care. One leaves off absolutely hungry & thirsty for his own *imprimatur* on the evidence he so glibly adduces & so ably marshals. I think very highly of all the latter part of the book, in every way but this,: & it is the opinion of all I have asked that the want of a summary & authoratative expression of opinion should not have been foregone, & can-not be forgiven— it is an abandonment of Lyells high position,— —he is like a King ducking down behind his throne, when obnoxious nobles crowd & frown. And we who quote him will have this thrown in our teeth ever after— In this Lyell & Stanley are in the same unenviable position, of reticence of their opinions on questions of the gravest import to the progress of Science.[14] To me & my papers he has done far more than mere justice—& I feel greatly gratified,[15] nay more, *touched*, by the evident kindly spirit that I can appreciate throughout.

I have a deal more to say, but no time to say it now. Ever yours affec | J D Hooker

Lyell is I suppose with you[16] if you *the least* care to, pray read him the above, I must tell him the same in pretty much the same words. & it is perhaps better that he should hear it in words not written under the vestment of a personal communication—if you do read it, please read this too.

DAR 101: 111–13

[1] The date is established by the reference to Lubbock 1863d (see n. 2, below), and by the relationship between this letter and the letter to J. D. Hooker, 5 March [1863]; the preceding Sunday was 1 March.

[2] John Lubbock delivered a lecture on ancient Swiss lake-habitations at the Royal Institution on 27 February 1863 (Lubbock 1863d; see also *Correspondence* vol. 10, letter from John Lubbock, 23 August 1862).

[3] In his paper, Lubbock argued that the lake-villages belonged to three periods, distinguishable on the basis of the materials from which the tools and other artefacts were made, namely, Stone Age, Bronze Age, and Iron Age (Lubbock 1862d, p. 29). Although this three-age system had been accepted by many archaeologists in Scandinavia and other parts of Europe since its invention in 1819, most of the British archaeological community criticised it; Lubbock belonged to a small group of younger geological archaeologists who adopted the system. See Van Riper 1993, pp. 41–2, 193–7.

[4] Charles Lyell discussed the three-age system and the ancient lake-habitations of Switzerland in chapter 2 of *Antiquity of man* (C. Lyell 1863a, pp. 8–32).

[5] Hooker served for many years as a scientific examiner for medical officers in the armed services (L. Huxley ed. 1918, 1: 387); he refers to examinations for admission to the Army Medical Service, held at Chelsea Hospital in February 1863 (see *Statistical, Sanitary, and Medical Reports* 5 (1865): 582–3, and letter from J. D. Hooker, [23 February 1863]). Hooker also refers to C. Lyell 1863a.

[6] CD had apparently sent Hooker Asa Gray's letter of 9 February 1863; this letter has not been found, but an indication of its contents is given by CD's reply (letter to Asa Gray, 20 March [1863]), and by the letter to H. W. Bates, 4 March [1863].

[7] Hooker and Gray held radically different views on the American Civil War, and had for some time tacitly agreed not to discuss the matter in their letters (see *Correspondence* vol. 10, letters from J. D. Hooker, [19 January 1862] and [14 December 1862], and letter from Asa Gray, 18 February 1862).

[8] The reference is to Joseph Trimble Rothrock, who had been a student of and an assistant to Gray at the Lawrence Scientific School, Harvard University, until the summer of 1862, when he enlisted in the Union army (see *Correspondence* vol. 10, letter from Asa Gray, 5 September 1862). The phrase used by Gray in his letter paraphrased a passage from Humphrey Mill's poem, *The Second Part of the Nights Search*, which reads: 'How those make wounds, that should apply the cure' (H. Mill 1646, p. 161).

[9] In a letter published in the *Athenæum* on 28 February 1863, p. 297, George Rolleston criticised at length three of the statements made in a letter by Richard Owen that appeared in the *Athenæum* on 21 February 1863, pp. 262–3. Owen's letter was a protest concerning the manner in which he felt his long-standing dispute with Thomas Henry Huxley and others, concerning the comparative anatomy of human and simian brains (the so-called 'hippocampus controversy'), had been misrepresented by Lyell in *Antiquity of man* (C. Lyell 1863a); Hooker had been fearful of the effect Owen's letter would have on Lyell (see letter from J. D. Hooker, [23 February 1863] and n. 3).

[10] In chapters 12–18 of *Antiquity of man* (C. Lyell 1863a, pp. 206–368), Lyell discussed recent research on glaciation and outlined a chronology of the Pleistocene glacial period. CD had expressed admiration for the same section of the book in his letter to J. D. Hooker, 24[–5] February [1863].

[11] Hooker refers to Lyell's critique, in C. Lyell 1863a, pp. 311–19, of Andrew Crombie Ramsay's attempt to account for the lakes of Europe and North America by reference to the erosive action of glaciers (Ramsay 1862).

[12] In the classification of the fossiliferous strata outlined in the introduction to *Antiquity of man* (C. Lyell 1863a, pp. 3–7), Lyell divided the post-Tertiary period into post-Pliocene (or lower post-Pliocene) and Recent (or upper post-Pliocene) divisions. Hooker apparently also refers to the extended discussion of prehistoric human remains, including ossiferous caves and stone implements, in chapters 2–11 of *Antiquity of man* (C. Lyell 1863a, pp. 8–205).

[13] Hooker refers to chapters 21 and 24 of *Antiquity of man* (C. Lyell 1863a, pp. 407–23 and 471–506), entitled 'On the origin of species by variation and natural selection' and 'Bearing of the doctrine of transmutation on the origin of man, and his place in the creation', respectively.

[14] The reference is to Arthur Penrhyn Stanley, who was regarded as the leader of a 'Rationalizing school' within the Broad Church party in the Church of England (Liddon 1893–7, 3: 334). When

the rationalist theological work *Essays and reviews* was projected in the late 1850s, Stanley declined to be involved with it, leading one historian to refer to the book as 'Hamlet without the prince' (Ellis 1980, pp. 11–12). Moreover, Stanley's review of *Essays and reviews* ([Stanley] 1861), which was widely expected to be a defence of the beleagured essayists, was 'painfully ambivalent' (Ellis 1980, p. 106). For the context of this debate, see *Correspondence* vol. 9, Appendix VI.

15 See, for example, the section entitled 'Dr. Hooker, on the theory of "Creation by variation" as applied to the vegetable kingdom' (C. Lyell 1863a, pp. 417–21), in which Lyell discussed Hooker's endorsement of natural selection in his *Introductory essay to the flora of Tasmania* (J. D. Hooker 1859), stating that 'no one was better qualified by observation and reflection to give an authoritative opinion on the question, whether the present vegetation of the globe is or is not in accordance with the theory which Mr. Darwin has proposed' (C. Lyell 1863a, p. 418).

16 Charles and Mary Elizabeth Lyell had intended to stay at Down House from 1 to 4 March, but as CD was ill with 'much sickness & weakness', he was forced to cancel the engagement (see letters to J. D. Hooker, 24[–5] February [1863] and 5 March [1863]).

From John Scott 3 March 1863

Botanic Gardens Edinburgh
March. 3^{d}. 1863.

Sir.

I am glad to hear that the variation of Imatophyllum, will be of service to you.[1] I was afraid that it might be of too short a standing for your noticing. It is certainly a very striking one, and may perhaps be the origin of a remarkably distinct race. I will be glad to afford every particular regarding it, you desire.

I have been endeavouring to get seeds of the *wild Maize*, which occurs on the Rocky Mountains; from description it appears to be remarkably distinct from any cultivated form.[2] Crosses with the latter might afford most interesting results: Do you regard *it* as the parent of the numerous forms now in cultivation? I suppose it has only as yet been found wild in the New World. I am daily expecting a selection of the most dissimilar varieties grown by Barr & Sugden.[3] So with those you have favoured me with, I will have a variety of subjects to work upon. I am very desirous, however, for the wild form, but I know not where to get it. I have searched the University Herbarium here, but it does not contain a single fertile specimen.

Are the *Passifloras* mentioned by you in Origin, invariably sterile when treat with 'own-pollen'; or is it a local occurrence?[4] *P. quadrangularis*, for example, fruits freely in some hot-houses, in others, though flowering most profusely it obstinately refuses to fruit. In our gardens a form perhaps a variety of the former called *Bonapartea* will not set a single fruit. It may, however, be otherwise in another locality: Will you be so kind as favour me with the names of any perfectly sterile species, with "own-pollen".? as I find that some of the species I had thought to be sterile are only locally so,: If *such* could be rendered fertile by application of pollen from other species, would they be of any service to you? I would much like a little information on these points; as I am working somewhat blindly: I have had yet little experience. I have the same difficulties with Lobelia:[5] I have made your desire known to Mr. McNab.[6] He informs me that he has already communicated all the information he could gather on weeping-trees; & inheritance of character from seeds, to *M. Neuman* of the Jardin des Plantes; who he supposes will shortly publish on the subject.[7] I do not think, however, from what I have heard him saying, that he has had much practical experience in these matters; his remarks, will I believe principally be founded on the experience of others. What the results may be I know not I am rather inclined to think they will be of a negative nature So far as I can ascertain—from the large experience of some of the nursery foremen here—the character rarely if ever—appears amongst the seedlings.

I did not mean to say that bisexual plants are *more variable* than unisexual, all that I intended, was, that the former presented a greater number of naturally established varieties, than the latter.[8] This I believe, to be a consequence of the more intimate relation of the sexual organs: as affording a better chance for *these* being imbued with *similar* formative tendencies, than occur in the unisexual arrangement. This opinion I hold partly on the grounds that similar variations will rarely appear in a number of individuals at any given time— much more frequently variations will be peculiar to individuals, or even portions of such will be originally representative: and thus in unisexual plants, the *embryo* must necessarily in general be the resultant of organs imbued with dissimilar formative tendencies—the normal specific, & a diverging form. This combination then causing as it does an immediate change in the conditions originating the latter forms and acting upon the most susceptible stages of the organism; will as I believe, cause an immediate reversion in the embryonic resultant, to the normal specifical form. This I think is fully supported by the results of crossing hybrids—incipient forms—with one or other of their original parents, established forms—as we see by a few successive crosses the other form entirely eliminated. Considering then the results from crossing of hybrids—which be it remembered are products of well established forms—what may we expect from such in the case of accidental variations *versus* the normal forms. Again, considering the truly casual agencies employed by Nature for effecting fertilisation in the vegetable kingdom; I really, candidly speaking, find great difficulty in understanding how species can possibly be manufactured by an innate, continuously watchful, selective principle. There seems to me a necessity for admitting the occasional actions of a superinduced power to explain the establishment of new forms of vegetative life. The struggle for existence, does not I think fully meet the difficulty—more especially as applied to annuals—inasmuch as the strongest individuals may be continually fertilised by the weaker, and thus constantly check the domination of their progeny.[9] Again supposing individual utility *alone*, as the great end striven after by Nature in the metamorphosis of organic beings; why should there necessarily be a dependence between the internal structure & external form?

The great morphological differences of the sexual forms of Catasetum, for example, seem to me—though I may be quite wrong—scarcely consistent with a mere regard to specific utility.[10] Is it not highly probable that in their native wilds, different insects will visit the different forms: and thus somewhat frustrate the plans of Nature. Does the occasional occurrence of the three sexual forms of Catasetum on one spike, not oppose the view that these forms were attained by a gradual modification? On the latter view, ought we not to have expected intermediate forms only? The metamorphosis you mention in Orchid Book on the authority of Mr. Rodgers is most remarkable;[11] more especially if we are right in regarding the hermaphrodite Myanthus as the original progenitor? Such an occurrence seemingly disproves what I hinted at in my last regarding Acropera, still no *variation* having ever been observed in the *latter* indicates a morphological status which may justify us in anticipating a homomorphic progeny.[12]

I have thus hurriedly and imperfectly attempted to lay before you a few of the difficulties, which have presented themselves to me, while considering species as products of Natural Selection; these & kindred points I have taken up in my paper.[13] But I know not whether I will yet have made my view sufficiently clear; I fail so completely in expressing myself: as I would like. I hope you will excuse me for troubling you with this subject as after all I may be quite wrong—though it has engaged me now for sometime.

In respect to the opinion that homomorphically fertilised Primulas, will produce their own-form only—I may remark that it occurred to me when I first considered the bearings of these plants upon my views. Since: I have examined the different forms very carefully, and I find in concordance with your observations—that the two forms are most decidedly permanent.[14] I have seen what, I suppose may be a similar case to that referred to by you on the authority of Mr. Wooler: viz, a long-styled variety of P. vulgaris, producing in early spring a number of short-styled flowers.[15] But—mark—there was no change in the position of the stamens: these as far as my observations go are always permanent. From this permanency then I at once concluded as above; having a firm conviction that established seed-variations, will always produce their own characters only, when kept pure: this I anticipate as the result of Vegetable Parthenogenesis—when satisfactorily established.[16] Some of the Cryptogams may be already mentioned as producing female forms only by parthenogenesis, e.g. certain Mosses & Chara. I thus regard Primula & Acropera as parallel cases: I would expect different results from Catasetum &c where individuals still produce the different forms. I do not believe that the relative lengths of stamens & pistils in the two forms of Primula could have been attained if these had been fertilised by *other* than own-form pollen. I am inclined to suppose that they even yet under cultivation, might be caused to revert by continuous crossing of the two forms. *P. Sinensis* is a case in point—which you first drew my attention to—being regularly raised from seed: and here we find pretty frequently a form with stamens & pistils equal.[17] These points, I likewise consider in my paper.[18] I may mention that I have raised long-styled plants of *P. cortusoides* only from own-form pollen. As I had little seed however, I raised only a few plants, consequently I could lay no particular stress on the experiment.

However, I thought I had evidence of such from varieties of P. vulgaris in the gardens here: there being beds of these plants—all of which are *long-styled*. I thought, therefore that they must have been propagated by seed. I have now asked Mr. McNab their history and he informs me that have been merely propagated by division. I have, however, a few seedlings of different forms of Primula all homomorphically fertilised—I expect some of these will be in flower shortly, and afford us some more information on this point.[19]

I am trying at present homomorphic crosses, with different coloured flowers of Primula vulgaris &c.[20] I am following out your hints likewise on decreased fertility between varieties presenting only variation in colour, it may perhaps be a more common occurrence than is supposed.[21] I will be much obliged if you favour me with a perusal of Asa Gray's 2d. review of Orchid Book.[22] I will return it immediately. I am busy at present with a *Bletia*;[23] I have at last been perfectly successful in my attempts to fertilise Gongora—it is by forced & unnatural work however—I have no less than fourteen capsules on a single spike swelling beautifully—I could almost have wished that it had been otherwise.[24] I will give you particulars when I write again, as I fear I will have already exhausted your time and patience.[25]

I remain | Sir | Yours very respectfully | J. Scott

DAR 108: 179

CD ANNOTATIONS
2.1 I have … Mountains; 2.2] *cross in margin, brown crayon*
2.1 Rocky Mountains; 2.2] *underl brown crayon*
2.8 I have … specimen. 2.9] *cross in margin, brown crayon*
3.8 If *such* … you? 3.9] *scored brown crayon*
3.18 So far … seedlings. 3.20] *scored brown crayon*
4.9 and thus … tendencies 4.10] *scored brown crayon*
4.16 as we … eliminated. 4.17] *scored brown crayon*
4.21 I really, … principle. 4.23] *scored brown crayon*
4.22 innate] *double underl brown crayon*
5.2 scarcely … utility. 5.3] *scored brown crayon*
7.2 I may … Mr. Wooler: 7.6] 'March 5th 1863' *added in margin, ink & del pencil*
7.6 viz, … permanent. 7.9] *scored brown crayon*; '*[*(Primula)*]*' *added in margin, pencil*
7.11 this I … Chara. 7.14] '*J. Scott.*' *added in margin, ink & del pencil*
7.22 I may … experiment. 7.25] *scored brown crayon, scoring del pencil*
8.5 I have, … point. 8.7] *double scored brown crayon, scoring del pencil*
9.2 I am … supposed. 9.4] *scored brown crayon*
9.6 I have … however 9.7] *scored brown crayon*

1 See letter to John Scott, 16 February [1863] and n. 6.

2 In his letters to Scott of 11 December [1862] and 19 December [1862] (*Correspondence* vol. 10), CD suggested that he repeat the experiments first conducted by Karl Friedrich von Gärtner on the degree of cross and hybrid sterility exhibited by differently coloured varieties of maize (see Gärtner 1844 and 1849). CD had subsequently sent, with his letter to Scott of 16 February [1863], seed of a cultivated variety of maize provided by Asa Gray.

3 Barr & Sugden was a firm of London nurserymen with premises at 12 King Street, Covent Garden.

4 Scott refers to CD's discussion in *Origin*, pp. 250–1, of plants that could be 'far more easily fertilised by the pollen of another and distinct species, than by their own pollen'. At CD's suggestion, Scott decided to carry out additional experiments on sterility and hybridisation in *Passiflora* (see *Correspondence* vol. 10, letter from John Scott, 17 December [1862]); he published his results in Scott 1864d.

[5] In his letter to Scott of 19 November [1862] (*Correspondence* vol. 10), CD mentioned that plants of the genus *Lobelia* were 'more easily fertilised' by the pollen of another species than by their own pollen. See also *Origin*, pp. 250–1.

[6] In his letter to Scott of 16 February [1863], CD asked Scott to pass on a request for information concerning the inheritance of the weeping habit in trees to James McNab, curator of the Royal Botanic Garden, Edinburgh.

[7] In *Variation* 2: 18–19, CD cited Verlot 1865 as the source for McNab's observations on the inheritance of the weeping character in birch and beech trees. The essay by Bernard Verlot, Louis Neumann's colleague at the Jardin des Plantes in Paris, won the prize in the open competition of the Société Impériale et Centrale d'Horticulture for 1862, and first appeared in the society's journal (Verlot 1864). There is an annotated copy of Verlot 1865 in the Darwin Library–CUL (see *Marginalia* 1: 817–19).

[8] In his letter to Scott of 21 January [1863], CD asked for clarification of Scott's views concerning the relationship between the form of reproduction and the heritability of variation in ferns. Scott's reply has not been found; however, it apparently included a more general discussion of the relationship between reproduction and inheritance in plants (see letter to John Scott, 16 February [1863] and n. 12). For an account of Scott's views on this subject, see the letter from John Scott, 16 January 1863, n. 11. By 'bisexual' Scott meant 'hermaphrodite' (that is, bearing both male and female reproductive parts in the same flower), and by 'unisexual' he meant 'diclinous' (that is, bearing male and female reproductive parts in different flowers, whether on the same or different plants).

[9] Scott first raised this objection to natural selection, based upon blending inheritance, in his letter to CD of 6 December [1862] (*Correspondence* vol. 10). For CD's response, see the letter to John Scott, 6 March 1863.

[10] In *Orchids*, pp. 236–47, CD established that the orchid species *Catasetum tridentatum* had hermaphrodite and female forms so different from the male as to have been assigned to different genera (*Myanthus barbatus* and *Monachanthus viridis*, respectively).

[11] Scott refers to the case, reported to CD by John Rogers and cited in *Orchids*, p. 236 n., of an orchid that one year produced flowers characteristic of *Myanthus barbatus*, and the following year produced those characteristic of *Catasetum tridentatum*. See *Correspondence* vol. 10, letter from John Rogers, 22 January 1862.

[12] See letter from John Scott, 18 February [1863].

[13] In a missing letter, Scott had evidently informed CD that he was preparing a paper on the relationship between the form of reproduction and the heritability of variation in plants (see letter to John Scott, 16 February [1863] and n. 12). This subject had been Scott's main focus for some time (see n. 8, above, and letter from John Scott, 16 January 1863 and n. 11); however, the planned paper was apparently never published.

[14] See letter from John Scott, 18 February [1863], and letter to John Scott, 20 [February 1863].

[15] CD cited William Alexander Wooler's case of polyanthuses that produced long-styled flowers late, but not early, in the season in 'Dimorphic condition in *Primula*', p. 79 (*Collected papers* 2: 47).

[16] On Scott's interest in vegetable parthenogenesis, see Scott 1862a and *Correspondence* vol. 10, letters from John Scott, 15 November [1862] and 17 December [1862]. CD was also interested in the analogy between parthenogenetic plants producing only female offspring and long-styled *Primula sinensis* producing only long-styled offspring when homomorphically fertilised (see *Correspondence* vol. 10, letter to John Scott, 11 December [1862], and this volume, letter to John Scott, 20 [February 1863]).

[17] See *Correspondence* vol. 10, letter to John Scott, 3 December [1862], and letter from John Scott, 6 December [1862].

[18] See n. 13, above.

[19] Scott experimented extensively with species of *Primula* during 1863, and, at CD's prompting, wrote a paper on the subject (Scott 1864a; see letters from John Scott, 21 May [1863], [3 June 1863], and 23 July [1863]).

[20] See letter from John Scott, 23 July [1863], and Scott 1864a, pp. 97–103.

21 Scott refers to CD's discussion in *Origin*, pp. 270–1, of Gärtner's experiments on maize, in which two differently coloured varieties of the same species were found to be infertile together, and his experiments on *Verbascum*, in which crosses between differently coloured varieties of the same or of different species were found to be less fertile than the parallel crosses between similarly coloured varieties. CD had encouraged Scott to attempt these and similar experiments in his letters of 19 November [1862] and 11 December [1862] (*Correspondence* vol. 10); he reiterated the point about differently *coloured* varieties of maize in his letters to Scott of 21 January [1863] and 16 February [1863].

22 A. Gray 1862a. See letter to John Scott, 16 February [1863] and n. 5.

23 See letter from John Scott, 21 March [1863].

24 See letter from John Scott, 6 January 1863, and letter to John Scott, 16 February [1863].

25 See letter from John Scott, 21 March [1863].

From Smith, Elder and Company[1] 3 March 1863

[65 Cornhill, London]
March 3rd. 1863

We beg to hand you acc sales of your works showing a balance due to you of £11 13 1 which we hope you will find correct.[2] In reference to your suggestion that we may make you an offer to purchase the remaining stock in hand of your Geology of S. America (76 Copies) & Volcanic Islands (99 Copies) we beg to say that the complete work comprising the whole 3 parts has been selling only at 10/6, or an average of 3/6 each part which price after all trade deductions leaves only about 2/– as the nett receipt for each volume sold— & the binding of each volume would cost 6d[3]

Considering therefore the uncertainty of our selling the whole of the stock of the works referred to,[4] in the ordinary course of trade, we are sorry we do not feel warranted in making you such an offer as we could have wished to make. If however you would like to dispose of the stock of both works for £5, we are willing to take them on our hands at that price— We delay remitting you a cheque for the balance due on the enclosed a/cts until we are favored with your reply

Contemporary copy
National Library of Scotland (MS 23181: 1)

1 The text of the letter is taken from a contemporary copy made by the company. A note on the copy records that the letter was assigned to 'Mr. Wooldridge'; this individual has not been further identified.

2 See letter to Smith, Elder and Company, 14 January [1863]. The balance represented CD's share from sales of *Coral reefs*, *Volcanic islands*, and *South America* (all published by Smith, Elder and Company). The exact nature of CD's agreement with the publisher is unclear, but see *Correspondence* vol. 4, letter to Smith, Elder and Company, [16 February 1849].

3 See letter to Smith, Elder and Company, 14 January [1863]. In 1851, Smith, Elder and Company reissued CD's three works bound as a single volume, with a new general title in place of the original title and half-title pages, and costing 10*s.* 6*d.* (*Geology of the 'Beagle'*); this reissue was evidently produced from the remaining copies of the original publications (see Freeman 1977). See also *Correspondence* vol. 3, letter to Smith, Elder and Company, 30 March [1846] and n. 2.

4 Smith, Elder and Company published a new edition of *Coral reefs* in 1874 (*Coral reefs* 2d ed.), and a combined edition of *Volcanic islands* and *South America* in 1876 (*Geological observations* 2d ed.).

To H. W. Bates 4 March [1863][1]

Down Bromley Kent
Mar 4—

Dear Bates

I heard from Asa Gray a week ago,[2] & must have the pleasure of sending you a copy of part of his letter. "The copy of Bates's paper addressed to me by the author has come.[3] I am by no means clear that I shall pass it on to Haldimann,[4] certainly not just yet. I have run thro' it. It is *fascinating* he at least is a full believer in Nat. Selection & the illustrations he brings to your aid are the prettiest possible.[5] I keep it in pickle along with Decandolle on oaks[6] & some other matters to make the subject of the evening for our Scientific Society which meets with me a fortnight hence,[7] when I hope to torment Agassiz delightfully,[8] and then I will myself draw up an abstract of the interesting matter for Silliman, but it is too late for the March number."[9]

I have been rather extra unwell for the last ten days so write this by dictation.[10] Don't trouble yourself to acknowledge this. I hope Book goes on well.—[11]

Yours sincerely | C. Darwin

About Agassiz is of course private.

LS(A)
Cleveland Health Sciences Library (Robert M. Stecher collection)

1 The year is established by the reference to A. Gray 1863a and 1863d (see n. 9, below).

2 Asa Gray's letter of 9 February 1863 has not been found; however, for an indication of its contents, see the letter to Asa Gray, 20 March [1863], and the letter from J. D. Hooker, [1 March 1863].

3 Bates had sent a copy of Bates 1861 to Gray in January, after CD had persuaded Gray to attempt to have it reviewed in the *American Journal of Science and Arts*, of which Gray was one of the contributing editors (see letter to H. W. Bates, 12 January [1863], and letter from H. W. Bates, 17 January [1863]).

4 Gray had initially intended to have Bates's paper reviewed by the American zoologist Samuel Steman Haldeman (see letter to H. W. Bates, 12 January [1863]); in the event, he reviewed the paper himself (A. Gray 1863a).

5 In his paper, Bates used the theory of natural selection to account for the phenomenon of mimicry in Amazonian butterflies, arguing that the case offered 'a most beautiful proof of the truth of the theory' (Bates 1861, p. 513).

6 In his paper proposing a revision of the oak family, Cupuliferae (A. de Candolle 1862a), Alphonse de Candolle discussed the question of the origin of species, arguing that evolution from a common ancestor was 'l'hypothèse la plus naturelle [the most natural hypothesis]' (p. 363), while withholding complete endorsement of the theory. Candolle also suggested that the theory of evolution was easier to reconcile with a belief in divine creation than was the theory of special creation, drawing particular attention to Gray's theistic interpretation of evolution (p. 61). See also *Correspondence* vol. 10, letter from Alphonse de Candolle, 13 June 1862 and n. 6.

7 The reference is to the Cambridge Scientific Club, a society formed chiefly of members of the science faculty at Harvard University. The club met twice monthly, with each member hosting the meeting in turn and presenting a paper to begin the discussion (see Dupree 1959, pp. 121–2).

8 Louis Agassiz was one of the leading opponents of CD's theory in the United States; he and Gray frequently sparred at the Cambridge Scientific Club, the Boston Society of Natural History, and the American Academy of Arts and Sciences (see Dupree 1959, pp. 284–6).

[9] Gray's reviews of A. de Candolle 1862a and Bates 1861 appeared in the May and September 1863 issues of the *American Journal of Science and Arts*, respectively (A. Gray 1863d and 1863a); the journal was commonly known as 'Silliman's journal' after its founder, Benjamin Silliman.

[10] The letter is written to this point in Emma Darwin's hand; the remainder is in CD's hand. Emma recorded in her diary (DAR 242) that after a fortnight of good health, CD became ill in the last week of February. She noted that he was 'faint in night', 'languid & heavy' every morning, and 'sick several times in course of week'. On 4 March she recorded: 'Ch. better but occasional sickness'.

[11] Bates's account of his eleven years as a naturalist in the Amazonian region of South America (Bates 1863) was published between 1 and 14 April 1863 (*Publishers' Circular* 26 (1863): 193).

From H. B. Dobell 5 March 1863

29 Duncan Terr
March 5 1863.

Dear Mr Darwin

I have delayed answering your letter until I could give you positive information.[1] After some trouble I found among my rough notes the quotation from White but with no other reference than "Library of Royal Medical Society Edinburgh", some years ago. The quotation is word for word the same as that given by Dr Carpenter p 480 of 1854 Edn. of his Comparative Physiology.[2] I sent a copy of it to my friend Dr John Brown of Edinburgh asking him to refer to the Library, if it still existed.[3] He has today answered me, & says "The volume is still in the Royal Med Library & the extract is correct— there is no further reference to the case. But if you would like to see it *in extenso* I can send you the volume in a week or two."

The author is the same Charles White who wrote the "Regular gradation in man & in different Animals & vegetables" which is in our Library at the Royal Med: Chir Society London'—& which I will lend you if you have not seen it.[4]

I am much gratified to learn that you think some of the arguments in my Lectures satisfactory.[5] I hope to hear that a further acquaintance with them has not altered that opinion. If I may venture to say so, I think the book should be read straight through to give it fair play, & that then it will be found that a simple course of argument has been carried on throughout. The medical reviewers dealt very hardly with the Lectures when they came out, & most of them denied that there was any argument in the book, & indeed chiefly aimed at making fun of it.

I believe that they could only come to such an opinion from absolute carelessness in not taking the trouble to find out the object of the book—for although I am deeply conscious of its great faults & with the ill success I have often had in making my ideas clear, I am conscious also that it cost me many years of work & that it was written with a simple desire to express to the profession what I believed to be valuable.

I am dear Sir | yrs very truly | Horace Dobell

If you would like to see the Memoir from Edinburgh, please say so.

Ch. Darwin Esq.

DAR 162: 188

[1] In his letter to Dobell of 16 February [1863], CD asked for details about a reference in Dobell 1861 to the work of Charles White on the regeneration of amputated supernumerary digits (White 1782, p. 338).

[2] The reference given in Carpenter 1854, p. 480 n., is to White's 'work on the "Regeneration of Animal and Vegetable Substances," 1785, p. 16'. This pagination relates to an offprint of White 1782 distributed to 'particular friends'; see Charles White, *On the regeneration of animal substances* (Warrington: printed by W. Eyres, 1785).

[3] The author and physician John Brown was an honorary member of the Royal Medical Society of Edinburgh (J. Gray 1952, p. 128).

[4] CD had read White 1799 in 1844 (see *Correspondence* vol. 4, Appendix IV, 119: 14a). Dobell refers to the Royal Medical and Chirurgical Society of London, of which he became a fellow in 1862 (Leyland ed. 1888, 1: 59).

[5] See letter to H. B. Dobell, 16 February [1863]. The reference is to Dobell 1861.

From Julius von Haast 5 March 1863

Lake Wanaka [New Zealand]
5 March 1863

My dear Mr Darwin!

Fearing that the parcels of letters which I sent from Lake Ohau and which contained also my letter to Dr Hooker with yours enclosed, has been lost, I enclose to you a copy of my last letter.—[1]

Coming just back from a most interesting journey across the Southern Alps to the West coast, I can only communicate to you, that I discovered a most wonderful pass, a break running N & S across the Alps being only 1600 to 1650′ above the sea level, which is the more remarkable as on both sides mountains of 10,000′ are covered with extensive glaciers.—[2] It was a very interesting although arduous task, large rivers & dense N.Z. forest all the way, and I shall do me the pleasure to send you as soon as I return to Christchurch a more detailed report in print which I wrote here in the field and which will give you an idea of the country traversed.—[3] How I lounge to come back to Christchurch, & study your work on the Orchideæ.—[4]

Believe me my dear Mr Darwin | very sincerely yours | Julius Haast.

[Enclosure]

(Copy)[5]

Lake Ohau. N.Z.
December 9th. 1862.

Dear Sir

Our common friend Dr J. D. Hooker tells me he did me the honour to send you my last letter & extracts and that you were interested in them, yet at the same time communicates to me that you wish for a specimen of our native rat & frog.[6] I had the pleasure to send you a month ago two productions of my pen (addressed to the British Museum) for the perusal of which I hope you will find a spare moment.[7] You will observe that I took the liberty to call one of our Alpine giants after you as a feeble tribute from the Southern hemisphere to the author of 'Origin of species'.[8]

I am very sorry that I have no specimen of the native rat, it being almost extinct, but I shall do my best to procure one for you; The frog exists only in one or two small creeks at Coromandel in the northern island, and I shall write instantly to one of my Auckland friends, Mr. Chs. Petschler to procure one, and to send it directly to you.[9] There are some more highly interesting animals in N.Z. quite unknown to science, as for instance a small quadruped in the rivers forming this lake, & as no doubt I shall be able to procure some, I shall do myself the honour to send them to you for examination & description.[10] Dr. Hooker tells me that you ask him if I had your 'Origin of Species', if ever in my life I could be induced to tell an untruth, it would be here the case, because I should consider it the highest compliment to receive a work like yours from the hand of the Author, and on returning to Ch. Church, I shall present my copy to our embryo of a library, so that I can then fairly say, I do not possess it.[11]

You will see in my address as Presite. of the Phil. Instte. of Canterbury, that I tried to explain in few words to the members & public at large, the great object of your work, so as to preserve them against the prejudices of bigoted people.[12] It will perhaps interest you when I tell you, that the 'Origin of Species' was my travelling companion during my last journey, in the N.Z. Alps, taking always a book with me, the careful study of which, the long evenings & rainy days in a tent afford the best opportunity, and I need scarcely say that very often I forgot hunger & fatigue, cold & storm in its perusal, & sometimes I was only roused from its study by the falling of an avalanche, or the howling of the storm.[13] If you wish me to make any observations on the subject of natural history, I shall be most happy to fulfil your desires. You will allow me to give you one instance which came under my knowlege, how animals in order to preserve their offspring, adapt themselves to circumstances: the 'Casarca Variegata', the 'Paradise duck' of the setlers, builds its nest along the bank of rivers, on the ground, but several instances have been observed at the Arowenui bush between Lyttelton and Timaru on the east coast of this island, that these webfooted birds when disturbed from their nests, have built new ones on the top of high trees, bringing afterwards their young ones on their backs down to the water. This occurrance has been observed by many respectable people so that there is not the least doubt about its truthfulness, & have not the deductions from such a change in the habit of an animal a very high bearing on the confirmation of your theories? Any parcel for me if entrusted to the care of Mr. J. Marshman our Provinl. Agent, 16 Charing Cross London, will reach me safely.[14] Having hurt my right hand slightly, by a fall with a horse, you will excuse my employing one of my travelling companions to write this. Hoping that you will find time to send me a few lines,

Believe me my dear Sir | very sincerely yours | Julius Haast

Chs. Darwin Esqr | ect FRS | London.

DAR 166: 1–2

CD ANNOTATIONS

Top of letter: 'Change in Habits in nest in Duck. | New Zealand Vertebrata.' *ink*

Enclosure

1.1 Our ... species'. 1.7] *crossed ink*

2.1 I am very sorry] *after opening square bracket, ink*

2.3 I shall write ... desires. 3.11] *crossed ink*

1 See enclosure and n. 5, below. Haast enclosed this letter and its enclosure with a letter to Joseph Dalton Hooker of 5 March 1863 that arrived in mid-June 1863 (see letter from J. D. Hooker, 19 June 1863 and n. 2).

2 Following the discovery of the rich Dunstan gold-fields in Otago province in August 1862, Haast, who was provincial geologist in the neighbouring Canterbury province, began an expedition through the Southern Alps of New Zealand's Middle Island (now South Island), with the aim of examining the rocks along the boundary line of the two provinces at the nearest point to the Otago gold-fields. In January 1863, Haast discovered a pass above Lake Wanaka (the 'Haast pass'), which allowed him to traverse the Southern Alps and descend to the west coast. See H. F. von Haast 1948, pp. 273–301.

3 Haast's provisional report to the superintendent of the province was dated 3 March 1863; it was sent ahead of him to Christchurch and was printed in the Christchurch *Press*, 1 April 1863, pp. 1–2, and 2 April 1863, pp. 2–3. Haast returned to Christchurch on 12 May 1863, and probably dispatched a copy of these newspaper reports in his letter to CD of 13 May 1863. See H. F. von Haast 1948, p. 289.

4 *Orchids* was published in May 1862. Hooker had told Haast about the impact of the book in a letter of 18 September 1862 (Alexander Turnbull Library, Wellington, New Zealand, MS.37.96).

5 Haast initially wrote this letter on 9 December 1862, evidently enclosing it with a letter to Hooker of 10 December 1862. Hooker received Haast's letter in mid-April 1863, but either he lost Haast's letter to CD, or Haast failed to enclose it as promised (see letters from J. D. Hooker, 20 April 1863 and [30 April 1863], and letter to J. D. Hooker, 23 April [1863]). For reasons of clarity, the copy of Haast's letter has been reproduced here, in addition to being reproduced in its proper chronological position (*Correspondence* vol. 10, letter from Julius von Haast, 9 December 1862).

6 See the enclosure to the letter from J. D. Hooker, 20 September 1862 (*Correspondence* vol. 10). See also *ibid.*, letter to J. D. Hooker, 21 [September 1862] and nn. 6 and 7.

7 J. F. J. von Haast 1862a and 1862b (see letter to Julius von Haast, 22 January 1863). There are annotated copies of these publications in the Darwin Pamphlet Collection–CUL.

8 J. F. J. von Haast 1862b, p. 127. Mount Darwin is at the northern end of the Malte Brun Range, in the Southern Alps of New Zealand. Haast explained his practice of naming mountains in a letter to William Jackson Hooker of 9 June 1862: 'When beginning with the survey of the Southern Alps of New Zealand, hitherto entirely unknown, I proposed myself to create a kind of Pantheon or Walhalla for my illustrious contemporaries amongst those never-trodden peaks and glaciers' (H. F. von Haast 1948, p. 213).

9 Charles Petschler was a merchant at Vulcan Lane, Auckland, New Zealand (*Chapman's New Zealand Almanac* 1860, 1862).

10 In J. F. J. von Haast 1861, p. 135, Haast stated: 'The native rat (Mus rattus) is the only known indigenous land quadruped'. However, in J. F. J. von Haast 1862a, p. 6, Haast discussed the possible existence of two further indigenous quadrupeds. One was a badger- or otter-like quadruped, 'called by the natives Kaureke', which he believed probably still existed in areas around the lakes and rivers of the Southern Alps. The second was a smaller, nocturnal quadruped, traces of which Haast had found 'in the river bed of the Hopkins, the stream which forms Lake Ohou [Ohau]'.

11 Haast refers to the Philosophical Institute of Canterbury, which he founded in the summer of 1862 (H. F. von Haast 1948, pp. 220–30).

12 In J. F. J. von Haast 1862a, p. 7, Haast presented his discussion of *Origin* as a 'tribute to its illustrious author', stating that this was 'the great work of the age' in natural history. While asserting that CD's theories were not altogether new, Haast claimed that CD's 'great merit' was that he had not only dealt with the subject in a 'true philosophical spirit' but had also collected 'a great mass of facts, which

throw new light on this inexhaustible source of inquiry'. After briefly summarising CD's theory, Haast indicated some points of significance raised in *Origin* for the study of the geology and palaeontology of New Zealand. See also letter to Julius von Haast, 22 January 1863.

13 As provincial geologist for the province of Canterbury, Haast had spent the period from January to May 1862 in the Southern Alps of New Zealand, carrying out the regular work of the geological survey, together with a search for gold-bearing deposits (H. F. von Haast 1948, pp. 199–219).

14 John Marshman was agent to the Canterbury Emigration Office, 16 Charing Cross, London (*Post Office London directory* 1863).

To J. D. Hooker 5 March [1863]

Down
March 5th

My dear Hooker

I have been having very bad 10 days with much sickness & weakness, & have been obliged to stop the Lyells.[1] It breaks my heart, but Emma says, I believe truly, that we must all go for two months to Malvern.[2] It is very provoking after London doing me so much good.[3] And so many experiments in prospect. I doubt whether I shall ever finish my book on Variation: I do not suppose I have done 3 months work during the last 9 months,[4] and poor little Horace is ailing much.[5] But it is no use complaining. One must grin & bear; but a grumble to you, my dear old friend, does one good.— A good severe fit of Eczema would do me good, & I have a touch this morning & consequently feel a little alive.[6] This might save me from Malvern. If not, we should go after Easter holidays early in April.[7] If I get any strength, & you could spare a Sunday, I shd very much like to see you here; but I doubt whether it wd. be worth *your* while.—[8]

A few words about the Stove Plants: they do so amuse me; I have crawled to see them 2 or 3 times. Will you correct & answer & return enclosed.[9] I have hunted in all my Books & cannot find these names, & I like much to know family. They nearly all look splendidly well; except nepenthes & like an ass, I knew it liked warmth & put it in the bed near pipe & now find it has been having 109° bottom Heat!!! One little plant has leaves browned. I have made list of plants, **165** in number!!![10] Do you not think you ought to be sent with Mr Gower[11] to the Police Court? and such interesting forms many of them to me.—

I am reading Wellwitschia:[12] what a wonderful plant it is; but the case requires more knowledge than I have fully to appreciate: those devilish ovules, embryos,, sacks & membranes drive my weakened brain half mad—

Cordial thanks for your deeply interesting letters about Lyell, Owen & co.[13] I cannot say how glad I am to hear that I have not been unjust about species-question towards Lyell.[14] I feared I had been unreasonable. When a bit stronger I must write to him. As for showing him your letter; I would as soon burn his house down.[15] I am too fearful & shall put the case much milder; but shall tell him that I am much disappointed. He has written to me, & thinks (perhaps truly) that his chapter will produce great effect on species question.—[16] He puts you with me on migration during mundane glacial period: what do you say to that? I fear you will not approve of this.[17] I have not yet seen Rolleston's letter. I shall be very glad if he exposes Owen's cloud of false words.—[18]

I have half a mind to get Owen's paper on Aye-Aye, in which Lyell tells me, that he claims whole credit of making out the derivation or origin of species; & if this is so, write a letter to Athenæum & show, what he has really done.—[19] It is in my Hist. Introduct. to 3[d] Edit of Origin; but I did not then point out the laugable definition he gives of "Creation"; after doubting whether certain species were "created".—[20]

I am so heartily glad to hear about Lubbock's lecture: I could not resist telling him by dictation in letter what you said.—[21]

Good Bye— I am weary.— I fear you will hardly read this. You really must not write so often. I cannot bear to add to your too hard work. Farewell | C. Darwin

Endorsement: '/63.'
DAR 115: 184

[1] Charles and Mary Elizabeth Lyell had planned to visit Down House from 1 to 4 March 1863 (see letter to J. D. Hooker, 24[–5] February [1863]). On CD's health, see the letter to H. W. Bates, 4 March [1863], n. 10.

[2] Emma Darwin. CD refers to James Manby Gully's hydropathic establishment in Great Malvern, Worcestershire, where Anne Elizabeth Darwin, CD's eldest daughter and favourite child, died of fever in 1851 (see *Correspondence* vol. 5).

[3] CD visited London from 4 to 14 February 1863 (see 'Journal' (Appendix II)).

[4] CD began drafting *Variation* in January 1860 (see *Correspondence* vol. 8, Appendix II); it was not published until 1868. Since June 1862, CD had drafted several chapters of *Variation* and had written 'Two forms in species of *Linum*' (see *Correspondence* vol. 10, Appendix II, and this volume, 'Journal' (Appendix II)).

[5] Horace Darwin. See also letter to G. V. Reed, 12 January 1863.

[6] CD suffered from eczema in June 1862, and had a further attack in October 1862 (see *Correspondence* vol. 10, letters to J. D. Hooker, 30 [June 1862] and 6 October [1862]).

[7] CD stayed with his family at Malvern Wells from 3 September to 12 or 13 October 1863 (see 'Journal' (Appendix II)).

[8] Hooker visited Down House on 22 March 1863 (see letter from J. D. Hooker, [24 March 1863]).

[9] The enclosure has not been found. In February, Hooker had sent CD plants from the Royal Botanic Gardens, Kew, for his new hothouse (see letter to J. D. Hooker, [21 February 1863]).

[10] See Appendix VI.

[11] William Hugh Gower, a foreman at the Royal Botanic Gardens, Kew, had assisted Hooker in selecting the plants sent for CD's hothouse (see letters to J. D. Hooker, 15 February [1863] and [21 February 1863]).

[12] The number of the *Transactions of the Linnean Society of London* in which J. D. Hooker 1863a appeared was published on 30 January 1863 (see Raphael 1970); CD's annotated copy is in the Darwin Library–CUL.

[13] CD refers to the letters from J. D. Hooker, [26 February 1863] and [1 March 1863], concerning reactions to Lyell's *Antiquity of man* (C. Lyell 1863a), especially those of Richard Owen and Hugh Falconer. See also letter from J. D. Hooker, [23 February 1863], and letter to J. D. Hooker, 24[–5] February [1863].

[14] See letter to J. D. Hooker, 24[–5] February [1863], and letter from J. D. Hooker, [1 March 1863].

[15] See letter from J. D. Hooker, [1 March 1863].

[16] The reference is to chapter 21 of *Antiquity of man* (C. Lyell 1863a, pp. 407–23), on 'the origin of species by variation and natural selection'. Lyell's letter has not been found; however, see the letters to Charles Lyell, 6 March [1863] and 17 March [1863].

17 The reference is to C. Lyell 1863a, p. 367, where Lyell referred to the attempt of 'Mr. Darwin and Dr. Hooker' to explain the distribution of numerous plants common to the northern and southern temperate zones in terms of migrations 'along mountain chains running from N. to S. during some of the colder phases of the glacial epoch'. CD and Hooker had consistently disagreed over the causes of the geographical distribution of plants and animals. In particular, Hooker could not accept the idea that tropical genera could be kept alive in 'so very cool a greenhouse' as would be required to allow temperate plants to cross the Equator (*Correspondence* vol. 6, letter from J. D. Hooker, 9 November 1856). Instead, he preferred to account for the distribution of plants and animals in terms of land-bridges and continental extensions occurring during the period, an explanation that CD rejected (see, for instance, *Correspondence* vol. 6, letter from J. D. Hooker, 4 August 1856). For CD and Hooker's continuing dialogue on this subject, see *Correspondence* vols. 9 and 10.

18 George Rolleston. See letter from J. D. Hooker, [1 March 1863] and n. 9.

19 Lyell's letter to CD has not been found. The reference is to Owen 1862c, pp. 89–97, where Owen discussed the question of the origin of species, committing himself to 'creation by law' while stating his 'ignorance of how such secondary causes may have operated', and refusing to endorse any of the mechanisms currently proposed (p. 96). In introducing the general principle of the natural origin of species, Owen mentioned Owen 1849 as the work in which he had first formally stated this position (Owen 1862c, pp. 90 n. – 91 n.); however, following the publication of *Origin* he had vehemently opposed CD's views (see *Correspondence* vols. 7 and 8). On Owen's views on evolution, see Rupke 1994, pp. 220–58. Having read the paper, CD told Hooker and Lyell that its wording gave him no scope for his proposed attack (see letter to J. D. Hooker, 17 March [1863], and letter to Charles Lyell, 17 March [1863]). There is an unannotated and unbound copy of the number of the *Transactions of the Zoological Society of London* in which Owen 1862c appeared in the Darwin Library–CUL.

20 CD refers to the 'Historical sketch of the recent progress of opinion on the origin of species', which was added to the third edition of *Origin* (*Origin* 3d ed., pp. xiii–xix). In detailing Owen's statements respecting the production of new species by natural means (pp. xvi–xvii), CD quoted several extracts from Owen 1858, and concluded:

> Farther on (p. xc.), after referring to geographical distribution, he adds, 'These phenomena shake our confidence in the conclusion that the Apteryx of New Zealand and the Red Grouse of England were distinct creations in and for those islands respectively. Always, also, it may be well to bear in mind that by the word "creation" the zoologist means "a process he knows not what."' He amplifies this idea by adding, that when such cases as that of the Red Grouse are 'enumerated by the zoologist as evidence of distinct creation of the bird in and for such islands, he chiefly expresses that he knows not how the Red Grouse came to be there, and there exclusively; signifying also by this mode of expressing such ignorance his belief, that both the bird and the islands owed their origin to a great first Creative Cause.'

In the fourth edition of *Origin*, pp. xvii–xviii, CD greatly enlarged his discussion of Owen's views, disputing Owen's claim to have promulgated a theory of evolution at an earlier date than had CD.

21 CD refers to Hooker's praise of John Lubbock's lecture on ancient Swiss lake-habitations, delivered at the Royal Institution on 27 February 1863 (Lubbock 1863d; see letter from J. D. Hooker, [1 March 1863]). CD's letter to Lubbock has not been found; however, see the letter from John Lubbock, 6 March 1863.

To Thomas Rivers 5 March [1863][1]

Down. | Bromley. | Kent. S.E.
March 5th

My dear Sir

I write one line just to thank you sincerely for additional information on weeping trees & for your *great* kindness in writing about the Sophora.[2] If you could remember, when you see the Weeping Elm in seed, & would send me a few seed; I would force them on, & see whether they weeped.—[3]

The double Peach is in beautiful flower & I do hope it may set some fruit. The almond,, alas, has not produced one flower; so I shall have to beg a specimen of a fresh fruit in summer for comparison with fruit of double Peach.[4]

I have been very unwell (& my hand is tremulous & worse than usual) for last 10 days & have very great fear I must knock off all work & go to Malvern for two months.[5] It breaks my heart, with so many things to do & observe.—

Depend on it that Lindley will never let your article pass unpublished.[6] I was at Kew about a fortnight ago & was mentioning your kindness, & Dr. Hooker burst out in admiration of your articles.—[7]

Believe me | Yours very sincerely | Ch. Darwin

Sotheby's, London (23–4 July 1987)

[1] The year is established by the relationship between this letter and the letter to Thomas Rivers, 1 February [1863], and by the reference to CD's visit to London (see n. 7, below).

[2] The letter from Rivers has not been found. CD wrote drafts of the chapters on inheritance for *Variation* (*Variation* 2: 1–84) between 23 January and 1 April 1863 (see 'Journal' (Appendix II)). In his letter to Rivers of 1 February [1863], CD asked for information concerning the character of seedlings from weeping trees; Rivers provided the information in his reply of [3 February 1863].

[3] See letter to Thomas Rivers, [14 February 1863] and n. 6.

[4] CD refers to plants sent to him by Rivers in January 1863 (see letters to Thomas Rivers, 11 January [1863], 15 January [1863], and 25 January [1863], and letter from Thomas Rivers, 21 January 1863). See also letter to Thomas Rivers, [9 May 1863], n. 4.

[5] CD refers to James Manby Gully's hydropathic establishment at Great Malvern, Worcestershire. On the state of CD's health, see also the letter to H. W. Bates, 4 March [1863], n. 10.

[6] John Lindley was editor of the horticultural section of the *Gardeners' Chronicle*; an article by Rivers on seedling raspberries and strawberries was published in the *Gardeners' Chronicle*, 14 March 1863, pp. 244–5.

[7] Joseph Dalton Hooker was assistant director of the Royal Botanic Gardens, Kew (R. Desmond 1994); CD visited Hooker there on 11 February 1863 (see 'Journal' (Appendix II)).

To Smith, Elder and Company 5 March [1863]

Down. | Bromley. | Kent. S.E.
March 5th

Dear Sir

I am much obliged for your letter & accounts.[1] As I am sure that you would offer the full amount of what is fair, & as it will save all future accounts I will gladly accept the 5£, & you can at your convenience send me a cheque for that & balance of £11:s13.d1. Will you be so good as to send me a proper form of receipt, to which I will affix a stamp & my signature.—

Please remember that you have to deduct small enclosed account.—[2]

One of my Boys has the common passion for collecting Postage stamps:[3] he tells me that you issue some peculiar kinds:[4] I know not in the least what they are & perhaps they are for India[5] (at least I have never met with them) & can only be sold in number; but if you have odd copies & could enclose one or two of each kind deducting amount from your cheque, I sh^d. be glad to please my Boy, but of course you must not think of this for a minute if in in anyway inconvenien⟨t.⟩

Pray believe me | Dear Sir | Yours very faithfully | Charles Darwin

Postmark: MR 5 63
National Library of Scotland (MS 23181: 8)

[1] Letter from Smith, Elder and Company, 3 March 1863.
[2] The enclosure has not been found. See letter to Smith, Elder and Company, 14 January [1863].
[3] Leonard Darwin.
[4] From 1857, firms were given the opportunity of having their names and addresses embossed on a circular collar placed around standard issue British postage stamps. Smith, Elder and Company was one of the first companies to make use of this facility, registering their collar in October 1857. It consisted of a plain circular band forming a complete ring around the stamp, with the name of the firm on a plain cartouche above the stamp, and the address on a similar cartouche below (Huggins 1970, pp. 167–70).
[5] Smith, Elder and Company were also East India agents (*Post Office London directory* 1863).

To H. B. Dobell 6 March [1863][1]

Down, Bromley, Kent, S.E.
March 6

My dear Sir

I am extremely much obliged to you for your great kindness and all the large amount of trouble which you have taken to oblige me.[2] Your information is sufficient: I feared that D^r. C. might not have given whole case.[3]

With very sincere thanks | Yours very faithfully | Ch. Darwin

Copy
DAR 143: 389

[1] The year is established by the relationship between this letter and the letter from H. B. Dobell, 5 March 1863.
[2] See letter from H. B. Dobell, 5 March 1863.
[3] See letter from H. B. Dobell, 5 March 1863 and n. 2. In *Variation* 2: 14 n., CD cited Carpenter 1854 in his discussion of this case of the regeneration of amputated supernumerary digits. There is a copy of Carpenter 1854 in the Darwin Library–CUL, in which the relevant passage is annotated (see *Marginalia* 1: 154–8).

From J. D. Hooker [6 March 1863][1]

Royal Gardens Kew
Thursday [Friday]

Dear Darwin

I am atrociously idle & prefer writing to you to anything else. What a bitter disappointment it must have been to put off the Lyells! but what could you do—pray God the Eczema has come out & relieved you ere this.[2]

Do not make "boiled greens" of your plants. Any of these tropical things that look sickly &c had better have a bell-glass over them, tilted at the bottom to let air in.[3]

Do not be in a hurry about repotting these tender little things, or you will lose them. The Sonerila likes a nice moist-warmth, but not too hot.—& perhaps screening from sun, but the art & mystery of *screening* is *utterly unintelligable to me*.

Falconer is working up to a state of savagery against Lyells book, & has arrived at a state of virtuous indignation about his treatment of Prestwich & Gunn's labors, which is the prelude to an onslaught about his own I expect in regard to the bone caves.[4] I hear that Lyell will answer Owen in tomorrows Athenæum—*if so I will send it* on by tomorrow's post, & you can return it, by Mondays. if done with[5]

I must read the Aye-Aye paper[6] but I hear that his onslaught on Nasmyth in 1851 (in I think the Med. Chirurg. review) is the masterpiece of Scientific vituperation & Billingsate, & well worth the perusal—[7] perhaps I shall get it & if so will let you know.

We had a good meeting at Linnean last night, & a very long paper by F. Smith on Wallaces Hymenopt insects *distribution*, & the very dullest thing I ever heard.[8] I do hope that Bates will write more & keep Entomology within the pale of Science—[9]

Wallace made a very few remarks worth all the paper.[10]

I doubt getting down on a Sunday to you—as I have promised to go to Lubbocks on the 21st. to meet Colenso![11]

You *must stick for a few months* to your Variation book & take the experiments mildly.[12]

I have 4 tubers of the Wild Potato— how shall I proceed with them?[13] My life is too great a worry to experiment properly. & I cannot bring mind or time to bear upon it. I do assure you that without joking Wedgewoods are an unspeakable relief to me—[14] I look over them every Sunday morning—& poke into all the little 2d. hand shops I pass in London seeking medallions. The prices of vases are quite incredible— I saw a lovely butter-boat, & was quite determined to go up to 30/ for it,—at the dirtiest little pig stye of a subterranean hole in the wall of a shop you ever were in,—the price was £25. All this amuses me vastly—& is an enjoyable contrast to grim science. No Lady enjoys **bonnets** more heartily!

Ever my dear Darwin | Yours affect | J D Hooker

DAR 101: 114–16

CD ANNOTATION

End of letter: 'Athenæum | Falconer | Wild Potato— | ['Treviranus'[15] *pencil del ink*] | Lyell | Huxley | Poplar' *pencil*

[1] The letter is dated by Hooker's reference to F. Smith 1863 (see n. 8, below); Hooker wrote 'Thursday' in error.

[2] Because of ill health, CD had been obliged to cancel the planned visit to Down of Charles and Mary Elizabeth Lyell; CD found that eczema relieved him of other symptoms (see letter to J. D. Hooker, 5 March [1863]).

[3] In his letter to Hooker of 5 March [1863], CD mentioned that he had had difficulty in cultivating some of the hothouse plants sent from the Royal Botanic Gardens, Kew.

[4] In a letter published in the *Athenæum*, 4 April 1863, pp. 459–60, Hugh Falconer objected to several aspects of Lyell's *Antiquity of man* (C. Lyell 1863a). His principal objection was that Lyell had failed properly to acknowledge the work by Joseph Prestwich on the age of the Quaternary deposits of the south of England, and his own work on ossiferous caves in which human artefacts had been found. Falconer also criticised Lyell for including, under Falconer's authority, but without his permission, a list of fossil species found on the Norfolk coast by John Gunn and others. For discussions of Falconer's attack on Lyell, see Bynum 1984, Grayson 1985, and L. G. Wilson 1996a.

[5] Hooker refers to Richard Owen's letter, published in the *Athenæum* on 21 February 1863, pp. 262–3, protesting about Lyell's treatment of him in *Antiquity of man* (C. Lyell 1863a); Lyell's reply was published in the *Athenæum* on 7 March 1863, pp. 331–2. See also letter to J. D. Hooker, 13 [March 1863].

[6] See letter to J. D. Hooker, 5 March [1863] and n. 19. The reference is to Owen 1862c.

[7] Billingsgate: 'Scurrilous vituperation, violent abuse' (*OED*). Hooker apparently refers to an anonymous review of Alexander Nasmyth's posthumous work on the development, structure, and diseases of the teeth (Nasmyth 1849), which appeared in the *British and Foreign Medico-Chirurgical Review* in April 1850 ([Owen] 1850b); the review harshly denounced Nasmyth's work. See also letter to J. D. Hooker, 3 January [1863] and n. 4.

[8] Frederick Smith's paper on the geographical distribution of the aculeate Hymenoptera collected by Alfred Russel Wallace in the Malay Archipelago (F. Smith 1863) was read at a meeting of the Linnean Society on 5 March 1863.

[9] Since his return from South America in 1859, Henry Walter Bates had published two papers on the insect fauna of the Amazon valley in the *Transactions of the Linnean Society* (Bates 1860 and 1861).

[10] Wallace's remarks are not recorded in the *Journal of the Proceedings of the Linnean Society*.

[11] In his letter to Hooker of 5 March [1863], CD expressed a hope that Hooker might visit Down, if he could 'spare a Sunday'. Hooker refers to the bishop of Natal, John William Colenso, the first part of whose book on the Pentateuch (Colenso 1862) had sparked religious controversy concerning rational biblical criticism. John Lubbock lived at Chislehurst, about five miles north of Down; Hooker visited CD from Lubbock's house on 22 March 1863 (see letters from J. D. Hooker, [15 March 1863] and [24 March 1863]).

[12] See letter to J. D. Hooker, 5 March [1863] and n. 4; the reference is to *Variation*.

[13] The specimens of South American wild potato were some of those brought back by Alfred Newton from his visit to the Americas (see letter to J. D. Hooker, 13 [March 1863], letter from J. D. Hooker, [15 March 1863], and letter from Alfred Newton, 21 March 1863). In November 1862, while preparing a draft of the part of *Variation* dealing with 'Facts of variation of Plants', CD had unsuccessfully sought 'odd *varieties*' of potato from Hooker, with the intention of growing a few plants of each for comparison (see *Correspondence* vol. 10, letter to J. D. Hooker, 27 [October 1862], letter from J. D. Hooker, 2 November 1862, and Appendix II).

[14] Hooker had begun to collect Wedgwood ware in 1862 (see *Correspondence* vol. 10, letter from J. D. Hooker, [27 or 28 December 1862]).

[15] Ludolph Christian Treviranus had sent CD two copies of the numbers of the *Botanische Zeitung* containing Treviranus 1863a, asking him to forward one set to Hooker (see letter from L. C. Treviranus, 12 February 1863, and letter to J. D. Hooker, 13 [March 1863]).

From John Lubbock 6 March 1863

15, Lombard Street. E.C.
6 M^{ch}./63

My dear M^{r}. Darwin

I have forwarded your proofs to Williams & Norgate & am very much obliged to you for the article which is just the thing for us.[1]

I was very sorry to hear that you were so unwell & much disappointed too on my own account. Will you please thank M^rs. Darwin for sending me report of your health & also for the extract from Hooker's letter.[2]

One would always be glad of his approval, but I was the more so having rather feared that I made a mess of the lecture.

I do hope you will soon be right again.

Yours affec^ly | John Lubbock

DAR 170: 38

[1] The reference is to CD's review of Bates 1861 ('Review of Bates on mimetic butterflies'), the manuscript of which CD had sent to Lubbock, one of the editors of the *Natural History Review* (see letter to John Lubbock, 4 January [1863]). The review appeared in the April 1863 number of the *Natural History Review*, which was published by Williams & Norgate.

[2] Emma Darwin's letter to Lubbock has not been found. Lubbock had intended to dine at Down House on 2 March 1863 (see letter from John Lubbock, 28 February 1863), but according to Emma's diary (DAR 242), CD became ill during the last week of February. CD dictated to Emma Joseph Dalton Hooker's favourable comments concerning Lubbock's lecture at the Royal Institution on 27 February 1863 (see letter from J. D. Hooker, [1 March 1863], and letter to J. D. Hooker, 5 March [1863]).

To Charles Lyell 6 March [1863]

Down
March 6th

My dear Lyell.

Putting you off was a great & bitter disappointment,[1] I sh^d. so have liked to have talked over many points & Owen's false letter.[2] But I had no choice, & I much fear that Emma is right & that I must knock off all work & all of us go to Malvern for two months.[3] I get on with nothing.— Thank you much for your interesting note.—[4] Keep Dana—[5]

I have been of course, deeply interested by your Book:[6] I have hardly any remarks worth sending, but will scribble a little on what most interested me. But I will first get out what I hate saying, viz that I have been greatly disappointed that you have not given judgment & spoken fairly out what you think about the derivation of Species.[7] I sh^d. have been contented if you had boldly said that species have not been separately created, & had thrown as much doubt as you like on how far variation & N. Selection suffices. I hope to Heaven I am wrong (& from what you say about Whewell it seems so) but I cannot see how your Chapters can do more good than an extraordinarily able review."[8] I think the Parthenon is right that you will leave the Public in a fog.[9] No doubt they may infer that as you give more space to myself, Wallace & Hooker than to Lamarck, you think more of us.[10] But I had always thought that your judgment would have been an epoch in the subject.[11] All that is over with me; & I will only think on the admirable skill with which you have selected the striking points & explained them.— No praise can be too strong, in my opinion, on that inimitable chapter on language in comparison with species.[12]

Now for a few trifling criticisms[13]

p. 187. Sentence beginning "my friend Mr Evans" not clear:[14]

p. 229 sentence "but long after that period" seems to me awkward & not clear.[15]

p. 231— "hence there will be passages" It is not clear whether you refer to glaciers or floating ice-bergs, if the latter might they not have been overturned?[16]

p. 257. "For Cap—Sullivan read Sulivan[17]

p. 264— I wonder you did not add a vivid sentence, which you could have done so well, at the end of Glen Roy, on our almost still seeing the glacier lakes.[18]

p. 278. the number of the p. itself at top is inverted—[19]

I think the glacier chapters are almost the best in the book.[20] The closing pages of Ch. XIV are magnificient, I can use no other term.[21] I think this discussion has interested me almost more than the antiquity of man. The gloss of novelty was worn off the latter, yet I have been *deeply* struck by the effect of the agglomerated evidence.

p. 294 might not you have added preservation of striæ under lakes & according to Smith of J. Hill under tidal fiords.[22]

p. 323 ought you not to have added a sentence or two on the glacial phenomena of S. America & N. Zealand?[23] Have you heard of Haast's statem^ts of pre-historic man in N. Zealand?[24] I was disappointed that you did not discuss the supposed warmer period after glacial period in N. America—

How admirable are your remarks on Ramsay's Theory![25]

p. 367. I do not know whether Hooker will approve of your coupling his name with mine on mundane glacial period He has often fought me strongly on it—[26] I feel a conviction that no view throws so much light on Geograph. Distrib: Hooker even in his late papers on Fernando Po & Cameroon M^ts does not allude to the coldness of the Tropics.[27] These latter cases seem to me to prove that Africa was then colder, & this was the only quarter of the globe where full proof was wanting. I think I convinced Hooker that the general phenomena are not explicable on the assumption that one Tropical region remained hot as a refuge for Tropical productions.[28]

p. 370. 5 lines from bottom Ought not "polished" to be added before "stone"? as it stands it is very puzzling.—[29]

p. 374— W^d not S. Africa be a better case with its elephants rhinos. Lions, Hippopotamus &c living in *swarms* formerly with savage man[30]

p. 379. I cannot see the force of your argument. Has not the Australian remained a savage to the present day?[31]

Ch. XXI I think it w^d have been advantageous to have enlarged a little more on Nat. Selec. explaining adaptations; at least this was the turning point with me.[32]

W^d it not have been well to have given a few more striking instances of rudiments?[33]

How admirably you treat the imperfection of Geolog. record in many places.[34]

p. 417 It is of little consequence, but Hooker published his Essay a month after the Origin; see my Hist. Introduct. I asked him.[35]

p. 421. Who is Sefström? neither Lubbock or I ever heard of him.[36]

p 446. 14 lines from bottom. A person might say that you thought that bats & rodents were not placentals. you show afterwards that this would be a mistake[37]

p. 447. The fewness of individuals in Islands I believe, as explained in Origin, w[d] account for the no great amount of transmutation therein.[38]

450. What a pity you give Owen's name of "macrurus" instead of von Meyer's proper name of Lithographicus.[39]

p. 469. Any one might argue from the middle paragraph that you were far from believing that man was descended from any animal.[40]

P. 497 Top of p. Compare mind of dog with its wild aboriginal.[41]

P. 500. I am surprized at what you say about man & Miocene strata considering great gap between man & other animals.—[42]

p. 505. Sentence at top of p. makes me groan.[43]

I suppose you could not lend me Owen's paper on the Aye Aye? Is the part very expensive? I am sorely tempted to expose in Athenæum what rubbish Owen has written on the subject.[44]

I know you will forgive me for writing with perfect freedom; for you must know how deeply I respect you, as my old honoured guide & master.— I heartily hope & expect that your Book will have gigantic circulation & may do in many ways as much good as it ought to do.—[45]

I am tired; so no more. I have written so briefly that you will have to guess my meaning. I fear my remarks are hardly worth sending— Farewell—with kindest remembrances to Lady Lyell | Ever yours | C. Darwin

LS(A)[46]
Endorsement: 'March 6. | 1863'
Postmark: MR 6 63
American Philosophical Society (289)

[1] Charles and Mary Elizabeth Lyell had planned to visit Down House from 1 to 4 March 1863, but CD cancelled the visit because of ill health (see letters to J. D. Hooker, 24[–5] February [1863] and 5 March [1863]).

[2] CD refers to Richard Owen's letter, published in the *Athenæum*, 21 February 1863, pp. 262–3, protesting about his treatment in Lyell's *Antiquity of man* (C. Lyell 1863a). For CD's view of Owen's criticisms, see letter to J. D. Hooker, 24[–5] February [1863], and letter to T. H. Huxley, 26 [February 1863].

[3] CD refers to James Manby Gully's hydropathic establishment at Great Malvern, Worcestershire.

[4] Lyell's letter has not been found; however, some indication of its contents is given by the letter to J. D. Hooker, 5 March [1863].

[5] CD may refer to the letter from J. D. Dana, 5 February 1863, which he apparently sent with his letter to Lyell of 17 [February 1863]. CD may also have sent a copy of Dana 1863c; however, there is a presentation copy of this work in the Darwin Pamphlet Collection–CUL.

[6] CD received a presentation copy of *Antiquity of man* (C. Lyell 1863a) on 4 February 1863 (see letter to Charles Lyell, 4 [February 1863]); there is an annotated presentation copy of this work in the Darwin Library–CUL (see *Marginalia* 1: 525–7).

[7] CD refers to Lyell's extensive but inconclusive discussion of theories of transmutation of species and organic progression in C. Lyell 1863a, pp. 385–506. On Lyell's unwillingness to commit himself to a belief in the transmutation of species, see Bartholomew 1973. On CD's reaction to Lyell's position,

see letter to Asa Gray, 23 February [1863], letter to J. D. Hooker, 24[–5] February [1863], and letter to T. H. Huxley, 26 [February 1863].

8 Lyell's letter has not been found, but see the letter to J. D. Hooker, 5 March [1863] and n. 16. The reference is to William Whewell, who consistently opposed theories of organic transmutation, including CD's theory (see Ruse 1991, and *Correspondence* vol. 8, letter from William Whewell, 2 January 1860). In a letter to Lyell of 28 February 1863 (Todhunter 1876, 2: 429–30), Whewell had written that he was 'deriving great pleasure' from reading C. Lyell 1863a, and that he was 'especially delighted' with Lyell's chapter comparing the development of languages and species (C. Lyell 1863a, pp. 454–70), which he considered 'admirably adapted to explain the difficulties and the solutions of the one theory by the other'. However, in the same letter, Whewell reiterated his earlier view that in the 'palætiological sciences' (that is, sciences of historical causes), 'no natural *beginning* is discoverable'.

9 The reference is to a review of C. Lyell 1863a published in the *Parthenon* on 21 February 1863, pp. 233–5. See letter to J. D. Hooker, 24[–5] February [1863], n. 10.

10 The references are to Alfred Russel Wallace, with whom CD announced the theory of natural selection (C. Darwin and Wallace 1858), Joseph Dalton Hooker, who was one of the first publicly to endorse natural selection (see J. D. Hooker 1859), and Jean Baptiste de Lamarck, whose evolutionary theory Lyell had opposed in his *Principles of geology* (C. Lyell 1830–3; see also Bartholomew 1973). In C. Lyell 1863a, Lyell discussed Lamarck's theory in a preliminary chapter on theories of progression and transmutation (pp. 385–406), and went on to describe CD's theory and discuss possible objections to it (pp. 407–53).

11 Lyell was a leading figure in the British scientific establishment, whose critical views of transmutation, first enunciated in *Principles of geology* (C. Lyell 1830–3), had played an important role in the British debate on the subject. Lyell had been CD's scientific mentor (see, for example, Browne 1995).

12 In the chapter 'Origin and development of languages and species compared' (C. Lyell 1863a, pp. 454–70), Lyell drew on the philological work of Friedrich Max Müller and others, to argue that many of the difficulties that were encountered by naturalists in attempting to explain the origin of species by modified descent were similar to those that had been solved by philologists with regard to the origin of languages. See also letter to Asa Gray, 23 February [1863] and n. 17.

13 In CD's copy of C. Lyell 1863a, which is in the Darwin Library–CUL, many of the passages in which CD suggested changes are annotated (see *Marginalia* 1: 525–7). Some indication of Lyell's reaction to CD's comments is given by his notes on a fragment of the envelope to this letter; Lyell wrote: 'Meridional belts of warm & cold will not explain all the phenomena of distribution of species. | 8. Selection explaining adaptations not enlarged enough | more examples of rudiments | 9. The fewness of individuals in islands prevents transmutation'. See nn. 26–8, 32–3, and 38, below. In addition, Lyell annotated with marginal crosses several of the points raised in CD's letter that correspond with corrections in the second edition of his book (C. Lyell 1863b; see nn. 14–17, 19, and 29, below). A discussion of the alterations made between the first, second, and third editions of *Antiquity of man* is given in Grayson 1985.

14 The reference is to John Evans. The sentence was reworded in the second edition (C. Lyell 1863b, p. 187).

15 The sentence was reworded in the second edition (C. Lyell 1863b, p. 229).

16 The passage was modified in the second edition (C. Lyell 1863b, p. 231).

17 Bartholomew James Sulivan. The name is corrected in the second edition (C. Lyell 1863b, p. 257).

18 No such addition was made to the second edition (C. Lyell 1863b, p. 264).

19 The number was corrected in the second edition (C. Lyell 1863b, p. 278).

20 In chapters 12–18 of *Antiquity of man* (C. Lyell 1863a, pp. 206–368), Lyell discussed recent researches on glaciation, and outlined a chronology of the Pleistocene glacial period. See also letter to J. D. Hooker, 24[–5] February [1863].

21 The last section of chapter 14 was entitled 'Time required for successive changes in physical geography in the Post-Pliocene period' (C. Lyell 1863a, pp. 284–9).

22 J. Smith 1848, p. 18. The Scottish geologist James Smith was known as 'Smith of Jordanhill'. No such addition was made to the second edition (C. Lyell 1863b, p. 294).

23 No such addition was made to the second edition (C. Lyell 1863b, p. 323).

24 In an address to the Philosophical Institute of Canterbury, New Zealand, Julius von Haast referred to reports of the discovery of pre-Maori stone implements in Wellington Province (J. F. J. von Haast 1862a, p. 7). Haast sent a copy of his address to CD in November 1862 (see *Correspondence* vol. 10, letter from Julius von Haast, 9 December 1862, and this volume, letter to Julius von Haast, 22 January 1863).

25 CD refers to Lyell's critique of Andrew Crombie Ramsay's controversial theory that many European and American rock-basins, now containing lakes, owed their origin to glacial erosion (see Ramsay 1862, and C. Lyell 1863a, pp. 311–19). Lyell was one of the leading critics of Ramsay's theory (see Davies 1969, pp. 305–6, and *Correspondence* vol. 10). See also letter from J. D. Hooker, [1 March 1863] and n. 11. Lyell marked this sentence with a pencilled bracket in the margin and the annotation 'D'.

26 Lyell referred to CD's and Hooker's belief in the trans-tropical migration of plant species along north-south aligned mountain chains during a global Pleistocene glacial period. For their differences on this subject, see the letter to J. D. Hooker, 5 March [1863], n. 17. The sentence remained unchanged in the second edition (C. Lyell 1863b, p. 367), but see n. 13, above.

27 CD considered that the recent discovery of temperate plants in the Cameroons Mountains and on Fernando Po provided additional evidence for his belief that temperate species had migrated into the tropical regions during a global Pleistocene glacial period. He refers to Hooker's paper on the plants of Fernando Po (J. D. Hooker 1861), and probably to an unpublished paper on the plants of the Cameroons Mountains, read before the Linnean Society of London on 5 June 1862 (see *Journal of the Proceedings of the Linnean Society of London (Botany)* 6 (1862): cvi; see also *Correspondence* vol. 10, letter from J. D. Hooker, 9 June 1862, and letter to J. D. Hooker, 11 June [1862]). Hooker's researches on the Cameroons flora were later published as J. D. Hooker 1863b.

28 See letter to J. D. Hooker, 17 March [1863], n. 17.

29 The sentence was reworded in the second edition (C. Lyell 1863b, p. 370).

30 CD refers to Lyell's argument that, given the difficulty of exterminating 'noxious' quadrupeds in modern times, 'even with the aid of fire-arms', it was reasonable to presume that the 'time demanded for the gradual dying out or extirpation of a large number of wild beasts which figure in the Post-pliocene strata, and are missing in the Recent fauna', was of 'protracted duration'; Lyell illustrated the point by reference to the tiger in India. The passage remained unchanged in the second edition (C. Lyell 1863b, p. 374).

31 In opposition to the commonly held view that humans had degenerated from a 'primæval state of superior intelligence', and that knowledge of the arts and sciences had been 'supernaturally communicated', Lyell argued that in such a case, given the 'improvable nature' of humans, the progress of civilisation would have been much more rapid than the record of human artefacts showed (C. Lyell 1863a, pp. 378–9). These paragraphs were not changed in the second edition (C. Lyell 1863b, pp. 378–9); Lyell pencilled a question-mark in the margin at this point in CD's letter.

32 Lyell outlined CD's theory in chapter 21 of *Antiquity of man* (C. Lyell 1863a, pp. 407–23). This discussion was not enlarged in the second edition (C. Lyell 1863b, pp. 407–23); however, Lyell marked this sentence of CD's letter with a pencilled brace in the margin and the annotation 'D'. See also n. 13, above.

33 Lyell made reference to only two examples of rudimentary organs (C. Lyell 1863a, p. 413). The paragraph remained unaltered in the second edition (C. Lyell 1863b, p. 413), but see n. 13, above.

34 References to the imperfection of the geological record appear throughout the volume, although there is a section dealing specifically with the subject in C. Lyell 1863a, pp. 448–53. CD's comment was probably prompted by the brief discussion on pp. 414–15, concerning the discovery of missing links.

35 Hooker's introductory essay to *Flora Tasmaniæ* (J. D. Hooker 1859), in which he endorsed CD's theory of evolution by natural selection, was published on 29 December 1859 (*Taxonomic literature*); *Origin* was published on 24 November 1859 (Freeman 1977). CD asked Hooker the date of publication of his essay in the letter to J. D. Hooker, 31 [January 1860] (*Correspondence* vol. 8); however, Hooker apparently did not reply to the query (see *Correspondence* vol. 8, letter to J. D. Hooker, 8 February [1860]). CD

detailed the order of publication in his 'sketch of the progress of opinion on the Origin of Species', which was first added as a preface to the American edition of *Origin*, published in July 1860 (*Origin* US ed., p. xi). In C. Lyell 1863a, p. 417, Lyell referred to Hooker's essay as having been published 'a few months before the appearance of the "Origin of Species"'. In the second edition of *Antiquity of man* the publication date of Hooker's essay was given as December 1859 (C. Lyell 1863b, p. 417). See also letter from J. D. Hooker, [15 March 1863], and letter to J. D. Hooker, 17 March [1863].

36 In the second edition Lyell changed 'Sefström' to 'Steenstrup' (C. Lyell 1863b, p. 421). Nils Gabriel Sefström was a Swedish chemist, physician, and natural historian (*SMK*); Japetus Steenstrup was a Danish zoologist (*DSB*). CD also refers to John Lubbock. No correspondence between him and CD on this subject has been found.

37 CD refers to Lyell's statement that the bats and rodents of Australia had not developed into the 'higher or placental type'; in the second edition Lyell changed this to read 'higher placental types' (C. Lyell 1863b, p. 446). Later in the same section, Lyell stated that the preoccupancy of Australia by marsupials might have 'checked the development of the placental rodents and cheiroptera' (C. Lyell 1863a, p. 448).

38 In a section headed 'Absence of Mammalia in islands considered in reference to transmutation' (C. Lyell 1863a, pp. 443–8), Lyell argued that one reason why the bats found in Madeira and the Canary Islands had not given rise, by transmutation, to other mammalian forms, was that there would be occasional crossing of the island bats with individuals arriving from the African continent. CD refers to the statement in *Origin*, p. 105, that 'fewness of individuals will greatly retard the production of new species through natural selection, by decreasing the chance of the appearance of favourable variations'. The passage remained unaltered in the second edition (C. Lyell 1863b, p. 447), but see n. 13, above.

39 The name *Archaeopteryx lithographica* was assigned by Hermann von Meyer on the basis of a fossil feather found in the lithographic slate quarries at Solenhofen, Germany (Meyer 1861). When a near-complete fossil skeleton of a bird-like reptile was discovered at Solenhofen shortly afterwards, Andreas Wagner gave it the name *Griphosaurus* (J. A. Wagner 1861b and 1862). The specimen was first described from personal observation by Owen (Owen 1862a), who gave it the name *Archeopteryx macrura*, arguing that its identity with Meyer's specimen could not be satisfactorily established (Owen 1862a, p. 33 n.). Lyell used the name *Archaeopteryx macrurus* in the second edition (C. Lyell 1863b, p. 450).

40 Lyell wrote (C. Lyell 1863a, p. 469):

> In our attempts to account for the origin of species, we find ourselves still sooner brought face to face with the working of a law of developement of so high an order as to stand nearly in the same relation as the Deity himself to man's finite understanding, a law capable of adding new and powerful causes, such as the moral and intellectual faculties of the human race, to a system of nature which had gone on for millions of years without the intervention of any analogous cause. If we confound 'Variation' or 'Natural Selection' with such creational laws, we deify secondary causes or immeasurably exaggerate their influence.

This paragraph remained unchanged in the second edition (C. Lyell 1863b, p. 469), but see letter to Charles Lyell, 12–13 March [1863] and n. 17.

41 CD refers to a quotation in C. Lyell 1863a, p. 497, from Sumner 1816, p. 19, in which it was argued that the power of 'progressive and improvable reason' was 'man's peculiar and exclusive endowment.' In his copy of C. Lyell 1863a, CD wrote in the margin, 'compared dog & wolf or jackall'; a second note refers to Rengger 1830, stating: 'Rengger says Monkeys are improvable' (see *Marginalia* 1: 525–7). The section remained unaltered in the second edition (C. Lyell 1863b, p. 497).

42 Following a discussion of fossil primates in European Miocene strata, Lyell wrote (C. Lyell 1863a, p. 500):

> But according to the doctrine of progression it is not in these miocene strata, but in those of pliocene and post-pliocene date, in more equatorial regions, that there will be the greatest chance of discovering hereafter some species more highly organised than the gorilla and chimpanzee.

The sentence was not changed in the second edition (C. Lyell 1863b, p. 500).

43 Lyell wrote (C. Lyell 1863a, pp. 504–5):

> If, in conformity with the theory of progression, we believe mankind to have risen slowly from a rude and humble starting point, such leaps may have successively introduced not only higher and higher forms and grades of intellect, but at a much remoter period may have cleared at one bound the space which separated the highest stage of the unprogressive intelligence of the inferior animals from the first and lowest form of improvable reason manifested by man.

In his copy of C. Lyell 1863a, CD underlined the words 'unprogressive' and 'improvable reason', scored the passage in the margin, and wrote 'oh' (see *Marginalia* 1: 525–7). The substance of the sentence was unchanged in the second edition (C. Lyell 1863b, pp. 504–5). See also letter from Charles Lyell, 11 March 1863 and n. 7.

44 CD refers to Owen 1862c, in which, Lyell had told him, Owen claimed the 'whole credit of making out the derivation or origin of species' (see letter to J. D. Hooker, 5 March [1863] and n. 19). The paper occupied an entire number of the *Transactions of the Zoological Society of London* (volume 5, number 2); according to the paper covers on CD's copies of the numbers, in the Darwin Library–CUL, it cost 18*s.* to fellows of the Zoological Society, and 25*s.* to the general public. CD was a fellow of the society (Freeman 1978); he started to subscribe to the society's transactions with this number of the journal. See also letter to Charles Lyell, 17 March [1863].

45 *Antiquity of man* was a great commercial success: at the publisher John Murray's trade sale in November 1862, 4000 copies of the book were sold to book-dealers (see *Correspondence* vol. 10, letter to J. D. Hooker, [10–]12 November [1862] and n. 25). A second edition was published in April 1863, only two months after the first, and by 19 May Lyell reported that 5000 copies had been sold (see K. M. Lyell ed. 1881, 2: 375). A third edition was published in November 1863 (Grayson 1985).

46 Emma Darwin acted as CD's amanuensis for the middle section of the letter, beginning with, 'Now for a few trifling criticisms', and ending with, 'what rubbish Owen has written on the subject'.

To John Scott 6 March 1863

Down Bromley Kent
Mar 6. 1863

Dear Sir

I have been unwell for 12 days, & must write more briefly even than usual.—[1] I used to think the husked Maize was wild & there is some evidence for **S.** America but I now hear from Asa Gray that it is very variable, I do not believe that it is the wild form.[2] I do not know where seed could be got.

I enclose information about Passiflora.[3] The experiment with P quandrangularis which fruits in some places & not in others, would be specially good to try with other pollens.[4] Thank you for the attempt to give me information on weeping trees.[5] I have got much from Mr Rivers.[6] I understand now about variability & bi-sexuality. I have been much interested by what you tell me on Primula, & hope to see all in detail in your paper.[7] I am *extremely* glad that you are experimenting on Primulas of different colours. I sent Asa Gray's Review by this morning's post.[8] I am very much interested about Gongora & fear more & more that I shall prove completely wrong about Acropera—[9]

I thank you for your criticisms on the Origin, which I have not time to discuss; but I cannot help doubting from your expression of an "**innate** . . . . . selective principle" whether you fully comprehend what is meant by Natural Selection.[10] Certainly when you speak of weaker (i.e less well adapted) forms crossing with the stronger, you take a widely different view from what I do on the struggle for existence; for such weaker forms could not exist except by the rarest chance.

With respect to utility reflect that $\frac{99}{100}$ths part of the structure of each being is due to inheritance of *formerly* useful structures. Pray read what I have said on "correlation":[11] Orchids ought to show us how ignorant we are of what is useful.[12] No doubt hundreds of cases cd be advanced of which no explanation c^{d} be offered; but I must stop. Your letter has interested me much—

I am very far from strong & have great fear that I must stop all work for a couple of months for entire rest & leave home.— It will be ruin to all my work.—

Pray believe me | Dear Sir | Yours very faithfully | Ch. Darwin

[Enclosure]

Passiflora

Lecoq De la Fécondation p. 70. says that *several* kinds cannot be fertilised by own pollen; but can be fertilised by that of other species; *names not given*.—[13]

(Transact. of Hort. Soc. of London Vol. 7. p. 95 Mr Mowbray I think (I see P. *racemosa* & *alata* set best by using pollen of one to other) gives names of species which he tried with same result.—)[14]

Bosse in German. Hort. Periodical (quoted by Gärtner Bastard s. 64) makes similar statements; & G. can confirm that P. racemosa can be more easily fertilised with pollen of P. cœrulea than with own;— &c &c—[15]

I have received 3 other private accounts of analogous cases; but cannot spare time to hunt them up; nor do I know whether names of species were given.[16]

I have found one case, viz that the Granadilla w^{d} never set with own pollen, but would with pollen of Passiflora Edulis. Mr Scott. Gardener to Sir G. Staunton[17]

Lobelia

Gärtner (s. 357) twice found that L. fulgens, though producing certainly good pollen, could not be fertilised by it, but c^{d} be fertilised by pollen of L. syphilitica & L. cardanalis.—[18]

Kölreuter (2^{d}. Fortsetz. & 3^{d}. Fort.) found that Verbascum phœniceum could be fertilised by 4 distinct species, but not by its own **apparently** good pollen.—[19]

LS(A)[20]
DAR 93: B66–8, B71

[1] Emma Darwin recorded in her diary (DAR 242) that after a fortnight of good health, CD became ill in the last week of February. She noted that he was: 'faint in night', 'languid & heavy' every morning, and 'sick several times in course of week'. On 4 March she recorded: 'Ch. better but occasional sickness'.

[2] See letter from John Scott, 3 March 1863. Asa Gray gave this information in a missing postscript to his letter to CD of 10 November 1862 (*Correspondence* vol. 10). See *Correspondence* vol. 10, letter to Asa Gray, 26[–7] November [1862] and n. 3.

[3] See enclosure.

[4] See letter from John Scott, 3 March 1863.

Charles Lyell
Portrait sketch by George Richmond, 1853.
(By courtesy of the National Portrait Gallery, London.)

Thomas Henry Huxley
Photograph by Ernest Edwards, *c.* 1863.
(From L. Reeve ed. 1863–6. By permission of the Syndics of Cambridge University Library.)

Hugh Falconer
Photograph by Ernest Edwards, *c.* 1863.
(From L. Reeve ed. 1863–6. By permission of the Syndics of Cambridge University Library.)

Richard Owen
Photograph by Ernest Edwards, *c.* 1863.
(From L. Reeve ed. 1863–6. By permission of the Syndics of Cambridge University Library.)

[5] See letter from John Scott, 3 March 1863 and nn. 6 and 7.

[6] The reference is to the Sawbridgeworth nurseryman, Thomas Rivers. CD wrote drafts of the chapters on inheritance for *Variation* (*Variation* 2: 1–84) between 23 January and 1 April 1863 (see 'Journal' (Appendix II)). His account of the weeping habit of trees draws heavily on information provided by Rivers. See also letters to Thomas Rivers, [14 February 1863] and 5 March [1863].

[7] Scott was preparing a paper on the relationship between the form of reproduction and the heritability of variation in plants; however, the paper was never published (see letter from John Scott, 3 March 1863 and n. 13). Scott experimented extensively with species of *Primula* in 1863 and, at CD's prompting, wrote a paper on the subject (Scott 1864a; see letters from John Scott, 21 May [1863], [3 June 1863], and 23 July [1863]).

[8] A. Gray 1862a. See letter from John Scott, 3 March 1863.

[9] On Scott's success in pollinating the orchid *Gongora atropurpurea*, and on its implications for CD's account of *Acropera* in *Orchids*, pp. 203–10, see letters from John Scott, 6 January 1863 and 3 March 1863, and letter to John Scott, 16 February [1863].

[10] See letter from John Scott, 3 March 1863.

[11] CD discussed what he called 'Correlation of growth' in *Origin*, pp. 143–50, stating that he meant by the expression 'that the whole organisation is so tied together during its growth and development, that when slight variations in any one part occur, and are accumulated through natural selection, other parts become modified' (p. 143).

[12] One of the objects of *Orchids* was to show that there were many 'contrivances by which Orchids are fertilised' (p. 1), the purposes of which naturalists had not previously known.

[13] Lecoq 1845, p. 70. There are annotated copies of the first and second editions of Henri Lecoq's *De la fecondation naturelle et artificielle des végétaux et de l'hybridation* (Lecoq 1845 and 1862) in the Darwin Library–CUL (see *Marginalia* 1: 495–7).

[14] William Mowbray's observations on *Passiflora* were communicated by letter to the secretary of the Horticultural Society of London in October 1824; the letter is summarised in the notices of communications to the society (*Transactions of the Horticultural Society of London* 7 (1830): 95–6).

[15] Bosse 1829, p. 431 and Gärtner 1849, pp. 64–5. There is an annotated copy of Gärtner 1849 in the Darwin Library–CUL.

[16] No letters on this subject have been found; however, see the account of cases of self-sterility in *Passiflora* given in *Variation* 2: 137–8.

[17] George Thomas Staunton and Alexander Scott. CD's source for this information has not been identified.

[18] Gärtner 1849, p. 357.

[19] Kölreuter 1761–6, 2: 9–40, 3: 2–5. There is an annotated copy of this work in the Darwin Library–CUL (see *Marginalia* 1: 458–71).

[20] Except for alterations, the letter is in Emma Darwin's hand up to 'Your letter has interested me much—'; the remainder of the letter and the enclosure are in CD's hand.

To W. D. Fox 9 March [1863]

Down Bromley Kent
March 9th

My dear Fox

I have just been quoting in my M.S. on your authority from an old letter that you crossed White Muscovy Drake with slate-coloured duck & that the young were always pied black & white like the common or aboriginal breed.—[1] I have many analogous facts in common Ducks, fowls, & pigeons & the case interests me much. Now can you tell me do the White Musk & Slate-coloured musk, (when not crossed) each breed quite or nearly true?

Again in same letter I have quoted that "12 white ewes of Mr Woodd's had 23 coal-black lambs by a Ram that had a small patch of black only".[2] Can you tell me what breed these ewes were? Was Ram of same breed? Do you know how big & where the black patch on ram was?—

I should almost expect that if two very different breeds of *white* sheep were crossed there would be some tendency to dusky lambs; & so with horns, if two hornless breeds were crossed; but I know not where to enquire. I am even inclined to *suspect* that there is a tendency to a return to primordial wildness in hybrids between two domestic species.—

I enjoyed much seeing you in London.—[3] My London trip did me some good; but I have since had a very bad fortnight; & Emma declares, I fear with truth, that we must all go for 6 or 8 weeks to Malvern.[4]

My dear old friend | Yours ever sincerely | C. Darwin

Endorsement: 'March 11. 1863'
Postmark: MR 10 63
Christ's College Library, Cambridge (Fox 138)

[1] The letter from Fox has not been found; however, see *Correspondence* vol. 6, letter to W. D. Fox, 3 January [1856], in which CD thanked Fox for 'particulars on the eggs and colour of the Muscovy Ducks & sheep' (see also n. 2, below). CD wrote a draft of the section of *Variation* dealing with inheritance (*Variation* 2: 1–84) between 23 January and 1 April 1863 (see 'Journal' (Appendix II)); the crosses conducted by Fox are briefly mentioned in *Variation* 2: 40.

[2] This information, incorporating corrections sent by Fox, is given in *Variation* 2: 30–1. The reference is to Charles Henry Lardner Woodd, Fox's brother-in-law.

[3] CD stayed at 6 Queen Anne Street, London, the home of his brother Erasmus Alvey Darwin, between 4 and 14 February 1863. He met Fox there on 13 February (see letter to W. D. Fox, [10 February 1863], and letter from W. D. Fox, [11 February 1863]).

[4] CD refers to James Manby Gully's hydropathic establishment in Great Malvern, Worcestershire. CD and Fox discussed the benefits of hydropathy and the treatment offered at Gully's establishment at length in 1849 (see *Correspondence* vol. 4); Fox had also undergone treatment there. On the state of CD's health, see also the letter to H. W. Bates, 4 March [1863], n. 10

From Smith, Elder and Company[1] 9 March 1863

[65 Cornhill, London]
9 March 1863

In reply to yr. favor of the 5th. inst we beg to hand you cheque for 16.13.1 being the Acct due to you after adding £5 for the purchase of stock on hand of yr. two works—viz 99 copies of *Volcanic Islands* & 76 copies of *South America*, & we also enclose a form of receipts for your signature agreeably with yr request—[2]

In reference to the postage stamps used by ourselves, we beg to say that they are not usually sold to the public, but we have much pleasure in requesting your son's acceptance of the enclosed set, as an addition to his collection.[3]

Contemporary copy
National Library of Scotland (MS 23181: 6v)

1 The text of the letter is taken from a contemporary copy made by the company. A note on the copy records that the letter was assigned to 'Mr. Wooldridge'; this individual has not been further identified.

2 See letter from Smith, Elder and Company, 3 March 1863. The references are to *Volcanic islands* and *South America*. CD recorded receipt of the cheque in his Account book–banking accounts on 14 March 1863.

3 See letter to Smith, Elder and Company, 5 March [1863] and n. 4. The reference is to Leonard Darwin.

To Smith, Elder and Company 10 March [1863][1]

Down Bromley | Kent
March 10th

Dear Sir

I enclose with thanks the receipt for your cheque.—[2] I am truly obliged for your most kind present of your own stamps, which have given unbounded satisfaction to my Boy.—[3]

Dear Sir | Yours faithfully & obliged | Ch. Darwin

National Library of Scotland (MS 23181: 13)

1 The year is established by the relationship between this letter and the preceding letter.

2 See preceding letter.

3 See preceding letter. The reference is to Leonard Darwin.

From Charles Lyell 11 March 1863

53 Harley Street:
March 11, 1863.

My dear Darwin,—

I see the 'Saturday Review' calls my book 'Lyell's Trilogy on the Antiquity of Man, Ice, and Darwin.'[1]

As to my having the authority you suppose to lead a public who up to this time have regarded me as the advocate of the other side (as in the 'Principles') you much overrate my influence.[2] In the new 'Year Book of Facts' for 1863, of Timbs, you will see my portrait, and a sketch of my career, and how I am the champion of anti-transmutation.[3] I find myself after reasoning through a whole chapter in favour of man's coming from the animals, relapsing to my old views whenever I read again a few pages of the 'Principles,' or yearn for fossil types of intermediate grade.[4] Truly I ought to be charitable to Sedgwick and others.[5] Hundreds who have bought my book in the hope that I should demolish heresy, will be awfully confounded and disappointed. As it is, they will at best say with Crawfurd, who still stands out, 'You have put the case with such moderation that one cannot complain.' But when he read Huxley, he was up in arms again.[6]

My feelings, however, more than any thought about policy or expediency, prevent me from dogmatising as to the descent of man from the brutes, which, though I am prepared to accept it, takes away much of the charm from my speculations on the past relating to such matters.

I cannot admit that my leap at p. 505, which makes you 'groan,' is more than a legitimate deduction from 'the thing that is' applied to 'the thing that has been,' as Asa Gray would say, and I have only put it moderately, and as a speculation.[7]

I cannot go Huxley's length in thinking that natural selection and variation account for so much,[8] and not so far as you, if I take some passages of your book separately.

I think the old 'creation' is almost as much required as ever, but of course it takes a new form if Lamarck's views improved by yours are adopted.[9]

What I am anxious to effect is to avoid positive inconsistencies in different parts of my book, owing probably to the old trains of thought, the old ruts, interfering with the new course.

But you ought to be satisfied, as I shall bring hundreds towards you, who if I treated the matter more dogmatically would have rebelled.

I have spoken out to the utmost extent of my tether, so far as my reason goes, and farther than my imagination and sentiment can follow, which I suppose has caused occasional incongruities.

Woodward is the best arguer I have met with against natural selection and variation. He puts conchological difficulties against it very forcibly. He is at the same time an out-and-out progressionist.[10]

I am glad that both you and Hooker like the 'ice' part of the Trilogy.[11] You are the first to allude to my remarks on Ramsay, who says 'I shall come round to his views in good time.'[12]

Falconer, whom I referred to oftener than to any other author, says I have not done justice to the part he took in resuscitating the cave question, and says he shall come out with a separate paper to prove this. I offered to alter anything in the new edition, but this he declined.[13] Pray write any criticism that occurs to you; you cannot put them too strongly or plainly.

Ever yours sincerely, | Charles Lyell.

Incomplete[14]
K. M. Lyell ed. 1881, 2: 362–4

[1] The anonymous review of C. Lyell 1863a in the *Saturday Review*, 7 March 1863, p. 311, stated that, strictly speaking, the work was 'a trilogy, the constituent elements of which should be headed respectively, Prehistoric Man, Ice, and Darwin.'

[2] CD was 'greatly disappointed' that Lyell had not written more positively in support of the transmutation of species in *Antiquity of man* (C. Lyell 1863a; see letter to Charles Lyell, 6 March [1863]). Lyell was a leading figure in the British scientific establishment, whose earlier critical views of transmutation, first enunciated in his *Principles of geology* (C. Lyell 1830–3), had played an important role in the British debate on the subject.

3 The *Year-book of facts in science and art* was an annual started in 1839 by John Timbs (*DNB*); the volume for 1863 began with a four-page biography of Lyell, including an engraved portrait as the frontispiece (*Year-book of facts in science and art* (1863): 3–6). The article stated (p. 6):

> With regard to 'the Origin of Species' . . . Sir Charles Lyell not only opposes this theory, but denies that in the history of the strata there is any evidence that the lowest forms of animals were created first. The only fact he admits favouring the hypothesis of development is the late appearence of man on the earth.

4 One of Lyell's chief concerns about transmutation theories had been their implications for the status of the human species (see L. G. Wilson ed. 1970, Bartholomew 1973, and *Correspondence* vols. 6–8). In the final chapter of *Antiquity of man* (C. Lyell 1863a, pp. 471–506), entitled 'Bearing of the doctrine of transmutation on the origin of man, and his place in the creation', Lyell relied heavily on quotations from other authors, leaving his own position on the subject unclear. See also n. 2, above.

5 Adam Sedgwick, one of CD's former mentors, was critical of *Origin* (see *Correspondence* vol. 7, letter from Adam Sedgwick, 24 November 1859).

6 The orientalist John Crawfurd, to whom CD sent a presentation copy of *Origin*, published one of the first negative reviews of the book ([Crawfurd] 1859; see *Correspondence* vol. 7, letter to Charles Lyell, 2 December [1859], and *Correspondence* vol. 8, Appendix III). Lyell discussed Crawfurd's views on the origin of languages in C. Lyell 1863a, pp. 455–6. The reference is to T. H. Huxley 1863b. Crawfurd discussed Lyell's *Antiquity of man* and Thomas Henry Huxley's *Evidence as to man's place in nature* in a paper read before the Ethnological Society on 14 April 1863 (Crawfurd 1863).

7 See letter to Charles Lyell, 6 March [1863] and n. 43. In a letter to J. D. Hooker of 9 March 1863, Lyell had written (K. M. Lyell ed. 1881, 2: 361):

> Asa Gray says that Lyell's doctrine is 'that the thing that is, is the thing that has been, and shall be.' Now if the thing that is, in the case of a man of genius born of ordinary parents and with ordinary brethren of the same parentage imply a slight leap, I do not see why Darwin should complain of my leap, given only as a speculation, for the highest unprogressive to the lowest progressive.

8 T. H. Huxley 1863b, pp. 105–8. See also K. M. Lyell ed. 1881, 2: 361.

9 The reference is to Jean Baptiste de Lamarck, whose evolutionary theory Lyell had opposed in his *Principles of geology* (C. Lyell 1830–3; see also Bartholomew 1973).

10 The reference is to Samuel Pickworth Woodward, whose opposition to species transmutation was well known to CD (see *Correspondence* vol. 6, letter to S. P. Woodward, 18 July 1856). Although Woodward did not review *Origin*, he had written critically about CD's theory on a number of occasions (see Woodward 1881, pp. 303–6). See also letter from S. P. Woodward, 5 June 1863 and n. 3. Most naturalists in mid-nineteenth-century Britain believed that the fossil record indicated a gradual progression from lower to higher forms of life, while at the same time rejecting species transmutation (see C. Lyell 1863a, pp. 395–406). Lyell had long been an outspoken critic of progressionism (see Bartholemew 1973).

11 See n. 1, above. See also letter to Charles Lyell, 6 March [1863] and n. 20.

12 See letter to Charles Lyell, 6 March [1863] and n. 25. The reference is to Andrew Crombie Ramsay.

13 In a letter published in the *Athenæum*, 4 April 1863, pp. 459–60, Hugh Falconer objected to several aspects of Lyell's *Antiquity of man* (C. Lyell 1863a). He claimed that Lyell had not properly acknowledged his role in 'the events which led to the re-agitation and proof of the question of primeval man', and particularly in the excavation of human artefacts from Pleistocene deposits in Brixham Cave in Devon. For discussions of Falconer's attack on Lyell, see Bynum 1984, Grayson 1985, and L. G. Wilson 1996a. Falconer never published the planned paper; however, he drafted a historical essay to affirm his role in the researches on the prehistory of the human species in 1863, intending it as an introduction to a book on the subject. The essay was published in Falconer 1868, 2: 570–600.

14 The original letter apparently included a postscript discussing various matters to which CD responded (see letter to Charles Lyell, 12–13 March [1863] and nn. 11 and 17).

To James Paget 11 March [1863]

Down. | Bromley. | Kent. S.E.
March 11th

Dear Paget

I shd. be very glad to insert in my Chapt. on inheritance the following sentence, as it comes in well after some other facts.[1] Do you object? And have I quoted the facts accurately?[2] If I receive *no answer* I will assume that it is accurate, & that I have your permission.—

"Many persons, as I hear from Mr Paget, have two or three hairs in their eye-brows (apparently corresponding with the vibrissœ of the lower animals) much longer than the others; and even so trifling a peculiarity as this runs in families."—

I wish I could hear of any case of inherited peculiarities in eye-lashes.—

Pray believe me | Dear Paget | Yours sincerely | Ch. Darwin

Endorsement: '1863'
Smithsonian Institution Libraries (MSS 405A, Special collections, Dibner)

1 CD wrote drafts of the chapters on inheritance for *Variation* (*Variation* 2: 1–84) between 23 January and 1 April 1863 (see 'Journal' (Appendix II)); the sentence appears in *Variation* 2: 8.

2 No earlier correspondence on this subject has been found; CD apparently discussed it with Paget while he was in London in February (see letter from James Paget, 7 February 1863). See also letter from James Paget, 16 March 1863.

From W. D. Fox 12 March [1863][1]

Delamere Ry | Northwich
March 12

My dear Darwin

I remember well what you mention about the slate coloured & White Muscovy Ducks.[2] The Slate coloured, I imagine, were the produce of some cross raised at Birmingham. I have never seen them since nor do I think I ever reared any, as I lost the Drake.

The white always breed true to colour, as far as my experience goes, & I keep a large flock of them—(I have about 20 now)—

With however this exception that they have a great tendency to a Black Topper on head (exactly as represented in Bewicks plate).[3]

Unless some considerable pains are taken in weeding out all these black topped Birds—the *pure white* would soon cease.

I never saw however a single black feather any where but on the top of Head.

The Black Lambs I also well remember.[4] They were Southdown Ewes. The Ram I never saw— He was provided by Giblitt the Great Butcher[5]—and I was told in answer to my enquiries that he was all but white, having a small patch only of Black—*I think* about the head. The flock was the queerest I ever saw, and *every* lamb being black in the lot, looks as if there was a positive law of some kind in the matter.

I will write to ask if Mr Woodds man[6] remembers where the Black was in Ram, but I fear it will not be satisfactorily answered.

The Ram appeared cross bred with a good deal of Southdown in him.

The only *pure* cross of Sheep I know (if one can use such a term)—are between pure Leicester & Southdown. I think it was the late Earl Morton[7] who had a beautiful flock of these. They had no horns—& partook of each strain.

I have a small flock of the black & white sheep—which keep very true to shape & colour. When crossed with Leicester they produce a race pretty nearly balanced—except that there is a great tendency to Black lambs. The only 3 born this year are black. Having once got the black strain in this way from a cross with the black & white & Leicester—it takes many generations—at least 6 or 7—I am told by a friend of mine—to get back the white colour—by continual crossing with a Leicester. I can only speak to the 3^d^ generation—which was as black as the first.[8] I have never seen or heard of *horned* produce from hornless crosses. I do not think they exist so in our English crosses. The Cheviot & Leicester are often crossed—& certainly have no horns— Neither have the Southdown & Leicester.

There seems little tendency in crosses of *Sheep* to return to primordial wildness—as the Irish have been almost renovated by the Leicester cross: and the Shropshire Down sheep are a greatly improved cross bred sheep.

I fear I have not added much to my former notes on these points.[9]

What I wrote at time, may be strictly depended upon— In this letter you will see what are surmises only.

I should be very glad to hear of your getting to Malvern for 6 or 8 weeks.[10] It is a glorious place for renovating the health in. If my time & means would allow of it, I sh^d^ go for a month every spring & be got into condition. I quite believe I owe my life to it in my last illness—and it was a curious illustration of *the System* versus the D^rs^. as Gully completely denied the existence of what I told him was my conceived idea of my wretchedness—but added "Well if it is so, the water cure will prove its existence—and so it did. I have never met with more than one similar case, which was most carefully treated according to rule, and the young man died of it.

I cannot tell you how glad I was to see you so well in London.[11] It was a great pleasure to see so many old & valued faces—& all looking so well. I very much want to see Susan and Caroline again.[12]

If you go to Malvern I shall try to come over & get a weeks bathing while you are there. I have some curious enquiries about you & your Book sometime. You ought to occupy a cage by the side of the Gorilla in the British Museum sometimes to let the Public see you.—

Kindest regards to Mrs Darwin.— Ever yours sincerely | W D Fox

DAR 164: 178

[1] The year is established by the relationship between this letter and the letter to W. D. Fox, 9 March [1863].

[2] See letter to W. D. Fox, 9 March [1863] and n. 1.
[3] Bewick 1826, 2: 317. Topping: 'the crest of a bird' (*OED*).
[4] See letter to W. D. Fox, 9 March [1863] and n. 2.
[5] The reference is possibly to William Giblitt, who had a butcher's shop at 110 Bond Street, London (*Post Office London directory* 1854).
[6] The reference is to Fox's brother-in-law, Charles Henry Lardner Woodd; his 'man' has not been identified.
[7] George Sholto Douglas.
[8] CD summarised Fox's information on the inheritance of colour in sheep in *Variation* 2: 30–1.
[9] The letter in which Fox originally provided information on these subjects has not been found; however, see the letter to W. D. Fox, 9 March [1863] and n. 1.
[10] See letter to W. D. Fox, 9 March [1863] and n. 4.
[11] CD stayed at 6 Queen Anne Street, London, the home of his brother Erasmus Alvey Darwin, between 4 and 14 February 1863. He met with Fox on 13 February (see letter to W. D. Fox, [10 February 1863], and letter from W. D. Fox, [11 February 1863]).
[12] The reference is to CD's sisters, Susan Elizabeth Darwin and Caroline Sarah Wedgwood. See also letter from W. D. Fox, 6 February [1863].

To Charles Lyell 12–13 March [1863]

Down Bromley Kent
12th

My dear Lyell

I thank you for your very interesting & kind, I may say charming letter.[1] I feared you might be huffed for a little time with me; I know some men would have been so.[2] I have hardly any more criticisms, anyhow worth writing. But I may mention that I felt a little surprise that old B. de Perthes was not rather more honourably mentioned.—[3] I would suggest whether you could not leave out some references to the "Principles";[4] one for the real Student is as good as a hundred; & it is rather irritating & gives feeling of incompleteness to general reader to be often referred to other book.— As you say that you have gone as far as you believe on species-question,[5] I have not a word to say; but I must feel convinced that **at times**, judging from conversation, expression, letters &c, you have as completely given up belief in immutability of specific forms, as I have done.— I must still think a clear expression from you, *if you could have given it*, would have been potent with the public, & all the more so, as you formerly held opposite opinion.—

The more I work the more satisfied I become with variation & n. selection; but that part of the case I look at as less important, though more interesting to me personally. As you ask for criticisms on this head (& believe me, I sh[d]. not have made them unasked) I may specify (p. 412, 413) that such words as M[r] D "labours to show"—"is believed by the author to throw light", would lead a common reader to think that you yourself do *not* at all agree, but merely think it fair to give my opinion.—[6]

Lastly, you refer repeatedly to my view as a modification of Lamarcks doctrine of development & progression;[7] if this is your deliberate opinion there is nothing to be said—; but it does not seem so to me; Plato, Buffon, my grandfather before Lamarck & others propounded the **obvious** view that if species were not created separately, they must have descended from other species:[8] & I can see nothing else in common between the Origin & Lamarck. I believe this way of putting the case is very injurious to its acceptance; as it implies necessary progression & closely connects Wallace's & my views with what I consider, after two deliberate readings, as a wretched book; & one from which (I well remember my surprise) I gained nothing.[9] But I know you rank it higher, which is curious, as it did not in the least shake your belief.— But enough & more than enough. Please remember you have brought it all down on yourself!!—

March 13th— I must add that Henrietta,[10] who is a first rate critic & to whom I had *not said a word* about Lamarck, last night said, "Is it fair that Sir C. Lyell always calls your theory a modification of Lamarcks? Why is it more a modification of his, than of any one's else?" Do not allude to this, for I do not suppose she would at all approve of my quoting her appropriate (in my opinion) criticism.— I have more trust in your judgment than in my own, so I hope you may be right, as far as mere policy is concerned, in your very gentle statement of your belief.—

Many thanks about Aye-Aye paper; I have written to Sir Henry.—[11]

I am very sorry to hear about Falconer's "reclamation";[12] I hate the very word, & have a sincere affection for him.— I have been much interested in Athenæum controversy:[13] how well Mr Flower puts in the way in which Owen has falsely dragged in size of brain.—[14] Your answer seemed to me very good; except I regretted your too great civility in saying that Owen must have *forgotten* what he said in Annals; it was a brazen lie, & ought not, I think, to have been treated so delicately.—[15]

Did you ever read anything so wretched as the Athenæum Reviews of you, & of Huxley especially Your *object* to make man old, & Huxley's *object* to degrade him. The wretched writer has not a glimpse what the discovery of scientific truth means.—[16] How splendid some pages are in Huxley; but I fear the book will not be popular.

I have just reread your letter; it seems that I mentioned the sentence "Should it ever become highly probable": I remember determining not to mention it; but it seems that I was overtaken.[17]

Ever my dear Lyell | Yours most truly | C. Darwin

I keep very queer in health, & we have resolved, eheu eheu, to start for Malvern after Easter holidays.—[18]

Endorsement: 'March 14 | 1863'
American Philosophical Society (290)

[1] Letter from Charles Lyell, 11 March 1863.

[2] CD refers to Lyell's reaction to the criticisms of C. Lyell 1863a that CD made in his letter of 6 March [1863]. In his letter to CD of 11 March 1863, Lyell stated: 'Pray write any criticism that occurs to you; you cannot put them too strongly or plainly.' CD's annotated copy of C. Lyell 1863a is in the Darwin Library–CUL (see *Marginalia* 1: 525–7).

[3] In the 1840s, Jacques Boucher de Perthes discovered shaped flints in Pleistocene deposits in the Somme valley, which, he argued in the first volume of *Antiquités celtiques et antédiluviennes* (Boucher de Perthes 1847–64), were human artefacts contemporaneous with the associated Pleistocene fossils. His explanation was initially widely rejected, but his findings were confirmed following visits to the location in 1858 and 1859 by Hugh Falconer and Joseph Prestwich. See Grayson 1983. Boucher de Perthes's research and collections are mentioned several times in C. Lyell 1863a; however, CD apparently considered that he was not given due credit for his priority in determining the prehistoric nature of the Somme artefacts.

[4] Lyell referred to several editions of his *Principles of geology* (C. Lyell 1830–3).

[5] See letter from Charles Lyell, 11 March 1863.

[6] In the second edition of *Antiquity of man*, 'labours to show' is changed to 'argues', and 'is believed by the author to throw light' is changed to 'would throw light' (C. Lyell 1863b, pp. 412, 413).

[7] CD refers to Jean Baptiste de Lamarck. See letter to Charles Lyell, 6 March [1863] and n. 10. See also letter from Charles Lyell, 11 March 1863.

[8] CD refers to Georges Louis Leclerc, comte de Buffon, and Erasmus Darwin, both of whom were mentioned in passing in CD's 'Historical sketch of the recent progress of opinion on the origin of species' (*Origin* 3d ed., pp. xiii–xiv n.) The reference to Plato is probably to *Timaeus*, 69B–C, a passage brought to his attention in a letter from G. H. K. Thwaites, 15 May 1862 (*Correspondence* vol. 10). See also *Correspondence* vol. 9, letter from R. E. Grant, 16 May 1861. In his 'Historical sketch', CD, like Lyell, treated Lamarck as 'the first man whose conclusions on this subject excited much attention' (*Origin* 3d ed., p. xiii).

[9] CD apparently refers to *Philosophie zoologique* (Lamarck 1809), which contains the most detailed exposition of Lamarck's transmutation theory; there is an annotated copy of the first volume of a later edition of this work (Lamarck 1830) in the Darwin Library–CUL (see *Marginalia* 1: 477–80). On 18 May 1839, CD noted in his reading notebook that he had read this work (see *Correspondence* vol. 4, Appendix IV, 119: 15a); there are frequent references to it in his transmutation notebooks from 1837 onwards (see *Notebooks*). CD also refers to Alfred Russel Wallace, with whom he announced the theory of natural selection in 1858 (C. Darwin and Wallace 1858). Unlike CD and Wallace's theory of transmutation, Lamarck's theory posited a natural tendency toward organic complexity (*DSB*). See also *Correspondence* vol. 3, letter to J. D. Hooker, [11 January 1844].

[10] Henrietta Emma Darwin.

[11] CD had asked whether he might borrow Lyell's copy of the part of the *Transactions of the Zoological Society of London* containing Owen 1862c (see letter to Charles Lyell, 6 March [1863] and n. 44). Lyell appears to have replied to this request in a missing portion of his letter to CD of 11 March 1863. CD himself bought a copy of the number, which is now in the Darwin Library–CUL. The reference may be to Henry Holland, although no such letter has been found.

[12] See letter from Charles Lyell, 11 March 1863 and n. 13.

[13] CD refers to the controversy following the publication of a letter by Richard Owen in the *Athenæum*, 21 February 1863, pp. 262–3. See letter from J. D. Hooker, [23 February 1863], n. 2. Some of Owen's charges were rebutted in a letter from George Rolleston, published in the *Athenæum*, 28 February 1863, p. 297. Lyell published a reply to Owen's letter in the *Athenæum*, 7 March 1863, pp. 331–2, which also contained a further letter from Owen. For CD's view of Owen's criticisms, see letter to J. D. Hooker, 24[–5] February [1863], and letter to T. H. Huxley, 26 [February 1863].

[14] In a second letter to the *Athenæum*, published with his first on 7 March 1863, pp. 331–2, Lyell enclosed an extract from a letter sent to him by William Henry Flower, which repudiated Owen's claims concerning disputed brain specimens. See also letter to J. D. Hooker, 24[–5] February [1863] and n. 25.

[15] The reference is to an article by Owen that appeared in the *Annals and Magazine of Natural History* (Owen 1861a; see *Athenæum*, 7 March 1863, p. 331). See also the following letter and n. 4.

[16] C. Lyell 1863a was reviewed in the *Athenæum*, 14 February 1863, pp. 219–21, and T. H. Huxley 1863b in the *Athenæum*, 28 February 1863, pp. 287–8. According to the publisher's marked copies of the *Athenæum* (City University Library, London) the reviews were written by John R. Leifchild.

[17] CD refers to C. Lyell 1863a, p. 469, which states:

> we ought by no means to undervalue the importance of the step which will have been made, should it ever become highly probable that the past changes of the organic world have been brought about by the subordinate agency of such causes as 'Variation' and 'Natural Selection.'

This passage is not mentioned in the extant text of the letter to Charles Lyell, 6 March [1863], although CD did criticise an earlier passage on the same page (see letter to Charles Lyell, 6 March [1863] and n. 40); it may have been one of the passages criticised in a letter from Joseph Dalton Hooker to Lyell (see the letter from J. D. Hooker, [15 March 1863] and n. 6). Lyell did not refer to the passage in his letter to CD of 11 March 1863, but it was probably discussed in a postscript to the latter that has not been found (see also n. 11, above). In the second edition (C. Lyell 1863b, p. 469), the phrase 'should it ever become highly probable' was altered to read 'should it hereafter become the generally received opinion of men of science (as I fully expect it will)'.

[18] Eheu: 'an interjection of pain or grief, *ah! alas!*' (Lewis and Short 1969). CD refers to James Manby Gully's hydropathic establishment in Great Malvern, Worcestershire.

To J. D. Hooker 13 [March 1863]

Down
13th

My dear Hooker

Treviranus has sent me *for you* copies of Bot. Zeitung, forwarded by this post, for which you won't care.—[1]

I sh^{d}. have thanked you sooner for Athenæum & very pleasant previous note,[2] but I have been busy & not a little uncomfortable, from frequent uneasy feeling of fulness, slight pain & tickling about the heart. But as I have no other symptoms of Heart complaint, I do not suppose it is affected. Were you not similarly plagued before you went to India?—[3] I was much interested by Athenæum; I was sorry that Lyell was so civil about that audacious lie of Owen's that there was not a word in Annals on Cerebellum & Cerebrum.—[4]

I have had a most kind & delightfully candid letter from Lyell, who says he spoke out as far as he believes.[5] I have no doubt his belief failed him as he wrote; for I feel sure that at times he no more believed in Creation than you or I.— I have grumbled a bit in answer to him, at his *always* classing my work as a modification of Lamarck's, which it is no more than of any author who did not believe in immutability of Species & did believe in descent.—[6] I am very sorry to hear from Lyell that Falconer is going to publish a formal reclamation of his own claims.[7] Did you ever read such a wretched review as that in Athenæum of Huxley's book?—[8] Some of the pages in the book struck me as magnificent.

Propagate the wild Potatoes in poorish soil most carefully: next year, if I am then up to any work, I would give anything for some tubers to test fertility with cultivated varieties.—[9]

It is cruel to think of it, but we must go to Malvern in middle of April,—it is ruin to me.—[10]

Will you make a very trifling observation for me on early fine day on any Poplars in garden (we have not one near here); viz whether visited by bees which you will easily see, & whether slight shake sends out cloud of pollen; for in latter case & if *not* visited by Bees, it is almost certain to be fertilised by wind, which I am curious about in relation to willows, which are fertilised by Bees.—

Farewell | Ever yours | C. Darwin

Endorsement: 'March/63'
DAR 115: 186

[1] Ludolph Christian Treviranus had sent CD two copies of the numbers of the *Botanische Zeitung* containing Treviranus 1863a, asking him to forward one set to Hooker (see letter from L. C. Treviranus, 12 February 1863). There are annotated copies of this work in the Darwin Pamphlet Collection–CUL.

[2] Letter from J. D. Hooker, [6 March 1863]. In his letter, Hooker promised to send CD his copy of the *Athenæum* for 7 March 1863, if it contained, as expected, Charles Lyell's reply to Richard Owen's critical remarks on C. Lyell 1863a; Lyell's letter was published on pp. 331–2.

[3] For a discussion of Hooker's health before and shortly after his departure in 1847 to explore the Himalayas, see the letter from J. D. Hooker, 20 February – 16 [March] 1848 (*Correspondence* vol. 4).

[4] In his letter to the *Athenæum*, 21 February 1863, pp. 262–3, Owen denied having discussed the relative backward extent of the cerebellum and cerebrum in Owen 1861a, as claimed in C. Lyell 1863a, p. 485. In his reply (*Athenæum*, 7 March 1863, pp. 331–2), Lyell quoted what Owen had written on the subject in that paper, commenting that Owen 'must surely have forgotten his own words'. See also preceding letter.

[5] See letter from Charles Lyell, 11 March 1863.

[6] See preceding letter.

[7] See letter from Charles Lyell, 11 March 1863 and n. 13. The reference is to Hugh Falconer.

[8] According to the publisher's marked copies of the journal (City University Library, London), the anonymous review of Thomas Henry Huxley's *Evidence as to man's place in nature* (T. H. Huxley 1863b) that appeared in the *Athenæum*, 28 February 1863, pp. 287–8, was by John R. Leifchild. There is an annotated copy of T. H. Huxley 1863b in the Darwin Library–CUL.

[9] Hooker had asked CD what he wanted done with a number of tubers of the South American wild potato that he had obtained (see letter from J. D. Hooker, [6 March 1863] and n. 13). Having read a report that some varieties of potato were sterile with their own pollen, but fertile with that from other varieties, CD was keen to have further experimental proof of the case (see *Correspondence* vol. 10, letters to John Scott, 11 December [1862] and 19 December [1862]).

[10] CD refers to James Manby Gully's hydropathic establishment in Great Malvern, Worcestershire, where Anne Elizabeth Darwin, CD's eldest daughter and favourite child, died in 1851 (see *Correspondence* vol. 5). CD stayed at near-by Malvern Wells with his family from 3 September to 12 or 13 October 1863 (see 'Journal' (Appendix II)).

From J. D. Hooker [15 March 1863][1]

Kew
Sunday

Dr. Darwin

I have written to Sclater who got me the 4 Potato tubers to get you some.[2] I have given 2 to our Pleasure Ground Keeper, & 2 to our Bot. Garden man, & shall watch results carefully.[3]

I will see to the Poplars, & report.[4]

I have been having a long correspondence with Lyell, & have given him quite as *deflagrating* a yarn as I sent you—& likened him to the Theologians!—adding that

I had always hitherto classed him as the sole sexagenarian philosopher who could change his opinion on good ground—[5] He proposes some alterations of the two obnoxious passages, which will at any rate *do justice* to the hypothesis as he states it,—which the former ones did not.[6] Lyell dwells, & with reason on the fact that he makes as many converts whether he withholds or gives his own opinion. I tell him *perhaps more*, as people like to draw their own inferences but that is not the particular point we as his friends now look to.

He has written to me also about the date of publication of the Australian Essay, as preceding your Origin,[7]—in this matter he has got into a fix by giving said Essay a prominence which in the history of the discussion it (& its author) do not deserve— I have such an extreme aversion to intrude myself personally into such matters, & such an abomination of reclamation, that I cannot set him right, even did the plan of his book *now* admit of his giving the Essay less prominence— As it is I am ashamed of seeing it paraded with an italicised heading, just as you & the "Origin" are.—& an importance given to its priority of publication which it never dreamt of claiming—[8] Had I really believed that your "Origin" would have been out so soon after it I really think I should have delayed the Fl. Tasmaniae, rather than antedated you—but though I knew you were actually printing the Origin, I knew how long it had been delayed,—I knew how uncertain your health was—, & I was working myself to death to get the Tasmania Flora & its (for me) gigantic Expenses off my hands.[9] As it is, Lyell seems to think me entitled to a goodly share of the credit, of *establishing* though not *originating*

1. because of your over-generous acknowledgement of assistance from me in the Origin.[10]

2. Because it was my making him eat the leek of variation, that so stupified his senses that he was enabled to swallow Origin, & apply selection (as gastric juice)[11]

3. Because I forced the card of non-reversion of varieties.[12]

4. Because I first applied many of your results to the class^{n}. & descrip^{n} of one Flora & country. in a way intelligeable to him.

5. Because he understood my arrangement of the subject better than yours—at least so he said—some 18 mo. ago

All this is no reason for putting me in the *same category with you as propounder* of the doctrine, which his work seems to me too much to do.— However I have not alluded to this subject to him—nor should I, if he had been as careful never to mention my name, as Huxley[13] would seem to be. not that he really is so in the least I am sure.

I am grieved about this Falconer affair, F. is so cruelly impracticable & churlish when his prejudices are touched or priority overlooked;[14] & he has a most mischievous backer in Dr Percy,[15] who he makes a great friend of. I really do not know the merits of the case, & am so dissatisfied myself with the confused & confusing elements of the first XII Chapters, that I can form no opinion whether he is right or wrong in saying that due credit is not given to himself Prestwich & J. Gunn.—[16] I do hope that he (F.), will not write a pamphlet & add another scandal to Science.[17] I have a great mind to see & talk to him, but he is the very devil to discuss a matter of the kind with.—

I have finished Lyell & am enchanted with the Glacial chapters, Language & the whole treatment of the origin & development subjects (with above qualifications.) it certainly is a grand book on the whole, & well worthy of Lyells Scientific Reputation. He never rises to the magnificence of Huxleys language, nor to the sublimity of some of the passages in H^s. little book on the Position of Man, which one can read 1000 times with fresh delight.[18] By the way surely the last few pages of Huxley are not clear, I do not see how he can logically throw away all the evidence of the two simioid human skulls as worth nothing, after his admissions regarding them.[19]

I go to Lubbocks for next Sunday & if I can will walk over to Down[20]

More Cameroon's Mt Plants are coming, which will enable me to complete my paper & discuss the cold period quoad tropical African Mts & Flora—[21] Were the Bees & combs worth anything?[22] Another man is getting you some in Luando. (Mr Monteiro)[23]

I have been plagued always at intervals ever since I was 18 with heart symptoms, but less this last 4 years (after a very bad bout) & latterly (2 years) hardly at all—dull pain over region—palpitations,—pain on left shoulder,—tingling in arms & fingers,—fainting feeling,—pricking of pins, & at other times worms crawling in region of heart—& so on.[24] I never could connect them with any physical or physiological condition, or use or abuse of functions, cerebral digestive or sexual—except that they were, if anything, worst when I was costive— All my friends relations & acquaintances have heart complaint! so think nothing of it— they were a horrible nuisance—especially the feeling of fainting—for which I have rushed to stimulants, more than once.

Ever Yours affection. | J D Hooker.

DAR 101: 117–20

[1] The date is established by the relationship between this letter, the preceding letter, and the letter to J. D. Hooker, 17 March [1863]; the intervening Sunday was 15 March 1863.

[2] Philip Lutley Sclater. See letter from J. D. Hooker, [6 March 1863], and the preceding letter.

[3] Alexander Williamson and John Smith.

[4] CD had asked Hooker to make observations on pollination mechanisms in poplars (see preceding letter).

[5] Hooker apparently refers to his letter to CD of [1 March 1863], in which he criticised Charles Lyell's refusal to publicly commit himself concerning CD's theory in *Antiquity of man* (C. Lyell 1863a), calling it an 'abandonment' of his 'high position' in science. The reference is to Lyell's private acceptance of transmutation, of which he had previously been one of the most influential critics in Britain (see Bartholomew 1973).

[6] The passages referred to in C. Lyell 1863a, which evidently related to CD's theory, have not been identified; however, some of the changes made by Lyell in the second and third editions of this work are noted in Grayson 1985. See also letter to Charles Lyell, 12–13 March [1863] and n. 17.

[7] Hooker's introductory essay to *Flora Tasmaniæ* (J. D. Hooker 1859) was published on 29 December 1859 (*Taxonomic literature*); *Origin* was published on 24 November 1859 (Freeman 1977). However, in C. Lyell 1863a, p. 417, Lyell referred to Hooker's essay having been published 'a few months before

the appearance of the "Origin of Species."' See also n. 9, below, and letter to Charles Lyell, 6 March [1863] and n. 35.

8 In chapter 21, 'On the origin of species by variation and natural selection' (C. Lyell 1863a, pp. 407–23), Lyell included a separate section, headed 'Dr. Hooker, on the theory of "Creation by variation" as applied to the vegetable kingdom', in which he discussed J. D. Hooker 1859.

9 Hooker was mistaken in his recollection (see n. 7, above, letter to J. D. Hooker, 17 March [1863] and n. 3, and letter from J. D. Hooker, [24 March 1863]). He was apparently misled by an erroneous statement on the first page of J. D. Hooker 1859, which stated that the essay was reprinted from the 'first volume of Dr. Hooker's "Flora of Tasmania," published in June, 1859'. Hooker's *Flora Tasmaniæ* (J. D. Hooker 1860a) was published in parts, but that containing the introductory essay was not published until 29 December 1859 (*Taxonomic literature*); the introductory essay itself was dated 4 November 1859. In subsequent editions of *Antiquity of man* the publication date of Hooker's essay was given as December 1859.

10 In C. Lyell 1863a, p. 417, Lyell referred to CD's description of Hooker as one who had, for fifteen years, aided him 'in every possible way by his large stores of knowledge and his excellent judgment' (*Origin*, p. 3).

11 Lyell argued that Hooker's extensive travels, his particular study of geographical variation, and his practical knowledge of classification, meant that there was no one 'better qualified by observation and reflection to give an authoritative opinion on the question, whether the present vegetation of the globe is or is not in accordance with the theory which Mr. Darwin has proposed' (C. Lyell 1863a, p. 418). The reference to leeks is an allusion to Shakespeare's *Henry V*, 5.1.1–59 (Wells and Taylor eds. 1988), in which Captain Fluellen forces Ensign Pistol to eat a leek as a penance for his mockery of the symbol of Welsh nationality.

12 In C. Lyell 1863a, p. 420, Lyell referred to Hooker's assertion in J. D. Hooker 1859, p. viii, that 'species which have remained immutable for many generations under cultivation, do at length commence to vary, and having once begun, are thereafter peculiarly prone to vary further'.

13 Thomas Henry Huxley.

14 See preceding letter and letter from Charles Lyell, 11 March 1863 and n. 13.

15 John Percy.

16 Joseph Prestwich and John Gunn. See letter from J. D. Hooker, [6 March 1863] and n. 4.

17 Hugh Falconer's response to what he considered to be Lyell's failure to give due acknowledgment in C. Lyell 1863a to his own work and that of Prestwich, was published in the *Athenæum*, 4 April 1863, pp. 459–60.

18 T. H. Huxley 1863b.

19 The third chapter of T. H. Huxley 1863b (pp. 119–59), entitled 'On some fossil remains of man', was confined to a discussion of the fragmentary fossils of human skulls found in caves in the Meuse valley, Belgium, in the 1830s, and in the Neanderthal valley, near Düsseldorf, Prussia, in 1857. After describing the specimens at length, and comparing them with modern skulls, Huxley concluded that they were truly human, and did not represent an approach to that 'lower pithecoid form' from which the human species had descended (pp. 155–9). Huxley used this conclusion to argue that it was necessary to 'extend by long epochs the most liberal estimate that has yet been made of the antiquity of Man'.

20 John Lubbock lived at Chislehurst, a village about five miles north of Down; Hooker visited CD from Lubbock's house on 22 March 1863 (see letter from J. D. Hooker, [24 March 1863]).

21 In 1861, Hooker had begun preparing a list of the plants collected in the Cameroons mountains and islands off the coast of West Africa by Gustav Mann. Hooker gave his first reports on the collection to the Linnean Society in March 1861 and June 1862 (J. D. Hooker 1861, *Journal of the Proceedings of the Linnean Society of London* (*Botany*) 6 (1862): cvi). Hooker's findings provided support for CD's view that temperate plants had migrated to tropical regions during a global glacial period (see *Correspondence* vol. 10, letter from J. D. Hooker, [5 May 1862], letter to J. D. Hooker, 9 May [1862] and n. 6, letter from J. D. Hooker, 9 June 1862, and *Origin*, 4th ed., p. 445). Hooker read a further paper on the

subject to the society on 5 November 1863 (J. D. Hooker 1863b). Hooker discussed the significance of his findings for CD's theory of migration during a global cold period in J. D. Hooker 1863b, p. 181.

22 See letter to J. D. Hooker, 17 March [1863]. Hooker had written to Mann, who was collecting in West Africa for the Royal Botanic Gardens, Kew, in August 1862, asking him for specimens of African bees and honeycombs to assist CD in writing his account of bees from different localities in *Variation* 1: 297–9 (see *Correspondence* vol. 10, letter from J. D. Hooker, 20 August 1862). See also letter to T. W. Woodbury, 15 March [1863], and letter from T. W. Woodbury, 17 March 1863.

23 Hooker had written in March 1862 to ask Joachim John Monteiro, a mining engineer and zoologist residing in Luanda, Angola, to try to obtain specimens for CD (see *Correspondence* vol. 10, letter from J. D. Hooker, 20 August 1862).

24 See preceding letter and n. 3.

From Charles Lyell 15 March 1863

53 Harley Street:
March 15, 1863.

My dear Darwin,—

Your letter will be very useful. I wish to get such passages so far in the Darwinian direction as not to be inconsistent with my general tone, and what Hooker calls some of my original arguments in favour of natural selection.[1] At the same time I am struck by the number of compliments, both in reviews and in conversation with the half-converted, which I receive, because I have left them to draw their own inferences, and have not told them dogmatically that they must turn round with me.[2] Hooker admits that in science people do not like to be told too plainly that they must believe, though in religion they wish to have it laid down for them. Yet he may be wrong, for if the 'Times' were to write for the next fortnight against the Southern States, and against the Poles,[3] nine-tenths of good society would whirl round, and the middle class which would stand firm would be able to do so partly because they read cheaper papers which are not interested in following the lead of the 'Times.' . . . .

I wish I deserved what you say about taking criticism kindly.[4] I often think I should be as touchy as anyone if the success of my works did not give me a constant opportunity of profiting immediately by every suggestion as to style and moral tone, and above all as to facts and logic.[5] Besides the increased responsibility which I incur by the trusting public, who before they had read a word induced the trade to bid for 3,850 copies, I have the prospect, if I improve my knowledge and my teaching, of future success in new editions with comparatively little labour.[6]

As to Lamarck I find that Grove, who has been reading him, is wonderfully struck with his book.[7] I remember that it was the conclusion he came to about man that fortified me thirty years ago against the great impression which his arguments at first made on my mind,[8] all the greater because Constant Prévost, a pupil of Cuvier's forty year ago, told me his conviction 'that Cuvier thought species not real, but that science could not advance without assuming that they were so.'[9] When I came to the conclusion that after all Lamarck was going to be shown to be right, that we must 'go the whole orang,' I re-read his book, and remembering when it was written, I felt I had done him injustice.[10]

Even as to man's gradual acquisition of more and more ideas, and then of speech slowly as the ideas multiplied, and then his persecution of the beings most nearly allied and competing with him—all this is very Darwinian.[11]

The substitution of the variety-making power for 'volition,' 'muscular action,' &c. (and in plants even volition was not called in) is in some respects only a change of names.[12] Call a new variety a new creation, one may say of the former as of the latter, what you say when you observe that the creationist explains nothing, and only affirms 'it is so because it is so.'

Lamarck's belief in the slow changes in the organic and inorganic world in the year 1800, was surely above the standard of his times, and he was right about progression in the main, though you have vastly advanced that doctrine.[13] As to Owen in his Aye Aye paper, he seems to me a disciple of Pouchet, who converted him at Rouen to 'spontaneous generation.'[14]

Have I not at p. 412 put the vast distinction between you and Lamarck as to 'necessary progression' strongly enough?[15]

Huxley's second thousand is going off well.[16] If he had leisure like you and me;—and the vigour and logic of the lectures, and his address to the Geological Society,[17] and half a dozen other recent works (letters to the 'Times' on Darwin, &c.),[18] been all in one book, what a position he would occupy! I entreated him not to undertake the 'Natural History Review' before it began. The responsibility all falls on the man of chief energy and talent; it is a quarterly mischief, and will end in knocking him up.[19]

I am sorry you have to go to Malvern.[20] The good of the water-cure is abstinence from work; a tour abroad would do it, I am persuaded, as effectually and more profitably.

I hope my long letter will not task you too much; when I sit down to write to you, I can never stop. Hooker, not having heard from you, is growing anxious, and hopes it is because you are corresponding with me and not because of serious ill-health.[21]

Ever affectionately yours, | Charles Lyell.

Incomplete[22]
K. M. Lyell ed. 1881, 2: 364–6

1 In his letter to Lyell of 12–13 March [1863], CD offered suggestions regarding the text of *Antiquity of man* (C. Lyell 1863a), particularly with respect to Lyell's treatment of natural selection. Joseph Dalton Hooker had sent Lyell a '*deflagrating* . . . yarn', concerning his failure publicly to endorse natural selection in the book (see preceding letter).

2 See letter from Charles Lyell, 11 March 1863 and nn. 2–6.

3 The references are to the Confederate forces in the American Civil War, and to the Polish uprising against Russian rule in 1863 (*EB*).

4 See letter to Charles Lyell, 12–13 March [1863].

5 Lyell refers to *Principles of geology* (C. Lyell 1830–3), which passed through nine editions between 1830 and 1853, and *Elements of geology* (C. Lyell 1838), which passed through five editions between 1838 and 1855.

[6] Lyell refers to the sale of copies of C. Lyell 1863a to book-dealers at John Murray's sale in November 1862 (see *Correspondence* vol. 10, letter to J. D. Hooker, [10–]12 November [1862] and n. 25). A second edition of the book was published in April 1863, only two months after the first, and by 19 May, Lyell reported that 5000 copies had been sold (K. M. Lyell ed. 1881, 2: 375). A third edition was published in November 1863, and a fourth in 1874. Some of the changes in the second and third editions are detailed in Grayson 1985.

[7] The references are to William Robert Grove and Lamarck 1809 (see letter to Charles Lyell, 12–13 March [1863] and n. 9).

[8] On Lyell's concerns about the implications of Jean Baptiste de Lamarck's theory of transmutation for the status of the human species, see Bartholomew 1973.

[9] The French geologist Louis-Constant Prévost studied medicine before turning to geology under the influence of Georges Cuvier; from 1823 to 1824, he and Lyell collaborated in comparing Secondary and Tertiary strata on either side of the English Channel (*DSB*).

[10] Lyell refers to his famous critical discussion of Lamarck's theory of the transmutation of species in *Principles of geology* (C. Lyell 1830–3, 2: 1–35).

[11] Lamarck 1809, 1: 349–57. Lamarck's account of the origin of the higher mental faculties was particularly innovative in that it posited the progressive development of such faculties, linked to the structural development of the nervous system (*DSB*).

[12] Lamarck's theory of species transmutation was based on two linked factors; namely, a natural tendency toward organic complexity, and the direct influence of the environment on the organism, causing heritable variations from this natural progression. In higher animals, however, Lamarck posited that there also existed a 'sentiment intérieur [inner feeling]', which, like the external environment, could affect the organism's form. Lamarck argued that this feeling, corresponding to agitations of the nervous fluid, resulted in muscular motion that, through repetition, could create new organs (*DSB*).

[13] Lamarck first publicly expounded his theory of transmutation in a preliminary discourse to his series of lectures on invertebrates at the Muséum d'Histoire Naturelle, Paris, in 1800; the discourse is reproduced in Lamarck 1801 (see also n. 12, above). In *Antiquity of man*, Lyell introduced the section relating to theories of species transmutation and progression with a preliminary chapter that first described Lamarck's theory, and then discussed subsequent debates (C. Lyell 1863a, pp. 385–406). Lyell had long been an outspoken critic of the progressivist palaeontology common in Britain during the mid-nineteenth century (see Bartholomew 1976). For CD's observations on progression, see the letter to Charles Lyell, 12–13 March [1863]; see also n. 15, below, and *Correspondence* vol. 3, letter to J. D. Hooker [11 January 1844].

[14] Lyell refers to Owen 1862c, in which Owen allegedly claimed the 'whole credit of making out the derivation or origin of species' (see letter to J. D. Hooker, 5 March [1863] and n. 19, and letters to Charles Lyell, 6 March [1863] and 12–13 March [1863]). In the paper, Owen committed himself to 'creation by law' while refusing to endorse any of the mechanisms currently proposed (Owen 1862c, p. 96). However, he gave most positive treatment to what he called the 'derivative hypothesis' of organisms, meaning the continual spontaneous generation and law-like transmutation of species, which, he considered, provided 'a kind of vantage-ground artificially raised to expand the view of the outlooker for the road to truth' (*ibid.*, p. 92 and n.). Félix Archimède Pouchet, director of the Muséum d'Histoire Naturelle at Rouen, was a prominent advocate of spontaneous generation and was involved in a major controversy on the subject with Louis Pasteur (see Farley 1977). Owen had given Pouchet's work qualified support on several occasions since 1859 (see Rupke 1994, pp. 239–40, 251–2). On Owen's views regarding spontaneous generation, see also Appendix VII.

[15] See letter to Charles Lyell, 12–13 March [1863], n. 9. In C. Lyell 1863a, p. 412, Lyell observed:

> One of the principal claims of Mr. Darwin's theory to acceptance is, that it enables us to dispense with a law of progression as a necessary accompaniment of variation. It will account equally well for what is called degradation, or retrograde movement towards a simpler structure

[16] T. H. Huxley 1863b. See letter from T. H. Huxley, 25 February 1863 and n. 4, and letter to T. H. Huxley, 26 [February 1863] and n. 6. See also letter to Charles Lyell, 12–13 March [1863].

[17] T. H. Huxley 1863a and 1862a.

[18] No letter from Thomas Henry Huxley to *The Times* relating to CD has been found; Lyell may refer to Huxley's anonymous review of *Origin*, 'The Darwinian hypothesis', which appeared in *The Times*, 26 November 1859, p. 8.

[19] In July 1860, Huxley had been offered 'effectual control' of the new series of the *Natural History Review*, due to start publication in January 1861, if he became one of the editors. Huxley had organised a 'commissariat' comprising eleven co-editors, but by June 1861 he was increasingly taking editorial responsibility on himself, writing: 'It is no use letting other people look after the journal. I find unless I revise every page of it, it goes wrong' (L. Huxley ed. 1900, 1: 209–10). CD had also warned Huxley that the editorship would consume much time that might otherwise have been spent on '*original* research' (see *Correspondence* vol. 8, letter to T. H. Huxley, 20 July [1860]).

[20] Lyell refers to James Manby Gully's hydropathic establishment at Great Malvern, Worcestershire (see letter to Charles Lyell, 12–13 March [1863]).

[21] CD wrote a letter to Hooker on 13 [March 1863], the first since his letter of 5 March [1863]; Lyell refers to CD's letters to him of 6 March [1863] and 12–13 March [1863].

[22] The manuscript of this letter has not been found; however, the ellipsis in the published version of the letter appears to indicate that a section of the original was omitted from the transcription. See also letter to Charles Lyell, 17 March [1863].

To [Thomas White Woodbury][1] 15 March [1863][2]

[Down]
March 15

[Suggests that Woodbury look at some bee and comb specimens received by CD from Africa.][3]

B. Altman & Co. (*New York Times*, 12 October 1975, p. 39)

[1] The recipient is identified by the relationship between this letter and the letter from T. W. Woodbury, 17 March 1863.

[2] The year is established by the relationship between this letter and the letter from T. W. Woodbury, 17 March 1863.

[3] The specimens were probably those obtained by Joseph Dalton Hooker from Gustav Mann, who was collecting botanical specimens in West Africa for the Royal Botanic Gardens, Kew (see letter from J. D. Hooker, [15 March 1863]). CD and Woodbury, who was one of the contributors to the beekeeping section of the *Journal of Horticulture and Cottage Gardener*, had corresponded in the summer of 1862, when CD was writing the section for *Variation* about bees from different localities (see *Correspondence* vol. 10).

To W. D. Fox 16 [March 1863]

Down
16th

My dear Fox

Very many thanks for your to me very interesting letter.—[1] I did not think it likely that you could add much.— if you should hear what Ram was I shd. like to add it.[2]

After writing I found sentence in your old letter, which I had overlooked saying that you believed white & slate Musks breed true, so I have put it in *cautiously*.—[3] These facts interest me greatly.

I suppose we must go to Malvern; but it breaks my heart.—[4] I am tired with a lot of letters, so goodBye with many thanks | Ever yours | C. Darwin

Endorsement: 'March 17/63'
Postmark: MR 16 63
Christ's College Library, Cambridge (Fox 137)

[1] Letter from W. D. Fox, 12 March [1863].

[2] See letter from W. D. Fox, 12 March [1863]. Fox's reply has not been found, but see the letter to W. D. Fox, 23 March [1863]. See also CD's description of the ram in *Variation* 2: 30.

[3] The letter from Fox to which CD refers has not been found; however, see the letter to W. D. Fox, 9 March [1863] and n. 1. The crosses conducted by Fox are briefly mentioned in *Variation* 2: 40.

[4] See letter to W. D. Fox, 9 March [1863], and letter from W. D. Fox, 12 March [1863]. CD refers to James Manby Gully's hydropathic establishment in Great Malvern, Worcestershire, where Anne Elizabeth Darwin, CD's eldest daughter and favourite child, died in 1851 (see *Correspondence* vol. 5).

From James Paget 16 March 1863

1 Harewood Place | Hanover Square | W.
March 16. 1863.

My dear Darwin

I send you, by this same post, a pamphlet with an account of an interesting case of malformations associated & inherited—[1] I ought not to have forgotten it when you asked me about such cases—[2]

And I enclose a paper which I turned-up a few days ago,[3] and which may serve as an illustration of a class of facts which one meets with in practice among diseases, and which probably have their parallels among 'malformations' and other varieties of form:—namely, that the members of one family are not so likely to have, all, the same disease as to have different diseases more or less nearly related to one another— In one family, as in this, are various diseases of the nervous system: in another, of the skin; in another, of the stomach & so on.— Since one might expect, as a general rule, that a progenitor having any well marked variety of form would have offspring in some of whom the same form would appear, but none of whom would present a variety of forms related to, yet different from, that which was peculiar in the progenitor— I do not know if this can suggest to you anything about related groups of forms; but I shall not have wasted much of your time even if all this is not worth consideration.

Of course, the sentence you forwarded to me was all right.[4]

Always truly your's | James Paget

Chas. Darwin Esqre.

DAR 174: 5

[1] The pamphlet has not been identified.
[2] CD apparently asked Paget for information on such cases while in London in February (see letter from James Paget, 7 February 1863, and letter to James Paget, 11 March [1863]).
[3] The paper has not been identified.
[4] See letter to James Paget, 11 March [1863].

To the secretary, Royal Prussian Academy of Sciences, Berlin 16 March 1863

Down. | Bromley. | Kent. S.E.
March 16th 1863

Honoured Sir!

I beg leave to acknowledge your letter, received this morning, in which you announce to me the distinguished honour conferred on me, by my election as a corresponding Member of the Royal Academy of Sciences of Berlin.—[1]

With my sincere thanks for this great honour | I beg leave to remain | Sir | Your most obedient servant | Charles Darwin

To the Secretary | Royal Acad. of Sciences

Archiv der Berlin-Brandenburgischen Akademie der Wissenschaften (II–III–120: 67)

[1] CD was elected a corresponding member of the Royal Prussian Academy of Sciences on 23 February 1863; the diploma sent to CD is reproduced in Appendix III. The letter referred to has not been found; however, it was probably written by either Christian Gottfried Ehrenberg or Johann Franz Encke, who, as the secretaries for the physical-mathematical section of the academy, were the joint signatories of the diploma. The academy comprised two sections: the physical–mathematical and the philosophical–historical; the former encompassed biological as well as physical sciences (Harnack 1900).

From Roland Trimen 16 March 1863

Colonial Office, | Cape Town.
16th. March, 1863.

My dear Sir,

I have received your kind letter of the 31st. January, and am very glad to hear that my Orchid sketches & obsns. interested you.[1] You are quite right in supposing that there is no movement of the pollinia in the species I have described; I imagine that such motion is chiefly found in those species which have a cap of membrane over the viscid matter of the pollinia discs in situ,—certainly none of the Cape species that I have seen (all of which have the discs naked) present any phenomenon of the kind.[2]

I have been very much surprised, since my attention has been turned to the subject, at never meeting with a single insect (save the fly I have mentioned in connection with the Stellenbosch *Disa* (?))[3] with a pollinium of a⟨ny⟩ kind attached to it. I have been collectin⟨g⟩ insects, especially Lepidoptera, ever since ⟨I have⟩ been out here,—and it certainly is ast⟨onishing⟩ that I have never fallen in with any ⟨pollinia-⟩laden moths or other insects.[4]

I am so little of a botanist ⟨ ⟩ that I really am not sure whether Orchideæ are propagated by the tubers from year to year:—is it so or is their increase only provided for by seed? In the latter case, many species would fare but badly, I should fancy.

I am fully sensible of the compliment you pay me in suggesting that I should write a paper on the fertilising contrivances in some genera of Cape Orchids for the Linnean or some other Society, but I am really not equal to it.[5] The subject is so new to me, & my botanical knowledge so very slight, that I have not a sufficient foundation whereon to construct a scientific paper. I far prefer to be permitted to send you any material I may obtain, for you to work up and elucidate in that able manner which is so thoroughly your own.[6] I trust that your half-expressed fears that you would not be able to publish again on the Orchid subject may prove unfounded,[7] and that I may be able to ⟨con⟩tribute something more to induce you to ⟨con⟩tinue yr. illustrations of a subject which ⟨ ⟩ to have hitherto been but "dimly ⟨ ⟩ ⟨an⟩d "little understood" by naturalists.[8]

⟨I am⟩ sorry to say my Nos. V & VI and ⟨ ⟩ ⟨Saty⟩rium are no longer in flower[9] I shall take care to send you some flowers next Spring, if I am still in this locality[10]

I have sent you with this, through my brother,[11] sketches of the structure of *Disa grandiflora* (the finest of S. African Orchids) and of *Herschelia caelestis*, Lindley, found last month on the mountains here.[12] I have tried to colour these, but, having unthinkingly made the outlines on glazed paper, have not done as well I might. My brother Henry (71, Guildford St., Russell Sqre. W.C.) is very fond of Botany, and has read your book[13] with much interest. I have therefore sent him these two drawings to inspect, and asked him to send them on to you.

You will, I am sure, be ready to make allowances for any roughness or irregularity of drawing in my sketches, when I tell you the disadvantages under which I work. To begin with, I am far from strong,—indeed, I came out here for the benefit of my health.[14] Living out of Cape Town, I have to leave Wynberg at ½ past 8 every morning & do not get back after office till 5 or 6 p.m. One is in a very unfit state for work requiring minuteness & attention after six hours office work & 2½ hours omnibus in this climate; and you can imagine how difficult it is to work with any satisfaction by fits & starts in the Office. What spare time I have, too, is very much occupied with entomology, and then, besides, I am told that I mustn't stoop more than I can help, & must take so much exercise per diem. Forgive these details; I cannot help grumbling sometimes, & thinking how delightful it would be to have all my time to give to natural history. I generally find some consolation in the reflection that very many others are in a similar position—obliged to follow distasteful work for a living.[15]

I am not quite clear as to the term 'rostellum'.[16] I don't know how much it is to be understood as including. The Cape Orchids do not appear to have that *special* rostellum you describe in many British species. If you write again, tell me, for instance, the limits of the rostellum in *Disa grandiflora*, if you are able to make it out from my drawings.[17]

I shall look out for *Bonatea*, but am afraid, from localities I have seen quoted, that the Genus does not occur *here*, but in the Eastern part of the Colony.[18]

I have ventured to enclose you my photograph,[19] as a prelude to my request that you will do me the kindness of sending me *yours*. It is true the advantage will be on my side in such an exchange, for Cape photographs are usually execrable, & never good. But the enclosed is the best I can get, & people here say (with one exception) that they "would know who it was"—an unusual admission as regards S. African cartes de visite.

I am, my dear Sir, | Yours very truly | Roland Trimen.

[Enclosure][20]

DISA GRANDIFLORA, LINN.

[watercolour sketches, reproduced facing p. 246]

A, Front view of flower, with posterior sepal almost wholly cut away.
B, Upper and front view of column, with labellum and basal portions of lateral sepals attached.
C, Front view of flower, with basal half of posterior sepal attached.
D, Side view of column, with petals and spur of posterior sepal attached
E, Do. do. with lateral petals removed.
F, Back and under do. do., with lateral petals attached.
G, Do. do. do. removed.
H, Pollinium attached to steel point (1, side, 2, upper view).
pp, lateral petals; l, labellum; sp, spur of posterior sepal, containing nectar.
dd, viscid discs of pollinia; s, stigma.

DAR 70: 180, DAR 178: 184

CD ANNOTATIONS
2.5 I have . . . insects. 2.6] *cross in margin, red crayon*
8.3 If . . . drawings. 8.5] *cross in margin, red crayon*
Top of letter: 'Oxalis seed' *pencil*
Enclosure:[21]
diagram A] *circled pencil*; 'B' *deleted pencil*; 'A' *pencil*
diagram E] *circled pencil*; 'E' *deleted pencil*; 'C' *pencil*
diagram F] *circled pencil*; 'F' *deleted pencil*; 'B' *pencil*
diagram H] *circled pencil*; 'H' *deleted pencil*; 'D' *pencil*

1 See letter to Roland Trimen, 31 January [1863] and n. 2.

2 See letter to Roland Trimen, 31 January [1863] and n. 6.

3 Trimen's initial letter to CD, which contained a manuscript account of his observations on South African orchids, has not been found (see letter to Roland Trimen, 31 January [1863] and n. 2). Some of Trimen's observations were published in Trimen 1863 (see n. 6, below), but that paper contains no mention of the observation given here; however, in 'Fertilization of orchids', p. 144, CD reported: 'Mr. Trimen informs me that he has seen a Dipterous insect, allied to *Bombylius*, frequenting the flowers [of *Disa grandiflora*]' (*Collected papers* 2: 141). Stellenbosch is a town in what was then Cape Colony (now part of South Africa), about thirty miles east of Cape Town (Heilprin and Heilprin eds. [1931]).

4 Trimen 1863, p. 147, notes the infrequency with which the pollinia were observed to have been removed from *Disa grandiflora*.

5 See letter to Roland Trimen, 31 January [1863].

6 In his letter to Trimen of 23 May [1863], CD reported that he had reworked Trimen's observations on *Disa grandiflora* as a paper for the Linnean Society (Trimen 1863). See also nn. 20 and 21, below.

7 See letter to Roland Trimen, 31 January [1863] and n. 10.

8 The source of these quotations, probably a review of *Orchids*, has not been identified.

9 In his letter of 31 January [1863], CD asked Trimen to send him flowers of several orchids previously described or drawn by Trimen.

10 See letter from Roland Trimen, 10, 13, and 18 October 1863.

11 Henry Trimen was a medical student at King's College, London (*DNB*).

12 See enclosure and nn. 20 and 21, below. The remaining enclosures have not been found. Trimen refers to John Lindley, who named the orchid *Herschelia caelestis*.

13 *Orchids*.

14 Trimen emigrated to Cape Colony in 1858 (*DSAB*).

15 Trimen worked in the Colonial Secretary's office for the Cape public service (*DSAB*).

16 Rostellum: a modified stigma, occurring in many orchids, which separates the stigmatic surface from the anthers in the column (a structure resulting from the fusion of the carpels and stamens) (see *Penguin dictionary of botany*). In *Orchids*, p. 6, CD wrote:

> Of the three pistils [carpels], which ought to be present, the stigma of the upper one has been modified into an extraordinary organ, called the *Rostellum*, which in many Orchids presents no resemblance to a true stigma. The rostellum either includes or is formed of viscid matter; and in very many Orchids the pollen-masses are firmly attached to a portion of its exterior membrane, which is removed, together with the pollen-masses, by insects.

17 Trimen refers to *Orchids*; however, the exact reference has not been found. See also letter to Roland Trimen, 23 May [1863].

18 See letter to Roland Trimen, 31 January [1863] and n. 12. Trimen later sent CD information on *Bonatea speciosa* (see *Correspondence* vol. 12, letter to Roland Trimen, 25 November 1864), and published a paper on the species (Trimen 1864). See also 'Fertilization of orchids', p. 157 (*Collected papers* 2: 154).

19 The photograph has not been found.

20 The enclosure, which includes watercolour sketches, is reproduced facing p. 246.

21 The enclosure was evidently sent by CD to the Linnean Society of London, together with a letter, which has not been found, giving instructions about the reproduction of four of the diagrams in Trimen 1863 (see n. 6, above); the four diagrams circled and lettered by CD appear as woodcuts in Trimen 1863, p. 145. Trimen's original labelling of these diagrams was altered, in an unknown hand, to a more sophisticated scheme used in publication. On the verso of the enclosure is an annotation, also in an unknown hand, which states: 'No. 1558. | Drawings on wood to be | made by M^r. Fitch. | A/c. for drawing & cutting | block to be sent to M^r. Darwin. | Letters of reference to be | engraved on the block.' The reference is to Walter Hood Fitch, a botanical artist at the Royal Botanic Gardens, Kew (R. Desmond 1994).

To J. D. Hooker 17 March [1863]

Down Bromley Kent
March 17th

My dear Hooker

What a candid honest fellow you are,—too candid & too honest. I do not believe one man in ten thousand would have thought & said what you say about your own work in your letter.[1] I told Lyell that nothing pleased me more in his work than the conspicuous position in which he very properly placed you.[2] About dates of your Essay & the Origin, I feared I might be thought to have told untruths, so I mentioned to Lyell that I had asked you (which I think you have forgotten) when I was writing my "historical sketch" the date of publication of your Essay & you wrote to me "December";[3] the Origin was published to world & every copy sold on Novr. 24th, but was finished ie last sheet corrected Oct 1. (& Oct 2d. I started for Ilkley), but was kept back by Murray for auction-sale; but private copies were distributed a good while before.[4] But all this is absolutely & wholly immaterial, excepting so far that anyone might think that Lyell has found out that I had misrepresented case.—[5]

I am really sorry for Lyell's troubles about so many claimants for notice: he has sent me the long P.S. addressed to you about Falconer;[6] I never heard such nonsense, as that of the monkey case.[7] Do see Falconer & see whether you can at all influence him, by saying what ill appearance Reclamations always have, & that the future historians of Science alone ought to settle such points.— It is wretched to see men fighting so for a little fame.—

I am so glad that you heartily admire parts of Huxley's book.[8] It *can* be only from brevity with which he treats species-question that he does not notice your great works: I do not remember that he even alludes to the grand subject of Geograph. Distribution.— The greatest blemish in my opinion in Lyell's work[9] (which I have said to no one) strikes me as a certain want of originality in the whole.— I have read Owen on Aye-Aye: it is nothing new: it gave me no scope for attacking him, & I had partly composed such *a good letter* (!);[10] I long to be in the same boat with all (except you) my friends ie at open war; but at same time I rejoice not to be annoyed at public quarrel, & it would annoy me much.

Thanks about Potatoes & *Poplars*.—[11] I was very glad of the Bee-combs; but they did not turn out anything specially interesting; & I am a fool to go on collecting materials for work, when I can clearly see that I shall never publish half my already half-worked out matter.—[12]

Thank you for telling me about your heart-symptoms, which are very like mine; but thank God I have not yet come to have "worms crawling over my heart"![13] This is first day I have had an hour's comfort.— If you can come over here on Sunday, we should indeed be delighted; but I shd. doubt it, as it is so far: I would send you back in carriage.—[14]

I am heartily glad to hear that you mean to think & write about mundane glacial period apropos to your grand Cameroon case.[15] How I wish I could have published my M.S in full on this subject, as the sketch in Origin does not do it justice.[16] Do not indulge in belief that any one continent could have remained a hot refuge for all tropical productions of world.[17] I have of late come to conclusion that there must have been former Tertiary or Secondary cold periods & migrations.[18]

You speak of Reversions in your letter:[19] I have been writing during last fortnight on this subject, i.e., on reversions to particular characters, & have got curious collection of facts & experiments. They have led me to view the whole case rather differently i.e. that the child *never* inherits from its grandfather or more distant ancestor, but that a crowd of characters lie latent in every living creature & parent.—[20]

Good Night | my dear old friend | C. Darwin

Endorsement: '/63.'
DAR 115: 187

1 See letter from J. D. Hooker, [15 March 1863].

2 See letter to Charles Lyell, 17 March [1863].

3 See letter from J. D. Hooker, [15 March 1863] and nn. 7 and 9. CD's historical preface, giving a 'sketch of the progress of opinion on the origin of species', was first added to the first American edition of *Origin*, which was published in July 1860 (see *Correspondence* vol. 8, Appendix IV). The sketch concluded (*Origin* US ed., p. xi):

> In November, 1859, the first edition of this work was published. In December, 1859, Dr. Hooker published his Introduction to the Tasmanian Flora: in the first part of this admirable essay he admits the truth of the descent and modification of species; and supports this doctrine by many original and valuable observations.

CD asked Hooker the date of publication of his essay (J. D. Hooker 1859) in the letter to J. D. Hooker, 31 [January 1860] (*Correspondence* vol. 8); however, no reply has been found (see *Correspondence* vol. 8, letter to J. D. Hooker, 8 February [1860]).

4 *Origin* was sold out at John Murray's trade sale on 24 November 1859 (see *Correspondence* vol. 7, letter to John Murray, 24 November [1859], and Appendix II). CD made an excursion to a hydropathic establishment in Ilkley, Yorkshire, between 2 October and 9 December 1859 (see *Correspondence* vol. 7, Appendix II). CD received his own copy of *Origin* early in November 1859, and his presentation copies were sent out shortly afterwards (see *Correspondence* vol. 7, letter to John Murray, [3 November 1859]). The earliest extant presentation letters are dated 11 November 1859, and the earliest letter of acknowledgment is the letter from Charles Kingsley, 18 November 1859 (*Correspondence* vol. 7).

5 See letter to Charles Lyell, 6 March [1863] and n. 35.

6 The reference is to Hugh Falconer, who was in dispute with Lyell over what he considered to be a lack of acknowledgment of his work in C. Lyell 1863a (see letter from Charles Lyell, 11 March 1863 and n. 13). The postscript has not been found; presumably it was enclosed with the letter from Charles Lyell, 15 March 1863. See also letter to Charles Lyell, 17 March [1863].

7 The reference has not been identified.

8 T. H. Huxley 1863b. See letter from J. D. Hooker, [15 March 1863].

9 C. Lyell 1863a.

10 Owen 1862c. See letter to J. D. Hooker, 5 March [1863] and n. 19.

11 See letter from J. D. Hooker, [15 March 1863] and nn. 2–4.

12 See letter from J. D. Hooker, [15 March 1863] and n. 22. CD made an extensive study of the cell-making instinct of bees in 1858, the results of which he never fully published (however, see *Origin* pp. 224–35); his notes on the subject are in DAR 48. See Prete 1990.

[13] See letter from J. D. Hooker, [15 March 1863].

[14] Hooker intended walking to Down House from John Lubbock's house in Chislehurst, Kent, on 22 March 1863; it is not clear whether Hooker made the journey on foot, but he subsequently walked the five miles back (see letter from J. D. Hooker, [24 March 1863]).

[15] See letter from J. D. Hooker, [15 March 1863] and n. 21.

[16] CD discussed his theory of the trans-tropical migration of temperate species during a global glacial period in *Origin*, pp. 365–82. Most of the draft chapter on geographical distribution written for CD's unpublished 'big book' on species was devoted to this subject (*Natural selection*, pp. 534–66); *Origin* was intended to be an 'abstract' of the 'big book' (see *Correspondence* vol. 7, letter to J. D. Hooker, 12 [October 1858]).

[17] CD and Hooker had persistently disagreed over the causes of the geographical distribution of plants and animals. See letter to J. D. Hooker, 5 March [1863], n. 17. In the fourth edition of *Origin*, pp. 450–1, CD added a new section to his discussion of what he called the 'Mundane glacial period', in which he stated: 'At one time I had hoped to find evidence that the tropics in some part of the world had escaped the chilling effects of the Glacial period, and had afforded a safe refuge for the suffering tropical productions.' He concluded, however, that such a discovery would be 'of no avail', since the tropical forms preserved in one such region 'could not have travelled to the other great tropical regions within so short a period as has elapsed since the Glacial epoch', and, moreover, the tropical species of the various regions were not uniform enough to have 'proceeded from any one harbour of refuge'. See also letter to Charles Lyell, 6 March [1863] and n. 27.

[18] See *Correspondence* vol. 10, letters to H. W. Bates, 4 May [1862] and n. 5, and 9 May 1862. In the fourth edition of *Origin*, p. 454, CD added a new passage to the conclusion of his discussion of the mundane glacial period, stating:

> It is extremely difficult to understand how a vast number of peculiar forms confined to the tropics could have been therein preserved during the coldest part of the Glacial period. The number of forms in Australia, which are related to European temperate forms, but which differ so greatly that it is impossible to believe that they could have been modified since the Glacial period, perhaps indicates some much more ancient cold period, even as far back as the miocene age, in accordance with the speculations of certain geologists.

[19] See letter from J. D. Hooker, [15 March 1863].

[20] CD wrote drafts of the chapters on inheritance for *Variation* (*Variation* 2: 1–84) between 23 January and 1 April 1863 (see 'Journal' (Appendix II)); he discussed 'reversion or atavism' in chapter 13 (*Variation* 2: 28–61). On the development of CD's 'provisional hypothesis' of inheritance ('pangenesis'), first published in *Variation* 2: 357–404, see Hodge 1985.

To *Journal of Horticulture and Cottage Gardener* [17–24 March 1863][1]

[Down]

Had Mr. Anderson asked me two days ago for any facts illustrative of his case of unopened flowers of Cattleya crispa and Dendrobium cretaceum producing seed-capsules, I could have given no sort of information;[2] nor can I now explain the fact. By an odd coincidence, yesterday I received a very interesting letter from Dr. Hermann Cruger, the Director of the Botanic Garden at Trinidad, who informs me that certain native species, and native species alone, of Cattleya, Epidendrum, and Schomburghkia, "are hardly ever known to open their flowers, but which nearly always set fruit."[3] In answer to Dr. Cruger, I have asked him to look at the seed or send me some, and inform me whether it appears good.[4]

Will Mr. Anderson have the kindness to send me a few seeds produced by his unopened flowers?[5]

I further asked Dr. Cruger whether these Orchids in their native haunts *never* open their flowers.[6] I can hardly believe that this can be the case, seeing how manifestly adapted the structure of their organs of fructification is to the action of insects. But it is known that several plants, such as Violets, Campanulas, Oxalis, &c., produce two kinds of flowers: one sort adapted for self-fertilisation, and the other sort for fertilisation by insect agency or other means.[7] In some cases the two kinds of flowers differ very little in structure; and it occurs to me as possible that something of this kind may occur with Orchids.

Dr. Cruger further informs me that with certain Orchids, as in those which do not open their flowers, the pollen-masses after a time become pulpy; and though remaining still *in situ*, emit their pollen-tubes, which reach the stigma, and thus cause fertilisation.[8]

An excellent observer, Mr. J. Scott, of the Royal Botanic Gardens of Edinburgh, will, I am sure, permit me to state that he has been making similar observations, and has seen the pollen-tubes emitted from the pollen-masses whilst still in their proper positions.[9]

These facts were all unknown to me when I published my small work on the Fertilisation of Orchids; but I ought, perhaps, to have anticipated their occurrence, for I saw the pollen-tubes emitted from the pollen within the anthers in the Bird's-nest Orchid, and likewise in *monstrous* flowers of the Man Orchis.[10] This latter fact seems related to Mr. Anderson's remark, that flowers of an imperfect character, wanting a petal or sepal, had a great tendency to produce seed-capsules.[11]

These curious observations by Dr. Cruger, Mr. Anderson, and Mr. Scott, convince me that I have in my work underrated the power of tropical Orchids occasionally to produce seed without the aid of insects; but I am not shaken in my belief that their structure is mainly related to insect agency.[12] With most British Orchids this conclusion may be looked on as established.

I will only add that since the publication of my work, a number of persons have set seed-capsules with various tropical Orchids.[13]

Charles Darwin, *Down*, *Bromley*, *Kent.*

Journal of Horticulture and Cottage Gardener n.s. 4 (1863): 237[14]

[1] The date range is established by the relationship between this letter and James Anderson's article in the *Journal of Horticulture*, 17 March 1863, pp. 206–8 (see n. 2, below), and by CD's reference to this letter in his letter to John Scott of 24 March [1863].

[2] In an article concerning orchid breeding, published in the *Journal of Horticulture*, 17 March 1863, pp. 206–8, James Anderson, who was a gardener in Uddingston, Scotland (R. Desmond 1994), asked if CD could tell him why some orchids, such as *Cattleya crispa* and *Dendrobium cretaceum*, formed apparently normal seed-capsules from apparently abortive, unopening flowers. CD's unbound annotated copy of this number of the *Journal of Horticulture* is in the Darwin Library–CUL; there are also notes on the article in DAR 70: 110–11.

[3] See letter from Hermann Crüger, 23 February 1863 and n. 5.

4 The letter to Crüger has not been found; however, see the letter from Hermann Crüger, 23 April 1863.

5 See letter from James Anderson, 1 April 1863.

6 See n. 4, above.

7 CD had long been interested by the occurrence in some plants of 'imperfect', unopening flowers in which self-pollination occurred (a phenomenon later called cleistogamy). His 'provisional conclusion' was that 'the final object of the imperfect flowers is to produce seed safely, *without any crossing* ... the perfect flowers being adapted for getting an occasional cross' (see *Correspondence* vol. 10, letter to Daniel Oliver, 12 [April 1862]).

8 CD cited this information in 'Fertilization of orchids', p. 152 (*Collected papers* 2: 149).

9 No letter from John Scott has been found recording observations of pollen-tubes emitted from pollen masses while still *in situ*; moreover, Scott denied having made such claims in his letter to CD of [1–11] April [1863]. See also letter to John Scott, 24 March [1863]. However, Scott had carried out, at CD's suggestion, numerous experiments on the emission of pollen-tubes from pollen masses placed on the rostellum of orchids (see, for example, letters from John Scott, 6 January 1863, 16 January 1863, and 21 March [1863]).

10 See *Orchids*, p. 324 n.

11 *Journal of Horticulture*, 17 March 1863, p. 107.

12 In *Orchids*, p. 1, CD stated that the object of the book was:

> to show that the contrivances by which Orchids are fertilised, are as varied and almost as perfect as any of the most beautiful adaptations in the animal kingdom; and, secondly, to show that these contrivances have for their main object the fertilisation of each flower by the pollen of another flower.

13 The only individual identified is Scott, who had succeeded in pollinating *Acropera loddigesii* and *Gongora atropurpurea* (see *Correspondence* vol. 10, letter from John Scott, 11 November 1862, and this volume, letter from John Scott, 3 March 1863).

14 There is an incomplete two-page draft of this letter at the American Philosophical Society, Philadelphia.

To Charles Lyell 17 March [1863]

Down Bromley Kent
March 17th

My dear Lyell.

I have been much interested by your letters & enclosure,[1] & thank you sincerely for giving me so much time, when you must be so busy.— What a curious letter from B. de P. He seems perfectly satisfied & must be a very amiable man.[2] I know something about his errors, & looked at his book many years ago, & am ashamed to think that I concluded the whole was rubbish![3] Yet he has done for man *something* like what Agassiz did for Glaciers.—[4] I am astounded & truly grieved at what I read in your letter to Hooker about Falconer: I never read anything like his conduct about the monkey-case![5]

With respect to your reference to Principles, it was in *early* part that I felt that others might feel bothered:[6] I did not mark passages so cannot find without re-reading, & I have no strength to spare.—

I cannot say that I agree with Hooker about the Public not liking to be told what to conclude, *if coming from one in your position.*[7] But I am heartily sorry that

I was led to make complaints, or something very like complaints, on manner in which you have treated subject; & still more so anything about myself.[8] I steadily *endeavour* never to forget my firm belief that no one can at all judge about his own work. As for Lamarck, as you have such a man as Grove with you, you are triumphant;[9] not that I can alter my opinion that to me it was an absolutely useless Book.—[10] Perhaps this was owing to my always searching books for facts; perhaps from knowing my Grandfather's earlier & identically the same speculation.—[11]

I will only further say that if I can analyse my own feelings (a very doubtful process) it as nearly as much for your sake, as for my own, that I so much wish that your state of belief could have permitted you to say boldly & distinctly out that species were not separately created. I have generally told you progress of opinion, as I have heard it on Species-question. A first-rate German naturalist (I now forget name!!) who has lately published grand folio has spoken out to the utmost extent on the "Origin".—[12] De Candolle in very good paper on Oaks goes, in Asa Gray's opinion, as far as he himself does;[13] but De Candolle in writing to me, says "we" "we" think this & that;[14] so that I infer he really goes to full extent with me; & tells me of French good Bot. Palæontologist (name forgotten, Count Laperda or Saperda or some such name) who writes to De Candolle that he is sure that my views will ultimately prevail.—[15]

But I did not intend to have written to all this. It satisfies me with the final result; but this result I begin to see will take 2 or 3 life-times. The entomologists alone are enough to keep subject back for $\frac{1}{2}$ a century.[16]

I really pity you having to balance the claims of so many eager aspirants for notice; it is clearly impossible to satisfy all. By the way I can see that Lubbock is not satisfied with notice of his Somme paper!![17] Certainly I was struck with the full & due honour you conferred on Falconer.—[18] I have just had note from Hooker;[19] I think he forgets that he told me himself date of publication of his Essay.[20] I am heartily glad that you have made him so conspicuous; he is so honest, so candid & so modest.[21] He tells me that he has got more plants from Cameroon mountains & that he will discuss mundane cold period.—[22]

I have read Owen's Aye-Aye:[23] I could find nothing to lay hold of, which in one sense I am very glad of, as I sh$^{d.}$ hate a controversy; but in another sense I am very sorry for, as I long to be in the same boat with all my friends; & I had written *so good a letter*(!) all ready, with a blank for his sentence claiming more than he had any right to; but I could pick out no such sentence. Hooker says he so despises him that he cannot hate him:[24] I do not know whether this a right frame of mind, but by Jove it is not my frame of mind.

I am heartily glad the Book is going off so well.—[25]

Ever yours | C. Darwin

I quite agree with what you say about Huxley & about the Review.[26] The Review, however, is excellent.— As I have spoken against entomologists; see in next number a notice by me on Bates' paper on mimetic resemblances;[27] it is, I think, worth your reading, & will give you cream of facts.

Endorsement: 'March—17. 1863'
American Philosophical Society (291)

[1] CD refers to the letter from Lyell of 15 March 1863, of which a portion is missing. He also mentions two items sent either with that letter, or possibly with another letter from Lyell that has not been found; namely, a letter from Jacques Boucher de Perthes to Lyell (see n. 2, below), and a letter from Lyell to Joseph Dalton Hooker, discussing the behaviour of Hugh Falconer (see n. 5, below).

[2] The letter from Boucher de Perthes has not been found (see n. 1, above). See letter to Charles Lyell, 12–13 March [1863] and n. 3.

[3] In the 1840s, Boucher de Perthes discovered shaped flints in Pleistocene deposits in the Somme valley, which he argued in the first volume of his *Antiquités celtiques et antédiluviennes* (Boucher de Perthes 1847–64) were human artefacts contemporaneous with the associated Pleistocene fossils. His explanation was initially widely rejected, but his findings were confirmed following visits to the location in 1858 and 1859 by Falconer and Joseph Prestwich. See Grayson 1983. The first volume of Boucher de Perthes 1847–64 is mentioned in CD's list of 'Books to be Read' for the period 1852–60 (*Correspondence* vol. 4, Appendix IV, *128: 163). On CD's reaction to Boucher de Perthes's work, see *Correspondence* vol. 7, letter to J. D. Hooker, 22 [June 1859].

[4] In the late 1830s, the Swiss-born geologist Louis Agassiz elaborated his belief that there had been a recent 'Ice Age' in the northern hemisphere, during which much of Europe had been covered by ice. Following Agassiz's visit to Britain in 1840, his views prompted a vigorous debate among British geologists, ultimately resulting in a widespread belief in the glacial origin of many geological and geomorphological phenomena (see *DSB*, and Davies 1969).

[5] Lyell's letter to Hooker has not been found. In his letter to Hooker of 17 March [1863], CD referred to its being 'a long P.S.' addressed to Hooker. The reference to the 'monkey case' has not been identified.

[6] In his letter to Lyell of 12–13 March [1863], CD suggested that C. Lyell 1863a contained too many references to the various editions of Lyell's *Principles of geology* (C. Lyell 1830–3); Lyell had apparently replied to this observation in the missing portion of his letter to CD of 15 March 1863.

[7] See letter from Charles Lyell, 15 March 1863.

[8] CD refers to his comments on Lyell's failure publicly to endorse evolution or natural selection in *Antiquity of man* (C. Lyell 1863a; see letters to Charles Lyell, 6 March [1863] and 12–13 March [1863]).

[9] See letter from Charles Lyell, 15 March 1863 and n. 7. The references are to Jean Baptiste de Lamarck and William Robert Grove.

[10] Lamarck 1809. See letter to Charles Lyell, 12–13 March [1863] and n. 9.

[11] CD's transmutation notebooks contain many references to Erasmus Darwin's *Zoonomia* (E. Darwin 1794–6), which outlined a theory of species transmutation, and to Lamarck 1809 (*Notebooks*). On CD's reactions to these two authors, see also *Autobiography*, p. 49.

[12] The reference is to Haeckel 1862, in which Ernst Haeckel first publicly endorsed CD's theory; there is a presentation copy of this work in the Darwin Library–Down. See also letter to Ernst Haeckel, 30 December [1863] – 3 January [1864].

[13] A. de Candolle 1862a. See letter from Asa Gray, 27 January 1863.

[14] The letter from Alphonse de Candolle has not been found, but see the letter to Alphonse de Candolle, 31 January [1863].

[15] Gaston de Saporta. See also letter to J. D. Hooker, 30 January [1863], letter to Alphonse de Candolle, 31 January [1863], and letter to Asa Gray, 31 May [1863].

[16] CD considered that his theory had been particularly 'attacked & reviled' by entomologists (see *Correspondence* vol. 8, letter to H. W. Bates, 22 November [1860]).

[17] CD refers to John Lubbock's paper reviewing the recent discoveries of prehistoric implements in the Somme valley with respect to the antiquity of the human species (Lubbock 1862c); Lubbock's paper was not cited in C. Lyell 1863a. No correspondence between CD and Lubbock on this point has been found.

[18] Falconer's researches were cited repeatedly in C. Lyell 1863a; however, Falconer considered that his work had not been given due credit by Lyell (see letter to J. D. Hooker, 17 March [1863] and n. 6).
[19] Letter from J. D. Hooker, [15 March 1863].
[20] J. D. Hooker 1859. See letter to Charles Lyell, 6 March [1863] and n. 35, and letter to J. D. Hooker, 17 March [1863] and nn. 3 and 4.
[21] In *Antiquity of man*, Lyell devoted a separate section to considering Hooker's views on the transmutation of species (C. Lyell 1863a, pp. 417–21). See also letter from J. D. Hooker, [15 March 1863], and letter to J. D. Hooker, 17 March [1863].
[22] See letter from J. D. Hooker, [15 March 1863] and n. 21.
[23] Owen 1862c. See letters to Charles Lyell, 6 March [1863] and n. 44, and 12–13 March [1863] and n. 11.
[24] See letter from J. D. Hooker, [23 February 1863].
[25] C. Lyell 1863a. See letter from Charles Lyell, 15 March 1863.
[26] CD refers to Thomas Henry Huxley and the *Natural History Review* (see letter from Charles Lyell, 15 March 1863 and n. 19).
[27] CD's anonymous review of Bates 1861 ('Review of Bates on mimetic butterflies') appeared in the April 1863 number of the *Natural History Review*.

From T. W. Woodbury 17 March 1863

Mount Radford, | *Exeter.*
17th March 1863

Dear Sir

I am exceedingly obliged to you for favouring me with specimens of bees and their comb from Africa.[1] I suppose the bees are *apis Adansonii* as they appear to tally with the description of that species in the Naturalist's Library.[2] They appear almost identical with *apis Ligustica* with the exception of being smaller. Very careful comparison of their comb with that of *mellifica* enables me to pronounce the cells exactly the same ⟨size⟩

How singular it appears that this small bee as well as the large species *apis testacea* do not vary one iota in the size of their cells from our European species! It would be very interesting to ascertain if the small species so common in India (*apis Indica*) also makes the same sized cell.

These bees being so small would not repay the trouble of attempting to import them,[3] but I still have a strong impression that the large Indian species (*apis dorsata*) would be very valuable. ⟨I⟩ dare say you may have seen my queries regarding them together with the answers in "The Field."[4] Little practical information has resulted, nor do I yet see my way to taking any steps for their importation. Have you noticed that the first four hives of *Ligustica* sent by me to Australia have all got there in safety?[5]

I have never heard if the piece of partially completed artificial comb proved at all interesting to you[6]

Yours greatly obliged | T W Woodbury

C. Darwin Esq

DAR 181: 150

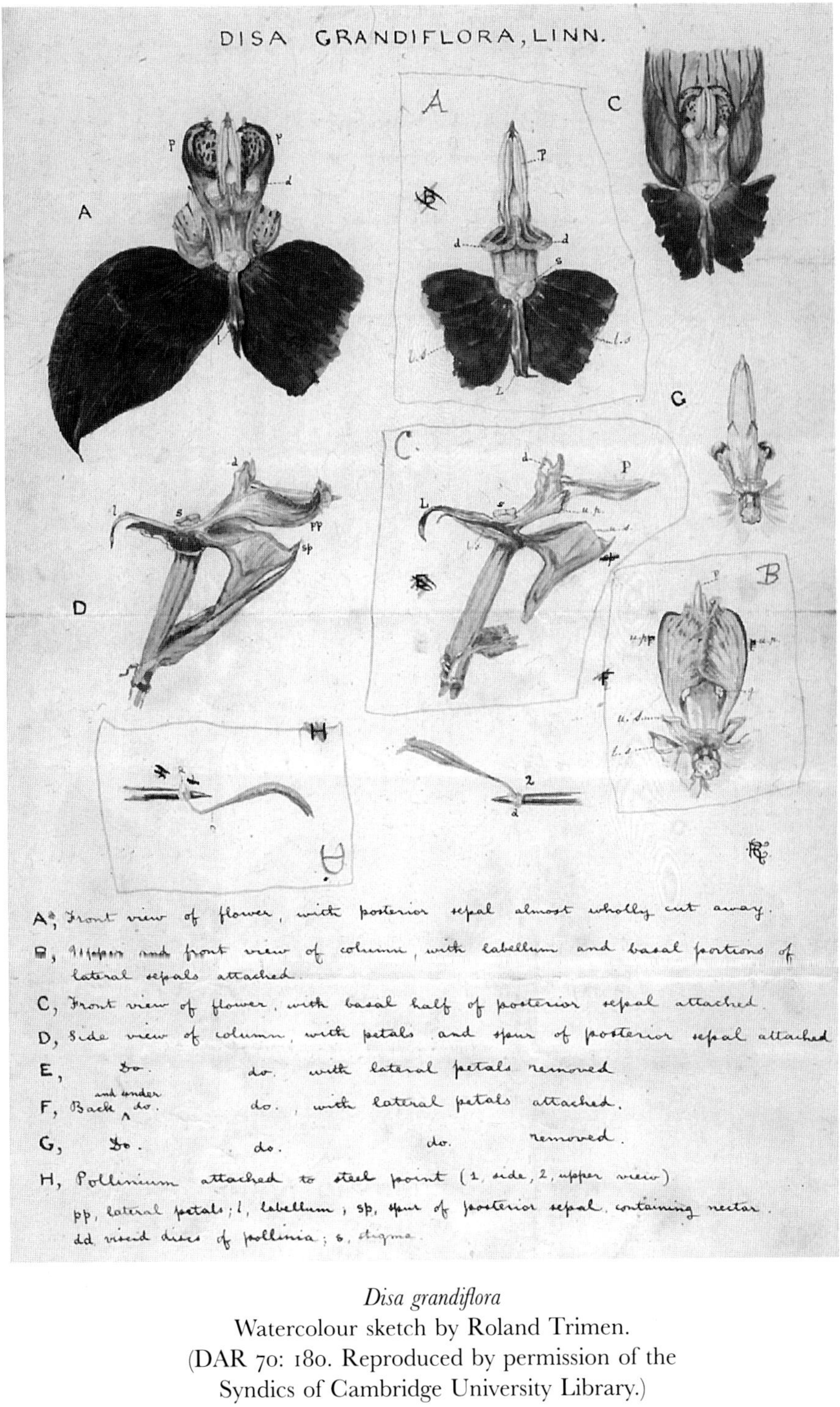

Disa grandiflora
Watercolour sketch by Roland Trimen.
(DAR 70: 180. Reproduced by permission of the Syndics of Cambridge University Library.)

Thomas Rivers
Photograph, before 1870.
(Courtesy of the Royal Horticultural Society, Lindley Library.)

1 See letter to T. W. Woodbury, 15 March [1863] and n. 3.
2 J. Duncan 1840, p. 274. J. Duncan 1840 was volume 18 in William Jardine's *Naturalist's Library*, published between 1833 and 1843.
3 In the summer of 1862, Woodbury had considered importing live bees from Africa, although with little hope that they would be 'extraordinary honey-gatherers' (see *Correspondence* vol. 10, letter from T. W. Woodbury, 9 August 1862).
4 *The Field* 21 (1863): 16, 60, 98, and 130.
5 At the International Exhibition in London in 1862, Edward Wilson, founder of the Acclimatisation Society of Victoria, ordered four hives of *Apis ligustica* through the London specialists George Neighbour & Sons. Woodbury, who had introduced the species to England, was consulted and asked to advise on packing the hives. The hives were shipped out of Southampton in September, and arrived in Australia in December, without the expected losses. See *Journal of Horticulture and Cottage Gardener* n.s. 3 (1862): 523 and n.s. 4 (1863): 235.
6 The reference is apparently to manufactured wax sheets with rhomboidal impressions, used by beekeepers to increase the yield of honey by reducing the amount of wax that hive-bees needed to produce in making their comb. The bees were found to excavate the artificial sheet further, and to use the wax thus obtained in the construction of the cells. The invention was introduced to Britain from Germany in 1862. See Neighbour 1865, pp. 70–1. Woodbury apparently sent CD the comb in December 1862 (see *Correspondence* vol. 10, letter to T. W. Woodbury, 7 December [1862]). CD had made a study of bee-cell construction in 1858, and discussed the subject in *Origin*, pp. 224–35, as part of the chapter on instinct, where he remarked (p. 224) upon the fact that 'a skilful workman, with fitting tools and measures, would find it very difficult to make cells of wax of the true form, though this is perfectly effected by a crowd of bees working in a dark hive'.

To T. W. Woodbury [after 17 March 1863][1]

I thought & still think that I wrote to thank you for the artificial comb, which *interested me much*; but if you did not get a letter, it must have been lost or I did not write it.—[2]

With apologies, Dear Sir, | & my best thanks, | Yours very faithfully, | C. Darwin

Incomplete
Eva Crane IBRA Library[3]

1 The letter is dated by its relationship to the preceding letter.
2 No such letter has been found. See preceding letter and n. 6.
3 A facsimile of this letter is reproduced in Brian and Crane 1959.

To Asa Gray 20 March [1863][1]

Down Bromley Kent
March 20th

My dear Gray.

I have very little to say, but will amuse myself by scribbling a few lines to thank you for information in last note of Feb. 9th[2] & to thank you in my dear little man's name for two precious stamps.[3] He told me with joyful triumph that in American stamps he equalled all other collections put together in the school. He exchanged a duplicate Blood's stamp for a whole lot of treasures.[4] He says there is an envelope of same value as the stamps you put on your letter, which would be of value to him.—

I have one request to make for myself, viz seed of Campanula perfoliata: I have tried in vain at Kew & elsewhere for some.—[5]

I am very glad you like Bates' paper;[6] I expect his Amazonian Travels will be good.[7] If you read Lyell's book, tell me what you think of it;[8] I (& Hooker) have told him that we regret much that he did not speak more boldly out about Species.[9] He answers that his belief in change fluctuates.—[10] His Book has made me reread your essay; & I admire it as much as ever.[11] What a dead hand you are in parrying a lounge & transfixing your adversary! You ask about Sprengels "Dichogamy":[12] he means by this a plant in which each flower first matures & *sheds* its pollen & then has its stigma mature; & much more rarely matures its stigma first & subsequently its pollen: so that these plants are in function monoœcious. I am sure his observations are to large extent correct, & the case is very common.[13] In the Primula-like cases the plants are in function Diœcious.—[14]

A couple of days ago I had an interesting letter from D^r. Cruger of Bot. Gardens of Trinidad,[15] & he tells me odd facts of native species (& only native species) of Cattleya &c which *never* open their flowers, & yet set seed-capsules. Happy man he has actually seen crowds of Bees flying round Catasetum with the pollinia sticking to their backs! I wrote to him to ask him to observe what insects did in flowers of Melastomaceæ;[16] he says not proper season yet, but that on one species a small Bee seemed busy about the horn-like appendages to the anthers. It will be too good luck if my study of the flowers in the green-house has led me to right interpretation of these queer appendages.[17] By the way, I have just built a hot-house & got some orchids, & it amuses me much.—[18] Some plants of Amsinckia spectabilis, at least the seed was so named (small dark orange flowers, elongated hairy leaves) have just begun to flower, & I find in two plants that stigma stands on exact level with anthers; hence I fear they cannot be dimorphic.—[19]

Your Mitchellas look healthy: I hope they will not flower very soon;[20] for my health (& that of my youngest Boy) has been of late so bad, that we have resolved all to go about middle of April for 6 or 8 weeks to Malvern for Water-Cure for me & change for my Boy.—[21] It breaks my heart: I shall never get my present Book on Variation under Domestication finished; yet it interests me much & I am now in middle of long chapter on Inheritance Reversion &c, giving results of my own & other Breeders' Experiments.—[22]

Good Night.— | My dear Gray | Yours most truly | Ch. Darwin

Many thanks for Pamphlet Chapters on History of war & newspaper just arrived.[23]

Gray Herbarium of Harvard University (58)

[1] The year is established by the relationship between this letter and the letter from Asa Gray, 13 April 1863.

[2] Gray's letter has not been found; however, a portion of the letter is quoted in CD's letter to H. W. Bates, 4 March [1863] (see n. 6, below).

[3] CD refers to his twelve-year-old son, Leonard Darwin; at CD's request, Gray had been sending United States postage stamps for Leonard's collection since the summer of 1862 (see *Correspondence* vol. 10).

[4] The reference is to a stamp issued by the Philadelphia carriers, D. O. Blood & Co. (see Sutton 1966, p. 41). See also letter to Asa Gray, 2 January [1863].

[5] *Campanula perfoliata* (also called *Specularia perfoliata*) is native to the eastern United States (Bailey and Bailey 1976). CD had been interested in the plant since learning from Gray that it bore unopened flowers in which self-pollination occurred, a phenomenon later known as cleistogamy (see, for example, *Correspondence* vol. 9, letter from Asa Gray, 11 October 1861, and *Correspondence* vol. 10, letters to Asa Gray, 10–20 June [1862] and 26[–7] November [1862], and letters from Asa Gray, 18–19 August 1862 and 29 December 1862).

[6] Gray evidently discussed Bates 1861 in a letter to CD written on 9 February 1863 that has not been found; CD quoted Gray's comments in a letter to Henry Walter Bates of 4 March [1863]. Bates sent a copy of his paper to Gray in January, after CD had persuaded Gray to attempt to have it reviewed in the *American Journal of Science and Arts*, of which he was one of the contributing editors (see letter to H. W. Bates, 12 January [1863], and letter from H. W. Bates, 17 January [1863]).

[7] Bates's account of his eleven years as a naturalist in the Amazon region of South America (Bates 1863) was published between 1 and 14 April 1863 (*Publishers' Circular* 26 (1863): 193).

[8] C. Lyell 1863a; see letter from Asa Gray, 20 April 1863.

[9] See letters to Charles Lyell, 6 March [1863] and 17 March [1863], and letter from J. D. Hooker, [15 March 1863].

[10] See letter from Charles Lyell, 11 March 1863. See also letter from Charles Lyell, 15 March 1863.

[11] A. Gray 1861a.

[12] See *Correspondence* vol. 10, letter from Asa Gray, 29 December 1862. The reference is to Sprengel 1793, in which the term 'dichogamie' was first used (Baillon *et al.* 1876–92).

[13] CD cited Sprengel 1793 extensively in *Orchids*, adding (p. 340 n.):

> I am aware that this author's curious work ... has often been spoken lightly of. No doubt he was an enthusiast, and probably carried some of his ideas to an extreme length. I feel sure, from my own observations, that his work contains a large body of truth.

Although little regarded by botanists, the work was notable for its statement of two related doctrines, namely, that flowers were generally adapted to be cross-pollinated, and that floral structures were often adapted for insect visitation (*DSB*). These doctrines related closely to CD's own belief that it was a 'law of nature ... that no organic being self-fertilises itself for an eternity of generations; but that a cross with another individual is occasionally ... indispensable' (*Origin*, p. 97).

[14] In his letter to Gray of 26[–7] November [1862] (*Correspondence* vol. 10), CD objected to Gray's use of the term 'Diœcio-dimorphism' to describe cases of flower dimorphism like those occurring in several species of *Primula*, on the grounds that he doubted the implied evolutionary transition between such dimorphism and dioeciousness. In his reply of 29 December 1862 (*Correspondence* vol. 10), Gray drew on CD's account of the functional similarity of dimorphic and dichogamous plants, suggesting that CD extend the meaning of the word 'dichogamy' to include such cases of dimorphism as that in *Primula*.

[15] Letter from Hermann Crüger, 23 February 1863.

[16] See letter to Hermann Crüger, 25 January [1863] and n. 2.

[17] See letter to Hermann Crüger, 25 January [1863] and n. 6.

[18] CD's new hothouse was completed by 15 February 1863 (see letter to J. D. Hooker, 15 February [1863] and n. 5); Joseph Dalton Hooker had sent CD orchids and other plants for cultivation from the Royal Botanic Gardens, Kew (see letter to J. D. Hooker, [21 February 1863]). See also Appendix VI.

[19] Gray had suggested that *Amsinckia spectabilis* might be dimorphic, a view CD endorsed, having seen dried specimens sent by Gray (see *Correspondence* vol. 9, letter from Asa Gray, 9 November 1861, and

letter to J. D. Hooker, 18 [December 1861]). CD subsequently grew plants from seed in order to experiment on the species (see *Correspondence* vol. 10, letter to Asa Gray, 15 March [1862], and this volume, letter to Asa Gray, 2 January [1863]). He recorded his observations on these plants, and on one belonging to his neighbour, George Henry Turnbull, in notes dated 23 March and 1 April 1863 (DAR 110: B2), concluding that the plant was not dimorphic, but that the length of the stigma was very variable, and the first-formed flowers tended to have stamens somewhat arrested in development. See also *Forms of flowers*, pp. 110–11.

20 Gray sent CD live specimens of the dimorphic plant *Mitchella repens* in December 1862, for use in crossing experiments (see *Correspondence* vol. 10, letter from Asa Gray, 9 December 1862, and this volume, letter to Asa Gray, 2 January [1863]). CD deferred his experiments until 1864, because the plants did not flower abundantly in 1863 (see letters to Asa Gray, 31 May [1863] and 26 June [1863], and *Forms of flowers*, p. 125).

21 CD had been suffering ill health since the end of February (see, for example, letter to J. D. Hooker, 5 March [1863]). Horace Darwin had been ill for much of 1862 (see letter from G. V. Reed, 12 January 1863 and n. 2); on 17 March 1863, Emma Darwin recorded in her diary (DAR 242) that he was again unwell. CD refers to James Manby Gully's hydropathic establishment in Great Malvern, Worcestershire.

22 According to his 'Journal' (see Appendix II), CD wrote a draft of the section of *Variation* dealing with inheritance (*Variation* 2: 1–84) between 23 January and 1 April 1863, noting: 'took me $6\frac{1}{2}$ weeks time lost by illness & London'. *Variation*, which CD began writing early in 1860, was not published until 1868.

23 The pamphlet has not been identified. Gray periodically sent CD American newspapers, although CD told Hooker that he never read them (see *Correspondence* vol. 10, letter to J. D. Hooker, 29 [December 1862]).

From Alfred Newton 21 March 1863

Elveden, Thetford
21 March 1863.

My dear Sir,

A few days ago I received from M^r. Sclater a note written to him by D^r. Hooker mentioning your wish to obtain a tuber of the Wild Potato—[1] I am very glad to say that it is in my power to supply you with two—the only two I have left— Had I known you were desirous of a specimen—I should certainly have placed the whole dozen that I brought home at your disposal. They were given to me by a M^r. Colchester who is a merchant at Valparaiso, and was my fellow passenger from the West Indies a short time ago—[2] He told me he dug them up just before he came away—I suppose in November last—

He said they were growing on low hills near the sea, and on ground which had never been disturbed, for I asked him as to the possibility of their having been descended from cultivated ancestors— This gentleman though no naturalist, is very intelligent in matters of natural history—and if you wanted a correspondent in that part of the world might prove very useful. He assured me however that these wild potatoes are very bitter to the taste—which does not accord with your experience of the Chiloë ones, as given in your 'Journal'—[3]

I hope shortly to send you the foot of a Red-legged Partridge to which an enormous ball of clay is attached, in illustration of what you have said as to Birds transporting seeds[4]—though in this case I am not sure that the bearer of the incumbrance could have ever flown with it.

I mean to show the specimen at a meeting of the Zoological Society—and will then, with the owner's permission, hand it over to you—[5] If this bird could have flown it could certainly have conveyed a very extensive Flora to a new island—the mass of clay being about 3 inches × $2\frac{1}{2}$ × 2.

I hope the potatoes will arrive safely, and I remain, Yours very truly | Alfred Newton

Charles Darwin Esq[re]. | F.R.S.

DAR 172: 39

[1] Learning that Joseph Dalton Hooker had received, through Philip Lutley Sclater, specimens of South American wild potato brought back by Newton from his visit to the Americas, CD expressed a desire to obtain such specimens in order to test their fertility with cultivated varieties (see letters from J. D. Hooker, [6 March 1863] and [15 March 1863], and letter to J. D. Hooker, 13 [March 1863]). Hooker subsequently wrote to Sclater, asking him to pass on his request to Newton (see the letter from Hooker to Sclater, dated [15 March 1863], and the covering letter from Sclater to Newton, dated 16 March 1863, in Cambridge University Library, Alfred Newton papers).

[2] Mr Colchester has not been further identified. Newton made a short visit to the West Indies in October and November 1862, returning to Britain in January 1863 (see Wollaston 1921, p. 75, and letter to Alfred Newton, 24 March [1863], n. 2).

[3] Newton apparently refers mistakenly to CD's description in *Journal of researches*, p. 347, of the wild potatoes he found in the Chonos Archipelago in January 1835. CD stated that the tubers, when cooked, were 'watery and insipid', and he contrasted this with the bitter taste others had reported of the wild potatoes growing on the nearby island of Chiloé.

[4] In the 1850s CD had made an extensive study of the geographical dispersal of seeds (see *Correspondence* vols. 5 and 6); the subject is discussed in *Origin*, pp. 356–65.

[5] Newton exhibited the foot of the partridge *Caccabis rufa* at a meeting of the Zoological Society of London on 21 April 1863; its owner was Henry Stevenson, secretary of the Norfolk and Norwich Museum (see Newton 1863). Newton made the specimen available to CD in October 1863 (see letter from Alfred Newton, 31 October 1863). In the fourth edition of *Origin*, p. 432, CD reported that the earth adhering to the foot of the specimen weighed $6\frac{1}{2}$ oz, and that, despite having been kept for three years, it yielded eighty-two plants when broken, watered, and placed under a bell glass.

From John Scott 21 March [1863][1]

Edinburgh
March 21st

Sir,

I am sorry to find from your last, that you are labouring under such a weak state of health.[2] I sincerely trust that a little relaxation will enable you to resume with renewed vigour, those arduous and varied works from which you will be so ill spared.

I thank you for your kind attention to my queries on Passiflora &c.[3] I will not trouble you now with any remarks on sterility of passion-flowers with own-pollen, further than merely stating, that I am inclined to believe—on grounds which if desirable ⟨I can⟩ communicate—you will find ⟨that⟩ —sterility—capable of a ⟨local⟩ application only. I know ⟨it to⟩ be so at least in species mentioned. Pray do not think me over-critical. I refer to it only in private, and would be the last publicly to question the validity of these long recorded statements; but being engaged in similar observations for you, I communicate *them* only as the results of my own *practical experiments*. Plants which with us may have exhibited various grades of sterility, may prove otherwise in another locality.[4]

I have now one or two singularly capricious cases of this phenomenon in species of Oncidium in store for you.[5]

I return by this evenings mail Asa Gray's Review, for which I am greatly obliged.[6] I may state that my observations on the capability of the rostellum for fertilisation are still entirely negative, & I have now performed not a few experiments.[7] In my examination, however, I frequently found short tubes protruding from grains, immersed in viscous matter of rostellum, but in no case did I find these insinuating themselves into the tissues of that organ Its viscous secretions, sufficed to generate their vital action, but its conductive prerogative seems lost; as in every case the tubes, were deflected along the anterior surface of the clinandrum towards the stigma. In *Laelia* I have now seen fertilisation thus accomplished, in *Bletia*, and *Cœlogyne* this was prevented by the plaister of Paris which covered the stigmas.[8] This organ, therefore, in these genera still retains its stigmatic function, though the stylar parts are unable to accomplish theirs. I am afraid, however, since I have read Asa Gray's observations, that I have misunderstood you, in supposing ⟨that the⟩ tubes would not only penetrate rostellum, but passing down amongst its vessels, likewise reach the ovary⟨.⟩

The capsules of Gongora atropurpurea are now swelling fast, and promise to be larger than that sent of Acropera—by the way, not a single seed has yet germinated.[9] I am at present working on *Gongora truncata* which presents even greater difficulties than the former: the stigmatic chamber being all but shut. This genus, with which I suppose *Acropera* is now associated, is indeed a most perplexing one from your point of view.[10] I will not, however encroach further at present, so with sincere wishes for a speedy recovery from your ill health,

I remain | Sir | Yours very respectfully | John Scott

DAR 177: 85

CD ANNOTATIONS
1.1 I am sorry ... locality 2.10] *crossed pencil*
4.2 I may state ... that organ 4.6] *scored brown crayon*
4.6 Its viscous ... vital action, 4.7] *reverse question mark added in margin, brown crayon*
5.2 by the way, ... but shut. 5.4] *scored red crayon*
Top of letter: 'Asa Gray | Gymnadenia' *pencil*

1 The year is established by the relationship between this letter and the letter to John Scott, 6 March 1863.

2 See letter to John Scott, 6 March 1863 and n. 1.

[3] See the enclosure to the letter to John Scott, 6 March 1863.

[4] See also letter from John Scott, 3 March 1863 and n. 4. Scott published his observations on sterility in self-pollinated *Passiflora* in Scott 1864d; they are summarised in *Variation* 2: 137–8.

[5] Scott's observations on this subject were published in Scott 1864b; CD cited them in *Cross and self fertilisation*, pp. 330–1.

[6] CD had lent Scott his copy of A. Gray 1862a (see letter to John Scott, 6 March 1863).

[7] In A. Gray 1862a, p. 426, Asa Gray reported that pollen falling onto certain viscid parts of the rostellum in the orchid *Gymnadenia tridentata*, readily sent pollen tubes into it, and that those parts of the rostellum appeared 'to act as stigmas'. In his letter to Scott of 3 December [1862] (*Correspondence* vol. 10), CD suggested that he might try a similar experiment with *Cattleya* 'or ... some allied orchis'. See also *Correspondence* vol. 10, letter from John Scott, 17 December [1862], and this volume, letters from John Scott, 6 January 1863 and 16 January 1863.

[8] In his letter to Scott of 21 January [1863], CD suggested that Scott apply plaster of Paris to the stigmas of orchid flowers in order to prevent pollen tubes from penetrating them and effecting fertilisation. Scott wished for a means to 'force the tubes either to penetrate the rostellum, or remain inert, and thus satisfactorily show its real functional condition' (see letter from John Scott, 16 January 1863).

[9] On *Gongora*, see letters from John Scott, 6 January 1863 and 3 March 1863. In his letter to CD of 18 February [1863], Scott sent CD a seed capsule of *Acropera loddigesii* that he had produced by artificial pollination, stating that he had retained some of the seeds to sow.

[10] See letter from John Scott, 6 January 1863 and nn. 3 and 4. In 'Fertilization of orchids', p. 153 (*Collected papers* 2: 150), CD noted that *Gongora* presented the same difficulty in pollination as *Acropera*, the entrance into the stigmatic chamber being 'so narrow that the pollen-masses cannot be inserted without the greatest difficulty'.

From Asa Gray 22–30 March 1863

Cambridge [Massachusetts]
22. March '63

My Dear Darwin

Argyle's article on the *Supernatural*—to which you called my attention a long while ago,[1] I never happened to see till to-day—when I have read it through. It is quite clever—not deep, but clear, and I think useful I see no occasion for finding fault with him—except in his attempts now and then to direct a little odium against you—which is unhandsome for his main points are those I hammered out in Atlantic. &c—indeed I see signs of his having read the same.[2] But it is hardly fair of him, after expressing his complete conviction that where the operation of natural causes can be clearly traced, the implication of design—upon its appropriate evidence—is not thereby rendered less certain or less convincing.—to go on to speak of derivation-doctrines in a way that implies the contrary.[3]

Of course we believers in *real design*, make the most of your frank and natural terms, 'contrivance, purpose", &c"—and pooh-pooh your endeavors to resolve such contrivances into necessary results of certain physical processes, and make fun of the race between long noses and long nectaries![4]

23^d March.

D^r. Wyman[5]—who is a sharp fellow tells me that—on the authority of the historian Prescott,—the Incas of Peru for—no one knows how long—married *their sisters*—to keep the perfect purity of the blood.[6] Quere. How did this strong case of close-breeding operate? Did they run out thereby? Wyman thinks there is no evidence of it.

If it is true—and the Incas stood it for a long course of generations, you must look to it—for it will bear hard against your theory of the necessity of crossing.[7]

If they run out, you will have a good case.

P.S. *30 March.*

I hear from Hooker that you are poorly,—am very sorry to hear it.—[8] hope it is something very temporary. I have been laid up two or three days with a sort of influenza and a bad throat,—now much better.

I have sent to Silliman extracts from Bates' paper, embracing almost all about mimetic analogy,—8 or 10 pages. My only fear is Silliman will demur to printing it.[9]

You think Lyell *too non-comittal* and *timid*. Well Huxley makes up for it, I should think![10]

Ever dear Darwin | Yours cordially | Asa Gray

DAR 165: 131

CD ANNOTATION
End of letter: 'Ohio—cousins'[11] *ink*

[1] In October 1862, George Douglas Campbell, eighth duke of Argyll, published an anonymous review entitled 'The supernatural' in the *Edinburgh Review*, in which he discussed *Orchids* together with five books on miracles and the supernatural ([Campbell] 1862). CD told Gray of the review in his letter of 23 November [1862] (*Correspondence* vol. 10), saying that while it was 'clever', it did not remove any of the 'difficulties of Theology'.

[2] The reference is to three articles on *Origin* published by Gray in 1860 (A. Gray 1860b), and reprinted as a pamphlet on CD's initiative (A. Gray 1861a); Campbell's name appears on the list of people to whom CD planned to send a copy of A. Gray 1861a (see *Correspondence* vol. 9, Appendix III). Like Gray, Campbell argued that the explanation of natural phenomena, including the origination of new species, in terms of material causes did not negate the widely held view that the design apparent in nature resulted from the intentions of a divine designer.

[3] In [Campbell] 1862, p. 388, Campbell stated that the 'various hypotheses of development', of which he considered CD's to be 'only a new and special version', were 'utterly destitute of proof' and involved 'such violations of, or departures from, all that we know of the existing order of things, as to deprive them absolutely of all scientific basis'. Moreover, while Campbell praised CD's teleological analysis of the floral structure of orchids, he ridiculed his attempt to explain how the structures had arisen by successive modifications (*ibid.*, pp. 391–5).

[4] Referring to CD's frequent use of words such as 'contrivance' and 'purpose' ([Campbell] 1862, pp. 292–3), Campbell commented: 'It is curious to observe the language which this most advanced disciple of pure naturalism instinctively uses when he has to describe the complicated structure of this curious order of plants.' Gray also refers to the fact that Campbell ridiculed CD's attempt to account for the great length of the nectary of the Madagascan orchid, *Angraecum sesquipedale*, by reference to a 'race in gaining length between the nectary of the Angræcum and the proboscis of certain moths'; in his account, Campbell had substituted the word 'nose' for 'proboscis' (see [Campbell] 1862, pp. 394–5, and *Orchids*, pp. 179–203).

5 The anatomist and ethnologist Jeffries Wyman was one of Gray's colleagues at Harvard University (*DSB*).

6 William Hickling Prescott. CD read Prescott's *History of the conquest of Peru* (Prescott 1847) in May 1850, describing it as 'goodish' (see *Correspondence* vol. 4, Appendix IV, 119: 22b).

7 Gray refers to CD's belief that it was a 'general law of nature . . . that no organic being self-fertilises itself for an eternity of generations; but that a cross with another individual is occasionally—perhaps at very long intervals—indispensable' (*Origin*, p. 97). See also *Correspondence* vol. 10, letter to Asa Gray, 26[–7] November [1862]. CD did not refer to Prescott's observation in print, but see the letter to Asa Gray, 20 April [1863].

8 In his letter to Joseph Dalton Hooker of 5 March [1863], CD reported that he had been particularly ill for ten days, with 'much sickness & weakness'.

9 Henry Walter Bates had sent Gray a copy of Bates 1861, in which he explained mimetic resemblances in South American butterflies in terms of natural selection, after CD had persuaded Gray to attempt to have it reviewed in the *American Journal of Science and Arts*, of which Gray was one of the contributing editors (see letter to H. W. Bates, 12 January [1863], and letter from H. W. Bates, 17 January [1863]). See also letter to H. W. Bates, 4 March [1863] and n. 4. The abstracts sent by Gray were so extensive that the journal's general editors, who included Benjamin Silliman and his son Benjamin Silliman Jr, insisted that they be held over for the July number of the journal; ultimately, Gray decided to write a review abstract of the paper instead (A. Gray 1863a), although this was not published until the September 1863 issue of the journal (see letters from Asa Gray, 13 April 1863, 7 July 1863, and 21 July 1863).

10 See letter to Asa Gray, 23 February [1863] and n. 16. The references are to Charles Lyell's *Antiquity of man* (C. Lyell 1863a), and Thomas Henry Huxley's *Evidence as to man's place in nature* (T. H. Huxley 1863b).

11 See letter to Asa Gray, 20 April [1863] and n. 12.

To W. D. Fox 23 March [1863][1]

Down. | *Bromley.* | *Kent. S.E.*
March 23^{d}

My dear Fox

I am particularly glad to have the authentic particulars & your own cases; both of which I shall give on your authority.[2] In the old note I, no doubt, read 23 for 13 lambs: at the time I supposed that the ewes had bred for two seasons.— The whole case seems a good one for showing preponderance of one colour over other.—[3]

Many thanks.—

We doubt a little about Malvern, for I am having an attack of Eczema on my face, which does me as much good as Gout does others; & which renders cold water to my face intolerable; so that I could not stand water-cure of any kind, as I found at Ilkley, when I had first attack of this horrid & blessed eczema.—[4]

Incomplete
American Philosophical Society (292)

1 The year is established by the relationship between this letter and the letters from W. D. Fox, 12 March [1863], and to W. D. Fox, 16 [March 1863].

[2] Fox's letter has not been found; however, see letters to W. D. Fox, 9 March [1863] and 16 [March 1863], and letter from W. D. Fox, 12 March [1863]. CD wrote a draft of the section of *Variation* dealing with inheritance (*Variation* 2: 1–84) between 23 January and 1 April 1863 (see 'Journal' (Appendix II)); the cases given by Fox are mentioned on his authority in *Variation* 2: 30–1 and 40.

[3] The letter from Fox to which CD refers has not been found; however, see the letter to W. D. Fox, 9 March [1863] and n. 1.

[4] CD refers to James Manby Gully's hydropathic establishment in Great Malvern, Worcestershire (see letter to W. D. Fox, 16 [March 1863] and n. 4), and to his visit to a hydropathic establishment in Ilkley, Yorkshire in 1859 (see *Correspondence* vol. 7). CD often found an attack of eczema to have an energising effect upon him (see, for example, letter to J. D. Hooker, 5 March [1863]).

To an editor[1] 24 March [1863?][2]

Down. | Bromley. | Kent. S.E.
March 24th

Private

Mr. Darwin takes the liberty to send by this post to the Editor a New Zealand newspaper for the very improbable chance of the Editor having some time spare space to reprint a Dialogue on Species. This Dialogue, written by some quite unknown to Mr. Darwin, is remarkable from its spirit & from giving so clear & accurate a view of Mr. Ds. theory.—[3] It is, also, remarkable from being published in a Colony exactly 12 years old, in which, it might have thought, only material interests would have been regarded.—[4]

Yours Obediently | Ch. Darwin

Canterbury Museum, Canterbury, New Zealand

[1] The editor and journal to which CD sent this letter have not been identified; however, in 1911, this letter was discovered among the papers of John Malcolm Forbes Ludlow, a Christian socialist with an extensive acquaintance (*DNB*; see 'Samuel Butler's lost dialogue', *Press*, 1 June 1912). See also n. 3, below.

[2] The year is conjectured from the reference to Samuel Butler's article (see n. 3, below).

[3] CD refers to Butler's anonymous article entitled 'Darwin on the origin of species: a dialogue', which was published in the *Press* of Christchurch, New Zealand, on 20 December 1862. The article, which was Butler's first published statement on evolution, presented a sympathetic exposition of CD's theory in dialogue form. It caused some controversy, and Butler wrote several letters to the newspaper in defence of his earlier piece, which are reproduced together with the original article in S. Butler 1923, pp. 188–207. CD had been sent a copy of the article anonymously, and thought that it might have been written by Julius von Haast (see letter to Julius von Haast, 18 July [1863]); Haast apparently informed CD of the true identity of the author (see letter from Emma Darwin to J. D. Hooker, [7 December 1863]). There is an annotated copy of the article in CD's Scrapbook of reviews (DAR 226.1: 131–2), identifying Butler as the author. The article was apparently not reprinted in any contemporary journal (see Harkness 1955).

[4] New Zealand became a British colony in November 1840, and was governed as a Crown Colony until 1852, when the New Zealand Constitution Act established a General Assembly. CD may refer to the formation in 1850 of the Canterbury Settlement (of which Christchurch was the main town) by the New Zealand Company. See Gardner 1992.

From J. D. Hooker [24 March 1863][1]

Royal Gardens Kew
Tuesday.

Dear Darwin

I have been looking at Poplars. Most of the Trees are so big, that one would require a ladder to examine.[2]

One nice young tree of P. canescens is covered with ♂ catkins, I find no honey secreted, any where, I see no bees or other insects about the tree, & I find the pollen to be shed in great quantity in the powdery form.

This is all I have made out as yet.

I forgot last Sunday to ask you about *Reversion*, I am so glad you have taken up the subject.[3] I do not understand you in saying that the child inherits nothing from its Grandparents except you mean (what I think you imply) that the Gndparents' properties, if developed by the grdchild, were latent in the parent—& which I entirely believe—[4] I wholly disbelieve Lyell's dogma that a talented child can be born of untalented parents.[5] In the case of every one of my Gdparents (Turners')[6] numerous grand children, I can descry some of his many wonderful qualities (bad good & indifferent), some patent in their parents, others latent *but producible* & I do not believe that one has a quality of its Grd parents, that was not transmitted through its parent.

I now well remember that the P.S. note to my Australian Essay was added to the 3d. Revise of the sheet, *after receiving* the "Origin" from you, & it was its actual distribution that prompted the reference in the P.S.—[7]

Falconer is *implacable* & takes a most violent & most prejudiced view of Lyell— I had at him in vain; & Lubbock did his very best too, seperately— F. was far too far gone to care whether Owen would enjoy it or not. From F.s tone it will be a very severe affair.[8]

Lubbock in N.H. Review, had in a note called attention to Lyell's "doing injustice" to Prestwich & Falconer.[9] I mollified the expression "injustice" in Lubbock's paper (which was friendly & apologetic) I am deeply sorry for it, but what can one do—? I do think Lyells first XII chapters a complete mess.[10]

Oliver asks if you have a spare copy of the *Linum* paper, he would be greatly obliged for one.[11]

I do not know what to think of Tropical plants during cold period— As to their living through it, it is an *impossibility*— I quite go along with you in suggesting as many Tertiary or Secondary cold periods of migrations as you please.—[12] But that such an Order as *Dipterocarpeæ*, whose species are all ultra tropical, all trees,—containing many diverse genera & species, should have survived a cold period, or been developed since, are equally preposterous surmises in *the present state of science*.

It was very kind of you to keep my medallion passion in mind, I do hope those you mentioned will turn up—[13] I should extremely like a cast of that of your Grandfather for our Museum here, & asked Woolner the Sculptor, who offered to have it done without any injury whatever, if you did not object to lend it for the purpose.—[14] It is immensely valuable, so I shall not be vexed if you decline.

Falconer was so superlatively jocund at Lubbocks, I did wish you had been there to hear.[15]

If ever you find any local application do good to your Eczema, pray let me know, as my Fathers is getting bad again.[16]

I had a very jolly walk back in 1h. 48″. to the house, & found the Busks & Tyndall arrived for dinner.[17]

I have not forgotten Edwardsia.[18]

I have a little girl down with bad sore throat not Diptheria thank God, something is evidently *going through* the family.[19]

Ever yours affec | J D Hooker

I hope I am not too severe on Lyells first Chapters—[20] the state of case is thus.—

I read first & second with delight, those reminded me of the Principles[21]—& after a long interval I skimmed III–X & was struck with the appearances of Lyells want of faith in all Prestwichs observations & facts—till ratified by his (Lyells) going down to spot & examining for himself—in short a decided *appearance* of suppression of credit due to Ps. originality & accuracy.[22]

I then reread all very carefully, but could hardly justify my last verdict, & got quite confused & am so still.

I quite agree with you as to want of originality of whole work, but I expected no originality.—

DAR 100: 154, DAR 101: 123–5

CD ANNOTATIONS

1.1 I have . . . examine. 1.2] *crossed brown crayon*
3.1 This is . . . yet.] *before closing square bracket, brown crayon*
4.1 *Reversion*,] *underl brown crayon*
4.1 I am so . . . indifferent), 4.8] *crossed pencil*
4.8 I do not . . . parent. 4.10] *scored brown crayon*
7.4 chapters . . . mess.] *underl brown crayon*
9.1 I do not . . . *impossibility*— 9.2] *scored brown crayon*
9.4 But that . . . trees, 9.5] *double scored brown crayon*
10.1 I do hope . . . up— 10.2] *scored brown crayon*
10.3 who offered . . . purpose.— 10.5] *double scored brown crayon*
12.1 If ever . . . again. 12.2] *double scored brown crayon*
18.2 & was struck . . . himself— 18.4] *double scored brown crayon*

[1] The date is established by the reference to Hooker's visit to Down on 22 March 1863 (see n. 3, below); the following Tuesday was 24 March 1863.

[2] CD asked Hooker for information on pollination in poplars in his letter of 13 [March 1863].

[3] See letter to J. D. Hooker, 17 March [1863] and n. 20. Hooker visited Down House on 22 March 1863, while on a visit to John Lubbock's house at Chislehurst, Kent (see letter from J. D. Hooker, [15 March 1863]).

[4] See letter to J. D. Hooker, 17 March [1863]. See also letter to J. D. Hooker, 26 [March 1863].

5 C. Lyell 1863a, pp. 504–5.

6 Hooker's maternal grandparents were the banker and botanist Dawson Turner, and Mary Turner.

7 In C. Lyell 1863a, p. 417, Charles Lyell had claimed that J. D. Hooker 1859 was published several months before *Origin*, whereas it was actually published a month after *Origin*. Hooker had also been confused over their respective dates of publication (see letter from J. D. Hooker, [15 March 1863] and n. 7, and letter to J. D. Hooker, 17 March [1863] and n. 3). Hooker ended his essay with a postscript, the last paragraph of which (J. D. Hooker 1859, p. cxxviii) read:

> I would further observe here, to avoid ambiguity, that my friend Mr. Darwin's just completed work 'On the Origin of Species by Natural Selection,' from the perusal of much of which in MS. I have profited so largely, had not appeared during the printing of this Essay, or I should have largely quoted it.

8 Hugh Falconer was angry about what he considered to be Lyell's failure properly to acknowledge his work, and that of Joseph Prestwich, in *Antiquity of man* (C. Lyell 1863a; see letters from J. D. Hooker, [6 March 1863] and [15 March 1863]). Hooker refers to conversations that took place at Lubbock's house on 22 March 1863 (see n. 3, above). Earlier in the year, Falconer had been involved in a vigorous dispute with Richard Owen, who had subsequently also been criticised in, and was critical of, Lyell's book (see, for example, letter to J. D. Hooker, 3 January [1863] and n. 2, and letter from J. D. Hooker, [23 February 1863] and n. 2).

9 Hooker refers to Lubbock's review of C. Lyell 1863a, which appeared in the number of the *Natural History Review* for April 1863 ([Lubbock 1863c]); Hooker must have seen the review in manuscript while visiting Lubbock (see n. 3, above). In the published version ([Lubbock] 1863c, p. 214), Lubbock stated: 'Though … we think that Sir Charles scarcely gives due prominence to the labours of his predecessors in these investigations, and especially to those of Dr. Falconer and Mr. Prestwich, we are quite satisfied that this is unintentional.'

10 C. Lyell 1863a. See also letter from J. D. Hooker, [1 March 1863] and n. 12.

11 CD's paper, 'Two forms in species of *Linum*', was read before the Linnean Society on 5 February 1863. It was published in the number of the *Journal of the Proceedings of the Linnean Society (Botany)* that was issued on 13 May 1863 (*General index to the Journal of the Linnean Society*, p. vi). However, CD obtained a number of offprints of the paper in mid-April for distribution to those who he considered would be interested; Daniel Oliver's name appears on CD's presentation list (see Appendix IV). Oliver had previously requested a copy of the paper in order to give an abstract of it in the *Natural History Review*, of which he was one of the editors (see letter from Daniel Oliver, 27 February 1863). He apparently wrote the abstract that was published in the July 1863 issue of the journal ([Oliver] 1863d; see letter to J. D. Hooker, 14 July [1863]).

12 See letter to J. D. Hooker, 17 March [1863] and nn. 17 and 18.

13 Hooker was an avid collector of Wedgwood ware (see letters from J. D. Hooker, 6 January 1863, [16 February 1863], and [6 March 1863]). Hooker and CD apparently discussed the subject during Hooker's visit to Down House of 22 March 1863.

14 Thomas Woolner. Hooker refers to a Wedgwood medallion of Erasmus Darwin and to the museum of the Royal Botanic Gardens, Kew.

15 See n. 3, above.

16 Both CD and William Jackson Hooker had been diagnosed as suffering from eczema in 1862 (see *Correspondence* vol. 10, letter from J. D. Hooker, 2 July 1862).

17 See n. 3, above. The references are to George and Ellen Busk, and to John Tyndall.

18 See letter to J. D. Hooker, 24[–5] February [1863] and n. 19. See also letter to Daniel Oliver, 24–5 March [1863].

19 The reference is to either Harriet Anne or Maria Elizabeth Hooker.

20 See n. 10, above.

21 C. Lyell 1830–3.

22 See n. 8, above.

To Alfred Newton 24 March [1863][1]

Down. | *Bromley.* | *Kent. S.E.*
March 24th

My dear Sir

I thank you sincerely for the potatoes, which arrived safely & are planted; they may come in very useful in crossing, as our cultivated potatoes present some odd peculiarities in their reproduction.—[2]

I shall be exceedingly glad of the foot of the Partridge, if you can get it for me: I will weigh the earth & then see, with due precautions, whether it contains any seeds capable of germination:[3] I presume the foot will turn out to be a wounded one. I suppose you feel sure it is not a stupid hoax.—

With my sincere thanks for your kindness— | My dear Sir | Yours sincerely | Ch. Darwin

CUL, Alfred Newton papers

1 The year is provided by the relationship between this letter and the letter from Alfred Newton, 21 March 1863.

2 See letter from Alfred Newton, 21 March 1863. See also letter to J. D. Hooker, 13 [March 1863], n. 9. Newton annotated this letter: '*Wild* Potatoes given me by Mr. Colchester who brought them from the W. coast of S. America & was my fellow passenger from St. Thos. to England in Jany./63.'

3 See letter from Alfred Newton, 21 March 1863 and nn. 4 and 5.

To Daniel Oliver 24–5 March [1863][1]

Down. | *Bromley.* | *Kent. S.E.*
March 24th

Dear Oliver

I remember looking in such Books as I had for explanation of the ovarium of Primula; which I found something like this,

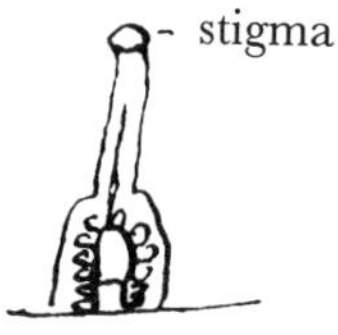

viz a knob in middle covered with ovules, which sent a thin stilyform process up centre of style. I could not make out from Books what this latter part was.[2] Last autumn I potted some Primroses; a long-styled plant is now in flower which in almost all its flowers has 3 rather short pistils with stigmas (5 stamens &c as usual); the three styles become a little foliaceous at base but are *not* united, so that by pulling them a little asunder without tearing you can see right into the ovarium with its ovules; & there is no central process from knob.— If capable of fertilisation the pollen tubes must go to base of knob & then turn up to ovules. I have put a few flowers in spirits. The case does not concern me;[3] & I mention it *solely* for *chance* of your caring for it, or for its throwing any light on morphology of ovarium in Primula.— It looks as if knob with ovules was prolongation of axis.

If you care for specimens per Railway or want me to ascertain whether these flowers can be fertilised, please write;[4] if not I shall understand that specimens may be thrown away.—

Please tell Hooker I sh^d. like to see flowers in tin-box of Edwardsia tetraptera;[5] & I sh^d like to see 2 or 3 leaves or bracts of Marcgravia, which Lindley speaks of as approaching in structure to pitchers of nepenthes.—[6] *Pods* of *Orchis*—[7] *Poplars*.—[8]

Dear Oliver | Yours very truly | C. Darwin

I think I will put pollen right on to ovules & cut off the stigmas & see if I can thus *directly* fertilise the ovules.—

P.S. 25th | Tell Hooker I have just received his interesting letter,[9] & will write soon. About Poplars suffice with particular thanks; perhaps he would look just once again for Bees on fine day, & gently again, shake branch.—

Case interests me on account of willows.

You shall surely have copy of *Linum* paper whenever I get copy—[10]

I am baddish today so no more—

Endorsement: '/63'
DAR 261 (DH/MS 10: 42)

1 The year is established by the relationship between this letter and the letter from J. D. Hooker, [24 March 1863].

2 CD refers to the tissue through which the pollen tubes pass from stigma to ovules, which was called 'conducting tissue' (see for example the letter from John Scott, 16 January 1863).

3 CD carried out crossing experiments with the plant; his notes, dated 24 March – 11 April 1863, are in DAR 108: 165. See also letter to Daniel Oliver, [12 April 1863] and n. 3.

4 See letter from Daniel Oliver, [26 March 1863].

5 See letter to J. D. Hooker, 24[–5] February [1863] and n. 19. See also letter from J. D. Hooker, [24 March 1863].

6 The reference has not been identified, but may be an imprecise allusion to Lindley 1853, p. 403. CD had been keen to obtain specimens of the pitcher-plant, *Nepenthes*, while preparing to stock his hothouse at the beginning of the year; he had obtained specimens from the Royal Botanic Gardens, Kew, through Joseph Dalton Hooker (see letters to J. D. Hooker, 13 January [1863] and 5 March [1863], and letter from J. D. Hooker, 15 January [1863]). CD was interested in *Nepenthes* as one of a few unrelated genera in which leaves were developed into pitcher-like structures; such cases of 'Similar & peculiar organs in beings far remote in the scale of nature' presented what he considered a difficulty for the theory of natural selection (*Natural selection*, pp. 374–5; see also *Notebooks*, p. 440; *Correspondence* vol. 2, letter to *Gardeners' Chronicle*, [late August 1843]; and *Correspondence* vol. 6, letter to J. D. Hooker, 16 [May 1857]).

7 CD probably refers to the seed-capsules of the orchid genus *Vanda* that he had been promised while on a visit to the Royal Botanic Gardens, Kew, in February; CD wished to compare a capsule produced artificially by John Scott from *Acropera loddigesii* with capsules from others of the Vandeae (see letter to John Scott, 16 February [1863] and n. 14).

8 See letter to J. D. Hooker, 13 [March 1863], and letter from J. D. Hooker, [15 March 1863]. See also letter from J. D. Hooker, [24 March 1863].

9 Letter from J. D. Hooker, [24 March 1863].

10 'Two forms in species of *Linum*'. See letter from J. D. Hooker, [24 March 1863] and n. 11.

To John Scott 24 March [1863][1]

Down Bromley Kent.
March 24th

Dear Sir

Your letter, as every one you have written, has greatly interested me.[2] If you can show that certain individual Passifloras, under certain known or unknown conditions of life, have stigmas capable of fertilisation by pollen from another species, or from another individual of its own species, yet not by its own individual pollen, (*its own individual pollen being proved to be good by its action on some other species*) you will add a case of **great** interest to me; & which in my opinion would be *quite* worth your publication. I always imagined that such recorded cases must be due to unnatural conditions of life; & think I said so in Origin.[3] I am not sure that I understand your result, whether it means what I have above obscurely expressed. If you can prove the above, do publish; but if you will not publish I earnestly beg you to let me have the facts in detail; but you *ought* to publish for I may not use facts for years.—[4]

I have been *much* interested by what you say on the Rostellum exciting pollen to protrude tubes;[5] but are you sure that the rostellum does excite them? would not tubes protrude if placed on part of column or base of petals &c., near to the stigma?— Please look at Cottage Gardener (or J. of Hort) to be published tomorrow week for letter of mine, in which I venture to quote you, & in which you will see curious fact about unopened orchid flowers setting seed in W. Indies.[6] Dr. Cruger attributes protrusion of tubes, to ants carrying stigmatic secretion to pollen; but this is mere hypothesis.[7] Remember pollen-tubes protrude within anther in Neottia nidus-avis—[8] I *did* think it possible or probable that perfect fertilisation might have been effected through rostellum.

What a curious case your Gongora must be:[9] could you spare me one of largest capsule: I want to estimate number of seed, & try my hand if I can make them grow. This, however, is foolish attempt; for Dr. Hooker who was here a day or two ago;[10] says they cannot at Calcutta, & yet imported species have seeded & have naturally spread on to the adjoining trees!

Dr. Cruger thinks I am wrong about Catasetum; but I cannot understand his letter:[11] he admits there are 3 forms in 2 species; & he speaks as if the sexes were separate in some & that others were hermaphrodite; but I cannot understand what he means.— He has seen lots of great Humble-bees buzzing about the flowers with the pollinia sticking to their backs! Happy man!!. I have the promise but not yet surety of some curious results with my homorphic seedling Cowslips: these have **not** followed rule of Chinese Primula; homorphic seedlings from short-styled parent have presented *both* forms, which disgusts me.—[12]

You will see that I am better; but still greatly fear that I must have compulsory holiday.

With sincere thanks & hearty admiration at your powers of observation.— Dear Sir | Yours sincerely | C. Darwin

My poor P. Scotica, look very sick, which you so kindly sent me.—[13]

DAR 93: B72–4

[1] The year is established by the reference to the letter from John Scott, 21 March [1863].
[2] Letter from John Scott, 21 March [1863].
[3] Following a discussion of genera, including *Passiflora*, in which some species had been found to be more easily fertilised by the pollen of another species than their own pollen (*Origin*, p. 251), CD wrote:

> Although the plants in these experiments appeared perfectly healthy, and although both the ovules and pollen of the same flower were perfectly good with respect to other species, yet as they were functionally imperfect in their mutual self-action, we must infer that the plants were in an unnatural state.

[4] Scott published the results of his experiments on *Passiflora* in Scott 1864d; CD cited Scott's paper in *Variation* 2: 137. Although CD wrote drafts of the chapters on inheritance for *Variation* (*Variation* 2: 1–84) between 23 January and 1 April 1863 (see 'Journal' (Appendix II)), the work was not published until 1868.
[5] See letter from John Scott, 21 March [1863] and nn. 7 and 8.
[6] See letter to *Journal of Horticulture and Cottage Gardener*, [17–24 March 1863].
[7] See letter from Hermann Crüger, 23 February 1863.
[8] See *Orchids* p. 324 n.
[9] See letter from John Scott, 21 March [1863].
[10] Joseph Dalton Hooker visited Down House on 22 March 1863 (see letter from J. D. Hooker, [24 March 1863]).
[11] Letter from Hermann Crüger, 23 February 1863.
[12] CD refers to seedlings raised from crossing experiments with cowslips (*Primula veris*) carried out in 1862 (see DAR 157a: 77 and DAR 108: 70; see also *Correspondence* vol. 10). These experiments are described in 'Illegitimate offspring of dimorphic and trimorphic plants', pp. 423–30, where CD notes that the 'short-styled form of the cowslip, when self-fertilized, does not transmit the same form nearly so truly as does that of *P. sinensis*.' With a single exception, CD had found that both long- and short-styled plants of *P. sinensis* (the Chinese primrose), produced only own-form plants when pollinated with own-form pollen (see letter to John Scott, 20 [February 1863]).
[13] Scott had sent CD specimens of *Primula scotica* in January (see letter from John Scott, 6 January 1863).

From Armand de Quatrefages[1] 25 March 1863

Paris
25 Mars 63

Monsieur et cher confrère

Au reçu de votre lettre je me suis occupé de votre commission.[2] J'ai écrit à un de mes amis dans le midi mais sa réponse n'était pas des plus satisfaisantes.[3] Heureusement j'ai pu la remettre en très bonnes mains. Mr Martins, professeur de Botanique et D^r^. du Jardin des Plantes de Montpellier, récemment nommé correspondant de notre académie a bien voulu s'en charger.[4] Il m'a promis de faire lui même et de faire faire par ses collegues de la Soc. d'agriculture une collection aussi complete que possible de toutes les races de vers à soie elevés aux environs de Montpellier.[5] Cette collection comprendra la chenille la chrysalide et le papillon—les deux premieres conservées dans l'alcool, le 3eme piqué comme à l'ordinaire. Le tout doit m'être envoyé, et je vous le ferai parvenir par le chemin de fer ou par toute autre voie que vous voudriez bien m'indiquer. De cette facon vous pourrez étudier par vous même les questions qui vous intéressent. Il m'a paru que ce serait le meilleur moyen de répondre à vos intentions.[6]

Je vous remercie des choses aimables que vous voulez bien me dire au sujet de mes *Metamorphoses*.[7] Un suffrage comme le vôtre en vaut à lui seul beaucoup d'autres.

J'ai été plus sensible que je ne saurais le dire à l'honneur que m'a fait votre Soc. d'Anthropologie en placant mon *Unité de l'espece humaine* presque en tête des travaux etrangers qu'elle se propose de traduire.[8] Vous faites certainement partie de cette Société et je suis heureux de penser que votre suffrage a pu être pour une part dans cette résolution de vos collègues.[9] Agréez en de nouveau mes remerciements.

Je suis en ce moment en grande lutte avec les Polygénistes de notre Société. Ils nient l'influence du milieu comme pouvant donner naissance à des races nouvelles. Mais je soutiens la variabilité de l'espèce sous cette influence.[10] Vous voyez, mon cher confrere, combien est vrai ce que je vous disais qu'entre nous il n'y a d'autre différence qu'une question de limite.[11] Mais il est bien étrange que ce soit au nom de la philosophie qu'on prétende nier l'action des agents extérieurs sur les êtres organisés et vivants. Lamarck, Geoffroy St-Hilaire ... seraient bien surpris de se voir de semblables disciples.[12]

Recevez Monsieur et cher confrere l'expression de mon affectueux devouement | De Quatrefages

DAR 175: 2

[1] For a translation of this letter, see Appendix I.

[2] CD's letter has not been found. In the summer of 1862, CD had requested information from Quatrefages on the varieties of the common silkworm (*Bombyx mori*), when writing the section of *Variation* dealing with this subject (*Variation* 1: 300–4; see *Correspondence* vol. 10, letter to Armand de Quatrefages, 11 July 1862 and n. 2). CD was particularly interested in the fact that, while the different varieties of silkworm varied greatly in the larval and pupal stages, they showed no constant differences in the adult stage. In a letter that is now missing, Quatrefages provided CD with information on this subject, which is quoted in *Variation* 1: 303. In a further letter to Quatrefages of 11 July 1862 (*Correspondence* vol. 10), CD asked him to send any 'fresh information' he might acquire in the ensuing year 'on the similarity of the moths of distinct races'. The season being too late to procure such information, Quatrefages asked CD to remind him of his request before the new season's eggs hatched the following April (see *Correspondence* vol. 10, letter from Armand de Quatrefages, [after 11 July 1862]).

[3] This individual has not been identified.

[4] Charles Frédéric Martins was elected as a corresponding member for the rural economy section of the Académie des Sciences, Paris, on 29 February 1863 (*Index biographique de l'académie des sciences*).

[5] The reference is probably to the Société d'agriculture of Montpellier, which was founded in 1807 (*Grande encyclopédie*).

[6] No such specimens are mentioned in *Variation* 1: 300–4, and it appears that they were never sent. Information from Martins is cited in *Variation* 1: 304 from an intermediate source (Godron 1859, p. 462); however, no information provided directly by Martins is included. An annotated copy of Godron 1859 is in the Darwin Library–CUL (see *Marginalia* 1: 331–5).

[7] CD's letter has not been found. Quatrefages promised to send CD a copy of Quatrefages 1862 in his letter of [after 11 July 1862] (*Correspondence* vol. 10); there is an annotated copy of this work in the Darwin Library–CUL (see *Marginalia* 1: 691).

[8] An annotated copy of Quatrefages 1861 is in the Darwin Library–CUL (see *Marginalia* 1: 693–4). Quatrefages refers to the announcement by the Anthropological Society of London, which was founded in 1863, that it hoped to publish an English translation of Quatrefages 1861. No copy of the first prospectus of the society has been found; however, in the 'Anthropological Review Advertising Sheet' issued with the first number of the society's *Anthropological Review* in May 1863, it was announced that the society intended to publish a series of works on anthropology, chiefly translations. Only one publication was described as 'in preparation'; Quatrefages 1861 headed a list of six other titles under the caption: 'translations will be recommended'. In subsequent prospectuses, the book was listed among those of which publication was contemplated, with George Frederick Rolph named as the prospective editor. However, the society's translation never appeared, and Quatrefages 1861 was not translated into English until 1879 (Quatrefages 1879). There is a copy of the 'Anthropological Review Advertising Sheet' in the unbound pamphlet collection in the Darwin Library–CUL.

[9] CD was an honorary member, although not an active one, of the Anthropological Society of London (see letter to Armand de Quatrefages, 27 March [1863]).

[10] Between February and August 1863, Quatrefages, who was president of the Société d'Anthropologie of Paris, engaged in extended discussions with the polygenists in the society over the question of the influence of the environment in modifying races. The substance of the discussions, together with Quatrefages's discourses on the subject on 5 March and 16 April 1863, are recorded in the *Bulletins de la Société d'Anthropologie de Paris* 4 (1863). See also Stocking 1968 and J. Harvey 1997, pp. 105–13.

[11] Quatrefages apparently made this observation in 1862 in a letter that has not been found (see *Correspondence* vol. 10, letter to Armand de Quatrefages, 11 July 1862, and letter from Armand de Quatrefages, [after 11 July 1862]).

[12] Quatrefages refers to Jean Baptiste de Lamarck and Etienne Geoffroy Saint-Hilaire, both of whom had adopted theories of species transmutation in which environmental influence played an important role (*DSB*). See also *Bulletins de la Société d'Anthropologie de Paris* 4 (1863): 149.

To J. D. Hooker 26 [March 1863]

Down
26th.

My dear Hooker

Is it worth while to send the Medallion by Railway? say the word & it shall be sent.—[1]

I consulted the great Mr Startin for my Eczema;[2] & the enclosed *local* applications are *certainly* to me very soothing. The muddy stuff must be shaken, a little poured out & smeared on part with broad camel-brush, & then mopped nearly dry with a bit of rag.— Startin said particularly go to Waugh 177 Regent St.—[3] Please return prescription.[4]

I hope & think that you are too severe on Lyell's early chapters: though so condensed & not well arranged, they seemed to me to convey with uncommon force the antiquity of Man, & that was his object. It did not occur to me, but I fear there is some truth in your criticism that nothing is to be trusted until he Lyell had observed it.—[5]

I am glad to see you stirred up about Tropical plants during Glacial period. Remember that I have many times sworn to you that they coexisted, so my dear fellow you must make them coexist.[6] I do not think greater coolness in a disturbed condition of things would be required than that zone of Himalaya in which you describe some Tropical & temperate forms commingling;[7] & as in lower part of Cameroons, & as Seemann describes on low mountains of Panama—[8] It is, as you say, absurd to suppose that such a genus as Dipterocarpea could have been developed since glacial era;[9] but do you feel so sure, as to oppose a large body of consideration on other side, that this genus could not have been slowly accustomed to a cooler climate. I see Lindley says it has not been brought to England[10] & so could not have been tried in greenhouse. Have you materials to know to what little height it ever ascends mountains of Java or Sumatra. It makes a mighty difference the whole area being cooled; & the area perhaps not being in all respects, such as dampness &c &c fitted for such temperate plants as could get in. But anyhow I am ready to swear again that Dipetrocarpea & any other genus you like to name did survive during a cooler period!

About Reversion you express just what I mean:[11] I somehow blundered & mentally took literally that the child inherited from his grandfather: this view of latency collects a lot of facts—both secondary sexual character in each individual—tendency of latent character to appear temporarily in youth—effect of crossing in educing latent character &c.— When one thinks of a latent character being handed down hidden for a thousand or ten-thousand generations & then suddenly appearing, one is quite bewildered at the host of characters written in invisible ink on the germ.— I have no evidence of the reversion of all characters in a variety.— I quite agree to what you say about genius; I told Lyell that passage made me groan.—[12]

What a pity about Falconer; how singular & how lamentable.—[13]

I am tired, so farewell | Ever my dear Hooker | Yours affect[ly] | C. Darwin

Remember Orchid pods.— I have a passion to grow the seeds (& other motives):[14] I have not a fact to go on, but have a notion (no, I have firm conviction!) that they are parasites in early youth on cryptogams!! Here is a fool's notion; I have some planted on sphagnum.—[15] Do any tropical lichens or mosses or European withstand heat grow on any trees in Hothouse at Kew; *if so for love of Heaven favour my madness* & have some scraped off & sent me. I am like a gambler, & love a wild experiment. It gives me great pleasure to fancy that I see radicles of orchis-seed penetrating the sphagnum; I know I shall not, & therefore shall not be disappointed.

Endorsement: 'March/63.'
DAR 115: 188

[1] In his letter to CD of [24 March 1863], Hooker asked if he could borrow a Wedgwood medallion of Erasmus Darwin so that a copy could be made for the museum at the Royal Botanic Gardens, Kew.

[2] James Startin was a leading skin specialist in London; he was the senior surgeon at the Hospital for Diseases of the Skin, Blackfriars, and ran a private consultancy at 3 Savile Row, Piccadilly (*Medical directory* 1863). An entry in CD's Classed account book (Down House MS) dated 24 August 1862 records payment of £1 7*s.* under the heading 'M[r]. Startin & Physic'; CD probably consulted Startin while in London in May 1862 (see *Correspondence* vol. 10, Appendix II, and letter to J. D. Hooker, 30 [June 1862]). In his letter to CD of [24 March 1863], Hooker inquired about treatments for eczema, from which his father, William Jackson Hooker, also suffered.

3 The reference is to the chemists George Waugh & Co. There is an entry in CD's Classed account book (Down House MS) under the classification 'Medical attendance', which records payment of £1 3*s.* to 'Waugh' on 11 January 1863.
4 The enclosure has not been found.
5 C. Lyell 1863a. See letter from J. D. Hooker, [24 March 1863].
6 See letter to J. D. Hooker, 17 March [1863] and n. 17, and letter from J. D. Hooker, [24 March 1863].
7 CD marked several passages in his copy of Hooker's *Himalayan journals* in which Hooker described the mixing of temperate and tropical plants in Himalayan localities (J. D. Hooker 1854b, 1: 109–10, 2: 18–19 and 319); CD's annotated copy of this work is in the Darwin Library–CUL (see *Marginalia* 1: 392–3).
8 Seemann 1852–7, p. 67.
9 Hooker referred to the 'Order . . . Dipterocarpeæ' (which approximates to the modern family Dipterocarpaceae) in his letter to CD of [24 March 1863]; in his letter of [28 March 1863], Hooker corrected CD's assumption that he had referred only to the genus *Dipterocarpus*.
10 The reference is apparently to John Lindley's statement in Lindley 1853, p. 393, that trees of this order were 'apparently unknown in Europe in a living state'.
11 See letter from J. D. Hooker, [24 March 1863].
12 See letter from J. D. Hooker, [24 March 1863] and n. 5. See also letter to Charles Lyell, 6 March [1863] and n. 43.
13 Hugh Falconer. See letter from J. D. Hooker, [24 March 1863] and n. 8.
14 CD probably refers to the seed-capsules of the orchid genus *Vanda* that he had been promised while on a visit to the Royal Botanic Gardens, Kew, in February 1863; CD wished to compare a capsule produced artificially by John Scott from *Acropera loddigesii* with capsules from others of the Vandeae (see letter to John Scott, 16 February [1863] and n. 15).
15 CD probably refers to some of the seeds from Scott's capsule of *Acropera loddigesii*; Scott's own attempts to germinate the seeds had so far proved ineffectual (see letter from John Scott, 21 March [1863]).

From Daniel Oliver [26 March 1863][1]

Royal Gardens Kew
Thursday

My dear Sir

I *shall be glad* of a sight of the monstrous ovaries of *Primula*.[2] The true character of the *free* placenta is not thoroughly understood—indeed Caspary[3] regards the 'pistil' as not homologous with the pistil (carpels) of other plants but as '*procarpels*' the true carpels being reduced to the funiculus & integuments of the ovules. Rather an absurd view.— But what we want is something better to connect this free placenta of *Primula* with the marginal placental^{n}. of other angiospermous ovaries.

I shall call at the New Conservatory this P.M. for flowers of *Edwardsia*[4]

Ever Sincerely yours | D Oliver

DAR 173: 18

1 The letter is dated by the relationship between this letter, the letter to Daniel Oliver, 24–5 March [1863], and the letter from Daniel Oliver, [27 March 1863]; the intervening Thursday was 26 March 1863.
2 See letter to Daniel Oliver, 24–5 March [1863].
3 Caspary 1861. Oliver summarised the findings of this paper in the October 1863 number of the *Natural History Review*, noting its conclusion that 'in *Primula*, the leaves, five in number, constituting the pistil,

are pro-carpels, the true carpels being each reduced to funiculus and integuments of as many ovules' (*Natural History Review* 3 (1863): 583).

4 See letter to Daniel Oliver, 24–5 March [1863] and n. 5.

From Edward Blyth 27 March 1863

White Hart Hotel, | *Bromley, Kent.*
March 27/63— | 4½ *p.m.*

My dear Sir,

Thus far on a flying visit to you; but I find that I cannot accomplish this today, as I am engaged to dine at the United University Club, Pall Mall, at 7 *p.m.* I had no idea that your residence is so far from Bromley.[1] I reached London on the evening of the 9th., just in time for the crowds on the Prince's Wedding day, which event I could have wished had happened on any other day.[2] But I found a refuge in the Somerset Hotel, Strand (next door to King's College), where I have been staying ever since, though seldom to be found there during the day. What with sight-seeing & visiting, my time has been pretty well taken up as yet; & about the end of this month I shall leave London for a few weeks, to pay a few visits in Sussex, Wilts, &c.[3] I am sorry that the fates have so ordained that I cannot see you today, but there is now no help for it; so kindly write me word upon what day of next week I am likely to find you at home, when I will leave town after breakfast—[4]

Trusting that your health is better than it was, I remain | Yours ever Sincerely, | E. Blyth—

DAR 160: 203

1 Down House was about six miles south of Bromley railway station (*Post Office directory of the six home counties* 1862).

2 Albert Edward, prince of Wales, married Alexandra, eldest daughter of Prince Christian of Denmark, on 10 March 1863 (*DNB*). Blyth, who had been curator of the museum of the Asiatic Society of Bengal since 1841, was forced to leave Calcutta for England at the end of 1862 as a consequence of ill health (see Blyth 1875, p. xii, and *Correspondence* vol. 10, letter from Edward Blyth, 23 November 1862). During the 1850s, Blyth had become one of CD's most important correspondents, providing him with a vast amount of information on the plants and animals of India (see *Correspondence* vols. 5–7).

3 Blyth had relations in Sussex, whom he visited in April 1863 (see letter from Edward Blyth, 7 April 1863 and n. 1). Later in the year, he visited Charles Edward Rendall of Brigmerston House, near Amesbury, Wiltshire (see the *Field* 21 (1863): 464 and *Post Office directory of Hampshire, Dorsetshire, and Wiltshire* 1859).

4 There is no evidence that Blyth visited CD at Down during 1863.

From Daniel Oliver [27 March 1863][1]

Royal Gardens Kew
Friday morng.

My dear Mr. Darwin

By this post the flowers of the *Edwardsia* are sent in small tin box. I enclose bracts of some *Marcgraaviaceae*.—[2]

I am busy upon West African *Amomums*,[3]—the genus furnishing "Grains of Paradise" &c. & characterised by anthers with *curious crest* variable in different species, as over 3-lobate crest

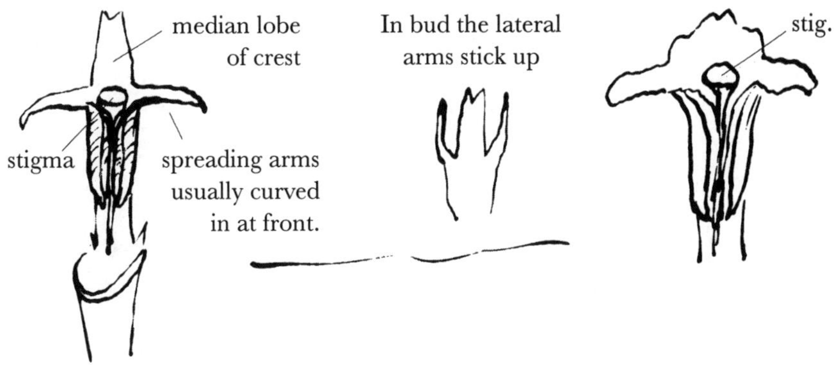

The flowers have each but one anther the cells of which embrace the top of style & the capsulate stigma projects at top. I can't guess how the crest-arms work; they are (the lateral ones) so usual & so frequently incurved, altering from bud to flower, that one cannot but think they are not useless but serve some thing.

Ever very sincerely yrs | D Oliver

DAR 173: 23

[1] The date is established by the relationship between this letter and the letters to Daniel Oliver, 24–5 March [1863] and 28 March [1863]; the intervening Friday was 27 March 1863.

[2] See letter to Daniel Oliver, 24–5 March [1863] and n. 6. See also letter to J. D. Hooker, 24[–5] February [1863], and letter from J. D. Hooker, [24 March 1863].

[3] A paper by Oliver and Daniel Hanbury, entitled 'On some new species of *Amomum* from West Africa', was read before the Linnean Society on 16 April 1863 (Oliver and Hanbury 1863).

To Armand de Quatrefages 27 March [1863][1]

Down. | Bromley Kent. S.E.
March 27th.

Dear Sir.

I thank you most cordially for all your great kindness, the trouble which you have taken and your most obliging note—[2] The specimens of which you have obtained the promise from M. Martins will be *most* interesting to me; & ultimately they shall be deposited in the British Museum.—[3] When I receive them I will write & thank M. Martins—[4] The best plan will be as you propose to send them by Railway

I had heard that your work was to be translated and I heard it with pleasure, but I can take no share of credit, for I am not an active, only a honorary member of the Society—[5] Since writing I have finished with extreme interest to the end, your admirable work on metamorphosis.[6] How well you are acquainted with the works of English Naturalists, and how generously you bestow honour on them.— Mr Lubbock is my neighbour & I have known him since he was a little boy—; he is in every way a thoroughly good man; as is my friend Huxley—[7] It gave me real pleasure to see you notice their works as you have done.

With sincere respect & cordial thanks, for your great kindness to me, I remain, Dear Sir | Yours truly obliged, | Charles Darwin.

Copy
DAR 147: 284

[1] The year is established by the relationship between this letter and the letter from Armand de Quatrefages, 25 March 1863.
[2] See letter from Armand de Quatrefages, 25 March 1863.
[3] See letter from Armand de Quatrefages, 25 March 1863 and nn. 4–6. There is no record that CD ever deposited such a collection in the British Museum (British Museum Zoological Accession Books, Natural History Museum). See also n. 4, below.
[4] No letter to Martins has been found.
[5] CD refers to Quatrefages 1861 and to the Anthropological Society of London (see letter from Armand de Quatrefages, 25 March 1863 and n. 8).
[6] CD's annotated copy of Quatrefages 1862 is in the Darwin Library–CUL (see *Marginalia* 1: 691).
[7] In his *Métamorphoses* (Quatrefages 1862), Quatrefages cited works by numerous British naturalists, including CD's study of cirripides (*Living Cirripedia* (1854); see Quatrefages 1862, p. 110 n.). Quatrefages repeatedly referred with praise to the work of Thomas Henry Huxley and John Lubbock (see, for example, Quatrefages 1862, pp. 165 n., 166, 170 n., 183, 189 n., 190 n., 258 n., 307 n.). He cited them extensively in his chapter entitled 'Théorie de la parthénogenèse', referring to them as 'deux des savants qui ont jeté le plus de jour sur cette question [two of the scholars who have thrown most light on the question]' (p. 281). Lubbock spent his youth at the family home of High Elms, about a mile and a half from Down House. CD encouraged his interest in science in the close personal relationship that grew between them.

From Joseph Wolstenholme to William Erasmus Darwin [27 March 1863?][1]

Christ's College,
Friday.

My dear Darwin,

I am very much ashamed of myself for the enormous time I have allowed to go over before answering your letter of last October term.[2] I was at the time grievously busy with preparing for my Senate House work,[3] but can offer no satisfactory excuse for the time elapsed since the end of January, when my labors in that respect ceased. I am not quite confident that I know sufficiently the working now-a-days of the two large colleges to offer an opinion worth very much. When I was an undergraduate, I have no doubt that a moderately good man got better trained in Mathematics at ⟨John's than⟩ at Trinity, while for the very best men there was then little to choose.[4] The first men at any college are nearly sure to be well cared for by both the college & private tutors. I rather fancy that the lecturing at John's now is much less conscientiously done than it was in my time, & I think that, in saying this, I am not merely judging from results though the last few years Triposes would certainly favor this opinion. There is however no doubt that the Math. Lecturing, & general management, at Trinity, has very greatly benefited by the great infusion of new blood into their staff the last few years. For Classical men, either moderately good, or first rate, I really think that we at Christ's may fairly claim to make more of our men than any other college. Shilleto[5] goes so far as to say, that on the average our men are a class higher than if they had gone elsewhere. And, though of course it shocks my modesty to say so, I think a good mathematician gets ⟨quite as⟩ well attended to here, in College lectures, as at any other place whatever, better, perhaps, than any but Caius.

I do not think there is anywhere any fear of damage to health by over-reading now-a-days. It is a tradition of past times altogether, & I am rather inclined to believe we have carried our reforms in that matter too far. Earnest students are far too rare, in any of the subjects we profess to teach, & the lovers of study for its own sake are extinct.

Trinity Hall has this year been blooming like an aloe, but I think the result rather accidental. They were the first to offer open scholarships, & their two men, who have certainly taken the two best degrees of the year, were their early harvest.[6] On the whole, if a large college be preferred, I should certainly now incline to Trinity, & of the small, I do not think an⟨y⟩ ⟨o⟩ught to be preferred to ours, unless very dubiously Caius, though against ⟨the⟩ race of men Caius breeds I have myself a prejudice, perhaps unjustifiable.

I am just packing up to be off to the Lakes tomorrow morning for a fortnight.[7] I wish you were with us again.[8] We are rather a large party, *[and]* I think mean to invade Wastdale on Sunday or Monday.

I am, | My dear Darwin | Yours faithfully, | Joseph Wolstenholme

DAR 181: 140

[1] The date is conjectured by reference to Wolstenholme's advice, which apparently concerned George Howard Darwin's choice of a Cambridge college in the winter of 1862–3 (see n. 2, below), by his mention of the 'time elapsed since the end of January', and by his reference to the end of the Cambridge University term and his plans for the vacation (see n. 7, below); in 1863, the university's Lent Term, which commenced in January, ended on Friday 27 March (*Cambridge University calendar* 1863).

[2] William graduated with a BA from Christ's College, Cambridge, in 1862, taking mathematical honours. He had apparently written to Wolstenholme, who was assistant tutor at Christ's College from 1852 to 1869 (*DNB*), to ask for advice on CD's behalf regarding George Darwin's choice of college at Cambridge University. George, who was seventeen years old, attended Clapham Grammar School in south-west London, where he had shown himself to be gifted in mathematics (see *Correspondence* vol. 10, letter to T. H. Huxley, 18 December [1862]). In a letter to William of [18 November 1862], which is in DAR 219.1: 67, Emma Darwin reported: 'G. to my surprize seems to wish to go to S[t] John's & there are such a number of scholarships that he is going to try for one & I should think he was likely enough to succeed'. In the event, George was not successful in obtaining a scholarship

at St John's in 1863, and remained at Clapham for an additional year (see letter to Edward Cresy, 13 May [1863]).

3 Wolstenholme probably refers to his preparing students for the written examinations of the Mathematical Tripos (final mathematical examination), which took place in the university's Senate House, early in January. In addition, Wolstenholme was one of the examiners for the Mathematical Tripos in 1863 (*Cambridge University calendar* 1863).

4 Wolstenholme matriculated at St John's College, Cambridge, in 1846; he graduated BA in 1850 as third wrangler (that is, third in the Mathematical Tripos) (*Alum. Cantab.*).

5 Richard Shilleto was a well-known Greek scholar (*Alum. Cantab.*).

6 In 1863, the senior wrangler was Robert Romer, from Trinity Hall. The next highest placed student from Trinity Hall in the Mathematical Tripos was Bryan Walker, the ninth wrangler, who was also eighth in the Classical Tripos for 1863, a considerable achievement (*Cambridge University calendar* 1863).

7 This was probably a reading party organised between the Lent and Easter terms. In 1863, the Lent Term ended on 27 March, and the Easter Term began on 10 April (*Cambridge University calendar* 1863).

8 William had accompanied Wolstenholme on a reading party in the Lake District during May or June 1859 (see *Correspondence* vol. 7, letter to W. E. Darwin, [5 May 1859] and n. 6).

From J. D. Hooker [28 March 1863][1]

Kew
Saturday

Dear Darwin

I send Athenæum by todays post, which please return. I thought you would like to see the first article![2]

I wish I could see any way of "ingenious wriggling" that would remove the crushing evidence in the shape of tropical forms—against tropical cold.[3] You have no idea of the magnitude of such a case as Dipterocarp**eæ** in Nat. Ord—not a mere genus, of 10 genera, & 112 species all from Ceylon, the Malayan Peninsula & Islands,—& of which a good 100 more species, & many more genera, are still to come from Borneo, Sumatra &c.— All are woody & for the larger proportion are huge timber trees— not one ascends at all to any height.—& analogous species to living are found in tertiary coal beds of Labuan &c.[4]

Zingiberaceæ & Marantaceæ are as bad cases in the Herbaceous way & there are many *many* others. I shall work out the Cameroon's case carefully & put the pros & cons in as strong a light as I can— I am very sorry for you, but what can I do or say!—[5]

A thousand thanks for your explanation about Reversion in which I am sure I shall go the whole hog with you; it is a subject on which I have a huge *latent* interest—[6] I am glad you pitched into Lyell about the passage regarding genius,—he will surely think we were in collusion! about his book.[7]

We went to Lord Dundreary last night & *roared*, he is a far more scientific character than I anticipated—not a broad comedy character, but very charming in his fatuity, nonchalance, ignorance, & his incoherent wit:[8] for the first time in my life

I fell in love with the character of "diverting vagabond" quite a new sensation—I assure you— he is not a puppy, like Albert Smith *was*.[9]

Please leave the Medallion any-where in town when you are sending up, & I will send for it.—[10]

Ever yours | J D Hooker

I have been awfully dissipated of late—out every night. We think of Weymouth & Guernsey on Thursday with 2 boys[11]

I wonder if your Willy would like to join us and botanize in Guernsey?— if you think so I will write to him[12]

We would cross from Southampton

N B. if the weather is bad we shall not cross, but take a cruise in New forest

DAR 101: 121–2

[1] The date is established by the relationship between this letter and the letters to J. D. Hooker, 26 [March 1863] and from J. D. Hooker, [31 March 1863]; the intervening Saturday was 28 March.

[2] Hooker refers to Richard Owen's anonymous review of Carpenter 1862, which appeared in the *Athenæum*, 28 March 1863, pp. 417–19 (for the attribution, see the publisher's marked copies of the journal at City University Library, London). The review, which is reproduced in Appendix VII, contained comments critical of CD's account of the origin of life and the origin of species (see letter to J. D. Hooker, [29 March 1863] and nn. 5 and 6).

[3] Hooker refers to CD's claim that the tropics must have been cool enough during the Pleistocene glacial period for temperate species to have migrated between the northern and southern hemispheres, a claim that Hooker disputed on the grounds that many tropical plants could not have survived in such conditions (see letter to J. D. Hooker, 26 [March 1863]). CD's capacity for 'ingenious wriggling' with regard to theoretical difficulties had been a joke between the two friends over a number of years (see, for example, *Correspondence* vol. 10, letter from J. D. Hooker, 7 November 1862 and n. 17).

[4] See letter to J. D. Hooker, 26 [March 1863] and n. 9. Labuan is an island off the north-west coast of Borneo.

[5] In the introduction to his flora of the temperate regions of the Cameroons mountains (J. D. Hooker 1863b, pp. 176–80), Hooker briefly characterised, under itemised headings, the extent of the overlap between the genera and species of the Cameroons mountains and those of various other temperate regions. He noted that the large proportion of plant genera in the Cameroons mountains that were also found in Europe was 'most singular', given the 'total isolation' of the regions by 'hot, low deserts', and he mentioned two hypotheses that could render the phenomenon explicable (p. 181). The first was CD's theory of migration during a global cold period, and the second was Hooker's preferred theory of transport by aerial currents and birds; however, Hooker did not clearly state which of the two theories he considered was favoured by the evidence. See also letter from J. D. Hooker, [15 March 1863] and n. 21.

[6] See letter to J. D. Hooker, 26 [March 1863].

[7] See letter to J. D. Hooker, 26 [March 1863] and n. 12.

[8] Hooker refers to Tom Taylor's comedy, *Our American cousin*, which had been running at the Haymarket Theatre, London, since 1861. The main feature of the play was Edward Askew Sothern's portrayal of the character Lord Dundreary, an English aristrocrat with a 'well-bred air married to a vacant stare', absurd mannerisms, twisted aphorisms, and inane lines. Sothern substantially rewrote the part for himself until it grew into a series of monologues and 'all but dwarfed the remainder of the play', creating one of the most celebrated comic parts in nineteenth-century theatre (Tolles 1940, pp. 173–80).

[9] Puppy: 'a vain, empty-headed, impertinent young man; a fop, a coxcomb' (*OED*). Between 1850 and 1860, the writer and dramatist Albert Richard Smith wrote and performed in a number of 'entertainments' in London, descriptive of his journeys in Asia (*DNB*).

[10] CD had agreed to send Hooker his Wedgwood medallion of Erasmus Darwin so that it could be copied for the museum at the Royal Botanic Gardens, Kew (see letter to J. D. Hooker, 26 [March 1863]).

[11] The reference is to Frances Harriet, William Henslow, and Charles Paget Hooker; Hooker's party travelled to Weymouth, Portland, Jersey, and Guernsey, returning home on 17 April 1863 (see letter from J. D. Hooker, 20 April 1863).

[12] CD's eldest son, William Erasmus Darwin, was a banker in Southampton. Since his removal to Southampton in 1861, CD had encouraged William to study botany, and had repeatedly asked him to carry out observations (see *Correspondence* vols. 9 and 10). In June 1862, William had visited Hooker at Kew (see *Correspondence* vol. 10, letter from J. D. Hooker, 19 [June 1862], and letter to J. D. Hooker, 23 June [1862]).

To Daniel Oliver 28 March [1863][1]

Down Bromley Kent
March 28th

Dear Oliver

You have indeed sent me some interesting specimens. Every part of the base of the flower of Edwardsia seems to secrete a surprising quantity of nectar: I thought from what Treviranus says it did not secrete till stamens or petals were broken off, which would have been a very surprising fact.—[2] What a very curious stamen you sent me: it would be fine sport to observe the living plant.[3] Can it serve as protection to stigma: in a Nierembergia, I saw the stamens formed into a sort cupola over the stigma; but I neglected to get the plant for observation. What singular bracts of the Marcgraviaceæ; if I can make anything of the Pitcher plant, which I do not suppose I shall, then Marcgraviæ would be interesting to observe.—[4]

I will send bottle with monstrous Primrose on Monday or Tuesday,[5] with the Medallion for Hooker of Dr. Darwin, (please tell him):[6] I have seen one flower with 4 pistils;[7] there is one in bottle one specimen with corolla removed. I remember I observed that 9 or 10 bundles of spiral vessels ran up in outer case of ovarium & up pistil, & none ran up the central needle, prolonged up the middle of pistil in growth from the placenta. I concluded at time that the single pistil is really ten pistils.— I had never heard of Caspary, but speculate whether what you say is the placenta, was not an inner whorl of pistils without stigmas.—[8]

With many thanks Yours very sincerely | C. Darwin

I have been looking at anther again, would it not force insects to approach only on one side & so rub against the side of anthers which dehisce?[9]

DAR 261 (DH/MS 10: 43)

[1] The year is established by the relationship between this letter and the preceding letter; see also n. 2, below.

[2] See letter to Daniel Oliver, 24–5 March [1863], and letter from Daniel Oliver, [27 March 1863]. See also letter to J. D. Hooker, 24[–5] February [1863] and n. 19. The reference is to Treviranus

1863a, p. 10; CD's copy of this paper is in the Darwin Library–CUL. A note in DAR 49: 84, dated 28 March 1863, records CD's observations on the specimens sent by Oliver:

> *Edwardsia tetraptera*— anther protruded from carina even in young flower & stigma— An *immense* quantity of nectar in drops ['nea' *del*] at bases of all petals & all stamens; this even in rather early flower hardly fully expanded— I scarcely ever saw so much.— When petals and stamens broken off ['not' *del*] no soon secretion from stumps.— I daresay old flower goes on secreting— What rubbish of old Treviranus.—

3 Oliver had apparently sent CD one of the anthers with a '*curious crest*' that were typical of species of *Amomum*, on which he was working (see letter from Daniel Oliver, [27 March 1863]).

4 See letter to Daniel Oliver, 24–5 March [1863] and n. 6. In 1864, CD obtained information about *Marcgravia* and *Nepenthes* (the pitcher plant) in the context of his work on climbing plants (see, for example, *Correspondence* vol. 12, letter to J. D. Hooker, 26[–7] March 1864, and letter from Richard Spruce, 29 July 1864).

5 See letter to Daniel Oliver, 24–5 March [1863], and letter from Daniel Oliver, [26 March 1863].

6 See preceding letter and n. 10. The reference is to Erasmus Darwin.

7 CD's notes recording this observation, dated 24 March 1863, are in DAR 108: 165 v.

8 See letter from Daniel Oliver, [26 March 1863] and n. 3. The reference is to Johann Xaver Robert Caspary.

9 CD refers to the specimen of *Amomum* (see n. 3, above).

From Armand de Quatrefages[1] [28 March –] 11 April 1863[2]

Monsieur et cher confrère

Comme je vous le disais je crois dans ma derniere lettre je soutiens en ce moment à la Société d'Anthropologie une lutte assez vive en faveur de la variabilité des espèces animales.[3] Un de mes confrères m'a défié de citer *une seule espèce* chez qui l'action des milieux eut produit des modifications *anatomiques*.[4] Il ne compte pour rien celles qui se rattachent à la peau, au tissu cellulaire ... Je vais repondre ⟨ ⟩ en insistant sur l'histoire du cheval.[5] Mais on ne s'arrêtera pas là et je voudrais placer sous les yeux de ces disciples de St Thomas un fait bien concluant[6]

J'ai songé aux *bœufs-dogues* que vous nous avez fait connaitre, aux bœufs Nãta ou Niata.[7] Vous en avez dites vous (Journ. of Research ...) une tête à Londres.[8] Pourriez vous en faire prendre une photographie et me l'envoyer?[9] Serait-il encore mieux possible d'en obtenir un moule pour les collections du Museum?[10] a-t-elle été publiée quelque part?[11] Si elle ne l'a pas été et que vous ou quelqu'un en Angleterre avez l'intention de le faire, je n'ai pas besoin de vous dire qu'aucune publication detaillée n'aura lieu en France et que je me bornerai à montrer soit le dessin soit le moule pour l'instruction de mes auditeurs et la conversion (si possible est) de mes contradicteurs.[12]

Pardon Monsieur et cher confrere de ce que mes demandes peuvent avoir d'indiscret et recevez de nouveau l'expression de mon bien affecteux dévouement | De Quatrefages

11 Avril 63

J'ai vu Mr Martins ⟨ ⟩ en ce moment à Paris pour la réunion des délégués des facultés de Province.[13] Je lui ai donné toutes les instructions, que mon habitude de ces petits details pouvaient me sugérer, pour que votre collection fut aussi bien faite que possible.[14] Je l'ai trouvé rempli d'excellentes intentions. Il a pris votre adresse et vous enverra directement les objets par le chemin de fer[15]

J'ai eu aussi ces jours derniers un veritable plaisir a causer longuement avec le Dr. Carpenter. ses travaux sur les Comatules m'ont très vivement interessé.[16] Merci de tout ce que vous me dites d'aimable ⟨a⟩u sujet de mes Métamorphoses[17]

DAR 175: 3

1 For a translation of this letter, see Appendix I.

2 The date range is established by the relationship between this letter and the letter to Armand de Quatrefages, 27 March [1863].

3 See letter from Armand de Quatrefages, 25 March 1863 and n. 10.

4 The reference is to André Sanson, and to a meeting of the Société d'Anthropologie held on 19 March 1863 (*Bulletins de la Société d'Anthropologie de Paris* 4 (1863): 159). For Quatrefages's responses to Sanson of 16 April and 16 July, see *Bulletins de la Société d'Anthropologie de Paris* 4 (1863): 213–23, and nn. 5 and 12, below. For Sanson's responses to Quatrefages of 7 May and 16 July, see *Bulletins de la Société d'Anthropologie de Paris* 4 (1863): 254–68, 374–5, and 380–3.

5 Quatrefages described the modification of breeds of horses through environmental influences at a meeting of the Société d'Anthropologie on 16 April 1863, citing the descent of the English racehorse from Arab stock as an example (*Bulletins de la Société d'Anthropologie de Paris* 4 (1863): 215–19). See also *Origin*, p. 35.

6 The reference is to the biblical narrative of the apostle Thomas's refusal to believe in the resurrection of Jesus Christ without direct experience (*EB*).

7 Niata: 'An abnormally small variety of cattle, found in South America' (*OED*). Quatrefages refers to *Journal of researches* 2d ed., pp. 145–6, where CD observed that the niata cattle 'appear externally to hold nearly the same relation to other cattle, which bull or pug dogs do to other dogs.'

8 *Journal of researches* 2d ed., p. 146. CD presented a skull of this breed from the *Beagle* collections to the Royal College of Surgeons of London in 1840 (see *Correspondence* vol. 4, letter to Francisco Javier Muñiz, 26 February 1847, n. 7).

9 CD sent photographs of the niata skull in May (see letter to Armand de Quatrefages, 14 May [1863], and letter from Armand de Quatrefages, 19 May [1863]).

10 Quatrefages was professor of the natural history of man at the Muséum d'Histoire Naturelle in Paris (*DSB*); apparently, no cast of the skull was sent (see letter to Armand de Quatrefages, 14 May [1863], and letter from Armand de Quatrefages, 19 May [1863]).

11 The skull is described in Owen [1853], 2: no. 3832; CD had also discussed it in his manuscript draft of *Variation* (see *Variation* 1: 89–91, and letter to Armand de Quatrefages, [14 April 1863]). See also *Origin* 6th ed., p. 177.

12 Quatrefages displayed the photographs at a meeting of the Société d'Anthropologie on 2 July 1863 (*Bulletins de la Société d'Anthropologie de Paris* 4 (1863): 350–2). See also letter from Armand de Quatrefages, 19 May [1863].

13 Charles Frédéric Martins was professor of botany at the Faculté de Montpellier (*Dictionnaire universel des contemporains*).

14 Quatrefages completed his own research on silkworms between 1857 and 1859 (see *DSB* and Quatrefages 1859).

15 Martins had agreed to send CD a collection of varieties of the common silkworm (*Bombyx mori*) at different stages in their life cycle; however, the collection was apparently never sent (see letter to Armand de Quatrefages, 27 March [1863] and n. 3).

16 William Benjamin Carpenter had travelled to Paris for recreational purposes during his Easter vacation as registrar of London University (Carpenter 1888, p. 84; *DNB*). However, he soon became

involved in the heated discussions that followed the discovery on 28 March 1863 of a fossilised human jaw at Moulin-Quignon in the Somme valley, discussions in which Quatrefages also played a leading role. See Carpenter's report of the discovery in the *Athenæum*, 18 April 1863, p. 523; see also Grayson 1983, pp. 213–17, and Van Riper 1993, pp. 134–39. Carpenter published his work on the echinoderm *Comatula* in Carpenter 1865.

17 See letter to Armand de Quatrefages, 27 March [1863]. The reference is to *Métamorphoses de l'homme et des animaux* (Quatrefages 1862), of which there is an lightly annotated copy in the Darwin Library–CUL (see *Marginalia* 1: 691).

To W. E. Darwin 29 [March–June 1863 or 1864?][1]

29th

My dear William.

I never saw such an organ as that which you draw;[2] I sh^d. very much like to see a specimen.— Does the cup secrete any viscid matter? Please send the spec. in tin cannister or box with a little damp blotting paper. Botanists call all out of the way organs "glands" & no doubt they would call this a gland.—

Your affect | C. Darwin

I do **not** want Oxalis.[3]

Are little insects ever caught in the cup??

DAR 210.6: 94

1 The date range is conjectured by the reference to *Oxalis* (see n. 3, below).

2 Neither the sketch, nor the letter with which it was enclosed, have been found. The organ has not been identified.

3 CD made a series of observations on *Oxalis*, beginning in April 1862 and ending in 1869; his experimental notes on this genus are in DAR 109: B3–27, B101, B108, and DAR 111: A44–48, B33–5, B43. CD's experiments on species of *Oxalis* native to Britain focused chiefly on *O. acetosella*, and were concentrated in the spring and summer of 1863 and 1864 (DAR 109: B3–7 and DAR 111: A44, A46–8). CD's interest in the occurrence of dimorphism in species of *Oxalis* began in 1861 (see *Correspondence* vol. 9, letters to George Bentham, 24 November [1861] and 30 November [1861]). He pursued investigations on heterostyly (see, for example, *Correspondence* vol. 10, letters to Daniel Oliver, 20 [April 1862], 24 April [1862], and this volume, letter to Roland Trimen, 23 May [1863]), and on cleistogamy (see, for example, letter to *Journal of Horticulture and Cottage Gardener*, [17–24 March 1863] and n. 7, and letter to Asa Gray, 31 May [1863]). CD published the results of his observations in *Forms of flowers*, pp. 169–183, 321–4.

To J. D. Hooker [29 March 1863]

Down
Sunday night

My dear Hooker

It is very good of you to think of William;[1] but I feel sure that he could not spare the time; barring the sea-sickness, I am sure he would have enjoyed it greatly. We will, however, tell him. If you go to Southampton & have spare time, beat up his quarters: he lodges at 1 Carlton Terrace; the little Bank is in heart of town.[2] I am glad you are so abandoned & dissipated.[3]

Many thanks for Athenæum, received this morning & to be returned tomorrow morning.[4] Who would have ever thought of the old stupid Athenæum taking to Oken-like transcendental philosophy written in Owenian style! It will be some time before we see "slime, snot or protolasm" (what an elegant writer) generating a new animal.[5] But I have long regretted that I truckled to public opinion & used Pentateuchal term of creation, by which I really meant "appeared" by some wholly unknown process.—[6] It is mere rubbish thinking, at present, of origin of life; one might as well think of origin of matter.—

The medallion will be sent with monstrous Primulas to Oliver.—[7]

I am unhappy about those horrid hot plants.[8] When one can say nothing it is best to declare solemnly "Magna est veritas et prevalebit".—[9]

I am sure there are cases of real tropical plants which to surprise of gardeners have been found to stand much cooler climates than could have been expected. Do think over such cases.— I wish Berberis Wallichii did not belong to that genus. Is Cassia partially a case. Think of Tiger— Think of Glacial Elephant & Rhinoceros, as far as genera are concerned.—

Goodnight— May you have a pleasant trip— goodnight | C. Darwin

Thwaites has sent me from Ceylon, I suppose through you, two splendid specimens of reciprocally dimorphic plants like Primula. One is Limnanthemum Indicum; & the other Sethia.—[10]

Endorsement: 'March 29/63.'
DAR 115: 189

1 In his letter to CD of [28 March 1863], Hooker suggested that William Erasmus Darwin accompany him to botanise on the island of Guernsey.

2 William was a partner in the Southampton and Hampshire Bank, 25 High Street, Southampton.

3 See letter from J. D. Hooker, [28 March 1863].

4 Hooker had sent CD his copy of the *Athenæum* of 28 March 1863 (see the letter from J. D. Hooker, [28 March 1863]), thinking CD would be interested to see an anonymous review of Carpenter 1862 (*Athenæum*, 28 March 1863, pp. 417–19). Hooker later discovered the review to be by Richard Owen (see letter from J. D. Hooker, [31 March 1863] and n. 2). Owen's review is reproduced in Appendix VII.

5 CD refers to the German naturalist and leading exponent of the romantic *Naturphilosophie*, Lorenz Oken, by whom Owen had been greatly influenced. In his review, Owen followed Oken in arguing that microscopic organisms like the Foraminifera were spontaneously generated on the beds of seas, lakes, and rivers, by the effect of a 'general polarizing force' on the 'slime' from dead and decaying organisms (*Athenæum*, 28 March 1863, p. 417; see also *DSB*, s.v. 'Oken, Lorenz', and Rupke 1994, pp. 175, 251–2, 309–10). Owen's reference to Foraminifera as 'Aggregates of slime, snot, or "protoplasm"' appeared on p. 418.

6 CD refers particularly to Owen's derisive statement that CD 'could only express' the creative force responsible for the origin of life 'in Pentateuchal terms as the primordial form "into which life was first breathed"!' (*Athenæum*, 28 March 1863, p. 419; see *Origin*, p. 484). Owen had sought to use this argument to impugn CD's commitment to naturalistic explanation on several occasions since 1859 (see Rupke 1994, pp. 239–40 and 251–2). CD omitted this phrase from the third edition of *Origin* (see Peckham 1959, p. 753). See letter to *Athenæum*, 18 April [1863].

7 See letter to Daniel Oliver, 28 March [1863].

8 See letter from J. D. Hooker, [28 March 1863] and nn. 3–5.

9 Magna est veritas et praevalebit: 'truth is great and will prevail' (*Chambers*); also given with 'prevalet' (prevails), in I Esd. (*Vulgate* III Ezra) 4:41.
10 See following letter.

To G. H. K. Thwaites 30 March [1863][1]

Down. | Bromley. | Kent. S.E.
March 30th

My dear Mr Thwaites

I thank you most cordially for the specimens which have interested me exceedingly.[2] The Sethia is a beautiful case of what I shall call "reciprocal dimorphism": unfortunately all the anthers were knocked off the long-styled, so I could not compare pollen, except from a bud in which pollen was not mature, & I dare not trust my measurements.[3] If you were to watch insects visiting these flowers I would wager that you would see body dusted by long stamens, & probosis by short anthers; & the object of unequal length of the alternate stamens is to get pollen on to proboscis from more than a pair of anthers; in short to rake them. I judge this from what I have seen in Lythrum.[4] Are there any Lythraceæ in Ceylon; you would be likely to find *three* forms, reciprocally related in sexual functions, in this order; & with two kinds of pollen in each flower. If you observe any more facts, I shd. be grateful for information & specimens. I have also been *very* glad to see the Limnanthemum: it, also, is a very pretty case.[5] I have got a hot-house: could I grow it? & if so, could you send me seed?[6]

Any flower with half anther of one colour & half of other colour (if not due to abortion of one set) probably indicates dimorphism or *tri*morphism.—[7]

I much wish to experiment on one plant in each order.— Sethia I presume is a tree.

I will soon send you a little paper on dimorphism of Linum.—[8]

Thank you for telling me about Discospermum; it is an important fact.—[9]

You have hit the nail on the head in supposing that I am interested about galls (& what curious ones those are that you have sent me!); but I unfortunately know hardly anything about them; nor do I know where I can find any full account. The subject is most curious, & I have often reflected on it, & at one time thought I would attempt a series of experiments. It is truly wonderful what a change a little poison or irritation has effected, & effected in so diversified a manner.—[10]

With respect to the curious specimens, which you so kindly have sent me of Gomphia & Lasemia, I do not know what to think.[11] Are they not the effect of insect puncture or disease, like those we see on our Birch-trees? I shd. fear to class them with what gardeners' call "sports" or bud variation. Can you illuminate me?

With cordial thanks | Dear Thwaites | Yours very sincerely | C. Darwin

American Philosophical Society (293)

[1] The year is established by the relationship between this letter and the letter from G. H. K. Thwaites, 17 February 1863.
[2] At CD's request, Thwaites had sent CD specimens of the dimorphic flowers of *Sethia acuminata* and *Limnanthemum indicum* with his letter to CD of 17 February 1863. See also preceding letter.
[3] A note in DAR 110: B28, dated 29 March 1863, records CD's observations on Thwaites's dried *Sethia* specimens. CD published his observations, including an estimate of the relative size of pollen from the two forms of flower, in *Forms of flowers*, p. 122.
[4] CD had begun experimenting on the trimorphic plant *Lythrum salicaria* in 1862 (see *Correspondence* vol. 10); he repeated this explanation of the purpose of the two lengths of stamens in *Sethia* in *Forms of flowers*, p. 122.
[5] CD's undated observational notes on these specimens are in DAR 110: B9–10. He published his observations in *Forms of flowers*, p. 116.
[6] Thwaites sent CD seed of *Limnanthemum* with his letter of 8 June 1863.
[7] CD had begun to consider the possibility that the occurrence of differently coloured anthers or pollen in the same or in different flowers of the same species might be 'a safe guide to dimorphism' (see *Correspondence* vol 10, letter to J. D. Hooker, 22 [August 1862] and n. 14).
[8] CD's paper, 'Two forms in species of *Linum*', was read before the Linnean Society on 5 February 1863. It was published in the number of the *Journal of the Proceedings of the Linnean Society (Botany)* that was issued on 13 May 1863 (*General index to the Journal of the Linnean Society*, p. vi). However, CD obtained a number of offprints of the paper in mid-April for distribution to those he thought would be interested; Thwaites's name appears on CD's presentation list (see Appendix IV).
[9] See letter from G. H. K. Thwaites, 17 February 1863 and n. 3.
[10] Thwaites apparently discussed this subject in the missing portion of his letter to CD of 17 February 1863. CD discussed galls in a section of *Variation* dealing with the 'direct and definite action of the external conditions of life' (*Variation* 2: 282–4). He observed of the specimens Thwaites sent him: 'some were as symmetrical as a composite flower when in bud, others smooth and spherical like a berry; some protected by long spines, others clothed with yellow wool formed of long cellular hairs, others with regularly tufted hairs' (*Variation* 2: 282).
[11] In his letter to Thwaites of 29 December [1862] (*Correspondence* vol. 10), CD had asked Thwaites for instances of what he termed 'bud-variations'. The part of Thwaites's reply dealing with this subject has not been found (see letter from G. H. K. Thwaites, 17 February 1863); however, it appears that Thwaites provided specimens of *Lesemia* and *Gomphia* exhibiting what he considered to be 'bud-variations'.

From J. D. Hooker [31 March 1863][1]

Kew
Tuesday.

Dear Darwin

One line to say that the article is surely Owens & none others.[2] I was certain of it on reading it, the expressions are identical with those of the Ed^r. Review,[3] & the whole method of traduction is typical of Owen & unapproached by any one else.— so are the longwinded phrases, & so is the unutterable rubbish of the doctrine. Good Lord *what* rubbish & what vulgarity. I think there must be truth in a report I once heard—that he tippled.— Your doubt caused me to hunt up Carpenter today,[4] who assures me it is so, both from his judgement, & from a conversation he had with Hepworth Dixon on Saturday about it—[5] he is going to answer it—mildly.[6]

Froude[7] told me today that Carlyle[8] is a violent anti-Darwinian—of course it is of no consequence, but I should not have expected it.

Ever Yours affec | J D Hooker

I have just got sent a splendid Wedgwood of Dr Johnson![9]

DAR 101: 126–7

[1] The date is established by the relationship between this letter and the letter to John Lubbock, 5 April 1863, and by the reference to the anonymous review of Carpenter 1862 in the *Athenæum* of 28 March 1863, pp. 417–19; the intervening Tuesday was 31 March.

[2] Hooker refers to Richard Owen's anonymous review of Carpenter 1862 in the *Athenæum*, 28 March 1863, pp. 417–19 (see letter to J. D. Hooker, [29 March 1863]). The attribution is confirmed by the publisher's marked copy of the *Athenæum* (City University Library, London). Owen's review is reproduced in Appendix VII.

[3] Hooker refers to Owen's anonymous review of *Origin* in the *Edinburgh Review* ([Owen] 1860a).

[4] William Benjamin Carpenter.

[5] William Hepworth Dixon was editor of the *Athenæum* (*DNB*).

[6] Part of Carpenter's response to Owen's review was published in the *Athenæum*, 18 April 1863, p. 525.

[7] James Anthony Froude.

[8] Thomas Carlyle.

[9] Samuel Johnson. Hooker was an enthusiastic collector of Wedgwood ware (see letters from J. D. Hooker, 6 January 1863, [16 February 1863], [6 March 1863], and [24 March 1863]).

From Richard Trevor Clarke [April? 1863][1]

Dear Mr Darwin

Enclosed are blossoms of the individual hautbois which I have used as female parent in my experiments.[2] The anthers are certainly only rudimentary. I may mention it is very possible that the microscope may shew some grains ⟨of⟩ pollen, other blossoms on the same spike and perhaps th⟨ese⟩ having been fertilized by myself.—

I am glad you received the plants safe.[3] I propose in future when any little curious fact turns up to send you a short memorandum, knowing that *working* men dont like too much correspondence

Very truly yours | R Trevor Clarke

⟨6⟩3 Eaton Square | for 10 days[4]

DAR 161: 164

[1] The conjectured date is the earliest possible for the flowering of strawberries subsequent to the letters from R. T. Clarke, [after 25 November 1862] and [after 27 November 1862] (*Correspondence* vol. 10). See n. 2, below.

[2] Clarke described his experiments with strawberry crosses in his letter to CD of [after 25 November 1862] (*Correspondence* vol. 10). See also *Variation* 1: 352.

[3] Clarke had sent strawberry plants to CD in November 1862 (see *Correspondence* vol. 10, letter from R. T. Clarke, [after 27 November 1862]). CD's reply has not been found.

[4] A fellow army officer of Clarke's, Major-general Alexander Murray Tulloch, lived at 63 Eaton Square, London (*Post Office London directory* 1863).

From James Anderson 1 April 1863

Meadow Bank, Uddingstone N.B.
1/4/63

Mr Darwin Dear Sir

I enclose for your inspection a few of the seeds of the abortive Cattleya crispa flower as well as a few of Lælia cinnabarina from a perfect flower without manipulation;[1] both of which have been ripe for the last two months. The Den. Cretaceum pods are yet unripe but when they are so I shall submit a few of the seeds for your inspection for I look upon them with more interest having watched the anomalous proceeding with more interest than the Cattleya flower,[2] some of the phases in the progress of which not being so closely watched, as it being the first instance of the kind that came under my cognisance I supposed it would be an entire failure. I have also sent pkts. to M[r] Gosse and I hope you will "report progress"[3]

I am exceedingly glad to embrace this opportunity of bearing testimony to your unwearied and painstaking abilities in going so minutely into what may appear at first a more difficult problem than what it really is.

I have great hopes of succeeding in rearing thousands of my "crosses" for I have now gone into the fertilisation of them with earnestness and from the results of practice I can with confidence reckon upon effecting a cross, if the pollen masses be taken off at the proper time and applied to a flower in a proper condition. We have hundreds of imported seedlings of a species of Rodriguesia that have come safe home on the blocks which they had germinated upon in Trinidad;[4] and I don't see why we should not be able to reproduce species and varieties here.[5]

I have perused your reply in J. of H with much pleasure and the more I know of the seed pods the more I am convinced that it is *almost* futile attempting to rear seedlings unless from flowers that have been *artificially* dealt with. I confess my inability in a dozen of trials with casual pods to produce a healthy germination. I, too, like M[r] Gosse, have had a very few swelling almost to the size of the "old" pin head but they too dissappeared. However, I should not like to be too dogmatical on this point.

I shall be glad to hear from you after you have analysed the chaff (?)

I am | D[r] Sir | Yours truly | Jas Anderson

P.S. By way of experiment, I gave a large plant of Phaius Grandiflorus a good 'shaking' to try if irritation would do anything in the way of inducing seed and I find some half dozen pods upon the plant. I was induced so to do as probably you may be aware that Erica elegans will not *cross* with any other species of heath neither will it set a *single pod* unless some artificial mode be adopted; but by shaking the plant the pollen will disperse and the plant thereupon will produce seed in abundance.[6]

J. A.

DAR 159: 59

[1] In the *Journal of Horticulture*, 17 March 1863, pp. 206–8, Anderson had asked CD to comment on cases like the 'abortive' flowers of the orchids *Cattleya crispa* and *Dendrobium cretaceum*, which did not open, yet still formed seed capsules. CD's reply was published in the *Journal of Horticulture*, 31 March 1863, p. 237; he asked Anderson to send him seeds produced by the unopened flowers (see letter to the *Journal of Horticulture and Cottage Gardener*, [17–24 March 1863]). In a note (DAR 70: 110), CD recorded:

> On or about March 18 —63 in Cottage Gardener Mr Anderson publishes account of certain orchids with abortive (unopened!) flowers setting seed-pods. He has sent me seeds *of such Cattleya crispa [*interl*] for examination, & not more [*above del* 'more'] than ['one in' *del*] $\frac{1}{100}$ or $\frac{1}{200}$ is anything [*above del* 'more'] than an [*above del* 'an'] empty testa; the few others have perfect embryo=nucleus— He sent me packet of seeds of Lælia cinnabrina, from plant which had not been manipulated (possibly visited by insect?) & here about $\frac{1}{4}$ were good.—

[2] See n. 1, above. Anderson sent CD one of the seed capsules produced by the *Dendrobium cretaceum* in June 1863 (see letter from James Anderson, 18 June 1863, 'Fertilization of orchids', p. 152 (*Collected papers* 2: 149), and *Orchids* 2d ed., p. 142).

[3] In the *Journal of Horticulture*, 24 March 1863, p. 217, Philip Henry Gosse published a response to Anderson's article, noting the viability of the *Odontoglossum* seed that Anderson had sent him. Since taking up residence in Devonshire, the zoologist Gosse had become interested in orchid cultivation and had 'formed a remarkable collection' (*DNB*).

[4] Since *Rodriguezia* species are epiphytic, the seedlings were probably growing on blocks of wood when transported.

[5] Anderson was gardener to Thomas Dawson at Meadowbank, Uddingston, Lanarkshire, where there was an orchid collection (*Orchid Review* (1899): 198; R. Desmond 1994).

[6] CD recorded this information on the orchid *Phaius grandifolius* in a note (DAR 70: 110); he referred to pollination in the species in *Orchids* 2d ed., p. 280.

From Arthur Rawson 1 April [1863][1]

Bromley Common
April 1

Dear M^r^. Darwin

I send you a little novelty in (what I suppose the literati call) morphology. I never observed a similar case. In the truss of primula accompanying this you will see that the crown flower though having only one corolla has two (apparently) perfect pistils and set of stamens. I don't know whether this may be a novelty to *you*, but it is to me.[2]

I ought to have written to you about the Gladiolus experiment, which I carried out strictly according to y^r^. letter,[3] and wh answered exactly as you expected, and according to my previous experience. The pollen of the same variety is *not* potent on itself, even when taken from one plant and applied to flowers on another.[4]

I have Ophrys aranifera in bloom, and shall soon have a good pot of Cypripedium pubescens— I shall be glad to *loan* them to you if you want to make any experiments.[5] I have just been reading your "orchis" book with great interest.[6]

Y^s^. very truly | A Rawson

Since writing the above, I have thought you may not be at home. The primula will be in bloom for some days. Will you send me a line if you wish to see it

DAR 176: 22

[1] The year is established by CD's reference to the loan of *Cypripedium pubescens* in his letter to Asa Gray of 20 April [1863].
[2] CD discussed double flowers in *Variation* 1: 167–8, 171–2, 200.
[3] CD's letters to Rawson have not been found. However, see the letter from Arthur Rawson, [6 April 1863], and n. 4, below.
[4] In *Variation* 2: 139–40, CD provided an account of the experiments carried out by Rawson, a 'skilful horticulturist', who worked with complex French varieties of hybrid *Gladiolus*. For other evidence of CD's interest in cross and hybrid sterility, see, for example, the following letter, and *Correspondence* vol. 10, Appendix VI.
[5] See letter from Arthur Rawson, [6 April 1863] and n. 2.
[6] *Orchids* was published on 15 May 1862 (Freeman 1977, p. 112).

From John Scott [1–11] April [1863][1]

Botanic Gardens
April

Sir.

You have correctly comprehended, what I imperfectly attempted to lay before you in my last: respecting the sterility of individual plants with own-pollen;[2] though, both pollen, and stigma were found perfectly capable of performing their functions, when treat reciprocally with another species. I have not as yet succeeded in proving it with different individuals of a species. I have a case or two ⟨*four lines excised*⟩ your cases in point of Passifloras;[3] but, I am looking forward for some interesting results from the experiments I will try this season. However I will give you illustrations of the above from my experiments on Orchids. I have made many experiments on our plants of *Oncidium sphacelatum* with own-pollen; but, in no case have I succeeded in fertilising a single capsule.[4] On one plant, I fertilised a hundred flowers and upwards, but every capsule aborted. I examined a number of these flowers, as they dropped off, and found in the style of each an *abundance* of *pollen-tubes*, which in most cases, I traced into the *ovary*. Thus the stigmatic function &c is perfect, abortion, apparently caused, by some inappreciable change, in sexual elements, preventing normal conjunction.

By applying the pollen of this species to *Oncidium ornithorynchum* I have got a nice plump capsule, though before I had *never* succeeded in fertilizing the latter; with '*own-pollen*'.— I have now a single capsule thus produced. My experiments with pollen from *O. ornithorynchum* or *sphacelatum* were *ultimately* unsuccessful; though *indications* were given, of the capsules thus treat *swelling*, which I *never* observed or very *slightly* with '*own pollen*'. I then tried pollen from O. *div*⟨*aricatum*⟩, from which **four capsules** are maturing. ⟨*two words excised*⟩ from *this* plant—**sphacelatum**—applied to ⟨*one line excised*⟩ effective ⟨*four lines excised*⟩ these.[5] There is something, ⟨*six words excised*⟩ *O. divaricatum*, at least with the plants in the Gardens here.[6] I know nothing of it in other places. My experiments upon individuals growing in baskets with '*own-pollen*' have been numerous, yet all without *exception* unsuccessful; the ovary scarcely ever give the *slightest indications* of being affected by the pollen. I have, however, succeeded, in *fertilising four flowers* with '*own-pollen*', upon a plant growing in a *pot*— In the latter state the plants are keeped much damper, and the roots less exposed, which *may have* some influence, especially in houses like those here in which a mixed collection of plants are grown. However, these capsules, are most *abnormal* looking productions, which I scarcely expect to contain a *single seed*.[7]

Though I have not in the above experiments succeeded in reciprocally crossing any two species, I have, nevertheless, I think satisfactorily shown an individual impotence of organs, capable of elimination by the action of other species. And this in a most complex and capricious manner, which might I think with illustrations you have already given, well convince your many caterers, that sterility is **not** a special endowment.[8] I have yet another case; and **this** of *reciprocal action*, with the *Maxillarias*, *atro-rubens* and *squallens*.[9] I have made *[many] fruitless* attempts to fertilise the former ⟨with⟩ 'own-pollen' I accordingly tried it with pollen from ⟨the l⟩atter; which has proved effective, as I have now ⟨several⟩ capsules maturing on *M. atro-rubens*, thus ferti⟨lised.⟩ Again, the functional efficiency of the latters pollen ⟨I⟩ have also proved by applying it to flowers of *M. squallens*, on which I have produced a capsule. I may state that *M. squallens*, produces capsules freely when fertilised with '*own pollen*'.

I have thus imperfectly given you all the information my experiments have yet afforded me on the above subject. The use to be made of it, I beg to leave entirely to you, and I shall certainly be highly gratified if it merits your recording.[10]

In respect to fertilising ovules by application of pollen to rostellum, permit me to make a few remarks. In my last I mentioned *certain genera* of Orchids, whose rostellum *still* exercised stigmatic function, as evidenced by inducing pollen-grains to emit tubes.[11] In making this statement, I did not at all *intend* its general application: it had reference to those genera mentioned,—*only*—in which I had *never* observed the pollen tubes produced *in situ*.[12] Your observations have long since convinced me that the power of inducing germination of the pollen-grains, is not a peculiar prerogative of the stigma. Nevertheless, this power of dependent germination is not at all a common pheno⟨menon as⟩ far as yet known, I think you will readily a⟨llow⟩, the great majority of plants have been found ⟨to prod⟩uce pollen-tubes only, when their pollen is in contact ⟨w⟩ith stigma or its secretions. It was therefore ⟨on this⟩ **consideration** that I mentioned the *above statement*. It is somewhat astonishing to see how much the degree of excitability of pollen-grains varies in species of a genus or even in individuals of the former. I am somehow inclined, however, to suppose that careful research and experiments will show that this difference in the degree of excitability is more **apparent** than **real**.

In respect to Dr Gray's observation, permit me to ask a single question.[13] Am I right in supposing that he found the tubes in the soft moist tip of rostellum *alone*?

or has he in reality traced them downwards? In his remarks, he speaks of the grains falling upon the naked cellular **tip** of the narrow process of rostellum, and upon the cellular sumit the process outside of each disc, and sending *tubes* down freely into their **substance,** I am somehow, inclined to think that the term "*substance*" has here a special reference to that portion of the rostellum mentioned—its cellular tip—though certainly it may possibly be intended to include the entire organ. As far as my observations go, the rostellum seems absolutely impervious by pollen-tubes; and certainly one might have expected to have seen, at least some, of the pollen-tubes protruding from pollen grains in contact with rostellum—mentioned in former letter[14]—penetrating the latter, if it had been possible, instead of being immediately deflected, and extending themselves upon its surface. I am now, indeed, ⟨more than⟩ ever inclined ⟨to a⟩ccept your f⟨irst su⟩ggestion on the absence of stigmatic utriculi— on a⟨ ⟩ ⟨ ⟩ tissue, which of course includes the former—probably preventing fertilisation.[15] Indeed if it proves otherwise it will be a fatal blow to our attempts at discovering physiological changes in the structure of the styles of sterile hybrids, or at least of giving *such* if discovered a *causal signification*.[16] But, independent of the above there are other difficulties, of an almost insuperable nature to contend with: for example, how is the *Catasetum* & *Acropera* cases to be reconciled with the preceeding view. You have shown that the structure of the female organs in the *former* is *much less abnormal* than in the *latter*; nevertheless, consider the different results of experiments on their fertilisation.[17] And even supposing, that Catasetum should ultimately yield a capsule to some assiduous & skilful operator, we will still have the anomalous *Acropera*! I candidly admit, that my experiments on the latter genus and Gongora—I have yet to state those on the latter, which are curious, I will do so in an early letter—lead me to doubt, (though perhaps, I may happily be wrong) the *absolute* sterility of *Catasetum*.[18] Dr. Hooker's experiments also as you will remember, in fertilising ovules of *Mecon⟨opsis* by⟩ direct application of the pollen, and thus entirely dispensing ⟨with⟩ stigma & style, present a grave difficulty.[19] ⟨*five words missing*⟩, such ⟨*three words missing*⟩ rapid extension of the tubes, as well as their passi⟨n⟩g through matter of a denser consistency will I am incline⟨d to⟩ think prevent any osmotic action between them and stigmatic utriculi. But I need not further particularize, as all are familiar to you, and I mention them merely as difficulties which I have felt in the following out of your suggestions. I will be glad, if time and health permit you, to hear whether or not, they are really such.

Since I last wrote you I have made an interesting and very opportune observation on *Epidendrum cochleatum*, which you may like to hear.[20] On a plant of the above I observed a somewhat abnormal looking flower, which was unusually long in expanding: this I found on examination to be due to complete adhesion of the outer segments of the perianth, which were much shorter than usual. Under the impression that there might possibly be some interesting abnormalities in the interior structure, I at last cut the flower for examination. I was disappointed in this, however, there being simply an abortion or nearly so of the inner circle of the perianth, and the column seemed perfectly normal. There, however appeared to be an unusual amount of viscous matter ab⟨ove the⟩ stig⟨ma⟩tic chamber, which covering ⟨in⟩ part the rostellum, *laterally* reached, the sides of an⟨thers⟩ *[illeg]* ⟨your⟩ ⟨ob⟩servations on *Neottia* and *Aceras* ⟨imme⟩diately induced a hope that probably this near approach of stigmatic secretions to the pollen-masses, might have excited the emission of pollen-tubes.[21] I accord⟨ing⟩ly placed sections of the column under the microscope and was much gratified to find that they had protruded; and this not only from below that *lateral portion* of the anther where the stigmatic secretions, all but touched the pollen-masses; but, likewise, from the *centre* extending themselves *over* the anterior surface of rostellum, downwards into the stigma. Thus the *simple rostellum* of Epidendrum has likewise afforded me no illustration of its penetrability by pollen-tubes. If I have an opportunity this season, I will make a few experimental observations on the British Gymnadenia, and see if I can discover anything analogous to Dr Gray's observation.[22]

I have seen your letter in Cottage Gardner.[23] I thank you most sincerely for the honour you have done me in referring to my observations.[24] Permit me to ask a single question or so in reference to Dr Cruger's observations on the pollen-masses of unopened flowers becoming ⟨pu⟩lpy, and then emitting their tubes.[25] Does he make no mention of the production of *fungi* upon this pulpy ⟨m⟩ass? This I ask in consequence of an ob⟨serva⟩ti⟨on⟩ upon polle⟨n m⟩asses of Blet⟨ia⟩ ⟨which⟩ assumed a somewhat pulpy form. I placed a little ⟨of⟩ th⟨is⟩ under the microscope, and was much pleased to observe, what I **then** fancied, an abundance of pollen-tubes. Great was my astonishment, however, on a careful examination under a higher power, to find that what I had fancied, pollen-tubes, were in reality the threads of some *fungus*. (Possibly—but I know nothing about fungi—a species of Aspergillus.) They protruded from the pulpy mass, in the form of elongated jointless filaments, some of which bore at their extremities processes from which neck-laces of spores originate. But for the *latter*, I candidly admit—though perhaps it is a most unpardonable admission—that I should have considered this as illustrative of the emission of tubes from pollen while still *in situ*. I have made similar observations on pollen from another flower, but after the most careful scrutiny I failed to observe a single pollen-tube, though I saw an abundance of fertile and sterile threads of a fungus.

Permit me to make a few remarks on Dr Cruger's suggestion that the emission of pollen-tubes-grains *in situ* is due to ants carrying stigmatic secretion to pollen,[26] and that of M^r^. Anderson's that flowers of an imperfect character have a great tendency to produce seed-capsules.[27] Dr Cruger, may possibly have based his suggestion on the observance of stigmatic secretions in contact with anthers. If it should be ⟨an observ⟩ation of this nature, my experiments on fertilising the rostellum, incline me to suppose, that he possibly mistakes the *effect* for the cause. I have already made—as you may remember—a somewhat similar mistake in Lælia.[28] I then stated that two lateral processes were developed by the stigmas which extended themselves upwards to rostellum, and thus connected the pollen-masses applied to that organ with the stigma. These processes are simply the hardened stigmatic secretions. In all my experiments on this point—excepting of course those in which plaister of Paris was applied to the stigma—I in general observed a superabundance of viscous matter exuded by the stigmas. But this I now believe does not take place until at least some of the pollen-tubes have penetrated the stigma: these somehow concentrate, assimilating or organising forces of the flower, and induce an increased activity in the secretion of viscous matter, which in most cases fills the stigmatic cavity, and even in that of Epidendrum, above referred to—reached the anther. Before the penetration of the stigma by the pollen-tubes the secretion of viscous matter is limited, in consequence of the assimilative forces of the flower being directed somewhat equally to its different parts, and we find that this balance is in *general* maintained in an unfertilised flower; its various parts decaying together. The penetration of the stigma by pollen-tubes however, turns the balance, the ⟨perianth⟩al parts quickly fade, and all is directed to the supply of the ovary. The mere circumstance then, of the viscous matter of the stigma reaching the anthers or pollen-masses—does not in my opinion, justify us in regarding it as necessary for the emission of tubes; it must rather in my opinion be regarded as a consequence of the *emission* and *penetration* of the stigma by pollen-tubes.

In respect to Mr Anderson's observations, or rather—as I see he limits it to flower wanting a sepal or petal—Dr Cruger's on unopened flowers so frequently setting fruit.[29] The greater **humidity**, and **equability** of temperature, consequent on such conditions, is I believe the *probable* cause for these abnormally conditioned flowers, so frequently fertilising themselves. A consideration of the extreme delicacy of the pollen-tubes must at once convince us of the greatly increased facility such conditions afford for accomplishing the above phenomenon, than obtained in the normal condition. Indeed; in my opinion, I do not think it improbable that the above phenomenon—the germination of pollen grains *in situ*—which in the majority of cases occurs *only* in *closed* flowers, is a purely accidental occurrence, having *no ultimate import* in the economy ⟨of⟩ the spec⟨ies.⟩ That in fact the latent vitality of the pollen grains is simply, and necessarily excited by the abnormal conditions of the flower. I certainly admit that those cases of *Viola* and *Oxalis*, on which the different flowers are produced at distinct periods, renders the above somewhat improbable.[30] But are you aware of the **strict observance** of this law in those cases? do they **not** also occur **together** upon the plant? the abnormal, self-fertilising flowers being arrested in their development. I think, but I will study the subject more carefully, that I have observed both flowers together on plants of *Oxalis Acetosella* in our hot-houses, in some of which it has become a great pest. In the cases of *Impatiens* and *Campanula* they seem to occur on plants at the same period and thus I think justify us in regarding the imperfect flowers, as simply due to a casual arrest in the development of the perianth, and having nothing whatever to do with special purposes in the economy of the species. That in fine we have no more *right* to look for such *ends* in this case—even though the ultimate end of vegetal life is perfectly accomplished—than we have in the case of *other* abnormalities in the flower, *regularly* reproduced—peloria, for example. I may be quite wrong in this view, nevertheless I cannot but think that we might w⟨i⟩th **equal justice** suppose that nature had some *ultimate end* to serve ⟨in the⟩ *latter* as well as the *former*. I have ventured these remarks, under the conviction that the inferences ⟨as w⟩ell as the demands, I have seen founded on this phenomenon, are alike unwarranted. I cannot see for example how it can possibly be maintained, that this continued self-fertilisation exercises no baneful influence on successive progeny. Now, inasmuch, as no single species, I think, can be instanced, as *absolutely* dependent on this exclusive mode of self-fertilisation: the "struggle for existence" between the offspring of the perfect and imperfect flowers, must necessarily be brought into action, and thus invalidates the objection, to your views, on the necessity of occasional crosses for the healthy continuation of species.[31] But I need not further follow out this subject, I have said sufficient to show you in what light I regard it. And if I am wrong, I have already said more than enough. I may remark that I have commenced experiments to elucidate, if possible the subject in Orchids, by **cementing** the **perianth** of **perfect** flowers shortly before their **expansion**. I will thus see if I can artificially produce the phenomenon.

The results with homomorphic Cowslips is a little provoking.[32] I now think, however, we ought to have anticipated these results, from experiments with such plants, as Cowslips, Primroses & Auricula &c. which present both forms. Inasmuch as the plant from which we save our seed might have heteromorphic parents, and thus would have a greater tendency to produce heteromorphic offspring. An exact experiment, would thus necessarily require a knowledge of a plants lineage for a number of generations. On a consideration of this I think we may look for satisfactory elucidations of this point from those long-cultivated exotic species, that present both forms in their native wilds; but of which *one* only has been introduced. These I look forward to with more hope; though certainly in the case of homomorphic crosses with Cowslips &c. we ought to have the majority of offspring presenting parent form.[33]

But I fear I will long ere this have exhausted your time and patience: Pray excuse me,

And now sincerely trusting that your health is still progressing favourably, I remain | Sir | Yours very respectfully | John Scott

I have just had a letter, in reply to one I sent, from one of the most experienced Auricula growers in Scotland.[34] He informs me that as soon as a long-styled plant presents itself it is immediately thrown out. And yet from seed saved from these **short-*styled*** plants, he states that *usually* **one-fourth** are **long-styled**. They certainly bear a *much* **higher proportion**, than I expected.[35] | J. S.

He also informs me that he never observed the **two forms** produced by *one* **plant**

Endorsement: 'April 1863'
DAR 108: 183, DAR 177: 86 (fragile)

CD ANNOTATIONS

2.8 There is . . . in a *pot*— 2.14] *scored brown crayon*
4.3 recording.] *before closing square bracket, brown crayon*
5.1 In respect] *after opening square bracket, brown crayon*
6.33 But I need . . . phenomenon. 10.42] *crossed ink*
11.1 The results . . . of which *one* 11.9] *crossed pencil*
11.9 only has been . . . John Scott 13.2] *crossed ink*
14.1 I have just . . . *one* **plant** 15.2] *scored brown crayon*
Top of letter: '*Sterility of Orchids* | John Scott.— | *Orchid Portfolio*— Rostellum' *ink*; 'Crüger' *blue crayon*
End of letter: 'April. 1863' *ink*

[1] The date range is established by the relationship between this letter, the letter to *Journal of Horticulture and Cottage Gardener*, [17–24 March 1863] (published in the issue for 31 March 1863), and the letter to John Scott, 12 April [1863].

[2] See letter from John Scott, 21 March [1863] and n. 4, and letter to John Scott, 24 March [1863] and n. 3.

[3] See letter from John Scott, 21 March [1863] and n. 4, and letter to John Scott, 24 March [1863].

[4] Scott's experiments in pollinating species of *Oncidium* with their own pollen are discussed in Scott 1863a, pp. 545–7. See also letter from John Scott, 21 March [1863] and n. 5.

[5] In his account of these experiments prepared for the Botanical Society of Edinburgh, Scott wrote that he found his most successful crossings to be those in which he pollinated six flowers of *Oncidium sphacelatum* with pollen from *O. divaricatum* var. *cupreum*; four fine capsules resulted. He also tried unsuccessfully to fertilise flowers of *O. altissimum* and *O. graminifolium* with pollen from *O. sphacelatum* (Scott 1863a, pp. 547–8).

[6] Scott was a foreman in the propagating department at the Royal Botanic Garden, Edinburgh (R. Desmond 1994).

[7] See Scott 1863a, pp. 548–9.

[8] Scott 1863a, p. 549. Scott discussed the function of sterility and its relation to CD's theory in Scott 1863a, pp. 543–4. CD discussed his opposition to the notion of sterility being a special endowment in *Origin*, pp. 260–3. See also letter from John Scott, 6 January 1863 and n. 10, and *Correspondence* vol. 10, Appendix VI.

[9] Scott's experiments with species of *Maxillaria* are described in Scott 1863a, p. 549.

[10] CD encouraged Scott to publish his results (see letter to John Scott, 12 April [1863]). In *Variation* 2: 133, CD discussed Scott's work with *Oncidium*, also stating that 'Mr. Scott found that *Maxillaria atro-rubens* was "totally insusceptible of fertilisation with its own pollen," but fertilised, and was fertilised by, a widely distinct species, viz. *M. squalens*'. See also *Orchids* 2d ed., p. 289.

[11] See letter from John Scott, 21 March [1863]. See also letter to John Scott, 24 March [1863], and letter to *Journal of Horticulture and Cottage Gardener*, [17–24 March 1863] and n. 9.

[12] See letter from John Scott, 21 March [1863] and nn. 7 and 8.

[13] Asa Gray had reported that in *Gymnadenia tridentata* pollen grains could fall onto 'the naked cellular tip of narrow process of the rostellum', germinate, and their pollen tubes would penetrate the rostellum (A. Gray 1862a, p. 426). After CD sent Scott a copy of A. Gray 1862a (see letter to John Scott, 6 March 1863), Scott discussed Gray's statement, noting that in his own experiments he had never observed pollen tubes 'insinuating themselves into the tissues' of the rostellum (see letter from John Scott, 21 March [1863]).

[14] See letter from John Scott, 21 March [1863].

[15] The reference is to *Orchids*, p. 311 (see letter from John Scott, 16 January 1863 and n. 6).

[16] See *Origin*, pp. 260–7.

[17] The role of the rostellum, stigma, and style in pollination appeared to be different in *Catasetum* and *Acropera* compared with other orchids (see *Orchids*, pp. 203–10, 231–48, and letter from John Scott, 18 February [1863] and nn. 4 and 5). See also letters to John Scott, 21 January [1863], 20 [February 1863], and 24 March [1863], and letter from John Scott, 3 March 1863.

18 Scott apparently refers to his pollination experiments with *Gongora truncata*, the difficulty and ultimate success of which he reported in his letter to CD of [after 12] April [1863]. Scott had previously told CD of his success in pollinating *G. atropurpurea* (see letter from John Scott, 3 March 1863).
19 The reference is to experiments described in J. D. Hooker 1854a. See also letter from John Scott, 6 January 1863 and n. 8.
20 CD compared the stigma and rostellum of *Epidendrum cochleatum* in *Orchids*, pp. 310–11, and *Orchids* 2d ed., pp. 249–50; Scott's specific observations were not mentioned.
21 See *Orchids*, p. 324 n., and letter to John Scott, 24 March [1863].
22 See n. 13, above.
23 CD's letter to the *Journal of Horticulture and Cottage Gardener* of [17–24 March 1863] appeared in the issue for 31 March 1863.
24 In his letter to the *Journal of Horticulture and Cottage Gardener* of [17–24 March 1863], CD referred to Scott as an 'excellent observer', and cited his observations of orchid pollen tubes.
25 See letter to *Journal of Horticulture and Cottage Gardener*, [17–24 March 1863]. See also letter from Hermann Crüger, 23 February 1863.
26 See letter to John Scott, 24 March [1863], and letter from Hermann Crüger, 23 February 1863 and n. 5.
27 See letter to *Journal of Horticulture and Cottage Gardener*, [17–24 March 1863], and James Anderson's letter in the *Journal of Horticulture*, 17 March 1863, pp. 206–8.
28 See letter from John Scott, 16 January 1863. See also letter from John Scott, 21 March [1863].
29 See letter to *Journal of Horticulture and Cottage Gardener*, [17–24 March 1863].
30 See letter to *Journal of Horticulture and Cottage Gardener*, [17–24 March 1863] and n. 7. CD began experimenting in 1862 with the unopening, 'imperfect' flowers of *Viola* and *Oxalis*; these appeared at different times of the year from the larger, opening flowers (see *Correspondence* vol. 10, letter to John Scott, 12 November [1862] and n. 5, and letter from John Scott, 15 November [1862]).
31 In the last paragraph of *Orchids*, p. 359, CD argued that nature 'abhors perpetual self-fertilisation'. See also *Origin*, pp. 96–101.
32 In his letter to John Scott of 24 March [1863], CD stated that cowslips raised from seed resulting from self-pollination of a short-styled parent yielded both short- and long-styled progeny; CD had expected short-styled seedlings only.
33 With CD's encouragement, Scott was already undertaking his own experiments with *Primula* species, including cowslips (see Scott 1864a, pp. 106–10).
34 Scott may refer to George Lightbody of Falkirk, who raised numerous varieties of *Primula auricula* (R. Desmond 1994).
35 Scott later published his own work with *Primula auricula* (see Scott 1864a, pp. 86–97).

To Charles Turner [1 April – 16 June 1863?][1]

Sir

I hope that you will excuse the liberty which I take in asking from you information on one small point, with permission to quote your name as my authority in a work which I am preparing for publication.[2] I have found that the varieties of several kinds of Hollyhock come true from seed;[3] now what I am anxious to know is, whether you find it necessary to grow the different varieties far apart from each other, in order to prevent crossing & to get true seed.— As I have repeatedly watched Bees carrying pollen from one variety to the other, I cannot understand how the seed can be true, unless the plants are grown very far apart: & yet I have been told that this plan is not always followed.—[4]

If you will have the kindness to answer me this question, I sh$^{d.}$ feel extremely much obliged. I should, also, be glad to know which varieties you have found come truest by seed.—[5]

Hoping that you will excuse the liberty which I have taken & grant this favour I beg leave to remain | Sir | Your obliged servant

Draft
DAR 96: 12

[1] The date is conjectured from the reference to CD's work on the draft manuscript of *Variation* (see n. 5, below). According to his 'Journal' (Appendix II), CD prepared a draft of the section on 'Crossing & Sterility' (*Variation* 2: 85–144) between 1 April and 16 June 1863.

[2] CD refers to *Variation*, which was eventually published in 1868.

[3] CD had been seeking information on hollyhock breeding for some time (see *Correspondence* vol. 2, letter from William Herbert to J. S. Henslow, 5 April 1839, and *Correspondence* vol. 5, letter to J. S. Henslow, 12 November 1855 and n. 2). See also *Natural selection*, pp. 65, 71, *Origin*, p. 271, and *Correspondence* vols. 8 and 9. See DAR 157a, p. 1, for CD's experimental notes on hollyhocks. CD had recently encouraged Charles William Crocker to undertake hollyhock experiments (see *Correspondence* vol. 10, letter from C. W. Crocker, [before 13 March 1862], and this volume, letter from C. W. Crocker, 1 May 1863).

[4] See *Correspondence* vol. 8, letter from William Masters, 8 May 1860, and *Variation* 2: 107–8.

[5] Turner's reply has not been found, but information from him on crossing in hollyhocks is given in *Variation* 2: 107–8.

To the chairman of Bromley District Highway Board from the parishioners of Down, Kent[1] 3 April 1863

[Document signed, being a testimony by the parishioners of Down, Kent, witnessing to the moral character and integrity of George Snow,[2] District Surveyor,[3] addressed to the Chairman of the Highway Board[4] and signed by nearly fifty local residents including 'Charles Darwin Down House'.]

Sotheby's, London (24 July 1995)

[1] District Highway boards were local administrative units responsible for the public highways. This new administrative system had just been introduced under the Highway Act of 1862. Each district comprised a number of parishes; the parish of Down was part of the Bromley highway district (*Return of highway districts*).

[2] George Snow was a coal dealer in Down who also ran a carrier service between Down and London (*Post Office directory of the six home counties* 1862).

[3] In addition to his activities as coal dealer and carrier (see n. 2, above), Snow may have been the parish highway surveyor or district highway surveyor under the administrative system prevailing before the introduction of the District Highway boards in 1862 (see n. 1, above). See *Highways (District Surveyors) Bill*, and Webb and Webb 1913, pp. 29–31. Snow was unsuccessful in his candidature for district surveyor; George Golding of Chislehurst was appointed instead (Minutes of the Bromley Highway Board, 847/HB/B/1, Bromley Central Library).

[4] At the first meeting of the Bromley District Highway Board, on 13 April 1863, John Farnaby Lennard was elected chairman (Minutes of the Bromley Highway Board, 847/HB/B/1, Bromley Central Library).

To John Lubbock 5 April [1863][1]

Down
April 5th

My dear Lubbock

You were so kind as to send to enquire after me the other day. I am better; a touch of that abominable yet blessed eczema did me as much good as a fit of gout. Whether my skin will allow me to try water-cure, I doubt; anyhow we shall not move till about the beginning of May.—[2] Could you come over here some day for a short talk; not that it is worth your while unless you could also take in High-Elms.—[3] This present week would not be quite so good, as our house is full of Boys & Relations.—[4] But whenever you can spare time & are so inclined, it will give me, as you know, real pleasure to see you.—

The last number of N. H. Review seems to me very good in many ways.—[5] I like particularly your Review of Lyell & agree to all you say.—[6] Whether Lyell will like it quite so much, especially about the B. of Oxford, may well be doubted.[7] But nothing you say can annoy him much; & it is all just. Your kick at Dr Cummings made me laugh.—[8] You are never weary of heaping honour on me.—[9]

Farewell & good night | Most truly yours | C. Darwin

I suppose you read that wonderful Review in Athenæum on Carpenter, which I almost suspected & which Hooker tells me is by Owen.—[10] What will the B. of Oxford say of slime &c generating new animals!—[11] I sometimes think Owen must be demented

DAR 263: 57

1 The year is established by the reference to Lubbock's review of Charles Lyell's *Antiquity of man* (C. Lyell 1863a; see n. 6, below).

2 CD had considered visiting James Manby Gully's hydropathic establishment in Great Malvern, Worcestershire (see letter to J. D. Hooker, 5 March [1863]).

3 In 1861, Lubbock moved from High Elms, the family home near Down, to Chislehurst, about five miles away (Hutchinson 1914, 1: 52).

4 Emma Darwin's diary (DAR 242) recorded the visits of a number of guests during the first week of April 1863. All the Darwin children, except for George, were home for Easter Sunday (5 April); CD's sister, Emily Catherine, was also visiting Down House. Emma Darwin's nephews, Laurence and Alfred Allen Wedgwood, arrived two days later when Edward Cresy also came for lunch. See also the letter from Emma Darwin to William Erasmus Darwin, [17 March 1863], in DAR 219.1: 71.

5 CD refers to the April number of the *Natural History Review*, of which Lubbock was one of the editors (see n. 6, below); CD's annotated copy is in the Darwin Library–CUL.

6 Lubbock reviewed *Antiquity of man* (C. Lyell 1863a) in the April number of the *Natural History Review* ([Lubbock] 1863c).

7 In his review of Lyell's *Antiquity of man* ([Lubbock] 1863c, p. 213), Lubbock quoted the sentence he thought was the strongest statement in the book concerning the origin of species (C. Lyell 1863a, p. 469):

> Yet we ought by no means to undervalue the importance of the step which will have been made, should it ever become highly probable that the past changes of the organic world have been brought about by the subordinate agency of such causes as 'Variation' and 'Natural Selection'.

Lubbock then stated that even the bishop of Oxford, Samuel Wilberforce, might agree to this sentence, 'but we can hardly believe that it expresses Sir Charles Lyell's real opinions, as the whole tenor of his argument is in favour of Mr. Darwin's theory' ([Lubbock] 1863c, p. 213). CD marked the quoted sentence in his annotated copy of C. Lyell 1863a, which is in the Darwin Library–CUL (see *Marginalia* 1: 525–7). See also letter to Charles Lyell, 12–13 March [1863] and n. 17. For Wilberforce's opposition to CD's theory, see [Wilberforce] 1860 and *Correspondence* vol. 8, Appendix VI.

8 The popular and controversial preacher John Cumming was the author of *Moses right and Bishop Colenso wrong* (Cumming 1863). In a postscript to his review, Lubbock criticised Cumming's misuse of geological authorities ([Lubbock] 1863c, p. 219); he charged Cumming with erroneously citing William Buckland, Edward Hitchcock, and Adam Sedgwick as believing that geological evidence supported the idea of a universal flood.

9 Although Lubbock stated that he wanted to confine his review of *Antiquity of man* to that part of the book concerned with prehistoric man ([Lubbock] 1863c, pp. 212–13), he referred to CD a number of times, concluding with a paragraph on natural selection in which he referred to 'the genius of Darwin', and describing CD as 'our illustrious countryman' (*ibid.*, pp. 218–19).

10 See Appendix VII for the review of Carpenter 1862 that appeared in the *Athenæum* on 28 March 1863, pp. 417–19. The publisher's marked copies of the *Athenæum* at the City University Library, London, confirm that Richard Owen was the reviewer. See letter to J. D. Hooker, [29 March 1863], and letter from J. D. Hooker, [31 March 1863].

11 Owen wrote that the mud or slime at the bottom of seas, lakes, and rivers contained a vital energy or force that 'condensed into a protoplasmic centre', eventually creating organisms 'of all low grades of organization' like the Foraminifera (*Athenæum*, 28 March 1863, p. 417, and Appendix VII). See also letter to J. D. Hooker, [29 March 1863] and n. 5. See also n. 7, above.

To Maxwell Tylden Masters 6 April [1863][1]

Down | Bromley. Kent. S.E.
April 6th.

My dear Sir,

I have been very glad to read your Paper on Peloria.—[2] For mere chance of following case being new, I send it.— A plant which I purchased as Corydalis tuberosa has, as you know one nectary, short, white, & without nectar; the pistil is bowed towards the true nectary; & the hood formed by the inner petals slips off towards opposite side (all adaptations to insect agency like many other pretty ones in this Family). Now on my plants there are several flowers (the fertility of which I well observe) with both nectaries equal & purple & secreting nectar; the pistil is straight, & the hood slips off either way— In short these flowers have exact structure of Dielytra & Adlumia— Seeing this I must look at the case as one of Reversion;[3] though it is one of the spreading of irregularity to two sides—

As Columbine has all petals &c. irregular & as Monks-Hood has two petals irregular, may not the case given by Serringe, & referred to you by you be looked at as Reversion to the Columbine State?[4] would it be too bold to suppose that some ancient Linaria or allied form, & some ancient Viola had all petals Spur shaped, & that all cases of "irregular peloria" in their genera are reversions to such imaginary ancient form?[5]

It seems to me in my ignorance, that it would be advantageous to consider the two forms of Peloria when occurring in the very same species as probably due to same general law,—viz one as reversion to very early state and the other as reversion to a later state than when *all* the petals were irregularly formed. This seems at least to me a priori a more probable view; than to look at one form of Peloria as due to Reversion & the other as something distinct—[6]

What do you think of this notion?—

My dear Sir | Yours sincerely | Ch Darwin.

Copy
DAR 146: 341

[1] The year is established by the relationship between this letter and the letter to M. T. Masters, [8–13 April 1863].

[2] Masters 1863.

[3] CD discussed *Corydalis* in the chapter of *Variation* that concerned reversion (*Variation* 2: 28–61); he had just completed a draft of this chapter, together with two others on inheritance (*Variation* 2: 1–84), on 1 April 1863 (see 'Journal' (Appendix II)). *Corydalis tuberosa* and Masters 1863 are discussed in *Variation* 2: 58–9. CD argued that the 'coadapted modifications' of two full nectaries functioning for 'insect agency' could not be attributed to chance, or to correlated variability; he suggested that such modifications should be attributed to 'reversion to a primordial condition of the species' (*Variation* 2: 59). See also n. 5, below. There are two notes in DAR 51 recording CD's observations on *C. tuberosa*, the first of which is dated 3 April 1863 (DAR 51: B16–17). *Dielytra* and *Adlumia*, like *Corydalis*, are genera in the family Fumariaceae.

[4] In his paper on peloric flowers (Masters 1863), Masters noted that in some garden varieties of columbine (*Aquilegia*), the irregularity disappeared almost entirely. He cited Nicolas Charles Seringe's description of a peloric flower in *Aconitum* (monks-hood) (Seringe 1823, 1: 132): 'all the sepals were helmet-shaped and all the petals presented that peculiar form, which, under ordinary circumstances, is assumed by two only, the other three being reduced to the condition of mere scales' (Masters 1863, p. 260). *Aquilegia* and *Aconitum* are genera in the family Ranunculaceae.

[5] CD was exploring the possibility of reversion in a species with several irregular parts of the flower to a form in which all parts were irregular (as in irregular peloria). In Masters 1863, Masters proposed the division of peloria, or regularity in normally irregular flowers (what the author called exceptions to the 'ordinary rule of floral structure', p. 258), into two categories. He suggested the term 'regular or congenital peloria' for cases where flowers 'contrary to their usual habit, retain throughout the whole of their growth their primordial regularity of form and equality of proportion'. The term 'irregular or acquired peloria' would designate flowers in which the 'irregularity of growth that ordinarily characterizes some portions of the corolla is manifested in all of them', resulting in a regular flower (*ibid.*, p. 260). See also *Variation* 2: 58–60.

[6] See *Variation* 2: 58–60.

From Arthur Rawson [6 April 1863][1]

Bromley Common
Monday night

Dear Mr. Darwin

I did not expect the arrival of any one for the Cypripedium quite so soon, but it is all right, and will bloom in a fortnight or three weeks.[2] It should be kept in a *cold* frame. Of course you may mutilate the flowers to any extent you please. The *plant* I should like back when you have done with it.

As to the Gladiolus experiment,—your statement is quite correct, and I tried the experiment carefully.[3] A. would *not* fertilize A. (I wish it would)—but B.C.D. would fertilize A. to any amount, and I have done this in many cases,—in fact I get all my seeds by thus crossing.[4] I find somewhat the same in Epimediums, but cannot speak so certainly.

Have you ever observed in Dianthus (or in other species, I have only in Dianthus) that in a pan of seedlings several will always come with *three* instead of *two* seed leaves, and that whenever this is so *three* true leaves follow? I have often noticed this, and intended, last year, to grow them separately, to see the result on the bloom, but something prevented me.

Once more. Have you ever studied "Dielytra spectabilis"?[5] I *cannot* cross it, or seed it, though I have tried this 6 years. If you can put me up to that, I shall be much obliged

Y^r. very truly | A Rawson

DAR 176: 23

[1] Dated by the relationship between this letter and the letter from Arthur Rawson, 1 April [1863], and by the date of flowering of Rawson's *Cypripedium pubescens* (see n. 2, below); in 1863, the first Monday following 1 April was 6 April.

[2] In his letter of 1 April [1863], Rawson offered CD a loan of the orchid *Cypripedium pubescens* in bloom 'to make any experiments' he wanted. CD had wished to examine additional specimens of *Cypripedium* after reading Asa Gray's observations on American species (A. Gray 1862a, pp. 427–8). Gray's observations were made partly in response to *Orchids*, pp. 274–5. See letter from Asa Gray, 2 January [1863] and n. 7. CD recorded his experiments with the flowers of the *C. pubescens* specimen sent by Rawson in DAR 70: 112–13; these notes are dated 'April 19'.

[3] CD's letters to Rawson have not been found; however, see n. 4, below, and the letter from Arthur Rawson, 1 April [1863] and n. 4.

[4] In *Variation* 2: 139–40, CD provided an account of Rawson's pollination experiments with varieties of a complex hybrid *Gladiolus*:

> Mr. Rawson, after repeated trials, found that none of the varieties would set seed with their own pollen, although taken from distinct plants of the same variety, which had, of course, been propagated by bulbs, but that they all seeded freely with pollen from any other variety.

[5] CD had studied *Dielytra spectabilis*, but primarily with regard to insect pollination (see *Correspondence* vol. 8, letter to J. D. Hooker, 26 April [1860], *Correspondence* vol. 10, letter to Daniel Oliver, 12 [April 1862], and *Variation* 2: 59).

From Edward Blyth 7 April 1863

The High Beeches, | near Crawley, Sussex[1]
April 7/63—

My Dear Sir,

Here I am, quietly and comfortably ensconsed for some time to come, with an occasional run up to London,—quite close upon Tilgate forest, the chief scene of Mantell's Researches;[2] and I find that my juvenile kinsfolk are in possession of *Iguanodon* teeth, & a fine Crocodilian bone, which I take to belong to the

Teleosaurus.[3] I have had no time, however, as yet for researches of any kind, & my only trip has been to Brighton Downs yesterday with my friends to see the Volunteer Review. Fortunately I was accompanied all day by Capt Hans Busk, who was the originator of the Volunteer movement, but who only attended yesterday as a spectator.[4] Of course it was much like other sham-fights, & there was a vast crowd of spectators— To me it is of course a treat to recognise all my old friends in the birds, plants, &c, & already I really sometimes feel as if I had never left England, almost.[5]

Bartlett told me that he would not trust your suggested experiments in other hands than his own, when he could find time to superintend them personally: otherwise they would be surely botched.[6] He shewed me some very curious hybrids. The mixed breed of the Ruddy Shieldrake or common 'Brahmini Goose' of India (*Casarca rutila*) with the European & Asiatic Shieldrake (*Tadorna vulpanser*) prove to be so like the Australian *Casarca* that the two are barely, if at all, distinguishable! They are in adjacent compartments, & I could not tell the one from the other! And yet there is also an Australian true *Tadorna*, & likewise I believe a *Casarca* peculiar to New Zealand,—also one that I know of only from S. Burmá, my *C. leucoptera*— The curious fact remains that by crossing the Eur-asian *Casarca rutila* & *Tadorna vulpanser*, you produce the exact likeness of the *Casarca* proper to Australia! The way the Feruginous Pochard has also been bred back from the mixed progeny of a *female Clangula vulgaris* is also worthy of notice.[7]

Yours ever sincerely, | E. Blyth

DAR 160: 205

[1] High Beeches, Slaugham, Sussex, was the seat of Robert Loder (*Post Office directory of the six home counties* 1862). Loder's relationship to Blyth has not been established.

[2] Gideon Algernon Mantell was well known for the numerous fossils he had collected in Tilgate Forest, Sussex (Cleevely and Chapman 1992).

[3] *Iguanodon* was a Cretaceous dinosaur from Tilgate Forest, first described by Mantell in 1825 (Mantell 1825); *Teleosaurus* was the name given by Etienne Geoffroy Saint-Hilaire (E. Geoffroy Saint-Hilaire 1825d) to the remains of a group of Jurassic crocodilian forms found in a limestone quarry in Normandy, France (see Buffetaut 1987, pp. 68–9, 80–3).

[4] The Volunteer Force was raised in 1859 as part of a movement to train military volunteers for countering the perceived threat of invasion by the French (*DNB*, *EB*). Hans Busk was Loder's brother-in-law (*Burke's peerage* 1980).

[5] Blyth had been appointed curator of the museum of the Asiatic Society of Bengal in 1841. He was forced to leave Calcutta for England at the end of 1862 as a consequence of ill health, arriving in March 1863 (see Blyth 1875, p. xii, and letter from Edward Blyth, 27 March 1863).

[6] Abraham Dee Bartlett was superintendent of the Zoological Society's gardens in Regent's Park, London (*Modern English biography*). He conducted experiments and made observations for CD on rabbits, geese, peacocks and wild cats (see *Variation* 1: 109–11, 114–15, 288, 290; 2: 150; and *Correspondence* vols. 8, 9, and 10). See also *Correspondence* vol. 13, letter to [A. D. Bartlett], 24 August [1860–3].

[7] CD discussed the chief domestic duck breeds, and their derivation from the common wild duck, in *Variation* 1: 277–87, but he did not mention the ducks or shelducks to which Blyth refers here. CD's interest in hybrid ducks had led him to keep some at Down House during the 1850s (see *Correspondence* vols. 5 and 6).

From John Lubbock 7 April 1863

15, Lombard Street. E.C.
7 April /63

My dear Mr. Darwin

I was very glad to hear that you were better again, & shall much like to come up & have a chat with you.[1]

After many alterations of plan I now propose to run down to Elgin on Saturday next to see those Scotch kjökken möddings.[2] I shall probably be away a week & when I come back will write & try to fix some day to come up & dine with you as you kindly propose.

I am glad you approve of our last Number, it pleased me more than most, as there was more variety.[3] It was very good of you to give us a help & without making a precedent of it, I hope you will consider us favourably another time.[4] I hope you don't think that Lyell could really be vexed by my article.[5] He certainly has not quite done justice to others & particularly to Prestwich[6] I don't altogether like Carpenter's letter in the Athenæum.[7] Surely he is mistaken in saying that you would expect Foraminifera to become any thing else.[8] I should have thot that you believed that once a For. always a For. Is it not so?

Yours affecly | John Lubbock

C Darwin Esq

DAR 170: 39

[1] In his letter of 5 April [1863], CD had invited Lubbock to visit 'for a short talk'.

[2] *Kjökkenmödding* is the Danish word for refuse-heap or kitchen midden; Lubbock had written about recent archaeological research on Danish kitchen middens in Lubbock 1861, and published an account of the Moray Firth shell mounds near Elgin in the July 1863 number of the *Natural History Review* (Lubbock 1863e).

[3] See letter to John Lubbock, 5 April [1863]. Lubbock was one of the editors of the *Natural History Review*.

[4] Lubbock refers to 'Review of Bates on mimetic butterflies', CD's anonymous contribution to the April 1863 number of the *Natural History Review*.

[5] Lubbock reviewed Charles Lyell's *Antiquity of man* (C. Lyell 1863a) in the April 1863 number of the *Natural History Review* ([Lubbock] 1863c). See letter from J. D. Hooker, [24 March 1863], and letter to John Lubbock, 5 April [1863] and n. 7. See also n. 6, below.

[6] See letter from J. D. Hooker, [24 March 1863] and n. 9. In a letter to the *Athenæum*, 4 April 1863, pp. 459–60, Hugh Falconer had accused Lyell of slighting the roles of Joseph Prestwich and himself in field investigations of human artefacts in France and England in C. Lyell 1863a (see letter to J. D. Hooker, [17 April 1863] and n. 4). For evaluations of this disagreement between Falconer and Lyell, see Bynum 1984 and L. G. Wilson 1996a.

[7] William Benjamin Carpenter had disputed the review of Carpenter 1862 written anonymously by Richard Owen in the *Athenæum*, 28 March 1863, pp. 417–19 (see Appendix VII). See also letter to J. D. Hooker, [17 April 1863], and letter to John Lubbock, 5 April [1863] and n. 10.

[8] See Appendix VII for Carpenter's letter to the *Athenæum*, 4 April 1863. Carpenter had stated in Carpenter 1862 that there had been no advance in the foraminiferous type since the palaeozoic period. In his letter to the *Athenæum*, 4 April 1863, p. 461, Carpenter wrote:

the present state of scientific evidence, instead of sanctioning the idea that the descendants of the primitive type or types of Foraminifera can ever rise to any higher grade, justifies the *anti-Darwinian* inference, that however widely they diverge from each other and from their originals, *they still remain Foraminifera.*

From H. W. Bates 8 April 1863

King Street | Leicester
8 April 1863

My Dear Mr Darwin

I have been very busy for the last 10 days with all the little details connected with a change of residence & am now in Leicester packing up my collections & library for removal to London.[1] This will account for my not writing before to thank you for the very beautiful review of my essay in the "Natural History review".[2] By this days post I send a copy of my book of travels for your acceptance.[3] I finished all my work connected with it 6 weeks ago & am now quite tired of it, the delays having been so numerous & tedious.[4] I found I had written for about 200 pages more than was necessary & so was obliged to leave out much matter including a long, unfinished account of origin of species by segregation of races in support of yr theory.

Mr Murray seems very slow to push my book, friends here complaining that it has never been properly advertised.[5] The expenses were very great (over 700£) & the price in consequence is fixed high.[6] Very few copies will be sent to reviewers so that there will be slow progress. Your most well-timed remarks at the end of your little review will be the best advertisement I have had or expect to have.[7]

Yours sincerely | H W Bates

I remain here till the 14th.

DAR 160: 74

1 On his return from South America in 1859, Bates had joined his brothers in the family hosiery business in Leicester, often visiting London for the sake of his scientific work. Bates and his wife, Sarah Ann, moved to London in April 1863 (Woodcock 1969, pp. 240, 255). See also letters from H. W. Bates, 24 January 1863 and 20 April 1863.

2 Bates refers to CD's anonymous review of Bates 1861, which appeared in the April 1863 number of the *Natural History Review* ('Review of Bates on mimetic butterflies').

3 Bates refers to his account of his travels in the Amazon River basin (Bates 1863). CD's annotated copy is in the Darwin Library–CUL (see *Marginalia* 1: 35–7). CD helped and encouraged Bates to write Bates 1863 (see *Correspondence* vols. 9 and 10, and Bates 1863, p. iv.)

4 See *Correspondence* vol. 10, letter to H. W. Bates, 15 December [1862], and this volume, letter from H. W. Bates, 17 January [1863].

5 *The naturalist on the river Amazons* (Bates 1863) was announced in John Murray's list of new works for April 1863 (*Publishers' Circular* 26 (1863): 172, 193). CD had encouraged Bates to publish with Murray, and had written to the latter in support of Bates's work (see, for example, *Correspondence* vol. 9, letter to H. W. Bates, 25 September [1861], and *Correspondence* vol. 10, letter to John Murray, 28 January [1862]).

[6] Bates 1863 was published in two volumes and sold at 28*s.* (*Publishers' Circular* 26 (1863): 172, 193).
[7] Concluding his review of Bates 1861 ('Review of Bates on mimetic butterflies', p. 224; *Collected papers* 2: 92) CD had written:

> We think so highly of the powers of observation and reasoning shown in this Memoir, that we rejoice to see by the advertisements that Mr. Bates will soon publish an account of his adventures and his observations in natural history, during his long sojourn in the magnificent valley of the Amazon. We believe that this work will be full of interest to every admirer of Nature.

To M. T. Masters [8–13 April 1863][1]

Down Bromley Kent

My dear Sir

I am much obliged for your note—[2] I send 2 spikes of the Corydalis; some of flowers are in nearly intermediate state.—[3] I daresay I drew false inference, in my own mind alone, from your division of Peloria into two classes.—[4]

I had used term of arrest of development;[5] but it seemed to me hardly applicable in such cases as this of Corydalis in which there is extra development of a part.—[6] It is no doubt very theoretical to call them Reversions.—[7]

My dear Sir | Yours sincerely | C. Darwin

Cleveland Health Sciences Library (Robert M. Stecher collection)

[1] The date range is established by the relationship between this letter, the letter to M. T. Masters, 6 April [1863], and the letter from M. T. Masters, 14 April 1863.
[2] The letter from Masters has not been found; however, see the letter to M. T. Masters, 6 April [1863].
[3] In his letter to Masters of 6 April [1863], CD mentioned two different flower structures in *Corydalis tuberosa*.
[4] See letter to Masters of 6 April [1863] and n. 5. The reference is to Masters 1863, p. 260.
[5] CD's previous use of this term has not been found. However, Masters cited *Origin*, p. 145 (Masters 1863, p. 260) where CD noted the abortion and also the shortening of nectaries in pelargoniums. In *Variation* 2: 58–9, CD referred to Masters's use of the term 'arrest of development' in regard to peloric flowers. For CD's 1862 and 1863 experimental notes on the occurrence of peloric flowers in pelargoniums, see DAR 51: B4–15.
[6] CD refers to the development of a second functioning nectary in some flowers of *Corydalis* (see letter to M. T. Masters, 6 April [1863]); he called this the 'redevelopment of a partially aborted organ' (*Variation* 2: 59).
[7] See letter to M. T. Masters, 6 April [1863] and n. 3.

To H. W. Bates 9 April [1863][1]

Down. | Bromley. | Kent. S.E.
April 9th.

Dear Bates

I have an hour ago received your note & the Book.[2] I thank you cordially for your kind present. I have cut the pages & looked at most of the illustrations, & it seems to me one of the best got up books I have ever seen. I shall read it slowly & am sure it will interest me greatly.[3] Whether it sells quickly or no, I feel a conviction that this publication will always be herafter a great satisfaction to you, when your severe labour will be forgotten. You will see it often alluded to in other works. I fear the great expence, which is not surprising, will lessen profits.[4] It is so clear that it is Murray's interest to sell the Book largely & as he has had large experience, I think you need not fear he will advertise to right degree.[5] Remember how expensive advertisements are. He is never a great advertiser; & I have fancied when he feels sure that the Book will sell he advertises little. I do not think he has advertised Lyell much, & I am sure he advertised the Origin extremely little.[6]

I cordially congratulate you on publishing such a work.— When settled in London & you have leisure send me a line to say what your prospects are.— Shall you be in Holles St?[7] I wrote there some little time since telling you of A. Grays "fascination" at your Butterfly paper.—[8]

With every good wish.— believe me yours sincerely | Ch. Darwin

Beinecke Rare Book and Manuscript Library, Yale University

1 The year is established by the relationship between this letter and the letter from H. W. Bates, 8 April 1863.

2 CD refers to the letter from Bates of 8 April 1863, and to Bates 1863.

3 CD's annotated copy of Bates 1863 is in the Darwin Library–CUL (see *Marginalia* 1: 35–7).

4 See letter from H. W. Bates, 8 April 1863 and n. 6.

5 Bates was concerned that the publisher, John Murray, did not seem to be advertising Bates 1863 very widely (see letter from H. W. Bates, 8 April 1863).

6 Murray had also published Charles Lyell's *Antiquity of man* (C. Lyell 1863a) and *Origin*.

7 In January 1863, Bates's address in London had been 10 Hollis Place (see letters from H. W. Bates, 17 January [1863] and 24 January 1863). After his marriage, Bates and his wife moved to 22 Harmood Street, off Haverstock Hill, Hampstead (Woodcock 1969, p. 255). See also letter from H. W. Bates, 8 April 1863 and n. 1.

8 Asa Gray's letter to CD of 9 February 1863, in which he complimented Bates's paper, has not been found, but see CD's letter to Bates of 4 March [1863]. See also letter to Asa Gray, 20 March [1863]. Gray published a review of Bates 1861 in the September 1863 number of the *American Journal of Science and Arts* (A. Gray 1863a).

From Asa Gray 11 April 1863

Cambridge, Mass.
April 11, 1863.

My Dear Darwin

You see that—at length—the thing is nearly done—and—to use the expression here rebeldom is "gone up" You have long seen, I suppose, that I was right in saying there was but one possible end to the war.— also that the continuance for a time or abolition of slavery depended simply on the rebels—that if the[y] obstinately and persistently resisted, slavery was thereby doomed.[1]

It has been a long, wearing and trying work. But the country has had the needed patience and nerve.— and the thing is done—once for all, at great cost, but to immen⟨se⟩ and enduring advantage.

You are the only Britisher I ever write to on this subject, and, in fact, for whose opinions about our country I care at all.[2]

So I hasten to rejoice with you over the beginning of the end.

How is your health? Ever, dear Darwin, | Yours cordially A. Gray

DAR 165: 132

[1] Gray refers to the American Civil War and to optimistic reports in northern newspapers of the Union navy's assault on Charleston harbour; the attack on 7 April 1863 was ultimately unsuccessful, but this intelligence had not yet reached the north (see the *New York Times*, 11 April 1863, pp. 1, 4, and McPherson 1988, p. 646). For Gray's views on the inability of the Confederacy to sustain conflict with the Union, and on the eventual end of slavery, see *Correspondence* vol. 9, letter from Asa Gray, 31 December 1861, and *Correspondence* vol. 10, letter from Asa Gray, [late June 1862].

[2] Gray and Joseph Dalton Hooker, for example, had ceased discussing the American Civil War in their letters, although CD often forwarded Gray's letters on the topic to Hooker (see *Correspondence* vol. 10, letter from J. D. Hooker, [14 December 1862], and this volume, letter from J. D. Hooker, [1 March 1863]).

To Daniel Oliver [12 April 1863][1]

Down Bromley Kent
Sunday

Dear Oliver

It is a shame to trouble, but will you tell me whether ovule of Primula is "anatropal" nearly as figured by Gray p. 123 Lessons of Botany.—or rather more tending to "amphitropal".—[2] I never looked at such a point before. Why I am curious to know is because I put pollen into ovarium of monstrous Primroses, & now after 16 days, & not before, (the length of time agrees with slowness of natural impregnation) I find abundance of pollen-tubes emitted, which cling firmly to ovules & I think I may confidently state penetrate the ovule; but here is odd thing they never once enter at (what I suppose to be) the "orifice" but generally at *c* the chalaza, when they fall.—[3]

Do you know how pollen-tubes go naturally in Primula? do they run down walls of ovarium & then turn up the placenta & so debouch near the "orifices" of ovules?

If you thought it worth while to examine ovules, I would see if there are more monstrous flowers & put pollen into ovarium & send you the flowers in 14 or 15 days afterwards. But it is rather troublesome, I would not do it unless you cared to examine the ovules. Like a foolish & idle man I have wasted a whole morning over them.—

When does Hooker return?[4]

Yours most sincerely | C. Darwin

One line in answer would suffice—

In 2 ovules there was an odd appearance as if the outer coat of ovule at the chalaza end (if I understand the ovule) had naturally opened, or withered, where most of the pollen tubes seemed to penetrate which made me at first think this was a widely open foramen.[5] I wonder whether the ovules would could be thus fertilised!

Endorsement: 'Apr. 1863'
DAR 261 (DH/MS 10: 46)

[1] The date is established by the relationship between this letter and the letter from Daniel Oliver, 14 April 1863, and by the observations on 'monstrous Primroses' (see n. 3, below); in 1863, 12 April was a Sunday.

[2] CD wondered whether *Primula* ovules were parallel to and fused with the stalk of the ovule ('anatropal'), or whether the ovule was positioned at a right angle to the stalk ('amphitropal') (see A. Gray 1857, p. 123, for diagrams). CD's lightly annotated copy of A. Gray 1857 is in the Darwin Library–CUL (see *Marginalia* 1: 347).

[3] CD's experimental notes (DAR 108: 165 v.) record that he began this experiment on 26 March 1863 and first observed pollen grains emitted on 11 April. Pollen-tubes in flowering plants normally enter the ovule through the micropyle or orifice. According to Francis Darwin (*ML* 2: 333), the observation that pollen-tubes might enter through the chalaza was first published by the Dutch botanist Melchior Treub (Treub 1891).

[4] Joseph Dalton Hooker was on a trip to the channel islands of Jersey and Guernsey (see letter from J. D. Hooker, 20 April 1863); Oliver was assistant to Hooker in the herbarium at the Royal Botanic Gardens, Kew (R. Desmond 1994).

[5] See also letter to Daniel Oliver, [after 14 April 1863].

To John Scott 12 April [1863][1]

Down Bromley Kent
Ap. 12th

My dear Sir

I really hardly know how to thank you enough for your very interesting letter.[2] I shall certainly use *all the facts* which you have given me (in a condensed form) on the sterility of orchids in the work which I am now slowly preparing for publication.—[3] But why do you not publish these facts in a separate little paper? they seem to me well worth it, & you really ought to get your name known.[4] I could equally well use them in my Book. I earnestly hope that you will experiment on Passiflora, & let me give your results.[5]

Dr. A. Grays observations were made loosely, he said in letter he would attend this summer further to the case, which clearly surprised him much.—[6] I will say nothing about the Rostellum, stigmatic utriculi, fertility of Acropera & Catasetum, for I am completely bewildered:[7] *it will rest with you to settle these points by your excellent observations & experiments.*— I must own I never could help doubting Dr. Hooker's case of the Poppy.—[8] You may like to hear what I have seen this morning: I found a Primrose plant with flowers having 3 pistils, which when pulled asunder without any tearing, allowed pollen to be placed on ovules. This I did with 3 flowers— pollen tubes did not protrude after several days. But this day, the sixteenth (N.B Primulas seem naturally slowly fertilised) I found many tubes protruded, & what is very odd they certainly seemed to have penetrated the coats of the ovules, but in no one instance the foramen of the ovule!![9] I mention this because it directly bears on your explanation of D^r. Cruger's case.—[10] I believe that your explanation is right; I sh^d. never have thought of it; yet this was stupid of me, for I remember thinking that the almost *closed* imperfect flowers of Viola & Oxalis were related to the protrusion of the pollen-tubes.[11] My case of the Aceras with aborted labellum squeezed against stigma supports your view.—[12] D^r. Cruger's notion about the ants was a simple conjecture.—[13] About crptogamic filaments rember D^r. C. says that the unopened flowers habitually set fruit—[14]

I think that you will change your view on the imperfect flowers of Viola & Oxalis;[15] I am now making a few observations on them: last year I observed a few intermediate forms.—[16]

I have asked everywhere for seed of Campanula perfoliata.[17] Can you get me any?

I am particularly obliged for your remark on Auriculas.—[18]

It seems to me a good experiment, that of preventing the orchid flowers opening.—[19] What an excellent & indefatigable observer you are.

with sincere respect | Yours very faithfully | Ch. Darwin

P.S. M^r. Anderson sent me some seeds of the "abortive" Cattleya crispa; & not one in 100 or 200 contained even a trace of nucleus; the testa being quite empty.—[20]

DAR 93: B59, B77–8

1 The year is established by the relationship between this letter and the letter from John Scott, [1–11] April [1863].

2 Letter from John Scott, [1–11] April [1863].

3 CD wrote a draft of the section of *Variation* on 'Crossing & Sterility' between 1 April and 16 June 1863 (*Variation* 2: 85–191; see 'Journal' (Appendix II)). Scott's observations on self-sterility and cross-fertility in *Oncidium* and *Maxillaria* were briefly discussed in *Variation* 2: 133, 164.

4 Scott detailed his observations on self-sterility and cross-fertilisation in the orchid genus *Oncidium* in a paper read before the Botanical Society of Edinburgh on 14 May 1863 (Scott 1863a). See also Scott 1864b.

5 CD had been encouraging Scott to experiment on *Passiflora* since March 1863 (see letter to John Scott, 24 March [1863] and nn. 3 and 4). Scott later published his experiments on *Passiflora* in Scott 1864d. CD referred to Scott's results in *Variation* 2: 137–8. See also *Cross and self fertilisation*, p. 330.

6 CD refers to Asa Gray's observations of pollination in *Gymnadenia tridentata*, made in July and August 1862 (see *Correspondence* vol. 10, letters from Asa Gray, 29 July 1862 and 18–19 August 1862), and to the letter from Asa Gray, 22 September 1862 (*ibid.*). See letter from John Scott, [1–11] April [1863] and n. 13.

7 See letter from John Scott, [1–11] April [1863] and nn. 16–17. CD subsequently discussed *Acropera* and *Catasetum* in 'Fertilization of orchids', pp. 153–4 (*Collected papers* 2: 150–1); see also *Orchids* 2d ed., pp. 247–57, for mention of these genera, and for further discussion of the rostellum.

8 The reference is to an article by Joseph Dalton Hooker on the possibility of artificially pollinating flowers of the blue poppy without the involvement of the stigma (J. D. Hooker 1854a). See letter from John Scott, [1–11] April 1863 and n. 19.

9 See preceding letter and n. 3, letter from Daniel Oliver, 14 April 1863, and letter to Daniel Oliver, [after 14 April 1863].

10 See letter from John Scott, [1–11] April [1863] and n. 25. Hermann Crüger had observed the emission of pollen-tubes while the pollinia remained *in situ* in unopened flowers of Epidendreae (see letter from Hermann Crüger, 23 February 1863).

11 See letter from John Scott, [1–11] April [1863] and n. 30.

12 In *Orchids*, p. 324 n., CD described 'monstrous' flowers of *Aceras* in which pollen germinated and formed pollen tubes while still in the anther. See also letter to John Scott, 24 March [1863], and letter from John Scott, [1–11] April [1863].

13 See letter from John Scott, [1–11] April [1863]. In his letter to CD of 23 February 1863, Crüger had suggested that the protrusion of pollen-tubes that he had observed in Epidendreae (see n. 10, above) was due to ants carrying stigmatic fluid to the pollen. See also letter to John Scott, 24 March [1863].

14 Scott had suggested (see letter from John Scott, [1–11] April [1863]) that the pollen-tubes described by Crüger as occurring in unopened orchid flowers (see letter to *Journal of Horticulture*, [17–24 March 1863]) could be fungal filaments; Scott had observed such filaments in the flowers of *Bletia*, initially mistaking them for pollen-tubes.

15 As he explained in his letter of [1–11] April [1863], Scott believed that the germination of pollen-grains *in situ* in cleistogamic flowers, like those of the imperfect forms of *Viola* and *Oxalis*, was a 'purely accidental occurrence, having *no ultimate import* in the economy ⟨of⟩ the spec⟨ies⟩'. For CD's study of the flowers, see n. 16, below, and letter to *Journal of Horticulture*, [17–24 March 1863] and n. 7. See also letter to Asa Gray, 31 May [1863].

16 For CD's observations of *Viola* and *Oxalis* in March and April 1863, see DAR 111: 6, 10, and 45. For CD's observations and experiments on the two genera in 1862, see *Correspondence* vol. 10. CD's published observations on cleistogamic flowers in these genera are in *Forms of flowers*, pp. 314–24.

17 See letter to Asa Gray, 20 March [1863] and n. 5. For CD's notes on the genus, see DAR 111: 29–32.

18 Scott had forwarded information to CD regarding the inheritance of style length in *Primula auricula* (see letter from John Scott, [1–11] April [1863]).

19 To investigate the cause of self-fertilisation in flowers that did not open, Scott had suggested cementing together the perianths of 'perfect' flowers shortly before their expansion (see letter from John Scott, [1–11] April [1863]).

20 See letter from James Anderson, 1 April 1863 and nn. 1 and 2.

From John Scott [after 12] April [1863][1]

Botanic Garden
April

Sir,

I duly received your Linum paper for which I am greatly obliged.[2] I have read it with great interest, and was quite unprepared for such remarkable results, as those you have so satisfactorily ⟨d⟩emonstrated. The fact that *they* ⟨ ⟩ have so long escaped observation ⟨ ⟩ ⟨show⟩s how much careful experimental observations are required.

I thank you most sincerely for the kind and friend advice you have given me in your last,[3] as well as for those complimentary remarks which you have frequently honoured me with: which gratify me in showing, that you do not regard me as an entirely unworthy correspondent. In regard to publishing more frequently, I may state that my chief objective was the idea that in my present position anything I might publish would be entirely disregarded.[4] I therefore, thought that ⟨I had⟩ better have a little patience, & see if I c⟨ould⟩ not succeed in getting some foreign appointment, where I would have a wider field for original observations, and likewise a better chance of their being attended to when published. I am anxious to remain where I am for this season, that I may work on to the best of my abilities the interesting experiments you have suggested to me:[5] after which I will be glad if I can [fortunately] succeed in getting a foreign or colonial appointment.[6]

At your suggest⟨ion I have⟩ drawn up a paper on the Ind⟨ ⟩ ⟨ ⟩ Sterility of Plants, which I will read at the ensuing meeting of the Botanical Society.[7] I am glad that I have seen your Linum paper before reading it: as I have stated that the sterility of the *Oncidium* was not simply due to the non-emission of pollen-tubes; having found an abundance of them in conducting tissues of styles.[8] You I observe have made similar observations on *Linum perenn⟨e⟩* & *Lythrum Salicaria*, which I will take the liberty of appending to a paper as a note[9]—contrasting them as you have done with the L. *grandiflorum* case.

Your experiment on the Primrose has interested me very much;[10] it is certainly very singular that the pollen-tubes should have penetrated the coats of the ovules. I have made of late a few experiments of a similar nature upon different plants; but with no success, all the capsules dropping off early. However, I neglected to see whether or not the pollen-grains had emitted tubes. I will look to this in my future experiments. I am sorry to say that I can get ⟨no⟩ seed of the *Campanula perfoliata* for you, or even ascertain where it is to be got![11] I have been trying to get plants of it for sometime back, that I might make a few observations on its imperfect flowers: but I have likewise failed in this.

I send off by mail a packet containing 4 ripe capsules of *Gongora atropurpurea*, and 2 unripe—of G. *truncata*.[12] Those each of the former are the largest & finest the plant has produced. I took the precaution to ⟨attach⟩ a small paper bag to each capsule as soon as it gave indication of dehiscing, as thus I might allow them to remain ⟨ ⟩ ⟨the plant⟩ till they were quite *ripe without any loss of seed*. Those of G. *truncata*, I have many seeds for comparison. I suppose they ⟨are⟩ nearly ful⟨l⟩ grown. I have had even greater difficulty in fertilising it than G. *atropurpurea*; its stigmatic orifice was in general so much contracted, that I could rarely force a pollen-mass into it. The *smaller* of the 2 capsules sent, will show you how I have succeeded in fertilising, when stigmatic orifice would not **admit** pollen-mass. I simply *cut* off the *clinandrum*, so as *fully* to *expose stigmatic chamber*, into which I immediately *insert* the *pollen*. Thus I accomplish fertilisation ⟨*by*⟩ *force*. It must certainly be a most ingenious insect that fertilises these plants in their native wilds, if they there present the same peculiarities they do with us! How delightful it would be to be there to watch the process: If more capsules of either are required, I will gladly furnish you with them.

Excuse the hurried scrawl.

I remain | Sir | Yours very respectfully | John Scott

P.S: I thought I would have got a few plants of *Primula Scotia* for you but I have failed, and we ⟨have⟩ only *three* plants in the Garden![13]

DAR 177: 87

CD ANNOTATIONS
0.3 Sir, . . . comparison. 5.6] *crossed pencil*
5.12 Thus . . . us! 5.14] *double scored red crayon*
8.1 P.S: . . . Garden! 8.2] *crossed pencil*
Top of letter 'Gongora' *ink, circled ink*

1 The date is established by the relationship between this letter, the preceding letter, and the letter to John Scott, 2 May [1863].
2 'Two forms in species of *Linum*'. Scott's name appeared on CD's presentation list for this paper (see Appendix IV).
3 See preceding letter.
4 Scott was foreman of the propagating department at the Royal Botanic Garden, Edinburgh (R. Desmond 1994).
5 CD had been encouraging Scott's experimental work since November 1862 (see *Correspondence* vol. 10). In his letter to Scott of 19 December [1862] (*ibid.*), CD sent a list of suggested experiments.
6 Scott became head of the herbarium at the botanic garden in Calcutta, India, in 1864 (Freeman, 1978).
7 Scott 1863a was read before the Botanical Society of Edinburgh on 14 May 1863.
8 Scott 1863a, p. 546.
9 See Scott 1863a, p. 547 n.
10 CD had described artificial fertilisation of ovules in a primrose (see preceding letter, letter from Daniel Oliver, 14 April 1863, and letter to Daniel Oliver, [after 14 April 1863]).
11 See preceding letter and n. 17.
12 Scott and CD had been discussing the pollination of *Gongora* and the related *Acropera* since November 1862 (see *Correspondence* vol. 10, letter from John Scott, 11 November 1862). In his letter of 3 March 1863, Scott noted that he had successfully pollinated *Gongora atropurpurea.* See also letter from John Scott, 21 March [1863], and letter to John Scott, 24 March [1863]. CD cited Scott's observations on the fertilisation of *Gongora* in *Orchids* 2d ed., p. 169.
13 CD had written that his *Primula scotica* specimens, which had been sent to him by Scott with his letter of 6 January 1863, looked 'very sick' (see letter to John Scott, 24 March [1863]).

From David Thomas Ansted 13 April 1863

Impington Hall | Cambridge.
13 April 1863

My dear Darwin

I am now back again and shall hope to hear from you and make some definite arrangement about the Bond.[1] I have written to Ransome[2] & shall probably see him in town on Thursday. Will you write to me *to the Athenæum* and let me know what you think right and fair.

always yours | D. T. Ansted

DAR 159: 74

1 In 1854, CD had invested in and made bonded loans to Ansted's Patent Siliceous Stone Company. CD had recently agreed to wait for the formation of Ansted's new company and the repayment of his loan (see CD Memorandum, 14 February 1863).
2 Ansted and Frederick Ransome were co-owners of the Patent Siliceous Stone Company (see CD's Investment book (Down House MS), entries for November 1854 and April 1855).

To William Henry Flower 13 April [1863][1]

Down. | Bromley. | Kent. S.E.
April 13th

My dear Sir

I have been so long familiar with your name, that I trust you will excuse me addressing you, as if we were acquainted.

Several years ago I presented to the College the skull of the Niata Ox.[2] This day I received a letter from M. Quatrefages, asking me, if I possibly could get for him a photograph of this skull, or a cast for the Museum.—[3]

Would it be possible to make so huge a cast & would it not be very expensive? Will you have the great kindness to assist M. Quatrefages & greatly oblige me.—

Do you know any photographer & could you get a photograph (I presume side & front view would be necessary) made.[4] If you will take this trouble I shd. be very much obliged. Perhaps, also, you can give me some information about a cast: if permission were granted & it could be made, I think I had better give M. Quatrefages some idea of cost. He could judge best after seeing Photographs.— I hope that you will forgive me thus troubling you. I would have come to London; but my health is so indifferent that this is no slight fatigue to me.—

My dear Sir | Yours faithfully | Ch. Darwin.

John Innes Institute

[1] The year is established by the relationship between this letter and the letter from W. H. Flower, 15 April 1863.

[2] CD presented a niata skull, collected on the *Beagle* voyage, to the Royal College of Surgeons in January 1840 (see letter from Armand de Quatrefages, [28 March –] 11 April 1863 and nn. 7 and 8). Flower was curator of the Hunterian Museum of the Royal College of Surgeons (*DNB*).

[3] See letter from Armand de Quatrefages, [28 March –] 11 April 1863. Quatrefages was professor of the natural history of man at the Muséum d'Histoire Naturelle in Paris (*DSB*). See also letter to Armand de Quatrefages, 14 May [1863].

[4] CD sent photographs of the niata skull to Quatrefages in May (see letter to Armand de Quatrefages, 14 May [1863], and letter from Armand de Quatrefages, 19 May [1863]).

From Asa Gray 13 April 1863

Cambridge, [Massachusetts]
April 13, 1863

My Dear Darwin,

Your latest to me is March 20.[1] I am very sorry that your health does not hold out as well as it ought.

To my disappointment all my extracts of Bates' paper on Mimetic Analogy are *laid over* to July no. of Sill. Journal,[2]—which you will not wonder at when I tell you that they have put in type at least 16 pages of mine for May no.—the larger part being review of A. DC. on *Species* derived from Oak-studies.[3]

⟨*half a page excised*⟩ am not surprised at his caution about adopting your theory, nor that his opinion about change—i.e. slow & gradual change should "fluctuate".[4]

Pray *do* finish & bring out your book on *Variation*. I have great expectations of it.[5]

I have to-day little to write, but cannot resist dropping you a line, though there be nothing in it.

Ever Yours | A. Gray

Incomplete
DAR 165: 133

CD ANNOTATIONS
On cover, recto: '(Plantago)[6] | angle of leaves | Bells— *[*Recurrent*]* angles'[7] *ink*
On cover, verso: 'Language | Lyell does not act as Judge, & this grieves me.—'[8] *ink, del ink*

1 Letter to Asa Gray, 20 March [1863].
2 Gray refers to his review of Henry Walter Bates's account of mimetic butterflies (Bates 1861), which eventually appeared in the September 1863 number of the *American Journal of Science and Arts* (A. Gray 1863a). The journal was commonly known as 'Silliman's journal', after its founder and editor Benjamin Silliman.
3 Gray's review of Alphonse de Candolle's 'Etude sur l'espèce á l'occasion d'une révision de la famille des Cupulifères' (A. de Candolle 1862a) appeared in the May 1863 number of the *American Journal of Science and Arts* (A. Gray 1863d).
4 Gray refers to Charles Lyell and his discussion of CD's theory in *Antiquity of man* (C. Lyell 1863a; see letter to Asa Gray, 20 March [1863]).
5 *Variation* was published in 1868.
6 See letter to Asa Gray, 11 May [1863] and n. 14.
7 See letter to Asa Gray, 11 May [1863] and nn. 8–11.
8 See letter to Asa Gray, 11 May [1863], and letter to J. D. Hooker, [9 May 1863].

From George Bentham [*c.* 14 April 1863][1]

25 Wilton Place | SW

My dear Darwin

Would you oblige me by telling me if you know of anything worth looking at for or against or upon your views of the Origin of species which may have appeared abroad[2]—besides what Asa Gray or A. De Candolle have written—[3] I do not mean mere reviews or analysis of your work such as those of the Revue des Deux Mondes[4] etc—but any original observations or rational discussion— Have the Germans taken up the subject? I believe Decaisne's memoir on the variation of species derived chiefly from the observation of cultivated fruits—and Naudin's on hybrids which contains I believe a great deal of important matter are neither of them yet published[5]

Yours very sincerely | George Bentham

DAR 160: 155

1 The date is established by the relationship between this letter and the letter to George Bentham, 15 April [1863].

[2] As president of the Linnean Society, Bentham was required to deliver the anniversary address to the society at a meeting on 25 May 1863; traditionally this address reviewed developments in natural history during the previous year. Bentham was planning to discuss the published reaction to CD's evolutionary theory (see Bentham 1863).

[3] A. Gray 1861a and A. de Candolle 1862a; CD's annotated copies of these works are in the Darwin Pamphlet Collection–CUL.

[4] Laugel 1860. There is an annotated copy of this work in the Darwin Pamphlet Collection–CUL. The *Revue des Deux Mondes* was a leading French literary and critical journal.

[5] The references are to Decaisne 1858–75 (published in 129 parts forming nine volumes), and to either Naudin 1863 or Naudin 1865 (see *Correspondence* vol. 10, letter from C. V. Naudin, 26 June 1862 and n. 4). Naudin 1863 was later also published as part of Naudin 1865; an annotated copy of the former is in the Darwin Pamphlet Collection–CUL, while an annotated copy of the latter is in the Darwin Library–CUL (see *Marginalia* 1: 638–9).

From M. T. Masters 14 April 1863

Rye Lane | Peckham. SE
April 14. 1863

My dear Sir

Many thanks for the specimens you were kind enough to forward to me[1] They singularly well illustrate the truth of the oft made remark that what is abnormal in one genus is the rule in another and have much significance when considered in relation to your views of inheritance and descent[2] I notice that the more perfectly developed flowers are placed toward the upper part of the cluster

The peculiar arrangement of the stamens in this family is exceedingly curious and not well-understood I think—[3]

Again thanking you for your kindness believe me dear Sir | Faithfully Y[rs]. | Maxwell. T. Masters.

Ch[as]. Darwin Esq

DAR 171: 69

[1] See letters to M. T. Masters, 6 April [1863] and [8–13 April 1863].

[2] CD discussed *Corydalis* in *Variation* 2: 58–9, as part of his chapter on reversion (*ibid.*, pp. 28–61). See also letter to M. T. Masters, 6 April [1863] and n. 3.

[3] In the Fumariaceae the six stamens are in two sets of three, on opposite sides of the flower, and are covered by the fused inner pair of petals. The three stamens in each set are often fused together at the base, with only the middle stamen having a complete anther (Ernst 1962).

From Daniel Oliver 14 April 1863

Royal Gardens Kew
14.IV.1863

My dear Sir

Let me *beg* you *never* to apologize. I feel so utterly unable to help you in any way unless it be about these little questions.[1]

The ovule of *Primula* is *amphitropal*, or what Agardh wd. call *apotropo-amphitropal* I suppose.[2]

Here it is.

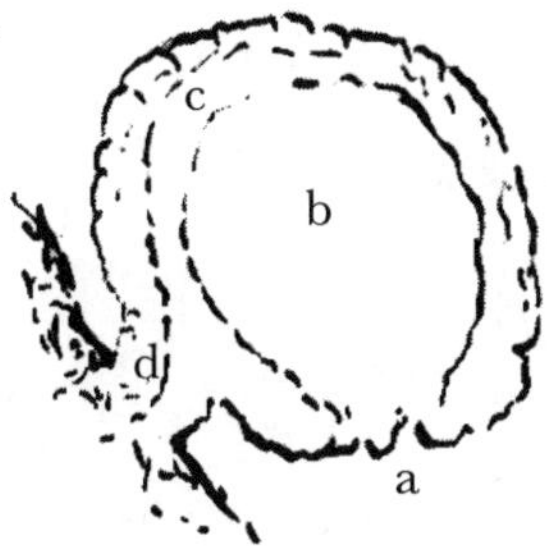

a. foramen or micropyle
b. Nucleus (Embr. sac.)
c. Chalaza
d. hilum

It is *all but* anatropal If the funicle were adnate quite down to the micropyle then it wd. be so.[3]

Hofmeister says Primulaceae differ from the great majority of micropetalous plants in having 2 coats to their ovules.[4] They have generally but 1 coat.

The inner is thick, the outer very thin. He says the exostome (opening through the *outer* coat) "liegt vom Endostom (openg. thro' inner coat) eine Strecke nach der Raphe hin entfernt; der Pollenschlauch hat von jenem zu diesem ein Stück zwischen beiden Eihüllen hinzukriechen"—[5] This is a very curious observation. The openings appear thus *not* to coincide.

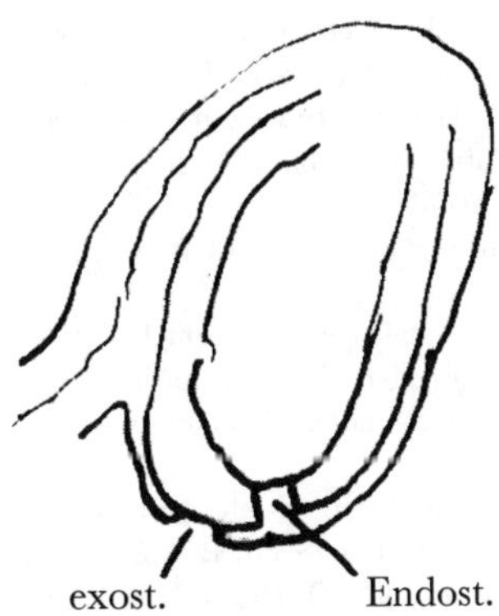

What you say is very curious about pollen-tubes penetrating ovules.[6] I do not remember any case of direct action except of course the normal place in Gymnosperms.

A good part of a notice of *Welwitschia* in N.H.R. I gave to the question of this Gymnospermy.—[7] I have made drawings from the very interesting *Primulas* you kindly sent,[8]—but they do not (seem to me to) furnish *decisive* evidence against Caspary.[9] I am not *sure* as to direction taken by the tubes but from what I see in an ovary now before me of *Primrose* think they do *not* go the round-about way down the sides of the cavity—& *up* the placenta, but that they strike right upon the top of it near the side of the spear-process from its centre. It must surely be so.

We expect D^r^. Hooker anytime.— Today I daresay.—[10] The opening at the Chalazal end is something astounding!—[11]

Ever very sincerely yours | D Oliver

I have re-opened this to say D^r^. Hooker is not coming until *Friday*. Lady H. told me a few days ago his tour had been quite successful so far.[12] I have looked further at ovary of Primrose & the tubes land right upon the placenta & do not trouble to go down & up again—at least I dont see them do so,—tho' they must creep down by each ovule to get to its micropyle—[13]

Yrs | D O

DAR 173: 21

[1] See letter to Daniel Oliver, [12 April 1863].

[2] In his letter to Oliver of [12 April 1863], CD asked about the position of the ovule in *Primula*; he wondered if it was 'amphitropal' or 'anatropal'. Jacob Georg Agardh was professor of botany at the University of Lund, Sweden (*SBL*). The use of the term 'apotropo-amphitropal' is not apparent in Agardh 1858, but see p. 331 and Tab. XXVII, fig. 1.

[3] See letter to Daniel Oliver, [12 April 1863] and n. 2.

[4] The reference is to Wilhelm Hofmeister and Hofmeister 1858, p. 199. Micropetalous: 'having very small petals' (*OED*).

[5] Hofmeister wrote that the exostome, the opening through the outer ovule coat, was 'situated a distance from the endostome in the direction of the raphe; the pollen tube has to creep between both ovular membranes from the former to the latter' (Hofmeister 1858, p. 119). The endostome was the opening in the inner ovule coat.

[6] CD had noted that pollen-tubes appeared to penetrate the ovule at the chalaza (see letter to Daniel Oliver, [12 April 1863]). See also letter to John Scott, 12 April [1863].

[7] [Oliver] 1863c, pp. 205–9, was a review of J. D. Hooker 1863a, and appeared in the April number of the *Natural History Review.*

[8] See letter to Daniel Oliver, 28 March [1863]; CD and Oliver were interested in the morphology of the ovary and its placenta (see also letter to Daniel Oliver, 24–5 March [1863], and letter from Daniel Oliver [26 March [1863]).

[9] Oliver refers to Caspary 1861 and Robert Caspary's notion of the relationship between the pistil and carpels (see letter from Daniel Oliver, [26 March 1863] and n. 3).

[10] In his letter to Oliver of [12 April 1863], CD asked when Joseph Dalton Hooker would be returning from the Channel Islands (see n. 12, below); Oliver assisted Hooker in the herbarium at the Royal Botanic Gardens, Kew, where Hooker was the assistant director (R. Desmond 1994).

[11] Oliver refers to CD's letter of [12 April 1863] in which he noted his observation of pollen-tubes penetrating the chalazal end of the ovule.

[12] Oliver refers to Hooker's mother, Maria Hooker. Hooker was on a trip to Dorset and the Channel Islands of Jersey and Guernsey (see letter from J. D. Hooker, 20 April 1863).

[13] See nn. 6 and 8, above.

To Armand de Quatrefages[1] [14 April 1863][2]

Ce nâta est pour vous, un cas excellent car la race est bien établie et doit tirer son origine de l'Amérique du Sud.[3] C'est une race très-singulière. Une courte description de la tête a été faite par le professeur Owen dans le *Catalogue descriptif de la collection ostéologique du collège des chirurgiens*, 1853 page 624. Vous y verrez que *la connexion des os* est modifiée, car le maxillaire ne s'unit pas aux os du nez.[4]

En examinant un grand nombre de squelettes de lapins, de canards, de poulets et de pigeons, j'ai rencontré plusieurs modifications remarquables dans les squelettes; mais comme elles ne sont pas encore publiées, vous ne pourriez en tirer aucun parti.[5] Le pigeon Powter (grosse-gorge ou boulans) descendant certainement du *Columba Livia*, est un cas intéressant car il a l'œsophage très-amplifié et modifié.[6]

Dans le premier chapitre de mon *Origine de l'Espèce*, vous trouverez un extrait des modifications remarquables chez les pigeons.[7]

Les poulets à crête, ou, comme nous les appelons, *polish fowls*, sont un excellent cas;[8] ils produisent parfaitement bien; autrefois, il n'y avait qu'un sexe dont le crâne fût affecté, maintenant les deux sexes sont affectés. J'insère ici une gravure sur bois très-bien faite.[9] J'ai examiné beaucoup de crânes. Il y a un changement étonnant de forme dans le *cerveau lui-même*, et dans *l'intérieur* du crâne, en outre du changement des os extérieurs.

Incomplete[10]
Bulletins de la Société d'Anthropologie de Paris 4 (1863): 378–9

[1] CD's original letter has not been found. For a translation of this text, see Appendix I. The extant portion of the letter-text was published in the *Bulletins de la Société d'Anthropologie de Paris* 4 (1863): 378–9. Quatrefages probably translated this portion for his presentation to the Société d'Anthropologie (see n. 2, below).

[2] The date was given by Quatrefages when he introduced the letter in a discussion at the Société d'Anthropologie on 16 July 1863 (*Bulletins de la Société d'Anthropologie de Paris* 4 (1863): 378).

[3] See letter from Armand de Quatrefages, [28 March –] 11 April 1863 and nn. 7, 8, and 9. CD thought that niata were a useful example of the environment's effect on anatomical form; the breed had been selected from cattle imported from Europe (*Variation* 1: 88–91).

[4] Niata were described in Owen [1853], 2: 3832 (see also *Variation* 1: 89–91). For Quatrefages's presentation of photographs of the niata skull at discussions of the Société d'Anthropologie de Paris, see the letter from Armand de Quatrefages, 19 May [1863] and n. 4.

[5] CD published this material in *Variation* 1: 103–304. *Variation* was not published until 1868; Quatrefages read this letter of CD's to the Société d'Anthropologie on 16 July 1863, and a report of William Bernhard Tegetmeier's presentation of a Polish or crested breed of fowl to the Zoological Society of London (*Bulletins de la Société d'Anthropologie de Paris* 4 (1863), pp. 378–80; see *Proceedings of the Zoological Society of London*, 25 November 1856, pp. 366–8). CD may have mentioned this Zoological Society report to Quatrefages in the missing portion of this letter; he cited his reliance on Tegetmeier's

information on poultry in *Variation* 1: 225 (see nn. 8 and 9, below). Quatrefages briefly mentioned examples of variation in poultry to the Société d'Anthropologie on 7 May and 2 July (see *Bulletins de la Société d'Anthropologie de Paris* 4 (1863): 273, 350–1).

6 An engraving and description of the Pouter pigeon is provided in *Variation* 1: 137–9.

7 *Origin*, pp. 20–8.

8 The Polish or crested breed of fowl and its sub-breeds are described in *Variation* 1: 227–30.

9 The enclosure has not been found, but a woodcut of the Polish fowl appears in *Variation* 1: 229.

10 The letter is considered incomplete since it does not include a salutation or valediction. In addition, Quatrefages probably translated only a portion of the letter from CD for his reading to the Société d'Anthropologie, and some further text may be missing (see n. 1, above).

From Robert Swinhoe 14 April 1863

18 Royal Avenue Terrace | Chelsea
14 April, 1863.

My dear Sir,

A strong case in point for you is that of the Squacco Herons of the Old World, which you may or may not have in mind.[1] It may not however have been brought before you before. It is the point of the various so-called species being almost identical in immature and winter dress, and yet perfectly distinct in their summer garb. The recognised species are the *Ardeola comata*, & of Africa and Southern Europe, the A. *leucoptera*, Boddaert of India, the *A. speciosa*. Horsf. of Java, and the A. *prasinosceles*, Swinhoe of China—[2] Those in their summer dress are quite dissimilar, but almost indistinguishable in the plumage of the winter— There is however in Africa and South Europe a bird the *A. ralloides*, and in India and its archipelago another the *A. malaccensis*, which retains the immature or winter plumage with but little alteration all the summer through The changes of plumage in this group due to whatever cause are worthy of record. I have mentioned them in my Formosan ornithology, which contains besides many other useful hints for you.[3] I shall be sure to supply you with a copy of this paper, as soon as it is all printed off.[4] I hope you will allow me to see you at your residence before I leave England.[5] There is however as yet plenty of time, for I do not intend returning till September.

I remain, | Your's very truly, | Robert Swinhoe

DAR 84.1: 18–19

CD ANNOTATIONS

1.1 A strong ... before. 1.3] *crossed pencil*

1.1 Squacco ... Old] *circled pencil & transposed to after* 'so-called species' 1.3

1.3 you before. ... point of] *del pencil*

1.3 the various] *after opening square bracket, pencil*

1.4 immature] *underl pencil*

1.4 winter] *underl pencil*

1.15 I hope ... September. 1.17] *crossed pencil*

Top of letter: 'On Sexual & immature dress, in Ardea' *ink*; 'This will work in well when I discuss the garb of young-Birds.'[6] *pencil*

[1] Swinhoe wrote to CD on 12 November 1862 and 2 December 1862 (see *Correspondence* vol. 10), but no intervening letters from CD have been found. CD discussed Squacco herons as an example of sexual selection in *Descent* 2: 190–2, citing Swinhoe (see n. 2, below).

[2] Pieter Boddaert named *Ardeola leucoptera*, but the species classification for *Ardeola* was under contention (see Blyth 1849, p. 281 and Swinhoe 1863d, p. 421); Thomas Horsfield named *A. speciosa* in 1821 (see Mayr and Cottrell eds. 1979, p. 218). Swinhoe described *A. prasinosceles*, as well as other Squacco heron species, in Swinhoe 1863a, pp. 421–2. This paper was reprinted with different pagination (Swinhoe [1863d]) and circulated with a set of Swinhoe's other 1863 publications (Swinhoe [1863b–g]; see letter from Robert Swinhoe, 29 July 1863 and n. 3). CD annotated the description of *A. prasinosceles* in his presentation copy of Swinhoe [1863b–g], which is in the Darwin Library–CUL (see *Marginalia* 1: 797). CD referred to Swinhoe's information on *Ardeola* in *Descent* 2: 190.

[3] See nn. 1 and 2, above. CD cited other information from Swinhoe [1863d] in *Descent* 2: 179, 190, 214–15.

[4] Swinhoe [1863b]; see n. 2, above.

[5] Swinhoe was British vice-consul in Formosa (now Taiwan) (P. B. Hall 1987). He had returned on leave to London in September 1862 (see *Correspondence* vol. 10, letter from Robert Swinhoe, 12 November 1862).

[6] CD discussed the immature plumage of birds, including *Ardea* and other herons, in relation to the plumage of both sexes when adult in *Descent* 2: 182–223.

To Daniel Oliver [after 14 April 1863][1]

Down

Dear Oliver

Many thanks about the Primula: I see that I was pretty right about the ovules.[2] I have been thinking that the apparent opening at the chalaza end must have been withering or perhaps gnawing by some very minute insects, as the ovarium is open at the upper end. If I have time I will have another look at pollen-tubes, as from what you say they ought to find their way to the microphyle. But ovules to me are far more troublesome to dissect than animal tissue; they are so soft, and muddy the water.

With many thanks | Yours very sincerely, | C. Darwin

Copy
DAR 147: 214

[1] The date is established by the relationship between this letter and the letter from Daniel Oliver, 14 April 1863.

[2] See letter to Daniel Oliver, [12 April 1863] and n. 3, and letter from Daniel Oliver, 14 April 1863 and nn. 6 and 11.

To D. T. Ansted 15 April 1863

Down
April. 15— 1863

My dear Ansted,

I still think that the offer which I made to Mr Ransome of receiving £100 in payment in full for the £250, was fair & liberal.—[1] My Solicitor, who knew all the circumstances, thought that I ought to claim the whole & remonstrated with me.—[2]

Mr Ransome has not paid me, as he asked for further delay owing to his expectation that a new company would be immediately formed.—[3]

As I before said I do not wish or intend to press my legal claim, & you can pay me the above sum of 100£, or part, or nothing, according as your circumstances permit.[4] On this point I cannot judge & you must decide for yourself.—

Yours sincerely | Ch Darwin

Draft
DAR 210.10: 25

[1] See CD memorandum, 14 February 1863, and letter from D. T. Ansted, 13 April 1863 and n. 1. No correspondence between CD and Frederick Ransome has been found.

[2] CD's solicitor was William Mackmurdo Hacon (Freeman 1978).

[3] Ransome's new company, the Patent Concrete Stone Company, went into production in East Greenwich in 1867 (*Engineering*, 28 June 1867, pp. 671–2). CD's bond, however, may have been settled before 1867 (see CD memorandum, 14 February 1863 and n. 2, and letters from D. T. Ansted, 17 April 1863 and 23 April 1863).

[4] In CD's Investment book (Down House MS), the last entry concerning Ansted and the Patent Siliceous Stone Company was dated May 1863; it reiterated the settlement recorded in this letter: 'I have cancelled Ansted [*after* 'R' *del*] Bond on understanding that if he can ever sell the shares, he will repay me'.

To George Bentham 15 April [1863][1]

Down. | Bromley. | Kent. S.E.
April 15th

My dear Bentham

I have run my eye over the Reviews & Notices on the Origin (amounting to about 90!) & very few are worth your notice.[2]

Pictet's is about the most valuable from the man's knowledge.[3]

Bronn wrote a final chapter, at my request, with excellent objections, in his German Edition, which I have tried partly to answer in the 3d. Edition of Origin p. 133–143[4]

There is a favourable & long Review in Revue Germanique, which is important from coming from so good observer, as Claparede.[5]

Much has been published in Germany but not worth your attention.[6]

Of English I suppose you know Hopkins in Fraser, one of most important.[7]

One of very best & partly Botanical is by Mr Maw in Zoologist.[8]

There is one by Dr. Dawson in Canadian Naturalist, which, however, seems to me poor.—[9]

There is good one in Calcutta Review.—[10]

Of course these are not all worth your looking at, but as Book Post is so cheap, & as it may save you trouble, I send the lot.[11]

Please be careful of them, & let me have them back, as soon as you have made any use of them

Believe me | Yours very sincerely | Ch. Darwin

Royal Botanic Gardens, Kew (Bentham correspondence 3: 700)

[1] The date is established by the relationship between this letter and the letter from George Bentham, 21 April 1863.

[2] In his letter to CD of [*c.* 14 April 1863] Bentham requested information on the published reaction to CD's theory for his 25 May 1863 anniversary address as president of the Linnean Society (Bentham 1863). CD's 'List of reviews' (DAR 262 (DH/MS* 8: 6–18)) served as an index to the reviews of his works. For a list of reviews of *Origin* published in 1859 and 1860, see *Correspondence* vol. 8, Appendix VII. See also letter from George Bentham, 21 April 1863.

[3] There is an annotated copy of Pictet de la Rive 1860 in the Darwin Pamphlet Collection–CUL.

[4] Heinrich Georg Bronn translated the first German edition of *Origin* (Bronn trans. 1860). CD refers to Bronn trans. 1860, pp. 495–520. There is an annotated copy in the Darwin Library–CUL (see *Marginalia* 1: 180–2).

[5] There is an annotated copy of Claparède 1861 in the Darwin Pamphlet Collection–CUL. Edouard Claparède was a Swiss naturalist and invertebrate zoologist.

[6] For discussions of German reactions to CD's theory, see the letter from Friedrich Rolle, 26 January 1863, and the letter to Friedrich Rolle, 30 January [1863]. See also Montgomery 1974, and Corsi and Weindling 1985.

[7] Hopkins 1860 was published in *Fraser's Magazine*. On his 'List of reviews' (see n. 2, above), CD included the statement, 'Hopkins in Fraser says my view differs in nothing from Lamarck' (DAR 262 (DH/MS* 8: 9)).

[8] The reference is to George Maw's review of *Origin* ([Maw] 1861). There is an annotated copy in the Darwin Pamphlet Collection–CUL.

[9] J. W. Dawson 1860. There is an annotated copy in the Darwin Pamphlet Collection–CUL.

[10] Anon. 1860, which was published in the *Calcutta Review*, may have been written by Edward Blyth (see *Correspondence* vol. 8, letter to J. D. Hooker, 30 May [1860] and n. 7). There is a copy of Anon. 1860 in the Darwin Pamphlet Collection–CUL.

[11] Of the reviews CD listed in this letter, Bentham made direct mention only of Pictet de la Rive 1860, Bronn trans. 1860, and Claparède 1861 (Bentham 1863, p. xvii).

From W. H. Flower 15 April 1863

Royal College of Surgeons of England, | *London,* | *(W.C.)*
15th. *day of* April *1863*

My dear Sir

I shall have great pleasure in complying with your request,[1] as far as I am able— In the first place I will take the skull to be photographed, as soon as I have obtained (which I have no doubt I shall) the permission of the Committee, and I will make enquiries as to the cost of a cast—[2]

I hope that some time when you are in town I may have the pleasure of showing you the recent additions to this Collection—to the value of which you have contributed so largely[3]

Believe me | very faithfully yours | W. H. Flower.

DAR 164: 137

[1] See letter to W. H. Flower, 13 April [1863].
[2] In his letter to Flower of 13 April [1863], CD had asked Flower if he could obtain a photograph or cast of a skull of the South American niata cattle breed from the Royal College of Surgeons for Armand de Quatrefages (see letter from Armand de Quatrefages, [28 March –] 11 April 1863). See also letter to Armand de Quatrefages, 14 May [1863], and letter from Armand de Quatrefages, 19 May [1863].
[3] CD's fossil vertebrates from the *Beagle* voyage had been given to the Hunterian Museum, where Flower was the curator (D. M. Porter 1985, pp. 997–9).

From William Hepworth Dixon 16 April 1863

Athenæum Office, | 20, Wellington Street, Strand, W.C.
April 16 *1863*

Dear Sir

The extreme pressure of Scientific Controversy enduces me to think that your note will appear with a clearer field & better effect next week[1]—in which opinion, I hope, when you see the Ath^{m}. of Saturday, you will agree.[2] Sir C Lyell takes nearly six columns to reply to D^{r}. Falconer,[3]—& D^{r}. Carpenter has a letter on the Strange discovery, so interesting to you, at abbeville.[4] Carpenter has been over to Abbeville & has seen the jaw & the bed in which it was discovered[5]

Very truly Yours | W. Hepworth Dixon

DAR 162: 186

[1] CD's letter to the *Athenæum*, 18 April [1863], appeared in the issue of 25 April; he evidently sent it to Dixon before 16 April. Dixon was editor of the *Athenæum.*
[2] Dixon refers to the issue of the *Athenæum* of 18 April 1863.
[3] Charles Lyell's letter to the *Athenæum*, 18 April 1863, pp. 523–5, was a response to Hugh Falconer's attack in the *Athenæum* of 4 April 1863, pp. 459–60. Falconer had criticised Lyell's *Antiquity of man* for not sufficiently acknowledging research by Joseph Prestwich and himself on the precise dating of sediments and caves where human relics had been found in France and Britain (C. Lyell 1863a). See letter from John Lubbock, 7 April 1863 and n. 6.
[4] William Benjamin Carpenter's letter to the *Athenæum*, 18 April 1863, p. 523, described the discovery on 28 March 1863 of a fossil human jawbone in the Moulin-Quignon gravel pit near Abbeville, France, by the archaeologist Jacques Boucher de Perthes.
[5] Carpenter examined the jaw himself on 13 April 1863; on the following day he visited the Moulin-Quignon gravel pit with Boucher de Perthes (*Athenæum*, 18 April 1863, p. 523). See also Grayson 1985, pp. 213–17 and Van Riper 1993, pp. 134–9.

From Isaac Anderson-Henry 17 April 1863

Hay Lodge, | Trinity, | Edinburgh.
April 17/63

My dear Sir

Your last Letter of 2^{d} February informed me, what I was truly sorry to learn, of your being obliged to be absent for a time from ill health.[1] I hope you are now restored & in the midst of your most interesting pursuits, advancing science by fresh discoveries & covering yourself with still brighter laurels. I too am, in my humble sphere, also at work. Today I have, as you suggested tried the experiment on 3 different plants of Primula polyanthus to test the result you accomplished by long and short anthers.[2] And, I think, when I come to report upon it, you will allow I have done it ample justice.[3] I have had the plants under glass for some 3 weeks.

I have also been testing Mr Cunninghams experiment by which the *Bryanthus erectus* was said to be produced—and also the fertility of that plant touched by the pollen of the alleged parents respectively.[4] I have the *Pods* of *M. Cærulea* (*Phyllodoce Cærulea*) offering to swell with the pollen of the *Rhodothamnus Chaemæcistus*;—& the *M. empetriformis* with the same pollen seems in a hopeful way— Those crosses *inverted* on the *Rhodothamnus* also bid fair But of all this again, & other things, I have on hand

Meantime I hasten to say that, in a Letter I have from Dr Jameson dated 7 March he says;—"I had the pleasure of receiving a Letter from Mr Darwin who has been for some time pursuing a course of investigation on the staminal organs of the Melastomaceæ; & is desirous of obtaining facts relative the fertilisation of the fruit. I shall attend particularly to that matter and communicate the result by an early opportunity"[5]

He further informs me that he had been asked by the President (of the Equator) to publish a Flora of those Regions, which he regrets his inability to accomplish for various reasons.—[6] Such a work by one so able to do it justice would have been a great gain to Botanical Science. It is a pity he cannot undertake it

My dear Sir would you forgive me asking, if you have one to spare, a *Carte* of yourself.— Next to the pleasure of seeing the original will be that of viewing the likeness of one I honour & admire so much. These things do little justice to those like me past the meridian of life— But I presume yours will show the features of a younger man than the enclosed, which I humbly take the liberty of offering for your acceptance.[7] Believe with best wishes | most faithfully yours | Is. anderson Henry

Charles Darwin Esq. F.R.S. | &c &c

DAR 159: 64

CD ANNOTATION
End of letter: 'Bates Book'[8] *pencil*

[1] See letter to Isaac Anderson-Henry, 2 February [1863] and n. 3.

[2] No letter from CD to Anderson-Henry suggesting experiments with *Primula* has been found. However, Anderson-Henry had previously mentioned his intention to experiment with *Linum* in regard to dimorphism (see letter from Isaac Anderson-Henry, 17 January 1863), and CD encouraged this in his letter to Anderson-Henry of 20 January [1863]. CD had sent Anderson-Henry a copy of his paper 'Dimorphic condition in *Primula*' in 1862 (see *Correspondence* vol. 10, Appendix III).

[3] See letter from Isaac Anderson-Henry, 7 May 1863.

[4] See letter from Isaac Anderson-Henry, 26–7 January 1863 and n. 14. The cross that supposedly resulted in *Bryanthus erectus* was reported to have been achieved by James Cunningham, a Scottish

nurseryman, by crossing *Phyllodoce* (*Menziesia*) *caerulea* with *Rhodothamnus chamaecistus* (*Gardeners' Chronicle and Agricultural Gazette*, 1 November 1851, p. 695).

5 William Jameson was the director of the Mint in Quito, Ecuador, and professor of botany and chemistry at the University of Quito; he had made extensive tours of South America collecting plants (*DNB*). CD was interested in botanical collectors who could make observations of the Melastomataceae (see letter to Hugh Falconer, 5 [and 6] January [1863] and n. 22). Anderson-Henry provided CD with Jameson's address (see letter from Isaac Anderson-Henry, 26–7 January 1863), but no correspondence with Jameson has been found.

6 Jameson published a two-volume work on the flora of Ecuador two years later (Jameson 1865). Gabriel García Moreno was professor of chemistry at the University of Quito, and president of Ecuador (*EB*). 'Ecuador' is the Spanish word for equator.

7 The photograph of Anderson-Henry has not been found.

8 In his reply to Anderson-Henry of 20 April 1863 (now missing), CD apparently recommended that he read Henry Walter Bates's *The naturalist on the river Amazons* (Bates 1863; see letter from Isaac Anderson-Henry, 24 April 1863).

From D. T. Ansted 17 April 1863

Athenæum Club | London
17 April 1863

My dear Darwin

I hoped to have seen Ransome to day in town but he has not called.[1] He tells me that he is getting on well with his Company but gives no details and I suppose as he has not yet paid you the £100 he has agreed to find he has not yet realised anything himself.[2]

I fully believe that his present work will be profitable. It is highly spoken of by practical as well as theoretical men and if any Company with a sound foundation can succeed in paying anybody but the stock brokers and solicitors (may I add managers!) I think this should.

I can quite see that your Solicitors would advise you that if a man is fool enough to run risks he must take the consequences—[3] I can also feel that in being contented to take a part of what you can legally claim you are performing a good natured & perhaps your lawyers may say a very unwise & unnecessary piece of Quixotism.

In my case however I have been so uniformly unlucky in every thing I ever undertook that had an appearance of business that I have almost lost the right of appealing for consideration. I must be utterly without judgment and suppose I ought to suffer. That I have for years suffered and shall for years to come continue to suffer is certain for everything without exception that I ever touched is either absolutely worthless or more frequently a terrible incumbrance that I cannot shake off

If in this case of Ransome's I can ever realise anything however small I will engage to give you the full benefit. If I do not I really dare not promise much. I know you have the power to put me to the greatest inconvenience & I ask you to let me enter into some arrangement that would involve your destroying the terrible *deed* If it can be so I shall be extremely grateful. Should I or you die unexpectedly this deed might be a very serious matter for me. Will you be contented to take my assurance in any form that you shall have a claim on any interest or capital arising from any Company connected with Ransome's patents in which I have shares? This I would give cheerfully. I would most willingly pay you money but I really have no available means.

Above all however I should like to feel that the deed was no longer in existence though I suppose it is too much to expect that it should be given up without something definite being paid for it.[4]

I am sure you will sympathise with my great trouble in this matter and tell me more clearly your ultimatum

Yours always | D. T. Ansted.

C Darwin Esq.

DAR 159: 75

[1] The reference is to Frederick Ransome, Ansted's partner in the Patent Siliceous Stone Company (see n. 2, below).

[2] See CD memorandum, 14 February 1863, letter from D. T. Ansted, 13 April 1863 and n. 1, and letter to D. T. Ansted, 15 April 1863.

[3] See letter to D. T. Ansted, 15 April 1863. CD's solicitor was William Mackmurdo Hacon (Freeman 1978).

[4] See letter to D. T. Ansted, 15 April 1863 and n. 4. The bond to which Ansted refers was returned to Ansted (see letter from D. T. Ansted, 23 April 1863).

To J. D. Hooker [17 April 1863]

Down
Friday Night

My dear Hooker

I have heard from Oliver that you will be now at Kew, & so I am going to amuse myself by scribbling a bit.[1] I hope you have thoroughily enjoyed your tour: I never in my life saw anything like the Spring flowers this year.— What a lot of interesting things have been lately published. I liked extremely your review of Decandolle—[2] What an awfully severe article that by Falconer on Lyell.[3] I am very sorry for it: I think Falconer on his side does not do justice to old Perthes & Schmerling.[4] This must have been *far* more odious to Lyell than Owen's article:[5] I shall be very curious to see how he answers it tomorrow (I have been compelled to take in Athenæum for a while.)[6] I am very sorry that Falconer sh^{d}. have written so spitefully, even if there is some truth in his accusations.[7] I was rather disappointed in Carpenter's letter:[8] no one could have given a better answer. But the chief object of his letter seems to me to be to show that though he has touched pitch he is no defiled.[9] No one would suppose he went so far as to believe all Birds came from one progenitor.[10] I have written a letter to Athenæum (the first & last time I shall take such a step) to say, under the cloak of attacking Heterogeny, a word in my own defence.[11] My letter is to appear next week, so Editor says;[12] & I mean to send new addition & quote Lyell's sentence in his 2^{d}. Edit; on the principle if one puffs oneself, one had better puff handsomely.[13] I have read 1st. vol. of Bates' book;[14] it is capital, & I think the best Nat. Hist. Travels ever published in England. He is bold about Species, & that d——d Athenæum coolly says he "bends his facts" for this purpose.—[15]

What an astounding production that of Owen's![16] it was rather silly in Carpenter to be vexed, as he clearly was, at Owen calling me his Master,[17] & how like Owen's clever malignity.— I fancy Bentham is going to discuss species; for he has asked me about Reviews of Origin, & I have sent him a bundle.—[18]

I stumbled on reference yesterday, "Gard. Chron. 1861. p. 74; Benguela mixture of Tropical & Temperate forms"—[19] I daresay you know what it is; I forget; but no harm in reminding you—

Now for one or two questions.—

(1) Can you tell me Planchon's address?[20]

(2) Remember Orchis pods.—[21]

(3) You sent me a pale pink Oxalis,[22] which can be known by crescentic purple mark on leaves, just as in Dutch Clover; it was called *O. Bowii*, but is quite different from plant sold under that name. Do you know whether this name is correct?[23] I rather want to know, as I find leaves are sensitive to touch; as they are I find in nearly all species of Oxalis, (whose leaves go to sleep) in a very slight degree, but plainer in this species.[24]

GoodNight— Tell me where you have been— goodnight | C. Darwin

Endorsement: 'April 17th/63'
DAR 115: 190

1 Hooker had been on a trip to Jersey and Guernsey from 2 to 17 April 1863 (see letter from Daniel Oliver, 14 April 1863, and letters from J. D. Hooker, [28 March 1863] and 20 April 1863). Daniel Oliver assisted Hooker in the herbarium at the Royal Botanic Gardens, Kew (R. Desmond 1994).

2 [J. D. Hooker] 1863c. Hooker was evidently the author of the review of A. de Candolle 1862a that appeared in the April 1863 number of the *Natural History Review*. CD's copies of the *Natural History Review* are in the Darwin Library–CUL; the review of A. de Candolle 1862a is annotated.

3 Hugh Falconer's letter in the *Athenæum*, 4 April 1863, pp. 459–60, accused Charles Lyell of giving insufficient credit to him and to the geologist Joseph Prestwich in the discussion of fossil human relics in C. Lyell 1863a (*Athenæum*, 4 April 1863, p. 460). See also letter from John Lubbock, 7 April 1863 and n. 6.

4 Jacques Boucher de Perthes was a French archaeologist; Philippe Charles Schmerling was a palaeontologist who described fossil human bones from Belgium in the 1830s (*BNB*).

5 Richard Owen's letter criticising C. Lyell 1863a appeared in the *Athenæum*, 21 February 1863, pp. 262–3.

6 CD refers to Lyell's letter in the *Athenæum*, 18 April 1863, pp. 523–5, which was written in response to Falconer's letter (see n. 3, above). See letter from W. H. Dixon, 16 April 1863. CD did not usually subscribe to the *Athenæum* (see letter to J. D. Hooker, [29 March 1863]), but had recently become interested in the coverage of controversies between scientists (see, for example, letter to J. D. Hooker, 13 [March 1863]).

7 See n. 3, above.

8 In the issue of the *Athenæum* for 4 April 1863, p. 461, William Benjamin Carpenter responded to Owen's anonymous review of his *Introduction to the study of the Foraminifera* (Carpenter 1862), which appeared in the *Athenæum*, 28 March 1863, pp. 417–19 (see Appendix VII).

[9] In his letter in the *Athenæum*, 4 April 1863, p. 461, Carpenter defended himself against the charge that he had accepted CD as his master and CD's hypothesis as his guide (see Appendix VII). See letter from John Lubbock, 7 April 1863 and nn. 7 and 8.

[10] In his anonymous review in the *Athenæum*, 28 March 1863, p. 418, Owen had accused Carpenter of arguing that all Foraminifera descended from one prototype. Carpenter replied (*Athenæum*, 4 April 1863, p. 461):

> After quoting my general conclusions as to the *probable* derivation of all the divergent forms of Foraminifera from a few family types, and the *possible* derivation even of these from a common original, your reviewer remarks: 'We here discern the influence of Mr. Darwin's volume on the mind of the writer'.

See Appendix VII. However, Owen had previously stated that he believed all birds had descended from one prototype (see *Correspondence* vol. 10, letter to Asa Gray, 23[–4] July [1862] and n. 18).

[11] See letter to *Athenæum*, 18 April [1863].

[12] See letter from W. H. Dixon, 16 April 1863. CD's letter appeared in the *Athenæum*, 25 April 1863, pp. 554–5.

[13] C. Lyell 1863b. See letter to *Athenæum*, 18 April [1863] and n. 11.

[14] Bates 1863.

[15] Bates 1863 was reviewed in the *Athenæum*, 11 April 1863, pp. 488–9. The anonymous review concluded:

> We have already stated that one of the principal objects of Mr. Bates's explorations was to gather facts 'towards solving the problem of the origin of species.' He thinks he has found such a solution in adopting Mr. Darwin's theory and making many of his facts bend to it.

In the publisher's marked copy of the *Athenæum* (City University Library, London), the review appears to be attributed to 'Seemann', probably the botanist Berthold Carl Seemann.

[16] See n. 8, above. See also letter to J. D. Hooker, [29 March 1863].

[17] See n. 9, above, and Appendix VII.

[18] See letter from George Bentham, [*c.* 14 April 1863], and letter to George Bentham, 15 April [1863]. CD had long been interested in George Bentham's reaction to *Origin* (see *Correspondence* vol. 7, letter from J. D. Hooker, [20 December 1859] and n. 3, and *Correspondence* vol. 8, letter to J. D. Hooker, 3 March [1860]).

[19] The reference is to Welwitsch 1861, which was quoted in the *Gardeners' Chronicle*, 26 January 1861, pp. 74–5; this account described the botany of Benguela (Angola) and other parts of west Africa. For CD's and Hooker's recent correspondence on the subject of migration during a global cold period, see letter to J. D. Hooker, 5 March [1863] and n. 17. CD had urged Hooker not to believe that 'any one continent could have remained a hot refuge for all tropical productions of world' (see letter to J. D. Hooker, 17 March [1863]). See also letter from J. D. Hooker, [24 March 1863], and letter to J. D. Hooker, 26 [March 1863]).

[20] Jules Emile Planchon had assisted in the herbarium of William Jackson Hooker, J. D. Hooker's father, at the Royal Botanic Gardens, Kew from 1844 to 1848 (R. Desmond 1994). CD intended to send him a presentation copy of 'Two forms in species of *Linum*' (see Appendix IV).

[21] See letter to J. D. Hooker, 26 [March 1863] and n. 14.

[22] This plant probably arrived with the shipment of plants sent to CD from the Royal Botanic Gardens, Kew, in February 1863, to stock CD's new hothouse. See letter to J. D. Hooker, [21 February 1863], and Appendix VI.

[23] CD mentioned heterostyly in '*Oxalis Bowei*', a wood sorrel, in his letter to the *Gardeners' Chronicle*, 11 August 1866, p. 756 (*Collected papers* 2: 132). In *Forms of flowers*, p. 179, CD used the spelling *O. Bowei* (the modern spelling is *O. bowiei*).

[24] This appears to be CD's first indication of an interest in nyctitropic movements (where particular positions in plant parts are assumed at night). Movements of the leaves in a number of species of *Oxalis* are discussed in *Movement in plants*, but *Oxalis bowiei* is not mentioned. There are experimental notes on the power of movement in *Oxalis*, dated 31 March – 9 April 1863, in DAR 66: 1 and 2.

To *Athenæum* 18 April [1863][1]

Down, Bromley, Kent,
April 18.

I hope that you will permit me to add a few remarks on Heterogeny, as the old doctrine of spontaneous generation is now called, to those given by Dr. Carpenter, who, however, is probably better fitted to discuss the question than any other man in England.[2] Your reviewer believes that certain lowly organized animals have been generated spontaneously—that is, without pre-existing parents—during each geological period in slimy ooze.[3] A mass of mud with matter decaying and undergoing complex chemical changes is a fine hiding-place for obscurity of ideas. But let us face the problem boldly. He who believes that organic beings have been produced during each geological period from dead matter must believe that the first being thus arose. There must have been a time when inorganic elements alone existed on our planet: let any assumptions be made, such as that the reeking atmosphere was charged with carbonic acid, nitrogenized compounds, phosphorus, &c. Now is there a fact, or a shadow of a fact, supporting the belief that these elements, without the presence of any organic compounds, and acted on only by known forces, could produce a living creature? At present it is to us a result absolutely inconceivable. Your reviewer sneers with justice at my use of the "Pentateuchal terms," "of one primordial form into which life was first breathed":[4] in a purely scientific work I ought perhaps not to have used such terms; but they well serve to confess that our ignorance is as profound on the origin of life as on the origin of force or matter. Your reviewer thinks that the weakness of my theory is demonstrated because existing Foraminifera are identical with those which lived at a very remote epoch. Most naturalists look at this fact as the simple result of descent by ordinary reproduction; in no way different, as Dr. Carpenter remarks,[5] except in the line of descent being longer, from that of the many shells common to the middle Tertiary and existing periods.

The view given by me on the origin or derivation of species, whatever its weaknesses may be, connects (as has been candidly admitted by some of its opponents, such as Pictet, Bronn, &c.)[6] by an intelligible thread of reasoning a multitude of facts: such as the formation of domestic races by man's selection,—the classification and affinities of all organic beings,—the innumerable gradations in structure and instincts,—the similarity of pattern in the hand, wing or paddle of animals of the same great class,—the existence of organs become rudimentary by disuse,—the similarity of an embryonic reptile, bird and mammal, with the retention of traces of an apparatus fitted for aquatic respiration; the retention in the young calf of incisor teeth in the upper jaw, &c.,—the distribution of animals and plants, and their mutual affinities within the same region,—their general geological succession, and the close relationship of the fossils in closely consecutive formations and within the same country; extinct marsupials having preceded living marsupials in Australia, and armadillo-like animals having preceded and generated armadilloes in South America,—and many other phenomena, such as the gradual extinction of old forms and their gradual replacement by new forms better fitted for their new conditions in the struggle for life. When the advocate of Heterogeny can thus connect large classes of facts, and not until then, he will have respectful and patient listeners.

Dr. Carpenter seems to think that the fact of Foraminifera not having advanced in organization from an extremely remote epoch to the present day is a strong objection to the views maintained by me.[7] But this objection is grounded on the belief—the prevalence of which seems due to the well-known doctrine of Lamarck[8]— that there is some necessary law of advancement, against which view I have often protested. Animals may even become degraded, if their simplified structure remains well fitted for their habits of life, as we see in certain parasitic crustaceans. I have attempted to show ('Origin,' 3rd edit. p. 135) that lowly-organized animals are best fitted for humble places in the economy of nature;[9] that an infusorial animalcule or an intestinal worm, for instance, would not be benefited by acquiring a highly complex structure. Therefore, it does not seem to me an objection of any force that certain groups of animals, such as the Foraminifera, have not advanced in organization. Why certain whole classes, or certain numbers of a class, have advanced and others have not, we cannot even conjecture. But as we do not know under what forms or how life originated in this world, it would be rash to assert that even such lowly endowed animals as the Foraminifera, with their beautiful shells as figured by Dr. Carpenter, have not in any degree advanced in organization. So little do we know of the conditions of life all around us, that we cannot say why one native weed or insect swarms in numbers, and another closely allied weed or insect is rare. Is it then possible that we should understand why one group of beings has risen in the scale of life during the long lapse of time, and another group has remained stationary? Sir C. Lyell, who has given so excellent a discussion on species in his great work on the 'Antiquity of Man,' has advanced a somewhat analogous objection, namely, that the mammals, such as seals or bats, which alone have been enabled to reach oceanic islands, have not been developed into various terrestrial forms, fitted to fill the unoccupied places in their new island-homes; but Sir Charles has partly answered his own objection.[10] Certainly I never anticipated that I should have had to encounter objections on the score that organic beings have not undergone a greater amount of change than that stamped in plain letters on almost every line of their structure. I cannot here resist expressing my satisfaction that Sir Charles Lyell, to whom I have for so many years looked up as my master in geology, has said (2nd edit. p. 469):— "Yet we ought by no means to undervalue the importance of the step which will have been made, should it hereafter become the generally received opinion of men of science (as I fully expect it will) that the past changes of the organic world have been brought about by the subordinate agency of such causes as Variation and Natural Selection."[11] The whole subject of the gradual modification of species is only now opening out. There surely is a grand future for

Natural History. Even the vital force may hereafter come within the grasp of modern science, its correlations with other forces have already been ably indicated by Dr. Carpenter in the *Philosophical Transactions*;[12] but the nature of life will not be seized on by assuming that Foraminifera are periodically generated from slime or ooze.

Charles Darwin.

Athenæum, 25 April 1863, pp. 554–5

[1] CD's letter appeared in the *Athenæum*, 25 April 1863, pp. 554–5 (*Collected papers* 2: 78–80). The letter from W. H. Dixon of 16 April 1863 indicates that CD wrote most of this letter before 16 April. He then added the portion discussing Charles Lyell on 18 April (see preceding letter, and n. 11, below).

[2] In the *Athenæum*, 4 April 1863, p. 461, William Benjamin Carpenter responded to Richard Owen's anonymous review of his *Introduction to the study of the Foraminifera* (Carpenter 1862), which appeared in the *Athenæum*, 28 March 1863, pp. 417–19. See Appendix VII.

[3] See Appendix VII. See also letter to J. D. Hooker, [29 March 1863] and n. 5, letter to John Lubbock, 5 April [1863] and n. 11, letter from John Lubbock, 7 April 1863 and nn. 7 and 8, and preceding letter.

[4] Owen's letter in the *Athenæum*, 28 March 1863, p. 418, is reproduced in Appendix VII. The phrase appears in *Origin*, p. 484. In the second edition of *Origin*, p. 484, CD changed the phrase to read: 'into which life was first breathed by the Creator'. This phrase was omitted from the third and subsequent editions (Peckham ed. 1959, p. 753). See also letter to J. D. Hooker, [29 March 1863] and n. 6.

[5] Carpenter's letter in the *Athenæum*, 4 April 1863, p. 461, is reproduced in Appendix VII.

[6] CD refers to François Jules Pictet de la Rive's review of *Origin* (Pictet de la Rive 1860) and Heinrich Georg Bronn's German translation of *Origin* (Bronn trans. 1860).

[7] Appendix VII. See letter from John Lubbock, 7 April 1863 and nn. 7 and 8.

[8] Jean Baptiste de Lamarck's theory of transmutation is outlined in Lamarck 1809 and 1815–22.

[9] CD's citation is contained in a section entitled 'On the degree to which organisation tends to advance' (*Origin* 3d ed., pp. 133–9).

[10] CD refers to Lyell's discussion in C. Lyell 1863a, pp. 443–8. There is an annotated copy of C. Lyell 1863a in the Darwin Library–CUL (see *Marginalia* 1: 525–7).

[11] C. Lyell 1863b, p. 469. The words in parentheses were introduced in the second edition of *Antiquity of man*, as a result of CD's criticisms (see letter to Charles Lyell, 12–13 March [1863] and n. 17). CD had received a copy of the second edition by 17 April, and decided to use the quote in his letter to the *Athenæum* (see letter to J. D. Hooker, [17 April 1863]). The second edition of *Antiquity of man* (C. Lyell 1863b) was published between 15 and 30 April 1863 (*Publishers' Circular* 26 (1863): 216).

[12] Carpenter 1850. CD's copies of the *Philosophical Transactions of the Royal Society* are in the Darwin Library–CUL; in them, Carpenter 1850 is lightly annotated.

To H. W. Bates 18 April [1863][1]

Down. | Bromley. | Kent. S.E.
April 18th

Dear Bates

I have finished vol. I.[2] My criticisms may be condensed into a single sentence, namely that it is the best book of Natural History Travels ever published in England. Your style seems to me admirable. Nothing can be better than the discussion on the struggle for existence & nothing better than the descriptions on the Forest scenery.[3]

It is a grand book, & whether or not it sells quickly it will last. You have spoken out boldly on Species; & boldness on this subject seems to get rarer & rarer.—[4] How beautifully illustrated it is.[5] The cut on the back is most tasteful.[6] I heartily congratulate you on its publication.—

The Athenæum was rather cold, as it always is, & insolent in highest degree about your bending facts.—[7] Have you seen the "Reader":[8] I can send it you, if you have not seen it.

I have got spare copies of my article in N. Hist Review,[9] which I will send you on Monday & you can use or burn them as you like.

I look forward with real pleasure to vol. 2.—

Yours very sincerely | Ch. Darwin

I will keep copies of Review till I know your London address.

Cleveland Health Sciences Library (Robert M. Stecher collection)

1 The year is established by the reference to Bates 1863.

2 Bates 1863. Bates's account of his eleven years in South America was published in two volumes; CD received a presentation copy (see letter to H. W. Bates, 9 April [1863]).

3 CD's annotated copy of Bates 1863 is in the Darwin Library–CUL (see *Marginalia* 1: 35–7). CD wrote notes on a sheet pasted into the back of the first volume of Bates 1863; these include a note relating to p. 55, 'competition in the Tropics', where Bates described (Bates 1863, 1: 54–5):

> the struggle which necessarily exists amongst vegetable forms in these crowded forests, where individual is competing with individual and species with species, all striving to reach light and air in order to unfold their leaves and perfect their organs of fructification.

The forest scenery of Pará is described on pages 44–9; see also pp. 290–3.

4 See, for example, Bates 1863, 1: 255–65, in which he discusses the origin of species and natural selection.

5 In the preface to *The naturalist on the river Amazons*, Bates acknowledged his indebtedness to Edward W. Robinson for the illustrations of smaller animals, and to Josef Wolf and Johann Baptist Zwecker for depicting most of the larger subjects and scenes (Bates 1863 1: vi). The first volume included eighteen illustrations.

6 There is no woodcut on the back of CD's existing copy. However, each of the two volumes of Bates 1863 bears on its front cover a gold-embossed illustration of a forest house surrounded by palms; standing in the foreground is a person holding a container on his or her head.

7 See letter to J. D. Hooker, [17 April 1863] and n. 15. The reference is to the *Athenæum*, 25 April 1863, p. 489.

8 A favourable review of Bates 1863 appeared in the *Reader*, 18 April 1863, p. 378.

9 CD's anonymous 'Review of Bates on mimetic butterflies' appeared in the April 1863 number of the *Natural History Review*.

To Charles Lyell 18 April [1863]

Down Bromley Kent
Ap. 18th

My dear Lyell

I was really quite sorry that you should have sent me a second copy of your valuable book.[1] But after a few hours my sorrow vanished for this reason.— I have written a letter to the Athenæum,[2] in order, under the cloak of attacking the monstrous article on Heterogeny to say a word for myself in answer to Carpenter,[3] & now I have inserted a few sentences in allusion to your analogous objection about Bats on islands,[4] & then with infinite slyness have quoted your amended sentence with your parenthesis ("as I fully believe");[5] I do not think you can be annoyed at my doing this, & you see that I am determined, as far as I can, that the Public shall see how far you go.—[6] This is the first time I have ever said a word for myself in any Journal, & it shall,, I think, be the last.— My letter is short & no great thing.—

I was extremely concerned to see Falconer's disrespectful & virulent letter.[7] I like extremely your answer, just read.[8] You take a lofty & dignified position, to which are so well entitled. I suspect that if you had inserted a few more superlatives in speaking of the several authors, there would have been none of this horrid noise.— No one, I am sure, who knows you, could doubt about your hearty sympathy with every one who makes any little advance in science.[9] I still well remember my surprise at the manner in which you listened to me in Hart St on my return from the Beagle's Voyage.— You did me a world of good.—[10] It is horridly vexatious that so frank & apparently aimiable a man as Falconer sh^d. have behaved so. Well it will all soon be forgotten.—

Is there any chance of Lady Lyell & you being able to come here for a little visit.[11] Emma was going to have written this very morning; but last night I had a furious attack of Eczema, which I suppose will take every bit of skin off my face, & as this generally makes me very unwell for many days, she did not write; but I have got better this evening, as this note, shows; & I hope I may be fairly well very soon; though far from as beautiful as usual.—

Farewell my good old master | C. Darwin

Have you seen Bates' Book it is **capital**.—[12]

Endorsement: '1863'
American Philosophical Society (294)

[1] The second edition of Lyell's *Antiquity of man* (C. Lyell 1863b) was published between 15 and 30 April 1863 (*Publishers' Circular* 26 (1863): 218).

[2] See letter to *Athenæum*, 18 April [1863].

[3] The 'monstrous article' to which CD refers was Richard Owen's anonymous review (*Athenæum*, 28 March 1863, pp. 417–19) of Carpenter 1862 (see Appendix VII). William Benjamin Carpenter responded in the *Athenæum*, 4 April 1863, p. 461. See letter to J. D. Hooker, [17 April 1863] and nn. 9–10.

[4] CD refers to the question regarding differences in the rate of change over time of particular species. This question was raised in *Antiquity of man* (C. Lyell 1863a), pp. 443–8, where Lyell discussed the absence of certain mammals on small and isolated islands (see letter to *Athenæum*, 18 April [1863]). Owen and Carpenter discussed the same question in their own recent letters to the *Athenæum* (see n. 3, above).

[5] Lyell actually wrote, '(as I fully expect it will)' (C. Lyell 1863b, p. 469; see letter to *Athenæum*, 18 April [1863] and n. 11).

[6] CD was concerned that Lyell was unwilling to endorse transmutation more wholeheartedly in print. For CD's disappointment regarding Lyell's position in C. Lyell 1863a, see, for example, the letter to Charles Lyell, 6 March [1863].

[7] See letter to J. D. Hooker, [17 April 1863] and n. 3. The reference is to Hugh Falconer's letter in the *Athenæum*, 4 April 1863, pp. 459–60.

[8] Lyell's letter was printed in the *Athenæum*, 18 April 1863, pp. 523–5. In it, Lyell pointed out that he had given both Falconer and Joseph Prestwich the opportunity to tell him which passages might require correction in the second edition, and that both had refused to do so. Lyell then proceeded to dispute Falconer's letter point by point, indicating those passages in which he cited the research of both Falconer and Prestwich.

[9] John Lubbock, like Falconer, thought that Lyell had not given enough credit to other researchers (see, for example, letter from John Lubbock, 7 April 1863 and n. 6). See also L. G. Wilson 1996a for correspondence during April 1863 between Lyell and William Pengelly regarding Falconer's concerns.

[10] In a letter to Caroline Darwin, [7 December 1836] (*Correspondence* vol. 1), CD wrote: 'I have seen Lyell, who was exceedingly friendly, & gave me much sensible advice and told me how he managed to make most of his time; an art which no one understands better.' In 1836, the Lyells lived at 16 Hart Street, Bloomsbury, London.

[11] Charles and Mary Elizabeth Lyell had planned to visit Down House from 1 to 4 March (see letter to J. D. Hooker, 24–5 February [1863]), but were unable to do so. Emma Darwin recorded in her diary that CD was ill on those days (DAR 242). There is no evidence that they visited Down House later in 1863.

[12] Bates 1863. See preceding letter.

From Philip Gidley King 19 April 1863

Goonoo Goonoo N.S.W
Ap^l 19/63.

My dear M^r. Darwin

You cannot think with what pleasure I received a note from you by the last Mail enclosing a Photograph of yourself—.[1] It was a Strange feeling that came over me as I identified one by one the *now* well remembered features of what you must permit me to call y^r. dear old face. And as I ga⟨ze⟩ upon your likeness r⟨emini⟩scences of my youth appear to come back to me.[2] I fancy I see you in your old corner behind the mizen-mast with neither room to bend or turn still pursuing your studies in Natural History with unwonted zeal, but I will not bore you with these reminiscences suffice it to say that few things have given me greater pleasure lately ⟨th⟩an the opportunity you have ⟨given⟩ me of reproducing in my minds eye the very expression of your eye & mouth.

I am so sorry to think you suffer so much from ill health but you are living on to a good age at any rate, and I trust the Scientific world will not lose you yet.

Accept my dear friend the best wishes of y^r. old Shipmate | Philip Gidley King

DAR 169: 27

[1] The letter to King has not been found.

[2] King had been a midshipman on the *Beagle* from 1831 to 1836. In 1836 he left the *Beagle* to join his father, Phillip Parker King, a retired naval officer who had emigrated to New South Wales, Australia. See *Correspondence* vol. 1.

From H. W. Bates 20 April 1863

22 Harmood Street | Haverstock hill N.W
April 20 1863

My Dear Mr Darwin

I have at length returned from Leicester & the above will be now my address for a long time.[1] Your last note to me at Leicester stated you had just received my book & were commencing to read.[2] I hope now you have finished & are ready to pronounce sentence.[3] With regard to the other matter you mention, namely my prospects;[4] I must be open with you as I have been heretofore. My total income is £123 £100 of which is allowed to me by my brothers on account of my withdrawing from their small business (which was not large enough to support 3 of us) & leaving £1000 of my capital on loan, the other £23 is interest on shares.[5]

Now I hope to add by scientific or literary work sufficient to enable me to live modestly. I should consider myself well off if I could gain an additional £150.

There is perhaps some slight chance of my getting in at British Museum which of course would fix me for life.[6] I can get plenty of Entomological work from private persons but it is of a tedious, mechanical nature & would prevent me from undertaking original researches.[7]

The house I have taken is a very *small one*, but retired & semi-rural Mrs B. is a plain domesticated woman so there you have it all.[8]

One of my reviewers "hopes that I have made my expedition answer in a pecuniary point of view"[9] It certainly did not answer for the total savings of my 11 years work did not exceed £800

Yours sincerely | H W Bates

I go to Murray today to ascertain the trade result of my book[10]

DAR 160: 75

[1] See letter from H. W. Bates, 8 April 1863 and n. 1, and letter to H. W. Bates, 9 April [1863] and n. 7.

[2] See letter to H. W. Bates, 9 April [1863]; the reference is to Bates's account of his eleven years in South America, *The naturalist on the river Amazons* (Bates 1863). Bates had evidently not yet received CD's letter of 18 April [1863], which had been sent to Bates's Leicester address.

[3] In his letter to Bates of 18 April [1863], CD said that he had read the first volume of Bates 1863, and his criticisms could be 'condensed into a single sentence, namely that it is the best book of Natural History Travels ever published in England'.

[4] See letter to H. W. Bates, 9 April [1863].

[5] Bates had been involved with his brothers, Frederick, John, and Samuel, in the family hosiery business in Leicester since returning from South America in 1859 (Woodcock 1969, pp. 16, 240).

[6] In 1862, Bates had hoped to be considered for a zoological assistantship in the natural history department of the British Museum (see *Correspondence* vol. 10, letter from H. W. Bates, 30 April 1862 and n. 10, letter to H. W. Bates, 4 May [1862], and Woodcock 1969, p. 250).

[7] Bates had difficulty finding acceptance within the London scientific establishment during the 1860s; for Bates's career after 1863, see Woodcock 1969, pp. 256–60.

[8] Bates married Sarah Ann Mason of Leicester on 19 January 1863 (see letter from H. W. Bates, 24 January 1863, and Woodcock 1969, pp. 253–5).

[9] The reviewer of Bates 1863 in the *Reader*, 18 April 1863, p. 378, expressed a hope that Bates had been able to pay the expenses of his expedition, as 'Mr. Wallace [Alfred Russel Wallace] was satisfied with the pecuniary result of his labours'.

[10] Bates 1863 had been published by John Murray. For information on the sales of Bates 1863, see the letter from H. W. Bates, 2 May [1863], and Bates 1892, p. lxx.

From H. B. Dobell 20 April 1863

29 Duncan Terr
April 20 1863.

My dear Sir

Knowing your interest in the subject of hereditary transmission, I hope you will not mind my troubling you by submitting to you the enclosed forms.[1]

There is so little acquaintance on the part of most persons with the common form in which a genealogical table is drawn up, that I find a great difficulty in getting full reports of families even when they are offered by medical friends. The usual plan being to mention only the individuals who have presented some peculiarity, and then the exact relationship between them is often difficult to get at; and all the collateral bearings of the case are lost.[2]

I am thinking, therefore, of having some forms & directions printed so that I can forward them to friends who are willing to report cases, and I have drawn up the specimen enclosed which appears to me to give all the assistance necessary. I think that I could thus get the materials from which to draw up a proper genealogical table of 3 or 4 generations.[3]

If you will be so good as to look at it & suggest any alterations which you think advisable I shall take it as a great favour, & I shall have great pleasure in furnishing you with any cases I may obtain which I think may interest you.

I hope you will pardon my troubling you in this matter, on the strength of so slight an acquaintance as ours, as I am very anxious to have your opinion before proceeding further in the matter.

I am my dear Sir | very truly yours | Horace Dobell.

Ch Darwin Esq | &c—

DAR 162: 189

[1] The enclosure has not been found. CD may have returned the forms to Dobell with his letter of 21 April [1863]; the revised form Dobell sent with his letter to CD of 12 May 1863 has not been found.

[2] Dobell was a physician interested in respiratory diseases and the inheritance of disease and morphological anomalies (see letter from James Paget, 7 February 1863 and n. 3).

[3] No copy of such a printed form has been found. However, in Dobell 1866, pp. 120–8, he analysed the genealogy of emphysema and bronchitis by tabulating the health of parents and siblings, suggesting (*ibid.* p. 120):

> the adoption of a similar form of table by other observers when registering family history will be found convenient, and calculated to prevent mistakes as to the real influence of the hereditary and collateral tendencies of diseases.

Dobell concluded that bronchitis, but not emphysema, was hereditary (*ibid.*, p. 126). See also letter to H. B. Dobell, 21 April [1863].

From Hugh Falconer 20 April [1863][1]

21 Park Crescent.
20th. April

My Dear Darwin

I have been on a trip to France—down the Loire examining the Faluns with Prestwich & Evans—and Galton.[2] I passed a day at Abbeville with Quatrefages—[3] who has no great affection for people on this side the channel—but makes a special exception in your favour. He desired me to offer his very best salutation to you and Huxley[4]—and I discharge my conscience by communicating it to you.

You will have seen—with grief I am sure—the sharp rencontre in the Athenæum between me and a great friend of yours. I will say no more[5]

Yours ever Sinly | H Falconer

PS. I was nearly forgetting the most important point. Bravard sent home from Brazil, to Paris, the head of a Mammal-fossil bigger than that of a Lama.[6] The Archæopteryx is a joke to it—for singularity.[7] The molar teeth nearest *Toxodon* essentially—but in form of crown, like those of a gigantic *Insectivore*, in contour of crown & overlapping. Upper incisors two—bigger than those of an Indian Rhinoceros—but in form & position like those of a Rodent. Brain case somewhat like that of a marsupial koala. Lower jaw with 4 incisors, the two central large, like the upper & 1 smaller on either side. Serres—by a happy inspiration proposed calling it *Mesotherium*—as being a common centre towards which all mammalia got happily confounded:[8] Bravard sent it home under the name of *Typotherium* as being the central type from which all mammals diverged.[9] Gratiolet told me he considered it as of the Toxodon family.[10] We shall probably soon have a cast of it in England. I have got Flower to apply for one[11]—Gratiolet having promised to assent. So much for Typotherium—being a hybrid between Toxodon & a kola in the first degree crossed in the Second with an Insectivore and ⟨a⟩ Rodent.

DAR 164: 14

[1] The year is established by the reference to letters in the *Athenæum* (see n. 5, below).

[2] Falconer refers to Joseph Prestwich and John Evans; he also refers to Douglas Strutt Galton, an engineer and public servant who was a close friend of Prestwich's, often accompanying him on excursions (G. A. Prestwich ed. 1899, pp. 148, 162, 196). This trip to France is referred to in Falconer *et al.* 1863, and Van Riper 1993, p. 135, in regard to the examination of the recently discovered human jaw at Moulin-Quignon, near Abbeville (see n. 3, below). A term originally introduced to English geology by Charles Lyell (C. Lyell 1833, 3: 203), the 'Faluns' are Miocene strata composed largely of Tertiary mollusca.

[3] Falconer examined the recently discovered Moulin-Quignon fossil jaw at Abbeville, and discussed it with Armand de Quatrefages on 15 April 1863 (see Falconer *et al.* 1863, pp. 424–5). See letter from Hugh Falconer, 24 April [1863] and n. 6.

[4] Thomas Henry Huxley. Quatrefages was a correspondent of CD's (see, for example, letter to Armand de Quatrefages, [14 April 1863]).

[5] The reference is to CD's friend and mentor Charles Lyell. Falconer's letter in the *Athenæum*, 4 April 1863, pp. 459–60, criticised C. Lyell 1863a (see letter to J. D. Hooker, [17 April 1863] and n. 3). CD also discussed the 'horrid' affair in his letter to Lyell of 18 April [1863]. See also letter to Hugh Falconer, 22 April [1863].

6 Auguste Bravard was a French palaeontologist living in South America. A decade earlier, CD had offered him advice on fossil hunting derived from his experience on the *Beagle* voyage (see *Correspondence* vol. 5, letter to G. R. Waterhouse, 8 September [1852]).

7 Falconer refers to *Archaeopteryx*, the fossil Jurassic bird recently discovered in the slate quarries of Solenhofen, Bavaria (see letter from Hugh Falconer, 3 January [1863]).

8 See Serres 1857, p. 961. Antoine Etienne Reynaud Augustin Serres propounded the view (sometimes called the Serres–Meckel Law) that there was only one underlying animal type and that, in the course of their development, the organs of the higher animals repeated the forms of the equivalent organs in lower organisms (*DSB*).

9 Both Serres's *Mesotherium* and Bravard's *Typotherium* have been used for this genus of Pleistocene notoungulates (hoofed mammals, largely from South America); *Mesotherium* is now the accepted name (Simpson 1980, p. 121). Serres later published a detailed description of the fossil (Serres 1867).

10 Louis Pierre Gratiolet. *Toxodon* is a genus of Pliocene and Pleistocene rhinoceros-like notoungulate mammals from Argentina (*EB*). CD had collected specimens of *Toxodon* on the *Beagle* voyage (see *Correspondence* vol. 2, and *South America*, pp. 180–1).

11 William Henry Flower was curator of the Hunterian Museum, Royal College of Surgeons (*DNB*).

To Asa Gray 20 April [1863][1]

Down Bromley Kent
April 20th

My dear Gray.—

Thanks for two notes,[2] one about the Duke of A. (did I tell you how capitally this was criticised in the Saturday Review?)[3] & the other note very savage against England:[4] I cannot help fearing that we shall drift into war:[5] what a curse it will be to us anyhow; for you seem to be getting to like war.— I wish I had known when I read the Correspondence that Mr Loring was your father-in-law:[6] I shd. have read it, if possible, with still greater interest. We must keep to science, I fear, for we both seem to be getting to think each other's country conduct worse & worse. But I shd. like to know whether General Butler is in your honest opinion a bad man.—[7]

Your remark from Wyman about the Incas came very appropriately for I am at present summing up all facts on this subject & give facts on both sides.—[8] I now regret that I have always intentionally evaded case of man; but have put in a note on such facts as I have heard of.—[9]

I sent off two days ago two copies of my Linum paper:[10] I hope the case may interest you; it has me. Amsinckia turns out with me only variability in length of pistil.—[11] I forgot to ask about Intermarriage; M. Devay says explicitly that state of Ohio, from evil shown by statistical returns, has legislated against cousins marrying: can this be true?[12]

Here is another heterogenous question: have you ever formed any theory, why in spire of leaves (I have been reading your most clear account in your Lessons)[13] the angles go $\frac{1}{2}$ $\frac{1}{3}$ $\frac{2}{5}$ $\frac{3}{8}$ &c— Why should there not be $\frac{1}{4}$ or $\frac{1}{5}$? It seems to me most marvellous— There must be some explanation. If you have theory, I know it would be too long to explain; but I shd like to hear whether you have. My good friend Falconer has been twitting me that these angles go by as fixed a law as that of Gravity & *never vary*.[14] I can fancy that packing of organs in very early bud may cause general alternation in the parts of the flower & consequent interruption in the spires.—

Speaking of Falconer, I was very sorry to see his letter in the Athenæum; so irreverent & virulent towards Lyell.[15] We have had lately sharp sparring in the Athenæum. Did you see the article on Heterogeny or Spontaneous generation, written I believe, certainly by Owen!! it was in Review on Carpenter, who seems to have been sillily vexed at Owen calling me Carpenter's master; it was like his clever malignity.[16] Under the cloak of a fling at Heterogeny I have sent a letter to Athenæum in defence of myself, & I take sly advantage to quote Lyells *amended* verdict on the Origin.—[17] I suppose my letter will appear next week:[18] it is no great thing.—

Be sure, if you have opportunity read Bates' Amazonia, because you will like & admire it.—[19]

I have got to say a few words on Orchids. I hope that you may have time to look at the Rostellum of Gymnadenia, this summer;[20] because a good observer in Bot. Garden at Edinburgh a Mr J. Scott has been experimenting on foreign genera, & he finds that the rostellum stimulates some kinds to protrude their tubes;[21] but that these tubes *never* penetrate the rostellum only creep along its surface to the stigma. He wrote to me to know whether I could give him any details on your observations.[22] By the way I lately found some primroses with 3 pistils, not united, so that I could peep into the ovarium: I put in pollen & afterwards found the tubes exserted & attached to & *apparently* penetrating the ovules; but never by the micropyle![23] I have now no doubt that you are perfectly right about fertilisation of Cypripedium;[24] a friend lent me a plant of C. pubescens.[25] I put excessively minute Bee (an Andrœna) into the Labellum & covered orifice with wet paper; but this precaution I afterwards found superfluous, for the edges all round of the orifice of labellum are folded over (just like insect traps for London kitchens) so that the Bee could not crawl out.— Well the bee immediately crawled out by one of the windows opposite the anthers, with his back towards the anther, against which he firmly pressed it, owing to the elastic wool opposite the anther. It was pretty to see under lens how the whole thorax & base of wings was smeared with pollen. I put him back into the labellum five times & five times I saw his back smeared. As you know he must pass under the stigma (with its spines directed towards the apex as you describe), for there is no other passage; & as I expected when I cut open the flower I found the stigma well smeared with pollen.—[26] It was beautiful.—

Good Night | My dear Gray | Ever yours cordially | Though an Englishman | Charles Darwin

Gray Herbarium of Harvard University (51)

[1] The year is established by the relationship between this letter and the letters from Asa Gray, 22–30 March 1863 and 11 April 1863.

[2] Letters from Asa Gray, 22–30 March 1863 and 11 April 1863.

[3] In his letter to CD of 22–30 March 1863, Gray had discussed the article by George Douglas Campbell, duke of Argyll, published in the October 1862 number of the *Edinburgh Review* ([Campbell] 1862). CD

refers to an article in the *Saturday Review*, 15 November 1862, pp. 589–90, that critically examined Campbell's review.

4 See letter from Asa Gray, 11 April 1863.

5 CD's fear of war may have been in part a response to *The Times*, 16 April 1863, p. 14, and 17 April 1863, p. 9, which reported a speech made in New York on 2 April by Benjamin Franklin Butler. Butler had advised retaliation against England for the actions of two ships, the *Florida* and *Alabama*, which had been built in England for the Confederacy. His solution was to sever relations with England; this included the curtailment of grain and gold shipments. Butler's behaviour and comments had already contributed to anti-Union sentiments in Britain (see Jenkins 1974–80, 2: 50–9). See also n. 7, below.

6 CD refers to the correspondence between Charles Greely Loring (father of Gray's wife, Jane Loring Gray), and Edwin Wilkins Field concerning the relations between Britain and the United States following the outbreak of the American Civil War (Loring 1862). In his letter to Gray of 23 February [1863], CD expressed his interest in the correspondence.

7 See n. 5, above. As military governor of occupied New Orleans, Butler had acquired the sobriquet 'Beast Butler' (McPherson 1988, pp. 551–2; Sifakis 1988).

8 See letter from Asa Gray, 22–30 March 1863 and nn. 5–7.

9 CD evidently refers to notes he was making for *Variation*. A brief reference to humans appeared in *Variation* 2: 122–4. CD worked on a draft of the chapter in which these pages appeared from 1 April to 16 June 1863 (see 'Journal' (Appendix II)). CD referred to 'the origin of man and his history' in one sentence at the conclusion of *Origin*, p. 488.

10 'Two forms in species of *Linum*' was read before the Linnean Society on 5 February 1863 and published in the second part of volume 7 of the *Journal of the Proceedings of the Linnean Society of London*. According to the *General index to the Journal of the Linnean Society*, this part was issued on 13 May 1863. Gray's name appears on CD's presentation list for this paper (see Appendix IV).

11 Having found from dried specimens sent by Gray that the pistil in one form of *Amsinckia spectabilis* was more than twice as long as in the other, with a corresponding difference in the lengths of the stamens, CD suggested that the species might be heterostyled ('Dimorphic condition in *Primula*', p. 95; *Collected papers* 2: 62). Since March 1862, he had been observing live specimens (see letter to Asa Gray, 2 January [1863] and n. 11). Experimental notes dated 23 March and 1 April 1863 record his growing conviction that *Amsinckia* was not dimorphic (see DAR 110: B2). In *Forms of flowers*, p. 110, CD reported that having raised 'many plants from seed', he 'soon became convinced that the whole case was one of mere variability'.

12 Devay 1862, p. 141. There is a lightly annotated copy of Francis Devay's book in the Darwin Library–CUL (see *Marginalia* 1: 194–5). In *Variation* 2: 122 n. 21, CD cited Devay on the prohibition by the legislature of Ohio of marriages between cousins and added that he had 'been assured, in answer to inquiries made in the United States,' that this statement was 'a mere fable'. See letter from Asa Gray, [10–16] June [1863].

13 A. Gray 1857, pp. 72–5. There is an annotated copy of this work in the Darwin Library–CUL (see *Marginalia* 1: 347).

14 Falconer 1863a, p. 80. See letter to J. D. Hooker, 13 January [1863] and n. 22, and letter from Daniel Oliver, 17 February 1863.

15 CD refers to Hugh Falconer's letter in the *Athenæum*, 4 April 1863, pp. 459–60, which was critical of Charles Lyell's *Antiquity of man* (C. Lyell 1863a). See letter to Charles Lyell, 18 April [1863], and letter to Hugh Falconer, 22 April [1863].

16 Richard Owen's anonymous review of Carpenter 1862 appeared in the *Athenæum*, 28 March 1863, pp. 417–19. William Benjamin Carpenter responded in the *Athenæum*, 4 April 1863, p. 461. Both articles are reproduced in Appendix VII.

17 See letter to the *Athenæum*, 18 April [1863]. The reference is to C. Lyell 1863b, p. 469. See also letter to J. D. Hooker, [17 April 1863], and letter to Charles Lyell, 18 April [1863].

18 CD's letter was published in the *Athenæum*, 25 April 1863, pp. 554–5 (see letter from W. H. Dixon, 16 April 1863).

19 CD refers to Henry Walter Bates's *The naturalist on the river Amazons* (Bates 1863).

[20] Gray had been puzzled by the type of self-pollination in *Gymnadenia* (see *Correspondence* vol. 10, letter from Asa Gray, 22 September 1862 and n. 13). He published comments on *Gymnadenia* in A. Gray 1863b.

[21] John Scott was foreman of the propagating department at the Royal Botanic Garden, Edinburgh (R. Desmond 1994). See letter from John Scott, [1–11] April [1863].

[22] See letter from John Scott, [1–11] April [1863] and n. 13.

[23] CD's notes on this experiment, dated 26 March – 11 April 1863, are in DAR 108: 165 v. See also letter to Daniel Oliver, [12 April 1863], and letter from Daniel Oliver, 14 April 1863.

[24] After reading Gray's observations of American species of *Cypripedium* in A. Gray 1862a, pp. 427–8, CD examined specimens and their pollination by bees. CD confessed that he had not thought of 'insects crawling into flower' (see *Correspondence* vol. 10, letters to Asa Gray, 10–20 June [1862] and this volume, letter to Asa Gray, 2 January [1863] and n. 7.

[25] The reference is to Arthur Rawson (see letters from Arthur Rawson, 1 April [1863] and [6 April 1863]).

[26] CD's note describing this experiment, dated 12 April [1863], is in DAR 70: 112–13.

From Asa Gray 20 April 1863

Cambridge. Mass.
April 20, 1863

My Dear Darwin

You asked me to tell you—when I had read it, what I thought of Sir Chas. Lyell's book.[1] I have only today finished the perusal of the copy he kindly sent me, i.e. all but half of the matter on Glacial period[2]—which I reserve till I can read it more attentively. Throughout it is a very interesting and to me a very satisfactory book. It is *three books*.— 1. a capital resumé and examination of what we know about the evidence of antiquity of man,[3]—no evidence we had not read of before,—but very clearly presented, of course.—[4] 2. a treatise on Glacial Per⟨iod.⟩[5] Out of this I have m⟨uch to⟩ learn, and must read it all again carefully. Of a part I have not yet cut the leaves.

3. On transmutation-matters.[6] That part of the book I can judge somewhat of, and I declare it *first-rate*. It is just about what I expected, and is characteristic of the man. I think that you, and Hooker, are unreasonable in complaining of Lyell that he does not come out "flat-footed" as we say, as an advocate of *natural-selection transmutation*.[7] For 1. it is evident that tho' inclined strongly towards ⟨it⟩ he is by no means satis⟨fied⟩ that nat. sel. will do ⟨all the⟩ work you put upon it. 2nd. He very plainly implies nearly all you would have him say. And, 3rd, he serves your cause (supposing it to be well-founded) quite as effectually,—perhaps, by his guarded position,[8]—by his keeping the position of a judge rather than of an advocate, and by considering still the case as not yet ripe for a decision.[9]

Very skillfully, too, has he presented the case of transmutation so as to commend it, as much as possible, to us orthodox people. (Huxley, I suppose, whose two books I have not seen, would put it in a way to frighten us off);[10] indeed I think he has shown remarkable judgment and taste, & will have much success in disarming prejudice— And this is all you could ask.

The Chapter on language makes the points I supposed would be made,[11] or some of them, but only *dips in*; leaving more to be said. But this is rather ticklish ground,—for, if we are not careful here, you would get the better of us in this field quoad *designs*.[12]

If I had got the book 3 or 4 weeks earlier, I should have worked in some notice of the last chapter into my review of De Candolle &c, on Species—in May no. of Sill. Journal.[13]

Now please, do not think of being ill this spring, and passing all your valuable time—wasting it—at water-cure.[14]

I have really—as you see, nothing special to write of this week, & no time to read what I have hurriedly penned. | Ever Yours cordially | Asa Gray

DAR 165: 134

1 Gray refers to Charles Lyell's *Antiquity of man* (C. Lyell 1863a). See letter to Asa Gray, 20 March [1863].

2 In chapters 12–18 of *Antiquity of man* (C. Lyell 1863a, pp. 203–368), Lyell discussed recent research on glaciation and outlined a chronology of the Pleistocene glacial period.

3 C. Lyell 1863a, pp. 8–205.

4 For a discussion of this aspect of *Antiquity of man*, see Van Riper 1993, pp. 139–42.

5 See n. 2, above.

6 C. Lyell 1863a, pp. 385–506.

7 See letters to Asa Gray, 23 February [1863] and 20 March [1863]. See also letter to J. D. Hooker, 24[–5] February [1863], letter from J. D. Hooker, [1 March 1863], and letter to Charles Lyell, 6 March [1863].

8 Lyell also suggested to CD that his moderate approach would bring hundreds towards CD's position, who, if he had treated the matter more dogmatically, 'would have rebelled' (see letter from Charles Lyell, 11 March 1863).

9 See CD's annotations to the letter from Asa Gray, 13 April 1863. See also letter to Asa Gray, 11 May [1863].

10 The references are to T. H. Huxley 1863a and 1863b.

11 The reference is to chapter 23 of *Antiquity of man*, entitled 'Origin and development of languages and species compared' (C. Lyell 1863a, pp. 454–70). Gray discussed the possible use of the analogy between the origin of species and the origin of languages in his letters to CD of 4 and 13 October 1862 and 24 November 1862 (*Correspondence* vol. 10).

12 Gray and CD had long debated the existence of a preordained design in nature (see *Correspondence* vols. 8, 9, and 10; see also A. Gray 1861). Gray had already suggested to CD that the analogy between the origin of species and languages might furnish conclusions 'adverse to *special design*' (see *Correspondence* vol. 10, letter from Asa Gray, 4 and 13 October 1862 and n. 3).

13 Gray's review of A. de Candolle 1862b (A. Gray 1863e) appeared in the May 1863 number of the *American Journal of Science and Arts*, which was known as 'Silliman's journal' after its founding editor, Benjamin Silliman.

14 See letter to Asa Gray, 20 March [1863] and n. 21.

From J. D. Hooker 20 April 1863

[Royal Gardens Kew]
April 20th/63

Dear Darwin

We returned on Friday having enjoyed the trip most amazingly— visiting Weymouth, Portland & its attractive establishment of rogues & villains, Jersey &

Guernsey.—[1] We had good weather, but the cool ocean kept down the temp. of the Islands, which are now a good 10 days *behind* England!

A thousand thanks for your letter.[2] I am grieved with Falconers letter, he has made out no case for himself or Prestwich, his vanity & presumption are overweening, & the last paragraphs depreciatory of Lyell as an original observer is as bad as any thing Owen ever writ in its way.[3] I am deeply sorry— Lyells answer, though not crushing is damning.[4] I like its tone temper & spirit— it is manly & temperate—well done in short.—& (were it possible) raises him in my eyes. What can Prestwich be thinking about to lend his weight to such an attack, by getting on F.s back, instead of fighting along side him, or by himself.—[5] The whole thing is bad

Carpenters love of self has cost him the loss of the best opportunity of smashing Owen that ever was offered to mortal man, & I cannot help thinking that Owen calculated it would be so!—[6] C might have smashed Owen like a Black beetle; but his motive was obviously to clear himself of the imputation of being dry-nursed by you,[7] & to let the world know that he had a Royal medal!.[8]

I have just received a bottle of cones of Welwitchia in spirits, these actually bristle with the projecting styliform processes of the useless ovules, all on the look out for pollen, of which they never get one grain, nor have the ovules a trace of embryo-sac.—[9] it confirms my conclusions. & is the most wonderful fact in vegetable embryology. only conceive *all* men having huge mammæ with no lactiferous glands whilst women (breeding) had microscopic ones, & yet the case would not be half so wonderful. There never was such a case of a retained organ whose function was gone—& the more I think of it the more I worship your theory.

I do not think Bentham will make much of the question of species, he only proposes to review the state of the question for Linnæan Anniversary.[10]

Have you ever read the "Admiral's daughter", one of "Two Old Mens tales" by the author of Emilia Wyndham.[11] I read it 25 years ago & was deeply interested. I picked it up the other day, & was almost equally affected: it is very powerful & very true. I can send it you (ends badly)

I see McCaul & Manchester have fairly ridden down poor Natal, about the Hare & I cannot but think that their answers will be considered triumphant by $\frac{9}{10}$ of readers.[12]

Please return Haasts letter at your leisure.[13]

My boy Willy[14] is at home now much the same as ever, very good & good natured, happy & amiable; but not strong & very nervous— his mind is an utter vaccuum, he cares for no one thing, attends to nothing, & regrets nothing. he is very young of his age. I do not think he made one single sensible observation throughout the journey— Charlie[15] is as much the other way, overflowing with animal spirits, never says or does a foolish thing, & is as amusing as a monkey & knowing as a Yankee he attracts a great deal of attention & would be spoiled were he less sensible, or were his mind less occupied— he walks 10 miles without knocking up, & is a keen observer and Collector— he must be very like what his grandfather Henslow was,[16] & I long to bring him to Down one day.

I am urging Oliver to take up Orchideæ,[17] & hope you will back me. I do not know an order in which there is more to do in every way, & we sadly want an orchidologist. Lindly returns from Vichy today—no better either as to memory or the use of his hands.[18]

I have notice of Benguela temp. & trop. form, it amounts to nothing per se (it was by Welwitsch).[19]—but it is important as backing up the Cameroons migrations & in reference to them. I am using it in the discussion.[20]

Planchon's address is Prof. Bot. Montpellier[21]

Your Oxalis is not in flower now with us, but is probably a Mexican species called *Deppei*, I have ordered an eye to be kept upon it.[22]

I send Orchis seeds, (do you want pods too?)[23]

Ever yours affec | J D Hooker

[Enclosure]

Lake Ohau N. Zealand
December 10th. 1862.

My dear D^{r}. Hooker

Having hurt my right hand slightly by a fall with a horse, I have to employ one of my travelling companions to write this.[24] I had the pleasure to receive here your letter of Sepr. 18th. which like all your communications has given me infinite pleasure.[25] I would have written to you sooner, had I not waited for the list of plants, promised to me in your letter of June.[26] In the mean time I had the pleasure to send you two memoirs which will I hope interest you.[27] You will find in them something more about the Physical geography of this island & on my return to Christchurch in April I shall begin to write a more extended report with maps & sections, illustrative of the Geology & physical geography of this part of New Zealand.[28]

You see I am again in the Alps, reexamining some spots where the inclemency of the weather did not allow me to remain long enough before, and then I shall continue my researches towards the sourses of lakes Wanaka & Hawea from the former reaching the west-coast to Jacksons Bay.[29] I found a few glacial shells, but they were so rotten that I could make nothing of them but I know a spot where there are more to be had, & I shall do my best to procure them. I discovered this latter spot only a few days ago, when on my road from Christchurch here. Amongst them a Mytilus in boulder clay, and if worth while, I shall send you my whole gathering, to have it examined.[30] It also struck me forcibly that our glaciers are enormous, when considering the small size of the island the comparatively low elevation of the mountains & the low latitude.

Have you not published your travels in India & by what publisher?[31] I should like very much to study your book if it is still to be procured. You are quite right in saying there are many reasons to deter you from coming to New Zealand, and I shall do my best to give you such accounts, that your interest in its physical geography & natural productions, will not be diminished. The other day I again studied your 'Flora' & observe that you describe the Discaria as being only a few feet high,[32] but this is not always the case as in the valley of the Alps & on the sides of the hills, I have seen it as high as 15 feet, having sometimes a knotty stem a foot in diameter. This spiney gentleman being called by the setlers 'Wild Irishman' with this gentle ascociate, aciphylla squarrosa, the 'bloody spaniard' of the setlers, forms some-times the only vegetation of these happy river beds, through which we had to wend our way, being hours in going one mile & emerging again on better ground with torn clothes & bleeding limbs.[33]

In coming here I made again a very striking observation viz. that travelling on the southern side of the Waitaki to this place & crossing hills about 2200 feet high, not a single species of Celmisia was to be found, and only a yellow flowering Sinecio with white leaves whilst on the upper parts of the Canterbury plains, & on the other side of the Pakaki river several species of the Celmisia are abundant, showing how local plants often are.

I have not yet seen Prof. Ramsay's paper, but as my last case of books arrived as I left town, I shall read it on my return.[34] The phenomenon that the Himalayan lakes lie only on the northern side whilst they are wanting on the southern, finds its counterpart in our Alps, but as the slopes of our Alps on their western side, are also narrow & steep, our lakes lie on the eastern, which have shallow & broad valleys; may this occurrance not offer us an easy explanation to this fact? A glacier on the steep side of a mountain will much sooner reach the foot, it cannot collect with others to form one large glacier, which by its compactness offers more resistance to the effect of the sun and atmospheric influences, in consequence the amount of detritus brought down, cannot be so great that it can form large terminal moraines, & the valley being steep, the rush of the waters will soon destroy & disperse this detritus, however large it may be and thus the first cause of the formation of lakes is destroyed.

I have not yet begun to study M^r. Darwin's new work,[35] which will be a great treat for me on my return; he is a man after my own heart & I do not recollect having enjoyed more intellectual pleasure, than when I began to peruse his wonderful work on the 'Origin of species', he has no more faithful disciple in the southern hemisphere than me & I have made it my duty to combat the popular prejudices against this work, & I dare say my voice has brought to the right path many an honest man who was seeking the truth on this important subject.[36] I consider it the highest compliment you could pay me, in sending my letter to M^r. Darwin & for which I thank you heartily.[37] I have written in the enclosed letter to him concerning the animals which he wishes to procure from N.Z.[38]

I am in hopes that the collection which M^r. Harris takes with him to Europe for you will reach you in time, so that you may use them for your supplement.[39] Already on the present journey I have collected a good many plants, some of which seem to me to be undiscribed, and this year I shall be in time in the Alps for the flowering season, so that I hope you will receive my new collection in time for your popular work on the N. Zealand Flora.[40] You will have heard from the General

Gover^t. of Auckland that the necessary money has been voted for the publication of this work,[41] altho' there was not the least doubt about it, our friend Monro[42] being speaker of the house and the majority of the members men of good education, M^r. Travers wrote separately to each member of our Province,[43] & I spoke to them personally before they left, all with one exception cordially entered into the subject, & the only gentleman who was opposed to it, said that the great expenses of the Maori difficulty did not allow them at present to think of science for that Session.[44] The consequence was that when I invited the most influential & best educated members of the community, to join me for the formation of the Philosophical Institute of Canterbury,[45] I purposely omitted him; Altho' as I heard afterwards he did vote for the publication of the work. Altho' he is one of our most influential men, this little lesson has done him good, and I could scarcely suppress a smile when he told me the other day, that in order to prepare himself to be elected a member of our Institute, he had bought some botanical handbooks to study the science.

I had the pleasure to receive the kind note of Sir W^m. Hooker,[46] for which you will thank him in my name

Our Provincial Agent M^r. John Marshman 16 Charing Cross London will be happy to forward anything, entrusted to his care for me.[47]

I see my letter gets so long that I think it wrong to intrude further on your valuable time and hoping that you will continue to let me hear sometimes from you.

Believe me my dear D^r. Hooker | Most sincerely & truly yours | Julius Haast

D^r. J. D. Hooker F.R.S. | Royal Gardens | Kew

DAR 101: 128–31; Royal Botanic Gardens, Kew (Director's correspondence 174 (New Zealand letters, 1854–1900): 281–2)

1 Between 2 and 17 April 1863, Hooker and his wife, Frances Harriet, travelled to the Dorset coast and to the Channel Islands of Jersey and Guernsey with their sons William Henslow Hooker and Charles Paget Hooker (see letter from J. D. Hooker, [28 March 1863]). Hooker refers to the convict prison on the Isle of Portland (*EB*).

2 Letter to J. D. Hooker, [17 April 1863].

3 Hooker refers to Hugh Falconer's letter in the *Athenæum*, 4 April 1863, pp. 459–60, which criticised Charles Lyell for not giving sufficient credit to his research and that of Joseph Prestwich in C. Lyell 1863a. See letter from John Lubbock, 7 April 1863 and n. 6. Hooker also refers to Richard Owen (see also n. 6, below).

4 Lyell answered Falconer's criticisms in the *Athenæum*, 18 April 1863, pp. 523–5.

5 Hooker refers to the last paragraph in Falconer's letter to the *Athenæum*, 4 April 1863, p. 460, which reads: 'I need hardly add, that I have been authorized by my friend, Mr. Prestwich, to make those statements contained in the above remarks which have reference to him.'

6 The reference is to Owen's anonymous review of Carpenter 1862, which appeared in the *Athenæum*, 28 March 1863, pp. 417–19, and to William Benjamin Carpenter's reply (*Athenæum*, 4 April 1863, p. 461). Both articles are reproduced in Appendix VII.

7 See letter to J. D. Hooker, [17 April 1863] and n. 9. See also letter to *Athenæum*, 18 April [1863], letter to Charles Lyell, 18 April [1863], and Appendix VII.

[8] See *Athenæum*, 4 April 1863, p. 461. The Royal Medals were instituted by George IV, and are awarded annually by the Royal Society of London for the two most important contributions to science published in the British dominions not more than ten years or less than one year before the date of the awards (*EB*).

[9] Hooker read an account of *Welwitschia* before the Linnean Society on 18 December 1862 and 16 January 1863; it was published in the society's *Transactions* (J. D. Hooker 1863a). He described a functionless male flower, which he called a hermaphrodite: six stamens, fused at the base, surrounded a sterile ovule with a large, tubular styliform body extending outwards (*ibid.*, pp. 21–4). Each such unit, of which there were a number, looked, with its associated bractlets, much like an angiosperm flower; such a structure is not otherwise known in gymnosperms. Hooker reasoned that *Welwitschia* was a transitional form between gymnosperms and angiosperms (*ibid.*, p. 38).

[10] See letter to J. D. Hooker, [17 April 1863] and n. 18. George Bentham had asked CD for information on published reactions to CD's theory to use in preparing his anniversary address as president of the Linnean Society, which was to be delivered on 25 May 1863 (Bentham 1863). See letters from George Bentham, [*c.* 14 April 1863] and 21 April 1863. Bentham's stated intention in regard to CD and *Origin* was to 'ascertain how far the discussion and verification of his views have proceeded since the publication of his work' (Bentham 1863, p. xvi).

[11] *Two old men's tales: the deformed, and the admiral's daughter* ([Marsh] 1834) and *Emilia Wyndham* ([Marsh] 1846) were written by Anne Marsh, later Marsh-Caldwell (*DNB*).

[12] Hooker refers to the controversy that followed the publication of the first two parts of *The Pentateuch and the book of Joshua critically examined* (Colenso 1862–79) by the liberal clergyman John William Colenso, bishop of Natal. In a speech delivered to the Church Missionary Society in Manchester on 17 March 1863, James Prince Lee, bishop of Manchester, was reported to have stated that anyone declaring one line of the sacred books to be unfaithful or untrustworthy destroyed the basis of Christian belief. In a letter published in *The Times*, 2 April 1863, p. 10, Colenso argued that nothing endangered the Christian tradition so much as such narrow rigidity; he cited a reference to the hare in Lev. 11:6, which he said did not meet the known facts of natural history and could not be read literally. Colenso's letter met with an extremely critical response from Joseph Benjamin M'Caul, who challenged his expertise in textual criticism (*The Times*, 4 April 1863, p. 9).

[13] See enclosure. Haast also wrote a letter to CD on 9 December 1862 (see enclosure to letter from Julius von Haast, 5 March 1863). See also letter from J. D. Hooker, [30 April 1863] and n. 2.

[14] William Henslow Hooker (see n. 1, above).

[15] Charles Paget Hooker.

[16] Hooker's father-in-law, John Stevens Henslow, had been professor of botany and CD's mentor at Cambridge; Henslow died in 1861 (see *Correspondence* vols. 1 and 9).

[17] Daniel Oliver assisted Hooker in the herbarium at the Royal Botanic Gardens, Kew, and was professor of botany at University College London (R. Desmond 1994).

[18] John Lindley was the leading orchid specialist in Britain. In his letter of 7 November 1862 (*Correspondence* vol. 10), Hooker informed CD that Lindley was suffering from a partial disablement of his arms, and that he feared Lindley was 'breaking up'. During the last four years of his life, Lindley reportedly suffered from 'gradual softening of the brain' (*DNB*).

[19] See letter to J. D. Hooker, [17 April 1863] and n. 19. The reference is to Welwitsch 1861.

[20] Hooker was writing a paper on the plants of the temperate regions of the Cameroons Mountains and the islands in the Bight of Benin (J. D. Hooker 1863b) that was pertinent to his continuing discussions with CD on plant and animal distribution (see letter from J. D. Hooker, 17 March [1863] and n. 17).

[21] In his letter to Hooker of [17 April 1863], CD had asked for the address of the French botanist Jules Emile Planchon.

[22] Immediately preceding this sentence, Hooker crossed out: 'Your *Oxalis* is one called *O. Deppii* omitted in Harveys Flora Capensis, & I suppose a var of *purpurata* [*after del* 'Bowiei, itself a'] of which O. *Bowiei* is another var.' Hooker refers to the first two volumes of Harvey and Sonder 1859–65. CD had questioned the name given to a wood sorrel that had been sent from the Royal Botanic Gardens,

Julius von Haast
(Frontispiece to H. F. von Haast,
The life and times of Sir Julius von Haast, 1948.
By permission of the Syndics of Cambridge University Library.)

Donald Beaton
(From *Cottage Gardener*, 9 March 1855, p. 154.
By permission of the Syndics of Cambridge University Library.)

Kew; he was interested in the nocturnal movements of its leaves (see letter to J. D. Hooker, [17 April 1863] and nn. 22–4).

23 See letter to J. D. Hooker, [17 April 1863], and letter to J. D. Hooker, 26 [March 1863] and n. 14.

24 Haast was on an expedition travelling across New Zealand's Middle Island (now South Island) from the east to the west coast. See J. F. J. von Haast 1879, pp. 44–58, and H. F. von Haast 1948, pp. 273–4 and map no. 4. See also Haast's letter to CD of 5 March 1863 and n. 2.

25 Hooker's letter to Haast of 18 September 1862 is in the Alexander Turnbull Library, Wellington, New Zealand, MS papers 37.96.

26 Haast refers to Hooker's letter of 2 July 1862, in which Hooker expressed enthusiasm about a plant collection that Haast had sent him. In Hooker's letter to Haast of 18 September 1862, he explained that he had not yet had time to finish identifying the plants. These letters are in the Alexander Turnbull Library, Wellington, New Zealand, MS papers 37.96.

27 Haast probably refers to J. F. J. von Haast 1862a and 1862b, which he also sent to CD (see letter to Julius von Haast, 22 January 1863).

28 Haast did not arrive at Christchurch until 12 May 1863, but he sent a preliminary report, dated 3 March, to the superintendent of Canterbury Province; the report was published in the Christchurch *Press*, 1 April 1863, pp. 1–2, and 2 April 1863, pp. 2–3. Haast also wrote a shorter report entitled 'Report, map, and section of the provincial geologist, shewing result of journey to the western part of the province', which he presented to the Provincial Council, Province of Canterbury, on 18 August 1863. The handwritten report and the map, but not the section, are in the National Archives, Christchurch (Archives Reference CH 287, CP 608a). See letter from Julius von Haast, 13 May 1863. A description of the expedition was later published in J. F. J. von Haast 1879, pp. 44–58.

29 See n. 24, above. Haast had explored some of the same areas in an expedition that took place during the first five months of 1862 (see J. F. J. von Haast 1879, pp. 18–44). Haast reached the west coast at the mouth of the Awarua River, about twenty miles north-east of Jackson's Bay, on 20 February 1863 (*ibid.*, p. 55).

30 Haast's interest in glacial shells was related to his proposal that portions of New Zealand had been submerged during the Pleistocene period; he thought that some of the surface geology would therefore be of marine origin (see J. F. J. von Haast 1862b, pp. 128–31). In 1864 Haast changed his mind, partly because the marine shells found inland were determined to be from Maori kitchen middens (see J. F. J. von Haast 1879, pp. 371–6, and H. F. von Haast 1948, pp. 268–9, 1023–37). Hooker and CD had taken a great interest in Haast's theory of New Zealand's submergence and later uplift (see, for example, letter to J. F. J. von Haast, 22 January 1863 and nn. 4 and 5, and *Origin* 4th ed., pp. 442–3). Hooker had asked Haast about further discoveries of glacial shells (see the letter from Hooker to Haast, 18 September 1862, in the Alexander Turnbull Library, Wellington, New Zealand, MS papers 37.96).

31 J. D. Hooker 1854b was published by John Murray of London.

32 Haast refers to *Flora Novæ-Zelandiæ* (J. D. Hooker 1853–5), 1: 46–7.

33 Haast mentioned his riverside encounter with *Discaria*, measuring '16 to 18 feet high', and '*Aciphylla Colensoi*, the Spaniard or Bayonet grass' in J. F. J. von Haast 1879, p. 25. See also H. F. von Haast 1948, pp. 41, 72 n., 188, 206, 296 and 464. In the first volume of Hooker's *Handbook of the New Zealand flora*, published in 1864, *Discaria* are described as 'spiny shrubs or small trees' (J. D. Hooker 1864–7, p. 43).

34 In his letter to Haast of 18 September 1862 (Alexander Turnbull Library, Wellington, New Zealand, MS papers 37.96), Hooker discussed Andrew Crombie Ramsay's paper on the glacial origin of lakes (Ramsay 1862). For CD's interest in the paper, see *Correspondence* vol. 10, letter to A. C. Ramsay, 5 September [1862].

35 Haast refers to *Orchids*.

36 In his address to the Philosophical Institute of Canterbury, Haast praised *Origin* (J. F. J. von Haast 1862a, p. 7). See letter from Julius von Haast, 5 March 1863 and n. 12.

37 Hooker sent Haast's letter of 9 June 1862 with his letter to CD of 20 September 1862 (*Correspondence* vol. 10).

38 The letter from Julius von Haast of 9 December 1862 (*Correspondence* vol. 10), is reproduced in this volume as an enclosure to the letter from Julius von Haast of 5 March 1863. For the confusion regarding the sending of this letter, see the enclosure to the letter from Julius von Haast, 5 March 1863 and n. 5.

39 Mr Harris has not been identified. For Haast's plans to have the collection transported to Hooker, and for a discussion of the collection, see the letter from Haast to Hooker of 10 August 1862, in the archives of the Royal Botanic Gardens, Kew (Director's Correspondence 174 (New Zealand letters, 1854–1900: 276)). Haast also mentioned the collection in a letter to Hooker of 9 June 1862 (see *Correspondence* vol. 10, enclosure to the letter from J. D. Hooker, 20 September 1862 and n. 7). Haast also refers to Hooker's forthcoming *Handbook of the New Zealand flora* (J. D. Hooker 1864–7), which supplemented J. D. Hooker 1853–5.

40 For a list of the plants collected by Haast on his 1862–3 expedition, see H. F. von Haast 1948, pp. 295–8. In the preface to J. D. Hooker, 1864–7, p. 12, Hooker paid tribute to the number and value of Haast's specimens.

41 Hooker's commission from the General Assembly of New Zealand to write a handbook of New Zealand flora (J. D. Hooker 1864–7) was discussed in his letter to CD of 6 January 1863.

42 David Monro was the speaker of New Zealand's House of Representatives from 1861 to 1870 (Scholefield 1950, p. 150).

43 Haast refers to William Thomas Locke Travers and to representatives of Canterbury province. Travers sent significant collections of New Zealand alpine plants to the Royal Botanic Gardens, Kew (J. D. Hooker, 1864–7, p. 12).

44 In late 1862, Canterbury province was represented by seven men (Scholefield 1950, pp. 154–67); the member of parliament opposed to funding Hooker's work has not been identified. The New Zealand Wars of 1845–72 were financed primarily by imperial resources in 1861 and 1862; provincial expenditures for the conflicts between the colonists and the Maori were generally not popular with the colonists, so the unidentified member held a minority opinion. See Belich 1986, pp. 76–80, 113–25.

45 See letter to Julius von Haast, 22 January 1863 and n. 2.

46 William Jackson Hooker.

47 John Marshman was the agent at the Canterbury Emigration Office (*Post Office London directory* 1863).

From George Bentham 21 April 1863

25 Wilton Place
April 21/63

My dear Darwin

I am afraid I have omitted thanking you for your great kindness in lending me the very things I wanted to see[1] I have been so very busy the last week the printer having almost got up to me in the flora Australiensis that I have not had a moment for anything else—[2] I now hope to send him the last words of the Preface tomorrow—and will then immediately go through these pamphlets and return them to you—[3] I trust you do not expect anything to come of it of value to you or any one else I do not feel up to a thorough discussion of your views founded upon such a vast amount of undeniable facts and wonderful observations which you have made. For though I do not go so far as Jos. Hooker in the thorough adoption of all your hypotheses, and though I fully agree with John Mill and other such experienced logicians in the appreciation of your arguments still I do not feel competent to enter into the lists and argue upon the very limited observations I have myself made.[4] All I want is to make myself as well acquainted as I can with the present state of the question with the view perhaps of saying a few words in my anniversary address.—[5] and this I knew I could not do without your assistance in telling me which of the unmanageable mass of criticisms it was worth looking into.

I have nearly got through Sir C. Lyell's book which is admirable as to the Geological as main portion—[6] I wish there had been a little more method in the early part—and I should rather he had left the brain question in Huxley's hands who is fully equal to holding his own with Owen[7]—but I cannot understand Dr Falconer's attack which seems to me quite unwarrantable.[8]

I will return the pamphlets as soon as possible—[9] may I keep them a fortnight?

Yours very sincerely | George Bentham

DAR 160: 156

[1] Bentham had asked CD for references to foreign publications that discussed CD's theory (see letter from George Bentham, [*c.* 14 April 1863] and n. 2, and letter to George Bentham, 15 April [1863]).

[2] The first volume of Bentham's *Flora Australiensis* (Bentham 1863–78) was published on 30 May 1863 (*Taxonomic literature*). It was printed by the firm of John Edward Taylor, 10 Little Queen Street, Lincoln's Inn Fields, London.

[3] See letter from George Bentham, 21 May 1863.

[4] The references are to Joseph Dalton Hooker and the philosopher John Stuart Mill. Bentham quoted Mill in his presidential address to the Linnean Society (Bentham 1863; see letter to George Bentham, 19 June [1863] and n. 9).

[5] Bentham devoted a considerable portion of his presidential address, delivered to the Linnean Society on 25 May 1863, to discussing the international reaction to CD's theory of the origin of species (Bentham 1863, pp. xi–xxiv). See preceding letter and n. 10.

[6] C. Lyell 1863a.

[7] Bentham refers to the debate between Thomas Henry Huxley and Richard Owen regarding differences between simian and human brains (see L. G. Wilson 1996b). Charles Lyell reviewed the state of the controversy in C. Lyell 1863a, pp. 480–93.

[8] Bentham refers to Hugh Falconer's letter in the *Athenæum*, 4 April 1863, pp. 459–60, which criticised Lyell for not giving sufficient credit to the research of Falconer and Joseph Prestwich in C. Lyell 1863a (see letter from John Lubbock, 7 April 1863 and n. 6, and letter to Charles Lyell, 18 April [1863] and nn. 7–9).

[9] See letter to George Bentham, 15 April [1863] and nn. 3–11.

To H. B. Dobell 21 April [1863][1]

Down. | Bromley. | Kent. S.E.
April 21st

My dear Sir

The importance of hereditary transmission can hardly be exaggerated under every point of view; & I sh^d.^ think it would be a considerable gain to have tables printed as you propose.—[2] I really hardly feel competent to make suggestions. But I would rather strongly advise that the relationship (if any) of the father & mother & of any subsequent marriage should be enquired for. To avoid errors, & as one man would seldom fill more than one or two papers, I should think it would be better to avoid *all* contractions & use "no issue" & "nothing known".— You do not ask for *age* in parents & children at which any peculiarity (not congenital) supervened: I would very strongly advise this to be made a point of high importance.—[3]

I am very much obliged for your kind offer of communicating any remarkable cases of inheritance to me:[4] I have to consider the subject only certain points of view & I hope & suppose in about a year I shall go to press with this portion of my subject.—[5]

If by any extraordinary chance you should stumble on any case of regrowth of amputated additonal digit, I sh^d. be grateful for information.—[6]

With many thanks pray believe me | Yours very faithfully | C. Darwin

P.S. Would it not be better to urge that Christian name be given in full & initial (when secrecy requisitc) of sirname alone; otherwise sex might be doubtful, & this of course w^d. be most serious doubt in all cases in which peculiarity clings to one sex?—

DAR 221.5: 6 (photocopy); Legends (catalogue 2, 1990)

[1] The year is established by the relationship between this letter and the letter from H. B. Dobell, 20 April 1863.

[2] See letter from H. B. Dobell, 20 April 1863 and n. 1. The drafts of forms that Dobell enclosed with that letter have not been found; CD may have returned them with this letter.

[3] For Dobell's publication of tabulated genealogical information, see Dobell 1863 and 1866, and letter from H. B. Dobell, 20 April 1863 and n. 3. Dobell 1880 includes a table (Table 1) entitled 'A contribution to the natural history of pulmonary consumption'; the last eight (of seventy-five) columns tabulate whether parents, siblings, uncles and aunts, and first cousins had consumption. In remarks on case histories, Dobell occasionally noted at what age these relations died of consumption, but included no further information on the onset of their disease (Dobell 1880, pp. 15–22).

[4] See letter from H. B. Dobell, 20 April 1863. Dobell had already presented CD with a copy of his book (Dobell 1862), and James Paget had sent Dobell's details of a case of inherited physical abnormality (see letter from James Paget, 7 February 1863).

[5] For CD's citations of Dobell in *Variation*, see letter from James Paget, 7 February 1863, n. 1. *Variation* was not published until 1868; CD worked on the chapters on inheritance (*Variation* 2: 1–84) from 23 January until 1 April 1863 (see 'Journal' (Appendix II)). These chapters included material on the inheritance of disease and physical abnormalities.

[6] See letter to H. B. Dobell, 16 February [1863] and n. 7.

From John Hutton Balfour 22 April 1863

27 Inverleith Row | Edinburgh
22 April 1863

My Dear Sir

Many thanks for your paper on Linum which is a most valuable contribution to science.[1]

One of my gardeners is prosecuting the subject of such fertilization vigorously.[2] He is much indebted to you for your kind encouragement.[3] He has just put into my hands a paper which I hope to read at next meeting of the Botanical Society.[4]

I am | Yours sincerely | J H Balfour

DAR 160: 32

[1] 'Two forms in species of *Linum*' was read before the Linnean Society on 5 February and published in the society's journal on 13 May 1863 (*General index to the Journal of the Linnean Society*, p. vi). Balfour's name appears on CD's presentation list for 'Two forms in species of *Linum*' (Appendix IV).

[2] The reference is to John Scott, keeper of the propagating department at the Royal Botanic Garden, Edinburgh, where Balfour was director (R. Desmond 1994).

[3] CD had become a mentor for Scott's experimental work (see, for example, letter from John Scott, [after 12] April [1863] and n. 5).

[4] Balfour refers to Scott's paper on orchid pollination, Scott 1863a, which Scott himself read before the Botanical Society of Edinburgh on 14 May 1863 (see letter from John Scott, 21 May [1863]).

To George Bentham 22 April [1863][1]

Down Bromley Kent
Ap. 22^d

My dear Bentham

Pray keep pamphlets just as long as you like.—[2]

I shall be greatly interested by reading whatever you may write on Species. But you must allow me to differ in toto from you when you say you are not up to treat the whole subject.[3] I cannot believe this, from what I hear of your Systematic knowledge & from what I have read of yours on affinities, stations, geograph. Distrib. &c.—[4] How well you could handle one great subject, that of affinities, in relation to descent & independent creation! But I can fully understand that you may be too much pressed for time to do anything.—

I am glad to see what you think on Falconer v. Lyell.[5] & quite agree with you.—

Yours very sincerely | Ch. Darwin

Royal Botanic Gardens, Kew (Bentham correspondence 3: 701)

[1] The year is established by the relationship between this letter and the letter from George Bentham, 21 April 1863.

[2] At Bentham's request, CD had sent Bentham copies of several reviews of *Origin* (see letters from George Bentham, [*c.* 14 April 1863] and 21 April 1863, and letter to George Bentham, 15 April [1863] and nn. 3–11). As president of the Linnean Society, Bentham was preparing a discussion of the reactions of the international scientific community to CD's theory in his anniversary address on 25 May 1863 (Bentham 1863).

[3] See letter from George Bentham, 21 April 1863.

[4] Bentham was a leading plant taxonomist (R. Desmond 1994). See also letter to J. D. Hooker, [17 April 1863], n. 18.

[5] Hugh Falconer and Charles Lyell. See letter from George Bentham, 21 April [1863] and n. 8.

To Hugh Falconer 22 April [1863][1]

Down,
April 22nd

My dear Falconer

It was very good of you to tell me about Quatrefages (who has been very good-natured towards me) and especially about the Brazilian beast.[2] I do hope a cast will be procured for this country. What an extraordinary creature it must be.[3] If it turns, out as is too probable, a comparatively modern cave-fossil, the fact will not be altogether so pleasant for me, seeing how intermediate it is. I must coolly assume it is a very old form & like one of the Ganoid fishes, which are so comparatively common in S. America,[4] that has lived almost to the present day. It is a converse case of unpleasantness to that of Plagioaulax.[5] Nevertheless so intermediate a form is very glorious. I have been, as you supposed, grieved as a man of science, and as a very sincere friend of Lyell and you, at the quarrel in the Athenæum.[6] But there is no need for me to say anything. With many thanks for your note

My dear Falconer | Yours sincerely | Ch. Darwin.

Copy
DAR 144: 31

[1] The year is established by the reference to letters in the *Athenæum* (see n. 6, below).

[2] Following his return from France, Falconer conveyed Armand de Quatrefages's 'very best salutation' to CD (see letter from Hugh Falconer, 20 April [1863]). He also informed CD of a large fossil mammal skull that the palaeontologist Auguste Bravard had sent to Paris from South America.

[3] Falconer described the head of Bravard's 'Mammal-fossil' (named *Typotherium*) as 'bigger than that of a Lama' and displaying features of a marsupial and rodent (see letter from Hugh Falconer, 20 April [1863]).

[4] The majority of Palaeozoic and Mesozoic fossil fishes were classified as members of the primitive order Ganoidei, as was the living South American lungfish (*Lepidosiren paradoxa*) (Owen 1861b, pp. 139–69; see especially p. 147). This order is no longer recognised in fish classifications, and the species are placed in other orders (Wheeler 1985).

[5] The Jurassic mammal *Plagiaulax* exhibited specialised dentition but was discovered in a stratum far more ancient (the Purbeck or upper Oolite) than those in which most other mammals had been found (see Falconer 1857b and C. Lyell 1857, pp. 18–23). Conversely, Bravard's mammal appeared to be a primitive form found in more recent, Tertiary strata. See *Correspondence* vol. 6, letter to J. D. Dana, 5 April [1857], and *Correspondence* vol. 10, letter to Hugh Falconer, 14 November [1862].

[6] Falconer and Charles Lyell had exchanged letters in the *Athenæum*, arguing about the attribution of credit for research reported in C. Lyell 1863a; see letter to Charles Lyell, 18 April [1863] and nn. 7–9.

From D. T. Ansted 23 April 1863

Athenæum Club
23 April 1863

My dear Darwin

I have only this day (on arriving in town) received your most kind note and enclosure of the Bond.[1] All I can say in reply is that whatever I can at any time obtain for the 250£ worth of shares shall be immediately paid over to you and that I will take care that if this should happen after my death you shall not be a loser.[2] I wish most sincerely that I could do more

I feel that you have acted throughout in the kindest & most friendly manner and as I should myself have desired to act under similar circumstances May you be rewarded—

I am sincerely sorry to hear that you are not well. I trust it is not serious or likely to be tedious.

Always yours most truly | D. T. Ansted.

DAR 159: 76

[1] The letter to Ansted has not been found. In 1854, CD invested in and made bonded loans to a company owned by Ansted and Frederick Ransome that had since failed. See CD memorandum, 14 February 1863, letter to D. T. Ansted, 15 April 1863, and letter from D. T. Ansted, 17 April 1863.

[2] In CD's Investment book (Down House MS), the last entry concerning Ansted and the Patent Siliceous Stone Company was dated May 1863: 'I have cancelled Ansted [*after* 'R' *del*] Bond on understanding that if he can ever sell the shares, he will repay me'.

From Hermann Crüger 23 April 1863

Trinidad
23 April | 63.

Dear Sir

I have received your kind note of 18 March.[1] I shall gladly make the observations you speak of & report when I shall have done so. I have continued to observe the bees which frequent the Heteronoma spoken of in my last.[2] There are two bees which visit this flower, one is a large spec, about the size of our European bee, but somewhat plumper, of a greyish color. The visitors come only for pollen, which they regularly get from the anthers by pressing them between their mandibles beginning from the base of the anther until they arrive at the top, when the clump of pollen is transferred in the usual manner to the legs. The bee takes its seat on the top of the bundles of anthers, of which in this plant there are two. The shorter of the stamens surround the pistil, & the pollen comes so in contact with the stigma by the instrumentality of the bee.

I have repeatedly examined the stamens & anthers of these flowers & could not find any the slightest wound or puncture, nor can I find any nectar. The other bee which visits these flowers is a small golden-green one, rather lively & difficult of approach, but I think I may affirm it also comes only for pollen. I shall send you some sketches, when I have a few more facts connected with this family.[3]

I am glad to see that my notes on orchideae interested you.[4] As I told you before, your book on Orchids has produced some doubts on the correctness of what I had noted down & partly sketched[5] I shall therefore begin the work again & send you notes, drawings, & specimens for your autopsy. The substance the humble-bees are so eager after in the Catasetum is some secretion in or near the inner surface of the labellum, but I must investigate this also anew.[6] For this purpose I have collected a lot of Catasetum tridentatum, & as soon as they flower I shall begin again. With regard to the orchids which rarely or never open, I must say that I have had a short time ago one of the Epidendrums of Lindleys folia Orchs.— No 60–68. (perhaps the very No 65. of which Ldl. says "the flowers never expand") fully open.[7] These have not set fruit, while commonly I do not remember to have noticed a flower which was not followed by fruit. The seeds of these plants I have examined but could not discover any difference between them & others.[8] I have never been able to make a single Orchid germinate here, but I shall try it again. Some species germinate here abundantly by themselves such as Oncidium luridum, ampliatum, Epidendrum rigidum, ramosum, fragrans.[9]

I have noted you other desiderata & shall attend to them. With regard to Ficus, I have spent some time on them when I occupied myself with the fecundation of plants. A short account of my studies on Ficus you will find in the Botanische Zeitung 1851, p 60.[10] I am firmly convinced that without insects Figs will rarely or never give seeds, but I shall try the experiment you suggest.—[11] I have your last two works, Origin of Species & the Orchids, & have read the Journal of researches often, if you send me the latter I shall be delighted to accept it as a keepsake from its celebrated author.—[12] I believe I wrote in my last, that this is our dry season & therefore little to be found in flower. The rains begin in the middle of May or in June. As soon as the several plants begin to flower you speak of I shall observe them carefully. I intend to observe Cassias & Mimosas as also some Solanums. Of Begonias we have only 4 species, three of them common enough about here, the fourth grows only at a certain elevation.[13] I am inclined to believe that they are fertilised by insects as foreigners even from neighbouring places produce no seed. I know that this is a dangerous test, but it is remarkable how many exotics while flowering most abundantly are shy seeders. Of course other explanations may be given, if a plant f[or] instance came from southern tropical countries, where the seasons are reversed some influence might be deduced from this circumstance. But this would not apply to cases where a plant only transplanted from Jamaica would not seed here. I do not remember a Vanda or any other Indian or African Orchid setting a single seed-vessel. And still insects commit sometimes errors, as it is perfectly possible that an humble bee may alight on a Dendrobium taking it for something else of this country.

If you have any thing else to be done in the sake of science & progress I repeat that I am at your command.

Believe me | dear Sir | very sincerely yours | H Crüger

DAR 161: 276, DAR 205.8: 68 (Letters)

CD ANNOTATIONS

3.1 I am . . . autopsy. 3.4] *crossed ink*
3.4 bees are . . . anew. 3.6] 'The substance the humble- | orchids' *added ink*
4.1 With . . . Solanums. 4.11] *crossed ink*
4.15 I know that . . . seeders. 4.16] *double scored ink*
4.20 I do . . . seed-vessel. 4.21] *double scored ink*
Top of letter: 'Dr Hermann Cruger' *ink*
End of letter: 'Sporting Plants.'[14] *ink, del ink*; 'April 23^{d}. 1863' *ink*

[1] CD's letter to Crüger has not been found.

[2] See letter from Hermann Crüger, 23 February 1863 and nn. 1 and 2.

[3] CD had hypothesised that flies or small wasps might puncture the horns on the stamens of Melastomataceae in order to collect the fluid within (see letter to Asa Gray, 19 January [1863]).

[4] CD's letter has not been found, but see the letter from Hermann Crüger, 23 February 1863.

[5] Crüger refers to *Orchids*, and to his observations of the supposedly distinct orchid species *Catasetum tridentatum* and *Myanthus barbatus* (see letter from Hermann Crüger, 23 February 1863 and n. 6).

[6] In 'Fertilization of orchids', p. 154 (*Collected papers* 2: 151), CD reported that Crüger had sent him specimens of bees that he had seen 'gnawing the inside of the labellum'. See *Correspondence* vol. 12, letter from Hermann Crüger, 21 January 1864, and Crüger 1864.

[7] Crüger refers to the section on Epidendrums in Lindley 1852–5. Numbers 60–8 are descriptions of nine species; number 65 is a description of *Epidendrum gravidum.*

[8] See Crüger 1864, *Orchids* 2d ed., pp. 147–8, and letter from Hermann Crüger, 23 February 1863 and n. 5. CD mentioned Crüger's self-pollinating orchids in his letter to the *Journal of Horticulture*, [31 March 1863]; he also discussed them with John Scott (see letter from John Scott, [1–11] April [1863]).

[9] CD was interested in attempting to germinate orchid seed (see following letter and n. 5), and had probably asked Crüger for information on the subject in the letter of 18 March 1863 that is now missing (see n. 1, above).

[10] Crüger 1851.

[11] CD evidently described an experiment in his missing letter to Crüger of 18 March 1863 (see n. 1, above), but see also *Correspondence* vol. 12, letter from Hermann Crüger, 21 January 1864.

[12] Crüger refers to *Origin*, *Orchids*, and *Journal of researches*. See letter to Hermann Crüger, 25 May [1863].

[13] CD had suggested crossing experiments on begonias to John Scott (see *Correspondence* vol. 10, letter to John Scott, 19 December [1862]), and may have asked Crüger about the fertility of begonias in Trinidad (see n. 1, above). For experimental notes on begonias dated 19 March and 6 April 1863, see DAR 49: 83.

[14] In his letter to Crüger of 25 May [1863], CD asked whether Crüger had any information on 'sports', or what CD called 'bud-variations'. For CD's interest in bud-variations, see letter to Asa Gray, 2 January [1863] and n. 18.

To J. D. Hooker 23 April [1863]

Down
Ap. 23^{d}

My dear Hooker

I return Haast's letter: he is a glorious species man.—[1] I am very sorry for your account of your Boy (but what a fine little fellow Charlie appears).[2] What anxiety our children give us: Horace has been very bad in his stomach again, & we have resolved to go on Monday to "Rev. C. Langton Hartfield Tonbridge Wells" & thence to Leith Hill for a fortnight's change, which indeed I much require.—[3] Very many thanks for gratifying my silly hobby-horse about orchid seeds & they are all planted.[4] If ever you have a very large pod of any foreign orchis, I sh$^{d.}$ like it to estimate, out of mere curiosity, total number of seed.—[5] One more thing, you told me of some plant, I think a Hot-House plant allied to Lythrum, with two coloured anthers; now I sh^{d} be really be *particularly* glad to examine fresh flowers of this plant this summer.—[6]

The more I think of Falconer's letter, the more grieved I am; he & Prestwich (the latter at least must owe much to the Principles) assume an absurdly unwarrantable position with respect to Lyell.[7] It is too bad to treat an old hero in Science thus.

I can see from note from Falconer (about a wonderful fossil Brazilian mammal, well called Meso- or Typo-therium) that he expects no sympathy from me.[8] He will end, I hope, by being sorry. Lyell lays himself open to a slap by saying that he would come to show his original observations; & then not distinctly doing so; he had better only have laid claim on this one point of man, to verification & compilation.[9] altogether I much like Lyell's letter. But all this squabbling will greatly sink Scientific men: I have seen sneers already in Times.[10]

I am heartily glad that your new specimens of Wellwitschia confirm your results: I suppose you are sure that these great "bristling" stigmas cannot serve to lift up or push out pollen, as in sterile florets of some Compositæ?[11] Have you ever looked at the common Ash? I find in my Field 3 classes of Trees; viz (1) females (seed-producers) with aborted anthers, & rarely with single good anther (2) Hermaphrodite (seed-producers) in my field few in number (3) Males "bristling" with good-sized pistils & stigmas, which soon drop off & with atrophied ovules: these male trees *apparently* do not produce a single seed; but I shall watch them—what a gradation![12]

Emma tells me that the "Admiral's Daughter" would not at all suit me,—too melancholy.—[13]

I am very glad you are stirring up Oliver for Orchideæ;[14] by Jove he would find it a job worthy his talents & I sh$^{d.}$ think most interesting. Do you ever see old Cottage Gardener; there is something odd about seeds of Orchids with little filament coming out, described by Gosse in last Wednesday's. paper.[15] If ever I have time I must have a look: one would suspect it was to attach seeds to bark.—

I am very glad you had such fine weather for your Trip.[16]

Farewell— | C. Darwin

Your plants in Hot-house are an everlasting amusement to me;[17] but heat is killing.—

Endorsement: '/63'
DAR 115: 191

[1] CD refers to Julius von Haast's letter to Hooker of 10 December 1862, in which he praised *Origin* (see enclosure to letter from J. D. Hooker, 20 April 1863).

[2] Hooker had included some candid remarks regarding his sons William Henslow Hooker and Charles Paget Hooker in his letter to CD of 20 April 1863.

[3] Emma Darwin recorded in her diary (DAR 242) that Horace Darwin, aged 11, was intermittently ill during 1863; entries on 17, 19, and 21 April noted his poor health. Hartfield Grove at Hartfield, Sussex, was the home of Charles Langton, widower of Emma Darwin's sister Charlotte. Leith Hill Place, near Dorking, Surrey, was the home of Josiah and Caroline Sarah Wedgwood, Emma's brother and CD's sister.

[4] See letter from J. D. Hooker, 20 April 1863 and n. 23.

[5] CD had recently counted seeds of another orchid pod, and did this for a number of plants (see letter to Hugh Falconer, 5 [and 6] January [1863], and letter to John Scott, 20 [February 1863] and n. 3).

[6] Hooker replied to CD's query regarding *Lagerstroemia* in his letter to CD of [24 May 1863].

[7] Hugh Falconer's letter in the *Athenæum*, 4 April 1863, pp. 459–60, criticised Charles Lyell for not giving sufficient credit to the work of Falconer and Joseph Prestwich in C. Lyell 1863a (see letter to Charles Lyell, 18 April [1863] and nn. 7–9, and letter to Asa Gray, 20 April [1863]). CD refers to the nine editions of Lyell's *Principles of geology* (C. Lyell 1830–3), Lyell's 'lifework', which established 'a new era in geology' (*DSB*).

[8] See letter from Hugh Falconer, 20 April [1863].

[9] Lyell published a letter replying to Falconer's (see *Athenæum*, 18 April 1863, pp. 523–5, and n. 7, above), in which he quoted (pp. 523–4) from Falconer's letter (see n. 7, above): 'I now invite Sir Charles Lyell, or any one else on his behalf, to point out a single new fact or one solitary special observation which he has contributed bearing on the immediate subject of the antiquity of man.' Lyell took up Falconer's challenge at the end of his letter (*ibid.*, pp. 524–5); he detailed his own work, which was largely on the geological background of human remains and relics at sites that had already been discovered and interpreted.

[10] The reference has not been identified.

[11] See letter from J. D. Hooker, 20 April 1863 and n. 9. CD mentioned this feature of Compositae in *Origin*, p. 452. See also *Natural selection*, p. 48.

[12] These observations of the common ash (*Fraxinus excelsior*) are reported in *Forms of flowers*, pp. 11–12. For CD's notes on the ash, dated 20 April 1863, see DAR 109: 51.

[13] [Marsh] 1834. See letter from J. D. Hooker, 20 April 1863 and n. 11.

[14] Daniel Oliver. In his letter to CD of 20 April 1863, Hooker mentioned that he was encouraging Oliver to concentrate on orchids and become Kew's 'orchidologist'.

[15] CD refers to Philip Henry Gosse's description of microscopic observations on orchid seeds that appeared in the *Journal of Horticulture and Cottage Gardener*, 21 April 1863, pp. 287–8.

[16] Hooker had just returned from a two-week trip to Dorset and to Jersey and Guernsey in the Channel Islands. See letter from J. D. Hooker, 20 April 1863.

[17] CD had recently had a hothouse built at Down House; Hooker contributed plants from the Royal Botanic Gardens, Kew (see Appendix VI).

From Isaac Anderson-Henry 24 April 1863

Hay Lodge, | Trinity, | Edinburgh.
April 24/63

My dear Sir

I embrace the first leisure to thank you for your very obliging note of the 20th Inst.[1] and the beautifully executed likeness of yourself so kindly accorded to me.[2] I have no Carte else I am so proud to possess

I was not a little struck to learn that they had got at Kew the supposed hybrid *Bryanthus* (*erectus*) *wild* from North America. Having a Letter from Dr Hooker by the same post which brought me yours, I have in my answer embraced the opportunity of asking him about it.[3] May it not be the B. *Gmelini* a true north American Species, not previously introduced and noticed by D Don in Edin: Phil: Journal 17 p 160.?[4] At same time there is, I must confess, a mystery about Mr Cunninghams *hybrid*, (*B. erectus*) His nephew now in a Nursery (Messrs Lawsons) hard by here, owned to me that his uncle was *not quite certain* of its parentage tho he believed it to be as given out,—viz a cross between the *Rhods. Chamæcistus* & a *Menziesia*.[5] I think I mentioned to you that I was at that time at work myself attempting a *mule* between these same things—but I made the *Rhodothamnus*, and not the *Menziesia* the seed bearer,—& did not succeed.[6] On hearing of Mr C's cross, I inverted the X, and made *Menziesia* (Phyllodoce) *Cærulea* the female, & the R Chamæcistus the male parent which succeeded and from the seeds I raised one or two plants, which I lost early by the incursion of a snail. I then held that those in that order must have been the parents of the *B. erectus*. I have this year wrought the *R. Chamæcistus* on both, vizt on the *Phyllodoce* (M) *Cærulea* and M. *empetriformis*. I am not sure that any one of the X[s] on *M. Cærulea* has taken—but all of those (4 or 5) on the *empetriformis* seem to have taken, & the pods are already unmistakeably swollen with the ripening seeds. This looks like confirmation of Mr C's selection and success. And I need not say that long experience (20 years of it) has made me most careful by previous *emasculation*, &[c], to secure against the interference of insects

Many many thanks for your most interesting Paper on the *Linum* family.[7] I wrought upon that tribe last year, and have, if true, one cross from it—viz[t]. *L. rubrum Grandiflorum* (annual) crossed on L. *album*, a perennial, the plants from which are now six inches high. But I was grievously puzzled & perplexed by the varying forms of the sexual organs on the different species,— in so much that I began to conclude from these and the want of success that Botanists had erred in the arrangement of the tribe, & that several species must be removed to another family. I wrought on a great many sp[s].— L. *trigynum* (with its 3 styles) L. *flavum*, L. *perenne*, L *Corymbiflorum*, yellow (a most obstinate thing which bore seed in no form), L. *album* a fine large white flowered perennial, L. *flavum* (which likewise bore no seed) and L. *rubrum Grandiflorum*— The L. *trigynum* let alone was the best seed bearer of them all; but I accomplished no certain cross upon it. I have the whole in array for further operations this year, & with the light you have thrown upon the *genus*, I will now resume my experiments with fresh avidity[8]

If I remember aright, I was perplexed in the same way in the *Phlox* family; with these short and long stamens—and perhaps it was from this cause & not falling on the corresponding styles, that I failed often, where I had reckoned success certain

And there are other tribes with which I have been grievously baffled, tho I cannot now recall them, for the like reason

Your *discovery* I am persuaded involves a law which pervades innumerable races.[9] It is akin to Newtons discovery of the law of Gravitation[10] The more I think of it, the more I am astounded by its importance & the further discoveries to which it will lead. I had long believed that different results were produceable by the different pairs of anthers on plants having these organs of unequal length,—e.g. the *Geraniaceæ*, the *Rhodoraceæ* &c &c Now, we must have regard to the styles—the length of them—tho I incline to the belief, that this law may be found limited mainly to tribes having the *anthers all* of *equal* lengths.

Believe me; you are yet but on the threshold,—tho you have unlocked the portal in which the greatest merit lies.

If I can aid in any way do kindly suggest to me Mean time I remain | very faithfully yours | Is. anderson Henry

I am keeping mind the experiment on the top flower of the Geranium[11]

Many thanks for your suggestion as to *Bates* Travels—[12] I devour all these Books of Travels whose writers speak at all intelligibly of Natural history But how few of them do?

P.S. On looking again, I find appearance of one pod of *Menziesia Cærulea* X *Rhod. Chamæcistus* having taken

We have formed a *Microscopic Club* here limited to 15 among whom are a good many Professors of the University. Dr Greville is our Chairman[13] Will the knowledge of the microscope aid me much in any experiments you would suggest?

Charles Darwin Esqr F.R.S. &c &c

DAR 159: 65

CD ANNOTATIONS
3.8 L. *trigynum* ... *Corymbiflorum*,] *cross in margin, ink*
4.1 I was ... stamens— 4.2] *cross in margin, ink*
6.7 I incline ... lengths. 6.8] *cross in margin, ink*
7.1 Believe ... lies. 7.2] *scored ink*
12.2 Will ... suggest? 12.3] *cross in margin, ink*

1 The letter to Anderson-Henry has not been found; it was a response to Anderson-Henry's letter of 17 April 1863.

2 In his letter of 17 April 1863, Anderson-Henry sent his own photograph and requested one from CD.

3 The reference is to Joseph Dalton Hooker, assistant director at the Royal Botanic Gardens, Kew (R. Desmond 1994). The supposed hybrid *Bryanthus erectus* was discussed in the letters from Isaac Anderson-Henry, 26–7 January 1863 and 17 April 1863. See also letter to Isaac Anderson-Henry, 2 February [1863] and n. 8.

4 *Bryanthus gmelini* was described by the botanist David Don (Don 1834); it is native to north-eastern Asia.

5 The references are to James Cunningham and to Peter Lawson & Sons, a firm with several nurseries in Edinburgh (*Post Office Edinburgh and Leith directory* 1863–4). Cunningham's nephew has not been identified. Cunningham was reported to have obtained *Bryanthus erectus* by crossing *Phyllodoce* (*Menziesia*) *caerulea* with *Rhodothamnus chamaecistus* (*Gardeners' Chronicle and Agricultural Gazette*, 1 November 1851, p. 695).

6 See letters from Isaac Anderson-Henry, 26–7 January 1863, and 31 January 1863 and n. 4.

7 'Two forms in species of *Linum*' was read before the Linnean Society on 5 February 1863. Anderson-Henry's name appeared on CD's presentation list for the *Linum* paper (see Appendix IV).

8 The crossing experiments with *Linum* were discussed in Anderson-Henry 1867b.

9 In 'Two forms in species of *Linum*', CD described dimorphism in some flax species, and demonstrated that each form was often far more fertile when crossed with another form, than when pollinated with own-form pollen.

10 Isaac Newton advanced his theory of gravitation in 1687 (*DSB*).

11 In his letter to Anderson-Henry of 20 January [1863], CD discussed his work with the central peloric flowers of *Pelargonium*, a genus of Geraniaceae. Anderson-Henry wrote that he would test the fertility of such flowers (see letter from Isaac Anderson-Henry, 26–7 January 1863).

12 Bates 1863; see CD's annotation to letter from Isaac Anderson-Henry, 17 April 1863.

13 Robert Kaye Greville was a botanist and entomologist, and honorary secretary of the Botanical Society of Edinburgh (*DNB*, *Medical directory* 1863). For the proliferation of Victorian microscope clubs and societies, see S. V. F. Butler *et al.* 1986, pp. 7–9, 11–16.

From Hugh Falconer 24 April [1863][1]

21 Park Crescent | N.W.
24 April

My Dear Darwin.

Lest I should have misled you through my banter about the *Insectivore* cross—in the *Typotherium*, I think it right to let you know, that I cited *insectivore* simply as a term of comparison to convey to you an idea of the outline of the crown of the molars.[2] In reality there is no insectivore affinity. The affinity is with *Toxodon*; the incisors of an overgrown *Rodent*; and the Cerebral box—of the skull shaped somewhat like some of the larger herbivorous *Marsupials*. There may be other odd points of detail. The Creature is so singular, that the names *Typotherium* and *Mesotherium* were thought of for it by the French.[3] I thought you would be interested about it—and I wrote to you. The skull was quite fresh looking. I suppose there is a Cast of it in England by this time, at the Coll. of S. Flower at my suggestion has written for it.[4]

Gratiolet—the greatest of living French Anatomists, told me, that *Glyptodon*, by its brain, is a true *Bradypus*[5] They have a magnificent entire skull of it—also—absolutely perfect

The fossil human jaw of Abbeville—is I believe a plant, got by by the workmen. I have written a letter about it.[6]

Yours Ever | H Falconer

Typotherium

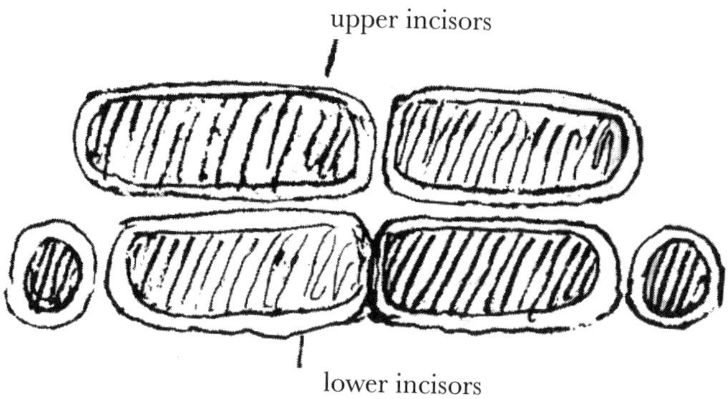

from rough memory

DAR 164: 15

[1] The year is established by reference to the fossil human jaw discovered at Moulin-Quignon near Abbeville, France (see n. 6, below).

[2] In his letter to CD of 20 April [1863], Falconer had likened the molar teeth of a fossil mammal skull from South America to those of a gigantic insectivore.

[3] See letter from Hugh Falconer, 20 April [1863] and nn. 8 and 9.

[4] William Henry Flower was curator of the Hunterian Museum of the Royal College of Surgeons (*DNB*).

[5] Louis Pierre Gratiolet. *Glyptodon* was a large Pleistocene armoured mammal, portions of which CD collected in South America (see *Correspondence* vol. 1, letter to Caroline Darwin, 24 October – 24 November [1832], *Correspondence* vol. 3, and *South America*, pp. 106–7); it evolved from armadillo-like ancestors and is related to the sloths, of which *Bradypus* is an extant genus.

[6] What appeared to be a fossil human jaw and tooth were found on 28 March 1863 by a workman at the Moulin-Quignon gravel pit near Abbeville, France. They were extracted by the archaeologist Jacques Boucher de Perthes and were examined by Falconer when he was in France earlier in the month (see letter from Hugh Falconer, 20 April [1863] and nn. 2 and 3). Initially impressed by its antiquity, Falconer examined a cast of the jaw on his return to London and subsequently grew suspicious of the fossil's authenticity (Carpenter 1863, Falconer *et al.* 1863, Van Riper 1993, pp. 135–6). Falconer's letter appeared in *The Times*, 25 April 1863, p. 14. See also Boylan 1979.

From George Maw 25 April 1863

Benthall Hall, | nr. Broseley.
25th. April 63.

Dear Sir

About four years ago my cousin[1] brought from Gibraltar a lump of stalactitic concretion containing bones— Sir C Lyell's book has interested me in examining it & I am strongly impressed with the idea that they are human.[2] a medical friend here[3] to whom I have shown the lump also thinks that the bones are human

My brother[4] who visited Gibraltar three years ago was shown by our friend there[5] (who is a medical man & who gave the lump above referred to to my cousin) another lump of rock containing what appeared to him to be a portion of a human skull & which I fancy must have come from the same locality as the mass of bones but at present I know nothing whatever of the circumstances under which either of the specimens were found indeed until Sir C. Lyell's book made me curious on the subject I had almost forgotten having heard about them—

I wrote to our friend at Gibraltar a few days ago asking him for the loan of the piece of skull & to collect for me any information he cd gather respecting the circumstances of its occurrence In the meantime it may perhaps be worth while to submit the lump of concrete to Sir C Lyell & if you would kindly furnish me with his address I shall be interested in placing it before him. as I am entirely unknown to Sir C Lyell perhaps you would not mind mentioning my having written to you on the subject of the bones when you next have occasion to communicate with him—

I am just now interested in the examination of some very fine sections of drift which are being opened out in a Railway Cutting at the Entrance of Coalbrook Dale. (1½ miles from here)—[6] The banks of the Severn the railway cutting & a gravel pit above it, afford an almost consecutive section or scores of sections of 160 to 170 feet in height & the top of this great drift bed is nearly 300 feet above the sea level. both the highest & lowest beds composing it are regular sea sand & well washed Pebbles & between them is a stratum of loam silt & Tough clay 50 or 60 feet thick interstratified with thin beds of gravel containing *drift*—coal Boulders & marine shells of which I have 7 or 8 species[7] I do not know much about them myself but fancy they are recent species— The great bulk of this drift bed & the strata of tough clays strike me as being very unusual

Believe me to remain D[r] Sir | very truly y[rs]. | Geo Maw.

C Darwin Esq.

DAR 171: 99

[1] This individual has not been identified.
[2] Charles Lyell discussed evidence of 'post-pliocene' human habitation on the Mediterranean coast in *Antiquity of man* (C. Lyell 1863a, pp. 174–8), but no finds from Gibraltar had at that time been reported (see *Correspondence* vol. 12, letter from J. D. Hooker, 26[–28] October 1864, Busk 1864, and Grayson 1983, p. 213).
[3] This individual has not been identified.
[4] Arthur Maw.
[5] This individual has not been identified.
[6] Maw lived in Shropshire near the Severn Gorge; he published an account of these drift deposits in Maw 1864.
[7] Maw listed these rocks and shells in Maw 1864, pp. 137 and 140.

To Lovell Augustus Reeve 25 April [1863][1]

Down. | *Bromley.* | *Kent. S.E.*
April 25th

My dear Sir

I am extremely much obliged for your very kind present of your really beautiful work on the British Land mollusks.[2] I am extremely glad to posses a copy & wish my knowledge rendered me more worthy of its possession.—

I shall take an early opportunity of carefully looking through it, for but lately I was wishing for information on two or three little points.—

With my best thanks | Pray believe me | Yours very faithfully | Charles Darwin

Liverpool Libraries and Information Services (HL AL)

[1] The year is established by the reference to L. A. Reeve 1863 (see n. 2, below).
[2] L. A. Reeve 1863. There is a lightly annotated copy of this work in the Darwin Library–Down (see *Marginalia* 1: 698).

From John James Aubertin 27 April 1863

São Paulo. Brazil—
April 27. 1863

My dear Mr Darwin—

I wonder whether you still recollect our joint sojourn at Ilkley Wells![1]—washing ourselves inside & out in order to wash the Satan of Illness out of us!—& seeking to purify our vile flesh by Water, as metals are purified by the more potent Element of Fire!—

If you do I hope you will accept of me the enclosed Vignette Portrait of myself,[2] trusting you will not find from it that I have within the past three years or more descended too rapidly into the vale of years!

Pray make my best remembrances to Mrs Darwin,[3] with best hopes also that you & all your family are well & flourishing!

I do not wake up our old acquaintance, short but pleasant, for the mere purpose of sending you my portrait: but I am sending also, all the way from Brazil to your quiet quarters at Down two lumps of stone, which I want you to see, as coming out of this Country: & not so only, but coming moreover out of the Tunnel we are blasting on our Great São Paulo Railway—[4] A friend who is going home takes the small box containing them to my brothers in London, by this mail of 9th May from Rio, & they will no doubt quickly forward the box to you, as I shall request them to do so—[5]

We have much curiosity among us here to know how the stone is to be strictly classified— Some declare it contains metal: & others that this metal is or ought to be *gold*! We are by no means however in search of the precious metal: but rather want to know whether the stone is likely to continue the features it at present presents, or whether it is likely to change its character for one which more readily responds to the Blast, before we get through our hill— There is a quantity of loose sort of wet schist outside: & gradually we have pushed forward into the character of stuff I now send you— The hill is a long sharp ridge, rising up into almost an edge at the Top—& the length of our Tunnel through it is somewhere about 500 yards— The Angles on both sides in upright lines must show something about an inclination of 1 in 5—but on one side this very sharp descent soon changes, & rather abruptly, into a far flatter one, from which we hope to find that the interior stone is more likely to prove a great thick upright wall than a broad flat bed: & that we shall not have to cut through hard stone for anything like the whole length of the 500 yards— We cannot yet tell anything of what the *exit* of the Tunnel is likely to be, as the form of the ground will oblige us to rely upon a very long heavy cutting, which of course has been perforce begun at its end: i.e about $\frac{3}{4}$ of a mile from what we propose should be the end of the Tunnel.[6]

In the course of our learned doubts hopes & fears it occurred to me that if I could send you a specimen stone or two you might feel interested in giving us a few hints upon the subject in general:[7] & I thought also my doing so might afford me a pleasant opportunity of re-opening our intercourse, though we are now rather farther apart than we were while conversing together at the Ilkley Fire-side, over the proof sheets of your *Naughty Book*—[8]

I have been writing thus far taking it for granted that you are aware of my being the Superintendent of this same Railway in Brazil, of which very likely you *are* aware—[9] We are pushing along very well, & hope we shall be able to complete all our works long within the time allotted to us—[10] I feel sure you have yourself visited Brazil, in your travels round the world: but I do not at all know whether you may have visited our Province of São Paulo.[11] You doubtless know the general character of the Brazilian Coast— I mean its great long, high, steep, wooded Serra! The barrier formed by this Phenomenon all along the Coast of this Province exhibits some of its most abrupt and obstinate difficulties: & the consequence is that we are all here much shut out of the civilized world: having nothing but monstrously bad roads, & mule-backs to get down to the coast with! Up the great Coast Serra our Ry comes by an incline of 1 foot in 10— We rise to the height of 2.600 in about five miles. We run up a great cheek of an immense buttress of the main ridge, not in zigzags: but in comparatively a straight line. Our incline is divided into four stages, each with its "landing" at the top of it: and on each of these "landings" is to be planted a stationary Engine, to draw up the carriages as far as itself by steel ropes: a new adjustment of the apparatus is to take place at each of these stages: & in all we shall have four stationary Engines.[12]

The work is a very venturesome one: but with Mr Fairbairn's & others' scientific applications I trust we shall surmount all difficulties—[13] The happy result of our exertions is a vital question for the Province which is pining for proper intercourse with mankind & commerce— The population is very scanty— We measure nearly $\frac{3}{4}$ of the area of France proper: & we have only 800.000 souls—blacks included! We are all very ignorant, & narrow-minded!— The Brazilians are of so weak & lazy a character that there seems almost more strength of character in many of the Blacks! In many of the Mulattos there assuredly is— Tho' I must confess that after contemplating the nature of the greater part of the Slaves here, I have a strong leaning to your doctrines in part: for I can scarcely hold to my old belief that the Negro can rank in common originin with the white, as a separate human race in the Animal Kingdom. Slavery here is very different from what it is in North America.[14] *There* the black is most rigidly separated, in all social points of view. Here, on the other hand this weak people let him in every where—even *petting* him—& the Mulatto has responded by taking, in more instances than merely a few, high, & even official position— A case in point indeed may be found in our very Provincial President here, who undoubtedly claims black blood, & is, in many respects, all the more active & intelligent therefrom![15] Coffee, as you may know is our chief produce here: I have been doing my best to get them to add Cotton—[16] Tobacco also might result in much gain— In short, our climate, & natural fertility with a little exertion would produce anything & everything. At St Paul's we are just under the Tropic of Capricorn, at about the altitude of Madrid— Corn & Cattle of all kinds might flourish: but all flour is imported, & almost all beef & mutton is horribly hard & uneatable— Chickens abound—& I wonder feathers are not growing out of some of us! Add to this eggs, eggs, eggs— But potted meats & hams & English beer, by myriads of Dozens come in now— Our Railway runs, or *is to* run some 85 miles inland, to a place called Jundiahy—a cold, ugly empty Town:[17] & we hope to extend it to Campinas—an equally ugly City some 30 miles onward, but situated in a more fertile district, & in a warmer soil— The Coffee Plantations begin to abound thence onwards— Sugar also is cultivated: but in these districts not well: more comes to you from the North. Some of our Coffee however, & some of our Tobacco ranks first in these productions in Brazil—

It is only a few mails ago that I took it into my head to write to D[r] Smith of Ilkley, sending him also a vignette, & recalling old scenes.[18] Among the number of these do your boys still remember the Lecture on the Mummy? beginning, if I rightly remember "Peace in the presence of the mighty Dead"—[19]

But how it ended I cannot venture to recall!— We got on there tolerably well altogether— The "course" did me a great deal of good: & I think it did you much good too. Our climate here is generally very hot at midday: but always cool at night & at morning—& I think it suits me very well altogether. Society there is just precisely & exactly *none*! I must by no means forget to ask after my friend Miss Butler![20] Pray make my kindest remembrances to her, with best hopes that she is quite well. She will not forget our jaunt to Harrowgate with the Miss Scotts.[21] Pray where are they? I suppose you have never repeated the dose at Ilkley, under "Wickers"—he of charade notoriety: & upon my honour a most admirable bath man![22] The beef & mutton at Ilkley were always first-class—& indeed all the eatables were the same— In drinkables there was (ex necessitate) no *very* great variety—

Any letter for me may be directed either to my brothers "Aubertin bros 6 G[t]. Winchester S[t]. Broad S[t]. E.C. to come in their post: or to the care of Messrs Baird Le Cocq & Co Rio de Janeiro—"[23] The Stones I send come from a range of hills some seven leagues up country from this City: or about 70 miles from the sea—

With just room for kind regards to M[rs] Darwin & yourself Believe me faithy yours | J. J. Aubertin.

DAR 159: 123

CD ANNOTATIONS

10.5 I must . . . well. 10.7] *cross in margin, brown crayon*
10.7 Scotts.] *underl brown crayon*
11.1 Any . . . post: 11.2] *scored brown crayon*

[1] CD had undergone hydropathic treatment at Ilkley Wells, Yorkshire, from 3 October to 9 December 1859 (see *Correspondence* vol. 7). See also nn. 8 and 18, below. Aubertin is not mentioned by CD in his letters from Ilkley.

[2] The enclosure has not been found.

[3] Emma Darwin.

[4] Aubertin was superintendent of the São Paulo railway from 1860 to 1869 (Graham 1968, p. 67). Eventually extending 139 kilometres from the port of Santos into the coffee districts of the province of São Paulo, the railway became the wealthiest in the country, and Britain's most profitable railway enterprise in Latin America during the nineteenth and early twentieth centuries (Graham 1968, pp. 60, 66). See also nn. 6, 10, and 17, below.

[5] Aubertin's brothers were merchants; their address was listed as Aubertin Brothers, 6 Great Winchester Street, London (*Post Office London directory* 1863).

[6] For the technical difficulties involved in building the São Paulo railway, see Graham 1968, pp. 63–4.

[7] See letter to J. J. Aubertin, 19 July 1863.

[8] *Origin* was published while CD was at Ilkley Wells in 1859, and he immediately began making corrections and additions for a second edition (see *Correspondence* vol. 7, and Freeman 1977, p. 75).

[9] See n. 4, above.
[10] The railway line opened in 1867.
[11] CD had never visited São Paulo, Brazil, but from 4 April to 5 July 1832, during the *Beagle* voyage, he spent time in the vicinity of Rio de Janeiro (*Journal of researches*, pp. 21–43).
[12] See n. 6, above.
[13] Aubertin apparently refers to the Scottish engineer William Fairbairn (*DNB*). Earlier in the century, Fairbairn had designed and constructed tubular railway bridges as well as locomotive engines (Pole 1877, pp. 197–213, 316–17, 320).
[14] CD wrote about slavery in Brazil during his *Beagle* voyage (see n. 11, above, and *Journal of researches* 2d ed., pp. 20, 24, 499).
[15] In 1863, Vicente Pires da Mota was president of the province of São Paulo (*Dicionário de história de São Paulo*).
[16] Aubertin was an active proponent of cotton culture in Brazil in the 1860s; in 1869 the Manchester Cotton Supply Association awarded him a gold medal for services rendered in creating a source of supply in Brazil (Graham 1968, pp. 67–8).
[17] See n. 4, above. The railway ran from the province's principal port, Santos, through its capital, São Paulo, to Jundiahy at the start of the coffee district (Graham 1968, pp. 60–6).
[18] Edmund Smith was the proprietor of Ilkley Wells hydropathic establishment (Metcalfe 1906, p. 107).
[19] Emma Darwin and the younger children joined CD at Ilkley Wells on 17 October 1859 (see *Correspondence* vol. 7, letter to W. D. Fox, [6 October 1859], and letter to W. E. Darwin, [14 October 1859]). In a letter to William Erasmus Darwin of [7 or 14 November 1859], Emma reported (DAR 210.6: 51):

> The other day the children had a charming evening at the Estab. A gent. gave a comic lecture on mummies (2 of the maids dressed up) & did it uncommonly well & afterwards they had a most ideotic game called the wild beast shew in which all the ladies yelled & barked & roared like mad.

[20] Mary Butler had been a patient at Ilkley Wells when CD and Aubertin were there in 1859 (*Correspondence* vol. 7). See *Correspondence* vol. 10, letter from Mary Butler, [before 25 December 1862].
[21] Neither the reference to Harrogate nor the Miss Scotts have been further identified.
[22] This individual has not been identified.
[23] See n. 5, above. The Rio de Janeiro company has not been identified.

From Edward Cresy 27 April 1863

Spring Gardens
27 April '63

My dear Sir

Many thanks for the copy of your paper on Linum which I should have acknowledged before but for pressure of business—[1] What a wonderful field for observation you continue to open up—infusing new life into the old dry bones of systematic botany— I can't think how it is the popularisers have not got hold of it yet, there seems such a fund of resource for them but I suppose they cant grasp the idea so as to feel their feet in using it—

I met with a curious passage in an old free thinker Francis Sanchez who lived in the latter part of the sixteenth century— the book is a rare one entitled "Quod nihil scitur" Frankfurt 1618—[2] Mackay the author of "Religious development of Greeks & Hebrews"[3] shewed it me among a lot of rare books he had been buying & I made a note of the following—

"In arboribus eadem mixtio cernitur et plantis aliis, ut in caulorapo malo persicis, amigdalo persicis et pluribus aliis quibus insitione media acquiritur natura inter insitum et id cui inseritur si denique addas mutationis specieum ut ex tritico saepe lolium et ex lolio triticum quandoque et ex secala avena fit"—[4]

I find I was right in telling you the London Chatham & Dover had withdrawn their Brighton line—[5] I have marked in pencil the two SE lines which are in progress New Cross to Dartford & New Cross to Tunbridge as well as London Chatham & Dover Sevenoaks branch on a map which I will forward you tomorrow[6]—together with the most recent map I can get of the lines round London— they are two long to go by post so I will send them by rail addressed to you at Bromley Station to be called for—

How did your son George get on at Cambridge[7] I heard he & young Carpmael went up together but did not see Carpmael this morning so have not heard the result—[8]

Yours very truly | E Cresy

C Darwin Esq—

DAR 161: 241

1 'Two forms in species of *Linum*' was read before the Linnean Society on 5 February 1863. Cresy's name appears on CD's presentation list for this paper (see Appendix IV).

2 Francisco Sanches (Franciscus Sanchez) wrote *Quod nihil scitur* [*That nothing is known*] in 1581 (Sanchez 1581), but his European reputation was established by the 1618 edition (Sanchez 1618). See Sanches 1988, p. 21.

3 The references are to Robert William Mackay and Mackay 1850.

4 In Sanches 1988, pp. 126–7 and 230, the passage is translated:

> We can see the same kind of mixing in trees and other plants, as for example in kohlrabi, in peaches and almonds, and several others, in which by means of grafting the plant acquires a nature intermediate between that of the graft and that of the root-stock, on to which it is grafted. If, again, you add changes in species, as darnel often comes from wheat and sometimes wheat from darnel, and wild-oats from rye

5 Cresy may be referring to a conversation that took place on 7 April 1863 when Cresy visited Down House for lunch (see Emma Darwin's diary (DAR 242)).

6 Cresy refers to railway lines from New Cross station in east London to several Kent villages; the fold-out map in *Bradshaw's railway guide* 1863 (see also *ibid.*, p. 28) displays a dotted line from London towards Brighton through Tunbridge Wells, to the Sussex village of Uckfield, indicating that the line was still under construction. The map sent to CD has not been found.

7 In 1863, George Howard Darwin competed unsuccessfully for an entrance scholarship at St John's College, Cambridge (*DNB*). See letter from George Howard Darwin to Emma Darwin, [15 April 1863], in DAR 251: 2232, and letter from Charles and Emma Darwin to W. E. Darwin, [4 May 1863]. See also letter to Edward Cresy, 13 May [1863].

8 The reference is to one of the sons of William Carpmael, a colleague of Cresy's at the Metropolitan Board of Works (*Minutes of Proceedings of the Institution of Civil Engineers* 30: 430). The son may have been Ernest Carpmael, who attended Clapham Grammar School, London, with George, and who evidently competed for a St John's College scholarship at the same time as George (*Alum. Cantab.*, *DNB*, *Eagle* 3: 356). See letter from George Howard Darwin to Emma Darwin, [15 April 1863], in DAR 251: 2232.

To ? 28 April [1863?][1]

[Down]

[referring to an exchange of photographs with Professor Edouard Claparède, the Swiss naturalist, 'for whom I feel the highest respect'.]

Christie's (6 August 1975, item 176)

[1] According to the catalogue description, the letter is dated 28 April. The year is conjectured by the relationship between this letter and an exchange of letters between CD and Claparède in 1862; the only extant letters between them are the letter to Edouard Claparède, [*c.* 16 April 1862], and the letter from Edouard Claparède, 6 September 1862 (*Correspondence* vol. 10).

To Andrew Crombie Ramsay 29 April [1863][1]

at Rev^{d}. C. Langton's | Hartfield | Tonbridge Wells[2]
April 29th

Dear Ramsay

I am extremely much obliged to you for sending me your Address.[3] I sh^{d}. think it would interest geologists; to me, as you may suppose, it is *deeply* interesting. Such discussions are the real method of discussing the modification of Species.[4] It is a great problem for the future how the geological record is so imperfect.—[5] I hope that you may be induced to go on with the series of Breaks.—[6] If you do, at the end, could you not give a table or diagram with names of Formations; & the Physical Breaks shown by *thicker* or *thinner* black lines—or interrupted lines.—or longer or shorter lines to show extent or degree of break & creeps. How grand it would be, if you could add to each physical Break, the percentage of Fossils which pass upwards; thus[7]

Wenlock			
______________	$\frac{13}{60}$	=	
Upper Llandovery	↑		
______________	$\frac{20}{28}$	=	
Lower Llandover	↑		
______________	$\frac{16}{500}$	=	·32
Caradoc	↑		

(Thickness of formation?)

I wish there were data to say how many thousand feet of fine sediment may be deposited with no change of species from top to bottom; or better still if we could say that after so many thousand feet, without a Break, some few, (some 2 or 3 per cent) of the species at base were *represented* by analogous species at top. But migration &c render this hope or wish Utopian.— I was glad to see you snub our good friend Huxley's geology: I was horrified at his view; it might to certain extent be tenable for strictly land productions.—[8]

I wish you would take opportunity & discuss "creeps" or overlappings of Formations.[9] Are the "creeps" over the land; & if so, why do we not often find, old sea-cliff overlapped?

Why not old river-beds &c overlapped? Are our sections not perfect enough to show these cases? Or have our chief great formations been all formed at mouths of great rivers, where estuary land is low? Or do large spaces of bottom of sea remain for ages uncovered, & are these sea-beds crept over? How with our Coal measures, & new Red Sandstone with Foot-prints; does a fault separate the sinking area from the stationary or rising land whence sediment is derived? Do give us a philosophical discussion on this whole subject.—

With sincere thanks for the great interest derived from your address, believe me Dear Ramsay | Your's very sincerely | Ch. Darwin

Excuse untidy paper.—

DAR 261 (DH/MS 9: 1)

1 The year is established by the reference to Ramsay 1863 (see n. 3, below).

2 Charles Langton, widower of Emma Darwin's sister Charlotte, lived at Hartfield Grove, Hartfield, Sussex. According to Emma Darwin's diary (DAR 242), the Darwins stayed at Hartfield between 27 April and 6 May 1863.

3 Ramsay delivered his presidential address to the Geological Society of London on 20 February 1863; it was printed in the *Quarterly Journal of the Geological Society* (Ramsay 1863).

4 In Ramsay 1863, the author related British palaeozoic stratigraphy to the fossils contained in the strata, mentioning several times CD's views on modified descent (*ibid.*, pp. xlviii, l, lii). Ramsay speculated (*ibid.*, p. l):

> Supposing, therefore, Mr. Darwin's hypothesis to be correct, it may be asked whether, under the changing conditions coincident with disturbance of strata, there may not have been influences at work that entailed a more rapid development of new species out of old, and of old species into new genera, than those that existed during an epoch when the conditions in a given area remained completely unchanged.

5 In his address, Ramsay discussed the incompleteness of the geological record (see, for example, Ramsay 1863, pp. l and lii). In his conclusion (*ibid.*, p. lii), he stated:

> Furthermore, many considerations, partly stated, lead me to suspect that we must look to the lapse of time unrepresented by strata, as the chief cause, or, rather, as the necessary accompaniment, of the influences that produced the great difference in species between any two formations one of which lies unconformably on the other, whether we adopt the old view of gradual extinction

and replacement by special creation, or Mr. Darwin's more philosophical argument of descent with modification.

[6] Ramsay's address was entitled 'Breaks in succession of the British Palæozoic strata' (Ramsay 1863). He defined 'breaks in succession' (*ibid.*, p. xxxvi) as:

> physical interruptions in stratification marked by the unconformity of an upper formation to one immediately underlying it, or, when such visible unconformity is wanting, by a sudden change in the fossils characteristic of the underlying and overlying formations.

At the end of his address, Ramsay noted that he had intended to apply similar evidence to breaks in succession in the 'Secondary and Tertiary formations' (Ramsay 1863, p. lii).

[7] CD's fractions in the diagram demonstrate the number of species that still survived in the overlying bed. For instance, of the approximately 500 species contained in the Caradoc sandstone, 16 were also found in the Lower Llandovery rocks; 3.2 per cent, therefore, 'pass upwards'. See Ramsay 1863, pp. xxxix–xlii, for a discussion of the South Wales strata that CD depicted and their fossils. Ramsay did not provide a diagram, but did include a table listing the principal species. In the text he stressed the significance of the 'doctrine of per-centages common to two or more formations', noting that it had destroyed the notions of catastrophism, and of life in any given period being suddenly extinguished 'by a distinct act of omnipotence' to be replaced by a 'universal new creation when the succeeding deposits began' (Ramsay 1863, p. xlix).

[8] In his address (Ramsay 1863, pp. li–lii), Ramsay criticised Thomas Henry Huxley's hypothesis that different fossil formations lying at great distances from one another might have existed at the same time owing to prolonged migration of species between continents: 'the process of transmission of a group from one area to another might be prolonged indefinitely, so that . . . a Silurian, a Devonian, and a Carboniferous fauna might all exist in different areas at the same time'. Huxley made his comments in his anniversary address to the Geological Society on 21 February 1862 (T. H. Huxley 1862a, p. xlvi). See also *Correspondence* vol. 10, letters to T. H. Huxley, 10 May [1862], 7 December [1862], and 18 December [1862].

[9] Ramsay referred to 'creeps' on one occasion during his address, noting that in 'Pembrokeshire the limestone does creep across the Old Red Sandstone in a manner suggestive of overlap rather than of break and unconformity' (Ramsay 1863, p. xliii).

To H. W. Bates 30 April [1863][1]

C. Darwin | at Rev^{d}. C. Langton's | Hartfield | Tonbridge Wells[2]
April 30th

Dear Bates

You will have received before this the note which I addressed to Leicester, after finishing vol. I, & you will have received copies of my little Review of your paper.[3] By the way I heard yesterday from Asa Gray that his article on same is delayed till next number in Silliman's Journal.—[4] I have now finished vol. 2. & my opinion remains the same; that you have written a truly admirable work, with capital original remarks, *first-rate* descriptions, & the whole in a style which could not be improved.[5] My family are now reading the book & admire it extremely; & as my wife remarks, it has so strong an air of truthfulness.—

I had a letter from a person the other day, unknown to you, full of praise of the book.—[6] I do hope it may get extensively heard of & circulated; but to a certain extent this, I think, always depends on chance.

I suppose the clicking noise of surprise made by the Indian, is that which end of tongue applied to palate of mouth & *suddenly* withdrawn makes?[7]

I have not written since receiving your note of April 20th, in which you confide in me & tell me your prospects—[8] I heartily wish they were better & especially more certain; but with your abilities & powers of writing it will be strange if you cannot add what little you require for your income. I am glad that you have got a retired & semi-rural situation.[9]

What a grand ending you give to your book contrasting civilisation & wild life![10] I quite regret that I have finished it: every evening it was a real treat to me to have my half hour in the grand Amazonian forest, & picture to myself your vivid descriptions. There are heaps of facts of value to me in a Nat. Hist. point of view— It is a great misfortune that you were prevented giving the discussion on species.—[11] But you will, I hope, be able to give your views & facts somewhere else.—

Once again I congratulate you & believe me | Yours very sincerely | C. Darwin

We shall stay here till Wednesday & then move for a week to

J. Wedgwood's Es^{qr}[12]
Leith Hill Place
Dorking
Surrey

We have come for change for my Boy & own health-sake[13]

Cleveland Health Sciences Library (Robert M. Stecher collection)

1 The year is established by the reference to Bates 1863 (see n. 3, below).

2 See preceding letter, n. 2. The reference is to Charles Langton.

3 See letter to H. W. Bates, 18 April [1863]. The references are to the first volume of Bates 1863 and to CD's review of Bates 1861 ('Review of Bates on mimetic butterflies').

4 See letter from Asa Gray, 13 April 1863. Gray's review (A. Gray 1863a) of Bates's account of mimetic butterflies (Bates 1861) appeared in the September number of the *American Journal of Science and Arts*, commonly known as 'Silliman's journal' after its founder, Benjamin Silliman.

5 CD refers to the second volume of Bates 1863. For CD's opinion of the first volume, see the letter to H. W. Bates, 18 April [1863].

6 This letter has not been identified.

7 CD refers to a discussion in Bates 1863, pp. 128–9. In CD's copy of Bates 1863, he noted (and crossed out) the following on a sheet pasted into the back of volume 2: 'close mouth & part front of tongue forming the palate & open mouth suddenly & this makes the click— open mouth sign of surprise' (see *Marginalia* 1: 35–7). CD's copy of Bates 1863 is in the Darwin Library–CUL.

8 See letter from H. W. Bates, 20 April 1863.

9 See letter from H. W. Bates, 20 April 1863.

10 In closing, Bates described his mixed feelings about leaving the Amazon, contrasting England with the tropics. Before concluding with his final view of the 'Great River', he wrote (Bates 1863, p. 417):

> The superiority of the bleak north to tropical regions however is only in their social aspect, for I hold to the opinion that although humanity can reach an advanced state of culture only by

battling with the inclemencies of nature in high latitudes, it is under the equator alone that the perfect race of the future will attain to complete fruition of man's beautiful heritage, the earth.

[11] See letter from H. W. Bates, 8 April 1863.

[12] Josiah Wedgwood III was CD's first cousin and brother-in-law. According to Emma Darwin's diary (DAR 242), the Darwins stayed with Charles Langton from 27 April until 6 May 1863, when they moved to Leith Hill Place.

[13] Both CD and Horace Darwin were ill at this time (see Emma Darwin's diary (DAR 242)).

From J. D. Hooker [30 April 1863][1]

Royal Gardens Kew
Thursday

My dear Darwin

I am puzzled & distressed at the loss of your letter from Haast.[2] I fancy that I do recollect it's coming in that I sent you, & I thought I had on its receipt put it out addressed to you to go by Post with a whole heap of other letters on the night of my return from Jersey.[3] My wife on the other hand saw no such letter when she stamped the batch. I had such a heap of letters to read through including *2* from Haast that I cannot speak positively of seeing the enclosure to you at all. My wife's not having seen it is odd—& makes me think that I may have confounded Haast's notice of its being enclosed, with my present idea that I saw itself I have hunted every where in vain for it, & am at my wits end about it. That it did not go to Post is certain, for I sent all letters through my wife. It is also odd that I should not have thought of enclosing it in a letter of my own to you, as I intended to write to you the first thing after my return.

I shall continue my search, meanwhile I see no way out of my perplexity.

Ever yours | Dumbfoundered | J D Hooker

P.S. | I am now going to Town about the R.S. Elections[4] have you any hints to give me.?

DAR 101: 132–3

[1] The date is established by the relationship between this letter, the letter from J. D. Hooker, 20 April 1863, and the letter to J. D. Hooker, 23 April [1863], and by the reference to the Royal Society elections (see n. 4, below).

[2] Hooker sent CD a letter from Julius von Haast to Hooker dated 10 December 1862 (see enclosure to letter from J. D. Hooker, 20 April 1863). In the letter Haast mentioned enclosing a letter to CD (dated 9 December 1862) and asked Hooker to forward it; either the letter was not enclosed by Haast, or Hooker misplaced it. CD eventually received a copy of the letter from Haast (see letter from Julius von Haast, 5 March 1863 and n. 5). Haast's letter of 9 December 1862 is printed in *Correspondence* vol. 10, and also in this volume, with his letter of 5 March 1863.

[3] Hooker had returned on 17 April 1863 from a trip to Dorset and the Channel Islands of Guernsey and Jersey (see letter from J. D. Hooker, 20 April 1863 and n. 1).

[4] The council of the Royal Society of London, of which Hooker was a member, drew up the shortlist of candidates for election to the fellowship at their meeting of 30 April 1863 (Royal Society, Council minutes, 30 April 1863).

From Charles William Crocker 1[–4] May 1863

Westgate, Chichester
May 1st. 63.

Dear Sir

I ought to have answered your kind letter before this but the excuse I have to make is one that will be accepted I know.[1] The day after I received your letter and the copy of the paper upon Linums,[2] I went out into the meadows for the first time for a long period. You know that I have promised myself to look up the Plantagos this season and I found that P. lanceolata was already coming into bloom.[3] The very first spike I examined, showed I fondly thought, that I was on the eve of a discovery. I tried all I could find in bloom and I fully believed that this species at least does not—cannot—be fertilised by its own pollen. You know how careful I am, and I was afraid to trust this first examination, and was waiting for better opportunities for confirming these observations. We have since had cold rains and very cold dry winds and these have checked the growth even of the weeds. I have therefore not carried my observations so far as I should otherwise have wished before mentioning them. But I must not delay my letter any longer and I cannot send it off without telling you what *I think* I have seen. The pubescent styles are protruded nearly or quite all the way up the spike before the stamens of the lower flowers even are ready to expand. By the time the lower flowers have hung out their anthers the styles are withered and useless (probably later in the season fertilised) half way up the spike at least.[4] How is fertilisation brought about? My idea is that the wind is the agent, although like yourself I dont give the wind credit for doing much in this line.[5] In the first place there is no coloured corolla, no odour, and (I think) no honey to attract insects. With all the time I have been able as yet to devote to this point I have seen only one single insect settle upon a spike of this plant. I watched him (he was a kind of fly—I am afraid of entomological terms) and he seemed to me to have settled there by accident perhaps merely to rest, for he did not set about examining flowers, he was simply resting himself—doing nothing—

Let us see what there is to favour the idea of the wind as being the agent. The long exserted styles are downy, ready to catch any pollen wafted to them; the anthers are hung out clear of the spike upon long thin filaments; the pollen is dry and non-cohesive in any way.— it is easily scattered from the large splitting anthers. So the matter stands at present, I think with me that you will say there is enough to induce one to hope for some facts worth having. Through the winter I have been looking up all my friends except those at Kew[6] and have collected above 30 species of the genus—hoping that something would turn up in regard to their structure and wondering what it would be. I expect mine is the largest set of these weeds in the country and should anybody get a view of my garden this year I shall be sure to get my German title back again—"The mad Englishman".[7] I am afraid the genus Bougueria is not in cultivation, most people would think it not worth growing. However I must if possible get hold of our British Littorella— Sir W. Hooker says in Brit Flora that it makes quite a turf on the margins of Highland lakes.[8] I have sent to a friend in Aberdeenshire— it grows too in our own country but rarely. I dont doubt that I shall get it. Both genera should be examined in connection with Plantago. they would throw side light upon the subject.[9] My wild specimen of P. lanceolata which last year produced a rosette of leaves instead of flower-spike (a foliaceous expansion of the bracts) has this season returned to the normal form.[10] I have however got seedlings from the few flowers produced among the leaves of that rosette and shall wait and see what they do.— Net like the sample you sent me will keep out the wind enough for practical purposes, dont you think it will?[11]

After no end of a lot of correspondence with all parts of the country I this winter succeeded in getting above two dozen genuine *single* Hollyhocks from a friend's friend in Yorkshire. They are doing well and are safe to flower— it remains however to see what the colours may be. Out of so many taken up at hazard we shall surely get a sufficient variety to suit our purpose.[12] With health there is little to fear on this point.

My Linum flavums have stood the winter well and though few in number will I hope be sufficient. Many thanks for your kind mention of my name in connection with this plant.[13]

Upon second thoughts I believe it would be worth while to let you have a list of my Plantagos, as possibly in your wanderings you may light upon one or two sp: which might be added to my collection, and I know you would not think me troublesome in asking it.

P. stricta. Schousb.[14]	P. pumila, L
mexicana, Link[15]	Hookeriana F. et Mey.[16]
arenaria W. et K.[17]	amplexicaulis. Cav.[18]
lagopus, L.[19]	microcephala, Poiret[20]
lusitanica L.	sibirica, Pl.[21]
Psyllum[22]	major
subulata, L. F.[23]	—"—monstrosa
Loeflingii, L.	—"—purpurea
maritima	intermedia
coronopus	lanceolata
serpentina	— " — var.
virginica, L.	Xorullensis, Kunth.[24]
Candollei, Rap.[25]	nitida
alpina	altissima ?
graminea	ciliata (or cynops?)

I am specially anxious to add all *vars* of Brit. sp: I can get hold of.

I found that Henderson of Pine Apple Place,[26] had got Peloria vars of Antirrhinum and Linaria vulgaris, so wrote and bought them with one or two Plantagos and other things, but although I did this 6 weeks ago I have not got them down yet.[27] I am afraid the order is so small they hardly think it worth doing. If I go to London this spring I shall go and fetch them — Out of 100 Primroses examined last week I found 44 long styled 56 short— the long ones differ much in length.— Out of 50 Cowslips—27 were long—23 short styled— Nobody has grown P. sinensis in quantity in this locality this year.—[28]

Believe me to remain, dear Sir, with the Greatest thanks for your kindness— | Yours very respectfully | Chas. W. Crocker

C. Darwin Esqr.

Three days have I see been lost (lost to science but used in money making) since I commenced this letter.[29]

DAR 110: 28, DAR 161: 260

CD ANNOTATION

6.5 Out of . . . styled— 6.7] *scored red crayon*

[1] The letter to Crocker has not been found.

[2] Crocker's name appears on CD's presentation list for 'Two forms in species of *Linum*'; CD distributed a number of offprints of the paper in mid-April (see Appendix IV).

[3] In a missing letter of November 1862, CD had directed Crocker's attention to dimorphism in plantains (see *Correspondence* vol. 10, letter from C. W. Crocker, 24 November 1862). Crocker subsequently carried out a number of experiments with *Plantago* (Crocker 1864); CD cited Crocker's observations on *P. lanceolata* in *Forms of flowers*, p. 306 n. (see also n. 4, below).

[4] In *Forms of flowers*, p. 307, CD observed that it was 'well known' that *Plantago lanceolata* was protogynous (that is, the female reproductive organs mature before the male). CD was interested in plants in which anthers and stigmas matured at different times in the same flower (dichogamy). Between 19 March and 24 August 1863, CD undertook a series of observations on dichogamy in a variety of plants, including this species; his notes are in DAR 49: 83–100 and DAR 109: A27–9.

[5] In 'Two forms in species of *Linum*', p. 78, CD stated: 'When insects are the agents of fertilization (and this is incomparably the more frequent case both with plants having separated sexes and with hermaphrodites), the wind plays no part, but we see an endless number of adaptations to ensure the safe transport of the pollen by the living workers' (*Collected papers* 2: 100–1).

[6] Crocker had worked as foreman of the propagating department at the Royal Botanic Gardens, Kew, from 1857 to 1863 (R. Desmond 1994).

[7] Prior to his appointment at Kew (see n. 6, above), Crocker had worked in Prussia as a gardener to Victoria, princess royal of England and crown princess of Prussia (*Journal of Horticulture* n.s. 14 (1868): 207).

[8] In the first edition of William Jackson Hooker's *British flora* (W. J. Hooker 1830), the description of the plantain shore-weed, *Littorella lacustris*, appeared on page 402. An eighth edition, written with George Arnott Walker Arnott, had been published in 1860 (W. J. Hooker and Arnott 1860).

[9] *Bougueria*, *Littorella*, and *Plantago* are all members of the Plantaginaceae (J. C. Willis 1973).

[10] See *Correspondence* vol. 10, letter from C. W. Crocker, 31 October 1862.

[11] The letter with which CD sent the netting has not been found; however, CD may have sent the sample in response to the letter from C. W. Crocker, 24 November 1862 (*Correspondence* vol. 10).

[12] In a missing letter, CD suggested that Crocker experiment with hollyhocks (see *Correspondence* vol. 10, letters from C. W. Crocker, [before 13 March 1862], 17 May 1862, and 24 November 1862). For an indication of the questions CD was pursuing with regard to hollyhocks, see *Correspondence* vol. 9, letter to C. W. Crocker, 18 May [1861]. See also this volume, letter to Charles Turner, [1 April – 16 June 1863?] and n. 3.

[13] In his discussion of *Linum flavum* in 'Two forms in species of *Linum*', p. 81, CD stated: 'I have not been able to try any experiments on this species; but a careful observer, Mr. W. C. Crocker, intends proving their reciprocal fertility next summer' (*Collected papers* 2: 104).

[14] Peder Kofod Anker Schousboe.
[15] Johann Heinrich Friedrich Link.
[16] Friedrich Ernst Ludwig von Fischer and Carl Anton von Meyer.
[17] Franz de Paula Adam Waldstein and Paul Kitaibel.
[18] Antonio José Cavanilles.
[19] Carolus Linnaeus (Carl von Linné (1707–78)).
[20] Jean Louis Marie Poiret.
[21] This may refer to Jules Emile Planchon; however, Planchon is not listed as the author of this species name in *Index Kewensis*.
[22] Crocker wrote 'Psyllum' in error for *Psyllium* (*Index Kewensis*).
[23] Linnaeus filius (Carl von Linné (1741–83), son of Carolus Linnaeus). However, Linnacus, not Linnaeus filius, is the author of the species name *Plantago subulata* (*Index Kewensis*).
[24] Carl Sigismund Kunth.
[25] Daniel Rapin.
[26] The reference is to John Andrew Henderson, proprietor of A. Henderson & Co. Nurseries, of 1 Pine-apple Place, Maida Vale and 1 Garden Road, St John's Wood, London (R. Desmond 1994, *Post Office London directory* 1863).
[27] Crocker discussed his crossing experiments with *Antirrhinum* in his letter to CD of 31 October 1862 (*Correspondence* vol. 10), and CD included the results in *Cross and self fertilisation*, p. 363. CD may have suggested that Crocker extend his investigation to include peloric flowers of this genus (see *Correspondence* vol. 10, letter from C. W. Crocker, 24 November 1862 and nn. 6 and 8), and of *Linaria vulgaris* (see *Correspondence* vol. 10, letter to M. T. Masters, 8 July [1862]). CD carried out a series of crossing experiments with peloric flowers of different varieties of *Pelargonium* in May 1862 and was seeking information on pelorism in other genera (see *Correspondence* vol. 10, and CD's experimental notes in DAR 51: B4–9, B12–13). See also this volume, letter to Isaac Anderson-Henry, 20 January [1863], and letter to M. T. Masters, 6 April [1863].
[28] Crocker began assisting CD in his observations of heterostyly in wild primroses in 1862, providing data on the frequency of long-, short-, and equal-styled forms (see *Correspondence* vol. 10, letters from C. W. Crocker, 17 May 1862 and 31 October 1862). CD published Crocker's observations of heterostyly in *Primula sinensis* in 'Illegitimate offspring of dimorphic and trimorphic plants', p. 415.
[29] Crocker worked as a cathedral official and journalist in Chichester (*Journal of Horticulture* n.s. 14 (1868): 206–7).

To Isaac Anderson-Henry 2 May [1863][1]

Hartfield.[2]
May 2nd. 1863.

My dear Sir,

I have left home for a little change and your kind letter has followed me.—[3] You think far too highly of my paper on Linum etc.[4] I agree that subject is only beginning. I write now partly to say that I am convinced (but evidence too long) that graduated length of stamens in *same* flower has no relation to powers of fertilization, *at least in most cases*.

B

A

For scarlet dwarf Pelargonium, if B represents the ordinary graduated stamens, you will find *occasionally* an *additional and abnormal* stamen (A) on opposite and lower side of flower. Now the pollen of this one occasional *short* stamen, I think, very likely would produce dwarf plants.— If you experiment on Pelargonium, I would suggest your looking out for this single stamen.[5]

I observed fluctuations in length of pistil in *Phloxes*, but thought it was mere variability.[6]

If you could raise a bed of seedling Phloxes of any species except P. Drummondii it would be highly desirable to see if two Forms are presented; and I should be very grateful for information and flowers for inspection. I cannot remember, but I know that I had some reason to look after *Phloxes*.—[7]

I am very glad to hear that you intend working at Linum.[8]

I do not know whether you have used microscopes much yet. It adds immensely to interest of all such work as ours, and is indeed indispensable for much work.— Experience, however, has fully convinced me that the use of the Compounds without the simple microscope is absolutely injurious to progress of Natural History (excepting, of course, with *Infusoria*).[9] I have, as yet, found no exception to the rule, that when a man has told me he works with Compounds *alone*, his work is valueless.

With my best thanks. | My dear Sir, | Yours very faithfully. | C. Darwin.

Copy
DAR 145: 3

[1] The text of the letter is taken from a copy made for Francis Darwin for his editions of CD's letters (*LL*, *ML*); part of the letter was published in *ML* 2: 298–9. The copyist dated the letter 2 May 1863; however, when Francis edited the transcription he added square brackets in pencil to indicate that the year had been added by the copyist and was not present in the original letter. The year is confirmed by the relationship between this letter and the letter from Isaac Anderson-Henry, 24 April 1863.

[2] Hartfield Grove, Hartfield, Sussex, was the home of Charles Langton, widower of Emma Darwin's sister Charlotte. According to Emma Darwin's diary (DAR 242), the Darwins stayed at Hartfield between 27 April and 6 May 1863.

[3] Letter from Isaac Anderson-Henry, 24 April 1863.

[4] Anderson-Henry's name appears on CD's presentation list for 'Two forms in species of *Linum*' (see Appendix IV).

[5] CD had suggested that Anderson-Henry make systematic experiments to test the reliability of Donald Beaton's statement (Beaton 1861) that the pollen from the two shortest stamens of scarlet *Pelargonium* produced dwarf plants (see letter to Isaac Anderson-Henry, 20 January [1863]).

[6] See letter from Isaac Anderson-Henry, 24 April 1863.

[7] CD's notes on 'Spring Phlox', dated 3 May 1863, are in DAR 49: 90. These observations formed part of a study on dichogamy (see preceding letter, n. 4). CD published his observations on *Phlox subulata* as part of a discussion of heterostyled dimorphic plants in *Forms of flowers*, pp. 119–20.

[8] See letter from Isaac Anderson-Henry, 24 April 1863.

[9] CD possessed both a simple microscope and a compound microscope, and had long held the view that these instruments should be used together in scientific work (see *Correspondence* vol. 4, letter to J. D. Hooker, [1 May 1847], and letter to Richard Owen, [26 March 1848], and *Correspondence* vol. 5, letter to J. D. Dana, 6 December [1853]). In the simple microscope, the magnified image of the object passed directly to the eye of the observer, but in the compound microscope, a first image

produced by the object-lens was focused within the tube of the microscope, and then magnified by one or more lenses forming the eye-piece (Beale 1861, p. 5). Although compound microscopes were capable of higher magnification, the direct image produced by the simple microscope made it easier to use for experimental manipulation or dissecting work. Work with the tiny unicellular *'Infusoria'*, however, would have required the resolving power of the compound microscope. CD used a simple microscope designed to his own specifications by the instrument makers Smith, Beck & Beck; the model was afterwards produced by the company for general sale as 'Darwin's Single Microscope' (see *Correspondence* vol. 4, letter to Richard Owen, [26 March 1848] and n. 2, and *Correspondence* vol. 10, letter to J. D. Hooker, 22 [August 1862] and n. 13). On CD's use of the microscope, see Burnett 1992.

From H. W. Bates 2 May [1863][1]

22 Harmood St. Haverstock Hill N.W.
May 2

My dear Mr Darwin

I have received both your letters containing remarks on my book & need not say how well satisfied I feel with them.[2] From all quarters, strangers & friends, the opinions are favourable. It is most curious; all find pleasure in the book, Darwinians, Calvanistic church ministers, Dissenting parsons, hard-headed men of business; women, old men & boys; philosophic naturalists & species grubbers. It makes one feel contented to be the author on the score of literary fame alone but more substantial rewards are not wanting for Mr Murray has given me a handsome sum (£250) although the edition (1250) is far from sold out.[3] He has behaved in the most friendly manner to me the friendliness visibly increasing every day.

I cannot forget that all this is due primarily to yourself who gave me so favourable an introduction to a publisher.[4] Not only that, if it had not been for your spurring me on I am quite sure the book would never have been written. As long as I live I shall remember these things & wish it were within the limits of etiquette to make them public. Murray said to me on Wednesday "I am very glad I am the publisher of the book".

Your former letter in which was a quotation from a letter of Asa Gray's gave me much pleasure[5] I find there is no end to the instruction that may be drawn from these cases of mocking butterflies; every time I reflect on them some new deduction flows forth, so that I think I have matter enough for another paper which might be illustrated by the same plates. Mr Wallace came last Sunday to spend the evening with me & after examining all my specimens of these mockers, came to the conclusion that all Nature does not furnish so plain & striking a case of the origin of species & of new & complex adaptations to new conditions, by the simple process of variation & Natural Selection.[6] But my first task will be to write a paper on the general subject of origin by segregation of local races.[7]

Believe me | Yours sincerely | H W Bates

DAR 160: 76

Henry Walter Bates. Adventure with curl-crested toucans.
(Frontispiece to H. W. Bates, *The naturalist on the river Amazons*, vol. 1, 1863.
By permission of the Syndics of Cambridge University Library.)

Isaac Anderson-Henry
(From *Gardeners' Chronicle*, 22 March 1873, p. 399.
By permission of the
Syndics of Cambridge University Library.)

Lydia Ernestine Becker
(From *The Graphic*, January 1874, p. 44. By permission
of the Syndics of Cambridge University Library.)

[1] The year is established by the reference to Bates 1863 (see n. 2, below).
[2] The references are to Bates's *The naturalist on the river Amazons* (Bates 1863) and to CD's letters to H. W. Bates, 18 April [1863] and 30 April [1863].
[3] Bates 1863 was published by CD's publisher, John Murray (see also n. 4, below).
[4] CD had encouraged Bates to write a book based on his experiences while collecting natural history specimens in South America from 1848 to 1859. CD also wrote to Murray on Bates's behalf, urging him to publish the book (see *Correspondence* vol. 10, letter to John Murray, 28 January [1862]).
[5] In his letter to Bates of 4 March [1863], CD wrote that Asa Gray had found Bates's paper on mimicry in butterflies (Bates 1861) '*fascinating*'. Gray's letter has not been found.
[6] Alfred Russel Wallace and Bates had collected together in South America from 1848 to 1850 (Woodcock 1969, pp. 32–3 and 67). Wallace's interest in Bates's study of mimicry is evident in the paper he read before the Zoological Society of London on 13 January 1863 (Wallace 1863a); in this paper, Wallace drew attention to examples of mimicry in birds (see letter from H. W. Bates, 24 January 1863 and n. 3). For the relationship between studies of mimicry and the theory of natural selection, see *Correspondence* vols. 9 and 10.
[7] Bates refers to an article entitled 'Geographical relations of species, and their varieties', which he had planned to write for the *Natural History Review*. However, he suspended work on this paper to undertake another commission, a monograph on the Mantidae, and does not appear to have returned to it (see letter from H. W. Bates, 29 September 1863 and nn. 7 and 9, and Bates 1892, pp. lxvii–lxx). See also Curle 1954, p. 26.

To John Scott 2 May [1863][1]

of Down, Bromley Kent [Hartfield][2]
May 2d.

My dear Sir

I have left home for a fortnight to see if I can with little hope improve my health—[3] The parcel of orchid pods, which you have so kindly sent me, have followed me.[4] I am sure you will forgive the liberty which I take in returning you the postage stamps.— I never heard of such a scheme as that you were compelled to practise to fertilise the Gongora![5] It is a most curious problem what plan nature follows. in this genus & Acropera.— Some day I will try & estimate how many seeds there are in Gongora.— I suppose & hope you have kept notes on **all** your observations on Orchids;[6] for with my broken health & many other subjects, I do not know whether I shall ever have time to publish again; though I have a large collection of notes & facts ready.—[7] I think you show your wisdom in not wishing to publish too soon; a young author who publishes every trifle gets sometimes unjustly to be disregarded.—[8]

I do not pretend to be much of a judge; but I can conscientiously say that I have never written one word to you on the merit of your letters that I do not fully believe in.—[9]

Please remember that I shd. very much wish for copy of your paper on sterility of individual orchids & on Drosera.—[10]

Thank you about Campanula perfoliata.— I have asked Asa Gray for seeds;[11] to whom I have mentioned your observations on Rostellum & asked him to look closely to case of Gymnadenia.—[12]

Let me hear about the sporting Imantophyllum if it flowers—[13]

Perhaps I have blundered about Primula; but certainly not about mere protrusion of pollen-tubes.—[14]

I have been idly watching Bees of several genera & Diptera fertilising O. morio at this place & it is a very pretty sight.— I have confirmed in several ways entire truth of my statement that there is no vestige of nectar in the spur; but the insects perforate the inner coat.[15] This seems to me a curious little fact, which none of my Reviewers have noticed.

with every good wish that I could anyhow aid you in any way.

Dear Sir | Yours sincerely | C. Darwin

DAR 93: B25–6

[1] The year is established by the reference to CD's absence from home (see n. 3, below).

[2] Hartfield Grove in Hartfield, Sussex, was the home of Charles Langton, widower of Emma Darwin's sister Charlotte (see n. 3, below).

[3] Emma Darwin recorded in her diary (DAR 242) that from 27 April to 6 May 1863, the Darwins stayed at Hartfield Grove (see n. 2, above), afterwards visiting Josiah Wedgwood III at Leith Hill Place, near Dorking, Surrey, between 6 and 13 May 1863.

[4] In his letter to CD of [after 12] April [1863], Scott mentioned that he was sending capsules of *Gongora atropurpurea* and *G. truncata*.

[5] Scott described how he had successfully pollinated *Gongora truncata* in his letter to CD of [after 12] April [1863]. Scott and CD had been discussing the difficult pollination and small stigmatic openings of *Gongora* and the related *Acropera* since November 1862 (see *Correspondence* vol. 10, letter from John Scott, 11 November 1862). See also this volume, letters from John Scott, 3 March 1863 and 21 March [1863], and letter to John Scott, 24 March [1863].

[6] Scott later followed up CD's plan to estimate the number of seeds produced (see *Correspondence* vol. 12, letter from John Scott, 28 March 1864); CD recorded his observations on *Gongora* and *Acropera* in *Variation* 2: 379 n.:

> Mr. J. Scott, of the Royal Botanic Garden of Edinburgh, calculated, in the same manner as I have done for some British orchids [*Orchids*, p. 344] . . . the number of seeds in a capsule of an Acropera, and found the number to be 371,250. Now this plant produces several flowers on a raceme and many racemes during a season. In an allied genus, Gongora, Mr. Scott has seen twenty capsules produced on a single raceme: ten such racemes on the Acropera would yield above seventy-four millions of seed.

[7] CD published the information he had accumulated on orchid reproduction subsequent to the appearance of *Orchids* in 1862, in 'Fertilization of orchids', published in 1869. A second edition of *Orchids* was published in 1877. See also *Variation* 1: 402–3, 2: 133–5.

[8] In his letter to CD of [after 12] April [1863], Scott stated that he had been reluctant to publish his observations on orchids, given his lowly status as foreman of the propagating department at the Royal Botanic Garden, Edinburgh. With CD's encouragement, Scott published his observations on sterility and cross-pollination in orchids in Scott 1863a and 1864b. Scott 1863a was read before the Botanical Society of Edinburgh on 14 May 1863.

[9] In his letter of [after 12] April [1863], Scott had thanked CD for the encouraging remarks in his letter to Scott of 12 April [1863].

[10] CD refers to Scott's draft paper on his experiments with species of *Oncidium* and *Maxillaria* (see letter from John Scott, [after 12] April [1863]); the paper was published as Scott 1863a (see n. 8, above). The second reference is to Scott's paper on *Drosera* and *Dionaea* (Scott 1862b), read before the Botanical Society of Edinburgh on 11 December 1862, abstracts of which appeared in the *Gardeners' Chronicle*,

10 January 1863, p. 30, the *Edinburgh New Philosophical Journal* n.s. 17 (1863): 317–18, and the society's *Transactions.*

11 See letter to Asa Gray, 20 March [1863]. CD had been trying to obtain seed of *Campanula perfoliata* (*Specularia perfoliata*) for some time, in connection with his experiments on self-pollination (see letter to John Scott, 12 April [1863] and n. 17). Scott had been unable to procure seed of this species, despite numerous inquiries (see letter from John Scott, [after 12] April [1863]).

12 See letter to Asa Gray, 20 April [1863]. For Scott's observations, see the letters from John Scott, 21 March [1863] and n. 7, and [1–11] April 1863 and n. 13. Asa Gray published his observations on *Gymnadenia tridentata* in A. Gray 1863b.

13 Scott sent CD information about a specimen of *Imatophyllum miniatum* in a missing letter (see letter to John Scott, 16 February [1863] and n. 6), in response to CD's request for examples of bud-variation; CD quoted the case in *Variation* 1: 385.

14 CD had observed pollen-tubes penetrating ovules in *Primula* through the chalaza. Pollen-tubes in flowering plants usually enter the ovule through the micropyle (see letter to Daniel Oliver, [12 April 1863] and n. 3, and letter to John Scott, 12 April [1863]). In his letter to CD of [after 12] April [1863], Scott expressed a wish to make further experiments on this point.

15 CD refers to his discussion of *Orchis morio* in *Orchids*, pp. 44–53. Despite having well-developed spur-like nectaries, *O. morio* did not appear to secrete nectar; CD's hypothesis was that in this, and in some other species of orchids, nectar was secreted into the intercellular spaces between the inner and outer membranes of the nectary. This required the insect to bore through the inner membrane of the nectary to reach the nectar. The observations on *O. morio* made by CD and George Howard Darwin at Hartfield at the beginning of May 1863 are recorded in DAR 70: 176. See also letter to Roland Trimen, 16 February [1863] and n. 4, and letter to J. D. Hooker, 24[–5] February [1863], n. 19.

From Charles and Emma Darwin to William Erasmus Darwin [4 May 1863][1]

Hartfield[2]
Monday

My dear W.

(Papa is dictating)

I am very glad to hear of the plant, but you are worse than a villain thus to tantalize me.[3] I hope it is not the Buck bean for that w^d not be new, tho I feel a special interest about it.[4] Compare the papillæ on the stigma. If possible compare & count the ovules in the two forms. Look for nectar. If pollen is very different & you have time you might see whether both kinds are on the stigma & both penetrated. If the case turns out good I shall insist on your sending a communication to Linn. Soc. thro' me.

You might begin thus "as the existence of two forms in the same species of plant has lately been shewn to be of functional importance I beg permission to lay the following case before the Linn. Soc"[5]

But after all perhaps the case will break down as so many have with me.

Now It's me—

I do hope poor little Evelyn will not be very ill.[6] Mrs Ashworth's death is most remarkable—[7] Your father remains so very languid & weak in the same way that he was after our London visit that we have almost given up all hopes of his being brisk enough to go to L. H. P. where we had intended driving to from here on

Wednesday[8] Any how Henrietta & Horace[9] will go. & we go home on Thursday, so you must not send off your mysterious flower without knowing where we are—

(Have you written about the photographs of your father to Cambridge as he wants to send one Please write if not.)[10]

Yesterday was a most lovely day & we (all but y^r^ father) walked over the common to look for buck bean but I suppose we are too early.[11] We have the pony here & it is a most charming country for riding. Hen. goes about by herself—

Edmund has invited the Josselinas to the boat races at Cambridge which pleases them very much. Uncle C. L. takes them & Aunt Eliz. Chaperones them.[12] Our visit here has not done Horace any good tho' he has a good deal of enjoyment but his stomach is worse than usual.

You have heard of G. leaving St John's he was so disgusted by the looks of his fellow students.[13] Mr Wrigley has been most amiable about it & I think the upshot will be that he will go for a year to a private tutor as he is very young.[14]

Goodbye my dear old man yours, | E. D.

DAR 219.1: 55

[1] Dated by the references to the Darwin family's visits to Hartfield Grove and Leith Hill Place (see nn. 2 and 8, below); the Monday before the Darwin family left Hartfield Grove was 4 May 1863.

[2] According to Emma Darwin's diary (DAR 242), the Darwin family stayed at Hartfield Grove in Hartfield, Sussex, the home of Charles Langton, between 27 April and 6 May 1863.

[3] The letter from William has not been found. William's observations related to a possibly dimorphic plant, later identified as *Pulmonaria angustifolia* (see following letter and n. 1). CD was interested in finding further examples of dimorphic species after discovering that dimorphic flowers in *Primula* and *Linum* were an adaptation for reciprocal pollination (see 'Dimorphic condition in *Primula*' and 'Two forms in species of *Linum*').

[4] CD refers to the mainly aquatic perennial *Menyanthes trifoliata*, in which the occurrence of dimorphism was an established fact (see *Correspondence* vol. 10, letter to C. C. Babington, 20 January [1862] and n. 3). CD was anxious to examine this species and had made attempts to obtain specimens in 1862 (see *ibid.*).

[5] William's observations on dimorphism in *Pulmonaria angustifolia* were not communicated to the Linnean Society; however, CD gave a detailed account of them in *Forms of flowers*, pp. 105–7.

[6] The reference is to Evelyn George Hammond Atherley, son of William's banking partner George Atherley ('Genealogical notes', Records research room, Southampton City Archives).

[7] Mrs Ashworth has not been further identified.

[8] Leith Hill Place, near Dorking, Surrey, was the home of Josiah Wedgwood III and his wife Caroline, CD's sister. According to Emma Darwin's diary (DAR 242), the Darwins stayed at Leith Hill Place from Wednesday 6 May 1863 until 13 May, when they returned to Down House. On Tuesday 5 May 1863, Emma recorded in her diary that CD had been 'poorly & languid all week'. CD had been in London for ten days in February (see 'Journal' (Appendix II)).

[9] Henrietta Emma and Horace Darwin.

[10] Emma apparently refers to the photograph of CD taken by William in April 1861 (see *Correspondence* vol. 9, frontispiece). In 1861, William was studying at Christ's College, Cambridge (*Alum. Cantab.*), and may have deposited the negative plate with a firm of photographers there. CD wanted a copy of the photograph to send to Roland Trimen (see letter to Roland Trimen, 23 May [1863]).

[11] On Sunday 3 May 1863, Emma Darwin recorded in her diary (DAR 242) that it had been a 'beautiful day' and that they had 'walked to Brook on common'. Buck bean (see n. 4, above) flowers between April and June.

[12] The 'Josselinas' was a family nickname for the daughters of Caroline and Josiah Wedgwood III (see *Correspondence* vol. 6, letter to W. E. Darwin, [17 February 1857] and n. 10). Sophy Wedgwood was 21 years old, Margaret 19, and Lucy 16. Emma also refers to Charles Langton, Sarah Elizabeth Wedgwood, and Charles Langton's son, Edmund Langton, who was an undergraduate at Trinity College, Cambridge (Freeman 1978, *Darwin pedigree*, *Admissions to Trinity College, Cambridge*).

[13] In 1863, George Howard Darwin competed unsuccessfully for an entrance scholarship at St John's College, Cambridge (*DNB*). See also letter to Edward Cresy, 13 May [1863].

[14] Alfred Wrigley was headmaster of Clapham Grammar School (*Alum. Cantab.*), where George had been a student since 1856 (*DNB*). It was subsequently decided that George should return to school for another year (see letter to Edward Cresy, 13 May [1863]).

From W. E. Darwin 4 May [1863]

Southampton.
May 4.

My Dear Father,

The flower is Anchusa officinalis, that is to say I am almost certain it that.[1]

I went over to the isle of Wight yesterday and gathered 52 plants and brought them back to look at them, and of these 52 plants 22 were Short styled 30 Long styled, and of the 30 Long styled 16 were *longest* styled, & 14 *long styled*, but I fancy this was mere variability in length of pistil.[2]

the difference between the long and short styled is that as well as the styles being different length the stamens tho' all the same length or nearly so are attached to *different places* on the Corolla.

I am not going to send you flowers till I know whether you are at Down or Hartfield. send me a line at once if you want them as they will be getting withered.[3]

The pollen of short-styled I think is the largest. The shortstyled seem (in the 50 examined) to have the smallest flowers, the long next and the longest finest of all but I think this must be chance[4] The shortstyled seem to have a greater tendency than others to become regularly 6 divisioned instead of 5. I have only had one short look at them this morning before breakfast, I shall look again tomorrow.

in Boragineæ the scales in mouth of Corolla are said to be aborted stamens and they have spiral vessels running up to them

Incomplete
DAR 110: 62

CD ANNOTATIONS

1.1 Anchusa officinalis,] *underl pencil*

Top of letter: 'W. says in 2^d letter[5] pollen of short-styled larger & more of it.—' *ink*; 'Pulmonaria' *pencil*; '1863' *pencil*; 'Anchusa' *pencil, circled & del pencil*

Top of second page: 'Anchusa' *pencil, circled pencil*

[1] In a letter that has not been found, William informed CD that he had been observing the possibly dimorphic flowers of a new species (see preceding letter). William's botanical notebook indicates

that the identification was changed from 'Anchusa officinalis' to 'Pulmonaria' (DAR 117: 67). CD cited William's observations in *Forms of flowers*, identifying the plant as *Pulmonaria angustifolia*, another member of the family Boraginaceae (*Forms of flowers*, pp. 105, 107).

2 William's observations on style lengths in *Pulmonaria angustifolia* collected on the Isle of Wight are mentioned briefly in *Forms of flowers*, p. 107 (see n. 1, above). William's botanical notebook contains notes on this species, beginning on 4 May 1863 (DAR 117: 67), and sketches of 'Anchusa officinalis' are preserved in his botanical sketchbooks (DAR 186: 43, p. 53). CD's brief pencil notes on the information sent by William are in DAR 110: 41–2.

3 Hartfield Grove in Hartfield, Sussex, was the home of Charles Langton (Freeman 1978). William sent the specimens of *Pulmonaria angustifolia* on 8 May 1863 (see letter from W. E. Darwin, 8 May [1863]).

4 William's observations on pollen and flower sizes in this species are mentioned briefly in *Forms of flowers*, pp. 105–6; his notes on flower sizes are in DAR 117: 67.

5 The first letter from William concerning 'Anchusa officinalis' has not been found (see n. 1, above).

To *Athenæum* 5 May [1863][1]

Down, Bromley, Kent, [Hartfield][2]
May 5.

I hope that you will grant me space to own that your Reviewer[3] is quite correct when he states that any theory of descent will connect, "by an intelligible thread of reasoning," the several generalizations before specified.[4] I ought to have made this admission expressly; with the reservation, however, that, as far as I can judge, no theory so well explains or connects these several generalizations (more especially the formation of domestic races in comparison with natural species, the principles of classification, embryonic resemblance, &c.) as the theory, or hypothesis, or guess, if the Reviewer so likes to call it, of Natural Selection. Nor has any other satisfactory explanation been ever offered of the almost perfect adaptation of all organic beings to each other, and to their physical conditions of life. Whether the naturalist believes in the views given by Lamarck, by Geoffroy St.-Hilaire, by the author of the 'Vestiges,' by Mr. Wallace and myself,[5] or in any other such view, signifies extremely little in comparison with the admission that species have descended from other species and have not been created immutable; for he who admits this as a great truth has a wide field opened to him for further inquiry. I believe, however, from what I see of the progress of opinion on the Continent, and in this country, that the theory of Natural Selection will ultimately be adopted, with, no doubt, many subordinate modifications and improvements.

Charles Darwin.

Athenæum, 9 May 1863, p. 617

1 The letter was published in the *Athenæum*, 9 May 1863.

2 According to Emma Darwin's diary (DAR 242), the Darwins stayed at Hartfield Grove in Hartfield, Sussex, the home of Charles Langton, between 27 April and 6 May 1863.

3 CD was responding to an anonymous letter by Richard Owen in the *Athenæum*, 2 May 1863, pp. 586–7 (see Appendix VII), which was critical of CD's letter to the *Athenæum* of 18 April [1863].

4 See the anonymous letter in the *Athenæum*, 2 May 1863, pp. 586–7 (Appendix VII). The publisher's marked copy of the *Athenæum* at the City University, London, reveals the author to have been Owen. Owen had taken the passage quoted from CD's letter to the *Athenæum* of 18 April [1863].

5 CD refers to Lamarck 1801, 1809, and 1815–22, E. Geoffroy Saint-Hilaire 1829, [Chambers] 1844, and C. Darwin and Wallace 1858. CD discussed the theories of transmutation that had preceded the announcement of his own theory in a historical preface to the third edition of *Origin*, pp. xiii–xix.

To W. E. Darwin [5 May 1863][1]

Hartfield.—[2]
Tuesday

My dear William

Hearty thanks for your letter about Anchusa.[3] The case interests me much; but there have been several statements published of dimorphism in the Boragineæ.—[4] I began to disbelieve,[5] for all have ⟨*third of a page excised*⟩ which occurs in Thyme & mint;[6] namely (1) form with long pistil & perfect anthers, (2) short-pistil & I think short stamen **with imperfect anthers** destitute of pollen: this second form bears seeds plentifully & must ⟨be fertilised by form (1).—⟩ ⟨*third of a page excised*⟩ probably is like Primula;—but I sh^d. anticipate that form (1) would be perfectly fertile with own pollen & the second form would require pollen from no^r I.— Do for Heaven-sake mark the place & get me seed in autumn; ⟨*four words excised*⟩ new case & would be well ⟨*third of a page excised*⟩ Down or go to Leith Hill, & we will write;[7] but as I have not the compound microscope, I cannot compare pollen; & 2 or 3 flowers would suffice just that I might glance at flowers & satisfy my curiosity.[8] ⟨*third of a page excised*⟩

Incomplete
DAR 210.6: 110

1 The date is established by the relationship between this letter and the letters from W. E. Darwin, 4 May [1863] and 8 May [1863]. In 1863, 5 May was a Tuesday.

2 According to Emma Darwin's diary (DAR 242), the Darwins stayed at Hartfield Grove in Hartfield, Sussex, the home of Charles Langton, between 27 April and 6 May 1863.

3 See letter from W. E. Darwin, 4 May [1863] and n. 1. The *Anchusa* was subsequently identified as *Pulmonaria angustifolia*.

4 CD's sources of information on dimorphism in Boraginaceae included Lecoq 1854–8: 7, 428–68, and Vaucher 1841: 3, 441–66, in which *Echium vulgare*, *Pulmonaria angustifolia*, and *Myosotis palustris* (now called *M. scorpioides*) were identified as dimorphic. CD read Lecoq 1854–8 in December 1861 and Vaucher 1841 in March 1862 (see *Correspondence* vols. 9 and 10). There are heavily annotated copies of both works in the Darwin Library–CUL (see *Marginalia* 1: 488–95, 812–15).

5 CD's own experiments with *Echium vulgare*, and several other members of the Boraginaceae, had led him to doubt suggestions that many species in this family exhibited heterostyly (see *Correspondence* vol. 10, letters to Asa Gray, 15 March [1862] and n. 11, and 9 August [1862] and n. 8). See also *Forms of flowers*, pp. 110–11.

6 Species of thyme and mint were taken by CD to be representative of a class of dimorphism that he later called 'gyno-dioecism', in which plants bear either hermaphrodite or female flowers (see *Correspondence* vol. 10, letter to Daniel Oliver, 12 [April 1862], and *Forms of flowers*, pp. 298–9). CD's experimental notes on *Mentha hirsuta*, dated 28 August 1862, are in DAR 109: A3. Notes on *Thymus*,

dated 29 May 1862 and 30 April 1863, are in DAR 109: A22 and A28. CD's Experimental notebook also contains several entries on *Thymus* dated between 5 June 1861 and June 1863 (DAR 157a: 72). CD frequently referred back to these notes when observing cases of gyno-dioecism in other species (see DAR 109). CD may have indicated in the missing portion of the letter that the *Anchusa* (actually *Pulmonaria angustifolia*) might also be gyno-dioecious. See also letter to W. E. Darwin, [10 May 1863] and n. 3. CD's experiments with *Echium vulgare* had established that some members of the Boraginaceae were gyno-dioecious (see *Correspondence* vol. 10, letter to Asa Gray, 9 August [1862]).

7 CD and Emma were undecided whether to go on from Hartfield to Leith Hill Place, near Dorking, Surrey (the home of Josiah Wedgwood III), or to return directly to Down (see letter from Charles and Emma Darwin to W. E. Darwin, [4 May 1863]). The Darwins stayed at Leith Hill Place from 6 to 13 May 1863 (see Emma Darwin's diary (DAR 242)).

8 William sent the specimens on 8 May 1863 (see letter from W. E. Darwin, 8 May [1863]).

From William Alexander Wooler 5 May 1863

Sadberge Hall | Darlington
5th May 1863

Dear Sir

The enclosed may interest you— They are from seed saved in 1860 from a cowslip and only from pips which I impregnated with the pollen of the polyanthus they were sown late in 1861 and have remained in the seed pans—in which the seed was sown undisturbed have been much neglected or otherwise no doubt they would have flowed larger— The ticket of the memo of the seed having been crossed remains in the pan they are growing in— These four are all that have flowed as yet— I should say there are about other 4 not yet flowered— Having been sowed late in 1861 did not bloom in 1862[1]

In haste yours very truly | W Wooler

C Darwin Esq | Bromley | Kent

DAR 181: 158

1 The experimental crosses that produced these seedlings are described in the letter from W. A. Wooler, 5 July 1862 (*Correspondence* vol. 10). CD presented Wooler with a copy of 'Dimorphic condition in *Primula*' (see *Correspondence* vol. 10, Appendix III) after corresponding with him on polyanthuses (see 'Dimorphic condition in *Primula*', p. 79 (*Collected papers* 2: 47); the correspondence has not been found). CD may also have corresponded with Wooler regarding the crossing experiments between polyanthus and the common cowslip (*Primula veris*) that he was undertaking at the end of April 1862 (DAR 157a: 76–7). The results were published in 'Three forms of *Lythrum salicaria*', p. 187 (*Collected papers* 2: 121).

From A. C. Ramsay 6 May 1863

DolauCothy, Llandielo
6 May 1863

My dear Sir

Having nothing to do I am so lazy that I have neglected for some days to answer your welcome letter.[1] In truth I have been here now 4 weeks trying to coax myself by idleness into my usual health. I have I hope partially succeeded but not thoroughly as yet. The prescription is however pleasant, viz a lovely place, old & well beloved friends a good library including plenty of novels, some pretty girls, my wife & one of my daughters,[2] good trout streams and a horse to ride on when I want it.

I am very glad that you approve of my address.[3] I supposed that if properly treated the subject would be sure to interest you both as a geologist & a Zoologist. Sir Charles Lyell writes me that he hopes I will continue the subject precisely as I began it if I continue President.[4] Such was my intention with possible additions in the summing up. Sir Charles wishes that experienced Surveyors all over the Continent would each do the same for their own Countries, but from what I have seen of Continental Surveys the mapping has not been done in sufficient detail to admit of it. Sella who reported to the Italian government on the Subject said to me that all the Continental surveys were done too *rapidly*.[5] For years I have had a diagram which I used in my lectures,[6] something like one that you suggest, but with a considerable difference, & I thought of using it in my next address if I can find time to correct it up to date.[7]

It is like this[8]

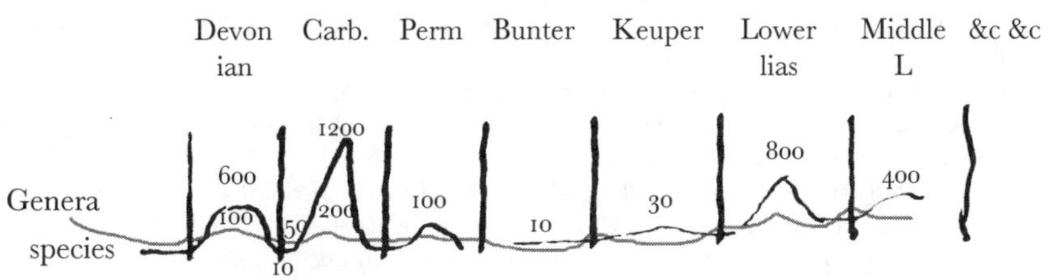

It shows the number of species in each formation, the number of genera, & also the number of genera & species that pass from one formation to another, & is very instructive.

Logan has a splendid Case in Canada of a cliff overlapped in the way you mention.[9] All the rocks concerned in the question are Old Silurian &c. I have no doubt he will send you a copy of the book when published.[10] I know of no positive case in England, but we have something like it too. "Creeps" are certainly often over the land.

You do me too much honour in supposing that I can give a satisfactory philosophical discussion on these & the other points you mention, but it must be done some day, & if I live & thrive perhaps I may try my hand at it.

Have you seen Geikie's new book on the Drift of Scotland?[11] It seems excellent.

This South Welsh land is full of Scratched stones & boulders. When I mapped it 20 years ago I knew the erratics when I saw them but I had then never seen a glacier & was not up to the *Scratched* stones.[12] There is a good field here still remaining for any one to work out in that way. For want of clear sections I cannot say whether the detritus is all or mostly *iceberg* work, or whether land ice did a deal of it. There are no *mountains* here, but all the country is hilly.

Ever sincly | And[rew] C Ramsay

DAR 176: 11

[1] Letter to A. C. Ramsay, 29 April [1863].

[2] The references are to Ramsay's friends John and Elizabeth Johnes of Dolaucothy, Llandilo, Carmarthenshire, Wales, and their two daughters, Charlotte Anna Maria and Elizabeth (*County families* 1860–76; Geikie 1895, pp. 76, 278, 340). Ramsay and his wife Louisa had four daughters and one son (*DNB*).

[3] See letter to A. C. Ramsay, 29 April [1863]. Ramsay gave his presidential address to the Geological Society of London on 20 February 1863, taking as his theme breaks in succession in the British Palaeozoic strata (Ramsay 1863). Ramsay concurred with CD's argument that the imperfection of the geological record explained the absence of transitional links between many species in successive geological formations (see *Correspondence* vol. 8, letter from A. C. Ramsay, 21 February 1860). In his address, Ramsay introduced the idea that the periods of time for which there was no stratigraphical evidence might be longer than the geological periods that were represented by stratified formations (see Geikie 1895, pp. 357–8).

[4] Ramsay was president of the Geological Society from 1862 to 1864 (*DNB*). CD had also encouraged Ramsay to continue his discussion with reference to breaks in succession in other strata (see letter to A. C. Ramsay, 29 April [1863]). See also n. 7, below. Ramsay's presidential address for 1864 discussed breaks in succession in the British Mesozoic strata (Ramsay 1864).

[5] Ramsay refers to the Italian geologist Quintino Sella (Sarjeant 1980–96).

[6] Ramsay probably refers to the annual series of lectures that he delivered at the Royal School of Mines, where he was professor of geology.

[7] In his letter to Ramsay of 29 April [1863], CD suggested that Ramsay publish a diagram of the physical breaks in the British geological record, indicating the percentage of fossils that continued to appear in the stratum above a break. Ramsay included tables showing the numbers and percentages of fossil species that passed from one formation to another in Mesozoic strata in his presidential address to the Geological Society in 1864 (Ramsay 1864).

[8] The lines on the graph represent the number of genera and species found in each geological formation and (on the intersections with the vertical lines) the respective number surviving from each formation to the next. For clarity, the genera line has been depicted here as a grey line to distinguish it from the species line. The annotation of the graph is only complete with reference to the Devonian and Carboniferous formations. In the Devonian formation, for example, the number of species rises to a peak of 600 and the number of genera rises to a peak of 100. In passing from the Devonian to the Carboniferous, 50 genera and 10 species are common to both formations. The number of species common to two formations is invariably smaller than the number of genera because in the intervening period many species have become extinct, while different species of the same genera have arisen. Ramsay eventually chose a tabular rather than a diagrammatic format to represent changes in the occurrence of fossil genera and species between geological formations (Ramsay 1864).

[9] The reference is to the Canadian geologist William Edmond Logan; the case referred to is described in [Logan] 1863, pp. 295–7. In his letter of 29 April [1863], CD suggested that Ramsay might discuss creeps (slow, more or less continuous movements of rocks under gravitational stresses), and asked: 'why do we not often find, old sea-cliff overlapped?'

[10] Ramsay refers to a report of the Geological Survey of Canada ([Logan] 1863) summarising the work of the survey since 1843. CD does not appear to have received a copy of this publication.

[11] Geikie 1863. Archibald Geikie was a colleague of Ramsay's on the Geological Survey of Great Britain.

12 After joining the staff of the Geological Survey of Great Britain in 1841, Ramsay spent the first four years mapping the geology of South Wales (*DNB*). In the 1840s, the role of glaciation in shaping the landscape of Britain was only just beginning to be appreciated (see Davies 1969, pp. 266–316). Ramsay subsequently conducted fieldwork that used scratched stones and boulders as evidence of past glaciation (Ramsay 1855).

From Isaac Anderson-Henry 7 May 1863

Hay Lodge, | Trinity, | Edinburgh.
May 7/63

My dear Sir

Your Letter of the 2d. reached me at my place in Perthshire from which I returned last night.[1] I cannot withdraw from the estimate I expressed of the importance and value of your discovery.—[2] You have let in the *first* ray of what I believe must burst in sunlight sooner or later

I have already, by the severest tests, proved you right among the *Primulas*.[3] I took three several plants of the Common garden *P. Polyanthus*—got them potted & put under Glass. I thereafter carefully labelled them, A, B, & C,—noting on the labels & in a Book, their respective properties. *A* indicated by *purple* thread had a *short style* and *long anthers*.— B, indicated by *yellow silk* had a *long style* and *Short anthers*.— C indicated by *red worsted*, like A, had a *short style* and *long anthers*.

Having secured against all interference by insects or otherwise I effected the experiment of *fertilisation* on 17 April last

On the plant *A* I did 2 flowers with the pollen of B., and one flower with its own pollen—carefully marking each with the appropriate thread.

On the plant B I did 4, two with the Pollen of *A* & two with its own pollen, marking as above

On the plant C I did one with the pollen of B. and one with its own pollen

Result

On *A*. the two done with the pollen of *B*. have already fine swollen seed vessels—the one done with its own pollen is a total failure

On B. all are failures—and the reason is this: The plant was sickly when I made the operation—and some days thereafter I found it nearly dead— I barely kept it alive by removal from a warm to a cold greenhouse

On C I did two flowers as I have said.— The one with the pollen of *B* has its seed pod swollen beautifully—the other, with its own, is a **total** failure

I have attempted a cross on *P Sinensis* with *long* style with the long anthered primrose known here as "Hose in hose" a mere form of *P. polyanthus* And tho' I cannot be confident it seems to have taken. The seed vessel at all events is swelling finely; but this, as you may have observed, is no sure test— And I confess I have my doubts—just because the cross is one of the severest—the *Sinensis* belonging, I hold, to a separate tho allied *genus*. Of this again

Meantime I regard the above results as clearly confirmatory of those of your published Experiments—and that the law you have discovered is established in nature

Thanks for your suggestions about the Pelargoniums—upon which I shall try my hand forthwith— So also of the *Linums* and *Phloxes*.[4] I had sown a potful of their seeds *last year* and wait their coming up. I have had my patience severely taxed in this way. I have got seeds to vegetate often, especially seeds from abroad after lying dormant for more than two years

In a Letter I had from Dr Hooker by the same post with yours— He says "The *Bryanthus* which Mr Darwin alludes to is a dried specimen from the M^ts. of California. If *B. erectus* is a hybrid (and no doubt you are correct) it is more probably a cross between 2 *Bryanthii* than with a *Rhododendron*."[5] All probable if Mr Cunningham had had the only two *Bryanthii then* (or perhaps *yet*) known, to make the X with—[6] But I presume they are yet to introduce. My Experiments this year may put this question to the test

Dr Hooker suggests to me "to try and cross American individuals of European plants with European Individuals. E.g. (he asks) would an American plant of *Caltha palustris* be as fertile with an English as with an American?" I must, in my manifold crosses have united *Americans*, as I have done innumerable *Asiatics*, with *Europeans*—eg among the Rhododendrons.—[7] As to the *fertility* of the hybrid, that may be another thing

I have got American *Brambles*, as well as Asiatic among which I have long resolved to try experiments to improve, if possible, the fruits of our European species—[8] And I am especially to try a union between the Blackberry & Raspberry—[9] And having my eye to your discovery, I am to apply *all* kinds of anthers, short and long, inner & outer, & in different states of maturity

Believe me ever | Yours very Sincerely | Is. anderson Henry

Thanks for your information about the Microscope.[10] But I had got one at a high price ere your Letter came, which however I required otherwise

I was told today that poor Donald Beaton was dying—at least, in a hopeless way[11]

Charles Darwin Esq F.R.S. | &c &c

DAR 159: 66

1 Letter to Isaac Anderson-Henry, 2 May [1863]. Anderson-Henry refers to his country estate, Woodend, in the parish of Madderty, Perthshire (*County families* 1878).

2 In his letter to CD of 24 April 1863, Anderson-Henry referred to CD's discovery that the presence of dimorphic flowers in a species was an adaptation for cross-pollination (see n. 3, below) as 'akin to Newtons discovery of the law of Gravitation'.

3 CD's first paper on dimorphic flowers, 'Dimorphic condition in *Primula*', was read before the Linnean Society in 1861 and was published in the society's journal the following year. CD sent Anderson-Henry a presentation copy of the paper (see *Correspondence* vol. 10, Appendix III).

4 CD had encouraged Anderson-Henry to perform experiments on *Pelargonium*, *Phlox*, and *Linum* (see letter to Isaac Anderson-Henry, 2 May [1863] and nn. 5, 7, and 8).

5 In his letters to CD of 26–7 January 1863, 17 April 1863, and 24 April 1863, Anderson-Henry referred to experiments he was making to ascertain the parent species of the hybrid *Bryanthus erectus* (*Phyllothamnus erectus*); he believed the parents to be *Rhodothamnus chamæcistus* and *Menziesia empetriformis*. CD may have mentioned to Anderson-Henry, in a missing letter, that there was a specimen of *B. erectus* at the Royal Botanic Gardens, Kew, of which Joseph Dalton Hooker was assistant director (see letter from Isaac Anderson-Henry, 24 April 1863).

6 The Edinburgh nurseryman James Cunningham claimed that he had produced *Bryanthus erectus* by crossing *Phyllodoce* (*Menziesia*) *caerulea* with *Rhodothamnus chamaecistus* (see letter from Isaac Anderson-Henry, 26–7 January 1863 and n. 14).

7 Many of Anderson-Henry's crossing experiments with rhododendrons are discussed in Anderson-Henry 1867a and 1867b.

8 Anderson-Henry reported that all his *Rubus* crosses attempted before 1867 had failed (Anderson-Henry 1867a), but he reported a successful cross in 1867 (Anderson-Henry 1867b).

9 See also letter from Isaac Anderson-Henry, 26–7 January 1863. Anderson-Henry's attempts to hybridise these two species failed (Anderson-Henry 1867a).

10 See letter to Isaac Anderson-Henry, 2 May [1863].

11 CD and the horticultural writer Donald Beaton had entered into a public correspondence in the *Journal of Horticulture* earlier in the year (see letters to *Journal of Horticulture and Cottage Gardener*, [before 27 January 1863] and [before 3 February 1863], and Appendix V). Beaton died in October 1863 (R. Desmond 1994).

From J. D. Hooker [7 May 1863][1]

Royal Gardens Kew
Thursday

My Dear Darwin

Enclosed came this morning from Gray without address, inside cover to me, & I read on till I found myself alluded to in the second page before discovering that it was not for me.[2]

There is a deal of gossip afloat about on many matters, you have no doubt seen the squib of "A hard case", also published in *part*, in "Public opinion"—[3] no one seems to know whose it is—certainly not Sir P. Egerton)[4] no copy came to me.

Falconer has his hands full & goes to Paris tomorrow to confront Quatrefage, Bouchet, & the chemists & anatomists who to a man say F. is wrong that both Flints & Jaw are ancient & *perfide Albion* at its old tricks of traduction.[5]

I met F. last night, he is beating up for allies to take over with him— I tell him he should go alone—it is his only chance of getting fair play— the more go the more opposition the more misunderstanding the more all that is bad.[6] I cannot abide this lugging of Science before the public in Times & Athenæum, & implore you my dear fellow not to do so again.[7] Owens answer to you is so triumphant in the eyes of the public[8] (who you wish to enlighten) as Manchesters is over Natal.[9] The only party that gains by these discussions is the proprietor of the paper, the only one that loses *every way*, is the maintainer of truth.

Science will be much more perfected if it keeps its discussions within its own circle.

I have not had time to look at Bates' book, which both Father & Mother pronounce capital.[10]

I am deep in Cameroons' cases which I shall discuss as best I can, the results are in many respects bamboozling, & I am as dissatisfied with migration by former cold, as by former elevation or by present agencies.[11]

I see from an allusion in the Athenæum that Herschell has a theory of Earth passing through *cold periods*—[12]

I am *out* almost every evening to my ennui & disgust & wish I were ill like you! We are all uncomonly well | Ever Yours affec | Jos D Hooker

DAR 101: 135–6

[1] The date is established by the reference to Richard Owen's anonymous letter in the *Athenæum*, 2 May 1863, pp. 586–7 (see n. 8, below), and by the relationship between this letter and the letter to J. D. Hooker, [9 May 1863]. The intervening Thursday was 7 May.

[2] See letter from Asa Gray, 20 April 1863. The letter was probably enclosed by Asa Gray with a letter to Hooker of 18 April 1863 (Royal Botanic Gardens, Kew (Asa Gray letters: 325)).

[3] Hooker refers to the anonymous satirical pamphlet, *A report of a sad case recently tried before the lord mayor, Owen versus Huxley* (Anon. 1863a), which was also published in *Public Opinion* 3 (1863): 497–8 (see Appendix VIII). There is an annotated copy of the pamphlet, which satirised recent scientific controversies, in the Darwin Pamphlet Collection–CUL.

[4] Philip de Malpas Grey-Egerton was the author of a squib on scientific controversies published anonymously in *Punch* in May 1861 (see *Correspondence* vol. 9, Appendix VIII).

[5] The reference is to Hugh Falconer and his attendance at a conference convened in Paris to decide on the authenticity of the flint tools and human jawbone discovered by the archaeologist Jacques Boucher de Perthes in a gravel pit at Moulin-Quignon, near Abbeville, France, in March 1863. Falconer had changed his original view that the remains were authentic after examining the structure of the jawbone itself, and that of a molar found with it (see letter from Hugh Falconer, 24 April [1863] and n. 6). The zoologist Armand de Quatrefages was a supporter of the case for the jawbone's antiquity. See also Grayson 1983, pp. 213–17, and Van Riper 1993, pp. 134–9. General opinion of the validity of the remains had divided along Anglo-French lines, with the British remaining sceptical. The quotation '*perfide Albion*' is from the French poet Augustin-Marie, marquis de Ximenès's poem 'L'Ere des Français' (*ODQ*): 'Attaquons dans ses eaux / La perfide Albion!' [Let us attack in her own waters perfidious Albion!]. Albion is an ancient name for the British Isles (*EB*).

[6] Falconer was accompanied by William Benjamin Carpenter, George Busk, and Joseph Prestwich (*Athenæum*, 23 May 1863, p. 682).

[7] Hooker refers to Falconer's letter in *The Times*, 25 April 1863, p. 14, concerning the Moulin-Quignon dispute (see letter from Hugh Falconer, 24 April [1863], n. 6), and to CD's letter to the *Athenæum* of 18 April [1863] (see also Appendix VII).

[8] Owen's anonymous letter in the *Athenæum*, 2 May 1863, pp. 586–7, was a critical response to CD's letter to the *Athenæum* of 18 April [1863] (see also Appendix VII).

[9] The reference is to the public criticism by James Prince Lee, bishop of Manchester, of the views espoused by John William Colenso, bishop of Natal, in the first two parts of Colenso 1862–79 (see letter from J. D. Hooker, 20 April 1863 and n. 12).

[10] The reference is to Bates 1863, and to Maria and William Jackson Hooker.

[11] Hooker was describing the plants collected by Gustav Mann between 1860 and December 1862 in the Cameroons Mountains and islands off the coast of West Africa (see letter from J. D. Hooker, [15 March 1863] and n. 21). Hooker's initial findings provided support for CD's view that temperate plants had migrated to tropical regions during the glacial period (see *Correspondence* vol. 10, letter from J. D. Hooker, [5 May 1862], letter to J. D. Hooker, 9 May [1862] and n. 6, and letter from J. D.

Hooker, 9 June 1862, and *Origin* 4th ed., p. 445). Hooker's final conclusion, given before the Linnean Society on 5 November 1863 (J. D. Hooker 1863b), was that the presence of European genera in the region's flora could be explained by two hypotheses: first, CD's migration theory, and second, the transportation of seeds by winds and birds (see also letter from J. D. Hooker, [28 March 1863] and n. 5).

12 In the 'Weekly Gossip' column in the *Athenæum*, 2 May 1863, p. 587, it was stated: 'It is already on record that Sir John Herschel considers the climate of the earth to be undergoing a change due to some cosmical cause'. John Frederick William Herschel had argued that changes in the ellipticity of the earth's orbit were causing a decline in the annual average receipt of solar radiation, thus effecting a gradual cooling of the earth's climate (Herschel 1830, pp. 145–8, Croll 1875, pp. 528–35).

To Charles Lyell [7 May 1863][1]

at J. Wedgwoods Esq[2] | Leith Hill Place | Dorking | Surrey
Thursday

My dear Lyell

When you have time please let me have an answer for M^r. Maw.—[3] We shall stay here till Tuesday morning.[4] What a poor letter the last one of Falconer's.[5] I was much pleased at the last sentence till I came to the last most unjust words; & never were words more unjust.—[6] That beggar Owen sneers well at me & with some justice. I was an ass to write to Athenæum.—[7]

You alluded in your previous note to Haughton; I received, but did not read, his paper, but shall do.—[8] I saw it was the same style of thing, which he has several times published before.—[9] He must be a desperately clever fellow; for, amongst other things, he has made very important discoveries in medicine, as he is an amateur attendant on Hospitals![10]

Ever yours very truly | C. Darwin

Asa Gray tells me he has written long Review on Decandolle for next Silliman & I *fancy* will discuss species question;[11] he has, also, written a Review on Bates' Butterflies for a succeeding number.—[12]

DAR 185: 46

1 The date is established by the relationship between this letter and the letter from George Maw, 25 April 1863, and by the address. According to Emma Darwin's diary (DAR 242), the Darwin family stayed at Leith Hill Place, near Dorking, Surrey, the home of Josiah Wedgwood III, from 6 to 13 May 1863; the only Thursday during this period was 7 May.

2 Josiah Wedgwood III was CD's cousin, Emma Darwin's brother, and the husband of CD's sister Caroline Sarah (*Darwin pedigree*).

3 See letter from George Maw, 25 April 1863.

4 According to Emma Darwin's diary (DAR 242), the Darwins returned home on Wednesday 13 May 1863.

5 The reference is to Hugh Falconer's letter published in the *Athenæum*, 2 May 1863, p. 586. The letter formed part of an ongoing controversy between Falconer and Lyell that began with Falconer's criticisms of *Antiquity of man* (C. Lyell 1863a) in the *Athenæum*, 4 April 1863, pp. 459–60, and continued with Lyell's response in the *Athenæum*, 18 April 1863, pp. 523–5. Falconer's main complaint was that Lyell had not given sufficient credit to his or Joseph Prestwich's contributions to the evidence of

human antiquity. See letters from J. D. Hooker, [23 February 1863] and n. 5, and [6 March 1863] and n. 4, and letter from John Lubbock, 7 April 1863 and n. 6.

6 The reference is to the last paragraph of Falconer's letter, which gave Lyell credit for his services to geology. Falconer's last sentence reads (*Athenæum*, 2 May 1863, p. 586): 'The day has gone by, when scientific works could be written in the style of Louis Quatorze: *La Géologie; c'est moi!—l'Ancienneté d'Homme; c'est moi aussi!*' (I am Geology!— I am also the Antiquity of Man!). Falconer was paraphrasing a quotation attributed to the French monarch Louis XIV: 'L'Etat, c'est moi.': 'The State! I am the State.' (Harbottle and Dalbiac 1901, p. 92).

7 CD refers to Richard Owen's anonymous letter in the *Athenæum*, 2 May 1863, pp. 586–7, criticising the views on the origin of species and spontaneous generation expressed in CD's letter to the *Athenæum* of 18 April [1863] (see Appendix VII). For CD's reply, see the letter to *Athenæum*, 5 May [1863].

8 The letter from Lyell has not been found. Samuel Haughton was professor of geology at Dublin University (*DNB*). There is a lightly annotated copy of S. Haughton 1863 in the Darwin Pamphlet Collection–CUL.

9 S. Haughton 1863 was particularly critical of CD's reasoning in *Origin* regarding the role of natural selection in explaining the geometrical configuration of the cells comprising honeycombs. Haughton made the bee cell a test case of the theory of natural selection. He published similar arguments against *Origin* in S. Haughton 1860a and [S. Haughton] 1860b. See Hull 1973, pp. 216–28.

10 Haughton had entered the University of Dublin's medical school in 1859; he graduated in 1862 and was appointed medical registrar of the school (*DNB*). Haughton had also recently published a paper on the action of poisons (S. Haughton 1862a).

11 See letter from Asa Gray, 13 April 1863. Asa Gray's review of A. de Candolle 1862a and 1862b appeared in the May number of the *American Journal of Science and Arts* (A. Gray 1863d and 1863e); the journal was commonly referred to as 'Silliman's journal' after its founder, Benjamin Silliman.

12 CD refers to Gray's review abstract of Bates 1862 (A. Gray 1863a), which appeared in the November issue of the *American Journal of Science and Arts*.

From W. E. Darwin 8 May [1863][1]

Southampton
May 8.

My Dear Father,

I sent off this morning a box containing two heads of Anchusa.[2] The pollen of short pistilled is decidedly the largest and there is most of it. I have entirely forgotten to look whether both bear seeds; but I will.[3] I am going to the isle of Wight on Sunday when I will mark some plants.[4]

I know whether you have examined Corydalis carefully; it seems almost as nicely arranged as Orchis.[5]

The two side petals form a hood over stigma enclosing it entirely, as soon as the pollen is mature it all falls into the cuplike-stigma when enclosed by the hood, then the back petal that is the one facing the nectary petal falls back and leaves the pistil just on the balance within the hood. This back petal prevents the hood coming off the stigma and it never seems to fall back till the stigma cup is full of pollen.

As soon as the back petal falls back, the least touch sends out the pistil with a bound. Carrying the pollen with it the stigma and style on bounding out fit exactly into the groove leading down to the nectary. But instead of the stigma going straight into the groove from a natural twist in the style it goes in side ways, so that as it buries itself at base of groove nearly all the pollen is pulled out by the friction and is left against side of groove in position for bees to take it, at the same time leaving stigma almost bare to receive other pollen.

So as to prevent the proboscis penetrating except at proper side (or at base?) of groove there is curve in side of groove which makes a sort of overhanging ledge.

Is this the same in fumitories?[6] As it is a beautiful arrangement making the pistil carrying the pollen in position for another flower.

I am glad to find you have managed to get to Leith hill.[7]

Your affect son | W. E Darwin

Tell me whether I am right about Corydalis

DAR 76: B188–90

CD ANNOTATIONS
1.1 I sent . . . plants. 1.4] *crossed ink*
End of letter: '(Wild white Corydalis near Southampton.)' *pencil, square brackets in MS*

1 The year is established by the relationship between this letter and the letter from W. E. Darwin, 4 May [1863].

2 The plant was subsequently identified as *Pulmonaria angustifolia* (see letter from W. E. Darwin, 4 May [1863], n. 1).

3 CD had asked William to make these observations on *Anchusa* (*Pulmonaria angustifolia*) to clarify the form of dimorphism exhibited by the flowers of this species (see letter to W. E. Darwin, [5 May 1863]).

4 William's observations on the sizes of pollen-grains in the different forms of flower in *Pulmonaria angustifolia* are discussed in *Forms of flowers*, p. 106; his original notes have not been found. According to his botanical notebook (DAR 117: 67), William visited the Isle of Wight to mark plants on 10 May 1863, but returned too late in the season to collect seed and had to repeat the work in 1864. CD's brief pencil notes on the information on this species sent by William in 1863 are in DAR 110: 41–2.

5 William's observational notes and sketches relating to *Corydalis claviculata* and *C. lutea* are in his botanical notebook (DAR 117: 61–2) and his botanical sketchbook (DAR 186: 43, pp. 48–9). See also letter to W. E. Darwin, [10 May 1863] and n. 4.

6 See letter to W. E. Darwin, [10 May 1863], n. 5. William's comparative observations on 'Fumitory' (*Fumaria*) are preserved in DAR 117: 63. Like *Corydalis*, *Fumaria* is a genus of the family Fumariaceae (J. C. Willis 1973).

7 Leith Hill Place, near Dorking, Surrey, was the home of Josiah Wedgwood III. According to Emma Darwin's diary (DAR 242), the Darwins stayed at Leith Hill Place from 6 to 13 May 1863.

From J. D. Hooker [8 May 1863][1]

Kew
Friday.

Dear Darwin

Mr Salwyn who has twice been to Guatemala & just returned—says he is game to go to Gallapagos if he had sufficient encouragement—(from scientific men in a scientific way)— he is a gentlemanly sort of fellow, but I believe sells his birds &c & has travelled as naturalist wholly.[2] I have offered to take all his plants, divide them into sets for him to sell for his own profit—& purchase the first set myself & describe them too.— A word of encouragement from you dwelling mainly on the *scientific interest* of the investigation might clinch it.— I told him I would write to you about him.

—Salwyn Esqr.
11. Hanover Terrace
Regents Park.

He can probably answer any questions about Guatemala annuals.— Why should he not go on to *Sandwich Isds*?
J D Hooker

DAR 101: 134

[1] The date is established by the relationship between this letter and the letter to Osbert Salvin, 11 [May 1863], and by the reference in the letter from Osbert Salvin, 12 May 1863, to Hooker's having suggested 'the other evening' that Salvin might travel to the Galápagos Islands. The Friday preceding 11 May 1863 was 8 May.

[2] After graduating from Trinity Hall, Cambridge, Osbert Salvin began his scientific travels with a natural history expedition to North Africa. He first travelled to Guatemala in 1857, and soon 'proved himself an unsurpassed collector'. Salvin, whose primary interests were ornithology and entomology, subsequently made several further expeditions to Guatemala, before returning to England in January 1863 (*DNB*).

From W. H. Flower 9 May 1863

Royal College of Surgeons of England, | *London,* | *(W.C.)*
9th. *day of* May *1863*

My dear Sir

I send you by this post, two copies of both views of the Niata Ox skull, presented by you to the College Museum.[1] I am sorry that there has been so long a delay, but the photographer was very dilatory in printing. Other copies can be had if required, or views in other positions taken.

I find that a cast will cost £3″3″0, though when the first is made, copies can be taken from the mould for a much smaller sum.[2]

Hoping that the photographs will prove satisfactory to M. Quatrefages | I remain | very faithfully yours | W. H. Flower

C. Darwin Esq | F.R.S. &c

DAR 164: 138

[1] In his letter to Flower of 13 April [1863], CD asked him to send a photograph of the niata skull that he had presented to the Hunterian Museum of the Royal College of Surgeons in 1840; Flower was curator of the museum. CD had been asked for the photograph by Armand de Quatrefages (see letter from Armand de Quatrefages, [28 March –] 11 April 1863). See also letter from W. H. Flower, 15 April 1863.

[2] In the letter to W. H. Flower, 13 April [1863], CD had also inquired into the costs of making a cast of the skull of the niata ox for Quatrefages.

To J. D. Hooker [9 May 1863][1]

Leith Hill Place[2]
Saturday

My dear Hooker.

You give good advice about not writing in newspapers;[3] I have been gnashing my teeth at my own folly; & this not caused by Owen's sneers, which were so good that I almost enjoyed them.[4] I have written once again to own to certain extent of truth in what he says;[5] & then if I am ever such a fool again have no mercy on me.— I enclose A. Gray's letter, as you might like to read all.[6] I quite disagree with what he says about Lyell acting as a Judge on Species; I complain that he has *not* acted as a judge; I sometimes wish he had pronounced dead against us rather than possessed such inability to decide.— [7]

I have read the Squib in Public Opinion:[8] it is capital; if there is more & you have copy, do lend it. It shows well that a scientific man had better be trampled in dirt than squabble.

Our outing has not done much for Horace or myself;[9] but I have been a bit better for the last 2 or 3 days.— I have been drawing diagrams, dissecting shoots & muddling my brain to a hopeless degree about the divergence of leaves & have, of course, utterly failed.[10] But I can see that the subject is most curious & indeed astonishing. I wish you or Oliver could give me reference to some paper by Asa Gray, of which you told me.—[11]

I am sure you will like Bates' book,[12] & it will be a rest & pleasure to you to read it.

It is a bad job that you can come to no even moderately clear conclusion about the Cameroons.[13] If the facts do not show it was migration during the Glacial period, so much the worse, as some one says, for the facts.—[14] About some of the same or allied species (in the case of Fernando Po) still existing in the Mauritius; do you think there can be some truth in what I say in Origin of the forms which become extinct on continents, still surviving on islands from less severe competition.—[15]

I was very sorry to see in Falconer's last letter, the parody of Louis' XIV words, applied to Lyell:[16] I have never seen any geological arrogance in Lyell.— I cannot think what he will do, now he has split with Owen & Falconer about naming mammals.—[17]

Goodnight— I long to be in my Hothouse & poking over my little experiments again; we return on Wednesday morning—[18] Goodnight | C. Darwin

That is a clever remark in Gray's letter about origin of language telling against each trifling variation being designed; Lyell shirked this point, which I urged him to grapple with.[19] I do not believe there are above half-a-dozen real downright believers in modification of Species in all England: certainly not more, who dare speak out.—

Darwin Hooker Huxley Wallace Lubbock Bates	} The only honest downright "flat-footed" (see A. Gray) men in all England !!![20]

Endorsement: 'May 11th/63.'
DAR 115: 192

[1] The date is established by the endorsement; the Saturday preceding 11 May 1863 was 9 May.
[2] According to Emma Darwin's diary (DAR 242), the Darwins stayed at Leith Hill Place, near Dorking, Surrey, home of Josiah Wedgwood III, from 6 to 13 May 1863.
[3] In his letter to CD of [7 May 1863], Hooker had implored him to stop writing letters to newspapers. CD had recently written to the *Athenæum* on two occasions (letters to *Athenæum*, 18 April [1863] and 5 May [1863]).
[4] CD refers to Richard Owen's anonymous letter in the *Athenæum*, 2 May 1863, pp. 586–7, which was a response to CD's letter to the *Athenæum* of 18 April [1863] (see Appendix VII).
[5] Letter to *Athenæum*, 5 May [1863].
[6] Letter from Asa Gray, 20 April 1863. Asa Gray had sent this letter in an envelope addressed to Hooker; Hooker read part of it before he realised it was for CD, and forwarded it to him (see letter from J. D. Hooker, [7 May 1863]).
[7] The reference is to Charles Lyell's treatment of transmutation theories and *Origin* in C. Lyell 1863a (see Bartholomew 1973, pp. 296–301). See also letter to Charles Lyell, 6 March [1863].
[8] The reference is to the squib satirising recent scientific controversies (Anon. 1863a), part of which appeared in *Public Opinion* 3 (1863): 497–8 (see Appendix VIII).
[9] According to Emma Darwin's diary (DAR 242), the Darwins stayed with relatives at Hartfield in Sussex from 27 April to 6 May, and at Leith Hill Place, near Dorking, Surrey, from 6 to 13 May 1863. The ill health of CD and his son Horace had prompted the trip (see letter to J. D. Hooker, 23 April [1863]).
[10] CD refers to his research on phyllotaxy. CD's interest in phyllotaxy was stimulated by a comment in Falconer 1863a (see *Correspondence* vol. 10, letter to Hugh Falconer, 1 October [1862], and this volume, letter to J. D. Hooker, 13 January 1863 and n. 22); CD wished to test Hugh Falconer's assertion that a law governed the evolution of leaves around the axis of a stem, producing constant arrangements (see Falconer 1863a, p. 80). See also letter to Daniel Oliver, 20 [February 1863]. There is a series of undated diagrams relating to phyllotaxy in DAR 51: 6–32.
[11] Daniel Oliver was Hooker's assistant in the herbarium and librarian at the Royal Botanic Gardens, Kew (R. Desmond 1994, *List of the Linnean Society of London* 1863). The reference is to A. Gray 1849; however, Hooker could not recall the article (see letter from J. D. Hooker, [13 May 1863]). Gray supplied the information in his letter to CD of [10–16] June [1863].
[12] Bates 1863.
[13] Hooker was writing a paper (J. D. Hooker 1863b) on a collection of plants from the Cameroons Mountains and was having difficulty explaining the geographical distribution of European genera in the region's flora; see letter from J. D. Hooker, [7 May 1863] and n. 11. See also n. 14, below.
[14] Possibly an allusion to a remark made by Erasmus Alvey Darwin, after reading *Origin* (*Correspondence* vol. 7, letter from E. A. Darwin, 23 November 1859): 'In fact the a priori reasoning is so entirely

satisfactory to me that if the facts wont fit in, why so much the worse for the facts is my feeling.' However, CD was hoping that Hooker's findings would corroborate his hypothesis that European plants had migrated to the tropics during the Pleistocene glacial period, and ascended the mountains with the return of a warmer climate (*Origin*, p. 378).

15 CD argued in *Origin*, pp. 106–7, that island plants and animals would face less competition and would consequently experience 'less modification and less extermination'. CD wondered whether this hypothesis would partially explain the affinity between the plants of Clarence Peak, Fernando Po, and the plants of Mauritius and other islands in the Indian Ocean that Hooker had noted in J. D. Hooker 1861, p. 3. See also letter from J. D. Hooker, [13 May 1863], and letter to J. D. Hooker, 15 and 22 May [1863]. This discussion formed part of an ongoing debate between CD and Hooker on the causal factors responsible for the geographical distribution of plants and animals (see, especially, *Correspondence* vols. 3 and 6).

16 See letter to Charles Lyell, [7 May 1863], n. 6. The reference is to Louis XIV of France.

17 Owen had been in dispute with Lyell since the publication of *Antiquity of man* (C. Lyell 1863a) in February 1863 (see letter to Charles Lyell, 4 [February 1863], n. 4, and letter from J. D. Hooker, [23 February 1863] and nn. 2–3). Falconer's disagreement with Lyell began in March (see letter to J. D. Hooker, 13 [March 1863]). Owen and Falconer were acknowledged authorities on mammalian comparative anatomy, and Lyell would probably have consulted them in the course of identifying mammalian fossils.

18 According to Emma Darwin's diary (DAR 242), the Darwins returned to Down on 13 May 1863.

19 The reference is to Gray's comments on C. Lyell 1863a, pp. 454–70, in his letter to CD of 20 April 1863. CD discussed variation and design in his letters to Charles Lyell, [1 August 1861] and 21 August [1861] (*Correspondence* vol. 9).

20 In his letter to CD of 20 April 1863, Gray had written: 'I think that you, and Hooker, are unreasonable in complaining of Lyell that he does not come out "flat-footed" as we say, as an advocate of *natural-selection transmutation*.' CD refers to Thomas Henry Huxley, Alfred Russel Wallace, John Lubbock, and Henry Walter Bates.

To Thomas Rivers [9 May 1863][1]

Leith Hill Place | Dorking
Saturday 10th

My dear Sir

I am very much obliged for the little peaches, which have been forwarded to me here,[2] where I am staying for a little change for health sake; but I shall return home in 3 or 4 days.— I am doubtful, whether the fruit will stick on the Chinese Double Peach which you gave me so kindly.—[3] Therefore I should esteem it a very great favour if you would send me, when ripe, a couple of the Double Chinese & Honey Peach.—[4] As for almonds I must buy varieties in the shell.—[5]

That was a curious monstrosity sent of the Wall-flower; the stamens seem converted into pistils with apparently good ovules.— I hope you will see whether it seeds: I shd. like to grow the seeds & see if so curious a monstrosity is hereditary.—[6]

My dear Sir | Yours very sincerely | C. Darwin

DAR 185: 84

1 The date is established by the relationship between this letter and the letter to Thomas Rivers, 11 January [1863] (see n. 2, below), and by the address. According to Emma Darwin's diary (DAR

242), the Darwins stayed at Leith Hill Place, near Dorking, Surrey, the home of Josiah Wedgwood III, between 6 and 13 May 1863. CD was mistaken as to the date; Saturday was 9 May.

2 No letter accompanying these specimens has been found. However, CD apparently refers to the fruit of a seedling peach tree that he had asked Rivers to send him in the summer (see letter to Thomas Rivers, 11 January [1863]).

3 Rivers, a nurseryman specialising in the cultivation of fruit trees, sent two fruit trees, including a 'double rose flowered chinese peach', for CD's greenhouse in January (see letter from Thomas Rivers, 21 January 1863, and letter to Thomas Rivers, 25 January [1863]).

4 CD wished to compare the stones of these two varieties of peach with almond stones to test Thomas Andrew Knight's conclusion (Knight 1817 and 1821) that the peach was a modified almond. See letters to Thomas Rivers, 11 January [1863] and nn. 7–10, 15 January [1863], 5 March [1863], and 17 August [1863]. CD acknowledged Rivers for providing specimens in *Variation* 1: 338–9. See also n. 5, below.

5 See n. 4, above. A set of engravings of peach and almond stones was included in *Variation* 1: 337, and featured five varieties of almond and the two varieties of peach referred to here. CD used the example to illustrate what could be produced 'by continued selection in two different lines on the same species' (see letter to Thomas Rivers, 15 January [1863]).

6 CD was interested in obtaining specimens of flowers different from the normal type in colour, form, size, or other character, to incorporate into the draft of the chapter on bud-variation (*Variation* 1: 373–411), which he had written in December 1862 (see *Correspondence* vol. 10, Appendix II). See *Correspondence* vol. 10, letter to Thomas Rivers, 28 December [1862], and this volume, letters to Thomas Rivers, 7 January [1863] and 1 February [1863].

To W. E. Darwin [10 May 1863][1]

Leith Hill
Sunday night

My dear William.—

I received Anchusa flowers safe but in broken box.[2] It is a splendid case, & I would not have missed seeing the flowers for anything. You must try & get me some seed; for it will be most necessary to test fertility; for the case will probably explain the state of Echium &c.—[3] Your observations on Corydalis seem *extremely* curious: the specimens which you sent, were rather withered (the plant withers so easily) & I did not examine them carefully.— I never saw anything like the very curious points which you describe in C. lutea; but I may have overlooked all.—[4]

Certainly it does not occur in Fumaria.— As I intend to discuss fertilisation of Fumariaceæ, I sh$^{d.}$ very much like some seed of your species.[5] Could you easily get some?— The seed in this family, I think, generally ripens soon.—

I sh^{d} like to cover up some plants, & watch insects at work.—

Hearty thanks for all your splendid facts.— We go home on Wednesday morning.—[6] Poor dear little Skimp keeps very indifferent; but he has some enjoyment, every day.—[7] I cannot boast much of myself. I have observed nothing curious. But George & I saw Bees of various species & genera fertilising Orchis morio.[8] Your Corydalis case must be somewhat like kidney Bean & the big Lathyrus.[9]

Goodnight, my dear old fellow. How does Mr Fold, Mr Fould Mr Foole, (what is his name?) get on in health?[10] Read Bates' Book; you will like it—[11] good night

| C. Darwin

DAR 210.6: 111

1 The date is established by the relationship between this letter and the letter from W. E. Darwin, 8 May [1863], and by the address. According to Emma Darwin's diary (DAR 242), the Darwins stayed at Leith Hill Place, near Dorking, Surrey, the home of Josiah Wedgwood III, between 6 and 13 May 1863. In 1863, 10 May fell on a Sunday.

2 See letter from W. E. Darwin, 8 May [1863]. The plant was subsequently identified as *Pulmonaria angustifolia* (see letter from W. E. Darwin, 4 May [1863], n. 1).

3 CD was trying to establish whether the 'Anchusa' (*Pulmonaria angustifolia*) was a heterostyled dimorphic plant. His experiments required plants raised from seed to test the fertility of the long- and short-styled forms: if one form was fully fertile only when pollinated with pollen from another form, then the species was heterostyled (see *Forms of flowers*, pp. 2–3). He initially thought that *Echium vulgare*, another member of the Boraginaceae, was heterostyled, but later concluded that this species, with aborted anthers in the long-styled form, was gyno-dioecious, namely, having a female and an ordinary hermaphrodite form (see *Correspondence* vol. 10, letter to Asa Gray, 9 August [1862], n. 8). CD evidently thought that *P. angustifolia* might exhibit a transitional stage between the heterostyled and gyno-dioecious conditions. See also *Correspondence* vol. 12, letter to W. E. Darwin, 14 May [1864]. For CD's conclusions on *P. angustifolia*, see *Forms of flowers*, pp. 105–10.

4 See letter from W. E. Darwin, 8 May [1863] and n. 5. As part of his investigations into the relationship between the structure of flowers and insect pollination, CD carried out a series of observations and experiments on *Corydalis lutea* (*Pseudofumaria lutea*) in 1858, 1859, and 1861; his notes are in DAR 76: 13, 15–17, and DAR 157a: 43. CD's continued interest in the species, a member of the Fumariaceae, is evident in his letters to Daniel Oliver of 12 [April 1862] and 15 April [1862] (*Correspondence* vol. 10). Further observations dated 22 May 1863, including notes on William's observations, are in DAR 76: 19. CD gave a summary of his observations on *C. lutea* in a section on 'Plants sterile without insect-aid' in *Cross and self fertilisation*, p. 359.

5 CD did not publish a description of pollination in the Fumariaceae until 1874, in a short letter to *Nature*, 6 April [1874] (*Calendar* no. 9393). CD began experiments on the pollination mechanisms of Fumariaceae in 1858, after Asa Gray argued that the family was one in which perpetual self-fertilisation occurred (see *Correspondence* vol. 10, letter to Daniel Oliver, 12 [April 1862] and nn. 7 and 8). CD's observational notes on species of *Fumaria*, dated 1858, 1861, and 1863, are in DAR 76: 13–21. CD summarised his findings in *Cross and self fertilisation*, p. 366; as with *Corydalis* (see n. 4, above), these observations formed part of a discussion of plants that were apparently fertile without insect aid.

6 According to Emma Darwin's diary (DAR 242), the Darwins returned to Down from Leith Hill Place on Wednesday 13 May 1863.

7 'Skimp' was a family nickname for Horace Darwin (Freeman 1978). Horace had been unwell since January 1862 (see *Correspondence* vol. 10).

8 George Howard Darwin. His and CD's observations of insects visiting *Orchis morio*, made at Hartfield from 1 to 3 May 1863, are in DAR 70: 176–7.

9 In the three plants to which CD refers, when an insect lands on the flower, the pistil protrudes and applies the stigma to the insect's body; pollen on the insect's body adheres to the stigma, and cross-pollination is thus effected. CD discussed pollination in the kidney bean (*Phaseolus vulgaris*) and in *Lathyrus grandiflorus* in the letter to the *Gardeners' Chronicle*, 18 October [1857] (*Correspondence* vol. 6), and in the letter to the *Gardeners' Chronicle*, [before 13 November 1858] (*Correspondence* vol. 7).

10 CD refers to Phillip Carteret Fall, formerly a partner in the Southampton and Hampshire Bank, who was friendly towards William when he joined the bank in 1861 (see *Correspondence* vol. 9, letter to W. E. Darwin, 17 [October 1861], and letter from W. E. Darwin, [17 November 1861]).

11 Bates 1863.

From Henry Fletcher Hance 10 May 1863

British Vice Consulate, | Whampoa,
10. May, 1863.

Dear Sir,

Knowing that you are engaged in researches on the influence of domestication on the external form and characteristics of animals, I take the liberty of inclosing a stuffed specimen of an albino sparrow from Japan, which I begged the other day from an acquaintance, in whose possession it had died.[1] I cannot say whether, the skeleton being absent, it is of the slightest use or interest to you, but I send it on the chance that it may prove so. I have seen several very curiously marked & colored Japanese sparrows, but they are very rare here, in the south of China, & I cannot procure any for transmission to you, nor do I think they would get home alive, unless very great care were taken of them, for they seem delicate in constitution, from what I can learn. M^r. Swinhoe[2] told me that the So. China sparrow is *Passer montanus*. I am no zoologist myself, & do not pretend to know whether the specimen sent & the others I have referred to are *sports* of that species or not, but the Japanese are such adepts at raising vegetable monsters that they are not unlikely to have outrun us in propagating & perpetuating animal races too.

It is not likely that you have heard my name, though I have for many years past worked steadily at the Flora of China; but a quiet student, who writes little, & is located in so distant & isolated a place can of course expect no other lot.[3]

I take advantage of this opportunity to say that an old friend of mine, Edw. Bradford, a *Linnaean*, formerly Princ: Med: Officer in the Hongkong Garrison, & now head of the medical staff at Sandhurst, who was stationed for some time at Dominica & Trinidad,[4] where he worked hard at Orchids, *con amore*, & with great skill too, & from whom I was so fortunate as to receive a number of specimens, wrote to me very lately, *à propos* of your suggestion (: Fertil. of Orch. p. 236:) that *Catasetum tridentatum* was ♂,[5] to say that he felt sure he had collected perfect fruit from it in Trinidad! This was about 17 years ago, but his memory was not treacherous, for I found on referring to my herbarium (which I did at his request) a fine full capsule, of which I have sent him a sketch—[6] Thus, your supposition is unfounded.[7] I have mentioned this, because I feel sure it is a fact which will greatly interest you; but, while doing so, I desire to explain in the most explicit manner that the discovery & all the credit pertaining thereto are *Mr. Bradford's*, **not** *mine*.

I am, | Dear Sir, | Faithfully Yours, | H. F. Hance.

My address is:

D^r. H. F. Hance,
H. B. M. Vice Consul,
Whampoa, | China.

or/to the care of

Mr. Franz Thimm,[8]
3 Brook Street,
Grosvenor Square.

DAR 166: 95

CD ANNOTATIONS
1.1 Knowing . . . lot. 2.3] *crossed pencil*
3.8 This was . . . sketch— 3.10] *double scored red crayon*
Top of letter: 'Catasetum seedling' *circled pencil*

1 Hance may have learned of CD's research from Robert Swinhoe (see n. 2, below).

2 Robert Swinhoe was an amateur naturalist, who had served with the British Foreign Office in the Far East since 1854, latterly as vice-consul in Formosa, now Taiwan. Swinhoe had corresponded with CD on several occasions (see *Correspondence* vol. 10, and this volume, letter from Robert Swinhoe, 14 April 1863).

3 Hance was the author of several articles on the plants of China (*Royal Society catalogue of scientific papers*) and, in 1871, produced 'Floræ Hongkongensis supplementum' (Hance 1871), a supplement to George Bentham's flora of Hong Kong (Bentham 1861).

4 Hance refers to the army surgeon Edward Bradford, who collected plants in Trinidad in 1845–6 (R. Desmond 1994). Bradford was the surgeon and deputy inspector-general of hospitals at the Royal Military College, Sandhurst; he was elected to a fellowship of the Linnean Society in 1859 (*List of the Linnean Society of London*).

5 In *Orchids*, pp. 236–48, CD argued that *Catasetum tridentatum* was the male, *Monachanthus viridis* the female, and *Myanthus barbatus* the hermaphrodite form of a single species.

6 The sketch by Hance was enclosed with the letter from Edward Bradford, 31 July 1863.

7 See n. 5, above. In *Orchids* 2d ed., p. 197 n., CD commented:

> Dr. Hance writes to me that he has in his collection a plant of *Catasetum tridentatum* from the West Indies bearing a fine capsule; but it does not appear to have been ascertained that this particular flower was that of Catasetum, and there is no great improbability in a single flower of Monachanthus being produced by a plant of Catasetum, as well as a whole scape, which we know has often occurred.

8 Franz Thimm was a London bookseller specialising in foreign literature (*Post Office London directory* 1863).

From George Howard Darwin [before 11 May 1863][1]

You will see in B that if this cylinder were squashed down flat the spire would become a circle, & then each of the buds would occupy one of the places marked 1, 2 . . . 21[2]

Thus Bud 2 would occupy the space betw: 13 & 14 Bud 3 betw: 5, 6; Bud 4 betw: 18, 19. &.c &c & thus every space would be filled but No bud would grow on the top of another. This may be all very old to you, but it was William's[3] suggestion to squash it down & see what happened.

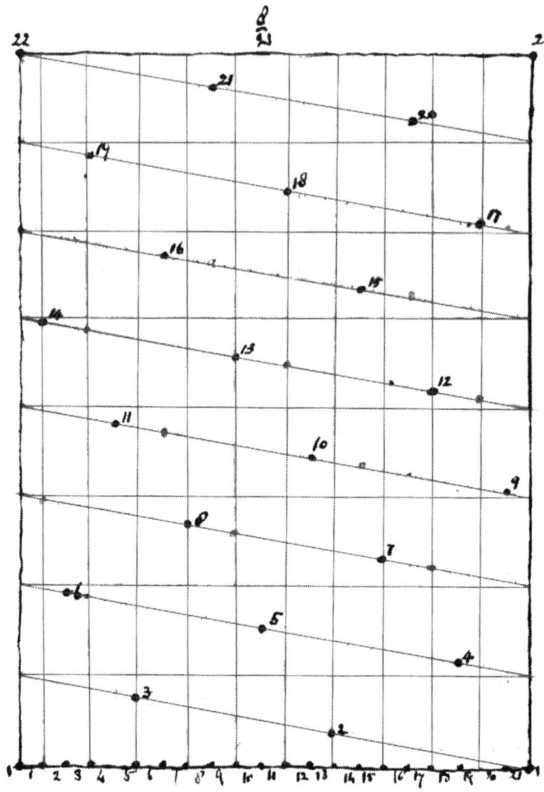

B^4

It strikes me as a very neat arrangement

Now if cylinder A were squashed, each bud would be pressed down by a bud above it & it would come to horrid grief

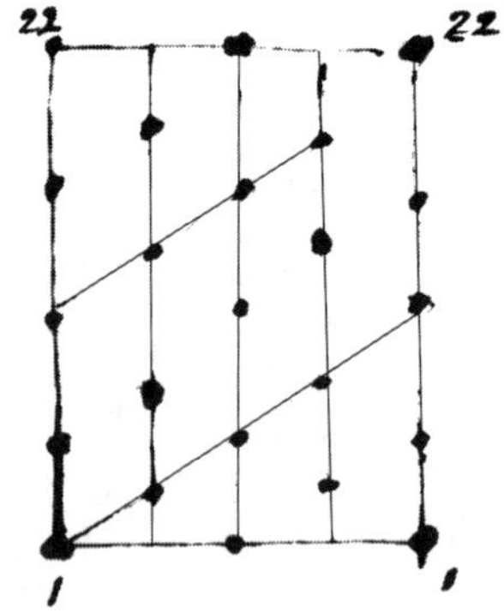

A

There are the same No: of leaves in each of these arrangements, if you count 1, & 1 the same & 22 & 22 the same in each.— Turn over.—

B

See how strangely one Secondary spire consists entirely of leaves with even numbers & the other with odd numbers.— The worst of it is that these things apply just as well to fractions wh: are not real.

The spires then are

1, 2, 3, ...	1 com: diff:
2, 4, 6, 8 . . .	2 com: diff:
1, 4, 7, 10	3 com: diff
1, 5, 9, 13 . . .	4 com: diff
1, 6, 11, 16 . . .	5 com: diff
{ 1, 7, 13 . . .	6 com diff }
but this is the same spire as	
1, 4, 7 &c	
1, 8, 15 . . .	7 com diff:
1, 9, 17—	8 com diff:

this is the same as 1, 5, 9, 13, 17 &.c
&.c &.c

So I suppose there are any no: of spires.[5]

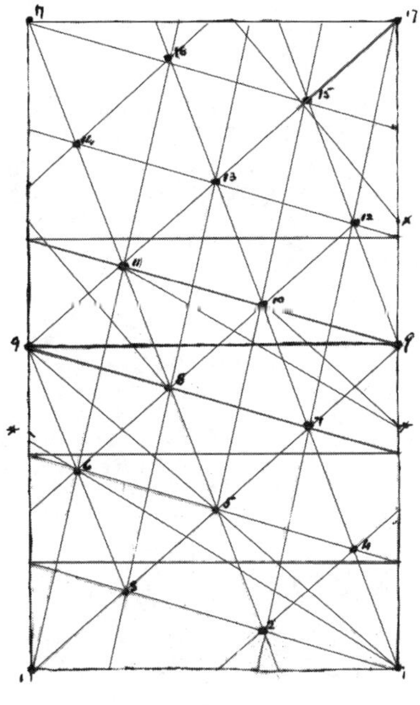

I've come to the conclusion that there are almost any number of secondary spires for instance 1, 4, 7, 10, 13 &.c

N.B. this is an arithmetical progression with 3 for common difference 1; 1+3=4; 4+3=7; 7+3=10; &.c

again 6, 8, 10, 12 &.c here 2 is com: diff:

Join 1 (on the left) 8 & 15 & this is another spire. 6

1, 8, 15, &c. here 7 is com: diff: Join 2 & 6 & the line emerges at the * & goes in on the other side at * & then by drawing a line parallel to the old line you pass thro' 10

thus another is 2, 6, 10

I dont quite understand your question about the cone,[6] but a piece of paper in this shape

if doubled round would form a *mathematical* cone with the top cut off.

If you look in the other diagram you will see that there you can make as many secondary spiries as you like

I've left all those lines in pencil so that you can rub them out.

Memorandum
DAR 51: 6–7

CD ANNOTATIONS
5.1 The spires then are] *cross brown crayon*
Above diagram B: *cross brown crayon*

[1] The date is established by the relationship between this memorandum and the following letter.

[2] George was assisting CD with his investigations on phyllotaxy by analysing the angles of divergence of the leaf buds in a spire (see following letter). For CD's research on phyllotaxy, see letter to Asa Gray, 20 April [1863], and letter to J. D. Hooker, [9 May 1863], n. 10. George's notes on phyllotaxy are in DAR 192: 1–7.

[3] William Erasmus Darwin.

[4] The diagram is reproduced at 66% of its original size.

[5] The following diagram is reproduced at 66% of its original size.

[6] CD's letter to George containing this information has not been found.

To Asa Gray 11 May [1863][1]

Down Bromley Kent [Leith Hill Place][2]
May 11th

My dear Gray

I have to thank you for 2 or 3 little notes.[3] The last I was glad to receive on Lyell, & will tell him, when I write, what you say on Species-portion.[4] I am pleased at it; but cannot quite agree. You speak of Lyell as a Judge; now what I complain of is that he declines to be Judge. It put me into despair, when I see such men as Lyell & you incapable (as you think) of deciding: I have sometimes almost wished that Lyell had pronounced against me.[5] When I say "me"; I mean only *change of species by descent*. That seems to me the turning point. Personally, of course, I care much about Natural Selection; but that seems to me utterly unimportant compared to question of *Creation* **or** *Modification*.— Like a huge Ass I have written two stupid letters to Athenæum:[6] the latter to above effect.— How clever & original & candid your remark about Language & Design.—[7]

Your little discussion on Angles of Divergence of leaves in a Spire has almost driven me mad.[8] My 2^d Boy George is a good mathematician, & when I showed him the fractions, he said they formed a converging series;[9] & I see when protracted, they do all crowd round one point.

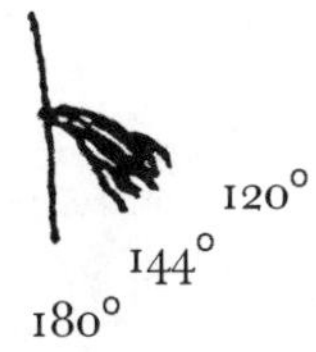

I have been drawing all the real angles & unreal angles on a spire, & I see the angles which do not occur in nature, are just as symmetrical in position as the real angles.[10] If you wish to save me from a miserable death, do tell me why the angles of $\frac{1}{2}$ $\frac{1}{3}$ $\frac{2}{5}$ $\frac{3}{8}$ &c series occur, & no other angles.—[11] It is enough to drive the quietest man mad.—

Did you & some mathematician publish some paper on subject; Hooker says you did.[12] Where is it? I have been visiting for a fortnight houses of relation to try to get a little health for my youngest Boy (the Natural Selection Hero) & self;[13] with very poor success. This has led me to muddle my brains over the angles of leaves.— Do you know of any plant in which angle is fluctuating or variable? I often bless science; for when observing I forget my discomfort & at no other time am I comfortable for two successive hours.— Remembering your statements I have been looking at Plantago lanceolata—[14] it is a **Female** Dichogam which is rather rare (ie pistil mature & fertilised *long* before anthers of same flower mature); fertilised by the wind; and a few plants have imperfect anthers, containing little pollen & a part of this imperfect.—

Euphorbia amygdaloides I find, is also a female dichogamous monoœcious plant, & which is diœcious in function at *any one period*.[15] But why I bother you with these trifling facts, I know not, except that I have nothing to do, & writing to you makes me forget my own odious self.—

Farewell my kind & good friend. If you can spare copy, send me one on De Candolle.[16]

Farewell— Yours most truly | C. Darwin

Gray Herbarium of Harvard University (59)

[1] The year is established by the relationship between this letter and the letter from Asa Gray, 20 April 1863.

[2] According to Emma Darwin's diary (DAR 242), the Darwins stayed at Leith Hill Place, near Dorking, Surrey, the home of Josiah Wedgwood III, between 6 and 13 May 1863.

[3] Letters from Asa Gray, 11 April 1863, 13 April 1863, and 20 April 1863.

[4] In his letter of 20 April 1863, Gray commented favourably on Charles Lyell's discussion of transmutation in C. Lyell 1863a.

[5] CD refers to C. Lyell 1863a, pp. 385–453; CD expressed his disappointment with Lyell's cautious tone regarding his views on transmutation in the letter to Asa Gray, 23 February [1863], the letter to J. D. Hooker, 24[–5] February [1863], and the letter to Charles Lyell, 6 March [1863]. See also letter from Charles Lyell, 11 March 1863 and n. 2, and Bartholomew 1973, pp. 296–301.

[6] Letters to *Athenæum*, 18 April [1863] and 5 May [1863]. See also Appendix VII.

[7] See letter from Asa Gray, 20 April 1863 and n. 11. See also letter to J. D. Hooker, [9 May 1863].

[8] CD had been reading Gray's account of the arrangement of leaves in a spire in *First lessons in Botany* (A. Gray 1857, pp. 72–5); an annotated copy of A. Gray 1857 is preserved in the Darwin Library–CUL (see *Marginalia* 1: 347). See also letter to Asa Gray, 20 April [1863]. In DAR 51: A30 there is a note, dated 10 May 1863, in which CD declared himself 'utterly confounded' by phyllotaxy.

[9] George Howard Darwin's notes on phyllotaxy are in DAR 192: 1–7. See also preceding memorandum.

[10] These diagrams are in DAR 51: A8–32.

[11] This is known as a Fibonacci series, in which any term is the sum of the two preceding terms (*Chambers*). CD had recently begun a series of observations on phyllotaxy (see letter to J. D. Hooker, [9 May 1863] and n. 10, and preceding memorandum). See also letter from Asa Gray, 26 May 1863.

[12] Joseph Dalton Hooker could not recall having made this statement (see letter from J. D. Hooker, [13 May 1863]); however, Gray had read a paper on phyllotaxy at the second meeting of the American Association for the Advancement of Science (A. Gray 1849). In the *Proceedings* of the association for 1849, Gray's paper preceded a paper discussing the fractions occuring in phyllotaxy given by the Perkins Professor of astronomy and mathematics at Harvard University, Benjamin Peirce (Peirce 1849). See letter from Asa Gray, [10–16] June [1863].

[13] According to Emma Darwin's diary (DAR 242), the Darwin family visited Hartfield Grove, Hartfield, Sussex and Leith Hill Place, near Dorking, Surrey, the homes of Charles Langton and Josiah Wedgwood III, from 27 April to 13 May 1863. Horace Darwin had been unwell since January 1862 (see *Correspondence* vol. 10). Horace is referred to as the 'Natural Selection Hero' in recognition of a remark he had made on the future evolution of adders, which applied the principles of natural selection (see *Correspondence* vol. 10, letter to Asa Gray, [3–]4 September [1862]).

[14] CD refers to Gray's observations on the pollination mechanisms of heterostyled species of *Plantago* in A. Gray 1862d, p. 419. There is an annotated copy of A. Gray 1862d in the Darwin Pamphlet Collection–CUL. CD had sought clarification of Gray's description of the apparent fertilisation of the short-stamened flowers while still closed (see *Correspondence* vol. 10, letter to Asa Gray, 26[–7] November [1862], and letter from Asa Gray, 29 December 1862), and made a series of observational notes on 'dichogamy' (the maturation of anthers and stigmas at different times in the same flower) between 19 March 1863 and 24 August 1863 (DAR 49: 83–100). CD's notes on *P. lanceolata*, dated 28 April and 9 May 1863, are in DAR 109: A27 and DAR 49: 91.

[15] CD's notes on *Euphorbia amygdaloides*, which formed part of his work on dimorphism and dichogamy (see n. 14, above) are in DAR 49: 91.

[16] Gray had recently reviewed two works by Alphonse de Candolle (A. de Candolle 1862a and 1862b) for the *American Journal of Science and Arts* (A. Gray 1863d and 1863e, respectively). CD probably refers to A. Gray 1863d, of which there is a heavily annotated copy in the Darwin Pamphlet Collection–CUL.

To Osbert Salvin 11 [May 1863][1]

Down Bromley | Kent [Leith Hill Place][2]
Monday 11th.

Dear Sir

Dr. Hooker has written to me to say that you would like to hear my opinion on the value of a collection made at the Galapagos.[3] Perhaps you may have read my Journal; & if so, you will have observed that all the productions are singularly unattractive in appearance, & this, no doubt, renders this collection less interesting.[4]

But under a purely scientific point of view, I think it would be scarcely possible to exaggerate the interest of a good collection of every species rigorously kept separate from each island. It would throw much light on variation (& as I believe on the origin of Species) & on geographical distribution.[5] No doubt many curious facts could be observed on the habits of the Birds & Reptiles. Probably there would be curious facts on the naturalisation & spreading of introduced plants & animals.—[6] Some of the islands were only just visited by the Beagle, & these would be well worth exploring.[7] The climate is perfectly healthy. The dismal scenery is like that of another world. I look back to my 6 weeks on these islands with extraordinary interest. But you must be prepared for mere amateurs in Natural History thinking little of any collection made there.

If you go, it would well deserve your attention to ascertain how the marine Amblyrhynchus breeds.[8] Pray attend to presence of sea-borne seeds in drift on the beaches exposed to prevailing currents.[9]

with most cordial wishes for your success if you start, & with sincere respect for your zeal, pray believe me Dear Sir | Yours very faithfully | C. Darwin

Sybil Rampen

1 The date is established by the relationship between this letter and the letter from Osbert Salvin, 12 May 1863.

2 According to Emma Darwin's diary (DAR 242), the Darwins stayed at Leith Hill Place, near Dorking, Surrey, the home of Josiah Wedgwood III, between 6 and 13 May 1863.

3 Joseph Dalton Hooker; see letter from J. D. Hooker, [8 May 1863].

4 CD visited the Galápagos archipelago for five weeks between 15 September and 20 October 1835, during his voyage on HMS *Beagle*. His account of the visit to the Galápagos Islands appeared in chapter 19 of the first edition of *Journal of researches* (pp. 453–78) and chapter 17 of the second edition (pp. 372–401).

5 The *Beagle* collections from the Galápagos islands indicated not only that the majority of the land birds were endemic to the archipelago as a whole, but also that some were endemic to specific islands. The results of the analyses of avian, mammalian, and botanical collections from the archipelago between 1837 and 1859 led CD to accord great importance to the role of geographical isolation in the production of new species (*Natural selection*, pp. 254–6, 273–4; *Origin*, pp. 107–8; Sulloway 1979 and 1984).

6 For a number of years CD had been collecting data on variation in plants and animals introduced into different environments (*Correspondence* vols. 5–7). CD believed insights would be gained into the processes of adaptation, variation, and natural selection as a result of examining the nature of plants or animals that had successfully replaced the endemic flora and fauna of any country. In particular, CD noted that naturalised plants exhibited a 'highly diversified nature' (*Origin*, pp. 114–16). See also letter to Julius von Haast, 22 January 1863.

7 For an itinerary of HMS *Beagle* between 15 September and 20 October 1835, see Freeman 1978, p. 147. See also *'Beagle' diary*, pp. 333–43.

8 CD refers to the marine iguana, which is endemic to the Galápagos Islands.

9 As part of his investigations into the geographical distribution of plants, CD had long been interested in the role of marine currents in seed dispersal, and had performed a series of experiments on the ability of seeds to survive immersion in sea-water (see *Correspondence* vol. 6, and *Origin*, pp. 358–61).

From H. B. Dobell 12 May 1863

29 Duncan Terr
May 12 1863.

Dear Mr Darwin

I have delayed writing that I might send you a copy of the "Table".[1] Allow me now to thank you very much for your kind & valuable suggestions for the improvement of the table—all of which you will find embodied in the enclosed form.[2]

I have this day had a letter from the country informing me of the birth of a child with two thumbs on one hand, & I am going to ask that one may be amputated & the result watched. I have just obtained the particulars of a case of a man with two thumbs on one hand (never amputated) who had 8 children neither of whom has any defect in the digits.[3]

I have just been informed that one child of the married cousins with deformed hands, is showing signs of disposition in the hands to assume the family deformity—I have not seen it yet but will report progress to you when I have.[4]

With many thanks | Believe me very truly, | Horace Dobell

Ch. Darwin Esq.

Darwin Pamphlet Collection–CUL (bound with G 395, Dobell 1862)

CD ANNOTATION
1.1 I have ... digits. 2.5] *crossed ink*

1 Dobell had drafted forms for recording genealogical information with the aim of assisting physicians and others in the collection of accurate data on the hereditary transmission of morphological characteristics and diseases (see letter from H. B. Dobell, 20 April 1863). CD suggested several amendments to the draft (see letter to H. B. Dobell, 21 April [1863]). Neither the original draft nor the amended version of the form has been found.

2 See letter to H. B. Dobell, 21 April [1863].

3 CD asked Dobell to send him information on the regrowth of supernumerary digits following amputation and on inherited characteristics in humans (see letter to H. B. Dobell, 21 April [1863]). CD was collecting information for the chapters on inheritance in *Variation* (*Variation* 2: 1–84).

4 Dobell refers to a case of inheritance of fingers with thickened joints in a family through five generations. In the fifth generation, the offspring from the marriage of two first cousins with the deformity did not initially appear to have inherited the defect. See also letter from James Paget, 7 February 1863, n. 1, and letter to H. B. Dobell, 16 February [1863] and n. 12. The case was documented in Dobell 1862. There is an annotated presentation copy of Dobell 1862 in the Darwin Pamphlet Collection–CUL.

To W. H. Flower 12 May [1863][1]

Leith Hill Place[2]
May 12th

My dear Sir

I am staying from home & your letter has followed me;[3] but I shall return home tomorrow & shall then, no doubt, find the Photographs. I am extremely much obliged to you for your kindness, & will inform M. Quatrefages that he owes the

Photograph to you. He evidently was very anxious for a copy. I will tell him about the Cast.—[4]

Will you be so kind as to inform me what I am indebted for the Photograph & I can repay you.—[5] Perhaps I will get copies for myself; if they will do for woodcuts; for I shall briefly describe this curious variety in my book on Variation under Domestication.[6]

Pray believe me, with very many thanks | My dear Sir | Yours very faithfully | Ch. Darwin

John Innes Institute

[1] The year is established by the relationship between this letter and the letter from W. H. Flower, 9 May 1863.

[2] According to Emma Darwin's diary (DAR 242), the Darwin family visited Leith Hill Place, near Dorking, Surrey, the home of Josiah Wedgwood III, from 6 to 13 May 1863.

[3] Letter from W. H. Flower, 9 May 1863.

[4] In his letter to CD of 9 May 1863, Flower wrote that the photographs of the niata ox skull at the Hunterian Museum that CD had requested for Armand de Quatrefages had been sent. CD had also asked Flower whether a cast of the skull might be made (see letter to W. H. Flower, 13 April [1863], and letter from W. H. Flower, 9 May 1863).

[5] According to CD's Classed account book (Down House MS), CD made a payment of 17*s.* for the photographs on 25 May 1863.

[6] Niata cattle were discussed in *Variation* 1: 89–91, 2: 66; however, CD did not include an illustration of the breed.

From John Edward Gray 12 May 1863

British Museum
12 May 63

My Dear Darwin

I wish to draw your attention to a fact that has recently developed itself to me

There are in view to be observed on the Bank of the Thames near Kew two most distinct forms of the Common Wild Chervil *Anthriscus Sylvestris* & I think you will very likely see the same near Down one is a larger bushy succulent pale green plant with thick stems with a few very large ridges on it, and rather larger flowers—

The other is a rigid hard slender comparatively slightly branched plant with thin stems with a number of small equal ridges a dark foliage with the leaves far apart & small flowers The Stems & leaves are purpurea but sometimes rarely dark green They grow on the same foot of soil The branches of the two varieties being often intermixed—so that it cannot arise from difference of soil or situation[1]

I have observed a some what similar variety in the Wild Wood anemones under the same circumstance that is grown side by side the one green with white flower the other with a dark foliage & purple flowers, & having narrower petals

Now my dear friend there is a Theme for you do consider why & how *two forms* of the same species should exist & grow abundantly (I could send you a barrow load of each) side by side on the same Soil & under the same circumstances

I think I have observed the same thing among other common Plants as the Ground Ivy & the Purple dead Nettle Lamium purpureum but I am not so sure of their being found exactly under the same circumstance as these are but will examine.

Ever yours sincerely | though not as convert[2] | J. E Gray

I cannot find that even the greatest of Species splitters has noticed this form of Anthriscus any more that they had noticed *anemone nemerosa purpurea*[3]

DAR 165: 208

[1] The plant described may have been *Anthriscus caucalis* (bur parsley) or *A. cerefolium* (garden chervil), both of which have smaller flowers than *A. sylvestris*, and both of which also occur in disturbed habitats (Stace 1991).

[2] On Gray's critical reception of *Origin*, see *Correspondence* vols. 7 and 8, and Gunther 1975, pp. 453–5.

[3] A purple form of *Anemone nemorosa*, the wood anemone.

To George Maw 12 May [1863][1]

of Down, Bromley, Kent [Leith Hill Place][2]
May 12th

Dear Sir

I have been visiting from home & your letter of April 25th did not reach me for some days.[3] I then wrote to Sir C. Lyell, & he was so busy that he was not able to answer me at once.—[4] He is much obliged to you, but he really does not know enough of the osteology of man to appreciate your specimen.—[5] He agrees with me that bones from Gibraltar must be of very doubtful age. Nevertheless, in my opinion, no skull of man ought to be neglected. If you get a skull or part of skull from Gibraltar, I would suggest that you had better write direct to Mr Busk, Secretary Linn. Soc. Burlington House or to Dr Falconer 21 Park Crescent, & state what the remains are & whether either one would like to examine them.[6] It would be a thousand pities if the specimens were wasted. You seem to have got a splendid section of the Drift.—[7] It would be highly adviseable for you to collect as many shells as possible & look out well for any bones.— Mr. Prestwich of the Geological Soc. Somerset House, would probably be interested by any considerable collection of shells from the Drift.—[8] Glacial shells have been found by me & others near Shrewsbury.[9]

Pray believe me | Dear Sir | Yours very faithfully | Ch. Darwin

Royal Horticultural Society, London

[1] The year is established by the relationship between this letter and the letter from George Maw, 25 April 1863.

[2] According to Emma Darwin's diary (DAR 242), the Darwins were at Leith Hill Place, near Dorking, Surrey, the home of Josiah Wedgwood III, from 6 to 13 May 1863.

[3] Letter from George Maw, 25 April 1863.

[4] Letter to Charles Lyell, [7 May 1863]; Lyell's reply has not been found.

[5] Maw had informed CD that he was in a position to obtain a specimen of a supposedly human skull from Gibraltar, and requested CD to ask Charles Lyell whether he was interested in examining it (see letter from George Maw, 25 April 1863).
[6] George Busk was the zoological secretary of the Linnean Society (*DNB*). Busk and Hugh Falconer were specialists in palaeontology, and had explored British and European caves in search of human fossil remains (*DSB*, Falconer 1868).
[7] See letter from George Maw, 25 April 1863.
[8] Joseph Prestwich was treasurer of the Geological Society of London (*Quarterly Journal of the Geological Society of London* 1863), and a leading expert on Tertiary geology. In 1862, Prestwich read two papers on the geology of drift deposits before the Royal Society of London; he completed the study, which included an analysis of the shells found in the deposits, towards the end of May 1863 (Prestwich 1862b, p. 247).
[9] In 'Ancient glaciers of Caernarvonshire', p. 185, CD briefly mentioned finding fragments of seashells in Shropshire and Staffordshire (*Collected papers* 1: 167).

From Osbert Salvin 12 May 1863

11, Hanover Terrace, Regent's Park, N.W.
12th. May 1863

Dear Sir,

I was very glad to receive your letter this morning giving me your ideas about a collection from the Galapagos Is.[1] Your Journal is one of the few books I carry abroad with me & your account of those Islands has long given me a wish to visit them.[2] I have but lately returned from a naturalizing tour with Mr. F. Godman in Guatemala & we are now engaged working up our collections or at least a part of them & intended to return to Costa Rica or S. America, or perhaps both, early next year.[3] Mentioning this the other evening to Dr Hooker he suggested the Galapagos as a point better worth attention and he so far convinced me that I think both Mr. Godman & myself have made up our minds to go there.[4]

Our object will be purely a Scientific one and as we look for no pecuniary return would be able to pay attention to all and every branch of Natural History no matter how dull or unattractive the species.

I would willingly start at once but our Central American collections will occupy some time but I reckon on being able to leave about the end of Feb. so as to get to the Islands before the rainy season sets in.—

There may be some things amongst our Central American Collections interesting to you, should you like to see them I should be most happy to shew them to you any day you might be in town.—

Believe me | Yours very faithfully | Osbert Salvin.

C. Darwin Esqre.

DAR 177: 17

[1] See letter to Osbert Salvin, 11 [May 1863].
[2] See letter to Osbert Salvin, 11 [May 1863] and n. 4. The reference is to *Journal of researches*, pp. 453–78.
[3] In 1861, Osbert Salvin and Frederick Du Cane Godman undertook an ornithological exploration of Guatemala; Salvin returned to Britain in January 1863 (*DNB*).

[4] Salvin's plan to return to South America in 1864 did not come to fruition: soon after his return to England he was induced to undertake the management of an ironworks in Rotherham, Yorkshire (*DNB*, Godman and Salvin eds. 1879–1915, 1: 6).

To Edward Cresy 13 May [1863][1]

Down, Bromley, Kent
May 13th.

Dear Cresy

I have been visiting for a fortnight for health sake and your pleasant note was forwarded to me;[2] but I delayed writing till my return home this afternoon when I found the maps.[3] We have all been studying them with much interest; and it was uncommonly kind in you to think of sending them.

My son George failed at St. John's; but the tutor gave a very good report of his examination.[4] He was unfortunate in not having questions in some of the higher branches of mathematics in which he is best up, as in the Differential Calculus.— But we have now resolved that he shall stay another year at school, as he is yet under 18.[5]

Our trip has not done my youngest boy or myself much good in health, I am sorry to say.—[6] In haste.

Yours very sincerely | C. Darwin

Copy
DAR 143: 323

[1] The year is established by the relationship between this letter and the letter from Edward Cresy, 27 April 1863.

[2] Letter from Edward Cresy, 27 April 1863. The Darwin family had been visiting relatives from 27 April to 13 May 1863 (see 'Journal' (Appendix II)).

[3] Cresy had sent CD maps of new railway lines in London and Kent (see letter from Edward Cresy, 27 April 1863). For a discussion of the development of the railway network in these regions see H. P. White 1982 and 1987.

[4] George Howard Darwin had entered a scholarship examination at St John's College, Cambridge (*DNB*). See letter from Edward Cresy, 27 April 1863 and n. 7.

[5] George was a pupil at Clapham Grammar School, Surrey (*DNB*).

[6] Horace Darwin accompanied his parents to Hartfield Grove, Hartfield, Sussex, and Leith Hill Place, near Dorking, Surrey, between 27 April and 13 May 1863; the ill health of Horace and CD had prompted the trip (see letter to J. D. Hooker, 23 April [1863], and Emma Darwin's diary (DAR 242)).

To H. B. Dobell 13 May [1863][1]

Down. | Bromley. | Kent. S.E.
May 13th

My dear Sir

The Table, as far as I can judge, seems excellent.—[2]

I shall be truly obliged for any further information with respect to the children of the Cousins; for the case surprised me much.—[3]

Pray believe me in Haste | Yours sincerely | Ch. Darwin

Paul C. Richards

[1] The year is established by the relationship between this letter and the letter from H. B. Dobell, 12 May 1863.
[2] See letter from H. B. Dobell, 12 May 1863. The table has not been found.
[3] See letter from H. B. Dobell, 12 May 1863 and n. 4.

From Julius von Haast 13 May 1863

Christchurch NZ.
May 13th 1863

Dear Sir!

Returning this moment from my summer campaign, I am delighted to find your very kindly note of the 22nd of January.[1] Many, many thanks! Those expressions of acknowledgment make me very happy & will give me new strength for my pursuits. The mail closes in half an hour, so I must be short. Inclosed an account of my last expedition.[2] You see, I have had a good result. With next mail I shall write more fully & give you some very interesting facts, but I would not let the mail part without a word of thanks.

Believe me my dear Sir | very faithfully yours | Julius Haast

Charles Darwin Esqre | FRS etc.
Down Bromley Kent SE.

DAR 166: 3

[1] Letter to Julius von Haast, 22 January 1863.
[2] The enclosure has not been found. Haast refers to his geological expedition across the Southern Alps of New Zealand's Middle Island (now South Island) to the west coast (see letter from Julius von Haast, 5 March 1863 and n. 2), and probably to the account of his expedition published in the Christchurch *Press*, 1 April 1863, pp. 1–2, and 2 April 1863, pp. 2–3. See also enclosure to letter from J. D. Hooker, 20 April 1863 and n. 28.

From J. D. Hooker [13 May 1863][1]

Kew
Wednesday.

Dr Darwin

Salwyn & his fellow traveller called today, very full of having heard from you, & quite bent upon Gallapagos.[2] I have tonight written to Consul at Guayaquil for information—[3] They talk of hiring a schooner & spending 10 months (I advise them to go on to Sandwich Islands), next March.

Thanks for A Grays letter.—[4] I quite agree with you that Lyell has in no way taken position of judge, but of advocate—[5] he is far too conscious of the merits of your case & too prudent of his own future fame not to have taken a side— Had he taken Judges seat I never should have blamed him—the fact is he is half-hearted but whole-headed— Do ask Gray to reconsider— I asked Bentham, who is *wholly* of our opinion— I think Bentham will make a very interesting discussion for the Linnæan anniversary on the *methods* of your adversaries.[6]

Your paper in Linn. Journal reads excellently well.[7]

I hope to get loan of copy of squib & will copy it.[8] Busk & Mrs B. have credit of it;[9] but I cannot believe it theirs, it is too clever & too impartial.

Bates book is charming—but I can see some ground for saying that it is *biassed* by Darwinism[10] It is a little evident to me, here & there, that his Darwinistic explanations of what he sees &c are after-thoughts— It is too bad to say that his facts are therefore twisted—but he says here & there, or leaves the impression of saying, in 1849, "We did so & so which is of such importance "au point de vue" of N. Selection or of Variation & N & S." whereas he never knew aught of these till 1859—[11] I express my meaning very clumsily—

I am very anxious to hear what you can say of divergence of leaves, PS. I was thinking of divergence of *species*— You mean Phyllotaxis[12] Oliver asks if you have seen Nageli on relations of leaves to vascular bundles of stem.[13] I have often tried to see something in it, but never could get the remotest glimpse & have been greatly disappointed in M^c^Cosh's & other books on the subject.[14] I remember no paper by A Gray[15]

I am groaning over Cameroons.[16] Mann has been again (4th time) up F. Po Peak, but got nothing new. he has been at deaths door 3 times & is on his way home.[17] We only found out last week that he was engaged to a very nice girl at Kew![18]

I have perfect faith in your doctrine of absence of competition favoring retention of continental forms on Islands, though how the devil one is to reconcile that with the extraordinary modification of other continental forms on same Islands passes my comprehension.[19] Except what you won't admit:— that they were common to continent & Island before disjunction of latter & the modification is of the *Continental* forms, the insular being the old original type— This is turning the tables on you with a vengeance but I will work it out in spite of you. Go to weep & howl[20]

The **Ferns** of Ascencion & St Helena are totally difft. from one another & from Cameroons, this is, or ought to be, a deathblow to all aerial migration, for ferns are notoriously widely dispersed & dispersable.— I wish I had never wasted a thought on the stupid subject

I feel quite sick & sorry when I think over this squabble of Falconer & Lyell[21] I am determined to bring them together if possible, & appeal to the magnanimity of each to allow me to do so.

I must send you a flower of *Clianthus Dampieri*, in which the pollen falls out of anthers into boatshaped pendulous keel, & rolls along the Kelson down towards the stigma—but it takes a Darwinian waggle of the keel to get it on stigma I expect.[22]

I shall be anxious to hear how the hot house plants get on.[23]

I quite expect that Bentham will speak *out* by & bye, & Oliver I am sure is far gone, but is bothered with external conditions.[24]

J. E. Gray spent $\frac{3}{4}$ hours abusing Bates to me the other day.—[25] I let him go on till he had made Bates out to be morally intellectually & physically, unutterably base & then I pitched into him hot & strong & made him eat **all** his assertions except that he had not collected 8000 new species (which I believe he has, but care nothing about.)—[26] He began by saying he had spent all his time in idleness & licentiousness amongst the natives on the Amazons & half an hour afterwards told me that it was the B.M. that had supported him, *all the 11 years, paying him £300–400 a year for the pick of his collections*!![27] I think I made him heartily ashamed of himself. I never heard such a slanderer in my whole life. I suppose it is because he so overdoes it that he makes so few real enemies thereby.

Bates is in a very difficult position & I am urging him to keep before the world in publishing & especially to take care not to quarrell with or show contempt for his brother Entomologists;[28] & to take their sneers & suspicions in perfect good part.— Poor fellow he dined with me the other day, & 2 days after wrote me that he had a hornets nest about his ears headed by J. E. Gray, who attacked him at the B.M.

I am certain if he only goes on quietly & goodnaturedly working hard & publishing such papers as he has in 3 years he will be the first living philosophical Entomologist & all the rest will be at his feet if he makes no enemies amongst them.

This is a cruel long letter to make you read—so I've done.

Ever yours affect | J D Hooker

DAR 101: 137–40

CD ANNOTATIONS
6.1 I am very . . . stem. 6.3] *cross in margin, pencil*
10.2 I am determined . . . do so. 10.3] *cross in right margin, pencil*
11.1 I must . . . keel, 11.2] *cross in right margin, pencil*
14.9 I suppose . . . thereby. 14.10] *cross in margin, pencil*
Top of letter: 'Primroses' *ink*

1 The date is established by the relationship between this letter, the letter to Osbert Salvin, 11 [May 1863], and the letter to J. D. Hooker, 15 and 22 May [1863]. The intervening Wednesday was 13 May.

2 See letter to Osbert Salvin, 11 [May 1863], and letter from Osbert Salvin, 10 May 1863. Osbert Salvin and Frederick Du Cane Godman had returned from a collecting expedition in Central America that began in the autumn of 1861 and ended in January 1863 (*DNB*). In his letter to CD of [8 May 1863], Hooker wrote of Salvin's interest in collecting natural history specimens in the Galápagos Islands and asked CD to write to encourage him to do so.

3 The chargé d'affaires and consul-general at Guayaquil in Ecuador was George Fagan (*British imperial calendar* 1863).

4 CD enclosed the letter from Asa Gray, 20 April 1863, with his letter to Hooker of [9 May 1863].

5 Charles Lyell; see letter from Asa Gray, 20 April 1863, and letter to J. D. Hooker, [9 May 1863].

6 George Bentham worked independently in the herbarium of the Royal Botanic Gardens, Kew, where Hooker was assistant director (R. Desmond 1995). Bentham, who was president of the Linnean Society, was preparing his anniversary address, to be read before the society on 25 May 1863; see letters from George Bentham, 21 April 1863 and 21 May 1863, and Bentham 1863.

7 'Two forms in species of *Linum*' was published in the *Journal of the Proceedings of the Linnean Society (Botany)* issued on 13 May 1863 (*General Index to the Journal of the Linnean Society*, p. vi).

[8] Anon. 1863a. See letter from J. D. Hooker, [7 May 1863] and n. 3, and Appendix VIII.

[9] George and Ellen Busk (see Appendix VIII).

[10] Bates 1863. See letter to J. D. Hooker, [17 April 1863] and n. 15.

[11] Bates 1863 was a narrative of Henry Walter Bates's travels in the Amazon between April 1848 and July 1859; *Origin* was published in November 1859.

[12] See letter to J. D. Hooker, [9 May 1863].

[13] The reference is to Daniel Oliver and the first part of Nägeli 1858–68, pp. 39–156. Oliver had been assisting CD with bibliographic references on phyllotaxy (see letter from Daniel Oliver, 17 February 1863, and letter to Daniel Oliver, 20 [February 1863]).

[14] Hooker refers to a series of articles on phyllotaxy by James McCosh (McCosh 1851, 1852, and 1854).

[15] See letter to J. D. Hooker, [9 May 1863] and n. 11. The reference is to Asa Gray.

[16] Hooker was writing a paper on the plants of the Cameroons Mountains that had been collected by the botanist Gustav Mann (J. D. Hooker 1863b). See letter from J. D. Hooker, [7 May 1863] and n. 11, and letter to J. D. Hooker, [9 May 1863].

[17] Hooker refers to Clarence Peak, Fernando Po; Mann's collections from this island mountain off the west coast of Africa were made between 1860 and 1863 (see letter from J. D. Hooker, [7 May 1863] and n. 11).

[18] Mann married Mary Anne Stovell on 12 November 1863 (Register of marriages, registration district of Chichester, 1863, no. 60 (General Register Office)).

[19] See letter to J. D. Hooker, [9 May 1863] and n. 15.

[20] Hooker refers to CD's discussion of the plants and animals of oceanic islands in *Origin*, pp. 388–406, in which CD argued that 'occasional means of transport' of species to islands, and the subsequent modification of those species, best explained the existing geographical distributions; he rejected explanations that invoked a former land connection between islands and the nearest continent. See also letter to J. D. Hooker, 15 and 22 May [1863]. CD and Hooker had frequently debated these theories on the origin of island plants and animals over many years (see, for example, *Correspondence* vol. 10, letter from J. D. Hooker, [30 December 1861 or 6 January 1862] and n. 7, and letter to J. D. Hooker, 4 November [1862] and n. 6).

[21] Hugh Falconer and Charles Lyell. See letter to Charles Lyell, [7 May 1863], nn. 5 and 6.

[22] See letter to J. D. Hooker, 15 and 22 May [1863] and n. 5.

[23] Hooker had sent CD a cart-load of plants from the Royal Botanic Gardens, Kew, for his new hothouse (see letter to J. D. Hooker, [21 February 1863]). See also Appendix VI.

[24] Bentham favourably discussed CD's theory of the origin of species by natural selection in his anniversary address to the Linnean Society (see n. 6, above, and Bentham 1863, p. xv–xvi). Hooker refers to CD's list of the supporters of his theory, those 'who dare speak out' (see letter to J. D. Hooker, [9 May 1863]).

[25] John Edward Gray was the keeper of the zoological department at the British Museum. Bates had worked on his South American collections at the British Museum since returning to England in 1859 (Woodcock 1969, p. 240), and had complained to CD that the naturalists at the museum did not appreciate his work (see *Correspondence* vol. 9, letter to H. W. Bates, 3 December [1861]). On Gray's character and antipathy towards evolutionary theory, see Gunther 1975, pp. 176–7 and 453–6.

[26] Bates claimed to have collected 14,712 species while in South America, over 8000 of them being new (Bates 1863, 1: v). See also Gunther 1975, p. 177.

[27] Bates had sent his specimens from South America to Samuel Stevens, his agent in London, who sold them for twenty per cent commission. Many were sold to the British Museum, which had agreed before Bates left in 1848 to purchase the rare insects (Woodcock 1969, pp. 30–1). See also letter from H. W. Bates, 20 April 1863.

[28] At this time Bates had no regular employment, but relied on a small private income (see letter from H. W. Bates, 20 April 1863). For the direction of Bates's career after 1863, see Woodcock 1969, pp. 255–61. Bates's evolutionary entomology was unpopular with the members of the Entomological Society of London, of whom CD wrote: 'No body of men were at first so much opposed to my views'

(see letter to Ernst Haeckel, 21 May [1867], *Calendar* no. 5544). See also n. 25, above, and letter to Charles Lyell, 17 March [1863] and n. 16.

To Richard Kippist 13 May [1863 or 1868][1]

Down. | Bromley. | Kent. S.E.
May 13th

My dear Sir

If "Audubon's Ornith. Biography" is in our Library, (5. vols.)[2]—will you be so kind as to have it ready for Carrier tomorrow Thursday morning.[3] I forget whether it is permitted to have 5 vols.; but if not, could you venture to break the rule & mention it to the Secretary, as I much want all vols. as I shall have to refer to them many times during next fortnight.—[4]

Yours very sincerely | C. Darwin

If Carrier does not bring any vols. I shall understand you have it not.—[5]

Harold B. Lee Library, Brigham Young University

1 The date range is conjectured from the stationery and from CD's request to consult Audubon 1831–[9] (see n. 2, below). The printed stationery on which the letter is written was in use between 1861 and 1869 (see Carroll ed. 1976, p. xxiii); 13 May fell on a Wednesday in only two years in that period, 1863 and 1868. Between 1 April and 16 June 1863 (see 'Journal' (Appendix II)), CD was writing a draft of the chapters on crossing and sterility for *Variation* (*Variation* 2: 85–191); Audubon 1831–[9] is cited in *Variation* 2: 154 and 157. Between 17 May and 26 December 1868, CD was working on the section of *Descent* (*Descent* 2: 38–238) that discussed sexual selection in birds (CD's journal (DAR 158)); Audubon 1831–[9] is cited extensively. CD made an unsuccessful attempt to purchase the book at the end of May 1868 (see letter from Williams & Norgate, 2 June 1868, *Calendar* no. 6222).

2 Audubon 1831–[9]. Fellows of the Linnean Society were allowed to borrow books from the society's library, with certain restrictions (*List of the Linnean Society of London* 1863); Kippist was the society's librarian (*DNB*). CD read Audubon 1831–[9] in December 1858 (*Correspondence* vol. 4, Appendix IV, 128: 22), and an abstract of the work is in DAR 71: 192–214. The annotated copy in the Darwin Library–CUL (see *Marginalia* 1: 21–3) appears to have been purchased after 1868 (see letter from Williams & Norgate, 2 June 1868, *Calendar* no. 6222).

3 George Snow of Down operated a carrier service between London and Down every Thursday (*Post Office directory of the six home counties* 1862).

4 George Busk was the zoological secretary of the Linnean Society (*DNB*).

5 There was a copy of Audubon 1831–[9] in the Linnean Society's library in 1866 (*Catalogue of the library of the Linnean Society*).

From George Chichester Oxenden 13 May 1863

Broome
May 13. 1863

Dear & kind Sir

I am just going abroad for some months—so that I shall lose the whole English-Orchid season—[1]

—If I meet with aught botanically-new to me, I will one day write you—

—Far as it is above *me*, your gift-book has *delighted* me incessantly—& I thank you for making it evident that Moths were not made & created solely to fatten "*Caprimulgi*"—[2]

—I hope your family sorrows are Overpast—[3]

Sincerely | G. Chichester Oxenden

DAR 173: 61

[1] During 1861 and 1862, Oxenden supplied CD with information about, and living specimens of, the rarer British orchids (see *Correspondence* vols. 9 and 10). CD acknowledged his assistance in *Orchids*, pp. 31–2.

[2] CD sent Oxenden a presentation copy of *Orchids* (see *Correspondence* vol. 10, Appendix IV). Oxenden probably refers to CD's observations on the pollination of orchids by moths, and his experiments indicating that moths possessed 'instinctive knowledge' (see *Orchids*, pp. 35–47). Caprimulgi are an order of birds that includes the nightjars.

[3] CD and members of his family had suffered ill health during the past year. Leonard and Emma Darwin fell ill with scarlet fever during the summer of 1862, and Horace Darwin had been ill since January 1862 (see *Correspondence* vol. 10). CD's health had been poor since the end of February 1863.

From Hugh Algernon Weddell 13 May 1863

Poitiers
May 13th. 1863

Dear Sir,

When your letter reached me I was suffering from a slight attack of congestion of the brain and was not able, on that account, to attend immediately to your request;[1] indeed it was only the day before yesterday that I was able to set out in search of your *Ophrys*, but regret now to say that it was in vain. *O apifera* has always been a rare plant in this neighbourhood, but this year more so, it appears, than usual, as not a single individual was to be found. If at any future period I am more lucky, I shall not fail to observe a sufficient number of them as respects their manner of fertilisation and make you acquainted with the results. Indeed I should have done it long since on my own account, had opportunity offered, for I was much struck by the difference of arrangement of pollinia in *O. apifera* & *O. arachnites*, when I read your able dissertation on the subject in the work you were so kind as to send me.[2]

Let me now thank you for your paper on "two forms in Linum".[3] This was quite new to me; but we are becoming acquainted with so many cases of this kind, that I have no doubt many others will still present themselves. You may have seen what I have said on different genera of *Rubiaceæ* where dimorphism of this class is observable (*Hist. nat. Quinq.*, pp. 21, 80 & 82; tab. 3 bis, 22 & 23)[4] I do not, at this moment, call to mind, any other of a remarkable kind, except in *Valerianæ*, cases of which I have figured in the 2 vol. of my flora of the Andes.[5] Many of the so said polygamous flowers in this genus, come, I suspect, under this head. A singular case of an hermaphrodite flower in which fecundation does not take place by the agency of its own stamens is met with in *Cynomorium*. This you will see a mention of in my paper on this plant, a copy of which I had the pleasure of forwarding to you through Baillière some time ago.—[6]

I remain, my dear Sir, Your's very sincerely | H A Weddell.

DAR 110: B60–1

CD ANNOTATIONS
1.1 When your ... Linum". 2.1] *crossed ink*
2.6 *Valerianæ*] *underl red crayon*
2.8 A singular ... Weddell. 3.1] *crossed ink*
Top of letter: 'Rubiaceæ' *blue crayon*

[1] The letter to Weddell has not been found.
[2] The arrangement of the pollinia in *Ophrys apifera* and *O. arachnites* was discussed in *Orchids*, pp. 63–73. CD showed that *O. apifera* was adapted for self-pollination, and *O. arachnites* for insect pollination, a functional difference that led him to conclude that *O. arachnites* was a separate species, and not a variety of *O. apifera* as some botanists had supposed. CD sent Weddell a copy of *Orchids* in 1862; for CD's presentation list for *Orchids*, see *Correspondence* vol. 10, Appendix IV.
[3] Weddell refers to 'Two forms in species of *Linum*'; Weddell's name is included on CD's presentation list for this paper (see Appendix IV).
[4] Weddell 1849. CD had been informed of Weddell's description of dimorphism in *Cinchona*, a member of the Rubiaceae, in 1861 (see *Correspondence* vol. 9, letter to J. D. Hooker, 25 November [1861] and n. 6).
[5] Weddell's observations on the Valerianaceae appear in Weddell 1855–7, 2: 17–34. CD had been informed of this work by Asa Gray (see *Correspondence* vol. 9, letter from Asa Gray, 31 December 1861).
[6] Weddell 1860. The copy presented to CD is in the Darwin Pamphlet Collection–CUL. Hippolyte Baillière was a firm of booksellers and publishers at 219 Regent Street, London, specialising in French medical and scientific texts (*Post Office London directory* 1863).

To James Digues La Touche 14 May [1863?][1]

Down, Bromley, Kent. S.E.
May 14th

Dear Sir

I am extremely much obliged for your very kind note and excellent drawing.[2] The case does interest me *much*.[3] Analogous cases have been recorded; but they are very rare. Your accurate account shows me that here, as (unfortunately) in almost every other case there is a doubt whether these parti-coloured fruits are really due to a bud formed at the point of junction of graft and stock, or whether it be not a variation analogous to a striped petal. In the case of peaches half nectarines, there is no reason to believe that the result is due to grafting; but in some other cases the evidence does point in this direction.[4]

With very sincere thanks for your kindness. Pray believe me, Dear Sir, In Haste | Yours very faithfully | Ch. Darwin

In one account of an apple half and half it is asserted that such fruit never yielded seed though other fruit on tree did yield seed.[5] I presume that it would be impossible for you to enquire from owner of Tree in Canada whether this was case.

Copy
DAR 146: 34

[1] The year is conjectured from the references to bud-variation and anomalous modes of reproduction and variation (see nn. 3–5, below). CD began writing a draft of a chapter on this topic for *Variation* on 21 December 1862 (see *Correspondence* vol. 10, Appendix II, and *Variation* 1: 373–411), and solicited examples from several correspondents. See, for example, letter to Asa Gray, 2 January [1863], letter to Thomas Rivers, 7 January [1863], and letter to John Scott, 21 January [1863].

[2] Neither the letter from La Touche nor the drawing has been found.

[3] CD cited this case in *Variation* 1: 392–3:

> The Rev. J. D. La Touche sent me a coloured drawing of an apple which he brought from Canada, of which half, surrounding and including the whole of the calyx and the insertion of the footstalk, is green, the other half being brown and of the nature of the *pomme gris* apple, with the line of separation between the two halves exactly defined. The tree was a grafted one, and Mr. La Touche thinks that the branch which bore this curious apple sprung from the point of junction of the graft and stock: had this fact been ascertained, the case would probably have come into the small class of graft-hybrids presently to be given. But the branch may have sprung from the stock, which no doubt was a seedling.

[4] CD cited cases of trees producing fruit that were half peach and half nectarine in *Variation* 1: 340–1.

[5] CD refers to a notice that appeared in the *Gardener's Magazine* 13 (1837): 230, concerning apples grown at the Château de Brequigny, France, that were red and acidic on one side, and yellow and sweet on the other. CD's annotated copy of the journal is in the Darwin Library–CUL. The case is described in *Variation* 1: 392.

To Armand de Quatrefages 14 May [1863]

Down. | *Bromley.* | *Kent. S.E.*
May 14th.—

Dear Sir

I send by this Post two Photographs (by Book Post) of the nãta Skull.[1] The delay in getting them made was caused by the Photographer.— Mr. Flower, the Curator of the Museum (whose name you will know in connection with the Brain discussion) has been very kind in aiding me for your sake.[2] He informs me that you could have a Cast made, but it would cost £3"s3"do, & there would be the expence of a large Box, & the cost of carriage.— Perhaps you will not think the skull worth all this expence; but if you do, Mr Flower will get one made & sent to the museum.—

With sincere respect | Dear Sir | Yours very faithfully | Ch. Darwin

Postmark: MY 14 63
Sotheby's, New York (1997)

[1] See letter from Armand de Quatrefages, [28 March –] 11 April 1863.

[2] See letter from W. H. Flower, 9 May 1863, and letter to W. H. Flower, 12 May [1863]. CD refers to the controversy that followed the publication of Owen 1857, in which Richard Owen argued that

humans should be classified as a distinct sub-class of mammals ('Archencephela'), because their brain anatomy differed markedly from that of apes. William Henry Flower showed that the anatomical criteria selected by Owen did not provide a reliable basis for a taxonomic distinction between humans and simians (Flower 1862). On Flower's contribution to the brain controversy, see Rupke 1994, p. 292, and L. G. Wilson 1996b, p. 201.

From La Société des Sciences Naturelles de Neuchâtel[1] 14 May 1863

College de Neuchatel,
le 14 Mai 1863

La Société des Sciences naturelles
de | Neuchatel | Suisse

Désirant s'associer les hommes qui, par leurs travaux & leur position, contribuent à réaliser le but qu'elle se propose, l'advancement des différentes parties de l'histoire de la nature, a élu dans sa séance du 7 Janvier comme son Membre correspondant | Monsieur Charles Darwin | et le prie d'accepter cette nomination comme un témoignage de sa haute considération[2]

Au nom de la Société, | Le Président | Louis Coulon
Le Vice Président | faisant fonction de Sécretaire
pour l'Étrangers | E. Desor

LS
DAR 230

1 For a translation of this letter, see Appendix I. This is a printed letter with the dates, signatures, and CD's name, added by hand.

2 CD's election as a corresponding member to the Société des Sciences Naturelles de Neuchâtel followed a proposal made by the vice-president Edouard Desor, at a meeting of the society on 5 December 1862, that the list of honorary members of the society be augmented to include 'several persons distinguished by their scientific merit'. Desor presented a list of candidates that included, in addition to CD, Charles Frédéric Martins, Ludwig Rütimeyer, and James Dwight Dana (*Bulletin Société des Sciences Naturelles de Neuchâtel* 6 (1864): 271). The society was founded in December 1832 to provide a forum for scientific discussion (*Histoire de l'instruction publique dans le Canton de Neuchâtel*, p. 708).

To J. D. Hooker 15 and 22 May [1863]

Down Bromley Kent.
May 15th

My dear Hooker

Your letter received this morning interested me more than even most of your letters;[1] & that is saying a good deal.—

I must scribble a little on several points. About Lyell & Species you put the *whole* case, I do believe, when you say that he is "half-hearted & whole-headed".[2] I wrote to A. Gray that when I saw such man as Lyell & he refuse to judge, it put me in despair; & that I sometimes thought I sh^d. prefer that Lyell had judged against modification of species, rather than profess inability to decide;[3] & I left him to apply this to himself.— I am heartily rejoiced to hear that you intend to try to bring L. & F. together again: but had you not better wait till they are a little cooled? you will do science a real good service. Falconer never forgave Lyell for taking the Purbeck bones from him & handing them over to Owen.—[4]

I was so glad to see the curious & most beautiful Clianthus (you beggar to make a man's wife & daughter laugh outrageously at him with your "wriggles"):[5] if you sh^d. look at flower again, I think from analogy & from what I saw of the packing of the pollen, that you will find that hairy pistil grows & brushes masses of pollen to apex of keel, & the insects, whilst sucking move the keel up & down & force pollen & stigma on to hairy abdomen: by thus gently working I brushed *all* the pollen out of keel by its apex.—

With respect to Island Floras, if I understand rightly, we differ almost solely how plants first got there:[6] I suppose that at long intervals, from as far back as later Tertiary periods, to the present time plants occasionally arrived (in some cases perhaps aided by different current from existing currents & by former islands) & that the old arrivals have survived little modified on the islands, but have been greatly modified or become extinct on the continents. If I understand, you believe that all islands were formerly united to continents & then received all their plants & none since; & that on the islands they have undergone less extinction & modification than on the continents.— The number of animal-forms on islands very closely allied to those on continents, with a few extremely distinct & anomalous, does not seem to me well to harmonise with your supposed view of *all* having formerly arrived or rather having been left together on the island.—

May 22^d. I have been very bad & chiefly confined to bed;[7] but will amuse myself by writing a little more to you. With respect to Bates & Wallace having distinct views on species during their Journey; what does astonish me is the extreme poverty of observation on this head in Wallace's book; with one discussion on very dissimilar Birds feeding alike showing, as it seemed to me, complete misunderstanding of the economy of nature.[8]

Will you ask Oliver for reference, if he can, to Nageli on relation of vessels & leaves;[9] it is just the point I was driven to.— I failed hopelessly; but found to my surprise that leaves arranged at angles, which do *not* naturally occur, stand in as perfect symmetry with other leaves (viz at intersection of the two diagonals of a rhomb, formed by joining the two two leaves above & two below) as do the leaves placed at the real angles from each other.[10] There is some curious law to be made out about these angles which all go on convergng to an imaginary angle.—

You have given excellent counsel to Bates & I hope he will follow it; what an old malignant fool Dr Grey is; but I never care an atom for his malignacy; it never makes me angry, & I believe your explanation is right; one gets used to it.[11]

Have you seen the Anthropological Review: there are some clever articles; with a fierce attack on Huxley.—[12]

Please remember & tell me name of Hot-House plant (& if possible send flower of) with two coloured anthers.—[13]

All my work is at wretched standstill, with everlasting sickness & devilish head-achs—

Goodnight | My dear old friend. | C. Darwin

The more I think of the "Sad Case" the cleverer it seems.—[14] Where is Huxley: is he alive? I sometimes think that my days of scientific work are very rapidly drawing to a close.— Goodnight dear old friend.—

Endorsement: '/63'
DAR 115: 193

[1] Letter from J. D. Hooker, [13 May 1863].

[2] CD refers to Charles Lyell's discussion of species transmutation in *Antiquity of man* (C. Lyell 1863a). See also letter to Charles Lyell, 6 March [1863] and n. 7, and letter from Charles Lyell, 11 March 1863 and n. 2.

[3] See letter to Asa Gray, 11 May [1863].

[4] Hugh Falconer and Richard Owen. Bones of early fossil mammals from the middle Purbeck beds at Durlstone Bay near Swanage, Dorset, were excavated in the 1850s, the first find being described by Owen (Owen 1854). In 1857, Lyell was instrumental in encouraging Samuel Husband Beckles to undertake further excavations of the beds (see C. Lyell 1857, pp. 13–14, and K. M. Lyell ed. 1881, 2: 238–9). Beckles sent the fossils directly to Lyell as they were unearthed, and Lyell transferred these and other finds from the area to Falconer, who made the initial determinations (see Falconer 1857b, p. 261, and *Correspondence* vol. 6, letter to J. D. Dana, 5 April [1857] and nn. 7–8). It was intended that when the collection was completed the Purbeck fossils would be transferred to Owen for description and publication (Falconer 1857b, p. 262). Lyell apparently transferred the fossils to Owen earlier than had been arranged. Owen published descriptions of the fossils in 1871 (Owen 1871).

[5] See letter from J. D. Hooker, [13 May 1863]. CD refers to *Clianthus dampieri*, an Australian legume, of which Hooker sent a specimen for CD to examine the pollination mechanism. Hooker wrote that he supposed that it required a 'Darwinian waggle of the keel' for the pollen to adhere to the stigma in this species. CD's 'wriggling' out of theoretical difficulties had been a joke between the two friends over a number of years (see, for example, letter from J. D. Hooker, [28 March 1863] and n. 3).

[6] See letter from J. D. Hooker, [13 May 1863] and n. 20.

[7] Emma Darwin's diary (DAR 242) for 1863 records: 'Ch. taken with pain in bowells' (15 May); 'very unwell' (16 May); 'Ch very unwell with pain' (17 May); 'pain left & sickness came on' (19 May); 'Ditto' (20 May); 'sick in mg' (21 May).

[8] Henry Walter Bates and Alfred Russel Wallace went on an expedition to South America in 1848 to gather species and collect natural history specimens; Wallace returned to England in 1852, after leaving Bates's expedition in 1850. Bates returned in 1859. Bates's narrative of his travels (Bates 1863) was notable for its use of Darwinian theories of natural selection (see letter to J. D. Hooker, [17 April 1863] and n. 15, and letter from J. D. Hooker, [13 May 1863]). CD also refers to Wallace 1853, pp. 83–5.

[9] See letter from J. D. Hooker, [13 May 1863] and n. 13. Daniel Oliver was librarian and assistant in the herbarium at the Royal Botanic Gardens, Kew, of which Hooker was assistant director (R. Desmond 1994, *List of the Linnean Society of London*). The reference is to the first part of Nägeli 1858–68, pp. 39–156.

[10] CD also discussed this observation with Asa Gray (see letter to Asa Gray, 11 May [1863]). CD had been pursuing research on phyllotaxy since February 1863; his experimental notes are in DAR 51: 1–32. See also memorandum from G. H. Darwin, [before 11 May 1863].

[11] See letter from J. D. Hooker, [13 May 1863] and n. 25. John Edward Gray was the keeper of the zoological collections at the British Museum.

[12] The first number of the *Anthropological Review* appeared in May 1863 and carried an anonymous review of Thomas Henry Huxley's *Man's place in nature* (T. H. Huxley 1863b), and a pseudonymous letter to the editor regarding Huxley's work from the physical anthropologist Charles Carter Blake (*Anthropological Review* 1 (1863): 111); both adversely criticised Huxley, and Blake's contribution was particularly hostile (*Anthropological Review* 1 (1863): 107–17, 153–62). The *Anthropological Review* was the journal of the Anthropological Society of London, which, as a body, tended towards anti-Darwinian and polygenist beliefs (see Stocking 1987, pp. 248–52).

[13] CD refers to *Lagerstroemia* (see letter from J. D. Hooker, [24 May 1863]). CD wanted to pursue his research on dimorphism by experimenting on a plant possessing two differently coloured sets of anthers, having hypothesised that the structure and colour of stamens might be a better guide to the occurrence of functional dimorphism than the length of stamens and pistils alone (see *Correspondence* vol. 10, letter to J. D. Hooker, 22 [August 1862] and n. 14, and letter to Daniel Oliver, 2 September [1862], and this volume, letter to J. D. Hooker, 23 April [1863]). CD's experimental notes on *L. indica*, dated August 1863, are in DAR 109: B116–17, and his findings are summarised in DAR 27.2: 17 v. These observations were published in *Forms of flowers*, pp. 167–8.

[14] The reference is to a squib (Anon. 1863a), part of which appeared in *Public Opinion* 3 (1863): 497–8 (see Appendix VIII).

From Isaac Anderson-Henry 16 May 1863

Hay Lodge, | Trinity, | Edinburgh.
May 16/63

My dear Sir

I hasten to say in reply to your welcome note[1] that the Sprig of *Linum* with bloom enclosed, is of all others I tried last Summer the most intractable of its race—the *L. luteum Corymbiflorum*. You are quite correct about its being *dimorphic*.[2] You have yours ahead of mine, which tho' in my Greenhouse, is only coming into bloom buds. It is a pity they won't all flower together. *L. flavum* with me will be over ere I have one to try with him, unless I get pollen off your *L. luteum*.[3]

Very faithfully yours | Is. Anderson Henry

I have just come in from emasculating among the rubus.[4] I saw at Dr Grevilles this day week what he told me was the old *haut bois* strawberry—[5] There seems to be a touch of the raspberry in its aspect—which may induce me to try a union there—I have several odd XX among strawberries to be tested this Summer—[6] *[illeg]*

Incomplete
DAR 110: 24

CD ANNOTATIONS

1.1 I hasten ... race 1.2] *crossed ink*
1.2 the *L. Luteum* ... *dimorphic*. 1.3] *enclosed in square brackets, ink*
1.3 You have ... him, 1.6] *crossed blue crayon*
1.6 unless ... *luteum*.] 'Have stigmas rotated in Fig?' *pencil*

[1] The letter to Anderson-Henry has not been found.

[2] CD classified *Linum corymbiferum* as heterostyled in *Forms of flowers*, p. 100.

[3] Anderson-Henry had offered to make crossing experiments with *Linum* in his letter to CD of 17 January 1863. CD hoped that Anderson-Henry's work would throw further light on the fertility of dimorphic

species of *Linum*, and he sent a copy of 'Two forms in species of *Linum*' in April 1863 to assist him (see letters from Isaac Anderson-Henry, 17 January 1863 and 24 April 1863, letter to Isaac Anderson-Henry, 20 January [1863], and Appendix IV).

4 Emasculation involves the removal of the anthers of a flower to prevent either self-pollination or the pollination of surrounding plants (*Penguin dictionary of botany*). See also n. 6, below.

5 Robert Kaye Greville, a botanist and entomologist, was honorary secretary of the Botanical Society of Edinburgh (R. Desmond 1994, *DNB*, *Medical directory* 1863).

6 Anderson-Henry had been attempting to hybridise raspberries and blackberries, and raspberries and strawberries (see letters from Isaac Anderson-Henry, 17 January 1863, 26–7 January 1863, and 7 May 1863 and n. 9).

From W. D. Fox [16–22 May 1863][1]

My dear Darwin

About two months since you wrote me you were suffering much—[2] I hope you are now quite yourself again. Do you ever see "the illustrated Times Newspaper". I was much amused by that of May 2–9–16 containing a quizz upon you, which I yesterday met with at my Neighbours.[3] If you have not seen this additional proof of Natural developement, I will send the papers to you. They will give you ten minutes amusement.

I hear Dr Gullys brain has quite broken down and disabled him from work.[4] He told me when I last saw him that it must do so soon.

He will be a great loss to a vast number of people. There is no one at Malvern fit to take up his place I fear. Do you ever see Dr Lane now? I think I saw an advertisement that he was practising near Kew— am I right.[5]

With kindest regards to Mrs Darwin[6] | Ever your aff Cousin | W D Fox

DAR 164: 175

1 The date range is established by the reference to the issues of the *Illustrated Times* for 2, 9, and 16 May 1863 (see n. 3, below), and by the relationship between this letter and the letter to W. D. Fox, 23 May [1863].

2 See letter to W. D. Fox, 23 March [1863].

3 Fox refers to three humorous cartoons and a poem, 'dedicated by natural selection to Dr. Charles Darwin' (*Illustrated Times*, 2 May 1863, p. 317, 9 May 1863, p. 333, and 16 May 1863, p. 348). These cartoons were the first instalments in a twenty-part series (see letter from Roland Trimen, 16, 17, and 19 July 1863 and n. 22).

4 James Manby Gully ran a hydropathic establishment at Great Malvern, Worcestershire, which both CD and Fox had attended (see *Correspondence* vols. 4–7 and Browne 1990). In 1863, Gully suffered from 'a severe (unspecified) illness', from which he recovered in June (Colp 1977, p. 217 n. 37). CD had been planning to visit Great Malvern again for hydropathic treatment (see letters to W. D. Fox, 9 March [1863] and 16 [March 1863]).

5 Edward Wickstead Lane had been proprietor of the Moor Park hydropathic establishment near Farnham, Surrey; CD had undergone hydropathic treatment there in 1857 (see *Correspondence* vol. 6). In 1860, CD attended Lane's new hydropathic establishment at Sudbrook Park, near Richmond, Surrey (see *Correspondence* vol. 8); Kew is about two miles north-east of Richmond.

6 Emma Darwin.

From Lydia Ernestine Becker 18 May 1863

Miss Becker presents her compliments to Mr. Darwin and takes the liberty of sending him the enclosed flowers of a variety of Lychnis dioica common in the woods here but which she has not observed elsewhere.[1] It has bisexual flowers and large dark purple anthers which give the plant a very striking handsome appearance. The same conspicuous anthers occur in flowers bearing stamens only of which one is sent. Miss Becker does not know whether such a variety as this would be interesting to Mr. Darwin, if not she must apologise for having troubled him.

Altham | Accrington | Lancashire
May 18th. 1863.

DAR 160: 108

CD ANNOTATION
2.1 Altham] 'Altham' *ink*

[1] Becker lived in Altham, Lancashire, between 1838 and 1865, during which time she developed an interest in botany (*DNB*). It was later established that these plants represented a diseased form of *Lychnis dioica* (*Silene dioica*) and not a new variety (see letter from L. E. Becker, 23–4 May [1863] and CD note, letter from L. E. Becker, 28 May [1863], and Becker 1869). When Becker published her observations she substituted the name *L. diurna*, a synonym for *L. dioica* (*Index Kewensis*).

From Armand de Quatrefages[1] 19 May [1863][2]

Paris
19 mai 6⟨3⟩

Monsieur et cher confrere

Mille remerciements pour votre photographie.[3] Elle a dépassé mon attente et je suis curieux de voir ce qu'en diront les partisans de la fixité absolue que j'ai à combattre en ce moment.[4] Je l'ai reçue il y a deja quelques jours et aurais voulu vous remercier sur le champ mais vous savez au milieu de quelles grandes affaires je me trouvais jeté. *Le procès de la machoire*, comme dit Mr Carpenter, me préoccupait entierement.[5] J'ai été profondément touché de la conduite de vos compatriotes dans cette circonstance. Il est impossible d'apporter dans l'examen d'une question très vivement controversée plus de franchise, de loyauté et de cordialité. J'espere qu'ils nous auront trouvés animés des mêmes sentiments. Je me suis fait un devoir et un plaisir à la fois d'exprimer hier à l'Académie tout ce que je pense à cet égard.[6]

J'ai depuis 4 jours sur mon bureau, plié à votre adresse, un exemplaire des deux photographies de la fameuse machoire en véritable étourdi je n'ai pas songé à la remettre à Mr Christy qui emportait le semblable pour Mr Falconer.[7] J'ai cru vous l'avoir deja envoyé mais ce ne sera qu'un court retard.

Adieu Monsieur et cher confrère et merci encore. Votre bien devoué | De Quatrefages

DAR 175: 1

[1] For a translation of this letter, see Appendix I.
[2] The year is established by the relationship between this letter and the letter to Armand de Quatrefages, 14 May [1863].
[3] CD sent Quatrefages two photographs of a skull of the niata breed of cattle. See letter to Armand de Quatrefages, 14 May [1863].
[4] Quatrefages was engaged in a debate at the Société d'Anthropologie de Paris with André Sanson, who challenged him to cite a single species in which the action of the environment had produced anatomical modifications (see letter from Armand de Quatrefages, [28 March –] 11 April 1863 and n. 4). Quatrefages responded with CD's example of the niata cattle (*Journal of researches* 2d ed., pp. 145–6) and, on 2 July 1863, exhibited the photographs that CD had sent to him. On 16 July 1863, Quatrefages continued to expand his case for the environmental modification of species by reading from *Journal of researches* 2d ed., pp. 145–6, and from the letter to Armand de Quatrefages, [14 April 1863] (*Bulletin Société d'Anthropologie de Paris* 4 (1863): 350–3 and 376–80). For Sanson's response to Quatrefages, see *Bulletin Société d'Anthropologie de Paris* 4 (1863): 374–5 and 380–3.
[5] William Benjamin Carpenter's phrase the 'trial of the jaw' was used to describe the inquiry into the authenticity of the fossil human jawbone and other remains discovered at Moulin-Quignon, near Abbeville, France, in March 1863 by Jacques Boucher de Perthes (Falconer *et al.* 1863, p. 423). See also n. 6, below. For a summary of this dispute, see the letter from J. D. Hooker, [7 May 1863], n. 5.
[6] A conference of French and British scientists on the Moulin-Quignon archaeological finds took place in Paris and Abbeville from 9 to 13 May 1863. Quatrefages paid tribute to the British scientists at the meeting of the Académie des Sciences in Paris on 18 May 1863 (*Comptes rendus hebdomadaires des séances de l'Académie des Sciences* 56 (1863): 933–5. The British scientists included Carpenter, Hugh Falconer, Joseph Prestwich, and George Busk. The French scientists included Quatrefages, Edouard Lartet, and Alphonse Milne Edwards. A summary of the conference proceedings was published in the *Athenæum*, 23 May 1863, p. 682. For a full report of the proceedings, see Falconer *et al.* 1863.
[7] Henry Christy was an English ethnologist, archaeologist, and geologist who worked closely with the French palaeontologist and archaeologist Lartet; together they excavated a number of European prehistoric sites in the 1860s (*DNB*). Lartet and Falconer were both involved in determining the authenticity of the Moulin-Quignon jawbone (see nn. 5 and 6, above).

From L. E. Becker 21 May [1863][1]

Altham | Accrington.
May 21

Sir

I have this day forwarded to you a small box containing plants of Lychnis which I hope will reach you tomorrow in tolerably good condition.[2] I have also enclosed a tin canister of flowers and I hope there will be pollen enough to enable you if you please to try the experiment of fertilising with it some of the plants grown in your neighbourhood. Since I wrote last[3] I took my note book into a different part of the wood and examined 122 plants of Lychnis with ⟨the⟩ following result—

Common form. males 41 females 23. hermaphrodite 29 males with dark stamens. 29 I pulled to pieces numbers of the flowers with long p⟨is⟩til but never found the stamens developed in them. The stamens are normally of medium length but in some plants they are very long, in others very short, the flowers on each plant are generally alike in this as in other respects.

With many apologies for troubling you by writing so much about these plants I remain Sir | Yours very respectfully | L. E. Becker

DAR 160: 106

CD ANNOTATIONS
2.1 Common form. . . . in them. 2.3] *scored ink*
2.3 in some . . . Sir 3.2] *scored ink*
End of letter: '(Counting all, except hermaphrodites, together we get proportion of Hermaphrodites.)' *ink*

[1] The year is established by the relationship between this letter and the letter from L. E. Becker, 18 May 1863.
[2] The reference is to a form of *Lychnis dioica* (*Silene dioica*) that Becker believed to be a new variety (see letter from L. E. Becker, 18 May 1863 and n. 1).
[3] See letter from L. E. Becker, 18 May 1863.

From George Bentham 21 May 1863

25 Wilton Place, | S.W.
May 21/63

My dear Darwin

I return you with a thousand thanks (p[r] Bookpost) your pamphlets—[1] I have read them all—with several others—till I got quite bewildered. My object was to say a few words in my address on the present state of the question and on some logical confusion in the arguments it has given rise to and my wish has been to do you full justice but not being gifted with your powers of expressing thoughts I fear I have only talked nonsense especially as age is beginning to tell upon me— you must look upon it however with an indulgent eye—and I sincerely trust you will find nothing in it really to annoy you.— I will send you a copy when printed[2]

I had intended saying something of what *I* considered the weak points of your hypothesis (not in its principle but in the generality of its application) but I found I could not give the necessary time and thought to it. There is one thing that I would wish you would further work out— What is that principle to a certain degree counteracting divergent variation which keeps certain species immutable for periods which further researches only lengthen— Not to speak of the Antiquity of Man How to account for the absolute identity (I mean identity within present small limits of variation) of species of plants of the temperate northern hemisphere and of Tasmania and the Victorian Alps when first discovered—plants that must have gone through so many thousand generations in both hemispheres unaltered[3] It is too much for me to suppose that Natural Selection has had no opportunity for acting upon these when others which appear to have been in similar circumstances have by her agency altered so much that the common origin of northern and southern representatives is difficult to recognise

The observed immutability of a large number of species (if taken within extended limits) which I had so long maintained continually haunts me although at the same time I feel the full force of your principle of Natural Selection the moment I divest it of that figurative personification which leads Asa Gray to say that your book on Orchids introduces the doctrine of final causes into the vegetable kingdom.[4] I feel that I am one of your converts but I cannot satisfy myself that I am right at all points, and therefore cannot go all lengths with you.

I agree with your notes on the merits of some of your reviewers except that I cannot see much to approve either in Hopkins or Maw.[5] I am amused by Huxley's publications, admire his raciness of style engrafted upon solidity of thought and correctness of views[6] I cordially agree in his opinions of your works but I have a great dislike to personal controversy and knock-down arguments. I have it is true been unable myself to refrain from a word or to on the unfairness of the Athenæum reviewer who quotes as authorities Pouchet and C° and ignores Pasteur but that is supposed to be anonymous and anonymous misrepresentations delivered ex cathedra ought I think to be put down by facts authenticated by a signature[7]

I shall endeavour as I get on with the Australian Flora to put together some notes on the comparative distribution of Australian British and other plants which I have specially worked upon—although Hooker has really done so much as to discourage one from pursuing the subject and some of my own ideas for instance as the connection between Europe and America through NE Asia have been taken up and worked out by more competent hands—so that I may after all be left to my old plodding task of systematic description which I have now carried on for nearly 40 years[8]

Ever yours most sincerely | George Bentham

DAR 160: 157

CD ANNOTATION

3.1 The observed . . . haunts me 3.2] *double scored brown crayon*

1 See letter to George Bentham, 15 April [1863] and nn. 3–5 and 7–10.

2 As president of the Linnean Society, Bentham was obliged to present an address to the society at its anniversary meeting, which in 1863 was held on 25 May. Bentham's address dealt in part with the reception of *Origin* and the influence of CD's hypotheses on the progress of natural history (Bentham 1863). Bentham sent CD a preprint copy of his address in June 1863 (see letter to George Bentham, 19 June [1863]; it was not published until 29 October 1863 (*General index to the Journal of the Linnean Society*).

3 Joseph Dalton Hooker had drawn attention to the number of European genera and species found in Tasmania and the Australian Alps, Victoria, in his introductory essay to the flora of Tasmania (J. D. Hooker 1859, pp. lxxxiv–lxxxv). Bentham had been preparing the first volume of his seven-volume *Flora Australiensis* (Bentham 1863–78; see n. 8, below), and thus had specialist knowledge of the distribution of European species in the region.

4 Bentham refers to *Origin* and to A. Gray 1862b, pp. 428–9. See also *Correspondence* vol. 10, letter to Asa Gray, 26[–7] November [1862], n. 32, and Bentham 1863, p. xvii.

5 Bentham refers to Hopkins 1860 and [Maw] 1861; see letter to George Bentham, 15 April [1863].

6 T. H. Huxley 1863a and 1863b.

7 See Bentham 1863, pp. xxv–xxvi. Bentham refers to Richard Owen's anonymous review of William Benjamin Carpenter's *Introduction to the study of Foraminifera* (Carpenter 1862), and his anonymous reply to CD's letter to the *Athenæum*, 18 April 1863, published in the *Athenæum* on 28 March 1863 and 2 May 1863, respectively (see Appendix VII). Owen commented favourably on the work of the French biologist Félix Archimède Pouchet (Pouchet 1858 and 1859), which purported to demonstrate the existence of spontaneous generation, or the appearance of living forms directly from non-living matter. Pouchet's work was refuted by the experiments of Louis Pasteur (Pasteur 1861), which, in showing that contaminated mercury in the experimental apparatus was the source of the micro-organisms detected by Pouchet, appeared to close the spontaneous generation debate in France (see Farley 1974).

8 The first volume of Bentham's *Flora Australiensis* (Bentham 1863–78) was published on 30 May 1863 (*Taxonomic literature*). Bentham refers to Hooker's essay on the flora of Australia (J. D. Hooker 1859; see n. 3, above), and to Asa Gray's study of the relationship between the plants of Japan and North America (A. Gray 1858–9). Gray's hypothesis was that an ancient connection between western North America and Eastern Asia enabled the passage of North American plants across Asia to Western Europe. The evidence in support of this hypothesis had recently been reviewed by Daniel Oliver (Oliver 1862c). See Bentham 1863, p. xxiii.

From John Scott 21 May [1863][1]

Botanic Gardens [Edinburgh]
May 21st.—

Sir,

I regret to find by your last that your health is still very weak.[2] I sincerely trust you have derived more benefit by your short absence from home than you anticipated. I was glad to hear that Orchid pods arrived safe.[3] I thank you for remittance of postage stamps; pray never mind anything of the kind in future; I am equally anxious with you that all should arrive safe, and it gratifies me much to have aught to send which is likely to interest you.

I thought you would be somewhat surprised when you heard the plan I had taken in certain cases to fertilise the Acroperas & Gongoras.[4] In some of my late experiments on *G. truncata*, I cut off not only the *clinandrum* but an eighth of an inch from the extremity of the stigma, before inserting pollinia; these capsules are now swelling rapidly. These plants must certainly present less contracted stigmatic chambers in their native haunts or at least secrete, as I formerly suggested, more viscid matter. But even granting this, I am quite at a loss to understand how the pollinia are applied to stigma! If ever I am fortunate enough to be amongst them in their native haunts, I shall certainly embrace every opportunity to satisfy myself as to Nature's *modus operandi*.

In respect to *Drosera* paper, I am sorry to say that it has not been published. Perhaps it might have found a place in the *Proceedings* of the *Bot. Soc. here*, if there had been no connection between the latter & the *Edin. New. Phil. Jour.*. But not being a paper adapted for the latter, it is consequently excluded from both—excepting the short notice you saw in *Gard. Chron.*—the one as respects this being simply a repetition of the other.[5] I will thus, I fear, be unable to send you a printed copy of it, as I have no acquaintance with any Editors likely to receive such contributions;

however, I will if you like send my manuscript, and you will then see if it contains aught worthy your notice.

I have read paper on *Sterility* of *Orchids*;[6] it appeared to be pretty well received by the Society in general: in the remarks upon it by Professors, Balfour, & Maclagan, no reference whatever was made to the opinions occasionally deduced from experiments.[7] I suppose this was due to the decided tendency I exhibited throughout to support your views.[8] The Edin. Phil. Jour. will be published in the beginning of July, & I will then send you a copy:[9] I will be impatient to have your opinion upon it. I have ventured in a a few preliminary remarks to refer to Prof. Huxley's "*absent link*" mentioned in "Mans Place";[10] and have there attempted to show that such cases of sterility as those stated in paper & others which I referred to in your writings connected indirectly the broken chain, and seemed to leave no difficulties to those of us who like Prof. Huxley needed only the above link to render your theory "*amply competent*" &c &c.[11]

By the way I enclose a copy of *note* in my paper on *closing of stigmatic orifice in Orchids*[12] I have drawn attention to an observation of yours in '*Orchid Fertilisation*" which has always struck me as very remarkable Indeed I have often wondered how it has escaped the attention of your numerous reviewers.[13] You will see that the phenomenon as presented by my experiments, is more strikingly adapted for the end in view, than in the *Bolbophyllum* case: fertilisation being dependent on external agents; Now it is evident that if stigmas close at definite periods, Nature's design may be frequently frustrated; whereas, when this is dependent on the mutual action of the organs which accomplish the latter, it is a most beautiful provision. Have you ever observed the phenomenon in any other Orchid yet? It seems most anomalous & exceptional; If you have not ⟨*several pages missing*⟩

am glad to say it has now done. I must state—a slight though in a certain sense an unimportant—peculiarity it has now presented. This is in regard to the style which varies a little in length: in some flowers it is exactly the same length as stamens, the latter reaching or nearly the mouth of corolla-tube: *these flowers produce nice plump capsules*. Other flowers again, have stigma protruded beyond the corolla-tube, and are accordingly perfectly sterile when the flower is protected. **Artificial** *fertilisation*, however, shows that they are *equally* as **fertile** as the *others*: irrespective then of this peculiarity—certainly important as affording an opportunity for crossing—as already stated, the difference is unimportant. I am trying this species also with others presenting the two forms. In my observations on *Primula*, I have found five distinct plants of *cultivated Auricula* with *stamens* & *pistils equal*; but these as in **case** alluded to by you in *Primula* paper;[14] have pollen-grains as small or even smaller than those from *normal short-styled forms*. It is different, however, with a *cultivated cowslip*, which I am at present experimenting upon, with *stamens* & *pistils equal*.[15] In this the *anthers* approach corolla mouth, *stigmas* reaching the **middle** of *anthers*. It is also producing *fine plump capsules*. I have examined *pollen-grains* & *stigmas* The former are as large as any I have yet observed on *short-styled* plants; stigmas *similar* in *form* to *those* of *long-styled* **plants**; **papillæ** *shorter* and more like those of *short-styled* forms

I am likewise trying results of *crosses between these* & the *long* & *short-styled forms*. I will of course detail more particularly my Primula work at some future time: it is not sufficiently advanced yet.[16] The **Passifloras** are now beginning to flower: so I will soon be able to commence work upon them.[17]

In conclusion allow me to present my most sincere thanks for the encouraging remarks you have been pleased to make on my occasional observations in your last.[18] I cannot, indeed, thank you as I would like to do, for the honour you have done me. Pray, excuse my inability and rest assured that the stimulus you have thus given shall be truly excitative, and make me strive more & more to improve those faculties, and to overcome other mental defects, or rather deficiencies, which a stern and inflexible Nature has been so niggardly in her bestowment. I used to entertain foolish ideas in respect to this: I thought in fact that Nature had formed me as it were in very mockery, endowed me with an insatiable desire for knowledge, & yet so constituted my mental nature, that it seemed all but unattainable. Now, however, I look upon myself as the subject of general, not *special* laws, so banishing these hard thoughts of Nature, my earnest endeavour is now to constantly strive to make the most of all.

I remain | Sir | Yours very respectfully | John Scott

[Enclosure][19]

⟨Mr. D⟩arwin records the following very remarkable ⟨observa⟩tion on the closing of the stigmatic orifice in *B. rhizopharæ*;— "Orchid fertilisation" p. 170.[20] "After the ⟨fl⟩ower has remained some time open, the sides of ⟨the⟩ oval orifice of the stigmatic chamber close in & shut it completely,—a fact which I have observed in no other Orchid, and which, I presume, is here related to the much exposed condition of ⟨the who⟩le flower." This case then of Mr. Darwin's ⟨differs⟩ from that above mentioned, and from all ⟨ot⟩hers which I have observed, in requiring no *external stimulus* for its excitation![21] *After* the application of pollen to the stigma, I have observed it in a number of species from different tribes; in *absence* of the pollen, however, it remained *open* to the *last*. There is thus in the closing as here manifested; an almost conscious sympathetic action—a case that the grand end of vegetal life is not too readily frustrated by an oversensitiveness of the stigmatic chamber; inasmuch as I find, that, when pollen is applied from widely distinct genera⟨, the⟩ orifice ⟨does not⟩ move. As illustrating this⟨, I⟩ applied pollinia from species of *Cypri⟨pedium⟩* and *Asclepias* to flowers of *Tric⟨hopilia tortilis*;⟩ nevertheless, though pollen tubes we⟨re emitted,⟩ the stigmatic orifice did not close, which ⟨it⟩ invariably does eighteen hours or so after the application of its own pollen. In respect ⟨to the⟩ suggestion of Mr. Darwin as to the special en⟨d⟩ subserved by the closing of the stigmatic chamber in *B. rhizophoræ*; I may remark that in m⟨any⟩ of ⟨the⟩ species exhibiting the phenomenon, ⟨it is evidently⟩ subservient to a highly beneficial end in affo⟨rd⟩ing more genial conditions for the perfect development of the pollen-tubes; than could otherwise be afforded in the naturally upturned and fully exposed condition of their stigma⟨s⟩ In others, however, it must be admitted that no such relation is evident: species whose stigmatic chamber closes being equally as well protected from physical injuries, as others in which it remains open. This, however, may be simply due to inheritance with modification; at least, it will be so regarded by those of us who believe that species are the modified descendants of previously existing species.

Incomplete
DAR 108: 181, DAR 177: 88

CD ANNOTATIONS
1.1 I regret . . . have not 5.11] *crossed pencil*
6.1 I must . . . *short-styled forms.* 6.13] *crossed red crayon*
6.10 I have found . . . *forms.* 6.13] *enclosed in square brackets, red crayon*
6.12 *Primula*] *underl red crayon*; 'see' *interl red crayon*
6.13 *normal*] *after interl pencil* 'short [stamens] of'
6.13 *short-styled*] 'Long' *added & del pencil, above* 'short'
7.1 I am likewise . . . them. 7.4] *crossed pencil*
8.1 In conclusion . . . Scott 9.1] *crossed ink*

[1] The year is established by the reference to Scott 1863a (see n. 6, below), which was delivered before the Botanical Society of Edinburgh on 14 May 1863.

[2] See letter to John Scott, 2 May [1863].

[3] See letter to John Scott, 2 May [1863] and n. 4.

[4] See letter to John Scott, 2 May [1863] and n. 5.

[5] Scott read a paper on the propagation and irritability of *Drosera* and *Dionaea* before the Botanical Society of Edinburgh on 11 December 1862; it was never published in full, but abstracts appeared in the *Transactions of the Botanical Society* [*of Edinburgh*] (Scott 1862b), the *Edinburgh New Philosophical Journal* n.s. 17 (1863): 317–18, and the *Gardeners' Chronicle*, 10 January 1863, p. 30.

[6] Scott read a paper entitled 'Experiments on the fertilisation of orchids in the Royal Botanic Garden of Edinburgh' before the Botanical Society of Edinburgh on 14 May 1863 (Scott 1863a). The paper was based on experiments described in the letter from John Scott, [1–11] April 1863. Scott sent CD an abstract of the paper with his letter of 28 May [1863] (see n. 9, below).

[7] John Hutton Balfour was professor of botany at Edinburgh University and one of the vice-presidents of the Botanical Society of Edinburgh (*DNB*, *Medical directory for Scotland* 1863); Andrew Douglas Maclagan was professor of medical jurisprudence and public health, and president of the Botanical Society of Edinburgh (*Edinburgh New Philosophical Journal* n.s. 18 (1863): 399; *Medical directory for Scotland* 1863). On Scott's experiments and the conclusions deduced from them, see n. 11, below.

[8] Balfour was known to have anti-Darwinian views (see letter from J. D. Hooker, 10 June 1863).

[9] An abstract of Scott 1863a appeared in the October 1863 issue of the *Edinburgh New Philosophical Journal*. Abstracts were also published in the *Edinburgh Evening Courant*, 28 May 1863, p. 8, and the *Gardeners' Chronicle*, 13 June 1863, p. 558. Scott sent a copy of the summary that appeared in the *Edinburgh Evening Courant* with his letter to CD of 28 May [1863]. See also letter from John Scott, 23 July [1863] and n. 19.

[10] In Thomas Henry Huxley's *Evidence as to man's place in nature* (T. H. Huxley 1863b, p. 107), it was argued that:

> our acceptance of the Darwinian hypothesis must be provisional so long as one link in the chain of evidence is wanting; and so long as all the animals and plants certainly produced by selective breeding from a common stock are fertile, and their progeny are fertile with one another, that link will be wanting. For, so long, selective breeding will not be proved to be competent to do all that is required of it to produce natural species.

Scott discussed this statement in Scott 1863a, pp. 543–5, arguing that *Origin* and 'Dimorphic condition in *Primula*' provided evidence that answered Huxley's objection. See n. 11, below.

11 Scott's paper (Scott 1863a) demonstrated the sterility of several species of *Oncidium* when flowers were pollinated with pollen from the same plant ('individual sterility'). The paper also demonstrated that the same plants could be successfully pollinated with pollen from a different species of *Oncidium*. Scott argued that as there were varying degrees of sterility in the vegetable kingdom, which did not necessarily conform to recognised systematic affinities, Huxley's use of sterility as the crucial test of origin of species by natural selection was flawed (Scott 1863a, pp. 543–5). For a discussion of the background to this debate, see also *Correspondence* vol. 10, Appendix VI.

12 The note was subsequently published in Scott 1863a, pp. 546–7 n. Scott's observations appeared to challenge CD's conclusions in *Orchids*, p. 170, regarding the closing of the stigmatic orifice in *Bolbophyllum rhizophorae* (see letter to John Scott, 25 and 28 May [1863] and n. 2).

13 For a list of reviews of *Orchids* published in 1862, see *Correspondence* vol. 10, Appendix VII.

14 Scott refers to CD's observations on an equal-styled specimen of *Primula auricula* in 'Dimorphic condition in *Primula*', p. 80 (*Collected papers* 2: 48).

15 Scott detailed his experiments and observations on this plant in Scott 1864a, pp. 92–7. See also letter from John Scott, [3 June 1863].

16 Scott's observations on and experiments with dimorphic and non-dimorphic species of *Primula* were published in Scott 1864a. See also letter to John Scott, 25 and 28 May [1863].

17 CD had suggested in 1862 that Scott experiment on *Passiflora* (see *Correspondence* vol. 10, letters to John Scott, 19 November [1862], and 11 December [1862] and n. 21). CD wanted to corroborate statements that some species of *Passiflora* could be pollinated more readily by different species than by their own pollen; Scott's results, which provided the experimental evidence, were published in Scott 1864d.

18 See letter to John Scott, 2 May [1863]. See also letter to John Scott, 12 April [1863].

19 See n. 12, above. The manuscript of the enclosure is damaged; missing words have been inserted from Scott 1863a, pp. 546–7 n.

20 *Orchids*, p. 170.

21 Scott refers to the case of *Oncidium sphacelatum* described in Scott 1863a, p. 545; he reported that he had observed its stigmatic orifice closing approximately twenty-four hours after the application of its own pollen.

To George Bentham 22 May [1863][1]

Down. | Bromley. | Kent. S.E.
May 22[d]

My dear Bentham

I am much obliged for your kind & interesting letter.[2] I have no fear of anything that a man like you, will say annoying me in the very least degree.[3] On the other hand any approval from one, whose judgment & knowledge I have for many years so sincerely respected, will gratify me much.— The objection, which you well put, of certain forms remaining unaltered through long time & space, is no doubt formidable in appearance & to a certain extent in realility according to my judgment.—[4] But does not the difficulty rest much on our silently assuming that we know more than we do? I have literally found nothing so difficult as to try & always remember our ignorance. I am never weary when walking in any new adjoining district or country of reflecting how absolutely ignorant we are why certain old plants are not there present, & other new ones are & others in different proportions. If we once fully feel this, then in judging the theory of natural selection, which implies that a form will remain unaltered unless some alteration be to its benefit, is it so very wonderful that some forms should change much slower & much less, & some few should have changed not at all under conditions which to us (who really know nothing what are the important conditions) seem very different.— Certainly a priori we might have anticipated that all the plants anciently introduced into Australia would have undergone some modification; but the fact that they have not been modified does not seem to me a difficulty of weight enough to shake a belief grounded on other arguments.— I have expressed myself miserably, but I am far from well today.—

I am very glad that you are going to allude to Pasteur; I was struck with infinite admiration at his work.—[5]

With cordial thanks, believe me Dear Bentham | Yours very sincerely | Ch. Darwin

In fact the belief in natural selection must at present be grounded entirely on general considerations. (1) on its being a vera causa, from the struggle for existence; & the certain geological fact that species do somehow change (2) from the analogy of change under domestication by man's selection. (3) & chiefly from this view connecting under an intelligible point of view a host of facts.—

When we descend to details, we can prove that no one species has changed: nor can we prove that the supposed changes are beneficial which is the groundwork of the theory. Nor can we explain why some species have changed & others have not. The latter case seems to me hardly more difficult to understand precisely & in detail than the former case of supposed change. Bronn may ask in vain the old creationist school & the new school why one mouse has longer ears than another mouse—& one plant more pointed leaves than another plant.—[6]

Royal Botanic Gardens, Kew (Bentham correspondence 3: 711–13)

[1] The year is established by the relationship between this letter and the letter from George Bentham, 21 May 1863.

[2] Letter from George Bentham, 21 May 1863.

[3] Bentham was preparing his anniversary address as president of the Linnean Society, to be delivered on 25 May 1863 (Bentham 1863). See letter from George Bentham, 21 May 1863.

[4] See letter from George Bentham, 21 May 1863.

[5] See letter from George Bentham, 21 May 1863 and n. 7. Although there are no publications by Louis Pasteur in the Darwin Library, CD was evidently familiar with his prize-winning essay, 'Mémoire sur les corpuscules organisés qui existent dans l'atmosphère' (Pasteur 1861). See *Correspondence* vol. 10, letter from Henry Holland, [*c.* April 1862] and n. 4.

[6] CD refers to the critical discussion that Heinrich Georg Bronn added at CD's suggestion (see *Correspondence* vol. 8, letter to H. G. Bronn, 4 February [1860]) to his German translation of *Origin* (Bronn trans. 1860, pp. 503–5). CD prepared a four-page list of Bronn's criticisms, which is appended to his copy of Bronn trans. 1860 in the Darwin Library–CUL (see *Correspondence* vol. 8, letter to H. G. Bronn, 5 October [1860], CD note, and *Marginalia* 1: 180–2). These criticisms were addressed in the third and later editions of *Origin*. For CD's treatment of this specific point, see Peckham ed. 1959, p. 232. See also *Correspondence* vol. 8, letter to H. G. Bronn, 5 October [1860], and letter to Charles Lyell, 8 October [1860].

From John Scott 22 May 1863

Botanic Gardens [Edinburgh]
May 22^{d}. 1863.

Sir.

Prof. Balfour has just offered me a situation in India;[1] regarding which I would very much like to have your advice. Before stating particulars in regard to this, however, I will now,—from the kind interest you have expressed in my behalf—state that I have not at all a pleasant situation under M^{r}. M^{c}.Nab here.[2] I formerly used to regard him, as, indeed, a friend, now however, I plainly see that he is doing as much as he rightly can to place obstacles in the way to my advancement. In illustration of this I may state the following. He ought to have allowed me to attend the Botanical Classes this season which he has not. Now; the time it is in, is only an hour daily, and there is not a single thing, in regard to my duties, which he could mention, to prevent my attendance.[3] There is this, however, he knows that I am interested in the place; engaged in numerous experiments which will induce me to submit to him, otherwise, I feel assured he would not have treated me thus, simply because I am of *service* to him, and might put him to a little inconvenience by leaving him. This and other little acts which I could mention if it were worth while, plainly show me his feelings towards me. I would much rather that I had not to mention these; but the kind interest you have manifested towards me has induced me to state a few particulars regarding my present situation. And I sincerely trust you will excuse me for so doing.

In respect to the Indian situation Prof. Balfour has given me the following information. Dr Anderson of Calcutta has written him that a few of his friends are desirous to get an intelligent young man to take charge of a Chinchona nursery, which they intend forming at Darjiling.[4] They purpose giving £.100 the first year and £.120 for the three subsequent years—as they wish to have an engagement of 4 years. On finding that, if I accepted it, I should have to leave here in latter end of August or beginning of September, I told Prof. Balfour that being engaged with a variety of very interesting experiments, I really could not think of entering into any such engagement this season: being anxious to get results.[5] He, however, remarked that I should think a little more over it before giving up what he thinks is a nice opening: he also said—for I suppose he understands that a number of my experiments have been proposed by you[6]—that I might be of greater service to you in that respect in the above place, and advised me to write you and mention the offer he had made.

I will therefore be greatly obliged if you will kindly favour me with your advice, for I do not know well what to do. In respect to the place itself, I really think it offers no great encouragement, the engagement is certainly long for the remuneration offered, more especially upon such an important product.[7] All things considered, I have certainly a strong inclination to give it up, even though by so doing, I should give a little offence to the Professor. Might I in the event of so doing take advantage of your kind offer in last letter,[8] and ask you if you could in anyway aid me in getting some foreign or colonial appointment. I should certainly regard it as a great favour

In the meantime | I remain | Sir | Yours very respectfully | J. Scott

DAR 177: 90

CD ANNOTATION
1.2 Before ... doing. 1.18] *crossed pencil*

[1] John Hutton Balfour was professor of botany at Edinburgh University and regius keeper of the Royal Botanic Garden, Edinburgh (*DNB*).

[2] James McNab was the curator of the Royal Botanic Garden, Edinburgh, where Scott was foreman of the propagating department (R. Desmond 1994).

[3] Balfour ran botany classes in a purpose-built classroom at the Royal Botanic Garden, Edinburgh; lectures for students of the University were held annually at 8 A.M. on weekdays from the beginning of May to the end of July (Fletcher and Brown 1970, pp. 141–2, 146; *Transactions and Proceedings of the Botanical Society* [*of Edinburgh*] 14 (1883): 160; Birse 1994, p. 65).

[4] Thomas Anderson was superintendent of the Calcutta botanic gardens (R. Desmond 1994). On the history of *Cinchona* plantations in Darjeeling, India, see the letter from J. D. Hooker, [23–7 May 1863], nn. 3 and 5.

[5] In 1863, Scott performed a series of crossing experiments with *Passiflora*, *Disemma*, and *Tacsonia*, with dimorphic and non-dimorphic species of *Primula*, and with species of *Oncidium* (see letters from John Scott, 3 March 1863, 21 March [1863], [1–11] April [1863], and 21 May [1863], and Scott 1864a, 1864b, and 1864d). These experiments were intended to throw light on three common problems: the relative fertility of crosses between individuals of the same and of different species; the relative fertility of crosses between different varieties of the same species; and the relative fertility of homomorphic and heteromorphic crosses in the same species. In a set of related experiments, Scott also attempted to make bi-generic crosses (see letter from John Scott, 6 January 1863 and n. 10).

[6] For CD's encouragement of Scott's experimental work, see, for example, *Correspondence* vol. 10, letters to John Scott, 19 November [1862] and 11 December [1862], and this volume, letters to John Scott, 12 April [1863] and 25 and 28 May [1863]. Scott's work was relevant to CD's work on the causes of cross and hybrid sterility (see *Correspondence* vol. 10, Appendix VI).

[7] *Cinchona* trees were valued for their bark, which was the source of quinine, an effective febrifuge or antipyretic medicine widely used in the tropics (see *Correspondence* vol. 10, letter to G. H. K. Thwaites, 15 June [1862], n. 4).

[8] See letter to John Scott, 2 May [1863].

From L. E. Becker 23–4 May [1863][1]

Altham | Accrington.
May 23rd.

Sir—

Allow me to thank you for your most kind and courteous reply to the communi-c⟨ation⟩ I ventured to make to you I am indeed grateful that you wish to investigate the structure and history of the curious Lychnis.[2] I hope in the course of next week to send you a small hamper containing roots which will probably continue to flower this season after they have been planted in your garden. On receipt of your second letter this morning I went into the wood and examined 137 plants of Lychnis,[3] taking them indiscriminately, though I do not pretend to have examined all I saw.

The plants were of four kinds, viz

1	Male flowers of the usual type small pale yellow anthers	56
2	Female flower do. with long spreading pistils	25
3	Hermaphrodites having large dark purple anthers and short straight upright p⟨istils⟩	31
4	Male flowers with stamens like those of the hermaphrodite.	25

I searched in vain for a female flower with pistils like the hermaphrodite or an hermaphrodite with pistils like the female flower. this suggests the query Can the long spreading pistils be an adaptation or a struggle to catch the pollen wafted from a distant flower, a provision needless in the hermaphrodite which has an abundant supply close at hand.

I tried in vain to classify the plants according to the ⟨length⟩ of the stamens this ⟨seems to⟩ vary indefinitely even in the same flower— And I think that when the plant has been gathered a few days, the stamens do grow rather long; you will readily sort the flowers I sen⟨t⟩ under the four heads I have enumerated. I have put in one or two of the common type for comparison with those which grow in your district. I have been given to understand that hermaphrodites occasionally appear in this Lychnis, but have never been able to make out whether they normally differ so much in the character of their stamens and pistils from the common form as do those found at Altham. I have never seen an hermaphrodite Lychnis diurna[4] except in the woods ⟨near⟩ here where it is, ⟨as⟩ you may gather from the numbers given above, very abundant, and so I have observed it several years in succession, I fancy it is hereditary. I will take care to collect as much se⟨ed⟩ as I can during the summer marking and noting the plants from which I obtain it, and transmit it to you at the end of the season in the capsules. I will also find out as much as possible of the distribution numbers and range of the variety. There is a practically unlimited supply of specimens here and I need hardly say that I shall be most happy to send you as many, and whenever you would like to have them. I hope it is not asking too much that y⟨ou⟩ will be good enough to favour me ⟨wi⟩th the result of your investigation and also let me know if you hear of the variety occurring in other places.

I am Sir | yours very respectfully | Lydia E. Becker

May. 24. I have kept the flowers fresh till today to avoid the detention in the post office over the Sunday. I hope they will arrive in good condition but I will enclose a tin can also with a bunch of the flowers in the packet of roots.

DAR 160: 107

CD ANNOTATIONS

2.2 Can the . . . hand. 2.5] *scored ink*

End of letter:

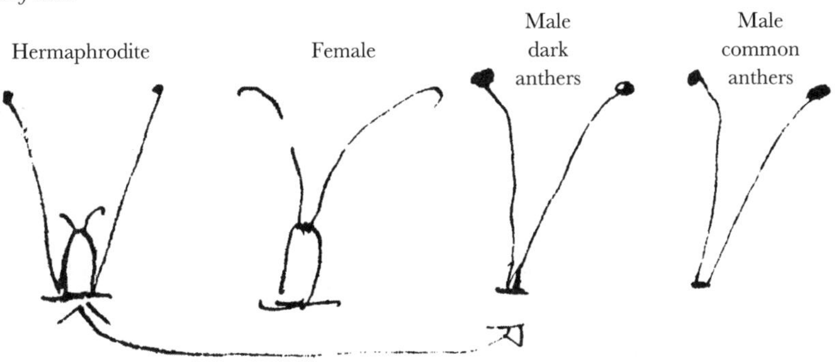

'White Lychnis = some Pods in separate paper' *ink*

CD note:
May 26th. The purple anthers a fungus— There is beautiful gradation in so-called hermaphrodite ovarium & stigma of all sizes & length—[5] In ['ma' *del*] several males a /minute/ point, or longish single green style, representing whole female organ.— In pure Female minute vestiges of stamen, some with minute double head, representing anthers.—[6]

[1] The year is established by the relationship between this letter and the letters from L. E. Becker, 18 May 1863 and 21 May [1863].

[2] The letter to Becker has not been found. The reference is to *Lychnis dioica* (*Silene dioica*) (see letters from L. E. Becker, 18 May 1863 and 21 May [1863]). CD's interest in information on *L. dioica* was probably partly related to his investigations into the relationship between differently coloured anthers and dimorphism. He had hypothesised that the structure and colour of stamens might be a better guide to the occurrence of functional dimorphism than the length of stamens and pistils alone (see *Correspondence* vol. 10, letter to J. D. Hooker, 22 [August 1862] and n. 14, and letter to Daniel Oliver, 2 September [1862] and n. 7, and this volume, letter to J. D. Hooker, 15 and 22 May [1863] and n. 13).

[3] The letter to Becker has not been found.

[4] *Lychnis diurna* was a synonym for *L. dioica* (*Index Kewensis*).

[5] CD's observations of the diseased specimens of *Lychnis dioica* led him to hypothesise that the pistil was developed in compensation for the pollen having been destroyed at an early period by the fungus (see Becker 1869).

[6] CD had long been interested in the rudimentary sexual organs found in the flowers of *Lychnis dioica* (see, for example, *Correspondence* vol. 9, letter to J. D. Hooker, 25 November [1861], and letters to Daniel Oliver, 30 November [1861], and 7 December [1861] and n. 5, and *Correspondence* vol. 10, letter to W. E. Darwin, 14 February [1862] and n. 6). In *Cross and self fertilisation*, pp. 410–11, CD used the rudimentary stamens and pistils found, respectively, in the female and male plants of this species to argue that some dioecious plants evolved from hermaphrodite species.

To W. D. Fox 23 May [1863]

Down. | Bromley. | Kent. S.E.
May 23d

My dear Fox

Many thanks for your kind note of enquiry.[1] I cannot say much for myself. We have lately been staying at Hartfield & Leith Hill Place (& I gave to Caroline your kind messages) for a fortnight;[2] but the change did me no good & I have been mostly in bed for the last week from my old enemy sickness. We gave up Malvern on account of my Eczema;[3] but it is all gone & perhaps after the holidays we may go there, unless I improve. Gully will be a great loss & I hardly know whom to consult there. I must be under some experienced man, for I could not stand much hard treatment. All this everlasting illness has stopped my work much.— I am glad you told me about Gully, for I had heard only a rumour.—[4]

Thanks about Illustrated Times, but I have seen it.[5] There has been a much better squib, on Owen & Huxley about the Brain; part of which appeared in Public Opinion.—[6]

I hope the world goes pretty well with you my old friend. I cannot say it does with me. Our youngest Boy is a regular invalid with severe indigestion, clearly inherited from me.[7]

Farewell | Yours ever truly | C. Darwin

Endorsement: 'Darwin May 26 | 1863'
Postmark: MY 24 63
Christ's College Library, Cambridge (Fox 139)

1 Letter from W. D. Fox, [16–22 May 1863].

2 See letters from W. D. Fox, 6 February [1863] and 12 March [1863]. According to Emma Darwin's diary (DAR 242), the Darwin family visited Hartfield Grove, Hartfield, Sussex, and Leith Hill Place, near Dorking, Surrey, the homes of Charles Langton and Joshiah Wedgwood III, respectively, from 27 April to 13 May 1863. Caroline Sarah Wedgwood was CD's sister and Fox's second cousin.

3 CD had recently had a recurrence of eczema; during eczema attacks, CD's general health improved (see letter to J. D. Hooker, 5 March [1863] and n. 6).

4 James Manby Gully was the proprietor of a hydropathic establishment in Malvern, Worcestershire (see letter from W. D. Fox, [16–22 May 1863] and n. 4). In the event, CD consulted James Smith Ayerst (see letter to W. D. Fox, 4 [September 1863]).

5 See letter from W. D. Fox, [16–22 May 1863] and n. 3. The reference is to the *Illustrated Times*, 2 May 1863, p. 317, 9 May 1863, p. 333, and 16 May 1863, p. 348.

6 The references are to Anon. 1863a, part of which was published in *Public Opinion* 3 (1863): 497–8, and to Richard Owen and Thomas Henry Huxley (see Appendix VIII).

7 Horace Darwin had been ill since January 1862 (see *Correspondence* vol. 10), and CD's health had been poor since the end of February 1863.

To J. D. Hooker 23 May [1863]

Down
May 23[d]

My dear Hooker

You can confer a real service on a good man, John Scott, the writer of the enclosed letter, by reading it & giving me your opinion.—[1] I assure John Scott is a truly remarkable man.— The part struck out is merely that he is not comfortable under M[r]. Macnab,[2] & this part must be considered as private. Now question is what think of you of the offer? Is expence of living high at Darjeeling? I may say it is healthy?— Will he find opportunity for experimental observations which are a passion with him.? It seems to me rather low pay. Will you advise me for him?—[3]

I shall say that as far as experiments in hand at Bot. Garden in Edinburgh are concerned, it would be a pity to hesitate to accept the offer.—[4] J. Scott is head of propagating department. I know you will not grudge aiding by your advice a good man.— I shall tell him that I have not slightest power to aid him in any way for appointment—

Yours affec[ly] | C. Darwin

I sh[d]. think voyage out & home ought to be paid for??

Endorsement: '/63'
DAR 115: 194

1 Letter from John Scott, 22 May 1863. CD was seeking advice on John Scott's behalf about a situation Scott had been offered in Darjeeling, India. See also following letter.

2 James McNab was curator of the Royal Botanic Garden, Edinburgh, and Scott's immediate superior.

3 Hooker had visited Darjeeling in April 1848 (J. D. Hooker 1854b, 1: 114–15). See also *Correspondence* vol. 4, letter from J. D. Hooker, 20 February – 16 [March] 1848.

4 See letter from John Scott, 22 May 1863 and n. 5.

From J. D. Hooker [23–7 May 1863][1]

D[r]. D.

I have written on enclosed my notions about Scotts plan which, if you think proper you can send to him.[2] I must add one or two lines in explanation

1) as to the plan, it is offered by a private company, & Cinchona is at present a *pure untried* speculation.[3]

2) Balfour is rather notorious for making bad appointments,[4] & I do not doubt that Anderson is in a fix—[5] he is not authorized to offer enough to get a good man from England, & so he goes to Scotland whence men are to be had cheaper—

3) I do not like Scotts quarrelling with M[c]Nab, of whom I never heard complaints as a bad master—[6] but I do know from sad experience that nothing is more difficult than to give good men good opportunities for improving themselves *in their own way in public establishments*.— Of all men in the world Gardeners (especially intelligent ones) are the most troublesome to deal with— give one the smallest advantage, or let it be seen that you think more of him than the others, & the "struggle for life" begins,—give them an inch for their *own* experiments, & they take an ell, & all sorts of jealousies spring up.

M[c]Nab at Edinburgh has to get his fair days work out of his men for fair days wages & lectures & experimenting are terrible excuses for all sorts of very unexpected & inexpedient delinquencies— As to a *head propagator* being away an hour daily, & that a fixed hour, I do not doubt it is most inexpedient—considering the enormous temptations to pilfer in that department: & that if one attends the lectures why should not all—& all their successors too.[7]

Then too much depends on the personal tact &c of the experimenter & his chief; experiments kill plants, as well as promote science!: in fact there are 2000 ways in which it interferes grievously with a public establishment except it is conducted under the personal superintendence of the head of the gardens.

Incomplete[8]
DAR 101: 141–2

[1] The date is established by the relationship between this letter, the preceding letter, and the letter to John Scott, 25 and 28 May [1863]. Since letters were collected from the Post Office at Down daily at 1:30 P.M., and an evening delivery to the Post Office at Kew at 7:30 P.M. was despatched immediately after arrival, it was possible for a letter from CD to be delivered to Hooker on the same day (*Post Office directory of the six home counties* 1862, *Post Office London directory* 1863).

[2] See letter from John Scott, 22 May 1863, and preceding letter. The enclosure has not been found; it was sent with the letter to John Scott, 25 and 28 May [1863].

[3] Attempts by the British to transfer Peruvian *Cinchona* for cultivation in India and Ceylon (Sri Lanka) began in 1860, the first plants arriving in the Nilgiri Hills in southern India in January 1861 (Markham 1880, pp. iii and 316). The establishment of plantations for *Cinchona* in Darjeeling between 1862 and 1864 is discussed in Markham 1880, pp. 389–90. See also n. 5, below.

[4] John Hutton Balfour. See letter from John Scott, 22 May 1863 and n. 1.

[5] Thomas Anderson was superintendent of the Calcutta botanic gardens (R. Desmond 1994). The new *Cinchona* plantations in Darjeeling were established under Anderson's direction; the first plants arrived there in June 1862 (Markham 1880, pp. 389–90; *Report on Cinchona cultivation in British Sikkim*).

[6] James McNab was the curator, and Scott the foreman of the propagating department, of the Royal Botanic Garden, Edinburgh. On Scott's dispute with McNab, see the letter from John Scott, 22 May 1863.

[7] Scott wished to attend a series of botanical lectures (see letter from John Scott, 22 May 1863 and n. 3).

[8] The letter appears to be a complete letter without a valediction or signature; it ends at the bottom of the third page with a blank verso. The valediction and signature may have been written on the missing enclosure (see n. 2, above).

To John Scott 23 May [1863][1]

Down. | Bromley. | Kent. S.E.
May 23[d]

Dear Sir

I have just received your two letters.[2] I am really incompetent from complete ignorance to advise you.[3] But I have done my best for I have written to D[r]. Hooker, who must often have sent out superintendents &c &c & who knows Darjeeling well, for his advice.—[4] He will do anything for me & as soon as I hear I will not lose a post & inform you.— Can you borrow his "Himalayan Journals", you will find much about Darjeeling.—[5] If the offer is fairly good (especially as you are rather uncomfortable in present position) I would not refuse on account of your experiments.[6] You will have plenty of opportunities.

I must not let you suppose that there is any chance of my aiding you to a situation abroad; my health is so broken that I rarely see a soul, & all such applications would be made to Kew or Edinburgh or Dublin—[7] I have mentioned your wish to D^{r}. Hooker; but rather suppose that he would feel bound to recommend anyone whom he had known at Kew.— I have only glanced at your scientific letter; but I see it contains much interesting to me.— Do not expect to interest many others equally; but rely firmly on it, that such experimental researches as yours will have a permanent value, & will be referred to years afterwards, when the endless flimsy papers of the present day are forgotten—

Yours faithfully. | C. Darwin

DAR 93: B15–16

[1] The year is established by the relationship between this letter and the letter from John Scott, 22 May 1863.

[2] Letters from John Scott, 21 May [1863] and 22 May 1863.

[3] In his letter to CD of 22 May 1863, Scott had asked CD's advice about a post he had been offered in Darjeeling, India.

[4] See letter to J. D. Hooker, 23 May [1863]. Joseph Dalton Hooker was assistant director at the Royal Botanic Gardens, Kew, and had travelled to Darjeeling in April 1848 (J. D. Hooker 1854b, 1: 114) to begin a three-year botanical exploration of the Himalayas.

[5] J. D. Hooker 1854b, 1: 113–176.

[6] See letter from John Scott, 22 May 1863 and n. 5. Scott was foreman in the propagating department of the Royal Botanic Garden, Edinburgh.

[7] CD refers to the Royal Botanic Gardens, Kew, the Royal Botanic Garden, Edinburgh, and the Botanic Garden, Glasnevin, Dublin.

To Roland Trimen 23 May [1863][1]

Down. | Bromley. | Kent. S.E.
May 23^{d}

My dear Sir

I have delayed thanking you for your note & photograph, as I had no photograph by me of myself.[2] I have never had a proper "carte" taken; but I enclose a photograph made of me by my son, which, I daresay will do as well.—[3]

Your accounts of the Disa & Herschelea are *excellent*. & your drawings first-rate.[4] I felt so sorry that such excellent work sh^{d}. remain locked up for an indefinite period in my portfolio, that you have made me break a solemn vow & I have drawn up from your notes (& selected 4 figures for woodcuts) an account for Linnean Soc.—[5] I have enlarged a little & explained & introduced a few remarks.— I hope the Soc^{y}. will publish the paper, & if so I will send you spare copies.—[6] The title is "On the Fertilisation of Disa grandiflora by Roland Trimen E^{qr} of the Colon. Off. C. Town: drawn up from notes & drawings sent to C. Darwin E^{qr}." I hope that you will approve of this, & not object to anything in the little paper.—

I am very sorry to hear so poor an account of your health & that you have so little time to spare for the exercise of your admirable powers of observation.—[7] I did not know all this; otherwise I sh$^{d.}$ not have thought of asking for plants.[8] Think not a moment more on subject.— Indeed I *ought* to work on other subjects.— Yet I am going to ask a favour, if you know any one who dabbles in Botany, viz for seed of any Cape Oxalis: several species present two forms, one with long pistil & short stamens, the other form with short pistil & longer stamens. It is of high interest to me to get seed of any such species.—[9]

To return to Orchids, I now believe that Hymenoptera & Diptera are generally the chief workers more than Lepidoptera.[10] With respect to the limits of Rostellum; it can in most cases be told only conjecturally: in Disa the 2 discs (& no part of caudicle of pollinia) & the part which connects the 2 discs with the medial upward central fold or ridge, & whole *face* of column down to the two confluent stigmas, may all be considered as the rostellum or modified third stigma.—[11]

With sincere thanks & every good wish, Believe me, my dear Sir | Yours sincerely
C. Darwin

Royal Entomological Society of London (Trimen papers, box 21: 56)

1 The year is established by the relationship between this letter and the letter from Roland Trimen, 16 March 1863.

2 Letter from Roland Trimen, 16 March 1863. Trimen's photograph has not been found.

3 William Erasmus Darwin's photograph of his father was taken in April 1861; the photograph is reproduced as the frontispiece to *Correspondence* vol. 9.

4 Trimen enclosed sketches of *Disa grandiflora* and *Herschelia caelestis* with his letter of 16 March 1863. The sketch of *Disa grandiflora* is reproduced facing p. 246. The sketch of *Herschelia caelestis* has not been found.

5 Trimen 1863. Trimen felt unequal to the task of writing up his observations on South African orchids for publication (see letter from Roland Trimen, 16 March 1863 and nn. 6 and 21).

6 Trimen 1863 was published in the *Journal of the Proceedings of the Linnean Society (Botany)* on 29 October 1863 (*General index to the Journal of the Linnean Society*).

7 See letter from Roland Trimen, 16 March 1863.

8 In his letter to Trimen of 31 January [1863], CD asked for specimens of the orchid flowers Trimen had described so that he might examine them.

9 See letter from Roland Trimen, 16, 17, and 19 July 1863. CD had been investigating dimorphism in species of *Oxalis* since 1861 (see *Correspondence* vols. 9 and 10).

10 In *Orchids*, p. 39, CD had indicated that moths were the most effective pollinators of orchids. CD's subsequent observations of the insect pollinators of orchids led him to conclude that species with 'moderately long' nectaries were pollinated by bees and flies ('Fertilization of orchids', p. 142; *Collected papers* 2: 139). See also letter to Roland Trimen, 31 January [1863].

11 See letter from Roland Trimen, 16 March 1863.

From J. D. Hooker [24 May 1863][1]

Kew
Sunday

D$^{r.}$ Darwin

I was aware how poorly you must be on receiving the sad case & Abbeville paper, for both of which much thanks.[2] No one yet knows who wrote said Case—

Busk & Egerton both deny it.[3] W^{m}. Gourlay F.L.S. is the man you ask about,[4] a most estimable fellow who did great good in Glasgow all ways, & died awfully of Fungus hæmatodes in the face! a dreadful death, & rapid.

I twice took *Clianthus*, to look at it, but had no time, & so sent it to you— not doubting that you would explain it readily & rightly, & that if I tried I should make a bungle of it.[5]

Thanks for your exposition of your Island views.[6] I think I understand them precisely— my difficulty in accepting them arises from the want of apparent accordance between the plants common to Isld & continent, & what I should have expected to be common. In other words migration inadequate to explain the presence of what is common to both, & the absence of what is absent in one. I am far from *believing* in ancient connection, all I hold is that in the present state of science it is to me the least difficult hypothesis—though a very bad one.[7] Cameroons Mts have much shaken my faith in our having any clue to ancient or modern migration as yet—[8] We want some new hypothesis, as novel as Nat. Selection, or Glacial Cold, & as stupendous as continental connection.

I do hope Godman & Salvin will stick to Gallapagos—[9] my fear is that G., a fine looking young man of means, will be bagged by some pretty girl before a year is over.

I had a preliminary talk with F. the other evening about a reconciliation with Lyell.[10] of course he was not *ready yet*, that I expected but I thought right to broach the subject before he was ready—it will probably be months yet. Poor Lyell seems perdu, no one sees him in London, & I am told by one that met him at Osborne, that he *button-holes* you about F.—as Wheatstone used about Brewster.[11]

I am going to re-read Wallaces book.[12] I do not remember that it interested me at all when it appeared. I wonder that Wallace does not fructify as Bates does.[13] Dr Gray is really not *malignant*, Owen is,—Gray has all the attributes of malignancy except malignance—[14] there then!—or rather, he talks like a malignant man without feeling in the least malignant.— I never knew Gray to do an action that sprung from an unkind motive or feeling. he abounds in all the active attributes of unkindness & malignancy without being either in heart.

I will get the Anthropolical.[15] I shall be intensely interested in your opinion of Benthams Address[16] It is neither judicial nor argumentative, only intended to show *his opinion* of the position of your hypothesis in regard to its acceptance, its influence & its future prospects, on scientific research— he has taken immense pains with it— of course he does not appreciate half your stand-points.

What a mess Falc. Busk, Carpenter & Prestwich have made of it![17] We had a "field night" at the Club last Thursday, & trotted them all out.[18] I cannot but think that Prestwich's position is most awkward— he had just claimed from Lyell the lion's share of authority in such matters & forthwith breaks down on a practical question.[19] I regard the position of all 4 as humiliating. Falconer is of his original opinion saving solely that no fraud was played, (how he reconciles this to his facts I cannot conceive). Busk believes a little more than F.— Carpenter more than either,

& P. is ready to believe any thing![20] Falc. assured us that his whole conversation with Lartet in the train from Paris to Moulin Quignon was, how so to word the Report as to give least umbrage to France's susceptibility![21] *Lubbock* tells me he is going next week to see for himself.[22]

My wife's knee is bad again but better— we go to the Nightingales for 3 days this week, & to Bury-hill next for as long or longer—[23] I do hate these sort of visits, but one has no business to cut old & kind friends, & certainly trotting about agrees with my wife, & it is very nice to see how people like her— How people with disagreeable wives can *visit* is a most fearful & wonderful thing.

I saw Huxley on Thursday looking quite well— his wife & children are at Felixstow.[24]

Palmerston has given my brother in Law Leonard a living of £600 & good house in Suffolk, near Halesworth.— his wife has had a very bad confinement indeed.[25]

I fancy I can feel the bad influence of Yankee affairs on A Gray's ordinary correspondence:[26] it makes him very bumptious scientifically!

I send you Nægeli's paper & I pity you,[27] but you are a hard headed man:—these subjects & papers floor me.— Please send it back when done with as it is Library copy

I will remember the 2-formed stamen plant (*Lagerstroemia*) when it flowers—[28]

I am glad that Lubbock is going to Abbeville— that young man *wants advice*— he is living too hard, a great deal, what with business Society & science—[29] I feel very strongly attached to him indeed. I think him the most faultless character I know, who is at the same time one of the best & most active & clever. Thank God there are some men worth living for, but really when one peeps beyond the immediate circle of ones friends one gets disgusted & disquited altogether.

Ever your affec^te grumbler | J D Hooker

DAR 101: 143–6

CD ANNOTATION
End of letter 'Kinglake | Nageli | Broom'[30] *pencil*

[1] The date is established by the reference to the meeting of the Philosophical Club of the Royal Society on 21 May 1863 (see n. 18, below); the following Sunday was 24 May.

[2] The references are to Anon. 1863a (see Appendix VIII), and to an unidentified article on the human jawbone and artefacts found at Moulin-Quignon near Abbeville, France (see n. 17, below). On CD's illness, see the letter to J. D. Hooker, 15 and 22 May [1863] and n. 7. CD apparently sent these papers on to Hooker before completing his letter to Hooker of 15 and 22 May [1863].

[3] George Busk and Philip de Malpas Grey-Egerton. See letters from J. D. Hooker, [7 May 1863] and [13 May 1863], and Appendix VIII.

[4] The reference is to the Scottish botanist William Gourlie. CD must have asked about Gourlie in a missing letter, possibly in a note accompanying the articles he had recently sent Hooker (see n. 2, above).

[5] Hooker had sent CD a flower of the legume *Clianthus dampieri* so that he could examine its pollination mechanism (see letter from J. D. Hooker, [13 May 1863], and letter to J. D. Hooker, 15 and 22 May [1863]).

6 See letter from J. D. Hooker, [13 May 1863] and n. 20, and letter to J. D. Hooker, 15 and 22 May [1863].

7 See letter to J. D. Hooker, 15 and 22 May [1863]. To account for the present distribution of animals and plants, Hooker was favourably disposed to explanations invoking the former existence of land-bridges, whereas CD had long been critical of land-bridge theories, preferring explanations based on trans-oceanic migration (see, for example, *Correspondence* vol. 6, letter from J. D. Hooker, 4 August 1856, *Correspondence* vol. 9, letter to J. D. Hooker, 28 [December 1861], *Correspondence* vol. 10, letter from J. D. Hooker, [30 December 1861 or 6 January 1862] and n. 7, and this volume, letter from J. D. Hooker, [13 May 1863]).

8 Hooker had been working on a collection of plants from the Cameroons Mountains, which extend from the mainland of west Africa to the coastal island of Fernando Po (J. D. Hooker 1863b). Hooker eventually concluded that the presence of European genera in the region's flora could be explained by two hypotheses, first, CD's theory that temperate plants had migrated to tropical regions during a global glacial period, and second, the transport of seeds by winds and birds (see letter from J. D. Hooker, [7 May 1863] and n. 11).

9 Frederick Du Cane Godman and Osbert Salvin had proposed making a collecting trip to the Galápagos Islands (see letter from J. D. Hooker, [8 May 1863]). CD had written to encourage them (see letter to Osbert Salvin, 11 [May 1863]).

10 Hugh Falconer and Charles Lyell had been involved in a public dispute since the beginning of April 1863 (see letter to Charles Lyell, [7 May 1863] and n. 5). Hooker was encouraging them to resolve their differences (see letter from J. D. Hooker, [13 May 1863]).

11 Lyell was staying at Osborne House, East Cowes, Isle of Wight, in early May 1863 (K. M. Lyell ed. 1881, 2: 366–72). Hooker also apparently refers to the antagonism nurtured by Charles Wheatstone towards David Brewster following their public dispute over the priority of invention of the stereoscope, conducted in the correspondence columns of *The Times* in 1856 (see Wade ed. 1983, pp. 2, 171–83).

12 Wallace 1853.

13 The reference is to Alfred Russel Wallace and Henry Walter Bates, both of whom had published books on their travels to Amazonian South America (Wallace 1853 and Bates 1863); CD considered Wallace's book to be inferior to Bates's as a work of natural history (see letter to J. D. Hooker, 15 and 22 May [1863]).

14 John Edward Gray and Richard Owen; CD had called Gray a 'malignant fool' in his letter to Hooker of 15 and 22 May [1863].

15 See letter to J. D. Hooker, 15 and 22 May [1863] and n. 12. Hooker refers to the first number of the *Anthropological Review*, published in May 1863.

16 Hooker refers to George Bentham's forthcoming anniversary address as president of the Linnean Society, which was to be delivered on 25 May 1863 (Bentham 1863). Bentham had evidently discussed his address, which assessed the impact of *Origin* on 'the science of life', with Hooker (see letter from J. D. Hooker, [13 May 1863]).

17 Falconer, George Busk, William Benjamin Carpenter, and Joseph Prestwich were the British members of the Anglo-French conference held at Paris and Abbeville to consider the authenticity of the flint tools and human jawbone discovered by the archaeologist Jacques Boucher de Perthes in the Moulin-Quignon gravel pit near Abbeville, France, in March 1863 (*Athenæum*, 23 May 1863, p. 682). See also letter from J. D. Hooker, [7 May 1863] and n. 5.

18 Hooker refers to the meeting of the Philosophical Club of the Royal Society, which took place once a month at 5 P.M. before the society meeting (Bonney 1919, p. 2). Falconer and Prestwich are recorded as having spoken at the May meeting, which Bonney dates 28 May (Bonney 1919, pp. 162–4), apparently in mistake for 21 May, the date of the society meeting (*Proceedings of the Royal Society of London* 12: 567).

19 During April and May 1863, Falconer had been pressing the claims of Prestwich as the foremost authority on the archaeological evidence for human antiquity, in opposition to the claims to authority he alleged had been made by Lyell in *Antiquity of man* (C. Lyell 1863a; see letter to Charles Lyell, [7 May

1863], nn. 5 and 6). Hooker probably refers to Prestwich's equivocation regarding the authenticity of the Moulin-Quignon jawbone and the associated handaxes (Bonney 1919, p. 164). See also Prestwich's summary of his findings in the *Athenæum*, 13 June 1863, pp. 779–80.

20 Carpenter and, initially, Falconer had lent their support to claims for the authenticity of the jaw found at Moulin-Quignon (see Falconer 1868, 2: 602, and *Athenæum*, 18 April 1863, p. 523). However, on closer examination of a molar belonging to the jaw, Falconer pronounced the find to be spurious (see *Athenæum*, 2 May 1863, p. 587, and letter from Hugh Falconer, 24 April [1863]). See nn. 17 and 19, above. Falconer and Busk finally concluded that the account of the jaw's discovery was authentic, but that the jaw was modern and not contemporaneous with the deposit in which it was found (Falconer *et al.* 1863, p. 462).

21 The palaeontologist Edouard Lartet had acted as an intermediary between the British and French palaeontologists concerning the authenticity of the Moulin-Quignon jaw, and was a member of the conference that took place in Abbeville and Paris between 9 and 13 May 1863 (Falconer *et al.* 1863). Opinions on the authenticity of the remains had divided along Anglo-French lines (see letter from J. D. Hooker, [7 May 1863] and n. 5).

22 John Lubbock's visit to the gravel pit at Moulin-Quignon was reported in the *Athenæum*, 6 June 1863, p. 748.

23 Lea Hurst, near Matlock, Derbyshire, was the home of William Edward Nightingale, Florence Nightingale's father (*DNB*). Bury Hill, near Dorking, Surrey, was the home of Arthur Kett Barclay (*Survey gazetteer of the British Isles*; *Post Office directory of the six home counties* 1862).

24 Hooker refers to Thomas Henry Huxley and his wife Henrietta Anne. Felixstowe, Suffolk, was a popular seaside resort (Walton 1983, pp. 57 and 187).

25 Hooker refers to Frances Harriet Hooker's brother Leonard Ramsay Henslow and his wife Susan Henslow. Henslow was appointed rector of St Mary Magdalene, Pulham, south Norfolk, by the prime minister, Henry John Temple, 3d Viscount Palmerston (*Alum. Cantab.*). The parish is nine miles north-west of Halesworth in north Suffolk.

26 Asa Gray continued to correspond with Hooker throughout the American Civil War (1861–5), but because of Hooker's lack of sympathy for the Union cause, they had agreed to stop discussing politics (see *Correspondence* vol. 9, letter from J. D. Hooker, [29 December 1861], and *Correspondence* vol. 10, letter from J. D. Hooker, [19 January 1862]).

27 The reference is to the first part of Nägeli 1858–68. CD wished to consult the work for information on phyllotaxy (see letter to J. D. Hooker, 15 and 22 May [1863]).

28 See letter to J. D. Hooker, 15 and 22 May [1863] and n. 13.

29 Lubbock, aged 29, helped to manage the family banking business, Robarts Lubbock and Co., and took an active part in the scientific community (*DNB*).

30 See letter to J. D. Hooker, 29 May [1863].

To Hermann Crüger 25 May [1863][1]

Down Bromley | Kent. S.E.
May 25.

Dear Sir

Various circumstances have delayed my thanks for your very interesting letter of Ap 23^{d}.—.[2] I thank you sincerely about the Melastomas:[3] it seems that my suspicion was quite unfounded yet it is just possible that bees may visit the flowers for pollen & suction.— I have wasted a fearful amount of time over this order.[4] There is something very odd about the difference in the two sets of stamens I first most strongly suspected that the plants on which I experimented were dimorphic; nor should I yet be surprised if this proved to be the case. I shall be very curious to know about the Catasetums & what attracts insects.[5] What a singular fact that of the orchids which did open their flowers setting seeds![6] A good observer Mr. Scott believes that when the flower is closed the pollen tubes come out of the anther & travel to the stigma.[7] You say you are going to try to make orchis seed germinate: Dr. Hooker tells me that they cannot succeed in Calcutta;[8] but that self-sown foreign plants appears on the surrounding trees!

I am very glad that you have not my Journal and I wrote a week ago to my publisher to send you a copy.—[9] I sent sometime ago a newspaper to you in which I alluded to some of your information.[10] I have also sent a copy of a little paper on the dimorphism of Linum:[11] This is a subject on which I am experimenting with *great* **interest**. Have you any analogous cases in the West Indies? You must kindly permit me to ask you questions: have you seen cases of what gardeners call "sports" but what I shall call "bud variation" (i.e. variation by buds & not by seed) in plants from warmer temperate regions cultivated in the West Indies.[12] I am collecting *all* cases. Sir R Schomburgk says that temperate plants as Dahlia, Roses &c cultivated in St Domingo are very liable to change in character & give off single shoots different from the mother plant.[13]

with cordial thanks for all your interesting information & great kindness | Dear Sir | Yours sincerely | Ch Darwin

Copy
DAR 143: 359

1 The year is established by the relationship between this letter and the letter from Hermann Crüger, 23 April 1863.

2 Letter from Hermann Crüger, 23 April 1863.

3 In his letter to Crüger of 25 January [1863], CD asked about the insects that visited the flowers of Melastomataceae. CD hypothesised that small flies or wasps were attracted by the fluid-filled horns of the anthers, which they penetrated with their probosces. Crüger assured him that the two species of bees seen visiting the flowers of *Heteronoma diversifolium* came only for pollen; however, he agreed to make further observations of other species of Melastomataceae (see letter from Hermann Crüger, 23 April 1863).

4 CD refers to the order Melastomaceae (Lindley 1853, p. 731), which corresponds approximately to the modern family Melastomataceae. CD had been working on the Melastomataceae since October 1861, suspecting that the family might exhibit a novel form of dimorphism (see letter to Hugh Falconer, 5 [and 6] January [1863] and n. 22). CD's experimental notes on Melastomataceae are in DAR 205.8.

5 In his letter to CD of 23 April 1863, Crüger wrote that he was going to investigate a secretion in the flower of the orchid *Catasetum tridentatum* that appeared to attract bees.

6 In his letter to CD of 23 April 1863, Crüger wrote of an *Epidendrum* species with flowers that usually did not open, though his specimen had 'fully open' flowers. Crüger reported: 'These have not set fruit, while commonly I do not remember to have noticed a flower which was not followed by fruit.'

7 No letter from John Scott recording these observations has been found; however, see letter to *Journal of Horticulture and Cottage Gardener*, [17–24 March 1863] and n. 9, and letter from John Scott, [1–11] April [1863].

8 Joseph Dalton Hooker gave CD this information when he visited Down House on 22 March 1863 (see letter to John Scott, 24 March [1863] and n. 10).

9 See letter from Hermann Crüger, 23 April 1863; the reference is to *Journal of researches* (1860). The letter to John Murray, CD's publisher, has not been found.

[10] CD included some information on orchid pollination reported by Crüger (see letter from Hermann Crüger, 23 February 1863 and n. 5) in the letter to the *Journal of Horticulture and Cottage Gardener*, [17–24 March 1863].

[11] 'Two forms in species of *Linum*'; Crüger's name appears on CD's presentation list for this paper (see Appendix IV).

[12] CD was seeking information on the question of whether plants introduced from different climates were particularly apt to produce bud-variations (see *Correspondence* vol. 10, letter to G. H. K. Thwaites, 29 December [1862]).

[13] Schomburgk 1857, p. 132.

To John Scott 25 and 28 May [1863][1]

Down Bromley Kent
May 25th

Dear Sir

Now for a few words on Science.— I do not think I could be mistaken about stigma of Bolbophyllum; I had the plant alive from Kew & watched many flowers.[2] That is a most remarkable observation on foreign pollen emitting tubes, but not causing orifice to close: it would have been interesting to have observed how close an alliance of form would have acted on orifice of stigma.—[3] It will probably be so many years, if ever, that I work up my observations on Drosera, that I will not trouble you to send your paper;[4] for I could not now find time to read it.—

If you have spare copy of your Orchid paper, please send it;[5] but do not get a copy of the Journal, for I can get one, & you must often want to buy Books.— Let me know when it is published.—

I have been glad to hear about Mercurialis;[6] but I will not accept your offer of seed on account of time, time, time, & weak health. For same reason I must give up Primula matter. What a wonderful indefatigable worker you are! You seem to have made a famous lot of interesting experiments. D. Beaton once wrote that no man could cross any species of Primula, you have apparently proved the contrary with a vengeance.—[7] Your numerous experiments seem very well selected, & you will exhaust the subject.— Now when you have completed your work, you should draw up a paper well worth publishing & give a list of all the dimorphic & non-dimorphic forms.[8] I can give you on authority of Prof. Treviranus in Bot. Zeitung case of P. longiflora non-dimorphic.—[9] I am surprised at your Cowslips in this state.[10] Is it a common yellow cowslip? I have seen Oxlips (which from some experiments I now look at as certainly natural hybrids) in same state.[11] If you think Bot. Soc. of Edinburgh would not do justice & publish your paper; send it to me to be communicated to the Linnean Socy.— I will delay my paper on successive dimorphic generations in Primula till yours appear; so as in no way to interfere with your paper.— Possibly my results may be hardly worth publishing; but I think they will; the seedlings from two successive homomorphic generations *seem* excessively sterile.—[12]

I will keep this letter till I hear from Dr. Hooker.—

I shall be very glad if you try Passiflora.—[13]

Your experiments on Primula seem so well chosen that whatever the result is; they will be of value. But always remember that not one naturalist out of a dozen cares for really philosophical experiments.

May 28th | I now enclose Dr. Hooker's answer which will console you for refusing the situation.—[14] Dr. H. permits me to forward his letter, but I think you had better not mention it to Dr. Balfour or Mr. Macnab, as it might make ill-will, his offering advice to an Edinburgh man.—[15]

Dr. Hooker in another note remarked that he had always heard a very good character for kindness in Mr. Macnab, so permit me to suggest to you to do all you can to please him. Dr. H. truly remarks that experiments take up much time, & that he knows himself that a superintendent is bound to see that the ordinary work is fully done.—[16] So pray do all you can to please & satisfy Mr. Macnab.

Dear Sir In Haste | Yours sincerely | C. Darwin

DAR 93: B41–4

1 The year is established by the relationship between this letter and the letters from John Scott, 21 May [1863] and 22 May 1863.

2 Scott had enclosed a note from his manuscript on orchid pollination (published as Scott 1863a) with his letter to CD of 21 May [1863]. Scott's study had led him to suspect that CD's observations on *Bolbophyllum rhizophorae* were anomalous. In *Orchids*, p. 170, CD had described how the stigma of *B. rhizophorae* would close after some time, even if not pollinated, something he had seen in no other orchid; the observation is repeated in *Orchids* 2d ed., p. 137. Joseph Dalton Hooker had sent CD a specimen from the Royal Botanic Gardens, Kew, in November 1861 (see *Correspondence* vol. 9, letters to J. D. Hooker, 5 [November 1861] and 25 November [1861]).

3 See enclosure to the letter from John Scott, 21 May [1863].

4 Scott read a paper, 'On the propagation and irritability of *Drosera* and *Dionaea*', before the Botanical Society of Edinburgh on 11 December 1862; only an abstract was published (Scott 1862b; see letter to John Scott, 2 May [1863] and n. 10, and letter from John Scott, 21 May [1863]). CD had carried out a series of experiments on *Drosera* and *Dionaea* between 1860 and 1862 (see *Correspondence* vols. 8–10); however, he decided to postpone this line of research until after *Variation* was completed (see *Correspondence* vol. 10, letter to Edward Cresy, 15 September [1862]). CD resumed his research on these plants in 1872, and *Insectivorous plants* was published in 1875; Scott's work on *Drosera* is cited on pp. 1–2 n. See also *Correspondence* vol. 10, letter to John Scott, 11 December [1862] and n. 10.

5 Scott sent CD an abstract of his paper on orchids with his letter of 28 May [1863]. The paper, which was read before the Botanical Society of Edinburgh on 14 May 1863, was later published in full as Scott 1863a. Scott sent a copy of Scott 1863a in July (see letter to John Scott, 1 and 3 August [1863]); it is in the Darwin Pamphlet Collection–CUL.

6 Scott apparently sent observations on *Mercurialis* in the missing section of his letter to CD of 21 May [1863].

7 Probably a reference to Donald Beaton's statement that polyanthuses (a member of the genus *Primula*) could not be crossed (*Cottage Gardener*, 5 June 1860, p. 150, and 3 July 1860, pp. 200–11). CD marked the statement in his unbound copy of the *Cottage Gardener*, which is in the Darwin Library–CUL. Beaton was a gardener and regular contributor to the journal (R. Desmond 1994). See also Appendix V. Scott discussed his crossing experiments with *Primula* in his letter to CD of 21 May [1863]. Scott had been corresponding with CD on *Primula* since November 1862 (see *Correspondence* vol. 10).

[8] CD communicated Scott's observations on dimorphic and 'non-dimorphic' species of *Primula* to the Linnean Society (Scott 1864a). The paper, which was read on 4 February 1864, contained lists of dimorphic, short-styled, long-styled, and 'non-dimorphic species' (Scott 1864a, p. 80).

[9] In his review of 'Dimorphic condition in *Primula*' (Treviranus 1863a, p. 4), Ludolph Christian Treviranus stated that *Primula longiflora* was non-dimorphic. Treviranus sent this review to CD in February (see letter from L. C. Treviranus, 12 February 1863). CD's annotated copy of Treviranus 1863a is in the Darwin Pamphlet Collection–CUL; at the head of the fourth page, CD noted that according to Treviranus, *P. longiflora* was found 'alone' and that it was 'nondimorphic' and 'shortstyled'. However, see the letter to John Scott, 25 [July 1863]. See also letter to Daniel Oliver, 20 [February 1863], and letter from Daniel Oliver, 27 February 1863.

[10] See letter from John Scott, 21 May [1863].

[11] The oxlips referred to are apparently those described in DAR 108: 24b. CD's experiments with cowslips and primroses (*Primula veris* and *P. vulgaris*) had led him to conclude that the common oxlip was a hybrid produced as a result of cross-pollination between these two species ('Dimorphic condition in *Primula*', pp. 93–4, *Collected papers* 2: 60–1). See also *Correspondence* vol. 10, letter to J. D. Hooker, 14 [October 1862] and n. 14, letter to John Scott, 3 December [1862] and nn. 13–15, and this volume, letter to John Scott, 25 [July 1863]. CD published his results, based on experiments performed between 1862 and 1867, in 'Specific difference in *Primula*', pp. 443–8. See also *Forms of flowers*, pp. 63–71.

[12] CD's experiments on the fertility of successive generations of *Primula* raised from homomorphic crosses were made with *P. sinensis*, *P. vulgaris (acaulis)*, and *P. veris (officinalis)*, and were carried out between 1862 and 1865. On the 1862 experiments, see *Correspondence* vol. 10, letter to Alphonse de Candolle, 17 June [1862] and n. 2, letter to J. D. Hooker, 23 June [1862] and n. 4, and letter to John Scott, 3 December [1862] and n. 6. CD's notes on homomorphic crosses of *P. sinensis* and *P. vulgaris*, dated March–June 1863, are in DAR 108: 50–5 and 165–7. CD eventually concluded that successive generations of long-styled and short-styled forms of these species, when pollinated with own-form pollen, showed variable changes in fertility, from slightly lessened fertility to absolute sterility. The results were published in 'Illegitimate offspring of dimorphic and trimorphic plants', pp. 410–37, which was read before the Linnean Society on 20 February 1868.

[13] Scott told CD of his intention to experiment on *Passiflora* in his letter of 21 May [1863]. CD, wishing to corroborate statements that some species of *Passiflora* could be fertilised more readily by different species than by their own pollen, suggested these experiments in 1862 (see *Correspondence* vol. 10, letters to John Scott, 19 November [1862] and 11 December [1862] and n. 21). Scott's experiments were published in Scott 1864d.

[14] The enclosure has not been found, but an indication of its contents is given in the letter from J. D. Hooker, [23–7 May 1863]. Scott had asked CD's advice on an appointment he had been offered in Darjeeling, India (see letter from John Scott, 22 May 1863, and letter to John Scott 23 May [1863]).

[15] John Hutton Balfour was the keeper of the Royal Botanic Garden, Edinburgh; James McNab was the curator (R. Desmond 1994).

[16] See letter from J. D. Hooker, [23–7 May 1863].

From Asa Gray 26 May 1863

Cambridge. Mass.
May 26 | 1863.

My Dear Darwin

Here I have a letter of yours a month old to acknowledge—a thing that rarely happens, you must allow.[1] On 3 successive post-days I have sat down for it, or endeavored to do so. But a regular whirl of business, of one sort or other has prevented

I counted on 2 or 3 hours this morning; but two of my neighbors have got squabbling over a Report I have to read at the Anniversary Meeting of the American Academy to-day,[2] and have wasted my time horribly, and distracted my brain, so that I am quite unfit to write to you. Yet I must drop a line, lest you should think me dead or worse.

From what you & Hooker & Boott had written about the sparring in the Athenæum I was anxious to see the nos.—but did not go to look it up at the Library, &c—as it comes to me thro' Boott, tho, tardily.[3] But the no. which contained Falconers attack on Lyell has failed. I suppose it may have gone down in the Anglo-Saxon.[4] And the nos. of April 18 & 25—reached me only 3 or 4 days ago!

I have read Lyell's reply to Falconer—which appears to dispose of him,— and Prestwich's letter, and Falconers rejoinder, I believe.[5]

This all seems to me a miserable business on the part of F. & P. I see no reasonable ground of excuse for their conduct. They appear childish.—

Your letter on *Heterogeny* is keen & good.[6] Owen's rejoinder ingenious.[7] But his dissent from your well-put claims of Nat. Sel. to attention & regard, is good for nothing *except* on the admission of the view that species are *somehow* derived genealogically—& this I judge from various of Owen's statements that he really in his heart believes to be the case,—and was (as I long ago intimated my suspicions) hunting about for some system of derivation, when your book came down upon him like a thunder-clap.[8]

Carpenter is weak: lacks *nerve* perhaps.—[9]

Wyman, here, is greatly pleased with Huxley's book on Man's place in Nature.[10] I have not even seen it.

Dana is dabbling in "*cephalization*".[11] I wish he could be *cephalized* more himself—at least could be made *harder-headed*. He is far too idealistic to ever make the naturalist he was intended for. He has capital points, but his head runs away with him.

Agassiz is writing very maundering geology & zoology, and worse botany (fossil) in the Atlantic Magazine.[12] He tells his readers that the *embryology* of trilobites (about which not a fact is existant) is better known than that of Crabs!![13] He "is joined to his idols",[14] and I have no expectation that he will ever be of any more direct use in nat. history. I hope better of his son,[15] who may do something.

Did you ever notice how prettily *Iris* is arranged for cross-fertilizing, by bees, &c.?

Your *Linum* paper has long been here—[16] But I have actually not had time to read it. I might have glanced at it. But I find it best to read only when I can do so with some attention.

You must had had Cypripediums out.[17]

C. arietinum—I had one blossom only—most of my stock was sent to you—is a very *clear* case—wholly confirming my notion—which is so obvious that nobody could fail to see it. The surface of the stigma is unusually bristly, I think, i.e—the papillæ longer than in the others—in proportion.—[18]

What you say from Edinb. will whet my curiosity to look into Gymnadenia 3-dentata.[19] If the case amounts to anything I will put buds & flowers into spirit for you.

On Cypripediums, you had better print a note, in Gard. Chronicle, when you have seen all the flowers you can.[20]

It *was* sly of you to cite Lyell, p. 469—[21] I had not seen this before, having only the 1st ed. A very considerable change from the first ed. and one that may well satisfy you.[22]

Your question Ohio l⟨aw⟩s of marriage I have forgotten to ask about, I will enquire.[23]

Phyllotaxis I have no notion in the world why the angular divergence should be of *that series* of nos. & not of others.[24] *Opposite* leaves give (decussating) the angle $\frac{1}{4}$. My puzzle has been to account for this system in *cycles* in leaves running into the system of decussating *whorls*, in flowers, (usually almost universally) You will see, the question by comparing in my Bot. Text. Book (not Lessons), pp. 236–237, with Chapt. V, sect. 1,—and you see I have drawn an illustration from it apropos to Falconer's remark.[25] But explaining the obscure by the obscure does not amount to much.

As to National affairs, how quarrelsome you English are. Here are we cooly & quietly occupied with our little affairs, never dreaming of harm from you,—and your people are trying their prettiest to pick a quarrel with us,—because we do what *Historicus* says the English have always done & will do again when the time comes, having Lord Stowell to back them![26] Tell me, who is *Historicus* in the Times.?[27] An able & most influential person, evidently.

The Government of England are now showing sense.[28] Do not wonder that some wild talk is given to the air in this rough country, after what you have heard in House of Commons, & read in Times, &c I am afraid we shall not like each other for a good while—the nations— But all shows I was right. We must carry out our little job, & hold the *U.S.* complete, & develop material strength at any cost,—or we could not live without eating more dirt than we like.

Boasting nonsense is pretty well knocked out of us, by severe discipline & sad reverses,—but the *determination* is stronger than ever.

Time up, & paper full. Forgive my maundering—& believe me to be | Ever Your affectionate | A Gray.

DAR 165: 135

[1] Letter to Asa Gray, 20 April [1863].

[2] The 'neighbors' have not been identified. Gray was corresponding secretary of the American Academy of Arts and Sciences; at the 26 May 1863 meeting, he read the annual report on behalf of the council (*Proceedings of the American Academy of Arts and Sciences* 6 (1866): 128–40). At this meeting, Gray was elected president of the academy.

[3] In his letter of 20 April [1863], CD referred to two disputes recently aired in the *Athenæum*: a letter highly critical of C. Lyell 1863a by Hugh Falconer (*Athenæum*, 4 April 1863, pp. 459–60), and a dispute

over heterogeny, or spontaneous generation, occasioned by Richard Owen's anonymous review of Carpenter 1862 in the *Athenæum*, 28 March 1863, pp. 417–19. For Falconer's dispute with Charles Lyell, see letter to Charles Lyell, 18 April [1863] and nn. 7–9. For an account of the dispute on spontaneous generation, see Appendix VII. Joseph Dalton Hooker and Francis Boott also corresponded regularly with Gray.

4 On 27 April 1863, the passenger steamship *Anglo Saxon* was wrecked on the rocks off Cape Race, Newfoundland, Canada, with the loss of almost 300 lives (*Annual Register* 1863, pt 2: 74–5).

5 Gray refers to Lyell's letter in the *Athenæum*, 18 April 1863, pp. 523–5, in reply to Falconer's criticisms of C. Lyell 1863a (see n. 3, above). Lyell's letter was answered by Joseph Prestwich (*Athenæum*, 25 April 1863, p. 555) and by Falconer (*Athenæum*, 2 May 1863, p. 586).

6 Letter to *Athenæum*, 18 April [1863]. See also Appendix VII.

7 Gray refers to Owen's anonymous letter to the *Athenæum*, 2 May 1863, pp. 586–7 (see Appendix VII).

8 Gray refers to his statement, in his review of *Origin* in the *Atlantic Monthly*, that Owen was 'apparently in travail with some transmutation theory of his own conceiving, which may yet see the light, although Darwin's came first to birth' ([A. Gray] 1860b, p. 115). For a discussion of Owen's views on evolution, prior to and immediately after the publication of *Origin*, see Rupke 1994, pp. 220–48.

9 The reference is to William Benjamin Carpenter's letter in the *Athenæum*, 4 April 1863, p. 461 (see Appendix VII), which replied to Owen's anonymous review of Carpenter 1862 (see n. 3, above).

10 The references are to Jeffries Wyman, Hersey Professor of anatomy at Harvard University, and to T. H. Huxley 1863b.

11 In January 1863, James Dwight Dana published a paper on zoological classification, using 'cephalization' as a fundamental principle of classification (Dana 1863c, p. 66). As both he and Gray were editors of the *American Journal of Science and Arts*, Gray may have known that Dana was working on a further paper on cephalisation, which was published in the journal in July 1863 (Dana 1863b). For CD's opinion of Dana 1863c and a definition of cephalisation, see the letter to Charles Lyell, 17 [February 1863].

12 Gray refers to a series of articles by Louis Agassiz on natural history that appeared in the *Atlantic Monthly* in 1862 and 1863; the articles, which continued into 1864, were based on notes taken by members of the audience at a popular series of lectures delivered by Agassiz at the Lowell Institute in Boston. The articles were afterwards collected and published as *Methods of study in natural history*, which Agassiz described as his 'opportunity to enter my earnest protest against the transmutation theory' (Agassiz 1863a, p. iii).

13 Agassiz 1863b, p. 468.

14 'Ephraim is joined to idols: let him alone' (Hos. 4:17). Agassiz adhered to a belief in special creation and was opposed to evolutionary theory (*DSB*). See also n. 12, above.

15 Alexander Agassiz was a zoologist, and since 1859 had been employed as an assistant in the United States Coastal Survey (*EB*)

16 An offprint of 'Two forms in species of *Linum*' was sent to Gray on 18 April 1863; Gray's name appears on CD's presentation list for the paper (see letter to Asa Gray, 20 April [1863], and Appendix IV).

17 Gray had sent CD live roots of the North American species of the orchid genus *Cypripedium* in December 1862 (see *Correspondence* vol. 10, letter from Asa Gray, 9 December 1862). There are notes on *C. acaule* 'sent by Asa Gray', dated 22 May 1863, in DAR 70: 121. See also n. 18, below.

18 Gray believed that if CD had the opportunity of examining North American species of *Cypripedium*, he would modify his conclusions on the manner in which pollination is effected in the genus. CD had suggested that the genus was pollinated by large insects inserting their probosces into either of the lateral orifices at the base of the labellum (*Orchids*, pp. 274–6). Gray, however, was convinced that pollination was effected by insects, probably flies, crawling bodily into the flower. Gray also pointed out that the rigid papillae on the stigma were well adapted to brushing the pollen from the insect's head or back (A. Gray 1862a, p. 428). See also *Correspondence* vol. 10, letters to Asa Gray, 10–20 June [1862] and n. 16, and 26[–7] November [1862].

19 In the letter to Asa Gray, 20 April [1863], CD informed Gray of John Scott's findings on the rostellum and the growth of pollen-tubes in some orchid genera observed at the Royal Botanic Garden, Edinburgh. See also the letter from John Scott, [1–11] April [1863]. CD and Gray discussed the structure and pollination mechanism of *Gymnadenia tridentata* on several occasions in 1862 (see *Correspondence* vol. 10), and Gray published his observations in A. Gray 1862a, p. 426, and A. Gray 1862c, p. 260 n. When Gray published further observations in the September 1863 issue of the *American Journal of Science and Arts* (A. Gray 1863b, pp. 293–4), he confirmed his earlier observation that the rostellum was penetrated by pollen-tubes. CD cited Gray's findings in 'Fertilization of orchids', p. 147 (*Collected papers* 2: 144).

20 After investigating Gray's observations on *Cypripedium* (see nn. 17 and 18, above), CD published a note on the genus in 'Fertilization of orchids', pp. 155–6 (*Collected papers* 2: 152–3), stating that Gray's hypothesis that the pollinating insects crawled into the flower through the overlapped edges of the labellum and out through the orifices near the anthers was correct. CD also included Gray's observations in *Orchids* 2d ed., p. 229. CD's observational notes on *C. pubescens* and *C. acaule*, dated April and May 1863, are in DAR 70: 112–13 and 121.

21 The reference is to C. Lyell 1863b, p. 469, which CD cited in his letter to the *Athenæum*, 18 April [1863]. See letter to Asa Gray, 20 April [1863] and Appendix VII.

22 In C. Lyell 1863a, p. 469, the passage read:

> Yet we ought by no means to undervalue the importance of the step which will have been made, should it ever become highly probable that the past changes of the organic world have been brought about by the subordinate agency of such causes as 'Variation' and 'Natural Selection.'

C. Lyell 1863a was published on 9 February 1863 (*Athenæum*, 7 February 1863, p. 176) and the second edition was published in April (see C. Lyell 1863b, p. vii, and letter to Charles Lyell, 18 April [1863]).

23 See letter to Asa Gray, 20 April [1863]; CD had asked whether the state of Ohio had legislated against the marriage of cousins.

24 CD had asked why the angles of leaves in a spiral up a stem were in the order of $\frac{1}{2}$, $\frac{1}{3}$, $\frac{2}{5}$, $\frac{3}{8}$, and so on; see letters to Asa Gray, 20 April [1863] and 11 May [1863].

25 The reference is to *Introduction to structural and systematic botany and vegetable physiology* (A. Gray 1858a, pp. 236–7 and 133–41), which formed the fifth revised edition of *The botanical text-book*. Hitherto, CD had been consulting *First lessons in botany* (A. Gray 1857) for information on phyllotaxy, following Hugh Falconer's claim that the angles of leaves in a spiral up a stem 'go by as fixed a law as that of Gravity & *never vary*' (see letter to Asa Gray, 20 April [1863] and n. 14, and Falconer 1863a, p. 80). In A. Gray 1858a, p. 236, Gray had remarked regarding phyllotaxy:

> in living bodies, endowed as they are with plasticity and a certain power of adaptation to circumstances, the positions assumed are not mathematically accurate; and the effects of unequal pressure in the bud in throwing the smaller parts more or less out of their normal position may be observed in almost any irregular flower.

26 The reference has not been identified. The maritime and international lawyer William Scott, from 1821 Baron Stowell, was a judge of the high court of admiralty between 1798 and 1828, and on many maritime points his judgments were the only source of law; they were often unfavourable to and unpopular with Americans (*DNB*).

27 From 1861, the lawyer and statesman William Vernon Harcourt wrote numerous letters to *The Times* under the name Historicus, defending the British policy of neutrality in the American Civil War on the basis of precedents in international law. In 1863, some of his letters to *The Times* were republished in book form ([Harcourt] 1863).

28 Gray probably refers to the British Government's decision that it was their duty as a neutral power to enforce clause 7 of the Foreign Enlistment Act, which made illegal the equipping of a ship for war in the service of a foreign belligerent (*Hansard's Parliamentary Debates* 3d ser. 170: 90–4, 703–59). It was claimed that enforcement of this law would be to the disadvantage of the Confederates (*The Times*, 27 April 1863, p. 8).

From John Scott 26 May [1863][1]

Botanic Gardens [Edinburgh]
May 26th.

Sir.

I have just received your letter.[2] I am sorry that I troubled you at all about the Darjeeling situation, as strangely enough, though the Professor asked me to consult you upon it, he has not allowed me time to hear from you before requiring a decisive reply.[3] The day following that on which I wrote you, he told me that Dr. Anderson expected to hear from him by this mail,[4] and desired me to state whether or not I would enter into any engagement: as he had another person in view if I failed to do so. I stated, that I certainly wished he could have given me a little more time; it being otherwise however, I could only repeat as before—that I would not at present enter into an engagement. He has therefore offered it to another who has accepted it.[5]

I may have erred in losing this opportunity of going abroad—as Prof. Balfour seems to think—in respect to my future; nevertheless, it is now at least consolatory, and will I doubt not be also retrospectively, to know that I have thus been afforded an opportunity to work out a series of interesting & instructive experiments.[6]

I thank you most sincerely for the kind and active interest you have manifested: and I beg you to excuse me for the unnecessary trouble I have put you to in respect to the above situation; for I certainly did not think that an answer would have been required of me till I had heard from you.

I will be glad to hear from you when convenient, about any of the more interesting questions of mine of the 21st.[7] I have written an abstract of paper on Sterility of Orchid for Edinburgh Evening Courant, which always notices the meetings of the Bot. Soc. I will send you a copy.[8]

In the meantime | I remain | Yours respectfully & obliged | J. Scott.

Thanks for mentioning me to Dr Hooker, though as you remark there would have been more hopes of his recommending me had I been at Kew.[9] J. S.

DAR 177: 91

[1] The year is established by the relationship between this letter, the letter from John Scott, 22 May 1863, and the letter to John Scott, 23 May [1863].

[2] Letter to John Scott, 23 May [1863].

[3] John Hutton Balfour, professor of botany at Edinburgh University and keeper of the Royal Botanic Garden, Edinburgh, had offered Scott a position with a proposed *Cinchona* nursery in Darjeeling, India (see letter from John Scott, 22 May 1863).

[4] Thomas Anderson was superintendent of the Calcutta botanic gardens (R. Desmond 1994).

[5] The reference may be to Mr Ryan, who, though not a trained gardener, was engaged to oversee the work of the native labourers in the *Cinchona* plantation at Darjeeling (*Report on Cinchona cultivation at Darjeeling*, p. 646).

[6] See letter from John Scott, 22 May 1863 and n. 5.

[7] Letter from John Scott, 21 May [1863].

[8] Scott read a paper entitled 'Experiments on the fertilisation of orchids in the Royal Botanic Garden of Edinburgh' before the Botanical Society of Edinburgh on 14 May 1863; a summary of the paper

was printed in the *Edinburgh Evening Courant*, 28 May 1863, p. 8. The paper was subsequently published in the *Transactions of the Botanical Society* [*of Edinburgh*] (Scott 1863a). There is an annotated copy of Scott 1863a in the Darwin Pamphlet Collection–CUL.

9 Joseph Dalton Hooker was assistant director of the Royal Botanic Gardens, Kew. See letter to J. D. Hooker, 23 May [1863], and letter to John Scott, 23 May [1863].

From John Thomas Austen 27 May 1863

Netherbury—
May 27—/63

My dear Sir,

I found your note on my return from Bromley on Monday.[1] I was late & could not in consequence write to you that Evening— yesterday I left early for this place having a long journey before me— I have written by this post to Dennen and asked for an explanation relative to the transaction you have referred to.[2] I do not understand it at all, it certainly looks as if he were engaged in some business which at any rate he ought not to meddle with. he ought not to have anything whatever to do with borrowing money at ten per Cent— The Bondsmen, he speaks of, are I suppose securities for the money advanced;[3] as to his own bond, it is worth nothing at all— I will desire him to write to me immediately and I will then let you know what he has said for himself; you must not however expect to hear from me before Sunday Morning, as we have no second Post from this place either out or in, and I believe you have no second d⟨eli⟩very at Down[4] I shall not get Dennen's answer before Friday morning— I am staying now at

Slape House
Netherbury
Beaminster
Dorset.

I saw your servant at Bromley on Monday; ⟨&⟩ am sorry to hear that you are not well;[5] if I had not been leaving home yesterday I would have rode over to Down & called on you, that we might have had some conversation on this matter. The affairs of the Bank are all right; no money goes through Dennen's hands.

I remain | Yrs very truly | John Thos. Austen.

C. Darwin Esq.
PS. I do not re⟨turn⟩ home before the end of the week after next, we remain here till ⟨Wedn⟩esday next—

DAR 159: 149

1 The letter to Austen has not been found.

2 CD had evidently written to inform Austen that George Dennen had been making unauthorised loans from the deposits of the Savings Bank, High Street, Bromley (see letter from J. T. Austen, 30 May 1863). Like CD, Austen, who was rector of West Wickham, Kent, was a trustee of the savings bank; Dennen was its secretary (see Horsburgh 1980, p. 303, and letter from J. T. Austen, 3 June 1863).

[3] Bondsman: one who becomes surety by bond (*OED*).

[4] Down received only one postal delivery daily, at 8:30 A.M. (*Post Office directory of the six home counties* 1862).

[5] Austen may refer to Joseph Parslow, the Darwins' butler (Freeman 1978).

From L. E. Becker 28 May [1863][1]

Altham | Accrington.
May 28

Sir—

At the risk of being troublesome I cannot refrain from expressing my thanks for your letter and the paper you have done me the honour to send me which I do indeed highly prize.[2] I had seen an abstract of it, and this induced me to send you the Lychnis.[3] I am of course much disappointed that it does not possess the interest you at first thought it might possibly have but I cannot regret having sent it as it has procured for me the pleasure and honour of the communications with which you have favoured me.[4] The resemblance of the black powder on the anthers to the smut of wheat had struck me but I never thought of the obvious inference that it was a similar cryptogamic growth. And there are one or two points which still excite my curiosity How is it that this parasite *invariably* attacks the hermaphrodites, and how does it get into the unopened buds? Can it be that the plant is weakened in the endeavour to perfect both stamens and pistils and so rendered liable to the attacks of disease. And does this presumed weakness in the plant account for the small size of the pistils?[5]

I found one solitary f⟨emale⟩ plant with small p⟨istils⟩ and upon watching it for three days to see if the pistils would grow, I cut off and send you the only flower expanded. Though the plant looks vigorous enough I suspect it is accidental weakness and not divergence of form which causes the peculiarity but I will try to save seed from it and also test the goodness of the seed of the hermaphrodites.

I am sorry to have troubled you with sending the roots now that they are not likely to prove interesting[6] indeed I hope they will not do mischief by spreading among other plants the insinuating parasite with which they ⟨ar⟩e charged, and I beg you to believe that I am not so unreasonable as to expect you to give yourself any further trouble on my account though I should be very glad if my difficulties could be satisfactorily explained.[7]

Again thanking you for your extreme courtesy and kindness—I am Sir—yours much obliged and very respectfully | L. E. Becker

DAR 160: 109

[1] The year is established by the relationship between this letter and the letters from L. E. Becker, 18 May 1863, 21 May [1863], and 23–4 May [1863].

[2] The letter to Becker has not been found. Becker probably refers to 'Two forms in species of *Linum*'; however, her name does not appear on CD's presentation list for the paper (see Appendix IV).

[3] The abstract has not been identified. Becker refers to *Lychnis dioica* (*Silene dioica*) (see letter from L. E. Becker, 18 May 1863).

[4] These letters to Becker have not been found. However, CD was evidently interested in the specimens of *Lychnis dioica* because he thought they might be dimorphic (see letters from L. E. Becker, 21 May [1863] and 23–4 May [1863]).

[5] In a missing letter, CD suggested that where the fungus had destroyed the pollen at an early stage, the pistil was developed in compensation (see letter from L. E. Becker, 23–4 May [1863], CD note, and Becker 1869). See also n. 7, below.

[6] See letter from L. E. Becker, 23–4 May [1863].

[7] Becker concluded that the fungus induced the development of stamens in female plants of *Lychnis dioica* (Becker 1869).

From John Scott 28 May [1863][1]

Botanic Gardens [Edinburgh]
May 28th.

Sir,

I enclose copy from the Edinburgh Courant, of notice of late meeting of the Bot. Soc.[2] The abstract of paper on Orchid I made at the request of the Secretary;[3] as I did it hurriedly with pencil, however, he has copied it very incorrectly; indeed in one place it is impossible to understand it at all. This I have attempted to correct in enclosed copy.— Similar mistakes will therefore be occurring in Gard. Chron.[4] It is very teasing.

You will perhaps be surprised at title now given. I think it is much less appropriate than the one I originally intended—Individual Sterility of Orchids: which clearly defined its nature.[5] Prof. Balfour proposed that now substituted: it has given it the appearance of a simple record of experiments destitute of any theoretical application.[6] The general reader at least will be somewhat surprised on passing from title to introductory remarks on Sterility! I was also anxious to emphasize the possibly individually limited nature of sterility manifested under my experience. Some may now be inclined to attribute to me a belief in the general sterility of their cultivated individual representatives; which is far from the truth. Indeed—as I have already stated—I am inclined to suppose that this will rarely if ever be the case—structural dimorphism of course excepted— certainly in cases of *Passiflora* & *Lobelia fulgens*, it has only a limited application. *Individual* plants of the latter, as well as the former, produce *seeds* with *own pollen*—while others are absolutely *self-sterile*.[7]

Perhaps farther illustrations may be afforded after the notice of my experiments in Gard. Chron. In the event of such it will be necessary to state the existence of an **individual** *sterility*; otherwise, they might be thought to disprove mine: as I have made no special allusion to it.

Will you kindly excuse the liberty I take, in asking you—time & health permitting —to reply and afford an explanation of the different results.[8]

In notice you will also find a curious case of variation—at least the writer seems to consider it such—[9] Perhaps, I am wrong, in regarding it as of a similar nature to the well-known *laburnum case*.[10] The specimens sent are much mutilated—from careless packing—which prevents satisfactory comparisons; otherwise we might have been better able to judge. Differing, however, as it does in so many characters from the supposed normal form—more evidence is certainly required before we can regard it as simply a "*bud-variation*".[11]

I remain | Sir | Yours very respectfully | J. Scott

DAR 177: 92

CD ANNOTATIONS

2.1 You will ... nature 2.3] *scored brown crayon*
2.12 *Individual* plants ... *self-sterile* 2.13] *scored brown crayon*
5.1 In notice ... variation—] *cross in margin, brown crayon*

1 The year is established by the reference to the *Edinburgh Evening Courant* of 28 May 1863 (see n. 2, below).

2 Scott refers to the *Edinburgh Evening Courant*, 28 May 1863, p. 8, which carried an article on the meeting of the Botanical Society of Edinburgh on 14 May 1863. See also n. 3, below.

3 Scott refers to an abstract of his paper on pollination experiments with orchids, which was read before the Botanical Society of Edinburgh on 14 May 1863. The paper was later published in full in the transactions of the society (Scott 1863a). There is a copy of Scott 1863a in the Darwin Pamphlet Collection–CUL. The honorary secretary of the Botanical Society of Edinburgh was Robert Kaye Greville (*Medical directory* 1863).

4 An abstract of Scott's paper on pollination experiments with orchids (see n. 3, above), also appeared in the *Gardeners' Chronicle*, 13 June 1863, p. 558. See also letter from John Scott, [3 June 1863].

5 Scott's paper was entitled 'Experiments on the fertilisation of orchids in the Royal Botanic Garden of Edinburgh' (*Edinburgh Evening Courant*, 28 May 1863, p. 8). The paper demonstrated the sterility of several species of *Oncidium* when flowers were pollinated with pollen from the same plant ('individual sterility'), and also showed that the same plants could be successfully fertilised when pollinated with pollen from a different species of *Oncidium*.

6 As regius keeper of the Royal Botanic Garden, Edinburgh, John Hutton Balfour was Scott's employer, as well as a leading figure in the Botanical Society of Edinburgh. Scott's main theoretical point was that as there were varying degrees of sterility in the vegetable kingdom, which did not necessarily conform to recognised systematic affinities, Thomas Henry Huxley's use of sterility as the crucial test of the origin of species by natural selection (T. H. Huxley 1863b, pp. 107–8) was flawed (Scott 1863a, pp. 543–5). For a discussion of the background to this debate, see also *Correspondence* vol. 10, Appendix VI.

7 Scott expressed similar reservations about the general applicability of observations documenting own-pollen sterility in *Passiflora* and *Lobelia* in his letters to CD of 3 March 1863 and 21 March [1863]. CD had encouraged Scott to experiment with *Passiflora* and *Lobelia* to provide data on plants more easily fertilised by the pollen of another species than by their own pollen (see *Correspondence* vol. 10, letters to John Scott, 19 November [1862] and 11 December [1862] and n. 21). He sent further information to assist Scott in his experiments on *Passiflora* in March 1863 (see letter from John Scott, 3 March 1863, and letter to John Scott, 6 March 1863). Scott published his observations on sterility in self-pollinated *Passiflora* in Scott 1864d; they are summarised in *Variation* 2: 137–8. See also letter to John Scott, 31 May [1863] and n. 7.

8 See letter to John Scott, 31 May [1863].

9 The reference is to a letter from William Bell, read at the meeting of the Botanical Society of Edinburgh, 14 May 1863, describing a plant of *Hibiscus tricuspis* that produced branches bearing leaves and flowers widely different from the normal forms (*Edinburgh Evening Courant*, 28 May 1863, p. 8).

10 *Cytisus adami*, Adam's laburnum, was thought to be a hybrid between *C. laburnum* and *C. purpureus*, two very distinct species. This tree was described in *Variation* 1: 387: 'Throughout Europe, in different

soils and under different climates, branches on this tree have repeatedly and suddenly reverted to both parent-species in their flowers and leaves'. CD did not regard this as a case of bud-variation (*Variation* 1: 389–91).

11 CD defined bud-variation as 'sudden changes in structure or appearance which occasionally occur in full-grown plants in their flower-buds or leaf-buds' (*Variation* 1: 373). CD did not include the case of *Hibiscus tricuspis* in the chapter on bud-variation in *Variation* (*ibid.*, pp. 373–411). CD had asked Scott to send examples of bud-variation (see letter to John Scott, 16 February [1863] and n. 7).

From J. T. Austen 29 May 1863

Netherbu⟨ry,⟩ Beaminster
Ma⟨y⟩ 29— 1863

Dear Sir

I have not heard from Mr Dennen this morning so that I shall not be able to let you k⟨n⟩ow what he says in reply to my enquiries before Tuesd⟨ay⟩[1]

I rem⟨ain⟩ | yours very truly | John Thos. Au⟨sten⟩

DAR 159: 150

1 See letter from J. T. Austen, 27 May 1863. The reference is to George Dennen.

From W. D. Fox 29 May [1863][1]

Delamere | Northwich
May 29

My dear Darwin

I am indeed sorry to receive such a poor account of you.[2] As Gully is not to be had[3]—I should recommend you a trip to Ben Rhydding near Ilkley in Yorkshire. Dr MacLeod is very careful and the place is a very nice one in many ways.[4] Why not go there for a fortnight *before* the holidays if you are so poorly—taking only Mrs Darwin[5] with you?

I fear your work will kill you at last. You want cessation from Brain work more than any thing else.

As A proof I think so I will not add more than our kind regards to Mrs Darwin & yourself | Ever yours truly | W D Fox

DAR 164: 179

1 The year is established by the relationship between this letter and the letter to W. D. Fox, 23 May [1863].

2 See letter to W. D. Fox, 23 May [1863].

3 James Manby Gully, physician and proprietor of a hydropathic establishment in Malvern, Worcestershire, was suffering from a severe illness (see letter from W. D. Fox, [16–22 May 1863] and n. 4).

4 William MacLeod was the senior physician at the hydropathic establishment at Ben Rhydding, near Otley, Yorkshire (*Medical directory* 1863).

5 Emma Darwin.

To J. D. Hooker 29 May [1863]

Down Bromley Kent
May 29th

My dear Hooker.

You have been uncommonly kind about Scott.[1] I forwarded your note,[2] but told him not to mention it; it will certainly be useful to him on any future occasion. For the same post that brought your note, brought one from him, with apologies & saying that Balfour had allowed him only 24 hours to decide, & being so pressed he had declined.[3] I sent him a preaching about Mr Macnab, & I daresay he is to blame, & all that you say about permitting men to experimentise is very true.[4] He certainly has tried a stupendous number. He told me formerly he did them out of work-hours.[5] He is evidently an eccentric man; an orphan of parents who were better off & who took to gardening as best chance of some science.[6] If he had leisure he would make a wonderful observer; to my judgment I have come across no one like him.—

Whether I have done right or wrong I know not, but I have urged him to publish on one little point, viz the impotence of certain orchids with their own pollen,[7] like Herbert's Crinum case.[8] When I get his paper in some newspaper I will send it you to read; if it seems at all worth while.[9] He has made lots of curious observations on pollen-tubes coming out of anthers in certain orchids & travelling all the way to the stigma.[10]

Many thanks for Nageli,[11] which shall soon be returned: owing to subject & the German I cannot understand all; but there are some very important statements, enough to convince me that Falconer's remark (which started me) on phyllotaxy being as fixed as law of gravity is much exaggerated:[12] there is, however, no explanation, as far as I can understand, why angles jump from one to another.— I suppose you have seen Asa Gray on A. Decandolle & on Falconer &c &c in Silliman;[13] he sent me 2 copies by mistake; but I presume you have one.— What a hard worker he is, & what a clever fellow: I wish this cursed war was over; if it were only to be at peace with him.—[14]

What a lot of gossip you told me in your last letter;[15] I am so heartily glad that you appreciate John Lubbock: I always fear that he will wear himself out.—[16]

If you have time & care to look at a pretty case of fertilisation in Leguminosæ, (far prettier than in that strange-looking Clianthus) look at common Broom.[17] It matures its pollen very early & you will see the pistil curled like a french horn out of the keel: I have seen two accounts that all this is for self-fertilisation.[18] All this is a mistake; if insects are excluded the pistil does not come out. When a bee alights to suck, the short anthers first protrude & rub & dust the under side of its thorax: then when the bee pushes harder or flies to an older flower, out flies the pistil & long anthers & strike (as as I have seen) the upper surface of thorax, where the stigma is well rubbed with its own pollen & that brought from other flowers; but here comes the pretty part, as soon as Bee flies away the pistil from its elasticity curls so much as to almost touch the shorter protruded anthers; & the stigma is now turned *upwards* & rubs against the *under* side of the thorax of the next bee dusted with pollen from many flowers. I never saw a clearer case of adaptation.— Subsequently the pistil curls uselessly still more.—[19]

Have you read Kinglake?[20] I am reading it with much amusement.

Good-night— | Ever yours | C. Darwin.

Endorsement: '/63.'
DAR 115: 195

[1] The reference is to John Scott (see letter to J. D. Hooker, 23 May [1863], and letter from J. D. Hooker, [23–7 May 1863]).

[2] The note has not been found; it was sent to CD with the letter from J. D. Hooker, [23–7 May 1863], and forwarded with the letter to John Scott, 25 and 28 May [1863]. The note gave Scott advice concerning the post he had been offered in Darjeeling, India (see letter from John Scott, 22 May 1863).

[3] See letter from John Scott, 26 May [1863]. The reference is to John Hutton Balfour.

[4] See letter to John Scott, 25 and 28 May [1863]. In his letter to CD of [23–7 May 1863], Hooker implied that the friction between Scott and James McNab, Scott's immediate superior, probably resulted from McNab's feeling that Scott's experiments were interfering with his job. Scott was foreman in the propagating department at the Royal Botanic Garden, Edinburgh.

[5] The letter in which Scott gave this information has not been found.

[6] See letter from John Scott, 6 January 1863 and nn. 23 and 24.

[7] Scott described these experiments in his letter to CD of [1–11] April [1863]; CD responded in his letter to Scott of 12 April [1863] with the suggestion that he publish his work. Scott read a paper on the experiments to the Botanical Society of Edinburgh on 14 May 1863; this was published in the society's transactions (Scott 1863a).

[8] The reference is to Herbert 1837, p. 32, where it was noted that various species of *Crinum* sent to Calcutta had crossed so freely that pure seed could not be obtained. CD used this case as an example of pollen from a distinct species obliterating over time the action of a plant's own pollen (*Cross and self fertilisation*, p. 395 n.).

[9] Scott had written an abstract of his paper on orchid sterility, read at the Botanical Society of Edinburgh (see n. 7, above), for the *Edinburgh Evening Courant* (see letter from John Scott, 26 May [1863]).

[10] See letters from John Scott, 21 March [1863] and [1–11] April [1863].

[11] The reference is to the first part of Nägeli 1858–68. See letter to J. D. Hooker, 15 and 22 May [1863], and letter from J. D. Hooker, [24 May 1863].

[12] The reference is to Hugh Falconer and his remark on phyllotaxy in Falconer 1863a, p. 80 (see letter to Daniel Oliver, 20 [February 1863], n. 2). CD was actively pursuing research on phyllotaxy during April and May 1863 (see letters to Asa Gray, 20 April [1863] and 11 May [1863], letter to J. D. Hooker, [9 May 1863], and memorandum from G. H. Darwin, [before 11 May 1863]). Asa Gray's work endorsed CD's conclusion that Falconer's claims regarding phyllotaxy were exaggerated (see letter from Asa Gray, 26 May 1863 and n. 25).

[13] A. Gray 1863d and 1863e. A. Gray 1863d was a review of A. de Candolle 1862a, and included a discussion of Falconer 1863a; A. Gray 1863e reviewed A. de Candolle 1863b. CD's annotated copies are in the Darwin Pamphlet Collection–CUL. The *American Journal of Science and Arts* was commonly known as 'Silliman's journal', after its founder Benjamin Silliman.

[14] The reference is to the American Civil War. The war introduced considerable tension into Anglo-American relations (see Ferris 1977 and *Correspondence* vols. 9 and 10), and this was reflected in Gray's correspondence with CD and Hooker. See, for example, *Correspondence* vol. 10, letter from J. D. Hooker, [19 January 1862], and this volume, letters to Asa Gray, 2 January [1863] and 23 February [1863], and letter from Asa Gray, 11 April 1863.

[15] Letter from J. D. Hooker, [24 May 1863].

[16] See letter from J. D. Hooker, [24 May 1863]. John Lubbock was the 'closest of CD's younger friends'

and had been a frequent visitor to Down House from childhood (Freeman 1978, p. 192), having grown up at High Elms, near Down.

17 CD refers to *Clianthus dampieri*, which Hooker suggested would be of interest with regard to the study of pollination mechanisms (see letter from J. D. Hooker, [13 May 1863], and letter to J. D. Hooker, 15 and 22 May [1863]), and to *Cytisus scoparius*, the common broom.

18 These references have not been identified.

19 CD's notes on the pollination of the common broom (*Cytisus scoparius*) are given in G. Henslow 1866 (*Collected papers* 2: 134) and *Cross and self fertilisation*, pp. 163–4. See also letter to Asa Gray, 31 May [1863].

20 CD refers to the first volume of Alexander William Kinglake's history of the Crimean War (Kinglake 1863–87).

From J. T. Austen 30 May 1863

Slape House | Netherbury— | Beaminster— Dorset
May 30— 1863

My dear Sir,

I ⟨inclose⟩ you Dennen's letter which I have just received.[1] I do so because it w^d^ give me a good deal of trouble to epitomize the contents of it, but I must request you to consider it *private*, for it is a stupid letter and if made use of would raise perhaps an *unfair* prejudice against him, it was written to me certainly in confidence— I have told him in reply that the case is simply this— The Savings Bank is founded to protect the savings of the poor, to induce & enable them to invest their savings where they will be secure instead of risking them with adventurers for the sake of hig⟨her⟩ interest, and therefore he o⟨ug⟩ht not to have made such a pr⟨op⟩osition to Percival—[2] I have not in⟨s⟩isted on knowing who the partie⟨s were⟩ to whom the money was to hav⟨e⟩ been lent, because perhaps it would be a breach of confidence on his part to disclose the name—

It appears to me that he has been only very stupid; perhaps you will kindly let me know whether you think the matter ought to be formally brought to the knowledge of the Trustees.[3] I would not bring it forward as a *charge* against the poor little man, but merely as a matter which ought to be noticed. I am almost of opinion that if you do not mind the trouble, it w^d^ be well for Dennen to call on you when convenient to you to see him and then you might obtain from him such explanation as you think it right he should give. He has certainly done the work of the Bank carefully and well. he appears to be trustworthy. by the system now adopted at the Bank no money passes through his hands, he does not in any way intervene between the Depositor & the Manager and he has in no instance that I am aware of sought to do so he understands accounts— any public en⟨qu⟩iry into this business would certainly so far prejudice him as to lose him support in any application for employment in a responsible situation. I therefore would not publish this affair without due consideration, & hope you will not do so—

I remain | my dear Sir | yrs very truly | John Thos. Austen.[4]

Chas. Darwin Esqre—

DAR 159: 131

[1] The letter from George Dennen to Austen has not been found. See letters from J. T. Austen, 27 May 1863 and 29 May 1863.
[2] This individual has not been identified.
[3] CD and Austen were trustees of the Savings Bank, High Street, Bromley (see letter from J. T. Austen, 3 June 1863). For an account of the bank's history, see Horsburgh 1980.
[4] Austen appended a postscript to this letter, but failed to enclose it. It was sent with the letter from J. T. Austen, 3 June 1863.

From Philip Henry Gosse 30 May 1863

Sandhurst. Torquay
May 30. 1863

My dear Sir

Will you kindly vouchsafe me a little word of help. With your charming book before me, I have been trying to fertilise the Orchids of my little collection, as they flower.[1] With some I succeed, with others there is difficulty. Let me tell you of the present "fix".

Stanhopea oculata opened four great blooms on Thursday: today they begin to flag, & I delay no longer to impregnate. I reach down your book, turn to your figure at p. 179,[2] & recognise the parts well enough. Then with a toothpick I lift the anther, & out come the pollinia very well depicted by you at p. 185 fig C;[3] except that in this species the poll. masses are much larger in proportion to the visc. disk. The disk is viscid enough, & I carry the whole on the toothpick. Now I want to find where to deposit it; I take for granted that it is in the hollow (marked a in my sketch,[4] wh. is the stigma. But there is no viscosity there, nor anywhere near, up or down, not the slightest; & I cannot get the pollinia to adhere.[5] How can this plant be fertilised? And how w^d. any insect do it? & what w^d. an insect be about, to touch the tip of this isolated projecting column? Supposing the great bee, or Scolia,[6] or what not, wants to get at the hollow hypochil[7] (tho' I don't find any honey there), he w^d. alight on the epichil[8] whose surface is already three fourths of an inch from the rostellum, & which being moveable w^d. bend away still further,—& creep between the horns of the mesochil:—[9] how thus could he touch the anther? & if he did, how c^d. he lodge the pollen on the stigma? & if he did, how c^d. it stick, seeing the place is not sticky?

Do resolve me these doubts; & believe me, | My dear Sir | Ever yours truly | P. H. Gosse

The *disk* at the end of the caudicle adheres to the stigma, but the *pollen masses* project, & won't touch it, tho' pressed against it with force.

Charles Darwin Esq^e.

DAR 165: 76

[1] Gosse refers to *Orchids*. Although primarily a zoologist, Gosse devoted much time during his later years to the study of orchids (R. Desmond 1994).

[2] Figure 22 (*Orchids*, p. 179) is a longitudinal section showing the structure of the column in the Vandeae. The column of an orchid flower is formed of three carpels, and generally four stamens, all confluent.

[3] Figure 23 (*Orchids*, p. 185) depicts the pollinia of four species of Vandeae; 'c' depicts the pollinium of *Stanhopea saccata* 'after depression'. It shows the two pollen masses, their mutual stipe, and the viscid disc at its base.

[4] The sketch has not been found.

[5] For pollination to be effected, the pollen masses or individual pollen grains must adhere to the stigma.

[6] *Scolia flavifrons*, the yellow-fronted digger wasp, is the largest European species of Hymenoptera (Grzimek ed. 1975, pp. 424–5).

[7] The hypochil is the basal part of the labellum, usually the largest of the three petals, of an orchid flower (Bailey and Bailey 1976).

[8] The epichil is the terminal part of the labellum of an orchid (Bailey and Bailey 1976).

[9] The mesochil is the intermediate part of the labellum in those orchids that have the labellum separated into three distinct parts. The rostellum is an extension of the upper edge of the stigma of an orchid, forming a flap of sterile tissue separating the stigmatic surface from the anthers in the column (Bailey and Bailey 1976).

To Asa Gray 31 May [1863][1]

Down Bromley Kent
May 31st

My dear Gray.

I was very glad to receive your Review of Decandolle a week ago.[2] It seems to me excellent & you speak out, I think, more plainly in favour of derivation of species, than hitherto, though doubtfully about natural selection.[3] Grant the first, I am easy about the second. Do you not consider such cases as all the Orchids next thing to a demonstration against Heer's view of species arising suddenly by monstrosities:[4] it is impossible to imagine so many coadaptations being formed all by a chance blow. Of course Creationists would cut the enigma.

What an indomitable worker you are! Why these Reviews, supposing I were to attempt them, would take me a month's work.[5] I have written twice to you not very long ago,[6] & sent 2 copies of my Linum paper;[7] but they & letter were sent about time of sailing of Anglo-Saxon, & were perhaps lost.[8] I only remember in my letter telling you how right you were about fertilisation of Cypripedium.[9] Of the species sent by you, C. acaule alone has flowered & has puzzled me. Mitchella, alas, does not look very healthy with all our care.[10] If you see & know Mr. Scudder please thank him particularly for his interesting paper on Pogonia, which I was very glad to read.—[11]

To return to your Review: I was very glad to see your Remarks in answer to Falconer on Phyllotaxy;[12] I infer you cannot explain why there are not intermediate angles. I have been looking at Nageli's work on this subject,[13] & am astonished to see that angle is not always the same in young shoot when leaf-buds are first distinguishable as in full-grown branch. This shows, I think, that there must be some potent cause for those angles which do occur: I daresay there is some explanation as simple as that for the angles of the Bees-cells.—[14]

You allude to Saporta's work;[15] Alp De Candolle sent me a copy of part of letter from him, in which he expressed strong belief that N. Selection would ultimately be triumphant in France, though now quite ignored.—[16]

I have nothing to tell you about my own doings: I work every day, that I can, on my big book & am now at all causes of sterility under domestication & cultivation.[17] I have got such an immense collection of facts, that the work though laborious & slow interests me, as I can generally come to some sort of conclusion. There never will be a man who will read my big book; it will be a sort of encyclopedia on special cases.—[18]

I have been looking again at the imperfect flowers of Oxalis & Viola: I was entirely wrong in supposing that in Oxalis the perfect flowers required insect-aid for fertilisation; so this view is knocked on the head. Viola, however, does require insects.[19] I must yet stick to my opinion that the imperfect flowers of Viola at least deserve to be ranked as something more than mere precocious flowers. In V. canina only 2 anthers are developed; the pollen-grains are smaller—the pistil widely different in shape; no nectar-appendages to the two fertile stamens & no spur.—[20] Remember, if you can get them, seed of Campanula perfoliata.—[21]

I suppose you are very busy, & I suppose the whirlwind of public affairs must waste much of your time. Do not think of writing to me unless you have any leisure; though a letter from you is always a real pleasure to me. I suppose there are few human beings in England who see so few persons out of their own family as I do.—

Good night. | Yours most sincerely | C. Darwin

I have been observing common Broom: hardly any orchid shows prettier adaptation to insects which are necessary for its fertilisation:— The *upper* & *lower* surface of thorax of Bees gets dusted with pollen, & first the stigma rubs the upper side of thorax & afterwards is rubbed by the lower side of thorax.—[22]

Gray Herbarium of Harvard University (84)

[1] The year is established by the reference to Gray's review of A. de Candolle 1862a (see n. 2, below), published in May 1863.

[2] Gray's review of A. de Candolle 1862a appeared in the *American Journal of Science and Arts* (A. Gray 1863d). There are annotated copies of A. de Candolle 1862a and A. Gray 1863d in the Darwin Pamphlet Collection–CUL.

[3] In his review, Gray argued that the views adopted by Alphonse de Candolle were representative of a general consensus among eminent naturalists, that is, that species had come into existence coincident in time and space with pre-existing and closely allied species, and that the distribution and origin of congeneric species had come to be 'regarded as something derivative' (A. Gray 1863d, pp. 437–8). On natural selection, Gray stated his opinion that: 'something more than natural selection is requisite to account for the orderly production and succession of species' (*ibid.*, p. 440).

[4] In his review of A. de Candolle 1862a, Gray noted that Candolle supported the concept of the 'derivation of species', but had been unclear as to whether he viewed that process as one of slow and very gradual change, following CD's view of the origin of species, or whether he believed, with the Swiss palaeobotanist Oswald Heer, that sudden, inexplicable change, and the production of monstrosities, were causal factors in the production of new species (A. Gray 1863d, p. 438). Heer's views on the origin of new species were developed within the context of his studies of Tertiary floras (Heer 1855–9).

[5] Gray was a regular reviewer for the *American Journal of Science and Arts*, a bimonthly publication.

[6] Letters to Asa Gray, 20 April [1863] and 11 May [1863].

[7] Gray acknowledged the receipt of 'Two forms in species of *Linum*' in his letter to CD of 26 May 1863. For CD's presentation list for this paper, see Appendix IV.

[8] See letter from Asa Gray, 26 May 1863, n. 4.

[9] See letter to Asa Gray, 20 April [1863] and n. 24.

[10] Gray had sent CD specimens of *Cypripedium acaule* and *Mitchella repens* to grow and examine for pollination mechanisms and heterostyly, respectively; CD received the plants at the end of December 1862 (see *Correspondence* vol. 10, letter from Asa Gray, 9 December 1862, and letter to J. D. Hooker, 29 [December 1862]). For CD's research on the genus *Cypripedium*, including *C. acaule*, see the letter from Asa Gray, 26 May 1863 and nn. 17, 18, and 20. CD had been interested in obtaining specimens of *Mitchella repens* since Gray suggested that the species would be a good case-history for research on dimorphism (see *Correspondence* vol. 9, letter from Asa Gray, 11 October 1861, and *Correspondence* vol. 10, letters to Asa Gray, 9 August [1862] and 6 November [1862]).

[11] The reference is to Scudder 1862c, which discussed the structure and manner of pollination of the orchid *Pogonia ophioglossoides*. CD cited Scudder 1862c in *Orchids* 2d ed., p. 86. Samuel Hubbard Scudder was Gray's colleague at Harvard University.

[12] In reviewing Hugh Falconer's assertion that phyllotaxy was governed by fixed physical laws (Falconer 1863a, p. 80), Gray pointed out that there were two distinct 'species' of phyllotaxy, alternate leaves and verticillate leaves, and that there could be a change from one to the other on the same stem. Gray commented: 'Here is a change from one fixed law to another, as unaccountable, if not as great, as from one specific form to another' (A. Gray 1863d, p. 440). See also letters to J. D. Hooker, [9 May 1863] and n. 10, and 29 May [1863] and n. 12, and letter from Asa Gray, 26 May 1863 and nn. 24 and 25.

[13] The reference is to the first part of Nägeli 1858–68. See letter from J. D. Hooker, [13 May 1863].

[14] CD believed the angles of the walls of bee-cells had been governed by natural selection for the purpose of economising wax (*Origin*, pp. 224–35).

[15] Gray discussed Saporta 1862a in A. Gray 1863d, pp. 440–1.

[16] In the letter to Alphonse de Candolle, 31 January [1863], CD acknowledged receiving news of the views of Gaston de Saporta. Neither the letter from Candolle, nor that from Saporta, has been found. Saporta was the principal founder of evolutionary palaeobotany in France (Tort 1996, pp. 3771–4). On the reception of *Origin* in France, see also Conry 1974, Farley 1974, Stebbins 1974, and Corsi and Weindling 1985.

[17] CD worked on a draft of the chapters for *Variation* on crossing and sterility (*Variation* 2: 85–191) between April and mid-June 1863 (see 'Journal' (Appendix II)).

[18] CD may refer to *Variation* (see n. 17, above), published in two volumes in 1868; it was the first part of a planned three-part 'big book' on the evolution of species (see *Origin*, p. 1, and *Correspondence* vol. 7, letter to T. H. Huxley, 16 December [1859], and letter to John Murray, 22 December [1859]). *Variation* was the only part of the 'big book' to be published during CD's lifetime; the remaining draft chapters were published posthumously as *Natural selection*.

[19] In the letter to the *Journal of Horticulture and Cottage Gardener*, [17–24 March 1863], CD had written:

> But it is known that several plants, such as Violets, Campanulas, Oxalis, &c., produce two kinds of flowers: one sort adapted for self-fertilisation, and the other sort for fertilisation by insect agency or other means.

CD made further observations on *Oxalis acetosella* between 31 March and May 1863 (see letter to John Scott, 12 April [1863], and experimental notes in DAR 111: 46 and DAR 109: B6). CD initially believed the 'perfect flowers' of the species to be heterostyled and hence adapted to cross-pollination by insects, but later concluded that the species was homostyled (see *Forms of flowers*, p. 182). CD's observations on *Viola* were published in *Forms of flowers*, pp. 315–21 and 336. CD had been experimenting on 'imperfect' (cleistogamic) flowers of *Oxalis* and *Viola* since 1862; he summarised his provisional conclusions on their function in the letter to Asa Gray, 26[–7] November [1862] (*Correspondence* vol. 10). There is a series of experimental notes on 'imperfect' flowers of *Viola* and *Oxalis*, dated between 15 April 1863 and 15 May 1863, in DAR 111: 8–11, 46.

[20] The flowers of *Viola canina* are described in *Forms of flowers*, pp. 314–17. CD used the case of 'imperfect' (cleistogamic) flowers in *Viola* as an example of how elements of flower structure, superfluous to the pollination mechanism, could gradually disappear (*Forms of flowers*, p. 336).

[21] CD had asked for seeds of this species, also called *Specularia perfoliata*, in connection with his research on cleistogamy (see letter to Asa Gray, 20 March [1863] and n. 5).

[22] CD's notes on the pollination mechanism of the common broom (*Cytisus scoparius*) are given in G. Henslow 1866 and *Cross and self fertilisation*, pp. 163–4. See also letter to J. D. Hooker, 29 May [1863].

To John Scott 31 May [1863][1]

Down
May 31st

Dear Sir

I am unwell & must write briefly.—

I am very much obliged for the Courant.—[2] The facts will be of *highest* use to me: I feel convinced that your paper will have permanent value.—[3] Your case seems excellently & carefully worked out.— I agree that alteration of Title was unfortunate; but after all title does not signify very much.—[4] So few have attended to such points, that I do not expect any criticisms: but if so, I shd. think you had much better reply; but I would if you wished it much.— I quite understand about the cases being *individual* sterility; so Gärtner states it was with him.—[5]

Would it be worth while to send a *corrected* copy of Courant to Gard. Chron.?[6] I did not know that you had tried Lobelia fulgens: can you give me any particulars, on number of plants & kinds used &c that I may quote, as in few days I shall be writing on this whole subject.—[7] No one will ever convince me that it is not a very important subject to Philosophical naturalists. The Hibiscus seems very curious case & I agree with your remarks.—[8]

You say you are glad of criticisms[9] (by the way avoid "former & latter," the reader is always forced to go back to look): I *think* you would have made case more striking if you had *first* showed that pollen of O. sphacelatum was good; secondly that ovules were capable of fertilisation, & lastly shown that the plant was impotent with own pollen.— "impotence of organs capable of elimination"—capable here strictly refers to organs, you mean to impotence— To *eliminate* impotence is a curious expression. it is removing a non-existence quality.— But style is a trifle compared with facts, & you are capable of writing well. I find it a good rule to *imagine* that I want to explain the case in as few & simple words as possible to one who knows nothing of subject.— I am tired— In my opinion you are an excellent observer.—

Yours very faithfully | C. Darwin

DAR 93: B47–8

[1] The year is established by the relationship between this letter and the letter from John Scott, 28 May [1863].

[2] See letter from John Scott, 28 May [1863]; Scott had sent a copy of a report of the meeting of the Botanical Society of Edinburgh on 14 May 1863 (*Edinburgh Evening Courant*, 28 May 1863, p. 8).

[3] The report in the *Edinburgh Evening Courant* (see n. 2, above) contained an abstract of Scott's paper on orchid pollination, which was subsequently published in full as Scott 1863a. See also letter from John Scott, 28 May [1863] and nn. 3–6.

[4] The title of Scott's paper on orchid pollination (see n. 3, above) had been changed from 'Individual sterility of orchids' to 'Experiments on the fertilisation of orchids in the Royal Botanic Garden of Edinburgh'. See letter from John Scott, 28 May [1863] and n. 5.

[5] See letter from John Scott, 28 May [1863]. CD refers to Gärtner 1849, pp. 64 and 357, which gave details of a number of crossing experiments with *Lobelia* and *Verbascum*. The experiments demonstrated the sterility of some individual plants when crossed with other plants of the same species; these same plants could be pollinated readily by pollen from a distinct species (see also *Variation* 2: 136–7).

[6] The abstract of Scott's paper on orchid pollination that appeared in the *Edinburgh Evening Courant* (see n. 3, above) was also due to be reprinted in the *Gardeners' Chronicle*. Scott had complained to CD that errors had been introduced into the abstract by Robert Kaye Greville, honorary secretary of the Botanical Society of Edinburgh (see letter from John Scott, 28 May [1863]). A revised version of the abstract appeared in *Gardeners' Chronicle*, 13 June 1863, p. 558.

[7] See letter from John Scott, 28 May [1863] and n. 7. CD was writing a draft chapter on crossing and sterility for *Variation* (see 'Journal' (Appendix II)). CD refers to a section that was published under the subtitle: 'On certain plants which either normally or abnormally are self-impotent, but are fertile, both on the male and the female side, when crossed with distinct individuals either of the same or another species' (*Variation* 2: 114, 131–40). CD cited Scott's experiments on orchid pollination and Gärtner's experiments on *Lobelia* and *Verbascum* (see n. 5, above), but did not mention Scott in connection with *L. fulgens*.

[8] See letter from John Scott, 28 May [1863] and nn. 9 and 10.

[9] CD's criticisms relate to the abstract of Scott's paper on orchid pollination that appeared in the *Edinburgh Evening Courant* (see n. 3, above).

To P. H. Gosse 2 June [1863][1]

Down. | *Bromley.* | *Kent. S.E.*
June 2^d

My dear Sir

It would give me real pleasure to resolve your doubts, but I cannot. I can give only suspicions & my grounds for them. I should think the non-viscidity of the stigmatic hollow was due to the plant not living under its natural conditions.[2] Please see what I have said on Acropera;[3] an excellent observer M^r. J. Scott of Bot. Garden Edinburgh finds all that I say accurate, but nothing daunted he with the knife enlarged orifice & forced in pollen-masses; or he simply stuck them into the contracted orifice *without coming into contact with the stigmatic surface*, which is hardly at all viscid; when lo & behold pollen-tubes were emitted & fine seed capsules obtained.[4] This was effected with Acropera Loddigesii; but I have no doubt that I have blundered badly about A. luteola.[5] I mention all this because as Mr Scott remarks, as the plant is in our Hot-House it is quite incredible it ever could be fertilised in its native land.[6] The whole case is an utter enigma to me.—[7]

Probably you are aware that there are cases (& it is one of the oddest facts in physiology) of plants which under culture have their sexual functions in so strange a condition, that though their pollen & ovules are in sound state & can fertilise & be fertilised by distinct but allied species; they cannot fertilise themselves.[8] Now Mr Scott has found this the case with certain Orchids, which again shows sexual disturbance. He has read paper at Bot. Soc. of Edinburgh & I daresay an abstract which I have seen will appear in Gard. Chron; but blunders have crept in in copying & parts are barely intelligible.[9] How insects act with your Stanhopea, I will not pretend to conjecture. In many cases, I believe the acutest man could not conjecture without seeing the insect at work. I could name common English plants in this predicament.— But the Musk orchis is case in point. Since publishing my son & self have watched the plant & seen the pollinia removed & where do you think they *invariably* adhere in dozens of specimens; always to the joint of femur with the trochanter of 1st. pair of legs & no where else.[10] When one sees such adaptation as this, it would be hopeless to conjecture on the Stanhopea till we know what insect visits it. I have fully proved that my strong suspicion was correct that with many of our English orchids no nectar is excreted, but that insects penetrate the tissues for it.[11] So I suspect it must be with many foreign species.

I forgot to say, that if you find you cannot fertilise any of your exotics take pollen from some allied form & it is quite probable that will succeed.—

Will you have the kindness to look occasionally at your Bee Ophrys near Torquay & see whether pollinia are ever removed. It is my greatest puzzle.[12] Please read what I have said on it & on O. arachnites;[13] I have since proved account of latter is correct.[14] I wish I could have given you better information.

My dear Sir | Yours sincerely | Ch. Darwin

P.S. | If the flowers of the Stanhopea are not too old—remove pollen-masses from their pedicels & stick them with a little liquid pure gum to stigmatic cavity. After the case of the Acropera, no one can dare positively say that they would not act.—

The Brotherton Collection, Leeds University Library

1 The year is established by the relationship between this letter and the letter from P. H. Gosse, 30 May 1863.

2 In his letter to CD of 30 May 1863, Gosse described the difficulties he had experienced in attempting to pollinate the orchid *Stanhopea oculata*, noting in particular that what he assumed to be the stigma was not viscous, so that pollinia would not adhere to it.

3 CD described the floral morphology of the orchid genus *Acropera* in *Orchids*, pp. 204–10, noting that, in the specimens he had seen, the orifice of the stigmatic chamber was so narrow that the pollen-masses could 'hardly be forced in' (p. 205). As with Gosse's *Stanhopea*, CD considered the possibility that this may have been due to unnatural conditions, but concluded that his experience of orchids grown in hothouses furnished 'no reason' to believe this to be the case (p. 209).

4 See, for example, *Correspondence* vol. 10, letter from John Scott, 11 November 1862, and this volume, letter from John Scott, 18 February [1863]. John Scott was foreman of the propagating department at the Royal Botanic Garden, Edinburgh (R. Desmond 1994).

5 In *Orchids*, p. 203 n., CD reported that *Acropera luteola* differed 'in little or in nothing from *A. Loddigesii*', and that it might be a variety of the same species. CD refers to his conclusion that there was a separation of the sexes in *A. luteola* (see *Orchids*, pp. 206–9), which he later described as 'a great error', Scott's results having convinced him that the species of this genus were hermaphrodite (see 'Fertilization of orchids', p. 153 (*Collected papers* 2: 150)).

6 In his letter to CD of [after 12] April [1863], Scott made an observation of this nature with respect to orchids of the genus *Gongora*, which had also been found to have a stigmatic chamber with a narrow orifice.

7 Before he concluded that there was separation of the sexes in *Acropera*, CD had considered the genus to be the 'opprobrium' of his work on the pollination mechanisms of orchids, since all the floral parts 'seemed determinately contrived that the plant should never be fertilised' (*Orchids*, p. 203). For CD's subsequent explanation of his observations, see the letter to Friedrich Hildebrand, 28 July [1863], and 'Fertilization of orchids', p. 153 (*Collected papers* 2: 150).

8 CD described cases of this sort in *Origin*, pp. 250–1.

9 CD refers to Scott 1863a, which was read before the Botanical Society of Edinburgh on 14 May 1863. An abstract of the paper was published in the *Edinburgh Evening Courant*, 28 May 1863, p. 8; Scott enclosed a corrected copy of the abstract with his letter to CD of 28 May [1863]. The abstract was also published in the *Gardeners' Chronicle*, 13 June 1863, p. 558.

10 CD refers to George Howard Darwin's observations of *Herminium monorchis*, made between 22 and 27 June 1862 (see CD's notes in DAR 70: 32–6), and later published in the German translation of *Orchids* (Bronn trans. 1862, pp. 47–8 n.) and 'Fertilization of orchids', pp. 145–6 (*Collected papers* 2: 142–3). *Orchids* was published in May 1862 (Freeman 1977, p. 112).

11 In *Orchids*, pp. 44–53, CD described the apparent failure of several species of *Orchis* to produce nectar, despite possessing nectaries. CD's hypothesis was that these plants did produce nectar, but secreted it into the intercellular spaces between the inner and outer membranes of the nectary, an arrangement requiring insects to spend more time in the flower in order to extract the nectar by boring through the inner membrane of the nectary. While staying at Hartfield Grove, Hartfield, Sussex, in May 1863, CD made additional observations of the nectaries of *Orchis morio* and *O. maculata*, and of the behaviour of insects visiting the flowers, which confirmed his earlier conclusions (see 'Fertilization of orchids', pp. 142–3 (*Collected papers* 2: 140), and the dated observational notes in DAR 70: 176–7). See also letter to John Scott, 2 May [1863] and n. 15.

12 CD described the bee ophrys (*Ophrys apifera*) in *Orchids*, pp. 63–72, stating that he found the case 'perplexing in an unparalleled degree', since the flowers appeared to have 'elaborate contrivances for directly opposed objects', that is, adaptations for both cross- and self-fertilisation. One of CD's objectives in *Orchids* was to demonstrate that cross-fertilisation was the 'main object' of the 'contrivances' by which orchids were pollinated (*Orchids*, p. 1), noting in his conclusion to the book that *O. apifera* was the only species in which he had observed 'special and perfectly efficient contrivances for self-fertilisation' (*Orchids*, p. 359). CD had carried out extensive but inconclusive observations designed to establish whether the pollinia were ever removed by insects in this species (*Orchids*, pp. 66–70). Gosse lived at Babbacombe Bay, near Torquay, Devon, where CD had made some of these observations (*Orchids*, p. 66).

13 In *Orchids*, pp. 72–3, CD noted that many botanists considered *Ophrys arachnites* to be a 'mere variety' of *O. apifera*; however, having observed that there was no adaptation for self-pollination in *O. arachnites*, CD argued that it should be considered a 'good species'.

14 Following the publication of *Orchids* in May 1862, CD continued to be interested in the functional differences between *Ophrys apifera* and *O. arachnites* (see *Correspondence* vol. 10). CD reported his additional observations on *O. arachnites* in 'Fertilization of orchids', p. 145 (*Collected papers* 2: 142).

From J. T. Austen 3 June 1863

Netherbury.
June 3—1863

My dear Sir,

I am sorry to tr⟨oub⟩le you with so much correspondence; the accompanying postscript I wrote with my last letter,[1] but by accident omitted putting it into the

⟨en⟩velope; it was only this morning that I became aware of the mistake— I have not heard a second time from Mr Dennen—[2] like yourself I feel very much perplexed, but a⟨m⟩ much obliged to you for havi⟨ng⟩ written to me ⟨on⟩ the subject[3] ⟨I⟩ shall return home next we⟨ek⟩ and will take an early ⟨oppor⟩tunity of seeing M^r^ De⟨v⟩as ⟨and⟩ mentioning this bu⟨si⟩ness to hi⟨m.⟩[4] I quite agree with ⟨yo⟩u th⟨at⟩ we must not conceal this Transaction from our fellow Trustees.[5] Deva⟨s is⟩ a man of business, and ⟨I th⟩ink the best person to speak to on the subject— I shall not speak to any of the Managers[6] it is the business of the Trustees rather than theirs— ⟨do⟩ not trouble yourself to ⟨wr⟩ite to me again; I leave this ⟨h⟩ouse the day after tomorrow ⟨an⟩d am not certain of my ⟨mo⟩vements for the next 10 days ⟨As so⟩on as I have seen Devas I ⟨will c⟩all on you; I should ⟨no⟩t be surprized if he ⟨adv⟩ises some official communication being made to the Trustees. I observed to you in my last letter that Dennen had wr⟨it⟩ten very stupidly. I really do not th⟨ink⟩ that he has been acting in any way dishonestly but I certainly do not at present see any *satisfactory* explanation of the Transaction. I expected he would have told me the *whole* story, but he has made a mystery of it— his account represents himself as concerned in the business, because his offer to be security to Percival[7] for the repayment of the loan is, as a matter of business ⟨adm⟩itting ⟨that⟩ he has a beneficial interest ⟨in⟩ borrowing the money— what ⟨worries⟩ me is that he brings forward ⟨this⟩ offer of his as *proof* & *eviden⟨ce* of⟩ his own honesty, of his having ⟨acted⟩ bonâ fide; in doing this he is either uncommonly silly or very artful; uncommonly silly 1st in proposing to give the ⟨secu⟩rity, 2^ndly^ in thinking that by doing so he credits the transaction and makes it appear a disinterested one on his part— if he *be* an interested party himself, of course he is ⟨v⟩ery artful in representing it as he has done. ⟨Th⟩e worst of it is that if he be *⟨on⟩ly silly*, the Trustees may ⟨be⟩ of opinion that it is reason ⟨for⟩ them not to have confidence ⟨in⟩ him.

I remain— | Yrs very sincerely | John Thos. Austen.

PS.[8] I should add that I have now written very plainly to Dennen. I have told him that people do not borrow money at 10 per cent unless there is great risk to the lender, and that any security offered under such circumstances is naturally supposed to be fictitious—that this business coming formally before the Trustees, he would be looked upon as engaged to persuade Percival to accept the security of two men of straw—that such a transaction made public would discredit the Trustees and the Savings Bank altogether.

June 3rd— 1863

PS. | I speaking of D⟨en⟩nen calling on you—I merely meant that he would do so if you *wished* it[9] I hope to be able to call on you DV[10] on Monday week, but fear it will not be sooner. I shall not I believe be at home before Friday the 12th.—

DAR 159: 131, 151

[1] Letter from J. T. Austen, 30 May 1863.

[2] See letter from J. T. Austen, 30 May 1863, and the first postscript to this letter. George Dennen was organising secretary of the Bromley Savings Bank (Horsburgh 1980, p. 303, and *Post Office directory of the six home counties* 1862); Austen and CD were trustees of the bank.
[3] The letter to Austen has not been found.
[4] Charles Frederick Devas was a prominent Bromley resident and magistrate (Horsburgh 1980).
[5] The remaining trustees of the Bromley Savings Bank have not been identified. The bank was under the control of a president, vice-president, and up to twenty trustees, together with a managing committee (Horsburgh 1980, p. 303).
[6] The managers of the Bromley Savings Bank have not been identified.
[7] This individual has not been identified, but see the letter from J. T. Austen, 30 May 1863.
[8] This postscript was originally written to be sent with the letter from J. T. Austen, 30 May 1863.
[9] See letter from J. T. Austen, 30 May 1863.
[10] D.V.: 'Deo volente' – 'God willing'.

From John Scott [3 June 1863][1]

Botanic Gardens [Edinburgh]
Wednesday.

Sir.

I duly received your letter of the 20th.[2] I am greatly indebted for trouble in my behalf: and favourable impression you have made upon Dr. Hooker; as shown by his kind & encouraging letter.[3] I can only express my most cordial thanks, and a hope that if through your influence with Dr. Hooker; he obtains for me a colonial appointment, you will have no cause to regret having exercised your influence in behalf of one, of whom you know so little. I will do my best to give satisfaction: In respect to Dr. Hookers letter permit me to ask you for a few hints—at some future time—as to mental qualifications required; that I may apply myself more directly to the cultivation of such. I am painfully alive to the defects of an imperfect education; circumstances, having compelled me to finish it just at that period, when I was beginning to acquire a taste for learning.[4]

Prof. Balfour seems to be not a little offended with me, for refusing the Darjeling situation.[5] I hope, however, he will soon forget it. I feel assured he would have given me an excellent recommendation to Dr. Anderson:[6] and I shall certainly make it my earnest endeavour to *merit* it no less in the future. I thank you for judicious counsel respecting duties to Mr. McNab.[7] I have always been very careful that my experiments do not in the least interfere with my other duties; and I am glad to say—that—so far as I am aware—he is perfectly satisfied with me in this respect, and would I believe readily recommend me. If ever he had given me the slightest grounds for supposing that experiments were interfering with my duties to him, I would at once have restricted them, but he has given me no cause for such suspicions. It was very natural, however, for Dr. Hooker to suppose such. I cannot say what is the exact cause of his colder and less encouraging treatment: but for some months back there has been clearly a great change, in bestowal of *voluntary* privileges. I was—and am—much grieved at observing this; as I previously thought that he had a kindly interest in my advancement. Rest assured, however, that though, unfortunately—working in perfect ignorance of the real cause of his changed demeanour, I will do my utmost to please him as to performance of duties. At present—I am sorry to say for him—I have more opportunities for doing so than ever: nearly all his duties in connection with Botanic Class having devolved upon me—he being unable to attend to them.[8]

I will have a care that neither Dr. Balfour nor M^r. M^c.Nab know anything respecting Dr. Hooker's letter: as I much fear your suspicion would otherwise be immediately verified.[9]

In conclusion, I would beg you to excuse this long detail of personal matters: they have interfered unpleasantly with our other correspondence. I would also again acknowledge how deeply I feel indebted for the kind and active interest you have taken in me. It would afford me a pleasure to be able to make you any return: for I have no right to trouble you with such matters. The experiments which you have already suggested, I will attend to: if there are any others which you would like me to try—if within my power—I will gladly undertake them.[10]

I will now turn to the scientific part of letter. In respect to closing of stigma in Orchids: I will yet attend to your hint, as to affinities required.[11] From observations on closing of orifice it occurred to me on reading Bolbolophyllum case, that foreign pollen might possibly induce it;[12] I had, however, at the time I performed experiments on this point, my Cypripedium Oncidium & the Trichopilia in flower: so that my observations are not at all numerous. I feel perfectly satisfied, however that they are correct: having applied pollen to a few flowers, and extracted a little of it a few days after, when I distinctly saw pollen-tubes. I did not observe—as I ought—the depth these penetrated: the flowers were allowed to remain until quite withered: orifice likewise remaining open.

I thank you for kind offer respecting Primula work. I will draw up a paper, when I have seen results of present experiments, and forward. I will be glad indeed if you consider it worthy of communicating to the Linn. Soc. I do not think such papers are much cared for here.[13] I am rather afraid now that the abstract of Orchid paper sent you, will be all that is published of it in Proceedings of Bot. Soc.[14] From the great interest & importance which your works & correspondence has attached to such questions in my mind, I have often wondered at the low estimate in which they appear to be held here. Your remarks in late letters enlighten me: and here of course anything that savours of the "*Origin*" is not at all palatable![15] Nevertheless, I, having a strong conviction of its truthful teachings, would not on account of such dogmatic prejudices—for I cannot think they are aught else—sacrifice the "integrity of my own mind". As Emerson has expressed it;— "Bravely let us speak the utmost syllable of our confession . . . . nor be ashamed of that divine idea which each of us represents"[16]— so would I did my qualifications permit.

Thanks for corroboration of *P. longiflora*, you will observe it with a note mentioned in list formerly transmitted as *non-dimorphic*.[17] *P. verticillata* is peculiar. In my list you will find it approaching the non-dimorphic state. The majority of flowers upon a plant which I am experimenting upon at present, have *Stamens* & *Styles equal*: both being included within corolla tubes: the others have *styles* slightly *exserted* beyond tubes. In a specimen which I have lately examined in the University Herbarium I find a truly *long-styled* form—anthers near the *middle* styles reaching the *mouth* of corolla-tube.[18]

P. floribunda appears to present similar peculiarities. In list you will find it referred to *short-styled* form, with note on the occurrence of a non-dimorphic form. This I have again found presented by another specimen, while another, has presented one with the *long*-styled condition. Of P. *floribunda* I have no plants for experiments, and of P. verticillata I have only the *non-dimorphic* form; which appears to be perfectly self-fertile.[19] I believe I have been successful in application of *its* pollen to long & short-styled *P. Auricula*.[20] I am extremely curious to see results of this from a case which I will presently refer to, and on account of which I should have liked much to have tested the reciprocal sexual relations of these apparently *trimorphic species*.

The sexual relations of the non-dimorphic Cowslip referred to in late letter, appear to be very different from what I expected![21] It seems to have undergone a special differentiation: being all but *inert* in its action as *male* on *long* & *short-styled* forms—capsules being very small—as *female*, pollen being applied from the *latter* form it promises to be more productive—capsules being larger. The size of respective capsules farther illustrates, I think a fact which you have already pointed out, viz that *short-styled* form is more feminine in nature &c &c.[22] Capsules on *short-styled* plants are larger than those on *long-styled*—both being fertilised with *non-dimorphic* form. Again those flowers on the *latter* fertilised with *long-styled* pollen have *larger* capsules than those fertilised with *short-styled*! I did not at all anticipate such results: and I am now sorry that I did not make more numerous experiments with these forms. They have surprised me much; indeed, perplexed me. Since you have mentioned case of *Linum Lewisii*—which is much like a case of *mere variability*—there will be no harm I think in giving the above case.[23] I am inclined to regard such cases as somewhat discordant with your views that these qualities have been *slowly* acquired. Analogous cases have been *already passed* by your *numerous critics* so I may be wrong.

Another curious fact which I have lately discovered respecting *non-dimorphic* Auriculas— will I think interest you. The few plants which I have observed in this state, agree with that mentioned by you in Primula paper, in having even smaller pollen-grains than the normal long-styled form: differing in this respect from the above Cowslip.[24] In crossing with the *long* & *short styled* forms, I find that with the *former* the capsules are *very small*: *larger* & *finer* with the *latter*: which is favourable to your opinion that it is a case of abnormal developement of anthers. The self-fertilised capsules on non-*dimorphic* form are **larger** than those produced by homomorphic crosses!

I am glad to find that the abstract of my Orchid paper interests you.[25] I am also greatly obliged for the few critical remarks you have made.[26] I now see that I would have rendered the case of *O. sphacelatum* much more striking if I had stated it as you have done. I wished to state it in as forcible a manner as possible but you see how signally I have failed! I have had so little experience. Your analysis of sentence on self-impotence, shows me how unfortunately I have expressed myself. It appears now so self-evident that I cannot help wondering how it escaped me! I thank you most sincerely for these criticisms: I will do my best to improve style, which is one of the greatest difficulties I have to contend with. I have no command of language, no power of expression; and I am afraid from the little knowledge I have of mental phenomena, that I will never overcome this; which I fear will eventually give me quite a distaste for writing. I often wonder at and envy you for that happy faculty of expressing the most recondite ideas in language which a child might almost understand

I have made no experiments on *Lobelia fulgens*. I forgot when I wrote last your experiments on this plant in "Origin" and was writing under the impression that Gartner regarded it as absolutely sterile with own-pollen. while a friend of mine had succeeded in seeding it by fertilising it with own-pollen.[27] I see, however, that I have misapprehended this case.

Are you making any experiments in *Linum Lewisii*? It will be interesting from above case of Cowslip to test the reciprocal sexual relations. We have plants in the Gardens; if you think it worth while, and are not already engaged with it, I will make a few experiments: if plants present the trimorphic condition.

You ask if non-dimorphic Cowslip be the common yellow?[28] It is a red variety of this approaching the form generally known as *rubra*.

I have not yet succeeded in experiments on self-fertilisation of Orchids, by preventing expansion of perianth.[29] I have seen pollen-tubes however developed by such means in a flower of *Bletia Tankervillia*: but all the others which I had thus treat dropped off, the capsules giving no indications of swelling. As there have been few species in flower, however, in the Gardens since the idea occurred to me, I have not been able to make my experiments: but I have no doubt that I will yet be able to demonstrate it.

I remain | Sir | Yours very respectfully | John Scott

P.S. I intended sending corrected copy of "Orchid paper to Gard. Chron.[30] Before doing so I went to Prof. Balfour thinking that he might not have sent the copy off. I was too late however, and, being busy he said he would look it over after, and see whether it would be worth while sending it. My astonishment was great next morning when his assistant told me that he considered the corrections unnecessary. I of course was silent; though I thought it strange that the Prof. could so express himself. I am pleased to find that you have a different opinion in respect to this. | J. Scott.

DAR 177: 93

CD ANNOTATIONS

1.7 In respect . . . such. 1.9] *scored brown crayon*

5.1 I will now . . . be wrong. 9.17] *crossed pencil*

1 The date is established by the relationship between this letter, the letter to John Scott, 31 May [1863], and the letter to J. D. Hooker, 8 [June 1863]; the Wednesday between 31 May and 8 June 1863 was 3 June.

2 Scott wrote '20th' in error; the reference is to the letter to John Scott, 25 and 28 May [1863] (see n. 3, below).

3 CD enclosed the missing portion of the letter from J. D. Hooker, [23–7 May 1863], with his letter to Scott of 25 and 28 May [1863].

4 See letter from John Scott, 6 January 1863 and n. 24.

5 John Hutton Balfour was keeper of the Royal Botanic Garden, Edinburgh, where Scott was foreman of the propagating department (R. Desmond 1994). Scott had declined a position in charge of a *Cinchona* plantation in India for which Balfour had offered to recommend him (see letters from John Scott, 22 May 1863 and 26 May [1863], and n. 6, below).

6 Thomas Anderson, superintendent of the Calcutta botanic garden, was a leading figure in a project to introduce species of *Cinchona* to India for the production of quinine (Balfour 1873, *DNB*). He planned to establish a plantation for the cultivation of *Cinchona calisaya* near Darjeeling, and had written to Balfour, his former professor at the University of Edinburgh, asking for recommendations for the post of plantation overseer (see letters from John Scott, 22 May 1863 and 26 May [1863]).

7 As curator of the Royal Botanic Garden, Edinburgh, James McNab was Scott's immediate superior. See letter from John Scott, 22 May 1863, letter from J. D. Hooker, [23–7 May 1863], and letter to John Scott, 25 and 28 May [1863].

8 See letter from John Scott, 22 May 1863, n. 3.

9 See letter to John Scott, 25 and 28 May [1863].

10 In December 1862, CD had, at Scott's request, suggested numerous experiments for Scott to attempt (see *Correspondence* vol. 10, letters from John Scott, 6 December [1862] and 17 December [1862], and letters to John Scott, 11 December [1862] and 19 December [1862]).

11 See letter to John Scott, 25 and 28 May [1863].

12 See the enclosure to the letter from John Scott, 21 May [1863], and letter to John Scott, 25 and 28 May [1863] and n. 2; Scott's observations were published in Scott 1863a, pp. 545–6 n.

13 In his letter to Scott of 25 and 28 May [1863], CD offered to communicate an account of Scott's work on dimorphism in *Primula* to the Linnean Society, when finished, if Scott considered that it would not be adequately noticed by the Botanical Society of Edinburgh. Scott's paper on *Primula* (Scott 1864a) was read before the Linnean Society on 4 February 1864, and published in the society's *Journal* for September 1864 (*General index to the Journal of the Linnean Society*).

14 Scott read his paper on the pollination of orchids before the Botanical Society of Edinburgh on 14 May 1863 (Scott 1863a); he enclosed an abstract of the paper, which appeared in the *Edinburgh Evening Courant* on 28 May 1863, p. 8, with his letter to CD of 28 May [1863]. Scott's paper was subsequently published in full in the *Transactions of the Botanical Society* [*of Edinburgh*] (Scott 1863a; see letter from John Scott, 16 June [1863]).

15 See letters to John Scott, 23 May [1863] and 25 and 28 May [1863].

16 Scott paraphrases Ralph Waldo Emerson's essay 'Self-reliance', where Emerson states: 'Nothing is at last sacred but the integrity of our own mind' (Emerson 1841, p. 50), and: 'Bravely let him speak the utmost syllable of his confession. We but half express ourselves, and are ashamed of that divine idea which each of us represents' (*ibid.*, p. 47).

17 See letter to John Scott, 25 and 28 May [1863]. Scott sent CD a list of dimorphic and non-dimorphic species of *Primula* at the end of 1862, but this list has not been found (see *Correspondence* vol. 10, letter to John Scott, 3 December [1862] and n. 2). A similar list appeared in Scott 1864a, p. 80.

18 In Scott 1864a, pp. 80 and 82, Scott treated *Primula verticillata* as non-dimorphic, but described the length of the style as 'slightly variable'.

19 In Scott 1864a, pp. 80–2, Scott treated *Primula floribunda* as a short-styled species, but explained that it occasionally presented individuals with a non-dimorphic structure.

20 See Scott 1864a, pp. 92–7.

21 See letter from John Scott, 21 May [1863], and Scott 1864a, pp. 105–10.
22 This is implied, although not explicitly stated, in 'Dimorphic condition in *Primula*', pp. 92–3 (*Collected papers* 2: 59–60).
23 In 'Two forms in species of *Linum*', pp. 82–3 (*Collected papers* 2: 104–5), CD described the occurrence in *Linum lewisii* of three forms of flower (short-styled, long-styled, and equal-styled) appearing on the same plant, a case that he believed to be 'unique'.
24 See 'Dimorphic condition in *Primula*', p. 80 (*Collected papers* 2: 48). Scott described his observations and experiments on the dimorphic *Primula auricula* in Scott 1864a, pp. 91–2.
25 See n. 14, above, and letter to John Scott, 31 May [1863].
26 See letter to John Scott, 31 May [1863].
27 See letter from John Scott, 28 May [1863], and letter to John Scott, 31 May [1863]. In *Origin*, p. 98, CD cited his observations of dichogamy in *Lobelia fulgens* as evidence that there were 'special contrivances' in many plants that prevented the stigma 'receiving pollen from its own flower'. See also *Correspondence* vol. 10, letter to John Scott, 3 December [1862]. Scott may have been confused by CD's additional reference in *Origin*, p. 250, to 'certain species of Lobelia' being 'far more easily fertilised by the pollen of another and distinct species, than by their own pollen'. See also *Correspondence* vol. 10, letter to John Scott, 19 November [1862]. Scott refers to Karl Friedrich von Gärtner.
28 See letter to John Scott, 25 and 28 May [1863].
29 See letter from John Scott, [1–11] April [1863].
30 See n. 14, above, and letter to John Scott, 31 May [1863]. The abstract of Scott 1863a published in the *Gardeners' Chronicle*, 13 June 1863, p. 558, has a number of minor differences from the abstract published in the *Edinburgh Evening Courant*, 28 May 1863, p. 8.

From P. H. Gosse 4 June 1863

Sandhurst. Torquay.
June 4. 1863.

My dear Sir,

I am exceedingly obliged for your kind & full reply.[1] Will the following additional facts throw any light on the matter?

The four flowers of Stanhopea oculata became thoroughly withered & flaccid by the 1st. Inst; the fourth day after opening; yet I allowed them to remain till this morning, when I cut off the raceme just before I received your letter. As one of the germens (& this one of those that I had tried to impregnate) came away with a touch, I took it as certain that no impregnation had taken place; & so threw the whole on the rubbish heap without further examination. But on reading your remarks I thought I wd. examine them again; chiefly to see if, by piercing the stigmatic surface which had been so perfectly dry, I could find any viscosity within. Looking first at one of those to wh. I had affixed the pollen-masses by means of their viscid disk, I was surprised to see that they were half imbedded in a mass of viscous fluid. The other wh. I had treated was in precisely the same condition; the viscum having exuded copiously, & oozing in a great globule, when I used pressure with my thumb & finger lower down the column. Let this then be fact the first; that though no viscum be visible at first on the stigma, it issues copiously *after the flower has faded*, from the interior, at the extreme point of the rostellum.

But secondly:— A day or two after my attempt at impregnation (wh. affected only *two* of the four flowers), I was surprised to see the pollinia of one of the *untouched* flowers adhering to the point of one of the ivory-like horns of the mesochil.[2] I wondered, but could not account for it, as I felt sure I had not accidentally detached & attached them in such a manner, while operating on the others. But just now, in my examination of the faded spike, I observe not only that the pollinia of *that* flower remain still on the tip of the horn,—but that *one of the horns of the other untouched flower* has rifled its own anther, & carries the pollinia in triumph on its point.

If this is accidental, it is surely a remarkable coincidence. But it suggests to me the following hypothesis. That the moveable lip of this curious flower, agitated by the wind brings the tips of the horns now & then into contact with the rostellum, so as to lift the anther, & carry away the pollinia by touching the viscid disk. That as soon as the viscum exudes from the stigmatic cavity & spreads over its surface, similar agitations of the lip would cause the pollinia to swing across the stigma, &, brushing the exuded globule of viscum, to adhere.

If this is tenable, here is a use for these extraordinary horns: tell me what you think of the thought. I regret that I was so hasty in cutting away the faded spikes; possibly, with a little more obstetric manipulation, or even an agitation of the flowers with my breath, I might have succeeded in impregnating, & in settling the point.

Tell me whether I may print this little correspondence in one of the Gardening Periodicals,[3] & believe me | Yours faithfully | P. H. Gosse

I will remember the Bee Ophrys.[4]

P.S. If my hypothesis should be correct, will it not show that Stanhopea affords another example of self-fertilisation?[5] For the horns of any blossom can rifle *only its own* anther, & can deposit on *only its own stigma*. But what an unexpected mode of proceeding! I inclose you one of the pollinia carried on the horn.

DAR 165: 77

[1] Letter to P. H. Gosse, 2 June [1863].
[2] Mesochil: 'the middle portion of the labellum of an orchid' (*OED*).
[3] Neither the correspondence, nor Gosse's observations on *Stanhopea oculeata*, appear to have been published (Freeman and Wertheimer 1980).
[4] See letter to P. H. Gosse, 2 June [1863].
[5] See letter to P. H. Gosse, 2 June [1863] and n. 12.

To P. H. Gosse 5 June [1863][1]

Down. | Bromley. | Kent. S.E.
June 5th

My dear Sir

If you could prove the truth of your hypothesis, it would be extremely curious & quite new.[2] It certainly seems very suspicious your having found the pollinium attached to the horns of the labellum so often. I am prepared to believe anything of these wonderful productions. But if I were in your place, I would wait till I could observe another spike & then you would, I have no doubt, definitely prove the case. Why I shd act so, is because I have so often noticed the pollinia removed in an unexpected manner: D[r]. Hooker published in Phil. Transact. that Listera ejected its pollinia to a distance, which is entire mistake.[3] The *conjecture* (& it was founded on nothing but despair) occurred to me that the vibrating Labellum in Acropera might remove the pollinia; but D[r]. Hooker tried on living plant & failed to make it act.—[4]

Nevertheless your case may prove quite true; the dried labellum seems very thin as if it had been flexible.— It is really a very curious case.— I have some Stanhopeas in my Stove (I know not what species) but I fear they will not flower this summer; should they do so, I will observe them & communicate the result to you.—[5]

If you thought fit to communicate your facts now to any periodical, it might induce others to observe;[6] but many persons are such bad observers that I doubt whether you would profit by it.—

I would suggest to you to get to know (if you do not already do so) the appearance of the viscid matter from the stigma, which abounds with isolated elongated cells, called by Brown utriculi:[7] these I find never present in viscid matter of Rostellum; & when these parts are close, it is important to distinguish them.— You could have thus probably told whether the fluid which exuded from your decaying flowers was a true stigmatic secretion.

I heartily hope your pretty little discovery will prove good & true.—

My dear Sir | Yours very sincerely | C. Darwin

Col. James Innes

[1] The year is established by the relationship between this letter and the preceding letter.

[2] See preceding letter.

[3] J. D. Hooker 1854c.

[4] No letter from Hooker containing this information has been found; however, see *Correspondence* vol. 9, letter to J. D. Hooker, 15 [October 1861].

[5] *Stanhopea* appears on CD's lists of hothouse plants (Appendix VI). CD did not succeed in pollinating *Stanhopea* until the following year (see *Correspondence* vol. 12, letters to J. D. Hooker, 22 October [1864] and 3 November [1864]). See also 'Fertilization of orchids', p. 153 (*Collected papers* 2: 150).

[6] See preceding letter and n. 3.

[7] Robert Brown 1833b, pp. 710–11; CD's abstract of this paper is in DAR 74: 169–74. There is also a presentation copy of a slightly modified preprint of this paper (Robert Brown 1831a) in the Darwin Pamphlet Collection–CUL.

From S. P. Woodward 5 June 1863

BM.
June 5/63

Dear Sir

I forward a note received this morning from an excellent & amiable friend of mine at Cirencester—[1] Tho' nothing new, you may be pleased to see what he writes. A week ago when he called here (& gave me some eggs of the *Teleosaurus*, from sub-aërial beds of the Bath oolite)[2] I told him of your book on Orchids—& as he was fonder of *watching* insects than of *sweeping* for them with a net—I posted a copy of my account of your researches—[3] He is a man of leisure—a good ornithologist &c & has a capital garden & greenhouse—[4]

In your last letter to me you expressed great & very natural indignation at the tricks played upon Prof. Buckman in the matter of his experiments on "Species" of plants—[5] But perhaps if you were of the age, & in the circumstances of those boys you would not wonder at what happened— When a conjurer says he will make a pancake in your hat, or develope a guinea-pig in it (without even the aid of "sarcode")[6]—a *boy* naturally feels inclined to take in the conjurer—if possible I don't care to scrawl all I have seen in this unlucky business—but you know To⟨mes⟩ ("the Bat"!) & he knows one of the delinquents.[7] The only point of any practical consequence is the *value* of certain observations—& I have seen reason to believe that they are—nil.

I have just been writing a long notice of Bates' capital book—but I can't see that he has done much to help the doctrine of "transmutation".[8]

I went to Sandhurst on Wed^{y}. & so escaped the great "jaw"[9]—at which every bad speaker felt bound to distinguish himself by talking (vulgo[10]) "rot" JC. Natal being present—[11] Mr J. Evans protested himself unconverted to the last.[12]

My brother Henry has just returned from a dredging excursion with Mr McAndrew, to Corunna bay—[13] They experienced several delays at Bilbao & Santander—& lost 2 dredges by getting among the rocks. But the only new data obtained (that I have learned) are additional instances of "peculiar Mediterranean" forms occurring on the Atlantic coast—

Dr Duncan's Coral paper strengthens my belief in the (general) diffusion of marine forms *Westward*, in the course of time—[14] Our *Miocene* may be Caribbean *Pliocene*—& Pacific *Holocene* (only I wouldn't coin a word to save my life)[15]

I took a slip of paper that I might not bore you, with much ado about nothing—Alas! for the best intentions—

Yours sincerely | S. P. Woodward

Chas. Darwin Esqre.

DAR 181: 155

1 The note has not been found. The reference may be to the educationalist Thomas Barwick Lloyd Baker of Hardwicke Court, near Cirencester, Gloucestershire (see n. 4, below).

2 *Teleosaurus* is a genus of crocodilian reptiles belonging to the oolitic epoch (a division of the Jurassic period) (*EB*).

3 Woodward apparently refers to the review of *Orchids* in the *Critic* 23 (1862): 10–12; according to his biographer, Woodward wrote most of the reviews of science books in this journal between 17 March 1860 and 1 August 1863 (see Woodward 1881, pp. 279 and 301 n.).

4 See n. 1, above. In 1846, Baker and Woodward were founding members of the Cotteswold Naturalists' Field Club in Cirencester (Woodward 1881, p. 288); Baker was one of the few early members of the club with an interest in ornithology (see *Proceedings of the Cotteswold Naturalists' Field Club* 1: 101).

5 The letter has not been found; it was probably a reply to Woodward's letter to CD of 14 February 1863, which discussed malicious interference by students in some of James Buckman's experiments at the Royal Agricultural College, Cirencester.

6 Sarcode: 'the protoplasm of animals' (*OED*).

7 Robert Fisher Tomes was the author of numerous papers on bats (*Royal Society catalogue of scientific papers*). See also letter from S. P. Woodward, 14 February 1863 and n. 10.

8 Woodward apparently refers to the review of Bates 1863 that appeared in the *Critic* 25 (1863): 356–8; see n. 3, above.

9 Woodward refers to the meeting of the Geological Society of London on 3 June 1863, at which Joseph Prestwich read a paper discussing a human jaw and teeth that had been found in March 1863 at Moulin-Quignon quarry, near Abbeville, France (Prestwich 1863). The jaw was believed by a number of French naturalists, including Jacques Boucher de Perthes and Armand de Quatrefages, to date from the post-Pliocene (Pleistocene) period. On examining flint implements found with the jaw, Prestwich initially concluded that they were forgeries, a finding that cast doubt on the authenticity of the jaw itself. However, after attending an international meeting in Paris to decide this question, Prestwich became convinced of the specimens' authenticity (see J. Prestwich 1863). Prestwich's associate, John Evans, remained unconvinced. See Grayson 1983, pp. 213–17, and Van Riper 1993, pp. 134–9.

10 Vulgo: 'commonly, popularly' (*OED*).

11 John William Colenso, bishop of Natal.

12 See n. 9, above.

13 Henry Woodward was a member of Robert M'Andrew's dredging expedition off the coast of Spain in May 1863 (M'Andrew and Woodward 1864).

14 On 6 May 1863, Peter Martin Duncan read the first part of his paper, 'On the fossil corals of the West Indian Islands' (P. M. Duncan 1863–7), before the Geological Society.

15 The word 'holocene' was coined in 1860 by Paul Gervais, to describe the period in which the most recent geological formations originated (*EB*).

To John Scott 6 June [1863][1]

Down Bromley Kent
June 6th.

Dear Sir

I fear that you think that I have done more than I have with respect to Dr. Hooker.[2] I did not feel that I had any right to ask him to remember you for a Colonial appointment: all that I have done is to *speak most highly* of your scientific merits.[3] Of course this *may* hereafter fructify.— I really think you cannot go on better for educational purposes than you are now doing, observing, thinking & some reading beat, in my opinion, all systematic education. Do not despair about your style: your letters are excellently written; your scientific style is a little too ambitious. I never study style: all that I do is to try to get subject clear as I can in my own head, & express it in the *commonest* language which occurs to me.— But I generally have to think a good deal before simplest arrangement & words occur to me.— Even with most of our best English writers, writing is slow work: it is a great evil but there is no help for it— I am *sure* you have no cause to despair.—

I hope & suppose your sending a paper to Linn. Socy will not offend your Edinburgh friends;[4] you might truly say that you sent the paper to me, & that (if it turns out so) I thought it worth communicating to Linn. Socy.—

I shall feel great interest in studying all your facts on Primula, when they are worked out & seed counted. Size of capsules is often very deceptive. I am astonished how you can find time to make so many experiments.

If you like to send me your paper tolerably well written, I would look it over & suggest any criticisms; but then this would cause you extra copying. Remember, however, that L^d. Brougham habitually wrote everything important three times over.—[5]

The cases of the Primulæ which lose by variation their dimorphic character seem to me very interesting. I find that the mid-styled (by variation) P. Sinensis is more fertile with own pollen even than a heteromorphic union![6]

If you have time it will be very good to experiment on Linum Lewisii: I wrote formerly to Asa Gray begging for seed.—[7] If you have time I think experiments on any peloric flowers would be useful.—[8]

I shall be sorry (& I am certain it is a mistake on part of the Society) if your Orchid paper is not printed in extenso.[9] I am now at work compiling all such cases, & shall give a very full abstract of your observations.[10] I hope to add in autumn some from you on Passiflora—[11] I would suggest to you the advantage at present of being very sparing in introducing theory in your papers (I formerly erred *much* in geology in that way):[12] *let theory guide your observations*, but till your reputation is well established be sparing in publishing theory. It makes persons doubt your observations.— How rarely R. Brown[13] ever indulged in theory, too seldom perhaps.— Do not work too hard, & do not be discouraged because your work is not appreciated by the majority.—

Dear Sir | Yours faithfully | C. Darwin

DAR 93: B38–40

[1] The year is established by the relationship between this letter and the letter from John Scott, [3 June 1863].

[2] Joseph Dalton Hooker. See letter from John Scott, [3 June 1863].

[3] See letter to J. D. Hooker, 23 May [1863]. See also letter to John Scott, 23 May [1863].

[4] Scott had been elected an associate of the Botanical Society of Edinburgh on 10 July 1862 (*Transactions and Proceedings of the Botanical Society* [*of Edinburgh*] 14 (1883): 160), and, to date, had read all his papers there (Scott 1862a, 1862b, 1862c, 1862d, and 1863a). However, CD had offered to communicate an account of Scott's work on dimorphism in *Primula* to the Linnean Society, when finished, if Scott thought it would not be adequately noticed by the Botanical Society of Edinburgh (Scott 1864a; see letter from John Scott, 3 March 1863 and n. 19).

[5] Henry Peter Brougham, Baron Brougham and Vaux.

[6] In the spring of 1862, having written 'Dimorphic condition in *Primula*', CD discovered an equal-styled form of *Primula sinensis*, and found that the flowers were 'abnormally fertile in comparison with those of ordinary long-styled plants when self-fertilized' ('Illegitimate offspring of dimorphic and trimorphic plants', p. 415). In 1863, CD carried out a similar experiment on some equal-styled plants of *P. sinensis*. The flowers produced a higher average number of seeds than CD had observed in any legitimate union, and CD considered that they had not only lost 'their proper dimorphic structure and peculiar functional power', but had acquired 'an abnormal grade of fertility' (*ibid.*, pp. 416–17).

[7] See letter to Asa Gray, 2 January [1863]. See also letter from John Scott, [26 July – 2 August 1863].

[8] See letter to John Scott, 20 [June 1863] and nn. 6–7.

[9] CD refers to Scott's paper on the pollination of orchids (Scott 1863a), which was read before the Botanical Society of Edinburgh on 14 May 1863. An abstract of the paper was published in the *Edinburgh Evening Courant*, 28 May 1863, p. 8; however, Scott was concerned that it would not be published in full in the society's *Transactions* (see letter from John Scott, [3 June 1863] and n. 14).

10 CD wrote a draft of the section of *Variation* on 'Crossing & Sterility' between 1 April and 16 June 1863 (*Variation* 2: 85–191; see 'Journal' (Appendix II)). Scott 1863a is discussed in *Variation* 2: 133.

11 In *Origin*, pp. 250–1, CD listed *Passiflora* among those genera in which individual plants had been found to be 'far more easily fertilised by the pollen of another and distinct species, than by their own pollen.' In his letter to Scott of 11 December [1862] (*Correspondence* vol. 10), CD suggested that Scott carry out further experiments on this subject, which Scott had indicated he would do (see *Correspondence* vol. 10, letter from John Scott, 17 December [1862], and this volume, letters from John Scott, [1–11] April [1863] and 21 May [1863]). Scott's article on the sterility and hybridisation of species of *Passiflora* was communicated to the Linnean Society by CD in 1864 (Scott 1864d); CD discussed this paper in *Variation* 2: 137–8.

12 In 1861, CD had concluded that one of his earliest geological papers, 'Parallel roads of Glen Roy', was 'one long gigantic blunder', and that his theory to account for the geological phenomena described in it was completely erroneous (see *Correspondence* vol. 9, letter to T. F. Jamieson, 6 September [1861]).

13 Robert Brown (1773–1858).

To J. D. Hooker 8 [June 1863][1]

Down
8th.

My dear Hooker.—

As you say you like to see Asa Gray's letter, I send enclosed de rebus omnibus:[2] it is atrociously written. The coolness of the poor dear man about England & U. States in the last two pages beats anything— Please return letter sometime.—

John Scott of Edinburgh is very grateful for your most kind letter of advice.[3] He says Balfour is offended with him for refusing the Darjeeling offer.[4] He tells me that he is sure that he has never lost any time over his experiments, & that it is not on this account that Mr. Macnab is offended with him & he cannot conjecture what the cause is.[5] He takes my preachment very well & declares he will do all he can to please him.— I send a slip containing some of his experiments, but it is very badly arranged & difficult to make out, & the poor fellow sees it now I have told him very plainly.[6]

I have told him to send me his paper on Primula crosses & if it seems good, I will communicate it to Linn. Soc; for such an abstract as this does not do justice to his work.[7]

As I have bothered you thus far, I will bother you a little more by enclosing a M.S. note of Scott's about closing of stigma in orchids, dependent on affinity of pollen & independent of protrusion of pollen-tubes.[8] This is not noticed in abstract, & he tells me he believes the Bot. Soc. will publish only the enclosed abstract.[9]

Forgive me | Yours affect | C. Darwin

Endorsement: 'June/62'[10]
DAR 115: 158

1 The date is established by the relationship between this letter and the letter from J. D. Hooker, 10 June 1863. See also n. 10, below.

2 De rebus omnibus: 'concerning everything'. CD refers to the letter from Asa Gray, 26 May 1863. See also letter from J. D. Hooker, [1 March 1863].

3 See letter from John Scott, [3 June 1863] and n. 3.

[4] See letter from John Scott, [3 June 1863] and n. 5; the reference is to John Hutton Balfour. See also letters to J. D. Hooker, 23 May [1863] and 29 May [1863].

[5] See letter from John Scott, [3 June 1863] and n. 7; the reference is to James MacNab. See also letter to J. D. Hooker, 29 May [1863].

[6] CD refers to a newspaper report, enclosed with Scott's letter to him of 28 May [1863], of a meeting of the Botanical Society of Edinburgh held on 14 May 1863, which included an account of Scott's paper entitled 'Experiments on the fertilisation of orchids, in the Royal Botanic Garden, Edinburgh' (*Edinburgh Evening Courant*, 28 May 1863, p. 8). For CD's comments on the paper and Scott's response, see the letter to John Scott, 31 May [1863], and the letter from John Scott, [3 June 1863]. See also n. 9, below.

[7] See letter to John Scott, 25 and 28 May [1863]. CD communicated Scott's paper to the Linnean Society in January 1864 (see *Correspondence* vol. 12, letter to J. D. Hooker, [25 January 1864]); it was read before the society on 4 February 1864 (Scott 1864a).

[8] See the enclosure to the letter from John Scott, 21 May [1863].

[9] See letter from John Scott, [3 June 1863]. Scott's paper was subsequently published in full in the *Transactions of the Botanical Society* [*of Edinburgh*] (Scott 1863a; see letter from John Scott, 16 June [1863]). The note referred to appeared in Scott 1863a, pp. 545–6 n.

[10] Hooker wrote '62' in error; see n. 1, above.

From G. H. K. Thwaites 8 June 1863

Peradenia, Ceylon
8 June 1863

My dear Darwin,

It gave me great delight to find from your last welcome letter that the specimens I sent you proved so interesting.[1] Be assured you shall be made acquainted with every fact that comes before me which seems to me to bear upon your most valuable & interesting researches. With this you will receive some seeds of the *Limnanthemum Indicum*, which you will probably be able to cultivate in a pan of water in your hothouse.[2]

Very many thanks for your little brochure on dimorphism in *Linum*.[3] It seems to me that there is a great analogy between the behaviours of the pollen grains in phænogams and what we observe in the cells of some of the lower Algæ (*Zygnema* for instance): for in one species of the last named genus we find a regular conjugation taking place between the cells of separate filaments

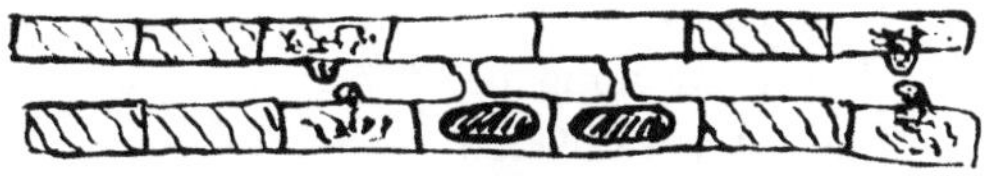

whilst in another species we have the adjoining cells of the same filament conjugating inter se, thus

—and, if my memory does not deceive me, there are cases where there is *apparently* no conjugation,

and I should be inclined to consider, under this view, that the differentiation you refer to in the second paragraph of p. 75 respecting *Linum grandiflorum*, is one of polarity rather than of a tendency to specific distinction;[4] but then again the question arises whether the process of specific differentiation is not coincident with modification of polarity or, what may be considered much the same thing, chemical affinity.

I enclose a letter I have just received from my very intelligent and accomplished correspondent, the Revd S. O. Glenie, Chaplain at Trincomalie, who with his good lady are quite enthusiastic in the study of botany, and with his letter is also the specimen of the flower of *Cassia Fistula* he refers to.[5]

I have requested him to keep his attention alive to cases of dimorphism in all the plants he examines. He takes great interest in your researches, but I don't at all know how far he goes with you.

Believe me always | truly & sincerely your's | G. H. K. Thwaites

P.S. Our dear friend Sir William Hooker has just sent me his Carte de Visite,[6] a most acceptable present, as you may believe; I need scarcely tell you how much I should be gratified with one of you, if you have had it taken.

[Enclosure]

Trincomale
5: June 1863

My dear Sir

I send in the accompanying packet a specimen of a lovely *Nauclea* I gathered this morning in some neglected ground near the Cuteherry where the tree (a small one) had doubtless been planted in better days. Is this the *Nauclea coadunata* **Roxb;**[7] DeC: 4.344 & Enum: 137.—[8] As far as I can gather from a careful examination the description of *C. coadunata*, (not a very full one by the bye) applies pretty well to my specimen but I do not feel at all certain—

The *Cassia Fistula* **L**[9] formed the subject of our study this morning, & a careful exhaustive study we made of it.

I enclose the stamens for your inspection premising that in 12 flowers at least I have found exactly the same appearance.

W. & A.: page 285 say the anthers have two clefts[10] Now I make out that the three (longer) lower, and the right and left of the the three (shorter) upper anthers have two clefts, while the middle one of the three lower and the four central anthers open by powers. This is a curious mingling of the two modes of discharging pollen, and worthy, I think, of notice.

I hope you will not look upon it as a trifle beneath notice, for I learn everyday the necessity of not passing over trifles—

Believe me | As Ever Yours sincerely | S. O. Glenie

Did you observe the sweet honey-like substance on the calyx of the Diospyros I forwarded a few days ago?—

DAR 178: 121–2

CD ANNOTATIONS

Enclosure:

1.1 I send . . . better days. 1.3] 'Gradation' *brown crayon, circled brown crayon*

4.3 the middle ... notice. 4.5] *scored brown crayon*
Top of enclosure: '*[*Inter*]*' *ink, del ink*; 'Gradation' *ink, circled ink*

[1] See letter to G. H. K. Thwaites, 30 March [1863] and n. 2.
[2] See letter to G. H. K. Thwaites, 30 March [1863].
[3] Thwaites's name appears on CD's presentation list for 'Two forms in species of *Linum*' (see Appendix IV).
[4] In 'Two forms in species of *Linum*', p. 75 (*Collected papers* 2: 98), CD stated:

> Taking fertility as the criterion of distinctness, it is no exaggeration to say that the pollen of the long-styled *Linum grandiflorum* (and conversely of the other form) has been differentiated, with respect to the stigmas of all the flowers of the same form, to a degree corresponding with that of distinct species of the same genus, or even of species of distinct genera.

[5] The references are to Samuel Owen Glenie, and possibly to his second wife Mary Elizabeth Louisa Glenie.
[6] William Jackson Hooker.
[7] William Roxburgh (*Taxonomic literature*).
[8] Candolle and Candolle 1824–73, 4: 344, and Thwaites 1858–64, p. 137.
[9] The 'L' denotes Carolus Linnaeus (*Taxonomic literature*).
[10] Wight and Arnott 1834.

From G. H. Darwin [9–15 June 1863][1]

I don't quite know what question to ask Wrigley, for I must have something definite.[2]

It seems as tho' there were an infinity of secondary spires— I don't quite understand your cone question, but it must come to the same as the other.[3]

How strange that the young shoot should jump from $\frac{3}{8}$ to $\frac{5}{13}$ & miss all intervening fractions. Though after all it is'nt much $\frac{3}{8} = \frac{39}{104}$ and $\frac{5}{13} = \frac{40}{104}$ only $\frac{1}{104}$ difference so there could'nt be many intervening fractions, except with very high denom^{rs}.—

I don't suppose those spires have anything to do with it, but it's very strange how they arise.—

As you know the holidays begin on next Tuesday week. As on that day the Oxford & Cambridge match is going on at Lord's I think I shall go there & come home in the evening, but the boys I suppose will come in the morning.[4] I'll let you know again before the time.

Next Tuesday there is a whole holiday & a cricket match old versus present members of the School & a good dinner in the evening.

L's letter is rather a curiosity in the way of spelling, but its because he does'nt look it over.[5]

I remain | Your affec: Son | G. H. Darwin

Incomplete
DAR 210.2: 1

CD ANNOTATION
2.2 other] *crossed pencil*; 'Cylinder' *pencil*

[1] The date range is established by the reference to the Oxford and Cambridge cricket match (see n. 4, below).

[2] Alfred Wrigley was the headmaster at Clapham Grammar School in south London, where George was a student; he was an accomplished mathematician (*Modern English biography*). CD had apparently asked George questions about the geometry of the positions of leaves on the stems of plants, either in a letter that has not been found, or in person; Emma Darwin's diary (DAR 242) records that George arrived at Down House on 6 June 1863. See also memorandum from G. H. Darwin, [before 11 May 1863]. CD's notes on phyllotaxy are in DAR 51: A6–32; there are also notes by George on this subject in DAR 192: 1–7.

[3] See memorandum from G. H. Darwin, [before 11 May 1863].

[4] George refers to the annual cricket match between the universities of Oxford and Cambridge, which took place at Lord's cricket ground in London on 22 and 23 June 1863 (*Cambridge University calendar*). Francis and Leonard Darwin were also students at Clapham Grammar School; Emma Darwin recorded 'boys from school' in her diary (DAR 242) on 23 June 1863.

[5] The letter from Leonard Darwin has not been found.

From Booth Bacon 10 June 1863

Penarth Harbour, Dock & Railway, | *Secretary's Office,* | *Cardiff*
10 June *1863*

Dear Sir.

I have received your Letter of 8th Inst relative to the Shares which I will place before the Directors on Monday next.[1]

With reference to the paying up the shares at once the Directors, I may inform you, are not receiving more now than 50 per cent of shares allotted—. they therefore would not receive more than that from you, say £22 per sh deposit & £28 per share in advance of calls.[2]

If this information should induce you in any way to modify your application of the 8th Inst. perhaps you will kindly inform me thereof before Monday.

Yours Truly | Booth Bacon | Secy

C. R. Darwin Esq
Down | Bromley | Kent.

DAR 160: 12

CD ANNOTATIONS
1.1 I have . . . Directors, 2.1] *crossed pencil*
2.3 say . . . Monday. 3.2] *crossed pencil*

[1] CD's letter to Booth Bacon has not been found. The Penarth Harbour, Dock & Railway company was established in 1856 to construct a railway line in Wales from the Taff Vale Railway to the river Ely in Glamorgan, and to convert part of the river into a tidal harbour (*Bradshaw's railway guide* 1863).

[2] CD's Investment book (Down House MS) records that, beginning in 1862, he invested heavily in the Penarth Harbour, Dock & Railway company; in 1863, he increased the value of his holding in the

venture from £3000 to £7000. On 22 June 1863, CD purchased thirty £100 five per cent preference shares, paying a deposit of £22 per share; on 20 July a further £32 was 'called up' on each of these shares.

From Asa Gray [10–16] June [1863][1]

Cambridge. [Mass.]
June ⟨ ⟩

My Dear Darwin,

I am kept distracting⟨ly⟩ busy. So look for nothing of any use from me yet a while

Your *Ohio* case of law against marrying of cousins, I put to my neighbor, Prof. Parsons, who had it looked up.[2] He tells me there is no *such law* at all on the Ohio Statute-books—nor is there a trace of any law on the subject to be found in the laws of any State in U.S. He doubts if there can really be any statistics which tell on the point, because 1. the marriage of 1st cousins is a rare thing in this country.— 2. the U.S. decennial Censuses do not afford any information on the matter.— 3. Nor any of the ⟨states'⟩ Census that he knows ⟨of.⟩

Campanula perfoliata *seed* I am trying to procure for you.[3]

Phloxes.[4] Last spring Dr. Torrey[5]—who has been examining *Polemoniaceæ*—told me that—the stamens remaining the same—some *species* have long styles, and other nearly related *species* short styles. I suggested dimorphism,—but he said they varied also in the no. of ovules. I have wished to look at Phloxes in this light—being confident that this means dimorphism.— But no wild ones grow in this region. Perhaps I can get seed from the West.—

Engelmann is the only reliable man,—probably because he is very busy.[6] I shall ask him. As I have lately been taking trouble for him, he will probably reciprocate.

Pray dont run mad over phyllotaxis! I can't save you, I am sure.[7]

George's 'Converging sines' is the same, perhaps, as what *Bravais* was after. His memoir may help you. (See Bot. Text Book, p. 141, par. 248)[8] Or, if you want something thoroughly mathematical, consult *Neumann*, of Berlin, in some paper—which I have no reference to.[9]

I have never meddled with this mathematical stuff. Hooker is quite mistaken.[10] But Prof. Peirce (our mathematical Prof. did so years ago, but *ran off into the sky*, dementedly—for which folly see Agassiz Contrib. vol. 1.—[11] It is not by me now.

—Oh, I see what Hooker was referring to. See the last pages of Proceedings of the American Association for the advancement of Science, 1849, vol. 2.[12]

You ask me for copy of article on *De Candolle*—on species.—[13] Why, I posted a copy to you—at the time, and you seem to refer to it. Do you mean to ask for a 2d copy. But my copies are exhausted quite.

If yours failed, I will try to get the sheets for you.— I dare say I can do so.

Never had I so much work to do in the University—lecturing & teaching all the while.[14] It quite uses me up.

I am sorry you do not give a better account of yourself. Be careful, and do not work too hard

Ever Yours | A. Gray

DAR 165: 136

[1] This letter was written before Gray received CD's letter of 31 May [1863], and was received by CD on or before 26 June (see letter to Asa Gray, 26 June [1863]); allowing ten to eleven days for the crossing of the Atlantic by mail packet, it was therefore probably written between 10 and 16 June 1863.

[2] See letter to Asa Gray, 20 April [1863]. Theophilus Parsons was Dane Professor of law at the Harvard Law School (*WWWA*).

[3] See letter to Asa Gray, 20 March [1863] and n. 5.

[4] CD thought that some species of *Phlox* might be dimorphic (see letter from Isaac Anderson-Henry, 24 April 1863, and letter to Isaac Anderson-Henry, 2 May [1863]); he probably asked Gray his opinion either in a missing postscript to his letter to Gray of 11 May [1863], or in a letter that has not been found.

[5] The botanist John Torrey was Gray's friend and mentor (Dupree 1959).

[6] The botanist George Engelmann was a very successful physician practising in St Louis, Missouri (*DAB*).

[7] See letter to Asa Gray, 11 May [1863] and nn. 8–11.

[8] Gray refers to George Howard Darwin, who was assisting his father in analysing the arrangement of leaves around the stems of plants (see letter to Asa Gray, 11 May [1863]). Gray discussed the work of 'M. Bravais' on phyllotaxy in A. Gray 1858a, p. 141; a specific publication is not cited, but the reference is evidently to the numerous papers on phyllotaxy published by Auguste and Louis Bravais in the 1830s (see *Royal Society catalogue of scientific papers* and *DSB*).

[9] Gray probably refers to Karl Friedrich Naumann, who published several papers on phyllotaxy, beginning with Naumann 1842 (*Royal Society catalogue of scientific papers*). See also Naumann 1845.

[10] See letter to Asa Gray, 11 May [1863] and n. 12; the reference is to Joseph Dalton Hooker.

[11] Benjamin Peirce was Perkins Professor of astronomy and mathematics at Harvard University (*DAB*). In the first volume of his *Contributions to the natural history of the United States of America* (Agassiz 1857–62, 1: 128–9), Louis Agassiz stated that Peirce had discovered 'the most perfect identity' between the laws regulating the arrangement of the leaves in plants, and the revolutions of the planets in the solar system.

[12] A. Gray 1849.

[13] Gray reviewed A. de Candolle 1862a in the May 1863 number of the *American Journal of Science and Arts* (A. Gray 1863d). See letters to Asa Gray, 11 May [1863] and 31 May [1863].

[14] Gray had been Fisher Professor of natural history at Harvard University since 1842 (Dupree 1959).

From J. D. Hooker 10 June 1863

Kew
June 10/63.

Dear Darwin

We have been visiting, eating, idling, riding & driving—& finish off next Saturday with the Clarke's at Bagshot park.[1]

Thanks for Scotts letter, he really must be a very superior man.[2] I know nothing of the present McNab, but as for Balfour, poor Scotts believing in aught you have writ is a *settler* in his eyes[3] You have a deal to answer for & may thank your stars that the days of torch & faggot are over, or some of your humble admirers would find out their **mistake**.— take that.

Balfour has the credit of ousting out of Edinburgh every man who shows any susceptibility for Botany.

If I hear of any thing at all likely to suit Scott I will bear him in mind. If he is a good cultivator the best place for him is a good liberal orchid growers garden like Rucker's—[4] It is a pity that he should throw away his papers on Bot. Soc. of Edinburgh, which has no status & no circulation of its Journals whatever.[5]

A Grays letter would be diverting were it not sad.—[6] What slaves men must be to *environment* that he should write & think so. How the deuce you can keep up the correspondence is a mystery to me— he & I would quarell over the 2^d^ letter we exchanged.[7]

What a capital letter Evan's is in Athenæum.[8]

Phyllotaxis is to me a most puzzling subject. I never get beyond the outline of the idea, I tried hard with $\alpha + \beta$[9]

Do you read Herbert Spencers First principles—[10] he asks me awfully hard questions in transcendental Botany.[11]

I am at Genera Plantarum—the only thing I am fit for—[12] I have stuck at Cameroons plants—& am hopeless & helpless—[13] Geog. Bot: must go to the dogs for me. I really cannot put a spoke in its wheel.

Ev yr aff | J D Hooker

DAR 101: 149–50

CD ANNOTATION[14]

End of letter: 'John Scott— Humble. | Herbert Spencer | Beer— Morphologie & Biologie of orchids.[15] | Bentham[16] | Haast letter—& Account—[17] | Visit here.— | Sneezing Pamphlet'[18] *pencil*

1 James and Barbara Clark lived at Bagshot Park in Surrey (*Post Office directory of the six home counties* 1862).

2 CD sent the enclosure to the letter from John Scott, 21 May [1863], with his letter to Hooker of 8 [June 1863].

3 Hooker refers to John Hutton Balfour, professor of botany, University of Edinburgh, and regius keeper of the Royal Botanic Garden, Edinburgh, and to James McNab, curator at the garden in succession to his father, William McNab (R. Desmond 1994). See letter to J. D. Hooker, 8 [June 1863].

4 Sigismund Rucker was a leading amateur horticulturalist and orchid-collector (*Gardeners' Chronicle* (1875), pt 2: 532–3; *Orchids*, p. 158 n.).

5 To date, Scott had read all his own papers before the Botanical Society of Edinburgh; they were all published, either in full or in abstract, in the *Transactions of the Botanical Society* [*of Edinburgh*] (Scott 1862a, 1862b, 1862c, 1862d, and 1863a). In his letter to Hooker of 8 [June 1863], CD mentioned his intention to communicate one of Scott's papers to the Linnean Society.

6 CD had enclosed the letter from Asa Gray, 26 May 1863, with his letter to Hooker of 8 [June 1863].

7 Hooker and Asa Gray held radically different views on the American Civil War, and had for some time tacitly agreed not to discuss the matter in their letters (see letter from J. D. Hooker, [1 March 1863] and n. 7).

8 Hooker refers to John Evans's letter in the *Athenæum*, 6 June 1863, pp. 747–8, in which Evans maintained that the human jaw and flint implements found in March 1863 at Moulin-Quignon quarry, near Abbeville, France, were not authentic. See Grayson 1983, pp. 213–17, and Van Riper 1993, pp. 134–9. See also letter from S. P. Woodward, 5 June 1863, n. 9.

[9] CD had become interested in phyllotaxy as a result of comments made on the subject with regard to natural selection in Falconer 1863a, p. 80 (see letter to J. D. Hooker, 13 January [1863], letter to Daniel Oliver, 20 [February 1863], and letter from Daniel Oliver, 17 February 1863). He carried out observations and measurements on the subject in May 1863, requesting references from Hooker (see letters to J. D. Hooker, [9 May 1863] and 29 May [1863], and letters from J. D. Hooker, [13 May 1863] and [24 May 1863]; see also CD's notes on this subject in DAR 51: 6–32). Hooker's comments were apparently prompted by Gray's remarks on the subject in his letter to CD of 26 May 1863 (see n. 6, above).

[10] Spencer 1860–2 constituted the first volume of a projected five-part 'System of philosophy' to which CD subscribed (see *Correspondence* vol. 8, letter to Herbert Spencer, 2 February [1860]); the volume presented Herbert Spencer's evolutionary philosophy in abstract terms, and the remaining four parts, published between 1863 and 1896 (*DNB*), dealt with its application to biology, psychology, sociology, and morality. There is a copy of Spencer 1860–2 in the Darwin Library–CUL; the last two numbers are uncut. See also n. 11, below.

[11] The second part of Spencer's 'System of philosophy', entitled *Principles of biology* (Spencer 1864–7), began publication in numbers in January 1863 (*ibid.*, p. [v]). CD's annotated copy of this work is in the Darwin Library–CUL (see *Marginalia* 1: 269–73).

[12] Bentham and Hooker 1862–83.

[13] Hooker was preparing an account of the plants collected in the Cameroons mountains by Gustav Mann (see, for example, letters from J. D. Hooker, [15 March 1863], [7 May 1863], [13 May 1863], and [24 May 1863]); his paper on the subject was read before the Linnean Society on 5 November 1863 (J. D. Hooker 1863b). CD was interested in the evidence these collections provided concerning the historical causes of the prevailing geographical distribution of plant species in tropical areas (see *Correspondence* vol. 10, letter to J. D. Hooker, 9 May [1862] and n. 6). However, Hooker had found that this case shook his confidence in the ability of naturalists to theorise successfully on the history of plant migrations (see letter from J. D. Hooker, [24 May 1863]). Hooker discussed the significance of his findings for CD's theory of migration during a global cold period in J. D. Hooker 1863b, p. 181.

[14] These annotations relate to subjects discussed by CD in his letter to Hooker of 23 [June 1863]. See also letter from J. D. Hooker, 19 June 1863.

[15] Beer 1863.

[16] Bentham 1863.

[17] The reference is to the letter from Julius von Haast, 5 March 1863, and probably to Haast's article in the Christchurch *Press*, 1 April 1863, pp. 1–2, and 2 April 1863, pp. 2–3 (see letter from J. D. Hooker, 19 June 1863, and letter to J. D. Hooker, 23 [June 1863]).

[18] Haliburton 1863.

To John Scott 11 June [1863][1]

Down
June 11th

Dear Sir

As the Abstract has not yet appeared in G. Chron. I sent your corrected copy to Dr. Hooker with the M.S. note on closing of stigma.—[2]

Dr. Hooker is evidently struck with your observations, & my sending these papers has produced the effect, which I hoped for as probable; for Dr. Hooker says "If I hear of anything at all likely to suit him, I will bear him in mind. If he is a good cultivator, the best place for him is in a good liberal orchid-growers garden like Mr Ruckers".[3] He adds "it is a pity he should throw away his papers on the Bot. Soc. of Edinburgh". This last sentence you must keep private; & indeed, I would advise you the whole.—

Dr. Hooker thinks that if it is known that you agree at all with my views on species, it is enough to make you unpopular in Edinburgh; so you had better keep your views pretty quiet.—

Dear Sir | Yours faithfully | Ch. Darwin

DAR 93: B22–3

[1] The year is established by the reference to the preceding letter.

[2] An abstract of Scott's paper on the pollination of orchids (Scott 1863a), read before the Botanical Society of Edinburgh on 14 May 1863, appeared in the *Gardeners' Chronicle*, 13 June 1863, p. 558. CD refers to a corrected copy of the abstract published in the *Edinburgh Evening Courant*, 28 May 1863, p. 8, which he had forwarded to Joseph Dalton Hooker, together with the enclosure to the letter from John Scott, 21 May [1863] (see letter to J. D. Hooker, 8 [June 1863]).

[3] Sigismund Rucker. See preceding letter and n. 4.

From John Scott 16 June [1863][1]

Botanic Gardens [Edinburgh]
June 16th.

Sir.

I feel quite ashamed with the trouble I am giving you.[2] I can only acknowledge my deep obligations, and express my cordial thanks.

I am glad to say that my Orchid paper is now in the press:[3] daily expecting the proofs. I will send you a copy when published: may I send one for Dr Hooker also? I would like to do so, now since you have done me the honour of forwarding abstract.[4] If you think I might do this, I will enclose a copy with yours, instead of sending it addressed to Dr Hooker.

It gratifies me very highly, to know that you and Dr Hooker are a little interested with my paper.[5] It is somewhat differently estimated here. But I care little for this; as I know well that its tendencies are a sufficient explanation. I have all along thought as you say Dr. Hooker thinks in respect to effects of its being known that I was favourable to your views of species.[6] I am quite sure it is a blemish in the eyes of my Edinburgh friends, nevertheless, it has not prevented me speaking the convictions of my mind. I regret that I have committed myself on this point in Orchid paper, *since* receiving your judicious caution on my introducing theory:[7] I see *now* that it will cause my observations to be doubted by many. This did not occur to me: I will remember and be guided by it in any future contributions which I may have occasion to make.

Thanks for Dr. Hooker's kind promise.[8] I will be glad indeed, to accept anything which he thinks likely to suit me. I am sorry that I have had so little experience as a cultivator.[9] I have somewhat neglected this, from a wish that I have long had to go abroad for a few years: and that on this account it would be of little importance. Dr. Hooker's suggestion makes me regret this: as nothing could afford me greater pleasure if I remain in this country than in a garden which contained a fine collection of Orchids.

I will indeed be glad if you can spare time to look over & correct my paper on Primulas.[10] It will be a pleasure for me to re-write it in accordance with your criticisms.

I will attend to experiments on *peloric* flowers.[11] I am only sure—of having plants—as yet of Gloxinias, & Antirrhinums; I suppose we have plants also of the peloric Geraniums but they are not yet in flower.

The few plants we have of *P. Scotica*, and *farinosa*. are now coming into flower. The former species is unfortunately so weak, that I fear my experiments upon it will be of little value.[12] I am busy with P. Sikkimensis at present, we had only the long-styled form in the Gardens here, but lately a friend has given me the use of a short-styled plant. I will thus be enabled to try the results of homomorphic & heteromorphic unions.[13]

I have been examining the structure of a few others of the genera of the Primulaceæ. Would it be worth while noticing their structure in paper?[14] I believe certain of the species of *Gregoriæ*—vitaliana for example—will be found to present similar sexual relations to the dimorphic Primulas. It presents both forms naturally: in our gardens I have seen as yet only the one form which produces *little* or no seed. I will yet perform a few careful experiments upon it.[15] Perhaps you can enlighten me on the genus *Cortusa*. I have examined a number of specimens of C. Matthioli from various parts: all present what one might regard as the long-styled form. Plants in the Gardens present this form also, and produce nice capsules containing good seed. I cannot as *yet* say as to proportion of *good* to *bad*; not having experimented upon the plants: the seeds produced being the result of natural fertilisation. It will be curious indeed perplexing in its bearing—should it prove perfectly fertile in this state??— one form only existing.[16] The allied genus *Androsace* has stamens & styles equal, as far as my examinations go.

Sir | Yours respectfully & obliged | J. Scott.

DAR 177: 94

CD ANNOTATIONS
2.2 for Dr Hooker] *underl red crayon*
6.2 Gloxinias,] *underl red crayon*
6.2 Antirrhinums] *underl red crayon*
8.3 *Gregoriæ*] *underl red crayon*

[1] The year is established by the relationship between this letter and the preceding letter.

[2] In his letter to CD of 22 May 1863, Scott requested advice about a position offered to him at a *Cinchona* nursery in India, and also solicited his aid in finding a better colonial appointment; CD had consulted Joseph Dalton Hooker (see letter to John Scott, 23 May [1863]). CD had also offered to communicate one of Scott's papers to the Linnean Society, advised Scott on his writing style, and sent Hooker an abstract of one of Scott's papers (see letters to John Scott, 25 and 28 May [1863] and 31 May [1863], and preceding letter).

[3] Scott 1863a. See letter from John Scott, [3 June 1863] and n. 14.

[4] See preceding letter and n. 2.

[5] See letters to John Scott, 31 May [1863] and 6 June [1863], and preceding letter.

[6] See preceding letter.
[7] See letter to John Scott, 6 June [1863].
[8] See preceding letter.
[9] Scott was foreman of the propagating department at the Royal Botanic Garden, Edinburgh (R. Desmond 1994).
[10] In his letter to Scott of 6 June [1863], CD offered to read and comment on the manuscript of Scott 1864a.
[11] See letter to John Scott, 6 June [1863].
[12] The results of Scott's experiments on *Primula farinosa* and *P. scotica* are given in Scott 1864a, pp. 114–17 and pp. 117–19, respectively. Scott had sent CD specimens of *P. scotica* and *P. farinosa* at the beginning of the year (see *Correspondence* vol. 10, letter from John Scott, 17 December [1862], and this volume, letter from John Scott, 6 January 1863).
[13] The results of Scott's experiments on *Primula sikkimensis* are given in Scott 1864a, pp. 110–11.
[14] Scott 1864a, pp. 78–86, discussed the structure of the reproductive organs in several genera of Primulaceae, including *Hottonia*, *Gregoria*, *Cortusa*, *Dodecatheon*, and *Soldanella*.
[15] Scott gave the results from his experiments on *Gregoria vitaliana* in Scott 1864a, pp. 83–4.
[16] Scott's experiments on *Cortusa matthioli* are described in Scott 1864a, p. 84.

From J. T. Austen 17 June 1863

West Wickham—
June 17 | 1863

My dear Sir,

I have seen Dennen who has thus explained the business of the negotiation for a loan of Percival's savings—[1] He says that his Brother R Dennen[2] owes money to a friend for which he is paying 10 per cent, that he is now in a positio⟨n⟩ to pay it off by Instalm⟨ents⟩ but the friend wishes th⟨e⟩ ⟨wh⟩ole t⟨o be⟩ repaid forthwith— it therefore occurred to him, G Dennen, that he could ⟨b⟩enefit both his Brother & Percival— he was certain his Brother could find securities, he had likewise a Policy on his life which would be given up— G Dennen offered 10 cent because his Brother was paying that rate of Interest. His Brother is laying by money ⟨to⟩ pay off the debt, but his ⟨fri⟩end did not wish⟨to r⟩eceive it by Instalments therefore G Dennen thought it would not be fair to propose to Percival to pay him by Instalments, as his Brother by Investing in a Deposit Bank would diminish the amount of Interest he now pays— He mentioned the subject to Percival without having spoken to his Brother, who he says did not approve of the terms— he considered the conversation to be preliminary only.

I am satisfied that he has never been before engaged in any similar transaction— he is very sorry that he acted so foolishly— I really believe he simply thought he was availing himself of an opportunity of benefitting an industrious man, but he has been a great fool— The Meeting will be called for Saturday week I believe— perhaps ⟨you⟩ will kindly in the mean time take the matter into your consideration— Dennen earnestly hopes there ⟨may⟩ be no exposure

yours very truly | John Thos. Austen.

C Dar⟨wi⟩n Esq

DAR 159: 132

[1] George Dennen was the secretary of the Bromley Savings Bank, of which CD and Austen were trustees (see letters from J. T. Austen, 27 May 1863, 29 May 1863, 30 May 1863, and 3 June 1863). Mr Percival has not been identified.

[2] Dennen's brother has not been identified.

From James Anderson 18 June 1863

Meadowbank | Uddingstone N.B
18th/June/63

Dear Sir

Enclosed is one of thc seed capsules of the Dendrobium cretacæum of which I wrote in the Journal for your inspection[1] It has taken fully a twelvemonths to mature. You may depend upon this act of fertilization being carried on within the floral envelope; and I should like to know the results of your investigation & opinion[2]

I remain | Dear Sir | very truly yours | Jas. Anderson

C. Darwin Esq

DAR 70: 181

CD note:[3]
Those seed produced by closed flowers were extraordinarily good; by searching I found a few empty testæ, & the nuclei of *a few [*after del* 'some'] appeared rather transparent. Closed flowers clearly admits, or rather favours, in this species, very perfect fertilisation; like imperfect flowers of viola &c. &c.—

[1] See letter from James Anderson, 1 April 1863 and n. 1. The reference is to a notice by Anderson in the *Journal of Horticulture*, 17 March 1863, pp. 206–8.

[2] No reply from CD has been found.

[3] The note is written on the verso of Anderson's letter. In 'Fertilization of orchids', p. 152 (*Collected papers* 2: 149), CD wrote:

> Mr. Anderson, a skilful cultivator of orchids in Scotland, informs me . . . that with him the flowers of *Dendrobium cretaceum* never expand, and yet produce capsules with plenty of seed, which, when examined by me, was found to be perfectly good. These orchids make a near approach to those dimorphic plants (as *Oxalis*, *Ononis*, and *Viola*) which habitually produce open and perfect, as well as closed and imperfect flowers.

To George Bentham 19 June [1863][1]

Down. | Bromley. | Kent. S.E.
June 19th

My dear Bentham

I have been extremely much pleased & interested by your address, which you kindly sent me.[2] It seems to me excellently done, with as much judicial calmness & impartiality as the Lord Chancellor could have shown. But whether the "immutable" gentlemen would agree with the impartiality may be doubted,—there is too much kindness shown towards me, Hooker & others, they might say.[3] Moreover I verily believe that your address, written as it is, will do more to shake the unshaken & bring on those leaning to our side, than anything written directly in favour of transmutation.[4] I can hardly tell why it is, but your address has pleased me as much as Lyell's book disappointed me,—that is the part on species, though so cleverly written.[5] I agree with all your remarks on the Reviewers. By the way, Lecoq is a believer in the change of species.—[6] I, for one, can conscientiously declare that I never feel surprised at anyone sticking to the belief of immutability; though I am often not a little surprised at the arguments advanced on this side.—[7] I remember too well my endless oscillations of doubt & difficulty. It is to me really laughable, when I think of the years which elapsed before I saw what I believe to be the explanation of some parts of the case: I believe it was 15 years after I begun before I saw the meaning & cause of the divergence of the descendants of any one pair.—[8]

You pay me some most elegant & pleasing compliments. I had not heard before that Mill had written anything on my treatment of the subject;[9] & this, as you may believe, is highly satisfactory to me; after the Bishop of Oxford having said he believed that the Origin "was the most illogical book ever published."—[10]

There is much in your address which has pleased me much, especially your remarks on various naturalists. I am so glad that you have alluded so honourably to Pasteur.[11] I have just read over this note; it does not express strongly enough the interest, which I have felt in reading your address.— You have done, I believe, a real good turn to the **right side**.

Believe me | Dear Bentham | Yours very sincerely | Ch. Darwin.

I have been rather bad for some days & was in bed nearly all yesterday; so my note is duller than it otherwise, perhaps, would have been.— Shall you think it a very odd request, if I ask you to show this note to Hooker, as he wanted to hear what I thought, & it will save me writing it again?—[12] I have written to you, exactly as I should have done to him—

Royal Botanic Gardens, Kew (Bentham correspondence 3: 709–10)

[1] The year is established by the reference to Bentham 1863 (see n. 2, below).

[2] CD refers to Bentham's anniversary address delivered to the Linnean Society on 25 May 1863 (Bentham 1863). The number of the *Journal of the Proceedings of the Linnean Society of London* containing Bentham's address was not issued until 29 October 1863 (*General index to the Journal of the Linnean Society*, p. vii); the preprint sent to CD by Bentham has not been found.

[3] In his address, Bentham sought to examine the present state of 'Biology', which he defined as 'the *science of life, i.e.* of the phenomena of life, in contradistinction to the description and classification of living beings' (Bentham 1863, p. xii). Bentham devoted most of his address to a review of the state of the literature on the origin of species (Bentham 1863, pp. xii–xxii), providing a detailed assessment of the scientific debate on the question since the publication of *Origin*, and referring to many reviews of *Origin* and to various other publications with a bearing on CD's theory, including J. D. Hooker 1859 and 1860b (see Bentham 1863, p. xx).

[4] In summing up the literature relating to *Origin*, Bentham suggested that the 'tide of opinion' among 'philosophical naturalists' was 'setting fast in favour of Mr. Darwin's hypothesis', and that few who

really considered the subject could deny that new species had been produced from a common ancestor by natural selection (Bentham 1863, pp. xxi–xxii). He continued:

> The great objections still urged are to the insufficiency of the data yet ascertained for the extension of the principle to all changes and to all species; and whilst many of Mr. Darwin's generalizations may be considered as adopted, there are others which many persons are disposed to refer for further proof, and many objects of research more or less relevant, indicated only by him, are still obscured from our view.

5 C. Lyell 1863a. See letter to Charles Lyell, 6 March [1863].

6 Henri Lecoq. See Bentham 1863, p. xiv.

7 In Bentham 1863, p. xiii, Bentham numbered himself among those who,

> through a long course of study of the phenomena of organic life, had been led more and more to believe in the immutability of species within certain limits, and have now felt their theories rudely shaken by the new light opened on the field by Mr. Darwin, but who cannot surrender at discretion so long as many important outworks remain contestable.

8 In his *Autobiography*, pp. 120–1, CD explained that, in developing his theory in the late 1830s and early 1840s, he had 'overlooked one problem of great importance', namely, 'the tendency in organic beings descended from the same stock to diverge in character as they become modified.' CD did not provide an explanation for this phenomenon until the mid-1850s, when he accounted for it in terms of the selective advantage of specialisation for closely related species living in the same territory. See *Origin*, pp. 111–26, and Ospovat 1981, pp. 170–90.

9 In his address, Bentham quoted John Stuart Mill's statement that, while not capable of proof, CD's theory was an 'unimpeachable example of a legitimate hypothesis', and that it constituted a 'wonderful feat of scientific knowledge and ingenuity' (J. S. Mill 1862, 2: 18 n.; see Bentham 1863, pp. xv–xvi).

10 In his letter to CD of 13–14 February [1860] (*Correspondence* vol. 8), Charles Lyell reported that the bishop of Oxford, Samuel Wilberforce, had declared *Origin* to be 'the most unphilosophical [book] he had ever read'.

11 In his paper, Bentham discussed the experiments detailed in Pasteur 1861 and 1862, as a result of which Louis Pasteur appeared to have completely refuted 'the idea of heterogeny in the production of microscopic animals or plants in organic matter in a state of fermentation' (Bentham 1863, p. xxv–xxvi). He also described the subsequent recognition accorded to Pasteur's work, before criticising the '*ex cathedrâ*' promulgation of 'a new form of spontaneous generation' in Richard Owen's anonymous review of Carpenter 1862, which appeared in the *Athenæum*, 28 March 1863, pp. 417–19 (see Bentham 1863, pp. xxvi–xxvii). See also letter to J. D. Hooker, 23 [June 1863] and n. 9, and Appendix VII.

12 Bentham worked with Joseph Dalton Hooker at the Royal Botanic Gardens, Kew. See letter from J. D. Hooker, [24 May 1863].

From J. D. Hooker 19 June 1863

[Royal Botanic Gardens, Kew]
June 19/63.

Dr. Darwin,

did you not ask me some weeks back about a letter from Haast to you—[1] I have just heard from him that some of his letters were lost before leaving N.Z.—& he sends me a copy of his former letter to me—& it encloses one for you, which I forwarded 2 days ago to Down.[2] I hope that it contains what you want— He has sent me a most interesting newspaper journal of his travels across the plains;[3] & Hector has sent by same mail, a more interesting but less successful account—[4] I will send both to you when my Father[5] (in Norfolk) returns them—together with Hector's letter.

How are you all?— we are all well—& settled at Kew for a long spell— though I think I shall steal a Sunday for Down ere long, & see you with my own eyes.

I forwarded to you today from Linn. Soc. a curious pamphlet about sneezing[6]

Ev yr aff | Jos D Hooker

DAR 101: 151

[1] No letter from CD to Hooker containing such an inquiry has been found, but see the letter from J. D. Hooker, [30 April 1863] and n. 2. The reference is to the letter from Julius von Haast, 9 December 1862 (*Correspondence* vol. 10), which is thought to have been enclosed with Haast's letter to Hooker of 10 December 1862 (Royal Botanic Gardens, Kew, Director's correspondence 174 (New Zealand letters, 1854–1900): 281–2).

[2] With a letter to Hooker of 5 March 1863, Haast enclosed a copy of his letter to Hooker of 10 December 1862 (see n. 1, above); with the same letter, Haast forwarded his letter to CD of 5 March 1863, which, in turn, enclosed a copy of Haast's letter to CD of 9 December 1862. Haast's letter to Hooker of 5 March 1863 has not been found, but details of its enclosures are given in Hooker's letter to Haast of 19 June 1863 (Alexander Turnbull Library, Wellington, New Zealand, MS. papers 37.96; see also H. F. von Haast 1948, p. 289).

[3] Hooker probably refers to articles in the Christchurch *Press*, 1 April 1863, pp. 1–2, and 2 April 1863, pp. 2–3; these articles described Haast's 1862–3 expedition in the Southern Alps of New Zealand's Middle Island (now South Island) (see the enclosure to the letter from J. D. Hooker, 20 April 1863 and n. 28; and Hooker's letter to Haast, dated 19 June 1863, in the Alexander Turnbull Library, Wellington, New Zealand, MS. papers 37.96; see also H. F. von Haast 1948, p. 289).

[4] James Hector was the provincial geologist in Otago province, New Zealand; between January and March 1863, he undertook an expedition to discover a practicable land route to the west coast from the eastern part of the province, only to be forced to turn back within sight of the coast, owing to a shortage of supplies (*DNZB*). The reference is apparently to an article entitled 'Dr. Hector's Expedition. Successful penetration to the West Coast to within sight of the sea, and eight miles of the shore', which appeared in the *Otago Daily Times*, 19 March 1863, p. 5.

[5] William Jackson Hooker.

[6] Hooker refers to a presentation copy of Haliburton 1863, which had been sent to CD at the Linnean Society. Following a general introduction, the pamphlet gave an account entitled 'The unity of origin of the human race, proved by the universality of certain superstitions connected with sneezing' (Haliburton 1863, pp. 9–14). CD's copy of the pamphlet is in the Darwin Pamphlet Collection–CUL.

To John Scott 20 [June 1863][1]

Down
20th

Dear Sir

I am poorly & must be brief— I am glad to hear about your orchid paper; by all means let me have copy for D[r]. Hooker.[2] It would be quite worth while to show that Gregoria is dimorphic & that Cortusa has the structure of one form alone.—[3] I can give you a M.S. short account of structure of pollen &c of Hottonia.—[4] You will thus make a fairly complete, & anyhow sufficient, account of Primulaceæ.—[5] I know nothing of the genera to which you refer.— I would make all this part brief.—

I shall be very glad if you have time to try peloric Gloxinias & Antirhinum; I have got both: I mean to try peloric on self & on common flower: & conversely with other forms; so that there will be 4 crosses.—[6] I doubt whether Pelargonium worth the labour; I have tried many crosses & can get no certain result—[7]

Please save every capsule of the self-fertile variety of red cowslip with equal anthers & stamens; I sh^d. like to count the seeds in each capsule, if you do not yourself do it—[8] In counting I find it safest plan to count all seeds, which are not *manifestly* bad; as this allows for partially unripe seed: of course it does not signify if same, as near as is possible, standard of goodness is used in all cases.—

Do not work too hard.—

Dear Sir | Yours faithfully | C. Darwin

DAR 93: B53–4

[1] The date is established by the relationship between this letter and the letter from John Scott, 16 June [1863].

[2] The references are to Scott 1863a and Joseph Dalton Hooker (see letter from John Scott, 16 June [1863]).

[3] See letter from John Scott, 16 June [1863] and nn. 14–16. The reference is to the paper Scott was preparing to write on the structure and functions of the reproductive organs in the Primulaceae (Scott 1864a).

[4] Scott gave CD's results in Scott 1864a, pp. 78–9. See also letter to John Scott, 2 July [1863].

[5] See letter from John Scott, 16 June [1863].

[6] CD had been interested in carrying out experiments on the fertility of peloric flowers for some time (see, for example, *Correspondence* vol. 10, letter to M. T. Masters, 8 July [1862]), and had ordered plants of peloric varieties of *Gloxinia* and *Antirrhinum* at the end of 1862 (see *ibid.*, letter to John Scott, 19 December [1862]). See also letter to Isaac Anderson-Henry, 20 January [1863], letter to J. D. Hooker, 15 February [1863], and Appendix VI. CD's experimental notes relating to peloric *Antirrhinum majus*, dated 26 June – 20 July 1863, are in DAR 51: B18–19. CD described his experiments with *A. majus* in *Variation* 2: 70–1 and 166–7.

[7] CD had carried out a number of crossing experiments with the normally sterile central peloric flowers of several varieties of Pelargonium in the summer of 1862 (see *Correspondence* vol. 10, letter to Asa Gray, 1 July [1862], and the experimental notes in DAR 51: B4–9, B12–13). There is a further set of CD's experimental notes on peloric Pelargoniums, dated 15 May – 20 July 1863, in DAR 51: B14–15. CD referred to his experiments on peloric Pelargoniums in *Variation* 2: 167.

[8] See letter from John Scott, 21 May [1863]. Scott presented his results from this experiment, including a tally of the number of seeds produced, in Scott 1864a, p. 106.

To Charles Thomas Whitley 20 June [1863][1]

Down. | Bromley. | Kent. S.E.
June 20th

My dear Whitley

I was sincerely glad to receive your note & have news of you;[2] for I often think of old days at Cambridge, when I took long walks with the expectant senior wrangler & when he instructed me in Engravings.—[3]

I thank you cordially for your kind invitation; but my health has been so bad for several years that I have been unable to attend the British Association, & have no hopes for this year.[4] In fact for the last two years I have led the life of a hermit, seeing no one & going nowhere; & doing nothing but two or three hours work daily on my good days at natural History. I am become that most wretched & despicable object, a confirmed valetudinarian. I have much, very much, to be thankful for in life; but everyone has his heavy drawbacks & my own health & even more that of my children is our sore drawback. For years we have had one or other of our children invalids. But I have said enough & more than enough.—

Again I thank you sincerely for your kind remembrance of me, & with hearty good wishes, remain | My dear Whitley | Yours very sincerely | Ch. Darwin

Shrewsbury School

1 The year is established by the reference to a meeting of the British Association for the Advancement of Science; the allusion is apparently to the meeting that took place in Newcastle in 1863 (see n. 4, below).

2 Whitley's letter has not been found.

3 CD and Whitley were contemporaries at Shrewsbury School and Cambridge University; Whitley was senior wrangler (that is, first in the final mathematical examination) in 1830 (*Alum. Cantab.*, *Correspondence* vol. 1).

4 It appears that Whitley had written to CD concerning the annual meeting of the British Association for the Advancement of Science for 1863 to be held in Newcastle-upon-Tyne from 26 August to 2 September (*Report of the 33d meeting of the British Association for the Advancement of Science*, p. 1); Whitley was vicar of Bedlington, Northumberland, about twelve miles north of Newcastle (*Alum. Cantab.*).

From William Allport Leighton [before 21 June 1863][1]

[Enclosure][2]

Days ago I was at Lee's[3] with our fellow Lambert,[4] &, as many species of Verbascum unknown to me were in blossom, I tried in all of them the phænomenon of their [irritability], with the greatest success in all of them. I remember having spoken to you about it; & as probably the English Verbasca are now in flower about Norwich I will be very glad if you observe it minutely. Strike with a stick & with some force, the stem of the plant under the branches 2 or 3 times & wait patiently for some minutes; you will see the limbus of the corolla move towards the centre, & afterwards the basis of the tubus slowly detach itself from the receptacle & fall—& immed$^{ly.}$ the calyx close itself with impatience & embrace the seeds & that will happen to all the open flowers one after another.[5] In Portugal[6] I had observed it in the Verbascum Thapsi—sinuatum— Blattaria—now at Lee's in many others & I wish I co$^{d.}$ observe all its species & the Celsiæ:— I know nothing analogous in other plants Next year this will be followed up, & made a Memoir for our Society.[7]

Abbé Correa to Sir JES. in lit July 4. 1795 see Life of Sir JES. 2.210[8]

DAR 108: 180

CD ANNOTATION

End of letter: 'Life of Sir J. E. Smith.' *ink*

[1] The date is established by the reference to CD's experimental notes on *Verbascum lychnitis* (see n. 5, below).

[2] The letter with which this enclosure was sent has not been found. The enclosure is a copy of part of a letter from Abade José Francisco Correa da Serra to James Edward Smith, transcribed by Leighton from P. Smith 1832, 2: 210–11.

[3] The reference is apparently to the Vineyard Nursery, Hammersmith, London, in which James Lee was a partner.

[4] Probably the botanist Aylmer Bourke Lambert, a friend of Smith's (*DNB*).

[5] CD attempted to repeat this experiment with *Verbascum lychnitis*, but without success. His experimental notes, dated 21 June 1863 (DAR 108: 5), state:

> Verbascum lychnitis, when struck does not cast corolla, nor do sepals close over ovarium. I believe flowers do not close. I believe V. phœniceum does not cast flowers [*interl*]; certainly sepals do not close over ovarium.—

[6] Correa da Serra was Portugese; he fled to Britain in 1795 to escape the inquisition (P. Smith 1832, 2: 198–224).

[7] Correa da Serra apparently refers to the Linnean Society, of which both he and Smith were fellows; however, he did not publish a paper on *Verbascum* in the society's transactions, and no paper by Correa da Serra on this subject has been found (see *Royal Society catalogue of scientific papers*, and Reuss 1801–21).

[8] P. Smith 1832, 2: 210–11 (see n. 2, above).

From Jacques Boucher de Perthes[1] 23 June 1863

Abbeville

23 juin 1863

Cher et honoré Monsieur

Votre lettre du 16 m'a causé une grande satisfaction et je vous en remercie.[2] Je tiendrai toujours en grand honneur de correspondre avec vous qui avez pris une si belle place dans la Science et dont les idées se rapprochent si bien avec les miennes

Je ne sais pas si vous avez eu connaissance de mon livre qui a paru en 1838 sous le titre *de la creation, essai sur la progression des etres*.[3] il a passé fort ignoré parce que la lecture n'en est pas amusante, mais tel qu'il est je vous prie de l'accepter. Je ne l'ai pas ici car imprimé a peu d'exemplaires. Il est devenu assez rare mais au moyen du bon ci joint vous pourrez par occasion le faire prendre à Paris. J'y joint le d^{er}. volume (de 1857 à 1860) de notre Societe d'Emulation ou est mon homme antediluvien.[4] Si vous avez deja ce volume il serait remplacé par celui de 1852 a 1857 il suffira que vous changiez ces dates sur le bon

Cette malheureuse machoire m'a donné bien de l'ennui, par suite des discussions a laquelle elle a donne lieu[5]— il en a été de meme il y a 28 ans quand j'ai montré mes premières haches.[6] Cependant alors comme aujourdhui j'étais sûr de ce que j'avançais dans ces questions je ne hasardais rien et quand j'affirme une chose c'est que je l'ai vue. J'ai dit il y a bien longtemps qu'on trouverait l'homme fossile et je repete ma prediction.[7] il en sera ici comme des haches que j'ai trouvé partout en Angleterre comme en France, enfin partout ou on en cherche.

Cet homme fossile est aussi là sous nos pieds, mais l'on n'en veut pas. Dès qu'ils se montrent on lui cherche querelles. À combien d'incroyables histoires celui de moulin quignon n'a-t-il pas donné lieu!—on a voulu qu'il y eut été enterré par des mistificateurs quand il a été reconnu que c'etait matériellement impossible, parce qu'il y avait trop de terrain à son exhumation pour rendre une fraude praticable, et d'ailleurs qu'au moins d'être aveugle ou insensé on reconnait toujours quand un terrain a été remué Le dernier des ouvriers même ne s'y trompe pas, on s'en est pris aux haches. On a dit qu'elles étaient fabriquées.— eh! bien l'un n'est pas plus vrai que l'autre— On a pu fabriquer des haches a St Acheul qui ressemblent assez à celles des tourbières qu'en effet on imite assez aisement, mais il n'en est pas ainsi a moulin quignon. Leur imitation demanderait beaucoup d'adresse de soin et de temps, du coup pour en tenir profits les ouvriers devraient les vendre cher, et le prix ordinaire en est de 25 centimes (cinq sous) j'ai voulu savoir à quoi m'en tenir. J'ai mis au concours la fabrication d'une hache, offrant 20 Francs à l'ouvrier qui en fabriquerait une ressemblant assez à celles de moulin quignon pour que l'on put s'y tromper. Apres beaucoup d'essais personne n'est parvenu à faire meme une hache passable.

Veuillez agréer cher monsieur l'expression de ma haute estime et de mon entier devouement. | J Boucher de Perthes

J'ai fait le bon pour les 3 derniers volumes des memoires de notre Société Veuillez les accepter tous les trois

A Monsieur Charles Darwin—

DAR 160: 257

[1] For a translation of this letter see Appendix I.

[2] CD's letter has not been found.

[3] The reference is to the first volume of Boucher de Perthes 1838–41, a philosophical work in which the author elaborates a theory of the universal evolution of nature, owing much to the romantic tradition (*DSB*).

[4] The enclosure has not been found. Boucher de Perthes refers to the volume of the *Mémoires de la Société Impériale d'Emulation d'Abbeville* for 1857–60, which contained Boucher de Perthes 1860a. There is an unannotated reprint of Boucher de Perthes 1860a (Boucher de Perthes 1860b) in the Darwin Pamphlet Collection–CUL.

[5] The reference is to the human jaw found in March 1863 at the Moulin-Quignon quarry, near Abbeville, France, together with teeth and some artefacts, and reported by Boucher de Perthes. The jaw was believed by Boucher de Perthes, Armand de Quatrefages, and other French naturalists to date from the Post-Pliocene (Pleistocene) period. However, on examining the finds in April 1863, Joseph Prestwich, John Evans, and Hugh Falconer concluded that they were forgeries. Between 9 and 13 May, a committee of French and English naturalists met in Paris to discuss the question, and concluded that the implements and jaw were authentic, although several scientists remained unconvinced. See Grayson 1983, pp. 213–17, and Van Riper 1993, pp. 134–9.

[6] Boucher de Perthes first became involved in the excavations for human artefacts in the Somme valley around 1837, although artefacts were found there as early as 1830 (Grayson 1983, pp. 118–20).

[7] See, for example, Boucher de Perthes 1847–64, 1: 245. See also Grayson 1983, pp. 128 and 213.

To J. D. Hooker 23 [June 1863]

Down
23d.

My dear Hooker

My eczema is well & consequently till it comes on again, I am languid & bedeviled & have[1] writing & hate everybody. No, that is not true for in my worst state I do not hate you; but I have not had spirit to thank you for two pleasant notes.—[2] John Scott of Edinburgh is very grateful for what you say about keeping him in mind.[3] I almost think he doubts his power to manage a great private establishment & leans to some foreign place; but I suspect that he is too modest.[4] He has asked me whether he might send you his orchid-paper when published; I told him by all means to send it—[5] About Haasts letter all right; you wrote some time ago saying that you thought you had lost one for me.—[6] I shall be very glad to see the account of his Explorations.[7] He seems a fine fellow. Thanks for sending Sneezing paper; how curious the case is, but how weak the explanation of origin at end.—[8]

The more I think of Bentham's address the more I like it. I quite enjoyed the snubs to Owen.—[9] You ask what I think of Herbert Spencer's great book: I never attempted to read any except last Part;[10] & that *greatly* disappointed me—all words & generalities, like Sir H. Holland's writings,[11] & I could grasp nothing clearly. But I suppose this is all my stupidity; as so many think so highly of this work.—

If Oliver knows "Beer's Morphologie und Biologie der Orchideen", I shd. like to know whether it would interest me: it is just published price 30s.—[12]

I shall enjoy extremely seeing you, if you can run down for a Sunday & I hope to God I may be decently well to enjoy it; but this gets rarer & rarer with me.

Farewell my old friend | C. Darwin

Endorsement: 'June | /63'
DAR 115: 196

[1] CD probably intended to write 'hate'. He believed that his attacks of eczema alleviated the symptoms of his chronic illness. See letter to J. D. Hooker, 5 March [1863].

[2] Letters from J. D. Hooker, 10 June 1863 and 19 June 1863.

[3] See letter to John Scott, 11 June [1963], and letter from John Scott, 16 June [1863].

[4] See letter from John Scott, 16 June [1863] and n. 9. See also letter from John Scott, [after 12] April [1863].

[5] Scott 1863a. See letter from John Scott, 16 June [1863], and letter to John Scott, 20 [June 1863].

[6] See letter from J. D. Hooker, [30 April 1863] and n. 2. Hooker had forwarded to CD the letter from Julius von Haast, 5 March 1863, which enclosed a copy of his letter to CD of 9 December 1862 (see letter from J. D. Hooker, 19 June 1863).

[7] See letter from J. D. Hooker, 19 June 1863 and n. 3.

[8] Haliburton 1863; see letter from J. D. Hooker, 19 June 1863 and n. 6. In his pamphlet, Robert Grant Haliburton explained the universality of certain superstitions respecting sneezing in terms of their having prevailed among a primeval human stock, from which all races had descended; he argued

that these primeval humans must have believed that the act of sneezing particularly exposed the soul 'to the influence of unseen enemies' (Haliburton 1863, p. 14).

9 CD refers to George Bentham's anniversary address to the Linnean Society (Bentham 1863), of which Bentham had sent him a preprint (see letter to George Bentham, 19 June [1863] and n. 2). In his address Bentham criticised the '*ex cathedrâ*' promulgation of 'a new form of spontaneous generation' in Richard Owen's anonymous review of Carpenter 1862, which appeared in the *Athenæum*, 28 March 1863, pp. 417–19 (see Bentham 1863, pp. xxvi–xxvii). See also letters to *Athenæum*, 18 April [1863] and 5 May [1863], and Appendix VII.

10 CD refers to Spencer 1860–2, the first volume of a projected five-part 'System of philosophy' which was issued to subscribers in parts (see letter from J. D. Hooker, 10 June 1863 and n. 10). There is a copy of this work in the Darwin Library–CUL; the final two numbers have their pages uncut.

11 Henry Holland was a leading London physician and a distant relative of CD's; there are a number of annotated copies of Holland's volumes of essays, some of which are presentation copies, in the Darwin Library–CUL and the Darwin Library–Down (see *Marginalia* 1: 385–8).

12 Daniel Oliver was assistant in the herbarium and librarian at the Royal Botanic Gardens, Kew (R. Desmond 1994, *List of the Linnean Society of London* 1863); CD refers to Beer 1863.

To Charles Loring Brace 24 June [1863][1]

Down. | Bromley. | Kent. S.E.
June 24th

Dear Sir

I am very much obliged to you for your kindness in sending me a copy of your new work on the "Races of the Old World".[2] I have not had time to read it all, but have looked through the latter part with interest. Your work must have been very laborious & I hope it will prove satisfactory.

I was interested by your remarks p. 388 on correlation of colour & constitution.[3] I have long thought this view probable; & you will be glad to hear, that about a year ago, with the assistance of the Director General of the medical department of the British army I circulated printed questions to the Surgeons of all Regiments serving in Tropical countries, asking them to tabulate the proportion of men who suffer from Tropical diseases in relation to the colour of their hair & skin.[4]

Whether I shall get any reliable data is of course very doubtful.—[5]

With my best thanks I beg leave to remain | Dear Sir | Yours very faithfully | Ch. Darwin

American Philosophical Society (310)

1 The year is established by the reference to Brace 1863 (see n. 2, below).

2 There is a presentation copy of Brace 1863 in the Darwin Library–Down (see *Marginalia* 1: 68).

3 CD discussed the correlation of skin and hair with 'constitutional peculiarities' in *Descent* 1: 242–5; however, Brace's book is not mentioned.

4 See *Correspondence* vol. 10, letter from George Busk, 1 April 1862, and letters from E. A. Parkes, 8 April 1862, 28 April 1862, and 29 June 1862. No copy of the printed memorandum has been found (see Freeman 1977, pp. 111–12); however, the text is given in *Descent* 1: 244–5 n. 48.

5 In *Descent* 1: 244 n. 48, CD reported that he had received no returns in response to his memorandum.

To J. D. Hooker 25 [June 1863]

Down Bromley Kent.
25th

My dear Hooker

I have been observing, *pretty* carefully, a little fact which has surprised me; & I want to know from you & Oliver[1] whether it seems new or odd to you; so just tell me whenever you write: it is very trifling fact so do not think of answering on purpose I have got a plant of Echinocystis lobata to observe the irritability of tendrils described by Asa Gray, & which, of course, is plain enough.[2] Having the plant in my study I have been surprised to find that the uppermost part of each branch, (ie the stem between the two uppermost leaves, excluding the growing tip) is *constantly* and slowly twisting round, making a circle in from 1½ to 2 hours: it will sometimes go round 2 or 3 times, & then at same rate untwists & twists in opposite direction. It generally rests half an hour before it retrogrades. The stem does not become permanently twisted. The stem beneath the twisting portion does not move in the least, though not tied. The movement goes on all day & all early night— It has no relation to light for the plant stands in my window & twists from the light just as quickly as towards it.—[3]

This may be common phenomenon for what I know; but it confounded me quite when I began to observe the irritability of the tendrils.— I do not say it is final cause, but the result is pretty for the plant every 1½ or 2 hours sweeps a circle, (according to length of bending shoot & length of tendril) of from 1 foot to 20 inches in diameter, & immediately that the tendril touches any object its sensitiveness causes it immediately to seize it. A clever gardener, my neighbour,[4] who saw the plant on my table last night, said "I believe, Sir, the tendrils can see, for wherever I put the plant, it finds out any stick near enough". I believe the above is the explanation, viz that it sweeps slowly round & round. The tendrils, have some sense, for they do not grasp each other when young.—

Yours affect | C. Darwin

Endorsement: 'June/63'
DAR 115: 197

[1] Daniel Oliver.

[2] CD refers to A. Gray 1858b, in which Asa Gray sought to corroborate from his own observations Hugo von Mohl's suggestion, first made in Mohl 1827, that the coiling of tendrils resulted from their being sensitive to touch. CD's annotated copy of the volume of the *Proceedings of the American Academy of Arts and Sciences* in which this paper appeared is in the Darwin Library–CUL. CD reread the article at the end of 1862, with the intention of trying 'a few experiments' on the subject, and Gray sent him seeds of *Echinocystis lobata* for this purpose (see *Correspondence* vol. 10, letter to Asa Gray, 6 November [1862], and letter from Asa Gray, 24 November 1862). CD's notes on his experiments with this species, dated 16 June – 29 July 1863, are in DAR 157.2: 29–51.

[3] In his *Autobiography*, p. 129, CD recalled that it was Gray's paper that led him to take up the subject of climbing plants more generally, and stated:

> He sent me seeds, and on raising some plants I was so much fascinated and perplexed by the revolving movements of the tendrils and stems, which movements are really very simple, though

appearing at first very complex, that I procured various other kinds of Climbing Plants, and studied the whole subject.

CD carried out numerous experiments on this subject in 1863 and 1864, and his paper entitled 'On the movements and habits of climbing plants' was read before the Linnean Society on 2 February 1865; his observations on *Echinocystis lobata* are given in 'Climbing plants', pp. 74–7. See also the observational and experimental notes in DAR 157.1 and 157.2.

[4] The reference is probably to John Horwood, gardener to CD's neighbour, George Henry Turnbull. Horwood had assisted CD with his botanical experiments over several years (see *Correspondence* vols. 9 and 10), and, earlier in 1863, had supervised the construction of CD's hothouse (see letter to G. H. Turnbull, [16? February 1863], and Appendix VI).

To Asa Gray 26 June [1863][1]

Down Bromley Kent
June 26th

My dear Gray.

I thank you for two pleasant letters; the former about Reviews in Athenæum & many other points, & the second shorter with answers to several of my questions.[2] In this latter you seem cruelly overworked. Although it is one of my pleasures to write to you & a very great pleasure to receive a letter from you; I earnestly beg you *never* to write to me when so busy; if I did not hear for six months or twelve months, I should understand the cause. Remember what a number of valuable & most interesting letters, I have received from you. So pray do not write unless you have a little leisure, which seems rare with you.—

I have little or nothing to say, for I see no one & hear from no one except dear Hooker.[3] But I must just run through some points in your letters. How curious the lie about Ohio marriages![4] I find it a dreadful evil in my compiling work, not knowing what to trust.[5] Many thanks for references about Phyllotaxis;[6] I have been half mad over it, but am having a lull: I have made no end of diagrams; but all my attempts have signally failed, as might have been expected. You will have seen by a later letter that I have received your Reviews on Decandolle.[7] I suppose you will have received Bentham's address, which has pleased me much, more than I understand why: it will do a world of good for *our* side.—[8] What you tell me about Phlox sounds very suspicious;[9] I have been looking at our one annual P. Drummondii & that is not dimorphic. Euonymus, I see, is dimorphic like Thyme, ie, hermaphrodite & female plants.[10] Mitchella has only two flower buds, alas! but we have just found out, why it is unhealthy, viz we gave the plants too much water.—[11]

The seeds of Sicyos did not germinate; & only one plant has come up of Echinocystis: I have been looking at its tendrils & seen with *great* interest their irritability; it is a very pretty little discovery of yours.[12] I am observing the plant in another respect, namely the incessant rotatory movement of the leading shoots, which bring the tendrils into contact with any body within a circle of a foot or 20 inches in diameter..— If I can make out anything clear about this movement, &

do not find that it is known, I will perhaps write a letter to you for the *chance* of its being worth inserting in Silliman or elsewhere.—[13]

Farewell. I have written this merely to tell you not to write, which proves that I am one of the best of human beings, | Farewell. C. Darwin

(I do not know who "Historicus" is.)[14]

Gray Herbarium of Harvard University (82)

[1] The year is established by the relationship between this letter and the letter from Asa Gray, 26 May 1863.

[2] Letters from Asa Gray, 26 May 1863 and [10–16] June [1863].

[3] Joseph Dalton Hooker.

[4] See letter from Asa Gray, [10–16] June [1863] and n. 2.

[5] CD refers to his work on the manuscript of *Variation*, which relied heavily on information gathered from a wide range of correspondents and published sources.

[6] See letter from Asa Gray, [10–16] June [1863].

[7] See letter from Asa Gray, [10–16] June [1863] and n. 13. CD refers to his letter to Gray of 31 May [1863] and to Gray's reviews of A. de Candolle 1862a and 1862b in the *American Journal of Science and Arts* (A. Gray 1863d and 1863e).

[8] The reference is to George Bentham's anniversary address to the Linnean Society, delivered on 25 May 1863, in which Bentham reviewed the scientific reaction to *Origin* (Bentham 1863). See also letter to George Bentham, 19 June [1863].

[9] See letter from Asa Gray, [10–16] June [1863] and n. 4.

[10] There are observational and experimental notes on *Euonymus*, dated 2 May, 4 June, and 20 July 1863, in DAR 109: 31–2; CD included the results from these experiments in *Forms of flowers*, pp. 290–1.

[11] At the end of 1862, Gray sent CD specimens of the dimorphic plant *Mitchella repens* for experimental purposes (see letter to Asa Gray, 2 January [1863] and n. 4). See also letter to Asa Gray, 31 May [1863].

[12] CD refers to A. Gray 1858b. See preceding letter and n. 2.

[13] CD refers to the *American Journal of Science and Arts*, commonly known as 'Silliman's journal' after its founder, Benjamin Silliman. CD worked extensively on climbing plants during 1863 and 1864, publishing a paper in the *Journal of the Proceedings of the Linnean Society (Botany)* ('Climbing plants'). See also preceding letter and n. 3.

[14] See letter from Asa Gray, 26 May 1863. From 1861, William Vernon Harcourt wrote many letters to *The Times* under the name Historicus, defending the British policy of neutrality in the American Civil War on the basis of precedents in international law. In 1863, some of his letters to *The Times* were published separately ([Harcourt] 1863). See Jenkins 1974–80.

To T. H. Huxley 27 June [1863][1]

Down. | Bromley. | Kent. S.E.
June 27th—

My dear Huxley

You will remember my telling you that Roget says a frog has a rudiment of a 6th toe.[2] I caught one yesterday (it is rarissima avis[3] here) & found (above what I suppose is the *great toe*) a projection clearly appearing like a rudiment of an extra digit on the hind feet. I removed skin, & removed the little projection without cutting it. It consists of transparent cartilage, of nearly (I cannot draw) form Fig. I.[4] This is crossed by lines which *may* represent articulations & then a rounded base which *may* represent the basal articulation. The organ has no muscles; but some are attached round the basal articulation & go to the next digit. Within the cartilage there are clusters of odd little bodies, like Fig II,[5] either cavities or hard particles (for I did not wish to cut open the specimen). Now from my entire ignorance of histology & embryology of vertebrates, I am dead stopped. Could you spare time to examine this cartilage & the enclosed little bodies; & see whether it is like what a rudiment of a bone might be expected to be, or like the first embryonic trace of a bone? Or whether it is merely thickened skin? Also whether the dark lines appear like first traces of articulations? The point is in itself, I think curious, & is to me *most* interesting in relation to strong inheritance & regrowth of extra digits &c &c?[6] The rudiment, without skin, is not much bigger than pin's head: I have wrapped it in tin-foil & put it, & the other hind-foot, in spirits, & could send it by Post in Box.

If you are too busy, to whom could I send it? Rolleston, I daresay, would examine it for me, or Mr Flower; but I do not know either personally?[7] As it is inner toe, I suppose I am right in thinking it an extra Great toe (if toe at all) & I sh^d.^ be glad of this, as I have been puzzled at the frequent doubling of the Great toe in Fowls.—[8] A newt has been seen with six toes.[9]

Let me hear, & tell me how Mrs. Huxley[10] & self are. *What are you doing now*??

I have never yet got hold of Edinburgh Review, in which I hear you are well abused.[11] By the way I heard lately from Asa Gray that Wyman was *delighted* at "Man's Place".—[12] I wonder who it is who pitches weakly, but virulently into you, in the Anthropological Review.[13] How quiet Owen seems;[14] I do at last begin to believe that he will ultimately fall in public estimation. What nonsense he wrote in Athenæum on Heterogeny![15] I saw in his Aye-Aye paper (I think) that he sneers at the manner in which he supposes that we should account for the structure of its limbs; and asks how we know that certain insects had increased in the Madagascan forests.[16] Would it not be a good rebuff to ask him how he knows there were trees at all on the treeless plains of La Plata for his Mylodons to tear down?[17] But I must stop, for if I once begin about that Devil there will be no end. I was disappointed in the part about species in Lyell.[18] You & Hooker[19] are the only two bold men.—

I have had a bad Spring & summer, almost constantly very unwell; but I am crawling on in my book on Variation under Domestication.[20]

Farewell my dear Huxley | Yours very sincerely | C. Darwin

It has only just occurred to me that I was very foolish not to look at bone in the real toes of the frog.

Imperial College of Science, Technology and Medicine Archives (Huxley 5: 225)

[1] The year is established by the relationship between this letter and the letter from T. H. Huxley, 2 July 1863.

[2] CD refers to Roget 1834, 1: 544, and to the letter to T. H. Huxley, 16 February [1863]. See also letter to T. H. Huxley, [after 16 February 1863].

[3] Rarissima avis: 'the rarest bird'; from 'rara avis', a rare bird or thing (*Chambers*).

[4] The diagram has not been found.
[5] The diagram has not been found.
[6] CD discussed polydactylism in man and other mammals in *Variation* 2: 12–17, suggesting that all such cases might be instances of 'reversion to an enormously remote, lowly-organised, and multidigitate progenitor'. In *Variation* 2: 16, CD noted that no 'mammal, bird, existing reptile, or amphibian' ordinarily possessed more than five digits, 'unless the tubercle on the hind feet of the toad and other tailless Batrachians be viewed as a digit'.
[7] George Rolleston was Linacre Professor of anatomy and physiology at the University of Oxford; William Henry Flower was curator of the Hunterian Museum at the Royal College of Surgeons (*DNB*).
[8] See *Variation* 2: 14
[9] See *Variation* 2: 14.
[10] Henrietta Anne Huxley.
[11] The reference is to Charles Carter Blake's anonymous review of T. H. Huxley 1863b in the April 1863 number of the *Edinburgh Review* ([Blake] 1863a); the attribution of authorship is given by the *Wellesley Index*.
[12] See letter from Asa Gray, 26 May 1863. CD refers to Jeffries Wyman and to T. H. Huxley 1863b.
[13] The first issue of the *Anthropological Review*, the journal of the newly formed Anthropological Society of London, appeared in May 1863. It contained an anonymous review of T. H. Huxley 1863b that, while critical of the book, was chiefly devoted to summarising its contents (Anon. 1863b). CD apparently refers to a letter that appeared in the same number under the title 'Man and beast', which attacked T. H. Huxley 1863b in more virulent terms (*Anthropological Review* 1 (1863): 153–62). The letter was signed '*Ανθρωπος*' (Anthropos), but the contents page for this volume of the journal subsequently identified its author as Blake. See also letter from T. H. Huxley, 2 July 1863. CD's annotated copy of the issue of the *Anthropological Review* containing these articles is in the Darwin Library–CUL.
[14] Richard Owen.
[15] CD refers to Owen's anonymous review of Carpenter 1862, published in the *Athenæum*, 28 March 1863, pp. 417–19. See also letters to *Athenæum*, 18 April [1863] and 5 May [1863], and Appendix VII.
[16] Owen 1862c, p. 95. There is an unannotated and unbound copy of the number of the *Transactions of the Zoological Society of London* in which Owen 1862c appeared in the Darwin Library–CUL.
[17] The reference is to Owen's monograph on the extinct giant sloth, *Mylodon robustus* (Owen 1842).
[18] C. Lyell 1863a. See letter to Charles Lyell, 6 March [1863].
[19] Joseph Dalton Hooker.
[20] CD was drafting the chapters on selection for *Variation* (*Variation* 2: 192–249; see 'Journal' (Appendix II)).

From J. D. Hooker [2]9 June 1863[1]

Kew
June 9/63.

My dear Darwin

Oliver[2] & I have little to say about your observation on the gyratory motion of the upper internode of *Echinocystis* except that it is most curious & important & goes a long way towards explaining the secondary causes which result in tendrils seeking & finding. indeed in so far as I know it is the *first* step in that direction—[3] I have set our people to watch various climbers with tendrils.

I am greatly relieved to find you like Bentham's address, I was very anxious about its result.[4] I wonder if Lyell will read it.[5]

We are overwhelmed with—& almost knocked up by visitors & visiting—of which there is no end & I have been atrociously idle of late— of course I have my Garden & Herb[m] work from 9 till 6, but my evening work, from which all my science *used* to be evolved is *no where* & I am disgusted with the extra part of the last 4 months— but then what am I to do?— unlike you, I have health & strength, & no large family connection with whom my children can mix, & I must keep up some society for them to mix in by & bye. added to which my position here demands large sacrifices of that sort. What a complicated machine life has become! I went to the Guards ball the other night & was deeply interested—[6] of course I knew so few people that I had abundant time & opportunity to roam about & observe & listen—admire & despise—the contrasts of old & young were ghastly— my God there were hideous old women in brides robes enough to keep you in night mares for a month of Sundays, & lovely girls enough to fill all the paradises of all the Turks.

The intellectual cut & exceeding handsomeness of both men and women was very satisfactory in the main, as was the cleanlyness & general health of the whole stock of high-bred humanity. To compare these with an equal number of the lower classes suggested many reflections—& strengthened me in my dogma that Brains × Beauty = Breeding + wealth.[7] I should extremely like to go to a similar selection in America, France or Austria; my impression is that the comparison would be ludicrous. Pray tell M[rs] Darwin & Henrietta[8] that I thought the Princess very charming looking, but neither pretty nor handsome, & that she was very badly dressed with a huge coronet of white roses enough to sink a ship— The Prince looked an utter nincompoop—& smiles with his lips only—a detestable savage feature.[9]

Have you read Goldwin Smiths "Empire"[10] I am much interested in it, probably most so because I had previously read only the abuse of his letters in the Times.[11] I think that after a *mild course of Darwinism* something might be made of him!

Ev yr affec & abandoned | Jos D Hooker

DAR 101: 147–8

CD ANNOTATION

Top of letter: 'Haast' *pencil*[12]

[1] Hooker wrote 'June 9' in error; the date is established by the relationship between this letter and the letter to J. D. Hooker, 25 [June 1863].

[2] Daniel Oliver.

[3] See letter to J. D. Hooker, 25 [June 1863] and nn. 2 and 3. CD had asked Hooker whether he or Oliver knew whether his observation of gyratory movements in *Echinocystis lobata* represented a new discovery. In 'Climbing plants', pp. 1–2, CD noted of his work on climbing plants in 1863 and 1864:

> My observations were more than half completed before I became aware that the surprising phenomenon of the spontaneous revolutions of the stems and tendrils of climbing plants had been long ago observed by Palm and by Hugo von Mohl, and had subsequently been the subject of two memoirs by Dutrochet.

The references are to Palm [1827?], Mohl 1827, Henfrey trans. 1852, pp. 147–58, and Dutrochet 1843 and 1844; there are annotated copies of Palm [1827?], Mohl 1827, and Henfrey trans. 1852 in the Darwin Library–CUL (see *Marginalia* 1: 589–94, 662–3). See also letters to J. D. Hooker, 12–13 August [1863] and 25 [August 1863], letter to Asa Gray, 4 August [1863], and letter from Asa Gray, 1 September 1863.

4 The reference is to George Bentham's anniversary address, delivered before the Linnean Society on 25 May 1863 (Bentham 1863; see letter to J. D. Hooker, 23 [June 1863] and n. 9). See also letters from J. D. Hooker, 20 April 1863 and [13 May 1863].

5 Charles Lyell. In his address, George Bentham included a survey of the scientific reaction to *Origin*, making reference to the 'eagerness' with which Lyell's judgment on CD's theory had been anticipated, and the 'great weight' that attached to his authority. He noted, however, that Lyell's opinion was 'not very easy to ascertain' from *Antiquity of man* (C. Lyell 1863a), and continued: 'the impression conveyed is, that he is generally convinced of the derivative origin of the present species, although he may leave it an open question whether there may not be exceptions, especially with regard to man' (Bentham 1863, p. xx). CD had been disappointed that Lyell had not endorsed natural selection more unambiguously in C. Lyell 1863a (see letter to Charles Lyell, 6 March [1863]).

6 The Guards' ball, described by *The Times* as '"the bright particular star" of this season', took place in London on the evening of 26 June 1863; among the many guests were foreign dignitaries, members of the royal household and the government, and representatives of many of the leading aristocratic families (*The Times*, 27 June 1863, p. 11).

7 Hooker refers to various remarks that he made to CD in 1862, on the development of an aristocracy being a consequence of natural selection (see *Correspondence* vol. 10). In his letter to CD of [23 March 1862] (*ibid.*), Hooker argued that the accumulation of 'Blood, Blunt [that is, wealth], Brains, [and] Beauty' by natural selection 'must culminate in an Aristocracy'.

8 Emma and Henrietta Emma Darwin.

9 The reference is to Alexandra, princess of Wales, and Albert Edward, prince of Wales, who were married on 10 March 1863 (*DNB*).

10 G. Smith 1863.

11 G. Smith 1863 comprised a series of letters on the British empire, contributed by Goldwin Smith to the *Daily News* in 1862 and 1863. In his letters, Smith argued for 'colonial emancipation', by which he meant that the self-governing colonies should be converted into independent states; his views attracted much attention, and caused widespread offence (*DNB*).

12 Julius von Haast. See letter to J. D. Hooker, 1 July [1863] and n. 2.

From W. B. Tegetmeier 29 June – 7 July 1863[1]

Muswell Hill | London. N.
July 7— 63.

My dear Sir

I thought that you might like to hear some of the results of the experiments that I have been trying, respecting the fertility of Mongrels.[2]

I have crossed three pair of Black Barbs and White Fantails, and am keeping the offspring for breeding next year.

The plumage has in all cases taken more after the male than the female parent.— In no case has there been any sign of want of fertility—[3]

In accordance with your suggestion I have also experimented on the Silk fowl and Spanish Cross with A very good pure bred Spanish Cock and two rather old Silk fowl hens.—[4] Only one of these laid. 9 eggs. They were left in the nest and she sat, producing 9 chicks, all black plumage, violet black skin.— Obviously identical with the Coq. Nêgre of Temminck—[5] None, even slightly, silky

I shall keep a cock and several pullets for breeding next season— Other fowls will breed for 5 or 6 seasons on end in breeding without a fresh cross, if these do not do the same it will be an indication of greater sterility in these mongrels than in pure breeds.[6]

There were some paragraphs in the Field the other day respecting all blue-eyed cats *not* being deaf, as I think you have stated.[7] If you have not seen the correspondence I will cut it out and forward it to you.

I forward you by this same post a number of the *Intellectual Observer* containing a short article of mine on Variation in Plumage at page 173—I have endeavoured to account for the circumstances that no small birds exist in a wild state with silky plumage but hardly expect that the idea has not occured to you.[8]

Trusting to hear that you are in good health | Believe me | Yours faithfully | W B Tegetmeier

C Darwin Esq

DAR 178: 58

[1] Tegetmeier originally dated the letter '29 June. 1863', but subsequently altered the date to 'July 7—63.'; the most likely explanation is that he began writing the letter on the former date, but was interrupted, resuming on the latter date.

[2] At the end of 1862, CD had suggested that Tegetmeier attempt to establish by experiment whether any existing breeds of pigeon had acquired 'accidentally any degree of sterility'; he also suggested a similar experiment with fowls (see *Correspondence* vol. 10, letter to W. B. Tegetmeier, 27 [December 1862]). See also letters to W. B. Tegetmeier, 26 [January 1863] and 19 February [1863], and letters from W. B. Tegetmeier, [after 26 January 1863] and 18 February 1863.

[3] Tegetmeier continued his experiments through two further breeding seasons (see letters from W. B. Tegetmeier, [before 21 December 1864] (*Correspondence* vol. 12) and 13 March 1865 (*Calendar* no. 4785)). CD reported Tegetmeier's findings with respect to the fertility *inter se* of different domestic varieties of pigeon in *Variation* 1: 192.

[4] See *Correspondence* vol. 10, letter to W. B. Tegetmeier, 27 [December 1862].

[5] Temminck 1813–15, 2: 253–5. See also Tegetmeier 1863b, p. 172.

[6] Tegetmeier published an account of these experiments in Tegetmeier 1867, p. 224; there is an annotated copy of this work in the Darwin Library–CUL (see *Marginalia* 1: 800–3). CD reported Tegetmeier's observations in *Variation* 1: 242.

[7] The reference is to *Origin*, p. 12, in which CD wrote: 'cats with blue eyes are invariably deaf'; see also *Origin*, p. 144. A series of letters disputing this assertion appeared in the *Field* 21 (1863): 416, 439, 524, and 547. In the third edition of *Origin*, CD altered the statement to read: 'cats with blue eyes are generally deaf' (*Origin* 3d ed., p. 12); in the fourth edition, published in 1866, he altered it again to read: 'cats which are entirely white and have blue eyes are generally deaf' (*Origin* 4th ed., p. 12). In discussing this phenomenon in *Variation* 2: 329, CD observed: 'I formerly thought that the rule was invariable, but I have heard of a few authentic exceptions'; however, the counter-examples cited are not those referred to here.

[8] Tegetmeier 1863b appeared in the April 1863 number of the *Intellectual Observer*. In his article, Tegetmeier demonstrated how a silky variety of fowl could be established by artificial selection from a naturally occurring variation. He then argued that no breed of small, silky fowl could survive in a state of nature, since this variation in plumage made fowls flightless, and it was 'obvious that a bird of moderate size, destitute of any natural means of defence, and incapable of escape by flight, could scarcely . . . avoid being destroyed by its natural carnivorous enemies' (*ibid.*, p. 173).

To J. D. Hooker 1 July [1863]

Down Bromley Kent
July 1st

My dear Hooker

We were all delighted at your account of Royalty & Aristocracy.[1] What an advantage it is to correspond with a man who habitually moves in the highest circles of fashion!!— When you can spare let me read N. Zealand newspapers about Haast.[2] Do not forget hot-house Lythrum-like flowers with 2 pollens.—[3]

I am sure I am not surprised that at this time of year, & with all your routine work, & most proper necessity of visiting (this is a sore point with me for our children's sake) that you have not much time for extra science: I have for years marvelled how on earth you have done such gigantic work, & retained a stomach & brains still good for anything.—

Thanks to you & Oliver about Tendrils; I thought all might have been known.[4] I have blundered in my last letter (I *invariably* do blunder at first), not about main fact of rotation, but about the means. It is a curiously complicated case & you must let me tell you a little about it. I find that when the shoot is **securely** lashed *close* above & below tendril, the tendril itself moves incessantly to & fro, but cannot make a circle. Secondly when tendrils are cut close off, so that neither their weight nor movement can affect the shoot itself, the shoot incessantly twists on its own axis (from 100° to 180°) from one side to the other. This twisting movement causes the whole shoot to move (about 30°) from side to side. Now from the correspondence in rate of movement of the tendril itself & of the gyratory movement of the shoot, I believe that after the tendril has moved one way & is ready to come back, the shoot twists & throws the tendril over to the opposite side, & then it travels back, & so performs a circle. In the pretty Cissus discolor which you gave me, the movement is entirely confined to the tendril itself, & it sweeps its circle in about 5 hours, day & night.— I hope I have not bored you with these details. Perhaps some day I will write a little paper on these movements—[5]

I have just read & laughed again over your splendid letter. I must get Goldwin Smith's book & read it.[6]

Ever yours affect | C. Darwin

Endorsement: '/63'
DAR 115: 198

1 See letter from J. D. Hooker, [2]9 June 1863 and nn. 6, 7, and 9.

2 In his letter to CD of 19 June 1863 Hooker promised to send CD a newspaper account of Julius von Haast's explorations in New Zealand. Haast evidently also sent the articles to CD, who received them later in July (see letter to Julius von Haast, 18 July [1863] and n. 4.

3 The reference is to *Lagerstroemia*; see letter to J. D. Hooker, 15 and 22 May [1863] and n. 13, and letter from J. D. Hooker, [24 May 1863].

4 Daniel Oliver. See letter to J. D. Hooker, 25 [June 1863], and letter from J. D. Hooker, [2]9 June 1863.

5 CD refers to his experiments on climbing plants, begun in June 1863. There are notes on CD's experiments with *Echinocystis lobata* (dated 16 June – 29 July 1863) in DAR 157.2: 29–51, and on his

experiments with *Cissus discolor* (dated 30 June – 18 July [1863]) in DAR 157.2: 55–6. CD carried out numerous experiments on this subject in 1863 and 1864, and his paper entitled 'Climbing plants' was read before the Linnean Society on 2 February 1865 (see also the observational and experimental notes in DAR 157.1 and 157.2); his observations on *E. lobata* and *C. discolor* are given in 'Climbing plants', pp. 74–7 and 83–4.

[6] G. Smith 1863; see letter from J. D. Hooker, [2]9 June 1863 and n. 11.

To *Gardeners' Chronicle* 2 July [1863]

A very slight shower, lasting hardly more than a minute, fell here this morning (July 2) about 10 o'clock. My wife gathering some flowers immediately afterwards noticed that the drops of water appeared yellowish, and that the white Roses were all spotted and stained. I did not hear of this circumstance till the evening; I then looked at several Roses and Syringas and found them much stained in spots. Between the petals of the double white Roses there were still drops of the dirty water: and this when put under the microscope showed numerous brown spherical bodies, $\frac{1}{1000}$ of an inch in diameter, and covered with short, conical transparent spines. There were other smaller, smooth, colourless sacs about $\frac{4}{7000}$ of an inch in diameter. I preserved a minute drop of the water beneath thin glass, cementing the edges, and next morning looked rather more carefully at it. I then observed that the water swarmed with elongated, moving atoms, only just visible with a quarter-inch object glass. Whether these inhabited the rain-drops, when they fell, I cannot of course say; but I suspect so, for the petals, now that they are nearly dry, seem stained with absolutely impalpable matter of the colour of rust of iron. This matter has chiefly collected, in the act of drying, on the edges of each spot. The Rev. M. J. Berkeley could tell us what the larger spherical bodies are which fell this day by myriads from the sky, carried up there, I presume, by some distant whirlwind.[1]

Gardeners' Chronicle and Agricultural Gazette, 18 July 1863, p. 675

[1] CD's letter was reproduced in the *Gardeners' Chronicle* with explanatory notes by Miles Joseph Berkeley, who had been a regular contributor to the journal since its commencement in 1841 (*DNB*). Berkeley thought that the 'larger bodies' described by CD were the 'pollen grains of some Thistle or Centaurea'. He also noted that the 'ferruginous spots on the white petals of Philadelphus' forwarded by CD consisted of 'siliceous particles', and 'multitudes of irregular bodies so minute as to present the Brownian molecular motion'.

From T. H. Huxley 2 July 1863

Jermyn St
July 2nd | 1863

My dear Darwin

I am horridly loth to say that I cannot do anything you want done—& partly for that reason & partly because we have been very busy here with some new arrangements during the last day or two, I did not at once reply to your note[1]

I am afraid, however, I cannot undertake any sort of new work— In spite of working like a horse (or if you prefer it like an ass) I find myself scandalously in arrear—and I shall get into terrible hot water if I do not clear off some things that have been hanging about me for months & years

If you will send me up the specimens, however, I will ask Flower[2] (whom I see constantly) to examine them for you— The examination will be no great trouble and I am ashamed to make a fuss about doing it—but I have sworn a big oath to take no fresh work, great or small, until certain things are done—

I wake up in the morning with somebody saying in my ear, A, is not done, & B, is not done & C is not done & D, is not done &c &c.—and a feeling like a fellow whose duns are all in the street waiting for him—

By the way you ask me what I am doing now—so I will just enumerate some of the A, B, & C; aforesaid—

A. Editing Lectures on Vertebrate skull & bringing them out in the Medical Times[3]

B. Editing & rewriting Lectures on Elementary Physiology—just delivered here & reported as I went along—[4]

C. Thinking of my course of 24 Lectures on the Mammalia at Coll. Surgeons in next spring & making investigations bearing on the same.[5]

D. Thinking of & working at a Manual of Comparative Anatomy (may it be d——d) which I have had in hand these seven years[6]

E. Getting heaps of remains of new Labyrinthodonts from the Glasgow coal field which have to be described[7]

F. working at a memoir on *Glyptodon* based on a new & almost entire specimen at the College of Surgeons[8]

G. preparing a new Decade upon Fossil fishes for this place[9]

H. knowing that I ought to have written long ago a description of a lot of most interesting Indian fossils sent to me by Oldham—[10]

I. being blown at by Hooker for doing nothing for the Natural History Review[11]

K. being bothered by sundry Editors—just to write articles "which you know you can knock off in a moment"

L. Consciousness of having left unwritten letters which ought to have been written long ago especially to C. Darwin—

M. General worry & botheration Ten or twelve people taking up my time all day— about their own affairs—

N. O. P. Q R. S T. U. V. W. X. Y. Z

Societies—

Clubs—

Dinners— Evening parties & all the apparatus for wasting time called 'Society'

Colenso-ism & botheration about Moses—[12]

Finally pestered to death in public & private because I am supposed to be what they call a "Darwinian"!

If that is not enough I could exhaust the Greek alphabet for heads in addition

I am very glad to hear that Wyman thinks well of my book—as he is very competent to judge—[13] I hear it is republished in America—but I suppose I shall get nothing out of it.—[14]

The man who does the virulence in the Edinburgh & Anthropological is a jackal of Owen's by the name of C. Carter Blake—[15] The same whom Falconer shewed up in the Elephant paper[16]

As a return for the impertinence against Rolleston & myself which that absurd Anthropological Society sanctioned—I sent them back the Diploma of Honorary Fellowship—by means of which after I had refused to join them in any other way they had lugged me in—[17]—

Owen is damning himself as fast as is good for us— The Aye-Aye wind up is the greatest bosh I ever read—[18]

Lyell is not as bold as I should have wished. But it is a great thing for a man past sixty to have eaten as much of his leek as he has[19]

My wife is better than I have seen her for years—but going to increase the population again in September I am sorry to say—[20] We have been grieved to get poor accounts of you occasionally I hope Mrs Darwin is well

Ever yours faithfully | T H Huxley

DAR 166: 298

[1] See letter to T. H. Huxley, 27 June [1863].

[2] William Henry Flower.

[3] T. H. Huxley 1863c. Huxley was appointed Hunterian Professor of comparative anatomy at the Royal College of Surgeons in 1863 (*DNB*). In the spring of 1863, he delivered two courses of lectures, one on the classification of animals and one on the vertebrate skull, both of which were published in the *Medical Times and Gazette*. The lectures were subsequently republished in book form (T. H. Huxley 1864).

[4] According to Leonard Huxley these lectures were delivered on Friday evenings from April to June at the Royal School of Mines, Jermyn Street, London, where Huxley was professor of natural history; Leonard Huxley also stated, incorrectly, that they were reported in the *Medical Times* (L. Huxley ed. 1900, 1: 246 n.). No published report of the lectures has been found; however, they subsequently formed the basis of T. H. Huxley 1866a.

[5] As Hunterian Professor of comparative anatomy at the Royal College of Surgeons, Huxley was obliged to deliver twenty-four lectures each year (see L. Huxley ed. 1900, 1: 236). Between 2 February and 26 March 1864, Huxley delivered twenty-four lectures on 'The structure and classification of the Mammalia' (see *Medical Times and Gazette* (1864), pt 1: 153 and pt 2: 145).

[6] In 1855, Huxley agreed to write a 'treatise on comparative anatomy for students' for the leading medical publisher, John Churchill (T. H. Huxley 1877, p. iii, and L. Huxley ed. 1900, 1: 85). The work finally appeared in two volumes as *A manual of the anatomy of vertebrated animals* and *A manual of the anatomy of invertebrated animals* (T. H. Huxley 1871 and 1877).

[7] As naturalist of the Geological Survey of Great Britain, Huxley was one of those responsible for cataloguing and describing the palaeontological specimens acquired by the Museum of Practical Geology (see Flett 1937, pp. 66–7; see also T. H. Huxley and Etheridge 1865). Huxley had recently described a fossil specimen of a newly discovered genus of labyrinthodont from the Lanarkshire coalfield in the *Quarterly Journal of the Geological Society* (T. H. Huxley 1862b).

[8] Huxley presented a 'brief preliminary notice' of a fossil specimen of the extinct edentate genus, *Glyptodon*, at a meeting of the Royal Society on 18 December 1862 (T. H. Huxley 1862c). Huxley

apparently refers to his full account of the specimen, which was read before the Royal Society on 28 January 1864, and published as T. H. Huxley 1864b.

9 T. H. Huxley 1866b. Since 1849, the Geological Survey had published an occasional series of 'Figures and descriptions of British organic remains'; each number contained ten plates of illustrations, and was thus called a 'decade' (Flett 1937, pp. 67–8). In 1861, Huxley contributed at length to a number on fish (T. H. Huxley 1861a, 1861b, and 1861c). He contributed an entire number on fish in 1866, and made a further contribution to the series in 1872 (T. H. Huxley 1866b and 1872).

10 T. H. Huxley 1865. Thomas Oldham was the director of the Geological Survey of India (Sarjeant 1980–96).

11 Joseph Dalton Hooker was one of the eleven editors of the *Natural History Review*, of which Huxley was the editor-in-chief (see letter from Charles Lyell, 15 March 1863, n. 19). Huxley apparently ceased to contribute to the organisation of the journal in July 1863, when the responsibility of the 'commissariat' was handed over to paid editors. Huxley observed: 'I did not foresee all this crush of work . . . when the *Review* was first started, or I should not have pledged myself to any share in supplying it' (L. Huxley ed. 1900, 1: 210; see also L. Huxley ed. 1918, 1: 413).

12 The reference is to John William Colenso, whose critical examination of the Old Testament had caused great controversy because of its denial that Moses was the author of the Pentateuch (Colenso 1862–79; see *DNB*). Huxley had become acquainted with Colenso, who was bishop of Natal in southern Africa, when Colenso attended some of his lectures at the Royal College of Surgeons in the spring of 1863 (see A. Desmond 1994–7, 1: 315–16).

13 See letter to T. H. Huxley, 27 June [1863] and n. 12. Huxley refers to Jeffries Wyman and to T. H. Huxley 1863b.

14 The first American edition of T. H. Huxley 1863b was published by D. Appleton & Co. of New York in May or early June 1863 (T. H. Huxley 1863d; see *American Publishers' Circular* 1 (1863): 43 and 192). At the time, there was no international copyright agreement, and American publishers often issued pirated editions of British books. D. Appleton & Co. was known for sometimes paying royalties to foreign authors, and had paid CD royalties on American editions of *Origin*; however, the company apparently issued pirated editions of both T. H. Huxley 1863a and 1863b (see L. Huxley ed. 1900, 1: 283–4, Tebbel 1972, pp. 291–3, A. Desmond 1994–7, 1: 311 and 323–4, and *Correspondence* vol. 8).

15 See letter to T. H. Huxley, 27 June [1863] and n. 13. Huxley refers to Richard Owen, to Charles Carter Blake's anonymous critical review of T. H. Huxley 1863b in the *Edinburgh Review* ([Blake] 1863a), and to Blake's pseudonymous letter on the subject in the *Anthropological Review* 1 (1863): 107–17. The attribution to Blake of the article in the *Edinburgh Review* is confirmed by the *Wellesley Index*.

16 Falconer 1863a, pp. 46–9. See also letter to J. D. Hooker, 3 January [1863] and n. 1.

17 Huxley apparently refers to the manner in which the account in T. H. Huxley 1863a of his long-standing dispute with Owen concerning the comparative anatomy of human and simian brains (the so-called 'hippocampus controversy') had been criticised in the *Anthropological Review* (Anon. 1863b, pp. 114–17). Although not specifically mentioned, George Rolleston had been one of Owen's main opponents on this subject (see, for example, L. G. Wilson 1996b). The *Anthropological Review* was produced by the Anthropological Society of London, which was founded, early in 1863, by James Hunt, in opposition to the Ethnological Society of London. Hunt and his society were explicitly opposed to CD's theory of natural selection. See Stocking 1987, pp. 245–54.

18 See letter to T. H. Huxley, 27 June [1863] and n. 16. Huxley refers to the conclusion of Owen's paper on the aye-aye, a Madagascan lemur, in which Owen reviewed the theories currently employed to account for the origin of new species (Owen 1862c, pp. 87–97). See also letter to J. D. Hooker, 17 March [1863].

19 See letter to T. H. Huxley, 27 June [1863] and n. 18. The reference is to C. Lyell 1863a. The phrase 'eat the leek' (Shakespeare, *Henry V*, 5.1.1–59 (Wells and Taylor eds. 1988)), denotes being 'compelled to take back one's words' (*Chambers*). Lyell, who had long been an outspoken opponent of transmutation theories, had significantly revised his views on the subject in the light of CD's theory.

[20] Henrietta Anne Huxley gave birth to a daughter, Nettie, on 21 September 1863 (A. Desmond 1994–7, 1: 322).

To John Scott 2 July [1863][1]

Down Bromley Kent
July 2d

Dear Sir

Many thanks for capsules.—[2] I would give table of Auricula, especially owing to enclosed extract which you can quote.[3] Your facts about varying fertility of the Primulas will be appreciated by but very few Botanists, but I feel sure that the day will come when they will be valued.—

By no means modify even in **slightest** degree any result.—[4] Accuracy is the soul of Natural History. It is hard to become accurate; he who modifies a hair's breadth will never be accurate. It is a golden rule, which I try to follow, to put every fact which is opposed to one's preconceived opinion in the strongest light.— Absolute accuracy is the hardest merit to attain & the highest merit.— Any deviation is ruin—

Sincere thanks for all your laborious trials on Passiflora.[5] I am very busy & have got two of my sons ill; I very much fear with scarlet-fever;[6] if so, no more work for me for some days or weeks.— I feel greatest interest about your Primula cases. I think it much better to count seed than to weigh; I wish I had never weighed: counting is more accurate, though so troublesome.—

Yours sincerely | C. Darwin

[Enclosure 1][7]

Prof. Treviranus says (Botanische Zeitung 1863 p. 4.) that according to Koch & Tausch all species of Primula, present two forms, except P. longiflora which is always short-styled.[8]

Prof. T. (idem p. 6.) seems to suspect that the long-styled P. auricula is sterile; for he says he found a plant of this long-styled form growing near a short-styled form, & the numerous flowers had not produced a single seed-capsule.

[Enclosure 2][9]

Hottonia

"Various authors have remarked that the Hottonia palustris presents two forms. Fresh Specimens sent to me from N. Wales were grandly dimorphic.[10] In the long-styled form, the pistil is more than twice as long as in the other form, & projects far out of the flower; the stigma is smaller & rougher than that of the short-styled: the anthers lie within the tube of the corolla. In the short-styled form, the anthers project far out of the flower & correspond in height with the projecting stigma of the other form; as does the short pistil with the short stamens of the long-styled form.— But the most remarkable difference is in the size of the pollen-grains:

those from the short-styled flowers, when distended with water, are $\frac{14}{14,000}$ of an inch in diameter; those from the long-styled (& therefore from the shorter stamens) are $\frac{9}{14,000}$ in diameter. Spheres differing in the proportion of 14 to 9 in diameter presented under the microscope a most remarkable contrast. The contents of the larger pollen-grains certainly seemed to be coarser-grained & of a browner colour.[11]

DAR 93: B79; Quentin Keynes

[1] The year is established by the reference to Scott 1864a, which was written in 1863 and read before the Linnean Society on 4 February 1864. See letter from John Scott, [3 June 1863].

[2] The seed capsules referred to were apparently sent in a letter to CD which has not been found (see n. 3, below). The specimens have not been identified, but probably resulted from Scott's experiments on either Primulaceae or orchids.

[3] See enclosure 1 and n. 7, below. In a letter which has not been found, Scott had apparently asked CD whether he should give the results from his crossing experiments with the two forms of *Primula auricula*. In Scott 1864a, pp. 86–90, Scott tabulated his results from homomorphic and heteromorphic crosses with *P. auricula*. He noted that, since CD had published evidence of the functional dimorphism of this species ('Dimorphic condition in *Primula*'), he had no intention of giving the results from these crosses in detail. He continued (Scott 1864a, pp. 86–7):

> My attention, however, has been lately directed to a tacit implication of the absolute sterility of the long-styled form. As this view is certainly erroneous, I will give, for the satisfaction of those who have had no means of testing the point at issue, the results of my own experience.

Scott's reference is to Treviranus 1863a, p. 6; see Scott 1864a, p. 90.

[4] Some indication of the nature of Scott's statement in the missing letter is given by Scott's letter to CD of 23 July [1863].

[5] Scott's letter has not been found, but see the letter to John Scott, 6 June [1863] and n. 11.

[6] Emma Darwin recorded in her diary (DAR 242) on 2 July 1863: 'boys with sore throat'. The reference may be to George Howard, Francis, or Horace Darwin; Leonard Darwin had contracted scarlet fever in 1862 (see *Correspondence* vol. 10).

[7] Enclosure 1 has been attached to this letter on the basis of Scott's reply, which establishes that CD sent notes on Treviranus 1863a with this letter (see letter from John Scott, 23 July [1863]).

[8] The references are to Treviranus 1863a, p. 4, Koch 1843–4, 2: 673, and Tausch 1821, p. 355. There is an annotated copy of Treviranus 1863a in the Darwin Pamphlet Collection–CUL.

[9] In his letter to Scott of 20 [June 1863], CD had offered to send Scott his observations on heterostyly in *Hottonia palustris*. See also Scott's letter to CD of 23 July [1863], which confirms that CD's notes were sent with this letter. Scott quoted this memorandum in Scott 1864a, pp. 78–9.

[10] The specimens of *Hottonia palustris* referred to were sent to CD from Llandudno, Carnarvonshire, by his niece Margaret Susan Wedgwood (see *Correspondence* vol. 10, letters from M. S. Wedgwood, [before 4 August 1862] and [6 August 1862], and letter to K. E. S., L. C., and M. S. Wedgwood, 4 [August 1862]). CD's original notes on the specimens are in DAR 110: A16.

[11] CD reported this observation in *Forms of flowers*, p. 51.

To T. H. Huxley 3 July [1863][1]

Down. | Bromley. | Kent. S.E.
July 3^d^

My dear Huxley

Very sincere thanks for your long letter.[2] I grieve that you are so fearfully hard worked. If Mr Flower will look at specimen, I shall be *particularly* obliged; pray tell him so.—[3] I send by this post the specimen in a Box.— Please tell Mr F. that the cleaned rudiment is in separate parcel of tin-foil within the larger: he must be careful in opening, as it is not bigger than pin's head. Just call his attention to points on which I am utterly ignorant.— viz whether this transparent (it was transparent before spirits) cartilage has character of embryonic bone?—what are little irregular embedded objects?—& whether the transverse dark lines can represent articulation? I rather fancy whole case will break down.—

With hearty thanks & hopes that you may get the thick of your work over | Ever yours | C. Darwin

Imperial College of Science, Technology and Medicine Archives (Huxley 5: 229)

1 The year is established by the relationship between this letter and the letter from T. H. Huxley, 2 July 1863.
2 See letter from T. H. Huxley, 2 July 1863.
3 William Henry Flower. See letter to T. H. Huxley, 27 June [1863] and letter from T. H. Huxley, 2 July 1863.

From Asa Gray to J. D. Hooker[1] 6 July 1863

Cambridge, Mass.
July 6, 1863.

My Dear Hooker

I have to thank you for a nice letter of June 12, and later, for the draft on your bankers for £72.12 for Parry.[2] The latter send profound thanks, and is quite set up by his venture on seeds. He ought to be (and is) very thankful to you.

A note from Bentham[3] came in your envelope. Let me answer one of his enquiries to you—who will let him know my answer: for I infer that this will not reach him at Kew until after his holiday, so I address to you

He asks whether the Major John Le Conte, Deceased, of Philadelphia, is the Entomologist and F.L.S.[4] Yes, doubtless, if a Fellow of long standing. His son John L. Le Conte is the more distinguished entomologist:[5] but he could have been chosen only a few years ago,—and I suppose you do not now elect Fellows in foreign parts. Whereas the old Major Le Conte was probably chosen about the same time with his contemporary Dr. Bigelow.[6]

Now I am to discourse, partly for Bentham, partly for you, anent Bentham's Anniversary address.[7]—which I like greatly, and yet have some flaws to pick at.— Here—far away from the great centres of scientific activity, we "are nothing if not critical".[8]

He rightly blames Huxley for his use of the term *Biology*.[9] I take the word to be essentially equivalent to Physiology and a better term. It certainly is not = Nat. History.

Darwin may well be pleased and satisfied at having made such an impression on Bentham. The latter's conclusion as to "the tide of opinion," on p. 13, I am pleased to see—;[10] both because it is *evidently true*, and because I had lately in Amer.

Journal. been saying the same.[11] Natural Selection he does not quite appreciate, I think; but few do.

The caution, in p. 10, is well enough. But few, I imagine, fit to handle such questions, ever take an *illustration* for an argument, at least for a proof.[12] A comparison is sometimes a capital argument for its purpose.— An analogous case may perfectly rebut an objection, tho' it establishes nothing for itself.

As to my Essay in Atlantic—which has been made so much of—Bentham seems to misapprend a little its object and the occasion.[13] It was not to "remove the prejudices excited by religious views" or to deprecate attack on Darwin in this regard, but to rebut and turn aside attacks which had been made on this ground, and partly by scientific men. To show the futility and unreasonableness of such attacks was only to remand the discussion to its proper court. "His opponents reply that his Natural Theology is not Religion" Bree, I suppose.[14] But the same line of argument holds as well. It is easy to show that the very same principles have been admitted in respect to astronomy & geology, and so may just as well be in Natural history.— But all this is of no account. But as a review of Darwin's Theory on scientific grounds, why did not B. mention my much earlier review, in Silliman's Journal,—perhaps the better of the two, and addressed to scientific men instead of *ad populum*.[15]

I am pleased to see him pitch into Owen for ill-treating Carpenter.[16] But, in the *Heterogeny* question, the prejudice (in which I share) against spontaneous generation, and too great deference to French Academy, has made him unjust to Wyman,—who is one of the most conscientious and accurate of observers.[17] Wyman will now probably publish the results of a lot of experiments, which he did not think were worth while.[18] As respects the heat employed to destroy germs, Wyman's experiments are better than Pasteur's. The latter subjected them to *dry heat* only, and to say 230° Fahr. But Wyman in the experiments under pressure, *boiled* at $250\frac{1}{2}$° and $307\frac{1}{2}$° Fahr.. You know how great the difference between the two modes.

I enclose you a brief mem. by Dr. Wyman.[19] By the way, when you want a new Foreign Member of Linn. Socy. you should think of Wyman.[20] He is pure gold.

As to canvassing for scientific honors—for which you blame A.DC,[21] one may be led into it, and I suspect it is not uncommon on the Continent. Last summer Ruprecht,—whom I know little of, wrote to ask me to furnish him with what I could to enable him to put me in nomination for a vacancy in Acad. Petro*[*ff*]*.[22] I declined to have anything to do or say in the matter, telling him to refer to any botanists he chose. Where upon I supposed the matter was dropped. But, last week, the diploma came to me.

I should have thought that they would have preferred Bentham or yourself. By the way, now that I am writing of myself I have, against my will been put into the chair of Amer. Acad. on the retirement of Dr. Bigelow. I could have succeeded in avoiding it had it not been that Agassiz and his small clique set to work with all their might against me,— —the result was a demonstration that made A open his eyes, and has, I hear, mortified him exceedingly.[23]

Bond is at home again, much pleased with his trip.[24] He says that the greatest thing to see, and the most satisfactory, on the other side of the water, is Kew Gardens & museum. We were pleased to see one who had lately seen you all, and brought such pleasant accounts of you and Mrs. Hooker, I should suppose that Mrs. Hooker is as well as possible.[25]

The sudden death of an aged and very dear Great Aunt of Mrs. Gray's took us to Connecticut last week, where we remained some while.[26] The good soul has bequeathed to my wife property which would amount in ordinary times to about £2000:—a great help to us. I now see my way clearer than ever to shaking off other things before long and taking myself to *Fl. N. Amer.*[27]

Flora Australiensis is a nice vol.[28] *Mesembryanthemum* was truly collected in California by Fremont[29] and by others—I think also by Coulter.[30]—said to be common on the coast. I always supposed it must have been introduced by the Spaniards.[31]

I note what you write of Seemann.[32] I never liked the fellow, and always had just as little as possible to do with him, while being civil distantly. I looked upon him as a sort of adventurer.— I shall keep very clear of him.

Ovules bearing pollen is a very curious business. But I do not see what use can be made of the case.

I take issue with you square, and say that the radicle *is* 1st internode. I support the proposition thus— 1. It is in the place of internode, i.e. below the first node; 2. It grows like an internode whenever (as is usual) it lengthens at all. 3. It produces roots like an internode, i.e. all internodes favorably placed, e.g. 2nd. 3rd. &c—

Royal Botanic Gardens, Kew (Asa Gray correspondence: 328–9)

1 Hooker enclosed this letter in a letter to CD of late July, which has not been found (see letter from J. D. Hooker, [21 July 1863], and letters to J. D. Hooker, 22 July [1863] and 26 [July 1863]).

2 Charles Christopher Parry.

3 George Bentham.

4 John Eatton Le Conte died on 21 November 1860, at Philadelphia, where he had been resident for ten or twelve years (*American Journal of Science and Arts* 2d ser. 31 (1861): 303, 463); his address was given as New York in the *List of the Linnean Society of London* 1863. Bentham was president of the Linnean Society.

5 John Lawrence Le Conte.

6 The American botanist Jacob Bigelow was elected a fellow of the Linnean Society in 1819; John Eatton Le Conte was elected a fellow in 1812 (*List of the Linnean Society of London* 1863).

7 Bentham 1863.

8 Shakespeare, *Othello*, 2.1.122 (Wells and Taylor eds. 1988).

9 Thomas Henry Huxley. In his anniversary address, delivered to the Linnean Society on 7 May 1863, Bentham questioned whether the term 'Biology' could be retained in the limited sense in which it had been originally proposed, 'that of the *science of life, i. e.* of the phenomena of life, in contradistinction to the description and classification of living beings', given that it had recently been used in a more general sense to refer to 'Zoology and Botany in all their branches' (Bentham 1863, p. xii). He continued: 'if Biology continues to be used by Professor Huxley and other distinguished public lecturers in the most general sense, it will be the one which will be definitively attached to it'.

10 Bentham devoted most of his address to a review of the state of the literature on the origin of species (Bentham 1863, pp. xii–xxii). In summing up the literature relating to *Origin*, he averred that the tide of opinion among 'philosophical naturalists' was 'setting fast in favour of Mr. Darwin's hypothesis',

and that few who really considered the subject would now deny that new species had been produced from a common ancestor by natural selection (Bentham 1863, pp. xxi–xxii; Gray evidently refers to the pagination of an offprint). For CD's reaction to Bentham's address, see the letter to George Bentham, 19 June [1863].

11 Gray probably refers to A. Gray 1863d, which was published in the May 1863 number of the *American Journal of Science and Arts*, and in which Gray discussed the 'general drift of opinion' on the question of the origin of species (A. Gray 1863d, p. 437).

12 In his address, Bentham argued that there was a 'general logical confusion' in most of the reviews of *Origin*, arising 'from that aptness of illustration and figurative language' in CD's book, which, Bentham considered, carried CD's supporters, and even sometimes CD himself, 'beyond strict logical bounds' (Bentham 1863, p. xviii). Among the figurative terms singled out as requiring 'the greatest caution in their use' were the phrases 'Struggle for Life' and 'Natural Selection'.

13 Gray refers to his three-part review of *Origin*, which appeared in 1860 in the *Atlantic Monthly* (A. Gray 1860b), and was reprinted as a pamphlet on CD's initiative (A. Gray 1861a; see *Correspondence* vol. 9, Appendix III). Bentham discussed Gray's 'remarkable dissertation' in Bentham 1863, p. xvii.

14 The reference is apparently to Bree 1860. Although Charles Robert Bree did not discuss Gray's natural theology in his book, he did argue that a developmental natural theology (like that espoused by Gray) was in conflict with revelation, and that a belief in the natural progressive development of species 'struck at the root of religion itself' (pp. 244–5).

15 *Ad populum*: 'to the people'. A. Gray 1860c was published in the *American Journal of Science and Arts*, commonly known as 'Silliman's Journal' after its founder, Benjamin Silliman.

16 Gray refers to Bentham's criticism, in Bentham 1863, pp. xxvi–xxvii, of Richard Owen's anonymous review of Carpenter 1862 in the *Athenæum*, 28 March 1863, pp. 417–19 (see Appendix VII).

17 In his address (Bentham 1863, pp. xxv–xxvi), Bentham criticised the work of Jeffries Wyman, Gray's colleague at Harvard University, who in 1862 began a series of experiments which paralleled those of Louis Pasteur, although with different results (Wyman 1862; see *DSB*). Bentham suggested that Pasteur's results (Pasteur 1861 and 1862) seemed to refute completely 'the idea of heterogeny in the production of microscopic animals or plants in organic matter in a state of fermentation' (Bentham 1863, p. xxv). The Académie des Sciences of Paris had awarded Pasteur the Alhumbert prize for his work on spontaneous generation in 1862 (see Farley 1977).

18 Wyman subsequently published a further paper detailing experiments relating to spontaneous generation (Wyman 1867).

19 The enclosure has not been identified.

20 Wyman was elected a foreign member of the Linnean Society in 1869; the number of foreign members was restricted to fifty (*List of the Linnean Society of London* 1870).

21 Alphonse de Candolle.

22 Franz Josef Ruprecht was director of the botanical museum of the Academy of Sciences of St Petersburg.

23 On Gray's election as president of the American Academy of Arts and Sciences in 1863, and the opposition of Louis Agassiz, see Dupree 1959, pp. 319–20.

24 The reference is probably to Gray's colleague at Harvard University, George Phillips Bond.

25 Frances Harriet Hooker.

26 Jane Loring Gray's great aunt, 'one of the country kin in Connecticut' (Dupree 1959, p. 332), has not been further identified.

27 In collaboration with John Torrey, Gray had first begun work on a planned flora of North America in 1833. A portion of the work was published between 1838 and 1843 (Torrey and Gray 1838–43), but it remained unfinished. Although Gray for many years planned to complete the work, he ultimately abandoned it, beginning to publish instead his *Synoptical flora of North America* (A. Gray 1878–84), which also remained unfinished (Dupree 1959, pp. 385–93).

28 Gray refers to the first volume of Bentham 1863–78.

29 John Charles Frémont.

30 Thomas Coulter.

31 In Torrey and Gray 1838–43, 1: 556, the authors stated that two or more species of *Mesembryanthemum* were reported to be naturalised in California, and that one of them was suspected to be native.

32 Berthold Carl Seemann.

From Asa Gray 7 July 1863

Cambridge. [Mass.]
July 7, 1863

Dear Darwin,

My last from you is May 31.[1]

I had arranged to reprint most of Bates on Mimetic analogy in Sill. Jour. but my long review of A. DC. crowded it out.[2] I then thought of a brief abstract, but have had no time to prepare it.[3] I wrote remarks & arranged long extracts of your Linum paper, and insisted on it for July no. of Sill. Jour.— But it too was laid over.[4]—not for any thing I had, for I have little in July no.

I like and agree to your remark that in Bates's Geographical varieties, etc. etc. we get about as near to seeing a species made as we are ever likely to get:[5]—and so believing I think your *gradual* way more likely than Heer's jumps.[6]

A propos to Heer, you ask me if it is not impossible to imagine so many & nice coadaptations as we see in Orchids *being formed all by a chance blow*.

I reply yes, perfectly impossible to imagine (and much the same by any number of chance blows).

So I turn the question back upon you, is not the fact that the co-adaptations are so nice" next to a demonstration against their having been formed by chance blows at all, one or many!

Here lives, I suppose the difference between us. When you bring me up to this point, I feel the *cold chill*.[7]

I have been doing nothing but attend to my daily work, & had got so fagged that I really thought I was about to have softening of the brain, or some other break down. But a week of respite—caused by the death of an aged relative of my wife's[8]—a dear old soul—taking us away from here perforce, has nearly set me up very nearly. And now after a week since comes my vacation, and we are off into the quiet country for 3 weeks.—

A little legacy of about £2000 to my wife comes in opportunely to relieve us of anxiety for the future. We have no children—(which I regret only that I have no son to send to the war),[9] and this with a little income—rather precarious, of about £200 a year would support us in our very simple way, if I were to throw up my place here. But I cannot do that yet.[10]

Hooker[11] writes that you have been ill. I am very sorry to hear it.

Look at *Impatiens* flowers. See if the most fertile "precociously fertilized" ones ever get crossed![12]

I have asked in three directions for seeds of the Specularia perfoliata.[13] Enclosed are depauperate specimens.

It is pretty to see Honey bees cross-fertilize Locust (*Robinia*) much as you say of Broom.[14] One of my students has been noticing the way bees act on Kalmia[15]

Now for my best thing for to-day.

An Orchid which I missed last year *Platanthera flava*, I knew would be curious, for I remembered a strong protuberance on base of labellum, on the median line. I have not time left to describe it now—having been sadly interrupted— But it is pretty.—equal to anything you have yet seen in British Orchids The process turns proboscis of insect either to right or left—where it will slip in to an imperfect ring (as seen from above) or deep groove (as seen from before)—in which lies the disk,—not flat but coiled up—ready to catch proboscis— It is like the eye of needle to re⟨ce⟩ive the thread.—[16]

Perhaps I will send you, or print—a sketch of the thing.

I am waiting for *Gymnadenia tridentata* to come on.[17]

But the post-hour has come, Good Bye | Ever | A Gray

DAR 165: 127, 137

[1] Letter to Asa Gray, 31 May [1863].

[2] Gray had planned to reprint extracts from Bates 1861 in the May 1863 issue of the *American Journal of Science and Arts*, commonly known as 'Silliman's Journal' after its founder Benjamin Silliman (see letter to H. W. Bates, 4 March [1863], and letter from Asa Gray, 22–30 March 1863). Gray also refers to his review of A. de Candolle 1862a (A. Gray 1863d). See also letter from Asa Gray, 13 April 1863.

[3] Gray prepared a review abstract of Bates 1861 for the September 1863 issue of the *American Journal of Science and Arts* (A. Gray 1863a; see also letter from Asa Gray, 21 July 1863).

[4] Gray's review abstract of 'Two forms in species of *Linum*' was published in the September 1863 issue of the *American Journal of Science and Arts* (A. Gray 1863c).

[5] In his review of Bates 1861 ('Review of Bates on mimetic butterflies', p. 223 (*Collected papers* 2: 92)) CD observed:

> It is hardly an exaggeration to say, that whilst reading and reflecting on the various facts given in this Memoir, we feel to be as near witnesses, as we can ever hope to be, of the creation of a new species on this earth.

[6] Oswald Heer. See letter to Asa Gray, 31 May [1863] and n. 4.

[7] Gray had expounded his interpretation of natural selection as consistent with design in a series of articles published in 1860 in the *Atlantic Monthly* (A. Gray 1860b). For Gray and CD's extended correspondence on the question of design in nature, see *Correspondence* vols. 8–10. See also Dupree 1959.

[8] The relative was a great aunt of Jane Loring Gray's, 'one of the country kin in Connecticut' (Dupree 1959, p. 332; see preceding letter); she has not been further identified.

[9] The reference is to the American Civil War.

[10] Gray did not retire as Fisher Professor of natural history at Harvard University until 1873 (*DAB*).

[11] Joseph Dalton Hooker.

[12] CD and Gray had discussed the occurrence of what Gray called 'precociously fertilized' flowers in *Impatiens* over a number of years (see *Correspondence* vols. 9 and 10). The reference is to small, unopening flowers in which self-pollination occurs (a phenomenon later known as 'cleistogamy'). CD discussed his experiments on the cleistogamic flowers of *Impatiens noli-me-tangere* in *Forms of flowers*, p. 328 and *Cross and self fertilisation*, p. 367.

[13] CD asked for seeds of the dimorphic plant *Specularia perfoliata* (which he called *Campanula perfoliata*) in his letters to Gray of 20 March [1863] and 31 May [1863]; it is a North American species that bears cleistogamous flowers.

[14] See letter to Asa Gray, 31 May [1863]

[15] The reference is apparently to William James Beal, a teacher in Union Springs, New York, who studied botany with Gray at Harvard University during his vacations (*DAB*). Beal published his observations of the role of bees in the pollination of *Kalmia* in Beal 1868, pp. 257–8.

[16] Gray gave an account of the pollination mechanism of *Platanthera flava* in A. Gray 1863b, pp. 292–3, which CD cited in 'Fertilization of orchids', p. 148 (*Collected papers* 2: 145).

[17] In the summer of 1862, Gray observed that *Gymnadenia tridentata* was adapted for self-pollination, reporting his results in A. Gray 1862a, p. 426, and A. Gray 1862c, p. 260 n. However, Gray introduced his more detailed account with the comment: 'we hesitate to bring forward our too scanty observations until another summer affords an opportunity to test them' (A. Gray 1862a, p. 426); he confirmed his observations in A. Gray 1863b, pp. 293–4. See also *Correspondence* vol. 10, letter to Asa Gray, 26[–7] November [1862] and n. 27.

From L. E. Becker 8 July [1863][1]

Altham | Accrington
July 8th.

Dear Sir

I send you some seed of Lychnis diurna gathered in Altham woods. You will probably not be surprised to learn that I have hitherto been unable to obtain a single capsule of seed from an hermaphrodite flower.[2] I have examined, I should think, hundreds of plants, and find, that invariably, the oldest capsules, corresponding in age to those which in the female plants, have already ripened their seed, are shrivelled up to nothing. Many of the younger capsules seem so fresh and healthy, (though none are so vigorous as on the female plants,) that I cannot help fancying they may come to something, but I have never discovered a good one at all approaching maturity. I will continue to watch the plants carefully, and should I obtain seed from the hermaphrodites I will not fail to send you some.

I was prevented by wet weather from going to the woods as soon as I wished to mark the plants, and those I at length ticketed have scarcely ripened their seed, but I entertain no shadow of a doubt, from the appearance of the capsules and the remains of the styles, that all the seed sent is from female plants. It was gathered in a wood abounding in hermaphrodites which seem to be nearly if not altogether barren an⟨d⟩ it may possibly interest you to try whether they will produce hermaphrodites when sown in another locality. I send it now, as I have an impression that in order to have flowers next spring, it is advisable to sow it so that they young plants may get well grown before winter. I observe a considerable variation in the form of the capsules, these are usually nearly globular, but some plants have them shorter others longer than the average. The elongation is occasionally carried so far as to seem like an imitation of the allied genus Cerastium I have put in separate papers a few capsules shewing the extreme forms, and I intend to try whether the peculiarities are hereditary, as well as constant in the individual, and can be perpetuated or increased by selection.

In your last letter you say that the pale colour of some of the flowers leads you to suppose that the white Lychnis may grow in the same wood, and that the hermaphrodites may be natural hybrids.[3] But in Professor Babington: "Manual" it is stated that both Lychnis diurna and vespertina vary in colour from red to white and from white to red.[4] I could have gathered a bunch in our woods, shaded through every tint from rich rose to pure white, which would have admirably illustrated this remark as regards diurna, but the whitest of them all did not make the slightest approach, in any other character than colour, to the genuine White Lychnis, which I have never seen growing in this district. And whenever I have been fortunate enough to meet with this very interesting and rather scarce plant, it did not grow in woods, but in open fields or hedges. Since I received your letter I have looked more particularly for it, but in vain.

It therefore seems as if we must look to another cause for the hermaphrodites, and there seems nothing but to attribute it to the parasite.[5] You say, adding the perfectly useless caution not to trust you on the point that the fungus may exist in the tissues of the plant and be propagated in the seed I could not, if I tried, help believing that this is so though I am at a loss to comprehend how the propagation is accomplished The female plants appear thoroughly free from the parasite while the infected plants seem incapable of producing seed, and their pollen you have found to be entirely destroyed by the fungus. Is it possible that the spores of the fungus may fall on the stigma and be conveyed along with the healthy pollen tube to the ovules? They must get into the tissues of the plant in some way as the fungus is found in the immature, unopened buds and I cannot do⟨ubt⟩ the correctness of your suspicion that it destroys the pollen and causes the development of the ovarium in some, though not in nearly all of the infected plants—

I hope you will be indulgent with me for presuming to write you so long a letter and believe me to be | yours most respectfully | L. E. Becker

DAR 160: 110

[1] The year is established by the relationship between this letter and the letter from L. E. Becker, 28 May [1863].

[2] See letter from L. E. Becker, 28 May [1863].

[3] CD's letter has not been found. For CD's interest in the relationships between the different species of *Lychnis*, and the possibility of several supposed species being mere varieties, see *Correspondence* vol. 5.

[4] Babington 1862, pp. 52–3.

[5] See letter from L. E. Becker, 28 May [1863]. Becker gave a more detailed explanation of these phenomena in Becker 1869. See also letter from L. E. Becker, 14 October 1869, *Calendar* no. 6937.

From George Sigerson 8 July 1863

82 Talbot St | Dublin,
July 8. 1863

Dear Sir,

It occurs to me that there is one part of my paper which I did not make sufficiently clear.[1] When it appeared in print I feared that my views were likely to be misunderstood through the faulty conciseness of my language. A specimen of orbicular leaf has been given, (Nelumbium speciosum) but I think that I should more fully have stated that it could by no means be supposed to realize the necessary condition of the orbicular.[2] The round leaf of the Nelumbium I take to be merely a 'segment', the limb of which is modified in a peculiar way, and the plane of the limb also changed with respect to its petiole. The Orbicular leaf which is ideal, must of course be supported by its axis, and therefore be a plant of itself. The round leaf given is merely to make more easy to be understood the views I put forth. The *disjecta membra* of the orbicular type I had intended to investigate throughout the simpler forms of vegetable life.[3] I was anxious to show, at once, that there is mutation in forms, and proved by actual facts that changes from comparatively simple to very complex forms do take place. Figures from 9 to 21 give examples which are evident proofs to any person, even though he be not a Botanist.[4] It was to these facts that I wished to draw your attention, and receive your opinion upon them. I thought they would naturally interest you, having read with so much pleasure your views on the 'Origin of Species' &c.

With respect to the angle of divergence to which you are so kind as to draw my attention, it does not encourage me to undertake the investigation to say that you have failed. I confess that as yet I have been able to pay little attention to the subject.[5] But, perhaps, you have remarked that I have perceived the angle to change for the leaflet coming off. (page 7, line 25, observations on change of direction in enlarging vein, in fig. 10, marked *g*.) It would be interesting to note the former angle, and the latter, comparing the difference with the progress of the leaflet in separating itself. This might give some clue. But, I do not wish to intrude crude speculations upon your attention.

At present, I am not in a position to pursue the subject, but I hope soon to be so; and, assuredly, the pleasure of acting upon your suggestion would equal to the pleasure of investigating an interesting point.

I have the honour to be, Dear Sir | Very truly Yours | Geo. Sigerson MD.

DAR 177: 162

[1] Sigerson 1863. No previous letters from Sigerson to CD have been found; however, there is a presentation copy of Sigerson 1863 in the Darwin Pamphlet Collection–CUL, the cover of which bears part of a May 1863 postmark.

[2] In his paper, Sigerson sought to account for phyllotaxy in terms of modification from a 'protomorphic leaf-type', which he believed to be the 'orbicular leaf' (Sigerson 1863, p. 1). Sigerson stated that each orbicular leaf was borne at the end of an independently arising stem, and that the phenomenon of phyllotaxy would be explained by the modification of this prototype such that the several radial ribs of the orbicular leaf were arranged separately along the stem (that is, separated by internodes), each forming the mid-rib of a lateral leaf. He thought the leaves of *Nelumbium speciosum* presented a 'good instance' of the orbicular form (see Sigerson 1863, p. 1; see also plate 1, figure 1).

[3] *Disjecta membra*: 'scattered remains' (*OED*). Sigerson stated in his paper that the 'prototype' of the orbicular leaf 'must be sought in a still simpler form', and that he believed it would be found 'in a cell' (Sigerson 1863, p. 1).

[4] Figures 9 to 21 were designed to illustrate the transition from a simple to a compound leaf by gradual stages; most of them depicted leaves of the cultivated raspberry (see Sigerson 1863, p. 6).

[5] CD's letter has not been found; however, CD had obviously made reference to his own work on phyllotaxy, carried out in May and June 1863, in which he sought to account for the series of fractions expressing the angles of divergence between successive leaves on the stems of plants. CD considered that he had 'signally failed' in his attempt to explain this phenomenon (see letter to Asa Gray, 26 June [1863]). See also letter to J. D. Hooker, 29 May [1863] and n. 12. The notes from CD's work on this subject are in DAR 51: 6–32.

To William Forsell Kirby 9 July [1863]

Down. | Bromley. | Kent. S.E.
July 9th

Dear Sir

I am very much obliged for the copy of your Paper on Geographical Distribution, but which I had already read in the Ent. Transacts.[1] I have been very much interested by it & it seems to me excellently done. There was of course much new to me; & I have been particularly struck by your observations on the Corsican sub-species & on those of N. America.—[2] It would be interesting if you could get a set of specimens from the Shetland Isld & see if you could observe the faintest trace of differentiation.—

with my best thanks, I remain | Dear Sir | Yours very faithfully | Ch. Darwin

Postmark: JY 10 63
American Philosophical Society (Getz 4237)

[1] Kirby 1863. CD's annotated copy of the volume of the *Transactions of the Entomological Society of London* in which this paper appeared is in the Darwin Library–Down; the part containing Kirby 1863 was issued in May 1863. In 1862, Kirby had written to CD telling him of his belief in the mutability of species; in reply, CD had expressed a hope that Kirby would some day 'write on variation in Butterflies' and express his beliefs 'on the subject of species' (see *Correspondence* vol. 10, letter to W. F. Kirby, 12 December [1862]). In his paper, Kirby discussed the geographical distribution of species and sub-species of butterflies in Europe on the basis of CD's theory (see especially Kirby 1863, pp. 481–2 and 491).

[2] CD refers to Kirby's comments on the role of geographical isolation in the production of new species and sub-species of butterflies in North America and Corsica, distinct from those in continental Europe (Kirby 1863, p. 491). Kirby observed:

> If geologists can tell us how long Corsica has been separated from the main land, we shall have some most reliable data by which to calculate the length of time that the formation of a species requires, for a large number of Corsican insects are already good species, and many more have become sub-species, which is the last step towards the formation of a new species.

To W. B. Tegetmeier 9 July [1863][1]

Down. | Bromley. | Kent. S.E.
July 9th

My dear Sir

I am very much obliged for your note[2] & am much pleased to hear that you are progressing with the breeding experiments.— I shall be very curious to hear next summer whether the crossed fowls are perfectly fertile.[3] I believe that I proposed to send you five guineas, which I now beg leave to enclose, as a sum which would probably repay you for actual cost of breeding the crossed fowls.— Will you kindly acknowledge its safe receipt?—[4]

I am much obliged for the "Intellectual Observer" which seems an excellent periodical.— I have read, as yet, only your interesting little paper:[5] I had not thought about silky plumage, but had said something like what you say about wingless birds in the Origin.[6]

I sh^d. be *very much* obliged if you would send me paragraphs about the Cats, & please say whether they are to be returned. I am sure that the statement is generally correct.—[7]

I hope you are well & no doubt very busy. I have been having a bad time for many months, but just now am better & am making rather quicker progress than hitherto with my book on "Variation under Domestication".— But I do not suppose I shall be ready for the M.S on fowls, with your corrections, for six months longer.—[8]

With very sincere thanks for all your kindness— | My dear sir | Yours sincerely | Ch. Darwin

The LuEsther T. Mertz Library at The New York Botanical Garden (Charles Finney Cox collection, Tegetmeier ser. 1: 28)

1 The year is established by the relationship between this letter and the letter from W. B. Tegetmeier, 29 June – 7 July 1863.
2 Letter from W. B. Tegetmeier, 29 June – 7 July 1863.
3 See the letter from W. B. Tegetmeier, 29 June – 7 July 1863 and n. 6.
4 See *Correspondence* vol. 10, letter to W. B. Tegetmeier, 27 [December 1862]. No note from Tegetmeier acknowledging receipt of the money has been found.
5 In his letter to CD of 29 June – 7 July 1863, Tegetmeier enclosed a copy of the April 1863 number of the *Intellectual Observer*, a monthly magazine of popular science edited by Henry James Slack, in which Tegetmeier 1863b was published.
6 See letter from W. B. Tegetmeier, 29 June – 7 July 1863 and n. 8. The reference is to *Origin*, pp. 134–5.
7 See letter from W. B. Tegetmeier, 29 June – 7 July 1863 and n. 7.
8 CD refers to his work on the draft manuscript of *Variation* (see 'Journal' (Appendix II)). CD had sent Tegetmeier the draft manuscript of his chapter on domestic fowls for comment in June 1861 (see *Correspondence* vol. 9, letter to W. B. Tegetmeier, 14 June [1861]). See also *Correspondence* vol. 10, letter to W. B. Tegetmeier, 27 [December 1862]. Tegetmeier did not return the manuscript with his comments until 1865 (see letter to W. B. Tegetmeier, [7 April 1865], *Calendar* no. 4806).

To W. H. Flower 11 July [1863][1]

Down. | *Bromley.* | *Kent. S.E.*
July 11th

My dear Sir

I am truly obliged for all the trouble which you have taken for me & for your very interesting note.—[2] I had only vaguely heard it said that Frogs had a rudiment of a sixth toe; had I known that such great men had looked to the point, I should not have dreamed of looking myself.[3] The rudiment sent to you was from a full-grown frog; so that if these bones are the two Cuneiforms they must, I sh^d. think, be considered to be in a rudimentary condition.— This afternoon my gardener[4] brought me some tadpoles, with the hind-legs alone developed, & I looked at the rudiment. At this age it certainly looks extremely like a digit, for the extremity is enlarged like that of the adjoining real toe, & the transverse articulation seems similar— I am sorry that the case is doubtful, for if these Batrachians had six toes, I certainly think it would have thrown light on the truly extraordinary strength of inheritance in Polydactylism in so many animals, & especially on the power of regrowth in amputated supernumery digits.[5]

With very sincere thanks | believe me | My dear Sir | Yours very faithfully | Ch. Darwin

John Innes Institute

[1] The year is established by the relationship between this letter and the letter to T. H. Huxley, 3 July [1863].

[2] The letter from Flower has not been found, but see n. 3, below.

[3] Having learned that Peter Mark Roget had stated that frogs have the rudiment of a sixth toe (Roget 1834, 1: 544), and having heard other such reports, CD had dissected a specimen, asking Thomas Henry Huxley to give his opinion of the dissection (see letters to T. H. Huxley, 16 February [1863] and 27 June [1863]). Huxley declined, but offered to ask Flower to answer CD's questions (see letter from T. H. Huxley, 2 July 1863, and letter to T. H. Huxley, 3 July [1863]). The 'great men' referred to in Flower's letter have not been identified; however, see *Variation* 2: 12–17.

[4] Henry Lettington.

[5] CD referred briefly to the possible occurrence in frogs of a rudimentary sixth toe, in his extensive discussion of polydactylism and inheritance in *Variation* 2: 12–17 (on p. 14).

From P. H. Gosse 13 July 1863

Sandhurst. Torquay
July 13. 1863.

My dear Sir

I had not forgotten your former request about Ophrys apif.[1] & having little time for searching, myself, I had put my young son into the service, who in his walks has collected me a few specimens.[2] The results are as follows:—

Of 16 plants, 32 flowers were open: of these N°. 1, with two fl. had the lower flower with one pollinium attached to stigma, the other poll. quite removed:—the upper flower had *both* poll. removed, but several of the cuneate pollen-masses were adherent to stigma. No trace of slugs or of gnawing visible on the plant.

N°. 2, one fl. open; both pollinia removed, & no trace left. Upper sepal *slightly* gnawed: no trace of slugs.

N°. 3— two fl. open, of which one had one pollinium removed: the other adherent to stigma

Thus out of the 32 flowers, two had lost both pollinia, & two had lost one each.[3]

I know not whether an experiment I tried has any value; it was to see if the wind were an agent in making the poll. come into contact with the stigma. Pollinia just escaped from anther cells, & still hanging unattached, I blew with my breath fitfully & strongly & in various directions; after many efforts I caused the poll. thus to adhere; & that in several instances.[4]

This is all I have been able to do in response to your first request.

On Saturday, after receiving your favour,[5] I immediately went down to Petit Tor, where on former occasions we have found the Bee Ophrys; & spent some time in searching; but perhaps from the flowering being nearly over, & so the plants being inconspicuous in the rank herbage,—I could find but *two* plants. Each of these had one well-swollen ovary, & one recently opened flower— In each case the former showed both poll. adherent to stigma; the latter had the poll. still in the anther cells.

I will, however, pursue the search further; seeking plants in the slopes you mention, & other known habitats.

Will you be so good as to tell me whether you have a good collection of Orchids, & whether you have much in flower just now.

Believe me, dear Sir | Yours very truly | P. H. Gosse

Ch. Darwin Esq[e]

DAR 165: 78

CD ANNOTATIONS
6.1 I know . . . request. 7.1] *crossed pencil*
9.1 I will . . . Gosse 11.1] *crossed pencil*
Top of letter: '*Bee 'Ophrys*' *ink*

1 CD had asked Gosse to observe whether insects ever removed pollinia from the bee ophrys, *Ophrys apifera* (see letter to P. H. Gosse, 2 June [1863] and nn. 12 and 13).

2 Edmund William Gosse was 13 years old.

3 Gosse's observations were not included in the revised account of *Ophrys apifera* that appeared in the second edition of *Orchids*, pp. 52–9; however, CD observed (p. 55):

> From what I had . . . seen of other Orchids, I was so much surprised at the self-fertilisation of this species, that I examined during many years, and asked others to examine, the state of the pollen-masses in many hundreds of flowers, collected in various parts of England.

For CD's own observations of this species in 1863, see the notes, dated 9 July – 25 August, in DAR 70: 53–5 and 57.

4 CD had described observations similar to these in *Orchids*, pp. 64–5.

5 CD's letter has not been found.

To J. D. Hooker 14 July [1863]

Down
July 14th

My dear Hooker

I am getting very much amused by my tendrils— it is just the sort of niggling work which suits me & takes up no time & rather rests me whilst writing.[1] So will you just think whether you know any plant, which you could give or lend me or I could buy with tendrils remarkable in any way, for development, for odd or peculiar structure or even for odd place in natural arrangement.—[2] I have seen or can see Cucurbitaceæ—Passion-flowers—Virginian creeper—Cissus discolor—common Pea & Everlasting pea.—[3] It is really curious the diversification of irritability (I do not mean the spontaneous movement, about which I wrote before & correctly as further observation shows);[4] for instance I find a slight pinch between thumb & finger at end of tendril of Cucurbitaceæ causes prompt movement, but a pinch excites no movement in Cissus.— The cause is that one side alone, (the concave) is irritable in former; wheres both sides are irritable in Cissus, so if you excite at same time both *opposite* sides there is no movement; but by touching with a pencil the two branches of tendril in any part whatever you cause movement towards that point; so that I can mould by mere touch the two branches into any shape I like

&c &c The peduncle of tendril is either not sensitive or sensitive only to *prolonged though slight* pressure &c &c.—[5]

If you can screw out a little time do come here for a Sunday, I shd. so like it, & I have been better of late & shd. stand some talking well.—

What a splendid number the last of N. Hist. Review.—[6] Capital, as they seemed to me, Botanical & Zoological papers.— The embryology of Echinodermata seemed capitally done.[7] I suppose I owe to Oliver the capital & clear article on Linum.[8]

GoodBye, it is awfully hot.— Ever yours affect— | C. Darwin

Endorsement: '/63.'
DAR 115: 200

1 CD refers to his experiments on climbing plants, begun in June 1862. See letters to J. D. Hooker, 25 [June 1863] and 1 July [1863].

2 On the back of CD's letter Hooker listed the following genera and family: 'Cobæa | Bignonia | Mutisia | Gloriosa | Flagellaria | Lygodium. | Smilax | Leguminosæ'. See also letter from J. D. Hooker, [21 July 1863].

3 CD's notes on his observations and experiments on climbing plants are in DAR 157.1 and DAR 157.2. These include notes on several members of the Cucurbitaceae, namely: *Echinocystis lobata*, dated 16 June – 24 May 1864 (DAR 157.2: 29–51 and 53); *Hanburya mexicana*, dated 12 October [1864] (DAR 157.2: 52); *Anguria warscewiczii*, dated 29 April 1864, and *Zanonia indica*, dated 11 April and 20 May [1864] (DAR 157.2: 54). There are also notes on several species of *Passiflora* (passion flower), dated 27 [July 1863] – 4 August [1864] (DAR 157.2: 69–77); *Ampelopsis hederacea* (Virginia creeper), dated 1 July – 16 August [1863] (DAR 157.2: 65–7); *Cissus discolor*, dated 30 June – 18 July [1863] (DAR 157.2: 55–6); *Pisum sativum* (common pea), dated 30 July – 23 August [1863] (DAR 157.2: 15–20); and *Lathyrus grandiflorus* (everlasting pea), dated 10–15 November [1863] (DAR 157.2: 22). These species are discussed in 'Climbing plants', pp. 65–7, 73–9, 83–7, and 89–91.

4 See letter to J. D. Hooker, 25 [June 1863].

5 CD's notes from these experiments, dated 30 June – 18 July [1863] are in DAR 157.2: 55–6. In his discussion of *Cissus discolor* in 'Climbing plants', CD stated (pp. 83–4):

> At the beginning of my work, and before examining this plant, I had observed only those tendrils which are sensitive on one side, and these when lightly pressed between the finger and thumb

become curved; but on thus pinching many times the tendrils of this *Cissus* no curvature ensued, and I was at first falsely led to infer that they were not at all sensitive to a touch.

6 CD refers to the July 1863 issue of the *Natural History Review*, of which Hooker was one of the editors; CD's unbound copy of this issue of the journal is in the Darwin Library–CUL.

7 Thomson 1863–4.

8 [Oliver] 1863d. Daniel Oliver, who was one of the botanical editors of the *Natural History Review*, had indicated that he wished to review 'Two forms in species of *Linum*' for the journal (see letter from Daniel Oliver, 27 February 1863; see also letter from J. D. Hooker, [24 March 1863] and n. 11).

From Friedrich Hildebrand 16 July 1863

Bonn
July 16th | 1863.

Honoured Sir,

induced by your work on the fertilisation of Orchids[1] I have made some interesting observations, not so much respecting the fertilisation by insects as the effects of the pollen applied to the stigma. Therefore I venture to send you an extract of a little treatise that is going to be printed in ⟨the⟩ Botanische Zeitung, and I shoul⟨d be⟩ very much obliged to you if ⟨you⟩ would read it in one of ⟨your⟩ botanical societies, or g⟨et it⟩ printed anywhere.—[2]

This last summer Professor Treviranus and myself have made observations on almost all the Orchids of our flora and can verify almost all you have said about them.[3] We observed also the selffertilisation of Ophrys apifera, I suppose Professor Treviranus will tell you more about it, as he would not believe at first in this selffertilisation.[4]

I am very much inclined to believe with you that there are certain Orchids as Acropera and Catasetum with their sexes separated, but after having read my observations you will agree with me that the absence or the incomplete state of the ovules ⟨can⟩not be called a proof for the ⟨ma⟩le sex of the plant.[5] I am sure ⟨you⟩ will find the ovules very ⟨incomp⟩letely developped in almost ⟨any tropi⟩cal Orchid you will look at.[6]

Your very interesting experiments on Linum perenne I have made with the same result.[7] I had a plant of the short-styled form in my room, a part of the flowers were fertilised with each other, another part each flower with itself, and of both kinds of fertilisation I got no capsule; a third part of flowers on the same plant I fertilised with pollen of the long styled form and got a good capsule of almost every flower.— With my experiments on Primula I have not come to an end as yet.[8]

I hope you will excuse my letter the more as I remember the kind answer I got on my former letter to you about the German translation of your Fertilisation of Orchids,[9] and remain dear Sir | yours | respectfully | F Hildebrand

DAR 166: 200

CD ANNOTATION
Top of letter: 'F. Hildebrand' *ink*

1 *Orchids*.

[2] The enclosure has not been found. The German version of Hildebrand's paper was published in the *Botanische Zeitung* on 30 October and 6 November 1863 (Hildebrand 1863a). In it Hildebrand discussed his observations of two independent effects of pollen in orchids, namely, the enlargement of the ovary and the fertilisation of the ovules. CD sought Daniel Oliver's advice about the suitability of the English abstract of Hildebrand's paper for the Linnean Society or the *Natural History Review* (see letter to Daniel Oliver, 18 July [1863], and letter from Daniel Oliver, 20 July 1863). It was ultimately published, through CD's intervention, in the September number of the *Annals and Magazine of Natural History* (Hildebrand 1863b; see letter to Friedrich Hildebrand, 28 July [1863]).

[3] On Ludolph Christian Treviranus's reaction to *Orchids*, see also the letter from L. C. Treviranus, 12 February 1863, Treviranus 1863a, and Junker 1989, pp. 143–4.

[4] CD described the bee ophrys (*Ophrys apifera*) in *Orchids*, pp. 63–72, stating that he found the plant 'perplexing in an unparalleled degree', since the flowers appeared to have 'elaborate contrivances for directly opposed objects', namely for both cross- and self-pollination. See also letter to P. H. Gosse, 2 June [1863] and n. 12. No letter from Treviranus on *O. apifera* has been found, but he questioned CD's findings in Treviranus 1863a, pp. 11–12, and later confirmed them in Treviranus 1863c. There are annotated copies of these papers in the Darwin Pamphlet Collection–CUL, and a manuscript translation of the latter in DAR 70.38. CD cited Treviranus's confirmation of his results in 'Fertilization of orchids', p. 145 (*Collected papers* 2: 142).

[5] In *Orchids*, CD identified four species of orchids that were dioecious, the male plants being known by the specific names *Acropera luteola*, *Catasetum saccatum*, *C. callosum*, and *C. tridentatum* (*Orchids*, pp. 203–48). One of CD's reasons for believing these plants to be male was that the ovules were in an atrophied condition (see *Orchids*, pp. 207–8 and 238). In his paper, Hildebrand observed that the ovules of recently expanded orchid flowers exhibited very varying degrees of development (Hildebrand 1863b, p. 171). John Scott had already convinced CD that *A. luteola* was hermaphrodite (see letter to P. H. Gosse, 2 June [1863]). In 'Fertilization of orchids', p. 153 (*Collected papers* 2: 150), CD accounted for his error in terms of his ignorance of the 'remarkable fact', established by Hildebrand, that 'in many orchids the ovules are not developed until several weeks or even months after the pollen-tubes have penetrated the stigma'.

[6] In his paper, Hildebrand observed that in some tropical orchids he had examined, he found no ovules in the ovary of the expanded flower, but nevertheless saw 'the enlargement of the ovarium after having applied the pollen to the stigma' (Hildebrand 1863b, p. 169).

[7] 'Two forms in species of *Linum*'.

[8] Hildebrand published his observations on dimorphism in *Linum perenne* and *Primula sinensis* in January 1864 (Hildebrand 1864), offering them as a confirmation of CD's views. CD had written on *P. sinensis* in his paper 'Dimorphic condition in *Primula*'.

[9] CD's letter to Hildebrand has not been found. Not realising that Heinrich Georg Bronn had already concluded his German translation of *Orchids* (Bronn trans. 1862) before his death on 5 July 1862, Hildebrand had offered to complete it (see *Correspondence* vol. 10, letter from Friedrich Hildebrand, 14 July 1862).

From Roland Trimen 16, 17, and 19 July 1863

Colonial Office, | Cape Town. | Cape of Good Hope.
16th. July, 1863.

My dear Sir,

I have to thank you for both your kind letters, respectively of Feb. 16th. and May 23rd., which arrived in due course (the latter to-day).[1] I was glad to see the copy of Professor Harvey's letter to you regarding my drawings, &c, of Orchids.[2] He naturally takes a botanist's view of the matter and looks to the discovery of new species rather than to the special line of research you have so ably opened up. Were all my time my own, I should gladly collect and dry botanical specimens; but as it is, I fear my collections in that way will not be extensive.

What you say about the position of the fertilising organs with reference to the labellum in *Satyrium* is very striking: the latter *do* seem on the wrong side.[3] When I first showed the common orange *Satyrium* to Dr. Pappe,[4] I spoke of the nectariferous portion of the flower as the "upper sepal" and was quite surprised to hear him pronounce it to be the labellum.

I have tried to induce several persons to collect Orchid roots for transmission to Kew.[5] One seems willing to try, but is rather puzzled how to discover the plants in the "veldt" when seeding-time is over. In fact, one ought to be resident on the spot to mark certain plants which might afterwards be dug up. I spoke to the Superintendent of the Botanic Gardens here on the subject, but was unfortunate in so applying, as that gentleman (naturally, it is said, of rather a crusty humour) seemed rather sore on the subject of Kew, asserting that he had on former occasions sent plants to the Royal Gardens, and not met with what he considered proper treatment.[6] But if I fail with the roots, you shall certainly have specimens of flowers in spirits.

Many thanks for your portrait: I hope you did not think me presuming too much, on so short an epistolary acquaintance, in asking for it.[7] But I have certainly the best of our exchange; mine was a very miserable affair, without *character* (as a naturalist would say). I shall endeavour to get some taken in the same manner as yours.

I rejoice that I have succeeded in making you break the "sullen vow" that you would publish no more on the Orchid subject.[8] You are pleased to be very complimentary on my performances, but I hereby declare that whatever merit they may possess is wholly due to your elaborate book, which opened up a terra incognita to me.[9] I think it very kind of you to publish the paper as coming from me, but am hardly satisifed as to the justice of my name being so prominently put forward.[10] I feel honoured by any remarks in amplification and explanation which you may have added, and shall be very glad of any copies of the paper which you may have to spare.

Like yourself, I found the subject most fascinating: there is a degree of mystery and wonder in this complete dependence of a whole group of plants upon a tribe of organic beings so remote from them as insects; and one can scarcely resist the idea that the strangely-formed and uniquely-developed Orchids are conscious and active agents in this constant mutually-advantageous intercourse. When I now look at a growing Orchid, I cannot help regarding it as a more *intellectual* being than other plants.

I intend to do my best to make further investigations as soon as I have got the second part of my Catalogue of the S. African Butterflies off my hands.[11] I have been a good deal encouraged of late in observing the growing interest in entomology in this part of the world, and am happy to think that I have not been unsuccessful in aiding the diffusion of a taste for collecting. My best correspondent is a gentleman in the Frontier Mounted Police, who is stationed in a thickly-wooded region of Kaffraria Proper.[12] He has sent down quite an extensive collection of Lepidoptera, and, being a good observer, numerous interesting notes on habits & localities.

In connection with the subject of Moths being able to perforate membranes with their proboscis,[13] I think it as well to mention the following. Since the last winter, which was excessively wet, certain Moths have been extraordinarily abundant, their larvæ causing serious ravages among the crops. The pest in this neighbourhood and over most of the Western Districts was *Heliothis armigera*, in the spring; the larvæ appearing in legions in the early summer (October to January). But in the middle of January, at Grahamstown, in the Eastern Province, a quite different moth was reported in the papers as doing much damage to *fruit*. I have since learnt from M^rs. Barber[14] (who is, I believe, a botanist, and a good correspondent of Harvey's) that this is a *fact*; and from specimens this lady forwarded, have ascertained the Moth to be *Achæa Chamæleon* Guén, a good-sized Quadrifid Noctua, excessively variable in colour & marking. I wrote to inquire as to particulars, asking whether the Moths ever sucked sound peaches, or if they did not rather confine themselves to those already injured by birds, remarking that I thought it impossible for the moths' trunk to pierce the tough skin. To this I received the following answer from M^rs. Barber:—"The moths certainly do puncture the skin of the peaches, commencing near the apex, where the peach first ripens, and puncturing it in patches, thus:—

and the small holes made by their proboscis can be distinctly seen. Much of the fruit destroyed by them was *very green*, for they came in such multitudes that "the famine was sore in the land" with them, and they were obliged to take what they could find." M^rs. B. had previously informed me that "in the evening you might see more than a dozen of them settled on one peach, draining the juices from it *till there was nothing left but the stone and the epidermis* of the fruit.". Does not this seem rather a strong case in favour of your belief (Fertil. of Orch. p. 50) that Moths perforate the lax inner membranes of the nectaries of Orchis to suck the fluid enclosed between it and the outer membrane? If Moths really possess such perforating power, in greater or less degree according to their size, one can still well believe that the *nocturnal* work of Orchid fertilisation, at any rate, is chiefly theirs,—as the chief difficulty with regard to their actually being the fertilising agents of the species with empty nectaries was their presumed want of this power.[15]

M^r. Bates very kindly sent me his interesting paper on the Amazonian Heliconidæ, which I like exceedingly.[16] I wrote to him by last mail in relation to certain points he advanced. It is interesting to observe how, one after another, naturalists eminent for thought or observatory power are forced into admitting the mutability of species.

With kind regards, believe me, dear Sir, | Yours very truly, | Roland Trimen.

P.S. (17th. July) With regard to *Oxalis*, I went to-day to the seed-house at the Gardens to ask if any seeds of the genus were to be had, but was unsuccessful, none being kept in stock.[17] I will endeavour to collect a few myself. A very common, large and handsome, bright-yellow species (*O. cernua*?), grows everywhere in and about the town. I gathered at random 10 flowers of this kind, and found that *3* of that number had the elongated pistils. It was quite pleasant to find my first handful so good a case in point. I have made a very rough, hurried sketch of the appearance of the column in both cases, which I enclose;[18] together with two flowers of the species, respectively presenting the variations you note. This species of *Oxalis* is very much frequented by some small *Staphylinus*-like insects, which are always moving about, and must be excellent conveyers of pollen in a flower whose petals so closely surround the reproductive organs. On one of the flowers gathered today, there was a strange-looking little blackish-grey *Aphis*, with a hump-back.

If you can find time, tell me what interesting discovery you have made with regard to this exchange of position between the stigmas and 5 higher anthers.[19]

I had a fine *Cypripedium* flower to examine the other day, and fully verified your lucid explanation of the method of fertilisation induced by its unique structure.[20] My specimen differed a good deal in detail from the species figured in your book, as you will see from the enclosed rough outline (lateral view) of column What is the use of the conspicuous ridged knob on the face of the "rudimentary shield-like anther?" I had no time to make a careful drawing.

*19*th.— Are you aware how very complete an insect-trap is an Asclepiadaceous plant, named "*Physianthus albens*" (in the Gardens here)? Last January, a gentleman told me that there was a plant "which caught butterflies" in the gardens, and I accordingly went to investigate. The plant is a creeper, purporting to come from Buenos Ayres. The cluster presented a singular spectacle, as from 30 to 40 specimens of Moths & Butterflies, besides other insects, were anchored fast, by the proboscis, to the dingy-white flowers, many of them fluttering vainly to escape. I at first imagined that it was some extremely viscid substance that kept them prisoners, but, on examination, attributed it rather to a mechanical force. The flower has a 5-partite corolla, which is tubular, the nectaries forming 5 swellings at the base. I omitted to make drawings at the time, but this fancy diagram will give you an idea of the process. Around the column (ovary?) are 5 pairs of plates, pp so arranged that the very narrow interstice between each two plates exactly commands, and is the only way of access to, each of the 5 nectary compartments,—the throat of the flower being so constricted that the column fills it. It appears to me that as soon as a proboscis is pushed into the slit between a pair of plates, the latter close on it after the fashion of a vice. In only

two instances did I succeed in pulling flower and insect apart without fracturing the proboscis, so firm is the hold of the plates. If you think the case is new or (or needs elucidation) I will endeavour, next summer, to draw up a paper on the subject, with proper illustrations, which I will send you.[21] The flowers are very strongly honey-scented. *Noctuæ* are the principal victims, but even powerful Sphingina (Chærocampa) are quite unable to escape when once trapped. I fear you will hardly thank me for so lengthy an epistle—but it is so pleasant, in a place where a naturalist is almost universally ridiculed or thought a trifler, to communicate one's ideas and observations to a mind so full of sympathy as your own. I sincerely trust that your health, which I am sorry to hear is delicate, is now more confirmed, and that the laborious works on which you are engaged are progressing satisfactorily.

I see some laughable "skits" upon your theory in the "Illustrated Times", consisting of a series of absurd (but clever) sketches illustrating the development of a man servant making off with a dish of eatables from a dog stealing some bones, and other equally sage instances.[22] I was glad to see some remarks of yours in the April "Athenæum" showing how different your theory essentially is from that of Lamarck.[23]

R. T.

Layard[24] tells me that he has seen a moth sucking at a sound peach, and believes that it pierced the skin. I have seen one statement (if not *two*) in the "Ent. Weekly Intelligencer", of Moths having refused to answer the invitation to "sugar" in consequence of their attachment to *blackberries*.[25]

[Enclosure]

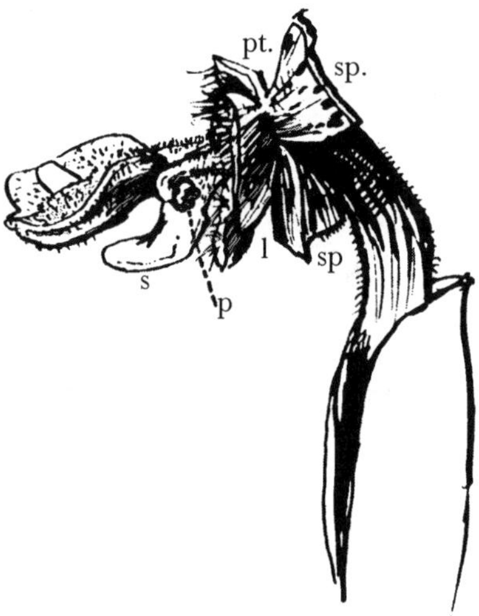

CYPREPEDIUM ?venustum

DAR 99: 13–16d

CD ANNOTATIONS
1.1 I have . . . extensive. 1.7] *crossed ink*
3.1 I have tried . . . localities. 7.9] *crossed ink*
8.2 Since the . . . January). 8.6] *crossed ink*

9.1 Mr. Bates . . . species. 9.5] *crossed ink*
11.1 P.S. . . . myself. 11.3] *crossed blue crayon*; 'Catching Physianthus Lepidopt' *blue crayon*
12.1 If you can . . . drawing. 13.6] *crossed ink*
14.25 *Noctuæ* . . . Lamarck. 15.6] *crossed ink*
Top of letter: 'Peach an Exotic' *ink, circled*

1 Letters to Roland Trimen, 16 February [1863] and 23 May [1863].
2 CD had enclosed a copy of the letter from W. H. Harvey, 3 February 1863, in his letter to Trimen of 16 February [1863].
3 See letter to Roland Trimen, 16 February [1863] and n. 3.
4 Karl Wilhelm Ludwig Pappe.
5 See letter to Roland Trimen, 16 February [1863].
6 The superintendent of the botanic gardens at Cape Town, James McGibbon, was a former gardener at the Royal Botanic Gardens, Kew (R. Desmond 1994).
7 See letter to Roland Trimen, 23 May [1863] and nn. 2 and 3.
8 See letter to Roland Trimen, 23 May [1863] and n. 5.
9 *Orchids*.
10 Trimen 1863. See letter to Roland Trimen, 23 May [1863] and n. 5.
11 Trimen 1862–6; the second part was published in 1866 (Trimen 1862–6, p. 353).
12 James Henry Bowker. Over the next four years, Bowker collected and transmitted to the South African museum, Cape Town, several thousand entomological specimens (Trimen 1862–6, p. 351). He subsequently collaborated with Trimen to publish a monograph on the extra-tropical South African butterflies (Trimen 1887–9).
13 This hypothesis was discussed in *Orchids*, pp. 48–53.
14 Mary Elizabeth Barber.
15 CD wrote a draft letter to the *Gardeners' Chronicle* detailing Trimen's observations, but the letter was not published (see letter to [*Gardeners' Chronicle*], [after 27 August 1863]. CD cited Trimen's observations in 'Fertilization of orchids', p. 143 (*Collected papers* 2: 140).
16 Bates 1861.
17 See letter to Roland Trimen, 23 May [1863] and n. 9.
18 The enclosure has not been found.
19 See letter to Roland Trimen, 23 May [1863] and n. 9.
20 See *Orchids*, pp. 270–6.
21 No paper by Trimen on this subject has been found.
22 Trimen refers to a series of twenty caricatures by Charles Henry Bennett, which were published in the London *Illustrated Times* between 2 May and 10 October 1863, under the title 'The origin of species, dedicated by natural selection to Dr. Charles Darwin' (*Illustrated Times* 2 (1863): 317, 333, 348, 364, 381, 396, 413, 429; and 3 (1863): 13, 29, 61, 77, 93, 125, 141, 157, 173, 205, 221, 237). The caricature he describes is one entitled 'Good dog', which appeared in the *Illustrated Times* for 16 May 1863.
23 Trimen refers to the letter to the *Athenæum*, 18 April [1863], and to Jean Baptiste de Lamarck.
24 Edgar Leopold Layard.
25 T[homas] L[innell], 'Failure of Sugar', *Entomologist's Weekly Intelligencer* 3 (1857–8): 51.

To Julius von Haast 18 July [1863][1]

Down. | Bromley. | Kent. S.E.
July 18th.

My dear Sir

I received about a fortnight ago a copy of your letter of Dec. 9th. & your note of March 5th,[2] & now I have received a brief & very kind note of May 12th.—[3]

with the latter came the Report of your expedition to the west coast.[4] I heartily congratulate you on your success. What fearful labour you have undergone & how well you must have managed the whole affair to have been so successful under such extreme difficulties.

On receiving your letter of December 9th (i.e. the copy of it) in which you say you would like a copy of the "Origin of species"; I wrote to my publisher to send one to you & I hope you will have received it before this reaches you.[5] I sh[d] be the most insensible & dull man if I were not highly pleased by what an observer like you says of the "Origin".— I have already expressed my sincere opinion on your admirable Address to your Society.[6] I wonder whether you were the Author of a very *amusing* & really excellently done Dialogue on Natural Selection, in a New Zealand paper, which was sent to me?—[7] I thank you for your information about the New Zealand Vertebrata:[8] I really think there is hardly a point in the world so interesting with respect to geographical distribution as New Zealand. I have been very glad to hear of the curious case of change in habits of nesting in the New Zealand Duck; & it is a case which I shall at some future time quote.

I am at present working very hard on Variation under Domestication & hope in six months to begin printing.[9] My weak health makes me live the life of a hermit & I really have no scientific news to communicate.

So you must, if you can, excuse this very dull letter, & believe me, my dear Sir, with sincere respect & my thanks | Yours very faithfully | Ch. Darwin

Alexander Turnbull Library, National Library of New Zealand (Haast Family papers, MS–Papers –0037–051)

[1] The year is established by the relationship between this letter and the letter from Julius von Haast, 5 March 1863.

[2] Haast had enclosed a copy of his letter to CD of 9 December 1862 (*Correspondence* vol. 10), the original of which he feared lost, with his letter to CD of 5 March 1863. The original of the letter from Julius von Haast, 9 December 1862 (*Correspondence* vol. 10) had indeed failed to reach CD (see letters from J. D. Hooker, [30 April 1863] and n. 2, and 19 June 1863 and n. 1).

[3] Letter from Julius von Haast, 13 May 1863.

[4] CD refers to Haast's geological expedition across the Southern Alps of New Zealand's Middle Island (now South Island) to the west coast (see letter from Julius von Haast, 5 March 1863, n. 2). The enclosure to the letter from Julius von Haast, 13 May 1863, has not been found; however, it was probably the account of Haast's expedition published in the Christchurch *Press*, 1 April 1863, pp. 1–2, and 2 April 1863, pp. 2–3.

[5] See the enclosure to the letter from Julius von Haast, 5 March 1863. The letter to John Murray, CD's publisher, has not been found.

[6] CD commented on Haast's address to the Philosophical Institute of Canterbury (J. F. J. von Haast 1862a) in his letter to Haast of 22 January 1863. See also enclosure to letter from Julius von Haast, 5 March 1863 and n. 12. There is an annotated copy of the address in the Darwin Pamphlet Collection–CUL.

[7] CD refers to Samuel Butler's anonymous article entitled 'Darwin on the origin of species: a dialogue', which was published in the Christchurch *Press*, on 20 December 1862 (see letter to an editor, 24 March [1863] and n. 3).

[8] See the enclosure to the letter from Julius von Haast, 5 March 1863.
[9] *Variation* was not published until 1868.

To Daniel Oliver 18 July [1863][1]

Down. | Bromley. | Kent. S.E.
July 18th

Dear Oliver

Would you have the kindness to look over the enclosed: it seems to me a valuable paper.[2] Dr. Hildebrand has sent it to me with the request to get it read before some Society, or get it published somehow.[3]

Now you will see that it is an abstract of a longer paper to be published in the Bot. Zeitung.[4] Therefore I fear it would not do for Linn. Soc. How is this; please consult Hooker?[5] If it will not do for Linn. Soc. would it do for the Nat. Hist. Review? & this is the reason I send it to you.—[6] If not fitted for N.H.R. I will send it to Annals of N. Hist. So at your leisure please let me hear as I must write to Dr. H..—[7] Dr. Hildebrand pleases me by telling me that he has been repeating my experiments on Linum & has observed fertilisation of all German orchids & finds my statements correct.— But he adds that he thinks I am probably wrong about sexes of Acropera & Catasetum & I very much fear, indeed feel almost sure, that I have blundered fearfully about Acropera, & probably about Catasetum, which is horrid to think of.—[8] I wish my Acropera would flower; I shall soon have two Catasetums in flower & must look to them.[9]

Dear Oliver | Yours very sincerely | C. Darwin

Tell Hooker I am madder than ever on Tendrils, & long to hear whether he has anything for me for observation[10]

DAR 261 (DH/MS 10: 51)

[1] The year is established by the relationship between this letter and the letter from Daniel Oliver, 20 July 1863.
[2] The enclosure has not been found, but see letter from Friedrich Hildebrand, 16 July 1863 and n. 2.
[3] See letter from Friedrich Hildebrand, 16 July 1863 and n. 2.
[4] Hildebrand 1863a.
[5] Joseph Dalton Hooker, a colleague of Oliver's at the Royal Botanic Gardens, Kew, was one of the vice-presidents of the Linnean Society.
[6] Oliver was one of the editors of the *Natural History Review.*
[7] The English abstract of Hildebrand's paper was ultimately published, through CD's intervention, in the September number of the *Annals and Magazine of Natural History* (Hildebrand 1863b; see letter to Friedrich Hildebrand, 28 July [1863]).
[8] See letter from Friedrich Hildebrand, 16 July 1863 and nn. 5 and 6.
[9] CD's notes of his observations and experiments on *Acropera luteola*, dated 29 August 1863, are in DAR 70: 114; he recorded successfully pollinating two flowers. See also 'Fertilization of orchids', p. 153 (*Collected papers* 2: 150).
[10] See letter to J. D. Hooker, 14 July [1863], and letter from J. D. Hooker, [21 July 1863].

To J. J. Aubertin 19 July 1863

Down. | Bromley. | Kent S.E.
July 19th./63.

My dear Mr. Aubertin.

I received your letter dated April 27th. many weeks ago, but only yesterday the stones—[1] I do not suppose anyone without examining the country could form a guess how thick the rocks, of which you sent specimens would run.— They form a very coarse micaceous schist, & I can well believe must be most difficult to tunnel through One may conjecture that the case is more hopeful than if the rock had been granite or porphyry— The rocks belong to the metamorphic series which I believe is developed to a larger extent in Brazil than in any other quarter of the world. I saw no other rocks besides these & Plutonic in Brazil & I examined a series collected along the whole coast from Rio to near the Plata—[2] I thank you for your very kind & long letter which interested me much; especially about the progress of the mulattoes. &c. I am very glad to hear that you are well & endure your life— I heard of your appointment shortly after you sailed—[3] I cannot give a very good account of myself, as I am constantly ailing— My eldest daughter thank God is well,[4] but one of my Boys is an Invalid[5] & we had fearful illness in our family last year.[6] Mrs. Darwin is well and begs to be kindly remembered to you— Owing to the state of our Family we have not seen anything of Miss Butler for a long time, but she has visited us here once or twice & was as pleasant as usual.[7] Of Miss Scott I know nothing.—[8]

I can tell you of only one other Ilkleyite, namely Mr Robinson; & he is married and has a living.—[9]

Will you grant me a little favour, I have a Boy, who remembers you, & is a passionate collector of postage stamps.—[10] He tells me that there are many kinds used in Brazil, now will you devote an envelope & tear off any old stamps which you may receive, & when you have got a few send them through your London Correspondent. You would give great pleasure, if you would take so much trouble

with every good wish for your success & health. | Pray believe me my dear Mr. Aubertin— | Very faithfully yours. | Ch Darwin.

Copy
DAR 143: 24

1 See letter from J. J. Aubertin, 27 April 1863. Aubertin was superintendent of a railway in Brazil and had sought CD's advice on tunneling through rocks of which he sent samples.

2 CD visited the Brazilian coast during the *Beagle* voyage in 1832 and in 1836 (see *'Beagle' diary*); his geological diary and notes from the voyage are in DAR 32–8, and his geological specimen books are in DAR 236.

3 CD refers to Aubertin's appointment in 1860 as superintendent of the São Paulo railway (Graham 1968, p. 67).

4 CD and Aubertin apparently met in the winter of 1859 while undergoing treatment at Edmund Smith's hydropathic establishment at Ilkley Wells, Yorkshire (see letter from J. J. Aubertin, 27 April 1863 and n. 1); at the time, CD had been concerned about the state of health of Henrietta Emma Darwin (see *Correspondence* vol. 7, letter to W. D. Fox, 23 September [1859]).

[5] CD refers to Horace Darwin, who had been chronically ill since the previous year (see letter from G. V. Reed, 12 January 1863 and n. 2).
[6] Leonard and Emma Darwin were both ill with scarlet fever in the summer of 1862 (see *Correspondence* vol. 10).
[7] Mary Butler had undergone treatment at Edmund Smith's hydropathic establishment at Ilkley Wells, Yorkshire, at the same time as CD and Aubertin (see letter from J. J. Aubertin, 27 April 1863 and n. 20). According to Emma Darwin's diary (DAR 242), Butler stayed at Down House from 25 April to 1 May 1860.
[8] Miss Scott has not been identified, but see the letter from J. J. Aubertin, 27 April 1863.
[9] Mr Robinson has not been identified.
[10] Leonard Darwin.

From Daniel Oliver 20 July 1863

Richmond, S.W.
20.VII./63

My dear Sir

I have been reading Dr. Hildebrand's paper & I think it scarcely suited to N.H.R.—[1] It may seem odd to talk of *anything* as scarcely suited when readers find we admit all manner of things suitable & unsuitable,—but we botanists have been for some time grumbling abt. the admission of papers wh. might suitably be read before Societies, &c. & we must consequently aim at least at some minimum of inconsistency. I believe Dr. Hooker to whom I have shewn the paper quite agrees.[2]

The subject is a very curious one.— I think Schacht in his Lehrbuch (ii. 373)[3] points out that some months elapse in the *Hazel* (*Corylus*) between the application of the pollen & formation of the ovules, & this may be likely to obtain in other *Corylaceae* some of which are slow in their reproduction work. I must refer to his book when I return to the Herbarium in the morning.[4]

I do not see that you need be so fearful about yr. *Catasetum*.[5]

I shd. think Naudin would be a likely man to suggest a few good Cucurbitaceae for yr. tendril-experiments.[6]

very sincerely yours | Danl. Oliver

Schacht says, l.c.[7]

"In vielen Fällen wird der obere Theil des Pollenschlauches, wenn das untere Ende sein Ziel erreicht hat, nicht mehr ernährt, er vertrocknet alsdann mit der Narbe; man findet in solchem Falle wohl in der Fruchtknotenhöhle oder im Staubwegcanal die Schläuche, aber man vermisst den Zusammenhang derselben mit dem Pollenkorn, dem sie vormals entsprungen sind, wodurch sich Rob. Brown[8] täuschen liess, indem er die Pollen-schläuche der Orchideen für Zellen des leitenden Gewebes erklärte. Dies gilt namentlich für diejenigen Pflanzen, bei welchen die Bestäubung & die Befruchtung der Zeit nach weit auseinander fallen, z.B. für die Haselnuss, Hainbuche & Erle, die im ersten Fruhjahr (die Haselnuss im Februar) bestäubt werden, zu einer Zeit, wo die beiden Samenknospen noch nicht vorhanden sind und erst im Sommer (die Haselnuss gegen Ende des Juni) zur Befruchtung kommen—"

If Hildebrand be going, as he says, to write about this in Bot. Zeit.[9] I really doubt the propriety of having this paper printed in England. Nothing is made clear as to specific action of pollen-tubes &c in causing swelling of ovary. Dr. Hooker tells me Mr. Smith[10] used to say—he caused some such swelling by rubbing them externally with **sand**!— D O.

DAR 173: 22

CD ANNOTATION
7.8 z.B. . . . kommen—" 7.12] *double scored ink*

[1] CD had sent Oliver a manuscript paper on orchid pollination, by Friedrich Hildebrand, for possible publication in the *Natural History Review*, of which Oliver was an editor (see letter to Daniel Oliver, 18 July [1863]).
[2] Joseph Dalton Hooker, Oliver's colleague at the Royal Botanic Gardens, Kew, was also an editor of the *Natural History Review*.
[3] Schacht 1856–9, 2: 373. See n. 7, below.
[4] Oliver was an assistant in the herbarium and librarian at the Royal Botanic Gardens, Kew (R. Desmond 1994, *List of the Linnean Society of London* 1863).
[5] See letter from Friedrich Hildebrand, 16 July 1863 and nn. 5 and 6, and letter to Daniel Oliver, 18 July [1863].
[6] See letter to Daniel Oliver, 18 July [1863] and n. 10. Charles Victor Naudin was a leading authority on the Cucurbitaceae (the gourd family); CD had already consulted him regarding the crossing of varieties of melon (see letter to C. V. Naudin, 7 February 1863). See also letter from J. D. Hooker, 24 January 1863 and nn. 1 and 3.
[7] l. c.: '*loco citato* in the place cited' (*Chambers*). Oliver refers to Schacht 1856–9, 2: 373. The passage translates:

> In many cases the upper part of the pollen tube ceases to be nourished when the lower end has reached its goal, whereupon it dries up with the stigma. In such cases one does indeed find the tubes in the ovarial cavity or in the stylar canal, but their connection with the pollen grain, from which they originated, is missing. Rob. Brown was misled by this as he declared the orchids' pollen tubes to be cells from the conducting tissue. This is particularly the case for those plants whose pollination and fertilization is far apart in time, such as hazel, hornbeam, and alder, which are pollinated early in the spring (February in the case of hazel), that is at a time when the two ovules are not yet present, and fertilised only in the summer (hazel at the end of June).

[8] Robert Brown.
[9] See letter to Daniel Oliver, 18 July [1863] and n. 4.
[10] The reference is apparently to John Smith, curator of the Royal Botanic Gardens, Kew (R. Desmond 1994).

From Daniel Oliver [after 20 July 1863][1]

In connection with Hildebrand's paper[2]—may be consulted also Duchartre, Sur un cas de grossissement, sans fécondation, des ovules de *Cycas revoluta*. Bull. Soc. Bot. ix. 531.[3]

The pollen of a *Ceratozamia* was placed upon the ovules.—
D. O.

Tuesday

Incomplete
DAR 173: 25

[1] The date is established by the relationship between this letter, the letter to Daniel Oliver, 18 July [1863], and the preceding letter.
[2] Hildebrand 1863b. See letter to Daniel Oliver, 18 July [1863].
[3] Duchartre 1862 described the enlargement of ovules under the influence of pollen, apparently without being fertilised. Similar observations were detailed in Hildebrand 1863b.

From Asa Gray 21 July 1863

Cambridge. [Mass.]
July 21, 1863

Dear Darwin.

Your latest is of 26th ult.[1] You need not *worry*! It never wearies nor bores me to write to you—in the off hand way I do.[2] I enjoy our correspondence too much to consent to curtail or interrupt it. I learn from you—here in this remote part of the world—a thousand things which I should not otherwise know at all. And you stimulate my mind far more than any one else, except, perhaps, Hooker.[3] So please do not make a *fuss*, but let me go on in my own fashion, and send me your fresh and stimulating letters—whenever you are in the mood for it.

I am now in my vacation, and already—having idled and dawdled a week or two, I am as well and hearty as possible, and in the best of spirits. We should leave home this week for 3 weeks run in the country. But the sickness of my wife's nephew, Lieut. Jackson of Mass. Cavalry[4] will keep us a while,—as tho' not alarming it might take a bad turn,—and so I may not be in the country for a week or two yet. We shall see.

So I must stop this idling and set to some of the work which has laid so long untouched around me. I am to-day fixing up an abstract of Bates's *Mimetic Analogy* paper, &c—for Sill. Journal.[5]

Get your *Mitchella* on a shady well-drained slope, and let it well alone, & it will thrive. I will send fresher seed of *Sicyos* & *Echinocystis*.[6] And if I can remember you shall have seeds of *Houstonia*.[7]

I have strong & fresh *Drose⟨ra⟩ rotundifolia*, & it will soon turn in its bristles & stick the viscid gland fast to a fly—binding him fast on all sides with Lilliputian cords.[8] But it is awfully slow about it,—say 3 or 4 hours. And the next day, the leaf sometimes becomes involute & folds over or curves around the insect. But what good. If the fly is not stuck fast in alighting, no movement takes place to hold him till he has got away if he ever could. However it is an indication of what is so effectually done in *Dionæa*.[9]

Rotary movement of end of tendril-bearing stems is common, is it not, and well known?[10]

Any notes you will give me to put in *Sill. Journal*, I shall ⟨alw⟩ays delight in.[11]

⟨I⟩ have been reading Owen's Aye-Aye paper.[12] Well, this is rich and cool! Did I not tell you in Atlantic long ago that Owen had a transmutation theory of his own.[13] It is your Hamlet, with the part of Hamlet left out.[14] But as you say now you don't so much insist on Nat. Selection if you can only have derivation of species,—and Owen goes in for derivation on the largest scale, you may as well lovingly embrace![15] Oh, it is rare fun. How I could now tease Agassiz, if I could see him,—only he is of late so cross and sore.[16]

He has been digging various pits for me, but has fallen himself into the pit that he digged. However, I am good-natured; and only laugh at him.

I have been so far disappointed in getting no *Gymnadenia tridentata*. But I still hope for it. I must have it, indeed.[17]

Benth.'s address is *good*—chiefly very good.[18] But he speaks of Wyman's paper without having duly considered it.[19] Wyman's experiments are far better than Pasteur's, & the results opposite![20]

A. Gray.

PS. | Papers just in—or rather telegrams—that you in London were daily awaiting & expecting the capture of Washington. etc.[21]—and speculating as to whether Jeff. Davis's envois from Washington might not be received at London—as a *fait accompli*.—[22] A good-deal of little-concealed joy. &c—

Oh foolish people! When will you see that there is only one end to all this:—and that the North never dreams of any other,—the complete putting down of the rebellion. And since 1863 began, it was clear that it would be attended with the annihilation of slavery.[23]

Time was when we should have highly valued English appreciation of the right cause.— We have now long ceased to care or think about it.

We only wish you had the city of New York. But the sympathizers with secession and riot there have done their worst—and lost their game.[24] The city of New York is the only part of our country which I am ashamed of. And the trouble there is that it is *not* American.

Enough— Good Bye | A. G.

DAR 165: 128, 138

CD ANNOTATION
6.1 *Rotary* ... known? 6.2] *scored brown crayon*

[1] Letter to Asa Gray, 26 June [1863].

[2] In his letter to Gray of 26 June [1863], CD expressed concern that Gray was overworked, and implored him not to write when busy.

[3] Joseph Dalton Hooker.

[4] Gray refers to Jane Loring Gray and to her nephew, Patrick Tracy Jackson III, who served as a first lieutenant in the Fifth Massachusetts Cavalry during the American Civil War (*Lamb's textile industries of the United States*, s. v. Jackson, Patrick Tracy IV).

[5] See letter from Asa Gray, 7 July 1863, nn. 2 and 3. The references are to Bates 1861, A. Gray 1863a, and the *American Journal of Science and Arts*, commonly known as 'Silliman's Journal' after its founder Benjamin Silliman.

[6] See letter to Asa Gray, 26 June [1863] and nn. 11 and 12.

[7] Gray had sent CD observations on and seed of the dimorphic species *Houstonia caerulea* in 1862 (see *Correspondence* vol. 10).

[8] The reference is to Jonathan Swift's *Gulliver's travels*, in which Gulliver was tied down to the ground

by six-inch-high Lilliputians ([Swift] 1726, 1: 7 *et seq.*). Since 1860, CD had corresponded with Gray concerning his numerous experiments on the sundew, *Drosera rotundifolia* (see *Correspondence* vols. 8–10).

9 CD had also corresponded with Gray concerning his experiments on *Dionaea muscipula* (Venus's fly-trap), in which the leaf movements that trap insects are relatively rapid (see *Correspondence* vols. 8 and 9).

10 See letter to Asa Gray, 26 June [1863] and nn. 12 and 13. See also letter from J. D. Hooker, [2]9 June 1863 and n. 3.

11 See letter to Asa Gray, 26 June [1863] and n. 13.

12 Owen 1862c.

13 Gray refers to his statement, in his review of *Origin* in the *Atlantic Monthly*, that Richard Owen was 'apparently in travail with some transmutation theory of his own conceiving, which may yet see the light, although Darwin's came first to the birth' ([A. Gray] 1860b, p. 115). In his monograph on the aye-aye (Owen 1862c, pp. 89–97), Owen discussed the question of the origin of species, committing himself to 'creation by law' while stating his 'ignorance of how such secondary causes may have operated' and refusing to endorse any of the mechanisms currently proposed (p. 96). For Owen's views on the origin of new species, see Rupke 1994, pp. 220–58.

14 Gray alludes to the common phrase 'Hamlet without the prince', which derives from an anecdote told by Walter Scott, concerning a play bill which is said to have announced a performance of William Shakespeare's play, *Hamlet*, in which the character of Hamlet, prince of Denmark, was to be omitted (*ODQ*, p. 560). The reference is to Owen's accepting a natural cause for the origin of new species, while rejecting natural selection.

15 See letter to Asa Gray, 31 May [1863]. For CD's reaction to Owen's claims, see the letters to J. D. Hooker, 5 March [1863] and 17 March [1863], and the letters to Charles Lyell, 6 March [1863] and 17 March [1863].

16 Gray refers to his Harvard colleague, Louis Agassiz, with whom he had a long-running dispute over transmutation. Gray had just been elected president of the American Academy of Arts and Sciences, against the wishes of Agassiz (Dupree 1959, pp. 319–20). See also letter from Asa Gray to J. D. Hooker, 6 July 1863.

17 See letter from Asa Gray, 7 July 1863 and n. 17.

18 Gray refers to George Bentham's anniversary address to the Linnean Society (Bentham 1863). See also letter from Asa Gray to J. D. Hooker, 6 July 1863 and n. 10.

19 The reference is to Wyman 1862 (see Bentham 1863, pp. xxv–xxvii). See letter from Asa Gray to J. D. Hooker, 6 July 1863 and n. 17.

20 Louis Pasteur. See letter from Asa Gray to J. D. Hooker, 6 July 1863 and nn. 17 and 18.

21 In the American Civil War, the previously successful Confederate campaign in Pennsylvania was brought to a dramatic end by the victory of the Union forces at the battle of Gettysburg, from 1 to 3 July 1863 (McPherson 1988, pp. 646–65).

22 Gray refers to the envoys of Jefferson Davis, president of the Confederate States of America.

23 The reference is to Abraham Lincoln's emancipation proclamation of 22 September 1862, which announced that from 1 January 1863 all slaves in the states still in rebellion against the United States would be freed (Denney 1992, pp. 248, 251).

24 Between 12 and 16 July 1863, there were extensive riots in New York City against conscription into the Union army; the majority of the rioters were immigrants from Ireland (McPherson 1988, pp. 609–10).

From Julius von Haast 21 July [– 7? August] 1863[1]

Christchurch Canterbury N Z.
21th July 1863

My dear Mr. Darwin!

I had the pleasure to write to you on the January last[2] and can not tell you, with what intense pleasure, I received your kind letter of the 22th of January.[3] If it is pleasant to hear words of encouragement, when living in the centre of civilisation, how much greater is the delight to receive them away from it and from such a man, as yourself, the noble champion of true philosophic enquiry.— But when I tell you, that your letter came into my possession just on returning from a very rough and toilesome journey across the Alps to the Westcoast, finding it with many others at an isolated sheepstation in the lower alpine regions, you may easily imagine, that this circumstance added not a little to my enjoyment.—[4]

I did not think, that my inaugural adress would interest any body except those few of my auditors, who really took an interest in scientific matters.—[5] Unfortunately the adress being too long, I had to curtail it and therefore many passages were omitted and amongst them one, which explained the enormous difference between your glorious researches and those of your predecessors in the same field. In one of my next papers, when noticing your magnificent work on Orchids, in which you have shown to many botanists, how to proceed, I shall return to the subject;[6] at the same time I shall endeavour to show not only, that the geological age of the world must be incalculable and elucidate the subject by describing the formation of our Alps, drawing some conclusions in looking at the enormous thickness of the strata, of which they are composed and their lithological character, but also in confirmation of your theory, I shall point out, that the old system of chronological sequence of the so called formations *all over the world*, has to be abandonned in a great degree.[7]

To this conclusion I came, when noticing some remarkable phenomena in connection with the fossiliferous strata of New Zealand. I see by an article in the Intellectual Observer[8] (: a splendid publication:) that Prof. Huxley has written a paper, treating of the same matter in the Quarterly Journal of the Geol. Society;[9] which unfortunately I have not yet seen, but which I expect every day. He seems to have come to the same conclusions.

I am just beginning to write my extended report on my researches of the last three years, which I hope will leave the printer's hands in about 12 months and which will be accompanied by maps, sections & sketches.[10] In it I shall give a great many details concerning the glacial period, which is developed in New Zealand in a most wonderful degree.[11]

During my last journey, I again observed very often the tracks of the supposed Quadruped, but although I had a capital dog with me, I never succeeded in obtaining the owners of the feet, which had imprinted them. It seems almost to me that the animal enters its hole from below the surface of the water like the otter.—[12] But I got a great many Kakapo's (Strigops) and as so very little is known of their natural history, I wrote a short paper on their habits, of which I send you enclose a copy, as it will be a long while before it will be printed.[13] It will without doubt interest you, as it bears on many points in connection with the "Origin of Species.[14] And the Solenhofen Bird—Amphibium, what a wonderful creature![15] The connecting links between the different species, genera and even order and classes *did* exist and will be found.

Your suggestion concerning the desirable record of the spreading of European weeds & insects in the Colony was a very good one, I have communicated it to the Philos. Institute & the members were extremely grateful to you.[16] Some of them have already offered to report and are occupied in collecting the necessary material

This subject had for a long time attracted my attention and I shall give you a few instances of the wonderful capabilities of European products, to settle themselves on antipodean ground & to destroy or drive away the indigenous inhabitants.[17] The Native (Maori) saying: "So as the white man's rat has driven away or killed our Kiore (native rat) the European housefly drives away our own (Bluebottle) the clover kills our fern (.Pteris esculenta) so will the Maori disappear before the white man" is a very true one. Yes, it is really wonderful to behold the botanical and zoological changes, which have taken place since first Captⁿ Cook visited the shores of New Zealand.[18] Some pigs which he and some other navigators after him left behind with the natives and the offspring of which in some cases became wild, have increased in such a way, that it is impossible to destroy them. There are large tracts of country, where they reign supreme, the soil looks as if ploughed hole near hole appearing from their burrowings. Some station owner of 100,000 acres have given contracts for killing them, (at 6 d. a tail) and as many as 22,000 on a single run have been killed by adventurous parties without any diminution being visible. Not only are they obnoxious in occupying the ground which the sheepfarmer needs for his flocks, but also in following assidiously the ewes, when lambing and eating the poor lambs, as soon as they make their appearance. They do not exist on the Western side of the Alps and only in the lower ground on the eastern side, where snow seldom falls, so that even the explorer has not the advantage of profiting by their existence. The boars are sometimes very large, covered with long black bristles and with enormous tusks, resembling closely the wild boars of the Ardennes & they are equally savage and courageous.

An other interesting fact is the appearance of the Norwegian rat. (At some spots the *English* rat is said to be abundant). It has thouroughly destroyed the native rat & is to be found everywhere even in the very heart of the Alps, growing to a very large size. The European house mouse follows it closely and what is surprising, where it makes its appearance it drives in a great degree the Norwegian rat away.— Amongst other quadrupeds cattle, dogs & cats are found in a wild state, although not abundant.

The European housefly is another importation Where it arrives, it expels the blue bottle fly, which seems to shun its company. But the spread of the European insect goes on very slowly, so that Settlers knowing their useful quality of their old home aquaintance have carried it in bottles and boxes to their new inland stations

Amongst the European plants the most prolifique is without doubt the Water-cress. Nobody who has not seen it, has any idea of its luxuriant growth. It has chocked up rivers, creeks & ditches so as to create great inconvenience to the Inhabitants. For instance the clearing of the river Avon, which meanders through Christchurch costs that city yearly many hundred pounds.

Docks, buckweeds & red Sorrel have spread everywhere even to the Alps, as far as cattle and sheep have travelled, and even higher up, they are found, the seeds without doubt being brought there by birds or winds. Several species of thistles are also spreading very fast and would soon overrun the whole country, as they have already done in some parts of Australia, had we not very stringent laws for their eradication.

You see there is ample room for research in New Zealand, also in this respect I am very anxious to see and study Sir Ch.s Lyell's new work on the evidence of antiquity of man, because it must be a masterwork, like everything this illustrious philosopher has written.[19]

As a feeble tribute from the Antipodes I may mention to you, that the Philosophical Institute of Canterbury at their last meeting have inanumously (:parsons ⟨no⟩t excepted) elected you an honorary member; you will probably receive the diploma with this mail.[20] In doing so, the members have honoured themselves and in associating the names of yourself and of our friend D.r Jos. Hooker[21] with the Institute, they have assured to their Society a powerful moral support.

In the Revue du Monde Coloniale of March 1863 (Paris) is an article: La science dans la nouvelle Zélande and in it a passage mentionning you, which I find so very appropriate that I venture to copy it for your perusal: "L'illustre Darwin a visité les rivages zélandais, alors que, comme un nouveau Platon, il preludait par de longs voyages à l'élaboration de ses savantes théories."[22]

But I see my letter becomes rather too long, therefore not to take any more of your valuable time, I shall conclude, but not without hoping, that you will again favor me with your letters, which are like an Oasis in this wilderness of money-making settlers.

Allow me to include my photograph[23] to enable you to know your correspondent by sight and may I ask you as a great favor not only to myself but to all your admirers, to send me yours in return, which I shall highly value.

Believe me my dear Sir with great respect yours very faithfully | Julius Haast.—

Ch.s Darwin Esq.re | FRS. etc. | Down, | Kent SE.

PS. | It is, I think, not necessary to tell you that you may make any use of my communications, how and when you like.—[24]

DAR 166: 4, 6; Darwin Pamphlet Collection–CUL (G304)

CD ANNOTATION

12.2 I am … (:parsons 13.2] *crossed pencil*

1 The date range is conjectured from the reference to CD's election as an honorary member of the Philosophical Institute of Canterbury, New Zealand (see n. 20, below). This letter was apparently sent in the same cover as the letter from Julius von Haast, 6 August 1863, which is folded in an identical manner.

2 No letter from Haast to CD of January 1863 has been found. Haast probably refers to his letter to CD of 9 December 1862 (*Correspondence* vol. 10), of which CD only received a copy in July 1863 (see letter to Julius von Haast, 18 July [1863] and n. 2).

3 Letter to Julius von Haast, 22 January 1863.

4 Haast refers to his geological expedition across the Southern Alps of New Zealand's Middle Island (now South Island) to the west coast (see letter from Julius von Haast, 5 March 1863, n. 2).

5 Haast refers to his inaugural address to the Philosophical Institute of Canterbury, New Zealand (J. F. J. von Haast 1862a). See letter to Julius von Haast, 22 January 1863 and nn. 2 and 3, and enclosure to letter from Julius von Haast, 5 March 1863 and n. 12.

6 No paper discussing *Orchids* is listed in the bibliography of Haast's publications in H. F. von Haast 1948, pp. 1088–1100.

7 Haast refers to CD's statement in *Origin*, p. 324, that when 'marine forms of life are spoken of as having changed simultaneously throughout the world', the expression should not be understood in any strict sense. In particular, CD noted that 'the most skilful naturalist would hardly be able to say whether the existing or the pleistocene inhabitants of Europe resembled most closely those of the southern hemisphere'. No publications on this subject by Haast have been identified.

8 The *Intellectual Observer* was a monthly popular science journal, which began publication in February 1862, edited by Henry James Slack.

9 Haast refers to Thomas Henry Huxley's anniversary address to the Geological Society of London (T. H. Huxley 1862a), which was briefly described by William Bernhard Tegetmeier in the March 1862 number of the *Intellectual Observer* (*Intellectual Observer* 1 (1862): 153), and discussed extensively in an article in the June 1862 number (Slack 1862). For CD's initial reaction to Huxley's 1862 address, see *Correspondence* vol. 10, letter to T. H. Huxley, 10 May [1862]; see also letters to T. H. Huxley, 7 December [1862] and 18 December [1862].

10 Haast refers to his geological work for the provincial government of Canterbury; he began working as a geologist in Canterbury province in November 1860, becoming provincial geologist in 1861 (*DNZB*). Although Haast published numerous papers and reports on the geology of Canterbury province (see H. F. von Haast 1948, pp. 1088–1100, and *Royal Society catalogue of scientific papers*), he did not publish an extended account until 1879 (J. F. J. von Haast 1879; see also letter from Julius von Haast, 27 September 1865, *Calendar* no. 4900). There is an unannotated copy of J. F. J. von Haast 1879 in the Darwin Library–Down.

11 For CD's interest in the evidence from New Zealand relative to the Pleistocene glacial period, see the letter to Julius von Haast, 22 January 1863 and n. 4. Haast discussed the glacial deposits of the Southern Alps of New Zealand in several of his papers (see H. F. von Haast 1948, pp. 1088–1100, and *Royal Society catalogue of scientific papers*).

12 See letter to Julius von Haast, 22 January 1863 and n. 6.

13 The manuscript copy of Haast's paper on the ground parrot, entitled 'Notes on the structure and habits of Strigops habroptilus read before the Philosophical Institute of Canterbury', is in the Darwin Pamphlet Collection–CUL. The paper was later published in German in the *Verhandlungen der Kaiserlich-Königlichen Zoologisch-Botanischen Gesellschaft in Wien* (J. F. J. von Haast 1863); an English translation appeared in *Ibis* in 1864 (J. F. J. von Haast 1864).

14 *Strigops habroptilus* possesses well-developed wings, but is flightless. In the manuscript copy of the paper he sent to CD (Darwin Pamphlet Collection–CUL, G304, pp. 9–10), Haast noted that an individual of the species frightened into falling from a tree did not open its wings to break its fall. He observed:

> is not this one instance more that all organic beings by adapting themselves to circumstances will insensibly change their habits & in consequence, their structure will also undergo great modification—the better to fit them for the struggle for existence, unless in the meantime they have been exterminated or have disappeared from other causes—?!— It would lead me too far were I to enter manfully into this important subject, to which the eminent labours of Darwin have directed the close attention of philosophers and naturalists—

15 *Archaeopteryx*. See letter to Julius von Haast, 22 January 1863.

16 See letter to Julius von Haast, 22 January 1863 and n. 9. Haast refers to the Philosophical Institute of Canterbury, New Zealand.

[17] CD sent Haast's letter to Joseph Dalton Hooker, who requested permission to publish Haast's observations on the naturalisation of species in New Zealand (see letter to J. D. Hooker, [30 October 1863], and letter from J. D. Hooker to Emma Darwin, 11 November 1863). A somewhat modified version of a portion of the letter was published in J. D. Hooker 1864, pp. 126–7.

[18] James Cook was the first to circumnavigate and accurately chart New Zealand; he first arrived there in 1769 (*DNB*).

[19] C. Lyell 1863a.

[20] According to the letter from the secretary of the Philosophical Institute of Canterbury, New Zealand, 14 September 1863, CD was elected an honorary member at a meeting held on 7 August 1863.

[21] Hooker was also elected an honorary member of the Philosophical Institute of Canterbury in 1863 (L. Huxley ed. 1918, 2: 508).

[22] Fonvielle 1863, p. 182. The passage translates: 'The illustrious Darwin visited the shores of New Zealand, at a time when, like a new Plato, he was developing through long voyages the elaboration of his learned theories.' Plato was reputed to have travelled widely before returning to Athens to found the Academy (*CBD*).

[23] The enclosure has not been found.

[24] See n. 17, above.

From J. D. Hooker [21 July 1863][1]

Kew
Tuesday.

Dear Darwin

Your observations on Tendrils &c are most curious & novel, & I am delighted that you are going on with them— you are "facile princeps" of observers.[2]

I am looking out some climbers that may serve your purpose & hope to send you

Bignonia
Cobæa
Gloriosa
Flagellaria

& perhaps others[3]

I owe you for 2 letters, & shall pay off soon, havi⟨ng⟩ a lot of gossip for you.[4] I am most anxious to get down for a Sunday, & shall do so by the earliest opportunity[5] My Father has been a month away, which has kept me very busy,—[6] London Society has been worse & really it demands serious consideration.[7] I cannot see my way to any mean course between dining out every-where & no where, without a system of prevarication that would be intolerable, & ⟨now⟩ that my Father never goes out I have double duty that way.— I must now get on with the N Z. Flora.[8]

Black (our Herb Curator)[9] is gone away on 6 months leave in very bad health— lungs affected—which throws an immense lot of work on me— happily Thomson is living at Kew & works all day at Herb. for love of the thing.[10] he says you should take note that in Cucurb. the tendril is a modified leaf, in Vines a shoot, (i.e. axis of growth.)

I will send Gray's letter to me tomorrow, he has come in to £2000 by death a relative of wifes.[11]

Ev yrs aff | J D Hooker

DAR 101: 152–3

CD ANNOTATION
Top of letter: '*Oliver* | *Limnothemium Indicus*' *ink*[12]

[1] The date is established by the relationship between this letter, the letter to J. D. Hooker, 14 July [1863], and the following letter; the intervening Tuesday was 21 July.
[2] *Facile princeps*: 'obviously pre-eminent' (*Chambers*). See letters to J. D. Hooker, 25 [June 1863], 1 July [1863], and 14 July [1863].
[3] See letter to J. D. Hooker, 14 July [1863] and n. 2.
[4] Letters to J. D. Hooker, 1 July [1863] and 14 July [1863].
[5] See letter to J. D. Hooker, 14 July [1863].
[6] Hooker was assistant director of the Royal Botanic Gardens, Kew, where his father, William Jackson Hooker, was director (R. Desmond 1994).
[7] See letter from J. D. Hooker, [2]9 June 1863.
[8] J. D. Hooker 1864–7. Hooker had been engaged by the colonial government of New Zealand to prepare a flora (J. D. Hooker 1864–7, p. 7*).
[9] Allan A. Black was curator of the herbarium at the Royal Botanic Gardens, Kew (R. Desmond 1994).
[10] Thomas Thomson had recently retired as superintendent of the Calcutta botanic garden and professor of botany at the Calcutta medical college (*DNB*). Hooker refers to the herbarium of the Royal Botanic Gardens, Kew.
[11] See letter from Asa Gray to J. D. Hooker, 6 July 1863, and letter from Asa Gray, 7 July 1863.
[12] See following letter. The reference is to Daniel Oliver.

To J. D. Hooker 22 July [1863]

Down Bromley Kent
July 22^d^

My dear Hooker

I shall be very glad of the plants you mention with *Tendrils*.—[1] I find now with 4 genera of climbing plants, which have no tendrils, that the upper free internode, night & day, sweeps a circle in 5 to 6 hours.—[2] See what amusement I owe to you, all these plants & Cissus are from Kew!—

Thank Thomson for message about Tendrils of Cucurbitaceæ:[3] I was thinking of asking & sh^d^. be particularly obliged sometime for answer.— I thought from position that these tendrils were branches; but I find A. Gray in his Book calls them *branches*,[4] & on looking at Vegetable Marrow, they did not seem to correspond so nicely with leaves as in my Echinocystis.— I sh^d^ be very glad to know whether Thomson spoke deliberately. As far as I have seen, **little as yet,** leaf-tendrils are sensitive but have not spontaneous movements, like tendrils of Cucurbitaceæ & Viniferæ.—[5]

Thank Oliver for note: I knew it was mere chance whether Hildebrand's paper would do for N.H.R.—[6] Hildebrand has two or three times been so obliging to me that I am bound to do what I can to gratify a little harmless vanity:[7] I have sent it to "Annals" & Editors must settle whether worth inserting.[8] In these Orchids the pollen-tubes must act like the spermathecas in insects—[9]

I am sorry you are so very busy so do not write any gossip for a long time, though I sh^d^. enjoy it; but there is one point on which I do much want information.

Thwaites has sent me seed of Limnothemium Indicum, which is grandly dimorphic, & he says sow it in "pan of water"; but I have no idea, how deep water ought to be & whether there ought to be Earth at bottom. Do you think Hugh Gower would know at all?[10]

I shall be glad to see Asa Gray's letter.[11] He tells me in a scrap about the £2000, which, I am heartily glad to hear of.[12] He tells me he has no children, which he regrets because he cannot send a son to the war! Did you ever hear the like.—

GoodBye— | Yours affect | C. Darwin

How opposite our troubles are about Society— you too much, I absolutely none.—[13]

Endorsement: '/63'
DAR 115: 199

[1] See preceding letter.

[2] CD first observed this phenomenon in June with *Echinocystis lobata*, a plant possessing tendrils (see letter to J. D. Hooker, 25 [June 1863] and n. 2). CD's experimental and observational notes on what he called 'Twiners' and 'Leaf-Climbers' are in DAR 157.1: 1–60 and 61–112 respectively. He apparently refers to his experiments with *Ceropegia gardnerii* (see the notes in DAR 157.1: 10–17, dated 21 July – 9 August 1863), *Stephanotis floribunda* (see the notes in DAR 157.1: 37, dated 14–15 July [1863]), and *Aristolochia gigas* (see the notes in DAR 157.1: 53, dated 22 July [1863]). These species, together with *Cissus discolor*, appear on the list of hothouse plants believed to be a record of those specimens sent to CD from the Royal Botanic Gardens, Kew, in February 1863 (see Appendix VI).

[3] Thomas Thomson. See preceding letter.

[4] A. Gray 1857, pp. 38–9. There is an annotated copy of this work in the Darwin Library–CUL (see *Marginalia* 1: 347).

[5] CD had made extensive observations of the spontaneous rotatory movements of the upper free internode in *Echinocystis lobata* (Cucurbitaceae), *Cissus discolor* (Vitaceae), and *Ampelopsis hederacea* (Vitaceae); see letter to J. D. Hooker, 14 July [1863], n. 3. (Viniferae was one of the synonyms for the vine family, Vitaceae; see Lindley 1853, p. 439). In a note dated 20 January 1863, CD observed that in *Bignonia speciosa*, the tendrils (which he considered to be modified leaves) were sensitive; however, he could find 'no spontaneous movement of tendril or internode' (DAR 157.1: 133). CD later concluded that nearly all 'twiners, leaf- and tendril-climbers' had 'the same remarkable power of spontaneously revolving' ('Climbing plants', p. 108).

[6] See letter from Daniel Oliver, 20 July 1863 and n. 1. The references are to Friedrich Hildebrand and to the *Natural History Review*.

[7] CD probably refers to the fact that Hildebrand had offered to complete Heinrich Georg Bronn's German translation of *Orchids* (Bronn trans. 1862), not realising that Bronn had already concluded it before his death on 5 July 1862 (see *Correspondence* vol. 10, letter from Friedrich Hildebrand, 14 July 1862).

[8] The editors of the *Annals and Magazine of Natural History* were Prideaux John Selby, Charles Cardale Babington, John Edward Gray, and William Francis. Neither CD's letter, nor the reply, has been found; however, see the letter to John Scott, 25 [July 1863], and the letter to Friedrich Hildebrand, 28 July [1863]. Hildebrand's paper was published in the September 1863 issue of the journal (Hildebrand 1863b).

[9] The spermatheca of insects is a receptacle for the reception and storage of spermatozoa. In his paper (Hildebrand 1863b), Hildebrand described his discovery that in many orchids the ovules were not developed until weeks or months after the pollen-tubes had penetrated the stigma; CD is suggesting that the pollen-tubes functioned as receptacles for the storage of the male gametes during that period.

[10] See letter from G. H. K. Thwaites, 8 June 1863. William Hugh Gower was foreman of the propagating department at the Royal Botanic Gardens, Kew.
[11] Letter from Asa Gray to J. D. Hooker, 6 July 1863. See preceding letter.
[12] See letter from Asa Gray, 7 July 1863, and preceding letter.
[13] See preceding letter.

From George Brettingham Sowerby Jr to Emma Darwin 22 July 1863

45 *Gt. Russell St. Bloomsbury, W.C.* | *London*
July 22 *1863*

Dear Mrs. Darwin

The cutting of the block will be—	1 · 0· 0
—The Drawing on Wood	· 10″—
Postage of Block	· 8
Loan of Trap	6
	£1″ 11″ 2

The cutting seems to me rather dear, but I hope it will answer your purpose[1]
Your very Obliged | G B Sowerby

DAR 157.2: 101

[1] The reference is to a woodcut of a steel trap for catching vermin which appeared in the pamphlet that Emma and CD had printed privately in August 1863 for distribution (*An appeal*, p. 2; see Appendix IX). There is an entry in CD's Account book–cash account (Down House MS), dated 24 July 1863, recording payment to Sowerby of £1 11*s.* 2*d.* George Brettingham Sowerby Jr. ran a business at 45 Great Russell Street, London, selling natural history specimens (*Post Office London directory* 1863). He was also a leading natural history artist, and had drawn the figures of fossil shells for *South America*, and the illustrations for *Living Cirripedia* (1851 and 1854), *Fossil Cirripedia* (1854), and *Orchids* (see *Correspondence* vols. 3, 4, 5, and 9).

From John Scott 23 July [1863][1]

Botanic Gardens Edinburgh
July. 23d.

Sir.

I duly received yours of the 2d.[2] I am greatly obliged for your account of the structure of Hottonia.[3] I will be able to had—if I had results—a few experiments on a *short-styled* form with *own-pollen*; and also with that from a *long-styled* form: having been favoured with a few flowers of the latter from a friend.[4] I can also add from observations on dried specimens of the North American *H. inflata*, as a **non-***dimorphic* species. The style is very short & reaches the anthers: pollen-grains very similar to those of long-styled H. palustris.[5] I should like much to hear whether the *H. sessiliflora* of Java is *dimorphic* or not—as it would enable me to include as

I believe all the representatives of the genus—but I cannot find a single specimen of it in any of the Herbaria to which I have access. The description of the species unfortunately takes no notice of either stamens or pistil.

I am much astonished with your extracts from Bot. Zeit. respecting the P. *longiflora*[6] I have had numerous specimens and never found an *approach* to the *short-styled* state: in those cases where styles & stamens differed in length I *invariably* found the *former organ projecting beyond the stamens*: anthers *always surrounding the mouth of tubes*. It is also quite a mistake of Prof. Treviranus to suppose that long-styled P. Auricula is sterile.[7] As bearing on this point however I mention in my paper the *long-styled* form of P. *denticulata*, from which I have never been able to get a **single** *seed* neither with own-pollen nor that of closely allied species.[8] I had no short-styled form of the species to try it with. Mr. McNab likewise informs me that though it has now been in cultivation in the Gardens for upwards of 18 years, yet he has never seen it produce a seed.[9]

—It is singular how Koch & Tausch could have hazarded such a statement respecting universal dimorphism in the species of Primula: for I cannot think if they had carefully examined the other forms could have escaped them.[10] Singularly enough we have also made directly opposite observations on P. *longiflora*![11] I am glad you gave me an authority to support me as to its being a *non-dimorphic* species.[12] Though I of course have the fine suite of specimens of this species in the Edinburgh University Herbarium to refer to in support of my statement.[13]

I am gradually getting results of my Primula work. Those from my crosses with **differently coloured** Primroses are most remarkable.[14] Thus from homomorphic crosses of P. *vulgaris* & var. *alba* I got an average of about **five** *seeds per capsule*: from the latter fertilised with *own-pollen* I had an average of **eleven**. But the most astonishing results were those from a **Red Primrose** & the common **yellow**, & *white*. *Reciprocal unions* between the former & the two latter have not yielded me a single seed. I also took the precaution to prove that that the plants of the **Red** & **White** varieties experimented upon were productive with own-pollen; as I find that many individuals of both varieties will not produce a single seed even after careful fertilisation. The less robust growing plants have alone proved productive with me; more especially in the case of the *Red*, with the *white* it is somewhat different. Both varieties have been cultivated in the Gardens here for many years yet Mr. McNab tells me that he has never known them to produce any seed! In my own mind I certainly have a conviction that I have here an illustration of the attainment of the *very zero* of *fertility between varieties of a species*; though I fear my experiments will not be sufficiently numerous to carry a like conviction to the minds of all. Indeed I had some little hesitation in bringing them forward in my paper at all; until I had further opportunities of repeating them. After a careful consideration of those I have already made, I think myself justified in stating them at least. Individuals of varieties perfectly sterile when crossed; yet productive at the same time with own-pollen, is certainly curious if it is not in reality due to the cause I suppose. As you have however so kindly offered to examine my paper;[15] I will leave it entirely in your hands, and you can form an idea from the nature of the evidence on which it is based, whether or not it would be advisable to record it along with my others.[16]

I have been most unfortunate with my unions of *Cowslips* & *Primroses*: my experimental plants were accidentally sent down to our Class-room in illustration of the Order; and best of the capsules were taken off.[17] I am very sorry for this as I fear I will be quite unable to give anything like a table of these unions.[18]

My Orchid paper is not published yet: they promised to bring out in the late no. of the Edin. New Phil. Jour. but it was pushed aside for others which they thought were more interesting.[19] It has certainly disappointed me very much. I have spoken to the publisher,[20] however, and he has promised to throw off a few copies of it in a short time; so I hope to be able to send you them sometime previous to the publication of the next Journal. By the way I have made out a few more cases of Individual Sterility in Orchids: I will state them in my next, in case you may need further illustrations.[21]

I know I was wrong in suggesting the slightest modification of results in my last.[22] Pray excuse me. I merely did it under the impression that otherwise, it might afford an argument for some of your numerous carping critics against the view you take, that for example the differentiation the male & female sexual e⟨le⟩ments of either form of Primulas have undergone with respect to their homomorphic action was a slowly acquired quality.[23] This I thought was opposed by the non-dimorphic cowslip. At least it appeared to me that any inclined to dispute your views might have instanced it as probably illustrative of the other forms having thus likewise suddenly attained their structural & functional peculiarities.[24] I however, may be quite wrong: anyhow, I will ever remember, & act in accordance with the judicious counsel you have given me.[25]

I regret to find by your last that you was afraid for an attack of scarlet-fever in your family.[26] I sincerely trust this has been awarded, and that all ere this are again enjoying that most invaluable of gifts.

Excuse my hurried scribble. | Very respectfully yours. | J. Scott

DAR 177: 95

CD ANNOTATIONS

4.18 After a . . . suppose. 4.21] *scored brown crayon*

5.1 I have been . . . Order; 5.3] *double scored brown crayon*

6.1 My Orchid . . . yet:] *two crosses in margin, pencil*

6.6 By the way . . . illustrations. 6.8] *cross in margin and scored, pencil*

1 The year is established by the reference to the letter to John Scott, 2 July [1863].

2 Letter to John Scott, 2 July [1863].

3 See the second enclosure to the letter to John Scott, 2 July [1863] and n. 9.

4 This individual has not been identified.

5 Scott reported the results of his experiments on *Hottonia*, including his own and CD's observations, in Scott 1864a, pp. 78–9. CD referred to Scott's experimental results in *Forms of flowers*, p. 51.

6 CD had told Scott of Ludolph Christian Treviranus's report that, according to Wilhelm Daniel Joseph Koch and Ignaz Friedrich Tausch, *Primula longiflora* was invariably short-styled (see the first enclosure

to the letter to John Scott, 2 July [1863] and n. 8.). The reference is to Treviranus 1863a, which was published in the *Botanische Zeitung*. Scott's own observation was that the styles of this species were either slightly longer than the stamens, or of an equal length (Scott 1864a, p. 81).

7 See the first enclosure to the letter to John Scott, 2 July [1863] and n. 3. The reference is to Treviranus 1863a, p. 6.

8 Scott 1864a, pp. 90–1.

9 James McNab was curator of the Royal Botanic Garden, Edinburgh.

10 See the first enclosure to the letter to John Scott, 2 July [1863] and n. 8. Scott refers to Koch 1843–4, 2: 673, and Tausch 1821, p. 355. Scott disputed their claim in Scott 1864a, p. 81.

11 See n. 6, above.

12 CD had previously told Scott that Treviranus claimed *Primula longiflora* was a non-dimorphic species (see letter to John Scott, 25 and 28 May [1863] and n. 9). The reference is to Treviranus 1863a, p. 5. See also letter to John Scott, 25 [July 1863] and n. 2.

13 Scott was foreman of the propagating department at the Royal Botanic Garden, Edinburgh.

14 See letter from John Scott, 3 March 1863. Scott reported on these crosses in Scott 1864a, pp. 97–103.

15 See letter to John Scott, 6 June [1863].

16 See letter to John Scott, 25 [July 1863] and n. 3.

17 There was a purpose-built classroom at the Royal Botanic Garden, Edinburgh, for the use of the keeper, John Hutton Balfour. Balfour gave classes in botany, apparently both for students of the University of Edinburgh, and for the staff of the botanic garden (see Fletcher and Brown 1970, pp. 141–3, 146; *Transactions and proceedings of the Botanical Society* [*of Edinburgh*] 14 (1863): 160).

18 In his paper (Scott 1864a, pp. 103–4), Scott presented results from crossing experiments carried out by CD with the common primrose (*Primula vulgaris*) and the cowslip (*P. veris*). See also letter to John Scott, 25 [July 1863].

19 In his letter to CD of 16 June [1863], Scott had reported that his paper on the pollination of orchids, read before the Botanical Society of Edinburgh on 14 May 1863, was in press. The quarterly *Edinburgh New Philosophical Journal* regularly published the proceedings of the Botanical Society of Edinburgh, including abstracts of some papers. An abstract of Scott's paper appeared in the October 1863 issue of the journal. However, the publication of the full version of the paper in the *Transactions of the Botanical Society* [*of Edinburgh*] apparently pre-dated the abstract (Scott 1863a; see letter to John Scott, 1 and 3 August [1863]).

20 The *Edinburgh New Philosophical Journal* was published in Edinburgh by Adam and Charles Black.

21 See letter from John Scott, [26 July – 2 August 1863].

22 Scott's letter has not been found, but see the letter to John Scott, 2 July [1863].

23 See 'Dimorphic condition in *Primula*'.

24 Scott refers to a non-dimorphic cowslip (*Primula veris*), which he had discovered among a number of seedling cowslips. He had been surprised to discover that it was remarkably sterile with both dimorphic forms, while being extremely self-fertile (see letters from John Scott, 21 May [1863] and [3 June 1863], and Scott 1864a, pp. 105–10).

25 See letter to John Scott, 2 July [1863].

26 See letter to John Scott, 2 July [1863] and n. 6.

To W. E. Darwin [25 July 1863][1]

Down
Saturday.

My dear William

There can be no doubt that the Anchusa would probably be a *very* important case for me.[2] But would it not be a frightful bore for you to get the plants? Could you not hire a man (& I would pay) at place with spade & make him do the digging

& packing for a dozen or score of plants would be heavy. It is a perennial, & though many would die, some would probably live.—[3]

The Boys return to school on the 12th of August.[4] I do hope that you may be able to come before then.—[5] It is a bad job Mrs. Atherley being sick.—[6] We have Uncle Eras. here,[7] & Strickland & Edmund have just gone.[8] We have had the Leith Hill people & altogether there has been a very jolly party.[9] Over & over again I have been wishing to see your dear old face here with the others. One day they all went to the Frys & had a gorgeous party with about 80 people chiefly from London & dancing on the Lawn & dinner in grand tent, Band, & ices &c &c.[10] Another evening they all went to the Bonham Carters for Crocket.—[11]

This evening we have a party of another kind viz 30 children from the Union for tea & play.[12] This evening Uncle Hensleigh & Fanny are coming.—[13]

George & all the Boys are very jolly. Do come as soon as you can.—

My hobby-horse at present is Tendrils; they are more sensitive to a touch than your finger; & wonderfully crafty & sagacious[14]

Good Bye my dearest William | Your affect. Father | C. Darwin

If you have time & inclination will you have a look at mid-styled Lythrum & see if you can really recognise difference in general appearance of the Plant. They are now just coming into full flower Perhaps difference is when capsules swollen.[15]

DAR 210.6: 112

1 The date is established by the reference to the departure of guests from Down House (see n. 8, below); 25 July 1863 was a Saturday.

2 CD is apparently responding to a letter from William that has not been found. In May 1863, William informed CD that he had discovered dimorphism in a plant that he identified as *Anchusa officinalis*; the species was subsequently identified as *Pulmonaria angustifolia*, another member of the Boraginaceae (see letter from W. E. Darwin, 4 May [1863] and n. 2). CD was anxious to investigate the occurrence of heterostyly in the Boraginaceae (see letters to W. E. Darwin, [5 May 1863] and [10 May 1863]).

3 In his letter to William of [10 May 1863], CD had requested seed of the plant, so that he could attempt crosses between the two forms in the following season. A note in William's botanical notebook states: 'on July 18. 63. I went to collect seeds and found all the flower stalks withered & seeds scattered' (DAR 117: 68). However, it appears that William managed to send CD seedlings, which formed the basis for experiments in 1864 (see *Forms of flowers*, pp. 104–10, and DAR 110: A41–55).

4 Leonard, Francis, and George Howard Darwin were pupils at Clapham Grammar School in South London (see *DNB* s.v. Darwin, Sir George Howard; F. Darwin 1920, p. 63; and letter from G. V. Reed, 12 January 1863 and nn. 1 and 2).

5 According to Emma Darwin's diary (DAR 242), William went to Down House on 1 August 1863.

6 Ellen Atherley was the wife of George Atherley, William's partner in the Southampton and Hampshire Bank.

7 According to Emma Darwin's diary (DAR 242), Erasmus Alvey Darwin arrived at Down House on 21 July 1863.

8 Sefton West Strickland was a friend of William's from his time at Cambridge University (Freeman 1978). CD probably refers also to Edmund Langton, Emma Darwin's nephew, who was a near-contemporary of Strickland's at Cambridge. They visited Down House from 22 to 25 July (see 'Journal' (Appendix II)).

9 CD refers to the family of Josiah Wedgwood III and Caroline Sarah Wedgwood, who lived at Leith Hill Place, near Dorking, Surrey. According to Emma Darwin's diary (DAR 242), the family stayed at Down House from 16 to 23 July, leaving Sophy Wedgwood behind on their departure.

10 The reference is apparently to James Thomas Fry, who lived at Baston, near the village of Hayes, about four miles north-west of Down (*Post Office directory of the six home counties* 1862).

11 Joanna Maria Bonham-Carter resided at Keston, about two miles north-west of Down. CD refers to croquet, which became popular in England at about this period (see *EB*).

12 The reference is apparently to children from the Bromley Union House, which was the local workhouse (working refuge for the indigent poor), located at Farnborough, about two miles north of Down (*Post Office directory of the six home counties*).

13 According to Emma Darwin's diary (DAR 242), Hensleigh and Frances Emma Elizabeth Wedgwood visited Down House from 25 to 28 July; they were accompanied by Eva Mackintosh, Frances Wedgwood's niece.

14 CD refers to his experiments on climbing plants, begun in June 1863 (see letter to J. D. Hooker, 25 [June 1863]). Many of the notes from these experiments are preserved in DAR 157.1 and 157.2.

15 In 1862, William reported to his father that he thought the general appearance of the mid-styled form of *Lythrum salicaria* was different from that of the other two forms; however, he had been unable to verify this at the end of the season, and subsequently doubted his assertion (see *Correspondence* vol. 10, letters to W. E. Darwin, [25 October 1862] and 30 [October 1862], and letter from W. E. Darwin, 28 October 1862). See also letter from W. E. Darwin, 21 August [1863].

To John Scott 25 [July 1863][1]

Down Bromley Kent

25th

Dear Sir

From what you say I looked again at Bot. Zeitung. Treviranus speaks of P. longiflora as *short*-styled, but this is evidently a slip of pen for further on, I see, he says the stigma always projects above anthers.[2] Your experiments on coloured Primroses will be most valuable, if proved true: I will advise to best of my power when I see the M.S. If evidence is not good, I would recommend you, for your reputation sake, to try them again.[3] It is not likely that you will be anticipated, & it is a great thing to fully establish, what in future time will be considered an important discovery (or rediscovery for no one has noticed Gärtners facts).[4] I will procure coloured primroses, for next spring, but you may rely I will not publish before you.—[5]

Do not work too hard to injure your health.—

I made some crosses between Primrose & Cowslip, & I send the results which you may use, if you like.[6] But *remember* that I am not quite certain that I *well* castrated the short-styled Primrose; I believe any castration would be superfluous; as I find all plants sterile when insects are excluded.—

Be sure save seed of the crossed differently coloured Primroses or Cowslips which produced least seed to test the fertility of the quasi-hybrids. seedlings.—[7] Gärtner found the common Primrose & Cowslip very difficult to cross, but he knew nothing on Dimorphism.—[8]

I am sorry about delay on your orchid paper:[9] I sh^{d}. be glad of abstract of your new observations on self-sterility in orchids, as I sh^{d}. probably use the new facts.[10]

There will be important paper in September in Annals & Mag. of N. Hist, on ovules of orchids being formed months after application of pollen, by Dr. F. Hildebrand of Bonn.[11]

In Haste | Yours sincerely | C. Darwin

Are you sure that your coloured Primroses are not descendants of coloured Polyanthus; so as to be hybrids.—[12]

This year I raised seedlings from wild oxlips, which had probably been naturally crossed with primroses; these seedlings closely resembled pure primroses, but the flower-peduncles rose from common scapes at the very crown of the plants, & so alone showed their parentage.[13]

DAR 93: B45–6, B69

[1] The date is established by the relationship between this letter and the letter from John Scott, 23 July [1863].

[2] See letter from John Scott, 23 July [1863] and nn. 6 and 12. The reference is to Treviranus 1863a, pp. 4 and 5. There are annotated copies of the numbers of the *Botanische Zeitung* in which the paper appeared in the Darwin Pamphlet Collection–CUL.

[3] See letter from John Scott, 23 July [1863]. See also letter to John Scott, 1 and 3 August [1863]. CD received the draft manuscript of Scott 1864a in September, and sent his comments to Scott in November; however, his detailed criticisms have not been found (see letter from Emma Darwin to John Scott, 24 September [1863], and letter to John Scott, 7 November [1863]; see also *Correspondence* vol. 12, letter from John Scott, 7 January [1864]). In introducing the results of his crosses between differently coloured varieties of the common primrose (*Primula vulgaris*), Scott observed: 'Certain of these are so remarkable, that I hesitate not a little in bringing them forward until I have had again an opportunity of repeating my experiments' (Scott 1864a, p. 97).

[4] CD refers to the experiments carried out on *Verbascum* and *Zea* by Karl Friedrich von Gärtner, described in Gärtner 1844 and Gärtner 1849, in which crosses between differently coloured varieties of the same species produced less seed than the parallel crosses between similarly coloured varieties. There are very heavily annotated copies of these works in the Darwin Library–CUL (see *Marginalia* 1: 248–98). CD discussed the experiments in *Origin* pp. 270–1; he viewed them as important in countering a presumption against natural selection based on the belief that there is 'some essential distinction between species and varieties' and that varieties 'cross with perfect facility, and yield perfectly fertile offspring' (*Origin*, p. 268). On the reception of Gärtner's work, see Graepel 1978.

[5] Scott sent CD seed from his red and white varieties of primrose (*Primula vulgaris*) with his letter of [26 July – 2 August 1863]. See also *Correspondence* vol. 12, letter from John Scott, 16 May [1864]. CD repeated Scott's experiments on a larger scale in 1865, with very different results (see letter to J. D. Hooker, 22 December [1865], *Calendar* no. 4953; DAR 108: 89–98; DAR 110: 2–4; 'Illegitimate offspring of dimorphic and trimorphic plants', pp. 420–3; and *Variation* 2: 109 n.).

[6] See letter from John Scott, 23 July [1863] and n. 17. The enclosure has not been found; however, there are notes dated 'July 1863' in DAR 108: 67–9, recording the results of CD's crossing experiments with cowslips and primroses. In addition, in reporting CD's results in Scott 1864a, pp. 103–4, Scott remarked in a footnote (p. 104 n.):

> Mr. Darwin, in his letter to me accompanying the above results, remarks that the seeds of this short-styled Primrose were very small; so that we may perhaps suspect a number of them unfit for germination.

[7] Scott did not discuss the fertility of the plants grown from crosses between the differently coloured varieties of primroses or cowslips in Scott 1864a.

[8] Gärtner 1849, p. 721. See also *Forms of flowers*, pp. 58–9.

[9] See letter from John Scott, 23 July [1863] and n. 19. The reference is to Scott 1863a.
[10] See letter from John Scott, 23 July [1863].
[11] CD refers to Hildebrand 1863b, which appeared in the September 1863 issue of *Annals and Magazine of Natural History* (see preceding letter). Friedrich Hildebrand was a university lecturer at Bonn (Junker 1989).
[12] The postscript, which is written on a separate piece of paper, has been assigned to this letter on the basis of its relationship to the letters from John Scott, 23 July [1863] and [26 July – 2 August [1863], and the letter to John Scott, 1 and 3 August [1863].
[13] CD recorded this observation in his Experimental notebook (DAR 157a, pp. 75–7) on 10 April 1863; he reported it in 'Specific difference in *Primula*', pp. 447–8. CD was interested in the commonly held view that common oxlips were the hybrid offspring of primroses and cowslips (see *Correspondence* vol. 10, letter to John Scott, 3 December [1862] and nn. 13–15, and this volume, letter to John Scott, 25 and 28 May [1863] and n. 11).

To J. D. Hooker 26 [July 1863][1]

Down
26th

My dear Hooker

Many thanks about Limnanthemium.[2]

What pleasant letters Asa Gray writes.[3] One might as well write to a madman as to him about the war.[4] It is a holy war & everything sh$^{d.}$ be sacrificed to it.— I never thought of Jesus & absence of family affection:— I hear Rénan has discussed this;—[5]

I see Henslow says tendrils of Cucurbitaceæ are stipules;[6] Gray branches,[7] & Thomson leaves—:[8] what is a poor devil to believe?

Have you by any chance seed of Lathyrus aphaca: it would be good for my purpose.—[9]

I am tired so good night— | C. D.

Remember "hot-house Lythrum-like flower."—[10] I am daily crossing my flowers of L. salicaria, & this reminds me.[11]

Endorsement: 'Aug$^{t.}$ | 63'
DAR 115: 203

[1] The month is established by the relationship between this letter and the letter to J. D. Hooker, 22 July [1863]; Hooker evidently endorsed the letter 'Aug$^{t.}$' in error.
[2] In his letter to Hooker of 22 July [1863], CD asked for advice about sowing seeds of *Limnanthemum indicum*; Hooker's reply has not been found.
[3] Hooker had evidently enclosed the letter from Asa Gray to J. D. Hooker, 6 July 1863, in his letter to CD (see the letter from J. D. Hooker, [21 July 1863], and the letter to J. D. Hooker, 22 July [1863]).
[4] In his letter, Hooker had apparently discussed CD's comments on Gray's attitude to the American Civil War (see letter to J. D. Hooker, 22 July [1863]).
[5] Renan 1863, p. 42.
[6] J. S. Henslow 1837, p. 72. There is an annotated copy of this work in the Darwin Library–CUL (see *Marginalia* 1: 369–71).
[7] A. Gray 1857, pp. 38–9. There is an annotated copy of this work in the Darwin Library–CUL (see *Marginalia* 1: 347).

[8] See letter from J. D. Hooker, [21 July 1863]. The reference is to Thomas Thomson.

[9] In the climbing plant, *Lathyrus aphaca*, the leaves generally develop into tendrils, their typical function being performed instead by large stipules; the plant is described in J. S. Henslow 1837, p. 72 (see n. 6, above). There are notes on this species, dated 19–29 October 1863, in DAR 157.2: 21; see also 'Climbing plants', p. 67.

[10] The reference is to *Lagerstroemia*, a member of the Lythraceae. See letter to J. D. Hooker, 15 and 22 May [1863], and letter from J. D. Hooker, [24 May 1863].

[11] In 1862, CD carried out ninety-four crosses between the three forms of flower in *Lythrum salicaria*, but decided that he needed to carry out further crosses in 1863 in order to be sure of his results (see *Correspondence* vol. 10, letter to J. D. Hooker, 27 [October 1862]). See the experimental notes in DAR 27.2: B39, B41, B43, and 'Three forms of *Lythrum salicaria*', pp. 180–4 (*Collected papers* 2: 115–18). In addition, CD carried out experiments in August 1863 using plants raised from the homomorphic crosses of the previous year (see DAR 109: B36–41, and 'Illegitimate offspring of dimorphic and trimorphic plants', pp. 394–410).

From John Scott [26 July – 2 August 1863][1]

[Royal Botanic Garden, Edinburgh]

of individual sterility in Orchids. I have *unfortunately failed* in giving illustrations from *genera* distinct from those mentioned in my paper.[2] This would have enhanced their interest; my experiments are so much limited here from want of subjects; our orchid collection being exceedingly poor.[3]

I can give you no information whatever as to origin of our *Red* & *White* Primroses.[4] I have asked several old and experienced growers as to this: I find that all agree in regarding them as varieties of the common yellow; but, none can affirm that he has raised either from a *perfectly pure* yellow primrose. It is possible they may have such an origin as you suspect. If I be permitted, however, to lay any stress upon my experiments, they appear to me somewhat opposed to this.[5] Thus though both varieties manifest curious idiosyncrasies in the production of seed when fertilised with own-pollen, individuals of both, nevertheless do as I have proved, occasionally produce seed. Hence if we admit them to be crosses; or hybrids between *cowslips* & *primroses*; ought we not naturally to look for an *increased fertility* from *conjunctions* with *either* of their presumed *parents*? This not being the case I had concluded before you had suggested the above that they were truly & simply **modified descendants** of the common yellow primrose. Since I have procured seeds from them, however, there are hopes for enlightenments on this. Perhaps you would like to try results. I therefore enclose a few seeds from both.[6] I am sorry that I had no more by me, having sown all as I gathered them to economise time in the elicitation of evidence. One of the packets of Red you will observe queried; this may chance to be from the *White* variety; having neglected to name when I gathered contents: you may rely on their being from one or other of same however. I have only a single capsule left—*certainly*—sowed from the White variety likewise enclosed: I will send young plants, however, if I succeed in raising them.

I enclose also a few self-fertilised seeds of my equal stamened & styled Cowslip: perhaps you will like to see their results likewise.[7] The great self-fertility of this plant conjoined with the high degree of sterility from unions with the other forms has surprised me much.

I am greatly obliged for your interesting tables on crosses between Cowslips & primroses.[8] I was much struck with results of the crosses between long-styled Cowslips & Primroses. These eight flowers *heteromorphicaly* crossed produced no seeds, while from the same number of **homomorphic** crosses I had three capsules, containing respectively 17, 10 & 6 seeds: they—the seeds—were all—seeds—smaller than usual. I have sown them, however & we will see how they germinate. Is it not singular how the homomorphic should so exceed the heteromorphic?[9] In your reciprocal cross—Primrose female—I see the heteromorphic cross is again abortive! Again how singularly productive your short-styled crosses are![10] The **average** production *per capsule* of my *heteromorphic* unions with **White** Primroses is about **21** *seeds*: the *homomorphic* about *13*. I here refer to unions with own-pollen—

I really do not think that there is any fear from imperfect castration when insects are guarded against. I have at least always found as you say when these plants were so protected that no seed was produced even when the anthers were all allowed to remain.[11]

The plants of *Linum Lewisii*, with us have all turned out L. *Libericum* which has provoked me much.[12] But, I have now got one plant which I am satisfied from description in Bot. Mag. is really L. Lewisii.[13] But though I have watched every flower, there is no variation, all are **short-styled**. I have just had a note from a friend,[14] however to call and see a plant of it which he informs me is producing in general *short*-styled flower, with others—occasionally sta. & sty. equal. This I will call and see. I trust it may prove so. I have looked over the specimens in the Herbaria here but they contain few specimens of it: they exhibit no traces however, of Planchon's trimorphic characteristics[15] Have you had any confirmatory evidence? Have you seen the *Linum Mongynum*, a *dimorphic* species? I find a long-styled plant of this singularly productive, and there is no *short-styled* plant near, indeed I have only seen dried specimens of the latter. The majority of the capsules produced by the former contained **10** apparently good seed—& this is the full number that it can possibly produce.

I am kept very busy at present; so pray excuse my hurried scrawl.

I remain | Sir | Yours very respectfully | J. Scott

Incomplete
DAR 177: 89

[1] The date range is established by the relationship between this letter and the letters to John Scott, 25 [July 1863], and 1 and 3 August [1863].

[2] In his letter to CD of 23 July [1863], Scott had offered to send CD details of examples of self-sterility in orchids, additional to those to be published in Scott 1863a. See also letters to John Scott, 25 [July 1863], and 1 and 3 August [1863]. The genera mentioned in Scott 1863a are *Oncidium* and *Maxillaria*.

[3] Scott was foreman of the propagating department at the Royal Botanic Garden, Edinburgh (R. Desmond 1994).
[4] See letter to John Scott, 25 [July 1863] and n. 12.
[5] In his paper, Scott noted: 'The pure descent of the red and white Primrose from the common yellow has been questioned, and a hybrid origin from the Cowslip and Primrose (*P. veris* and *P. vulgaris*) ascribed to them' (Scott 1864a, p. 98 n.). He argued that his experimental results 'negatived' such a view, and expanded the argument given here.
[6] See letter to John Scott, 25 [July 1863] and n. 5.
[7] See letter to John Scott, 20 [June 1863] and n. 8. Scott described his experiments with this plant in Scott 1864a, pp. 105–10. CD raised plants from the seeds sent him by Scott, repeating Scott's experiments in 1865 and 1866 with contradictory results (see DAR 108: 99–124 and 'Illegitimate offspring of dimorphic and trimorphic plants', pp. 426–30). See also DAR 78: 4 v and 65–8, and *Cross and self fertilisation*, pp. 222–5.
[8] See letter to John Scott, 25 [July 1863] and n. 6.
[9] Scott refers to the results of crosses between a long-styled cowslip (*Primula veris*) and long- and short-styled primroses (*P. vulgaris*). His other experimental plants were accidentally damaged (see Scott 1864a, p. 103, and letter from John Scott, 23 July [1863]).
[10] CD's results are reported in Scott 1864a, p. 103.
[11] See letter to John Scott, 25 [July 1863].
[12] See letter from John Scott, [3 June 1863], and letter to John Scott, 6 June [1863].
[13] The reference has not been identified; neither *Linum lewisii* nor *L. liburnicum* were mentioned in the *Botanical Magazine* (see Chittenden 1956).
[14] This individual has not been identified.
[15] Scott refers to the statement by Jules Emile Planchon, reported by CD in 'Two forms in species of *Linum*', pp. 82–3 (*Collected papers* 2: 104–5), that *Linum lewisii* was trimorphic (Planchon 1847–8, p. 175).

To Friedrich Hildebrand 28 July [1863][1]

Down. | Bromley. | Kent. S.E.
July 28th.

Dear and Respected Sir

Your paper on Orchids seems to me extremely curious & valuable, & has interested me greatly.[2] But I grieve to say that I doubt whether I shall succeed in getting it read or published in England. Our Societies have a fixed rule not to allow a paper to be read, which will be published elsewhere.— I sent your paper to one of our best Journals: the Editor wrote to me that he thought it very valuable, but would not publish it because it was to appear in Bot. Zeitung.[3] I have now sent it to another Journal & you may rely I will do what I can; but our Journals choose to publish original articles or to make their own abstracts & Reviews.—[4] Your paper has explained much to me, that I could not understand. I sometimes saw the ovules & sometimes I could not, & in this latter case I attributed the fact to my want of experience & skill. I saw the ovules best in the hot-house Goodyera discolor from Brazil. Other facts, besides what you tell me, make me fearful that I have made a great mistake about Acropera and Catasetum; but I shall soon have some flowers of latter genus open in my own hot-house.—[5]

And now let me thank you for your very kind letter. I am extremely much pleased to hear that you have been looking at the manner of fertilisation of your native Orchids, & still more pleased to hear that you have been experimenting on Linum.[6] I much hope that you may publish the results of these experiments; because I was told that the most eminent French Botanists in Paris said that my paper on Primula was the work of imagination, & that the case was so improbable they did not believe in my results.—[7]

With my thanks & sincere respect | I remain Dear Sir, | Yours very faithfully | Ch. Darwin

P.S. I have just heard that the Editor of the Annals & Mag. of N. History (which is a good periodical) will very gladly publish your "interesting paper".— in the September number.[8]

Courtesy of Eilo Hildebrand (photocopy)

1 The year is established by the reference to Hildebrand 1863b.
2 CD refers to an English abstract of Hildebrand 1863a. See letter from Friedrich Hildebrand, 16 July 1863 and n. 2.
3 See letter from Daniel Oliver, 20 July 1863. In his letter to Oliver of 18 July [1863], CD had asked whether the abstract of Hildebrand's paper might be published in the *Natural History Review*, of which Oliver was one of the botanical editors.
4 CD had sent the English abstract of Hildebrand's paper to *Annals and Magazine of Natural History* (see letter to J. D. Hooker, 22 July [1863] and n. 8). See also n. 8, below.
5 See letter from Friedrich Hildebrand, 16 July 1863 and n. 5.
6 See letter from Friedrich Hildebrand, 16 July 1863 and nn. 7 and 8.
7 'Dimorphic condition in *Primula*'. The reference has not been traced, but see also *Correspondence* vol. 10, letter to Asa Gray, 23[–4] July [1862]. CD cited the results of Hildebrand's experiments on *Primula* (Hildebrand 1864) in 'Three forms of *Lythrum salicaria*', p. 180 n. (*Collected papers* 2: 129–30, n. 9).
8 The letter has not been found. The English abstract of Hildebrand's paper appeared in the September issue of the *Annals and Magazine of Natural History* (Hildebrand 1863b). The editors were Prideaux John Selby, Charles Cardale Babington, John Edward Gray, and William Francis.

From Robert Swinhoe 29 July 1863

Oriental Club. | Hanover Square.
29 July, 1863.

My dear Sir,

I have just received your's of yesterday, and am very sorry to hear that you are again suffering, and that I cannot have the pleasure of meeting you again before I leave.[1] I trust however to see you at Newcastle.[2] I shall not neglect to send you copies of all the papers I have published in England in connection with Natural History & Formosa,[3] and if you are interested on Chinese topics, (which are normally distasteful to the British mind), I will present you with a copy of my work on the China war.[4]

You are doubtless aware of the cosimilarity of tendency to change of the Japanese as compared with the Chinese birds with the British birds as compared with those of the European continent. The causes are probably due to the influence on the respective atmospheres produced by the Gulf streams of the Pacific and the Atlantic.

Whatever the causes the facts are pretty patent. The differences in plumage of several Japanese birds from allied forms on the main has been often noted, but the strongest in point is in the case of the Wagtails. You are acquainted with the instance of the Grey-backed Wagtail (*Motacilla alba*, L) in England acquiring a black back—the *M. Yarellii* Gould.[5] In eastern Asia the same thing occurs in other forms of this group. The grey-backed wagtail of Eastern Siberia, Amoorland, and China, in Japan acquires a black back. The black-backed race was procured and noted by Pallas as a Kamtschatkan variety of *M. albeola* (Zoograph. Ross. Asiat. Vol I. p. 507),[6] and has been wrongly referred by modern authors to *M. lugubris* of western Siberia. In Formosa however where the influence of the gulf stream is strongly felt in the prevalence of rainy weather, the grey-backed is the resident form; though in winter the black-back also occurs, as it does during that season on the China Coast. There is a marked distinction between the two birds, sufficient to distinguish them at first sight; but in my large series of specimens I have one undoubted grey-back with marks here & there of black on its back.

The *Budytes* group in the east are affected in a similar way to what prevails in Europe. In China, south of Shanghai, the Yellow Wagtail moults in summer into the plumage of typical *B. flavas*, L. North of Shanghai it adopts the plumage of *B. cinereocapilla*. In India & Siam that of *B. melanocephala*. In Japan that of *B. rayi* of Great Britain. In Formosa similar to *B. rayi* but with brown instead of yellow cheeks.— In Java & the Archipelago with black cheeks—*B. melanotis*. Thermally considered the lines of latitude coincide of the areas of the respective forms in the east & west.— *Climate* in these cases must therefore have something, if not all, to do with the cosimilarity of change. But there are many other like subjects, which will doubtless prove of interest to you have been considered at some length in the different papers I have written, & the shortest way therefore to communicate them to you will be to send you the papers, which I will not fail to do, as soon as they come to hand.

Meanwhile I remain, with hopes for your speedy recovery to health & strength. | Your's ever truly, | Robert Swinhoe

P.S. I trust you will have no objection to send me a Photograph of yourself in return for mine which I now enclose.[7] I would esteem it as a great favour. Kindly sign your name at the foot of the *Carte*. R S.

DAR 47: 176–7

CD ANNOTATIONS
1.1 I have ... Newcastle. 1.3] *crossed pencil*
5.1 P.S. ... R S. 5.3] *crossed pencil*
Top of letter: 'Direct Action of Conditions very remarkable' *pencil*

[1] CD's letter has not been found; however, see the letter from Robert Swinhoe, 14 April 1863 and n. 5.

[2] In 1863, the annual meeting of the British Association for the Advancement of Science was held in Newcastle-upon-Tyne from 26 August to 2 September (*Report of the 33d meeting of the British Association*

for the Advancement of Science, held at Newcastle-upon-Tyne, p. l). Swinhoe gave a paper at the meeting entitled 'Ethnology of the island of Formosa' (see Swinhoe [1863b] and [1863c]).

3 Swinhoe was an amateur naturalist who had served with the British Foreign Office in the Far East since 1854, latterly as Vice-Consul of Formosa (now called Taiwan). He returned to London on leave in September 1862, bringing with him a large collection of specimens. During his leave, he lectured widely on his collections, publishing many articles before his return to Formosa late in 1863 or early in 1864 (P. B. Hall 1987 and *Royal Society catalogue of scientific papers*). Swinhoe issued reprints of many of his papers in 1863; these were apparently circulated in two bound collections, copies of which are in the Darwin Library and Darwin Pamphlet Collection–CUL, and in the British Library (*BLC*). There are copies of Swinhoe [1863b–g], bound in a publisher's case binding with the title on the spine 'Formosa | R. Swinhoe', in the Darwin Library–CUL; the volume is inscribed to CD, and is annotated (see *Marginalia* 1: 797). There are copies of Swinhoe [1863h–o], stitched into a single paper wrapper, in the Darwin Pamphlet Collection–CUL; the collection is inscribed to CD, and is annotated.

4 There is a presentation copy of Swinhoe 1861 in the Darwin Library–Down.

5 John Gould identified the British pied wagtail as a separate species, which he named *Motacilla yarelli* (Gould 1837); it is now regarded as a sub-species of the species found on the European continent, and is called *Motacilla alba yarelli* (Sibley and Monroe 1990–3).

6 Pallas 1811–31.

7 The enclosure has not been found.

To G. H. K. Thwaites 29 July [1863][1]

Down. | Bromley. | Kent. S.E.
July 29th

My dear Thwaites

I thank you sincerely for the seed of Limnanthemium, which I have planted, & if the plants will flower, they will be a treasure to me.[2]

The view on the conjugation of Algæ is familiar to me from a pamphet which you sent me many years ago, & which records a remarkable double embryo in Fuchsia to which I have lately had to refer.[3] This view of Algæ prefiguring the higher stages of reproduction has since reading your paper always seemed probable to me.—

I was glad to see the Cassia & to hear of the two forms of anthers in the same flower.[4] This class of gradation always seems to me very interesting. What a curious flower that of the Cassia. Mr Glennie ought to watch it & see how insects visit it; I should expect that the pistil would change its position during inflorescence.[5] You ask for my "carte de visite"; but I have never been done; I can only send you a rather poor small photograph made by my son.—[6]

With very sincere thanks & good wishes, | believe me | Yours very sincerely | Ch. Darwin

American Philosophical Society (295)

1 The year is established by the relationship between this letter and the letter from G. H. K. Thwaites, 8 June 1863.

2 Thwaites had sent CD seed of *Limnanthemum indicum*, a species with dimorphic flowers (see letter from G. H. K. Thwaites, 8 June 1863).

[3] CD refers to the letter from G. H. K. Thwaites, 8 June 1863, and to Thwaites 1848, which suggested parallels between reproduction in lower and higher plants. There is an annotated presentation copy of Thwaites 1848 in the Darwin Pamphlet Collection–CUL. CD cited Thwaites's account of a double embryo in *Fuchsia coccinea* (Thwaites 1848, pp. 163–4) in *Variation* 1: 391; he prepared a draft of the chapter of *Variation* in which this reference appears between 21 December 1862 and 23 January 1863 (see *Correspondence* vol. 10, and this volume, 'Journal' (Appendix II)).

[4] See the enclosure to the letter from G. H. K. Thwaites, 8 June 1863. In his letter of 8 June 1863 Thwaites sent CD a letter that he had received from Samuel Owen Glenie; Glenie's letter contained specimens and descriptions of stamens from a flower of *Cassia*. CD had already learned from Joseph Dalton Hooker that species of *Cassia* had two forms of anther in the same flower (see *Correspondence* vol. 10, letter from J. D. Hooker, [12 October 1862], and letter to J. D. Hooker, 14 [October 1862]).

[5] CD thought that the occurrence of differently coloured anthers in the same flower might be a guide to functional dimorphism, representing a gradation towards structural dimorphism (see *Correspondence* vol. 10, letter to J. D. Hooker, 22 [August 1862] and n. 14). Moreover, having noticed in some of the Melastomataceae that the pistils changed their positions as the flowers became older, he considered that this structural change might be accompanied by a functional change with respect to the pollen from the two sets of anthers (see, for instance, *Correspondence* vol. 10, letter to Asa Gray, 16 February [1862], and letter to J. D. Hooker, 15 [May 1862]).

[6] CD apparently refers to the photograph taken by William Erasmus Darwin in April 1861, which is reproduced as the frontispiece to *Correspondence* vol. 9 (see *ibid.*, letter to Asa Gray, 11 April [1861] and n. 19).

From L. E. Becker 31 July [1863][1]

Altham | Accrington
July 31st.

Dear Sir

After repeated and fruitless search I at last discovered one of my marked plants of Lychnis which had ripened its seeds—[2] I was so much surprised to see this solitary exception to the otherwise universal rule of sterility in the hermaphrodites that I suspected myself of error in marking it—but after soaking one of the withered flowers in water for a day or two I got out the remains of a stamen skin which I have enclosed and which seems to place beyond a doubt that in this one instance an hermaphrodite plant has ripened its seeds ⟨I hav⟩e kept a few of the ⟨seeds⟩ to try if they will ⟨ger⟩minate and enclose the rest of the plant just as I picked it. It has occurred to me that the failure of the hermaphrodites to produce seeds may be owing not solely to imperfection in the ovarium but to difficulty in getting fertised, its own pollen being destroyed. The stigmas are very minute, wholly concealed within the calyx tube, and thickly covered with the mass of diseased stamens, so unless visited by an insect of an unusually pushing disposition it seems impossible that they ever should be fertilised. I tried to effect this artificially in some of the plants in my garden and have not yet had time to be quite sure that no results will follow though as yet there is no difference perceptible between the flowers operated on and the others.

Your great kindness in sending me your pamphlet[3] must be my excuse for venturing to remark that I am watching with great interest a pot of scarlet Linum which I procured soon after reading it. The plants are blooming freely but all that have come out hitherto have been long styled flowers Some of the styles are much shorter than others but all present the same characters.

I have two other pots of seeds of my own sowing and perhaps some of these plants will shew the other form when they come into flower

I remain Sir, | with much respect | yours sincerely | L E Becker

DAR 160: 111

[1] The year is established by the relationship between this letter and the letter from L. E. Becker, 8 July [1863].

[2] See letter from L. E. Becker, 8 July [1863]. Becker reported her observations on *Lychnis diurna* in Becker 1869.

[3] 'Two forms in species of Linum'.

From Edward Bradford 31 July 1863

Roy⟨al Military College | Sandhurst⟩

Dear Sir.

The interest which I had ⟨derived⟩ from your recent work on the ferti⟨l⟩isation of Orchids[1] will I hope excuse my intrusion on your notice.

Some years since while stationed in Trinidad I gave much attention to plants of this family of which I collected and I may say domesticated about 80 species—[2]

One of the most common of these is the plant which is called Catasetum tridentatum— This I learn from your work you have been led to conclude is a male plant only—[3]

Now I am very confident that I never met with any flower having the forms called Monachanthus or Myanthus in the island But recollecting that in my collection there was a dried specimen of what I considered a ripe fruit of Catasetum tridentatum I wrote to my friend D^r. Ha⟨n⟩ce of the British Consulate in China ⟨into⟩ whose ⟨hands they⟩ had passed, and he has ⟨sent the⟩ enclosed sketch of it. You may perhaps think it worth your notice—[4]

I must admit that I did not see the flower of this specimen but seen in its dried state it seems to me to shew enough of the withered antennæ to prove it to be a fertile Catasetum.

I submitted about 2 years since to the Linnean Soc^y. a paper descriptive of a large number of West Indian orchids I was unfortunately unable sufficiently to revise it to fit it for publication—[5]

Had I when collecting possessed the light which your researches have cast on the subject, it might have been otherwise—

I remain Dear Sir | Very faithfully Yours | Edward Bradford.

31 July 1863.

Charles Darwin Esqr. F.R.S | &c &c &c

[Enclosure]

Extract from D^r. Hance's le⟨tter dated⟩ 2⟨ ⟩ Ap⟨ril⟩

"You are perfectly right about Catasetum tridentatum— The specimen in my herbarium labe⟨lled⟩ as gathered by you in Trinidad in August 1845 has a large ripe fruit on it quite full of sacciform seeds— The accompanying sketch is a faithful representation of the fruit in question".

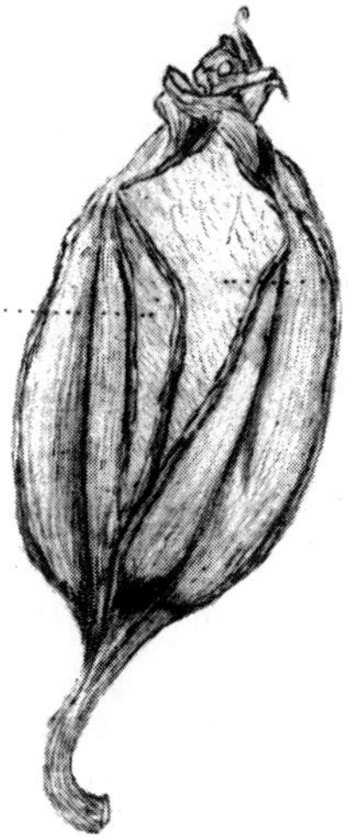

Fruit of Catasetum tridentatum[6]

DAR 160: 274, DAR 166: 95/1

CD ANNOTATIONS
1.1 The interest . . . species— 2.2] *crossed pencil*
4.1 Now . . . island 4.2] *double scored pencil*
4.1 confident] 'confident' *added pencil*
Top of letter: 'Catasetum' *blue crayon*
Verso of last page: '(Catasetum)' *pencil*

[1] *Orchids*.

[2] Bradford was an army surgeon; he collected plants, especially orchids, while in Trinidad during 1845 and 1846 (R. Desmond 1994).

[3] In *Orchids*, pp. 178–208, CD argued that the supposed species *Catasetum tridentatum*, *Monachanthus viridis*, and *Myanthus barbatus*, were in fact the male, female, and hermaphrodite forms respectively of a single species.

[4] See the enclosure. The reference is to Henry Fletcher Hance. See also letter from H. F. Hance, 10 May 1863.

[5] Bradford refers to his paper entitled 'Notes and descriptions of orchidaceous plants from Trinidad', which was read before the Linnean Society on 21 June 1860 (*Journal of the Proceedings of the Linnean Society* 5 (1861): xxxvii).

[6] The sketch is reproduced at 60 per cent of its original size.

From J. D. Hooker [31 July 1863][1]

Kew
Friday.

Dear Darwin

A small parcel of 3 or 4 tendrilliferous plants go to you today.[2] I wish there were more— with them is a flowering plant of the Lagerstrœmia with its 2 forms of stamen—[3]

All are tropical except the *Cissus antarctica*—[4] They go by Rail today & will no doubt arrive at Bromley tomorrow (Saturday).

I hoped to have gone to Down on tomorrow night for Sunday, but Boott is *very* ill, & must go & see him— he is I hope better but in a very hazardous state—the lungs again.[5]

When I go to you may I bring Willy with me? he will be no plague to any one & can amuse himself now.[6]

You will find a paper on Tendrils of Cucurbit. in Ann. Sc. Nat. by Naudin who I think proves them to be foliar in the opinion of most readers, but I have no opinion of my own[7]

Have you seen Decaisne on Pear vars.?—no great novelty but authoritative, & hence useful.—[8] He says that Larkspurs' *fertilize before Expansion*—[9]

Wollaston was here the other day & spent 2 days very pleasant, looking *dreadfully* ill.[10] he snubs Bates & *all that kind of work* (yours of course included) consumedly—like a good thoroughly consistent out & out uncompromising species-monger & typic of an *Entomologist*.[11]

Oliver goes to country tomorrow[12] my Father is away & I am "Head-clerk, Cook & bottle-washer" to the Establishment.[13]

I cannot get to Down for at least 3 weeks I fear.

Examinations come thick & fast for first 3 weeks of August[14]

Incomplete
DAR 101: 154–5

CD ANNOTATION
Top of letter: 'Catasetum tridentatum | Book' *ink*

[1] The date is established by the relationship between this letter and the letters to J. D. Hooker, 26 [July 1863] and 3 August [1863]; the intervening Friday was 31 July.

[2] See letter to J. D. Hooker, 14 July [1863], and letter from J. D. Hooker, [21 July 1863].

[3] See letter to J. D. Hooker, 26 [July 1863] and n. 10.

[4] There are experimental notes on *Cissus antarctica*, dated 14 August 1863, in DAR 157.2: 57.

[5] The botanist Francis Boott, who had long suffered from poor health, died on 25 December 1863; his cause of death was reportedly 'disease of the right lung, induced by pneumonia' (*DNB*).

[6] William Henslow Hooker.

[7] Naudin 1855. See letter to J. D. Hooker, 26 [July 1863].

[8] Decaisne 1863. Francis Darwin's manuscript index of CD's collection of offprints refers to an article by Decaisne on 'variation of fruit trees' (DAR 252); however, there is no offprint corresponding to this description in the Darwin Pamphlet Collection–CUL. See also n. 9, below. CD referred to Decaisne 1863 in *Variation* 1: 350.

[9] Decaisne 1863, pp. 10–11. CD disputed Joseph Decaisne's statements on this point in *Cross and self fertilisation*, p. 129 n., apparently giving the pagination of an offprint of the paper. See also n. 8, above.
[10] Thomas Vernon Wollaston.
[11] Like many entomologists, Wollaston was an outspoken critic of CD's theory (see Westwood 1860 and *Correspondence* vol. 8). Henry Walter Bates, by contrast, had endorsed natural selection, arguing that mimicry in Amazonian butterflies offered 'a most beautiful proof of the truth of the theory' (Bates 1861, p. 513). See also letter from J. D. Hooker, [13 May 1863] and n. 28.
[12] Daniel Oliver was assistant in the herbarium and librarian at the Royal Botanic Gardens, Kew (R. Desmond 1994, *List of the Linnean Society of London* 1863).
[13] Hooker was assistant director of the Royal Botanic Gardens, Kew; his father, William Jackson Hooker, was director.
[14] Hooker probably refers to examinations at the University of London, where he was examiner in botany (see *Correspondence* vol. 10, letter from J. D. Hooker, [24 July 1862] and n. 4). He was also examiner for the Society of Apothecaries' prize in botany, and a botanical examiner for medical officers in the armed services and the East India Company (L. Huxley ed. 1918, 1: 385–7 and 537).

From Henrietta Emma Darwin [1 August 1863][1]

A young cat of about the age of 10 months when put upon a soft cloak or gown used to take a piece in her mouth & then pound with her claws out & purr.[2]

Memorandum
DAR 189: 9

CD ANNOTATIONS
gown] 'generally' *interl after, ink*
purr.] '& close her eyes & look happy.— Explain the pounding.— No mistake.' *added ink* | '(& Purr)' *added pencil* | 'Aug. 1.— 1863 | Expression'[3] *added ink*

[1] The date is established by CD's annotations to the letter.
[2] Henrietta made a series of observations on behaviour in cats; there is a memorandum in DAR 189: 7, annotated 28 March 1862, in which Henrietta recorded her observations on the cries of cats and their meaning.
[3] CD evidently preserved these observations for use in *Expression*, pp. 126–9. *Expression* was published by John Murray in 1872.

To John Scott 1 and 3 August [1863][1]

Down. | *Bromley.* | *Kent. S.E.*
Aug 1st.

Dear Sir

I am rather overwhelmed with letters to answer, but I must thank you for the copy of your Orchid paper.—[2] I can with entire truth say it has *greatly* interested me.— Every naturalist who cares for anything besides naming species, will, I believe, be interested by it.— I am pleased by the notice you take of my papers; but you take almost too much notice of them.[3] My sole criticism is, as you know, that some parts might have been written in a[4] simpler style,[5] like your letters. I shall hereafter read your paper to Dr. Hooker,[6] so that he shall see it; but for about 3 weeks he is terribly overworked owing to the absence and illness of several of the officers at

Kew.[7] I congratulated you on the printing of your Paper and now must wish you good night for I am very tired.

Dear Sir | Yours faithfully | C. Darwin.

Aug 3^d. P.S.—[8] | I have kept back my note, owing to receipt of yours,[9] which as usual contained *much* to me very valuable matter.— I shall give all your new facts on sterility of orchids; as you have established the case far better than anyone.—[10]

Your argument on coloured primroses not being hybrids seems to me good, & ought to be just mentioned in your paper—[11] It is a good thought of yours to sow seed of the coloured primroses & so perhaps discover parentage. I begin to think more & more that perhaps you had better keep back this subject for separate paper.[12]

Many thanks for seed of the red Cowslip— it is a very remarkable case.—[13]

I suppose I understand rightly that the seed of white & red Primrose are self-fertilised seed—if I am wrong please tell me.—[14]

I am very glad that I sent the result of crosses of Primrose & Cowslip:[15] I doubted whether worth sending; as I thought the non result of the heteromorphic union of cowslip & primrose was mere accident.—[16] I wish I had made more crosses, but I was then so busy.—

I have no more evidence about L. Lewisii: Asa Gray considers it only as var.—[17] L. corymbiferum is dimorphic.— I was right about fertility of L. flavum.—[18] I *believe* I once saw a lot of L. monogynum & they were all same form. *Can seed* of L. *monogynum be purchased??* What you say about long-styled is curious.[19]

Dear Sir | Yours faithfully & obliged | C. Darwin

[Enclosure][20]

Possibly Dr Hildebrand of Bonn would notice your paper in Bot. Zeitung— if you have *spare* copy & like; I will forward one to him—[21]

DAR 93: B24, B27–8, B70; DAR 147: 455

1 The year is established by the reference to Scott 1863a; see n. 2, below.

2 The reference is to Scott 1863a, which was published in the *Transactions of the Botanical Society* [*of Edinburgh*]. There is a copy of Scott 1863a in the Darwin Pamphlet Collection–CUL. Scott had originally expected that a full version of the paper would be published in the *Edinburgh New Philosophical Journal* in July (see letters from John Scott, 21 May [1863] and 23 July [1863]); however, the journal only published an abstract.

3 In Scott 1863a, pp. 544–5, Scott prefaced the discussion of his experiments on *Oncidium* with a commentary on the theoretical points raised by *Origin* and 'Dimorphic species in *Primula*'. See also letter from John Scott, 21 May [1863] and nn. 10–11.

4 The original manuscript of the portion of the letter written on 1 August 1863 is missing from this point, and the text is taken from a copy made for Francis Darwin in DAR 147: 455. The copyist apparently mistranscribed 'read' for 'send', and 'congratulated' for 'congratulate'.

5 See *Correspondence* vol. 10, letter to John Scott, 11 December [1862].

6 Joseph Dalton Hooker.

7 See letters from J. D. Hooker, [21 July 1863] and [31 July 1863]. Hooker was assistant director at the Royal Botanic Gardens, Kew (R. Desmond 1994).

8 The text from this point is transcribed from the original manuscript in DAR 93: B27–8.

9 Letter from John Scott, [26 July – 2 August 1863].

10 CD cited Scott 1863a on the self-sterility of species of *Oncidium* and species of *Maxillaria* in *Variation* 2: 133, *Cross and self fertilisation*, pp. 330–1, and *Orchids* 2d ed., p. 289.

11 See letter from John Scott, [26 July – 2 August 1863]. CD refers to Scott's argument against *Primula vulgaris* var. *rubra* and *P. vulgaris* var. *alba* being hybrids of the cowslip and primrose (*P. veris* and *P. vulgaris*). Scott rejected this suggestion on the grounds that while he found that both the purple and white varieties of *P. vulgaris* were perfectly fertile *inter se*, crossing with the common yellow primrose resulted in decreased fertility; he argued that the purple and white varieties of *P. vulgaris* were therefore modified descendants of the yellow primrose. See also letter from John Scott, 23 July [1863], and Scott 1864a, pp. 98–9 n. In his letter of 25 [July 1863] CD urged Scott to repeat the experiments. CD discussed Scott's experiments with differently coloured varieties of primrose in *Variation* 2: 109, n. 25, 'Illegitimate offspring of dimorphic and trimorphic plants', p. 420, and *Forms of flowers*, pp. 224–5. CD reported that he had been unable to corroborate Scott's results.

12 Scott published his observations on the Primulaceae in Scott 1864a; there is an annotated copy of the paper in the Darwin Pamphlet Collection–CUL.

13 Scott discussed the equal-stamened and -styled cowslip in his letter of 21 May [1863], and enclosed the seed with his letter of [26 July – 2 August 1863]. CD discussed Scott's crossing experiments with a red equal-styled *Primula veris* in *Variation*, 2: 109 n., 'Illegitimate offspring of dimorphic and trimorphic plants', pp. 426–7, and *Forms of flowers*, pp. 234–5. CD's subsequent experiments with the seedlings confirmed the high degree of self-fertility of this form of *P. veris*, but did not corroborate Scott's finding that the form exhibited a high degree of sterility when crossed with other forms. See also n. 11, above.

14 See letter from John Scott, [26 July – 2 August 1863]. Scott's reply has not been found, but in *Variation* 2: 21, CD stated that the seeds of the purple (red) variety came from self-pollinated plants. See also *Correspondence* vol. 12, letter from John Scott, 16 May [1864].

15 CD sent the results of his crosses of *Primula vulgaris* and *P. veris* with the letter to John Scott, 25 [July 1863], and Scott published them in his paper on the Primulaceae (Scott 1864a, p. 103). CD undertook a series of crossing experiments with primroses and cowslips from April 1862 to 1867 (see DAR 157a, pp. 76–7, and DAR 108). Some of the results of the heteromorphic crosses made before August 1863 are recorded in DAR 108: 67, 69–70. CD published his conclusions in 'Specific difference in *Primula*', pp. 439–40.

16 Scott's experiments confirmed the observation that heteromorphic crosses between *Primula vulgaris* and *P. veris* were infertile (see letter from John Scott, [26 July – 2 August 1863], and n. 15, above).

17 CD was seeking confirmation of Jules Emile Planchon's statement that *Linum lewisii* was trimorphic (Planchon 1847–8, p. 175). CD cited this statement in 'Two forms in species of *Linum*', pp. 82–3 (*Collected papers* 2: 104–5). The occurrence in *L. lewisii* of three forms of flower (short-styled, long-styled, and equal-styled) all appearing on the same individual plant was regarded by CD as 'unique', and he was seeking to obtain specimens for further experiments (see letters to Asa Gray, 2 January [1863], and 19 January [1863], letter to John Scott, 6 June [1863], and letter from John Scott, [26 July – 2 August 1863]). Asa Gray thought that *L. lewisii* was merely a variety of *L. perenne* (see letter from Asa Gray, 27 January 1863). See also *Forms of flowers*, p. 101.

18 Apparently a reference to the reciprocal infertility of the two forms of *Linum flavum* (see 'Two forms in species of *Linum*', p. 81 (*Collected papers* 2: 104)). See also *Correspondence* vol. 10, letter to Daniel Oliver, 13 October [1862] and n. 3.

19 Scott reported a seed-forming long-styled *L. monogynum* when no short-styled form was nearby (see letter from John Scott, [26 July – 2 August 1863]).

20 The enclosure has been attached to this letter on the basis of its relationship to the letter from John Scott, 21 September [1863] (see n. 21, below).

21 Friedrich Hildebrand was a lecturer in botany at the University of Bonn (Correns 1916, Junker 1989). He was one of the first botanists in Germany to publish in support of CD's theory (Hildebrand 1861);

in 1862 he wrote to CD offering to help produce a German edition of *Orchids* (see *Correspondence* vol. 10, letter from Friedrich Hildebrand, 14 July 1862). Hildebrand regularly contributed to *Botanische Zeitung*, a weekly botanical magazine published in Leipzig (Correns 1916, *Royal Society catalogue of scientific papers*). In his letter of 21 September [1863], Scott stated that he had sent two copies of Scott 1863a 'a few weeks ago'; however, the paper was not reviewed in *Botanische Zeitung*.

To L. E. Becker 2 August [1863][1]

Down. | *Bromley.* | *Kent. S.E.*
Augt 2^{d}

Dear Madam.

I am very much obliged for the Lychnis seed, which shall be sown immediately;[2] & I will try, as you have done, next year to fertilise the little ovaria.—

It is odd about the Linum; I have never failed to find the two forms, though they certainly resemble each other closely.—[3]

In Haste, | Dear Madam | Your sincerely obliged | Ch. Darwin

Charterhouse Archives (ACC 013)

1 The year is established by the relationship between this letter and the letter from L. E. Becker, 31 July [1863].

2 Becker enclosed a hermaphrodite plant of the usually dioecious *Lychnis diurna* (*L. dioica*), bearing a seed capsule, with her letter of 31 July [1863].

3 Becker reported having a pot of scarlet *Linum*, all the seedlings of which had the long-styled flower form (see letter from L. E. Becker, 31 July [1863]).

To J. D. Hooker 3 August [1863]

Down
Aug 3^{d}

My dear Hooker

The plants arrived on Saturday.[1] The whole lot (far more than 3 or 4) are really priceless treasures; I hardly know which I am most curious about. It is very odd that gardeners sh^{d} not have noticed the movement of the shoots of simple climbing plants.[2] The Apocyneæ, which you gave me go on swinging great shoots, 18 inches long, round & round day & night. What a curious rhythmical contraction or expansion there must be; for there is no twisting of the shoot.[3] The common garden pea, both shoot & tendrils are in constant movement.[4] What a mistake the old one was that animals moved & plants did not.— I long to look at pollen of Lagerstrœmia;[5] I have just finished crossing, marking &c &c &c. 134 flowers of Lythrum; so I will work out that case well, with the 96 crosses made last year.[6]

How desperately hard worked you must now be. Do not come here till you can do it with smooth conscience & by all means bring Willy.[7] It is, however, probable we may be away for about a fortnight towards end of this month.—[8] I shall so like to see you.—

I am very sorry to hear about kind-heart Dr Boott.[9]

I fear I am an incubus to you; for I enclose 3 questions, which **at any time** I sh^{d}. be grateful for answer—

Ever yours affect^{y}.— | C. Darwin

Reference to Decaisne on Pear-trees: is in it that he mentions Larkspurs?[10]

I have "Henslow's Botany"[11];— "Asa Gray's Lessons."[12]; "Aug. St. Hilaires Leçons de Botanique":[13] can you name any good book with miscellaneous information for me to get. Did not Henfrey translate some German Book on Bot. Phys.—[14]

I have had lots of letters on Catasetum tridentatum seeding:[15] I have good reason to suspect there are two species under this name. Hence it w^{d}. be important for me to examine same plant that I had before. Can you lend me a plant? or is this too unreasonable? I wish to try to fertilise it.—

Endorsement: '/63'
DAR 115: 201

1 See letter from J. D. Hooker, [31 July 1863]; Hooker wrote that he was sending '3 or 4 tendrilliferous plants'.

2 CD had begun making observations on the movements of climbing plants (see letters to J. D. Hooker, 25 [June 1863], 1 July [1863], and 22 July [1863], and letter to Asa Gray, 26 June [1863]). A long paper by CD, 'Climbing plants', was read before the Linnean Society on 2 February 1865 and published in a double number of the society's journal in June. Longmans and Williams & Norgate also ran off-prints of the paper from the type set and issued *Climbing plants* as a commercial venture (see Freeman 1977, pp. 116–18).

3 CD included measurements of stem movement for *Dipladenia crassinoda* and *D. urophylla* (Apocynaceae) in 'Climbing plants', p. 16.

4 CD discussed the movements of *Pisum sativum* in 'Climbing plants', pp. 65–7. His experimental notes on this species are in DAR 157.2: 15–20.

5 Hooker had sent CD a dimorphic specimen of a flowering *Lagerstroemia* (see letter from J. D. Hooker, [31 July 1863]). See also letter to J. D. Hooker, 15 and 22 May [1863], and letter from J. D. Hooker, [24 May 1863].

6 CD became interested in *Lythrum* after reading that *L. salicaria* and *L. thymifolia* were trimorphic (see *Correspondence* vol. 9, letter to J. D. Hooker, 18 [December 1861]). He began experiments on *L. salicaria* at the end of July 1862 and, after making the 94, not 96, crosses referred to, concluded that he would have to make a further 106 crosses before he could publish his results (see *Correspondence* vol. 10, letters to Asa Gray, 9 August [1862] and [3–]4 September [1862], and letter to J. D. Hooker, 27 [October 1862] and n. 12). See also this volume, letter to J. D. Hooker, 26 [July 1863] and n. 11. CD's experimental notes on *Lythrum*, dated 1862–4, are in DAR 27: B1–57, and DAR 109: B36–51. CD's paper on *Lythrum*, 'Three forms of *Lythrum salicaria*' was read before the Linnean Society on 16 June 1864.

7 See letter from J. D. Hooker, [31 July 1863]; the reference is to Hooker's eldest son, William Henslow Hooker. There is no evidence that Hooker visited Down later in 1863.

8 CD was expecting to visit a hydropathic establishment in Malvern, Worcestershire, if his ill health continued (see letter to W. D. Fox, 23 May [1863], and letter to Charles Lyell, 14 August [1863]). According to Emma Darwin's diary (DAR 242), CD was at Malvern Wells between 3 September and 12 or 13 October 1863. See also 'Journal' (Appendix II).

9 See letter from J. D. Hooker, [31 July 1863]. Francis Boott died of a lung disease on 25 December 1863 (*DNB*).

10 See letter from J. D. Hooker, [31 July 1863] and n. 9. In Decaisne 1863, pp. 10–11, Decaisne claimed that the flowers of *Delphinium* (larkspur) self-pollinate in the bud and do not normally intercross.

11 The reference is to John Stevens Henslow's *Descriptive and physiological botany* (J. S. Henslow 1837). CD's annotated copy is in the Darwin Library–CUL (see *Marginalia* 1: 369–71).

12 CD refers to Asa Gray's *First lessons in botany and vegetable physiology* (A. Gray 1857). CD's annotated copy is in the Darwin Library–CUL (see *Marginalia* 1: 347).

13 Auguste de Saint-Hilaire's *Leçons de botanique* (Saint-Hilaire 1841). CD's annotated copy is in the Darwin Library–CUL (see *Marginalia* 1: 733–8).

14 The reference appears to be to *Principles of the anatomy and physiology of the vegetable cell* (Henfrey trans. 1852), Arthur Henfrey's translation of Hugo von Mohl's *Grundzüge der Anatomie und Physiologie der vegetabilischen Zelle* (Mohl 1851). There is an annotated copy of Henfrey trans. 1852 in the Darwin Library–CUL (see *Marginalia* 1: 589–90).

15 See letter from H. F. Hance, 10 May 1863 and nn. 5 and 7, and letter from Edward Bradford, 31 July 1863. For CD's discussion of the information that *Catasetum tridentatum*, though a male plant, occasionally produced seed-capsules, see 'Fertilization of orchids', p. 154 (*Collected papers* 2: 151), and *Orchids*, 2d ed., p. 197 n.

To Alphonse de Candolle 4 August [1863]

Down. | *Bromley.* | *Kent. S.E.*
Aug. 4th

My dear Sir

Will you permit me to beg you to interest any young Botanist at Geneva to observe one point for me. But the possibility rests on Lythrum Thymifolia or Hyssipifolia growing near you & being still in flower. Vaucher says (Hist. Phys. Tom. 2. p. 371)[1] that these two species are dimorphic like Primula. I have examined several dried specimens of L. hyssipifolia & can see no trace of this fact.[2] If either species present any difference in length of pistil & stamens, it is most likely that three forms should be presented. Could you get this point looked at for me; & if possible a few flowers of the two or three forms sent me in a letter.[3] I may add that in some species of Lythrum *two* of the forms have the pistil hidden within the calyx.

I have been trying most laborious experiments on L. salicaria, & am extremely anxious to hear about & see if possible, L. thymifolia.—[4]

I trust to your kindness to forgive me for troubling you & remain, my dear Sir, with sincere respect | Yours very faithfully | Ch. Darwin

Endorsement: '4 aout 63'
La Fondation Augustin de Candolle

1 Vaucher 1841, 2: 371. There is an annotated copy in the Darwin Library–CUL (see *Marginalia* 1: 812–5). For CD's interest in *Lythrum*, see preceding letter and n. 6.

2 CD's observational notes on *L. hyssopifolia* are in DAR 27: B41a–44 and DAR 109: B92. CD evidently thought that *L. hyssopifolia* growing wild in Switzerland might differ from the specimens he had examined.

3 Candolle's reply has not been found.

4 See preceding letter and n. 6. *Lythrum salicaria*, *L. thymifolia*, and *L. hyssopifolia* are discussed in 'Three forms of *Lythrum salicaria*'. CD concluded that *L. thymifolia* was non-dimorphic; Candolle was not cited ('Three forms of *Lythrum salicaria*', p. 191 (*Collected papers* 2: 124)).

To Asa Gray 4 August [1863][1]

Down Bromley Kent
Aug 4th

My dear Gray.

Thanks for two letters of July 7th & 21st,[2] full of interest to me. I am heartily glad to hear that you are so thoroughily refreshed by your short holiday; for you seemed to be almost worn out by work.[3] It is capital news about your Legacy;[4] I only wish it was five or ten times as much, & then you could give up all your time to new work. How goodnatured you are to my little man about stamps: those enclosed in two last letters were ineffable treasures.[5] Many thanks for Specularia;[6] but, good Heavens, I fancied it was a specimen to show me the flowers & I put it in warm, not boiling, water: I then discovered my mistake & hope I have not killed the seeds. Was not that a misfortune?— With your previous letter came a most obliging & kind note from Mr Brace;[7] if you have any communication with him, pray thank him; I will not write, as I had *previously* written to thank him for his kind present of his book.[8] What a pleasant book his was on Hungary.—[9]

After I had written to you Versus Heer's doctrine of sudden changes, I suspected what you would say:[10] what, I think, ought to give you the severest "cold chill" is the case of Pouter, Fantail-pigeons &c: were not these variations accidental as far as the purpose man has put them to?[11] Lyell said he would grapple with this, but I suspect he found it would be most prudent to shirk the question.[12] In my present book I have been comparing variation to the shapes of stones fallen from a cliff, & natural or artificial selection to the architect;[13] but I cannot at all work a metaphor like you do.— That seems a very pretty case of the orchid with prominence on labellum.—[14]

I have lots of Hobby-horses at present, fertility of peloric flowers & especially of "Homorphic" seedlings, which I suspect will throw much theoretical light on Hybrids.—[15] I have worked Lythrum like a Trojan & have just finished 134 crosses, no slight labour; but the case seems to me worth any labour, for I declare I think it about the oddest case of reproduction ever noticed.—a triple marriage between three hermaphrodite.—[16] I am so glad your Nesæas, after a sickly infancy are growing splendidly![17] But my present chief Hobby-horse I owe to you, viz tendrils;[18] their irritability is beautiful, as beautiful in all its modifications as anything in orchids. About the *spontaneous* movement (independent of touch) of the tendrils & upper internodes I am rather taken aback by your saying "is it not well known?".[19] I can find nothing in any book which I have: neither Hooker nor Oliver knew anything of these movements.[20] The spontaneous movement of the tendrils is independent of the movement of the movement of the upper internodes, but both work harmoniously together in sweeping a circle for the tendrils to grasp a stick. So with all climbing plants (without tendrils) as yet examined, the upper internodes go on night & day sweeping a circle in one fixed direction. It is surprising to watch the Apocyneæ with shoots 18 long, beyond the supporting stick, steadily searching for something to climb up. When the shoot meets a stick, the motion at that point is arrested, but in the upper part is continued, so that the climbing of all plants yet examined is the simple result of the spontaneous circulatory movement of the upper internodes.— Pray tell me whether anything has been published on this subject: I hate publishing what is old; but I shall hardly regret my work, if it is old, as it has much amused me.— If I do publish my paper, it will be too long to send you; for I am (or shall) examine very many plants.—[21]

I am very glad you are going to review Bates' paper & I must say I enjoy anything which riles Agassiz.[22] He seems to grow bigoted with increasing years; I once saw him years ago, & was charmed with him.[23]

Depend on it, you are unjust on the merits of my beloved Drosera:[24] it is a wonderful plant, or rather a most sagacious animal. I will stick up for Drosera to the day of my death. Heaven knows whether I shall ever publish my pile of experiments on it.—[25]

How profoundly interesting American new is— I declare almost more so even to us than Crimean new.—[26] Do not hate poor old England too much. Anyhow she is the mother of fine children all over the world. I declare no man could have tried to wish more sincerely for the north that I have done.— My reason tells me that perhaps it would be best,—of course best if it would end Slavery, but I cannot pump up enthusiasm. The boasting of your newspapers & of your little men, & the abuse of England, and the treatment of the free coloured population, and the not freeing Maryland slaves stops all my enthusiasm.[27] If all the States were like New England the case would be different.— I find a man cannot hope by intention. You will think me a wretched outcast. Farewell & do not hate me much. What devils the low Irish have proved themselves in New York.[28] If you conquer the South you will have an Ireland fastened to your tail.—

Good night & Farewell | C. Darwin

Gray Herbarium of Harvard University (83)

1 The year is established by the relationship between this letter and the letters from Asa Gray, 7 July 1863 and 21 July 1863.

2 Letters from Asa Gray, 7 July 1863 and 21 July 1863.

3 See letter from Asa Gray, 21 July 1863.

4 See letter from Asa Gray, 7 July 1863 and n. 8. Gray's wife, Jane Loring Gray, had inherited £2,000.

5 Letters from Asa Gray, 7 July 1863 and 21 July 1863. Gray had enclosed United States stamps for Leonard Darwin, an avid stamp collector.

6 See letter from Asa Gray, 7 July 1863 and n. 13.

7 The letter from Charles Loring Brace has not been found. Brace was a nephew of Gray's wife, Jane Loring Gray (Dupree 1959, p. 192).

8 CD refers to Brace 1863 and the letter to C. L. Brace, 24 June [1863].

9 Brace 1852. CD read Brace 1852 in December 1854 (*Correspondence* vol. 4, Appendix IV, 128: 9).

10 See letter to Asa Gray, 31 May [1863] and n. 4, and letter from Asa Gray, 7 July 1863 and n. 7; CD refers to Oswald Heer and his theory that sudden changes and the production of monstrosities were causal factors in the emergence of new species.

[11] CD had discussed variation and selection in domestic pigeons in *Origin*, pp. 20–29, and provided an extended discussion of this subject in *Variation* (*Variation* 1: 131–224). He selected this example because the accidental variations in pigeons had been 'extraordinarily great' (*Variation* 1: 131).

[12] See *Correspondence* vol. 9, letters to Charles Lyell, [1 August 1861], 13 [August 1861], and 21 August [1861]; Charles Lyell's replies, which may have included the statement referred to, have not been found. CD had evidently hoped that Lyell would discuss the role of accidental variation versus providential design in *Antiquity of man* (C. Lyell 1863a). On Lyell's unwillingness to commit himself fully to a belief in CD's theory of the origin of species, see Bartholomew 1973.

[13] CD was writing a draft of *Variation*, which formed the first published part of a planned three-part work on species (*Variation* 1: 3–14; see also *Correspondence* vol. 10, letter from Edouard Claparède, 6 September 1862, n. 9); the architect simile appears in *Variation* 2: 248–9 and 430.

[14] See letter from Asa Gray, 7 July 1863; the reference is to *Platanthera flava*.

[15] CD discussed the fertility of peloric flowers in *Variation* 2: 166–7. 'Homomorphic seedlings' were seedlings produced by artificially pollinating dimorphic or trimorphic plants with pollen taken from stamens not corresponding in height with the pistil. CD termed the result an 'illegitimate union', since it was the product of an only partially fertile cross. CD believed that there was the 'closest identity' in character and behaviour between illegitimate plants and hybrids (see *Variation* 2: 181–5). CD published a full account of his experiments with homomorphic seedlings in 'Illegitimate offspring of dimorphic and trimorphic plants'.

[16] In *Lythrum salicaria* three distinct forms occur: short-, mid-, and long-styled. Each of these forms is an hermaphrodite, each is distinct from the other two forms in its pistils, and each form has two sets of stamens differing from each other in appearance, and at different levels in the flower from the style. In all there are three kinds of stamen, short-, mid-, and long-length. There are therefore 18 possible combinations of crosses of pollen and stigmas. CD thought that these crosses should be repeated 20 times to guarantee reliable results, but completed his experiments having made 233 crosses. ('Three forms of *Lythrum salicaria*', pp. 169–76, 179–80 (*Collected papers* 2: 106–7, 113–14).) See also letter to J. D. Hooker, 3 August [1863] and n. 6.

[17] Gray sent specimens of *Nesaea verticillata* with his letter of 27 October 1862 (*Correspondence* vol. 10). Like *Lythrum salicaria*, *Nesaea verticillata*, also a member of the Lythraceae, presents three forms ('Three forms of *Lythrum salicaria*', p. 191 (*Collected papers* 2: 125)).

[18] CD developed an interest in tendrils after reading A. Gray 1858b, and began making observations on tendrils and plant movements by observing a seedling of *Echinocystis lobata* grown from a seed sent to him by Gray (see letter to J. D. Hooker, 25 [June 1863] and nn. 2 and 3, and letter to Asa Gray, 26 June [1863]).

[19] See letter from Asa Gray, 21 July 1863.

[20] See letter from J. D. Hooker, [2]9 June 1863. CD refers to Joseph Dalton Hooker and Daniel Oliver at the Royal Botanic Gardens, Kew.

[21] CD reported observations on over 100 plant species in 'Climbing plants', which was read before the Linnean Society on 2 February 1865.

[22] See letter from Asa Gray, 21 July 1863. Gray reviewed Bates 1861 in the issue of the *American Journal of Science and Arts* for September 1863 (A. Gray 1863a). CD refers to Louis Agassiz, an opponent of the theory of natural selection and Gray's colleague at Harvard University (*DSB*).

[23] CD may be referring to the 1846 meeting of the British Association for the Advancement of Science at Southampton, which he attended, and during which Agassiz spoke to a meeting of the Ray Society (see *Correspondence* vol. 4, letter to J. E. Gray, 18 December 1847, n. 5).

[24] See letter from Asa Gray, 21 July 1863.

[25] Since 1860, CD had carried out numerous experiments on the sundew, *Drosera rotundifolia* (see *Correspondence* vols. 8–10). *Insectivorous plants*, in which CD reported his findings on the species, was not published until 1875.

[26] CD refers to the reporting of the American Civil War: the English newspapers were predicting that Washington, D.C., was about to fall following the invasion of Maryland and Pennsylvania by the

Confederate army (see letter from Asa Gray, 21 July 1863). CD also probably refers to the recent revival of interest in the history of the Crimean War (1853–6) following the publication of the first volume of Alexander William Kinglake's account of the invasion of the Crimea (Kinglake 1863–87). Kinglake's book was described as 'probably the most notable book of 1863', and rapidly passed through four editions (*Annual register* 1863, 1: 347) after receiving widespread newspaper coverage.

27 President Abraham Lincoln's Emancipation Proclamation, which came into effect on 1 January 1863, freed the slaves only in states rebelling against the Union. Since Maryland had not seceded from the Union, slaves were not freed in that state (Denney 1992, pp. 248, 251, 284–7).

28 CD refers to the prominent role played by Irish immigrants in the riots in New York City in July 1863 against the drafting of men into the Union army; the riots resulted in the deaths of at least 105 people (McPherson 1988, pp. 609–10).

From Julius von Haast 6 August 1863[1]

Christchurch Canterbury NZ.
Aug 6. 1863.

My dear Dr Darwin!

Our mutual friend Dr Hooker tells me, that he could not deliver to you my enclosed letter of December,[2] because it miscarried in some way, fortunately I sent a copy of it in March, so that you will now be aware of its contents.[3] A few days ago I had a letter from Mr. Petschler of Auckland, who thinks, he will very soon be able to obtain the frog in question.[4] I had the pleasure of sending you by the bookpost a map of the Province of Canterbury, the first product of our new lithographic-press.[5] I have marked in different colours my various journeys; from the △ upwards the country was unexplored and unsurveyed so that I had to construct also the topographical map.

I also marked Dr Hector's (the Otago geologist) line of journey,[6] who in crossing the Alps near the boundary of both Provinces, came upon a river flowing in a northerly direction, and thus came into our Province. When within 8 or ten miles of the sea he had to return, from want of Provisions. Another journey is also marked, that of the late Mr Whitcombe a very accomplished surveyor,[7] who was drowned, begin of Mai, at the mouth of the R. Teramakau (Westcoast). Unfortunately his field book was lost and the only thing found on the body, was the map on which the course of the Okitiki was traced, which has been copied in this map.

In looking over the map, you will without doubt observe, that your name adorns one of our snowy giants.[8] It was the only feeble tribute which one of your zealous adherents could offer you.—

Believe me ever | most faithfully yours | Julius Haast.

DAR 166: 5

1 This letter was evidently posted with the letter from Julius von Haast, 21 July [– 7? August] 1863 (see letter from Julius von Haast, 21 July [– 7? August] 1863, n. 1).

2 Letter from Julius von Haast, 9 December 1862 (*Correspondence* vol. 10). See letter from J. D. Hooker, 19 June 1863, and letter from Julius von Haast, 5 March 1863.

3 See letter from Julius von Haast, 5 March 1863.

[4] At CD's request, Haast had written to Charles Petschler, asking him to obtain a specimen of the native New Zealand frog (see *Correspondence* vol. 10, letter from Julius von Haast, 9 December 1862); Charles Petschler was a merchant with premises at Vulcan Lane, Auckland, New Zealand (*Chapman's New Zealand Almanac* 1862). On CD's interest in the native New Zealand frog, see *Correspondence* vol. 10, letter to J. D. Hooker, 21 [September 1862] and n. 7.

[5] The map has not been found.

[6] James Hector served as the provincial geologist for Otago, New Zealand (*DNZB*). Haast refers to Hector's expedition along the Matukituki River in February and March 1863. CD had received a newspaper report of the expedition through Hooker (see letter from J. D. Hooker, 19 June 1863 and n. 3).

[7] John Henry Whitcombe was the provincial surveyor for Canterbury, New Zealand (*DNZB*).

[8] Mount Darwin is in the Southern Alps of New Zealand (J. F. J. von Haast 1862b, p. 127).

From Hermann Crüger 8 August 1863

Trinidad
8 August | 63.

Dear Sir

Your kind note of 23[d] May has duly come to hand as also the Journal of Researches & the Paper on Linum for which I beg to tender my best thanks.[1] Although my time has been of late much taken up with matters of business, I have at last sight of the researches you recommended me.[2] The experiment on Ficus is in progress & I shall report later on it.[3] With Melastomas I am not much more advanced but your supposition that insects visit them for nectar also is apparently correct.[4] The pollen-seekers are however by far the earliest risers & I think the act of fecundation is performed by them. All Melastomaceae I am acquainted with, or which I have noticed lately, are abundantly visited by insects, so much so that in the forest you often "hear" them in flower before you see them. I am unable to detect any dimorphism, but in the genus Spennera & Nepsera one can distinguish varieties, as in many Species of the arborescent Melastomaceae.

I have found lately a new instance of a plant which to all appearance cannot fertilize itself & which is abundantly provided with a substance attracting insects. This is Narantea, a remarkable Marcgraviacea, & very beautiful creeper, native of our woods. As we cultivate it here in the Garden I have it constantly under observation while it flowers.[5] The whole anthers, which separate from the filament, with the included pollen, or perhaps the pollen alone form a viscid extremely sticking mass, soluble with difficulty in water. The stigma reaches maturity only after the anthers. But the pitchers into which the Bracteae of this flower are converted contain a large quantity of syrup, clear as crystal, often 5–6 drops. This is only secreted when the flowers open, the Bracteae are empty up to that moment. Insects are very fond of this syrup & as the opening of the generally pendulous bract is towards the flower, they naturally alight on the latter & proceed from there to drink the syrup. Marcgravia has powdery pollen, & only a few pitchers to a whole umbel of flowers, the pitchers however contain nectar.[6]

I am curious to examine Ruyschia, but that is a rare plant here. It will perhaps be useless to look amongst these strongly characterized plants for immediate striking varieties, but a Nature which creates itself difficulties in order to overcome them as it were for sport, has been for me always a very sterile idea. The clue that your theory gives to these phenomena is much more satisfactory.[7] Have you directed your attention to the genus Mentha? I am afraid that the Mints generally are not easy to make experiments with, but I believe that Salvia also shows sometimes dimorphism.— With regard to sports in cultivated plants here I cannot agree with Schomburgk.[8] By neglect dahlias & such degenerate, ie double & fine flowers are no more produced, but otherwise I can see nothing particular. That plants vary here perhaps more than in temperate zones appears to be correct & in this respect I wish to draw your attention to the genus Capsicum. The innumerable varieties of the species, varieties which in their turn become at once true from seed, are something very astonishing. I should think it could be proved there, if one had the time, how quick a variety takes all the characters of a species.[9] As generations of plants succeed each other much quicker here, the process of transmutation must go on, caeteris paribus,[10] much more rapidly. If I had more time & means at my command I should very much like to experimentalize in these matters but it is impossible under present circumstances to do anything.—

Another subject to which I shall draw your attention, (although it may not be necessary) is that of the periods in which certain cultivated plants begin to vary. It is not my original idea, and is I believe to be found in older writers,[11] but I believe I should have been struck with it. You know that for instance the old Caladium bicolor has been cultivated in Europe for nearly a Century now. Yet all the varieties, as far as I can see, are of the last ten or 20 years. Some of them are supposed to be species, but where is the proof.

The most wonderful part of it is however, that here, where the plant has been cultivated for ornament & for superstitious purposes,[12] & where it has become wild, the varieties have made their appearance also, & some of the prettiest are to be found in uncultivated spots, where certainly seeds of the introduced ones have not reached. Of course I do not allude to the new Colocasias, & such, but only to the Caladiums. I believe that many more cases will be found in this matter, & not always to be explained by the increased attention people have paid to plants at certain periods.—.

The Catasetums which I am cultivating, begin to show flower stalks now, & in a month or so, I shall study the matter anew.[13] Of late I have often observed our larger Orchids Such as Stanhopeas, Gongoras, Coryanthes, & in all of them insects are certainly very busy, but in some the nectar-seeking ones are certainly not those which occasion fecundation. This is particularly the case, as far as my observation goes with Gongora & Coryanthes.[14] But then it must be admitted that the insect world of cultivated grounds like this garden is in many respects different from that of the high woods where these orchids prosper.

Besides there are whole tribes of nocturnal insects, which escape our observation, particularly those of the forests. We have here for instance a tribe of nocturnal or vespertini Wasps, which I believe visit flowers like many of their relations. It is extremely likely that many of these visit the stigmas of Orchids, for the sweet fluid found there. I think it is not sufficiently known or kept in mind, that the tissues which compose the stigma of many Orchids, are affected at the critical period in a manner analogous to that in ripening fruits, ie. that part of the cellulose & starch is used up to form sugar, by which process the cells cease at the same time to cohere, & form a sort of pulp, which is very attractive to insects. It is well known that pollen tubes are emitted best artificially in syrup.—

Excuse the length of this letter, I shall report soon on the Catasetum etc, & send you specimens in spirit.[15]

Very sincerely yours | H Crüger.

DAR 161: 277, 277/1

CD ANNOTATIONS
2.1 I have found . . . contain nectar. 2.14] *enclosed in square brackets, brown crayon*
2.3 This is Narantea . . . it flowers. 2.5] *scored brown crayon*
3.1 I am curious . . . vary. 4.2] *crossed pencil*
5.1 The most wonderful . . .] *after opening square bracket, brown crayon*
5.1 The most . . . periods.— 5.8] *crossed pencil*

CD note:[16]
Melastomas also *visited by [*interl*] nectar-seeking insects,—& swarm with insects. | Dichogamy [*brown crayon*] | Narantea (one of Marcgravaciæ) Bracts so placed to draw insects to flowers. Insects very busy about larger orchids, but not right for

1 In the letter to Crüger of 25 May [1863], CD indicated that he had sent a copy of *Journal of researches* (1860), and an off-print copy of 'Two forms in species of *Linum*'. For CD's presentation list for 'Two forms in species of *Linum*', see Appendix IV.

2 In a letter to Crüger of 18 March 1863, which has not been found, CD suggested a series of observations that Crüger might make on plants in Trinidad, including *Ficus*. See letter from Hermann Crüger, 23 April 1863.

3 See this volume, letter from Hermann Crüger, 23 April 1863, and *Correspondence* vol. 12, letter from Hermann Crüger, 21 January 1864.

4 See letter to Hermann Crüger, 25 January [1863] and nn. 2 and 6, and letter to Asa Gray, 10 January [1863].

5 Crüger was director of the botanic garden in Trinidad (R. Desmond 1994).

6 Crüger's observations on the excretion of a sweet fluid by the bracts of the Marcgraviaceae were reported in *Cross and self fertilisation*, pp. 404–5.

7 Like *Norantea*, *Ruyschia* was a member of the Marcgraviaceae. Crüger probably refers to CD's theory that floral structures were adapted to ensure occasional cross-pollination (see, for example, *Origin*, pp. 96–7, *Orchids*, p. 1, 'Two forms in species of *Linum*', pp. 82–3 (*Collected papers* 2: 105). For CD's comments on floral structures in the Marcgraviaceae, see *Cross and self fertilisation*, pp. 404–5.

8 In his letter to Crüger of 25 May [1863], CD had asked whether Crüger could confirm Robert Hermann Schomburgk's observation that plants such as dahlias and roses when cultivated in the tropics were likely to form sports (Schomburgk 1857, p. 132). CD was seeking information on the question of whether introduced plants from different climates were particularly apt to produce bud-variations (see also *Correspondence* vol. 10, letter to G. H. K. Thwaites, 29 December [1862]). CD eventually concluded that change of climate was not a necessary factor in the production of bud-variations (*Variation* 1: 408).

[9] CD omitted the genus *Capsicum* from his discussion of variation in cultivated plants because botanists were 'not agreed which kinds ought to rank as species and which as varieties', and the wild parent-species were unknown (*Variation* 1: 371).
[10] *Caeteris paribus*: 'other things being equal'.
[11] The reference has not been identified. CD took up this point briefly in *Variation*, under the heading: 'On the accumulative action of changed conditions of life' (*Variation* 2: 261–3). CD argued that while no effect was produced on a species until it had been exposed to cultivation or changed conditions for several generations, once a species had begun to vary, the greater was its tendency to vary further.
[12] No information has been found on the ritualistic use of *Caladium bicolor* in Trinidad. However, further south, the Amerindians of French Guiana used this species in shamanistic initiations and to propitiate spirits (Grenand *et al.* 1987, pp. 133, 135).
[13] See letters from Hermann Crüger, 23 February 1863 and n. 6, and 23 April 1863. Crüger reported his observations on *Catasetum* in his letter of 21 January 1864 (*Correspondence* vol. 12). See also Crüger 1864, pp. 127–9; 'Fertilization of orchids', p. 154 (*Collected papers* 2: 151); and *Orchids*, 2d ed., pp. 205–6.
[14] Crüger published his observations on the insect pollination of *Catasetum tridentatum*, *Coryanthes macrantha*, *Stanhopea grandiflora*, and *Gongora maculata* in Crüger 1864, pp. 127–31. CD cited Crüger's work in 'Fertilization of orchids', pp. 154, 157 (*Collected papers* 2: 151, 153–4), and *Orchids* 2d ed., pp. 168, 171, 175, 200.
[15] Crüger sent specimens of *Catasetum* with his letter of 21 January 1864 (*Correspondence* vol. 12).
[16] CD's note is on a separate piece of paper in DAR 161.2: 277.

To J. D. Hooker 12–13 August [1863][1]

Down
Aug. 13th.

My dear Hooker

Many thanks for Decaisne returned by this post.[2] I have been glad to see it, though, as you say, it contains nothing new.[3] I remember formerly he was much inclined to look at pears as descended from several aboriginal forms.[4] As for the Larkspurs, I shall be astonished if he prove right.[5] I will at once try them on a small scale. It will be marvellous to me that a flower shd. be so elaborately adapted to be sucked by bees, & not profit by its own structure.—[6]

I go on working at climbers & tendrils much to my own amusement & I have now on the table a Ceropegia which has caught firm hold of a stick, 26 inches distant from itself & centre of rotation.[7] It troubles me much not knowing what is known:[8] thus today I see that the peduncle of young leaves of Tropæolum Canariense are all very sensitive to touch, & as the top of shoot rotates, the peduncles are brought into contact with sticks & cling to them; which to me is new method of climbing.[9] Virginian creeper is the oddest of climbers; contact promptly causing a cellular cushion to grow out on the back & sides of the point, & this exudes a resinous cement, or at least a cement, which essential oils & no other reagent act on.[10] But good Lord, how I do maunder on to you about my hobby-horses. Your new plants are not yet recovered enough to observe.—[11]

I wrote the above yesterday & I now find out that Dutrochet has observed the spontaneous movements of the internodes of climbers & tendril-bearing plants.[12] So that I have thrown away some good time; but I hardly regret it, for I have had some good sport. Possibly I may have observed enough new to draw up short paper, but I must get his papers in the Comtes Rendus.[13]

I wish some Botanist, like Oliver, would publish a full book on Bot. Phys & miscellaneous matter, with lots of references. It would be awfully laborious, but I sh^d. think very useful.[14]

There is going to be a marriage in our family, between my sister, Catherine who is only a year younger than I am, & Mr Langton, a widower of my wifes sister.[15]

Good night: I hope you will soon be rather less busy— goodnight | Yours affect.— | C. Darwin

Remember the medallion of D^r. Darwin by Wedgwood, which you had for a cast; do not let it be lost[16]

Endorsement: '/63.'
DAR 115: 202

1 The year is confirmed by the relationship between this letter and the letter to J. D. Hooker, 3 August [1863]. CD changed the date from '12' to '13'.

2 Decaisne 1863. See letter to J. D. Hooker, 3 August [1863] and n. 10.

3 Letter from J. D. Hooker, [31 July 1863].

4 CD presumably refers to Decaisne 1858–75, which CD placed on his list of 'books to be read' between 1858 and 1860 (see *Correspondence* vol. 4, Appendix IV, *128: 169). CD discussed Joseph Decaisne's revised views on pear varieties (Decaisne 1863) in *Variation* 1: 350.

5 The reference is to Decaisne's statement that the flowers of *Delphinium* (larkspur) self-pollinated in the bud and did not normally intercross (Decaisne 1863, pp. 10–11).

6 In *Variation* 2: 21, CD reported that experimental crosses in annual larkspurs (*Delphinium consolida*) showed that insect pollination was necessary for this species to achieve its full fertility.

7 CD's experimental notes on *Ceropegia*, dated from 21 July to 10 August 1863, are in DAR 157.1: 10–17. His observations on *Ceropegia* were published in 'Climbing plants', pp. 4–5, 12–13, 16.

8 CD had unsuccessfully sought information from Hooker and Daniel Oliver on scientific literature relating to climbing plants (see letter to J. D. Hooker, 25 [June 1863], and letter from J. D. Hooker, [2]9 June 1863 and n. 3). CD was unaware of much of the research that had been done on this subject until receiving the letter from Asa Gray, 1 September 1863. See also n. 12, below.

9 This species was not discussed in 'Climbing plants'.

10 CD's experimental notes on the attachment of disks in Virginia creeper, *Parthenocissus quinquefolia*, (which CD called *Ampelopsis hederacea*), dated August [1863], are in DAR 157.2: 66–7. CD published his observations on this species in 'Climbing plants', pp. 84–7.

11 Hooker sent CD several 'tendrilliferous' plants in July (see letter from J. D. Hooker, [31 July 1863]).

12 CD refers to René Joachim Henri Dutrochet and to Dutrochet 1843 and 1844 ('Climbing plants', p. 2 n.).

13 Dutrochet 1843 and 1844 appeared in *Comptes rendus hebdomadaires des séances de l'Academie des Sciences*.

14 In addition to being an assistant in the herbarium and the librarian at the Royal Botanic Gardens, Kew, Daniel Oliver was professor of botany at University College, London (R. Desmond 1994, *List of the Linnean Society of London*). He also compiled botanical bibliographies for the *Natural History Review*. Oliver's knowledge of European botanical literature had prompted CD to describe him as 'omniscient' (*Correspondence* vol. 10, letter to J. D. Hooker, 11 June [1862]). Oliver did not publish an advanced text-book on physiological botany.

15 Emily Catherine Darwin, known as Catherine, was 53; on 8 October 1863 she married Charles Langton, widower of Charlotte Wedgwood (*Emma Darwin* (1915), 2: 180–1).

16 CD had lent Hooker a Wedgwood medallion of Erasmus Darwin (see letter from J. D. Hooker, [24 March 1863]).

To Charles Lyell 14 August [1863][1]

Down. | Bromley. | Kent. S.E.
Aug. 14th

My dear Lyell

I was very glad to get your letter & congratulate you on finding the Arctic shells: I looked long there in vain. It is really a grand case.—[2]

Hunt sent me a copy of his paper,[3] & I was much struck with it; but Dana made some good remarks on caution necessary.[4] By the way I fear poor Dana will never do much more work. I heard from him 2 or 3 months ago, & he said that he was forced to be very cautious in all mental exertion.[5]

I am very glad to hear that the Antiquity goes on selling so well;[6] but if it gets another edition of "Origin" I shall not be grateful; for I dread the very thought of that job.[7]

Have you seen Benthams remarks on Species in his address to Linn. Soc: they have pleased me more than anything I have read for some time.[8]

I have no news, for I have seen not a soul for months & have had a bad spring & summer, but have managed to do a good deal of work. Emma is threatening me to take me to Malvern & perhaps I shall be compelled, but it is a horrid waste of time.[9]

You must have enjoyed N. Wales, I shd. think; it is to me a most glorious country.[10]

I do not know whether you will have heard of the approaching marriage of my sister Catherine to Mr. Langton, who married Emma's sister: I believe that the marriage will answer well to both.—[11]

If you have not read Bates' book; I think it would interest you.[12] He is second only to Humboldt in describing a Tropical forest.[13] Talking of reading I have never yet got the Edinburgh, in which I suppose you are cut up.—[14]

With kind remembrance to Lady Lyell—[15] Ever yours | C. Darwin

P.S. I answered the Manchester Pigeon man for you.—[16]

Endorsement: '1863'
American Philosophical Society (296)

[1] The year is confirmed by the reference to the forthcoming marriage of Emily Catherine Darwin and Charles Langton; see n. 11, below.

[2] The letter from Lyell has not been found; however, see K. M. Lyell ed. 1881, 2: 379. In the letter to J. D. Hooker, 25 [August 1863], CD mentioned that Lyell had informed him that he had found 'Trimmers Arctic shells on Moel Tryfan'. Joshua Trimmer reported finding broken fragments of the shells of marine molluscs on Moel Tryfan, and argued that their presence indicated that they were deposited when the summit of Moel Tryfan was submerged beneath the sea (Trimmer 1831). CD had pursued fieldwork on Moel Tryfan in 1842 but had not found shells ('Ancient glaciers of Caernarvonshire', pp. 184–5 (*Collected papers* 1: 167)).

[3] CD refers to the American military engineer Edward Bissell Hunt, and his paper in the *American Journal of Science and Arts* on the Florida coral reef (Hunt 1863); this offprint is in the Darwin Pamphlet Collection–CUL.

[4] James Dwight Dana was one of the editors of the *American Journal of Science and Arts*; Dana appended long editorial footnotes to Hunt 1863, drawing attention to what he and CD had written on the

formation of coral reefs in Dana 1851–2 and 1853, and in *Coral reefs*. *Coral reefs* was published in 1842. See also *Correspondence*, vol. 2. Dana's main reservations related to Hunt's estimates of the chronology of coral reef formation, which he argued would be complicated by changes of sea-level during the post-Tertiary period (Hunt 1863, p. 209 n.).

5 CD appears to refer to correspondence that followed his letter to J. D. Dana of 20 February [1863]; no letters have been found. However, see letter from J. D. Dana, 5 February 1863 and n. 3.

6 The reference is to *Antiquity of man*; the first edition appeared in February 1863 (C. Lyell 1863a), the second in April (C. Lyell 1863b), and the third in November (C. Lyell 1863c). See Grayson 1985, p. 106.

7 The third edition of *Origin* was published in 1861, and the fourth in 1866 (Freeman 1977, pp. 85–6).

8 George Bentham delivered his presidential address to the Linnean Society on 25 May 1863 (Bentham 1863). See also letter to George Bentham, 19 June [1863] and n. 2.

9 CD intended to take a course of hydropathic treatment at Malvern, Worcestershire, if his health did not improve (see letter to W. D. Fox, 23 May [1863]). According to Emma Darwin's diary (DAR 242), CD was at Malvern Wells between 3 September and 12 or 13 October 1863.

10 CD visited north Wales on eight occasions between 1818 and 1842 (Freeman 1978).

11 On 8 October 1863, Emily Catherine Darwin married Charles Langton, widower of Charlotte Wedgwood (*Emma Darwin* (1915), 2: 180–1).

12 Henry Walter Bates's account of his eleven years as a naturalist in the Amazonian region of South America (Bates 1863) was published in April 1863 (*Publishers' Circular* 26 (1863): 193).

13 CD refers to Alexander von Humboldt and the descriptions of tropical forests in *Personal narrative of travels to the equinoctial regions of the new continent during the years 1799–1804* (Humboldt and Bonpland 1822). CD's annotated copy is in the Darwin Library–CUL. On the inspiration CD derived from reading Humboldt, see *Correspondence* vols. 1–3, and *Autobiography*, pp. 67–68.

14 The *Wellesley Index* identifies James David Forbes as the author of the review of *Antiquity of man* (C. Lyell 1863a) in the July issue of the *Edinburgh Review* ([J. D. Forbes] 1863).

15 Mary Elizabeth Lyell.

16 This individual has not been identified.

From Robert Swinhoe 14 August 1863

18 Royal Avenue Terrace. | Chelsea.
14 Aug[t]. 1863.

My dear Sir,

Many thanks for the *vignette* you have sent me. I do not think it quite does you justice.[1]

I have been lately spending a few days with an Uncle, a retired Indian officer, who has been a great hunter & sportsman in his day;[2] and in conversation with him I learnt two facts which I think will interest you. The first is that on one occasion he shot a female *Flamingo* in India, the legs of which were thickly covered with small barnacles (*Balanus*).

Now it is a known fact that the Flamingo sits on her nest astraddle, and probably in this instance the birds legs were left during incubation soaking in salt water during which time the barnacles had time to adhere to them.

The other fact is with reference to the Wild Ass (*Asinus* sp—) of Kutch.[3] There is a story prevailing among the natives that the old male *bites* off the *testes* of all the young males he can overpower. Sir W. Harris shot many wild asses while in Kutch,[4] and he several times met with males which had been gelded & with every appearance about the parts of the testes having been roughly torn away, which to his mind confirmed the stories of the natives. This fact my Uncle heard from Sir W. Harris himself, and I think it is trustworthy, though it sounds much like an exaggerated eastern tale—

I remain, | Your's very sincerely | Robert Swinhoe

Chas Darwin, Esqre

DAR 205.2 (Letters): 253

CD ANNOTATION

Top of letter: '18' *brown crayon*; '(good as queer case)' *pencil*

1 In the postscript to his letter of 29 July 1863, Robert Swinhoe requested a photograph of CD. The letter to Swinhoe enclosing CD's photograph has not been found. Presumably CD sent him a copy of the photograph taken by William Erasmus Darwin in 1861 (see, for example, letter to J. D. Dana, 7 January [1863]).

2 Probably Samuel Swinhoe, who retired from the Bengal Army in 1855 (*Army list* 1858).

3 Kutch is an area in the north of the province of Gujarat on the north-west coast of India (*Edinburgh world atlas*).

4 William Cornwallis Harris served as a major in the Bombay engineers (*DNB*).

To *Gardeners' Chronicle* [before 15 August 1863][1]

In a hard gravel walk close to my house, my gardener[2] and myself distinctly remember, about five or six years ago, two little rosettes of purplish leaves pushing their way up. We neither of us could imagine what they were; they were soon trampled down and apparently killed. But this spring they have re-appeared in exactly the same spot, and were protected. They have now flowered and prove to be Epipactis latifolia. This Orchid, though by no means a rare plant, I have never seen in this neighbourhood, and have heard only once of its having been found in a wood about a mile and a half distant. The gravel walk was made 20 years ago; and before that time the spot existed as a little-used carriage drive; and about 25 or 26 years ago it was a pasture field. How could this Epipactis, which is so rare a plant here, have come to this spot? The root stock seems to have lain dormant under the gravel for the last five or six years. Could a seed have been blown here from a distance and have germinated during some season when the walk was neglected? The tall stems growing up in the midst of the bare gravel surface present an odd appearance, and the case seems to me a singular one.[3]

Charles Darwin,

Down, Bromley, Kent.

Gardeners' Chronicle and Agricultural Gazette, 15 August 1863, p. 773

1 The letter was published in the issue of the *Gardeners' Chronicle* for 15 August 1863 (see also *Collected papers* 2: 83).

[2] Henry Lettington.

[3] CD published further observations on these plants in 'Fertilization of orchids', p. 150 (*Collected papers* 2: 146–7), and in *Orchids* 2d ed., pp. 101–2. There are observational notes on *Epipactis latifolia* in DAR 70: 56, dated 13 August 1863.

To Thomas Rivers 17 August [1863][1]

Down. | Bromley. | Kent. S.E.
Aug 17th

My dear Sir

The almond tree which you gave me has produced no fruit; but the Double Chinese Peach has 3 fruit.[2] Now, if you have any bearing Almond trees, would you have kindness, when fruit ripe to send me 2 or 3; & if you have any other **odd** peach 2 or 3 to compare the stones.—[3] The best address is

"C. Darwin"
"care of Down Postman"
per Rail "Bromley"
"Kent"

I presume you have no more facts for me to quote on inheritance of weeping trees; you said, if you had time, you could prove the weeping elm.—[4]

Now I am going to beg another favour. Have you yourself seen any kind of modification in fruit or foliage any fruit-tree from being grafted. Andrew Knight & others assert that slight differences do result;[5] others positively deny this. Can you give me any certain facts?[6]

Lastly I ask on mere improbable chance of the point having occurred to you: viz do you believe & can you any how account for the positive statements of several good pears (& several varieties of wheat) having been discovered as seedlings in hedges & woods. On mere doctrine of chances, this might perhaps have occasionally occurred; but the occurrence seems too frequent for mere chance. I have not yet collected such recorded cases, as I have met with, but I know that I have several & in a day or two shall collect them.[7]

Pray excuse me thus troubling you, & believe me, My dear Sir | Yours sincerely | Ch. Darwin

DAR 185: 85

[1] The year is established by the relationship between this letter and the letter from Thomas Rivers, 21 January 1863.

[2] See letter from Thomas Rivers, 21 January 1863, and letters to Thomas Rivers, 25 January [1863], 5 March [1863], and [9 May 1863].

[3] See letter to Thomas Rivers, [9 May 1863] and n. 4. CD acknowledged Rivers for providing specimens of peach and almond stones in *Variation* 1: 338–9.

[4] This comment may have been in the part of the letter from Thomas Rivers, [3 February 1863], that is now missing. CD cited observations made by Rivers on seedlings of varieties of weeping elm in

Variation 2: 19. See also letters to Thomas Rivers, [14 February 1863] and nn. 3 and 6, and 5 March [1863].

5 The reference is to Thomas Andrew Knight and to Knight 1815. CD discussed Knight's conclusions on the effects of grafting in *Variation* 1: 387, and 2: 278 n. 29.

6 No letter from Rivers in answer to this query has been found; however, Rivers evidently provided examples of cases in which grafting had produced alterations in the stock plants (see *Variation* 1: 394).

7 Examples of new varieties of fruit-trees and wheat occurring in woodland and wasteland, including one provided by Rivers, are given in *Variation* 2: 260–1. This information appeared in the section of *Variation* entitled 'Causes of variability' (*Variation* 2: 250–70); CD had completed the draft of the preceding section, entitled 'Selection', on 20 July 1863 (see 'Journal' (Appendix II)).

From W. E. Darwin 21 August [1863][1]

Southampton.
Au 21

My Dear Father,—

I have signed the six orders for Dividends & forwarded them,[2] also sent Aunt Catherines letter to Uncle Ras—[3]

Thanks for Mama's nice long letter[4] it has given me more news than I have had for a long time.

I am reading hard away at Mülder Organic Chemistry & find it very interesting tho' rather long.[5]

On examining the Lythrums they turned out to be nothing, tho' it was not a fair trial.[6] Mr. Waterhouse Hawkins gave a good lecture last week at the Hartley upon Unity of Vertebrate structure but no "origin" turned up[7]

Dr. Lankaster is going to lecture on Monday which I shall go and hear[8]

Your affect son | W E Darwin

I am glad to hear you are going to have Dr B—? down[9]

DAR 162: 95

CD ANNOTATIONS

1.1 I have ... long. 3.2] *crossed pencil*

5.1 Dr. Lankaster ... down 7.1] *crossed pencil*

1 The year is established by the relationship between this letter and the letter to W. E. Darwin, [25 July 1863].

2 These orders for dividends have not been traced. CD's dividend receipts during late August and September 1863 were principally from railway shares and are recorded in CD's Account book–banking accounts (Down House MS).

3 Emily Catherine Darwin (known as Catherine) and Erasmus Alvey Darwin were CD's sister and brother. Catherine Darwin's letter has not been found; the letter may have referred to her forthcoming marriage to CD's brother-in-law Charles Langton, and its postponement owing to her illness (see letter to J. D. Hooker, 25 [August 1863]).

4 The letter from Emma Darwin to W. E. Darwin has not been found.

5 The reference is to Mulder 1849.

6 In the letter to W. E. Darwin, [25 July 1863], CD asked William to make some observations on the mid-styled form of *Lythrum salicaria*.

[7] William refers to the artist, anatomist, and popular natural history author and lecturer, Benjamin Waterhouse Hawkins, and to the Hartley Institution, Southampton (*Men and women of the time* 1865, *Post Office directory of Hampshire, Dorsetshire, and Wiltshire* 1867). The Hartley Institution was a literary and scientific institution, opened in Southampton in October 1862 to 'promote the study and advancement of the sciences of Natural History, Astronomy, Antiquities, Classical and Oriental Literature in the town'. Between October 1862 and September 1863 use of the lecture hall was granted to various local societies (Patterson 1962, pp. 12, 21–2).

[8] Edwin Lankester, surgeon, physician, and professor of natural history at New College, London, was also a popular author and lecturer in natural history (*DNB*, *Medical directory* 1863).

[9] The reference is to the London physician William Brinton; Brinton attended CD in November and December 1863 (CD's Classed account books (Down House MS)). See also letter from George Busk, [*c.* 27 August 1863], and letters to J. D. Hooker, [13 November 1863] and 5 [December 1863]. Brinton was a specialist in stomach disorders at St Thomas's Hospital, London (*Medical directory* 1863).

From John Goodsir 21 August [1863][1]

South Cottage. Wardie— | Edinburgh
Aug^st. 21th—

Dear Sir.

As I am living at present in the country, your letter of the 18th reached me too late for an answer by return of post.[2]

I will most willingly examine the slide; or if not giving you too much trouble, a small quantity of the fluid with the flocculent & tenacious matter sent in a tube or small phial.[3]

The spherical bodies are probably the cells of *Torula*, or spores of *Penicillium*.[4] If *Sarcina* be present, it will be at once detected by its square form and peculiar segmentation. *Sarcina* and *Torula* often occur together.[5]

Sarcina has been the object of much observation in this country and on the continent. M^r. Busk has paid much attention to the subject.[6] D^r. Jenner,[7] I believe recommends the use of Hyposulphite of Soda in those cases of irritable stomach in which *Sarcina* occurs. If your medical adviser has no objection you might try Creosote. In the case in which *Sarcina* was first detected, one drop of Creosote was taken at bed-time, and afterwards two drops in the forenoon, and two drops at bedtime with complete success—

You will find this in full detail, with a chemical analysis of the fluid ejected from the stomach, in the Edin. Medical and Surgical Journal—April 1842—page 430—[8] I am sorry I have no separate copy to send to you. I will transmit, if you desire it, a list of references to authorities on the subject.

Here I must in⟨fo⟩rm you that the ⟨cur⟩rent opinion of ⟨t⟩he medical profession appears to be that these vegetable forms are not immediate agents in the morbid action. *Sarcina* has been found in the healthy stomach, and in other parts of the body. A perverted and weakened action of the stomach would appear to supply the conditions for the rapid multiplication of these vegetable forms. But if not the cause or source of your distress, they may, assuming them to be present, very much increase your discomfort—

I think therefore you do well to inquire into this matter, and to make use of such means, as without taxing your strength, may relieve your more urgent symptoms.

Accept of this expression of my sympathy, and believe me | very faithfully yours John Goodsir.

Charles Darwin Esq

DAR 165: 73

CD ANNOTATION

Top of letter: 'Mr W. mentioned Dr. Harley of U. College—'[9] *ink, circled ink*

[1] The year is established by the relationship between this letter and the letter to J. D. Hooker, 25 [August 1863].

[2] The letter to Goodsir has not been found.

[3] Goodsir was professor of anatomy at the University of Edinburgh; CD had evidently written regarding the cause of his vomiting (see also letters from John Goodsir, 27 August [1863] and 28 August [1863]). In the middle of August 1863, CD began vomiting every morning (see letter to J. D. Hooker, 25 [August 1863]). For a summary account of CD's health during 1863, see Bowlby 1990, pp. 370–4.

[4] *Torula* and *Penicillium* were both genera of fungi; in the early nineteenth century, *Torula* was used as the generic term for yeasts (Barnett *et al.* 1990, p. 20).

[5] *Sarcina* is a genus of acid-tolerant bacteria; *S. ventriculi* is common in human intestinal tracts. *Torula* was a genus of fungi, now represented by several modern genera (see n. 4, above).

[6] George Busk was a leading figure in the Royal College of Surgeons (Plarr 1930); while a naval surgeon on the hospital ship at Greenwich, he published a paper on the occurrence of *Sarcina* in the human stomach (Busk 1842).

[7] The London physician William Jenner was professor of medicine at University College and practised at the University College Hospital (*Medical directory* 1863); CD consulted Jenner in 1864 (CD's Classed account books (Down House MS)).

[8] Goodsir 1842.

[9] Apparently a reference to the University College Hospital physician George Harley, a liver specialist. Harley was also professor of medical jurisprudence and a lecturer on physiology and histology at University College, London (*Medical directory* 1863). 'Mr W.' may be the family dentist Alfred J. Woodhouse, whom George had visited on 18 August 1863 (Emma Darwin's diary (DAR 242)).

From Hugh Falconer 24 August [1863][1]

21 Park Crescent N.W.
24 Augt.

My Dear Darwin

I learn from your Brother that you are up to the eyes in tendrils[2]—and I think you were disposed to be rather angry with me about my *raid*, in the Spring.[3] But I am not going to let myself be dropt out of your acquaintance—on either account.

About this time last year, I sent you the M.S. of remarks on the persistence of specific characters in the Elephants—founded upon mammoth molar anterior to the Boulder clay.[4]

I have now got another still stranger case—out of the Pliocene Fauna of the "Forest Bed" of the Norfolk Coast—namely a genus described by Owen as extinct—yet the form in reality an *existing* species frequenting rivers in Russia.[5] It is an insectivore The fact was made out by Lartet—who was lately over on a visit to me,[6] and he communicated it to Owen. I carefully compared, the old & the new specimens. In my opinion they are of the *same species*—unchanged—absolutely identical.

Has Suess sent you his brochure—"Über die Verschiedenheit &c".?[7] He there formulates Heer and me for the following generalization p. 25.[8]

"That the time during which a new species is formed, is (as a rule) very short in comparison with the time during which it persistently presents the same peculiar specific characters".

The "Forest-Bed" Species gives wonderful Countenance to this heretical generalization. But your fertility of resource, in the way of explanation, will I have no doubt, make light work in demolishing it.[9]

My Dear Darwin | Yours Ever Truly | H Falconer

P.S. Lartet has not yet published his rectification.[10] I have therefore only mentioned the case to you in a general way. But if you desire it I will send you all particulars—for demolition hereafter.

DAR 164: 16

CD ANNOTATIONS
1.1 I learn . . . clay. 2.3] *crossed ink*
Top of letter: '*21*'[11] *brown crayon*

[1] The year is established by the relationship between this letter and the letter from Hugh Falconer, 29 August 1863.

[2] Erasmus Alvey Darwin. CD made numerous observations on the movements of plant tendrils during the summer of 1863 (see letters to J. D. Hooker, 25 [June 1863] and nn. 2 and 3, 1 July [1863], 14 July [1863], 3 August [1863], and 12–13 August [1863], and letters to Asa Gray, 26 June [1863], and 4 August [1863] and n. 18).

[3] Falconer refers to his letter concerning *Antiquity of man* (C. Lyell 1863a) in the *Athenæum*, 4 April 1863, pp. 459–60 (see letter to J. D. Hooker, [17 April 1863] and n. 3, and letter to Charles Lyell, 18 April [1863]).

[4] Falconer sent CD a draft of a portion of Falconer 1863a with his letter of 24–7 September [1862] (*Correspondence* vol. 10). See also letter to Hugh Falconer, 1 October [1862] (*ibid.*), and this volume, letter from Hugh Falconer, 3 January [1863] and nn. 15 and 16.

[5] The reference is to Richard Owen. The fossil species referred to has not been identified (see n. 9, below). The 'Forest Bed' of the Norfolk coast is described in C. Lyell 1863a, pp. 212–16.

[6] Edouard Lartet was a French lawyer who, since the 1850s, had devoted his time to palaeontological research (*DSB*).

[7] Falconer refers to the Austrian palaeontologist Eduard Suess and to Suess 1863.

[8] Falconer refers to the Swiss palaeobotanist, Oswald Heer. Falconer's copy of Suess's 'Über die Verschiedenheit und die Aufeinanderfolge der tertiären Landfaunen in der Niederung von Wien' ('On the diversity and the succession of Tertiary land faunas in the lowland of Vienna': Suess 1863) may have been a preprint copy; it was published in part 1 of volume 47 of *Sitzungsberichte der Kaiserlichen Akademie der Wissenschaften in Wien*, pp. 306–31. The paragraph to which Falconer refers is on page 330. On Heer's theory of species change and its reception, see following letter and n. 5.

[9] In Falconer 1863a, p. 80, Falconer had put forward the evidence of the persistence of specific characters in fossil elephants as an argument against natural selection. CD responded to this point, after reading the argument in draft (see n. 4, above), by noting that the proboscideans were verging

on extinction, and that it was the flourishing members of each order that generally changed and gave rise to new species. CD also wrote that he did not believe that there was 'some unknown law of evolution' by which species necessarily changed (*Correspondence* vol. 10, letter to Hugh Falconer, 1 October [1862]).

10 Lartet does not appear to have published a 'rectification' concerning the Pliocene fossil insectivore.

11 The number of CD's portfolio of notes on palaeontology and extinction.

To Hugh Falconer [25–6 August 1863][1]

Down

thank you for telling me about the pliocene mammal,[2] which is very remarkable; but has not Owen stated that the pliocene badger is identical with the recent?[3] Such a case does indeed well show the stupendous duration of the same form. I have not heard of Suess' pamphet,[4] and should much like to learn the title, if it can be procured; but I am on different subjects just at present. I should rather like to see it rendered highly probable that the process of formation of a new species was short compared to its duration; that is if the process was allowed to be slow and long: the idea is new to me.— Heer's view that new species are suddenly formed like monsters, I feel a conviction from many reasons is false.—[5] Whenever I come to London I have several little things to ask about; but when that will be I do not know. I have had a bad summer with much sickness of late, and we are all going to Malvern for a month, and start in a week's time.[6] I have managed to do this summer a fair share of work and have been greatly interested by the spontaneous movements and irritability of tendrils and twining plants; but only a little of my work is new.—[7] Do you remember telling me that I ought to study phyllotaxy;[8] well I have often wished you at the bottom of the sea; for I could not resist, and I muddled my brains with diagrams etc. and specimens and made out, as might have been expected, nothing. Those angles are a most wonderful problem and I wish I could see someone give a rational explanation of them.[9]

I am tired, so good night.

My dear Friend, | Yours very Sincerely. | Ch. Darwin.

Copy incomplete[10]
DAR 144: 32

1 The date range is established by the relationship between this letter, the preceding letter, and the letter from Hugh Falconer, 29 August 1863, and by the reference to CD's visit to Malvern (see n. 6, below). Francis Darwin dated this letter 26 August [1863] (*LL* 3: 51), but there is no corroborating evidence to support this.

2 See preceding letter.

3 CD refers to Richard Owen and to Owen 1846b, p. 109, where he discussed the badger *Meles taxus*. CD was presumably referring Falconer to an analogous case of the persistence of specific characters in a mammal species from the Pliocene to the present (see preceding letter). There is an annotated copy of Owen 1846b in the Darwin Library–CUL (see *Marginalia* 1: 649–52).

4 The references are to Eduard Suess and to Suess 1863. In Suess 1863, Suess had linked Falconer's views on species with those of Oswald Heer (see preceding letter, and n. 5, below).

5 Working mostly with Tertiary plants and insects, the Swiss palaeontologist Oswald Heer maintained that species were generally constant, but that during occasional periods of creation, existing types underwent abrupt variation and gave rise to new species (see Heer 1860, p. 56). CD's annotated copy of Heer 1860 is in the Darwin Library–CUL (see *Marginalia* 1: 363–6). By 1863, Heer's view of new species formation was being presented in the international literature as a rival to CD's theory of slow evolution by natural selection (see A. Gray 1863d, p. 438). For other comments on Heer's theory of species change, see letters to Asa Gray, 31 May [1863] and n. 4, and 7 July 1863.

6 On 1 September 1863, CD went to London with his daughter Henrietta Emma Darwin, and the rest of the family departed for Malvern Wells. According to Emma Darwin's diary (DAR 242), CD and Henrietta travelled from London to Malvern Wells on 3 September 1863.

7 See letter to J. D. Hooker, 12–13 August [1863], and following letter and n. 5.

8 It is not clear when Falconer made the suggestion that CD study phyllotaxy; however, a manuscript paper sent by Falconer in September 1862 in which he commented on the theory of natural selection directed CD's attention to phyllotaxy (see *Correspondence* vol. 10, letter to Hugh Falconer, 1 October [1862], and this volume, letter to J. D. Hooker, 29 May [1863]). Falconer's paper was published in the *Natural History Review* in January 1863 (Falconer 1863a), his comments on phyllotaxy appearing on page 80. CD began preliminary research on phyllotaxy in February, and made a series of detailed observations in May and June 1863 (see, for example, letter to Daniel Oliver, 20 [February 1863], letter to J. D. Hooker, [9 May 1863], and letters to Asa Gray, 11 May [1863] and 26 June [1863]). CD's phyllotaxy notes are in DAR 51: 6–32; there are also notes by George Darwin in DAR 192: 1–7.

9 See memorandum from G. H. Darwin, [before 11 May 1863], letters to Asa Gray, 11 May [1863] and 31 May [1863], and letter to J. D. Hooker, 29 May [1863].

10 The opening paragraph of the letter, which is now missing, dealt with CD's opinion of Falconer's criticism of Charles Lyell's *Antiquity of man* (C. Lyell 1863a) in the *Athenæum*. See letter from Hugh Falconer, 29 August 1863 and n. 10.

To J. D. Hooker 25 [August 1863][1]

Down Bromley Kent
25th

My dear Hooker

I have made up my mind to go on Thursday week with all our household for a month to Malvern.[2] I have had a deal of sickness of late; every morning for a fortnight. I have been in communication with Prof. Goodsir of Edinburgh, as I find what I suppose are vegetable cells in the limpid fluid which I throw up, & on my return I must consult someone skilled in such cases.—[3] Goodsir thinks this is not cause, but consequence of enfeebled stomach.

Whenever you write please tell me the **Vol. Title** of Journal, **Year**, & page of your paper on the "Climate &c" of the Himalaya.[4] And I ask you, whether you ought not to be crucified alive for sending out a valuable pamphlet with no means of giving a reference? Though Dutrochet has published the cream of my work, I have been going on at Tendrils &c; for the subject has interested me much, & Dutrochet left something undone.[5] So do not forget me, if you notice at Kew any plant with *odd* tendrils. Those of Bignonia unguis are very peculiar; as are those of Smilax aspera, which latter have quite stumped me.—[6]

I had the other day a little note from Lyell, who has found Trimmers arctic shells on Moel Tryfan.—[7] He goes to Newcastle;[8] I suppose you will not have time. I have heard from no one else.—

I suppose I told you that my sister Catherine was going to marry our brother-in-law, Langton;[9] well only a few days before, she caught the Scarlet-Fever so badly that she has been in some risk of her life.—

When we return from Malvern you must try & spare a Sunday; it is so very long since I have seen you.[10] Do not write till you have something approaching to leisure.

Goodnight my dear old friend | C. Darwin

DAR 115: 204

[1] The date is established by CD's reference to his planned visit to Malvern. CD travelled to Malvern Wells, Worcestershire, on 3 September 1863 (see Emma Darwin's diary (DAR 242) and n. 2, below).

[2] CD stayed at Malvern Wells, Worcestershire, for six weeks beginning 3 September 1863, returning to Down on 14 October 1863. Emma Darwin travelled there in advance, arriving on 1 September 1863 to secure lodgings for the family (Emma Darwin's diary (DAR 242)). During his stay, CD underwent treatment at James Smith Ayerst's hydropathic establishment at Old Well House, Malvern Wells (see letter to W. D. Fox, 4 [September 1863]).

[3] John Goodsir was professor of anatomy at the University of Edinburgh. See letter from John Goodsir, 21 August [1863].

[4] CD refers to 'On the climate and vegetation of the temperate and cold regions of East Nepal and the Sikkim-Himalaya Mountains', published in the *Journal of the Horticultural Society* (J. D. Hooker 1852); there is a presentation copy of the pamphlet in the Darwin Pamphlet Collection–CUL.

[5] The reference is to the French plant physiologist René Joachim Henri Dutrochet and to Dutrochet 1843 and 1844 (see letter to J. D. Hooker, 12–13 August [1863]). CD had begun making observations on climbing plants in June 1863 (see letter to J. D. Hooker, 25 [June 1863], and letter to Asa Gray, 26 June [1863]). Hooker sent living specimens of climbing plants from the Royal Botanic Gardens, Kew, at the end of July 1863 (see letter from J. D. Hooker, [31 July 1863]).

[6] CD's observational notes on *Bignonia unguis*, dated between 23 and 31 August 1863, are in DAR 157.1: 118, 120. CD's observations on the tendrils of *B. unguis* were later published in 'Climbing plants', pp. 50–1; *Smilax aspera* var. *maculata* is discussed on pp. 68–70.

[7] The references are to Charles Lyell and Joshua Trimmer. The letter from Lyell has not been found; however, see the letter to Charles Lyell, 14 August [1863].

[8] In 1863, the British Association for the Advancement of Science held its annual meeting in Newcastle-upon-Tyne between 26 August and 2 September (*Annual register* 1863 (pt 2), pp. 131–9).

[9] See letter to J. D. Hooker, 12–13 August [1863]. Emily Catherine Darwin and Charles Langton were married on 8 October 1863 (*Emma Darwin* (1915), 2: 180–1).

[10] There is no evidence that Hooker visited CD later in 1863.

From J. D. Hooker 26 August 1863

Kew
Aug 26th /63.

Dear Darwin.

I address this on the bare chance of your being alive, for it is a good epoch since I have heard of you; all my own fault. as I owe you two—[1]

I had a good fortnights very hard work at examinations,[2] & have been enlivening my dull life by picking up Wedgewoods again with redoubled vigor[3]—& doggedly plodding on at the New Zealand Flora,[4] what will I hope yield some very curious results in distribution about which we will have many a yarn yet. Black my Herbarium factotum is away ill in Scotland,[5] & Oliver away for his holiday too,[6] which throws a deal of work on my narrowing shoulders. Then my wife has been much away nursing her Aunt at Cheltenham who is dying by inches *precisely* as Henslow did, & at the same age! (62)—or very near it:[7] & I daily expect to be called down to the funeral.

Your Medallion of Dr. D. is all safe, I will send it you as soon as I get it back, Woolner has had some casts of it taken for me, it is still with him.[8]

Planchon writes in extraordinary excitement about your Linum experiments, & wants to devote his life to such researches; but of course will do nothing of the sort.[9]

Boott is better, but not out of his room, he has grown morbid & unreasonable & will not take any advice from Drs or aught else, poor Mrs & Miss Boott are at their wits ends about him.[10]

How go on the tendrils,?[11] I wish I had some such engrossing pursuit & leisure to follow it up.

I am sure I do not know when I shall get down to Down, at present I see no prospect of it for some weeks;[12] I expect I shall have to go & visit Chatsworth & some other Great Gardens apropos of our cultivation here.[13] Haast has sent a nice map of part of the Glacial ranges of N. Zeald[14] shall I send it on to you?

I have read Jamesons paper with great pleasure.[15] I am sure most of my terraces in Himalaya are cuttings away of the vast flowing of glacial ? detritus that once filled the valleys 800 feet deep, & still fills the valleys of the drier country of Tibet.[16] Others are however glacial-lacustrine as I have hinted & as Jameson holds for Glen Roy

Ev Yrs affec | J D Hooker

DAR 101: 157–8

[1] Hooker refers to CD's letters to him of 3 August [1863] and 12–13 August [1863].

[2] Hooker probably refers to examinations at the University of London, where he was examiner in botany (see *Correspondence* vol. 10, letter from J. D. Hooker, [24 July 1862] and n. 4). He was also examiner for the Society of Apothecaries' prize in botany, and a botanical examiner for medical officers in the armed services and the East India Company (L. Huxley ed. 1918, 1: 385–7 and 537).

[3] Hooker was an avid collector of Wedgwood ware (see *Correspondence* vol. 10, letter from J. D. Hooker, [27 or 28 December 1862], and this volume, letter from J. D. Hooker, 6 January 1863).

[4] J. D. Hooker 1864–7.

[5] Allan A. Black was curator of the herbarium at the Royal Botanic Gardens, Kew (R. Desmond 1994).

[6] Daniel Oliver was assistant in the herbarium and librarian at the Royal Botanic Gardens, Kew (R. Desmond 1994, *List of the Linnean Society of London*).

[7] The references are to Hooker's wife, Frances Harriet Hooker, her aunt, Anne Frances Henslow, and her father, John Stevens Henslow. Anne Henslow died in Cheltenham, Gloucestershire, in September 1863 (see letter from J. D. Hooker, 15 September 1863).

[8] Hooker had borrowed a Wedgwood portrait medallion of Erasmus Darwin (see letter to J. D. Hooker, 12–13 August [1863]). Thomas Woolner was a sculptor (*DNB*).

[9] Jules Emile Planchon was professor of botany at the University of Montpellier; he had been an assistant in the herbarium at the Royal Botanic Gardens, Kew, between 1844 and 1848 (R. Desmond 1994). Hooker refers to the letter from Planchon to Hooker of 20 August 1863 (Director's correspondence, vol. 131, doc. 1482, Library and Archives, Royal Botanic Gardens, Kew).

[10] The reference is to the physician and botanist Francis Boott, who was dying of lung disease (*DNB*), and to his wife Mary Boott. Boott had four daughters (Census returns 1841 (Public Record Office HO 107/673/14)); the particular daughter has not been identified.

[11] CD had written of his interest in the movements of tendrils in climbing plants in the letters to J. D. Hooker, 3 August [1863] and 12–13 August [1863].

[12] Hooker visited CD at Down on 22 March 1863 (see letter from J. D. Hooker, [24 March 1863]); there is no evidence that he did so again in 1863.

[13] Chatsworth in Derbyshire was the estate of William Cavendish, the seventh duke of Devonshire (*DNB*, *EB*). See also letter to J. D. Hooker, [28 August 1863], n. 9.

[14] Julius von Haast had sent Hooker a map entitled: 'The Southern Alps of the Middle Island of New Zealand'. The map, drawn by Haast, is in the general collections of the library and archives of the Royal Botanic Gardens, Kew.

[15] The reference is to Thomas Francis Jamieson's paper on the geology of Glen Roy in Lochaber, Scotland (Jamieson 1863). Jamieson believed that during a great 'Ice Age', ice-flows trapped a series of lakes in the glen and that the 'parallel roads' or terraces of Glen Roy represented the shorelines of three of these former lakes (see also, letter from T. F. Jamieson, 28 January 1863 and n. 1, and *Correspondence* vol. 9, Appendix IX).

[16] Hooker described terrace formation, including glacial-lacustrine terraces, in his *Himalayan journals* (J. D. Hooker 1854b, 1: 234–5, 242–3, 269–70, and 2: 119).

From George Busk [*c.* 27 August 1863][1]

Roberts Lodgings | Maentwrog | N. Wales

My dear Darwin

The above address will account for my delay in not answering your letter sooner, which I received this morning.—[2] I am very sorry to hear that you are suffering so much. The particular form of complaint you describe, is what the doctors term 'Waterbrash".[3] It is obviously a diseased secretion of the stomach, but of its true nature I am not aware that anything very certain is known, as it accompanies various kinds of general derangement. The treatment usually recommended is light bitters with subnitrate of bismuth & hydrocyanic acid & I think it not improbable unless, as is most probable, you have already tried it that some such combination might do you good. But I think a useful adjunct under any circumstances would be to give the stomach something to do, very early in the morning, & before the vomiting usually occurs. Perhaps a cup of milk & water with a few grains of calcined Magnesia, say 10 or 15 taken early in the morning with a little biscuit or dry toast, might persuade the stomach to limit itself to its proper function. I have enclosed a prescription for the other medicine which you can try or not as you think proper.[4]

With respect to the flocculent matter you mention as contained in the ejecta I should not be at all surprized to hear that Goodsir detects his Sarcina in it.[5] That organism however, whatever it may be is I believe extremely common, though under certain circumstances it becomes so much increased in quantity as to cause the idea that it is the cause of the disordered function.

I do not think that this is so, & in minute quantity I do not myself consider its presence a sign of any importance whatever. All that I have been led to conclude respecting it when more copious than usual, is that for some reason or another the contents of the stomach are delayed longer than usual in that organ, I imagine that your stomach is weak in its muscular power & perhaps in consequence very dilatable, & that it does not push on its contents as rapidly as it ought to do— whether there is any cause of obstruction besides this, I cannot say, but think, that reason is alone sufficient—

I think however that with this change of Symptoms as it would seem to be, you had better ask some advice,— as regards the man to apply to—I think you would get a good opinion from Dr Brinton who has paid much attention more especially to diseases of the stomach & is a sensible & prudent man at any rate not likely to do you any harm—[6]

I have been down here with my family for about a week, but until today we have nothing but rain, though it seems now inclined to take up.—

Believe me | Yours very truly | Geo Busk

DAR 160: 378

[1] The date is established by the relationship between this letter and the letters from John Goodsir, 21 August [1863] and 27 August [1863]; see also n. 2, below. Presumably CD wrote to Busk after receiving Goodsir's first letter, which would have taken two days to arrive at Down, on or around 23 August 1863. CD's letter to Busk had to be forwarded from London, which suggests that it would not have been received before 25 August 1863. CD had not received Busk's letter by 28 August 1863, or he would have mentioned it in the letter to J. D. Hooker, [28 August 1863]. Therefore, the best estimate for the date of Busk's letter is 27 August.

[2] Busk lived at 15 Harley Street, West London (*Post office London directory* 1863). The letter to Busk has not been found; however, CD mentioned that he had written to Busk in the letter to J. D. Hooker, [28 August 1863].

[3] 'Waterbrash' was a condition involving the eructation of dilute acid from the stomach causing a burning sensation at the back of the throat, or 'heartburn' (*Butterworths medical dictionary*).

[4] Busk's prescription has not been found.

[5] The reference is to the Scottish surgeon and professor of anatomy at the University of Edinburgh, John Goodsir. See letter from John Goodsir, 21 August [1863].

[6] William Brinton was a physician at St Thomas's Hospital, London, and a specialist in stomach disorders (*Physicians*). According to Emma Darwin's diary (DAR 242), Brinton attended CD at Down on 3 November 1863.

From John Goodsir 27 August [1863][1]

South Cottage. Wardie | Edinburgh.
Augt 27.

My dear Sir.

I have obd. examined the slide—[2] I have obd. no *Sarcinæ* on it. The spherical bodies are *Torulæ* in various stages of development.[3] The minute black atoms are evidently the result of decomposition, probably of biliary matter. I find also epithelial scales from the mouth, single ⟨and⟩ coherent, generally in the vicinity or attached to small masses of a filamentous aspect, and probably also from the mouth. There is nothing on the slide referable to the food.

You will bear in mind, that *Torula* is generally present in acid matter ejected from the stomach.[4]

The Carb. of Potash has no doubt acted beneficially;[5] and I trust you will now derive benefit from your visit to Malvern—[6]

I shall most willingly repeat the examination should the vomiting recur. Hoping that it will not—I remain | My dear Sir | Yours most truly | John Goodsir.

Charles Darwin Esq.

DAR 165: 74

[1] The year is established by the relationship between this letter and the letter from John Goodsir, 21 August [1863].

[2] CD had sent Goodsir a microscope slide and phial of his vomit so that it could be examined for organisms that might be causing his persistent vomiting; see letters from John Goodsir, 21 August [1863], and 28 August [1863]. No letter to Goodsir accompanying the slide has been found.

[3] See letter from John Goodsir, 21 August [1863] and n. 4.

[4] Under highly acidic stomach conditions, a characteristic microflora consisting of yeasts (formerly designated by the generic term *Torula*) and bacteria may develop (Brock *et al.* 1994, p. 403).

[5] Potassium carbonate (carbonate of potash) was used as an alkaline remedy for stomach upsets (Paris 1820, p. 337).

[6] CD was planning to visit a hydropathic establishment in Malvern, Worcestershire (see letter to W. D. Fox, 23 May [1863], and letter to J. D. Hooker, 25 [August 1863]).

From J. D. Hooker [27 August 1863][1]

[Royal Gardens Kew]
Thursday

Thanks for yours of this morning.[2] I am distressed to hear of your being again ill. Busk has the most fertile brain of any man I know in regard of all such matters as your stomach—& I really think it might be worth while to consult him.[3]

My Climate Essay was *published* in Journ. Hort. Soc. Lond. VII 69 (1852).[4]

Ev yrs | J D Hooker

DAR 101: 156

CD ANNOTATION

End of letter: 50 *ink*
69
—
for 119
—

[1] Dated by the relationship between this letter and the letter to J. D. Hooker, 25 [August 1863], and the letter from J. D. Hooker, 26 August 1863. In 1863, 27 August was a Thursday.

[2] Letter to J. D. Hooker, 25 [August 1863].

[3] Hooker refers to the surgeon and anatomist George Busk. CD had already written to Busk (see letter from George Busk, [*c.* 27 August 1863] and n. 1).

[4] J. D. Hooker 1852. See letter to J. D. Hooker, 25 [August 1863].

To Roland Trimen 27 August [1863][1]

Down. | Bromley. | Kent. S.E.
Aug 27th

My dear Sir

I am very much obliged for your very pleasant letter.[2] You have hit upon the right case in Oxalis, & seeds will really be a treasure to me.[3] I have posted a paper for you on the dimorphism of Linum which if you will read, you will see why I am anxious for Oxalis[4] I have a more curious case unpublished;[5] but the whole class of facts strike me as very surprising. You may rely on my statements for they have been verifyed. Linum perenne agrees with your Oxalis.[6] I am also very glad indeed to hear about the Peaches,[7]—the more so as it is an exotic in S. Africa.— I am going in a weeks time to Malvern for a month to try & get a little strength, & when there I will probably draw up a notice for Gardeners' Chronicle on your peach case.—[8]

I daily expect proofs of your paper on Disa;[9] a rough woodcut is made.— [10] You must not waste time in sending me many specimens of Orchids in spirits, for I declare I do not know whether I shall ever have time to work up mass of new matter already collected on Orchids. It is capital sport to observe & a horrid bore to publish.—

It pleases me to read your admiration on my beloved orchids—[11] I quite agree they are intellectual beings! By the bye, I believe I have blundered on Cypripedium:[12] Asa Gray suggested that small insects enter by the toe & crawl out by the lateral windows.—[13] I put in a small bee & it did so & came out with its back smeared with pollen; I caught him & put him in again, & again he crawled out by the window: I cut open the flower & found the stigma smeared with pollen!—

Read Bates Travels[14] they will,, I am sure, interest you.— With respect to Physianthus, I do not know whether fact is known; but I think it would be well worth investigating.[15] It is certain that the Asclepiadæ require insect aid for fertilisation. The pollen-masses are wonderfully like those of Orchids. You ought to read R. Browns admirable paper on Asclepias in Transact. Linnean Soc. about 15 or 20 years ago.[16] In the Apocyneæ, (which are allied to the Asclepiadæ) there is a genus, which catches Diptera by the hundred; I have a plant but cannot make it flourish, as I have always wished to investigate the case.[17] It is said that the Diptera are caught by the wedge-shaped spaces between filaments of anthers. But I suspect that plant somehow profits or requires visits of insects. You ought to try whether Physianthus will seed if insects are excluded by a net.— I have seen Hymenoptera from N. America with numbers of pollen-masses of some Asclepias sticking to their tarsi; & the pollen-masses are thus dragged over the stigmas.— R. Brown's paper has beautiful illustrations.—

This is a disjointed, dull letter, but I have been working all day with very little strength.—

With every good wish & sincere thanks | Pray believe me | My dear Sir | Yours sincerely | Ch. Darwin

Endorsement: '1863'
Royal Entomological Society of London (Trimen papers, box 21: 57)

1 The year is confirmed by the relationship between this letter and the letter from Roland Trimen, 16, 17, and 19 July 1863.

2 Letter from Roland Trimen, 16, 17, and 19 July 1863.

3 See letter from Roland Trimen, 16, 17, and 19 July 1863; Trimen had said that he would collect seeds of dimorphic species of *Oxalis* for CD. CD had been investigating dimorphism in species of *Oxalis* since 1861 (see *Correspondence* vols. 9 and 10).

4 'Two forms in species of *Linum*'; for CD's presentation list for this paper, see Appendix IV.

5 CD refers to the existence of trimorphic flowers in species of *Lythrum* (see letter to J. D. Hooker, 3 August [1863] and n. 6). See also letter to Alphonse de Candolle, 4 August [1863], and letter to Asa Gray, 4 August [1863]. CD described his experiments on *Lythrum* in 'Three forms of *Lythrum salicaria*', which was read before the Linnean Society on 16 June 1864 and published in *Journal of the Proceedings of the Linnean Society (Botany)* in December 1864 (*General index to the Journal of the Linnean Society*).

6 Trimen had sent CD sketches of the pistils of the two flower forms and the flowers of what he took to be *Oxalis cernua* with his letter to CD of 16, 17, and 19 July 1863. CD had discussed *Linum perenne* in 'Two forms in species of *Linum*', pp. 75–81 (*Collected papers* 2: 98–103).

7 Trimen had written that moths often injured peaches in South Africa by penetrating the skins with their probosces; this was consistent with CD's hypothesis that moth pollinators penetrated the inner walls of the spurs of certain species of orchids (see letter from Roland Trimen, 16, 17, and 19 July 1863 and n. 13, and *Orchids*, pp. 48–53). See also n. 8, below.

8 CD wrote a draft letter detailing Trimen's observations, but the letter was not published (see following letter). CD cited Trimen's observations in 'Fertilization of orchids', p. 143 (*Collected papers* 2: 140), and *Orchids* 2d ed., p. 40.

9 Trimen 1863. See letter to Roland Trimen, 23 May [1863] and nn. 5 and 6.

10 Trimen 1863, p. 145, carried a woodcut comprising four diagrams illustrating the structure of *Disa grandiflora*. The woodcut was made from Trimen's sketches (see letter from Roland Trimen, 16 March 1863 and nn. 20 and 21, and letter to Roland Trimen, 23 May [1863]). These watercolour sketches are reproduced facing p. 246.

11 See letter from Roland Trimen, 16, 17, and 19 July 1863.

12 In his letter of 16, 17, and 19 July 1863, Trimen remarked that he had verified CD's explanation of pollination in this orchid genus. CD described pollination in *Cypripedium* in *Orchids*, pp. 270–6, where he hypothesised that the pollinating insect would insert its proboscis at the base of the labellum directly over an anther.

13 See *Correspondence* vol. 10, letter to Asa Gray, 10–20 June [1862] and n. 16. Asa Gray's observations on pollination mechanisms in *Cypripedium* were published in A. Gray 1862b. See also this volume, letter to Asa Gray, 20 April [1863] and n. 24, and letter from Asa Gray, 26 May 1863 and nn. 17, 18, and 20. CD modified his account of pollination in this genus in 'Fertilization of orchids', pp. 155–6 (*Collected papers* 2: 152–3), and *Orchids* 2d ed., pp. 229–31.

14 Bates 1863.

15 See letter from Roland Trimen, 16, 17, and 19 July 1863 and n. 21; Trimen had reported that the milkweed *Physianthus albens*, a member of the Asclepiadaceae, trapped insects that landed on its flowers.

16 The references are to Robert Brown and to Robert Brown 1833b.

17 The reference is to *Apocynum androsaemifolium*, which CD's father, Robert Waring Darwin, had grown.

See *Correspondence* vol. 8, letters to Daniel Oliver, 16 November [1860] and [21 November 1860], and letter from Daniel Oliver, 23 November 1860. CD did not discuss this species in *Insectivorous plants*.

To [*Gardeners' Chronicle*] [after 27 August 1863][1]

Peaches perforated & sucked by moths.[2]— Have any of your readers seen moths or butterflies sucking peaches, plums or other fruit, of which the skin was not broken?

A well-known entomologist, Mr Roland Trimen, writes to me from the Cape of Good Hope, that a moth, the [*Achaea chamaeleon*], has lately swarmed in the province of Natal;[3] & after advancing some indirect evidence that moths are capable of perforating with their delicate proboses the skin of the peach, he sends me the following extract, together with a sketch of a perforated peach, from a letter written to him by an accomplished lady & botanist.[4]

Draft, incomplete
DAR 70: 172

[1] The date is established by the relationship between this letter and the preceding letter. See n. 2, below.

[2] In the preceding letter, CD wrote to Roland Trimen of his intention to send a letter to the *Gardeners' Chronicle* on this subject. He later wrote to Trimen that he had started this task but had not finished it owing to illness (see letter to Roland Trimen, 25 November [1863]).

[3] Letter from Roland Trimen, 16, 17, and 19 July 1863. Natal is a province on the north-east coast of South Africa. However, CD was mistaken; the observations were made in Grahamstown, Cape Province, South Africa.

[4] See letter from Roland Trimen, 16, 17, and 19 July 1863. The reference is to Mary Elizabeth Barber.

From John Goodsir 28 August [1863][1]

South Cottage— Wardie. [Edinburgh]
Augt. 28.

My dear Sir,

I enclose a slip of Litmus paper, from which you will observe that the fluid contained in the phial you have sent to me, is acid—[2]

A great part of the tenacious mass contained in the phial consists of bundles of that filamentous structure, to which I alluded in my account of your slide; but which I could ⟨not⟩ define or account ⟨for the⟩ putrifactive chang⟨es that⟩ had taken place. These bundles of minute filaments, appear to me to be referable to some form of *Leptothrix*; probably *L buccalis*, which is an Alga found on the mucous membrane of the Tongue and mouth, but accumulating occasionally in the stomach and bowels.[3]

I see these bundles of filaments on my slides ⟨along⟩ with numerous epi⟨theliu⟩m scales from the ⟨mouth⟩— These scales are seen enve⟨loped by⟩ molecular ma⟨tter and⟩ debris from w⟨hich⟩ bundles of filaments ⟨ ⟩cied

I enclose a rough sketch of a bundle attached to molecular matter investing an epithelium scale; a⟨nd⟩ of a group of *Torula* cell⟨s,⟩ two buccal or saliva cells, and the minute cells, probably the spores of *Leptothrix*.[4]

As I have already stated—medical opinion ⟨*2 words*⟩ to any ⟨*2 words*⟩ as to the ⟨ ⟩ of decreasing those parasites—[5]

Believe me— | my dear Sir— | sincerely yours | John Goodsir.

Charles Darwin Esq.

DAR 165: 75

[1] The year is established by the relationship between this letter and the letter from John Goodsir, 21 August [1863].
[2] See letters from John Goodsir, 21 August [1863] and 27 August [1863]. CD had consulted Goodsir for a diagnosis of the cause of his persistent vomiting. No letter to Goodsir accompanying the phial of vomit has been found.
[3] *Leptothrix buccalis* (now named *Leptotrichia buccalis*) is a bacterium (*Bergey's manual of systematic bacteriology*).
[4] These sketches have not been found.
[5] Probably a reference to Goodsir's letter to CD of 21 August [1863], paragraph 6.

To J. D. Hooker [28 August 1863][1]

Down
Friday night.

My dear Hooker

The very sight of your handwriting always does me good.—[2] You poor miserable wretch you must groan when you see mine, for I think I have always some request. Do not return medallion of Dr. Darwin, but keep it till you come here on our return—[3] I start next Wednesday or Thursday; perhaps staying in London for a day to see Busk if there.[4] From what Goodsir said I had already written to consult Busk.[5] but was glad to be confirmed by your advice.[6]

About New Zealand, at last I am coming round & admit it must have been connected with some Terra firma; but I will die rather than admit Australia.[7] How I wish mountains of New Caledonia were well worked.—

Never mind about New Zealand map.—[8]

I think you will enjoy a tour in the great gardens: I was speechless with admiration, years ago, at Chatsworth.—[9]

I have only one request (wonderful moderation): will cuttings strike, or can suckers be separated of a palm (Calamus??) with shoots ending in hooks.[10] If possible, on my return in 5 or 6 weeks I shd. very much like a young plant, *with growing internodes*, to look for movements.—[11] Probably this is not possible. Ask Hugh Gower.—[12]

Goodnight | Your troublesome old friend | C. D.

I am very sorry to hear about Miss Henslow.—[13]

DAR 115: 205

1 Dated by the relationship between this letter and the letter from J. D. Hooker, 26 August 1863. In 1863, 28 August was a Friday.
2 Letters from J. D. Hooker, 26 August 1863 and [27 August 1863].
3 Hooker had borrowed a Wedgwood portrait medallion of Erasmus Darwin (see letter to J. D. Hooker, 12–13 August [1863], and letter from J. D. Hooker, 26 August 1863). There is no evidence that Hooker visited Down after CD's return from Malvern Wells, Worcestershire (see n. 4, below).
4 CD travelled to London on 1 September 1863 with Henrietta Emma Darwin; they travelled on to Malvern Wells, Worcestershire, on 3 September 1863, returning to Down on 14 October 1863 (Emma Darwin's diary (DAR 242)). George Busk was probably in Wales at this time (see letter from George Busk, [*c.* 27 August 1863]).
5 John Goodsir. See letter from John Goodsir, 21 August [1863]. The letter to Busk has not been found.
6 See letter from J. D. Hooker, [27 August 1863].
7 CD and Hooker had frequently debated competing theories of the origin of island plant and animal assemblages over the years (see *Correspondence* vols. 5 and 6, and *Correspondence* vol. 10, letter from J. D. Hooker, [30 December 1861 or 6 January 1862] and n. 7, and this volume, letter to J. D. Hooker, 15 and 22 May [1863]). Hooker was favourably disposed to the idea of land-bridges to account for the present distribution of animals and plants, whereas CD believed that there was insufficient geological evidence in most cases, and preferred explanations based on trans-oceanic migration. In *Origin*, pp. 389–94, 399, CD evidently considered New Zealand always to have been an island. In the fourth edition of *Origin*, published in 1866, CD suggested that New Zealand and New Caledonia had once been attached to Australia (*Origin*, 4th ed., p. 466). See also Peckham ed. 1959, pp. 621–2, and *Correspondence* vol. 10, letter from William Branwhite Clarke, 16 January 1862 and nn. 13 and 14.
8 See letter from J. D. Hooker, 26 August 1863 and n. 14.
9 Chatsworth in Derbyshire was the estate of William Cavendish, the seventh duke of Devonshire, and was renowned for its great conservatory. Designed by Joseph Paxton, and covering an acre, the conservatory was completed in 1840 (*EB*). CD visited Chatsworth in 1845 and was impressed by the large collection of tropical plants (see *Correspondence* vol. 3, letter to Charles Lyell, 8 October [1845], and letter to J. D. Hooker, 28 October [1845]).
10 CD refers to *Calamus*, a genus of climbing palms native to humid forests in the tropics. Few species of palm can be propagated from suckers or by division (Bailey and Bailey 1976, p. 814).
11 Hooker promised in his letter of 15 September 1863 to send young *Calamus* plants. Since June 1863, CD had been making a series of observations on the movements of climbing plants (see letter to J. D. Hooker, 3 August [1863], n. 2).
12 William Hugh Gower was a foreman at the Royal Botanic Gardens, Kew (R. Desmond 1994).
13 The reference is to Anne Frances Henslow, Frances Harriet Hooker's aunt, who was ill (see letter from J. D. Hooker, 26 August 1863).

From Hugh Falconer 29 August 1863

British Association for the Advancement of Science | Meeting 1863. | Newcastle upon Tyne
29 Aug^{t} *1863*

My Dear Darwin,

Thanks for your note.[1] I am distressed to learn from it that you have suffered so much during the Summer. Your pluck—to hold out with such indomitable perseverance, with your researches—under such adverse conditions, is truly admirable. Long may it last.

With regard to the value of the pliocene species—identical with a living one—to be of any value the instances must be sound, and unquestionable: and if a sufficient number of good cases is adduced, there is a very wholesome basis for a generalization. With regard to the Badger, I cannot say.[2] As soon as I get back to London I will send you Suess paper.[3] I was not a little surprized to find myself so precisely *formulated* for so decided a Conclusion.[4] Had I not thought that Suess must have sent it to you, I should have forwarded it to you. He deals directly with natural selection, under its German designation[5]

Armstrong, gave the best address I ever listened to here:[6] i.e. limited in the main to a single point upon which he was strong—and touching lightly and modestly on other matters. I send you a copy.[7] You will have already seen it in the Times[8]—but a detached copy may be convenient, as he took you gently in hand—with high appreciation and much modesty.[9]

I will not touch on the remark in the opening part of your note—except to say that *nothing in the course of my life* ever gave me so much *pain*, as the reclamation. It was no spurt of sudden feeling. But . . . . I will not trust myself to say more on paper, about it.[10]

My Dear Darwin | Yours very Sin[ly] | H Falconer

DAR 164: 17

1 Letter to Hugh Falconer, [25–6 August 1863].

2 See letter from Hugh Falconer, 24 August [1863], and letter to Hugh Falconer, [25–6 August 1863] and n. 3. Falconer was interested in the stability of specific characteristics over long periods of geological time (see Falconer 1863a, p. 80).

3 Falconer refers to Eduard Suess and to Suess 1863 (see letter from Hugh Falconer, 24 August [1863], and letter to Hugh Falconer, [25–6 August 1863]).

4 See n. 3, above.

5 Suess summarised the conclusions of a number of palaeontologists on the subject of variation in Tertiary land fauna. His article concluded with a discussion of natural selection ('natürliche Auswahl') that was critical of Falconer's comments on natural selection in Falconer 1863a (Suess 1863, pp. 330–1). With some reservations, Suess was in agreement with Falconer's belief that the period during which new species emerge is very brief compared with the period over which a species occurs with constant characteristics. Suess emphasised the importance of external conditions in generating new species, but maintained that there was as yet no satisfactory explanation of the forces which come into play in that process, whether by 'natural selection' or by other processes.

6 The industrialist William George Armstrong was president of the British Association for the Advancement of Science in 1863; the presidential address (Armstrong 1863) was delivered on the evening of 26 August (*Report of the thirty-third meeting of the British Association for the Advancement of Science, held at Newcastle-upon-Tyne*). Armstrong discussed the progress of science and industry, and dealt particularly with the history of coal formations, coal consumption, the mining industry, and improvements in the efficiency of energy use.

7 The copy of Armstrong 1863 has not been found.

8 *The Times*, 27 August 1863, p. 7.

9 In the conclusion to his address, Armstrong briefly referred to *Origin*, and to the doctrine of natural selection (Armstrong 1863, p. lxiii):

> The novelty of this ingenious theory, the eminence of its author, and his masterly treatment of the subject have perhaps combined to excite more enthusiasm in its favour than is consistent with that dispassionate spirit which it is so necessary to preserve in the pursuit of truth. Mr. Darwin's views have not passed unchallenged, and the arguments both for and against have been urged with great vigour by the supporters and opponents of the theory. . . . In the present instance we

may without difficulty suppose it to have been part of the great scheme of creation that natural selection should be permitted to determine variations amounting even to specific differences where those differences were matters of degree; but when natural selection is adduced as a cause adequate to explain the production of a new organ not provided for in original creation, the hypothesis must appear, to common apprehensions, to be pushed beyond the limits of reasonable conjecture.

10 The first part of the letter to Hugh Falconer, [25–6 August 1863], is missing. Falconer refers to his letter in the *Athenæum*, 4 April 1863, pp. 459–60, which was highly critical of Charles Lyell's *Antiquity of man* (C. Lyell 1863a). See letter from W. H. Dixon, 16 April 1863 and n. 3, and letter to Charles Lyell, 18 April [1863].

From John Brodie Innes 29 August [1863][1]

Milton Brodie | Forres.
29th. August.

Dear Darwin,

We often speak and still more frequently think of you and yours, but my friend Stephens is too much occupied with parochial and family matters to send me many Gazettes about friends though he does write fully of all important parish matters,[2] so I have not had particulars of you for an age and have determined to address a line to yourself to ask how you all are, How your son likes his Banking Employment[3] I hope he will find it as profitable as the Lubbocks seem to do and not so conducive to gout as I hear it (or something else) is with them.[4] The idea of a cottage shew of garden produce was a capital one, and from the report seems to have been quite a success.[5] I suggested an addition of cottagers flowers to enliven the scene and encourage the ornamental, but it was not in time to be carried out this year.

I have to congratulate you on an accession to your honours, that your natural history researches should have caused a number of the lower animals to be called by your name was in the natural course but I did not look for a Duke Darwinii but there he is.[6] By the way I hear he is very ill, if he dies you must get Owen to stuff him for you.[7] I do not think I have any natural history for you, but that I lately saw a second instance of a donkey without a stripe.[8] What do you think of the toads? My friend and neighbour Sir A Gordon Cumming has written about these being found in the cuttings of the new railway in large numbers.[9] I cannot find that there is any good evidence of their being found actually in the stone, but it seems certain that they turned up, no one seems to know whence in very queer places, and though one here and there might be found in some deep hole with very little opening for air and food, it is odd that so many should appear, and I am sorry that no really scientific observer should have taken pains to make out when they really lived. I believe Dr. Innes of Forres or Dr. Gordon of Birnie would have been trustworthy but I think they both laugh at the notion and did not care to investigate.[10]

You and I have often had small arguments about rating for schools[11] If you were here I think you would come over to my side. We pay in this parish about double what the Downe School costs, the education is not good and the behaviour of the children and moral conduct of the adult poor is very sad indeed. The abuse has crept in to make the schoolmaster's post a step to the office of preacher in the Establishment,[12] so the master who gets probably £120 and a house puts in an inefficient substitute and goes off to Edinburgh to keep terms The result of all this is that some clever boys get on and learn enough to go to college or get good situations and the general run who can only be at school a short time learn little. The moral condition is bad enough, and illegitimate children are in swarms and don't they lie and cheat with a vengeance!

As to ourselves I have not much to tell you. M^rs. Innes[13] is much as usual, not even as much up to walking as she was but seldom really poorly. She dines out at rare intervals, which is a gain— I dare say you heard we had a small run to Kent in the spring It was a fine contradiction of your theory that all goes well when the wife is master of events. I thought M^rs. Innes would want to see an invalide sister[14] and offered to take her up and make arrangements for her to be in England for two months which she firmly declined then suddenly got a worse report of her sister and must go at a couple of days notice when I could make no comfortable arrangement, could only stay a few days, and was obliged to come back leaving all undone I specially wished, chief to see you and other friends at Downe. However happily the sister is better and I was able to give Johnny[15] a week in Paris which pleased him. He works on pretty steadily with his tutor is vastly grown in stature and breadth and has been quite well since we have been here. I should have preferred his going on at School but think he suffers less than many boys would from being at home

I wish you would come down here and take a look at the red sandstone, and other interests of the North—

Our kindest regards to M^rs. Darwin and your family—

Believe me Dear Darwin | Yours faithfully | J Brodie Innes—

DAR 167: 11

[1] The year is established by the relationship between this letter and the letter to J. B. Innes, 1 September [1863].

[2] Thomas Sellwood Stephens was the curate of St Mary's church, Down; Innes was the perpetual curate (*Clergy list*). Having inherited an estate, Innes moved to Forres, Scotland, in 1862 (see *Correspondence* vol. 10). Innes continued to be the non-resident incumbent of Down until 1869 (*Crockford's* 1894, Freeman 1978).

[3] In 1861, William Erasmus Darwin became a partner in the Southampton and Hampshire Bank, Southampton (see *Correspondence* vol. 9).

[4] The Down House grounds bordered the High Elms estate, seat of the Lubbocks; John William Lubbock and his son, John Lubbock, were partners in the banking house Robarts, Lubbock & Co., Lombard Street, London (*Post Office London directory* 1863, Freeman 1978).

[5] The correspondence concerning the show of garden produce has not been found, but Innes presumably refers to the activities of one of Down's local societies. CD played an active part in parish affairs (see Moore 1985, p. 469).

[6] The reference is to George Douglas Campbell, the eighth duke of Argyll; Campbell had been nicknamed the 'Darwinian Duke' by the newspapers following a speech he gave to the Highland and

Agricultural Society of Scotland at the beginning of August (see letter from J. B. Innes, 4 September [1863] and n. 3).

7 Although the reference is to G. D. Campbell, the duke of Argyll, Innes was mistaken; Campbell was not ill (see letter from J. B. Innes, 4 September [1863]). Innes also refers to Campbell's friend Richard Owen (Rupke 1994, p. 213); Owen was superintendent of the natural history departments at the British Museum (*DNB*, *DSB*).

8 See *Correspondence* vol. 8, letter to John Innes, 11 September [1860]. CD was interested in the inheritance of shoulder stripes in horses and asses, and had asked several of his correspondents to report any observations they had made on this subject (see *Correspondence* vols. 6–8). CD considered the presence of stripes to be a reversion to a primitive characteristic of an ancient progenitor (see *Origin*, pp. 163–7, and *Variation* 1: 55–64 and 2: 41–3; however Innes's observations are not cited).

9 The reference is to Alexander Penrose Gordon Cumming of Altyre (*County families*). In April and May 1863, Cumming sent letters to the *Elgin Courier* and the *Forres Gazette* reporting that he had observed live toads deep in the strata exposed by the cuttings of the Inverness and Perth Railway, which was being constructed near Altyre. The story was reprinted in the *Scotsman* and *The Times* (see *The Times*, 16 April 1863, p. 7, and 25 May 1863, p. 9). The myth that toads or frogs contemporaneous with coal or rock formations had been exhumed alive in modern times was prevalent in popular natural history books and periodicals of this period (see Barber 1980, p. 18).

10 Innes refers to the Forres surgeon John George Innes (*Medical directory* 1863) and to the botanist and geologist George Gordon (R. Desmond 1994).

11 No correspondence between CD and Innes on this subject has been found.

12 The reference is to the Church of Scotland, a state establishment, and one of the three churches which formed the main divisions of Scottish presbyterianism (Cameron *et al.* 1993).

13 Eliza Mary Brodie Innes.

14 Innes refers to Mrs Swan (see *Correspondence* vol. 12, letter from J. B. Innes to Emma Darwin, 23 January [1864]).

15 John William Brodie Innes.

From Asa Gray 1 September 1863[1]

Cambridge, Mass.
Sept. 1, 1863

My Dear Darwin,

Your fine long letter of Aug. 4th. reached me up in the country—in my native region—in the centre of the State of New York rusticating, and enjoying ourselves mightily.[2] We were among the people of a thriving region—a well-to-do set.— no *poverty* near us for miles & miles, i.e—no hardship, except any that a drunken laborer might bring on his family; and I longed to take you out with us in our drives that you might see a happy and comfortable country—more and more so every year,—and perhaps a larger ratio of the population refined to a reasonable degree in feeling and life than I know of in any other part of the world.

Well, I am at home again as well as possible, and as hearty as an ox,—thoroughly well, and all disagreeable feelings about my head quite forgotten. I hope it may keep so.

I hope to get other *Specularia* seeds for you.[3]

I will *consider* about fantastic variation of pigeons.[4] I see afar trouble enough ahead quoad *design* in nature but have managed to keep off the chilliness by giving the knotty questions a rather wide birth. If I rather avoid, I cannot ignore the difficulties—ahead. But if I adopt your view bodily⟨,⟩ can you promise me any less difficulties?

—If your *Lythrum*-paper shall be at all equal in interest to that on *Linum* it will be a gem.[5]

As to *tendrils*, What are Hooker & Oliver (the latter a Professor too) about, and where have they lived not to know anything of them?[6] Every body must have seen, in *Cucurbitaceæ* & *Passiflora*, tendrils *reaching out straight* for a certain time, and then, if they reach nothing, coiling up from the end.— Also the *sweeping of* stems. But I think I never noticed that till it was pointed out in print. You ask me to tell you where anything is published on this subject. I had indicated it well enough in the little note of mine on the *visible coiling*—which excited your interest. See the *first 3 sentences*.[7] And consult *Mohl* (who is worth all Germany besides) *On the Vegetable Cell*, transl. by Henfrey, and published by Van Voorst, 1852, pp. 156–158, also 151. &c,—and be thankful to me for having instigated (in 1851) Van Voorst to get Henfrey to translate this little book.—tho you English pay no attention to it, when you have got it.[8]

No do not abandon this subject, for it will be fruitful in your hands.

You should go on and connect this with something *analogous* in the spiral twist of the wood & bark of Coniferous trees—and others— It is very marked in *Thuja occidentatis*. The twist here is to the right (of the observer). But in noticing a hundred trunks or so, while up in the country, I found 4 or 5 which twisted in the opposite direction.

Next week I will send you sheets containing abstract of Bates' mimetic analogy, & your Linum paper.[9]

As to Agassiz, you must not infer that I have any real ill-will toward him. I have long been an impartial and perhaps useful friend to him.[10] But he is not a person to be satisfied long with any one who acts as well as thinks independently. So, a good while ago, he undertook to put me down, and has once or twice repeated the attempt. He has not yet succeeded. All the revenge I take is to worry and tease him upon occasion. But he has many excellent points and great (morphological) ability, and I am truly sorry he does not make a career at all up to his early promise.

Ever your cordial A. Gray

P.S.[11] Three nos. of Boston newspaper, recently sent you, two by this mail (in which my good beau-père is again "*spiking* the English")[12] please to forward to Reuben Harvey, Esq, Limerick, Ireland.[13]

Parsons (the first) I think puts the case neater, and in much fewer words.[14]

You are quite out in supposing that hatred of England is increasing, or that there is the least desire to meddle with you, except in self defence[15]

My own feelings were very sensitive at first, because I expected better things, and I then deferred much to British opinion. I now do neither, and nothing strikes me more than the smallness of mind and largeness of gullibility of the British people, as far as I can judge from their press, (weeklies, quarterlies, & Times). But

I do not suppose you will fight us because you dislike us. And so conversely. I suppose I do not see the papers which so abuse England, tho, I read influential & respectable papers. But from what I do see, I think we receive far more abuse and misrepresentation, and unfair usage than we give.

As to the course of the war and policy of our country as to slavery, some day when you turn back to some early letter of mine you will see that I was a fairly good prophet.—that the South might have *delayed* the abolition of slavery by giving up early in the conflict,—but that every month of continued resistance hastened and ensured the downfall of slavery[16] That is now *doomed*, and sure near to rapid death, quick in some places—slower in others, but sure.—

Ill usage of *negroes*—who make such good soldiers—will soon be unheard of—except with Irish. It will take some generations of American life to breed out the barbarism they bring to the country.

Good bye— Ever Yours | AG.

DAR 157.2: 108; DAR 165: 139, 140

CD ANNOTATION

6.4 But I think . . . promise. 10.7] *crossed pencil*

1 This letter was apparently sent through Joseph Dalton Hooker, who forwarded it to CD (see letter from J. D. Hooker, 15 September 1863 and n. 3).

2 Letter to Asa Gray, 4 August [1863]. The Grays had been in Sauquoit, New York, visiting Gray's mother (Dupree 1959, p. 308).

3 Gray had sent CD impoverished seeding specimens of the dimorphic plant *Specularia perfoliata* with his letter of 7 July 1863, but CD thought it likely that he had inadvertently killed the seeds on arrival (see letter to Asa Gray, 4 August [1863]).

4 See letter to Asa Gray, 4 August [1863] and nn. 10 and 11.

5 See letter to Asa Gray, 4 August [1863]. CD's paper entitled 'Three forms of *Lythrum salicaria*' was read before the Linnean Society on 16 June 1864. Gray also refers to 'Two forms in species of *Linum*', a copy of which CD had sent Gray in April 1863 (see Appendix IV).

6 In his letter to Gray of 4 August [1863], CD had reported that neither Joseph Dalton Hooker nor Daniel Oliver had been able to give him any information regarding the spontaneous rotatory movements of the tendrils and upper internodes of climbing plants.

7 CD's interest in climbing plants had been stimulated by his reading of A. Gray 1858b (see letter to J. D. Hooker, 25 [June 1863] and nn. 2–3). The paper begins (A. Gray 1858b, p. 98):

> As much as twenty years ago, Mohl suggested that the coiling of tendrils 'resulted from an *irritability* excited by contact.' In 1850 he remarked that this view has had no particular approval to boast of, yet that nothing better has been put in its place. And in another paragraph of his admirable little treatise on the Vegetable Cell (contributed to Wagner's *Cyclopedia of Physiology*), he briefly says: 'In my opinion, a dull irritability exists in the stems of twining plants and in tendrils.'

The references are to Mohl 1827, pp. 63–72 and Mohl 1853, pp. 303 and 307; see also Henfrey trans. 1852, pp. 151 and 156. There is an annotated copy of Mohl 1827 in the Darwin Library–CUL (see *Marginalia* 1: 590–4). Hugo von Mohl's article on the vegetable cell (Mohl 1853) was originally published in the fourth volume of Rudolph Wagner's *Handwörterbuch der Physiologie* (R. Wagner ed. 1842–53). From Gray's reference to Mohl's article appearing in 1850, and from the existence of an offprint of the article dated 1851 (see *Taxonomic literature*), it appears that the fourth volume of Wagner's *Handwörterbuch der Physiologie* was issued in parts over several years.

8 Mohl 1851 was translated into English by Arthur Henfrey, and published by the London natural history publisher, John Van Voorst (Henfrey trans. 1852). Gray visited Britain during 1850 and 1851 (Dupree 1959, pp. 190–3). There is a copy of Henfrey trans. 1852 in the Darwin Library–CUL; CD made annotations on the pages cited by Gray (see *Marginalia* 1: 589–90).

9 Gray published review abstracts of Bates 1861 and of 'Two forms in species of *Linum*' in the September 1863 issue of the *American Journal of Science and Arts* (Gray 1863a and 1863c respectively).

10 See letter from Asa Gray, 21 July 1863, and letter to Asa Gray, 4 August [1863]. Gray refers to Louis Agassiz, his colleague at Harvard University, with whom he had a long-standing dispute concerning CD's theory and other scientific and political issues (see Dupree 1959, pp. 264–306 and 313–31).

11 The postscript, which is written on a separate sheet of paper, has been assigned to this letter on the basis that both contain responses to the letter to Asa Gray, 4 August [1863]. In addition, the postscript bears a London postmark of 14 September, which is consistent with its having been sent from the United States in the first few days of September.

12 Gray probably refers to five articles written by his father-in-law, Charles Greely Loring, for the *Boston Daily Advertiser*, and published on 20, 22, 25, 27, and 29 August 1863. These and six subsequent articles by Loring were later published as a pamphlet (Loring 1863) entitled *Neutral relations of England and the United States.* Earlier in the year, Gray had sent CD a copy of Loring's *Correspondence on the present relations between Great Britain and the United States of America* (Loring 1862; see letter to Asa Gray, 23 February [1863]).

13 The prefatory note to Loring 1863, which presumably included the articles Gray refers to (see n. 12, above), explained that all the compiled articles 'originated in an undertaking to reply to a letter from a highly respected correspondent in Ireland'.

14 The reference is apparently to Gray's colleague at Harvard University, Theophilus Parsons, who was an ardent exponent of the Union cause in the American Civil War (*DAB*); the newspaper article has not been identified.

15 See letter to Asa Gray, 4 August [1863].

16 See *Correspondence* vol. 9, letter from Asa Gray, 31 December 1861.

To J. B. Innes 1 September [1863]

Down. | Bromley. | Kent. S.E.
Sept. 1st

Dear Innes

I was very glad to get your kind & pleasant letter, with a good account of your son & a fair account of Mrs. Innes.[1]

I have had a bad summer, with my stomach as bad as it well can be & tomorrow we start for a month at Malvern. My wife has gone first with half the party to look for a house.[2]

We are in the same predicament, as you were, about our youngest boy, Horace who is too delicate as yet for school, & a great evil it is.—[3] He learns nothing & what on earth we are to do about him I am sure I do not know.—

You know what a hermit's life I lead & it has been worse than ever this summer; & I have not even seen John Lubbock for months.[4] I think everything goes on well & smoothly in the village.[5]

Today poor old Spearing was buried & I think before a year or two is over young Mr Smith will follow all his family to the grave.—[6] I am much surprised at what you say about your poor; I thought they were the best educated & moral people in Europe.—[7]

I go on working at Natural History & have taken a good deal to Botany & it is all that I am good for; I can just do an hour or two's work, when I can do nothing else. By the way there is one sentence in your letter that utterly puzzles me, about a "Duke Darwinii" who is going to die & ought to be stuffed![8]

I hope that you have had a day's shooting over your own land & brought home a good bag. What a glorious day I used to think it.[9] I have heard guns going off at a great rate all round here today & it has made me think of old times.—

Your old Terrier Tartar is still alive & well, but awfully venerable in all his movements.[10]

And now I will wish you good night & with kind remembrances to M[rs] Innes, believe me | Dear Innes | Yours very sincerely | Ch. Darwin

I do not believe a word about the toad-stories; though they must get into very queer places to cause so many strange stories.—[11]

Endorsement: '1st Sept. 63'
American Philosophical Society (Getz 4287)

[1] See letter from J. B. Innes, 29 August [1863]. The references are to Eliza Mary and John William Brodie Innes.

[2] Emma Darwin travelled to Malvern, Worcestershire, on 1 September 1863, and rented a house at Malvern Wells; CD and Henrietta Emma Darwin joined her there from London on 3 September 1863 (see Emma Darwin's diary (DAR 242) and letter to W. D. Fox, 4 [September 1863]). CD was to visit James Smith Ayerst's hydropathic establishment in Malvern Wells (see letter to William Darwin Fox, 4 [September 1863]).

[3] Horace Darwin had been ill since 1862, and was tutored privately from October to December 1862 (see letter to G. V. Reed, 12 January 1863 and n. 1). On Horace's continuing ill health see Emma Darwin's diary (DAR 242). Innes's son, John William Brodie Innes, apparently suffered delicate health for a number of years (see *Correspondence* vol. 8, letter to John Innes, 18 July [1860], and *Correspondence* vol. 9, letter to John Innes, 19 December [1861]).

[4] John Lubbock had moved from High Elms, near Down, to Chislehurst, a village about five miles north of Down, in August 1861 (John Lubbock's diary (British Library, Add. Ms. 62679: 64 r.)).

[5] Innes had been the Anglican incumbent of Down since 1846; he moved to Milton Brodie, near Forres, Morayshire, Scotland, in 1862, having inherited an estate there, and employed a curate to undertake his parochial duties in Down (see *Correspondence* vol. 10, letter from J. B. Innes, 2 January [1862] and n. 1).

[6] The references are apparently to William Spearing, an agricultural labourer resident at Down, and John Smith, a farmer of Down Court, a short distance from Down House (Census returns 1861 (Public Record Office, RG9/462: 71, 78)). Smith's father, John, died in the early 1850s; his mother, Sarah, and brother, Josiah, had apparently died since the 1861 census (Census returns 1861 and 1871 (Public Record Office, RG9/462: 78 and RG10/875: 35v) and *Post Office directory of the six home counties* 1851 and 1855).

[7] See letter from J. B. Innes, 29 August [1863].

[8] See letters from J. B. Innes, 29 August [1863] and 4 September [1863].

[9] CD refers to the start of the partridge-shooting season on 1 September (see *British Imperial calendar* 1863, p. 8). CD had been fond of shooting game-birds in his youth (see *Correspondence* vol. 1 and *Autobiography*, pp. 54–5).

[10] Tartar had apparently been left with the Darwin family, along with Innes's son's dog Quiz, when Innes moved to Scotland in 1862 (see *Correspondence* vol. 10, letter to J. B. Innes, 1 May [1862]).

[11] See letter from J. B. Innes, 29 August [1863] and n. 9.

From J. J. Aubertin 3 September 1863

St Paul's
Sep 3. 1863

My dear Mr Darwin—

I am much obliged to you for your letter of the 19th July—[1] Our geological aspirants here are quite satisfied in finding your report of the stones I sent you from our tunnel confirmatory, more or less, of their own ventured opinions![2]

This is indeed a strange country geologically as well as æthnologically! Everything in Creation has a real meaning I suppose! & so I suppose Brazil & the Brazilians must have! There is not much more variety in their Postage Stamps than there is in their pursuits, arts, & acquirements! But such as there are to be found I have much pleasure in enclosing for your boy, & wish they were prettier![3] The numbers on them represent the "reis"[4]—one thousand of which (according to "Cocker")[5] make the mil-reis— 1#000.

The value of this coin, at Mint Par, being (as perhaps you already know) 27^{d}— That is the simplest division of money here, but there are a hundred & fifty other clumsy calculations, as bad as our own—

The House of Commons, I see, has voted the Decimal & Metric Systems, but is far from obtaining the satisfaction of all!—[6] I suppose there are objections to every system—& lately I have heard the proposal to make "8" the Basis, as giving even halves & quarters &c—which of course "10" does not. There are a good many more, I fancy, troubled with having nothing to count than there are who are troubled to count what they have! At all events, in general matters of life, each of us pretty well knows how to state the account in his own favour!

I am very sorry to hear of such illness in your family, but trust you may now all be pretty well, yourself included.[7] You have, I understand had a very beautiful summer in England, with every prospect of a most abundant harvest, which I hope has been by this time tolerably well realized.—

We have the small-pox very badly here, which makes terrible looking objects of some of the poor Blacks! I understand it has been an Epidemic too, in England this year!—[8] I was re-vaccinated the other day, but it resulted ⟨in a⟩ *false* pustule!

Believe me very faithfully Yours | J. J. Aubertin.

DAR 159: 124

1 Letter to J. J. Aubertin, 19 July 1863.

2 See letter from J. J. Aubertin, 27 April 1863, and letter to J. J. Aubertin, 19 July 1863. Aubertin was superintendent of a railway in Brazil and had sought CD's advice on tunneling through the rocks of which he sent samples.

3 CD had requested Aubertin to send postage stamps for Leonard Darwin's collection.

4 Reis: 'A Portugese money of account of very small value . . . of which one thousand form a Milreis' (*OED*).

5 According to Cocker: 'reliably, correctly' (*Brewer*). Edward Cocker was a seventeenth-century arithmetician, whose popular arithmetical text-book went through over a hundred editions (*DNB*).

6 The reference is to a bill for a decimal system for British weights and measures, and for making them correspond with the units of the metric system, that was introduced in the House of Commons by William Ewart in May 1863, and received its second reading (during which the principle of the bill was approved) on 1 July 1863 (*Hansard's Parliamentary Debates* 3d ser. 172 (1863): 43). However, since the bill did not have government support, it did not pass through the various stages required for it to become law before the end of the parliamentary session, and thus lapsed. With government support, Ewart successfully introduced a more limited bill in the following parliamentary session, which legalised the use of metric terms in contracts (Connor 1987, *EB*).

7 See letter to J. J. Aubertin, 19 July 1863 and nn. 5 and 6.

8 An epidemic of smallpox in London, beginning in the middle of 1862 and culminating in the second quarter of 1863, claimed 2281 lives by the end of 1863. Outbreaks also occurred in several English provincial towns (*Sixth report of the medical officer of the Privy Council*, pp. 82, 105–8, 157–9, and 187–9).

To Hugh Falconer 4 [September 1863][1]

Malvern Wells
Friday 4th.

My dear Falconer

I write merely to give you my address;[2] but please do *not* lend me Suess, unless you can spare it for a month, as I have no brains or German Dictionary to read anything at all difficult.[3]

I see Carter Blake was well pitched into at Newcastle.[4] Many thanks for Armstrong's admirable speech.[5]

Your very stupid friend | C. Darwin

Copy
DAR 144: 33

1 The date is established by the relationship between this letter and the letters from Hugh Falconer, 29 August 1863 and 10 September 1863.

2 According to Emma Darwin's diary (DAR 242), CD stayed at Malvern Wells, Worcestershire, from 3 September to 12 or 13 October 1863. See also following letter.

3 Suess 1863. See letter from Hugh Falconer, 29 August 1863.

4 The reference is apparently to the discussion of Charles Carter Blake's paper entitled 'On some points in the cranioscopy of South American nations' (Blake 1863b) at the meeting of the British Association for the Advancement of Science in Newcastle-upon-Tyne on 29 August 1863. In *The Times*, 1 September 1863, p. 5, it was reported that 'the ethnologists got up a warm debate' regarding Blake's paper, and that John Crawfurd 'denied altogether that cranioscopy afforded any sound knowledge of the affinity of races'. The debate reflected a schism within Victorian anthropological science. Cranioscopy, the measurement of skull anatomy, had become an important tool in the work of members of the Anthropological Society seeking to establish that humanity was composed of discrete and permanently unalterable races. Members of the Ethnological Society, like Crawfurd, believed in human biological unity and were deeply opposed to the views propounded by Blake and other leading figures of the Anthropological Society (see Rainger 1978, pp. 56–60; Stepan 1982, pp. 44–6; Stocking 1987, pp. 248–54). For Falconer's opinion of Blake, see the letter from Hugh Falconer, 8 January [1863] and n. 4.

5 Falconer had sent CD an offprint of William George Armstrong's presidential address to the British Association, later published as Armstrong 1863 (see letter from Hugh Falconer, 29 August 1863 and n. 6).

To W. D. Fox 4 [September 1863]

Villa nuova | Malvern Wells
Friday 4th.

My dear Fox

I have had a bad amount of sickness of late & came here yesterday; & as I could get no country in Grt. Malvern we are here & I have put myself under Dr. Ayrehurst, though very sorry not to be under Dr. Gully.[1] Emma came here a day or two first & took this house.[2]

And now I come to the painful subject which makes me write at once. Emma went yesterday to the church-yard & found the gravestone of our poor child Anne gone.[3] The Sexton declared he remembered it, & searched well for it & came to the conclusion that it has disappeared.[4] He says the churchyard a few years ago, was much altered & we suppose that the stone was then stolen.

Now some years ago, you with your usual kindness visited the grave & sent us an account.[5] Can you tell what year this was? I was so ill at the time & Emma hourly expecting her confinement that I went home & did not see the grave.[6] It is not likely, but will you tell us what you can remember about the kind of stone & where it stood; I think you said there was a little tree planted. We want, of course, to put another stone. I know your great & true kindness will forgive this trouble.

Your affect | C. Darwin

Endorsement: 'Sep | /63'
Postmark: SP 4 63
Christ's College Library, Cambridge (Fox 140)

1 According to Emma Darwin's diary (DAR 242), CD arrived in Malvern Wells, Worcestershire, on 3 September 1863. CD had originally intended to be treated by James Manby Gully, who practised at nearby Great Malvern; however, Fox informed him that Gully was seriously ill and unable to practise (see letter from W. D. Fox, [16–22 May 1863], and letter to W. D. Fox, 23 May [1863]). James Smith Ayerst's hydropathic establishment at Malvern Wells was reportedly opened 'in conjunction at first with Dr. Gully', although it was publicised in Ayerst's name alone (*Medical directory* 1858–68; *Post Office directory of Birmingham* 1860 and 1864; Metcalfe 1906, p. 94). Gully discontinued his own hydropathic establishment at Great Malvern in the late 1850s, although he continued to practise from his residence (*Post Office directory of Birmingham* 1854 and 1860; B. S. Smith 1978, p. 211). He visited CD at some time during his stay at Ayerst's establishment (see letters from Emma Darwin to W. D. Fox, [29 September 1863] and 8 December [1863]).

2 According to her diary (DAR 242), Emma Darwin travelled to Great Malvern on 1 September 1863.

3 Anne Elizabeth, the Darwins' eldest daughter, died at Great Malvern on 23 April 1851 (see *Correspondence* vol. 5).

4 This individual has not been identified.

5 Fox's letter has not been found, but see *Correspondence* vol. 6, letter to W. D. Fox, 3 October [1856]. See also letter from W. D. Fox, 7 September [1863].

6 CD had left Great Malvern on 24 April 1851, before Anne's burial, which was arranged by Frances Emma Elizabeth Wedgwood (see *Correspondence* vol. 5, letter to F. M. Wedgwood, 25 April [1851]); there is a sketch showing the location of Anne's grave in the Priory churchyard at Great Malvern in DAR 210.13: 39. Emma Darwin was at the time pregnant with Horace, who was born on 13 May 1851.

From J. B. Innes 4 September [1863][1]

Milton Brodie | Forres
4th Sept

Dear Darwin,

I must write a line to thank you for your letter and enlighten you as to what seems an enigma.[2] The Duke of Argyle lately made a speech at an agricultural meeting wherein he made large reference to your theory of the origin of species by artificial selection, and improved the occasion by shewing how new and valuable varieties of stock had been bred exhorting to further efforts. Hence some of the papers have named him the Darwinian Duke.[3]

I heard somehow, before it got public that one of our Dukes was very ill, somehow I mistook which, and thought it Argyle whereas it is Athol,[4] so I made a rare confusion, which I thus interpret.

I hope your trip to Malvern will do you much good.[5] You had better have come North and next year should we be here I hope you will be tempted to come in a body. There is a fine hydropathic establishment now being built on a beautiful spot at Forres and with our famous climate the vicinity of the sea and good opportunities for excursions to many interesting places I hope the conviction that it is *the* place will come on you.[6] I am afraid I can hardly tempt you and Mrs. Darwin so far towards the North pole now, but I shall be very glad if I can.

I do not shoot much.[7] the harvest is not in yet and the birds are small. In this country there is not much lowland shooting before Octr. I persecute rabbits as enemies and knock over a bird occasionally now—

I am sorry to hear Smith is ill.[8] If he should die I should like to get Down Hall for a parsonage but likely the property may go to some one as unwilling to sell as he is.[9]

The Scotch people have continued to call themselves the most moral in Europe till they not only seem to believe it but persuade others who don't know them.[10] The proportion of bastards is larger than in any country in Europe, and the morals in this respect tally exactly. They are certainly far from honest, but they are all as full of pious talk as an English Dissenter. What can I say more?

Our kindest regards to all yours | Faithfully yours | J Brodie Innes

DAR 167: 12

[1] The year is established by the reference to the letter to J. B. Innes, 1 September [1863].

[2] See letter to J. B. Innes, 1 September [1863].

[3] See letter from J. B. Innes, 29 August [1863], and letter to J. B. Innes, 1 September [1863]. The reference is to George Douglas Campbell, eighth duke of Argyll, who chaired the banquet at the Highland and Agricultural Society's show held at Kelso, Roxburghshire, Scotland, on 4 August 1863. In his speech at the dinner, Campbell heralded agriculture as a 'science', and celebrated 'the power of man . . . over the animal and over the vegetable world in creating almost, as it were, new species for his own benefit and his own use' (*Scotsman*, 5 August 1863, p. 6). He continued:

> I will not tell you . . . that that power is an unlimited power; but this I will tell you, that it is a power of which we do not know the limits, because we have never arrived at them and probably

> never shall. . . . It is not properly a creative power, and yet it produces results which are very closely allied to creation, and perhaps are difficult to be separated from it.

He proceeded to state that no book that he had ever read contained 'more fertile principles for the progress of agriculture' than *Origin*, and he referred favourably to CD's theory of natural selection, drawing parallels between it and the artificial selection familiar to agriculturalists. The newspaper references to the 'Darwinian Duke' have not been identified.

4 George Augustus Frederick John Murray, sixth duke of Atholl, died of cancer in January 1864 (*Complete peerage*).

5 CD and his family had gone to Malvern Wells, Worcestershire, so that CD could undergo a course of treatment at James Smith Ayerst's hydropathic establishment (see 'Journal' (Appendix II)).

6 Innes refers to the Cluny Hill hydropathic establishment near Forres, Morayshire, the building of which commenced in 1863. In 1864, Alexander Monro was appointed medical superintendent of the new establishment, which was formally opened in August 1865. See Metcalfe 1906, pp. 158–64, and Douglas 1934, pp. 347–8.

7 See letter to J. B. Innes, 1 September [1863] and n. 9.

8 See letter to J. B. Innes, 1 September [1863] and n. 6. The reference is to John Smith.

9 Smith lived at Down Court, Down, Kent; he apparently also owned and let Down Hall, which was nearby, beside the parish church (Census returns 1861 (Public Record Office, RG9/462: 78), *Post Office directory of the six home counties* 1862). For some time before his removal to Milton Brodie, near Forres, Scotland, in 1862, Innes had been looking for a suitable property in Down to use as a parsonage (see *Correspondence* vol. 8, letters to John Innes, 18 July [1860] and 6 September [1860], and Moore 1985, p. 469). He had subsequently acquired a property in Down for the use of his curate, Thomas Sellwood Stephens (*Post Office directory of the six home counties* 1862).

10 See letter from J. B. Innes, 29 August [1863], and letter to J. B. Innes, 1 September [1863].

From John Lubbock 4 September 1863

15, Lombard Street. E.C.
4 September 1863

My dear Mr. Darwin

It is an age since I have heard of or seen you. Is it any use asking whether you will come & see us in any way? Either (which you know we should like best) staying a few days with us, or coming to dine or lunch?[1] or if you cannot do any of these are you well enough to see me? I know you will not hesitate to say so if you do not feel up to it.

I am just back from a trip to Norway, Sweden, & Denmark, which we have enjoyed very much.[2] Have you heard that they have made me President of the Ethnological Society? are you a member, if not I wish you would think of it.[3]

But I will not attempt to tell you the news as I shall hope to see you before long.

With my kind regards to Mrs Darwin believe me, dear Mr Darwin, | Yours affectly | John Lubbock

DAR 170: 40

1 Lubbock lived at Chislehurst, a village about five miles north of Down (*Post Office directory of the six home counties* 1862). The address given is that of Robarts, Lubbock & Co., the banking firm in which Lubbock was a partner (*Post Office London directory* 1863); Lubbock travelled into London daily (Hutchinson 1914, 1: 52).

[2] According to his diary (British Library, Add. Ms. 62679: 64 v.), Lubbock travelled in Norway, Sweden, and Denmark with his wife, Ellen Frances, and a friend, between 7 July and 1 September 1863. During the tour they visited some of the Danish kitchen-middens, or prehistoric refuse heaps, about which Lubbock had previously published a paper (Lubbock 1861; see Hutchinson 1914, 1: 61).

[3] Lubbock was elected president of the Ethnological Society at the anniversary meeting on 19 May 1863 (*Athenæum*, 23 May 1863, p. 686). CD was made an honorary fellow of the Ethnological Society in 1861 (see *Correspondence* vol. 9, letter to James Hunt, 28 May [1861]).

From Emma Darwin to W. D. Fox [6–27 September 1863][1]

Malvern Wells
Sunday.

Dear Mr Fox

I have been sending out a good many of these papers & have been very much surprized & pleased to find that a good many persons Squires Ladies & MPs are interested in the matter & anxious to find a remedy.[2]

Charles recommends me to send one to you & to beg you to send one to Mr Wilmot (Sir F's son in law)[3] I send one or two more in hopes that you or Mrs Fox[4] may approve of them & use them in some way.

With my very kind remembrances to her I am yours | very sincerely | E. Darwin

Several have suggested offering a prize for a humane trap—[5]

Christ's College Library, Cambridge (Fox 142a)

[1] The date range is established by the relationship between this letter and the letter from Emma Darwin to W. D. Fox, [29 September 1863], and by the address. According to her diary (DAR 242), Emma arrived in Malvern Wells, Worcestershire, on 1 September 1863; the first Sunday after her arrival was 6 September.

[2] The reference is to *An appeal*, a four-page circular concerning the cruelty of using steel traps for catching vermin, which Emma and CD had privately printed for distribution in August (see letter from G. B. Sowerby Jr to Emma Darwin, 22 July 1863, and Appendix IX). None of the replies to the circular mentioned here have been found.

[3] Emma refers to Francis Sacheverel Darwin and to his son-in-law, Edward Woollett Wilmot, who was CD's first and Fox's second cousin by marriage (*Darwin pedigree*).

[4] Ellen Sophia Fox.

[5] See letter from Emma Darwin to W. D. Fox, 8 December [1863] and n. 4, and Appendix IX.

From Edward Levett Darwin 7 September 1863

Derby
Sept 7. 1863

My dear Cousin

I am very glad to find it *was* my Cousin I was writing to, for altho our name is not a common one yet there are I know other Darwins.[1] Your Wife sent me 2/1 for a copy of the Book and so I have 2^{d} to return you.[2] The Books will be ready to-morrow & I will send you one. I am sure it would be a great thing if the necessity for that trap did not exist but I do not see my way.[3] You will find I allude to the difficulty of catching large vermin in my other book.[4] I am on the move for the next six weeks and we go to-morrow to my Brother in law Marcus Huish of Castle Donington where we stay a fortnight.[5] We have let our House for the 2 months to our Cousins Boteler Bristowe.[6]

I hope that you will find Malvern agree with you[7] I never was there but once when passing through to Worcester. It seemed a very delightful place as far as I remember. I am extremely glad that the "Manual" has been the means of an introduction in an *epistolary* way and sincerely trust that some day or other it will result in being *personal*.

Yours very sincerely | Edw[d] L Darwin

Chr. Darwin Esq | F.R.S.

DAR 99: 17–18

[1] Edward Darwin was CD's first cousin, son of his father's half-brother, Francis Sacheverel Darwin (*Darwin pedigree*). No other correspondence between CD and Edward Darwin has been found.

[2] Emma Darwin had apparently sent for a copy of the fourth, 'considerably enlarged' edition of Edward Darwin's *Game preserver's manual*, which was published in 1863 (E. L. Darwin 1863).

[3] Emma Darwin had apparently sent Edward Darwin a copy of *An appeal*, a four-page circular concerning the cruelty of using steel traps for catching vermin, which she and CD had privately printed for distribution in August (see letter from G. B. Sowerby Jr to Emma Darwin, 22 July 1863, and Appendix IX). See also preceding letter. The two parts of Edward Darwin's *Game preserver's manual* dealt with the prevention of poaching and the 'destruction of vermin'; in the latter part he stated that, of the traps in common use, 'the steel trap stands, as it always must, pre-eminent' ([E. L. Darwin] 1859, p. 22).

[4] The different editions of the *Game preserver's manual* are the only works by Edward Darwin listed in the *NSTC*. The reference may be to the first edition of the work, in which the author detailed the particular procedures necessary for trapping foxes ([E. L. Darwin] 1859, pp. 25–6).

[5] Marcus Huish, of Castle Donnington, Leicestershire, was married to Edward Darwin's sister, Frances Sarah (*Darwin pedigree*).

[6] The reference is to Samuel Boteler and Albertine Bristowe of Beesthorpe Hall, Nottinghamshire; Samuel Boteler Bristowe was CD's and Edward Darwin's second cousin, once removed (*Darwin pedigree*).

[7] CD and his family had gone to Malvern Wells, Worcestershire, so that CD could undergo a course of treatment at James Smith Ayerst's hydropathic establishment (see 'Journal' (Appendix II)).

From W. D. Fox 7 September [1863][1]

Delamere Rectory | Northwich
Sep 7

My dear Darwin

I have not been able to answer your letter[2] before in consequence of an attack of fever which has kept me in bed the last 5 days—& I am at present about as strong as a drowning kitten— I hope to be about again soon. I wish I was at Malvern under Gully.[3]

What you mention about your dear childs grave stone is to me, almost incredible.[4] Is it possible you have overlooked it in consequence of its being buried up by surrounding trees & shrubs, as I can easily imagine it might be by this time.

It is very difficult to describe an exact locality—but either my wife[5] or I could go to the spot, almost blindfold. A line drawn in line with East end of Church to the road—would nearly cut it I think.

Again—entering the Yard from the lower Gate—& following the path to the main entrance of Church—it is almost 1 third of way—no great distance to right (as to go to the Church)—among se- several tombs which have shrubs & trees thickly planted round them—& even when I last saw it—a good deal covering up this stone.

I am sure it was there in June 1859.[6] It was a good strong upright stone—and I remember well "To a good & dear child".[7]

If it has been taken away it is one of the most audacious and abominable villainies I ever heard of—& without object—as the digging up—carrying off & reworking the stone would cost more than its worth.

I must not write more— Pray rest yourself & give Malvern a fair chance— Our kindest regards to Mrs Darwin | Yours ever | W D Fox

DAR 164: 180

1 The year is established by the relationship between this letter and the letter to W. D. Fox, 4 [September 1863].

2 Letter to W. D. Fox, 4 [September 1863].

3 CD was undergoing treatment at James Smith Ayerst's hydropathic establishment at Malvern Wells, Worcestershire (see letter to W. D. Fox, 4 [September 1863] and n. 1). Both CD and Fox had earlier undergone treatment at James Manby Gully's hydropathic establishment at nearby Great Malvern (*Correspondence* vol. 4).

4 In his letter of 4 [September 1863], CD told Fox that Emma Darwin had been unable to find the gravestone of their daughter Anne Elizabeth, who had died at Great Malvern in 1851, and that they feared it had been stolen.

5 Ellen Sophia Fox.

6 No reference has been found to Fox's visit to Great Malvern in 1859; however, see the letter to W. D. Fox, 4 [September 1863] and n. 5.

7 A photograph of the grave appears opposite p. 631.

From W. F. Kirby 8 September [1863]

29 Nelson Terrace | New Hampstead Road | NW
Sept 8

Dear Sir

I must apologize for not answering your letter of July 9th long before, but I was unable to attend the August meeting of the Entomological Society, and consequently never received it till yesterday.[1] I am very glad you approve of my paper, but it is a very imperfect sketch for want of more extensive materials.[2] There is a slight error in my remarks on the genus *Papilio*, as the number of described species is (in round numbers) not 300 but 400.[3] I think Mr Wallace once remarked to me that the depth of the sea between Corsica and Sardinia and the main land of Italy proved that they had long been separated.[4] Mr Doubleday[5] tells me that *Coenonympha Typhon* (which

I regard as a distinct species from *C. Davus*, though he does not) gradually becomes paler and paler as it advances north till in Scandinavia and the northern islands of Scotland it is almost white and immaculate forming the variety *Isis* of Thunberg.[6] And there is a whitish *Coenonympha* nearly related to this in the British Museum from California. *C Davus* (or rather *C Typhon* I imagine, which is the commoner form) occurs in Amoorland.

On referring to Menetries I find that two or three forms of *Isis* occur in Amoorland, and after examining a large number of specimens, he is of opinion that it is only a northern form of *Davus* (*Typhon*)[7] I think Mr Doubleday once told me that *Isis* occurred in Iceland, but I am not certain. The most perplexing question in the geographical distribution of butterflies is how to account for that of *Pyrameis Atalanta* and *Callirhoe*. I suppose that while *Atalanta*, a species of gigantic power like all of its genus was establishing itself on the continent of Europe and Africa, the Canaries were separated from the mainland so as to prevent its crossing over and crushing out *Callirhoe*. I had suggested this in my paper, but struck it out at the suggestion of Mr Bates who kindly looked it over before I read it, and thought the hypothesis too fanciful.[8] Still this would not account for the occurrence of Atalanta in North America, as I have no evidence of its having occurred east of Asia Minor, and if it occurred in Siberia, such a conspicuous insect must have found its way into some of the excellent Russian works on the Lepidoptera of Northern Asia. To be sure Cramer says that it occurs in China,[9] but if it occurs in Northern Asia at all, it must be very rare, and no dependence whatever can be placed in localities given for insects by the old authors, some of whom used the phrase "ab Indiis" as equivalent to "exotic"[10]

With every wish for the success of your series of works on species,[11] | Believe me, dear Sir | Yours very truly | W. F. Kirby

PS Mr F. Smith[12] will tell you there is a humble-bee (Bombus Smithianus) peculiar to the northern Islands of Scotland I forget which

Postmark: London. W.C. 6 SP 9 63
DAR 205.3 (Letters): 280

CD ANNOTATION
On cover: 'Kirby. Geograph Distribution of ['In' *del*] Butterflies | Possibly useful' *ink*

1 CD's letter to Kirby of 9 July [1863] was addressed to him at the Entomological Society of London; Kirby was an assistant in the zoological department of the British Museum. Kirby refers to the meeting of the Entomological Society on 3 August 1863; he apparently collected the letter at the following meeting, on 7 September (*Transactions of the Entomological Society of London* 3d ser. 1 (1862–4), Proceedings, pp. 164 and 167).

2 Kirby 1863.

3 Kirby 1863, p. 484.

4 See letter to W. F. Kirby, 9 July [1863] and n. 2. The reference is to Alfred Russel Wallace.

5 Henry Doubleday.

6 Carl Peter Thunberg named this variety in 1791 (*Index animalium*).

7 Ménétriés 1859, pp. 43–4.

[8] Henry Walter Bates. Kirby discussed the distribution of *Pyrameis atalanta* and *P. callirhoe* in Kirby 1863, p. 488.
[9] Cramer *et al.* 1779–91, 1: 132.
[10] *Ab Indiis*: from the Indies.
[11] CD had described *Origin* as an abstract of a larger work he was preparing on natural selection (*Origin*, p. 1). He intended to publish the larger work in three volumes (see *Correspondence* vol. 7, letter to T. H. Huxley, 16 December [1859], and letter to John Murray, 22 December [1859]); however, the only part of the projected work to appear in his lifetime was *Variation*, which was published in 1868.
[12] Frederick Smith was a colleague of Kirby's at the British Museum.

To John Price [8 September – 13 October 1863][1]

Malvern. Wells
Tuesday

Dear Price,

I have been ill, & am very weak. I do read your numbers—& last night I read Heureka.[2]

It seems to me as good as, perhaps better, than, the other articles.—

I am astonished at your vivid recollection of old times.[3]

I wonder you do not give us more about the wilder parts of North Wales, which I admire with the fervour of a Welchman.[4]

I sincerely wish you all success & that you may improve & please many readers—

I can write no more.—

Yours very sincerely | Ch. Darwin.

Copy
DAR 147: 273

[1] The date range is established by the reference to Price 1863–4, and by CD's address. According to Emma Darwin's diary (DAR 242), CD stayed in Malvern Wells, Worcestershire, from 3 September to 12 or 13 October 1863; the first Tuesday after his arrival was 8 September.
[2] The reference is to Price's miscellaneous work, *Old Price's remains* (Price 1863–4), published in monthly parts between April 1863 and March 1864. CD refers to a humorous article in the number for September 1863, entitled '*Ε῾ΥΡΗΚΑ, Ε῾ΥΡΗΚΑ!*' (Eureka, eureka!) (*ibid.*, pp. 277–82). CD's copy of the work is in the Darwin Library–CUL.
[3] Price was a childhood friend of CD's; they were at Shrewsbury School together (see *Correspondence* vol. 1).
[4] Shropshire, of which CD's home town Shrewsbury is the county town, borders on north Wales, and CD made several trips there as a boy (see *Correspondence* vol. 1 , 'Journal' (Appendix I)); he continued to visit Wales while he was at Cambridge and afterwards (see *Correspondence* vol. 2, 'Journal' (Appendix II)).

From Hugh Falconer 10 September 1863

Geological Society Somerset House, | W.C.
10th. Septr. 1863

My Dear Darwin

I sent about a week ago Suess's Essay to my niece, in Edinburgh to translate for me,[1] a most ignoble confession on the part of the Foreign Secy Geol. Society.[2]

I told her as soon as the translation was finished to send the original to you to Malvern Wells.[3]

Yours Ever | H Falconer

I start tomorrow morning for Auvergne—for a months run—on duty—old bone department.[4]

DAR 164: 18

[1] Suess 1863; see letter to Hugh Falconer, 4 [September 1863]. On 5 September 1863, Falconer sent CD's letter of 4 [September 1863] to his niece, Grace Anne McCall, commenting: 'Your note of yesterday acknowledging the pamphlet by Suess, has just come to hand. I enclose a note from Darwin which will amuse you. You will see by it, that the said pamphlet is in request' (Falconer Museum, Forres, ZF/97). Falconer again wrote to his niece on 20 September: 'Remember that Charles Darwin has written to me for the German pamphlet and that I told him you would forward it to him' (Falconer Museum, Forres, ZF/110).

[2] Falconer was elected foreign secretary of the Geological Society of London on 20 February 1863 (*Quarterly Journal of the Geological Society of London* 19 (1863): xviii).

[3] CD was undergoing treatment at James Smith Ayerst's hydropathic establishment at Malvern Wells, Worcestershire (see 'Journal' (Appendix II)).

[4] Falconer refers to his ongoing research on the palaeontological evidence relating to prehistoric humans, and on Pleistocene mammals (*DSB*). Falconer had arranged to visit the Auvergne with Edouard Lartet, who organised visits to palaeontological collections and meetings with palaeontologists (see the letter from Lartet to Falconer of 3 September 1863 in the Falconer Museum, Forres, ZF/300). Falconer travelled to Le Puy, in the Velay region, and to Clermont, the capital of the Auvergne, between 14 and 20 September, and also visited museums in Paris and Chartres, returning to London on 23 September (see the letters from Falconer to Grace McCall, 15 September 1863 and 20 September 1863, in the Falconer Museum, Forres, ZF/98 and 110, and Falconer 1868, 2: 479).

From the secretary of the Philosophical Institute of Canterbury, New Zealand 14 September 1863[1]

Philosophical Institute of Canterbury,
New Zealand

Christchurch,
14 Septr. 1863.

I have the honor to inform you, that at the Ordinary Meeting of the Philosophical Institute of Canterbury, held on the 7th of August last—you were unanimously—elected an Honorary Member of the Society.[2]

I have the honour to be, | Your most obedient Servant, | E. C. J. Stevens Hon. Secretary.[3]

Charles Darwin Esquire. F.R.S. | &.c. &.c. &.c. | Down, Bromley | Kent

LS
DAR 230

[1] This is a printed letter with the dates, signature, name and address of addressee, and the word 'unanimously', added by hand.

[2] See also letter from Julius von Haast, 21 July [– 7? August] 1863 and n. 20.

[3] Edward Cephas John Stevens.

From J. D. Hooker 15 September 1863

Kew
Sept 15/63

Dear old Darwin

I really have nothing worth writing about, but obey the Spirit's motion to creep towards you with a letter— Thwaites of Ceylon seems to suffer as you do from Nettle rash, swelling up terribly— the slightest acid brings it on with him.[1]

I shall take care that you have a *Calamus* (climbing Palm) when you are ready for it.[2]

I sent you a letter from A Gray two days ago.[3]

Miss Henslow' died last week at Cheltenham & I spent the week there being her chief executor I had a lot of matters to attend to besides burying her.[4]

I am working away at the New Zealand Flora—[5] Your admission of a previous continental extension of that Island excites all my attention—& I am most anxious to know what you have to say on the subject. Now that you have so taken it up I expect that great light will be thrown on the whole subject of Southern distribution—what I have longed to see some master hand at work upon.[6]

I am still extremely puzzled at the prodigious & undue amount of variation amongst new Zealand plants;—whole genera *more* undifferentiated by far than Rosa & Rubus with us, i.e. the genus presenting far greater diversity of form & yet the species all as intimately connected by intermediates. This may be due to there being plenty of space for preservation of intermediates: Can this be accounted for by the land rising, & increased space thus being afforded— We (you & I) clash a little here Extreme diversity of form should according to you, follow on much destruction of individuals.—[7] According to me if you give space enough you will have as much variation as if you killed off in a smaller area, & New Zealand looks like a case in point.— The quantity of marvellous odd form of Veronica, e.g. from the Middle Island[8] is quite extraordinary & yet however odd, you can trace their origin in slight characters of previously discovered forms. An enormous proportion of the N.Z. Flora is made up of endemic species of a few rather widely distributed genera—more so than any other country I know of— I am getting intensely interested in the work.—but "heu me miserum"[9] I am only half through the descriptive work! And now in my old age I am threatened with having to do Flora Indica, which I had abandoned all intention of taking up long ago.[10] The Indian Govt. talk of commissioning me to do it.[11] Pay would tempt me, but only because it would hold out a prospect of early retirement from the struggle of scientific work for one's livelyhood—& shaking the dust off my feet at the Govt & Kew Gardens—but for God's sake let this go no further— I regard succession to my fathers place with horror.[12] Not that a better scientific place exists in the world *except my own*. I am beginning too to hate the οἱ πολλοι [13] of Science— Huxley, Lubbock[14] & half a dozen others are enough for me, of the workers, outside my own immediate pale which includes only yourself Bentham Oliver & Thomson.[15] As to Murchisonian science[16] & all that sort of thing, like K.C B^s.[17] it makes me sick to read his science at the New Castle meeting.[18]

Lyell I have not seen even for 6 months.—nor the Horners nor Falconer,—[19] perhaps when the Phil. Club meets again I shall feel a little more charitable & loving to my betters—[20] meanwhile I am sunning myself in the vacation of scientific Society as far as I am concerned, & besotting over Wedgewood ware.[21]

On Monday I have to go down to Chatsworth[22] to inspect the plant cultivation there & to Biddulph Grange (the Batem⟨an⟩ place)[23] & shall probably break away to Stafford & see the potteries[24] (don't sneer) but if you know any one there that has interest to show me over a good pottery establishment ⟨send⟩ me a line care of Jas ⟨Bateman⟩ Esqre. Biddulph Gra⟨nge⟩ Cheshire.—where I s⟨hall be⟩ Monday to Wednes⟨day⟩ ⟨Let⟩ me know how you a⟨ll are⟩ getting on.— Henrietta[25] ⟨ ⟩ & all

I saw Huxley the other da⟨y⟩ looking well, you heard I suppose ⟨of⟩ the awful death of his brother.—[26] The N.H. Review is made over to Sclater & Oliver as Editors ⟨in C⟩hief.—& the thing is to be made more of a Review of next year.[27]

We are all extremely well— My wife sends her love to you & Mrs Darwin

Ever dear Darwin | Yr aff | J D Hooker

DAR 101: 163–6

CD ANNOTATIONS

1.1 I really . . . burying her 4.2] *crossed pencil*

Top of letter: 'New Zealand | Character of Species—Varying' *pencil*

[1] George Henry Kendrick Thwaites was superintendent of the Botanic Gardens, Peradeniya, Ceylon (R. Desmond 1994).

[2] See letter to J. D. Hooker, [28 August 1863] and n. 10.

[3] The reference is apparently to the letter from Asa Gray, 1 September 1863, which Asa Gray had presumably enclosed with his letter to Hooker of the same date (Asa Gray letters: 332, Royal Botanic Gardens, Kew). Gray often included letters to CD in packets sent to Hooker, presumably to save on postage. Gray addressed his letter to CD at Down, and Hooker apparently omitted to correct the address, since the letter bears a London western district postmark for 14 September (presumably Kew), a London south-eastern district postmark for 15 September (presumably Down), as well as Malvern and Malvern Wells postmarks for 16 September.

[4] The reference is to Hooker's aunt by marriage, Anne Frances Henslow, who died at Cheltenham, Gloucestershire, on 30 August 1863 (*Gentleman's Magazine* n.s. 15 (1863): 520).

[5] J. D. Hooker 1864–7.

[6] See letter to J. D. Hooker, [28 August 1863] and n. 7. See also letter to J. D. Hooker, [22–3 November 1863].

[7] Hooker refers to CD's doctrine of the 'divergence of character', according to which 'the more diversified the descendants from any one species become in structure, constitution, and habits, by so much will they be better enabled to seize on many and widely diversified places in the polity of nature, and so be enabled to increase in numbers' (*Origin*, p. 112). CD argued that this process of divergence of species involved a great deal of extinction, such that 'all the intermediate forms between . . . the less and more improved state of a species, as well as the original parent species itself, will generally tend to become extinct' (*Origin*, p. 121). On CD's doctrine of divergence, see Ospovat 1981, pp. 170–209, and Browne 1983, pp. 210–20.

John Goodsir. Oil on canvas.
(Courtesy of the Royal College of Surgeons of Edinburgh.)

William Brinton
Photograph by C. T. Newcombe, before 1862.
(Courtesy of the Wellcome Institute Library, London.)

Gravestone of Anne Elizabeth Darwin,
Priory churchyard, Great Malvern.
Photograph by James Moore.

[8] 'Middle Island' is now called the South Island of New Zealand.

[9] *Heu me miserum*: Alas! wretched me!

[10] Hooker had projected a *Flora Indica* with Thomas Thomson during his Indian botanical expedition of 1847–50. He and Thomson completed one volume of the planned work in 1855 (J. D. Hooker and Thomson 1855). However, with Thomson's return to India in 1854 and Hooker's appointment as assistant director of the Royal Botanic Gardens, Kew, in 1855, and in the absence of the anticipated financial assistance from the East India Company, the work was laid aside. See L. Huxley ed. 1918, 1: 339–42 and 354–62.

[11] Starting in 1861, Hooker had sought the assistance of his friend Thomas Anderson in bringing a proposal before the Indian government to fund his *Flora Indica*. The Indian government officially sanctioned the *Flora* on 10 November 1863, but Hooker did not resume work on the project until 1870, when he began to prepare a more circumscribed work entitled *Flora of British India* (J. D. Hooker 1875–97). See L. Huxley ed. 1918, 1: 354–62 and 2: 12–17.

[12] Hooker was assistant director of the Royal Botanic Gardens, Kew, where his father, William Jackson Hooker, was director (R. Desmond 1994).

[13] οἱ πολλοι: (hoi polloi) 'the many: the rabble' (*Chambers*).

[14] Thomas Henry Huxley and John Lubbock.

[15] Daniel Oliver was Hooker's colleague at the Royal Botanic Gardens, Kew, where Thomas Thomson and George Bentham also worked in the herbarium (see letter from J. D. Hooker, [21 July 1863], and *DNB* s.v. Bentham, George).

[16] Hooker refers to the geologist, Roderick Impey Murchison, who was adept at using social influence to further his scientific ambitions (*DNB*, *DSB*, and Stafford 1989). Hooker particularly objected to Murchison's tendency to lionise travellers before fully establishing the value of their reports, and he regarded the Royal Geographical Society, of which Murchison was a former president and prominent member, as a 'sort of seton upon science' which drew 'all odium for scientific lion-hunting, toadying and tuft-hunting away from the Linnean, Royal and Geological [Societies]—only that the latter are too fond of following in [the] wake!' (L. Huxley ed. 1918, 1: 407).

[17] Murchison had been made a Knight Commander of the British Empire (KCB) in 1863 (*DNB*).

[18] The reference is apparently to Murchison's inaugural address as president of the geographical section of the British Association for the Advancement of Science at the annual meeting held in Newcastle-upon-Tyne from 26 August to 2 September 1863 (Murchison 1863a). In his address, Murchison described the development of British geography over the previous twenty-five years, dwelling particularly on the progress of colonial exploration. Extracts from the address appeared in *The Times*, 28 August 1863, p. 7, and the *Athenæum*, 5 September 1863, p. 311. An editorial discussion of the speech in *The Times*, 29 August 1863, p. 8, began: 'Sir Roderick Murchison's address to the British Association describing the progress of geographical discovery is almost too triumphant. There will be very soon little left to discover.' Murchison also read a geological paper at the meeting (Murchison 1863b), which was reported in *The Times*, 3 September, p. 8, and the *Athenæum*, 26 September 1863, pp. 403–4.

[19] Hooker refers to Charles Lyell, Leonard Horner and his family, and Hugh Falconer.

[20] The Philosophical Club of the Royal Society next met on 29 October 1863 (Bonney 1919, p. 164).

[21] Hooker had recently become an avid collector of Wedgwood ware (see *Correspondence* vol. 10, letter from J. D. Hooker, [27 or 28 December 1862], and this volume, letter from J. D. Hooker, 6 January 1863).

[22] Hooker refers to Chatsworth House, Derbyshire, home of William Cavendish, seventh duke of Devonshire, which possessed 'one of the most magnificent gardens in Britain' (Hadfield *et al.* 1980, p. 98). See also letter from J. D. Hooker, 26 August 1863.

[23] The botanist and horticulturalist James Bateman had created 'one of the most remarkable gardens in England' at Biddulph Grange, north Staffordshire (Hadfield *et al.* 1980, p. 30). See also letter from J. D. Hooker, 26 August 1863.

[24] Hooker refers to the district of north Staffordshire known as 'the Potteries', which was the principal site of the English china and earthenware industries (*EB*). The Wedgwood works were at Etruria,

near Hanley, one of the principal towns of the Potteries; Biddulph Grange is approximately seven miles north of Hanley.

25 Henrietta Emma Darwin. The Darwin family were staying in Malvern Wells, Worcestershire, where CD was undergoing treatment at James Smith Ayerst's hydropathic establishment (see letter to W. D. Fox, 4 [September 1863]).

26 George Knight Huxley had died suddenly, early in August 1863 (L. Huxley ed. 1900, 1: 248).

27 Since the start of the new series in 1861, the *Natural History Review* had been edited by a 'commissariat' of eleven co-editors, with T. H. Huxley as *de facto* editor-in-chief. However, Huxley relinquished this role in July 1863 because of the pressure of work. See L. Huxley ed. 1900, 1: 209–10. Philip Lutley Sclater and Daniel Oliver were co-editors of the new series of the journal, which contained a mixture of reviews and original articles.

From Edward Blyth 21 September 1863

19 Essex St, Strand—
Septr. 21/63—

My dear Sir,

I have the pleasure to send you some evidently original observations on the variation of our native ferns, by some practical fern-grower,[1] which you may consider worth finding place for in your series of exemplifications.[2]

You have probably not heard of me for some time. I am now fixed in England, having obtained a small annual pension, sufficient to live upon;[3] and just now I am trying to find some place of abode at Kew or in its vicinity, where I may have the advantage of proximity to the B. Gn.,[4] & be away from the turmoil of London, which I abominate. Eventually I shall probably settle at a greater distance from the metropolis, in some quiet place by the sea-side.

Trusting that your health and strength are such as to enable you to enjoy existence and to continue your valuable la⟨bors and⟩ researches, | I remain | Ever Sincerely Yours, | E Blyth.

DAR 160: 206

1 The enclosure was apparently a clipping headed 'The fern mania', now in DAR 205.10: 3, which was taken from an unidentified newspaper. The text of the clipping had itself been extracted from an article entitled 'British ferns and their varieties', which appeared in the *Gardener's Weekly Magazine and Floricultural Cabinet* on 5 September 1863 (*Gardener's Weekly Magazine and Floricultural Cabinet* 5 (1863): 286–8; quotation on p. 288); the article was unsigned, but the journal was conducted by Shirley Hibberd. The author stated: 'It is a fact of no less interest to the philosopher speculating on the origin of species than to the fern-grower, who cares not a jot about that grave question, that to form a complete collection of British ferns and their varieties is simply impossible.'

2 Blyth apparently refers to CD's statement, in the introduction to *Origin*, that the work was an abstract of a larger one that he had been preparing, and that he hoped to publish a work giving 'all the facts, with references,' on which his conclusions had been grounded (*Origin*, p. 2).

3 Blyth returned to England from India, where he had been curator of the museum of the Asiatic Society of Bengal, early in 1863 (see letter from Edward Blyth, 27 March 1863 and n. 2). He was granted a pension of £150 by the British government in 1863 (*Proceedings of the Asiatic Society of Bengal* 33 (1864): 73). See also *Correspondence* vol. 10, letter from Edward Blyth, 23 November 1862 and nn. 1, 2, and 4.

[4] The reference is to the Royal Botanic Gardens, Kew.

From John Scott 21 September [1863][1]

Edinburgh | Botanic Gardens
Sep^{t}. 21st.

Sir.

I have now finished my work on Primulas for this season.[2] The greater part of it you will find in my Manuscript which I send off by to-days mail.[3] You kindly promised to review it for me; and should it prove worthy, lay it before the Linnæan Society.[4] I shall therefore be glad—if you can spare time to look it over—to re-write it in accordance with your criticisms. I am sorry that I had no ⟨opportunity to⟩ consult as to arrang⟨ement⟩ as I fear that my want of ⟨order⟩ will give you much unnecessary trouble.

I sent you a few weeks ago two copies of my paper on Sterility of Orchids.[5] I hope they came duly to hand.

In your last letter you asked about seeds of the *Linum monogynum*.[6] I have had a long-styled plant from a friend: it has now *four fine plump capsules* upon it, which I will send you when ripe.[7]

I will send you shortly results of my work on Peloric Antirrhinums,[8] and Verbascums.[9] The latter promise ⟨to be very⟩ interesting, they have afforded ⟨evidence of individual⟩ ⟨s⟩terility, both male ⟨and femal⟩e proving good by the action ⟨of different⟩ ⟨spe⟩cies

It will be later in the season before I can send results on the Passifloras & Maize.[10] The former promise to be of great interest.

Some months ago I mentioned that I had observed *female flowers* on the *terminal male spikes* of Maize.[11] If you would *now* like to see such a spike I can send you one.

I shall be glad to hear what you think of my Primula work when you have a little time to spare.[12] And for the present | I remain | Yours respectfully & obliged | J. Scott

DAR 177I 96

CD ANNOTATION
Top of first page: 'Typical form Acropera'

[1] The year is established by the relationship between this letter and the following letter.
[2] See letters from John Scott, [3 June 1863] and 23 July [1863].
[3] Scott refers to the manuscript of Scott 1864a.
[4] See letters to John Scott, 25 and 28 May [1863] and 6 June [1863]. CD communicated Scott 1864a to the Linnean Society; it was read on 4 February 1864.
[5] Scott 1863a. CD's copy of this paper is in the Darwin Pamphlet Collection–CUL. Scott had sent a second copy of the paper for CD to give to Joseph Dalton Hooker (see letter from John Scott, 16 June [1863], and letter to John Scott, 20 [June 1863]).
[6] See letter to John Scott, 1 and 3 August [1863].
[7] See *Correspondence* vol. 12, letter from John Scott, [13 January 1864].

[8] CD had suggested that Scott conduct crossing experiments with peloric plants (see *Correspondence* vol. 10, letter to John Scott, 19 December [1862]; see also this volume, letters from John Scott, 6 January 1863 and 16 June [1863], and letters to John Scott, 6 June [1863] and 20 [June 1863]). Scott sent CD the results from his experiments in 1864, but they differed from CD's own, and were not included in *Variation* (see *Correspondence* vol. 12, letters from John Scott, 19 March 1864, 10 June [1864], and 20 June [1864], and *Variation* 2: 166–7).

[9] At CD's suggestion, Scott had undertaken to repeat some of the crossing experiments of Karl Friedrich von Gärtner with differently coloured varieties of *Verbascum* (see *Correspondence* vol. 10, letters to John Scott, 19 November [1862], 11 December [1862], and 19 December [1862], and letter from John Scott, 17 December [1862]). Scott published his results from these experiments in Scott 1867. See also letter to John Scott, 7 November [1863].

[10] CD had encouraged Scott to carry out crossing experiments on *Zea mays* (maize) and species of *Passiflora* similar to those discussed in *Origin*, pp. 250–1 and 270 (see *Correspondence* vol. 10, letters to John Scott, 11 December [1862] and 19 December [1862], and letter from John Scott, 17 December [1862]). See also this volume, enclosure to letter to John Scott, 6 March 1863, letters from John Scott, 21 March [1863] and n. 4, and 21 May [1863] and n. 17, and letters to John Scott, 24 March [1863] and 7 November [1863]. Scott published the results of his experiments with *Passiflora* in Scott 1864d.

[11] See *Correspondence* vol. 10, letter to John Scott, 19 November [1862], and letter from John Scott, [20 November – 2 December 1862]. See also Scott 1863a.

[12] See letter to John Scott, 7 November [1863].

From Emma Darwin to John Scott 23 September [1863][1]

Malvern Wells
Sep 23.

Sir

Mr C. Darwin requests me to thank you for your letter[2] & to say that he is too unwell to write or even to read any thing.[3] He begs you will not send the Primula Paper until after his return home when he will write himself—[4]

I am dear Sir | Yours truly | Emma Darwin

DAR 93: B1–2

[1] The year is established by the address; according to Emma Darwin's diary (DAR 242), CD stayed at Malvern Wells, Worcestershire, from 3 September to 12 or 13 October 1863.

[2] Preceding letter.

[3] According to Emma Darwin's diary (DAR 242), CD was repeatedly sick in the period between 20 and 23 September 1863. See also letter from Emma Darwin to W. D. Fox, [29 September 1863].

[4] The reference is to the manuscript of Scott 1864a; see preceding letter.

From Joanna B. Horner 24 September 1863

Hamilton House. Felixstowe
24 Sepr 1863.

My dear Mr Darwin

I blame myself much for not having fulfilled my intention sooner to have written to thank you for your very kind and prompt reply to me about the Pigeons.[1] Lady Napier was so very much obliged, as you just gave her the information she required.[2]

I have been spending about a fortnight here with Leonora and her little girls,[3] and find it a nice healthy place for children, though as we are 13 miles from a Railroad it is in a very primitive state, and very dead, without much pretty scenery. The sea is always however a pleasant object, and we search for shells among the crag cliffs.

I hope that you and M[rs] Darwin and all your circle have enjoyed Malvern and benefitted by it—as I think you were going there when I heard from you.[4] We return to the Bunbury's house Barton Hall in a few days where we have left Papa and Susan[5] We have good accounts from Scotland where are still both houses of Lyell but we are looking forward to gathering in London early next month.[6]

With our united best regards to all your party Believe me dear Mr Darwin with many thanks for the trouble you took to answer my questions yours very sincerely Joanna B. Horner

DAR 166.2: 269

[1] Neither Horner's earlier letter nor CD's reply has been found.

[2] The reference may be to Frances Napier, whose brother-in-law, William Francis Patrick Napier, had been known to the Horner family (K. M. Lyell ed. 1890, 2: 48).

[3] Leonora Pertz was one of Joanna's five sisters (Freeman 1978); she was on a visit to England from Germany with her daughters, Annie and Dora (K. M. Lyell ed. 1890, 2: 359).

[4] CD, Emma Darwin, and their family had gone to Malvern Wells, Worcestershire, at the beginning of September so that CD could undergo a course of treatment at James Smith Ayerst's hydropathic establishment (see 'Journal' (Appendix II)).

[5] Joanna refers to her father, Leonard, and her unmarried sister, Susan (Freeman 1978). Another sister, Frances, and her husband, Charles James Fox Bunbury, lived at Barton Hall, Great Barton, near Bury St Edmunds, Suffolk (*Post Office directory of Cambridge, Norfolk, and Suffolk* 1865). The Horners had apparently gone to stay at Barton Hall on 18 August 1863 (K. M. Lyell ed. 1890, 2: 360).

[6] Joanna's eldest sister, Mary Elizabeth, was the wife of Charles Lyell; another sister, Katherine Murray, had married Lyell's younger brother, Henry (Freeman 1978). Joanna apparently refers to Kinnordy, Kirriemuir, Angus, the Lyell family seat; both Charles and Henry Lyell also had houses in London (*Post Office London directory* 1863). Various members of the Lyell and Horner families met in London for Leonard Lyell's birthday party on 21 October 1863 (Ames ed. 1909, 1: 446–7).

From Emma Darwin to John Scott 24 September [1863][1]

Malvern Wells
Sep. 24

Dear Sir

Mr Darwin desires me to say that your M.S. is safely arrived & as there is no chance of his being able to study it for several weeks would you like it returned to you or shall he keep it?[2]

I am dear Sir | yours truly | E. Darwin

Mr Darwin desires me to thank you for the papers on Sterility of Orchids.[3] He sent one to the Botanische Zeitung & one to D[r] Hooker, but he has not heard from D[r] Hooker since.[4]

The first thing Mr D can apply his mind to shall be your paper but I am afraid that it must be some time first.[5]

DAR 93: B3–4

[1] The year is established by the address; according to Emma Darwin's diary (DAR 242), CD stayed at Malvern Wells, Worcestershire, from 3 September to 12 or 13 October 1863.
[2] The reference is to the manuscript of Scott 1864a, which Scott had sent to CD for comment (see letter from John Scott, 21 September [1863]).
[3] Scott 1863a; see letter from John Scott, 21 September [1863].
[4] The reference is to Joseph Dalton Hooker; see letter from John Scott, 21 September [1863] and n. 5.
[5] See letter to John Scott, 7 November [1863].

From G. H. K. Thwaites 24 September 1863

Peradenia, Ceylon
24th Sept 1863

My dear Darwin,

Very very many thanks for the capital photograph of your excellent self;[1] I value it greatly; it recalls to my memory so vividly and pleasantly the delightful day I spent with you and our dear friends Hooker and Harvey at Nuneham many years ago.[2]

I am glad you were pleased with the flowers of the *Cassia*:[3] my friend Mr. Glenie is so delighted at having been able to communicate something to interest you that he promises to observe carefully the changes that take place in the flowers of the different species of Cassia during the time they are expanded.[4] With reference to another species of *Cassia* (C. Roxburghii, DC.)[5] he writes me

> "We have been examining the C. Roxburghii of late carefully, and find in it, as well as in C. Fistula, a peculiarity though not of the same marked nature, in the anthers.— The 3 upper appear to be abortive;—the middle 4 open by pores only; the lower 3 (the longest) open by pores and by clefts at the base of the lobes"

I will collect for you and send you as soon as I can, flowers of all the *Cassias* and other *Cæsalpinieæ* I can get here. It seems likely to me that the careful study of the differences *inter se* of the closely allied forms or species of one group, and the comparison of these with corresponding differences in other groups, will be one of the modes of iluminating the law which has operated in bringing about these differences, whether it be that of natural selection or of a law operating with it or independently of it; for such a vast period of time seeming to be a necessary element in the process, there would appear to be slight hope of direct experiment (in selection) leading to any result beyond what we see in the races of dogs, or of other domestic animals producing fertile crosses. But do these fertile crosses between very dissimilar races shew any tendency to degenerate or disappear after a certain number of generations? The mixed race here, between the Portuguese and

Cinghalese, are not a healthy set of people judging from those I have had anything to do with, and they certainly will not compare in symmetry or good looks with either of the original races.

Believe me always | My dear Darwin | Most sincerely yours | G. H. K. Thwaites.

DAR 48: 74

CD ANNOTATIONS

1.1 Very . . . ago. 1.4] *crossed pencil*

On cover: '**Grdation** | **Cassia**' *pencil*[6]

[1] At Thwaites's request, CD had sent a photograph of himself with his letter to Thwaites of 29 July [1863].

[2] Joseph Dalton Hooker and William Henry Harvey. CD and Thwaites first met at the annual meeting of the British Association for the Advancement of Science held in Oxford in 1847 (see *Correspondence* vol. 5, letter to G. H. K. Thwaites, 10 December 1855); during the meeting an excursion was made to Nuneham, a beauty spot near Oxford (see *Correspondence* vol. 4, letter to J. S. Henslow, [26 September 1849]).

[3] See letter from G. H. K. Thwaites, 8 June 1863, and letter to G. H. K. Thwaites, 29 July [1863].

[4] The flower of *Cassia fistula* enclosed in the letter from G. H. K. Thwaites, 8 June 1863, had been collected by Samuel Owen Glenie; Thwaites also sent CD Glenie's covering letter describing the specimen.

[5] Thwaites enclosed with this letter a packet containing a pressed flower, labelled 'Cassia Roxburgii, DC.'. The flower is in DAR 48: 74. 'DC' is the botanical abbreviation for Augustin Pyramus de Candolle, who named *Cassia roxburghii* in 1825 (Candolle and Candolle 1824–73, 2: 489).

[6] See the letter to G. H. K. Thwaites, 29 July [1863] and n. 5.

From John Scott to Emma Darwin 25 September [1863][1]

Edinburgh | Botanic Gardens
Sepr 25th.

Madam.

I beg to acknowledge the receipt of your notes of the 23^{d} & 24th. ult.[2] I deeply regret to find by them that M^{r}. Darwin is again labouring under such a weak state of health. I hope & trust that you may be deceived in your anticipations of a tardy recovery, and that you may soon have the satisfaction of seeing him again in the full enjoyment of health.

Please convey to M^{r}. Darwin my best thanks for the trouble he has given himself with my paper, and also my gratefulness for the honour he has done me in considering it at all worthy of his notice.[3]

There is no occasion, whatever, for your returning my M.S. on account of M^{r}. Darwin's weak state of health. Presuming that he had been enjoying his usual state of health, I knew full well that his time was too much engaged to permit for a moment the idea that he would at once examine my paper. I merely sent it that it might await his convenience.

Please retain it there until M^{r}. Darwin is again well, and perhaps when he has a little time to spare, he will glance over it & see whether or not it is worthy of his study.— It is far too kind of him thinking for a moment of making my paper his first study—time will then be too precious for his wasting it on aught of mine.

And now with kindest wishes for Mr. Darwins speedy recovery, and cordial thanks for the trouble you have taken in writing me | I remain | Madam | Yours very respectfully | John Scott.

DAR 177: 97

[1] The year is established by the relationship between this letter and the letters from Emma Darwin to John Scott, 23 September [1863] and 24 September [1863].

[2] Letters from Emma Darwin to John Scott, 23 September [1863] and 24 September [1863].

[3] Scott refers to the manuscript of Scott 1864a, which he had sent to CD for comment (see letter from John Scott, 21 September [1863]).

From W. B. Tegetmeier [*c.* 26 September 1863][1]

Dear Sir

Excuse this hurried note but I thought you would like to see the Enclosed from last weeks Field Newspaper.

Yours most truly | W B Tegetmeier

Do not trouble to acknowledge receipt.

[Enclosure][2]

A fact for Mr Darwin—zebra-striped ass.—[3] There is a grey coloured female jackass, about 15 years old, in the possession of a fruiterer in this neighbourhood, which has the four legs striped like a zebra up to the knees and hocks respectively. The stripes grow fainter behind; on the forelegs anteriorly, though not complete as in the zebra, they are very distinct and symmetrical. The muzzle and jaw, half way up to the eye, are black, and there is a broad black band at the base of each ear. All the history I could obtain about the animal was that "it came out of Suffolk," and that it was "a rare good 'un"—this last exclamation being produced by an offer on my part to buy the skin of the animal, should it ever die, a fact which is, I suppose, somewhat problematical, as one never sees a dead ass. I opine, from the above exclamation and other "hints," that its owner would rather I did not wait for its decease, but is quite ready to part with it now for "a consideration."—C. R. Bree, M.D. (Colchester)[4]

DAR 178: 59

[1] The date is established by the date of publication of the enclosed abstract from the *Field* newspaper (see n. 2, below).

[2] The enclosure, which appeared in the *Field*, 19 September 1863, p. 298, is with Tegetmeier's letter in DAR 178: 59.

[3] CD was interested in the inheritance of shoulder stripes in horses and asses, and had asked several of his correspondents to report any observations they had made on this subject (see *Correspondence*

vols. 6–8); he considered the presence of stripes to be a reversion to a primitive characteristic of an ancient progenitor (see *Origin*, pp. 163–7, and *Variation* 1: 55–64; 2: 41–3).

[4] The zoologist Charles Robert Bree was a surgeon at the Essex and Colchester Hospital (*Medical directory*). There is no record of any correspondence between CD and Bree on the inheritance of shoulder stripes in asses; Bree was not cited in *Variation*. Bree had been a strong opponent of *Origin* (see Bree 1860 and *Correspondence* vol. 8).

From A. R. Wallace 26 September 1863

5, Westbourne Grove Terrace, W.
Sept[r]. 26th. 1863

My dear Mr Darwin

I enclose you some flowers of a *Melastoma* just received from a friend at Singapore—[1] Unfortunately he gives me very little information about them except that "*in every case they were swarming with ants*" Perhaps by examining the flowers you can find out something.[2]

My friend Mr. Tristram[3] informed me the other day of an interesting fact on acclimatization of plants similar to that of the *rhododendrons* mentioned by D[r]. Hooker.[4] I note the particulars on the opposite page.

I have seen quite a number of *striped horses* in London— At least 4 or 5 Cab horses striped on the legs all more or less *clay coloured*, & lately a pony, with strong dorsal stripe, *two shoulder* stripes, & bands on fore legs.[5]

I hope you are now better in health & that we may soon hope to have your volume on "Domestication &c."[6]

I have the bees comb of oval cells promised from two friends in the East but there seems some difficulty in getting it.[7]

With best wishes I remain | My dear Mr Darwin | Yours very faithfully | Alfred R. Wallace—

C. Darwin Esq.

"W. E. Surtees Esq. (of Seaton Carew, Durham),[8] had a quantity of furze killed by the frost (Xmas 1860) at an estate of his in Devonshire, *all except a small patch which he had raised himself from seed from the neighbourhood of Aberdeen*"

Communicated by Mr. Surtees to the Rev[d]. H. B. Tristram, Greatham Vicarage, Durham

DAR 47: 146–7

CD ANNOTATION

8.1 "W. E. Surtees … *Aberdeen*" 8.3] 'Acclimatisation under Nature' *added pencil*

[1] This individual has not been identified.

[2] CD was interested in the pollination mechanisms of the Melastomataceae, a family in which many genera have two sets of stamens of different length within the same flower; he suspected that they might exhibit functional dimorphism (see *Correspondence* vol. 10). He had sought information earlier in the year from a number of correspondents, apparently including Wallace, on the identification of

insects frequenting flowers of species of Melastomataceae (see letter from A. R. Wallace, [23 January 1863?] and n. 2). In *Melastoma* the two sets of stamens are of very unequal length (J. C. Willis 1973).

3 Henry Baker Tristram.

4 The reference is to Joseph Dalton Hooker and to J. D. Hooker 1853–5, p. xi and n. CD cited Hooker's observations in *Origin*, p. 140, stating:

> we have evidence, in the case of some few plants, of their becoming, to a certain extent, naturally habituated to different temperatures, or becoming acclimatised: thus the pines and rhododendrons, raised from seed collected by Dr. Hooker from trees growing at different heights on the Himalaya, were found in this country to possess different constitutional powers of resisting cold.

5 In *Origin*, pp. 163–7, CD explained the occasional appearance of stripes in horses and several other equine species in terms of reversion to an ancestral type; he added further examples to subsequent editions (see Peckham ed. 1959, pp. 313–7). CD also discussed the phenomenon in *Variation* 1: 56–61 and 2: 351.

6 *Variation* was not published until 1868.

7 No correspondence discussing this honeycomb has been found. In *Origin*, pp. 224–35, CD had sought to account for the cell-making instinct of hive bees in terms of natural selection, describing the gradation in form between the cells made by different species of bee. See Prete 1990.

8 William Edward Surtees had residences at Seaton Carew, West Hartlepool, Durham, and Tainfield House, Taunton, Somerset (*Men-at-the-bar*).

From J. D. Hooker [28 September 1863][1]

Dear dear friend

My darling little 2d. girl died here an hour ago,[2] & I think of you more in my grief, than of any other friend.[3] Some obstruction of the bowels carried her off after a few hours *alarming* illness—with all the symptoms of strangulated Hernia.

Your affectionate | J D Hooker

Kew Monday.

DAR 101: 159

1 The date is established by the reference to the death of Maria Elizabeth Hooker (see n. 2, below).

2 Maria, the Hookers' second daughter, died on 28 September 1863, aged 6 (L. Huxley ed. 1918, 2: 61). Hooker's reaction to her death is discussed in Jalland 1996, pp. 350–1.

3 Hooker apparently refers to CD's extreme grief following the death in 1851 of his eldest daughter, Anne Elizabeth (Annie), at the age of 10 (see *Correspondence* vol. 5).

From Louis Agassiz 29 September 1863

Cambridge, [Massachusetts]
Sept. 29, 1863

My dear Sir,

Allow me to introduce to you Mr. J. P. Lesley,[1] the most accomplished ⟨g⟩eologist of the United States & to take this opportunity to express to you my kindest regards.

Very truly yours | Ls Agassiz

Charles Darwin Esq.

DAR 159: 9

[1] J. Peter Lesley was professor of mining at the University of Pennsylvania. In October 1863 he began a three-month visit to Europe on behalf of the Pennsylvania Railroad Company, in order to study methods of hardening the surface of rails (Ames ed. 1909, 1: 436). Lesley reported having met CD at Charles Lyell's house in London on 21 October 1863, describing him as 'a gentle shy person of 55+ years of age' (*ibid.* 1: 447). However, Emma Darwin's diary (DAR 242) makes no reference to CD's having gone to London on that date, and records that he had a 'bad day' with 'sickness'.

From H. W. Bates 29 September 1863

22 Harmood St. Haverstock Hill. N.W.
29 Sept 1863

My dear Mr Darwin

I am very much concerned to hear of your increased illness; but I expected it having been told by various persons throughout the summer that you were not so well this year.[1] Let us hope that Malvern wells will give you a little relief so as to enable you to work cheerfully & correspond with your friends again. If it would not be annoying I should like to send you a letter with Natural History chit chat now & then, regardless of your answering me punctually.

Many thanks for the loan of Asa Gray's reviews.[2] Really my little paper in the Linnean has been greatly honoured to be reviewed by Darwin & Asa Gray.[3] How capitally Gray has done it. I did not think it possible that the complex details of variation &c which I gave could be so briefly & luminously brought to a focus. I think I must write to him to thank him.

The review in the Times of my book has caused quite a commotion.[4] I consider it the best that has yet been written. It is also of great general importance because it is a public concession on the part of the highest literary tribunal, of the claims of philosophical natural history to the attention of the public. My old Father happens to be on a visit to me & the review came very apropos, causing great elation in our little family circle.[5] My Father is an old man of business who thinks everything right that is said by the Times; & who begins now to see that his son really has written a goodish book.

The longest review that has yet appeared is in the Revue des deux mondes by Forgues.[6] It is also most excellently done & I think shows a closer examination & higher appreciation of the book than any that has yet appeared in England except that of the Times.

You ask me what I am doing. I have been commissioned by Mr W. Wilson Saunders to write a monograph of the Mantidæ (a remarkable family of Insects).[7] The work is to appear in the Ray Society series, in 4to illustrated by 20 plates by Westwood.[8] This has occupied me the last 4 months & will continue to occupy me for 18 months longer. Mr Saunders pays me (moderately) for the work & leaves me all the credit. This work leaves me time for other things, such as short articles

& I had commenced one on the whole subject of local variation, intending to incorporate details of facts of new varieties interbreeding with counterparts, which you require.[9] But the monograph is so much pleasanter work just now that I have laid this paper aside for the present.

I will not write more just now | Yours sincerely | H W Bates

DAR 160: 77

[1] CD's letter to Bates has not been found; however, see n. 2, below. CD and his family had gone to Malvern Wells, Worcestershire, at the start of September, so that CD could undergo hydropathic treatment (see letter to W. D. Fox, 4 [September 1863]).

[2] In his letter to CD of 1 September 1863, Asa Gray had promised to send sheets from the September 1863 number of the *American Journal of Science and Arts* containing his review abstracts of Bates 1861 and 'Two forms in species of *Linum*' (A. Gray 1863a and 1863c) the following week. CD had evidently forwarded the sheets to Bates.

[3] CD reviewed Bates 1861 in the April 1863 number of the *Natural History Review* ('Review of Bates on mimetic butterflies').

[4] Bates 1863 was favourably reviewed in *The Times*, 24 September 1863, p. 8.

[5] Bates's father, Henry, was a hosiery manufacturer in Leicester (*DSB*); Bates had moved to London in April 1863 (see letter from H. W. Bates, 8 April 1863).

[6] Bates 1863 was reviewed by the French man of letters, Paul-Emile Daurand Forgues, in the 1 August 1863 issue of the *Revue des Deux Mondes* (Forgues 1863).

[7] The reference is to William Wilson Saunders, a wealthy insurance broker with an interest in entomology (*DNB*). Bates continued to work on the monograph until at least 1864, when he was appointed assistant secretary of the Royal Geographical Society (Bates 1892, pp. lxxiii–lxxv); however, the work was never published (Curle 1954, p. 26).

[8] The references are to John Obadiah Westwood, a skilled entomological artist who supplemented his small private income by his drawings (*DNB*), and to the Ray Society. The Ray Society was established in 1844 to publish important works of natural history that were unlikely to prove commercially profitable (Curle 1954, p. 2).

[9] Bates had intended to include a long discussion of the 'origin of species by segregation of races' in Bates 1863, but there had been insufficient space (see letter from H. W. Bates, 8 April 1863). He consequently engaged to write an article entitled 'Geographical relations of species, and their varieties' for the *Natural History Review* (see letter from H. W. Bates, 2 May [1863] and n. 7, and Bates 1892, pp. lxvii–lxx); however, the paper was never completed (*Royal Society catalogue of scientific papers*). In his letter to Bates of 25 November [1862] (*Correspondence* vol. 10), CD had asked Bates to include in Bates 1863, or elsewhere, all his facts 'about similar varieties pairing', including how many he had caught, and how many he now had in his collection. CD considered this point 'very important' because of its bearing on the process of speciation (see *Correspondence* vol. 10, Appendix VI). See also *Correspondence* vol. 10, letter to H. W. Bates, 15 December [1862].

From Emma Darwin to W. D. Fox [29 September 1863]

Malvern Wells
Tuesday.

Dear Mr Fox

I am writing instead of Charles to thank you for your precise answer to his enquiry.[1] I am glad to say that by the help of your directions & the lady at whose house our poor Annie lodged we have found the tomb stone.[2] It is very much covered with trees & looks so green & old I am sure I looked at it many times thinking it quite out of the question that should be it. Also the iron palisades are gone, at least both the sexton[3] & lady thought there had been rails round it, but that does not signify. This has been a great relief.

Charles has been quite ill last week but for the last 5 days he has decidedly improved, but I expect his recovery will be very slow.[4]

We like Dr Ayerst tho' he has not the influence of Dr Gully. Dr G. it is hopeless to try to see tho' I must say he has been to see Ch. twice & he quite approves of his treatment.[5] He takes 2 or 3 wet rubbings in the day & small walks in the garden, but he is weak— Our stay here has however been of real use to our sick boy & put us on a better system with him.[6] Ch. appetite is so good I think he must get strength soon & he has struggled on for 5 days without sickness.

I sent you a paper concocted between Ch. & myself wisely directed to Delamere *Droitwich* & so I will put in another in hopes you may employ it in some way.[7] I have met with a good deal of encouragement & I see it today in the Worcester paper endorsed by the Member of Parliament[8]

Will you remember us both very kindly to Mrs Fox[9] & | believe me | yours very sincerely | E. Darwin

Postmark: SP 29 63
Christ's College Library, Cambridge (Fox 141)

1 See letter from W. D. Fox, 7 September [1863].

2 Emma had been unable to locate in the Priory churchyard, Great Malvern, the grave of her daughter Anne Elizabeth (Annie), who died in 1851 (see letter to W. D. Fox, 4 [September 1863]). At CD's request, Fox, who visited the grave in 1856 and 1859, had sent specific directions for finding it (see letter to W. D. Fox, 4 [September 1863], and letter from W. D. Fox, 7 September [1863]). The reference is to Eliza Partington, the lodging-house keeper who ran Montreal House, where Annie had apparently lodged (*Correspondence* vol. 5, letter to E. A. Darwin, 19 April 1851; *Post Office directory of Birmingham* 1850 and 1864).

3 This individual has not been identified.

4 In her diary (DAR 242), Emma Darwin recorded that CD was sick every day from 20 to 23 September, but that he was better over the succeeding five days, while suffering intermittently from flatulence and 'head swimming'.

5 James Manby Gully, under whose care both CD and Fox had been treated at Great Malvern on previous occasions, had been seriously ill (see Browne 1990 and letter from W. D. Fox, [16–22 May 1863]). As a result, CD was receiving treatment in Malvern Wells at the hydropathic establishment run by James Smith Ayerst, an associate of Gully's (see letter to W. D. Fox, 4 [September 1863] and n. 1).

6 According to Emma Darwin's diary (DAR 242), Horace Darwin, who had been ill intermittently during 1863, began hydropathic treatment on 15 September 1863.

7 See letter from Emma Darwin to W. D. Fox, [6–27 September 1863]. The enclosure referred to is reproduced in Appendix IX. According to CD's Address book (Down House MS), Fox's address was 'Delamere R[ectory]. Northwich'; see also *Post Office directory of Cheshire* 1864.

8 The newspaper has not been identified. In 1863, the members of parliament for Worcester were Richard Padmore and Osman Ricardo (*Dod's parliamentary companion* 1862, p. 136).

9 Ellen Sophia Fox.

To A. R. Wallace [29? September 1863][1]

Malvern Wells
Tuesday

My dear Mr. Wallace

Your kindness is never failing.[2] I got worse & worse at home & was sick every day for two months; so came here, where I suddenly broke down & could do nothing; but I hope I am now very slowly recovering, but am very weak.—[3]

Sincere thanks about Melastomas:[4] these flowers have baffled me & I have caused several friends much useless labour; though, Heaven knows, I have thrown away time enough on them myself.—[5]

The gorze case is very valuable & I will quote it, as I presume I may.—[6]

I was very glad to see in the Reader, that you have been giving a grand paper, (as I infer from remarks in discussion) on Geographical distribution.[7]

I am very weak, so will say no more | Yours very sincerely | C. Darwin

British Library (Add. MS. 46434: 36–7b)

[1] The date is conjectured from the relationship between this letter and the letter from A. R. Wallace, 26 September 1863; the first Tuesday after that date was 29 September.

[2] See letter from A. R. Wallace, 26 September 1863.

[3] CD stayed at Malvern Wells, Worcestershire, from 3 September to 12 or 13 October 1863, in order to undergo treatment at James Smith Ayerst's hydropathic establishment (see Emma Darwin's diary (DAR 242) and letter to J. B. Innes, 1 September [1863]).

[4] See letter from A. R. Wallace, 26 September 1863 and n. 2. CD never published his research on the Melastomataceae (see *Cross and self fertilisation*, p. 298 n., and *ML* 2: 292–302); his working notes on the subject are in DAR 205.8.

[5] See, for example, letter to Hugh Falconer, 5 [and 6] January [1863] and nn. 22–3, letter from John Scott, 6 January 1863, letter to H. W. Bates, 12 January [1863], letter to Asa Gray, 19 January [1863], and letter to Hermann Crüger, 25 January [1863].

[6] See letter from A. R. Wallace, 26 September 1863. CD discussed the acclimatisation of domesticated varieties in *Variation* 2: 305–15; however, he did not cite the case supplied by Wallace in *Variation*, or in any of his subsequent publications.

[7] CD refers to a report in the *Reader* 2 (1863): 352–3, describing Wallace's paper, 'On the geographical distribution of animal life' (Wallace 1863c), which was read on 31 August 1863 at the annual meeting of the British Association for the Advancement of Science in Newcastle-upon-Tyne. The report made no reference to the discussion of Wallace's paper; however, it was followed by a report of Henry Baker Tristram's paper, 'On the variation of species as pointing to Western Asia as the centre of the palæarctic area of creation', in which Tristram referred to Wallace's 'magnificent epitome of all that science had yet attained in respect to the variations and limits of the different groups'. There are annotated copies of both reports in DAR 205.10: 2.

From J. D. Hooker 1 October 1863

Kew
Oct 1. 1863

Dear old Darwin

I have just buried my darling little girl & read your kind note.[1] I tried hard to make no difference between her & the other children, but she was my very own, the flower of my flock in everyone's eyes, the companion of my walks, the first of my children who has shown any love for Music and flowers, & the sweetest tempered affectionate little thing that ever I knew. it will be long before I cease to hear her voice in my ears—or feel her little hand stealing into mine, by the fire side & in the Garden.— wherever I go she is there—

The Funeral service had no more effect on me than on her— the association with her personally snapped as the ceremonial left my door, & oddly enough I felt nothing at seeing the little white coffin go into the vault, my mind was wandering amongst sweeter memories elsewhere.

And now I can calmly think of what sorrows I am spared— Her's was no contagious disease threatening the whole family for weeks afterwards.— she suffered comparatively little—& above all do I rejoice, that she was yet so young & happy, that death did not enter her little head during her illness—& I was spared the agony of seeing my darling pass through the "Valley of the shadow of death"—[2] Then too strangely enough I never knew she was dying till 3 minutes before the breath left her body For 3 hours I was blind to every one of those symptoms of rapidly approaching dissolution that every nurse knows & every novelist describes—& I have seen myself so often.[3] The doctor came in just 3 minutes before she died & told me to my horror she was dying. I knew the extreme danger, but assumed she had many hours to live. The retrospect of that last night is thus in some respects comforting in others hideous, & I can still feel the cold shudder that every misinterpreted symptom still sent through me, during that long night of agony & suspense.

My wife is still very ill, with rheumatism & neuralgia, weakness & sorrow—& but for the presence of her friend Miss Hawthorn[4] I do not know what I should have done—during the last 5 days.

I do hope to be able to move her by Monday to my Uncle Gunn's in Norfolk.[5] poor thing she has now no Uncle or Aunts house or other of the many resorts she once had to fall back upon.[6] Since we married her family has foundered like an ill equipped fleet in an open roadstead.

And now I must go about my work till Monday—& have heaps to do till then.

I had hoped that you were back at Down & better— Kindly let Henrietta write & tell me how you all are.—[7] your last bulletin was a very sad one— my address will be

Revd. J. Gunn.
Irstead Neattishead[8]
Norwich

Ever yrs affec | J D Hooker

DAR 101: 160–2

[1] CD's note, in response to the letter from J. D. Hooker, [28 September 1863], has not been found. Maria Elizabeth Hooker died on 28 September 1863, aged 6 (Allan 1967, 'Hooker pedigree'). See also Jalland 1996, pp. 350–1.

[2] *Psalms* 23:4.

[3] Hooker had been medically trained and served as a naval surgeon between 1839 and 1843 (*DNB*).

[4] Miss Hawthorn has not been identified. She may be Grace Hawthorn, sister of Sarah Hawthorn, who married Leonard Jenyns in 1862 (*DNB* s.v. Blomefield [Jenyns], Leonard).

[5] John Gunn, the husband of Hooker's maternal aunt Harriet, was vicar of Barton Turf, and rector of Irstead in Norfolk (*Clergy list*, L. Huxley ed. 1918, 1: 18–19).

[6] Frances Harriet Hooker's aunt, Anne Frances Henslow, died in September 1863 (see letters from J. D. Hooker, 26 August 1863 and 15 September 1863). Her father, John Stevens Henslow, died in 1861 (see *Correspondence* vol. 9).

[7] Henrietta Emma Darwin.

[8] Irstead and Neatishead are villages in Norfolk, ten miles north-east of Norwich (*Survey gazetteer of the British Isles*).

To J. D. Hooker [4 October 1863][1]

Malvern Wells[2]
Sunday

My dear old Friend.

Your note is most pathetic[3] I understand well your words: 'wherever I go, she is there".— I am so deeply glad that she did not suffer so much, as I feared was inevitable. This was to us with poor Annie the one great comfort.—[4] Trust to me that time will do wonders, & without causing forgetfuless of your darling.

I am very weak & can write little.— My nervous system has failed & I am kept going only by repeated doses of brandy; but I am certainly better, much, & sickness stopped.—

God Bless you my best of friends.— Yours affect | C. Darwin

P.S. | I must add that I shall be grateful for a line whenever you are inclined to write.

My head swims badly so no more.—

DAR 115: 206

[1] The date is established by the relationship between this letter and the preceding letter. In 1863, 4 October was a Sunday.

[2] CD was resident in Malvern Wells, Worcestershire, while undergoing treatment at James Smith Ayerst's hydropathic establishment (see 'Journal' (Appendix II)).

[3] Preceding letter. Hooker had written concerning the death of his six-year-old daughter, Maria Elizabeth Hooker.

[4] Anne Elizabeth Darwin, the Darwins' eldest daughter, had died aged 10 in 1851 (see *Correspondence* vol. 5).

From Roland Trimen 10, 13, and 18 October 1863

Colonial Office, | Cape Town.
10th. October, 1863.

My dear Sir,

I have today received your very welcome letter of August 27th.,[1] with the paper on *Linum*, for which I am very thankful.[2] I have read the paper with much interest, but have not yet thoroughly studied it. How very astonishing are the marked differences in individuals of one and the same species in the growth and position of the eminently important sexual organs: and, still more, the wonderful fact that the pollen of either form is wholly (or very nearly) ineffectual in fertilising an individual presenting the same modification. To me, the dimorphism presented by Linum and Oxalis seems more difficult to account for than the strangest Orchid modification,[3] because of the extraordinary fact that the same great law in constant operation has produced exactly opposite structural modifications in different individuals of the same species,—and that *not* a plant with separated sexes. By what remarkable process are such conflicting results, so evenly balanced, of natural selection to be accounted for?

The very common White Oxalis (n°. 1) of which I enclose a specimen is often pink or pink-tinged, and, like the yellow one I sent in my last,[4] is dimorphic. I gathered 37 flowers at random here and there, and found on examination that in *10 only the stigmas were uppermost*; in the remaining 27 the anthers held the superior position.

I was very much delighted to find two other species of *Oxalis* which are *trimorphic*. Of the first of these (unfortunately I have preserved no examples) I only gathered 13 flowers (they are *white, with narrow crimson edgings to each petal*, and, never opening widely, the crimson lines have a beautiful half-spiral effect);—

proportion of 3 forms as follows:

Stigmas	lowermost,	3	The leaves of this
::	central,	7	*Oxalis* are linear or
::	uppermost,	3	nearly so.
		13.	

The other, of which I enclose 2 specimens of each form, is *pinkish-red, with a wide yellow centre* (also a very common species). I found the following to be the relative proportion of the three forms:[5]

Stigmas	lowermost,	21	
::	central,	12	N°. 3.
::	uppermost,	10	
		43.	

I am glad the "peach-puncturing" is likely to be of use, & hope that your renewed strength at Malvern will enable you to give publicity to the evidence it seems to afford in favour of the Moth agency in fertilising the "Scheinsaftblumen".[6]

With regard to *Cypripedium* (though it is very likely Asa Gray may be right in his suggestion) I venture to observe that I don't think your *putting in* a small bee quite a fair experiment in re the plant's fertilisation, because it does not appear that the said bee would voluntarily have acted as he twice did under exceptional circumstances.[7]

I have read a good part of Bates' book, and think it very good,[8] especially his remarks on the universal predominance of an unmistakeable sylvan type in the whole fauna of Equatorial America.[9] His remarks on Lepidoptera are especially interesting to me.[10]

*13*th.— By this mail I received a very kind and interesting letter from Mr. Bates,[11] chiefly referring to a few observations I ventured to make on his striking & able paper on the *Heliconidæ* and their mimickers.[12] I had some difficulty in regarding the Nymphalideous butterflies, with their more or less atrophied fore-legs, as the supreme types or highest forms of Lepidopterous life, but Mr. Bates has most simply explained it to me by the remark that it appears to him "that the more *aerial* a Lepidopteron becomes, the more perfect it is, and the atrophy of the legs is a sign of less terrestrial habits, &c." He has kindly offered me a series of Amazonian Rhopalocera, in return for which I hope to supply some gaps in his South African representation of the group

By Mr. Adamson,[13] the Chief Clerk in the Office where I serve, I have sent you (addressed to care of Mr. Stevens)[14] a bottle of spirit in which are a few Orchid flowers in different stages of growth. I have a few more bottled for you, which I hope to have the pleasure of sending before long, although you warn me in your last letter that you have little hope of being able to work up the accumulation of new Orchid material.[15] Still, in case you ever *should* have a little leisure, I should be sorry to have missed the chance of affording you some aid, be it ever so little.

I hope you will pardon me for asking you to do me the favour of posting the enclosed note to Mr. Bates. Under the altered postal arrangements, I find my English correspondence somewhat expensive, and yet I do not like to forego the convenience of the regularity ensured by the monthly line of Steamers[16]

This morning, and often recently, I have been trying to find some *Oxalis* seed, but without success. The plants of the yellow species I sent you are plentiful enough,— their flowers are all gone, but I can find no seeds. I found many flowerstalks, the ends of which were brown and shrivelled, from which ends it appeared as if the capsule & part of stalk immediately below it had wasted away. Is there anything unusual in the seeding of Oxalids, or am I merely unfortunate or stupid?

Is not the case of a *mule* casting a foal very rare? I believe I can furnish you with some authentic details of such a case which recently occurred in this Colony. The foal is now in the Museum,[17] and is a weird-looking object enough. The sire of this still-born oddity is stated by the donor to have been a donkey. Layard has a letter containing particulars,[18] of which I have no doubt I could send you a copy, if it would be of any use.

Wishing you renewed strength for the prosecution of your researches, | I remain, my dear Sir, | Yours very truly, | Roland Trimen.

P.S. (18th.)— I leave tomorrow on a week's trip to a part of the country which I have not yet visited. I trust to fall in with some local forms of both Insects & Plants.

DAR 109: B122–3

CD ANNOTATIONS
0.2 1863.] *underl red crayon*
1.1 I have . . . for? 1.13] *crossed blue crayon*
2.1 (N°. 1)] *underl blue crayon*
3.14 N°. 3.] *underl red crayon*
4.1 I am . . . circumstances. 5.5] *crossed red crayon*
4.1 I am . . . material. 8.6] *crossed blue crayon*
9.1 I hope . . . Plants. 13.2] *crossed blue crayon*

1 Letter to Roland Trimen, 27 August [1863].
2 'Two forms in species of *Linum*'; for CD's presentation list for this paper, see Appendix IV.
3 CD and Trimen had been discussing dimorphism in *Oxalis* and *Linum*. See letter from Roland Trimen, 16, 17, and 19 July 1863, and letters to Roland Trimen, 23 May [1863], and 27 August [1863] and nn. 3 and 6.
4 Trimen sent CD specimens and a sketch of dimorphic *O. cernua* with his letter of 16, 17, and 19 July 1863. The specimens Trimen mentions in this letter have not been found.
5 CD quoted Trimen's observations in *Forms of flowers*, p. 169.
6 'Scheinsaftblumen' are sham nectar-producers, or flowers with false nectaries (*Orchids*, pp. 45–50). See letter from Roland Trimen, 16, 17 and 19 July 1863, and letter to Roland Trimen, 27 August [1863] and nn. 7 and 8. See also letter to [Gardeners' Chronicle], [after 27 August 1863].
7 See letter to Roland Trimen, 27 August [1863] and nn. 12–13. These experiments led CD to revise his conclusions regarding pollination in *Cypripedium* (see 'Fertilization of orchids', pp. 155–6 (*Collected papers* 2: 152–3)).
8 Trimen refers to Henry Walter Bates's *The naturalist on the river Amazons* (Bates 1863).
9 Presumably a reference to Bates 1863, 1: 48, 71–2, and 108.
10 Trimen was a specialist in Lepidoptera (*DSAB*).
11 Henry Walter Bates.
12 Bates 1861.
13 L. Adamson was chief clerk in the Colonial Office at the Cape of Good Hope (*Colonial Office list*).
14 Samuel Stevens was a natural history agent at 24 Bloomsbury Street, London (*List of the Linnean Society of London*, *Post Office London directory* 1863). The specimens have not been found.
15 Letter to Roland Trimen, 27 August [1863].
16 The rates of postage from Cape Town had probably increased considerably; the rates of postage for mail from London to Cape Town had doubled in April 1863 (*British postal guide* 1863). The official mail service from Cape Town was available monthly; mail could also be carried by private ship, but the availability of this means of transport was more uncertain (*ibid.*).
17 Trimen refers to the South African museum in Cape Town (see also n. 18, below).
18 The reference is to Edgar Leopold Layard who, in addition to his duties as a civil servant, was the founder and curator of the South African museum in Cape Town (*DSAB*, *Modern English biography*). Layard had corresponded with CD in 1856 (see *Correspondence* vol. 6).

From John Lubbock 14 October 1863

15, Lombard Street. E.C.
14 Oct 1863

My dear M^r. Darwin

I have undertaken to write a little notice for the N.H.R. of Huxley's Lectures to working Men on O. of S.[1]

If I remember right the Record was particularly ridiculous about you[2] Can you give me a reference to the article or by chance lend me the thing itself. Any hints or references to any particularly good or bad point in any review of you would be thankfully accepted.

I hope that you are better, but heard a poor account of you yesterday from Hooker.[3] You have of course heard of the loss which he, poor fellow, has had;[4] we are going together to St. Acheul on Saturday,[5] but I expect to be back on Monday.

I am longing for a good talk with you, & am dear Mr Darwin with kind regards to Mrs Darwin & all your family,

Yours affec | John Lubbock

C Darwin Esq

I send this to Down, having forgotten your present address[6]

DAR 170: 41

[1] The notice of Thomas Henry Huxley's six lectures to working men (T. H. Huxley 1863a) appeared in the January 1864 number of the *Natural History Review* ([Lubbock] 1864).

[2] Lubbock refers to the anonymous review of *Origin* in the tri-weekly, evangelical *Record* on 12 December 1860, p. 4. The readership of the *Record* is described in the *Newspaper press directory* 1863. There is an annotated copy of the review in the 'Scrapbook of reviews' (DAR 226.1: 139).

[3] See letter to Joseph Dalton Hooker, [4 October 1863].

[4] Hooker's six-year-old daughter, Maria Elizabeth, had died on 28 September 1863 (see letters from J. D. Hooker, [28 September 1863] and 1 October 1863).

[5] A site near Amiens, France, where prehistoric flint implements were found. See also letter from J. D. Hooker, 23 October 1863 and n. 4.

[6] According to Emma Darwin's diary (DAR 242), CD had returned to Down from James Smith Ayerst's hydropathic establishment in Malvern Wells, Worcestershire, via London, on 14 October 1863.

From H. E. Darwin to Thomas Warner[1] 14 October [1863][2]

Down, Bromley | Kent S.E.
Oct. 14th

Sir,

Mr. Darwin is too unwell to write himself but will be happy to have his name put down for Mr. Buckman's testimonial.[3]

your obedient servant | H. E. Darwin

Mr. Darwin encloses a cheque for 2.2.0[4]

Mrs Carole Cockett

[1] Possibly Thomas Warner of Gosditch Street, Cirencester, Gloucestershire (*Post Office directory of Gloucestershire, with Bath, Bristol, Herefordshire, and Shropshire* 1870). A Thomas Warner is listed as an original shareholder in the Royal Agricultural College, Cirencester, founded in 1845 (Sayce 1992, p. 351). See also n. 3, below.

[2] The year is established by reference to CD's Account book–cash accounts (Down House MS). See n. 4, below.

[3] James Buckman retired from his position as professor of geology and botany at the Royal Agricultural College, Cirencester, Gloucestershire, in 1863; upon resigning his appointment, Buckman 'received two valuable testimonals, one from the inhabitants of Cirencester and his scientific friends, and the

other from his pupils' (*Men and women of the time* 1868). Though only one letter to Buckman is extant (*Correspondence* vol. 6, letter to James Buckman, 4 October [1857]), CD evidently took an interest in Buckman's research (see letter from S. P. Woodward, 14 February 1863, n. 7).

4 CD's Account book–cash accounts (Down House MS) for 14 October 1863 records the payment of £2 2*s*. to Thomas Warner.

From J. D. Hooker 23 October 1863

Coltishall Norwich
October 23/63.

My dear Darwin.

Since writing last,[1] I have heard, through Lubbock,[2] a rather better account of you—& sincerely hope that it may soon be confirmed by yourself. I have been wandering about, seeking rest & finding much— first I went to my Uncle Gunn's with wife & 2 youngest;[3] & I had a day at the Norfolk Cliffs with him— Then I returned to Kew & started last Saturday with Lubbock, Gunn, Thomson, & a Mr Jansen (friend of Lubbock) to Amiens & Abbeville,[4] where I had look at the gravel pits, & much discussion thereupon. . Of course I am no authority on such a subject, but like most ignorant men must have an opinion of my own, & that the most confounded one, of simply being dissatisfied with the explanations of others. The most mysterious thing of all is the great numbers of worked flints found, & the vast area over which they are found. Considering the "imperfection of the Geological Record" it is obvious that these implements must be in the ground by tens of millions[5]— they are found over areas of very many square miles;— at various elevations in the valleys, from the present river-beds to several hundred feet above it, in beds *capping* the hills;—& at all depths in 〈 〉 that are 30 ft thick! 〈There〉 are several unsatisfactory points in the published explanations &c.[6]

1. The so called Loess or Superficial deposit, is, I think, clearly nothing but the altered condition of the exposed beds, due to ærial action.

2. There is in most of the beds both in the Somme & Oise, a curious admixture, of very rounded pebbles & angular flints, *without intermediate forms*: seeming every where, as if a detritus of angular flints had been rolled up with Seaside flint pebbles.

3. It is difficult to suppose 〈that〉 the angular & rounded flints have be〈en〉 carried by ice action to th〈eir〉 present position & the 〈 〉 worked flints (intermixed with them) *not so*— No〈thing〉 that I could see or hear of at 〈all coun〉tenanced the assumption that angular flints & pebbles were transported, & the worked flints simply dropped in during the deposition of the beds.

4. The theory of the accumulation of any of the beds by periodic overflow of large rivers, seems to me simply preposterous.[7]

The so called Loess is *full* of angular flints, weighing from an oz to a pound, besides round pebbles & sandstone boulders, all of which nothing, but violent current (*if of water*) could have transported, & to suppose that such rivers covered uniformly so many thousand square miles of 〈h〉illy country is to me quite 〈in〉conceivable— such deposit 〈re〉present the denudation of 〈m〉ountains many thousands feet high—& I cannot believe that the Oise & Somme of former days could have had anything material to do with it. Supposing that the so called Loess had been a fine silt—I doubt whether even then it would have been accountable for any river deposits— I doubt rivers making such deposits by annual floods at all—*Except the land be sinking*.

A system of vast Lakes & morasses with lofty mountains, torrents & ice action would do much to explain the whole deposit,—but where are the ancient boundaries of such lakes?— Next best would be estuarial ice deposit along a sinking coast.

We returned on Tuesday night after having visited Creil, & Chauny on the Oise; Amiens Abbeville St Acheul &c on the Somme. On Thursday I came here to my sisters,[8] & this morning went to the gravel pits hard by. The similarity of the phenomena is so striking that it is impossible to suppose that the two countries were dissimilar, when the hatchets were deposited in Suffolk, Norfolk & Picardy, & I have a great notion that these post-tertiary beds do not represent so many climates as is supposed.— I must send you a wonderful sketch just received from Hector,[9] of the ancient terraced glacial detritus filling the broad N. Zealand valleys, at *only 1000* ft above the level of the sea—the exact counterparts of my sketches of Himalayan valleys at ⟨ ⟩[10] Now this, in N.Z; is in the ⟨ ⟩ & elevation of Palms, tree Ferns, & Dracænas!— Were Elephants & Co now to inhabit such a valley, they would be liable to be swept by torrents into this glacial detritus—& be regarded by future Geologists as evidently glacial animals; whilst the vegetable remains hard by would be referred to quite a different climate, & hence a difft geological period!

Torell's account of the *glacial marine* animals, inhabiting a fresh-water lake in Sweden fa⟨r⟩ south of the latitude where th⟨ese⟩ animals now inhabit ⟨ ⟩ sea,[11] is another confoundi⟨ng⟩ circumstance— . & Lubbock brought me 2 marine Fuci, ⟨from⟩ a perfectly fresh Lagoon, wh⟨ich⟩ ⟨ ⟩ the rocks.

⟨Tomorrow⟩ I go to the cliffs at Bacton & Happisburgh.

My wife is much better—[12] we go on Monday to Yarmouth, & I return to Kew on Wednesday of next week

I am very well, but it will be long before I get over this craving for my child;[13] or the bitterness of that last night. To nurse grief I hold is a deadly sin, but I shall never cease to wish my child back in my arms, as long as I live.

I have written you a most presumptuous yarn. It is delightful to have a good fellow to say what ⟨he⟩ likes to!!!— I would not have ⟨blas⟩phemed modern Geology so ⟨to Pr⟩estwich *Lyell* or Falconer ⟨ever in⟩ the world.[14] My wife sends ⟨her⟩ love

Ever yours affec | J D Hooker

P.S. Just heard that Willy[15] is sent back to Kew convalescent of scarlet fever; we had not been told he was ill! My Mother[16] has taken him in for a week, after that I must take him somewhere.

DAR 101: 167–70

[1] Letter from J. D. Hooker, 1 October 1863.

2 John Lubbock.
3 The palaeontologist and geologist John Gunn was married to Maria Hooker's sister Harriet, making him Hooker's uncle by marriage (Allan 1967, 'Hooker pedigree'). Hooker also refers to his wife, Frances Harriet Hooker, Charles Paget Hooker, aged 8, and Brian Harvey Hodgson Hooker, aged 3 (*ibid.*).
4 The references are to John Lubbock, John Gunn, the surgeon and botanist Thomas Thomson (R. Desmond 1994), and probably Frederick Halsey Jansen, a London solicitor (*Law list* 1863). Abbeville, near Amiens, in the Somme valley, France, was a renowned site of prehistoric human artefacts (Van Riper 1993). See also letter from John Lubbock, 14 October 1863.
5 An allusion to chapter 9 of *Origin*, which was entitled: 'On the Imperfection of the Geological Record'.
6 The principal study of the formation of the Somme valley deposits was made by Joseph Prestwich, presented in a paper read before the Royal Society of London on 27 March 1862 (J. Prestwich 1862a), and later published in revised form in the *Philosophical Transactions of the Royal Society of London* (J. Prestwich 1862b). Charles Lyell had also included a section on the geological structure of these deposits and their associated artefacts in *Antiquity of man* (C. Lyell 1863a, pp. 106–47).
7 The reference is to the theory expounded in Prestwich 1862b, pp. 286–98.
8 Hooker's sister Elizabeth was married to the surgeon Thomas Robert Evans of Coltishall, Norfolk (Allan 1967, 'Hooker pedigree').
9 James Hector was the provincial geologist of Otago, New Zealand (*DNZB*).
10 Presumably a reference to Hooker's sketches of the terraced moraines in the Yangma valley of Nepal published in *Himalayan journals* (J. D. Hooker 1854, 1: 242, and end of volume).
11 The reference is to the Swedish geologist and zoologist Otto Martin Torell and to Torell 1859.
12 Frances Harriet Hooker (see letter from J. D. Hooker, 1 October 1863).
13 Maria Elizabeth Hooker had died aged 6 on 28 September 1863 (see letters from J. D. Hooker, [28 September 1863] and 1 October 1863).
14 Joseph Prestwich, Charles Lyell, and Hugh Falconer.
15 William Henslow Hooker, the Hookers' eldest child (Allan 1967).
16 Maria Hooker.

From H. W. Bates 24 October 1863

22. Harmood St Haverstock Hill. N.W.
24 Oct 1863

My dear Mr Darwin

I just send you a line to say that I posted the tract containing Asa Gray's criticisms immediately after receiving your note.[1]

The sad news conveyed in your letter has caused great sorrow to myself & those of our friends to whom I have communicated it. Of course much correspondence is out of the question until you are better therefore several small matters I wished to mention must remain for another time

Yours sincerely | H W Bates

DAR 160: 78

1 The letter to Bates has not been found. Asa Gray reviewed Bates 1861 and 'Two forms in species of *Linum*' in the September issue of the *American Journal of Science and Arts* (A. Gray 1863a and 1863c). CD had sent Bates copies of the reviews. See the letter from Asa Gray, 1 September 1863, and the letter from H. W. Bates, 29 September 1863.

From John Lubbock to Emma Darwin 24 October 1863

15, Lombard Street. E.C.
24 Oct. 1863

My dear Mrs. Darwin

I am indeed grieved to hear so bad an account of Mr. Darwins health, but hope that he may possibly be better now that he has returned home.[1]

Your kind note reached me just as I was starting for a short visit to the Valley of the Somme.[2]

The article in the Record, was the one I wanted;[3] it shall be returned to you ere long.

Please give my kindest regards to Mr. Darwin & believe me | Yours very sincerely John Lubbock

DAR 170: 42

[1] The letter from Emma Darwin to Lubbock has not been found. According to Emma Darwin's diary (DAR 242), CD returned to Down from Malvern Wells, Worcestershire, via London, on 14 October 1863.

[2] See letter from John Lubbock, 14 October 1863, and letter from J. D. Hooker, 23 October 1863.

[3] See letter from John Lubbock, 14 October 1863 and n. 2.

From Emma Darwin to W. E. Darwin [28 October 1863][1]

Down
Wednesday.

My dear William

I believe it is months since I have written to you but I have had a good ⟨d⟩eal of writing on my hands.

On Saturday Mr Engle—went up to consult Dr Brinton (phys. to Guy's H.) on your father's behalf.[2] He brought back a prescrip. & after he has tried it some little time we shall have Dr. B. down to see him.[3] He has a great rash come out on his shoulders partly encouraged by a compress & I believe it will be of use to him. He has walked a little further every day & yesterday accomplished twice round the sand walk—

He got so weak at Malvern after you left us he could not walk a step but from one room to another.[4]

I forgot to tell you that Mrs Acland called after you left.[5] I thought her pleasing & pretty but Hen. wd not allow the prettiness.[6] I went to Bromley yesterday in Reeves's fly as Billy has entirely struck work & lay flat down in the yard the last time he was requested to draw.[7] I heard at Nash's of the melancholy death of poor Frances Wells whom you thought you saw at Malvern[8] She was dressed & going out to dinner when she was suddenly taken ill & died in an hour's time. It must have been the heart I suppose. I picked up At Eliz. at the station & she is come home as fresh as a lark, but they had 16 rainy days which is not what one goes abroad for.[9]

Lizzy comes on Saturday.[10] Last evening there was a brilliant fire to the South. We settled it to be that timber yard & carpenter's near the Westerham road but it turned out to be much further off a farm near Tatsfield & a very extensive fire I hope the poor man is insured.[11]

Oddly enough early this m^{g} before light two stacks at Mr Solomon's were burnt down— I don't think it can be an incendiary. Mr S. is insured.[12] Hen. goes next week to Everleigh for a week— Hope[13] assures her she will find it very dull & I think it is possible, but young people like a new place. Mr Huxley writes to Papa that I am to treat him like Vivien did Merlin & shut him up in an oak, he then apologizes that Vivien is not a very proper person, but as Papa does not read Tennyson concludes it does not signify.[14] They have been coming to fisticuffs in Australia about Darwin & Huxley &c—[15]

Horace is about as usual & is gone on the pony this m^{g} with Duberry to look after the fire.[16] He does a little reading every day.

Goodbye my dear old man | E. D.

DAR 219. 1: 78

[1] The date is established by the reference to Emma's trip to collect Sarah Elizabeth Wedgwood from Bromley station; see n. 9, below. According to Emma Darwin's diary (DAR 242), Sarah Elizabeth Wedgwood arrived on 27 October 1863. The following Wednesday was 28 October.

[2] The references are to the local surgeon Stephen Paul Engleheart (*Post Office directory of the six home counties* 1862), and to the London physician William Brinton, a specialist in stomach disorders (see letter from George Busk, [*c.* 27 August 1863]). Brinton was a physician at St Thomas's Hospital, London, and does not appear to have been on the staff of Guy's Hospital, London (*Medical directory* 1863).

[3] William Brinton attended CD at Down in November and December 1863 (see letter from W. E. Darwin, 21 August [1863], n. 9).

[4] CD underwent treatment at James Smith Ayerst's hydropathic establishment in Malvern Wells, Worcestershire, between 3 September and 12 or 13 October 1863. According to Emma Darwin's diary (DAR 242), William arrived at Malvern Wells on 13 September and left on 25 September 1863.

[5] Mrs Acland was probably the wife of Lawnford Acland of Langdowne, Hythe, near Southampton (letter from Emma Darwin to William Erasmus Darwin, [after 7 October 1861], in DAR 210.6: 76; *Post Office directory of Hampshire, Dorsetshire, and Wiltshire* 1867–75).

[6] Henrietta Emma Darwin was William's sister.

[7] William Reaves (or Reeves) was a blacksmith and farrier in Down, Kent (Census Returns 1861 (Public Record Office RG9/462: 71), *Post Office directory of the six home counties* 1862); Down was five miles from Bromley. 'Billy' was the name of a horse (see letter from Henrietta Emma Darwin to William Erasmus Darwin, [February 1864?], in DAR 210.6: 115).

[8] Probably a reference to the shop Nash & Lukey, linendrapers, silk merchants, and milliners, which was situated in the High Street, Bromley (*Post Office directory of the six home counties* 1862). Frances Wells has not been identified.

[9] Sarah Elizabeth Wedgwood was Emma Darwin's sister (*Darwin pedigree*).

[10] Elizabeth Darwin was William's sister (Freeman 1978). Elizabeth may have been returning from school; she was sent away to school with Miss Buob at the beginning of the year (see letters from Emma Darwin to William Erasmus Darwin, 29 October 1862 and [15 April 1863], in DAR 219.1: 63 and 73). CD's Classed account book (Down House MS) records two payments to Miss Buob in April and August 1863.

11 Tatsfield is a village eight miles south-east of Croydon, three miles south-west of Down (*Survey gazetteer of the British Isles*).

12 Mr Solomon has not been further identified.

13 Hope Elizabeth Wedgwood was CD's first cousin once removed (Freeman 1978).

14 The letter from Thomas Henry Huxley has not been found. The reference is to Alfred Tennyson's poem 'Vivien', from *Idylls of the king* (Tennyson 1859). Vivien was a seductress.

15 The controversy which followed the publication of *Origin* in Australia (Mozley 1967) had presumably been reawakened by the publication of Huxley's *Evidence as to man's place in nature* (T. H. Huxley 1863b). Huxley may have told CD of the Australian climate of opinion in a missing letter (see n. 14, above). The majority of Australia's leading scientists were opposed on religious grounds to CD's theory of evolution by natural selection (see Mozley 1967, Moyal 1986, pp. 144–7, 184, Butcher 1988, pp. 141, 143).

16 Horace Darwin was William's brother. The Duberry family lived in Down, Kent. Richard Duberry and his son George were agricultural labourers; Duberry's eldest son, Mark, was a carter (Census returns 1861 (Public Record Office, RG9/462: 71)).

To J. D. Hooker [30 October 1863]

[Down]

My dear old friend.

I must just have pleasure of saying this.

Yours affect^{y}. | C. Darwin

I have a long letter from Dr Haast: I do not know whether you would care to see it. It has something on spreading of European plants.[1]

Endorsement: 'Oct 30/63.'
DAR 115: 207

1 The letter from Julius von Haast, 21 July [– 7? August] 1863, contained information on the spread of European plants and animals in New Zealand. See letter from Julius von Haast, 21 July [– 7? August] 1863 and nn. 16 and 17, and letter from J. D. Hooker to Emma Darwin, 11 November 1863 and nn. 3–5.

From Alfred Newton 31 October 1863

Magdalene College, | Cambridge.
31 Oct. 1863.

My dear Sir,

I have the pleasure to enclose you a copy of my remarks on the Red-legged Partridge's foot concerning which I before wrote to you—[1] The specimen I have left at the Zoological Society's House (11 Hanover Square) and it will be delivered to any messenger you may send for it. I regret that so long a delay has taken place in this matter—which I fear must be fatal to the vitality of any seeds enclosed in the ball of Earth—[2] However this was not my fault but that of the wood-cutter, who I began to think would never finish the engraving.

M^{r}. Buckland I see has suggested in the 'Field' that the conglomeration has been caused by a tumour on the bird's foot, resulting from a wound—[3] I should be glad to hear, whether you find this to be the case on examination.[4]

When I exhibited the specimen at the Zool. Soc$^{y's}$ meeting, Dr Gray would have it, that the ball of earth had not been *gradually* aggregated—[5]

This notion is I think erroneous and I trust you will notice what I have said in my remarks on this point, and bear them in mind when examining the specimen.[6]

With sincere respect | I remain, yours very truly | Alfred Newton

C. Darwin Esqre

DAR 172: 40

[1] The reference is to Newton 1863; see letter from Alfred Newton, 21 March 1863 and n. 5. Part of the enclosure is in DAR 205.9.3: 366 and is annotated in CD's hand. See also nn. 2 and 4, below.

[2] Newton refers to a specimen of a foot of a red-legged partridge (*Caccabis rufa*) with a ball of clay attached. CD grew eighty-two plants from the seeds embedded in this specimen and reported the results in the fourth edition of *Origin*, p. 432. These experiments supplemented CD's study of the geographical dispersal of seeds made in the 1850s (see letter from Alfred Newton, 21 March 1863 and n. 4). CD's annotations to the enclosure (see n. 1, above) indicate that the seeds were planted on 13 November 1863. CD discussed the results in his letter to J. D. Hooker, 5 December [1863], and in the letter from Emma Darwin to J. D. Hooker, [7 December 1863]. See also *Correspondence* vol. 12, letter to J. D. Hooker, 26[–7] March 1864, and letter to Alfred Newton, 29 March [1864].

[3] The reference is to remarks on the specimen by Francis Trevelyan Buckland (*Field*, 16 October 1863, p. 368); Buckland was a staff writer for the *Field* (*DNB*). See also *Correspondence* vol. 12, letter from Alfred Newton, 2 April 1864, and n. 4, below.

[4] CD's notes indicate that he examined the leg on 13 November 1863 (DAR 205.9 (3): 366). His observations read: 'Earth broken open; no concentric arrangement— Whole tarsus crooked enormously enlarged & one toe cut off— probably viscous secretion. Three Oats were extracted from larger fragments Leg clay weighed about $\frac{1}{3}$ of oz.— Seeds planted Nov. 13th/63/'. See also *Correspondence* vol. 12, letter to Alfred Newton, 29 March [1864].

[5] Newton exhibited the foot of the partridge at a meeting of the Zoological Society of London on 21 April 1863 (see Newton 1863). The reference is to John Edward Gray, who was keeper of the zoological collections at the British Museum. See also *Correspondence* vol. 12, letter to Alfred Newton, 29 March [1864], and letter from Alfred Newton, 2 April 1864.

[6] Newton gave his reasons for concluding that the ball of clay had accumulated gradually in Newton 1863, pp. 128–9.

From J. D. Hooker [1 or 3 November 1863][1]

Kew

⟨ ⟩u⟨ ⟩[2]

My dear Darwin

Mrs Darwins very ⟨ ⟩ letter & your scrap are s⟨ilver⟩ linings to my black cloud.[3] I do trust that you are to have a little rest from your intolerable illness. I will see about the prickly Palm (Calamus) at once & send it to Down postman.[4]

I should extremely like ⟨to⟩ see Haast's letter,[5] as I am quite excited about ⟨ ⟩ Geology, in one letter ⟨ ⟩ mentions, besides ⟨ ⟩ *cuts* in the valley ⟨ ⟩ read like Gl⟨ ⟩ ⟨ ⟩ on Hectors sketches—[6] ⟨ ⟩ *[illeg]* in Otago, which ⟨close⟩ly resemble those on ⟨the up⟩per Himalayan valleys.[7] ⟨I h⟩ave been along the Norfolk Coast again with Gunn,[8] from Bacton to Happisburgh, & cannot but think all the beds there, from the "forest bed" up to Trimmer's warp, are terms of one series, & the said warp, ⟨wh⟩ere present, is a myth, only ⟨the⟩ effect of atmosphere &c &c on ⟨ ⟩ part of upper bed.[9] ⟨The simi⟩larity of the whole ⟨ ⟩ those of Picardy is ⟨ ⟩ & suggestive[10] I ⟨ ⟩ ⟨sa⟩w Ramsay[11] at ⟨ ⟩ seemed to agree with me about all being terms of one series & not due to alternations of climates & periods. Surely Prestwich's distinctions are too fine-drawn:[12] he is a wonderful close accurate observer certainly

Willy goes on well,[13] but has given the Scarlatina to my Mother who is very ill with it,[14] but as there are no bad symptoms we are not alarmed. She took it from me 35 years ago![15] My wife & other children are still away at Yarmouth where I have ordered them to remain & where they will be now I suppose for a month at least,[16] & I have turned nurse in my old age, as my father & mother live all alone.[17] No doubt it is good for me, but it is a sad mess altogether, & I do not cotton to my condition at all. I bought a whole lot of Wedgwoods some time ago & have not the heart to open them yet:[18] by the bye I wish Gladstone would hold his tongue & not raise the price of them with his nonsense.—[19] What a clever dog he is, but he seems to me to be but a small minded man in many matters.

What do you think we should do about the Poles, to go to war about them seems absurd, but surely we have behaved ⟨as lo⟩west sneaks, we incurred ⟨res⟩ponsibilities by the Treaty ⟨o⟩f Vienna & now ignore them. I for one wish the French may go at the Russians.[20]—not but what I suppose the Poles are a miserable race who will never survive the struggle for life if not crossed with some better breed.— If the French do go ahead we must follow.

What does Asa Gray say now? I think Lairds position is a most dishonorable one to a British merchant & Legislator.[21] he should make a clean breast of it & cry "peccavi",[22] or justify himself if he can do so. This squabble between Lord Russell & Sir James Hudson is a pitiful affair,[23] I supp⟨ose⟩ both are wrong—at leas⟨t⟩ ⟨ ⟩

Thanks many for Huxleys v⟨ery⟩ amusing & clever letter & its enclosure.[24]

Ramsay told us at Phil. Club that they had found old moraines in Mts of Roxburghshire, at about 12–1500 ft elevation I think.[25]

Now you are not to answer this discussion of mine; but thank Mrs Darwin heartily for letter.—[26]

Ever dear old Darwin | Yours | J D Hooker

What a horrible business our destruction of that Japan town is.—[27]

DAR 101: 173–5
Endorsement: 'Nov 63'

[1] The date is established by the endorsement and by the relationship between this letter and the letter to J. D. Hooker, 10 [November 1863]. Hooker dated the letter either Sunday or Tuesday (see n. 2, below), making the most likely date 1 or 3 November 1863.

[2] Only the second letter, 'u', of this word is legible, suggesting that Hooker wrote either Sunday or Tuesday.

[3] Emma Darwin's letter to Hooker is not included in the collection of J. D. Hooker correspondence at the Royal Botanic Gardens, Kew. For CD's letter, see the letter to J. D. Hooker, [30 October 1863].

4 CD discussed this climbing palm in *Origin* 6th ed., p. 158. See letters to J. D. Hooker, 3 August [1863], n. 2, and [28 August 1863] and n. 10, and letter from J. D. Hooker, 15 September 1863. See also CD's note in DAR 157.1: 66.

5 Hooker refers to the letter from Julius von Haast, 21 July [– 7? August] 1863 (see letter to J. D. Hooker, [30 October 1863]).

6 James Hector; see the letter from J. D. Hooker, 23 October 1863 and n. 9.

7 In his letter of 23 October 1863, Hooker discussed the geology of New Zealand, having received Hector's sketches; he was struck by the similarity of the effects of glacial action in New Zealand to what he had observed on his Himalayan travels (see J. D. Hooker 1854b). Hooker exhibited photographs of Hector's sketches at the 29 October 1863 meeting of the Philosophical Club of the Royal Society (Bonney 1919, p. 165).

8 John Gunn. Hooker had also visited the Norfolk coast during the previous month (see letter from J. D. Hooker, 23 October 1863).

9 Hooker describes an approximately three-mile stretch of a north-east Norfolk coastal cliff that included a lower stratum of an ancient forest buried in loam and lignite (see C. Lyell 1863a, pp. 212–16). Joshua Trimmer had named what he thought was an upper alluvial deposit of surface soil 'Warp of the drift, or erratic Warp', since it appeared to fill cavities in the glacial drift (see Trimmer 1850, p. 23). Hooker evidently attributed this stratum to weathering of the underlying deposit.

10 Hooker had recently visited the archaeological sites near Amiens and Abbeville in the French province of Picardy, where he was able to observe the geology of the Somme valley (see letter from J. D. Hooker, 23 October 1863 and n. 4).

11 Andrew Crombie Ramsay.

12 Hooker refers to Joseph Prestwich's conclusions that particular Pleistocene deposits in north-western France and eastern England were due to intense and periodic flooding of large, ancient rivers, and to the associated action of river ice; Prestwich believed the resulting scouring and deposition were correlated with significant changes in climate (see J. Prestwich 1862a and 1862b, pp. 286–98). There is a copy of J. Prestwich 1862b in the Darwin Library–CUL. See also letter from J. D. Hooker, 23 October 1863 and n. 6.

13 William Henslow Hooker was Hooker's eldest son.

14 Maria Hooker.

15 Hooker contracted a mild case of scarlet fever at the age of six (see L. Huxley 1918, 1: 5).

16 Hooker refers to his wife, Frances Harriet Hooker, and their two youngest children, Charles Paget Hooker and Brian Harvey Hodgson Hooker (Allan 1967).

17 Maria and William Jackson Hooker.

18 Hooker had started collecting Wedgwood ware in 1862 (see *Correspondence* vol. 10, letter from J. D. Hooker, [27 or 28 December 1862]).

19 Hooker refers to William Ewart Gladstone's address delivered at the laying of the foundation stone of the Wedgwood Institute, which was built in Burslem, Staffordshire, in honour of Josiah Wedgwood I. The address extolled Wedgwood's industry, and the beauty and quality of the ware (see *The Times*, 27 October 1863, pp. 5–6, and Gladstone 1863).

20 In 1863, Poland attempted to achieve independence from Russia by a prolonged insurrection. Under the Congress of Vienna (1814–15), Great Britain and other western European nations had advocated limiting Russia's control and incorporating Poland as a separate kingdom with the capability eventually of establishing its own constitution. In his 5 November 1863 address to the new legislature, the French Emperor, Napoleon III, was expected to announce his intention to intervene on the Polish side in the war with Russia (*EB*; *The Times*, 2 November 1863, p. 6, 5 November 1863, p. 6).

21 The reference is to the industrialist and MP for Birkenhead, John Laird (*DNB*). His shipbuilding firm, Laird Brothers, had built the *Alabama*, which was sold to the American Confederacy in 1862 and played an active role in the Civil War as a commerce raider (Jenkins 1974–80, 2: 121–2). During October 1863 the firm was also completing two armour-plated vessels carrying gun turrets and a seven-foot iron spike for piercing wooden ships below the waterline. When the authorities seized the

two so-called 'Laird rams' on 27 October, stating that they violated British neutrality, Laird denied that the ships were to be sold to the Confederacy; he also continued to evade the question of his earlier intentions for the *Alabama* (*The Times*, 27 October 1863, p. 7, 28 October 1863, p. 8, and 29 October 1863, p. 9). See Jenkins 1974–80, 2: 289–306. See also letters from Asa Gray, 26 May 1863 and n. 27, and 23 November 1863 and nn. 8 and 9.

22 *Peccavi*: 'I have sinned'; an acknowledgment or confession of guilt (*OED*).

23 The references are to Lord John Russell, the foreign secretary, and Sir James Hudson, the former British envoy at Turin, who had resigned in April 1863. Though the two men were not themselves openly quarrelling, a public outcry supporting Hudson raised the question whether he had been forced from office by Russell (*The Times*, 25 August 1863, 24 October 1863, p. 8, 28 October 1863, p. 5, 29 October 1863, p. 8; *DNB*).

24 Thomas Henry Huxley's letter has not been found; however, see the letter from Emma Darwin to W. E. Darwin, [28 October 1863].

25 Hooker refers to Andrew Crombie Ramsay's statement at the 29 October 1863 meeting of the Philosophical Club of the Royal Society that members of the Geological Society had discovered and mapped terminal moraines in the south of Scotland (Bonney 1919, pp. 164–5).

26 See n. 3, above.

27 On 15 and 16 August 1863 seven British navy warships bombarded Kagoshima (the capital of Satsuma province) causing an estimated £1,000,000 of damage; the action was in response to repeated attacks on British people and property and to demands for indemnities. In Britain, this attack was hotly debated and widely condemned (see Fox 1969, pp. 97–119, particularly pp. 114–19).

From Emma Darwin to Alfred Newton 4 November [1863]

Down. | Bromley. | Kent. S. E.
Nov 4

Sir

Mr Darwin begs me to thank you for your note & Remarks on the partridge's leg.[1] He has been so ill as not to be able to attend to any thing & he still cannot even write a note,[2] but he will send [for] the specimen as soon as he can & examine it with care. He begs to be very kindly remembered to you.

I am dear Sir | yours truly | Emma Darwin

Endorsement: 'Nov. $\frac{4}{6}$/63.'[3]
Cambridge University Library (Alfred Newton papers)

1 See letter from Alfred Newton, 31 October 1863. See also letter to J. D. Hooker, 5 [December 1863], and *Correspondence* vol. 12, letter to J. D. Hooker, 26–[7] March 1864.

2 The entry in Emma Darwin's diary (DAR 242) for 4 November 1863 is 'in bed all day'.

3 Newton's endorsement may include the day he received the letter, 6 November 1863.

From John Lubbock to Emma Darwin 6 November 1863

15, Lombard Street. E.C.[1]
6 Nov. 1863

My dear Mrs. Darwin

I return with many thanks the extract from the Record which you were kind enough to lend me.[2]

I should be very much obliged if you would drop me one line to say how M^r. Darwin is now.

Please give him my love & Believe me | Yours very truly | John Lubbock

DAR 170: 43

[1] The address of the London bank Robarts, Lubbock & Co., of which Lubbock was a partner.

[2] Lubbock refers to the review of *Origin* in the *Record* of 12 December 1860, p. 4, which he had borrowed from CD (see letter from John Lubbock to Emma Darwin, 24 October 1863).

To John Scott 7 November [1863][1]

Down. | *Bromley.* | *Kent. S.E.*
Nov 7.

Dear Sir

Every day that I c^d do any thing, I have read a few pages of y^r paper & have now finished it & return it registered.[2] It has interested me deeply, & is I am sure, an excellent memoir. It is well arranged & in most parts well written. In the proof sheets you can correct a little with advantage I have suggested a few alterations in pencil for yr consideration, & have put in here & there a slip of paper. There will be no occasion to re-write the paper, only, if you agree with me, to alter a few pages. When finished return it me & I will with the highest satisfaction communicate it to the Linn. Soc.[3] I sh^d be proud to be the author of the paper. I shall not have caused much delay as the 1st meeting of the Soc. was on Nov 5.

When yr primula paper is finished if you are so inclined I sh^d like to hear *briefly* about y^r Verbascum & Passiflora experiments.[4] I tried Verbascum & have got the pods but do not know when I shall be able to see to the results.[5] This subject might make another paper for you. I may add that Acropera Luteola was fertilized by me & had produced 2 fine pods.[6]

I congratulate you on your excellent paper & believe me My dear Sir | yours very sincerely | Charles Darwin

P.S. In summary to P. paper can you conjecture what is the typical or parental form i.e. equal long or short-styled[7]

LS[8]
DAR 93: B5–6

[1] The year is established by the relationship between this letter and the letter from John Scott to Emma Darwin, 25 September [1863], and by the references to Scott 1864a (see nn. 2 and 3, below).

[2] CD refers to the manuscript of Scott 1864a, which Scott had sent to CD for his comments (see letter from John Scott, 21 September [1863]).

[3] Scott 1864a was read before the Linnean Society on 4 February 1864.

[4] Scott and CD had been discussing *Verbascum* and *Passiflora* since November 1862 (*Correspondence* vol. 10, and this volume, letters to John Scott, 6 March 1863 and 2 July [1863], and letter from John Scott, 21 March [1863]). See also *Correspondence* vol. 12, letter from John Scott, 19 March 1864. Scott's observations on *Passiflora* were published in Scott 1864d, and those on *Verbascum* in Scott 1867. There is an annotated copy of Scott 1867 in the Darwin Pamphlet Collection–CUL.

[5] For CD's experimental notes on *Verbascum lychnitis*, dated 21 June 1863, see DAR 108: 5–6. CD published his work on *Verbascum* in 'Specific difference in *Primula*', pp. 451–4.

[6] In *Orchids*, pp. 203–10, CD had concluded that *Acropera* was dioecious, and that the specimens he had examined were male. However, Scott had successfully pollinated two flowers of *A. loddigesii* (see *Correspondence* vol. 10, letter from John Scott, 11 November 1862). CD continued to puzzle over *Acropera* in 1863 (see, especially, letter to P. H. Gosse, 2 June [1863]); for CD's eventual successful pollination of *Acropera*, see the experimental note, dated 29 August 1863, in DAR 70: 114.

[7] See Scott 1864a, p. 126. See also letter from Emma Darwin to John Scott, 19 November [1863] and n. 4.

[8] The letter is written in Emma Darwin's hand.

From E. A. Darwin 9 November [1863][1]

9 Nov

Dear Charles.

M[r] Conway the Yankee about whom I formerly spoke to you has written to inquire if you are well enough to seee him as he would call on say Friday if you would see him.[2] I have written to him to say it is very doubtful.[3]

D[r] Carpenter showed me the extract from Mills Logic which he read when he argued for your having the Copley Medal.[4] Have you seen it? I hope you are getti⟨ng⟩ better— the number of inquiries I receive is infinite

ever yours | E Darwin

DAR 105: B13–14

[1] The year is established by the relationship between this letter and the following letter.

[2] The letter referred to has not been found. Moncure Daniel Conway was an American preacher and author with strong anti-slavery views; in 1863, he moved from Boston to London (*DAB*).

[3] See following letter.

[4] In the fifth edition of John Stuart Mill's *System of logic*, Mill discussed the proper use of scientific hypotheses, noting that 'Mr. Darwin's remarkable speculation on the origin of species is another unimpeachable example of a legitimate hypothesis' (J. S. Mill 1862, 2: 18 n.). Arguing that CD had not violated the 'rules of Induction' since he did not pretend his doctrine was proven, Mill wrote (*ibid.*):

> He has opened a path of inquiry full of promise, the results of which none can foresee. And is it not a wonderful feat of scientific ingenuity to have rendered so bold a suggestion, which the first impulse of everyone was to reject at once, admissible and discussable, even as a conjecture?

See also *Correspondence* vol. 9, letter from Henry Fawcett, 16 July [1861] and n. 3, and Hull 1973, pp. 27–8. CD was considered for the Royal Society's Copley Medal in 1863, but on 5 November the medal was awarded to Adam Sedgwick; William Benjamin Carpenter was a member of the society's council and seconded John Lubbock's nomination of CD for the medal (Royal Society, Council minutes, 5 November 1863). The Copley Medal was awarded, generally annually, by the council of the Royal Society, for outstanding 'philosophical research' (*Record of the Royal Society*, Appendix IV).

From E. A. Darwin to Moncure Daniel Conway 9 November 1863

6 Queen Annes
9 Nov 63

Dear M[r] Conway

My Brother I am sorry to say has been seriously unwell for a considerable time & I feel very doubtful if he can see any one.[1]

I have written to him & am sure he will be very glad to see you if able

I will let you know as soon as I hear & give you the necessary directions for getting there.[2]

Believe me | Yours very truly | Eras. Darwin

M. D. Conway Esq

Columbia University, Rare Books and Manuscript Library (Moncure D. Conway papers)

[1] See preceding letter.
[2] See letter from E. A. Darwin to M. D. Conway, 11 November 1863.

From George Clendon Jr 10 November 1863

Provost Marshal's Office, | 16th Cong. District, N.Y. | Plattsburgh, November 10th *1863.*

My dear Sir

The following supposed cause of the paucity of fossil remains of "intermediate forms of life" does not I think occur in your work on the "Origin of Species"[1] It is perhaps worthy of consideration

Let us imagine a continent of large area that has maintained its configuration of hill and valley for a long period of time. It will be peopled by living forms adapted to the various stations Variation accompanied by natural selection will make but little change from the fact that each plant and animal by the previous action of these causes is adapted to its location. This state of things may continue for a long period, and the remains of plants and animals become fossil constitute a geological formation

A great change in the physical geography of the continent is now supposed. Islands are formed by the subsidence of the neighbouring land or a mountain chain is upheaved. Variation and selection are now in full operation, new forms of life are produced to conform to the changes of their habitation. When this is accomplished, the new order of things may remain for a long period constituting another geological formation

In this way we shall find fossil remains of two formations differing materially from each other, and from the period of disturbance being shorter than the period of rest shall not be likely to find many intermediate forms. It is not necessary to imagine a great catastrophe or sudden upheaval, All that is requisite is that the period of disturbance be short compared with the period of rest and I know of no geological facts opposed to this view.

I am Sir Very Respectfully | Your Obedient Servant | George Clendon Junr

Chas Darwin Esq MLGS[2]

DAR 47: 178

CD ANNOTATION
End of letter: '*Paleontology*' pencil

[1] Chapters 9 and 10 of *Origin*, pp. 279–345, were entitled 'On the imperfection of the geological record' and 'On the geological succession of organic beings'. Clendon's suggestions, which apply to these chapters, were not mentioned in later editions of *Origin*, nor has a reply to Clendon's letter been found.

[2] MLGS: member of the Geological Society of London.

From Friedrich Hildebrand 10 November 1863

Bonn
Novbr 10th | 1863

Dear and respected Sir

I must beg your pardon that I have not answered your very kind letter before this, but I was waiting for my paper on Orchids being printed in the Botanische Zeitung to send you a copy;[1] I hope that I shall be able to do it in the next days, you will see that it does not contain much more as you know as yet from my last letter.[2] And now let me thank you that you have taken so much pain to get my paper printed, I have seen it in the Annals and Mag. of n. hist. and was very glad of it; I feel very much obliged to you.[3] I am glad being able to do you a service in publishing my experiments on Linum and Primula, but I can speak only about L. perenne and P. sinensis, beacause my experiments on P. officinalis and elatior were destroyed by childeren in the botanical garden.[4] On Linum I spoke a few words on the meeting of German naturalists in Stettin.[5]

In your paper on Primula you say, that you never placed pollen of the same flower on its own stigma[6] but I thought this manner of fertilisation to be of interest too, and I had in this way six unions instead of your four. The result was that the two heteromorphic unions gave most seeds, namely each capsule 43 good seeds on an average; the two homomorphic unions, if the flowers were crossed, gave 18 seeds on an average per capsule, the flowers selffertilised only 13. In the first four unions I got good capsules of every fertilised flower, only in the last two I did not always succeed in the fertilisation.— You see that, though I have counted the seeds, not weighed, I have come to the same results as you.[7] But I must say that I cannot agree with you about the pollen of P. sinensis, for I found the pollen grains of the long styled form almost in the same manner differing from that of the short styled as in P. officinalis and elatior; between the small transparent grains of the short-styled form there were only a few of larger seize and opace.—[8]

Perhaps it will interest you to hear that I have made other experiments to answer the question to what form the descendants of the two forms belong. From seeds that I had got last year by your four unions I got about 60 young plants and these were the interesting results: Of the descendants from long-styled plants fertilised with their own form pollen most, but not all, were again long styled; the short-styled form fertilised with its own-form pollen produced in the greatest part short styled plants; if the short styled form was fertilised with the long-styled or the long styled with the short styled nearly the half of the young plants was short-styled the other half long-styled.—[9]

Coming to a close I must say that if there is anything, where I can serve you, perhaps with microscopical researches, I should be very glad if you would tell me about it, the winter especially leaves me a great deal of time.

Professor Treviranus wanted me to send you his respects and remembrances.[10] Some time ago I heard that you were not quite well but I hope that this letter will find you in good health and I remain most respected Sir | yours | sincerely | Hildebrand

DAR 166: 201

[1] See letter to Friedrich Hildebrand, 28 July [1863]. The German version of Hildebrand's paper on orchids (Hildebrand 1863a) was published in the issues of *Botanische Zeitung* for 30 October and 6 November 1863; there is an annotated copy of Hildebrand 1863a in the Darwin Pamphlet Collection–CUL.

[2] In his letter to CD of 16 July 1863, Hildebrand included an English abstract of Hildebrand 1863a.

[3] At Hildebrand's request, CD asked Daniel Oliver and Joseph Dalton Hooker about the possibility of publishing an abbreviated version of Hildebrand 1863a in an English journal (see n. 2, above, letter to Daniel Oliver, 18 July [1863], and letter to J. D. Hooker, 22 July [1863]). Hildebrand's paper, entitled 'On the impregnation in orchids as a proof of the two different effects of the pollen' appeared in the September 1863 issue of *Annals and Magazine of Natural History* (Hildebrand 1863b). See letter to Friedrich Hildebrand, 28 July [1863].

[4] See letter to Friedrich Hildebrand, 28 July [1863]. Hildebrand published his experiments on *Linum* and *Primula sinensis* in Hildebrand 1864.

[5] Hildebrand's statement that he had confirmed CD's results with *Linum perenne* by experiment is briefly reported in *Amtlicher Bericht über die 38. Versammlung Deutscher Naturforscher und Ärzte in Stettin im September 1863*.

[6] 'Dimorphic condition in *Primula*', p. 87 (*Collected papers* 2: 55).

[7] In 'Dimorphic condition in *Primula*', pp. 86–8 (*Collected papers* 2: 54–6), CD described experiments that tested the potency of the pollen of the different forms of *Primula veris*, *P. sinensis*, and *P. auricula*. To avoid the possible negative effects of close interbreeding, CD never placed the pollen of a flower on its own stigma. Hildebrand published his observations in Hildebrand 1864; CD commented on Hildebrand's results in 'Illegitimate offspring of dimorphic and trimorphic plants', pp. 411 n. and 413 n., and in *Forms of flowers*, pp. 40–3, 192 n., and 217 n.

[8] While CD observed little difference in the pollen grains of the two forms of *P. sinensis* ('Dimorphic condition in *Primula*', pp. 80–1 (*Collected papers* 2: 48)), Hildebrand wrote (Hildebrand 1864a, p. 2) that the pollen grains of the long-styled plants were smaller and more transparent than those of the short-styled form. In this letter, Hildebrand writes that it was the short-styled plants that yielded small and transparent pollen grains. By 'seize and opace' Hildebrand presumably meant size and opacity. CD noted the published information in *Forms of flowers*, p. 38 n. See also 'Illegitimate offspring of dimorphic and trimorphic plants', p. 416.

[9] For Hildebrand's data, see Hildebrand 1864, pp. 4–5.

[10] Ludolph Christian Treviranus had been Hildebrand's teacher in Bonn (Correns 1916, p. 29).

To J. D. Hooker 10 [November 1863]

Down
10th

My dear old Friend.

I was so glad to hear of your French tour & that you are determined & are able to exert yourself.—[1] What an addition of trouble your case of S. Fever.—[2]

I now send Haast:[3] I have tried many days to write to you, but could not.— Dr. Brinton has been here, (recommended by Busk)[4] he does not believe my Brain or heart are primarily affected; but I have been so steadily going down hill, I cannot help doubting whether I can ever crawl a little up hill again. Unless I can, enough to work a little, I hope my life may be very short; for to lie on sofa all day & do nothing, but give trouble to the best & kindest of wives & good dear children is dreadful. I shd. much like to write on N. Zealand distribution & superficial deposits &c (I have been rereading your old letters) but am too weak.[5]

God Bless you my best of Friends. | Yours affect. | C. Darwin

Haast sent me a Report of some kind & large map.—[6]

Endorsement: '/Nov 1863.'
DAR 115: 208

1 See letter from J. D. Hooker, 23 October 1863.

2 Hooker's son, William Henslow Hooker, and mother, Maria Hooker, were both ill with scarlet fever (see letter from J. D. Hooker, [1 or 3 November 1863]).

3 Letter from Julius von Haast, 21 July [– 7? August] 1863 (see letter from J. D. Hooker, [1 or 3 November 1863] and n. 5).

4 William Brinton was recommended to CD in the letter from George Busk, [c. 27 August 1863]. According to Emma Darwin's diary (DAR 242), Brinton visited CD on 3 November 1863.

5 See letter to J. D. Hooker, [28 August 1863] and n. 7, and letter from J. D. Hooker, 15 September 1863.

6 The map and report have not been found in the Darwin Archive–CUL. CD may refer to the map Haast mentioned in his letter to CD of 6 August 1863. Alternatively, Haast may have sent copies of a handwritten report and map that he presented with a geological section to the Provincial Council, Province of Canterbury, on 18 August 1863 (Archives Reference CH 287, CP 608a, National Archives, Christchurch). See also letter from Julius von Haast, 21 July [– 7? August] 1863 and n. 10.

From E. A. Darwin to M. D. Conway 11 November 1863

6 Queen Anne St | W
11 Nov 63

Dear Mr Conway

I am sorry to say that I have heard from Mrs Darwin that my Brother is no better & that he is forbidden by his physician to see any one.[1] He begs me to say how sorry he is that he cannot have the pleasure of making your acquaintance.

Believe me | Yours very truly | Eras. Darwin

Columbia University, Rare Book and Manuscript Library (Moncure D. Conway papers)

[1] See letter from E. A. Darwin, 9 November [1863], and letter from E. A. Darwin to M. D. Conway, 9 November 1863, and following letter.

From E. A. Darwin to Emma Darwin 11 November [1863][1]

11 Nov.

Dear Emma

Thanks for your letter.[2] I thought Mr Conway would be better satisfied if I wrote to inquire tho' I had'nt an idea that Charles would be able to see him.[3] I have very little to say about the Copley Medal.[4] The numbers were 8 to 10 for Charles, but the Cambridge men mustered very strongly for Sedgwick.[5] I did not know enough to ask Carpenter any questions.[6] He also mentioned that Benthams address to the Linnæan was on Origin & very good.[7] Lyell was talking again how much he wished the Dogs to be published & Huxley the same, & that even Hooker grudged Charles employing his time on plants instead of animals.[8]

Yours affectionately | E. D

DAR 105: B116–17

[1] The year is established by the reference to the Copley Medal (see n. 5, below).

[2] The letter has not been found.

[3] See preceding letter and n. 1.

[4] The reference is to the Royal Society's Copley Medal (see n. 5, below, and letter from E. A. Darwin, 9 November [1863]).

[5] On 5 November 1863, the council of the Royal Society resolved to award the Copley Medal to Adam Sedgwick, professor of geology at the University of Cambridge. According to the council minutes, only seventeen, rather than eighteen, of the twenty-one council members were present at the meeting. Warren De la Rue, Philip de Malpas Grey-Egerton, Robert Godwin-Austen, and Joseph Henry Green were absent; however, it was not unusual for all but De la Rue to miss meetings (see Royal Society Council minutes, 1863, and Lyons 1944, pp. 272–5). See also letter from Edward Sabine to John Phillips, 12 November 1863 and n. 4. Four members of the Council of the Royal Society were directly associated with Cambridge University in 1863: Charles Cardale Babington, William Hallowes Miller, George Gabriel Stokes, and Robert Willis (*DNB*).

[6] William Benjamin Carpenter was a member of the council of the Royal Society (see letter from E. A. Darwin, 9 November [1863] and n. 4). Carpenter and John Lubbock had nominated CD for the Copley medal (see Royal Society Council minutes, 11 June 1863).

[7] The reference is to George Bentham's presidential address to the Linnean Society of London on 25 May 1863, in which Bentham discussed the reception of *Origin* by the international scientific community (Bentham 1863). For CD's reaction to the address, see the letter to George Bentham, 19 June [1863] and the letter to J. D. Hooker, 23 [June 1863].

[8] The references are to Charles Lyell, Thomas Henry Huxley, and Joseph Dalton Hooker. In working on his 'big book' on species, of which *Origin* was an abstract (see letter to Asa Gray, 31 May [1863], n. 18), CD investigated variation in domestic dogs (see *Correspondence* vols. 5–9). CD and Lyell discussed dog ancestry in 1859 (see *Correspondence* vol. 7). Since *Origin* appeared in 1859, CD had published a book on orchid pollination (*Orchids*) and papers on polymorphism in *Primula* ('Dimorphic condition in *Primula*') and *Linum* ('Two forms in species of *Linum*'). The first part of the planned three-part 'big book' was published in 1868 as *Variation*; the first chapter was a discussion of variation in domesticated breeds of dogs and cats (*Variation* 1: 15–48).

From J. D. Hooker to Emma Darwin 11 November 1863

Kew
Nov 11/63.

My dear Mrs Darwin

Will you kindly te⟨ll⟩ me whether I ought ⟨to⟩ write to Mr Darwin whil⟨st⟩ he is so ill—promising th⟨at⟩ my letters require no answering—[1] Do you know that it is such a real pleasure to me to write to him, that I feel it a positive privation not to—

I was very glad to get his letter,[2] this morning, but he must not try to write to me.

Please tell him that I ⟨ha⟩ve prepared a mere notice of ⟨the⟩ introduced plants of ⟨New⟩ Zeald: from letters of Mr Travers, for Nat Hist Review—[3] would he object to my adding Haast's notes on the introduced Animals? I see Haast says that Mr Darwin may make "any use of his communications how & when you like."[4]— I should also like to read the letter to the Phil. Club.[5]

My wife & 2 children are still in Norfolk, & will be till end of month.[6] Willy has been scoured with oil & soft soap & sent to Hastings.[7] My Mother still keeps her bed, but is well recovering.[8]

With most sincere regards | Ever ty yrs | J D Hooker

DAR 101: 171–2

[1] No reply from Emma Darwin has been found, but see CD's letter to J. D. Hooker, [13 November 1863].

[2] Letter to J. D. Hooker, 10 [November 1863].

[3] Hooker's account (J. D. Hooker 1864) was published in the January 1864 issue of the *Natural History Review*. Hooker refers to the botanist and ornithologist William Thomas Locke Travers (R. Desmond 1994). Travers was quoted at length in J. D. Hooker 1864, p. 124. Hooker was well aware of CD's interest in the introduction of plants and animals to new environments (see J. D. Hooker 1864, p. 125, and letter to Julius von Haast, 22 January 1863 and n. 9). See also letter to Osbert Salvin, 11 [May 1863] and n. 6.

[4] CD sent the letter from Julius von Haast of 21 July [– 7? August] 1863 with his letter to J. D. Hooker of 10 [November 1863]. See J. D. Hooker 1864, pp. 126–7, for his inclusion of Haast's information.

[5] Hooker refers to the letter from Julius von Haast of 21 July [– 7? August] 1863 and to the Philosophical Club of the Royal Society; however, he does not appear to have read the letter to the club. See Bonney 1919.

[6] Hooker refers to Frances Harriet Hooker and their two youngest children, Charles Paget Hooker and Brian Harvey Hodgson Hooker (Allan 1967).

[7] William Henslow Hooker, Hooker's eldest son, had had scarlet fever (see letter from J. D. Hooker, [1 or 3 November 1863]). Eleanor Jane Jacobson, Hooker's aunt, lived in Hastings, Sussex (Allan 1967).

[8] Maria Hooker had contracted scarlet fever from William Henslow Hooker (see letter from J. D. Hooker, [1 or 3 November 1863]).

From Edward Sabine to John Phillips 12 November 1863

The Royal Society, | Burlington House, London. W.
Nov. 12 1863.

Dear Phillips

I thank you heartily for your improvements, all of which I have adopted and will with your leave send you the part which relates to Sedgwick *in type*, that you may have a second opportunity of *touching up*.[1] I am endeavoring to have the whole address ready for the printer by the 19th.; so that if possible I may have a proof on Saturday the 22d: and a revised sheet on Saturday the 29th. *at latest*.

Sedgwick will attend himself which will be a great pleasure to us all here. Murchison will *not* be present; which is so far well that we shall avoid embarrassment;—to say the least.[2] Will *you* not come up to sit by your friend at dinner. The Society meets, as you know, at 4 PM. It would be a most gratifying circumstance to Sedgwick, and we, of the council, should feel the benefit of such an evidence of your thorough approval of the Award.[3]

It may be that I do not partake sufficiently, by reason of my age, of the spirit of the opinions which appear to have taken a strong hold of the younger Geologists and Zoologists. With all respect for Darwin's great services, and recognising that his recent work on Orchids must be classed amongst these, I cannot see without extreme concern the efforts of a very strong party to obtain the award of the Copley Medal to him expressly on the ground of his conclusions as to the "Origin of Species".[4] A more decided interference that I desire to exercise a second time with the thorough independence of the votes of individual members of the Council, was required to prevent such an award being made in this year instead of to Sedgwick.[5] We may not have so good an alternative next year;[6] and I cannot but deem it a great misfortune to the RS. that we cannot prevail on the older, & may I say more soberminded of the naturalists to give us their aid in the Council. It is the *only* source from which I have any *present* anticipation of trouble.

always truly yours | Edward Sabine

American Philosophical Society (Misc. MS collection)

[1] Sabine evidently sent Phillips drafts of the anniversary address that he was to give on 30 November 1863 as the President of the Royal Society (Sabine 1863). The anniversary meeting was to include the presentation of medals; see Sabine 1863, pp. 31–5, for the presentation of the Copley Medal to Adam Sedgwick. The medal was awarded (Sabine 1863, p. 31) for Sedgwick's

> observations and discoveries in the Geology of the Palæozoic Series of Rocks, and more especially for his determination of the characters of the Devonian System, by observations of the order and superposition of the Killas Rocks and their Fossils, in Devonshire.

For a discussion of the significance of medals in Victorian science, particularly the Royal Medal, see MacLeod 1971.

[2] Sedgwick and Roderick Impey Murchison had disagreed over the classification of British Palaeozoic rocks since the 1840s; the two men, once collaborators, were no longer on speaking terms (see Secord 1986, pp. 262–8). Murchison had publicly defended his classification as recently as 1861, when he gave the presidential address to the geology section of the thirty-first meeting of the British Association for the Advancement of Science in Manchester (Murchison 1861). Murchison had received the Copley Medal in 1849.

[3] The council of the Royal Society, composed of twenty-one members including Sabine, voted on recipients of medals (*Record of the Royal Society of London*, pp. 227–8). Phillips was professor of geology at Oxford University, and had been president of the Geological Society in 1859 and 1860 (*DNB*). For Phillips's recollections of his friendship with Sedgwick, see Phillips 1873. See also Secord 1986, pp. 295–6.

[4] Sabine refers to *Orchids* and *Origin*. At the 11 June 1863 meeting of the Royal Society council, John Lubbock nominated CD for the Copley Medal, and William Benjamin Carpenter seconded the nomination. Two others were also nominated at that meeting (Henri Victor Regnault and August Wilhelm von Hofmann), but it was not until the following meeting on 18 June 1863 that Sedgwick was nominated by Richard Owen; Philip de Malpas Grey-Egerton seconded the nomination. The other members of the council of the Royal Society in 1863 were Charles Cardale Babington, George Bowdler Buckton, Warren De la Rue, Douglas Strutt Galton, Robert Godwin-Austen, Joseph Henry Green, Joseph Dalton Hooker, James Clerk Maxwell, William Allen Miller, William Hallowes Miller, William Sharpey, Henry John Stephen Smith, George Gabriel Stokes, James Joseph Sylvester, Charles Wheatstone, and Robert Willis. See Royal Society, Council minutes, 6 November 1862, 11 June, 18 June, and 5 November 1863. See also letter from E. A. Darwin to Emma Darwin, 11 November [1863] and n. 5.

[5] Sabine wrote 'that' before 'I desire' when presumably he meant 'than'. See also letter from E. A. Darwin to Emma Darwin, 11 November [1863]. For a discussion of Sabine's presidency of the Royal Society, see Hall 1984, pp. 104–9.

[6] CD received the Copley Medal in 1864. See, for example, *Correspondence* vol. 12, letter from Edward Sabine, 3 November 1864, and letter from Charles Lyell, 4 November 1864.

To J. D. Hooker [13 November 1863]

[Down]

My dear Hooker.

A letter from you is always one of my greatest pleasures.—[1] I enclose Haast, which I did not suppose you w$^{d.}$ care to see.—[2] There was a printed address to the Phil. Soc. of Canterbury published some 6 or 12 months ago by Haast.—[3] Was it this you wished to see?

I escaped vomiting yesterday & have felt these 2 days better. D$^{r.}$ Brinton tells me that a little head-work not bad.[4] Dr Gully writes again this morning to do nothing for 6 months.[5] I presume I must observe & judge for self.— I am glad to hear about Willy.—[6] Do not, my dear friend, write when much overworked. | C. D.

Of course use any & all Haast's facts[7]

Who reviewed Lyell in Edinburgh in July last??? I inferred it was a real geologist, but one at least 100 years old.[8] over

I daresay you have no other tendril-bearing plants.—[9] over

Endorsement: 'Nov 13/63'
DAR 115: 209

[1] See letter from J. D. Hooker to Emma Darwin, 11 November 1863.

[2] Following the interest Hooker had expressed in the letter from Julius von Haast of 21 July [– 7? August] 1863 (see letter from J. D. Hooker to Emma Darwin, 11 November 1863), CD may have decided to send also the letter from Julius von Haast of 6 August 1863, or the map and report that CD mentioned in the postscript to his letter to J. D. Hooker of 10 [November 1863].

[3] CD refers to J. F. J. von Haast 1862a. There is a copy in the Darwin Pamphlet Collection–CUL.

[4] Emma Darwin wrote in her diary (DAR 242) 'good' for 12 November and 'fine day' for 13 November; she also wrote that CD had been visited by William Brinton on 3 November 1863.

[5] The letter from James Manby Gully, who had treated CD at his Great Malvern hydropathic establishment in 1849, has not been found.
[6] William Henslow Hooker was recovering from a bout of scarlet fever; see letter from J. D. Hooker, [1 or 3 November 1863], and letter from J. D. Hooker to Emma Darwin, 11 November 1863.
[7] Hooker had asked CD if he could use the information Haast included in his letter of 21 July [– 7? August] 1863 (see letter from J. D. Hooker to Emma Darwin, 11 November 1863 and n. 4.
[8] CD refers to the review of *Antiquity of man* (C. Lyell 1863a) in the *Edinburgh Review* ([J. D. Forbes] 1863); the author of the review was James David Forbes (*Wellesley Index*).
[9] In July, Hooker sent CD some tendril-bearing plants from the Royal Botanic Gardens, Kew, for his research (see letter from J. D. Hooker, [31 July 1863] and n. 2, and letter to J. D. Hooker, 3 August [1863]).

To J. D. Hooker 16 [November 1863]

Down
16th

My dear H.

My Father had some small vases, black & brown, given him by Jos. Wedgwood, & which my Father valued much.[1] I, of course, do not know whether they are really good; but I have begged one from my sister Susan[2] & she will send it in about a week's time to Kew.[3]— I shall be so pleased, if it turns out good; but that must be a mere chance.—

Ever my dear H. | C. D.

I have now been six days with no vomiting!!! & my head feels more solid.—

I see you have been at alpine W. African Plants[4]

Endorsement: '/Novr. 63'
DAR 115: 210

[1] CD refers to Robert Waring Darwin, and to either Josiah Wedgwood I, master-potter and founder of Wedgwood & Sons, or his son, Josiah Wedgwood II.
[2] Susan Elizabeth Darwin lived at The Mount, Shrewsbury, the family home.
[3] Hooker, assistant director at the Royal Botanic Gardens, Kew, was an avid collector of Wedgwood ware (see letter from J. D. Hooker, [1 or 3 November 1863] and nn. 18 and 19).
[4] CD probably refers to the report of J. D. Hooker 1863b in the *Gardeners' Chronicle*, 14 November 1863, pp. 1085–6, which was read to the Linnean Society on 5 November 1863. CD's annotated copies of the *Gardeners' Chronicle* are in the Cory Library, Cambridge Botanic Garden.

From Emma Darwin to John Scott 19 November [1863][1]

Down. | Bromley. | Kent. S.E.
Nov 19.

Dear Sir

My husband wishes me to write & thank you for your letter[2] He has not been so well as he was & cannot answer it himself & for the same reason he begs you will not send the specimen.[3] He quite agrees "about reversion & it was that made him ask you about the parental form"—[4]

The discovery about crossing Cryptogams must be very interesting[5]

With Mr Darwin's kind regards I am | yours very truly | E. Darwin

DAR 93: B31

[1] The year is established by the relationship between this letter and the letter to John Scott, 7 November [1863] (see n. 4, below).

[2] The letter from Scott has not been found.

[3] The letter mentioning the specimen has not been found, but Scott may have again referred to a ripe capsule of *Linum monogynum* (see letter from John Scott, 21 September [1863]). He eventually sent seeds of this species to CD in January 1864 (see *Correspondence* vol. 12, letter from John Scott, [13 January 1864]).

[4] After reading the manuscript of Scott's *Primula* paper (Scott 1864a), CD asked him to conjecture in the summary what he thought the ancestral form of the genus was (see letter to John Scott, 7 November [1863]). Scott's reply has not been found; however, in the last two sentences of Scott 1864a (p. 126), Scott referred to his experiments with a non-dimorphic form of the cowslip, *P. veris* (*ibid.*, pp. 105–8), that he believed had gradually derived from the more common dimorphic forms. He took this as evidence that the original progenitor had been a non-dimorphic form that had also gradually given rise to a 'dimorphically characterized race'. The non-dimorphic form thus presented a case of reversion. CD noted Scott's conclusion in 'Illegitimate offspring of dimorphic and trimorphic plants', p. 435, where he also referred to his own discussion of reversion in *Variation* (2: 28–61). He had completed a draft of the chapter on reversion and two others on inheritance on 1 April 1863 (see 'Journal' (Appendix II)).

[5] The letter in which Scott sent this information has not been found. Scott had been working on the sexuality of cryptogams; his observations were later published as Scott 1864c (see *Correspondence* vol. 12, letter from John Scott, 12 [February 1864]). In Scott 1864c, p. 195, Scott described the growth of a plant that he believed was a successful cross between two species of the clubmoss *Selaginella*.

From Emma Darwin to Friedrich Hildebrand 20 November [1863][1]

Down. | Bromley. | Kent. S.E.
Nov 20

Sir

My husband wishes me to write & thank you for your very kind letter & to say that he is so unwell as to be unable to write or attend to any scientific subject.[2] He desires also to send his respectful remembrances to Professor Treviranus & to thank him for the paper on Orchids he so kindly sent & to say how much he w[d] have liked to have written to him on the subject but he is not able.[3]

Mr Darwin wishes to ask if Orchis Pyramydalis grows in your neighbourhood he w[d] very much like to see a fresh specimen of it next summer as he is convinced the structure of the stigma is different from what it is here[4]

With Mr Darwin's kind regards I am | Sir yours very truly | Emma Darwin

Courtesy of Eilo Hildebrand (photocopy)

[1] The year is established by the relationship between this letter and the letter from Friedrich Hildebrand, 10 November 1863.
[2] See letter from Friedrich Hildebrand, 10 November 1863.
[3] The references are to Ludolph Christian Treviranus and to Treviranus 1863c, which was published in the 7 August 1863 number of *Botanische Zeitung*. There is a lightly annotated copy of Treviranus 1863c in the Darwin Pamphlet Collection–CUL, and a manuscript translation in DAR 70: 38–52. The most recent extant correspondence between CD and Treviranus is the letter from L. C. Treviranus, 12 February 1863, in which he mentioned sending Treviranus 1863a. See also letter from Friedrich Hildebrand, 16 July 1863.
[4] In Treviranus 1863c, p. 243, Treviranus discussed what he thought were two errors in CD's diagram of *Orchis pyramidalis* (*Orchids*, p. 22). CD wrote in 'Fertilization of orchids', p. 141 (*Collected Papers* 2: 139), that Treviranus confirmed his 'observations on this remarkable species', but differed from him in 'one or two minor points'. See also *Correspondence* vol. 12, letter from Friedrich Hildebrand, 21 June 1864, and letter to Friedrich Hildebrand, 25 June [1864].

From Robert Monsey Rolfe [20 November 1863][1]

Holwood | Bromley
⟨ ⟩

⟨My d⟩ear Sir

My contribution to your Downe Charities is paid about this time of the year—& as we may perhaps be going to pay some visits next week, I may as well send you a cheque while the matter is in my mind—[2]

I hope that you & your family are all pretty well— Mrs Norris[3] a⟨n⟩ ⟨*one or two words*⟩ Cranworth who is wi⟨th us⟩ gives us an unsatisfactory acc^t^ of Sir H Holland—[4] He was ill last week & after getting much better seems to have imprudently exposed himself which has brought on a sharp attack of Gout or something of the sort—

Yours very truly | Cranworth[5]

Chas Darwin Esqre

DAR 161.2: 230

[1] The date is established by entries in CD's Account books–cash and bank accounts, and by the Down Coal and Clothing Club account book (Down House MS); 1863 is the only year in which Rolfe's name (Lord Cranworth) appears both in CD's Account books and as a subscriber to the Down Coal and Clothing Club. Also, all years but 1863 are already accounted for by an annual letter enclosing his donation (see, for example, *Correspondence* vol. 10, letter from R. M. Rolfe, 28 November 1862, and n. 2, below).
[2] As treasurer of the Down Friendly Club and the Down Coal and Clothing Club, CD distributed donations, and allotted funds to a day school and a Sunday school in Down (Freeman 1978; see *Correspondence* vol. 7, letter to John Lubbock, 14 December [1859]). In 1863, CD recorded Rolfe's annual donation in his Account book–banking account on 20 November for £10 10*s*. He recorded the following for 21 November in his Account book–cash accounts: 'L. Cranworth Charities. School 7.7. Friendly Club 1.1. Coal Club. 2.2.0'.
[3] Mrs Norris has not been identified.
[4] The London physician Henry Holland was a distant relative and a correspondent of CD's (Freeman 1978).

[5] Rolfe was created Baron Cranworth in 1850; his country residence was a mile and a half north of Down House (*DNB*).

From Emma Darwin to Patrick Matthew 21 November [1863][1]

Down. | Bromley. | Kent. S.E.
Nov 21.

Dear Sir

Mr Darwin begs me to thank you warmly for your letter which has interested him very much.[2] I am sorry to say that he is so unwell as not to be able to write himself.

With regard to Natural Selection he says that he is not staggered by your striking remarks. He is more faithful to your own original child than you are yourself.[3] He says you will understand what he means by the following metaphor.

Fragments of rock fallen from a lofty precipice assume an infinitude of shapes— these shapes being due to the nature of the rock, the law of gravity &c— by merely selecting the well-shaped stones & rejecting the ill-shaped an architect (called Nat. Selection could make many & various noble buildings.[4]

Mr Darwin is much obliged to you for sending him your photograph.[5] He wishes he could send you as good a one of himself. The enclosed was a good likeness taken by his eldest son but the impression is faint.[6]

You express yourself kindly interested about his family. We have 5 sons & 2 daughters, of these 2 only are grown up.[7] Mr Darwin was very ill 2 months ago & his recovery is very slow, so that I am afraid it will be long before he can attend to any scientific subject.

Dear Sir | yours truly | E. Darwin

National Library of Scotland (Acc. 10963)

[1] The year is established by the references to CD's illness, and by a similar description of CD's 'metaphor' having been sent to Asa Gray earlier in 1863 (see letter to Asa Gray, 4 August [1863], and n. 4, below).

[2] The letter from Matthew has not been found. CD and Matthew also corresponded in 1862 (see *Correspondence* vol. 10, letter to Patrick Matthew, 13 June [1862] and letter from Patrick Matthew, 3 December 1862).

[3] Matthew claimed that he anticipated CD's concept of natural selection in his work on naval timber and arboriculture (Matthew 1831). CD conceded Matthew's priority in correspondence (see *Correspondence* vol. 8, letter to *Gardeners' Chronicle*, [13 April 1860], and letter to Joseph Dalton Hooker, 13 [April 1860]). CD also discussed Matthew's views in the 'Historical sketch' that prefaced the third edition of *Origin*, pp. xiv–xv. See also Dempster 1996.

[4] Darwin's metaphor appears in *Variation* 2: 248–9 and 430. See also letter to Asa Gray, 4 August [1863].

[5] The photograph has not been found.

[6] CD's photograph was taken by William Erasmus Darwin in April 1861; the photograph is reproduced as the frontispiece to *Correspondence* vol. 9.

[7] The references are to William and Henrietta Emma Darwin, who were 23 and 20 years old respectively; George Howard was 18, Elizabeth was 16, Francis was 15, Leonard was 13, and Horace was 12 (Freeman 1978).

To J. D. Hooker [22–3 November 1863][1]

Down
Sunday & Monday.

My dear old Friend—

Since writing I have had much sickness & am weak & must be brief— [2] I have been delighted by your long letter & have had it read thrice;[3] but *in truth you must not write so much as you are so overworked*.—

Pray thank Mr Gower for the excellent plants which will amuse me nicely on the table by my sofa—[4] You have *not* sent Asa Gray; I **must** see his praise of you—[5] It is charming your being thus converted to the Northern side.[6] I never heard of the dear old woman with her Mesopotamia—[7] The enclosed curious letter (please return it) is worth reading—[8] You will agree with what he says on n. selection; I do not quite.—

I have got your note this morning—[9] we think (*& will enquire*) that the Vases must have come from the Fr. Wedgwoods of Etruria.—[10] Your Zeal has stimulated me to send some old letters to Miss Metyard for her life.— [11] Whoever sent the vases, I do rejoice you are pleased: it cuts us both that we have nothing to give.— Thwaites letter & enclosure on Cassia not worth Linn. Soc—[12]

If you know, *sometime* tell me who reviewed Lyell in Edinburgh.— [13] I hope you can read this; I can thus write lying down.—

A German has sent me pamphet, (which I have not read) on the Darwinsche theorie applied to Language with a diagram like mine with names of Languages instead of mere letters.—[14] The more I look at Plants the higher they rise in my mind: really the tendril-bearers are higher organised, as far as adapted sensitivity goes, than the lower animals. Does Flagellaria, (are you sure?) use leaves like Tendrils?[15]

Excuse my jumping from subject to subject.—

I shall be curious to read Oliver's paper:[16] I have often admired the curling of valves.— I suppose O. knows Wymans paper in (I think) Proc. Boston Soc. of bursting of gourd of Echinocystis—[17] I suppose Oliver read Bot. Zeitung: ask him to tell me, if John Scotts paper on sterility of Orchids with own pollen is noticed that I may tell the poor fellow:[18] I want much to encourage him. He is a splendid worker & has paper for Linn. Soc. that interests at least me greatly.—[19]

I have been just looking again at your former letter.[20] How well I remember your feeling when we lost Annie,[21] that it was my greatest comfort that I had never spoken a harsh word to her. Your grief has made me shed a few tears over our poor darling; but believe me that these tears have lost that unutterable bitterness of former days.—

I have no new ideas on New Zealand.—[22]

I agree with what you say on Sedgwick & medal:[23] he has done good descriptive work—prepared way for Murchison[24] & discovered, but not explained, cleavage.[25] Hardly enough for Copley Medal. I was told, (but ought not to have heard) that I was put in opposition to poor old Sedgwick;[26] with Owen[27] to lead against me I shd. have no chance whatever—

I am so glad to hear about your Willie.—[28] Henrietta is much pleased at your invitation,[29] which I have no doubt she will *gladly* accept in part, but will write again.

GoodBye I am tired. | C. Darwin

Endorsement: 'Nov 23/63'
DAR 115: 211

[1] The date is established by the relationship between this letter and the letter to J. D. Hooker, 16 [November 1863]; in 1863, 22 and 23 November fell on the Sunday and Monday after that date.

[2] Letter to J. D. Hooker, 16 [November 1863]. See Emma Darwin's diary (DAR 242) for her daily record of CD's symptoms during the week after 16 November; on 22 November, CD had a 'v.g. day'.

[3] The letter from Hooker has not been found.

[4] William Hugh Gower was a foreman at the Royal Botanic Gardens, Kew (R. Desmond 1994). In his letter to Hooker of [13 November 1863], CD had asked if he had any other tendril-bearing plants. Hooker had sent several tendril-bearing plants in July (see letter from J. D. Hooker, [31 July 1863] and n. 2). A note dated 19, 20, and 21 November records the movements of *Hoya carnosa* past objects in CD's study (see DAR 157.1: 39v.).

[5] See enclosure to letter to J. D. Hooker, 27 [November 1863]. In the enclosure Asa Gray commented on Hooker's account of *Welwitschia* (J. D. Hooker 1863a). A review of J. D. Hooker 1863a by Gray appeared in the November issue of the *American Journal of Science and Arts* (A. Gray 1863f).

[6] CD refers to the Union side in the American Civil War, which had begun in 1861. Because of Hooker's view of the Civil War, he and Gray had stopped discussing politics in their letters (see letter from J. D. Hooker, [1 March 1863]). See also L. Huxley 1918, pp. 39–44.

[7] Mesopotamia: 'a word which is long, pleasant-sounding, and incomprehensible; used allusively for something which gives irrational or inexplicable comfort or satisfaction to the hearer' (*OED*).

[8] This may be a reference to a letter from Patrick Matthew; the letter has not been found but see preceding letter and n. 3.

[9] The note has not been found.

[10] See letters to J. D. Hooker, 16 [November 1863], 27 [November 1863], and 5 [December 1863]. Francis Wedgwood was CD's first cousin and brother-in-law; Etruria was the site of the Wedgwood pottery works in Staffordshire.

[11] The writer Eliza Meteyard was working on a life of Josiah Wedgwood I (Meteyard 1865–6; *DNB*).

[12] Hooker evidently wondered if information in the letter from George Henry Kendrick Thwaites, 8 June 1863, and the enclosed notes on *Cassia* from Samuel Owen Glenie, might be read as a paper to the Linnean Society. See also letter from G. H. K. Thwaites, 24 September 1863.

[13] CD refers to the review of Charles Lyell's *Antiquity of man* (C. Lyell 1863a) in the *Edinburgh Review* ([J. D. Forbes] 1863). The attribution to James David Forbes is confirmed by the *Wellesley Index.*

[14] Schleicher 1863. There is a lightly annotated copy of August Schleicher's pamphlet in the Darwin Pamphlet Collection–CUL. CD refers to the illustration in *Origin* (facing p. 117) representing the divergence of forms through time.

[15] CD discussed *Flagellaria* in *Climbing plants*, p. 46.

[16] Daniel Oliver read a paper on the dehiscence of *Pentaclethra macrophylla* capsules before the Linnean Society on 19 November 1863 (Oliver 1863e). Oliver discussed this research in his letter to CD of 27 November 1863.

[17] CD refers to Jeffries Wyman and Wyman 1854, which was published in the *Proceedings of the American Academy of Arts and Sciences* (see letter to Daniel Oliver, [before 27 November 1863]).

[18] CD had offered to send a copy of Scott 1863a to Friedrich Hildebrand in the hope that it would be reviewed in *Botanische Zeitung*, but it was not noticed or reviewed there (see letter to John Scott, 1 and 3 August [1863]).

[19] Scott 1864a.

[20] Letter from J. D. Hooker, 1 October 1863.
[21] The Darwin's eldest daughter, Anne Elizabeth, died in 1851 when she was 10 years old.
[22] See letter to J. D. Hooker, 10 [November 1863] and n. 5.
[23] Hooker's letter has not been found. CD was considered for the Royal Society's Copley Medal in 1863 but on 5 November, Adam Sedgwick was chosen for the award (Royal Society, Council minutes, 5 November 1863). See letter from E. A. Darwin to Emma Darwin, 11 November [1863].
[24] CD evidently refers to Adam Sedgwick's stratigraphic mapping, based primarily on structural geology, which preceded Roderick Impey Murchison's emphasis on fossils, in addition to structure, for classifying strata (see Secord 1986).
[25] See Sedgwick 1835, and *Correspondence* vol. 3, letter to Daniel Sharpe, [1 November 1846].
[26] See letter from E. A. Darwin, 9 November [1863], letter from E. A. Darwin to Emma Darwin, 11 November [1863], and n. 23, above.
[27] Richard Owen was a member of the Council of the Royal Society; he had opposed CD's theory since the publication of *Origin* (see, for example, [Owen] 1860a, and letter from Edward Sabine to John Phillips, 12 November 1863 and n. 4). See also letter from E. A. Darwin to Emma Darwin, 11 November [1863] and n. 5.
[28] William Henslow Hooker had been suffering from scarlet fever, but was evidently recovering (see letter to J. D. Hooker, [13 November 1863] and n. 6).
[29] The invitation to Henrietta Emma Darwin has not been traced.

From Asa Gray 23 November 1863

Cambridge, Mass.
Nov. 23/63

My Dear Darwin

I wish to thank Mrs. Darwin most heartily for her kind note of the 29th. ult.[1]

I feared you had been ill, and I think I had long ago some intimation to that effect, in some indirect way, but I hardly believed it, as just before I had received one of your usual bright letters.[2] Hooker not having written since the death of his daughter,[3] and Boott having been very ill,[4] all my sources of intelligence were cut off.

It is sad to hear that you are obliged to suspend all your work for half a year. Pray disappoint the physicians and get all right again sooner. For there is a deal for you to do. But after all, we must be patient. It is far more important that you get well surely than speedily. And pray, somehow, continue to let me know how you are getting on.

I have no scientific news for you. I have been busy at very humdrum botanical work. I read lately, with *gusto*, Wallace's *exposé* of the Dublin man (Haughton) on Bee-cells, &c—[5]

The next-best thing, of late, is the *exposé* of Lindsay & Geo. *Saunders* (the Confederates) by *Historicus*.[6]

I trust Historicus' previous letters, in which he shows—about the same time my father-in-law's articles on the subject reached England[7]—that it is the duty of a country to see that armed or war vessels are not fitted out, quite irrespective of all municipal law, have produced their proper effect.[8] Something has produced a great effect, and a great change in the idea of what it was incumbent on the government to do; and nothing can be more satisfactory than the views now taken:[9] and the effect here is excellent. For we are sure that when the right notions once get a lodgement, as they have, England will faithfully carry them through.

Lawyers whom I know here were confident how the law would ultimately be laid down by your courts,[10]—but we greatly feared it would be done only after a few more such vessels had got to sea. All will go well now.

The newspaper I occasionally send you is a fair specimen of the *influential* part of the press here.

Such articles as the *Times* likes to cite have far less effect here than you suppose in the determination of events.[11]

The result of the autumn elections will let you know[12]—what has all along been clear to me—that the *North* has no notion whatever of l⟨e⟩aving its work till it is done.

Thanks for the article in the Reader[13]—true enough, no doubt, sensible & dull.

The best thing I have read of late was the showing up of that consummate humbug (as for many years I have regarded him) Brougham, in the Spectator.[14]

The confidence that every month's prolongation of the contest makes the destruction of slavery surer, quite reconciles to the cost, the loss of life even, and even to the blunders and shortcomings!

Agassiz reports that a letter he has from Faraday pronounces that "*Darwinism*" has about had its day in England.[15] But when, at length, A. showed me Faraday's note, I could not see that it said or implied anything of the sort. It is amusing to see how *worried* Agassiz is.

By the way, now that you are in for "reading of the very lightest sort" let some one read to you Agassiz on *Glaciers* in the *Atlantic* Mag. for November.[16] It will not strain your brain, tho' it may your diaphragm to read how the hairy elephants and the *bears* &c were luxuriating in a tropical climate, with their thick coats on to keep them comfortable, when, all of a sudden it changed to bitter winter,—so suddenly that they could not run away, nor even rot, but were *frozen up* before there was time for either!

Pray set Wallace[17] upon these articles, and upon A's book just issued, made up of an earlier series of such articles: "Method of Study" (forsooth), and which A. has sent over to European *savans* as an important Contribution to Science.[18] It is announced, you may see in the preface, that this little book completely does away with all "*Darwinism*"[19]

Thanks for the little paper on Disa.[20]

These stamps—of a sort now outlawed here—being clean, may please Master Darwin, or be useful for his exchanges.[21]

Ever, dear Darwin— Your sincere | A. Gray

DAR 165: 141

[1] Emma Darwin's letter has not been found.

[2] The most recent extant letter from CD is the letter to Asa Gray, 4 August [1863], but Gray responded to this with his letter to CD of 1 September 1863.

[3] Maria Elizabeth, Joseph Dalton Hooker's daughter, had died aged 6 on 28 September 1863 (see letter from J. D. Hooker, [28 September 1863]).

[4] Francis Boott, an American-born physician and botanist who lived in London, had been ill for some time (see letter from J. D. Hooker, [31 July 1863]).

[5] The appendix of Samuel Haughton's paper (S. Haughton 1862b), 'On the form of the cells made by various wasps and by the honey bee; with an appendix on the origin of species', was a slightly shorter version of Haughton's earlier review of *Origin* ([S. Haughton] 1860b). A lightly annotated reprint of S. Haughton 1862b (S. Haughton 1863) is in the Darwin Pamphlet Collection–CUL. Alfred Russel Wallace's review of S. Haughton 1862b appeared in the October number of the *Annals and Magazine of Natural History* (Wallace 1863b). In his review, Wallace defended CD's argument in *Origin* (3d ed., pp. 254–56) that the complex structure of hive cells, and the more economical use of wax by the hive bee in relation to the humblebee, provided evidence of the hive bee's development through natural selection (Wallace 1863, pp. 305–7). Wallace noted that many of Haughton's statements were 'often so deficient in clearness as to suggest the idea that "The Origin of Species" has been but superficially studied by him' (p. 304). See CD's note on Wallace 1863b in DAR 205.11: 126. See also Hull 1973, pp. 216–228.

[6] Under the pen-name Historicus, the lawyer William Vernon Harcourt published letters in *The Times* on the American Civil War and Britain's neutrality policy (see letter to Asa Gray, 26 June [1863] and n. 14). In a series of letters and exchanges with George Nicholas Sanders and William Schaw Lindsay in October and November 1863, Historicus published and discussed correspondence revealing plans made with the American Confederacy for constructing six iron-clad ships in England. These letters exposed the role of the Confederate agent, Sanders, and the shipping magnate, Lindsay. Lindsay, a leading figure in the pro-Confederate lobby, and MP for Sunderland, also owned one of the two firms that were to build the ships. See, especially, *The Times*, 23 October 1863, p. 7, and 28 October 1863, p. 9. See also Loring 1864, and Jenkins 1974–80, 2: 202, 209–34.

[7] Gray may be referring to the issues of the *Boston Daily Advertiser* which included articles by his father-in-law, Charles Greely Loring; in August and September Gray had sent several newspapers containing Loring's articles to CD (see letter from Asa Gray, 1 September 1863 and n. 12, and Loring 1863). These would have reached CD at about the same time as two letters from Historicus appeared in *The Times* discussing the actions Britain should take as a neutral country when its shipbuilders built vessels for the American Confederacy (see *The Times*, 11 September 1863, p. 7, and 18 September 1863, p. 7, and n. 6, above).

[8] The legality of building, arming, and equipping ships to be used in warfare against a country with which Britain was at peace was in question during 1862 and 1863, particularly after depredations against the Union by the British-built ship *Alabama* (see letter from J. D. Hooker, [1 or 3 November 1863], n. 21). Historicus had become an authority on the technicalities of the Foreign Enlistment Act, which made illegal the equipping of a ship for war in the service of a foreign belligerant, and was listened to by the foreign secretary (see letter from Asa Gray, 26 May 1863 and n. 27, and Jenkins 1974–80, 2: 264). On 27 October 1863, the British authorities confiscated two iron-clad vessels as they neared completion at the Birkenhead shipyards; it was believed that these so-called 'Laird's rams' were intended for the American Confederacy (see letter from J. D. Hooker, [1 or 3 November 1863] and n. 21, and *The Times*, 29 October 1863, p. 9). Gray may also be referring to newspaper reports of the detention of the *Alexandra*, another British-built Confederate ship (see, for example, *The Times*, 6 November 1863, p. 11, and 7 November 1863, p. 8).

[9] The seizure of John Laird's two ships (see n. 8, above) reflected the views of the British authorities on the illegality of selling ships to the Confederacy for warfare; editorials in *The Times*, in addition to Historicus's views on the interpretation of the Foreign Enlistment Act, clearly supported this action (see *The Times*, 28 October 1863, p. 8, 4 November 1863, p. 6, 7 November 1863, p. 8).

[10] See n. 9, above. Gray may also be referring to the *Alexandra* case (see n. 8, above).

[11] One of the newspapers Gray sent to CD was the *Boston Daily Advertiser* (see *Correspondence* vol. 10, letter from Francis Boott, 26 December 1862 and n. 7, and n. 7, above). During the month of October

1863, *The Times* quoted newspapers from cities including New York, Baltimore, and Washington, but rarely Boston, near where Gray resided.

12 Several state elections took place on 13 October 1863, including gubernatorial contests in Ohio and Pennsylvania; the outcomes favoured the Republicans, and were regarded as a demonstration of growing support for Abraham Lincoln's emancipation proclamation (McPherson pp. 684–8).

13 The article in the *Reader* had not been identified.

14 Gray refers to the article 'Lord Brougham on America', *Spectator*, 10 October 1863, pp. 2596–7. The article examined and dismissed Henry Peter Brougham's arguments supporting the secession of the southern states from the Union; these arguments were advanced in his presidential address to the National Association for the Promotion of Social Science in Edinburgh (Brougham 1863).

15 Gray refers to Louis Agassiz and Michael Faraday. Agassiz had long opposed CD's theory of species transmutation (see *Correspondence* vol. 8, Dupree 1959, pp. 216–32, and n. 19, below). For an interpretation of Faraday's views on *Origin*, see Cantor 1991, pp. 145–6.

16 Gray refers to Agassiz's article in the November issue of the *Atlantic Monthly* (Agassiz 1863b).

17 Alfred Russel Wallace.

18 Gray refers to *Methods of study in natural history* (Agassiz 1863a), which comprised a series of articles on natural history, based on popular lectures, that appeared in the *Atlantic Monthly* in 1862 and 1863 (see letter from Asa Gray, 26 May 1863 and n. 12).

19 In the preface to Agassiz 1863a, dated 22 August 1863, Agassiz noted that his indirect intention in the volume was to enter his 'earnest protest against the transmutation theory' (*ibid.*, p. iii).

20 Trimen 1863. For CD's role in the publication of this paper, see letter to Roland Trimen, 23 May [1863].

21 Gray often sent stamps for Leonard Darwin's collection (see letter to Asa Gray, 2 January [1863] and n. 15).

To Roland Trimen 25 November [1863][1]

Down. | *Bromley.* | *Kent. S.E.*
Nov. 25

My dear Sir

I have been laid on the shelf for nearly three months, & am ordered to do nothing for 6 months by my doctors.[2] To write this is against rules.— Many thanks for specimens of orchids & for your kind letter.[3] I dare not look at Oxalis flowers. I regret much that you cannot get seed, especially of your trimorphic flowers. Most species of Oxalis shed their seed by a spurt & the capsules are sensitive to a touch.[4]

Could you employ anyone to dig up the bulbs of the 2 or 3 forms & allow me to pay, i.e. if they are bulb-bearers.

The last job I began & broke down was a letter to G. Chronicle on your Peach case.—[5] I must write no more.— I live in *hopes* some day to be able to work a very little more, but it will be long before I can.—

Sincere thanks for your very kind letter.

Yours very sincerely | C. Darwin

I forwarded letter to Bates.[6] Pray use me as often as you like.—

Royal Entomological Society of London (Trimen papers, box 21: 58)

1 The year is established by the relationship between this letter and the letter from Roland Trimen, 10, 13, and 18 October 1863.

2 See letter to J. D. Hooker, [13 November 1863].

[3] In his letter of 10, 13, and 18 October 1863, Trimen noted that he had separately sent several orchid flowers in bottles of spirit.
[4] In his letter of 10, 13, and 18 October 1863, Trimen enclosed dried specimens of dimorphic *Oxalis*; he was puzzled as to why he could find no seed for either the dimorphic or the trimorphic plants.
[5] In the letter from Roland Trimen, 16, 17, and 19 July 1863, Trimen reported a case of moths penetrating the skins of peaches in order to obtain their juices. Trimen felt that this helped strengthen CD's hypothesis in *Orchids*, pp. 45–52, that moths visit the flowers of some orchid species to puncture the lining of false nectaries for the fluid; these flowers contain nectar between the inner and outer membranes of the nectary, rather than in the nectary itself. See the letter to the [*Gardeners' Chronicle*], [after 27 August 1863] for CD's draft summary of the case, which he intended to publish in order to elicit information on similar cases. The letter was never published, but Trimen's information was included in 'Fertilization of orchids', pp. 142–3 (*Collected papers* 2: 140).
[6] Trimen had enclosed a letter to Henry Walter Bates with his letter to CD of 10, 13, and 18 October 1863.

To Daniel Oliver [before 27 November 1863][1]

[Down]

Dear Oliver—

I think Wymans very short notice in Proc. American Acad. of Arts & Sciences. Vol. III (1852–1857) p. 167, worth your looking at.—[2] I am proud *for once* to give you a reference.[3]

I am bad so no more.—

I am so sorry to hear about Hooker's boy with scarlatina.—[4]

Yours very sincerely | C. Darwin

DAR 261 (DH/MS 10: 53)

[1] The date is established by the relationship between this letter and the letter from Daniel Oliver, 27 November 1863.
[2] CD refers to Jeffries Wyman and Wyman's abstract on contractility in the capsules of two plants (Wyman 1854). Oliver copied Wyman 1854 on the back of this note from CD; CD's interest was related to the paper Oliver read before the Linnean Society on 19 November 1863 (Oliver 1863e). The paper was published in the society's transactions; CD is acknowledged for providing the reference to Wyman 1854 (Oliver 1863e, p. 419).
[3] Oliver, an assistant in the herbarium and librarian of the Royal Botanic Gardens, Kew, often sent CD botanical information and references to articles (see, for example, letters from Daniel Oliver, 22 January 1863 and 17 February 1863).
[4] Joseph Dalton Hooker's son William Henslow Hooker had been suffering from scarlet fever (see letter from J. D. Hooker, 23 October 1863, and the following letter).

To J. D. Hooker 27 [November 1863]

Down
27th

My dear Hooker

I grieve to hear about the Scarlet-Fever:[1] my poor dear old friend you are most unfortunate. The tide must turn soon.— Be careful about secondary symptoms &

do not expose your Boy *to a chill*. We think Leonard's fearful illness was thus caused, last summer.—[2]

I write now chiefly to say vases did not come from Etruria & therefore must from my elder sister Susan.[3] The day after she must have sent them, she was struck down with fever, believed to be Scarlet & communicated apparently by a shawl sent by my sister Catherine[4] & worn once by her.—

Susan is now very ill; when better I will enquire, & forward your note in which you seem so pleased by Vases,[5] & this will be best thanks you can send.—

I return Asa Gray's letter & have been very glad to see it.[6] It makes me, however, rather envious that I have not myself knowledge enough to thoroughily appreciate your grand paper.—[7]

Farewell my good fellow: I shall be glad to hear sometime about your Boy, whom you love so.— Much love much trial, but what an utter desert is life without love.

God Bless you | C. D.

[Enclosure]

Cambridge.
Oct. 25: 1863

My Dear Hooker,

A letter from Boott to his sister (in Boott's own handwriting which I was so glad to see again)[8] which was sent to me yesterday, told us—before we saw the Gard. Chronicle, of your and Mrs. Hooker's great and sudden sorrow, and deeply do we sympathise with you both in your affliction.[9] I suppose the departed one was named both after your mother and your sisters.[10] All your children have been born since I saw you,[11] but your accounts of them, from time to time, have kept up our interest in them.

Your photograph is before me, and reproaches me with not having thanked you for the gift.—if indeed it be, as I fear, that I have not replied to yours of Aug. 26–Sept. 4th. At first I was not well satisfied with the photographer's work. He has allowed your head to take a position which brings out a furrow on your cheek which is yet out of place, I presume, in the original. But, through longer familiarity I am getting to like the picture better. Many thanks for it.

We are delighted to know that dear Boott is really convalescent and gaining strength steadily, if slowly.

I must have written to you early in Sept. to say how much my wife had gained by our holiday.[12] The gain she held well until 3 weeks ago—when a piece of overdoing—caused, I fear by a rather unexpected houseful of visitors, has put her down again as bad as ever.—confined to her room for the present, and with the prospect of being an invalid all winter![13] Her illness, though not very serious, perhaps, has damped our spirits, and has kept me from doing very much.

I am well in the midst of a study of the *Astragali* of N. Amer.,[14] and before I print the results of my revision of the materials which our herbaria furnish, I may ask a question or two, as to those of Hook. Fl. Bor—Am.[15]

At present I am at a total loss only in regard to *A. diaphanus*, Dougl. in Hook. Fl. Bor. Am.[16]

It is in none of our herbaria—(i.e. Douglas's specimens),[17] nor have we anything I can identify with it. It is hardly likely you can help me to know it.

I must hereafter look over your remarks about Compositæ in your last letter. I shall be pleased if you can keep up my *Mniopsis*, which when I gave the name I thought a famous good genus. But, as you say, the Gnaphaliad genera must be done all over before we can get satisfaction—probably small satisfaction then.[18]

If I were younger—or even as I am, if I had not a weary load of work upon my shoulders I would "come over, and work up Compositæ" for your Gen. Pl.[19] Some parts of the job I think I could do fairly well.

I had been told I could have little room in Nov. no. of Sill. Jour. and so had not attacked Welwitschia. But being lately told I could fill 4 pages I wrote and sent on a notice of your memoir[20]—I trust an appreciating one, though hasty and imperfect.

I consider the memoir *grand*—and that it does you very great credit. The clearness with which you have put every thing is unrivalled, perhaps even by any thing you ever did before. I do not believe that there is any monograph which exceeds it. You had a first rate opportunity, and I think *you have improved it.*

I lately find that Seemann had taken it into his head to reprint my review of *DC. Memoirs* et *Souvenirs* in his Journal.—[21] That, as he likes, But he has cut out here and there as he liked.— which he was also welcome to do if he indicated the *lacunæ*,—as he did not.

But what vexes and astonishes me is that he had the impudence to assert that it was *communicated with corrections by the author.*—which surely is not true in any sense.[22] I must rap his knuckles for that piece of impudence.[23] Then he will probably leave me alone, which is all I want.

Ever, dear friend, Your affectionate and sympathising A. Gray

Endorsement: 'Nov/63.'
DAR 115: 212; Royal Botanic Gardens, Kew (Asa Gray correspondence: 333)

1 The letter from Hooker has not been found. William Henslow Hooker had contracted scarlet fever in October but had been recovering (see letter from J. D. Hooker, 23 October 1863 and letter to J. D. Hooker, [22–3 November 1863]). William may have suffered a relapse as CD's son Leonard had the previous summer (see n. 2, below).

2 Leonard Darwin had been ill with scarlet fever during the summer of 1862, suffering two relapses (see *Correspondence* vol. 10, letters from J. D. Hooker, [24 July 1862] and n. 11, and letter to A. R. Wallace, 20 August [1862]).

3 See letters to J. D. Hooker, 16 [November 1863] and 5 [December 1863]. The reference is to Susan Elizabeth Darwin.

4 Emily Catherine Langton.

5 Hooker's note has not been found.

6 See enclosure. Gray's letter, requested in the letter to J. D. Hooker, [22–3 November 1863], must have been sent to CD with Hooker's missing letter (see n. 1, above).

7 CD refers to Hooker's study of *Welwitschia* (J. D. Hooker 1863a).

8 Gray refers to Francis Boott, who was dying of lung disease (see letter from J. D. Hooker, 26 August 1863). Boott's sister has not been identified.

9 Maria Elizabeth, the six-year-old daughter of Hooker and his wife, Frances Harriet, died on 28 September 1863 (see letter from J. D. Hooker, [28 September 1863]). A notice of her death in the *Gardeners' Chronicle* has not been found.

10 Hooker's mother was named Maria. Gray also refers to Hooker's sisters, Maria McGilvray and Elizabeth Lombe.

11 Hooker's eldest child, William Henslow Hooker, was born in 1853, two years after Gray had last been in England (Dupree 1959). The Hookers had since had four more children: Harriet Anne, Charles Paget, Maria Elizabeth, and Brian Harvey Hodgson.

12 Gray and his wife, Jane Loring Gray, had been on holiday in New York State (see letter from Asa Gray, 1 September 1863 and n. 2).

13 In the 1850s and 1860s, Jane Loring Gray suffered chronic ill health (Dupree 1959, p. 182).

14 Gray refers to his paper published in the 11 November 1863 issue of the *Proceedings of the American Academy of Arts and Sciences* entitled 'Revision and arrangement (mainly by the fruit) of the North American species of *Astragalus* and *Oxytropis*' (A. Gray 1863h).

15 Gray refers to William Jackson Hooker's *Flora Boreali-Americana; or the botany of the northern parts of British America* (W. J. Hooker 1840).

16 *Astralagus diaphanus* was described in W. J. Hooker 1840, 1: 151.

17 Gray refers to the plant collector David Douglas.

18 Gray probably meant to write *Mniodes*, rather than *Mniopsis*. He had listed *Mniodes*, pp. 138–9, as a section of the genus *Antennaria* in the Gnaphaliad alliance of the family Compositae (A. Gray 1861b). Hooker and George Bentham raised *Mniodes* to the generic level in Bentham and Hooker 1862–83, 2: 301.

19 Gray refers to George Bentham's and Hooker's *Genera Plantarum* (Bentham and Hooker 1862–83). See n. 18, above.

20 Gray's review of Hooker's monograph on *Welwitschia* (J. D. Hooker 1863a) was published in the *American Journal of Science and Arts* (A. Gray 1863f). This was commonly known as 'Silliman's Journal', after its founder Benjamin Silliman.

21 Gray's review of Alphonse de Candolle's *Mémoires et souvenirs* (A. de Candolle 1862c) originally appeared in the January 1863 issue of the *American Journal of Science and Arts* (A. Gray 1863g). Berthold Carl Seemann began editing the *Journal of Botany, British and Foreign* in 1863; the first volume reprinted Gray's review, in part, on pp. 107–120. See also n. 22, below. There is a copy of A. de Candolle 1862c in the Darwin Library–Down.

22 The version of A. Gray 1863g that appeared in Seemann's *Journal of Botany, British and Foreign* 1 (1863): 107–120, was followed on p. 120 by the statement: 'Abridged from the American Journal of Science and Art, Second Series, with corrections by the Author'.

23 Gray expressed his objection to Seemann's reprint of Gray 1863g in a note in the November 1863 issue of the *American Journal of Science and Arts* (2d ser. vol. 36), p. 434.

From Daniel Oliver 27 November 1863

Royal Gardens Kew
27. Nov. 1863

My dear Sir

It was very kind of you to trouble to send the reference to Wyman's paper.[1] I understood it was in Boston Journal.[2] I must look it up next time I go in to the Societies.[3]

The hygroscopic bundles in pods of *Pentaclethra* undergo wonderful amount of extension when moistened—increasing about 16 per Cent. It is curious that this extension is in direction of the *length* of the cells of the contracting & expanding tissue. In Common Pea, *Lathyrus* &c. the contraction on drying—causing curling up of valves is in direction of *breadth* of the cells. In wood the principal contraction is also transverse.[4] It is curious to cut with scissors the strips marked from pod of sweet Pea **dry**, & put *one* of them in water a few minutes & ⟨then⟩ compare. The directio⟨n of⟩ cells of contracting ⟨tissue⟩ is indicated on inner ⟨surface⟩ of the valves ⟨by the⟩ fine striation. The p⟨ods⟩ sh[d]. be cut to equal le⟨ngth⟩[5]

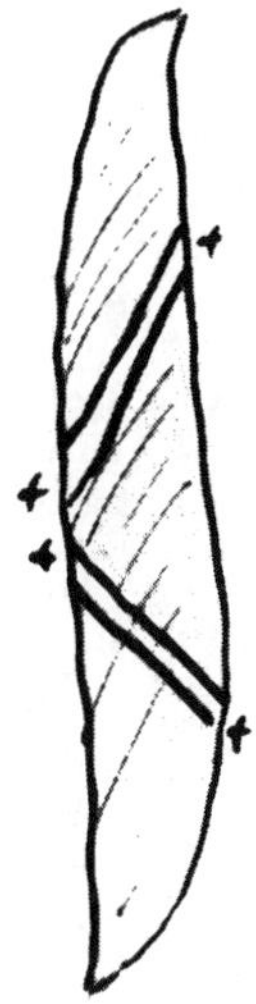

I see v. Mohl is wr⟨iting⟩ in Bot. Zeit. on the s⟨mall⟩ hermetic flowers of *Oxali⟨s⟩* ⟨ ⟩[6]

He has made more ⟨ ⟩ obs. than we previously ⟨ ⟩ but seems to attach u⟨ndue⟩ importance to the phen⟨omenon⟩[7]

If there were any p⟨ods⟩ with these flowers **only** ⟨ ⟩ might be ground for such.

However he makes out that *Fumaria* is necessarily a self-fertilizing plant & makes a little flourish about it which by the way I think you have made some observations wh. may enable you to refute him.[8]

I forget whether you said you had *actually* found bees carrying pollen of Fumitories.[9] But it is too bad of me troubling you about things while so poorly.

Very sincerely yours | Dan[l]. Oliver

I tho't of noticing Mohl in N.H.R. & would refer to y[r]. obs. on Fumitory if worth while?[10]

Incomplete[11]
DAR 173: 24

CD ANNOTATION
1.1 It was ... equal le⟨ngth⟩ 2.10] *crossed pencil*

[1] See letter to Daniel Oliver, [before 27 November 1863]. Oliver's interest in Wyman 1854 was related to a paper he read on the mode of dehiscence of the pods of *Pentaclethra macrophylla*, a legume from Fernando Po, before the Linnean Society of London on 19 November 1863 (Oliver 1863e).

[2] In the letter to J. D. Hooker, [22–3 November 1863], CD stated that Wyman 1854 had been published in the *Proceedings of the Boston Society of Natural History* rather than in the *Proceedings of the American Academy of Arts and Sciences*.

[3] Oliver was a fellow of both the Linnean Society and the Royal Society (R. Desmond 1994).

[4] Oliver wrote this sentence in the left hand margin of the second page of the letter.

[5] Oliver described this experiment on *Lathyrus* in Oliver 1863e, p. 419.

[6] Hugo von Mohl discussed *Oxalis acetosella* in Mohl 1863, pp. 321–2, 326, 327. There is an annotated copy of Mohl 1863 in the Darwin Pamphlet Collection–CUL.

[7] In [Oliver] 1864b, pp. 243–4 (see n. 10, below), Oliver noted the observations that Mohl 1863 had added to what was already known of the small, unopening flowers (later called cleistogamic); Oliver referred back to his own discussion ([Oliver] 1862b, pp. 238–43) of *Oxalis acetosella* and several other plants bearing cleistogamic flowers. In reviewing Mohl's discussion of the unopening flowers, Oliver

wrote ([Oliver] 1864b, pp. 245–6) that Mohl:

> ascribes an undue importance to them, especially when he proceeds to couple with them the self-fertilising homomorphic flowers ... and opposes them collectively to the alleged 'law of nature,' that both in the vegetable and animal kingdoms* there shall be an occasional intercross of distinct individuals.

Oliver's asterisk referred to *Origin*, p. 101, where CD stated that 'both in the vegetable and animal kingdoms, an occasional intercross with a distinct individual is a law of nature'.

8 Mohl referred to the family Fumariaceae as necessarily self-fertilising, without specifying the genus *Fumaria* (Mohl 1863, p. 325). He believed that Fumariaceae provided a counter-example to CD's statement in *Orchids*, p. 359: 'Nature thus tells us, in the most emphatic manner, that she abhors perpetual self-fertilisation'. CD had observed insects pollinating other genera of Fumariaceae but had never observed pollination of *Fumaria*; however, the structure of the *Fumaria* flowers convinced him that they were occasionally pollinated by diurnal insects (see *Correspondence* vol. 9, letter to Daniel Oliver, 9 April [1861], *Correspondence* vol. 10, letter to Daniel Oliver, 12 [April 1862], and *Cross and self fertilisation*, p. 366; see also letter to *Nature*, 6 April 1874 (*Calendar* no. 9393, *Collected papers* 2: 182–3)). CD's botanical notes on Fumariaceae, dated from 1857 to 1863, are in DAR 76: B13–21.

9 CD had observed bees inside flowers of a plant that he had thought was *Fumaria* but that Oliver later identified as *Corydalis* (also a member of the Fumariaceae) (see *Correspondence* vol. 9, letter to Daniel Oliver, 9 April [1861] and DAR 76: B17).

10 A review by Oliver of Mohl 1863 and seven other works on dimorphic flowers also published in 1863 appeared in the April 1864 issue of the *Natural History Review* ([Oliver] 1864b; see also n. 7, above). A note ([Oliver] 1864b, p. 243 n.) stated that the papers reviewed were 'suggested by the researches of Mr. Darwin'. See also n. 8, above, and *Correspondence* vol. 12, letter to J. D. Hooker, 5 April [1864].

11 A strip of paper evidently containing a note was cut from the margin of the fourth page of the letter. Some words on the third page were also excised as a result.

To Daniel Oliver 28 [November 1863][1]

Down
28th

Dear Oliver

Your note has interested me much.—[2] But I cannot look to notes on Fumitories, have not near strength enough.[3] These plants are fertile in large degree without insect but are all (or nearly all) manifestly adapted to visits of insects,[4] which favour & increase their fertility— But I must write no more.—

I shall be very curious to read sometime Hugo von Mohl on the little flowers.[5] I did some work on them & on violets this summer.[6]

You may rely that perfect flowers of violets except V. tricolor are fertile **only** when visited by insects: I marked flowers visited by Bees & prevented Bees visiting others &c—[7] The imperfect flowers are of course fertile without insect-visits—

Yours very sincerely | C. Darwin

How curious about the pods.—[8]

The hydrostatic movement must be sometime like that of the pollinia in orchids.—[9]

DAR 261 (DH/MS 10: 54)

[1] The date is established by the relationship between this letter and the preceding letter.
[2] Preceding letter.
[3] See preceding letter and nn. 8–10. CD's botanical notes on Fumariaceae, dated from 1857 to 1863, are in DAR 76: B13–21.
[4] Oliver quoted this sentence in his review of works on dimorphic flowers ([Oliver] 1864b, p. 246). For CD's observations on Fumariaceae, see the preceding letter and n. 8.
[5] CD began observing the small, unopening flowers (later called cleistogamic) of *Oxalis* and *Viola* during the spring of 1862 (see, for example, *Correspondence* vol. 10, letter to Asa Gray, 10–20 June [1862] and n. 24). He made additional observations in 1863 (see n. 6, below). CD commented on Mohl 1863 in his manuscript of 'Three forms of *Lythrum salicaria*' (DAR 27.2: 44a–b), with a shorter reference to his observations only appearing in the published version, p. 196 (*Collected papers* 2: 128). In *Forms of flowers* he also referred to Mohl's observations on cleistogamy (pp. 314, 320–1, 330), only disagreeing with Mohl's insistence that the open flowers on cleistogamous plants were sterile. There is an annotated copy of Mohl 1863 in the Darwin Pamphlet Collection–CUL.
[6] For CD's 1863 notes on *Oxalis acetosella*, see DAR 109: B6–7 and DAR 111: A46. For his 1863 notes on *Viola*, see DAR 111: A5–12. See also letter to Asa Gray, 31 May [1863] and n. 19.
[7] Some of CD's observations on *Viola tricolor* are preserved in DAR 111: A27. See also *Cross and self fertilisation*, pp. 123–8.
[8] See preceding letter.
[9] In *Orchids*, pp. 335–9, CD described the movement of some orchid pollinia, and the disc to which they are attached, when exposed alternately to air and water.

To J. D. Hooker 5 [December 1863]

Down
Saturday 5th

My dear Hooker

I am very glad that this will reach you at Kew & that Mrs Hooker will soon be with you.[1] You will then get rest & I do hope some lull in anxiety & fear.[2] Nothing is so dreadful in this life as fear: it still sickens me when I cannot help remembering some of the many illnesses our children have endured.—[3] My Father, who was a sceptical man, was convinced that he had distinctly traced several cases of scarlet-fever to handling *letters* from convalescents [4]—

The vases did come from my sister Susan;[5] she is recovering & was **much** pleased to hear that you liked them: I have now sent one of your notes to her, in which you speak of them as "enchanting" &c.—[6]

I have had a bad spell, vomiting every day for eleven days & some days many times after every meal.[7] It is astonishing the degree to which I keep up some strength. Dr Brinton[8] was here 2 days ago & says he sees no reason I may not recover my former degree of health.— I sh^d like to live to do a little more work & often I feel sure I shall & then again I feel that my tether is run out.—

Your Hastings note, my dear old fellow, was a Copley medal to me & more than a Copley medal;[9] not but what I know well that you overrate what I have been able to do. Now that I am disabled, I feel more than ever what a pleasure observing & making out little difficulties is.—

By the way here is a very little fact which may interest you.— A partridge's foot, is described in Proc. Zoolog. Soc. with huge ball of earth attached to it, as hard as rock—[10] Bird killed in 1860— Leg has been sent me & I find it diseased & no doubt exudation caused earth to accumulate: now already 32 plants have come up from this ball of earth.—[11]

Many thanks about Edinburgh R.—[12] Do not send Hochstetter.—[13]

By Jove I must write no more— good Bye, my best of friends. | C. Darwin

Remember me most kindly to Huxley[14] when you see him What a *capital* paper the "Reader" has become[15]

There is an Italian Edit. of Origin preparing!!![16] This makes fifth foreign Edit, ie in five foreign countries.[17] Owen will not be right in telling Longmans that Book w^{d} be utterly forgotten in ten years—[18]

Hurrah!

Endorsement: '(Dec/63)'
DAR 115: 213

[1] The letter from Hooker conveying this information has not been found. Frances Harriet Hooker had been staying with relatives in Norfolk since the burial of their daughter, Maria Elizabeth Hooker (see letters from J. D. Hooker, 1 October 1863 and 23 October 1863).

[2] Hooker's son, William Henslow Hooker, and mother, Maria Hooker, were both ill with scarlet fever (see letters from J. D. Hooker, 23 October 1863 and [1 or 3 November 1863], and letter to J. D. Hooker, 27 [November 1863]).

[3] Recent serious family illnesses included Charles Waring Darwin's fatal attack of scarlet fever and Henrietta Emma's attack of diphtheria in 1858, and Leonard Darwin's attack of scarlet fever in 1862; Horace Darwin also had a serious illness throughout the early part of 1862 (see *Correspondence* vols. 7 and 10).

[4] Robert Waring Darwin was a physician (Freeman 1978). See also letter to J. D. Hooker, 27 [November 1863].

[5] CD refers to his sister Susan Elizabeth Darwin, who lived at The Mount, Shrewsbury, and to some Wedgwood vases that she had sent to Hooker (see letters to J. D. Hooker, 16 [November 1863], [22–3 November 1863], and 27 [November 1863]).

[6] Hooker's note has not been found.

[7] According to Emma Darwin's diary (DAR 242), CD had been vomiting almost daily from mid-November 1863.

[8] William Brinton was a physician at St Thomas's Hospital, London, and a specialist in stomach disorders (*Physicians*).

[9] The letter from Hooker has not been found. The Copley Medal was awarded by the council of the Royal Society, usually annually, for outstanding 'philosophical research' (*Record of the Royal Society*, Appendix IV). Adam Sedgwick was awarded the Copley Medal in 1863; CD was an unsuccessful candidate (see letter from E. A. Darwin, 9 November [1863], letter from E. A. Darwin to Emma Darwin, 11 November [1863], and letter to J. D. Hooker, [22–3 November 1863]).

[10] Newton 1863. See letter from Alfred Newton, 31 October 1863, and letter from Emma Darwin to Alfred Newton, 4 November [1863].

[11] See letter from Alfred Newton, 31 October 1863, nn. 2 and 4.

[12] The reference is to the *Edinburgh Review* (see letter to J. D. Hooker, [22–3 November 1863], and n. 13).

[13] Presumably this is a reference to Hochstetter 1863.

[14] Thomas Henry Huxley.

[15] The *Reader* was a weekly journal reviewing literary and scientific works. The journal, which was issued for the first time on Saturday 3 January 1863, was liberal, religiously neutral, and had a circulation

of about 1,000 (Ellegård 1990, Sullivan ed. 1984). CD began subscribing to the *Reader* by April 1863 (see letter to H. W. Bates, 18 April [1863]).

16 An Italian edition of *Origin*, translated by Giovanni Canestrini and Leonardo Salimbeni, was published in Modena by the firm of Nicola Zanichelli in 1864 (Freeman 1977, p. 105).

17 Several editions of *Origin* had appeared in the United States (Freeman 1977, p. 85), and translations had been published in Germany (Bronn trans. 1860), the Netherlands (Winkler trans. 1860), and France (Royer trans. 1862). There are copies of Bronn trans. 1860 and Winkler trans. 1860 in the Darwin Library–CUL; Bronn trans. 1860 is annotated (see *Marginalia* 1: 180–1). CD's copy of Royer trans. 1862 has not been found but a lightly annotated copy of the preface is in the Darwin Pamphlet Collection–CUL.

18 The occasion on which Richard Owen made this remark has not been identified; CD mentioned Owen's comment in letters to J. D. Hooker, 3 March [1860] and [2 July 1860] (*Correspondence* vol. 8). The London publishing house Longman, Green, Longman, Roberts & Green was commonly known as 'Longmans' (P. A. H. Brown 1982).

From Emma Darwin to J. D. Hooker [7 December 1863][1]

Down
Monday

Dear Dr Hooker

I cannot give a very good account of Charles.[2] He has frequent attacks of sickness but recovers from them in a wonderful manner & they are often with very little distress. The stomach retains the food in a surprizing manner which accounts for his not getting thin. His medical men speak confidently of his regaining his former state of health.[3]

He desires me to say that your letters always give him the *greatest pleasure* tho' he grudges you the time you employ in them. The partridge's foot has now produced 54 plants which he hopes may stick in your throat.[4] "Do not think me bigotted for I fully admit an immense residuum of unexplained facts". What is Galton's book on Domest. Animals?[5] He sends you a letter of Asa Gray which tho' not very interesting you may like to see.[6] Also 2 squibs by the Author of the Dialogue in the New Zealand paper on Origin.[7] He is a Mr Butler Grandson of the old master of Shrewsbury C.'s schoolmaster.[8]

Dr Haast says they have made a noise but he says himself he did not think them fit for publication but the Editor put them in.[9]

I am so glad you will soon have Charley at home safe & sound & that you have got Mrs Hooker & the rest[10] Will you give her my love & thank her for her note to me & for her help in my trap affair which has succeeded better than I expected & is drawing to a conclusion,[11]

I am dear Dr Hooker | yours very sincerely | E. D.

DAR 115: 215

1 The date is conjectured by the relationship between this letter and the letter to J. D. Hooker, 5 [December 1863], and the entries in Emma Darwin's diary (DAR 242). See n. 2, below. In 1863, 7 December was a Monday.

2 Emma Darwin's diary entries on 5 and 6 December 1863 record that CD was vomiting four times a day (DAR 242). See also letter to J. D. Hooker, 5 [December 1863] and n. 7.
3 See letter to J. D. Hooker, 5 [December 1863] and n. 8.
4 See letter to J. D. Hooker, 5 [December 1863]. CD viewed this case of the transport of seeds by a bird as further evidence in support of his favoured explanation of the geographical distribution of island plants, namely, occasional oceanic and wind transport (see letter from Alfred Newton, 31 October 1863 and nn. 2 and 5). Hooker favoured explanations that involved the existence of land-bridges in past geological eras (see letter from J. D. Hooker, [13 May 1863] and n. 20, and letters to J. D. Hooker, 15 and 22 May [1863], and [28 August 1863]).
5 On 22 December 1863, Francis Galton read a paper on the domestication of animals before the Ethnological Society of London (Galton 1863). Galton circulated a privately printed preprint of the paper, which Hooker may have referred to in a missing letter, in advance of the lecture. There is an annotated copy of the preprint in the Darwin Pamphlet Collection–CUL.
6 The letter from Asa Gray has not been found.
7 CD refers to Samuel Butler (1835–1902), who wrote an anonymous article entitled 'Darwin on the Origin of Species: a dialogue', which was published in the *Press* of Christchurch, New Zealand, on 20 December 1862 (see letter to an editor, 24 March [1863?] and n. 3). The squibs were published in the *Press* on 13 June 1863, p. 1, and 15 September 1863, p. 2. There are copies in the 'Scrapbook of reviews' in the Darwin Archive–CUL (DAR 226.1: 130 and DAR 226.1: 133–4, respectively).
8 Samuel Butler (1774–1839) was headmaster at Shrewsbury School, 1798–1836; CD attended Shrewsbury between 1818 and 1825 (Freeman 1978).
9 The letter from Julius von Haast that may have accompanied the squibs from the Christchurch *Press* (see n. 7, above) has not been found.
10 Frances Harriet Hooker and the two youngest Hookers, Charles Paget Hooker and Brian Harvey Hodgson Hooker, had been staying in Norfolk because William Henslow Hooker, the Hookers' eldest son, had scarlet fever (see letter from J. D. Hooker, [1 or 3 November 1863], and letter from J. D. Hooker to Emma Darwin, 11 November 1863). CD's comment suggests that Charles Paget Hooker had also fallen ill.
11 Following the publication and distribution of an appeal against the use of steel traps to control vermin, Emma organised the raising of subscriptions to fund a competition for the design of a humane trap under the auspices of the Royal Society for the Prevention of Cruelty to Animals (see letter from Emma Darwin to W. D. Fox, [6–27 September 1863], and Appendix IX).

From Emma Darwin to W. D. Fox 8 December [1863]

Down. | Bromley. | Kent. S.E.
Dec 8

Dear Mr Fox

Thank you very much for your letter.[1] I must own it has discouraged me a good deal as to the necessity of steel traps.[2] I believe you are the only person in England who has energy & humanity to visit the traps at midnight & Charles tells me you once carried a toad over the Menai Bridge for fear he should come to a bad end—[3]

I want to ask you to do me a favour. My game keeper friend[4] here who is so smooth spoken I suspect he chiefly tells me what he thinks I want to hear, tells me that the animals almost always break the leg in their first frantic efforts to escape—is this true according to your experience?[5] But I want you to be so kind as to have a trap wound round with cloth or flannel over the iron teeth so as to see whether it will save much suffering to have the leg only pinched instead of torn with the teeth. I thought if it is true that the leg is broken the animal would suffer nearly as much in whatever way it is held fast.

I am going to send out papers for subscriptions & when I have got as many names as I can I must put it into the hands of the Soc. in Pall Mall, as the awarding the prize will be a difficult matter.[6] I send you a paper to shew what sort of a beginning we have made & I am more grateful for two 5s/ subscriptions than one at £1, for I think the number of names who take an interest in the attempt is what chiefly signifies[7]

I can give rather a better account of Charles this week but he is never long without attacks of sickness. The water cure seemed to do him harm & Dr Gully quite owned that he was not strong enough to bear it[8] Since we came home he has consulted Dr Brinton of Guy's Hospital & I hope mineral acid is doing him some good[9].

He is wonderfully cheerful when not positively uncomfortable He does not feel the least temptation to disobey orders about working for he feels quite incapable of doing any thing.

His good symptoms are losing *no* flesh & having a good appetite so that I fully hope that in time he will regain his usual standard of health which is not saying much for him. He desires me to remember him most kindly to you & with my very kind remembrances to Mrs Fox[10] I am dear Mr Fox yours very sincerely | E. Darwin

I will trouble you with the new paper when it is printed.[11] I wish you had told us how you are yourself

Postmark: DE 8 63
Christ's College Library, Cambridge (Fox 142)

[1] The letter from Fox has not been found.

[2] Presumably Fox had remarked on the pamphlet prepared by CD and Emma to protest against the use of steel traps by gamekeepers (*An Appeal*; see Appendix IX); the pamphlet was included with the letter from Emma Darwin to W. D. Fox, [29 September 1863]. See also preceding letter, n. 11.

[3] The Menai Bridge spans the Menai Straits to connect Anglesey to Caernarvonshire, Wales. No other reference to Fox's actions has been found.

[4] This individual has not been identified. Two resident gamekeepers were recorded in Down, Kent, in 1861: John Higwood and William Folgate (Census returns 1861 (Public Record Office, RG9/462: 73, 77)).

[5] No answer from Fox has been found.

[6] Emma refers to the Royal Society for the Prevention of Cruelty to Animals, which had offices at 12 Pall Mall, London (*Post Office London directory* 1863). There is an entry in CD's Account books–banking accounts (Down House MS) recording the receipt of a subscription towards funding a prize for the design of a humane trap for the destruction of vermin (see Appendix IX).

[7] The enclosure has not been found.

[8] CD underwent treatment at James Smith Ayerst's hydropathic establishment in Malvern Wells, Worcestershire, between 3 September and 12 or 13 October 1863 (see 'Journal' (Appendix II)). CD had formerly received hydropathic treatment from James Manby Gully at Great Malvern, Worcestershire, but was treated on this occasion by Ayerst, presumably on the recommendation of Gully, owing to Gully's own ill health (see letter from W. D. Fox, [16–22 May 1863], and letters to W. D. Fox, 23 May [1863], and 4 [September 1863] and n. 1).

9 Emma was mistaken; William Brinton was a physician at St Thomas's Hospital, London; no position at Guy's Hospital is listed in the *Medical directory*. The mineral acid referred to may have been hydrocyanic acid (see letter from George Busk, [*c.* 27 August 1863]).

10 Ellen Sophia Fox.

11 Presumably a reference to Emma's proposed circular soliciting donations for a prize to be offered for the design of a humane vermin trap (see n. 6, above).

From Emma Darwin to Julius von Haast 12 December [1863][1]

Down. | Bromley. | Kent. S.E.
Dec 12

Sir

I write by my husband's desire as he is too unwell to be able to answer your letter himself, to thank you for your kind & interesting letter.[2] He was much interested by your account of the marvellous ground parrot & also by the Reports & Facts on naturalisation of animals in New Zealand.[3] He has communicated them to D[r] Hooker who is also writing on the subject.[4]

Mr Darwin as he is not able to do so himself begs you will present his thanks for the honour of being elected to the Philosophical Institute of Canterbury.[5]

I am dear Sir | yours truly | Emma Darwin

I am sorry to say that Mr Darwin has been very unwell for some months.

Alexander Turnbull Library, National Library of New Zealand (Haast family papers, MS-Papers -0037-051)

1 The year is established by the relationship between this letter and the letter from Julius von Haast, 21 July [– 7? August] 1863.

2 Letter from Julius von Haast, 21 July [– 7? August] 1863.

3 See letter from Julius von Haast, 21 July [– 7? August] and nn. 13, 14, and 17.

4 See letters to J. D. Hooker, 10 [November 1863] and [13 November 1863], and letter from J. D. Hooker to Emma Darwin, 11 November 1863. Joseph Dalton Hooker incorporated Haast's information into an article published in the *Natural History Review* (J. D. Hooker 1864).

5 CD was elected an honorary member of the Philosophical Institute of Canterbury, New Zealand, on 7 August 1863 (see letter from the secretary of the Philosophical Institute of Canterbury, New Zealand, 14 September 1863). Haast initiated the founding of the Institute in 1862 and was the first president (H. F. von Haast 1948, p. 221).

From Emma Darwin to John Murray [before 17 December 1863][1]

Down Bromley | Kent.

Dear Sir

Mr Darwin who is not well enough to write himself desires me to ask you to be so kind as to send a copy of the last Edition of the Origin of Species to M[r] Quaritch Piccadilly with the enclosed note to inform him where it is to be sent.[2]

I am yours obediently | Emma Darwin

Mr Darwin desires me to say that as you have never hesitated to authorize a foreign translation he has taken upon himself to authorise a translation into Italian without consulting you[3]

John Murray Archive (Darwin 128–9)

[1] The date is established by the relationship between this letter and the letter to J. D. Hooker, 5 [December 1863], in which CD told Joseph Dalton Hooker about the new Italian edition of *Origin*, and by the recipient's annotation to the top of the letter: 'Copy sent to M[r] Quaritch Dec. 17'.

[2] The reference is to the third edition of *Origin*, which was published by John Murray in 1861 (Freeman 1977, p. 85). Bernard Quaritch was a book dealer with premises at 15 Piccadilly, London (*Post Office London directory* 1863). The note accompanying this letter has not been found.

[3] CD's letter authorising an Italian translation of *Origin* has not been found. The first Italian translation was published in Modena by the firm of Nicola Zanichelli, translated by Giovanni Canestrini and Leonardo Salimbeni, in 1864 (Freeman 1977, p. 105). The copy of the third edition of *Origin* (see n. 2, above) may have been for Canestrini and Salimbeni to use for their translation.

From Emma Darwin to Alphonse de Candolle 17 December [1863]

Down. | *Bromley.* | *Kent. S.E.*
Dec 17.

Dear Sir

Mr Darwin wishes me to write for him as he is too unwell to be able to write himself. He begs me to thank you for your letter & very kind present of the pamphlet.[1] I am sorry to say that he has been very unwell for 3 months & though he is making slow progress I am afraid it will be long before he can apply himself to his usual pursuits.

I am with his kind remembrances | yours very truly | Emma Darwin

Endorsement: 'Mad. Darwin | 17 dec. 63'
La Fondation Augustin de Candolle

[1] The reference has not been identified; the pamphlet may have related to CD's inquiry about *Lythrum* (see letter to Alphonse de Candolle, 4 August [1863]).

From J. B. Innes 17 December [1863][1]

Milton Brodie
17th Dec[r].

Dear Darwin,

I sent this to Malvern duly and you see the fate it has met with.[2] I send it off again as I write so little it is a pity to lose what I do—

And as you are in some perplexity about your son[3] let me tell you that the Rev[d]. I Gresson who is a most charming fellow, and who I hear has a very nice sister is just starting a small establishment for young boys before they go to public schools at Worthing.[4] You will know all about the sanitary condition of Worthing.[5] If you want such a place it may be worth while to ask about this I can answer for Gresson being a most agreeable man with boys, as well as grown folk He has been second master at Bradfield for many years and Stevens can you more about him.[6]

Our best regards | Faithfully yours | J Brodie Innes.

DAR 167: 13

[1] The year is established by the relationship between this letter and the letter to J. B. Innes, 1 September [1863].

[2] CD underwent treatment at James Smith Ayerst's hydropathic establishment in Malvern Wells, Worcestershire, between 3 September and 12 or 13 October 1863 (see 'Journal' (Appendix II)). Innes presumably refers to his letter to CD of 4 September [1863], which was apparently returned to Innes from Malvern after some delay and sent to CD with this letter.

[3] See letter to J. B. Innes, 1 September [1863] and n. 3. The reference is to Horace Darwin, whose ill health had led to the interruption of his education with the Reverend George Varenne Reed in January 1863 (see letter from G. V. Reed, 12 January 1863). Horace resumed his studies with Reed in April 1864 (CD's classed account books (Down House MS)). See also *Correspondence* vol. 12, letter from G. V. Reed, 15 August 1864.

[4] Innes refers to John George Gresson, who was a master at St Andrew's College, Bradfield, Berkshire (see n. 6, below). Gresson's sister has not been further identified. Gresson opened West Mansion Preparatory School at Heene, Worthing, Sussex, in 1864 (*Clergy list* 1863–90, *Crockford's clerical directory* 1889). This was one of many private educational establishments opened up in seaside resorts in Sussex during the second half of the nineteenth century (Walton 1983, p. 97).

[5] Worthing, in common with many other seaside towns, was a popular place of resort for convalescents (see Walton 1983, p. 109, N. Wood 1912, p. 161). However, Innes may be alluding to CD's possible knowledge of Edward Cresy's report to the General Board of Health on the sanitary condition of Worthing (Cresy 1850); Cresy was a surveyor and civil engineer, and CD's neighbour. CD had assisted Cresy in obtaining the appointment of superintending inspector to the General Board of Health in 1848 (see *Correspondence* vol. 4). CD had also received some information about Worthing from Joseph Dalton Hooker, who stayed there with his wife Frances Harriet Hooker in 1861 (see *Correspondence* vol. 9, letter to J. D. Hooker, 13 July [1861]).

[6] St Andrew's College, Bradfield, Berkshire, was a private school founded for the education of sons of clergymen and gentlemen (Blackie 1976, p. 1). Gresson had been a master at St Andrew's College since 1855 (see *Crockford's clerical directory* 1889, and Blackie 1976, p. 37). Thomas Stevens was the Warden of St Andrew's College (Blackie 1976, p. 225).

From Emma Darwin to J. D. Hooker 26 December [1863]

Down. | Bromley. | Kent. S.E.
Dec 26

Dear D[r] Hooker

Charles begs me to tell you that nothing w[d] please him more than your having a bust of him & if he should get well enough to sit to Mr Woolner he will be sure to let you know but at present I am sorry to say that he could not even sit for

5 minutes.[1] One day is a little better & one a little worse but I cannot say that he makes progress at present. He stays in his bed room & gets frequently in & out of bed & occasionally goes down stairs for a very short time, but he can only stand very short visits even of the boys.[2] When not very uncomfortable his spirits are wonderfully good, but I am afraid he may remain just as he is very long before there is a struggle in his constitution & that the sickness is conquered.

With Charles & my love to Mrs Hooker[3] & you | I am yours very sincerely | E. D.

About America I think the slaves are gradually getting freed & that is what I chiefly care for.[4] The Times evidently thinks that is to be deplored,[5] but I think all England has to read up Olmsted's works[6] again & get up its Uncle Tom again[7]

Endorsement: '/63'
DAR 115: 214

1 The reference is to the sculptor Thomas Woolner. The letter in which Hooker made this request has not been found. CD did not sit for Woolner until 1868 (see letter to J. D. Hooker, 26 November [1868] (*Calendar*, no. 6476)).

2 William Erasmus, George Howard, Francis, Leonard, and Horace Darwin.

3 Frances Harriet Hooker.

4 President Abraham Lincoln's Emancipation Proclamation, which came into effect on 1 January 1863, freed the slaves in those states in rebellion against the Union. The Union forces effectively became armies of liberation (McPherson 1988, pp. 557–8).

5 *The Times* had maintained a hostile tone towards the Union cause in its coverage of the American Civil War. CD and Emma had long been unhappy with the newspaper's treatment of the war with regard to slavery (see *Correspondence* vol. 10, letter to Asa Gray, 16 October [1862], and this volume, letter to Asa Gray, 23 February [1863] and n. 22).

6 Frederick Law Olmsted wrote a number of accounts of the social and economic conditions of the southern states of America: *A journey in the seaboard slave states* (Olmsted 1856), *A journey through Texas* (Olmsted 1857), and *A journey in the back country* (Olmsted 1860); these accounts were republished as a two-volume work entitled *Journeys and explorations in the cotton kingdom* (Olmsted 1861). CD had read Olmsted 1856 and 1857 (see *Correspondence* vol. 4, Appendix IV, 128: 23 and 128: 25, and *Correspondence* vol. 9, letter to Asa Gray, 17 September [1861]).

7 A reference to Harriet Elizabeth Beecher Stowe's novel entitled *Uncle Tom's cabin* (Stowe 1852). Stowe's novel, which was popular in Britain, was an influential indictment of slavery (McPherson 1988, pp. 38–9, 88–9).

From Emma Darwin to George Maw[1] 28 December 1863

Down Bromley | Kent
Dec 28 | 1863.

Sir

Mr Darwin being too unwell to write himself begs me to tell you that he has had the greatest satisfaction in signing the paper & forwarding it as you wish— [2]

I am Sir | Yours very truly | Emma Darwin

Royal Horticultural Society (Lindley library, box 69, c 16)

[1] The attribution is based on the archival history of the letter, namely, that it was deposited in the collections of the Royal Horticultural Society, London, with the George Maw papers, and on the relationship between this letter and the following letter.
[2] See following letter.

To the Geological Society of London [1] [*c.* 28 December 1863][2]

George Maw FLS FSA FRHS— VP of the Severn Valley Field Club member of the R Agricultural College[3] & member of the Council of the British Archæological Association residing at Benthall Hall near Broseley—[4] being desirous of admission into the GEOLOGICAL SOCIETY OF LONDON, WE, the undersigned, recommend him as a proper person to become a FELLOW thereof.

Jan.2.1864[5]

Charles Darwin[6]
W. S. Symonds of Pendock[7]
J. Gwyn Jeffreys[8]
Edwd Meryon[9]
Jos. Prestwich[10]
} From Personal knowledge

Proposed
Notice of Ballot

{ Elected
No.

Geological Society Archives (F1/6)

[1] The proposal form for George Maw's election as fellow of the Geological Society was presumably sent by CD at about the same time as the preceding letter. The proposal was formally registered on 6 January 1864, the notice of ballot was given on 20 January, and Maw was elected on 24 February 1864 (Geological Society Archives, F1/6).
[2] The date is established by the relationship between this letter and the preceding letter (see n. 1, above).
[3] In 1860 Maw was elected fellow of the Linnean Society and of the Society of Antiquaries (*List of the Linnean Society of London* 1860, *List of the Society of Antiquaries of London* 1861). The document also refers to Maw's fellowship of the Royal Historical Society and vice-presidency of the Severn Valley Naturalists' Field Club (see also n. 6, below). Maw was a member of the Royal Agricultural College, Cirencester, Gloucestershire; the college awarded him a diploma with honours in 1852 (Bathurst and Kinch eds. 1898, p. 178).
[4] Broseley was a village in Shropshire, approximately fourteen miles south-east of Shrewsbury (*Survey gazetteer of the British Isles*).
[5] The date apparently refers to the date of the last signature, and must have been added after CD had signed the fellowship proposal.
[6] See preceding letter. Maw initiated a correspondence with CD in March 1861, and sent CD a presentation copy of his review of *Origin* in July 1861 ([Maw] 1861). See *Correspondence* vol. 9, letter

to George Maw, 13 July [1861]. See also *Correspondence* vol. 10, and this volume, letter to George Bentham, 15 April [1863]. In February 1863, Maw had been instrumental in CD's acceptance of an honorary membership of the Severn Valley Naturalists' Field Club (see letters from George Maw, 19 February 1863 and 25 February 1863).

7 William Samuel Symonds. Symonds was rector of Pendock, Worcestershire (*DNB*). This signature and the signatures which follow (see nn. 8–10, below) were presumably added after CD had signed the fellowship proposal.

8 John Gwyn Jeffreys.

9 Edward Meryon.

10 Joseph Prestwich.

To Ernst Haeckel 30 December [1863] – 3 January [1864][1]

Down. | Bromley. | Kent. S.E.
Dec 30

My dear Sir

Your letters are always most kind.[2] I shall be proud to receive your monograph on the Radiolariæ which I saw & much admired in London;[3] but I know so little of the group that in truth I am not worthy of your present. I sent off a few days since to Jena a copy of a paper by me, which if you have time I think is worth your reading.[4]

My health continues very weak & I have no strength to spare, so I am sure you will excuse the brevity of this note.

With very sincere respect believe me my dear Sir | Yours truly obliged | Charles Darwin

P S. Jan 3.

By a mistake this letter was put in a wrong envelope & you will have have received another note addressed to a friend which I fear will have surprized you.[5]

Please to tear it up. As there has been so much delay owing to my unfortunate mistake, I have directed this to Jena instead of to Berlin[6]

LS
Ernst-Haeckel-Haus (Bestand A/ Abt. 1: 1028/1)

1 The years are established by the relationship between this letter and the letter from Ernst Haeckel, 2 January 1864 (*Correspondence* vol. 12).

2 The letter from Ernst Haeckel has not been found.

3 Haeckel 1862. CD visited London several times in 1863 (see 'Journal' (Appendix II)), and may have seen Thomas Henry Huxley's copy of Haeckel's book, which Huxley had obtained in October 1862 (see L. Huxley ed. 1900, 1: 237).

4 CD may have sent Haeckel a copy of 'Two forms in species of *Linum*'.

5 CD had mistakenly sent the letter for Ernst Haeckel to Hugh Falconer, and vice versa. See following letter, and *Correspondence* vol. 12, letter from Ernst Haeckel, 2 January 1864. The letter intended for Falconer has not been found.

6 Haeckel must have told CD in a missing letter of his plans to move from Berlin to Jena in January 1864; Haeckel was professor extraordinarius of zoology at the University of Jena (*DSB*, *NDB*). See also *Correspondence* vol. 12, letter from Ernst Haeckel, 2 January 1864.

From Hugh Falconer 31 December [1863][1]

21 Park Crescent
31 Dec.r

My Dear Darwin.

A happy new year to you and yours—with all the best wishes of the Season.

On opening your cover, I find that the wrong letter has been sent to me[2]—and that mine has probably gone to your Radiolariæ friend—who I trust is not the Pres. R.S![3]

I return the former enclosed.

Yours Ever Sinly | H Falconer

DAR 164: 25

[1] The year is established by the relationship between this letter and the preceding letter.

[2] See preceding letter and n. 5.

[3] See preceding letter and n. 5. The letter intended for Falconer has not been found. Falconer probably refers to Edward Sabine, who was president of the Royal Society of London, and instrumental in the choice of Adam Sedgwick for the award of the society's Copley Medal in 1863, in preference to CD (see letter from Edward Sabine to John Phillips, 12 November 1863 and nn. 4 and 5).

APPENDIX I
Translations of letters

From Henri Louis Frédéric de Saussure[1] 17 January 1863

Geneva.
17 Jan 63.

Sir

I have indeed received your letter of 8th Jan. and would like to thank you for it.[2]

The work on *Mexico* will contain some geology and therefore might eventually be of interest to you.[3] As for the one on birds, there are some copies of it at Rothchild's in Paris (14 rue de Buci)[4] It is almost sold out.

I am currently working on various projects, but there is one which has been occupying me for a long time and which will be of interest to you. It concerns the *descent* of genera, species and social behaviour in wasps.[5] In this group there are peculiar transitions and specific and obvious behavioural changes.

I had thought that my map of Mexico would be of interest at this time and have deposited some copies with Stanford, *6 Charing Cross*.[6] But it seems that in England no one cares much for that country, as the map is selling badly, even though it is very beautiful and has been put together with great care, and adds a great many details to the geography of the area.

I remain, my dear sir, most faithfully yours | H. de Saussure

May I ask you to forward the enclosed letter to the British Museum.[7] There is no hurry and your work probably takes you to that establishment.

DAR 177: 40

1 For a transcription of this letter in the original French, see p. 54.

2 CD's letter has not been found.

3 The reference is to the first volume of Saussure 1858–71, a memoir on the natural history of Mexico, the Antilles, and the United States.

4 Saussure 1858. CD's lightly annotated copy of Saussure 1858 is in the Darwin Pamphlet Collection–CUL. Saussure refers to the Paris publisher, Jules Rothschild.

5 Saussure had already published a major study on wasps (Saussure 1852–8); his synopsis of American wasps formed the fourth part of Saussure 1858–71.

6 Saussure 1862. The London bookseller, stationer, and map engraver, Edward Stanford, was an agent for ordnance maps, geological survey maps, and Admiralty charts (*Post Office London directory* 1861).

7 The enclosed letter has not been found.

From Friedrich Rolle[1] 26 January 1863

Bad Homburg bei Frankfurt am Main
26th. January 1863.

Most honoured Sir,

Please excuse me for not replying earlier to your kind letter of last Autumn.[2] First I wanted to wait for the conclusion of my little work "Darwin's Lehre".[3]

I was greatly pleased to see that my work met with your approval and I hope that in the concluding part you will come across more matters that will attract your attention. You will find how much physiology can be gleaned from palaeontology.

I am basically a geologist and a palaeontologist. My main task in recent years was to establish the genealogy of the European marine fauna and its derivation from tertiary fossils. I have published only little on this subject, because last spring, before completing these studies I left the Kaiserl. Königl. Hofmineralien-Cabinet in Vienna, where for five years I had been first assistant and later adjunct.[4] Since then I have settled in my home town of Homburg as a dealer in natural history objects and hope in this manner to be able to earn an independent living.

At that time I turned to the publisher Herr Suchsland in Frankfurt, to obtain the commission for a scientific work.[5] Herr Suchsland had read about your doctrine in periodicals and expressed great interest in it. Therefore he commissioned me to write a commentary upon it.

Thus the task to broaden and spread your theory has fallen to me. The most active supporter of your views in Vienna is actually my friend Doctor Gustav Jaeger, a native Swabian, founder and now director of the Zoological Garden in Vienna.[6]

Gustav Jaeger and I already shared your views even before we became acquainted with your work. We were thinking of writing a joint work on the genealogy of animals; Jaeger was to work on the zoological and physiological part, I on the geological and palaeontological part. But we did not succeed in fully realising our intentions.[7]

When your work appeared, Jaeger undertook to explain the theory in Vienna in public lectures and journals. A certain Herr Pelzeln presented himself as his opponent in Vienna.[8] But public opinion spoke for Jaeger. Herr Pelzeln never once found a place for his objections. Thus he had his reply published privately as a pamphlet (August von Pelzeln, Bemerkungen gegen Darwin's Theorie vom Ursprung der Spezies. Wien bei Pichler 1861. 8º. 17 pag.)[9]

However, this pamphlet is so poorly, so absurdly and at the same time so fanatically written, that it is best not to mention it at all, and Jaeger has not replied to it.

At that time Jaeger also gave a lecture at the annual meeting of the German ornithologists in Stuttgart; Professor Blasius (of Braunschweig) and Professor Altum (of Münster) spoke in opposition.[10]

Since then Jaeger has done no more about it, because he was too occupied with the founding of the Vienna Zoological Garden. However, he intends to resume his efforts soon, and I am endeavouring to urge him to do so.

Our German Professors are partly your opponents, partly they say nothing.

Professor Quenstedt of Tübingen has always had similar views. But he works without method and makes many mistakes, and therefore has never found general acceptance in Germany.[11]

D^r. Weinland of Frankfurt (a former assistant to Agassiz) and D^r. Hensel of Berlin have pronounced themselves in favour of your views.[12] Weinland has strongly supported my work. I do not know Hensel personally; you will find some important observations in Hensel's paper on Hipparion in the Proceedings of the Berlin academy.[13] He remarks that the horse is distant from the Pachyderms. As an embryo, the horse already exhibits the abnormal pedal structure. However, there are monstrous structures in the horse in which the normal structure of the pachyderm foot is visible. This is an echo of the descent of the horse from the Pachyderms. The fossil genus Hipparion confirms this and provides a distinct transition from the horse to the rhinoceros, hippopotamus, and other pachyderms. Hensel says also that rudimentary organs cannot be explained naturally by any researcher with his feet on the ground except by descent.[14]

I would be happy if I could be of any service to you. I am especially willing to help with Tertiary molluscs and other palaeontological matters.

Please forgive me for replying in German, but I have so little command of English that you would not understand me if I were to write in it.

I am, dear Sir, | your obedient servant, D^r. Friedrich Rolle

DAR 176: 201

1 For a transcription of this letter in the original German, see pp. 86–7.

2 Letter to Friedrich Rolle, 17 October [1862] (*Correspondence* vol. 10).

3 Rolle 1863 was published in instalments, the first and second of which had been sent to CD before 17 October 1862. The publication of the third and fourth parts was announced on 19 January 1863 (*Börsenblatt für den Deutschen Buchhandel* 30 (1863): 113). In translation, its title read: 'Charles Darwin's theory of the origin of species in the plant and animal kingdoms and its application to the history of creation'. In the foreword to the book (Rolle 1863, p. [ii]), Rolle referred to CD's approval of his efforts (see *Correspondence* vol. 10, letter to Friedrich Rolle, 17 October [1862] and n. 3). CD's annotated copies of all four parts of Rolle 1863 are in the Darwin Pamphlet Collection–CUL.

4 For a description of Rolle's life in Vienna at the Kaiserlich-königliche Hofmineraliencabinet (Royal Imperial Cabinet of Mineralogy), and for a discussion of his reasons for leaving academic life and moving to Homburg in March 1862, see Martin and Uschmann 1969, pp. 18–26.

5 The publisher and bookseller Friedrich Emil Suchsland was head of the Frankfurt firm Johann Christian Hermann'sche Verlagsbuchhandlung (Martin and Uschmann 1969).

6 Gustav Jäger announced his support for transmutation and CD's theory in lectures given at the annual meeting of the Deutsche Ornithologen-Gesellschaft in Stuttgart on 18 September 1860 (see n. 10, below), and at the Verein zur Verbreitung naturwissenschaftlicher Kenntnisse in Vienna on 10 and 15 December 1860 (Jäger 1862; see n. 8, below). Jäger was one of the directors of the zoological garden in Vienna, which opened on 25 May 1863 (see Weinreich 1993, pp. 48–64).

7 *Zoologische Briefe* (Jäger 1864) included a chapter in which Jäger argued in support of the transmutation of species theory; the author claimed that this was written before the publication of *Origin* in 1859 (Jäger 1864, pp. 36–56). There is a copy of Jäger 1864 in the Darwin Pamphlet Collection–CUL. For a discussion of Jäger's reaction to *Origin*, see Weinreich 1993, pp. 36–47.

[8] Rolle refers to the confrontation between August Pelzel von Pelzeln and Jäger at the meeting of the Zoologisch-botanische Gesellschaft in Vienna on 5 December 1860. Pelzeln had given a lecture in which he criticised CD's theory, and which was later published as Pelzeln 1861; CD's annotated copy of this work is in the Darwin Pamphlet Collection–CUL. According to the report in the *Verhandlungen der kaiserlich-königlichen zoologisch-botanischen Gesellschaft in Wien* 10 (1860): 98, Jäger strongly opposed Pelzeln's view and offered to speak in support of CD's theory (see Jäger 1862; see also n. 7, above).

[9] Pelzeln 1861. See n. 8, above.

[10] Jäger's lecture in support of CD's theory was given at the annual meeting of the Deutsche Ornithologen-Gesellschaft on 18 September 1860. In the subsequent discussion, Bernhard Altum and Johann Heinrich Blasius maintained that CD's theory was a purely speculative and unsubstantiated hypothesis (see *Bericht über die XIII. Versammlung der Deutschen Ornithologen-Gesellschaft zu Stuttgart vom 17. bis 20. September 1860*, pp. 6, 40–6). In the late 1860s, Altum, a clergyman and lecturer in zoology, became known as one of the most uncompromising opponents of the theory of evolution in Germany (see Stresemann 1975, pp. 273–4, 328–32).

[11] Friedrich August Quenstedt had been Rolle's tutor at the University of Tübingen. In 1856, he published a popular book in which he made a number of remarks in favour of biological evolution and common descent (see Quenstedt 1856, pp. 227–9).

[12] The references are to David Friedrich Weinland and Reinhold Friedrich Hensel. Weinland, a lecturer in zoology at the Senckenberg Museum, Frankfurt, had been an assistant to Louis Agassiz in Cambridge, Massachusetts, between 1855 and 1858 (*Zoologischer Beobachter* 57 (1916): 160). In February 1861, he published a short but favourable review of *Origin* (Weinland 1861). Hensel had been a tutor in natural history at the University of Berlin until 1860; three years later, he was commissioned to conduct zoological research in southern Brazil (*Leopoldina* 18 (1882): 19–21).

[13] Hensel 1860.

[14] Hensel mentioned CD's theory briefly when he considered the importance of rudimentary organs and maintained that biological traits were gradually changing over time (see Hensel 1860, pp. 69–70).

To Charles Victor Naudin[1] 7 February 1863

7 February 1863

Dear Sir,

An absence from home has prevented me from offering you my sincere thanks for the trouble that you took in writing your letter of February 2, so full of such valuable information.[2]

It pleases me greatly to have your opinion (as far as you are able to express it) on the crossing of varieties of melons; and I will use it in quotations.[3]

Recently I made extensive use of your admirable study of the Cucurbitacae,[4] as much as a source of quotations as for my own information. I thank you sincerely for your intention of sending me your latest paper.[5] I look forward with great interest to your work on hybridisation.[6]

Please accept my cordial thanks for your great kindness and believe me, with much respect, your most obedient servant.

Charles Darwin[7]

Progressus rei botanicæ 4 (1913): 94

[1] For a transcription of this letter in the original French, see pp. 119–20.

[2] Naudin's letter has not been found. CD was in London from 4 to 14 February 1863, staying at Erasmus Alvey Darwin's house at 6 Queen Anne Street (see 'Journal' (Appendix II)).

[3] CD had sent a memorandum to Naudin requesting information on crossing varieties of melon (see *Correspondence* vol. 10, letter to J. D. Hooker, 24 December [1862], and this volume, letter from J. D. Hooker, 24 January 1863). CD cited Naudin's study of varieties of melon (Naudin 1859) in *Variation* 1: 359–60. CD's notes on Naudin 1859 are preserved in DAR 205.7 (2): 146. CD cited additional information supplied by Naudin in *Variation* 2: 108.

[4] CD refers to Naudin 1856 and 1862. In discussing the variations of cucurbitaceous plants in *Variation* 1: 357–60, CD quoted extensively from Naudin's experiments on the genus *Cucurbita* (Naudin 1856). CD's notes on Naudin's article are in DAR 205.7 (2): 141. CD wrote a draft of this chapter between October and December 1862 (see *Correspondence* vol. 10, Appendix II); it dealt with variation in fruits, ornamental trees, and flowers.

[5] Naudin may have referred to his forthcoming publication on hybridity (Naudin 1863). There is an annotated copy in the Darwin Pamphlet Collection–CUL.

[6] Naudin's researches on the hybridisation of plants were published in Naudin 1863 and 1865, of which there are annotated copies in the Darwin Library–CUL. For Naudin 1865, see *Marginalia* 1: 638–9. CD's opinion of Naudin's work is discussed in *Correspondence* vol. 10, letter to J. D. Hooker, 30 [June 1862], and letter to Asa Gray, 1 July [1862].

[7] A footnote in Blaringhem 1913, p. 94, notes that only the signature of this letter is in CD's hand.

From Camille Dareste[1] 8 February 1863

Lille
8 February 1863

Sir

Although I have not the honour of being known to you, I wish nevertheless to express the satisfaction I felt in reading your admirable book on the origin of species. I have been engaged in the study of the natural sciences for twenty years; and I had long ago through my own observations and reflections come to consider the idea of the fixity of species, so generally accepted by naturalists, as a view completely in contradiction with the facts. But though I was unable to content myself with the existing theories, I nevertheless did not see how they could be replaced; and I found myself vacillating between doctrines which were no longer acceptable and the greater uncertainty concerning the doctrines that ought to replace them. For me your book has caused a light to shine in the midst of darkness, by showing me where to look for the guiding principle that had hitherto been lacking in my reflections on the nature of species.

On this subject, I take the liberty of sending you a small work which I have just published.[2] It is a memoir that received a prize a month ago from the Académie des Sciences of Paris.[3] If you will take the trouble to peruse these pamphlets, you will see that I have endeavoured to solve by the experimental method some questions wholly comparable to those you deal with in your book.[4] After some very lengthy experiments I have indeed succeeded in producing, as did Geoffroy Saint-Hilaire thirty years ago, a great number of artificial monstrosities;[5] and I hope that in pursuing these studies I shall one day be able to explain the formation of a great number of anomalies This work, if I am able to proceed with it, will I hope one day enable me to apply the experimental method to the investigation of a branch of science where you have won yourself such a prominent position.

I regret, Sir, being obliged to write to you in French: although I know enough English to have been able to read your book before it was translated,[6] I am not sufficiently conversant with your tongue to speak it, and even less so to write it.

I remain, with profound respect | Your very humble and very obedient servant Camille Dareste | Professor at the Faculty of Sciences of Lille

Mr Charles Darwin Esq.

DAR 162: 42

[1] For a transcription of this letter in the original French, see pp. 121–2.

[2] Dareste 1863. There is a lightly annotated copy of this work, in which Dareste provides an account of monstrosities artificially induced in chicken embryos, in the Darwin Pamphlet Collection–CUL.

[3] Dareste refers to the Alhumbert Prize (*Comptes rendus hebdomadaires des séances de l'Académie des Sciences* 55: 979–82).

[4] Dareste's experiments on hen's eggs were published in two complementary pamphlets as Dareste 1862 and 1863. The wording of this letter, and the letter to Camille Dareste, 16 February [1863], suggests that Dareste only sent the second pamphlet; however, Dareste appears to assume that CD also had access to the earlier pamphlet. There are annotated copies of Dareste 1862 and 1863 in the Darwin Pamphlet Collection–CUL. CD discussed the implications of embryology for the theory of natural selection in *Origin*, raising the question of the relationship of embryos to external conditions (*Origin*, pp. 439–50). CD cited Dareste's experiments in *Variation* to illustrate the point that external conditions could be the direct cause of modifications in the structure of organisms (see *Variation* 2: 289).

[5] Etienne Geoffroy Saint-Hilaire and his son, Isidore Geoffroy Saint-Hilaire, pioneered the scientific study of monsters (see E. Geoffroy Saint-Hilaire 1825a, 1825b, 1825c, 1826, 1827a, and 1827b, and I. Geoffroy Saint-Hilaire 1832–7). By artificially generating monstrosities in embryos, they hoped to throw light on the laws governing the production of monsters and new species (see E. Geoffroy Saint-Hilaire 1825a and I. Geoffroy Saint-Hilaire 1832–7, 3: 500–6).

[6] A French translation of *Origin* (Royer trans. 1862) was published on 31 May 1862 by Guillaumin et Cie (*Journal Générale de l'Imprimerie et de la Librarie* 2d ser. 6 (pt 3): 341).

From Armand de Quatrefages[1] 25 March 1863

Paris

25 March 63

Sir and dear colleague

As soon as I received your letter I concerned myself with your request.[2] I wrote to a friend of mine in the South of France, but his reply was not entirely satisfactory.[3] Fortunately, I have been able to place the matter in very capable hands. M. Martins, professor of botany and director of the Jardin des Plantes at Montpellier, recently elected a corresponding member of our academy, has agreed to take care of the matter.[4] He promised me that with his colleagues at the Soc. d'Agriculture, he would personally make as complete a collection as possible of all species of silkworm bred in the vicinity of Montpellier.[5] This collection will include the caterpillar, the chrysalis, and the butterfly—the first two preserved in alcohol, the third mounted in the usual way. It should all be sent to me, and I will have it sent to you by train or any other means that you care to specify.[6] In this way you will be able to study for yourself the questions that interest you. It seemed to me that this would be the best way to meet your requirements.

I thank you for the pleasant things which you are kind enough to say on the subject of my *Métamorphoses*.[7] Approval from someone like you is, on its own, worth more than that of many others put together.

I was more touched than I can say by the great honour that your Soc. of Anthropology has done me by placing my *Unité de l'espèce humaine* almost at the top of the list of foreign works that it intends to translate.[8] You are of course a member of that society and it pleases me to think that your support might have played a part in the decision of your colleagues.[9] Please accept my thanks once again.

I am at present engaged in a mighty struggle with the polygenists in our Society. They deny the effect of the environment in the creation of new races. But I maintain the variability of species under this influence.[10] You see, my dear colleague, the truth of what I said to you, that the only difference between us is one of degree.[11] But it is quite strange that it should be in the name of philosophy that the action of external influences on organised and living beings is being denied. Lamarck, Geoffroy S[t]-Hilaire . . . would be quite surprised to see themselves such disciples.[12]

Please accept, Sir and dear colleague, my most affectionate regards.

De Quatrefages

DAR 175: 2

[1] For a transcription of this letter in the original French, see pp. 263–4.

[2] CD's letter has not been found. In the summer of 1862, CD had requested information from Quatrefages on the varieties of the common silkworm (*Bombyx mori*), when writing the section of *Variation* dealing with this subject (*Variation* 1: 300–4; see *Correspondence* vol. 10, letter to Armand de Quatrefages, 11 July 1862 and n. 2). CD was particularly interested in the fact that, while the different varieties of silkworm varied greatly in the larval and pupal stages, they showed no constant differences in the adult stage. In a letter that is now missing, Quatrefages provided CD with information on this subject, which is quoted in *Variation* 1: 303. In a further letter to Quatrefages of 11 July 1862 (*Correspondence* vol. 10), CD asked him to send any 'fresh information' he might acquire in the ensuing year 'on the similarity of the moths of distinct races'. The season being too late to procure such information, Quatrefages asked CD to remind him of his request before the new season's eggs hatched the following April (see *Correspondence* vol. 10, letter from Armand de Quatrefages, [after 11 July 1862]).

[3] This individual has not been identified.

[4] Charles Frédéric Martins was elected as a corresponding member for the rural economy section of the Académie des Sciences, Paris, on 29 February 1863 (*Index biographique de l'académie des sciences*).

[5] The reference is probably to the Société d'agriculture of Montpellier, which was founded in 1807 (*Grande encyclopédie*).

[6] No such specimens are mentioned in *Variation* 1: 300–4, and it appears that they were never sent. Information from Martins is cited in *Variation* 1: 304 from an intermediate source (Godron 1859, p. 462); however, no information provided directly by Martins is included. An annotated copy of Godron 1859 is in the Darwin Library–CUL (see *Marginalia* 1: 331–5).

[7] CD's letter has not been found. Quatrefages promised to send CD a copy of Quatrefages 1862 in his letter of [after 11 July 1862] (*Correspondence* vol. 10); there is an annotated copy of this work in the Darwin Library–CUL (see *Marginalia* 1: 691).

[8] An annotated copy of Quatrefages 1861 is in the Darwin Library–CUL (see *Marginalia* 1: 693–4). Quatrefages refers to the announcement by the Anthropological Society of London, which was founded in 1863, that it hoped to publish an English translation of Quatrefages 1861. No copy of the first prospectus of the society has been found; however, in the 'Anthropological Review Advertising Sheet' issued with the first number of the society's *Anthropological Review* in May 1863, it was announced that the society intended to publish a series of works on anthropology, chiefly translations. Only one publication was described as 'in preparation'; Quatrefages 1861 headed a list of six other titles under the caption: 'translations will be recommended'. In subsequent prospectuses, the book was listed among those of which publication was contemplated, with George Frederick Rolph named as the prospective editor. However, the society's translation never appeared, and Quatrefages 1861 was not translated into English until 1879 (Quatrefages 1879). There is a copy of the 'Anthropological Review Advertising Sheet' in the unbound pamphlet collection in the Darwin Library–CUL.

[9] CD was an honorary member, although not an active one, of the Anthropological Society of London (see letter to Armand de Quatrefages, 27 March [1863]).

[10] Between February and August 1863, Quatrefages, who was president of the Société d'Anthropologie of Paris, engaged in extended discussions with the polygenists in the society over the question of the influence of the environment in modifying races. The substance of the discussions, together with Quatrefages's discourses on the subject on 5 March and 16 April 1863, are recorded in the *Bulletins de la Société d'Anthropologie de Paris* 4 (1863). See also Stocking 1968 and J. Harvey 1997, pp. 105–13.

[11] Quatrefages apparently made this observation in 1862 in a letter that has not been found (see *Correspondence* vol. 10, letter to Armand de Quatrefages, 11 July 1862, and letter from Armand de Quatrefages, [after 11 July 1862]).

[12] Quatrefages refers to Jean Baptiste de Lamarck and Etienne Geoffroy Saint-Hilaire, both of whom had adopted theories of species transmutation in which environmental influence played an important role (*DSB*). See also *Bulletins de la Société d'Anthropologie de Paris* 4 (1863): 149.

From Armand de Quatrefages[1] [28 March] – 11 April 1863[2]

Sir and dear colleague,

As I think I told you in my last letter, I am engaged at present in a rather lively battle at the Société d'Anthropologie on behalf of the variability of animal species.[3] One of my colleagues has challenged me to cite a *single species* in which the action of the environment has produced *anatomical* modifications.[4] He does not take into account those which concern the skin, the cellular tissue … I am about to reply ⟨ ⟩ by insisting on the history of the horse.[5] But this will not be the end of it, and I should like to produce before the eyes of these disciples of St Thomas one decisive fact.[6]

I have considered the *pug-cattle* which you brought to our attention, the Nãta or Niata cattle.[7] You say (Journ. of Research …) that you have the head of one in London.[8] Could you have a photograph of it taken and sent to me?[9] Even better—would it be possible to have a cast made for the museum collections?[10] Has it been published anywhere?[11] If it has not and either you or somebody else in England intends to do so, it goes without saying that no detailed publication will be produced in France, and I will restrict myself to showing either the drawing or the cast for the instruction of my students and the conversion (if it be possible) of my opponents.[12]

Forgive me Sir and dear colleague for any indiscretion in my questions and please accept again the expression of my affectionate devotion | De Quatrefages

11 April 63

I saw Mr Martins who is at present in Paris for the meeting of the delegates of the provincial faculties.[13] I gave him all the guidance that my experience of these small details could suggest,[14] so that your collection should be as well arranged as possible. I found him full of excellent intentions. He took your address and will send you the objects directly by train.[15]

A few days ago I also had the great pleasure of talking at length to D^r^. Carpenter; I have found his work on Comatulae of very great interest.[16] Thank you for all the kind things you said about my Métamorphoses.[17]

DAR 175: 3

[1] For a transcription of this letter in the original French, see pp. 275–6.

[2] The date range is established by the relationship between this letter and the letter to Armand de Quatrefages, 27 March [1863].

[3] See letter from Armand de Quatrefages, 25 March 1863 and n. 10.

[4] The reference is to André Sanson, and to a meeting of the Société d'Anthropologie held on 19 March 1863 (*Bulletins de la Société d'Anthropologie de Paris* 4 (1863): 159). For Quatrefages's responses to Sanson of 16 April and 16 July, see *Bulletins de la Société d'Anthropologie de Paris* 4 (1863): 213–23, and nn. 5 and 12, below. For Sanson's responses to Quatrefages of 7 May and 16 July, see *Bulletins de la Société d'Anthropologie de Paris* 4 (1863): 254–68, 374–5, and 380–3.

[5] Quatrefages described the modification of breeds of horses through environmental influences at a meeting of the Société d'Anthropologie on 16 April 1863, citing the descent of the English racehorse from Arab stock as an example (*Bulletins de la Société d'Anthropologie de Paris* 4 (1863): 215–19). See also *Origin*, p. 35.

[6] The reference is to the biblical narrative of the apostle Thomas's refusal to believe in the resurrection of Jesus Christ without direct experience (*EB*).

[7] Niata: 'An abnormally small variety of cattle, found in South America' (*OED*). Quatrefages refers to *Journal of researches* 2d ed., pp. 145–6, where CD observed that the niata cattle 'appear externally to hold nearly the same relation to other cattle, which bull or pug dogs do to other dogs.'

[8] *Journal of researches* 2d ed., p. 146. CD presented a skull of this breed from the *Beagle* collections to the Royal College of Surgeons of London in 1840 (see *Correspondence* vol. 4, letter to Francisco Javier Muñiz, 26 February 1847, n. 7).

[9] CD sent photographs of the niata skull in May (see letter to Armand de Quatrefages, 14 May [1863], and letter from Armand de Quatrefages, 19 May [1863]).

[10] Quatrefages was professor of the natural history of man at the Muséum d'Histoire Naturelle in Paris (*DSB*); apparently, no cast of the skull was sent (see letter to Armand de Quatrefages, 14 May [1863], and letter from Armand de Quatrefages, 19 May [1863]).

[11] The skull is described in Owen [1853], 2: no. 3832; CD had also discussed it in his manuscript draft of *Variation* (see *Variation* 1: 89–91, and letter to Armand de Quatrefages, [14 April 1863]). See also *Origin* 6th ed., p. 177.

[12] Quatrefages displayed the photographs at a meeting of the Société d'Anthropologie on 2 July 1863 (*Bulletins de la Société d'Anthropologie de Paris* 4 (1863): 350–2). See also letter from Armand de Quatrefages, 19 May [1863].

[13] Charles Frédéric Martins was professor of botany at the Faculté de Montpellier (*Dictionnaire universel des contemporains*).

[14] Quatrefages completed his own research on silkworms between 1857 and 1859 (see *DSB* and Quatrefages 1859).

[15] Martins had agreed to send CD a collection of varieties of the common silkworm (*Bombyx mori*) at different stages in their life cycle; however, the collection was apparently never sent (see letter to Armand de Quatrefages, 27 March [1863] and n. 3).

[16] William Benjamin Carpenter had travelled to Paris for recreational purposes during his Easter vacation as registrar of London University (Carpenter 1888, p. 84; *DNB*). However, he soon became involved in the heated discussions that followed the discovery on 28 March 1863 of a fossilised human jaw at Moulin-Quignon in the Somme valley, discussions in which Quatrefages also played a leading role. See Carpenter's report of the discovery in the *Athenæum*, 18 April 1863, p. 523; see also Grayson 1983, pp. 213–17, and Van Riper 1993, pp. 134–39. Carpenter published his work on the echinoderm *Comatula* in Carpenter 1865.

[17] See letter to Armand de Quatrefages, 27 March [1863]. The reference is to *Métamorphoses de l'homme et des animaux* (Quatrefages 1862), of which there is an lightly annotated copy in the Darwin Library–CUL (see *Marginalia* 1: 691).

To Armand de Quatrefages [1] [14 April 1863][2]

This nãta is an excellent case for you because the race is well established and must derive its origin from South America.[3] It is a very singular race. A short description of the head was made by Professor Owen in the *Descriptive catalogue of the osteological collection of the College of Surgeons*, 1853 page 624. You will see there that the *connection of the bones* is modified, since the maxillary is not fused to the bones of the nose.[4]

In examining a great number of skeletons of rabbits, ducks, poultry, and pigeons, I have observed several remarkable modifications in the skeletons, but since these are not yet published, they would not be of any use to you.[5] The Powter pigeon (inflated gullet or boulans) which undoubtedly descends from *Columba Livia*, is an interesting case since it has a very enlarged and modified oesophagus.[6]

In the first chapter of my *Origin of species* you will find a section about remarkable modifications in pigeons.[7]

The crested fowls, or, as we call them, *polish fowls* are an excellent case;[8] they reproduce perfectly well; in the past there was only one sex in which the skull was affected; now both sexes are affected. I am enclosing a very well drawn woodcut.[9] I have examined a great many skulls. There is an astonishing change of form in the *brain itself* and in the *interior* of the skull, besides the change in the external bones.

Incomplete[10]
Bulletins de la Société d'Anthropologie de Paris 4 (1863): 378–9

[1] CD's original letter has not been found. For a transcription of this letter in French see p. 313. The extant portion of the letter-text was published in the *Bulletins de la Société d'Anthropologie de Paris* 4 (1863): 378–9. Quatrefages probably translated this portion for his presentation to the Société d'Anthropologie (see n. 2, below).

[2] The date was given by Quatrefages when he introduced the letter in a discussion at the Société d'Anthropologie on 16 July 1863 (*Bulletins de la Société d'Anthropologie de Paris* 4 (1863): 378).

3 See letter from Armand de Quatrefages, [28 March –] 11 April 1863 and nn. 7, 8, and 9. CD thought that niata were a useful example of the environment's effect on anatomical form; the breed had been selected from cattle imported from Europe (*Variation* 1: 88–91).

4 Niata were described in Owen [1853], 2: 3832 (see also *Variation* 1: 89–91). For Quatrefages's presentation of photographs of the niata skull at discussions of the Société d'Anthropologie de Paris, see the letter from Armand de Quatrefages, 19 May [1863] and n. 4.

5 CD published this material in *Variation* 1: 103–304. *Variation* was not published until 1868; Quatrefages read this letter of CD's to the Société d'Anthropologie on 16 July 1863, and a report of William Bernhard Tegetmeier's presentation of a Polish or crested breed of fowl to the Zoological Society of London (*Bulletins de la Société d'Anthropologie de Paris* 4 (1863), pp. 378–80; see *Proceedings of the Zoological Society of London*, 25 November 1856, pp. 366–8). CD may have mentioned this Zoological Society report to Quatrefages in the missing portion of this letter; he cited his reliance on Tegetmeier's information on poultry in *Variation* 1: 225 (see nn. 8 and 9, below). Quatrefages briefly mentioned examples of variation in poultry to the Société d'Anthropologie on 7 May and 2 July (see *Bulletins de la Société d'Anthropologie de Paris* 4 (1863): 273, 350–1).

6 An engraving and description of the Pouter pigeon is provided in *Variation* 1: 137–9.

7 *Origin*, pp. 20–8.

8 The Polish or crested breed of fowl and its sub-breeds are described in *Variation* 1: 227–30.

9 The enclosure has not been found, but a woodcut of the Polish fowl appears in *Variation* 1: 229.

10 The letter is considered incomplete since it does not include a salutation or valediction. In addition, Quatrefages probably translated only a portion of the letter from CD for his reading to the Société d'Anthropologie, and some further text may be missing (see n. 1, above).

From La Société des Sciences Naturelles de Neuchâtel[1] 14 May 1863

College de Neuchatel
14 May 1863

La Société des Sciences naturelles
de | Neuchatel | Suisse

Desiring to be associated with those men who by their work and position contribute to the realisation of its proposed aim, the advancement of different aspects of natural history, has elected at its meeting of 7 January as a corresponding member | Mr Charles Darwin | and begs him to accept this nomination as a token of its high regard.[2]

In the name of the society, | the president | Louis Coulon
the vice president | fulfilling the function of foreign secretary | E. Desor

LS
DAR 230

1 For a transcription of this letter in the original French, see p. 419. This is a printed letter with the dates, signatures, and CD's name, added by hand.

2 CD's election as a corresponding member to the Société des Sciences Naturelles de Neuchâtel followed a proposal made by the vice-president Edouard Desor, at a meeting of the society on 5 December 1862, that the list of honorary members of the society be augmented to include 'several persons distinguished by their scientific merit'. Desor presented a list of candidates that included, in addition to CD, Charles Frédéric Martins, Ludwig Rütimeyer, and James Dwight Dana (*Bulletin Société des*

Sciences Naturelles de Neuchâtel 6 (1864): 271). The society was founded in December 1832 to provide a forum for scientific discussion (*Histoire de l'instruction publique dans le Canton de Neuchâtel*, p. 708).

From Armand de Quatrefages[1] 19 May [1863][2]

Paris
19 May 6⟨3⟩

Sir and dear colleague,

A thousand thanks for your photograph.[3] It has exceeded my expectation and I am curious to see what the adherents of absolute fixity, with whom I am currently fighting, will say about it.[4] I received it some days ago and would have liked to thank you immediately, but you know of the great affair into the midst of which I have been thrown. I have been totally preoccupied with *The trial of the jaw*, as Mr Carpenter calls it.[5] I have been deeply touched by the behaviour of your countrymen in these circumstances. It is impossible to bring to the examination of such a controversial question more candour, loyalty, and cordiality. I hope they will find us animated by the same sentiments. I found it both a duty and a pleasure to express to the Academy yesterday everything I feel in this regard.[6]

For 4 days I have had on my desk, addressed to you, a copy of two photographs of the famous jaw through absent-mindedness I forgot to give it to Mr Christy, who is taking a similar copy to Mr Falconer.[7] I thought I had already sent it to you, but this will mean only a short delay.

Farewell, dear Sir and colleague and thank you again | Yours most sincerely | De Quatrefages

DAR 175: 1

1 For a transcription of this letter in its original French, see p. 424.

2 The year is established by the relationship between this letter and the letter to Armand de Quatrefages, 14 May [1863].

3 CD sent Quatrefages two photographs of a skull of the niata breed of cattle. See letter to Armand de Quatrefages, 14 May [1863].

4 Quatrefages was engaged in a debate at the Société d'Anthropologie de Paris with André Sanson, who challenged him to cite a single species in which the action of the environment had produced anatomical modifications (see letter from Armand de Quatrefages, [28 March –] 11 April 1863 and n. 4). Quatrefages responded with CD's example of the niata cattle (*Journal of researches* 2d ed., pp. 145–6) and, on 2 July 1863, exhibited the photographs that CD had sent to him. On 16 July 1863, Quatrefages continued to expand his case for the environmental modification of species by reading from *Journal of researches* 2d ed., pp. 145–6, and from the letter to Armand de Quatrefages, [14 April 1863] (*Bulletin Société d'Anthropologie de Paris* 4 (1863): 350–3 and 376–80). For Sanson's response to Quatrefages, see *Bulletin Société d'Anthropologie de Paris* 4 (1863): 374–5 and 380–3.

5 William Benjamin Carpenter's phrase the 'trial of the jaw' was used to describe the inquiry into the authenticity of the fossil human jawbone and other remains discovered at Moulin-Quignon, near Abbeville, France, in March 1863 by Jacques Boucher de Perthes (Falconer *et al.* 1863, p. 423). See also n. 6, below. For a summary of this dispute, see the letter from J. D. Hooker, [7 May 1863], n. 5.

6 A conference of French and British scientists on the Moulin-Quignon archaeological finds took place in Paris and Abbeville from 9 to 13 May 1863. Quatrefages paid tribute to the British scientists at the

meeting of the Académie des Sciences in Paris on 18 May 1863 (*Comptes rendus hebdomadaires des séances de l'Académie des Sciences* 56 (1863): 933–5. The British scientists included Carpenter, Hugh Falconer, Joseph Prestwich, and George Busk. The French scientists included Quatrefages, Edouard Lartet, and Alphonse Milne Edwards. A summary of the conference proceedings was published in the *Athenæum*, 23 May 1863, p. 682. For a full report of the proceedings, see Falconer *et al.* 1863.

7 Henry Christy was an English ethnologist, archaeologist, and geologist who worked closely with the French palaeontologist and archaeologist Lartet; together they excavated a number of European prehistoric sites in the 1860s (*DNB*). Lartet and Falconer were both involved in determining the authenticity of the Moulin-Quignon jawbone (see nn. 5 and 6, above).

From Jacques Boucher de Perthes[1] 23 June 1863

Abbeville
23 June 1863

My dear and honoured Sir

Your letter of the 16th has given me great satisfaction and I thank you for it.[2] I will always hold it a great honour to be in correspondence with you who have obtained such a fine place in Science and whose ideas approach so closely to my own.

I do not know if you were aware of my book which appeared in 1838 under the title *de la creation, essai sur la progression des etres*.[3] it has been mostly ignored, because it did not read as an amusing book, but such as it is—I ask you to accept it. I do not have it here because few copies were printed. It has become rather rare but by means of this enclosed voucher you will be able at some stage to get someone to collect it in Paris. I also enclose the latest volume (from 1857 to 1860) of our Société d'Emulation, in which my antediluvian man appeared.[4] If you already have this volume it can be replaced by that of 1852 to 1857. All you have to do is change the dates on the voucher.

That unfortunate jaw has caused me much trouble following the discussions to which it has given rise.[5] It was the same 28 years ago when I exhibited my first axes.[6] Nevertheless then as today I was certain of whatever I proposed on these matters I guessed at nothing and when I assert something it is because I have seen it. I said a very long time ago that fossil man would be found and I repeat my prediction.[7] it will be the same as those axes that I found everywhere in England as in France, in short everywhere one looks for them. Fossil man is also there under our feet, but he is not wanted. As soon as a specimen is discovered, people seek to quarrel with it. The Moulin Quignon find has given rise to so many incredible stories!— people wanted it to have been buried by hoaxers although it was acknowledged that this was physically impossible because there was too much earth at its exhumation to render a fraud practicable, and moreover, unless one is blind or senseless, one can always see when a site has been disturbed Even the least experienced workman is not fooled, so then the axes became a target. It was said that they were forgeries— well! one is no truer than the other— It was possible to manufacture some axes at St Acheul which are fairly similar to those found in peat bogs, which are indeed fairly easily imitated, but this is not the case at Moulin Quignon. Their imitation would require a great deal of skill care and time, so that in order to make a profit the workmen would have to sell them for a great deal of money, and the normal price is 25 centimes (five sous). I wanted to be absolutely sure of this. I set up a competition in which I offered 20 Francs to the workman who could make one similar enough to those of Moulin Quignon to fool people. After very many tries, no one has come near to making even a passable axe.

Please accept, my dear sir, the expression of my high esteem and complete devotion. | J. Boucher de Perthes

I made out the voucher for the last 3 volumes of the memoirs of our society Please accept all three.

To Mr Charles Darwin—

DAR 160: 257

[1] For a transcription of this letter in its original French, see pp. 502–3.

[2] CD's letter has not been found.

[3] The reference is to the first volume of Boucher de Perthes 1838–41, a philosophical work in which the author elaborates a theory of the universal evolution of nature, owing much to the romantic tradition (*DSB*).

[4] The enclosure has not been found. Boucher de Perthes refers to the volume of the *Mémoires de la Société Impériale d'Emulation d'Abbeville* for 1857–60, which contained Boucher de Perthes 1860a. There is an unannotated reprint of Boucher de Perthes 1860a (Boucher de Perthes 1860b) in the Darwin Pamphlet Collection–CUL.

[5] The reference is to the human jaw found in March 1863 at the Moulin-Quignon quarry, near Abbeville, France, together with teeth and some artefacts, and reported by Boucher de Perthes. The jaw was believed by Boucher de Perthes, Armand de Quatrefages, and other French naturalists to date from the Post-Pliocene (Pleistocene) period. However, on examining the finds in April 1863, Joseph Prestwich, John Evans, and Hugh Falconer concluded that they were forgeries. Between 9 and 13 May, a committee of French and English naturalists met in Paris to discuss the question, and concluded that the implements and jaw were authentic, although several scientists remained unconvinced. See Grayson 1983, pp. 213–17, and Van Riper 1993, pp. 134–9.

[6] Boucher de Perthes first became involved in the excavations for human artefacts in the Somme valley around 1837, although artefacts were found there as early as 1830 (Grayson 1983, pp. 118–20).

[7] See, for example, Boucher de Perthes 1847–64, 1: 245. See also Grayson 1983, pp. 128 and 213.

APPENDIX II
Chronology 1863

This appendix contains a transcription of Darwin's 'Journal' for the year 1863. Darwin commenced his 'Journal' in August 1838 and continued to maintain it until December 1881. In this small notebook, measuring 3 inches by $4\frac{1}{2}$ inches, Darwin recorded the periods he was away from home, the progress and publication of his work, and important events in his family life.

The version published by Gavin de Beer as 'Darwin's Journal' (1959) was edited before the original 'Journal' had been found and relied upon a transcription made by an unknown copyist. The original, now in the Darwin Archive in Cambridge University Library (DAR 158), reveals that the copyist did not clearly distinguish between the various types of entries contained in the transcription.

From 1845 onward, Darwin recorded anything that pertained to his work (including his illnesses, since these accounted for time lost from work) on the left-hand pages of the 'Journal', while the periods he was away from home, and family events, were noted on the right-hand pages. In order to show clearly Darwin's deliberate separation of the types of entries he made in his 'Journal', the transcription has the left- and right-hand pages labelled.

All alterations, interlineations, additions, and the use of a different ink or pencil have been noted. In addition, the editors have inserted additional information relevant to CD's correspondence throughout this transcription of the 'Journal' for 1863. These interpolations are enclosed in square brackets to distinguish them from Darwin's own entries, the source of the information (either letters in the *Correspondence* or in the Darwin Archive, Cambridge University Library, or Emma Darwin's diary (DAR 242)) being given in footnotes.

[Left]

1863

[21 December 1862. Bud-variation.][1]

Jan 23^{d} Chapt. on Inheritance.[2] finished April 1st— took me $6\frac{1}{2}$ weeks time lost by illness & London—[3]

April 1st Ch. on Crossing & Sterility;[4] finished June 16th.— took me 8 weeks; time lost by Hartfield & illness &c.—[5]

June 16th.[6] Selection.—[7] July 20th finished

[Right][8]
1863
Feb 4 to 14th Stayed with Erasmus[9]
[8 February. Visited Thomas Henry Huxley.][10]
[11 February. Visited Joseph Dalton Hooker at Royal Botanic Gardens, Kew.][11]
[13 February. William Darwin Fox came to visit at Erasmus Alvey Darwin's house.][12]
[22 February. John Lubbock dined.][13]
[22 March. J. D. Hooker visited.][14]
[7 April. Edward Cresy lunched.][15]
April 27th to May 13th Hartfield & Leith Hill Place[16]
[16–23 July. Josiah Wedgwood III and family visited.][17]
[21–8? July. Erasmus Alvey Darwin visited.][18]
[22–5 July. Sefton West Strickland and Edmund Langton visited.][19]
[25–8 July. Hensleigh and Frances Emma Elizabeth Wedgwood, and Eva Mackintosh visited.][20]
[1 August. William came.][21]
[2 August. Edward Cresy came.][22]
Sept. 2^{d} Malvern Wells;[23] returned Oct. 14th— Ill, there.— Ill to end of year
[3 November. Dr Brinton came.][24]
[3 December. Dr. Brinton came.][25]

1 CD was preparing a draft of the manuscript of *Variation*, which was not published until 1868. He continued working on bud-variation in January 1863; this material eventually formed chapter 11 of *Variation* (see *Correspondence* vol. 10, Appendix II).

2 CD refers to material that eventually appeared in chapters 12 to 14 of *Variation* (*Variation* 2: 1–84).

3 CD was unwell from the start of the year (see letter to Asa Gray, 2 January [1863]), and stayed in London from 4 to 14 February, as he put it 'chiefly to see if change will do me good' (letter to J. D. Hooker, 30 January [1863]). Emma Darwin recorded in her diary (DAR 242) that, after a fortnight of good health, CD became ill in the last week of February, and he was ill several times during March.

4 CD refers to material that eventually appeared in chapters 15 to 19 of *Variation* (*Variation* 2: 85–191).

5 See n. 16, below.

6 '7' *over* '6'.

7 CD refers to material that eventually appeared in chapters 20 and 21 of *Variation* (*Variation* 2: 192–249).

8 At the bottom of the right-hand page, CD added three entries in pencil. Two of these related to visits by the physician William Jenner, on 10 April and 22 May; the third recorded, for the date 13 April, 'lost much', apparently referring to CD's weight. No reference to Jenner attending CD appears either in the correspondence, or in Emma Darwin's diary (DAR 242), before 1864; moreover, there is no record of any payment to Jenner before 1864 in CD's Classed account book (Down House MS). Visits by Jenner to CD are recorded in Emma Darwin's diary (DAR 242) for 10 April and 22 May 1864, and her diary also reports that CD weighed himself on 12 April 1864. This evidence suggests that CD added these three entries to the page of his journal for 1863 in mistake for the page for 1864; in consequence, they have not been transcribed here, but will appear in the chronology for 1864 (*Correspondence* vol. 12, Appendix II).

9 Erasmus Alvey Darwin lived at 6 Queen Anne Street, London.

10 See letter to T. H. Huxley, [8 February 1863].

11 Emma Darwin's diary (DAR 242) records a visit to Kew.

12 See letter from W. D. Fox, [11 February 1863].

[13] Emma Darwin's diary (DAR 242).

[14] See letter from J. D. Hooker, [24 March 1863].

[15] Emma Darwin's diary (DAR 242).

[16] CD and his family travelled to Hartfield Grove, Hartfield, Sussex, home of CD's brother-in-law, Charles Langton, on 27 April; they travelled on to Leith Hill Place, near Dorking, Surrey, home of Emma Darwin's brother, Josiah Wedgwood III, on 6 May, returning home on 13 May (see Emma Darwin's diary (DAR 242) and letter to J. D. Hooker, 23 April [1863]).

[17] Emma Darwin's diary (DAR 242).

[18] Emma Darwin's diary (DAR 242) records Erasmus's arrival at Down House on 21 July. He was still there on 25 July (see letter to W. E. Darwin, [25 July 1863]) and was probably among those recorded as returning to London in Emma's diary for 28 July.

[19] Emma Darwin's diary (DAR 242) and letter from Emma Darwin to William Erasmus Darwin, [14 July 1863], in DAR 219.1: 75.

[20] Emma Darwin's diary (DAR 242) and letter to W. E. Darwin, [25 July 1863].

[21] Emma Darwin's diary (DAR 242).

[22] Emma Darwin's diary (DAR 242).

[23] CD stayed at a rented house, Villa Nuova, while undergoing treatment at James Smith Ayerst's hydropathic establishment at Malvern Wells, Worcestershire (see letter to W. D. Fox, 4 [September 1863]). According to Emma Darwin's diary (DAR 242), CD travelled to Malvern Wells via London, arriving on 3 September; he left Malvern Wells for London on either 12 or 13 October, returning home on 14 October.

[24] Emma Darwin's diary (DAR 242). The reference is to William Brinton.

[25] See letter to J. D. Hooker, 5 [December 1863].

APPENDIX III
Diploma presented to Charles Darwin

In 1863, Darwin received the following diploma from the Royal Prussian Academy of Sciences in Berlin. Although not a letter in the conventional sense, it represents significant communication between Darwin and the Academy, and the citations in such diplomas often provide valuable indications of those aspects of Darwin's work that were considered worthy of honour. In view of this, it has been included here, together with a translation.

From the Royal Prussian Academy of Sciences, Berlin[1] 26 February 1863

REGIA SCIENTIARUM ACADEMIA BORUSSICA
Virum eruditissimum suisque titulis condecorandum

CAROLUM DARWIN

in Epistolarum de re litteraria commercium, cum classe potissimum physica et mathematica cooptat, et, quoties aliquid augendis et promovendis studiis optimis quibusque inserviens aut ab ipso repertum aut aliunde cognitum habebit, ut illud secum humaniter communicet, invitat.

Datum Berolini d. 26 Febr: a. MDCCCLXIII.

Classis physicae et mathematicae Secretarii:
Joannes Franciscus Encke[2]
Christianus Godofr. Ehrenberg[3]

[Translation]

THE ROYAL PRUSSIAN ACADEMY OF SCIENCES
Elects the most learned man, who is to be graced with its titles,

CHARLES DARWIN

to a correspondence on matters of learning, especially in the department of physics and mathematics, and, whenever he either himself discovers, or learns from elsewhere, of something serving the expansion and promotion of the best studies, it invites him kindly to communicate the same to itself.

Despatched at Berlin, 26 Febr: 1863

The secretaries of the department of physics and mathematics:
Johann Franz Encke[2]
Christian Gottfried Ehrenberg[3]

DAR 229: 50a

[1] The Royal Prussian Academy of Sciences in Berlin was founded in 1700. The list of corresponding members was limited to 200. For an account of the academy and its history, see Harnack 1900.

[2] Johann Franz Encke was professor of astronomy at the academy and at the University of Berlin, and director of the Berlin observatory (*DSB*).

[3] The comparative anatomist and microscopist Christian Gottfried Ehrenberg was professor of medicine at the University of Berlin (*DSB*). He corresponded with CD between 1844 and 1846 (see *Correspondence* vol. 3).

APPENDIX IV
Presentation list for 'Two forms in species of *Linum*'

Darwin's studies of dimorphism began in 1861 with a series of experimental crosses of the long- and short-styled forms of *Primula* (*Correspondence* vols. 9 and 10). These experiments marked the beginning of a sustained research project on dimorphism in many other plant genera, including *Linum*. Darwin sought to demonstrate that the existence of long- and short-styled forms was not mere variability, but an adaptation to ensure cross-pollination between the different forms. He sought to test his hypothesis by investigating the relative fertility of homomorphic and heteromorphic crosses (see *Correspondence* vol. 10, letter to J. D. Hooker, 22 [August 1862] and n. 16); these experiments reflected Darwin's continuing interest in the adaptation of flowers to particular modes of pollination, and were an extended investigation of his conclusion in *Orchids*, p. 359, that 'Nature ... abhors perpetual self-fertilisation'.

Darwin began experimenting on the functional meaning of the two forms in species of *Linum* in September 1861 ('Two forms in species of *Linum*', p. 70 (*Collected papers* 2: 93–4); see also *Correspondence* vol. 9, letter to Asa Gray, 16 September [1861]); however, the bulk of the research was conducted in the summer of 1862 ('Two forms in species of *Linum*', p. 71 (*Collected papers* 2: 95); see also *Correspondence* vol. 10, letter to Asa Gray, 14 July [1862], and letter to J. D. Hooker, 22 [August 1862]). Darwin's experimental notes on *Linum* are no longer extant. The paper was written between 11 and 21 December 1862 (see *Correspondence* vol. 10, Appendix II). In his letter to Asa Gray of 2 January [1863], Darwin described his study of dimorphism in *Linum* as a 'much better case than Primula'. The paper was due to be read before the Linnean Society on 5 February 1863, and Darwin travelled to London the preceding day to read it himself; however, illness prevented him from attending the meeting (see letter from W. B. Tegetmeier, 18 February 1863, and letter to W. B. Tegetmeier, 19 February [1863]). The paper was published in the botanical section of the society's *Journal of Proceedings*, in a number issued to members on 13 May 1863 (*General index to the Journal of the Linnean Society*, p. vi), and Darwin ordered extra offprint copies from the society's printers; there is an entry in his Classed account book (Down House MS) for 18 February 1864 recording a payment of 19*s.* 6*d.* to Taylor & Francis (printers of the *Journal of Proceedings*) representing the cost of approximately thirty additional copies. These he began to distribute soon after the middle of April (see, for example, letter to Asa Gray, 20 April [1863]).

In his botanical work, Darwin relied heavily on the help of individuals from whom he could obtain specimens, notes, observations, or practical or bibliographic information. Consequently, the names on the presentation list include a number of gardeners, horticulturists, and plant collectors throughout the world; also included are people likely to review his work, family members and friends, and British and foreign scientists he wished to keep apprised of his research. Many of his closest friends and colleagues, for example, Joseph Dalton Hooker and Thomas Henry Huxley, do not appear on the list: as fellows of the Linnean Society, they would have received the paper with their copies of the society's journal. Not all of the individuals to whom Darwin sent a copy of the paper appear on this list: Roland Trimen was sent a copy later in the year (see letter to Roland Trimen, 27 August [1863]), and Fritz Müller received one in November 1865 (see letter from Fritz Müller, 5 November 1865 (*Calendar* no. 4929a)).

The manuscript list is in DAR 210.11: 23, and is reproduced below as close to its original format as possible, with all alterations, interlineations and additions noted. The symbol ✘ denotes a cross in ink. All other marks and alterations are recorded in footnotes in the positions in which they occur, and are in ink unless otherwise stated.

The list is followed by a biographical register, arranged alphabetically, identifying each recipient and, where possible, indicating why Darwin would have sent the individuals concerned a presentation copy of 'Two forms in species of *Linum*'. Individuals not otherwise mentioned in the volume are only identified here and are not included in the main Biographical register and index to correspondents.

Linum

✘ I.. Henderson Henry[1] Hay Lodge. Trinity Edinbg[2]
✘ J. Scott Bot. G. Edinbg
✘ W. C. Crocker South St Chichester.[3]
✘ Th Rivers. Sawbridgeworth Herts—
✘ J. Salter Versailles now Hammersmith
✘ W. Turner[4] Slough
✘ Naudin 3d France
✘ Heer 3d viâ France Hofmeister[5]
✘ Alphe De Candolle
✘ Planchon— Weddell ✘ Austria 4d.[6]
✘ Quatrefages.
✘ Mr Hugh Gower
✘✘ Asa Gray
✘ A. R Wallace ?[7]
✘ Prof Treviranus. Bonn. *[Preface]* 3d.
✘ Dr Hermann Cruger. Bot. Garden Trinidad 3d[8]
✘ J. Anderson Esq Meadow Bank. Uddingstone
✘ Prof. Alex Braun— Berlin.— 3d. via France
✘ Oliver wants copy

✗ A. Rawson 24[9]
✗ Berkeley
✗ D[r] Brown Sequard 25 Cavendish Sq
✗ Sir H. Holland
✗ William Darwin
✗ Editors of Journ of Hort[10]
✗ E. Cresy
Babington[12]
✗ Dickie
✗ Balfour
✗ Egerton Sir P.
✗ Leighton
✗ J. R. Greene
Claparede[16]
✗ H C. Watson
✗ Duke of Argyll.[17]
A. G. More. Bembridge[18]
✗ Thwaites

Jamieson of Quito[11]
✗ Dana
A. M[13]
✗ M. Professor Charles Martins[14]
Montpellier
Hérault
France[15]

1 CD misspelled Isaac Anderson-Henry's name.
2 'Edinbg' interlined.
3 'South St' interlined.
4 Charles Turner.
5 Written in pencil and deleted in ink.
6 Deleted.
7 '?' deleted.
8 'Bot. Garden' interlined.
9 Written in brown crayon.
10 The editors of the *Journal of Horticulture* were George William Johnson and Robert Hogg.
11 Deleted. CD misspelled William Jameson's name.
12 Deleted.
13 Probably Andrew Murray.
14 'Professor' interlined.
15 'Montpellier . . . France' deleted.
16 Deleted.
17 George Douglas Campbell, eighth duke of Argyll.
18 Deleted.

Biographical register of recipients of presentation copies of 'Two forms in species of *Linum*'

Anderson, James (1831/2–99). Scottish gardener. Gardener to iron and shipping agent Thomas Dawson at Uddingston in Lanarkshire; later a landscape gardener in Glasgow and Manchester. (R. Desmond 1994.) Anderson provided CD with information on and specimens of *Cattleya crispa* and *Dendrobium cretaceum* (see 'Fertilization of orchids', p. 152 (*Collected papers* 2: 149); see also letters from James Anderson, 1 April 1863 and 18 June 1863).

Anderson-Henry, Isaac (1800–84). Scottish solicitor and horticulturist. Bred new and rare plant varieties in his garden at Hay Lodge, Edinburgh (see Anderson[-Henry] 1853 and Anderson-Henry 1863). Anderson-Henry had an established interest in *Linum* and intended to undertake further experiments (see letter from Isaac Anderson-Henry, 17 January 1863); his name also appears on the presentation list for 'Dimorphic condition in *Primula*' (see *Correspondence* vol. 10, Appendix III).

Babington, Charles Cardale (1808–95). Botanist. In 1861, succeeded John Stevens Henslow as professor of botany at the University of Cambridge; an expert on plant taxonomy. (*DNB*, *DSB*.) Babington provided CD with seed for his study of polymorphism (see *Correspondence* vol. 10, letters to C. C. Babington, 20 January [1862], 1 February [1862], and 2 September [1862], and letters from C. C. Babington, 17 January 1862, 30 January 1862, and 16 October 1862); CD sent Babington copies of 'Dimorphic condition in *Primula*' and *Orchids* (see *Correspondence* vol. 10, Appendixes III and IV).

Balfour, John Hutton (1808–84). Physician and botanist. Professor of medicine and botany at the University of Edinburgh and Regius keeper of the botanic garden from 1845. (*DNB*, *DSB*, *Medical directory for Scotland* 1863.) CD sent Balfour copies of *Orchids* and 'Dimorphic condition in *Primula*' (see *Correspondence* vol. 10, Appendixes III and IV).

Berkeley, Miles Joseph (1803–89). Clergyman and botanist. Perpetual curate of Apethorpe and Wood Newton, Northamptonshire, 1833–68. An expert on British fungi, he was also an authority on British algae and mosses. (*DNB*, *DSB*.) Berkeley was a regular contributor to the *Gardeners' Chronicle*, and CD cited several of his observations on pollination and crossing varieties in *Variation* (see, for example, *Variation* 1: 397 and 400). In 1862, he published a highly favourable review of *Orchids* in the *London Review* ([Berkeley] 1862).

Braun, Alexander Carl Heinrich (1805–77). German botanist. Professor of botany and director of the botanic garden at the University of Berlin since 1851. (*DSB*, *NDB*.) He was deeply influenced by *Naturphilosophie*, and his main studies were on plant morphology, physiology, and the structure of cryptogamic plants. CD had sent Braun a presentation copy of *Orchids* (see *Correspondence* vol. 10, Appendix IV).

Brown-Séquard, Charles Edouard (1817–94). French physiologist. Pioneer in endocrinological and neurological research. Practised medicine in the United States and France before moving to London in 1858. (*DBF*, *DSB*, *London and provincial medical directory* 1863.) Brown-Séquard was the editor of the *Journal de la Physiologie de l'Homme et des Animaux*, and CD had encouraged him to review the French translation of *Origin* (Royer 1862; see *Correspondence* vol. 10, letters to C. E. Brown-Séquard, 2 January [1862] and 16 April [1862], and letter from

Brown-Séquard, Charles Edouard, cont.
C. E. Brown-Séquard, 13 January 1862). Brown-Séquard's name also appears on the presentation lists for 'Dimorphic condition in *Primula*' and *Orchids* (see *Correspondence* vol. 10, Appendixes III and IV).

Campbell, George Douglas, 8th duke of Argyll (1823–1900). Whig politician and author of works on science, religion, and politics. A defender of the concept of design in nature. (*DNB*.) Campbell discussed *Origin* in his presidential address to the Royal Society of Edinburgh in December 1860 (Campbell 1860) and, in 1862, reviewed *Orchids* for the *Edinburgh Review* (Campbell 1862); CD sent him presentation copies of Asa Gray's pamphlet on natural selection and natural theology (A. Gray 1861; see *Correspondence* vol. 9, Appendix III), and *Orchids* (see *Correspondence* vol. 10, Appendix IV).

Candolle, Alphonse de (1806–93). Swiss botanist. Professor of botany and director of the botanic gardens, Geneva, 1835–50. (*DSB*.) CD had been greatly interested by Candolle's work on the geographical distribution of plants (A. de Candolle 1855), and had sent him presentation copies of *Origin* (see *Correspondence* vol. 8, Appendix III) and Asa Gray's pamphlet on natural selection and natural theology (A. Gray 1861) (see *Correspondence* vol. 9, Appendix III). Candolle had recently sent CD his monograph on oaks and related genera (see letter to Alphonse de Candolle, 14 January [1863] and A. de Candolle 1862a). Candolle's name appears on the presentation lists for 'Dimorphic condition in *Primula*' and *Orchids* (see *Correspondence* vol. 10, Appendixes III and IV).

Claparède, Jean Louis René Antoine Edouard (Edouard) (1832–71). Swiss zoologist. Professor of comparative anatomy, Academy of Geneva, 1862. (*Dictionnaire historique & biographique de la Suisse*.) Claparède published a favourable review of *Origin* in 1861 (see letter to George Bentham, 15 April [1863]), and in 1862 assisted with the French translation of *Origin* by Clémence Auguste Royer (see *Correspondence* vol. 10, letter from Edouard Claparède, 6 September 1862). CD sent Claparède a copy of *Orchids* (see *Correspondence* vol. 10, Appendix IV).

Cresy, Edward (1824–70). Surveyor and civil engineer. Principal assistant clerk at the Metropolitan Board of Works. Founder member of the Geologists' Association, 1858 (*Annual Report of the Geologists' Association* 1859–70, *DNB* s.v. Cresy, Edward (1792–1858)). Cresy, a neighbour and family friend of the Darwins, assisted CD with his experiments on insectivorous plants in the autumn of 1860 (see *Correspondence* vol. 8, letters from A. W. von Hofmann to Edward Cresy, 13 October 1860 and 27 October 1860, and letters to Edward Cresy, 14 October [1860] and 2 November [1860]). CD sent Cresy copies of the second and third editions of *Origin* (see *Correspondence* vol. 8, Appendix III, and *Correspondence* vol. 9, Appendix VII), and his name appears on the presentation lists for 'Dimorphic condition in *Primula*' and *Orchids* (see *Correspondence* vol. 10, Appendixes III and IV).

Crocker, Charles William (1832–68). Foreman of the propagating department, Royal Botanic Gardens, Kew, 1857–61/2. (*Correspondence* vol. 10, letter to J. D. Hooker, [before 15 February 1862], letter from J. D. Hooker, [before

15 February 1862], and letter from C. W. Crocker, 17 February 1862; R. Desmond 1994.) Crocker had carried out experiments for CD on crossing varieties and pollination (see *Correspondence* vol. 9, letters to C. W. Crocker, 18 May [1861] and 1 June [1861], letter to Daniel Oliver, 27 May [1861], and letter to J. D. Hooker, 11 October [1861]). By 1862 Crocker was growing a number of plants for experiments including *Linum grandiflorum* and *L. flavum*, and CD may have discussed the progress of his own research with him (see *Correspondence* vol. 10, letter from C. W. Crocker, 31 October 1862; see also 'Two forms in species of *Linum*', p. 81 (*Collected papers* 2: 104)). Crocker's name appears on the presentation lists for 'Dimorphic condition in *Primula*' and *Orchids* (see *Correspondence* vol. 10, Appendixes III and IV).

Crüger, Hermann (1818–64). German pharmacist and botanist. Government botanist and director of the botanic garden in Trinidad from 1857. (R. Desmond 1994, S[chlechtendal] 1864.) Crüger provided CD with information on orchids and melastomads (see letters to Hermann Crüger, 25 January [1863] and 25 May [1863], and letters from Hermann Crüger, 23 February 1863 and 23 April 1863).

Dana, James Dwight (1813–95). American geologist and zoologist. Professor of natural history, Yale University, 1856–64; of geology and mineralogy, 1864–90. An editor of the *American Journal of Science and Arts* from 1846. (*DAB*, *DSB*.) CD sent Dana a presentation copy of *Origin* in 1859 (see *Correspondence* vol. 8, Appendix III), and in 1863 Dana sent CD a copy of his *Manual of geology* (Dana 1863a).

Darwin, William Erasmus (1839–1914). CD's eldest child. Banker. (*Alum. Cantab.*) William carried out botanical experiments for CD, who encouraged his interest in botany (see *Correspondence* vol. 9, letter to J. D. Hooker, 19 June [1861], and *Correspondence* vol. 10, letters to W. E. Darwin, 14 February [1862] and 4 [July 1862]). William's name appears on the presentation lists for *Origin* (see *Correspondence* vol. 8, Appendix III), 'Dimorphic condition in *Primula*' and *Orchids* (see *Correspondence* vol. 10, Appendixes III and IV).

Dickie, George (1812–82). Scottish botanist. Professor of botany, University of Aberdeen, 1860–77. Specialist on marine algae. (R. Desmond 1994, *DNB*.) Dickie had supplied CD with orchid specimens and made observations on his behalf (see *Correspondence* vol. 9, letter to George Dickie, [5 July 1861], *Correspondence* vol. 10, letter from George Dickie, 30 May 1862, and *Orchids*, p. 152). Dickie's name is on the presentation lists for 'Dimorphic condition in *Primula*' and *Orchids* (see *Correspondence* vol. 10, Appendixes III and IV).

Egerton, Philip de Malpas Grey-, 10th baronet (1806–81). Palaeontologist. Specialised in fossil fish. Tory MP for South Cheshire, 1835–68. (*British imperial calendar* 1863, *DNB*, Sarjeant 1980.) Egerton was a prominent member of several London scientific societies and was a long-standing acquaintance of CD's. His name appears on the presentation list for 'Dimorphic condition in *Primula*' (see *Correspondence* vol. 10, Appendix III).

Gower, William Hugh (1835–94). Nurseryman. Foreman at the Royal Botanic Gardens, Kew. (R. Desmond 1994.) Gower had provided CD with information

Gower, William Hugh, cont.
about pollination in *Victoria regia*, the giant waterlily at Kew (see *Correspondence* vol. 9, letter from W. H. Gower, 23 November 1861), and selected specimens for CD's botanical research (see *Correspondence* vol. 9, letter to J. D. Hooker, [30 and 31 January 1861], *Correspondence* vol. 10, letter to J. D. Hooker, 30 [June 1862], and letter from Daniel Oliver, 28 July 1862, and this volume, letter to J. D. Hooker, 15 February [1863]). Gower's name also appears on the presentation list for 'Dimorphic condition in *Primula*' (see *Correspondence* vol. 10, Appendix III).

Gray, Asa (1810–88). American botanist. Fisher Professor of natural history at Harvard University since 1842. (*DAB*, *DSB*.) A regular correspondent, collaborator and adviser to CD; superintended the production of the American edition of *Origin* (see *Correspondence* vol. 8), and assisted CD in his investigation of the occurrence of dimorphism in plants (see *Correspondence* vols. 9 and 10). Gray had received presentation copies of *Origin* (see *Correspondence* vol. 8, Appendix III, and *Correspondence* vol. 9, Appendix VII), 'Dimorphic condition in *Primula*' and *Orchids* (see *Correspondence* vol. 10, Appendixes III and IV).

Greene, Joseph Reay (*fl.* 1857–77). Irish zoologist. Professor of natural history, Queen's College, Cork, 1858–77. Specialist on the Hydrozoa and the fauna of the south and west coasts of Ireland. One of the editors of *Natural History Review*. (*Catalogue of graduates, University of Dublin*; Praeger 1949.) Greene's name appears on the presentation list for *Orchids* (see *Correspondence* vol. 10, Appendix IV).

Heer, Oswald (1809–83). Swiss botanist, palaeontologist, and entomologist. An expert on Tertiary flora. Lecturer on botany, University of Zürich, 1834–5; director of the botanic garden, 1834; associate professor, 1835–52; professor of botany and entomology, 1852–83. (*DSB*, *NDB*.) CD had been greatly interested in Heer's work on the geographical distribution of fossil plants, and in his theory of the sudden production of new species as described in Heer 1855–9 (see *Correspondence* vols. 9 and 10, and this volume, letter to Asa Gray, 31 May [1863]). Heer's name appears on the presentation list for *Origin* (see *Correspondence* vol. 8, Appendix III).

Hofmeister, Wilhelm Friedrich Benedikt (1824–77). German botanist. Expert on the embryology of flowering plants. Appointed professor of botany and director of the botanic garden at the University of Heidelberg in 1863. (*ADB*, *DSB*, Goebel 1926, *NDB*.) In his letter to CD, 14 April 1863, Daniel Oliver referred to Hofmeister as an authority on the physiology of the Primulaceae.

Hogg, Robert (1818–97). Scottish horticulturalist. From 1855, co-editor of the *Cottage Gardener* (the *Journal of Horticulture and Cottage Gardener* from 1861). (R. Desmond 1994.) CD frequently published observational notes and requests for assistance in the journal (see *Correspondence* vols. 9–11).

Holland, Henry, 1st baronet (1788–1873). Physician. Distant relative of the Darwins and Wedgwoods. Physician in ordinary to Prince Albert, 1840, and to Queen Victoria, 1852. President of the Royal Institution for many years. (*DNB*, *Physicians*.) CD consulted Holland on matters of family health, and sent him presentation copies of *Origin* (see *Correspondence* vol. 8, Appendix III), *Orchids*, and 'Dimorphic condition in *Primula*' (see *Correspondence* vol. 10, Appendixes III and IV).

Jameson, William (1796–1873). Scottish botanist. Director of the mint in Ecuador and professor of chemistry and botany at the University of Quito. (*DNB*.) Jameson collected plants in South America; CD wrote to him for information about Melastomataceae (see letters from Isaac Anderson-Henry, 26–7 January 1863 and 17 April 1863).

Johnson, George William (1802–86). Co-editor of the *Cottage Gardener*, which he founded in 1848 (the *Journal of Horticulture and Cottage Gardener* from 1861). (R. Desmond 1994.) CD frequently published observational notes and requests for information and assistance in the journal (see *Correspondence* vols. 9–11).

Leighton, William Allport (1805–89). Botanist, clergyman, and antiquary. School-fellow of CD's in Shrewsbury, 1817. Curate of St Giles's, Shrewsbury, 1846–8. Published a flora of Shropshire (1841). (*Alum. Cantab.*, R. Desmond 1994, *DNB*, Freeman 1978.) Leighton supplied CD with specimens for his experiments on dimorphism, and his name is on the presentation list for 'Dimorphic condition in *Primula*' (see *Correspondence* vol. 10, letter to W. A. Leighton, 26 November [1862] and Appendix III).

Martins, Charles Frédéric (1806–89). French meteorologist and botanist. Professor of botany, University of Montpellier, 1846. (*NBU*, *Dictionnaire universel des contemporains*.) CD hoped to receive specimens of silk worms from Martins (see letter from J. L. A. Quatrefages, 25 March 1863).

More, Alexander Goodman (1830–95). Naturalist. Lived at Bembridge on the Isle of Wight before moving in the 1860s to Ireland, where he made a study of the island's flora and worked in the Dublin Natural History Museum. (R. Desmond 1994, Gilbert 1977, Moffat ed. 1898, *Natural Science* 6 (1895): 351.) More provided CD with several specimens of orchids, and was sent a presentation copy of *Orchids* in 1862 (see *Orchids*, pp. 67, 95 n., 99 n., and 101 n., *Correspondence* vol. 9, and *Correspondence* vol. 10, Appendix IV).

Murray, Andrew (1812–78). Scottish botanist and entomologist. Assistant secretary to the Royal Horticultural Society of London from 1860. In entomology, specialised in the Coleoptera; in botany, in the Coniferae. (R. Desmond 1994.) Murray published a critical review of *Origin* (Murray 1860). CD sent Murray presentation copies of *Origin* in 1861 (see *Correspondence* vol. 9, Appendix VII) and 'Dimorphic condition in *Primula*' in 1862 (see *Correspondence* vol. 10, Appendix III).

Naudin, Charles Victor (1815–99). French botanist. Aide-naturaliste at the Muséum d'Histoire Naturelle from 1854. Experimented widely on plants, particularly on acclimatisation and hybridisation. (*DSB*.) Naudin provided CD with information on crossing varieties and hybrids (see letter from J. D. Hooker, 24 January 1863, and letter to C. V. Naudin, 7 February 1863). Naudin's name appears on the presentation lists for *Orchids* and 'Dimorphic condition in *Primula*' (see *Correspondence* vol. 10, Appendixes III and IV).

Oliver, Daniel (1830–1916). Botanist. Assistant in the herbarium and librarian at the Royal Botanic Gardens, Kew. Professor of botany at University College London from 1861. (R. Desmond 1994, *List of the Linnean Society of London* 1859–91.) Oliver regularly provided specimens, corroborated observations, and undertook bibliographic searches for CD (see, for example, *Correspondence* vol. 10, letters to Daniel Oliver, [before 11 June 1862] and 13 October [1862], and letters to J. D. Hooker, 11 June [1862] and 6 October [1862]). Oliver wrote reviews of both 'Dimorphic condition in *Primula*' ([Oliver] 1862b) and 'Two forms in species of *Linum*' ([Oliver] 1863d) for the *Natural History Review*.

Planchon, Jules Emile (1823–88). French botanist. Professor of botany at the faculty of sciences and director of the school of pharmacy at Montpellier since 1853. (R. Desmond 1994, *Grande encyclopédie*.) CD cited Planchon's monograph on *Linum* (Planchon 1847–8) in 'Two forms in species of *Linum*', p. 81 (*Collected papers* 2: 103–4), and in *Forms of flowers*, p. 100, where he referred to Planchon as the only botanist to have inferred that 'heterostylism might have some important functional bearing'.

Quatrefages de Bréau, Jean Louis Armand de (Armand de Quatrefages) (1810–92). French physician, zoologist, and anthropologist. Professor of the natural history of man at the Muséum d'Histoire Naturelle since 1855. (*DSB*.) CD and Quatrefages were regular correspondents and frequently exchanged copies of their respective publications (see *Correspondence* vols. 5, 7–9, and this volume, letters from Armand de Quatrefages, 29 March 1863 and 11 April 1863). Quatrefages's name appears on the presentation list for *Origin* (see *Correspondence* vol. 8, Appendix III).

Rawson, Arthur (1818–91). Perpetual curate of Trinity Church, Bromley, Kent, 1843–82. A keen gardener who bred florists' flowers, particularly pelargoniums. (*Alum. Cantab.*, R. Desmond 1994.) Rawson corresponded with CD on the subject of hybridisation, and sent him a number of specimens (see letters from Arthur Rawson, 1 April [1863] and [6 April 1863]; *Variation* 2: 139–40). Rawson's name appears on the presentation list for 'Dimorphic condition in *Primula*' (see *Correspondence* vol. 10, Appendix III).

Rivers, Thomas (1798–1877). Nurseryman in Sawbridgeworth, Hertfordshire. Author of several works on roses and fruit-growing and a regular contributor to gardening journals. (R. Desmond 1994, *DNB*.) Rivers provided CD with information on variation and persistency of character in fruit-trees and roses (see letters from Thomas Rivers, 21 January 1863, 26 January 1863, and 30 January 1863, and *Variation* 1: 332–84 *passim*), and on inheritance in seedling trees (see letter from Thomas Rivers, [3 February 1863], and *Variation* 2: 18–19).

Salter, John (1798–1874). Nurseryman and florist. Conducted a nursery for the sale of English flowers at Versailles, 1838–48. Established a florist shop specialising in chrysanthemums in Williams Street, Hammersmith, *circa* 1863. (R. Desmond 1994, *Gardener's year-book* 1864, *Post office directory of the six home counties* 1862–6.) A regular contributor to gardening journals, Salter provided CD with information on pelargoniums, chrysanthemums, and species of *Phlox* (see letter to John Scott, 21 January [1863], and *Variation* 1: 378–84 *passim*).

Scott, John (1836–80). Scottish botanist. Gardener at Chatsworth House, Derbyshire, before becoming foreman of the propagating department at the Royal Botanic Garden, Edinburgh, in 1859. (R. Desmond 1994, Freeman 1978, *Transactions of the Botanical Society* [*of Edinburgh*] 14 (1883): 160–1.) CD discussed his research on *Linum* in correspondence with Scott (see *Correspondence* vol. 10, letters to John Scott, 3 December [1862] and 11 December [1862], and this volume, letter from John Scott, 21 January 1863). Scott was engaged in a study of dimorphism in *Primula* (see Scott 1864a).

Thwaites, George Henry Kendrick (1811–82). Botanist and entomologist. Director of the Peradeniya botanic gardens in Ceylon from 1857. (R. Desmond 1994, *DNB*.) Thwaites corresponded with CD on the subject of dimorphism, and sent specimens and conducted experiments and observations on CD's behalf (see *Correspondence* vol. 10, letter to G. H. K. Thwaites, 15 June 1862, and letter from G. H. K. Thwaites, 20 June [1862]; see also *Forms of flowers*, pp. 116 and 286). Thwaites's name appears on the presentation list for *Origin* (see *Correspondence* vol. 8, Appendix III).

Treviranus, Ludolph Christian (1779–1864). German botanist. Professor of botany at the University of Bonn from 1830. (*ADB*, *DSB*.) Treviranus's review of *Origin* from a botanical perspective (Treviranus 1861) was the first to appear in Germany; he had also intended to translate *Orchids* (see *Correspondence* vol. 10, letter from H. G. Bronn, 21 June 1862), and later wrote a review (Treviranus 1863b) of Heinrich Georg Bronn's translation (Bronn trans. 1862). See also letter from L. C. Treviranus, 12 February 1863.

Turner, Charles (1818–85). Nurseryman and florist. Proprietor of the Royal Nurseries in Slough and Salt Hill, Buckinghamshire. Edited the *Florist*, 1851–60. (R. Desmond 1994, *Gardener's year-book* 1864.) Provided CD with information on crossing hollyhocks (see letter to Charles Turner, [22 April – 16 June 1863], and *Variation* 2: 107–8).

Wallace, Alfred Russel (1823–1913). Naturalist, lecturer, and author. Collected natural history specimens in the Amazon, 1848–52, and in the Malay Archipelago from 1854. Independently formulated a theory of natural selection in 1858. Returned to England in 1862. (*DNB*, *DSB*.) Wallace's name appears on the presentation lists for *Origin* (see *Correspondence* vol. 8, Appendix III) and *Orchids* (see *Correspondence* vol. 10, Appendix IV).

Watson, Hewett Cottrell (1804–81). Botanist, phytogeographer, and phrenologist. Published guides to the distribution of British plants. (R. Desmond 1994, *DNB*, *DSB*.) A supporter of the idea of the progressive development of plant species, Watson corresponded with CD over several years about the geographical distribution of species and varieties (see *Correspondence* vols. 6–10). His name

Watson, Hewett Cottrell, cont.
appears on the presentation lists for *Origin* (see *Correspondence* vol. 8, Appendix III, and *Correspondence* vol. 9, Appendix VII), 'Dimorphic condition in *Primula*', and *Orchids* (see *Correspondence* vol. 10, Appendixes III and IV).

Weddell, Hugh Algernon (1819–77). British-born botanist and physician who worked in France. Specialist on South American flora. (R. Desmond 1994.) In 1861, Asa Gray informed CD that Weddell had observed dimorphism in *Impatiens* and *Cinchona* (see *Correspondence* vol. 9, letter from Asa Gray, 11 October 1861, and letter to J. D. Hooker, 25 November [1861]). CD sent Weddell presentation copies of *Origin* (see *Correspondence* vol. 8, Appendix III), *Orchids*, and 'Dimorphic condition in *Primula*' (see *Correspondence* vol. 10, Appendixes III and IV).

APPENDIX V
Donald Beaton's responses to Charles Darwin's letters to the *Journal of Horticulture*

On 27 January 1863, the *Journal of Horticulture and Cottage Gardener* published a letter from Darwin discussing experiments made by the German physician and botanist Karl Friedrich von Gärtner. Gärtner's experiments indicated that pollen from one variety of pea could exert a direct influence on the colour and form of the seeds of another variety of pea. Darwin's letter solicited further information on analogous cases from Donald Beaton, a regular contributor to the journal (see letter to *Journal of Horticulture and Cottage Gardener* [before 27 January 1863] and n. 4). Between 21 December 1862 and January 1863 Darwin had been preparing a draft of a chapter on bud-variation for *Variation* (see 'Journal' (Appendix II)). This chapter incorporated a section on the 'direct or immediate action of the Male Element on the Mother Form' (*Variation* 1: 397–403), and included a discussion of Gärtner's observations (see Gärtner 1849, pp. 80–7). Beaton's response, which appeared in the issue of the *Journal of Horticulture* for 27 January 1863, flatly contradicted Darwin's statement that the direct influence of pollen of one species on the seed and fruit of another species was 'well-ascertained'; he was particularly scathing about Darwin's reliance on the experimental work of Gärtner to substantiate the statement. Darwin replied in his letter to the *Journal of Horticulture and Cottage Gardener* [before 3 February 1863], both to defend his original statement, and to defend the reputation of Gärtner as a reliable experimentalist.

Darwin's original request for information from Beaton could be seen as a straightforward attempt to collect further examples of a phenomenon he had been writing about from a horticultural journal that he used regularly. Darwin had been taking the *Cottage Gardener*, which became the *Journal of Horticulture, Cottage Gardener, and Country Gentleman* in 1861, since about 1860. The journal proved so suited to his needs that he decided to buy back-issues (see 'CD's Reading notebooks' (DAR 128: 153) and *Correspondence* vol. 4, Appendix IV); he regarded it as a better journal than the *Gardeners' Chronicle*, and used it to provide material for his work on the variation of plants and animals under domestication (see *Correspondence* vol. 9, letter to J. D. Hooker, 23 [April 1861] and n. 23). Darwin's copies of the *Cottage Gardener* and the *Journal of Horticulture*, an imperfect set running from 1855 to 1865, are in the Darwin Library–CUL. He numbered the items of interest to him in a sequence from 1 to 92, keyed to his own index, which was kept with item 92. The copies, particularly some of Beaton's articles, are heavily annotated, suggesting both that the series was frequently used by Darwin and that he was familiar with Beaton's contributions.

Beaton had been a regular contributor to the journal, which appeared weekly, since 1848. As Beaton was a celebrated gardener, an expert on crossing garden flowers, who took a keen interest in botanical science, it is not surprising that he should have engaged Darwin's attention. Darwin wrote to Joseph Dalton Hooker in May 1861 that he had been going through the *Cottage Gardener* chiefly on account of Beaton's articles, adding: 'he strikes me as a clever, but d—d cock-sure man', who 'has made me indignant by the way he speaks of Gärtner, evidently knowing nothing of his work.—' (see *Correspondence* vol. 9, letter to J. D. Hooker, 14 May [1861]).

Darwin's correspondence suggests that his letter to the *Journal of Horticulture* in January 1863 represented more than a simple request for factual information. In 1861 he told Hooker: '[Beaton] alludes to me occasionally & I cannot tell with what spirit. He speaks of "*this* M^r. Darwin" in one place, as if I were a very noxious animal'; and: 'I mean to try & pump him in Cot. Gard. & shall perhaps defend Gartner.—' (see *Correspondence* vol. 9, letter to J. D. Hooker, 14 May [1861]). Darwin was evidently irritated by the fact that Beaton did 'not much venerate botanical authorities' (*Correspondence* vol. 9, letter to *Journal of Horticulture*, [before 9 July 1861]). Maintaining Gärtner's reputation as a reliable observer and experimentalist was important to Darwin. Anxious to demonstrate that sterility was not a 'universal & infallible criterion of species' (*Correspondence* vol. 8, letter to Charles Lyell, 12 [February 1860]), he had invested Gärtner's crossing experiments with great theoretical significance (see *Correspondence* vol. 10, Appendix VI). Among the most important of these experiments were those published in Gärtner 1844 and 1849, demonstrating that the offspring of similarly coloured varieties of the same or different species of *Verbascum* were less sterile when crossed than were those of differently coloured varieties. In *Origin*, pp. 270–1, Darwin had cited these experiments as evidence that there was no essential distinction between species and varieties, but that there was a gradation of sterility in nature.

On the day Darwin wrote to Hooker about Beaton, the first of Darwin's attempts to 'pump' Beaton appeared in the *Journal of Horticulture*: a polite request asking Beaton to state whether varieties of the same species of composite plants frequently cross by insect agency or by other means (*Correspondence* vol. 9, letter to *Journal of Horticulture*, [before 14 May 1861]). Beaton's reply, which was also published in the issue of 14 May 1861, confirmed Darwin's initial impression that Beaton was 'not to be trusted' (*Correspondence* vol. 9, letter to J. D. Hooker, 18 [May 1861]):

> He does not well know his own subject of crossing. He has learnt largely from Herbert, but exaggerates all Herberts opinions grossly. Says not one plant in a thousand is ever fertilised by its own pollen! gives some exceptions & these are false & so on.— Coolly assumes that all Botanists have blundered & that it is an entire mistake that pollen-grains emit a tube which penetrates the stigma!!!—

Darwin continued his dialogue with Beaton in his letters to the *Journal of Horticulture*, [17 May 1861] and [before 9 July 1861] (*Correspondence* vol. 9). In these letters

Darwin cast doubt on the accuracy of many of Beaton's published observations and statements, drawing attention to inconsistencies. In private correspondence, Darwin sought corroboration of Beaton's statements from other experienced plant hybridisers, and was quick to point out cases where experimental observations did not correspond with Beaton's assertions (see letter to Isaac Anderson-Henry, 20 January [1863], and letter to John Scott, 25 and 28 May [1863] and n. 7).

Darwin's opinion of Beaton was at variance with that of others. William Herbert, an undisputed authority on plant hybridisation, held a high opinion of Beaton, describing him as 'a most intelligent gardener' and naming the genus *Beatonia* in his honour (Herbert 1846, p. 97; R. Desmond 1994). John Claudius Loudon, editor of the *Gardener's Magazine*, the journal in which Beaton began his writing career, described him as the author of a 'most original paper' and a man worthy of a 'first-rate situation' (*Gardener's Magazine* 13 (1837): 205). Maxwell Tylden Masters, editor of the *Gardeners' Chronicle*, stated that Beaton had a 'first-rate reputation as a practical gardener, and was esteemed for his shrewdness and humour' (*ML* 2: 269 n.). By the 1850s and 1860s, with his vast experience of the horticultural scene in Britain, Beaton was widely regarded as a leading authority in the field. Even Darwin, although not without irony, referred to him as 'the great M^r^. Beaton' (*Correspondence* vol. 9, letter to Charles William Crocker, 18 May [1861]). As his obituarist in the *Journal of Horticulture* pointed out, Beaton had been 'in the van of English horticulture' for 'upwards of thirty years' and for many years he was (*Journal of Horticulture*, n.s. 5 (1863): 349):

> the leader of that branch of it which more immediately concerns the flower garden. To Mr. Beaton we are mainly indebted for the direction that has been given to the modern style of English flower gardening, . . . and it is generally allowed that through his articles, as published periodically in the pages of this Journal, his fine taste and skill in the harmonising of colours have exercised an influence which has operated in all the best garden establishments in the country. . . . Although he has not taken a position among botanists, he was no mean proficient in that science, and there is, perhaps, not another example on record in the history of our British gardening of one who applied that science more skilfully and beneficially to the practice of horticulture.

On 3 February 1863 Beaton wrote an apology to Darwin for having contradicted him. It was to be the last item Beaton wrote for publication; he suffered a stroke soon afterwards, and following a second attack of paralysis, died in October 1863. There is no extant private correspondence between Beaton and Darwin, although Darwin appears to have sent Beaton a presentation copy of 'Dimorphic condition in *Primula*' (see *Correspondence* vol. 10, Appendix III). As a supplement to the correspondence about Beaton's articles, Beaton's responses to Darwin's letters to the *Journal of Horticulture* in 1863 are printed below, in chronological order of their publication.

Beaton's first response to the letter from Charles Darwin to the *Journal of Horticulture and Cottage Gardener*, [before 27 January 1863][1]

In answering Mr. Darwin's question, allow me, first, to clear myself of any participation in his opening remark, that "Few facts in vegetable physiology are more remarkable than the well-ascertained influence of the pollen of one species or variety on the seed and fruit of another species or variety while still attached to the female plant."[2] Gärtner never proved that—he only asserted it; and when he was pushed to the proof, he lowered his sails, made a second edition of his great work, and confessed many of his errors.[3]

The most practical cross-breeder who has yet appeared has stated "Gärtner's report of the cross-bred seed he *has obtained*, to be nothing but a mere enumeration of the crosses he *has tried to obtain*."[4] And with regard to very many of the cases of impregnation mentioned by Gärtner, he, the said cross-breeder, otherwise Dr. Herbert, "utterly repudiates the probability of such impregnation;"[5] and well he might.

It was not Gärtner, but Dr. Wiegman, in 1823, who first said he found the Pea changed colour from being planted along with Vicia sativa, or common Vetch;[6] and Gärtner, two years subsequently, said he caused the same change by means of the pollen;[7] and our Mr. Knight was somewhat smitten with that doctrine.[8]

I had a commission to work over, again and again, every experiment mentioned by Gärtner, Wiegman, and Lageret, and I found over and over again each experiment was without a base.[9] Others proved the same, but it remained for the late Professor Henslow to prove by scientific investigation that the pollen has no visible effect on the seed impregnated;[10] and no cross-breeder of any practice in England at the present day would like to have his name associated with that of Gärtner, for or against any exploit in crossing. Nevertheless, I am firmly of the opinion that Gärtner was right in his belief of the *way* impregnation is effected.[11]

I forget the plant I mentioned last year as having the penduncle, or stalk, of the flower affected by pollen.[12] Of all the plants I recollect having mentioned, for the last twelve months, there are only five genera that are affected that way—Corbularia and Hermione among the Narcissi, and Erodium, Pelargonium, and Geranium, in that class of plants. The best generic distinction for upholding Corbularia and Hermione as distinct from the true Narcissi, is that the peduncle rises slowly from a horizontal to an upright posture as the seeds ripen in the pod. The Erodiums and a section of Pelargoniums, with, I believe, all the true Geraniums, have the peduncle affected differently from the great mass of Pelargoniums.

It is probable, however, that Mr. Darwin has been thinking of what was said on the different stages of impregnation at page 330, of Vol. XXVII., the part for this time last year.[13] And I think Gärtner's idea of how the pollen acts is proved both by the Erodiums,[14] including Pelargonium, which is not a natural genus, and the true Geraniums, inasmuch as the footstalk of each flower in all that mass of species, and in their endless varieties, is the first part that is affected by the pollen. Indeed, I am quite certain of that, for I had at least ten thousand proofs of it. But let me explain.

Take any of the true species out of all the Geraniaceæ, cut off the stigma before the pollen can intrude, and the peduncle of that flower will retain its original or natural posture till it decays. A Tom Thumb, or any of the bedders, will prove that very shortly. Take another flower in the same truss, and cross it with the pollen of another species which you know will not cross with it, and the peduncle will turn to the opposite of its natural posture, and will never regain the true position, but die ultimately, as the process of fertilisation went no farther than to affect the footstalk. The next stage is the quickening of the pod, the next that of the lobes of the seeds, and all this may be, and yet no life be given to the embryo of the seeds.

Early next May any one may influence a hundred flowers of the Scarlet Defiance Geranium as far as the footstalk of its flowers, but no farther. Every flower of that one kind which is crossed by its own pollen, or by foreign pollen, will, in eight or ten hours, reverse the posture of the footstalks; but the seedpod is not reached by the contents of the pollen, and the consequence is no seed and no rising of the peduncle. There are many seedlings which will prove the same as Defiance at the end of October, and through November.

Notwithstanding these facts, one can conceive a state of climate which would render the effect of that same pollen active through the three or four stages of impregnation, and produce fertile seed from Defiance. Indeed, the pollen of some kinds or varieties, which will only affect the footstalk very early and very late in the season, will effect a cross in the middle of July; and I have obtained, and I believe I am the only one who has done so, a real cross from Scarlet Defiance, which is over fifteen years old; but I may be mistaken. The case of Cybister, or "The Tumbler" Nosegay Geranium, is a different sport altogether, the truss comes naturally in a reversed position, and takes from three weeks to a month to make a right round change of position.—D. Beaton.

Journal of Horticulture n.s. 4 (1863): 70–1; Darwin Library–CUL

CD ANNOTATIONS
1.5 Gärtner] *after double quotation marks, pencil*
1.5 asserted it;] *before double quotation marks, pencil*
1.6 second edition of his great work] *underl pencil*
1.7 errors.] *before double quotation marks, pencil*
3.1 It was not] *after double quotation marks, pencil*
4.5 and no cross-breeder . . . in crossing 4.7] *scored, pencil*
7.1 Take any . . . seeds. 7.9] *scored, pencil*

[1] This is a reply to CD's letter to the *Journal of Horticulture and Cottage Gardener*, [before 27 January 1863]; the exchange appeared in the issue of the *Journal of Horticulture* for 27 January 1863, pp. 70–1, under the heading 'Influence of pollen on the appearance of seed'; CD's annotated copy of the journal is in the Darwin Library–CUL.

[2] See letter to the *Journal of Horticulture and Cottage Gardener*, [before 27 January 1863].

[3] Beaton refers to the work of the German botanist, Karl Friedrich von Gärtner; however, it is not clear to which of Gärtner's publications he refers. Beaton did not read German and appears to have been familiar with Gärtner's work through the writings of others, particularly William Herbert (see nn. 4, 5, 6, and 7, below); in his final response to CD's letters (letters to the *Journal of Horticulture and Cottage Gardener*, [before 27 January 1863] and [before 3 February 1863]), he stated: 'I only wish I could read . . . [Gärtner] in the original, or that he had a good practical English reviewer instead of a highly scientific one.' In Gärtner's first report of experimental results, published in a paper in *Naturwissenschaftliche Abhandlungen: Herausgegeben von einer Gesellschaft in Würtemberg* (Gärtner 1826), he explicitly stated (p. 61) that pollen from one species or variety applied to the mother-plant of another species or variety had no direct influence on the seeds; however, in 1849, in his last publication, *Versuche und Beobachtungen über die Bastarderzeugung im Pflanzenreich* (Gärtner 1849), an account of twenty-five years of experimental research, Gärtner reversed his view, having repeated the trials, and conceded that in fact some seeds did change in colour and form (see n. 7, below). Gärtner 1849 was a corrected and extended edition of a treatise on hybridisation honoured by the Royal Dutch Academy of Sciences in 1837 (Gärtner 1838). For an account of the revisions of Gärtner's 'great work', see the foreword to Gärtner 1849, pp. iii–xii.

[4] Beaton refers to William Herbert's discussion of Gärtner 1826 in Herbert 1837, pp. 348–52; the passage quoted appears on p. 349. Herbert 1837 is a treatise on the classification of the Amaryllidaceae; however, a discussion on hybridisation is appended, entitled 'On crosses and hybrid intermixtures in vegetables' (pp. 335–80).

[5] Herbert 1837, p. 351. Herbert was referring to Gärtner's claim to have successfully crossed *Datura laevis* and *Nicotiana rustica*, announced in Gärtner 1826, p. 58.

[6] In his account of his hybridisation research, Arend Friedrich Wiegmann stated that, as a result of cross-pollination between peas and vetches, the leguminous seeds unmistakably changed colour and form (Wiegmann 1828, p. 23). These experimental trials were conducted in Braunschweig in the summer of 1823 (Wiegmann 1828, pp. 14, 15). Beaton's information probably comes from Herbert 1837, p. 352, where it is stated that Wiegmann's trials were conducted in 1823 (see nn. 3 and 4, above).

[7] Beaton is mistaken (see n. 3, above). In Gärtner's first report of his experimental research, conducted in Calw in Würtemberg in the summer of 1825 (Gärtner 1826, p. 36, Herbert 1837, p. 348), Gärtner stated (Gärtner 1826, p. 61):

> Der Einfluss des fremden Pollens ändert demnach nichts in den der Mutterpflanze eigenthümlichen Formen und äusserlichen Eigenschaften der Früchte, der Saamen und selbst des Embryo; er erzeugt in diesem letzteren nur die Fähigkeit, durch das Keimen und die fernere Entwicklung der neuen Pflanze eine aus beyden concurrirenden Arten innig gemischte Form der Theile hervorzubringen. [The impact of foreign pollen thus changes nothing in the shape and appearance of the fruit, the seed or even the embryo, all of which remain as is common in the mother plant. All the seed does is to create in the embryo the capability of developing a new plant which is a mix of the two competing species, by way of germination and further growth.]

However, in Gärtner 1849, he conceded that in fact some seeds did change when differently coloured varieties of garden peas were crossed, as Wiegmann had described (Gärtner 1849, p. 84). Gärtner did not accept Wiegmann's statement that peas and vetches could be crossed. See also CD's letter to the *Journal of Horticulture and Cottage Gardener*, [before 3 February 1863].

[8] Thomas Andrew Knight discussed the subject of crossbreeding and the influence of pollen on the seed coats and cotyledons of peas in Knight 1823.

[9] The references are to the work of the French agronomist and horticulturalist Augustin Sageret (Sageret 1826), Wiegmann (see n. 6, above), and Gärtner (see n. 3, above). Beaton refers to the experiments he conducted in the 1830s while employed by William Gordon of Haffield House, Herefordshire. The results of this research do not appear to have been published; however, in a brief autobiographical sketch, Beaton reported that Gordon 'was a practical botanist, a great lover of experiments', and that under his patronage Beaton 'crossed all sorts of plants' and one year 'had more than an acre of cross seedlings' (*Cottage Gardener* 13 (1854): 157).

[10] The reference has not been identified. Until his death in 1861, John Stevens Henslow had been professor of botany at the University of Cambridge (*DNB*).

[11] Gärtner did not provide a detailed account of the processes involved in pollination and fertilisation; in the foreword to *Beiträge zur Kenntniss der Befruchtung durch der vollkommeneren Gewächse*, he apologised for his failure, as a result of the poor state of his eyesight, to conduct thorough anatomical examinations using a microscope (Gärtner 1844, p. vi). See also n. 13, below.

[12] In his letter to the *Journal of Horticulture and Cottage Gardener*, [before 27 January 1863], CD asked Beaton if he would repeat a statement concerning the effect of applying the pollen of one species of plant on the position of the footstalk of the seed-capsule of a different species. The passage referred to appeared in an article on crossing flowers in the *Cottage Gardener* in 1860. Beaton stated that by crossing *Imatophyllum miniatum* (the pods of which stand as erect as its umbellate flowers) with *I. cyrtaniflorum* (the pods of which hang down like its flowers), the pods of the former become as pendant as those of the latter (*Cottage Gardener* 24 (1860): 254).

[13] In his article in the *Journal of Horticulture* n.s. 2 (1862): 329–30, 'The process of fertilisation', Beaton argued that, though we are apt to think of it as one event, there are 'more stages than one in the process of fertilisation'.

[14] See nn. 11 and 13, above.

Beaton's second response to the letter from Charles Darwin to the *Journal of Horticulture and Cottage Gardener*, [before 27 January 1863][1]

I wrote the answer to Mr. Darwin so hurriedly,[2] that I made a mistake or two, which might lead readers into greater errors. Thus, "I had a commission to work over, again and again, every experiment (for changing the colour of Peas) mentioned by Gärtner and Weigmann," and the absurd assertion of "Sageret," about crossing between the Cabbage and Horseradish.[3]

I did not try over again all the experiments they put on record; at least, not under that commission. That was as far back as 1833, 1834, and 1835,[4] and the conclusion was, that Dr. Weigmann was not even aware that the garden Pea could not be crossed by any other Pea or plant whatever, without artificial means, and very likely there is no want of such opinions at the present day.

From thirty to forty trials in each of those years, Gärtner's assertion that he changed the colour of a Pea by pollen was proved to be wrong, and yet he very probably had a different-coloured Pea in the mother pod as he says;[5] and if so, that was caused by a natural sport, more than one-half, if not every one, of our present race of Peas being natural sports induced by cultivation probably. But, in some cases, the change would appear to be induced by some chemical constituent of the soil in which the plants are grown, for I know a soil which will change the bean of the Scarlet Runner to a jet black three times out of five sowings, and, no doubt, many gardeners have noticed the same effect in this and in other seedlings of their own rearing.

Again, when I said Pelargonium "is not a natural genus," I meant not botanically, but for the purposes of the hybridiser, in its strict meaning—that is to say, that all the species of Pelargonium could not cross together without destroying Erodium, because there is a section of Pelargoniums in which the species are affected in the peduncle by the pollen, exactly as the species of Erodium, and the species of Erodium are as differently affected in the peduncle from those of Pelargonium as the affection itself is different from the more usual course of nature. In the "usual course" impregnation would seem to be only one single process, which it certainly is not in many plants.

To explain the affection to people who know little of such subjects, let me say that a truss of such flowers is like the foot of a Game cock; the claws are the peduncles, or stalks, which support the individual flowers in a head, truss, or bunch of flowers. When the pollen affects one of such claws, that claw is, as it were, paralysed, and is drawn-in under the foot of the bird, the nail on the claw being the flower, but it is then the seed part. Now, if the effect of the pollen reached the young seed, or ovum, and put life into it, the claw would, in time, be restored very gradually from the seeming paralysis to its natural position, which position would be gained just as the seed was fit to sow, but not just quite ripe for harvest. It is a very pretty phenomenon; but being as common to gardeners as covering cold pits, few of them care aught about it.

That way of drawing-in the claws under the foot one after the other as soon as each is affected by the pollen is peculiar to about two-thirds of the species of Pelargonium only. The other third of the species are differently affected, and in the same way as all Erodiums are. With these, the Erodiums, the joint of the leg above the claws, or what you might call the ankle, would seem to be out of joint, and to pull up the claws, one at a time, straight up against the leg; and by the time the seed would be ripe each claw slackens from the rigid strain, and finally turns up to take the original spread-out form as the seeds become ripe. But without the pollen, the claws or peduncles would never change out of the original and natural direction.

Now, would it not seem the oddest of the doings of Nature to undo the joint of the footstalk of a flower as the first result from the effect of the pollen, if that effect merely passed through the style to the seed in a long tube, as it is said to do? But when you have five hundred wild species, and a multitude of seedling flowers, which go directly to prove the footstalk out of joint, or affected by pollen, although the seeds of none of that vast number of plants had been touched, or affected at all, the mind of man could hardly conceive a greater error in natural history than the way they say the pollen reaches the seed; and yet nine-tenths, if not all, the learned professors of botany of this generation believe that way firmly.

Lastly, when I said I obtained a real cross from "Scarlet Defiance, which is over fifteen years old; but I may be mistaken," I ought to have said that I believed Defiance was fifteen years old, though I might be mistaken about its age, not mistaken as to the cross, as the sentence might be taken to imply.

I would ask, in return, Has any gardener obtained a real cross from this Defiance during the last fifteen years, and if so, what is the name of the seedling and who let it out? It may be of some use to cross-breeders to know the reason why I ask this.

It took me eight years to obtain that cross, and yet they say I ought to know all about such ways. Some years I touched over a hundred of its flowers without a fertile result, and yet I was quite sure the pollen took a certain degree of effect; for the footstalk of a great number of the flowers touched relapsed to the perpendicular downward position, but some did not, because the pollen I made use of was foreign to the kind.

If a plant is absolutely barren the footstalk cannot be thus affected; and when we say a plant is barren because we cannot force a seedling from it, we may be saying what is not the fact. Of one thing, I think, we may be sure in a seedling of Geraniaceæ, if the peduncle yields to the effect of pollen—namely, that that seedling is not naturally barren, though we may fail to force it to seed. And that is what I wish that cross-breeders should bear in mind, and therefore endeavour by a change of treatment—such as a warmer or a colder climate, a different soil, and a differently aged plant, with different degrees of health and strength in the plants—to obtain the desired cross.

Even should all attempts fail us, we may rest assured the cross is obtainable in some other locality or country differing from our own.

Then, if that be true of Geraniaceæ, and I can see no cause to doubt it, there can hardly be a question about its not being an isolated fact in one order of plants only; although we cannot discern it in those orders where the process of fecundation is positive, or effected at one stage of the process.

I said, long since, that the surest way to effect a difficult cross in Geraniaceæ was to subject the mother, previous to attempting impregnation, to a sudden check, and that was how I overcame the inertness of Scarlet Defiance, although I failed for seven years by that move; but another move seems to have been equally necessary. The plant, for the eighteen months preceding the period of that seeding, was kept at, or as near, the starvation-point as could be without killing it; the poorest sandy soil, and the smallest pot to cramp the roots in, with no more water than to lift the leaves after they all flagged, was the treatment repeated through two summers.

The fact is, that plant was a giant sport, and it was necessary to reduce the giant to the level of its ordinary kindred before it could seed by the pollen of any one of them; and I have found it necessary to reduce it considerably in strength before I could seed it by its own pollen, which makes me anxious to learn if it has been found more easy to effect by any one else whose soil and treatment might be very different from mine.

I am supposed, by many, to know more about crossing than many others, which obliges me to be so much the more particular about everything I have to say on the subject. But every one of our readers who has a turn for crossing knows just as much about it as I do, for I know nothing of it which is not down in these volumes, and some may know much more of it; for all the cross-breeders over the country have, each one, their own peculiar breeds to work on, and, of course, every one knows something, practically, which no one else can know so well for the want of practice in that particular breed. One thing, however, we all know, and that is, how very little the best of us really does know about it. Depend upon it, there are scores who have never crossed twenty kinds of plants in all their practice, who can tell you twice as much about it as the most practical of us who have been at it all our gardening lifetime. They can do every conceivable turn in it without going out of doors; we, very little indeed, even if we devote our whole time to it. But the greatest difference between them and us is, that they can account for every effect produced, which enables them to foretel events in crossing; while we cannot reach farther than our experience places before our eyes. D. Beaton.

Journal of Horticulture n.s. 4 (1863): 88–9; Darwin Library–CUL

CD ANNOTATIONS
6.1 That way of . . . may be mistaken," 8.2] *scored, pencil*
15.1 The fact . . . from mine. 15.6] *scored, pencil*

1 This is Beaton's second response to CD's letter to the *Journal of Horticulture and Cottage Gardener*, [before 27 January 1863]; it was published in the issue of 3 February 1863, pp. 88–9, under the heading 'Cross-breeding geraniums'; CD's annotated copy of the journal is in the Darwin Library–CUL.
2 See Donald Beaton's first response to CD's letter to the *Journal of Horticulture and Cottage Gardener*, [before 27 January 1863], above.
3 The references are to Gärtner 1826, Wiegmann 1828, and Sageret 1826, p. 297. Beaton's information about hybridisation trials on the continent appears to come primarily from William Herbert (see Donald Beaton's first response to CD's letter to the *Journal of Horticulture and Cottage Gardener*, [before 27 January 1863] and nn. 3, 4, 5, 6, and 7, and his final response, below). The work of Gärtner, Wiegmann, and Sageret is discussed in Herbert 1837, pp. 348–54
4 See n. 3, above. Between 1829 and 1837, Beaton was employed by William Gordon of Haffield House, near Ledbury, Herefordshire (see Beaton's first response and n. 9, above).
5 It is not clear to which of Gärtner's publications Beaton refers (see Beaton's first response and nn. 3 and 7, above).

Beaton's final response to the letters from Charles Darwin to the *Journal of Horticulture and Cottage Gardener*, [before 27 January 1863] and [before 3 February 1863][1]

The above[2] is a most interesting fact, bearing directly on the question of changing the colour of the garden Pea by the influence of the pollen;[3] and the two cases, as far as I know, stand isolated, the one in the vegetable and the other in the animal kingdom. The other fact adduced goes directly against the inference he draws from it, for to bring it within the analogy it is not the offspring of the white woman that would or should be the dark-skinned, but the white woman's own skin. It is not the offspring of the crossed Pea or second generation which the believers in Gärtner's cross acknowledge, but the Pea itself.[4] The Pea was crossed, and the cross changed that Pea, not the offspring of the Pea, although we may infer the offspring to be of the same colour as the parent. By that analogy it would be necessary to adopt the notion that the male influence in the animal kingdom changed the colour of the skin or coat of the female, not the skin or coat of their offspring, or second generation. Now, if we analyse the more plausible side of the question which is thus raised—the egg changed by the influence of the male, we shall be able, without any stretch of idea, to dispose of it as easily as of the other. The egg in the feathered race represents by analogy the pea-pod, not the body of the Pea; and to have a bearing on the question in hand, Gärtner's experiment would need to change the colour of the pod, not of the Pea. But the fact is, we can trace no direct analogy between the modes of fecundation in the two kingdoms; all we know is, the effect of the mode is the direct opposite in the two. Repeated approaches of consanguinity is a well-known cause of degeneracy in the animal kingdom, and the surest way of improving races in the vegetable kingdom—that is, improving flowers and fruits; and yet when we push the inquiry or the mode of improving to its ultimate, we find, or, at least, we have already found, that these improvements have been acquired at the expense of the health of that strain or race ere we reached the limits of the ultimatum or final stage of improvement, as florists could tell us from their records of their different "worts," every one of which has been getting less and more less healthy, and by consequence more difficult to preserve, from the first cross by in-and-in breeding.

This seems the best place for me to make a suitable apology to Mr. Darwin, and to ask a thousand pardons for my seeming contradiction; but I had not the slightest idea even of contradicting him, much less of discourtesy. I wrote hurriedly and very earnestly for a particular reason—well knowing that every word Mr. Darwin says carries great influence with it to the ends of the earth. I knew Gärtner was the least reliable of all the old authors; that he published the greatest absurdities about it, above all who wrote before him; that his first edition was the text-book of all who wrote upon crossing without experience in it themselves;[5] and that it would be a thousand pities if Mr. Darwin lent the power of his authority to Gärtner's speculations, which have not advanced the art one single degree. It cannot possibly be an attack on an author for another to expose errors he might have fallen into. I have no idea, or any wish, but to speak the truth only about this author, when I affirm that nine-tenths of his nine thousand recorded experiments are nothing better than idle speculations.[6] I am satisfied there are not nine of them out of nine hundred which would come exactly alike next year or next trial the same season, and such might be recorded a thousand-fold without helping the art of crossing to advance one single step. As the work of an amiable naturalist, and as pleasant reading, I have no doubt but Gärtner will be read as long as Sir Walter Scott.[7] I only wish I could read him in the original, or that he had a good practical English reviewer instead of a highly scientific one.—[8] D. Beaton.

Journal of Horticulture n.s. 5 (1863): 415–16

1 This was the last item written by Beaton for publication in the *Journal of Horticulture*; it appeared under the heading 'The late Mr. Donald Beaton' in the *Journal of Horticulture*, 24 November 1863, pp. 415–16). Beaton died on 31 October 1863. Robert Hogg, one of the co-editors of the *Journal of Horticulture*, became the executor of Beaton's estate; he found this 'very interesting relique' among his deceased friend's papers, apparently the last contribution Beaton had the power of writing: 'It was written in answer to a note forwarded to him. Paralysis seized him before he had returned it to us'. The note was from a reader of the *Journal of Horticulture* (see n. 2, below).

2 Beaton refers to a note he had received from a reader of the *Journal of Horticulture* (see n. 1, above). The note was reproduced above Beaton's letter (*Journal of Horticulture* n.s. 5 (1863): 415):

> I never cross-bred a plant intentionally in my life, so I am not about mounting a pillar like Simon Stylites, and proclaiming myself an authority, but I have been a cross-breeder of poultry, and have observed facts there which make me hesitate from agreeing that it is 'absurd' to think that the skins of Peas may be altered in colour by the pollen which fertilised them. For instance: Cochin-China hens lay buff-coloured eggs, but if they are associated with a Dorking or Spanish cock the shells of their eggs frequently, after a time, are laid nearly white. Again, another fact we all have observed, a white woman has white children, but if her husband is a black man her offspring becomes dark-skinned.— *February 3rd*, 1863.

3 See CD's letters to the *Journal of Horticulture and Cottage Gardener*, [before 27 January 1863] and [before 3 February 1863]. See also Beaton's responses, above.

4 Gärtner 1849, p. 84. See also Donald Beaton's first response and nn. 3 and 7, above.

5 Beaton may refer to Gärtner 1826 or 1838.

6 Beaton refers to Gärtner 1849. In the letter to the *Journal of Horticulture and Cottage Gardener*, [before 3 February 1863], CD referred to Gärtner's 'great and last work', which he said contained 'in 790 closely-printed pages the detailed results of nine thousand distinct experiments in crossing, together with admirable observations on the whole subject of hybridisation'. Gärtner 1849 was a corrected and extended edition of a treatise on hybridisation honoured by the Royal Dutch Academy of Sciences in 1837 (Gärtner 1838).

7 The reference is to the Scottish poet and novelist Walter Scott, author of several celebrated books of poetry and the 'Waverley novels' (*DNB*).

8 Probably a reference to William Herbert's discussion of Gärtner's work in Herbert 1837, pp. 348–52 (see Beaton's first response and nn. 3 and 4, above).

APPENDIX VI
Darwin's lists of hothouse plants

Towards the end of 1862, Darwin resolved to build a small hothouse at Down House, for 'experimental purposes' (see *Correspondence* vol. 10, letter to J. D. Hooker, 24 December [1862], and this volume, letter to Thomas Rivers, 15 January 1863). The decision was evidently prompted by his growing engagement in botanical experimentation, and the building of the hothouse early in 1863 marked something of a milestone in Darwin's botanical work, since it greatly increased the range of plants that he could keep for scientific investigations. Of course, Darwin had been engaged in botanical experiments over many years, and there had been a greenhouse, evidently built at least in part for his botanical work, at Down House since the winter of 1855–6 (see CD's Classed account book (Down House MS) and *Correspondence* vol. 5, letter to J. D. Hooker, 19 April [1855]). Darwin became increasingly involved in botanical experiments in the years after the publication of *Origin*, and made more frequent references in correspondence to his use of the greenhouse for experiments (see *Correspondence* vols. 8–10). Though his greenhouse was probably heated to some extent, Darwin found himself on several occasions in need of what he called a 'hothouse', meaning a construction suitable for tropical plants. In 1861 and 1862, while preparing *Orchids*, he was allowed 'free use' of the hothouses and some of the tropical plants of his friend and neighbour George Henry Turnbull, and he relied again on Turnbull's hothouses in his experiments, begun in 1861, on the Melastomataceae, a family of tropical and sub-tropical plants (see *Orchids*, p. 158 n., and *Correspondence* vols. 9 and 10). However, he found it increasingly necessary to ask the botanists at the Royal Botanic Gardens, Kew, and private individuals with collections of hothouse plants to make observations and even experiments on his behalf.

Darwin's decision to build a hothouse was thus obviously prompted by his increasing use of plants for a variety of experimental purposes. The particular spur seems to have been his desire to grow the tropical plant *Oxalis sensitiva*, in order to investigate the effects of various substances on its sensitivity to touch (see *Correspondence* vol. 10, letter to J. D. Hooker, 12 [December 1862] and n. 13). Initially, Darwin purchased for this purpose a glass plant-case, heated by means of a dish that had to be filled with hot water twice daily. However, either because he found this to be unsatisfactory, or because he simply decided to work on a larger scale, he soon determined to build a hothouse as well. He was persuaded to do so by John Horwood, the gardener of his neighbour George Turnbull; Horwood had assisted him in the use of his employer's hothouses over the previous two years. In a letter of 24 December [1862] (*Correspondence* vol. 10) Darwin told Hooker:

> I have almost resolved to build a small hot-house: my neighbours really first-rate gardener has suggested it & offered to make me plans & see that it is well done, & he is a really a clever fellow, who wins lots of prizes & is very observant. He believes that we sh[d] succeed with a little patience; it will be grand amusement for me to experiment with plants.—

According to John Claudius Loudon's *Encyclopedia of gardening* (Loudon 1835), a copy of which Darwin signed in 1841 (see the copy in the Darwin Library–CUL), the word 'hothouse' had two meanings. It was applied generally to all forms of protected plant accommodation, including frames, glass cases, greenhouses, orangeries, conservatories, dry-stoves, and moist- or bark-stoves (p. 1012). More particularly, it was synonymous with 'bark-stove', a construction designed to house those plants that required 'the highest degree of heat' (p. 1100). The latter was the sense in which Darwin used the word.

The building of the hothouse was begun in mid-January, and completed by mid-February (see letters to J. D. Hooker, 13 January [1863] and 15 February [1863]). It was built by William Ledger, a builder from the nearby village of Hayes, and cost a total of £85 11*s*. 1*d*.; this included £22 5*s*. for Horwood, who superintended the operation, and £9 15*s*. 10*d*. for fittings purchased in July (see CD's Classed account book and his Account book–cash accounts (Down House MS)). When it was completed, Darwin told Turnbull that without Horwood's aid he would 'never have had spirit to undertake it', and that, if he had had, he would 'probably have made a mess of it' (letter to G. H. Turnbull, [16? February 1863]).

Even before work on the hothouse started, however, Darwin began making preparations to stock it with plants for use in a wide variety of experiments. He told Hooker that he was 'looking with much pleasure at catalogues to see what plants to get', adding 'I shall keep to curious & experimental plants' (letter to J. D. Hooker, 13 January [1863]). Darwin apparently refers to the catalogues of a local nurseryman, John Cattell of Westerham, with whom he had dealt over many years. In his letter to Hooker, Darwin mentioned that he hoped to be able to borrow 'some few orchids' from the Royal Botanic Gardens, Kew, dreading the 'awful sums' that he imagined they would cost to buy. Hooker's response was unequivocal: 'You will give me *deadly* offence if you do not send me your Catalogue of the plants you want before going to Nurserymen' (letter from J. D. Hooker, [15 January 1863]). Darwin agreed to send Hooker his list of 'experimental' plants, while observing: 'anything ornamental[,] which, however, I shall avoid[,] of course I must not have from Kew' (letter to J. D. Hooker, 30 January [1863]).

Darwin probably gave his list of plants to Hooker when he visited Kew on 11 February (see Emma Darwin's diary (DAR 242)). Four days later, the construction was completed and he was impatiently waiting to hear what plants Hooker could let him have, telling him: 'I long to stock it, just like a school-boy' (letter to J. D. Hooker, 15 February [1863]). On 20 February, the plants from Kew had arrived.

Darwin was delighted, telling Hooker: 'I am fairly astounded at their number! why my hot-house is almost full!— I have not yet even looked out their names; but I can see several things which I wished for, but which I did not like to ask for' (letter to J. D. Hooker, [21 February 1863]). He had, he confessed to Hooker, 'stewed so long admiring the plants' that he developed a headache. What he did not confess to Hooker was that he had also been in a state of 'agitation' because his gardener, Henry Lettington, had potted up the temperamental and valuable tropical orchids in 'common earth'. Horwood had had to be sent for 'instanter', 'to undo them & pot them as they like' in a particular mixture of moss, peat, and charcoal (see the letter from Henrietta Emma Darwin to William Erasmus Darwin, [22 February 1863] in DAR 210.6: 109). There were other teething problems, and Hooker had to warn Darwin not to make 'boiled greens' of his plants, proffering further advice on cultivation (see letter from J. D. Hooker, [6 March 1863]).

Darwin derived enormous pleasure from his hothouse plants, telling Hooker: 'Henrietta & I go & gloat over them; but we privately confessed to each other, that if they were not our own, perhaps we sh^d. not see such transcendent beauty in each leaf' (letter to J. D. Hooker, 24[–5] February [1863]). Darwin's aesthetic appreciation of the tropical plants was probably partly due to the recollections they inspired of his own travels in the tropics. Even before he left on the *Beagle* voyage, Darwin used the hothouses in the Cambridge University botanic garden to envision the tropics (see *Correspondence* vol. 1, letter to Caroline Darwin, [28 April 1831]), and when, on the *Beagle*, he heard about the building of a hothouse at his father's house in Shrewsbury, he commented: 'how when I return I shall enjoy seeing some of my old friends again' (*Correspondence* vol. 1, letter to Catherine Darwin, May–June [1832]). Years later, the great hothouses at Chatsworth House, Derbyshire, made him 'thrill with delight at old recollections' of the tropics (*Correspondence* vol. 3, letter to Charles Lyell, 8 October [1845]).

Having indulged his senses, Darwin soon began the more serious work of listing the plants, seeking to identify the families to which they belonged. In his letter to Hooker of 5 March [1863], he announced that the plants from Kew were '**165** in number!!!', continuing: 'Do you not think you ought to be sent with Mr Gower to the Police Court?' (William Hugh Gower, superintendent of the propagating department at Kew, had helped select the plants for Darwin). Hooker had also sent seeds, and was attempting to obtain others from Paris, so that Darwin felt he had to reiterate: 'All that were on list were for experiments, which seem to me really worth trial' (letter to J. D. Hooker, 21 February [1863]).

Darwin's hothouse became an important focus for his experiments. By the spring of 1864, he was thinking of expansion, telling Hooker: 'I am so magnificent that I am thinking of building a large greenhouse & turning the present green house into a cool Stove [that is, cool hothouse]' (*Correspondence* vol. 12, letter to J. D. Hooker, 26[–7] March 1864). The plan was quickly set in motion, and once again Darwin relied on Horwood to superintend the work, while William Ledger did the building. By August 1864, he had spent £126 10*s.* on the new heated greenhouse, and on iron pipes to convert the original greenhouse into a cool hothouse (Classed account book

(Down House MS)). Subsequent references make clear that the new greenhouse comprised two halves kept at different temperatures, and in 1869 Darwin told the botanist William Chester Tait that he had '4 houses of different temperatures' (letter to W. C. Tait, 12 and 16 March [1869], *Calendar* no. 6661).

Darwin's greenhouse complex, comprising these four distinct sections, was a series of lean-to constructions, built inside the boundary wall of Darwin's long kitchen garden, at the end nearest the house (see plan, opposite). Inspection of the surviving brickwork of the original base confirms that the complex resulted from three separate building phases, although the superstructure was replaced after Darwin's death, and one section of the 1864 greenhouse was subsequently demolished. The appearance of the complex at about the time of Darwin's death, showing the four distinct sections, is represented in the illustration reproduced as the frontispiece to this volume, from the *Century Illustrated Monthly Magazine* for January 1883. The drawing provides an interior view, looking away from the house, apparently from within the original greenhouse (later the cool hothouse). Thus, as far as can be determined, the section in the foreground, with pipes clearly visible, is the hothouse of 1863.

Over many years, the greenhouse complex was an important site of Darwin's research, complementing his study, in which he carried out many botanical experiments (see, for example, A. de Candolle 1882, p. 495). The greenhouses were, according to Francis Darwin, the first port of call on his father's midday walk, when he would look 'at any germinating seeds or experimental plants which required a casual examination' (*LL* 1: 114). Even when he was ill, Darwin would make the effort to walk a distance that he thought 'must be more than 100 yards' to the greenhouses (*Correspondence* vol. 12, letter to J. D. Hooker, [25 January 1864]).

In view of the importance of Darwin's hothouse plants in his experimental work, it is appropriate to reproduce here two lists of hothouse plants drawn up by Darwin; these lists are in DAR 255: 8 and DAR 255: 2–5. The first is a list that Darwin apparently drew up from the catalogue of John Cattell, from whom he had bought plants for his garden over many years. Darwin's Classed account book gives an entry under 'Science', dated 28 March 1863, for five guineas' worth of plants bought from Cattell, suggesting that he bought at least some plants for the new hothouse from Cattell at this time. (Usually, purchases of plants from Cattell were listed in the account book under the heading 'Garden'.) The heading to the list suggests that it is a copy of a list drawn up for the nurseryman Hugh Low, of Clapton, London, for the purpose of obtaining comparative prices. It was certainly not the list that Darwin gave to Hooker, since Hooker 'marked out' on that list the plants he could not supply (see letter from J. D. Hooker, [16 February 1863]). However, it can be dated with some certainty to the period when the first hothouse was built by at least two details. Firstly, Darwin specifically mentioned 'Gloxinia droopy & upright' both in this list and in his letter to J. D. Hooker, 15 February [1863]. Secondly, he mentioned in this list, as an afterthought, the species *Edwardsia tetraptera*, citing Treviranus 1863a, which he received in mid-February (see letter from L. C. Treviranus, 12 February 1863).

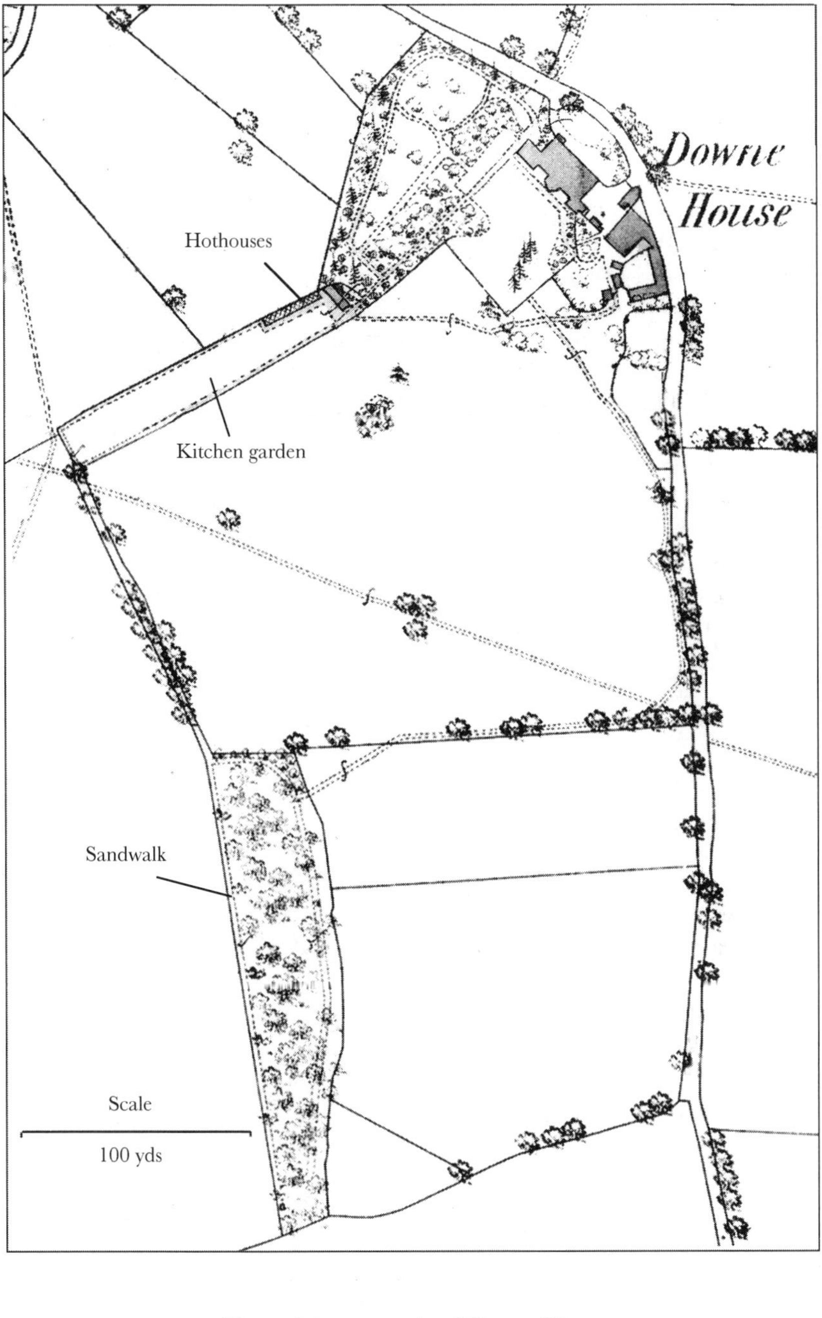

Plan of the grounds of Down House
from the 1868–9 25″ Ordnance Survey map of Kent, sheet XXVIII

The second list is headed 'Stove Plants', and itemises more than 130 species and their associated orders or families. This list is on four sheets of identical paper, bearing the watermark date 1860. This may be the list that Darwin made of the plants sent to him by Hooker (see letter to J. D. Hooker, 5 March [1863]), since many of the species listed were referred to in subsequent correspondence as having been sent to Darwin from Kew. Darwin said in the letter to Hooker of 5 March [1863] that he had received 165 plants from Kew; however, the numerical discrepancy may be accounted for in a number of ways. There may, for example, have been an additional sheet which is now missing, or Darwin's '165' may have referred to the number of individual plants rather than the number of species. Perhaps most likely is that several species were grouped under a single heading in the list, such as 'Melastomaceæ (Sin[g]apore)' or 'Catasetum species?'. However, even if the list is not that referred to by Darwin, it provides important evidence of the species that he had been anxious to obtain at this time, and as such is a useful indication of his experimental concerns.

The manuscript lists are reproduced below as closely as possible to the original format, with all alterations, interlineations, and additions noted. The symbol ✓ denotes a tick marked in pencil. The footnotes are intended to clarify the transcription and, where possible, to provide the correct nineteenth-century spellings of the species, genera, families or orders listed. Corrected spellings are taken from Lindley 1853, *Index Kewensis*, the *Gray Herbarium index*, and *Index Londinensis.*

Hugh Low & Co.[1]— Cattell[2] will get me at prices below without carriage

✓Asclepias currasavica[3]	1.6
—— rosea	do
✓Seed of common annual Mimosa	
Canna Warscewisii—[4]	2.6
✓Gloxinia droopy & upright	1.6—to 3.6.
Malpighia urens	5
Pentas rosea	1.6
Poinsettia pulcherrima	3.6
✓Cyanophyllum magnificum	7.6— 21
—— speciosa[5]	do. do.
✓Melastoma atropurpurea[6]	2.6
—— trinerva[7]	2.6.
Maranta ——	2.6
Ruellia aurantiaca	2.6
✓—— maculata	1.6.
✓Allamanda & Dipladenia	2.6 or 3.6.

Get Edwardsia tetraptera mentioned by Treviranus Honey.[8]

Acropera Loddigesii.—	10.6	Catasetidæ	
Catasetum cristatum	do	Clowesia	
—— Russelianum[9]	do.	Cyrtopodium.	
		——	
✓Gongora atropurpurea	5	✓Cyrtopodium Andersonii	10.6
✓—— maculata	5	—— punctata[10]	10.6
Mormodes auraticum[11]	21	{ Anoectochilus argenteus	5^{s}.[12]
✓Stanhopea grandiflora	7.6.	{ —— pictus	8^{s}.[13]
		{ Neotteæ	

Disa
Vinca-like Plant
Drosera[14]
Sarracenia— Nepenthes
Asclepias— species
Poinsettia
Catasetidæ
Melastoma with good sized flowers— a Bush.
Cyanophyllum—
Bolbophyllum[15]
Cypripedium barbatum
moveable Labellum. (Bateman[16])
Stilidyum adnatum.[17] 1.6. Low
Brugmansea—[18] flowers turn up in Moonlight
{ Bryanthus erectus hybrid form
{ Menziesia cærulea empetriformis[19]
{ Rhodathamnus chæmæcistus[20]

DAR 255: 8

1 Hugh Low & Co. was a firm of nurserymen, seedsmen, and florists with premises at Clapton, London (*Post Office London directory* 1863).
2 John Cattell was a florist, nurseryman and seedsman with premises at Westerham, Kent (*Post Office directory of the six home counties* 1862).
3 Asclepias curassavica.
4 Canna Warszewiczii.
5 'speciosa' deleted in pencil.
6 This name has not been found in botanical reference works.
7 Melastoma trinerve or M. trinervium.
8 This sentence added in pencil. The reference is to Ludolph Christian Treviranus and to Treviranus 1863a, p. 10. See also letter to J. D. Hooker, 24[–5] February [1863] and n. 19.
9 Catasetum Russellianum.
10 '—— punctata' deleted in pencil. CD misspelled Cyrtopodium punctatum.
11 Mormodes aurantiaca
12 'Anoectochilus argenteus 5^{s}.' deleted in ink.
13 '—— pictus 8^{s}.' deleted in ink.
14 'Drosera' added in pencil.
15 Bulbophyllum.

[16] The reference is to James Bateman, an orchid specialist (R. Desmond 1994).
[17] Stylidium adnatum.
[18] Brugmansia.
[19] 'cærulea' deleted in pencil, 'empetriformis' added in pencil.
[20] Rhodothamnus chamaecistus.

Stove Plants

Cordilyne heliconæfolia[1]	Asphodeleæ
Marcgravia umbellata	Marcgraviaceæ. Guttiferæ
Aristolochia Thwaitesii	
Rivinia humuli[2]	Phytolacceæ[3]
Artanthe mollicoma	Piperaceæ
Myroxylon cleriesii[4]	Flacourtaceæ[5]— Violales
Passiflora punctata	
P. Hartwissiana[6]	
— cærulea racemosa	
— hybrida floribunda	
Chirita Walkerii[7]	Cyrtandraceæ— Gesneraceæ
Lopezia axilare[8]	Onagreæ[9]
Leea coccinea	Amelidæ[10]— Santalaceæ
Alloplectis chrysanthum[11]	Gesneraceæ
Saccharum officinarum	Gramineæ
Acropera luteola	*Orchids*
—— Loddigesii	
Gongora maculata	
Catasetum purum	
Bolbophyllum barbigerum[12] major	
—— careyanum[13]	
Angræcum eburneum	
Cyrtopodium punctatum	
Coffea Bengalensis	Cinchonaceæ
Nepenthes distillatoria	
Desmodium gyrans	Leguminosæ
Oxalis sensitiva: O Braziliensis[14]	
O. crenata. O Bowii.[15] O. cluphurifolia[16]	
Caladium Houlletii: Chemtini:[17] verschaffeltii	Aroidæ[18]

C. hæmatostigma.
Gloriosa Leopoldi[19]
Gloxinia atro-violacea[20]— Laura erect bits[21]
carthusiana[22]

Costus Malortianus[23]		Scitamineæ
Lasiandra Fontanesiana		Melastomaceæ
Noryska Chinensis[24]		Hypericineæ
Ixora coccinea		Rubiaceæ
—— Javania[25]		"
Stemonacanthus macrophyllus		Acanthaceæ
Capparis ferruginei[26]		Capparidæ[27]
Sericographis squarrosa		Acanthaceæ
Hamelia patens		Rubiaceæ
Eucharis grandiflora		Ameryllidæ[28]
Urostigma rubiginosum		Figs
Cyanophyllum magnificum	Mr Low[29] | of	Melastomaceæ
Daphne Indica rubra odorata	Clapton	
Ruellia maculata		Acanthaceæ
Dipladenia micro? *urophylla*[30]		Apocyneæ
Brugmansia arborea		
Nepenthes ampullacea		
Sarracenia viriolaris[31]		
Catasetum species?		
Gongora ?		
Cyrtopodium ?		
Stanhopea ?		

Jasminium pauciflorum[32]	
Meyenia erecta	Solanaceæ Cestrineæ
Abutilon Duc de Melekoff	=Sida Malvaceæ
Higgensia argyroneurra[33]	Cinchonaceæ
—— pyrophylla[34]	"
—— discolor[35]	
Columnea scandens	Scrophulariæ— S. America
—— Schiedeana	"
Freycinettia imbricata[36]	Bromeliaceæ
Ficus barbata	
Tecoma undulata	Bignoniaceæ— Personatæ
—— capensis	
Phyllarthron comorense	Jus. Masc.[37] Bignoniaceæ
Plumbago capensis	

Franciscea latifolia	Brazil — Rhinantheæ[38] Scrophulariaceæ
Pandanus graminifolius	
Cyphomandra betacea	Solanaceæ
Cryptostegia grandiflora	Asclepiadæ[39] Ind. Or.[40]
Bexia Madgascariensis[41]	Aquifoliaceæ
Thryallis brachystachys[42]	Malpighiaceæ Brazil
Gardenia florida	Gardeniaceæ or Rubiaceæ Ind. Or. China
Bougainvillea speciosa[43]	Nyctagineæ
Torenia pucherrima[44]	Scrophulariaceæ
Hoya carnosa	Asclepiadæ[45]
Clusia incarnata[46]	Guttiferæ
Dipeteracanthus Herbstii[47]	=Ruellia Acanthaceæ
Allamanda Schotti[48]	Apocyneæ
—— nerifolia[49]	
Pentas carnea	Rubiaceæ
Chrysophyllum albidum	Sapoteæ[50]
Aristolochia gigas	
Lagerstroemia regina[51]	Lythraceæ &c Malabar
—— indica	
Pleroma heteromalla[52]	Lythorariæ—[53] Melastomaceæ
Justicea carnea[54] splendens[55]	Scophuseriaceæ[56]
" " diclyptera[57] speciosa	
" " hyssopifolia	
Roxburgia[58]	Aroidæ[59] Ind. Or.
Dieffenbachia seguine picta	= Caladium
Pogostemon patchouli[60]	Labiatæ
Xylophylla speciosa	Euphorbiaceæ —Rutaceæ Euphorb. Jamaica
	do
Pointsettia pucherrima[61]	
Dracæna terminalis	Asphodeleæ
vivipara[62]	
—— fragrans	
Sphærostema marmorata[63]	Schizandraceæ— near Menispermales
Chiococca racemosa	Rubiaceæ Ind. Occ.[64]
Stylidium graminifolia[65]	Campanulaceæ— Stylideæ
Ficus aspra[66]	
Clidemia crenata	= Melastoma Brazil
Adhadota cydonifolia[67]	Acanthaceæ,— Justiceæ[68]

Vinca rosa[69]	alba Apocyneæ
Stephanotis floribunda	= Ceropegia— Apocyneæ— Asclepiadæ[70]
Cyrtandra Dohliana[71]	Scrophulariæ
Melastomaceæ (Sinapore)[72]	
Vinca rosea alba	Apocyneæ
Plumbago zeylarnica[73]	Plumbagineæ
Sonerilla margaritacea[74]	Lythraceæ or Melastomaceæ
Scutellaria trianea[75]	Labiatæ
Anthurium cartilagineum	Aroidæ[76] = *P*othos
—— podophyllum	
—— cardifolium[77]	
Ceropegia gardnerii[78]	Apocyneæ— Asclepiadæ[79]
Croton longifolium[80]	Euphorbiæ
—— variegata[81]	
Aschynanthus lianus[82]	Bignoniaceæ
—— Teysmanianus[83] minialis[84]	
Clavija undulata	Myrsineæ. S. Am.
Euphorbia lophogona	
Malpighia punicifolia	Malpighiaceæ
—— aquifolia	
Arduina grandiflora	Apocyneæ
Melastoma macrocarpa[85]	Melastomaceæ
Cissus discolor	Viniferæ
Andropogon podophyllum[86]	Gramineæ
Elaeis guiniensis[87]	Palmæ
Coelibogyne ilicifolia[88]	Euphorbiæ
Impatiens Hookeriana	Balsamineæ

DAR 255: 2–5

1 Cordyline heliconiaefolia
2 Rivina humilis
3 'Phytolacceæ' preceded by 'o' added in pencil.
4 This name has not been found in botanical reference works.
5 Flacourtiaceae.
6 Passiflora Hartwiesiana.
7 Chirita Walkeri.
8 Lopezia axillaris.
9 Onagrae.
10 Ampelidae.
11 Alloplectus chrysanthus.
12 Bulbophyllum barbigerum.
13 Bulbophyllum Careyanum.
14 Oxalis brasiliensis.
15 Oxalis Bowiei.

[16] Oxalis bupleurifolia.
[17] Caladium Chantinii.
[18] Aroideae.
[19] Gloriosa Leopoldii.
[20] This name has not been found in botanical reference works.
[21] 'erect' interlined in ink.
[22] This name has not been found in botanical reference works.
[23] Costus Malortieanus.
[24] Norysca chinensis.
[25] Ixora javanica.
[26] Capparis ferruginea.
[27] Capparideae.
[28] Amaryllideae.
[29] Hugh Low was the proprietor of Hugh Low & Co., nurserymen, with premises at Clapton, London. After Low's death in 1863 the firm was conducted by his son, Stuart Henry Low (R. Desmond 1994).
[30] 'micro?' interlined in ink.
[31] Sarracenia variolaris.
[32] Jasminum pauciflorum.
[33] Higginsia argyroneura.
[34] Higginsia pyrophylla.
[35] Higginsia discolor.
[36] Freycinetia imbricata.
[37] 'Jus. Masc.': Jussieu. Mascarene Islands. The reference is to Antoine Laurent de Jussieu, who described the family Bignoniaceae ('Bignoniæ') in 1789 (Jussieu 1789, p. 137).
[38] Rhinanthideae.
[39] Asclepiadeae.
[40] 'Ind. Or.': India Orientalis (East Indies).
[41] Brexia madagascariensis.
[42] An 'a' is superimposed in pencil on the final 's' in 'brachystachys'. CD mispelled Thryallis brachystachis.
[43] Bougainvillaea speciosa.
[44] Torenia pulcherrima, but the name has not been found in botanical reference works.
[45] Asclepiadeae.
[46] This name has not been found in botanical reference works.
[47] Dipteracanthus Herbstii.
[48] Allamanda Schottii.
[49] Allamanda neriifolia.
[50] Sapotae.
[51] Lagerstroemia reginae.
[52] Pleroma heteromallum.
[53] Lythrarieae.
[54] Justicia carnea.
[55] This name has not been found botanical reference works.
[56] Scrophulariaceae.
[57] Justicia dicliptera.
[58] Roxburghia.
[59] Aroideae.
[60] Pogostemon Patchouly.
[61] Poinsettia pulcherrima.
[62] This name has not been found in botanical reference works.
[63] Sphaerostema marmoratum.

[64] 'Ind. Occ.': India Occidentalis (West Indies).
[65] Stylidium graminifolium.
[66] Ficus aspera.
[67] Adhatoda cydoniaefolia.
[68] Justiciadae.
[69] Vinca rosea.
[70] Asclepiadeae.
[71] This name has not been found in botanical reference works.
[72] Singapore.
[73] Plumbago zeylanica.
[74] Sonerila margaritacea.
[75] Scutellaria Trianaei.
[76] Aroideae.
[77] Anthurium cordifolium.
[78] Ceropegia Gardneri.
[79] Asclepiadeae.
[80] Croton longifolius.
[81] Croton variegatus.
[82] Aeschynanthus. Species of the genus are tropical vines; CD mispelled 'lianas'.
[83] Aeschynanthus Teysmanniana.
[84] Possibly Aeschynanthus miniata.
[85] Melastoma macrocarpum.
[86] This name has not been found in botanical reference works.
[87] Elaeis guineensis.
[88] Coelebogyne ilicifolia.

APPENDIX VII

Notices in the *Athenæum* on William Benjamin Carpenter's *Introduction to the study of Foraminifera* and the origin of species

On 28 March 1863, the *Athenæum* carried, as the leading article of its literature section, an anonymous review of William Benjamin Carpenter's *Introduction to the study of Foraminifera*, published by the Ray Society at the end of 1862. The publisher's marked copy of the *Athenæum* at the City University, London, reveals the author to have been Richard Owen. The identity of the reviewer was immediately suspected by Darwin and his circle: Joseph Dalton Hooker posted a copy of the 28 March issue to Darwin (letter from J. D. Hooker, [28 March 1863]), who detected 'Oken-like transcendental philosophy written in Owenian style!' (letter to J. D. Hooker, [29 March 1863]). Hooker made inquiries and concluded, from the style of writing, from the resemblance of passages to Owen's critical review of *Origin* ([Owen] 1860a), and from a reported conversation with the editor of the *Athenæum*, William Hepworth Dixon, that the article was 'Owens & none others'. Hooker's overall assessment of the review was damning: 'Good Lord *what* rubbish & what vulgarity. I think there must be truth in a report I have heard—that he tippled' (letter from J. D. Hooker, [31 March 1863]).

Owen's review of *Introduction to the study of Foraminifera* was critical of Carpenter's general conclusion that the divergent forms of Foraminifera were probably derived from a few family types, and that these, too, might possibly be derived from a common ancestor. Owen seized on these conclusions as evidence that Carpenter's study was influenced, if not distorted, by the 'Darwinian hypothesis'. Carpenter had given Darwin qualified support in two reviews of *Origin* ([Carpenter] 1860a and 1860b), and this led him to be identified as a supporter of Darwin's thesis (*Correspondence* vol. 8, letter to J. D. Hooker, 3 March [1860]). Owen then proceeded to address the problem of the origin of the 'father of all Foraminifera'; he accused Carpenter of assigning the beginning of such an organism to an occult, creative force, and compared this notion to Darwin's use of 'Pentateuchal terms' in *Origin* to explain the beginning of life (see *Origin*, p. 484). Owen, by contrast, advocated the 'Lamarckian principle of the heterogenous production of the primitive types of organisms'. Heterogeny, or spontaneous generation, was the starting-point of Jean Baptiste de Lamarck's theory of natural progression, according to which unorganised matter was transformed into primitive forms of life by subtle fluids, particularly electricity and heat (Burkhardt 1977, pp. 151–7, Farley 1977, pp. 40–2). Owen used this review, like his review of *Origin* ([Owen] 1860a, p. 514), as an opportunity to 'out-naturalize Darwin' (Gillespie 1979, p. 93, Rupke 1994, p. 240).

Carpenter had been working on the Foraminifera, a group of mostly microscopic, single-celled animals (rhizopods), since the late 1840s (Janal 1988), and was awarded a medal by the Royal Society in recognition of his work in 1861. Stung by the imputation that Darwin was his 'master', and that his study of Foraminifera had slavishly followed Darwinian principles, he wrote to the *Athenæum* on 30 March 1863 to assert his intellectual independence. Darwin and his supporters were disappointed by Carpenter's response, seeing it as a lost opportunity to 'smash' Owen. Darwin thought 'the chief object . . . of his letter seems to me to be to show that though he has touched pitch he is no[t] defiled' (letter to J. D. Hooker, [17 April 1863]). The exchange prompted Darwin's first venture into the newspaper columns in defence of *Origin* (letter to *Athenæum* 18 April [1863]), 'under the cloak of attacking Heterogeny' (letter to J. D. Hooker, [17 April 1863]). Darwin used this opportunity to quote the new edition of *Antiquity of man* (C. Lyell 1863b), and claim Charles Lyell's authoritative support for *Origin*: 'if one puffs oneself, one had better puff handsomely' (letter to J. D. Hooker, [17 April 1863]; see also letter to Charles Lyell, 18 April [1863]). Owen replied anonymously to Darwin in a letter published in the *Athenæum* on 2 May 1863. The letter aimed to further discredit Darwin's account of the origin of life, and to reassert the superiority of the doctrine of heterogeny, in a discussion endorsing Félix-Archimède Pouchet's work on heterogeny (Pouchet 1859). Hooker, unaware that Darwin had already written a second letter on 5 May [1863] to the *Athenæum*, was dismayed (letter from J. D. Hooker, [7 May 1863]):

> I cannot abide this lugging of Science before the public . . . & implore you my dear fellow not to do so again. Owens answer to you is so triumphant in the eyes of the public (who you wish to enlighten) as Manchesters is over Natal. The only party that gains by these discussions is the proprietor of the paper, the only one that loses *every way*, is the maintainer of truth.

Darwin regretted his actions: 'That beggar Owen sneers well at me & with some justice. I was an ass to write to Athenæum' (letter to Charles Lyell, [7 May 1863]).

As a supplement to the correspondence about Owen's anonymous review of Carpenter, and the exchanges which ensued, Owen's original review, Carpenter's reply, and Owen's anonymous response to Darwin's letter to the *Athenæum* of 18 April [1863], are printed below, in chronological order of their publication.

Athenæum, 28 March 1863, pp. 417–19

Introduction to the Study of the Foraminifera.
By William B. Carpenter, M.D.; assisted by William K. Parker, Esq. and T. Rupert Jones, Esq. (Hardwicke.)

Water, containing earth in solution or suspension, gives to the atoms of such earth a freedom of moving according to the impulse of the pervading polaric forces. These may be affected by the reaction of the earthy atoms themselves as aggregation proceeds; and both the course of motion and the resultant form of the crystal or other aggregate may be modified according to the nature of the fluid field of such aggregation. This fact has been lucidly demonstrated by Rainey's beautiful experiments on the artificial production of organic conditions of aggregation of carbonate of lime, inducing the same departure from crystalline shape as when the atoms aggregate in the vital production of shell and coral.

When the fluid is pure water the atoms of the earth are aggregated, under polaric guidance, into crystalline forms, distinct for each chemical kind of earth, and affording, by the angles and other properties of the crystal, characters of the "species" of the mineralogist. When the fluid is slimy, the operation of the polarizing force is modified, and apparently, though not really, impeded, and the character of a crystal is proportionally lost: the modification of, and departure from, angular or crystalline forms being greater as the slime is thicker; and especially when the atoms of such are in the condition, through condensation, to act and react upon the surrounding slime-suspending fluid, so as to attract or draw in outside atoms of slime, and to be attracted by or drawn out to such particles of assimilable slime.

The exuviations, ejections, and decay of organisms of all kinds and grades now supply to seas, lakes, rivers and other aggregates of water, the raw material for the re-commencement of those modes of action of the all-pervading force which are termed "vital."

This raw material abounds in, and indeed partly gives its character to, the kind of bed of sea or lake which is called ooze or "mud," the nature and varieties of which are manifold. At whatever depth of the ocean such "mud-bed" has been reached, the contained slime has been observed, when duly scrutinized, to be manifesting the vital form of force, and modifying the shapes in which the atoms of earth,—flinty or limy,—may be aggregated within the vitally acting slime.

The bulk of the present folio of 319 pages is devoted to the description, and its 22 plates and 47 woodcuts to the delineation, of the various leading forms in which the particles of carbonate of lime are aggregated and arranged within the portions of slimy matter, called, according to the prevalent character of their lime-aggregates, "Foraminifera." The organic basis or field of those aggregates has not been distinguished from that in which particles of silex may be precipitated, either in the form of fine needles, as in sponges, or of delicate cases, as in Polycystina; nor from that substance in which, probably from the want of the raw material in the fluid field of its existence, neither flint nor chalk is deposited, as *e.g.* in Actinophrys and Amœba; and such common organic basis has received the name of "sarcode" or rudimentary flesh, and of "protoplasm" as being that "in which every form of animal" (and plant?) "structure has its origin, and from which it is evolved by a process of gradual differentiation." Under the latter term the author defines this substance as being "composed of an albuminoid base, with oil particles in a state of very fine division diffused through it; it is tenacious, extensible and contractile; it is diaphanous, reflecting light more than water, but less than oil; and it is dissolved by alkalies, rendered transparent by acetic acid, and dyed brown by iodine. In the midst of this substance are usually to be seen *vacuoles*, or cavities containing a thinner fluid, which is often coloured; these are extremely variable, both in number and size, and their deficiency in any definite limiting walls is rendered obvious by their not unfrequent coalescence" (p. 13).

Under what modification or combination of the general polarizing force, the slime of mud or ooze is first condensed into a protoplasmic centre of such low vital force is still undetermined; but, once begun, and growing as such, it successively puts forth other portions or centres; and the "Foraminifera afford one of the best examples which the animal kingdom can present of that 'vegetative' or 'irrelative' repetition which is an essential feature of all low grades of organization." The primitive protoplasmic centre of a potential foraminifer, however, contains lime-water in much greater proportion than either of the substances defined by the author as composing it; it combines the calcium of such water with carbonic acid, and precipitates it on its exterior as a thin porous crust. The sarcode or protoplasmic matter can extend itself, like threads, from the pores of the crust or shell, and hence the name "foraminifer." It is an "organism" without organs; and manifests life at its lowest grade. Budding out a second protoplasmic centre, this forms a second calcareous porous case, which may or may not be detached from the first; most commonly it is not, and the manifold variety of the Foraminifera depends upon the direction of the budding-forth, and the shape of the buds or "sarcode" centres; whilst their beauty is mainly due to the repetition of the same structure on a regular plan. Some have a rectilinear, more have a curvilinear axis of growth; and the latter, being spiral, may be either flat, like the shell of the nautilus, or may coil round a lengthened axis, like that of a whelk. But between the rectilinear and curvilinear forms there is a complete gradation.

When the varying conditions of the ocean-beds, which are the seat of the foraminiferous mode of resumption of vital force, are considered,—either in regard to distance from the surface, decreasing from depths exceeding two miles, as reached by the soundings of the Atlantic Telegraph Company, and of 1,600 fathoms as plumbed by Wallich between Cape Farewell and the Island of Rockall, to the shallows of tidal shores,—or in respect of the constituents of the sands and muds in which the slime-elements become aggregated into protoplasmic centres,—corresponding modifications of the forms of deposit of carbonate of lime in and upon such centres might be expected,—are, indeed, inevitable under such low conditions of a centre of vital power, with corresponding weakness of resistance of surrounding influences. Accordingly, a result of the summary of the observed forms of foraminiferous shells in Dr. Carpenter's "Introduction" is, that a definition of a species is impossible, in the terms at least to which organisms with definite organs lend themselves; and that the like "ordinary method of definition" of the genus is "impracticable." Upwards of twenty pages are, nevertheless, devoted to the attempt to classify the zoologically unclassifiable Foraminifera. The efforts of the mineralogist towards a like systematic end are painful enough, and meet with as inadequate a reward; but the conditions of the aggregation of mineral atoms, under the guidance of polar force, in the purer fresh waters, are more definite and less varied than those in the complex solution which forms the ocean.

We have supplemented the author's definition of the protoplasmic base or seat of the chalk-precipitate by the quantity of "lime-water" which fills out the delicate, filmy, albuminous lime. Our meaning is this: every river pours a solution of lime into the sea; but pure water only evaporates from its surface. All sea-water, therefore, holds (with many other salts) a proportion of lime in solution; and if this caustic earth were not disposed of, the effect of evaporation would, in time, unfit the sea for the purposes or as a medium of life. Amongst the instruments by which the soluble caustic lime is precipitated in the condition of a mild, insoluble carbonate of lime, the Foraminifera are the most active and the most widely diffused. In the deep-sea soundings to which reference has been made "the surface layer of the ooze was formed of living shells; whilst its principal mass consisted of the exuviæ of preceding generations."

To Ehrenberg we are indebted for the capital discovery of the large proportion in which the shells of Foraminifera enter, either entire or comminuted, into the formation of chalk ('Ueber die Bildung der Kreidefelsen und der Kreidemergel durch unsichtbare Organismen,' in the *Transactions* of the Royal Academy of Berlin for 1838; and 'Ueber noch jetzt zalreich lebende Thierarten der Kreidelbildung und den Organismen der Polythalamien,' *Ibid.* 1839). When this constitution of the downs and more notable cliffs of Albion is duly examined, it is found to bear a close resemblance to that of the mud which has been dredged up from a great tract of the bed of the Atlantic, and which was composed of little else than the calcareous shells of Globigerinæ and other foraminiferous forms, with the silicious shields of the allied Polycystineæ.

If the finer part of a portion of chalk be washed away, the remaining sediment consists almost entirely of foraminiferous shells. They have also been found in similarly upheaved beds of older oceans, down to the Lower Silurian, in the instances of calcareous rocks of a consistency to be washed; and even in hard limestones and marbles foraminiferous shells can be detected, by the microscope, in polished sections and in thin slices laid upon glass.

Another kind of evidence of the share taken by the polythalamous Foraminifera in the formation of the earth's crust has been made known to us also by the original and persevering labours of the great micrographer of Berlin, in his discovery that their shells occasionally become infiltrated by a solution of silicate of iron, which fills by precipitation not only the chambers, but the pores and co-extended canals, even to their minutest ramifications; so that when a shell thus infiltrated undergoes decomposition, a perfect internal cast of the original protoplasm and its root-like extensions remains:—

> "Of such casts it has been shown by Prof. Ehrenberg that the green-sands which present themselves in various geological formations, from the Silurian system upwards, are in great part composed; and his discovery has a two-fold value, as, on the one hand, it places before us far more exact representations

> of the configuration of the animal body, and of the connexions of its different parts, than we could obtain even from living specimens by dissolving away their shells with acid (its several portions being disposed to heap themselves together in a mass when they lose the support of the calcareous skeleton), whilst it also enables us to identify with great certainty the types of Foraminifera by which these casts were originally formed, notwithstanding the entire destruction of the shells."

The meaning which the author attaches to the term "type" may be illustrated by his characters of the family of Lagenidæ. The shell "substance" is hyaline or vitreous, and very hard, perforated by tubes of extreme fineness. The "septal" aperture is small, circular and central, more or less prominent, sometimes produced into a siphon-like tube.

The generic name Lagena is retained for the Foraminifera which form a single-chambered shell, globose, with a prolonged neck and terminal aperture. But from such "typical" Lagenæ the deviations are numerous—*e.g.*, into an ovate form, pointed at the hinder end, or so narrowed as to become "fusiform," &c., or with introversion of the flask-shaped neck, which then gives occasion for another generic name—*e.g.*, Entosalenia, &c.

If the protoplasm, operating under the conditions which lead it to precipitate its chalk in the texture and form of Lagena, be so favourably placed in regard to the acquisition and assimilation of fresh slime as to bud out successive portions thereof, an elongated vitreous shell is formed of a series of segments joined end to end in a straight line, sometimes with progressive increase, giving the entire shell a conical form.—

> "From the rectilineal Nodosaria, through the gently-curved Dentalina and the more strongly-curved Marginula, we are led, in a series so perfectly gradational as to forbid lines of demarcation from being anywhere drawn across it to the spiral Cristallaria, which may be regarded as the highest manifestation of the lagenoid type. From this central stem we find a set of offsets more or less divergent, presenting modifications of the typical characters on which generic distinctions have been founded, but which we cannot recognize as of higher than varietal importance."

Nevertheless—and, indeed, of necessity in respect to finite capacities of comprehending the manifold results of the conditions of creative force—the author proceeds to arrange about a "type form" a second group of genera, forming the family Globigerinida; and a third group around another "type," forming the family Nummulinida.

This latter includes the largest and most highly-developed kinds of Foraminifera, the existing members of which are almost exclusively confined to tropical or subtropical regions. Under analogous influences must have been produced those giants among Foraminifera which multiplied during that early tertiary period which has been called, after them, "nummulitic." Their shells form, literally, mountain masses, once in the depths of ocean, now upraised to the surface. The "nummulitic" limestones are mere aggregates of such organic products, the interspaces of the entire shells being occupied by smaller specimens of the same kinds or by fragments of the larger ones. These rocks occur in Southern Europe, Northern Africa, India, Jamaica. The "Great Pyramid" of Egypt is built of the stone formed of the kind of Foraminifer called *Nummulites nummularia*, from its resemblance to a coin.

The above-recited three "types" or "families" of Foraminifera are grouped together under the "sub-order" Perforata, the shell having the tubular apertures for the exit of the filaments of protoplasm called "pseudopodia." They may be regarded as the "typical" Foraminifera. All the kinds have calcareous shells, and are marine.

Aggregates of slime, snot, or "protoplasm," however, occur, which, putting forth their pseudopodial extensions at one point only of the surface, cover the rest with a case which is imperforate. In the Miliolida, which are and were marine, the shell is of porcellanous texture. In the Lituolida, some kinds of which live in shallow seas or brackish water, the shell is "arenaceous," or composed of foreign particles cemented together upon the outside of the protoplasm. In the Gromida, of which many inhabit fresh water, the shell is membranous or "chitinoid."

The freshwater Amœbidæ resemble the protoplasmal basis of the "perforate" Foraminifera; but the conditions for obtaining the lime are poor or absent in the sphere of their existence. The truly vital essential part of all these minute and beautiful organisms is alike. Since the great step which Dujardin ('Observations sur les Rhizopodes et les Infusoires,' *Comptes Rendus de l'Acad. des Sciences*, 1835) made in the right appreciation of their nature, observers have progressively concurred in interpreting the results of their scrutiny more or less in accordance with the following expressions of our author:—

> "The physiologist has here a case in which those vital operations which he is accustomed to see carried on by an elaborate apparatus are performed without any special instruments whatever,—a little particle of apparently homogeneous jelly changing itself into a greater variety of forms than the fabled Proteus; laying hold of its food without members, swallowing it without a mouth, appropriating its nutritious material without absorbent vessels or a circulating system, moving from place to place without muscles, feeling (if it has any power to do so) without nerves, propagating itself without genital apparatus; and not only this, but in many instances forming shelly coverings of a symmetry and complexity not surpassed by those of any testaceous animals." (p. vii.)

The large experience of the Foraminifera which Dr. Carpenter has accumulated from his own studies, derivative and direct, and from those of his coadjutors in the present work, Messrs. Parker and Rupert Jones, has led him to the conclusion that—

> "the ordinary notion of *species* as assemblages of individuals marked out from each other by definite characters that have been genetically transmitted from original prototypes similarly distinguished is quite inapplicable to this group.

* * The only natural classification of the vast aggregate of diversified forms which this group contains, will be one which ranges them according to their direction and degree of divergence from a small number of principal family-types. * * The subordinate groupings of genera and species which may be adopted for the convenience of description and nomenclature, must be regarded merely as assemblages of forms characterized by the nature and degree of the modifications of the original type, which they may have respectively acquired in the course of genetic descent from a common ancestry. * * Even in regard to these family types, it may fairly be questioned whether analogical evidence does not rather favour the idea of their derivation from a common original than that of their primitive distinctness." (p. xi.)

We here discern the influence of Mr. Darwin's volume, 'On the Origin of Species,' on the mind of the writer, the essentially distinctive character of which work is the attribution of all animal species, as descendants "from at most only four or five progenitors, and plants from an equal or lesser number,"—the author adding—"Analogy would lead me one step further, namely, to the belief that all animals and plants have descended from some one prototype." The resemblance is so close between the terms in which Dr. Carpenter repeats his main conclusion as to the origin of the acrite Foraminifera, and those in which Mr. Darwin expresses his conclusions in regard to the origin of all the living beings of this planet, that we are led to question whether a like influence may not have affected Dr. Carpenter's conclusions and expressions as to the nature of the species and genera of antecedent investigators and classifyers of the Foraminifera.

With regard to the highly interesting and instructive facts, of the identity of some (Ehrenberg determined it in ten) of the so regarded species of fossil Foraminifera of the chalk with still existing species, and of the identity of more of the Foraminiferous Nodosariæ of the Triassic, Permian and Carboniferous strata, with existing forms, Dr. Carpenter explains it, agreeably with the Darwinian hypothesis, as a case "of the genetic continuity between the Foraminifera of successive geological periods; graduating backward to the period when they began to descend from a common original," or, to use the Master's words, "from some one primordial form, into which life was first breathed." (*Darwin*, p. 484.)

According to this hypothesis we are compelled to conceive, or try to conceive, a period in our planet's history in which the conditions of life, or the circumstances tending to change the modes of operation of all-pervading force in the vital direction, were uniform, or at least that the uniformity which led to the impossibility of the existence of more than one "original form" had only been broken so far as to allow of the variation of such form into those of a few progenitors of succeeding classes of beings; the grade of differentiation of such progenitors from the primordial one being such as might be manifested by an Amœba or whatever Dr. Carpenter may conceive to have been the original "father of all Foraminifera."

The inevitable corollary of the Darwinian hypothesis seems to us to demonstrate its weakness; and to show by contrast the superiority of the Lamarckian principle of the heterogenous production of the primitive types of organisms. Agreeably with this principle, we conceive that "particles of apparently homogeneous jelly" are now, as of old, being aggregated through the operation of existing interchangeable modes of force, under special conditions, capable of causing, for example, the special characters of the Foraminifera.

The difference of atomic condition between a speck of jelly exercising the attraction of cohesion and one supplementing such attraction by the force of more intimate atomic union, call it incorporation or assimilation, may not be greater than the difference of atomic condition in a non-magnetic and a magnetic piece of iron: it is certainly very analogous to it. The latter draws to itself contiguous pieces of iron, which stick to the hard surface and cannot get in; the looser suspended atoms of the jelly-speck, under the assimilative or organic state of force, allow the attracted particles to permeate the mass, there to adhere atomically to their like, in short to be assimilated; the unassimilatable particles give way and are pushed out.

Every observer of the movements of such a sarcodal aggregate as Amœba, or Actinophrys, may recognize enough of the above-indicated correspondence with magnetic attraction and repulsion to place afresh before his judgment, for unbiassed consideration, the question whether such properties of the jelly-mass compel the invocation of a miraculous "inbreathing of life" for their initiation, and of generation exclusively for their continued display. Analogy of the phenomena of convertible modes of operation of polaric force leads us to conclude that the difference between snot and sarcode, between passive and active states of mucus, is due to a conversion of attractive or other mode of polar force into the mode called "assimilative," "organic," or "vital." We believe, to use Darwin's common expression, that such conversion occurs under the conditions favouring it; that such conditions exist now as of old, and that they have operated in the production of Foraminifera in the ooze or mud of sea-beds from the Silurian period, or the remotest one for the possibility of such conditions of vivifying jelly-specks, to the present day.

Dr. Carpenter's studies of the porous chalk-deposits in and on these jelly-specks appear, however, to have induced him to adopt a different conclusion; to assign the beginning of such organism to an occult cause, to an operation of force or to a concurrence of forces which have now no place in Nature,—a creative force in fact, which Darwin could only express in Pentateuchal terms as the primordial form "into which life was first breathed"! Such primordial form of Foraminifera, once created, with its power of growing by assimilative atomic attraction and of propagating by pushing out and detaching particles of its substance, increased and multiplied. The foraminiferous offspring in time spread into positions and fell under conditions which, by reaction on the primordial organic properties, modified diversely the resultant conditions of the foraminiferous body and shell. The "Foraminifera of successive geological periods" have all descended from the first-created little particle of homogeneous jelly, having the properties above quoted from p. vii; they are related thereto by genetic continuity; in the author's words and type, they are "the *direct lineal descendants* of the Foraminifera of very ancient geological periods." The evidence of this astounding assumption he regards to be "as complete as the nature of the case admits," that is to say, as it possibly can be, or, in fewer words, to be demonstrated! Accordingly, we find writers who accept the dictum, and *in verba magistri* use the language of the Darwinian hypothesis, remarking, "not without astonishment, that scarcely any of the species of Foraminifera met with in the secondary strata have become extinct." "Nodosariæ of the trias have transmitted descendants without loss of any of their essentially specific features to the present day." (*Phil. Trans.*, 1860, p. 582.)

Observation has, however, some value, though it may, or rather must, in natural objects, fall short of the extent and perfection of its subject. For example, the earliest or oldest strata with evidence of life present such evidence under a rather considerable variety of grades and conditions, agreeing, however, in a general way with the variety of states arising from different chemical constitutions of sedimentary rocks, from different depths in the sea, and other conditions likely to influence the state and form of the co-existing organisms of the period. The uniform condition of the surface of the earth at the period when the organic form of force was first capable of energizing upon such surface, which uniformity must be assumed at the period when the Darwinian primordial form of all organisms, and the Carpenterian primordial form of all Foraminifera existed in solitary state, is a condition of the earth's surface of which no geological observations whatever afford the slightest foundation or reason for assuming.

As, according to John Mill, the aim of the physical philosopher is to determine "what are the fewest and simplest assumptions, which being granted, the whole existing order of nature would result" ('Logic,' 3rd edit., vol. i., p. 327); so we hold, that the aim of the philosophic naturalist is to determine what are the assumptions having the best ground of acceptability from observation and analogy, on which organic forms and conditions, through known natural powers, might arise, propagating modifiable offspring concurrently with the continued operation of heterogenous productive force, which would people the globe, in the long course of its life-periods, with the various grades and forms constituting its present Fauna and Flora. We hold it to be contrary to the spirit of Mill, and of every searcher after truth, to prepose to the mind any quality of "number of primitive types," whether small or large, as a guiding idea in this high aim.

Reverting, then, to the question of the origin, or coming into existence, of a "particle of seemingly homogeneous jelly," with the powers above cited from p. vii. of the work under review: the assumption of an occult, miraculous, creative cause operating once for all, with the inevitable corollary of uniformity of vital conditions of existence during the hypothetical, uniform, primitive and primordial state of the earth's inhabitants, appears to us to be a less "simple assumption" than the following, viz., that at the period when life became possible on the earth's surface the conditions were sufficiently varied to permit the conversion of the general polaric into a special organic mode of force to operate under circumstances resulting in a variety of the simplest forms of life, such as "monad," "mucor," "amœba," "lichen-spore," &c., and that such conditions have continued to operate in the heterogenous production of "organisms without organs" to the present day.

The atoms, attracting and attracted, causing the pushing out and drawing in of the "sarcode," intus-suscepted at any part of the surface and assimilated, augment the mass, which buds out; in some the bud falls off, in others it coheres, and may form a polythalamous or dendritic whole. So in the crystalline salt, the atoms, attracted to the surface, cohere according to the law of its crystallization and form, sometimes as beautiful dendritic wholes, as in Infusoria, zoophytes and plants. In all these cases the single cell and the primitive crystal continue to be independently and "heterogenously" brought into being.

When observation shows that a particular form of Foraminifer occurs in fossiliferous rocks of all ascertained geological periods, we deduce from the fact that the conditions of the heterogenous creation of such form have been the same. Instead of concluding that the particular form or kind of Nodosaria of to-day is the great-great-&c.-grandchild of the Nodosaria of the carboniferous period, itself the lineal offspring of the occultly-created primordial Foraminifer, we deem it the "simpler" assumption, that the conditions of the "ooze" which led to the "fusiline" form Nodosaria coming into being when coal-plants flourished, operate in like manner in producing that particular form of shell-making jelly at the present day. In like manner we regard the valuable labours of Messrs. Parker and Rupert Jones 'On the Nomenclature of the Foraminifera' as a suggestive body of proof of the degree in which uniform conditions of the development of Foraminifera, characterized and named "specifically," have operated during long geological periods on this planet.

Agreeably with this "heterogenous" hypothesis is the great and significant fact of there having been "no advance in the foraminiferous type from the Palæozoic period to the present time." Most adverse is this fact to the hypothesis that one primordial form was first, and once for all, created, with a power of producing progeny susceptible of variation and improvement to the illimitable degree postulated by Darwin. It is expressly admitted by his disciple, the author of the work under notice, that "there is no indication of any tendency to elevation towards a higher type" in the foraminiferous Fauna of any period; but the significancy of this fact seems to be lost on Dr. Carpenter in reference to the most important, indeed supreme, deduction from the numerous particulars which he has ably and instructively set forth in his 'Introduction to the Study of the Foraminifera.'

In regard to the history of these most interesting and beautiful, though for the most part minute, forms of Protozoa, the salient features which, in zoology, will be remembered gratefully and with admiration of the originality, acuteness and perseverance of the observers, are—the determination of the true nature of the Foraminifera and their kinship to Amœba and other freshwater rhizopods by Dujardin,—and the discovery of their antiquity and important share in forming the earth's crust by Ehrenberg.

To the general reader there is no work on the Foraminifera which will prove more instructive and satisfactory than the richly-illustrated volume which Dr. Carpenter has brought out, with the aid of Messrs. Parker and Rupert Jones, under the auspices of the Ray Society. To the scientific reader, especially looking to the bearing of the recorded facts and observations on the generalizations of highest interest at the present phase of his science, the result, we regret to state, is such as the author himself seems to recognize when, with becoming modesty, he confesses that "he cannot venture to anticipate that his work will prove altogether satisfactory."

Athenæum, 4 April 1863, p. 461
Dr. Carpenter and his reviewer.

University of London, March 30, 1863

Whilst thanking you for the honour conferred upon my 'Introduction to the Study of the Foraminifera' by the elaborate review of it contained in your last number, I must beg to be allowed to correct some misapprehensions into which the author of that article has fallen.—

1. After quoting my general conclusions as to the *probable* derivation of all the divergent forms of Foraminifera from a few family types, and the *possible* derivation even of these from a common original, your reviewer remarks: "We here discern the influence of Mr. Darwin's volume on the mind of the writer"; and he is further "led to question whether a like influence may not have affected Dr. C's conclusions and expressions as to the nature of the species and genera of antecedent investigators and classifiers of Foraminifera." So far is this from being the fact, that these conclusions had been arrived at by my coadjutors and myself before the publication of Mr. Darwin's 'Origin of Species,' and in utter ignorance of his views; and any one who may take trouble to refer to my Address as President of the Microscopical Society in 1855, and to the first of my memoirs on this group in the *Philosophical Transactions* for 1856, will see that even at that period I was far on the road to them. The question of the variability of species is one to which my attention was very early directed by Dr. Pritchard; during more than a quarter of a century I have taken every opportunity of gaining information in regard to it from zoologists, botanists and palæontologists; and the expectation of finding in the group of Foraminifera an entirely new and valuable body of material for the prosecution of this inquiry, was one of my chief reasons for applying myself to the systematic study of it.

2. In the succeeding paragraph your reviewer cites me as representing the identity in the forms of certain fossil and existing Foraminifera, "agreeably with the Darwinian hypothesis, as a case 'of the genetic continuity between the Foraminifera of successive geological periods, graduating backwards to the period when they began to descend from a common original.' " I should be glad to be informed of the page of my book from which this professed citation is taken; I am unable to identify it with anything I have written. That I do regard the repetition of identical forms through a succession of geological epochs as indicative of continuous genetic descent is most true; and I have always understood this, so far from being "an astounding hypothesis," to be the doctrine current among our most esteemed palæontologists. Surely the derivation of a certain number of the Mollusks at present inhabiting the

Mediterranean, by direct continuity of descent from the identical types whose shells are entombed in the Tertiary formations of its shores, is a fact as well established as the derivation of the existing races of Men from those which peopled the globe during the pre-historic period. And I am at a loss to see what other evidence of genetic connexion Palæontology can ever supply, than that afforded by continuity, either of identical forms, or of forms undergoing a modification so gradational as to exclude the idea of new specific creations. If your reviewer prefers to suppose that new types of Foraminifera originate from time to time out of the "ooze," under the influence of "polar forces," he has, of course, a right to his opinion; though by most naturalists such "spontaneous generation" of rotalines and nummulites will be regarded as a far more "astounding hypothesis" than the one for which it is offered as a substitute. But I hold that mine is the more scientific, as being conformable to the fact that Foraminifera do propagate their kind with more or less of modification; whilst his is not supported by any evidence that rotalines or nummulites ever originate spontaneously, either in "ooze" or anywhere else.

3. Under the influence of his foregone conclusion that I have accepted Mr. Darwin as my master and his hypothesis as my guide, your reviewer represents me as blind to the significance of the general fact stated by me, that "there has been no advance in the foraminiferous type from the palæozoic period to the present time." But for such a foregone conclusion, he would have recognized in this statement the expression of my conviction that the present state of scientific evidence, instead of sanctioning the idea that the descendants of the primitive type or types of Foraminifera can ever rise to any higher grade, justifies the *anti-Darwinian* inference, that however widely they diverge from each other and from their originals, *they still remain Foraminifera.*

I cannot but regret that in his anxiety to warn your readers against the heterodox tendencies of my treatise, your reviewer should have passed by the question whether the classification proposed by my coadjutors and myself as a substitute for the artificial system of M. D'Orbigny, previously in vogue, is really founded on natural principles. The fact that the Royal Society has honoured my individual labours in this field by the award of one of the royal medals in 1861, might have been expected to secure for them a trial before any scientific tribunal upon some fairer issue than their supposed tendency to Darwinism.

William B. Carpenter.

Athenæum, 2 May 1863, pp. 586–7
Origin of species.

Mr. Darwin's notes on the modification of species, given in our impression of last week, call for a few observations. We differ from such a naturalist as Mr. Darwin with respect; but the science which he cultivates is only in its infancy; and from the conflict of opinion we shall confidently expect the truth to appear.

The "similiarity of pattern in the construction of the limbs of the great class of Vertebrata," with the like evidences of unity of plan in the organization of the Articulata; the resemblance of the different vertebrate classes to each other in their embryo state, and of all classes in their ovum state—and most of the other generalizations cited by Mr. Darwin, have been established by careful and sufficient induction, and are the well-earned fruits of such labour. They may seem to be explained or to be "connected by an intelligible thread of reasoning" on the supposition that one species engenders another, or on the hypothesis of the genetic relationship of all species with a common ancestor. Lamarck adduced the facts of the change of size and shape of parts by exercise and by disuse to explain the passage of species to species, viewing them as transitory conditions of life; but, according to observation, the Lamarckian conditions of change are too limited to originate a species. Mr. Darwin assigns the tendency to variation in individuals as sufficing in time, and with concurrent external conditions, to bring about all the various forms and grades of vegetable and animal life; but observation teaches the limits to which variation extends in regard to the genetic powers and characters of a species. Neither this, nor any better attempt to explain the transmutation of species, would entitle the guesser to appropriate to himself the application of the genetic relationship of species to a better comprehension of the cause of unity of plan, and of other established truths in biological science. The conformity of type in recent and extinct animals of circumscribed localities would be equally intelligible on the assumption of species being the offspring of species, whether under Lamarck's, or Geoffroy St.-Hilaire's, or 'Vestiges',' or Darwin's views of the way in which such genetic relation was brought about. Observation has only, however, made us acquainted with the power of the individual to engender others of its own kind. But to arrogate for a hypothesis of the *modus operandi* of "transmutation" all the consequences and applications of the old and often-mooted notion of the engendering of species in the abstract, is a large assumption, and lets a new light into the character of the mind that indulges in it. The different generalizations cited by Mr. Darwin as being "connected by an intelligible thread of reasoning" exclusively through his attempt to explain specific transmutations, are, in fact, related to it in this wise: that they have prepared the minds of naturalists for a better reception of such attempts to explain the way of the origin of species from species.

Our application of Heterogeny to the production of Foraminifera pre-supposed, as in Prof. Pouchet's exposition of the theory in the production of infusoria, the presence of binary and ternary compounds, which are due as a rule, to living acts; but the results of organic chemistry show that life is not, or may not be, essential to the production of organic compounds. Howsoever produced, or originally created, the organic matter is widely diffused over earth and sea, and the conditions of its assumption of the sarcodal or monadic forms and acts of life are a legitimate subject of inquiry, in no degree exploded by the inadequacy of such conditions to the production of a shell-fish or insect. Prof. Pouchet has prosecuted the research chiefly by the way of exclusion,—by his most ingenious contrivances and combinations to prevent or test the presence of the hypothetical atmospheric spores and germs to which Mr. Darwin with other homogenists would ascribe the appearance of living monads in infusions or sediments of organic substances. Mr. Darwin condemns the observer and expounder of the heterogenous operations in such ooze as finding therein a fitting hiding-place for the obscurity of his ideas; but the unprejudiced student of Prof. Pouchet's 'Hétérogénie' will acquit him of being open to that sneer. This work, of 672 pages, was published in the same year (1859) as Mr. Darwin's 'On the Origin of Species': to which it offers a striking contrast in the absence of mere speculative views and of expressions of feelings of belief and conviction, devoid of the grounds or proofs of such convictions. A conscientious history and close analysis of the course of human thought on Hétérogénie are premised; the rest of the work is a model of calm, clear and close philosophical experimental research and exposition. The 'Hétérogénie' resembles the 'Origin of Species' in this respect, that it aims, in the chapter devoted to the variability and temporary nature of species, to refute the arguments of Cuvier, De Blainville, Dumeril, and Lyell (p. 510), for the fixity of species. Prof. Pouchet discusses the three predicaments; viz., the origin of each species by creation as such,—their successive appearance through secondary laws,—their derivation from one unique, primitive, and only created species. This last hypothesis Prof. Pouchet rejects (p. 503) and Mr. Darwin adopts—each independently of the other.

A finite intelligence cannot ascend to the first beginning of organic matter, or of the living being, or of the elements which compose a complex crystal. The prevalent belief rests on the acceptance of the revelation of the creative acts; and these the Pentateuch teaches are of two kinds—"the waters brought forth abundantly the moving creature that hath life,"—"the earth brought forth grass," &c.; but of man it is written, He "breathed into his nostrils the breath of life."

We deem the recent evidence adduced by Pouchet, Joly, Musset, Schauffhausen, Mantegazza of Pavia, Wyman (of Boston, U.S.), sufficient to show that, in reference to the origin of the lowest forms of life, the first expression best accords with the little which it has been given to us to know on this subject.

APPENDIX VIII
The 'squib' on scientific controversies

On 2 May 1863, *Public Opinion*, a general weekly periodical, published a 'squib' satirising recent disputes between prominent men of science, particularly those between Thomas Henry Huxley and Richard Owen concerning human and simian brain anatomy. Other scientists involved in disputes mentioned in the squib were Gideon Algernon Mantell, Charles Lyell, Hugh Falconer, Joseph Prestwich, John Edward Gray, and Darwin. Rumours circulated about the authorship of the squib: those most frequently named as possible authors were Philip de Malpas Grey-Egerton, and George and Ellen Busk; however, all three denied involvement (see letter from J. D. Hooker, [24 May 1863]). Joseph Dalton Hooker clearly did not believe rumours that Egerton was the author, and was also sceptical of the suggestion that his friends, the Busks, were responsible: 'it is too clever and too impartial' (letters from J. D. Hooker, [7 May 1863] and 13 May 1863).

Egerton had already achieved notoriety with the satirical poem 'Monkeyana', signed 'Gorilla', which was published in *Punch* in May 1861 (*Correspondence* vol. 9, Appendix VIII). Like the squib transcribed below, 'Monkeyana' gave a humorous slant to the scientific controversies that followed the publication of *Origin*, particularly those concerning human descent. Several of the verses of 'Monkeyana' were devoted to the controversy between the two chief protagonists in the squib, Owen and Huxley. As a member of several London scientific societies, and having palaeontological interests, Egerton was well placed to write on the in-fighting that was fracturing the scientific community in 1863.

George Busk was similarly well qualified to comment on the increasingly public displays of disharmony amongst the scientific elite. A specialist in palaeontology, Busk was active in a number of London scientific institutions, including the Linnean Society, the Royal Society, and the Zoological Society. Busk and his wife Ellen, who was a freethinker with an active interest in natural science, were also close friends of 'Tom Huxley'. With Darwin, Busk had been one of the small group invited to sign the memorial addressed to Frederick Temple, archbishop of Canterbury, in defence of the liberal theology of *Essays and reviews* (*Correspondence* vol. 9, Appendix VI). It may be significant, therefore, that the anonymous author of the squib refers to the controversial liberal clergyman John William Colenso, bishop of Natal, as one of Huxley's 'set'.

A later source attributes the squib to a provincial surgeon, Dr George Pycroft of Exeter, Devon (G. A. Prestwich 1899, p. 180). Pycroft's public and scientific life centred chiefly on Devon, and he was not a member of London's scientific society (*Modern English biography*). If this attribution is correct, it suggests that knowledge of the controversies described had extended far beyond the small scientific coterie concerned.

Hooker drew Darwin's attention to the squib in his letter of [7 May 1863]. Darwin replied saying: 'I have read the Squib in Public Opinion: it is capital; if there is more & you have a copy, do lend it. It shows well that a scientific man had better be trampled in dirt than squabble' (letter to J. D. Hooker, [9 May 1863]). Later Darwin commented: 'The more I think of the "Sad Case" the cleverer it seems' (letter to J. D. Hooker, 15 and 22 May [1863]). *Public Opinion* reproduced only part of the squib, which had been printed anonymously in April 1863 (G. A. Prestwich 1899, pp. 180–1). It is not known how Darwin subsequently came to possess the copy of the full version of the squib that is in the Darwin Pamphlet Collection–CUL. This version, together with Darwin's annotations, is transcribed below.

A Report of a sad case Recently tried before the Lord Mayor,
Owen *versus* Huxley,
In which will be found fully given the Merits of the great Recent bone case.
London. 1863.

A sad case.
Mansion House.—April 23, 1863.
(Before the Lord Mayor.)

T. H. Huxley, well known about the town in connection with monkeys, and Richard Owen, in the old bone and bird-stuffing line, were charged by Policeman X with causing a disturbance in the streets.

The prisoners exchanged glances of such a character that it was thought prudent to keep them separated in the dock. then *Policeman X*, being sworn, stated as follows:—My attention was called to the prisoners by a crowd of persons, who seemed much excited—they appeared to take sides, and some were for Owen and some for Huxley. On coming near I saw Huxley snapping his fingers at Owen, and telling him he was only a little better than an ape; he seemed dreadful angry, and would have done Owen some bodily harm if I had not been near. He told Owen he had quite as much brains as he had, and he called him some awful names. Must I repeat the bad words your worship?

Lord Mayor—Certainly. You *must* state what he said.

Policeman X—Well, your Worship, Huxley called Owen a lying Orthognathous Brachycephalic Bimanous Pithecus; and Owen told him he was nothing else but a thorough Archencephalic Primate.

Lord Mayor—Are you sure you heard this awful language?

Policeman X—Yes, your worship, and some more I could not exactly understand.

Lord Mayor—Did you see any violence used?

Policeman X—Yes, your Worship. Huxley had got a beast of a monkey, and he tried to make it tread on Owen's heels—and said 'twas his grandfather—and like him—and just the same breed and all that; and some gentlemen cheered and said "Bravo."

Lord Mayor—Did you see the man Huxley actually put the monkey on the other prisoner—was there no interval between them?

Policeman X—He put the beast so near as ever he could; he tried to make him go quite close but he could not, and he kept singing out, "Look at'em, a'nt they like as peas?"

Lord Mayor—Did Owen appear much annoyed by this outrage?

Policeman X—He behaved uncommon plucky, though his heart seemed broke. He tried to give Huxley as good as he gave, but he could not, and some people cried "Shame," and "he's had enough," and so on. Never saw a man so mauled before. 'Twas the monkey that worrited him, and Huxley's crying out, "There they are—bone for bone, tooth for tooth, foot for foot, and their brains one as good as t'other."

Lord Mayor—That was certainly a great insult.

Huxley—So they are, my lord, I can show——

Here a scene of indescrible confusion occurred. Owen loudly contradicted Huxley; the lie was given from one to the other; each tried to talk the other down; the order 'Silence' was unheeded; and for a time nothing could be heard but intemperate language, mingled with shouts of "Posterior Cornu," "Hippocampus," "Third Lobe," &c., &c. When order was restored, the Lord Mayor stated that, in all his experience, he had never witnessed such virulent animosity among costermongers.

The Lord Mayor here asked whether either party were known to the police?

Policeman X—Huxley, your Worship, I take to be a young hand, but very vicious; but Owen I have seen before. He got into trouble with an old bone man, called Mantell, who never could be off complaining as Owen prigged his bones. People did say that the old man never got over it, and Owen worritted him to death; but I don't think it was so bad as that. Hears as Owen takes the chair at a crib in Bloomsbury. I don't think it be a harmonic meeting altogether. And Huxley hangs out in Jermyn Street.

Lord Mayor—Do you know any of their associates?

Policeman X—I have heard that Hooker, who travels in the green and vegetable line, pats Huxley on the back a good deal; and Lyell, the resurrectionist, and some others, who keep dark at present, are pals of Huxley's.

Lord Mayor—Lyell, Lyell; surely I have heard that name before.

Policeman X—Very like you may, your Worship; there's a fight getting up between him and Falconer, the old bone-man, with Prestwitch, the gravel sifter, for backer.

Owen—He's as bad as any of 'em, my lord. I thought he was a friend of mine, but he's been saying things of me as I don't like; but I'll be even with him some day.

Lord Mayor—Silence! Have you seen the prisoners in the company of any ticket-of-leave men?

Policeman X—No, your Worship; but from information I have received, I believe Huxley is one of the same set with John William Natal, or some such name, for he is one of those chaps as has got a lot of aliases, who has lately returned from abroad. John's been kicking up a pretty row—he has.

Lord Mayor—I desire you to bring him before me if you detect him in creating any disturbances.

Policeman X—Oh! your Worship, there's plenty trying to catch him; but he's so artful they can't trap him no how. They wanted to take his ticket from him, but they could not; then they tried to coax him to give it up, but he would not; not he. You see when he was across the water, he took to the bush and got in with the savages, and tried to come over them, but one of they Kaffirs gave him such a topper that he's never been the same man since.

Lord Mayor—You have not seen them together?

Policeman X—No, your Worship; but I believe they are both tarred with the same brush.

As there appeared to be no case against Owen, he was allowed to be sworn. Hereupon, Huxley demanded to be sworn likewise, but Owen objected, declaring it was impossible to swear a man who did not believe in anything, and Huxley declared it was equally impossible to swear Owen. Owen, however, was directed to take the book in his right hand, whereupon, Huxley vociferated, "He does not know a hand from a foot." An angry altercation ensued between the parties, amidst the din of which the words, "peronœus longus," "moveable toe," "thumb," "astragalus," and "short flexor," could be distinguished. The Mayor addressed both parties, and declared such violent conduct was scarcely human, at which Huxley laughed, and Owen looked grave. He then gave his evidence as follows:—

I knew the prisoner in former years. We were both in the same business, and I looked upon him as a quiet, well-meaning man. But since he has risen in the world, he has become highly dangerous, so much so, that I am willing to believe his conduct proceeds from diseased brain.

Here the Mayor called upon Dick Owen to come at once to the point.

Owen proceeded—For the last two years, my life has been a burden to me. That fellow Huxley has got new pals, Charlie Darwin, the pigeon fancier, and Rollstone, and others of that awful lot, and he waylays me in public, and throws dirt at me. Indeed, he has hit me very much about the head; very hard indeed; and he tries to make believe that I don't know my trade, and that he can teach me; and he tries to make me ridiculous in the eyes of the public, and I can't bear it. And lately I went down to Cambridge, and who should I see there but that Tom Huxley and his low set, and they all attacked me at once—(Here the Mayor directed the witness to keep to the point.)

Owen continued—I could live well enough if you could only keep that beastly monkey away from me, and make Huxley hold his tongue about comparing our brains: indeed, continued Owen, how would you like to be told in public that physically, morally, and intellectually, you were only a little better than a gorilla.

Huxley was now called upon, and said as follows:

Me and Dick is in the same line—old bones, bird skins, offal, and what not.

The Mayor—Do you mean the marine store line?

Huxley—No, your worship; that's Bowerbank and Woodward's business. Well, as I was saying, we was in the same line, and comfortable as long as Dick Owen was top sawyer, and could keep over my head, and throw his dust down in my eyes. There was only two or three in our trade, and it was not very profitable; but that was no reason why I should be called a liar by an improved gorilla, like that fellow.

[Here the Mayor cautioned the prisoner.]

Well, in my business I put up monkeys, and the last monkey I put up was Dick Owen's.

[Here the Mayor declared, on the repetition of such language, he would at once commit Huxley.]

Well, as I was saying, Owen and me is in the same trade; and we both cuts up monkeys, and I finds something in the brains of 'em. Hallo! says I, here's a hippocampus. No, there ain't, says Owen. Look here, says I. I can't see it, says he, and he sets to werriting and haggling about it, and goes and tells everybody, as what I finds ain't there, and what he finds is, and that's what no tradesman will stand. So when we meets, we has words. He will stick to his story, your worship, he won't be right himself, nor let anybody else be right. As to this here monkey business, I can't help the brutes treading on his heels. If he was to go forward more, why you'd see he'd be further off from the beast; but he's one of these here stand-still Tories, what they call the orthodox lot, as never moves forward. If he'll keep his tongue in his head, why I'll keep mine, but he shan't have the last word, or my name's not Tom Huxley.[1]

A gentleman, who we afterwards learnt was Mr. Bull, here stepped forward, and presented his card. On the Lord Mayor asking if he knew anything bearing on the case, he said he was sorry to say he knew all the parties named, and with pain must give his testimony to the bad feeling existing among them all. It was not only the case with the two prisoners, but with all their associates. There were no two of them who agreed—no one had a good word to say of the other. The whole neighbourhood was unsettled by their disputes; Huxley quarrelled with Owen, Owen with Darwin, Lyell with Owen, Falconer and Prestwich with Lyell, and Gray the menagerie man with everybody. He had pleasure, however, in stating that Darwin was the quietest of the set. They were always picking bones with each other and fighting over their gains. If either of the gravel sifters or stone breakers found anything, he was obliged to conceal it immediately, or one of the old bone collectors would be sure to appropriate it first and deny the theft afterwards, and the consequent wrangling and disputes were as endless as they were wearisome.

Lord Mayor—Probably the clergyman of the parish might exert some influence over them.

The gentleman smiled, shook his head, and stated he regretted to say that no class of men paid so little attention to the opinions of the clergy as that to which these unhappy men belonged.

The Lord Mayor thanked Mr. Bull for his evidence, and proceeded to sum up in a good-humoured strain. He said it would be absurd to make a tragedy of the matter; at the same time the public had a right to claim protection from annoyance. It appeared to him to be simply an exemplification of the proverb, "two of a trade never agree." He did not think any punishment was called for at his hands—first, because no penalty could put a stop to rivalries which seemed to be peculiar to the character of man; and secondly, because he knew of no form of punishment which could reform offenders so incorrigible. He thought if the parties would listen to a little friendly advice, they would themselves see the propriety of altering their conduct. He put it to Owen, first, whether it was worth his while to feel bitterness at being compared to an ape, and whether he could not better prove his dissimilarity by the practice of kindness, gentleness, forbearance, and humility,—qualities peculiar to man—than by giving up those qualities, and attempting to prove his manhood simply from his anatomical structure; also, whether the practice of grasping at every other person's possession was not more ape-like than human. And he put it to Huxley, whether it really *was* truth, pure and simple, he was fighting for; whether it was not more, or at least partly, for the purpose of exposing a weak point in his rival. However, he would not be too severe upon either of them. He would advise them in future to be friends—to assist each other in their work—to be helpers not rivals; and he would recommend them to associate in a social manner—to follow the example of good Edward Forbes, who used to call his fellow tradesmen together after their work was over to the Red Lion, and indulge in a little harmony; and he undertook to say, that if they were jovial, open-hearted, merry, and wise, as he was; if they would meet at some house of call, say once a week, and have their glass and their pipe together, in fact, continued the Mayor, good-humouredly, relax a little, unbend the bow, and cultivate the feelings that warmed the hearts of Tam O'Shanter, and Sonter Johnny, they would soon cease to vex themselves, and disturb the public. At the same time, it was his duty to warn them that a persistance in their present bitter course, and in giving a loose rein to passion, must eventually bring them both to the gallows.

At the conclusion of the Mayor's address, there was loud applause in the court, which was immediately checked.

Policeman X. wished to know if Huxley's monkey was to be restored to him.

Lord Mayor—Certainly, there was no law to enable him to detain it. The parties then left the dock with an expression on their countenances, which foreboded mischief. No sooner did they reach the street than an altercation ensued, and the loud voices of Huxley and Owen could be heard in the court. Policeman X. asked if he should go out to put a stop to the riot, but the Lord Mayor decided it were best not to notice it, the time and attention of the country could not be occupied with such frivolous disputes. At the same time he stated that if they were brought before him again he certainly should commit them both to Earlswood. As our reporter was passing Jermyn-street in the evening, he recognized Huxley vilifying Owen before a large assembly, holding over his head a board, on which was painted a series of monkey skeletons, beginning with the Gibbon, and ending with man, the words—"I'll let him know his place in nature," were heard, but as the assemblage consisted of working men, and as they were very orderly they were not interfered with by the police.

DAR Pamphlet Collection–CUL (R: 90)

[1] The version published in *Public Opinion*, 2 May 1863, pp. 497–8, ends here.

CD ANNOTATIONS
Top of document: 'C. Darwin' *ink*
p. 770 T. H. Huxley, . . . streets.] *square bracket in left margin, ink*
p. 771 Lord Mayor—Did Owen . . . outrage?] *angle bracket in left margin, ink*
p. 771 *Policeman X*—He behaved . . . as t'other."] *square bracket in left margin, ink*
pp. 771–2 The Lord Mayor here asked . . . same man since.] *enclosed in angle brackets, ink*
p. 772 As there appeared . . . demanded] *square bracket in left margin, ink*
p. 772 Hereupon, Huxley] *after closing square bracket, blue crayon*
p. 772 *Owen* proceeded . . . , and throws] *enclosed in square brackets, ink*
p. 772 Indeed, he has hit me] *after closing square bracket, ink*
p. 772 And lately . . . attacked me at once] *enclosed in square brackets, ink*
p. 773 *Huxley* was now called upon . . . profitable;] *enclosed in square brackets, ink*
p. 773 Well, as I was saying, Owen] *after opening angle bracket, ink*
p. 773 As to this here monkey business,] *after closing square bracket, ink*
pp. 773–4 A gentleman, who . . . belonged.] *enclosed in angle brackets, left margin, ink*
p. 773 A gentleman, who . . . influence over them.] *enclosed in angle brackets in right margin, ink*

APPENDIX IX
An appeal

This appendix reproduces a four-page pamphlet, entitled 'An Appeal', which Emma and Charles Darwin jointly composed (see letter from Emma Darwin to W. D. Fox, [29 September 1863]). The pamphlet, which protested against the cruelty of steel vermin-traps, was privately printed in July, and Emma organised the distribution of the pamphlet in August and September 1863 (see letter from G. B. Sowerby Jr to Emma Darwin, 22 July 1863 and n. 1, and letters from Emma Darwin to W. D. Fox, [6–27 September 1863], 29 September [1863], and 8 December [1863]). Emma sent a copy to the *Bromley Record*, which published the pamphlet on 1 September 1863, 'in the hope that it may have the desired effect of diminishing cruelty' (*Bromley Record*, 1 September 1863, p. 168). In addition, a version of the pamphlet was published in the *Gardeners' Chronicle* on 29 August 1863, pp. 821–2, under the title 'Vermin and traps' (*Calendar* no. 4282). The wording of the letter to the *Gardeners' Chronicle* is almost identical to the wording of the pamphlet, but the letter did not include the woodcut of the steel vermin-trap, and was signed 'C. D.'

The Darwins' 'Appeal' appears to have been inspired by a series of letters published in the London daily newspaper the *Morning Star* in 1862 (see *Animal World*, 1 January 1870, p. 75, 1 February 1872, p. 66). The letters, sent by the writer and historian William Howitt, drew attention to the scale of trapping on a game-preserving estate in North Wales, and stressed the infrequency with which the traps were visited and the hours of suffering that trapped animals experienced. Howitt focused on the injuries inflicted by the traps, and the large variety of mammals and birds trapped by them, including 'vermin' – rabbits, hawks, jays, crows, stoats, weasels, martens, polecats, foxes, and cats – and accidental victims such as pheasants and dogs. Howitt's letters, and other campaign literature, made the gamekeeper the embodiment of cruelty. In common with the Darwins' 'Appeal', the humane-trap campaigners assumed that the gentlemen employing the keepers were unwitting participants in animal torture, whose finer sensibilities would revolt against the practice if they were confronted with the evidence (Royal Society for the Prevention of Cruelty to Animals Annual Report, 1864, p. 32; *Animal World*, 1 February 1872, p. 66, 1 April 1872, pp. 99–100, 1 April 1874, p. 56).

Charles and Emma distributed the 'Appeal' to their own acquaintances, and to 'a good many persons Squires Ladies & MPs' (see letter from Emma Darwin to W. D. Fox, [6–27 September 1863], and letter from Emma Darwin to J. D. Hooker, [7 December 1863]). Although none of the replies to the circular have survived, it is clear that Emma had more success with the campaign than she expected (see the letter from Emma Darwin to William Erasmus Darwin, [2 September 1863] (DAR 219.1: 77), and this volume, letter to W. D. Fox, [29 September 1863]). Several responses suggested that a prize should be offered for a humane trap, and Emma accordingly sent out papers for subscriptions towards a prize to be awarded under the auspices of the Royal Society for the Prevention of Cruelty to Animals (see CD's Classed account book (Down House MS), 20 August 1863, recording a payment of £2 11*s.* 6*d.* for distributing the 'cruelty pamphlet', and letter from Emma Darwin to W. D. Fox, 8 December [1863]).

There is no surviving record of the subscription papers, or of Emma's correspondence with the RSPCA; however, the RSPCA Annual Report for 1864 records that 'a benevolent lady, intensely interested in the matter, induced several noblemen and country gentlemen to subscribe an amount, which she kindly placed at the disposal of your Committee' (p. 32). Sufficient funds were raised to fund a £50 prize for the design of a trap serving the purposes of game preservers 'without inflicting torture'. The scheme received the support of 'many of the leading sportsmen of the country', and Francis Trevelyan Buckland, well known to sportsmen through his regular columns in the *Field* newspaper, was a prominent supporter of the campaign. An exhibition of more than 100 humane traps, submitted in competition for the prize, was held at the Royal Horticultural Gardens, South Kensington, in June 1864 (*The Times*, 27 May 1864, p. 11, and 28 May 1864, p. 14). However, despite an optimistic start, the competition failed to produce a design that was portable, cheap, and effective. The judges appointed by the RSPCA awarded three competitors a total of £10 in recognition of the merit of their designs, but the remaining £40 was reserved in the hope of finding a better design in the future (RSPCA Annual Report 1865, p. 20). The competition was held again in 1865 and 1866, but still no single design fitted the requirements (Moss 1961, pp. 146–7). During the early 1870s the RSPCA continued to campaign against steel traps through the columns of its monthly paper *Animal World*, and prominently linked Charles Darwin's name to the offer of a prize for a humane trap (*Animal World*, 1 January 1870, p. 75). The prize was still available in 1876, when the RSPCA made a further attempt to encourage the invention of a suitable device (RSPCA General Committee minutes, 22 May 1876). Altogether the campaign had little direct effect (Moss 1961, pp. 146–7, *Emma Darwin* 2: 200).

Although the RSPCA considered in 1864 that many game preservers had 'already discontinued the use of the old steel trap', and that others were 'awakening to its barbarity' (RSPCA Annual Report 1864, p. 32), the use of the steel trap was defended by other game preservers as a practical necessity, and some maintained that it involved no more cruelty than the possible alternatives (see letter from E. L. Darwin, 7 September 1863, letter from Emma Darwin to W. D. Fox, 8 December [1863], and *Animal World*, 1 February 1872, pp. 65–6). It is clear that the use of steel traps for vermin control continued to be widespread: one estimate suggested that at least 10,000 animals were caught in steel traps in England and Wales every night (*Animal World*, 1 January 1870, p. 75). Under these circumstances, Emma was prompted to make a further public appeal against the use of steel traps in 1877 (*Spectator*, 6 January 1877, p. 15, and *Emma Darwin* 2: 200–1).

While Emma Darwin appears to have taken much of the initiative in the organisation of the campaign, Charles Darwin evidently shared her abhorrence of cruelty to animals. His sensitivity towards the suffering of animals is recorded in his autobiography, a sensitivity he attributed to the influence of his sisters during his boyhood (*Autobiography*, pp. 26–7). As an adult, he took pains to prevent cruelty to domestic animals, reporting a neighbouring farmer to the RSPCA in 1852 for working horses with sore necks (see letter from Emma Darwin to William Erasmus Darwin, [23 April 1852] (DAR 219.1: 4)); he threatened to report a similar case of cruelty in 1866 (see letter to [Local landowner], [1866], *Calendar* no. 4963). He regularly subscribed to the RSPCA, payments being recorded from 1854 to 1861, in 1863 and 1864, from 1871 to 1875, and in 1878 and 1880 (CD's Classed account books (Down House MS)). In *Descent* 1: 101, he argued that sympathy for animals was one of the noblest moral qualities possessed by human civilisation. However, Darwin was not against blood-sports: his autobiography makes clear the intense pleasure he took in shooting, both as a boy and as a young naturalist (*Autobiography*, pp. 44, 54–5, 78, and *Correspondence* vol. 2, letter to W. D. Fox, 28 August [1837]). Later he gradually gave up shooting, but did not censure those of his friends and family who enjoyed the sport (see *Autobiography*, pp. 78–9, *Correspondence* vol. 7, letter to W. E. Darwin, 22 [September 1858], and this volume, letter to J. B. Innes, 1 September [1863]).

The 'Appeal' should be set in the context of two historical movements in Victorian Britain that were to some extent in conflict: first, the extension of game preserving, and second, the growth of a humane movement opposed to the deliberate infliction of suffering on animals. The improvement of shotguns, particularly the introduction of the cartridge in 1808 and the breech-loading shotgun in 1861, transformed the traditional sport of shooting partridges, pheasants, and other game-birds (Carr 1981, pp. 475–6, Mackenzie 1988, p. 18, Allen 1994, p. 127). The transformation was encapsulated in the 'battue', in which large numbers of birds were driven from bushes by beaters towards a point where sportsmen waited to shoot them; the size of 'bags' (kills) at shooting events increased accordingly. This practice was well established by the 1830s, and was extended rapidly over the course of the century. Shooting game was the most widespread and rapidly growing country pursuit of the landed classes (Thompson 1981, p. 459). On country estates game-birds began to be reared on a large scale for this sport, and the topography of rural England was transformed as coverts, or 'game preserves', were set aside to provide habitat for them (Mackenzie 1988, p. 26). The extension of game preserves brought with it increased problems of pest control: the number of foxes and other predators had to be controlled to protect the game-birds, and the new coverts provided habitat for a growing rabbit population, which had to be controlled to reduce the depredation of farmers' crops on neighbouring land. As a result of the need to protect game and control vermin, there was a sixty per cent rise in the number of gamekeepers in England between the 1860s and the early 1900s; Kent was one of the most densely 'keepered' counties (Carr 1981, p. 484, Thompson 1981, pp. 460–1). The extension of the use of steel vermin-traps was a result of these developments.

The early nineteenth century also saw the growth of a significant humane movement, arousing the sensitivities of sections of the upper and middle classes to the sufferings of animals. For these groups, kindness to animals had become an integral part of what it meant to be morally distinguished and civilised (Turner 1980, pp. 17–45, Thomas 1984, pp. 149–50, 300–3, Ritvo 1990, pp. 126–135). As a result of systematic campaigning, legislation was passed in 1822, 1835, and 1849 (see nn. 1 and 5, below) to prevent cruelty to farm animals, and to animals involved in the blood sports of the 'lower orders'. In the 1860s, the humane movement extended its sympathy also toward wild animals, a transition which was marked by campaigning and legislation to prevent the wholesale slaughter of categories of wild birds by sportsmen (Sheail 1976, pp. 22–7, Allen 1994, p. 177), and which some scholars have attributed to a growing 'biocentricism' following the publication of *Origin* (Turner 1980, pp. 60–2, 124–128, Worster 1985, pp. 179–80, 184–7).

An appeal

It is a common observation that cases of brutality to horses, asses, and other large quadrupeds, are much less frequently witnessed now than they were some time ago. This is no doubt owing to the general increase of humanity, and to these animals being now under the protection of the law.[1]

An English gentleman would not himself give a moment's unnecessary pain to any living creature, and would instinctively exert himself to put an end to any suffering before his eyes; yet it is a fact that every game preserver in this country sanctions a system which consigns thousands of animals to acute agony, probably of eight or ten hours duration, before it is ended by death. I allude to the setting of steel traps for catching vermin.

Some women may never have seen a trap, and therefore I give a wood-cut of one.[2]

[Reproduced by kind permission of the Master and Fellows of Christ's College, Cambridge]

The iron teeth shut together with so strong a spring, that a pencil which I inserted was cracked and deeply indented by the violence of the blow. The grip must be close enough not to allow of the escape of a small animal, such as a stoat or a magpie; and therefore when a cat or a rabbit is caught, the limb is cut to the bone and crushed. A humane game-keeper said to me, "I know what they must feel, as I have had my finger caught."[3] The smaller animals are often so fortunate as to be killed at once. If we attempt to realise the sufferings of a cat, or other animal when caught, we must fancy what it would be to have a limb crushed during a whole long night, between the iron teeth of a trap, and with the agony increased by constant attempts to escape. Few men could endure to watch for five minutes, an animal struggling in a trap with a crushed and torn limb; yet on all the well-preserved estates throughout the kingdom, animals thus linger every night; and where game-keepers are not humane, or have grown callous to the suffering constantly passing under their eyes, they have been known by an eye-witness to leave the traps unvisited for twenty-four or even thirty-six hours. Such neglect as this is no doubt rare; but traps are often forgotten; and there are few game-keepers who will leave their beds on a cold winter's morning, one hour earlier, to put an end to the pain of an animal which is safely in their power.

I subjoin the account of the appearance of a rabbit caught in a trap, given by a gentleman, who, last summer witnessed the painful sight many times.[4]

> "I know of no sight more sorrowful than that of these unoffending animals as they are seen in the torture grip of these traps. They sit drawn up into a little heap, as if collecting all their force of endurance to support the agony; some sit in a half torpid state induced by intense suffering. Most young ones are found dead after some hours of it, but others as you approach, start up, struggle violently to escape, and shriek pitiably, from terror and the pangs occasioned by their struggles."

We naturally feel more compassion for a timid and harmless animal, such as a rabbit, than for vermin, but the actual agony must be the same in all cases. It is scarcely possible to exaggerate the suffering thus endured from fear, from acute pain, maddened by thirst, and by vain attempts to escape.

Bull-baiting and cock-fighting have rightly been put down by law;[5] I hope it may never be said that the members of the British Parliament will not make laws to protect animals if such laws should in any way interfere with their own sports.

Some who reflect upon this subject for the first time will wonder how such cruelty can have been permitted to continue in these days of civilisation; and no doubt if men of education saw with their own eyes what takes place under their sanction, the system would have been put an end to long ago.

We shall be told that setting steel traps is the only way to preserve game, but we cannot believe that Englishmen when their attention is once drawn to the case, will let even this motive weigh against so fearful an amount of cruelty.

The writer of these remarks will be grateful for any suggestions, addressed to A. B.,[6] *Mr. Strong, Printer, Bromley, Kent.*[7] or to Mrs C. Darwin | Down| Bromley| Kent.[8]

Christ's College Library, Cambridge (Fox 142)

[1] The Act to prevent the cruel and improper treatment of cattle, 1822, prohibited the ill-treatment of horses, asses, sheep, and cattle, excluding bulls (*Statutes, public and general*, 3 Geo. IV c. 71). On attitudes to cruelty to animals in nineteenth-century England, see Turner 1980.

[2] This sentence, and the woodcut of the vermin-trap, did not appear in the version published in the *Gardeners' Chronicle*, 29 August 1863, pp. 821–2 (*Calendar* no. 4282). The woodcut was arranged through George Brettingham Sowerby Jr (see letter from G. B. Sowerby Jr to Emma Darwin, 22 July 1863 and n. 1).

[3] This individual has not been identified; however, see letter from Emma Darwin to W. D. Fox, 8 December [1863]. Only two resident gamekeepers were recorded in Down, Kent, in 1861: John Higwood, under-gamekeeper, age 34, and William Folgate, gamekeeper, age 50 (Census returns 1861 (Public Record Office, RG9/462: 73, 77)).

[4] The reference is to William Howitt; the quotation is taken from Howitt's letter to the *Morning Star*, 8 December 1862 (see *Animal World*, 1 January 1870, p. 75).

[5] Animal baiting, including bull-baiting and cock-fighting, was made an offence by the Act to consolidate and amend the several laws relating to the cruel and improper treatment of animals, 1835 (*Statutes, public and general*, 5 & 6 Will. IV c. 59), and the Act for the more effectual prevention of cruelty to animals, 1849 (*Statutes, public and general*, 12 & 13 Vict. c. 92).

[6] The 'Appeal' was signed 'C. D.' in the version published in the *Gardeners' Chronicle*, 29 August 1863, pp. 821–2 (*Calendar* no. 4282).

[7] Edward Strong was a printer and stationer with premises on High Street, Bromley (*Post Office directory of the six home counties* 1863).

[8] The closing words, printed here in roman type, were added in Emma Darwin's hand at the end of one of the two copies of the pamphlet among the W. D. Fox papers in Christ's College Library, Cambridge.

MANUSCRIPT ALTERATIONS AND COMMENTS

The alteration notes and comments are keyed to the letter texts by paragraph and line numbers. The precise section of the letter text to which the note applies precedes the square bracket. The changes recorded are those made to the manuscript by CD; readers should consult the Note on editorial policy in the front matter for details of editorial practice and intent. Superscript numerals are used where necessary to distinguish between occurrences of the same letter in a word. The following terms are used in the notes as here defined:

del	deleted
illeg	illegible
interl	interlined, i.e., inserted between existing text lines
omitted	omitted by the editors to clarify the transcription
over	written over, i.e., superimposed

To Asa Gray 2 January [1863]
1.8 live] *after del* 'a'
2.3 (to anthers) 2.4] *interl*
2.5 Mitchellas] *altered from* 'Mitchella'
2.6 these] *altered from* 'this'
2.6 two] *interl*
2.6 Families] *altered from* 'Family'
8.1 flowers of] *interl*
8.1 well-colored] *interl*

To Joseph Dalton Hooker 3 January [1863]
1.1 do] *interl*
1.2 in N.H. Review] *interl*
1.5 Lyell,] *comma over* '—'
1.7 mind,] *before omitted comma*
1.7 as I] *after del* 'I'

To John Lubbock 4 January [1863]
1.5 & I will do my best. 1.6] *interl*
2.1 What] 'W' *over illeg*

To James Dwight Dana 7 January [1863]
1.7 for the Society] *interl*
2.3 his Book] *above del* 'it'

To Thomas Rivers 7 January [1863]
1.3 Knight's] *altered from* 'Knighst'
3.1 1786] *altered from* '1886'
3.10 comparatively] *above del* 'commonly'
4.4 Has ... moss-rose?] *added*
6.1 very] *interl*
6.5 in the Peach] *interl*
6.6 apparently] *interl*
6.7 Pears or] 'o' *over* 'b'

To Thomas Henry Huxley 10 [January 1863]
1.5 me] *interl*
1.13 any] *interl*
2.1 over] 'ov' *over* 'of'
2.1 (but I now doubt)] *interl; square brackets in MS*
2.5 mammals] *above del* 'animals'
2.13 or more] *interl*
2.13 perfectly] *after del* 'almost'
4.2 has] *after del comma*
4.3 do,] *interl*
4.6 your] *interl*
4.8 waste] 'was' *over* 'wai'

To Thomas Rivers 11 January [1863]
2.2 has] *interl*
5.6 your] *above del* 'the'
5.7 almond;] ';' *altered from* ':'
7.2 chance] *after del* 'hope'

To Henry Walter Bates 12 January [1863]
1.3 by Book Post] *interl*

4.3 state;] ';' *altered from* ':'

To J. D. Hooker 13 January [1863]
1.1 have] *after del* 'hope'
1.3 qualities] *after del* 'disea'
1.5 begins] *after del* 'tak'
3.2 this] *over* 'it'
3.2 letter,] *interl*
3.5 any] *interl*
5.2 Andrew Murray knows.—] *added*
7.1 How] *above del* 'How'
7.3 1853] *interl*
8.1 & allies,] *interl*
8.2 has] *above del* 'wor'
8.4 or struggle] *after del* 'of'
9.3 only] *interl*
9.4 your Hedysarum, Mimosa 9.5] *interl*
9.5 daresay] 'd' *over* 'D'
10.3 when here] *interl*

To Alphonse de Candolle 14 January [1863]
1.5 I noted . . . achenia &c &c.—] *added in margin*
2.7 real] *interl*
2.10 will] *after del* 'has'

To Smith, Elder & Company 14 January [1863]
1.1 unpaid] *interl*
1.3 now] *interl*

To Thomas Rivers 15 January [1863]
2.2 & *fruit*] *interl*
2.3 not rob you & would] *interl*
3.1 immediately] *interl*

To Thomas Rivers 17 [January 1863]
1.2 from a Peach] *interl*

To Asa Gray 19 January [1863]
1.8 next] *before omitted point*
1.9 fruit of] *interl*
1.11 of Orchids] *interl*
2.7 Monochætum,] [1]'o' *interl*
2.9 or horns] *interl*

To Daniel Oliver 20 [January 1863]
4.3 (& I then] *parenthesis over comma*

To John Scott 21 January [1863]
1.4 only] *interl*
1.5 Rostellum & in stigma of] *interl*
1.7 (& what] *parenthesis over comma*
1.8 anxiously] *after del* 'always'
1.8 had] *added*
1.8 & still hope to do] *interl*
1.9 or] *after del* 'of'
2.2 quite] *interl*
2.2 new;] ';' *over* ':'
2.3 small flower of] *interl*
2.6 gum] *above del* 'gum'
2.9 on Linum] *interl*
2.12 the] *interl*
2.12 -masses] *interl*
3.4 lower ends of] *interl*
4.5 buds] *above del* 'bulbs'
6.1 Am] *over illeg*
8.2 seeded] *interl*
9.1 Yesterday] 'Y' *over illeg*
9.1 of Hay Lodge, Trinity 9.2] *interl*
9.6 & asking for information from] *above del* 'to'
11.7 but] *after del* 'except'

To H. W. Bates 26 January [1863]
2.4 &] *over* 'but'

To William Bernhard Tegetmeier 26 [January 1863]
3.1 white] *interl*

To J. D. Hooker 30 January [1863]
4.2 of Good Hope] *interl*
4.3 more] *after del* 'th'
4.6 experimental:] ':' *over comma*
4.6 which,] *after del illeg*
4.6 which, . . . avoid] *interl*
4.7 from Kew] *interl*
4.9 distinct flower,] 'distinct' *interl*
6.1 it announced] *above del* 'announced'
7.2 old] *interl*

To Alphonse de Candolle 31 January [1863]
1.5 nearly] *interl*
1.6 (not very . . . greatly)] *square brackets in MS*

To Roland Trimen 31 January [1863]
1.7 are,] *interl*
1.10 high] *interl*
1.21 the] *interl*
1.22 & discover . . . insects.—] *added*
1.27 &] *over* '.—'
2.3 or very] *after del* 'fo'
2.4 in each great class.] *interl*

To Thomas Rivers 1 February [1863]
1.3 when opened] *interl*

1.7 on lower] *after del comma*
1.8 inch.!] '!' *over* '—'
3.2 inexplicable] *after del* 'of'
3.3 weeping] *interl*

To John Joseph Briggs 2 February [1863]
1.2 tells] 't' *over* 's'
1.8 of fin 1.9] *interl*

To Charles Lyell 4 [February 1863]
1.3 very soon] *interl*

To T. H. Huxley [8 February 1863]
1.1 blundering; & 1.2] ';' *over* ':'; '&' *added*
1.2 generally] *interl*
1.3 do??] '??' *over* '.—'
1.4 cases] *after del* 'pe'
2.1 solely] *interl*

To William Darwin Fox [10 February 1863]
1.4 (*possibly*, ... Saturday—) 1.5] *square brackets in MS*
2.5 relations,] *interl*
2.5 at a time] *interl*

To George Gabriel Stokes [12 February 1863?]
2.2 calculating] *above del* 'putting'
3.1 for myself] *interl*
3.1 sentence] *interl*
3.2 & figures] *above del* 'ne'

CD memorandum 14 February 1863
1.1 of Prof. ... 1855] *interl*
1.3 dispose] *after del illeg*

To Thomas Rivers [14 February 1863]
3.2 & "natural selection".] *interl*

To J. D. Hooker 15 February [1863]
2.7 before] *after del* 'by'
3.1 & of course ... cold, 3.2] *interl*
3.3 at Garden 3.4] *interl*
4.1 (of same species)] *parentheses over commas*
4.1 of *erect*] 'of' *interl*
4.1 of *drooping*] 'of' *interl*
5.1 overrated] 'over' *above del* 'under'
6.1 old] *over* 'od'

To Horace Benge Dobell 16 February [1863]
1.6 on] *after del* '"'
2.6 would you 2.7] 'you' *interl*
2.10 be] *over* 'been'

To T. H. Huxley 16 February [1863]
1.1 once] *interl*
1.5 the wonderfully] 'the' *interl*
1.6 May I say] *altered from* 'I may say'
1.7 by] *after del* 'into'
1.7 very] *after del* 'almost'
1.9 one of] *interl*
2.2 good] *interl*

To John Scott 16 February [1863]
2.4 now] *interl in CD's hand*
7.1 kinds] *interl*
7.1 Small red ... the latter. 7.2] *added*

To Roland Trimen 16 February [1863]
1.2 orchids] *interl*
1.4 pollen-] 'e' *over* 'i'
2.12 on two sides] *interl*

To Charles Lyell 17 [February 1863]
1.4 grown] *interl*
1.5 as] *over illeg*
1.6 analogous] *interl*
1.6 (for they ... to mouth) 1.7] *square brackets in MS*

To W. B. Tegetmeier 19 February [1863]
1.3 nests;] *after del* 'litters'
2.2 mentioned] *above del* 'selected'
3.3 (for instance ... Tumblers)] *parentheses over commas*

To J. D. Dana 20 February [1863]
1.1 Man] 'M' *over* 'm'
2.2 separation] *altered from* 'separate'
2.2 of countries] 'of' *interl*
2.3 really now] *interl*
2.4 I admit] 'I' *over illeg*
2.11 As] *altered from* 'as'
2.16 would have been] *interl*
3.5 with teeth] *interl*
3.5 which] *after del* 'whi'
3.10 Rhinoceroses,] *altered from* 'Rhinoscerose,'

To Daniel Oliver 20 [February 1863]
1.4 points] *above del* 'cases'
1.6 enlighten me] *before omitted point*
1.7 that of] *interl*
2.4 he] *over* 'his'
3.4 sexual] *interl*
3.4 all three] *interl*
3.4 together] *after del* 'all'
4.1 the Orchids] *interl*

To John Scott 20 [February 1863]
1.2 not] *interl*
1.2 seed] *interl*
2.3 from variable size & transparency] *interl*
3.6 still] *after del* 'do'
3.10 ie produced . . . 2 generations] *interl*
3.11 a single] 'a' *interl*
4.1 I cannot . . . yet.—] *added*

To J. D. Hooker [21 February 1863]
1.2 number!] '!' *over full stop*
4.1 Bell-glasses] *interl*

To Asa Gray 23 February [1863]
1.3 mutual] *interl*
2.3 attachment.] *before omitted point*
3.1 parts of] *interl*
3.9 he] *interl*
3.11 form] *after del illeg*
3.12 lately] *interl*
5.5 you conquer] 'you' *interl*
5.5 the border] 'the' *interl*
5.6 For] *interl*

To John Lubbock 23 [February 1863]
3.1 with you] *interl*

To J. D. Hooker 24[–5] February [1863]
1.2 tomorrow;] ';' *over* ':'
1.5 the best] 'the' *interl*
1.7 of the evidence] *interl*
1.10 species;] ';' *over* ':'
2.1 discussion] *interl*
2.3 nearly as follows 2.4] *interl*
2.11 openly] 'op' *over illeg*
2.12 that man] 'that' *over* 'then'
2.13 proper] *after del* 'fa'
2.15 (but] *parenthesis over comma*
2.15 but then] *after del* 'but'
2.17 that] *added*
2.18 a certain] 'a' *interl*
4.1 & Acropera &c.] *interl*
4.3 but] *over* '&'
4.5 Have you . . . observe it.— 4.6] *added*
5.1 by their tips] *interl*
6.4 that] *interl*
7.6 of Species,] *interl*
7.8 (which . . . for me)] *parentheses over commas*
8.3 Owen's] *above del* 'his'
8.8 Owen] 'O' *over* 'o'
8.11 ; it was . . . approbation.] *interl*
8.14 a single] 'a' *interl*
9.2 already] *interl*
10.1 present] *interl*

To T. H. Huxley [before 25 February 1863]
3.2 in his Strawberry Book] *interl*

To T. H. Huxley 26 [February 1863]
1.7 as compared . . . monkeys] *interl*
3.3 the highest 3.4] *interl*
4.6 will] *after del* 'whether'

To George Maw 28 February [1863]
3.1 on Coal] *after del* 'are'

To H. W. Bates 4 March [1863]
4.1 Agassiz] *after del* 'Assi'

To J. D. Hooker 5 March [1863]
1.2 the] *interl*
1.5 ever] *interl*
2.1 they] *altered from* 'They'
2.2 & answer] *interl*
2.3 nearly] *interl*
2.5 now] *interl*
2.7 with] *interl*
5.2 or origin] *interl*
5.2 &] *after del* 'by descent—'
5.4 of Origin] *interl*
5.5 gives] *after del* 'there'

To Thomas Rivers 5 March [1863]
1.3 & would] '&' *over* 'I'
1.3 seed; I 1.4] 'seed' *interl*
2.1 double] *above del* 'dwarf'
3.1 (& my hand . . . usual)] *added*
3.3 heart, with] ', with' *over* '.—'

To Smith, Elder & Company 5 March [1863]
1.2 the full amount of] *interl*
1.3 gladly] *interl*
3.1 has] *altered from* 'have'
3.3 (at least . . . them)] *interl*

To Charles Lyell 6 March [1863]
1.3 all of us] *interl*
2.6 have not] 'have' *above del* 'are'
9.1 itself at top] *interl in CD's hand*
10.4 *deeply*] *underl in CD's hand*
14.5 These latter cases seem] *altered in CD's hand from* 'This case seems'
14.6 full] *interl in CD's hand*
15.2 as it . . . puzzling.—] *interl in CD's hand*

16.1 Lions, Hippopotamus 16.2] *interl in CD's hand*
16.2 living] *interl in CD's hand*
21.2 Origin;] *altered in CD's hand from* 'Origin.'
21.2 see my Hist. Introduct. I asked him.] *added in CD's hand*; 'him.' *interl*

To John Scott 6 March 1863
1.1 even] *interl in CD's hand*
4.3 "correlation"] *quotation marks added in CD's hand*
7.1 *several*] *above del* 'many'
7.2 be fertilised by that of] *interl*
7.2 other] *after del* 'by'
8.1 (I see . . . other) 8.2] *interl; square brackets in MS*
8.1 (Transact. . . . same result.—) 8.3] *square brackets in MS*
9.1 in German. Hort. Periodical] *interl*
9.2 similar] *above del* 'same'
9.3 own;—] ';' *over* ':'
9.3 &c &c—] *below del* 'but may be female again'
10.1 I have] *after del* 'I think Sabine gives case in Hort. Transactions of Passiflora.—'; 'no' *circled & added over deletion*
11.1 I have . . . Staunton 11.2] *added*
12.1 certainly] *interl*

To W. D. Fox 9 March [1863]
3.5 domestic] *above del* 'tamed'

To Charles Lyell 12–13 March [1863]
2.2 part of the case] *interl*
2.2 , though] *over* '.—'
2.3 me,] *interl*
2.6 at all] *interl*
3.3 before Lamarck 3.4] *interl*
3.5 they] *over* '&'
3.6 in common] *after del caret*
3.6 between the Origin &] *interl*
3.6 Lamarck] *after del* 'with'
3.8 Wallace's] 'W' *over* 'm'
3.9 as] *interl*
3.9 (I well remember my surprise)] *parentheses over commas*
3.10 rank] *altered from* 'ranked'
3.11 but enough] 'e' *over* 'E'
4.5 approve] *after del* 'he'
4.5 (in my opinion)] *parentheses over commas*
4.7 your very] 'your' *interl*
6.3 falsely] *interl*
7.3 has] *above del* 'does'
7.3 discovery of] *interl*

To J. D. Hooker 13 [March 1863]
2.7 Cerebellum &] *interl*
3.3 Creation] 'C' *over* 'c'
3.6 & did believe in descent] *interl*
4.1 Propagate] 'P' *over* 'K'
6.2 which you will easily see, 6.3] *interl*
6.4 to] *interl*

To W. D. Fox 16 [March 1863]
2.1 sentence] *above del* 'P.S'
3.1 Malvern;] ';' *over* '.—'

To J. D. Hooker 17 March [1863]
1.8 & every copy sold] *interl*
1.9 ie last sheet corrected] *interl*
1.9 (& Oct 2[d]. . . . Ilkley)] *interl*
3.4 (which I . . . no one) 3.5] *parentheses over commas*
3.8 (except you)] *interl*
4.1 very] *interl*
5.3 on Sunday 5.4] *interl*
7.3 & experiments.] *interl*
7.3 the whole] *interl*
7.4 inherits from] *above del* 'resembles'
7.5 & parent. 7.6] *interl*

To Charles Lyell 17 March [1863]
1.5 for man] *interl*
1.6 at] *after del* 'about'
3.8 Book] 'B' *over illeg*
4.2 nearly] 'n' *over illeg*
4.5 on Species-question.] *interl*
4.5 now] *interl*
4.7 on Oaks] *interl*
4.9 with me] *interl*
5.2 The entomologists . . . century. 5.3] *interl*
6.5 himself] *interl*
6.5 date] *after del* 'that'
6.7 from] *over* 'on'
7.2 very sorry 7.3] 'very' *interl*

To Asa Gray 20 March [1863]
2.2 in vain] *interl*
3.4 in change] *interl*
3.7 & *sheds*] *interl*
4.8 right] *interl*
4.12 stigma] *after del* 'pistil'
5.3 to Malvern for] *interl*
7.1 Chapters on History of war] *added*

To W. D. Fox 23 March [1863]
1.3 lambs] *interl*
3.4 &] *over* 'e'

To an editor 24 March [1863]
1.4 & accurate 1.5] *interl*

To Daniel Oliver 24–5 March [1863]
1.4 from Books] *interl*
1.5 long] *above del* 'short'
1.6 rather short] *interl*
1.6 with stigmas] *interl*
1.7 the] *over* 'three'
1.7 a little] *interl*
1.7 *not*] *interl*
1.8 without tearing] *interl*
1.13 It looks . . . of axis.] *interl*
2.1 per Railway] *interl*
2.1 these] *interl*
2.2 not] *interl*
3.1 tetraptera;] *interl*
3.2 2 or 3] *interl*
3.2 or bracts] *interl pencil*
6.1 25th] *interl*
6.2 About] *interl*
6.2 with particular thanks] *interl*
7.1 Case . . . of willows.] *interl*

To John Scott 24 March [1863]
1.7 that such] *after del* 'it was'
1.7 recorded] *interl*
1.10 you will] *interl*
1.10 not publish] 'publish' *interl*
1.11 for I may . . . for years.] *interl*
3.3 is] *over* 'if'
3.5 the adjoining] 'the' *added*
4.3 that] *interl*
4.3 others were] 'were' *interl*
4.5 the promise] 'the' *interl*
4.5 but not yet surety 4.6] *interl*
4.7 parent have 4.8] *interl*
4.8 presented] *altered from* 'present'

To J. D. Hooker 26 [March 1863]
1.1 &] *over* 'it'
2.4 particularly] *interl*
3.1 early] 'e' *over* 'E'
4.5 describe] 'd' *over illeg*
4.11 little] *interl*
4.14 such] *after del* 't'
5.2 the] *interl*
5.3 secondary] *interl*
8.2 a notion] 'a' *added*
8.2 (no, I have firm conviction!)] *interl*; *square brackets in MS*
8.4 or European withstand heat 8.5] *interl*
8.5 in Hot-house] *interl*

To Daniel Oliver 28 March [1863]
1.2 of Edwardsia] *interl*
1.6 Nierembergia] *above del* 'Nerima'
1.6 I saw] *after del* 'or such such name'
2.3 one specimen] *interl*
2.5 ran] *interl*
2.5 the middle of pistil] *interl*

To J. D. Hooker [29 March 1863]
1.2 barring] *after del* '&'
1.2 have enjoyed] 'have' *interl*
2.4 writer] *before del quotation mark*
5.1 real] *interl*
5.4 Glacial] *above del* 'Arctic'

To George Henry Kendrick Thwaites 30 March [1863]
1.7 of the alternate stamens] *interl*
1.10 in sexual functions,] *interl*
1.11 more] *above del* 'unus'
1.14 seed?] '?' *over* ':'
2.1 Any flower . . . *tri*morphism.— 2.2] *added*
2.1 anther] *after del* 'stam'

To Charles Turner [1 April – 16 June 1863?]
1.1 asking] *above del* 'begging'
1.1 from you] *after del* 'a small'
1.4 of several kinds of Hollyhock] *interl*
1.4 Hollyhock] *after del* 'of the'
1.4 true] *after del* 'fr'
1.6 to prevent crossing &] *interl*
1.6 As] *altered from* 'as'
1.7 Bees] *after del* 'the'
1.8 have been told that 1.9] *above del* 'believe'
2.1 answer] *above del* 'answer'
2.1 feel] *below del* 'be'
2.2 which varieties] *above del* 'whether'
2.2 come] *after del* 'many of the varieties'
3.1 & grant this favour] *interl*

To Maxwell Tylden Masters [8–13 April 1863]
1.2 false] 'fa' *over illeg*

To H. W. Bates 9 April [1863]
1.4 a conviction] 'a' *interl*
1.8 large] *altered from* 'largely'

To Daniel Oliver [12 April 1863]
1.2 or rather . . . "amphitropal".— 1.3] *interl*
1.5 days, &] '&' *over illeg*

1.8 (what I suppose to be)] *parentheses over commas*
1.8 "orifice"] *after del* 'for'
1.8 generally] *after del* 'at'
2.1 pollen-tubes] *after del* 'Pri'
7.2 (if I understand the ovule)] *interl*
7.4 could] *altered from* 'Could'

To John Scott 12 April [1863]
1.2 (in a condensed form)] *parentheses over commas*
1.5 well worth it] 'well' *interl*
2.2 this summer] *interl*
2.2 further] *after del* 'to'
2.7 plant] *interl*
2.11 the ovules] 'the' *interl*
2.14 I sh^d^.] 'I' *over* '&'
2.15 the almost] *interl*
2.16 protrusion . . . pollen-tubes.] *above del* 'flowers being closed.'
2.19 habitually] *interl*
5.1 your] *interl*

To David Thomas Ansted 15 April 1863
1.1 that] *interl*
3.1 not] *interl*
3.1 can] *above del* 'must'
3.2 sum of 100£,] *interl*
3.3 judge] *above del* 'decide'
3.3 decide] *after del* 'fix'
4.1 Yours] *after del* 'Believe me'

To George Bentham 15 April [1863]
3.1 in his] 'in' *above del* 'to'
5.1 but] *above del* '&'
7.1 & partly Botanical] *interl*
10.2 save] *above del* 'safe'

To J. D. Hooker [17 April 1863]
1.4 lately] *interl*
1.10 his] *interl*
1.14 last] *before omitted point*
7.1 known] *below del* 'told'
7.4 are I find] 'I find' *interl*
7.5 a] *interl*

To H. W. Bates 18 April [1863]
3.2 on Monday] *interl*

To Charles Lyell 18 April [1863]
1.4 myself] 'l' *over* 'f'
1.5 analogous] *interl*
1.8 this] *over illeg*
1.9 is] *after del* 'if'
1.9 the first] 'the' *added*
1.10 My letter . . . thing.— 1.11] *added*
3.4 for] *over illeg*

To Asa Gray 20 April [1863]
1.1 (did] *parenthesis over colon*
2.1 from] *over* 'on'
3.3 M. Devay] *above del* 'Devay'
3.4 evil shown by] *interl*
4.6 go by] *above del* 'are'
4.7 of organs] *interl*
4.8 alternation] *above del* 'alteration'
4.8 parts of the] *interl*
5.2 towards Lyell.] *interl*
7.3 a Mr J. Scott] *interl*
7.3 genera] *above del* 'species'
7.7 some primroses] *altered from* 'a primrose'
7.13 all round] *interl*
7.13 orifice of labellum are 7.14] *above del* 'opening is'
7.14 the Bee 7.15] 'the' *interl*
7.17 firmly] *interl*
7.22 well] *interl*

To H. B. Dobell 21 April [1863]
1.4 (if any)] *parentheses over commas*
1.4 & of any subsequent marriage 1.5] *interl*
1.6 or two] *added*
1.6 papers] *altered from* 'paper'
1.7 contractions] *after del* 'corre'
1.7 nothing] *altered from* 'not'
1.8 (not congenital)] *parentheses over commas*
5.1 in full] *interl*
5.2 alone] *interl*
5.3 in all] 'in' *over* '.—'

To George Bentham 22 April [1863]
3.1 & quite agree with you.—] *added*; '& quite' *over* '—'

To J. D. Hooker 23 April [1863]
1.4 on Monday] *interl*
1.8 total] *interl*
1.9 some] *interl*
1.9 a Hot-House plant] *interl*
2.4 mammal] *after del* 'fo'
2.6 saying] *after del* 'alluding'
2.7 show] *interl*
2.8 of man,] *interl*
3.1 of Wellwitschia] *interl*
3.2 "bristling"] *interl*
3.3 sterile] *above del* 'very'

3.4 (1)] *interl*
3.5 (seed-producers)] *interl*
3.6 (seed-producers)] *interl*
3.6 in number] *interl*
3.6 "bristling"] *after del* 'with'
3.8 male] *interl*
5.3 of Orchids] *interl*
5.4 paper.] *added*

To Andrew Crombie Ramsay 29 April [1863]
1.6 the end] 'the' *interl*
1.7 Physical] *interl*
1.8 & creeps.] *interl*
1.9 physical] *interl*
1.15 (Thickness of formation?)] *square brackets in MS*
2.3 say] *after del* 'tell'
2.5 our] *after del* 'at'
2.6 his] *interl*
3.7 sea-beds] *interl*
3.9 stationary or rising] *interl*
3.10 whole] *interl*
4.1 the great] 'the' *interl*

To H. W. Bates 30 April [1863]
3.2 of mouth] *interl*
5.1 you give] 'you' *over* 'to'
8.1 have] *interl*

To John Scott 2 May [1863]
1.6 in this] 'in' *over* '—'
4.1 Thank] *altered from* 'Thanks'
4.1 for seeds;] *interl*
7.1 of several genera & Diptera] *interl*
7.2 a very] 'a' *interl*
7.2 in several ways] *interl*

To W. E. Darwin [5 May 1863]
1.3 & mint 1.4] *interl*
1.7 form] *after del* 'one'
1.7 (1)] *interl*
1.8 second] *above del* 'other'
1.9 new case & would be 1.10] *interl*
1.12 flowers] *interl*

To Charles Lyell [7 May 1863]
0.1 at] *before omitted point*
2.2 do] *over illeg*

To J. D. Hooker [9 May 1863]
1.4 in what he says] *interl*
7.2 ; we return on Wednesday morning] *interl*

To Thomas Rivers [9 May 1863]
1.5 Double] *interl*

To W. E. Darwin [10 May 1863]
1.5 sent,] *after del* 'were'
5.1 Mr Fould] *interl*
5.1 (what] *parenthesis over comma*

To Asa Gray 11 May [1863]
1.4 Judge.] 'Judge.' *altered from* 'judge.'
2.7 angles of 2.8] *interl*
3.8 which is rather rare] *interl*
4.1 amygdaloides] *interl*

To Osbert Salvin 11 [May 1863]
2.2 of every species] *interl*
2.3 on the] *after del closing parenthesis*

To William Henry Flower 12 May [1863]
2.2 if] *after del* 'f'

To George Maw 12 May [1863]
1.9 would like] 'would' *interl*

To Richard Kippist 13 May [1863 or 1868]
1.3 the rule] 'the' *interl*

To J. D. Hooker 15 and 22 May [1863]
2.1 & Species] *interl*
2.4 Lyell had] *interl*
2.8 cooled?] '?' *over* ':'
3.1 (you beggar . . . "wriggles") 3.2] *square brackets in MS*
3.3 the packing] 'the' *interl*
3.4 the pollen] 'the' *interl*
3.5 move the] *above del* 'work'
4.2 from] *interl*
4.3 to the present time] *interl*
4.4 from existing currents] *interl*
4.4 by former] 'by' *interl*
4.5 arrivals] *after del* 'plants'
4.6 the continents] 'the' *interl*
4.6 all] *above del* 'the'
4.7 to continents] *interl*
4.9 animal-] *interl*
4.10 , with a few . . . anomalous,] *interl*
4.11 your supposed] *above del* 'this'
4.11 of] *over* '.—'
4.11 *all*] *above del* '*all*'
5.4 very] *interl*
9.1 Hot-House] *interl*

To George Bentham 22 May [1863]
1.4 you] *interl*
1.15 few] *interl*
1.18 the fact] *after del* 'we certainly cannot tell why'
5.2 supposed] *interl*
5.4 case] *interl*
5.5 supposed] *interl*
5.6 school] *interl*

To W. D. Fox 23 May [1863]
2.2 on] *after del* 'chiefly'

To J. D. Hooker 23 May [1863]
1.4 Mr.] *after del* 'J'
1.4 part] *interl*
4.1 out &] *interl*

To John Scott 23 May [1863]
2.8 to] *interl*

To Roland Trimen 23 May [1863]
2.6 title] *after del illeg*
4.1 now] *interl*
4.4 2 discs] '2' *interl*
4.5 two confluent] *interl*

To John Scott 25 and 28 May [1863]
3.7 should] *above del* 'would'
3.15 in Primula] *interl*
3.17 seedlings] *after del* 'twice crossed'
6.1 on Primula] *interl*

To J. D. Hooker 29 May [1863]
2.3 in some newspaper] *interl*
5.6 rub & dust] *interl above del* 'rub'
5.7 to] *after del* 'out'
5.7 & long anthers 5.8] *interl*
5.8 (as] *parenthesis over comma*
5.8 surface] *above del* 'part'
5.8 the stigma 5.9] 'the' *interl*
5.12 the thorax] 'the' *interl*
5.12 of the next bee] *interl*
5.14 Subsequently ... still more.—] *added*

To Asa Gray 31 May [1863]
1.7 Of course ... enigma.] *added*
2.2 to you] *interl*
3.4 when ... distinguishable 3.5] *interl*
5.6 cases.] *above del* 'facts.'
6.4 of Viola at least 6.5] *interl*
6.7 nectar-] *interl*
6.8 get them,] *interl*
9.2 which are ... fertilisation:] *interl*
9.3 of Bees] *interl*
9.3 upper side] *above del* 'top'

To John Scott 31 May [1863]
2.2 permanent] *before omitted point*
3.1 Chron.?] *over* '—'
3.3 on number ... used &c] *interl*
3.4 No] 'N' *over* 'I'
4.3 showed that] *above del* 'proved'
4.6 to impotence] 'to' *interl*
4.7 it is ... quality.] *interl*
4.8 with] *interl above del* 'to'

To Philip Henry Gosse 2 June [1863]
1.6 he simply] 'he' *interl*
2.13 with the 2.14] *above del* '&'
4.1 near Torquay] *interl*
4.4 is correct] 'is' *interl*
6.3 the case] 'the' *interl*
6.3 positively] *interl*

To P. H. Gosse 5 June [1863]
2.2 a very] *interl*
2.3 but] *over* '&'
4.5 probably] *interl*

To John Scott 6 June [1863]
1.7 is a] 'is' *interl*
1.8 as I can] *interl*
1.11 most of] *interl*
6.1 very] *interl*
6.2 formerly] *interl*
7.4 from you] *interl*
7.5 at present] *interl*
7.6 in geology] *interl*
7.7 theory] *interl*
7.8 rarely] *above del* 'rarly'

To J. D. Hooker 8 [June 1863]
1.1 de rebus] *after del* 'about'
2.7 I have told him 2.8] *interl*
3.2 for] *above del* 'but'
3.2 does] *after del* 'gives'

To John Scott 11 June [1863]
2.4 in] *interl*

To George Bentham 19 June [1863]
1.16 the explanation] 'the' *interl*
2.4 the Origin] *above del* 'it'

To John Scott 20 [June 1863]
2.1 very] *interl*
3.1 self-fertile] *interl*
3.3 not *manifestly* 3.4] 'not' *interl*
3.5 , as near as is possible,] *interl*

To J. D. Hooker 23 [June 1863]
1.10 a] *added*
2.5 is] *interl*
3.2 30^s] '3' *over* '1'

To Charles Loring Brace 24 June [1863]
1.3 but] 'b' *over* '&'
2.2 be] *after del* 'per'
2.4 of the] *before omitted point*

To J. D. Hooker 25 [June 1863]
1.3 it is . . . purpose 1.4] *interl*
1.7 stem] *above del* 'upper space'
1.8 and] *above del* 'the stem'
1.10 It generally . . . retrogrades.] *interl*
1.11 does] 'do' *over* 'is'
2.5 the] *over* 'it'
2.5 tendril] *interl*
2.8 it] *above del* 'they'
2.8 finds] *altered from* 'find'

To Asa Gray 26 June [1863]
1.7 I have] *after del* 'which'
1.7 from] *over illeg*
2.13 alas!] '!' *over* ':'
3.5 20] *above del illeg*
3.6 diameter.] *above del* 'circumference'
5.1 (I . . . is.)] *square brackets in MS*

To T. H. Huxley 27 June [1863]
1.4 little] *interl*
1.5 (I cannot draw)] *interl*
1.8 Within] 'W' *over* 'T'
1.12 the enclosed] 'the' *interl*
1.14 Or whether . . . skin?] *interl*; '?' *over full stop*
1.14 appear] *after del illeg*
2.5 A newt . . . six toes.] *interl*
4.1 Review,] *interl*
4.5 ultimately] *interl*

To J. D. Hooker 1 July [1863]
1.1 Royalty] *interl*
1.1 Aristocracy] 'A' *altered from* 'a'
1.4 about Haast.] *interl*
2.3 extra] *interl*
3.3 of] *after del* 'abo'
3.4 a little] *interl*
3.5 the] *over* 'it'
3.5 to & fro,] *interl*
3.7 the shoot incessantly] 'the' *over* 'this'; 'shoot' *interl*
3.7 on its own axis] *interl*
3.8 (from] *after del* 'from'
3.8 from 100°] *above del* 'through 90'
3.8 180°] *above del* 'about 120° or 130°'
3.9 the correspondence] 'the' *interl*
3.10 & of] 'of' *over* 'in'

To John Scott 2 July [1863]
2.4 preconceived] *after del* 'con'
2.5 Any . . . ruin— 2.6] *interl*
5.2 & Tausch . . . except] *interl*
5.2 which] *added*
6.1 p. 6.] *interl*
6.1 seems] *after del* 'suspects'
6.2 long-styled] *interl*
7.1 *Hottonia*] *after del* 'See'
7.3 Fresh] *interl*
7.5 flower] *above del* 'corolla'
7.5 that of] *above del* 'in'
7.7 projecting] *interl*
7.8 with] *after del* 'with'
7.10 $\frac{14}{14,000}$] *after del* '$\frac{1}{1000}$'
7.12 Spheres] *above del* 'Spheres'
7.12 in diameter] *interl*

To T. H. Huxley 3 July [1863]
1.4 is] *added*
1.7 cartilage] *interl*

To William Forsell Kirby 9 July [1863]
1.3 it seems] 'it' *interl*

To W. H. Flower 11 July [1863]
1.11 have] *interl*
1.13 supernumery] *after del* 'add'

To J. D. Hooker 14 July [1863]
1.9 between thumb & finger 1.10] *interl*
1.13 at same time] *interl*
1.15 mould by mere touch] *above del* 'mold'

To Julius von Haast 18 July [1863]
1.6 difficulties] *altered from* 'difficulty'
2.3 one] *before omitted point*
2.4 were] *altered from* 'was'
2.4 not] *before omitted point*
2.7 Dialogue] *above del* 'article'

To Daniel Oliver 18 July [1863]
1.2 it to] 'it' *interl*
2.5 as I must write to Dr. H. 2.6] *interl*
2.7 fertilisation of] *interl*

To J. D. Hooker 22 July [1863]
1.2 4] *over* '3'
2.7 Viniferæ] *altered from* 'Vitiferæ'
3.2 do] *over* 'for'
3.2 Hildebrand] *altered from* 'He'
4.1 long] *after del* 'f'
4.5 Earth] 'E' *over illeg*
4.5 think] *altered from* 'thing'
7.1 are] *interl*

To W. E. Darwin [25 July 1863]
1.1 probably] *interl*
2.3 have just] 'have' *before omitted point*

To John Scott 25 [July 1863]
1.1 at] *before omitted point*
1.5 the M.S.] *above del* 'them.'
1.5 for] *over* 'to'
1.7 fully establish,] *above del* 'make'
4.1 differently] *interl*
4.2 seedlings.] *interl*
4.3 very] *interl*
5.2 in orchids] *interl*
9.4 alone] *interl*

To J. D. Hooker 26 [July 1863]
2.1 Asa] *altered from* 'Asy'
2.1 a madman] 'a' *interl*
6.1 -like] *interl*

To Friedrich Hildebrand 28 July [1863]
1.5 very] *interl*
2.2 manner of] *interl*

To G. H. K. Thwaites 29 July [1863]
1.2 if] *interl*
2.3 to which . . . refer.] *interl*

To John Scott 1 and 3 August [1863]
7.2 non] *interl*
8.4 about long-styled] *interl*

To J. D. Hooker 3 August [1863]
2.3 I shall so like to see you.— 2.4] *added*
8.1 good] *interl*

To Alphonse de Candolle 4 August [1863]
1.5 of L. hyssipifolia] *interl*
2.2 about] *interl*

To Asa Gray 4 August [1863]
2.1 After] *after del* 'When'
2.1 to you] *interl*
2.7 to the architect;] 'to' *interl*
3.10 it] *after del* 'not'
3.11 neither] 'ei' *over illeg*
3.13 movement of the upper] 'movement of the' *interl*
3.19 in] *interl*
3.19 part] *interl*
3.24 or] *over* '&'
6.6 of your little] 'of your' *interl*

To J. D. Hooker 12–13 August [1863]
0.2 13] *over* '12'
1.5 elaborately] *interl*
2.4 young] *interl*
2.5 very] *added*
5.1 Catherine] *interl*
7.2 Remember . . . be lost] *added*

To Charles Lyell 14 August [1863]
3.1 on] *over illeg*
6.1 think] *after omitted point*

To Thomas Rivers 17 August [1863]
1.2 bearing] *interl*
1.3 fruit] *interl*
3.2 fruit or foliage] *interl*
3.2 -tree] *added*
3.3 assert] 'a' *over illeg*
4.7 & in a day . . . them.] *added*
5.1 you,] *after del comma*

To J. D. Hooker 25 [August 1863]
1.4 are] *interl*
2.1 of Journal] *interl*
2.6 something] *above del* 'a good deal'
2.6 at] *over* 'in'
2.6 at Kew] *interl*

To Roland Trimen 27 August [1863]
1.3 of Linum] *interl*
3.5 in] *interl*
4.11 are excluded] 'are' *interl*

To [*Gardeners' Chronicle*] [after 27 August 1863]
2.4 delicate] *interl*
2.4 proboses] 'e' *over* 'i'

2.6 to him] *interl*

To J. D. Hooker [28 August 1863]
1.3 return] *above del* 'return'
1.5 had] *altered from* 'have'
1.6 but] *over* '—'
1.6 but . . . your advice.] *added*
4.1 great] *altered from* 'grat'
5.1 strike] *altered from* 'strke'
5.4 is] *interl*

To John Brodie Innes 1 September [1863]
4.1 has] 'h' *over* 'w'

To W. D. Fox 4 [September 1863]
2.5 &] *added*
3.4 kind of] *interl*

To Alfred Russel Wallace [29? September 1863]
4.2 discussion)] *parenthesis over comma*

To J. D. Hooker [4 October 1863]
1.3 with poor Annie] *interl*

To J. D. Hooker [13 November 1863]
4.1 ???] *over* '.—'

To J. D. Hooker 16 [November 1863]
1.1 small] *interl*

To J. D. Hooker [22–3 November 1863]
1.2 long] 'g' *over* 'ge'
1.2 *in truth . . . overworked.*— 1.3] *double scored, left margin*
5.2 theorie] 'orie' *over* 'rie'
7.2 (I think] *parenthesis over comma*
7.3 I suppose] *above del* 'Does'
7.6 &] *over* '.—'
8.1 former] *interl*

To Roland Trimen 25 November [1863]
3.3 before] *over* '.—'

To J. D. Hooker 27 [November 1863]
2.2 must have] *interl*

To Daniel Oliver 28 [November 1863]
3.1 except V. tricolor] *interl*
3.3 The imperfect . . . insect-visits—] *added*

To J. D. Hooker 5 [December 1863]
8.1 see] *after del* 'visit'
9.2 foreign] *interl*

BIOGRAPHICAL REGISTER

This list includes all correspondents and all persons mentioned in the letters and notes that the editors have been able to identify. Dates of letters to and from correspondents are given in chronological order. Letters to correspondents are listed in roman type; letters from correspondents in italic type; third-party letters are listed in italic type with the name of the recipient given in parentheses.

Acland, Lawford. Of Langdown Lawn, Hythe, near Southampton. (Letter from Emma Darwin to William Erasmus Darwin, [after 7 October 1861] in DAR 210.6: 76; *Post Office directory of Hampshire, Dorsetshire, and Wiltshire* 1867 and 1875.)

Adamson, L. (*fl.* 1860s). Colonial office clerk, *circa* 1850–71. Chief clerk, Colonial Office, Cape of Good Hope, 1865–71. (*Colonial Office list* 1869–71.)

Agardh, Jacob Georg (1813–1901). Swedish botanist. University of Lund, 1847; professor of botany, 1854–79. Established the Lund Botanical Gardens. Published widely, primarily on marine algae. Member of Swedish parliament, 1867–72. (*DSB*, *SBL*.)

Agassiz, Alexander (1835–1910). Swiss-born zoologist, oceanographer, and mining engineer. Son of Louis Agassiz. Emigrated to the United States, 1849. Joined the US survey of Washington Territory boundaries, 1859. Superintendent, Calumet copper mine, Michigan, 1867; later president of the corporation. Curator of the Harvard Museum of Comparative Zoology, 1874. (*DAB*.)

Agassiz, Jean Louis Rodolphe (Louis) (1807–73). Swiss-born zoologist. Professor of natural history, Neuchâtel, 1832–46. Emigrated to the United States, 1846. Professor of natural history, Harvard University, 1848–73. Established the Museum of Comparative Zoology at Harvard in 1859. Foreign member, Royal Society of London, 1838. (*DAB*, *DSB*, *Record of the Royal Society of London*.)
29 September 1863

Albert Edward, prince of Wales. *See* Edward VII.

Alexandra, princess of Wales (1844–1925). Eldest daughter of Prince Christian of Schleswig-Holstein-Sonderburg-Glücksburg, later Christian IX of Denmark. Married Albert Edward, prince of Wales, later King Edward VII, in 1863. (*DNB*.)

Altum, Bernhard (1824–1900). German clergyman and zoologist. Specialised in ornithology and entomology. Lecturer in zoology, University of Munster, 1859. Professor of zoology, Forstakademie, Eberswalde, 1869. (*NDB*; Stresemann 1975, pp. 328–32.)

Anderson, James (1824–93). Naval officer. Entered the merchant navy in 1840. Entered service with the Cunard line in 1851. Commanded the steamer *Great*

Eastern during the laying of the Atlantic telegraph, 1865–6. Managing director, Eastern Telegraph Company, 1866–93. Fellow of the Royal Geographical Society, 1870. Knighted, 1866. (*Modern English biography.*)

Anderson, James (1831/2–99). Scottish gardener; orchid specialist. Gardener to the Glasgow iron and shipping agent Thomas Dawson, at Meadowbank, Uddingston, Lanarkshire. Following Dawson's death, worked as a landscape gardener in Glasgow and Manchester. Founding member of the Manchester and North of England Orchid Society. Editor of the *Northern Gardener.* (R. Desmond 1994, *Orchid Review* 7 (1899): 198, *Post Office Glasgow directory* 1863–4.)

Anderson, Thomas (1832–70). Scottish physician and botanist. Entered the Bengal medical service in 1854. Superintendent, Calcutta botanic garden, 1861–8; managed the forestry department, Bengal, 1864–9; retired because of ill health. Instituted experiments that led to the successful cultivation of *Cinchona* in India. (R. Desmond 1992, pp. 100–1; R. Desmond 1994; *DSB.*)

Anderson-Henry, Isaac (1800–84). Scottish lawyer and horticulturalist. Practised as a solicitor in Edinburgh from 1834 or earlier. Retired from practice in 1861 when his wife succeeded to the estates of Woodend, Perthshire, and, as a condition of the entail, took the additional name Henry. Established a garden in Edinburgh known for its many rare plants. Interested in plant hybridisation and acclimatisation. Member of various scientific societies in London and Edinburgh. President of the Botanical Society of Edinburgh, 1867–8. (*County families* 1878; *Curtis's botanical magazine dedications*, pp. 175–6; *Gardeners' Chronicle*, 22 March 1873, p. 399; *Post Office Edinburgh directory* 1834–64; *Proceedings of the Linnean Society of London* (1886–7), pp. 42–4.)

17 January 1863, 20 January [1863], *26–7 January 1863*, *31 January 1863*, 2 February [1863], *14 February 1863*, *17 April 1863*, *24 April 1863*, 2 May [1863], *7 May 1863*, *16 May 1863*

Ansted, David Thomas (1814–80). Geologist. Professor of geology, King's College, London, 1840–5. Assistant secretary to the Geological Society of London, 1844–7; editor of the society's journal from 1844. From 1845, lecturer on geology at the military college at Addiscombe, Surrey, and professor of geology at the College of Civil Engineers, Putney. Consulting geologist and mining engineer from *circa* 1850. FRS 1844. (*DNB*, *Geological Magazine* n.s. 7 (1880): 336, *Record of the Royal Society of London*, Sarjeant 1980–96.)

13 April 1863, 15 April 1863, *17 April 1863*, *23 April 1863*

D. Appleton & Co. New York publishing house from 1831. Founded by Daniel Appleton (1785–1849). His son William Henry Appleton (1814–99) was taken into partnership in 1838. American publishers of works by CD, Thomas Henry Huxley, and Herbert Spencer. (*DAB.*)

Argyll, duke of. *See* Campbell, George Douglas.

Armstrong, William George, 1st Baron Armstrong of Cragside (1810–1900). Inventor and industrialist. Pioneer in the fields of hydraulics and hydro-electricity. Inventor of the Armstrong gun. Knighted, 1859; created Baron Armstrong, 1887.

Armstrong, William George, cont.
President of the British Association for the Advancement of Science, 1863. FRS 1846. (*DNB*, *Report of the 33d meeting of the British Association for the Advancement of Science, held at Newcastle-upon-Tyne.*)

Arnott, George Arnott Walker (1799–1868). Scottish botanist. Worked with William Jackson Hooker in Glasgow, 1830–40. Regius professor of botany, University of Glasgow, 1845–68. (R. Desmond 1994, *DNB.*)

Atherley, Ellen. Daughter of Arthur Frederick. Married George Atherley, William Erasmus Darwin's partner in the Southampton and Hampshire Bank, Southampton, in 1851. (Genealogical notes, Records research room, Southampton City Archives.)

Atherley, Evelyn George Hammond (b. 1852). Army officer. Son of George Atherley. Lieutenant, Royal Horse Guards, 1875; captain, 1883; major, 1889; retired in 1895. (*Army list*; Genealogical notes, Records research room, Southampton City Archives.)

Atherley, George (1818–83). Banker. William Erasmus Darwin's partner in the Southampton and Hampshire Bank, Southampton. (*Alum. Cantab.*; *Banking almanac* 1862; Genealogical notes, Records research room, Southampton City Archives.)

Atholl, duke of. *See* Murray, George Augustus Frederick John.

Aubertin Brothers. Merchants with premises at 6 Great Winchester Street, London. Conducted by the brothers of John James Aubertin. (*Post Office London directory* 1863.)

Aubertin, John James (1818/19–1900). Traveller and writer. Superintendent of the São Paolo Railway, Brazil, 1860–9. Promoted the cultivation of cotton in Brazil; awarded a gold medal by the Manchester Cotton Supply Association in 1869. Met CD at Ilkley Wells hydropathic establishment in Yorkshire in 1859. (Graham 1968, pp. 67–8; letter from J. J. Aubertin, 27 April 1863; *Modern English biography.*)
27 April 1863, 19 July 1863, *3 September 1863*

Audubon, John James La Forest (John James) (1785–1851). American ornithologist and illustrator. His published illustrations, such as those in *The birds of America* (1827–38), introduced him to the scientific community in Europe. (*DAB*, *DSB.*)

Austen, John Thomas (1793/4–1876). Clergyman. Vicar of Aldworth, Berkshire, 1832–48; rector of West Wickham, Kent, 1848–76; honorary canon of Canterbury, 1873–6. (*Alum. Cantab.*)

Ayerst, James Smith (1824/5–84). Physician. Assistant surgeon, Royal Navy, 1847–55. Physician, Great Malvern, Worcestershire, 1856–7. Ran a hydropathic establishment at Old Well House, Malvern Wells, 1857–68, initially in conjunction with James Manby Gully. Practised homoeopathy and hygienics. Physician, Torquay, Devon, 1868–84. (*Medical directory* 1856–85; Metcalfe 1906, p. 94; *Navy List* 1847–55.)

Babington, Charles Cardale (1808–95). Botanist, entomologist, and archaeologist. Involved in natural history activities at Cambridge for more than forty years; an expert on plant taxonomy. A founding member of the Cambridge Entomological Society and the Cambridge Antiquarian Society. Editor, *Annals and Magazine of Natural History*, from 1842. Chairman, Cambrian Archaeological Association, 1855–85. Professor of botany, Cambridge University, 1861–95. FRS 1851. (*DNB*, *DSB*.)

Bacon, Booth (*fl.* 1860–84). Secretary of the Penarth Harbour, Dock and Railway company from *circa* 1860. From 1862, CD invested heavily in the company. (*Bradshaw's railway guide* 1859–85, CD's Investment book (Down House MS).)
10 June 1863

Bacon, Francis, Baron Verulam and Viscount St Albans (1561–1626). Lawyer, statesman and philosopher. Lord Chancellor, 1618–21. Created Baron Verulam, 1618; Viscount St Albans, 1621. (*DNB*, *DSB*.)

Baer, Karl Ernst von (1792–1876). Estonian zoologist and embryologist. Professor of anatomy at Königsberg University, 1819; professor of zoology, 1826–34. Professor of zoology at the Academy of Sciences, St Petersburg, 1834–67. Demonstrated the existence of the mammalian egg, 1826. Propounded the influential view that embryological development proceeds from the general to the specific. Foreign member, Royal Society of London, 1854. (*DSB*, *NDB*, *Record of the Royal Society of London*.)

Baillière, Hippolyte (1808/9–67). Bookseller and publisher. Moved to London *circa* 1827 and established a business selling French medical texts. Collected books on behalf of the Royal College of Surgeons. (*Modern English biography*.)

Baillon, Ernest Henri (1827–95). French botanist. Professor of medical natural history, and of hygiene and natural history applied to industry, Ecole Centrale des Arts et Manufactures, Paris, 1863–95. Editor of the botanical journal, *Adansonia*, 1860–70. (*DBF*.)

Baker, Thomas Barwick Lloyd (1807–86). Educationalist and social reformer. Resided at Hardwicke Court, Gloucestershire. Officer in the Gloucestershire yeoman cavalry, 1835–65. Instrumental in the founding of the Social Science Congress and a leading figure in the reform school movement; established the Hardwicke Reform School in 1852. Author of an ornithological index (1835). Founding member of the Cotteswold Naturalists' Field Club, and first president, 1846–59. (*CDEL*, *DNB*, Gebhard trans. [1878], *Modern English biography*, *Proceedings of the Cotteswold Naturalists' Field Club* vols. 1–4.)

Balfour, John Hutton (1808–84). Scottish physician and botanist. Professor of botany, Glasgow University, 1841–5. Professor of botany, University of Edinburgh, and Regius keeper of the Royal Botanic Garden, Edinburgh, 1845–79. Founding member of the Botanical Society of Edinburgh, 1836. Founder of the Edinburgh Botanical Club, 1838. Co-editor of the *Edinburgh New Philosophical Journal*. FRS 1856. (*DNB*, *DSB*.)
22 April 1863

Barber, Mary Elizabeth (1818–99). British-born naturalist, artist, and writer in South Africa. Sister of James Henry Bowker. Married Frederick William Barber, a chemist, in 1845. Studied birds, moths, reptiles, and plants, and corresponded with a number of leading scientists, providing them with specimens and drawings. (R. Desmond 1994, *DSAB*.)

Barclay, Arthur Kett (1806–69). Brewer. Head of Barclay, Perkins & Co., brewers. Built an observatory at Bury Hill near Dorking, Surrey, 1848. Paralysed, 1855. Grandson of Robert Barclay (1751–1830), who established a large collection of exotic plants in the gardens at Bury Hill. FRS 1852. (*Burke's landed gentry*; R. Desmond 1994 s.v. Barclay, Robert; *Modern English biography*.)

Barr & Sugden. Nursery established by Peter Barr and Edward Sugden in 1861, with premises at 12 King Street, Covent Garden, London. (R. Desmond 1994, *Post Office London directory* 1863.)

Bartlett, Abraham Dee (1812–97). Zoologist. Taxidermist, 1834–52. Superintendent of the natural history department, Crystal Palace, 1852–9; of the Zoological Society's gardens, Regent's Park, 1859–97. (*Modern English biography*.)

Bateman, James (1812–97). Industrialist, banker, landowner, and horticulturalist. Built Biddulph Grange in north Staffordshire and established a well-known garden there. Cultivated tropical plants, especially orchids. Acquired orchids in Central and South America, and published important works on the Orchidaceae. President of the North Staffordshire Field Club, 1865–70. FRS 1838. (R. Desmond 1994, *DNB*, Hadfield *et al.* 1980.)

Bates, Frederick (1829–1903). Brewer and naturalist. Younger brother of Henry Walter Bates. Managing director of a brewing company in Leicester. A keen entomologist, particularly interested in the Heteromera, publishing many papers and forming an extensive collection. Also formed collections of Cicindelidae and British Coleoptera. Cultivated orchids for some years. (*Entomologist's Monthly Magazine* 39 (1903): 286–7; Gilbert 1977.)

Bates, Henry (d. 1870). Hosiery manufacturer in Leicester. Father of Henry Walter Bates. (*DNB* s.v. Bates, Henry Walter.)

Bates, Henry Walter (1825–92). Entomologist. Undertook a joint expedition to the Amazon with Alfred Russel Wallace, 1848–50; continued to explore the area alone until 1859. Provided the first comprehensive scientific explanation for the phenomenon subsequently known as Batesian mimicry. Published an account of his travels, *The naturalist on the River Amazons*, in 1863. Assistant secretary, Royal Geographical Society of London, 1864–92. President, Entomological Society of London, 1868, 1869 and 1878. FRS 1881. (*DNB*, *DSB*.)

12 January [1863], *17 January [1863]*, *24 January 1863*, 26 January [1863], 4 March [1863], *8 April 1863*, 9 April [1863], 18 April [1863], *20 April 1863*, 30 April [1863], *2 May [1863]*, *29 September 1863*, *24 October 1863*

Bates, John. Younger brother of Henry Walter Bates. (Woodcock 1969.)

Bates, Samuel. Younger brother of Henry Walter Bates. (Woodcock 1969.)

Bates, Sarah Ann. Daughter of a Leicester butcher. As Sarah Ann Mason, married Henry Walter Bates in 1863. (*DSB* s.v. Bates, Henry Walter; Woodcock 1969, pp. 253–5.)

Beal, William James (1833–1924). American botanist and teacher. School-teacher in Union Springs, New York, 1859–1868; studied under Louis Agassiz and Asa Gray at Harvard University. Lectured in botany at the University of Chicago, 1868–71. Taught at Michigan Agricultural College, 1871–1910. (*DAB*.)

Beaton, Donald (1802–63). Scottish gardener. An expert on bedding schemes and hybridisation. Trained in the gardens at Beaufort Castle, Inverness-shire. Gardener to William Gordon-Cumming at his estate in Altyre, at the Dickson & Turnbul nursery in Perth, and at the Caledonian Horticultural Society in Edinburgh. Head gardener and estate manager to William Gordon of Haffield House, Herefordshire, 1829–37; gardener to Thomas Harris of Kingsbury, Kilburn, London, then head gardener to William Fowle Middleton of Shrubland Park, Suffolk, from 1840. A regular contributor to the *Gardener's Magazine* and the *Cottage Gardener*. (*Cottage Gardener*, 28 November 1854, pp. 153–8; R. Desmond 1994; Hadfield *et al.* 1980.)

Becker, Lydia Ernestine (1827–90). Leading member of the women's suffrage movement, botanist, and astronomer. Published *Botany for novices* (1864); awarded a Horticultural Society Gold Medal, 1865. Founder and president of the Manchester Ladies' Literary Society, 1867. Secretary to the Manchester Women's Suffrage Committee from 1867, and later to the Manchester National Society for Women's Suffrage. Member of the Manchester School Board, 1870. Editor of and regular contributor to the *Women's Suffrage Journal* from 1870. Secretary to the London Central Committee for Women's Suffrage from 1880. (Blackburn 1902, R. Desmond 1994, *DNB*, *Macmillan dictionary of women's biography*.)
18 May 1863, 21 May [1863], 23–4 May [1863], 28 May [1863], 8 July [1863], 31 July [1863], 2 August [1863]

Beckles, Samuel Husband (1814–90). Barrister and palaeontologist. Discovered in the Purbeck beds the oldest known mammalian fossils. FRS 1859. (*Modern English biography*, Sarjeant 1980–96.)

Bell, William (*c.* 1833–1916). Scottish-born gardener. Gardener, Royal Botanic Garden, Edinburgh. Head gardener, Saharanpur Botanic Garden, India, 1862. Returned to Edinburgh in 1869. (R. Desmond 1994.)

Bennett, Charles Henry (1829–67). Illustrator and author. Wrote and illustrated children's books and provided illustrations for other publications. Published a series of caricatures entitled 'The origin of species, dedicated by natural selection to Dr. Charles Darwin'. (*DNB*, *Illustrated Times* 2 (1863): 317, 333, 348, 364, 381, 369, 413, and 429; 3 (1863): 13, 29, 61, 77, 93, 125, 141, 157, 173, 205, 221, and 237.)

Bennett, John Joseph (1801–75). Botanist. Assistant keeper of the Banksian herbarium and library at the British Museum, 1827–58; keeper, 1858–70. Secretary of the Linnean Society of London, 1840–60. FRS 1841. (*DNB*.)

Bentham, George (1800–84). Botanist. Donated his botanical library and collections to the Royal Botanic Gardens, Kew, in 1854, and was provided with facilities there for his research from 1861. President of the Linnean Society of London, 1861–74. Published *Genera plantarum* (1862–83) with Joseph Dalton Hooker. FRS 1862. (*DNB, DSB.*)

*16 January 1863, [*c. *14 April 1863]*, 15 April [1863], *21 April 1863*, 22 April [1863], *21 May 1863*, 22 May [1863], 19 June [1863]

Berkeley, Miles Joseph (1803–89). Clergyman and botanist. Perpetual curate of Apethorpe and Wood Newton, Northamptonshire, 1833–68. Vicar of Sibbertoft, Northamptonshire, from 1868. Editor of the *Journal of the Royal Horticultural Society*, 1866–77. An expert on British fungi, one of the collections he described being from CD's *Beagle* voyage. Awarded the Royal Society's Royal Medal, 1863. FRS 1879. (*DNB, DSB.*)

Bewick, Thomas (1753–1828). Illustrator and naturalist. Produced and illustrated numerous natural history books. (*DNB.*)

Bigelow, Jacob (1786–1879). American botanist and physician. Professor of materia medica, Harvard Medical School, 1815–55; Rumford Professor of the application of science to the useful arts, 1816–27. President, American Academy of Arts and Sciences, 1847–63. Published extensively on botany, medicine, and technology. (*DAB.*)

Adam & Charles Black. Edinburgh publishing house; from 1834, a partnership between Adam Black (1784–1874) and his nephew Charles Black (1807–54). Publishers of the *Encyclopedia Britannica* and the *Edinburgh Review.* Acquired the rights to Walter Scott's Waverley novels in 1851. (Anon. 1957; *DNB* s.v. Black, Adam.)

Black, Allan A. (1832–65). Scottish-born botanist. First curator of the herbarium at the Royal Botanic Gardens, Kew, 1853–64. Superintendent, Botanic Garden, Bangalore, India, 1864. (R. Desmond 1994.)

Blake, Charles Carter (b. *c.* 1840, d. after 1887). Anthropologist, palaeontologist, and comparative anatomist. Lecturer on zoology, London Institution, 1862–3. Founding member of the Anthropological Society of London, 1863; assistant secretary, *circa* 1863–6; curator and librarian, *circa* 1863–7. Lecturer in comparative anatomy and zoology, Westminster Hospital School of Medicine, 1869–81. Founding member and honorary foreign secretary of the London Anthropological Society, 1873–5; joint editor of *Anthropologia*, 1873–5. (*CDEL*; correspondence with Richard Owen (Natural History Museum); *Geologist* 5 (1862): 323; Gorman 1910; *List of the Geological Society of London* 1865–8; *Medical directory* 1866–81; Officers and council of the Anthropological Society of London for 1865, and for 1869; *Memoirs read before the Anthropological Society of London* 1 (1863–4), 3 (1867–9); Subscription records, Royal Anthropological Institute; *Wellesley index.*)

Blasius, Johann Heinrich (1809–70). German ornithologist and zoologist. Professor of natural history, Collegium Carolinum, Brunswick, 1836–70. Founder and director of the natural history museum in Brunswick from 1859 or later. (*ADB, DBE, NDB.*)

D. O. Blood & Co. Carriers in Philadelphia, USA. Issued stamps for local use between 1841 and *circa* 1860. (Sutton 1966, pp. 41, 231.)

Blyth, Edward (1810–73). Zoologist. Pharmacist in Tooting, London. Wrote and edited zoological works under the pseudonym Zoophilus. Curator of the museum of the Asiatic Society of Bengal, Calcutta, India, 1841–62. Returned to Britain in 1863, where he continued to write on zoology and on the question of the origin of species. (*DNB*, *DSB*.)
27 March 1863, 7 April 1863, 21 September 1863

Boddaert, Pieter (b. *c.* 1733, d. after 1786). Dutch physician and naturalist. MD, University of Utrecht, 1764; lecturer on natural history, University of Utrecht. Published on zoology and dentistry. (*Dutch medical biography*.)

Bolton & Barnitt. Importers and manufacturers of chemicals and of photographic materials and equipment, with offices at 146 Holborn Bars, London. (*Post Office London directory* 1863.)

Bond, George Phillips (1825–65). American astronomer. Assistant observer, Harvard College Observatory, 1845–59; director, 1859–65. Credited with the discovery of Hyperion, the eighth satellite of Saturn, and with establishing the nature of Saturn's rings. (*DAB*.)

Bonham-Carter, Joanna Maria (1791–1884). Widow of John Bonham-Carter (died 1838), Whig MP for Portsmouth, 1812–38. Family friend of the Darwins. Resided at Ravensbourne Lodge, Keston, two miles north-west of Down, Kent, from 1853; moved to Keston House by 1862. (Bonham-Carter 1960, *Burke's landed gentry* 1965, *Dod's parliamentary companion*, *Post Office directory of the six home counties* 1862.)

Bonpland, Aimé Jacques Alexandre (Aimé) (1773–1858). French physician, traveller, and botanist. Accompanied Alexander von Humboldt on his South American travels, 1799–1804. Curator of the garden at the royal château of Malmaison, 1809. Returned to South America, 1817. (*DBF*, *EB*.)

Boott, Francis (1792–1863). American physician and botanist resident in London. Lecturer on botany, Webb Street School of Medicine, London, 1825. Conducted a successful medical practice in London, 1825–32. Secretary, Linnean Society of London, 1832–9; treasurer, 1856–61. (R. Desmond 1994, *DNB*.)
23 January 1863

Boott, Mary (b. 1794/5). Daughter of the botanist Lucy Hardcastle of Derby (*fl.* 1760s–1830s), and wife of Francis Boott. (Census returns 1861 (Public Record Office RG9/173: 123); R. Desmond 1994 s.v. Hardcastle, Lucy; *DNB* s.v. Boott, Francis; *Proceedings of the Linnean Society* 1863–4, pp. xxiii–xxvii.)

Bosse, Julius Friedrich Wilhelm (1788–1864). German landscape gardener. Curator of the ducal gardens at Oldenburg, Germany. (Friedl *et al.* 1992, pp. 88–9.)

Boucher de Crèvecoeur de Perthes, Jacques (Jacques Boucher de Perthes) (1788–1868). French customs official and archaeologist. Director of customs, Abbeville, France, from 1825. President of the Société d'Emulation, Abbeville.

Boucher de Perthes, Jacques, cont.
Published on prehistoric archaeology and palaeontology. Discovered controversial evidence of early humans in the Somme River valley gravels. (*DSB*.)
23 June 1863

Bowker, James Henry (1822–1900). South African farmer, soldier, civil servant, and naturalist. Brother of Mary Elizabeth Barber. Inspector in the Frontier Armed and Mounted Police, Cape Colony, 1855; commandant, 1858; commanding officer, 1870. High commissioner's agent in Basutoland, 1868. Chief commissioner at the diamond fields of Griqualand West, after leading the expedition that secured their annexation to the Cape Colony in 1871. Major contributor of specimens to the South African museum's collection of Lepidoptera. Collaborated with Roland Trimen to publish *South-African butterflies: a monograph of the extra-tropical species* (Trimen 1887–9). (*DSAB*, Trimen 1887–9.)

Brace, Charles Loring (1826–90). American philanthropist, author, and ethnologist. After studying theology at Yale University, toured Europe, 1850–1. One of the founders of the New York City Children's Aid Society, 1853, and subsequently director. Published a small number of ethnological treatises. (*DAB*.)
24 June [1863]

Bradford, Edward (1802–88). Army surgeon and botanist. Honorary deputy inspector-general of hospitals, 1858; deputy inspector of hospitals, 1870. Surgeon at the Royal Military College, Sandhurst, 1860–7. Honorary surgeon to Queen Victoria, 1859–88. Collected plants in Trinidad, 1845–6. (*Army list* 1860–7, R. Desmond 1994, *Medical directory* 1889, *Modern English biography*.)
31 July 1863

Braun, Alexander Carl Heinrich (1805–77). German botanist. Professor of botany, Freiburg, 1846–50. Professor of botany and director of the botanic garden, University of Berlin, 1851–77. Brother-in-law of Louis Agassiz. Deeply influenced by *Naturphilosophie*; studied plant morphology. Established the doctrine of spiral phyllotaxy. (*DSB*, *NDB*.)

Bravais, Auguste (1811–63). French naval officer, botanist, and physicist. Took part in a number of voyages of scientific exploration, 1832–40. Taught at the Faculté des Sciences, Lyons, 1841–5. Professor of physics at the Ecole Polytechnique, 1845–56. Pioneer in the field of crystallography. (*DBF*, *DSB*.)

Bravais, Louis. French botanist and medical practitioner. Practised medicine in Annonay. Undertook research into plant organography with his brother Auguste Bravais. (Bravais and Bravais 1838; *DSB* s.v. Bravais, Auguste.)

Bravard, Pierre Joseph Auguste (Auguste) (1803–61). French cartographer and palaeontologist. Emigrated to Argentina in 1852. Appointed director of the national museum at Parana and inspector-general of Argentinian mines. (*DBF*.)

Bree, Charles Robert (1811–86). Medical practitioner and zoologist. Medical officer in the Polish army, 1831; qualified to practise medicine in Britain, 1833. MD, Edinburgh, 1859. General practitioner, Colchester, Essex, and physician, Essex and Colchester Hospital, 1860–82. Entomological editor of the *Naturalist*,

1858. Published articles critical of CD's theory of the origin of species. (*CDEL*, Freeman 1978, *Medical directory* 1847–87.)

Brewster, David (1781–1868). Scottish physicist who specialised in optics. Invented the kaleidoscope in 1816. Edited the *Edinburgh encyclopædia*, 1807–30. Assisted in the establishment of the British Association for the Advancement of Science in 1831. Principal of the United College of St Salvator and St Leonard, St Andrews University, 1838–59. Vice-chancellor, Edinburgh University, 1860–68. Knighted, 1832. FRS 1815. (*DNB*, *DSB*.)

Bridgman, William Kencely (1812–84). Dentist and naturalist. Practised as a surgeon-dentist in Norwich. (*NUC*, *Post Office directory of Cambridge, Norfolk, and Suffolk* 1865.)

Briggs, John Joseph (1819–76). Farmer, naturalist, and topographer. Regular contributor to journals such as the *Field*, the *Zoologist*, the *Reliquary*, the *Derby Reporter*, and the *Leicestershire Guardian*. (*DNB*.)
2 February [1863], 16 February [1863]

Brinton, William (1823–67). Physician. Specialised in stomach disorders; author of *Diseases of the stomach, with an introduction on its anatomy and physiology* (1859). Treated CD in 1863. Member of the faculty at the medical school, King's College, London, 1847–53. Physician, Royal Free Hospital, 1852–60. Lecturer, medical school, St Thomas's Hospital, 1853; physician, 1860–4. FRS 1864. (CD's Classed account books (Down House MS), *DNB*, *Medical directory*, *Physicians*.)

Bristowe, Albertine (b. 1835). Daughter of Jean Jacques Lavit of Paris. Married Samuel Boteler Bristowe in 1856. (*Darwin pedigree*.)

Bristowe, Samuel Boteler (1822–97). Lawyer and politician. CD's second cousin once removed. Called to the bar, 1848. Fellow of Trinity Hall, Cambridge, 1849–56. Recorder of Newark, 1869–70. Liberal MP for Newark-on-Trent, 1870–80. County court judge, 1880–91. (*Alum. Cantab.*, *Darwin pedigree*.)

Brongniart, Adolphe Théodore (1801–76). French palaeobotanist, plant anatomist, and taxonomist. One of the founders of the *Annales des Sciences Naturelles*, 1824. Professor of botany, Muséum d'Histoire Naturelle, from 1833. (*DSB*.)

Bronn, Heinrich Georg (1800–62). German palaeontologist. Professor of natural science at Heidelberg University, 1833. Translated and superintended the first German editions of *Origin* (1860) and *Orchids* (1862). (*DSB*, *NDB*.)

Brougham, Henry Peter, 1st Baron Brougham and Vaux (1778–1868). Scottish-born statesman, lawyer, reformer, and educationalist. A founder of the *Edinburgh Review*, 1802, the Society for the Diffusion of Useful Knowledge, 1825, and London University, 1828. Prolific author. Lord chancellor, 1830–4. President of the Social Science Association, 1857 and 1860–5. Elected chancellor of Edinburgh University, 1859. Created Baron Brougham and Vaux, 1830. FRS 1803. (*DNB*, *Record of the Royal Society of London*.)

Brown, Edwin (1818/9–76). Bank manager and naturalist. Employee of the Burton, Uttoxeter, and Ashbourne Union Bank in Burton-upon-Trent, Staffordshire, 1834; manager, 1865. Member of the Entomological Society of London, 1849;

Brown, Edwin, cont.
fellow of the Geological Society of London and the Royal Geographical Society; member of the Botanical Society of London. A life-long friend and patron of Henry Walter Bates; Bates's letters to him from the Amazon were published in the *Zoologist*, 1849–58. (*Banking almanac* 1865; Bates 1892, p. 20; R. Desmond 1994; *Entomologist* 9 (1876): 240; *Entomologist's Monthly Magazine* 13 (1876–7): 116–7, 257–8.)
14 February 1863

Brown, John (1810–82). Scottish physician and essayist. Practised medicine in Edinburgh, 1833–82. (*DNB*.)

Brown, Robert (1773–1858). Scottish botanist. Naturalist to the expedition surveying the coast of Australia, 1801–5; published descriptions of the plants he collected. Librarian to the Linnean Society of London, 1806–22; to Sir Joseph Banks, 1810–20. Continued as curator of Banks's collections after his death in 1820 and negotiated their transfer to the British Museum in 1827. Keeper of the botanical collections, British Museum, 1827–58. FRS 1811. (*DNB*, *DSB*.)

Brown, Robert (1842–95). Botanist, geographer, and journalist. Studied at the Universities of Edinburgh, Leyden, Copenhagen, and Rostock. Travelled widely throughout mainland Europe, Greenland, and North and South America, 1861–3. Botanist to the expedition to British Columbia, 1863. Commander of the exploration of Vancouver Island, 1864. Made a study of the glaciation of Greenland in 1867. Lectured in natural history, School of Arts and Heriot-Watt College, Edinburgh, from 1869. Moved to London in 1876 and wrote for a number of popular science publications. (*DNB*.)

Brown, Robert N. (d. *c.* 1862). Scottish-born gardener. Pupil of John Hutton Balfour at Edinburgh University. Superintendent of the Agri-horticultural Society Garden, Madras, India, 1857–62. (R. Desmond 1994.)

Brown-Séquard, Charles Edouard (1817–94). French physiologist. Practised medicine in France and the United States before accepting a professorship at Virginia Medical College, Richmond, in 1854. Returned to Paris in 1855; moved to England in 1858. Physician, National Hospital for the Paralysed and Epileptics, 1860–3. Professor of physiology and pathology, Harvard Medical School, 1864–7. Professor of medicine, Collège de France, 1878–94. Conducted pioneering research in neurology and endocrinology. FRS 1860. (*DBF*, *DSB*, *Record of the Royal Society of London*.)

Buckland, Francis Trevelyan (1826–80). Naturalist, popular science-writer and surgeon. Son of William Buckland. Trained and practised medicine at St George's hospital, London, 1848–53. Assistant surgeon in the Life Guards, 1854. Staff writer on the *Field*, 1856–65. Established an exhibit on pisciculture at the South Kensington Museum, *circa* 1865, and launched a weekly journal, *Land and Water*, in 1866. Inspector of salmon fisheries from 1867. (*DNB*.)
26 January [1863], *[before 1 February 1863]*, 1 February [1863]

Buckland, William (1784–1856). Geologist and clergyman. Professor of mineralogy, Oxford University, 1813; reader in geology, 1819–49. President of the Geological Society of London, 1824–5 and 1840–1. Dean of Westminster from 1845. FRS 1818. (*DNB*, *DSB*.)

Buckle, Henry Thomas (1821–62). Historian. Heir to a London shipowner. Travelled widely and published a number of volumes on the history of English and European civilisation. (*DNB*.)

Buckman, James (1814–84). Geologist and agricultural botanist. Curator of the Birmingham Philosophical Institute, 1842–7. Professor of geology and botany at the Royal Agricultural College, Cirencester, 1847–63. Farmed according to scientific principles in Dorset, 1863–84. (*DNB*, Sarjeant 1980–96.)

Buckton, George Bowdler (1818–1905). Chemist and entomologist. Assistant to August Wilhelm von Hofmann at the Royal College of Chemistry, London. Wrote papers on chemical subjects until 1865, when he moved to Haslemere, Surrey, and began researches that resulted in a series of monographs on British Homoptera. FRS 1867. (*Entomologist's monthly magazine* 41 (1905): 282, Gilbert 1977, *Nature* 72 (1905): 587–8.)

Buffon, Georges Louis Leclerc, comte de (1707–88). French naturalist, philosopher, and mathematician. Keeper, Jardin du Roi, 1739–88. His theory of transmutation is outlined in *Histoire naturelle* (1749–1804). FRS 1739. (*DBF*, *DSB*, *Record of the Royal Society of London*.)

Bunbury, Charles James Fox, 8th baronet (1809–86). Botanist. Collected plants in South America, 1833–4; in South Africa, 1838–9. Accompanied Charles Lyell to Madeira in 1853. Justice of the peace and deputy lieutenant for Suffolk; high sheriff, 1868. Succeeded to the baronetcy, 1860. FRS 1851. (*County families* 1875, R. Desmond 1994, Sarjeant 1980–96.)

Bunbury, Frances Joanna (1814–94). Daughter of Leonard Horner. Married Charles James Fox Bunbury in 1844. Published memorials of her husband (1890–3) and his life and letters (1894). (*Burke's peerage* 1863 and 1980, *CDEL*, *County families* 1875.)

Buob, Miss (*fl.* 1863–7). Schoolmistress. Ran a private school at 2 Courtland Terrace, Kensington, London, *circa* 1862 – *circa* 1867. (*Correspondence* vol. 10, letter to W. E. Darwin, 30 [October 1862]; letter from Emma Darwin to William Erasmus Darwin, [29 October 1862], in DAR 219.1: 63; *Post Office London directory* 1862–8.)

Burdach, Karl Friedrich (1776–1847). German physiologist. Practised medicine, 1799–1811. Professor of anatomy, physiology, and forensic medicine, University of Dorpat, 1811–13; professor of anatomy, Königsberg, 1813. Director of the Königliche Anatomische Anstalt, Königsberg, 1817. Influenced by *Naturphilosophie*. Made significant contributions to embryology and neuroanatomy. (*ADB*, *DSB*.)

Busk, Ellen. Daughter of Jacob Hans Busk. Married her cousin, George Busk, in 1843. (*DSB* s.v. Busk, George.)

Busk, George (1807–86). Russian-born naval surgeon and naturalist. Served on the hospital ship at Greenwich, 1832–55. Retired from medical practice, 1855. Member of several scientific societies, president of the Microscopical Society, 1848–9; of the Anthropological Institute, 1873–4. Zoological secretary of the Linnean Society of London, 1857–68. Hunterian Professor of comparative anatomy, Royal College of Surgeons of England, 1856–9; council member, 1863; member of board of examiners, 1868; president, 1871. Specialised in palaeontology and in the study of Bryozoa. FRS 1850. (*DNB*, *DSB*, Plarr 1930.)
[c. 27 August 1863]

Busk, Hans (1815–82). Lawyer. Founder of volunteer movement and writer on military topics. In 1858 restored vitality to the Victoria Rifles, which became a senior battalion of Britain's mid-nineteenth-century volunteer movement. Published treatises and gave lectures on volunteer corps, the rifle, and naval construction. High sheriff of Radnorshire, 1847. (*Burke's landed gentry*, *DNB*.)

Butler, Benjamin Franklin (1818–93). American lawyer, soldier, and politician. Practised law in Massachusetts, 1840–93. Democratic member of the Massachusetts House of Representatives, 1853; of the Senate, 1859. Commander of Union troops during the American Civil War. Major-general of volunteers, 1861; military governor of New Orleans, 1862. Congressman, 1866–75 and 1878. Governor of Massachusetts, 1882–3. (*DAB*.)

Butler, Mary (d. 1865?). Patient at Moor Park and Ilkley Wells hydropathic establishments. Visited the Darwins at Down. Probably sister of Richard Butler, vicar of Trim, Ireland, 1819–62 (*Modern English biography*). (*Burke's landed gentry* 1879; *Correspondence* vol. 7, letters to Mary Butler, 20 February [1859] and 11 September [1859], and letter to John Murray, 14 November [1859].)

Butler, Samuel (1774–1839). Educationalist and clergyman. Headmaster, Shrewsbury School, 1798–1836. Bishop of Lichfield and Coventry, 1836; of Lichfield, 1836–9. (*DNB*.)

Butler, Samuel (1835–1902). Author and artist. Grandson of Samuel Butler (1774–1839). Emigrated to New Zealand in 1859; returned to Britain after publishing an account of his time farming in the colony, *A first year in Canterbury settlement* (1863). Published books on art, music, literature, and philosophy, including the novels *Erewhon* (1872) and *The way of all flesh* (1903); published a two-volume life of his grandfather, *Life of Samuel Butler, Bishop of Lichfield* (1896). Became a critic of Darwinism from the 1870s. (*Autobiography*, pp. 167–219, *DNB*.)

Cairnes, John Elliot (1823–75). Irish political economist. Whately Professor of political economy, Trinity College, Dublin, 1856–61; professor of political economy and jurisprudence, Queen's College, Galway, 1859; professor of political economy, University College, London, 1866. Published influential works in support of the Union in the American Civil War. (*DNB*.)

Campbell, George Douglas, 8th duke of Argyll (1823–1900). Scottish statesman and author of works on science, religion, and politics. A defender of the concept of design in nature. Chancellor of St Andrews University, 1851. President of the Royal Society of Edinburgh, 1861. Privy seal, 1853–5, 1859–60, 1860–6, and 1880–1; postmaster-general, 1855–8 and 1860; secretary of state for India, 1868–74. Succeeded to dukedom in 1847. FRS 1851. (*DNB.*)

Candolle, Alphonse de (1806–93). Swiss botanist, lawyer, and politician. Active in the administration of the city of Geneva until 1860. Responsible for the introduction of postage stamps to Switzerland. Professor of botany and director of the botanic gardens, Geneva, from 1835. Concentrated on his own research after 1850. Son of Augustin Pyramus de Candolle. Foreign member, Royal Society of London, 1869. (*DSB*, *Record of the Royal Society of London.*)
14 January [1863], 31 January [1863], 4 August [1863]

Candolle, Augustin Pyramus de (1778–1841). Swiss botanist. In Paris, 1796–1808; professor of botany, Ecole de Médecine and Faculté des Sciences, Montpellier, 1808–16; professor of natural history, Academy of Geneva, 1816–35. Foreign member, Royal Society of London, 1822. (*DSB*, *Record of the Royal Society of London.*)

Canestrini, Giovanni (1835–1900). Italian zoologist. Translated the first Italian edition of *Origin* (1864) in association with Leonardo Salimbeni. Professor of zoology, University of Modena, 1862–9; University of Padua, 1869–1900. (*Enciclopedia Italiana*; Freeman 1977, p. 105; Pancaldi 1991, pp. 77–106.)

Carlyle, Thomas (1795–1881). Essayist and historian. (*DNB.*)

Carpenter, William Benjamin (1813–85). Physician and naturalist. Fullerian Professor of physiology at the Royal Institution of Great Britain, 1844–56; professor of forensic medicine, University College, London; physiology lecturer, London Hospital, 1845–56. Registrar of the University of London, 1856–79. Founding member of the Marine Biological Association. FRS 1844. (*DNB*, *DSB.*)

Carpmael, Ernest (1844–1921). Barrister; contemporary of George and Francis Darwin at Clapham Grammar School and Cambridge University. Son of William Carpmael. Formally admitted as a member of St John's College, Cambridge, and elected to an Open Exhibition, 1863. Called to the bar, 1869. (*Alum. Cantab.*; Biographical notes, St John's College, Cambridge; *The Eagle* 3: 356.)

Carpmael, William (1804–67). Engineer. Served on the council of the Institute of Civil Engineers. Member of the Institute of Mechanical Engineers. Member of the Metropolitan Board of Works, 1855–67. (*Minutes of Proceedings of the Institution of Civil Engineers* 30: 430; *Proceedings of the Institution of Mechanical Engineers* (1868): 14–15.)

Caspary, Johann Xaver Robert (Robert) (1818–87). German botanist. Director, Bonn herbarium, 1856. Professor of botany and director of the botanic gardens at the University of Königsberg from 1858. Specialised in water plants. (*ADB.*)

Cattell, John. Florist, nurseryman, and seedsman in Westerham, Kent. (R. Desmond 1994, *Post Office directory of the six home counties* 1859–66.)

Cavanilles, Antonio José (1745–1804). Spanish clergyman and botanist. In Paris, 1777–81. Director of the botanical gardens in Madrid, 1801–4. (*Taxonomic literature.*)

Cavendish, William, 7th duke of Devonshire and 2d earl of Burlington (1808–91). Statesman and benefactor of scientific and industrial enterprises. Whig MP for Cambridge University, 1829–31; for Malton, Yorkshire, 1831; for Derbyshire, 1831; for North Derbyshire, 1832–4. Succeeded to the earldom, 1834; to the dukedom, 1858. Chancellor of London University, 1836–56; of Cambridge University, 1861–91. President of the Royal Agricultural Society, 1870. Privy councillor, 1876. Knighted, 1858. FRS 1829. (*DNB*, *Record of the Royal Society of London.*)

Christ, Konrad Hermann Heinrich (Konrad Hermann) (1833–1933). Swiss lawyer, botanist, and plant geographer. Professor of botany, University of Basel. (*DBE*, *Taxonomic literature.*)

Christy, Henry (1810–65). Ethnologist, archaeologist, geologist, and banker. Partner in the London banking house Christy & Co.; director of the London Joint-Stock Bank. Travelled for ethnological purposes in the East, 1850; in Scandinavia, 1852–3; in Canada, the United States, Cuba, and Mexico, 1856–7. Member of the Geological Society of London, 1858; conducted excavations of caves in the Vézère valley in southern France with Edward Lartet in the 1860s. (*DNB.*)

Churchill, John Spriggs Morss (John) (1801–75). Bookseller and medical publisher. Originally a retailer of medical texts; became a publisher in the 1830s. Published *The Lancet*, 1842–7; the *Medical Times* from 1850; the *Medical Times and Gazette* from 1852. Engaged solely in publishing from 1854. Settled in Pembridge Square, Bayswater, London in 1862. Retired, 1870. (*DNB.*)

Claparède, Jean Louis René Antoine Edouard (Edouard) (1832–71). Swiss naturalist and invertebrate zoologist. Professor of zoology and comparative anatomy, Academy of Geneva, 1862. Specialised in invertebrate anatomy, histology, embryology, and evolution. One of the first Swiss naturalists to endorse CD's theory. (*Dictionnaire historique & biographique de la Suisse*, Gilbert 1977.)

Clark, Barbara. Daughter of the Reverend John Stephen. Married the physician James Clark. (*DNB* s.v. Clark, James.)

Clark, James, 1st baronet (1788–1870). Scottish naval surgeon and physician. Naval surgeon, 1809–15. Practised as a physician in Rome, 1819–26; in London, 1826–60. Physician to Prince Leopold, the duchess of Kent, Queen Victoria, and Prince Albert. Member of the senate of London University, 1838–65; served on the general medical council, 1858–60. Created baronet, 1837. FRS 1832. (*DNB*, *Modern English biography.*)

Clarke, Richard Trevor (1813–97). Army officer and horticulturalist. Major in the Northampton and Rutland Infantry Militia, 1862. Bred nearly thirty new varieties of begonias and many new strains of cotton. Awarded a gold medal by the Cotton Supply Association of Manchester. Member of the Royal Horticultural Society; served on the council and scientific committee for many years; awarded the society's Veitchian medal, 1894. (*Army list*; R. Desmond 1994; *Gardeners' Chronicle*, 17 April 1897, p. 263.)

[April? 1863]

Clarke, William Branwhite (1798–1878). Clergyman and geologist. Emigrated to Australia in 1839, where he carried out extensive geological surveys. Discovered gold in New South Wales in 1841. Incumbent of Willoughby parish, Sydney, New South Wales, 1847–70. FRS 1876. (*DNB*, *DSB*.)

Clendon, George Jr (b. 1826/7). American military officer. Provost-marshal of the sixteenth congressional district, New York, 1863–5. Commissioned captain in the Union army, 1861; major, 1862. Wounded in action in Virginia, 1862; discharged for disability from wounds, 1863. (Phisterer 1912, p. 1988; *War of the rebellion* 5 (ser. 3): 531, 895.)
10 November 1863

Cobbett, William (1762–1835). Reform campaigner, journalist, essayist, and agriculturalist. Published reference works, newspapers, and political journals, including, from 1802, *Cobbett's Weekly Political Register.* Established experimental seed-farms in Kensington and Guildford. MP for Oldham, 1832–5. (Craig 1989, *DNB*.)

Cocker, Edward (1631–75). Arithmetician, teacher, and author. Ran a school for writing and arithmetic near St Paul's Cathedral, London, from 1664. Wrote a number of widely used arithmetical textbooks. (*DNB*.)

Colenso, John William (1814–83). Clergyman, mathematician, and missionary. Fellow, St John's College, Cambridge, 1837; college tutor, 1842–6. Mathematical tutor, Harrow, 1839–42. Vicar of Forncett St Mary, Norfolk, 1846–53. Bishop of Natal, South Africa, 1853–83. Published *Critical examination of the Pentateuch* (1862–79), in which he argued that these books were post-exile forgeries. There was an attempt to remove him from the bishopric in 1863, but his possession of the see was confirmed by the law courts in 1866. (*DNB*.)

Conway, Moncure Daniel (1832–1907). American preacher and author. Pastor of the Unitarian Church, Washington, until 1856, when he was dismissed because of his outspoken anti-slavery views. Preacher at the First Congregational Church, Cincinnati, until 1862, when he became editor of the Boston anti-slavery newspaper *Commonwealth.* Moved to Britain in 1863 and, in addition to writing and public speaking, became pastor at the South Place Chapel, Finsbury, London, until 1884, and 1892–7. (*DAB*.)

Cook, James (1728–79). Naval officer, navigator, explorer, and marine surveyor. Commander of several voyages of discovery. Circumnavigated the world, 1768–71 and 1772–5. FRS 1776. (*DNB*, *DSB*, *Record of the Royal Society of London*.)

Cooke, Benjamin (*fl.* 1735–1812). Of Newport, Isle of Wight. Published more than twelve papers, particularily on botanical subjects, mostly in the *Philosophical Transactions of the Royal Society*. FRS 1735. (*Philosophical Transactions* 40 (1738): 379–82, 43 (1744–5): 525–6; *Record of the Royal Society of London*; Watt 1824.)

Cooper, Susan Augusta Fenimore (1813–94). American author. Daughter of the novelist James Fenimore Cooper (*DAB*); acted as his amanuensis until his death in 1850. Wrote and edited books on the countryside, of which the most successful was *Journal of a naturalist in the United States* (also published as *Rural hours*) (London, 1850). (*DAB*, *BLC*, *NUC*.)

Correa da Serra, José Francisco (1750–1823). Portuguese priest and naturalist. Studied for the priesthood at Naples and Rome. Returned to Portugal in 1778 and helped establish the Royal Academy of Sciences, Lisbon, in 1779. His scientific enquiries made him suspect to the Catholic establishment and he fled Portugal in 1795, going first to London and then to Paris; emigrated to the United States in 1812. Returned to Portugal in 1821. FRS 1796. (Davis 1955, *O grande livro dos Portugueses*, s.v. Correia da Serra, Abade José Francisco; *Record of the Royal Society of London.*)

Coulon, Paul Louis de (Louis) (1804–94). Swiss naturalist. Friend of Louis Agassiz, whom he helped to establish at Neuchâtel. Director of the Musée d'Histoire Naturelle at Neuchâtel, founded by his father in 1837, until his own death. President of the Societé Neuchâteloise des Sciences Naturelles for more than fifty years. (*Dictionnaire historique & biographique de la Suisse.*)

Coulter, Thomas (1793–1843). Irish physician and botanist. Physician to Real del Monte Company in Mexico, 1824 – *circa* 1829. Collected plants in Mexico, California, and Arizona, *circa* 1831–3. Returned to Ireland in 1834 and was appointed curator of the herbarium at Trinity College, Dublin. (R. Desmond 1994, Hadfield *et al.* 1980.)

Cramer, Pieter (d. 1779). Dutch entomologist. (Gilbert 1977, *IBN.*)

Cranworth, Lord. *See* Rolfe, Robert Monsey.

Crawfurd, John (1783–1868). Scottish-born physician, diplomat, and orientalist. Joined the East India Company medical service in 1803. Held several civil and political posts in Java, India, Siam, Cochin China, Singapore, and Burma. Returned to England in 1827; promoted the study of Indo-China and campaigned on behalf of the Singapore trading community against the East India Company. FRS 1818. (*DNB*, *Record of the Royal Society of London*, Turnbull 1989, pp. 25–30.)

Cresy, Edward (1824–70). Surveyor and civil engineer, resident at Ham Moor, Black Boy Bridge, Chertsey, Surrey. Son of Edward Cresy (1792–1858), the architect and civil engineer who advised CD about the purchase and improvement of Down House. Worked as an architectural draftsman in his father's office as a young man. Assisted his father in preparing his *Cyclopædia of civil engineering* in 1845. Assistant surveyor under the Commissioners of Sewers, 1849; afterwards engineer. Principal assistant clerk at the Metropolitan Board of Works, 1859. Architect to the fire brigade, 1866. Founder member of the Geologists' Association, 1858; president, 1864–5; vice-president, 1865–70. (*Annual Report of the Geologists' Assocation* 1859–70; Census returns 1861 (Public Record Office RG9/422: 118a); *DNB* s.v. Cresy, Edward (1792–1858); *Engineer* 30 (1870): 409.)
27 April 1863, 13 May [1863]

Crocker, Charles William (1832–68). Spent time in Germany as a gardener at the summer residence of Victoria, crown princess of Prussia, near Berlin, before becoming foreman of the propagating department at the Royal Botanic Gardens, Kew, 1857–61/2. Moved to his native Chichester where he worked as a journalist and cathedral official. (*Correspondence* vol. 10, letter to J. D. Hooker, [before 15 February 1862], and letter from J. D. Hooker, [before 15 February 1862]; R. Desmond 1994; *Journal of Horticulture* n.s. 14 (1868): 206–7.)

1[–4] May 1863

Crotch, William Duppa (1831/2–1903). Entomologist and zoologist. Specialised in the Lepidoptera, Coleoptera, and Hemiptera. Travelled with his brother George Robert Crotch (1841–74) on several collecting expeditions. Married a Swede and settled in Scandinavia, where he studied lemmings. (*Entomologist's Monthly Magazine* 39 (1903): 256, Gilbert 1977.)

Crüger, Hermann (1818–64). German pharmacist and botanist. Apothecary in Trinidad in the West Indies from 1841; government botanist and director of the botanic garden, Trinidad, from 1857. Collected plants in Jamaica, Trinidad and Venezuela. (R. Desmond 1994, S[chlechtenda]l 1864.)

25 January [1863], *23 February 1863*, *23 April 1863*, 25 May [1863], *8 August 1863*

Cumming, Alexander Penrose Gordon, 3d baronet (1816–66). Scottish army officer. Captain in the Seventy-first Highlanders and Fourth Light Dragoons; major in the Inverness and Elgin Militia. Succeeded to the baronetcy in 1854. Justice of the peace and deputy lieutenant for Elgin. (*County families.*)

Cumming, John (1807–81). Scottish clergyman and apiarian. Prominent in numerous ecclesiastical and theological debates. Educated in Aberdeen before becoming a preacher at the National Scottish Church at Crown Court, Covent Garden, London. Widely known for his writings on prophecy. Wrote in *The Times* as 'Beemaster'. (*DNB.*)

Cunningham, James (1784–1851). Scottish nurseryman. Apprenticed as a gardener in Scotland; afterwards engaged at the Royal Botanic Gardens, Kew; returned to Scotland to establish a nursery at Comley Bank, Edinburgh, in 1810. Nurseryman and florist to Queen Victoria in Scotland, 1837. First to take up the hybridisation of rhododendrons in Scotland. (R. Desmond 1994; *Gardeners' Chronicle and Agricultural Gazette*, 1 November 1851, p. 695.)

Cunningham & Fraser. Nursery established by James Cunningham at Comley Bank, Edinburgh, *circa* 1845. (Bean 1970–88.)

Cuvier, Jean Léopold Nicolas Frédéric (Georges) (1769–1832). French systematist, comparative anatomist, palaeontologist, and administrator. Professor of natural history, Collège de France, 1800–32; professor of comparative anatomy, Muséum d'Histoire Naturelle, 1802–32. Permanent secretary to the Académie des Sciences from 1803. Foreign member, Royal Society of London, 1806. (*DBF*, *DSB.*)

Dana, James Dwight (1813–95). American geologist and zoologist. Naturalist with the United States expedition to the Pacific, 1838–42; wrote reports on the geology, zoophytes, and Crustacea. One of the editors of the *American Journal of Science* from 1846. Professor of natural history, Yale University, 1856–64; professor of geology and mineralogy, 1864–90. Foreign member, Royal Society of London, 1884. (*DAB*, *DSB.*)

7 January [1863], *5 February 1863*, 20 February [1863]

Dareste, Gabriel Madeleine Camille (Camille) (1822–99). French zoologist. A specialist in experimental embryology. Doctor of medicine, 1847. Doctor of science, 1851. Taught natural history at various provincial institutions. Professor of zoology, University of Lille, 1864–72. Professor of ichthyology and herpetology, Muséum d'Histoire Naturelle, Paris, 1872. Director of the laboratory of teratology, later attached to l'Ecole des Hautes-Etudes, 1875. Awarded the grand prize in physiology by the Académie des Sciences for *Recherches sur la production artificielle de monstruosités* (1877). (*DBF*, *Dictionnaire universel des contemporains.*)
8 February 1863, 16 February [1863]

Darwin, Anne Elizabeth (Annie) (1841–51). CD's daughter. (*Darwin pedigree.*)

Darwin, Caroline Sarah. *See* Wedgwood, Caroline Sarah.

Darwin, Charles Waring (1856–8). Youngest child of CD. Died of scarlet fever. (*Darwin pedigree.*)

Darwin, Edward Levett (b. 1821). CD's first cousin. Son of Francis Sacheverel Darwin. Captain in the Second Derby militia from 1856. Lived in Buxton, Derbyshire. (*Army List* 1863, *Darwin pedigree.*)
7 September 1863

Darwin, Elizabeth (Bessy) (1847–1926). CD's daughter. (*Darwin pedigree*, Freeman 1978)

Darwin, Emily Catherine (Catherine). See Langton, Emily Catherine.

Darwin, Emma (1808–96). Youngest daughter of Josiah Wedgwood II. Married CD, her cousin, in 1839. (*Emma Darwin* (1904) and (1915).)
[4 May 1863] (W. E. Darwin), *[6–27 September 1863]* (W. D. Fox), *23 September [1863]* (John Scott), *24 September [1863]* (John Scott), *[29 September 1863]* (W. D. Fox), *[28 October 1863]* (W. E. Darwin), *4 November [1863]* (Alfred Newton), *19 November [1863]* (John Scott), *20 November [1863]* (Friedrich Hildebrand), *21 November [1863]* (Patrick Matthew), *[7 December 1863]* (J. D. Hooker), *8 December [1863]* (W. D. Fox), *12 December [1863?]* (Julius von Haast), *[before 17 December 1863]* (John Murray), *17 December [1863]* (Alphonse de Candolle), *26 December [1863]* (J. D. Hooker), *28 December 1863* (George Maw)

Darwin, Erasmus (1731–1802). CD's grandfather. Physician, botanist, and poet. Advanced an evolutionary theory similar to that subsequently propounded by Jean Baptiste de Lamarck. FRS 1761. (*DNB*, *DSB.*)

Darwin, Erasmus Alvey (1804–81). CD's brother. Attended Shrewsbury School, 1815–22. Matriculated at Christ's College, Cambridge, 1822; Edinburgh University, 1825–6. Qualified in medicine but never practised. Lived in London from 1829. (*Alum. Cantab.*, Freeman 1978.)

Darwin, Frances Sarah (1822–81). Daughter of Francis Sacheverel Darwin. Married Gustavus Barton of Congleton, Cheshire, in 1845. Widowed in 1846; married Marcus Huish of Castle Donnington, Leicestershire, in 1849. (*Darwin pedigree.*)

Darwin, Francis (1848–1925). CD's son. Botanist. BA, Trinity College, Cambridge, 1870. Qualified as a physician but did not practise. Collaborated with

CD on several botanical projects, 1875–82. Lecturer in botany, Cambridge University, 1884; reader, 1888–1904. Edited CD's letters. President of the British Association for the Advancement of Science, 1908. Knighted, 1913. FRS 1882. (*DNB*, *DSB*.)

Darwin, Francis Sacheverel (1786–1859). Son of Erasmus Darwin by his second wife, Elizabeth Collier Pole (*Darwin pedigree*). Justice of the peace and deputy lieutenant of Derbyshire. Knighted, 1820. (*Alum. Cantab.*)

Darwin, George Howard (1845–1912). CD's son. Mathematician. BA, Trinity College, Cambridge, 1868; fellow, 1868–78. Called to the bar in 1872 but did not practise. Plumian Professor of astronomy and experimental philosophy, Cambridge University, 1883–1912. President of the British Association for the Advancement of Science, 1905. Knighted, 1905. FRS 1879. (*DNB*, *DSB*.)
[before 11 May 1863], *[9–15 June 1863]*

Darwin, Henrietta Emma (1843–1927). CD's daughter. Married Richard Buckley Litchfield (*Alum. Cantab.*) in 1871. Assisted CD with some of his work. Edited *Emma Darwin* (1904) and (1915). (*Burke's landed gentry* 1952, Freeman 1978.)
[1 August 1863], *14 October [1863]* (Thomas Warner)

Darwin, Horace (1851–1928). CD's son. Civil engineer. BA, Trinity College, Cambridge, 1874. Apprenticed to an engineering firm in Kent; returned to Cambridge in 1875 to design and make scientific instruments. Founder and director of the Cambridge Scientific Instrument Company. Mayor of Cambridge, 1896–7. Knighted, 1918. FRS 1903. (*Alum. Cantab.*, *DNB*.)

Darwin, Leonard (1850–1943). CD's son. Military engineer. Attended the Royal Military Academy, Woolwich. Commissioned in the Royal Engineers, 1870; major, 1889. Served on several scientific expeditions, including those for the observation of the transit of Venus in 1874 and 1882. Instructor in chemistry and photography, School of Military Engineering, Chatham, 1877–82. Intelligence service, War Office, 1885–90. Liberal Unionist MP, Lichfield division of Staffordshire, 1892–5. President, Royal Geographical Society of London, 1908–11; Eugenics Education Society, 1911–28. Chairman, Bedford College, London University, 1913–20. (Keynes 1943, Sarjeant 1980–96, *WWW*.)

Darwin, Robert Waring (1766–1848). CD's father. Physician. Had a large practice in Shrewsbury and resided at The Mount, which he built *circa* 1796–8. Son of Erasmus Darwin and his first wife, Mary Howard. Married Susannah, daughter of Josiah Wedgwood I, in 1796. FRS 1788. (Freeman 1978.)

Darwin, Susan Elizabeth (1803–66). CD's sister. Lived at The Mount, Shrewsbury, the family home, until her death. (*Darwin pedigree*, Freeman 1978.)

Darwin, William Erasmus (1839–1914). CD's eldest son. Banker. Attended Rugby School. BA, Christ's College, Cambridge, 1862. Partner in Southampton and Hampshire Bank, Southampton, 1861. Chairman of the Southampton Water Company. Amateur photographer. (*Alum. Cantab.*, F. Darwin 1914.)
21 [January 1863], *9 November [1863]*, *9 November 1863* (D. M. Conway), *11 November 1863* (D. M. Conway), *11 November [1863]* (Emma Darwin)

Davis, Jefferson (1808–89). American statesman. President of the Confederate States of America. As senator, announced the secession of Mississippi from the Union in 1861; became president of the Confederacy in the same year. Remained president until the fall of the Confederacy and his own capture by unionists in 1865. Advocated an independent Southern American nation. (*DAB.*)

Dawson, John William (1820–99). Canadian geologist and educationalist. Investigated the geology of the maritime provinces with Charles Lyell in 1842 and 1852. Superintendent of education for common schools in Nova Scotia, 1850. Appointed principal and professor of geology at McGill University, 1855. Knighted, 1884. FRS 1862. (*DNB*, *DSB.*)

Dawson, Thomas. Glasgow shipping agent. Resided at Meadowbank, Uddingston, Lanarkshire. (R. Desmond 1994 s.v. Anderson, James; *Post Office Glasgow directory* 1863–4.)

Decaisne, Joseph (1807–82). French botanist. Gardener at the Jardin des Plantes, Paris, 1824. Professor of agricultural statistics, Collège de France, 1848. Professor of plant cultivation, Muséum d'Histoire Naturelle, 1850. Foreign member, Royal Society of London, 1877. (*DBF*, *NBU.*)

De la Rue, Warren (1815–89). Printer, astronomer, and chemist. Educated in Paris. Entered his father's printing firm, De la Rue & Sons. Published researches on practical chemistry and astronomy which he conducted at his private laboratory and observatory. Noted for his work on celestial photography. President of the Royal Astronomical Society, 1864–6; of the Chemical Society, 1867–9 and 1879–80. FRS 1850. (*DNB*, *DSB.*)

Dennen, George (*fl.* 1860s). Secretary of the Savings Bank, High Street, Bromley, Kent, and, from 1866, manager of the Bromley branch of the London and County Joint Stock Banking Company. (*Banking almanac*, *Post Office directory of the six home counties.*)

Desor, Pierre Jean Edouard (Edouard) (1811–82). German-born geologist. Secretary and assistant to Louis Agassiz in Neuchâtel from 1837. Joined Agassiz in America in 1848. Returned to Switzerland in 1852. Professor of geology at Neuchâtel. Founder and first president of the Société d'Histoire de Neuchâtel, 1866. President of the International Congress of Anthropology and Prehistoric Archaeology, 1866. Presided over the council of the Academy of Neuchâtel. Member of the Commission for the Geological Map of Switzerland. Involved in both regional and national government. (*Bulletin Société des Sciences Naturelles de Neuchâtel* 7 (1864–7): 153, 260; *Dictionnaire historique & biographique de la Suisse.*)

Devas, Charles Frederick (1825/6–96). Bromley resident, active in local government. Justice of the peace for Kent. Chairman of the Bromley Local Board from 1867. Resided at Bromley Lodge from 1855; moved to Pickhurst Manor in 1870. (*Alum. Oxon.*; Horsburgh 1980, pp. 258, 423; *Post Office directory of the six home counties* 1862–82; *Registrum Orielense.*)

Devay, Francis Marie Antoine (Francis) (1813–63). French medical practitioner. Professor of clinical medicine, Ecole de Médicine de Lyon, 1854. (*DBF.*)

Devonshire, duke of. *See* Cavendish, William.

Dickie, George (1812–82). Scottish botanist. Lecturer in botany, King's College, Aberdeen, 1839–49. Professor of natural history, Queen's College, Belfast, 1849–60. Professor of botany, Aberdeen University, 1860–77. Specialised in the study of marine algae; described many of the collections sent to the Royal Botanic Gardens, Kew. FRS 1881. (R. Desmond 1994, *DNB*.)

Dixon, William Hepworth (1821–79). Historian, traveller, man of letters. Published lives of William Penn (1851), Robert Blake (1852), and Francis Bacon (1862). Editor of the *Athenaeum*, 1853–69. Justice of the peace for Middlesex and Westminster, 1869. Member of the London School Board, 1870. Fellow of the Society of Antiquaries and the Royal Geographical Society of London. (*DNB*.)
16 April 1863

Dobell, Horace Benge (1827–1917). Physician and medical author. Entered general practice in London, 1849. Qualified as a physician and entered practice as a consultant, 1856. Physician to the Royal Hospital for Diseases of the Chest, London, 1859; honorary consulting physician, 1875. Physician to the Royal Albert Orphan Asylum, 1864. Fellow of the Royal Medical and Chirurgical Society of London, 1862. (*British Medical Journal*, 10 March 1917, p. 351; Leyland ed. 1888.)
16 February [1863], *5 March 1863*, 6 March [1863], *20 April 1863*, 21 April [1863], *12 May 1863*

Don, David (1800–41). Botanist. Professor of botany, King's College, London, 1836–41. Librarian, Linnean Society of London, 1822–41. (R. Desmond 1994, *DNB*.)

Donald, Alexander (b. 1812/13). Scottish gardener. Head gardener to Isaac Anderson-Henry at Hay Lodge, Edinburgh. (Census returns 1871 (General Register Office for Scotland 692/1/19/119).)

Doubleday, Henry (1808–75). Entomologist and ornithologist; grocer, ironmonger, and insurance agent in Epping, Essex. Noted for his work on the systematics of Lepidoptera. (R. Desmond 1994, *DNB*, *Post Office directory of the six home counties* 1862.)

Douglas, David (1799–1834). Scottish gardener and plant collector. Collected plants in North America and the Pacific islands during six expeditions. Introduced over two hundred new species. (R. Desmond 1994.)

Douglas, George Sholto, 17th earl of Morton (1789–1858). Diplomat and statesman. Retired from service, 1825. Succeeded to earldom, 1827; a representative peer of Scotland, 1830–58. Lieutenant-colonel, Midlothian yeomanry cavalry, 1843–4; vice-lieutenant of Midlothian, 1854–8. (*Modern English biography*.)

Duberry, George (b. 1847/8). Agricultural labourer. Resident of Down, Kent. (Census returns 1861 (Public Record Office RG9/462: 71).)

Duberry, Mark (b. 1844/5). Carter. Resident of Down, Kent. (Census returns 1861 (Public Record Office RG9/462: 71).)

Duberry, Richard (b. 1811/12). Agricultural labourer. Resident of Down, Kent. (Census returns 1861 (Public Record Office RG9/462: 71).)

Duchartre, Pierre Etienne Simon (1811–94). French botanist. Specialised in plant physiology. Professor of botany, Agricultural Institute, Versailles, 1849–52; professor of botany, Faculty of Sciences, Paris, 1861. (Barnhart 1965, *DBF.*)

Duchesne, Antoine Nicolas (1747–1827). French botanist. Professor of natural history, Ecole Centrale, Versailles, 1794. (*DBF.*)

Duncan, Peter Martin (1821–91). Physician and geologist. Physician, Essex and Colchester Hospital, 1848–59; consultant physician, county asylum and Oldham Club. Practised at Blackheath from 1860. Professor of geology, King's College, London, 1870. Secretary of the Geological Society of London, 1864–70; president, 1876–8. Specialist on corals and echinoids. FRS 1868. (*DNB*, *Medical directory* 1849–76.)

Dutrochet, René Joachim Henri (1776–1847). Physician and natural scientist. French military medical officer, 1808–9. Corresponding member of the Académie des Sciences, 1819; full member, 1831. Wrote works on plant physiology; noted for his research on osmosis and diffusion. (*DBF*, *DSB.*)

Edward VII (1841–1910). Eldest son of Queen Victoria and Prince Albert. King of Great Britain and Ireland and of the British dominions beyond the seas, and emperor of India, from 1901. As Albert Edward, prince of Wales, married Princess Alexandra of Schleswig-Holstein-Sonderburg-Glücksburg on 10 March 1863. FRS 1863. (*DNB.*)

Edwards, Alphonse Milne (1835–85). French naturalist. Trained as a physician. Assistant to his father, the naturalist Henri Milne Edwards. Professor, school of pharmacy, Paris, 1865. Professor of zoology, Muséum d'Histoire Naturelle, Paris, 1876, in succession to his father; director, 1891. (*Dictionnaire universel des contemporains.*)

Egerton, Francis Henry, 8th earl of Bridgewater (1756–1829). Scholar and patron of learning. His literary works, many on his notable ancestors, were mostly printed for private circulation. Bequeathed 8000 guineas in his will to commission works illustrating the 'goodness of God as manifested in the Creation'; the money to be divided among eight persons. The resulting essays have become known as the Bridgewater treatises. FRS 1781. (*DNB*, Topham 1993.)

Egerton, Philip de Malpas Grey-, 10th baronet (1806–81). Palaeontologist. Of Oulton Park, Cheshire. Tory MP for South Cheshire, 1835–68; for West Cheshire, 1868–81. Specialised in fossil fish. FRS 1831. (*DNB*, Sarjeant 1980–96.)

Ehrenberg, Christian Gottfried (1795–1876). German zoologist, comparative anatomist, and microscopist. Wrote extensively on infusoria and the development of coral reefs. Professor of medicine at Berlin University, 1839. Foreign member, Royal Society of London, 1837. (*DSB*, *NDB.*)

Emerson, Ralph Waldo (1803–82). American clergyman, essayist, and poet. (*DAB.*)

Encke, Johann Franz (1791–1865). German astronomer. Appointed professor of astronomy, Academy of Sciences, Berlin, and director of the Berlin observatory, 1825; professor, Berlin University, 1844. Retired, 1862. (*DSB.*)

Engelmann, Georg (George) (1809–84). German-born physician, botanist, and meteorologist. Studied medicine at Heidelberg, Berlin, and Würzburg. Moved to Paris in 1832 and, later in the same year, to the United States. Practised medicine in St Louis, Missouri. One of the founders of the St Louis Academy of Science, 1856. (*DAB*.)

Engleheart, Stephen Paul (1831/2–85). Surgeon. Member of the Royal College of Surgeons of England, 1859; licentiate of the Royal College of Surgeons of Edinburgh, 1860. Surgeon in Down, Kent, 1861–70. Medical officer, Second District, Bromley Union, 1863–70; divisional surgeon of police, 1863–70. Resident in Shelton, Norfolk, 1870–81; in Old Calabar, Nigeria, 1882–5. (*Medical directory* 1861–86, *Post Office directory of the six home counties* 1862.)

Esar-haddon (d. 668 B.C.). Assyrian king. Son of Sennacherib. Came to the throne 680 B.C. Father of Assur-bani-pal and Samas-sum-yukin. (*EB*.)

Etheridge, Robert (1819–1903). Palaeontologist. Curator, Bristol Philosophical Institution, 1850–7. Assistant palaeontologist to the Geological Survey of Great Britain, 1857; palaeontologist, 1863. Assistant keeper in geology, British Museum, 1881–91. Vice-president of the Royal Society of London. FRS 1871. (*DNB*.)

Evans, Elizabeth (1820–98). Joseph Dalton Hooker's sister. Married the surgeon Thomas Robert Evans in 1853. (Allan 1967 s.v. 'Hooker pedigree'.)

Evans, John (1823–1908). Paper manufacturer in Hemel Hempstead, Hertfordshire; archaeologist, geologist, and numismatist. In 1859, his study of chipped flints helped to establish the antiquity of humans in western Europe. Published an important paper on the fossil bird, *Archeopteryx*, in 1865. Developed a theory of evolution with regard to coins, 1849–50, and later applied natural selection to numismatics. Active member of many archaeological, scientific, and industrial societies. Vice-president of the Royal Society of London from 1876; treasurer, 1878–89. Honorary secretary of the Geological Society of London, 1866–74; president 1874–6. Knighted, 1892. FRS 1864. (*DNB*.)

Evans, Thomas Robert (1821–1902). Surgeon at Coltishall, Norfolk. Married Elizabeth Hooker, Joseph Dalton Hooker's sister, in 1853. (Allan 1967 s.v. 'Hooker pedigree'; *Medical directory*.)

Ewart, William (1798–1869). Politician. Radical MP for Bletchingley, Surrey, 1828–30; for Liverpool, 1830–7; for Wigan, 1839; for Dumfries district, 1841–68. Supported repeal of the corn laws, abolition of the death penalty for a range of offences, electoral reform, university reform, and the establishment of free public libraries. (*DNB*.)

Fagan, George. Diplomat. Chargé d'affaires and consul-general at Guayaquil in Ecuador. (*British imperial calendar* 1863.)

Fairbairn, William, 1st baronet (1789–1874). Scottish engineer and shipbuilder. Associate of George Stevenson. Patented a new method for the manufacture of wrought-iron girders. President of the Institution of Mechanical Engineers, 1854; of the British Association for the Advancement of Science, 1861. Created baronet, 1869. FRS 1850. (*DNB*.)

Falconer, Hugh (1808–65). Palaeontologist and botanist. Superintendent of the botanic garden, Saharanpur, India, 1832–42. Superintended the arrangement of Indian fossils for the British Museum, 1844. Superintendent of the Calcutta botanic garden and professor of botany, Calcutta Medical College, 1848–55. Retired owing to ill health and returned to Britain in 1855; pursued palaeontological research while travelling in southern Europe. Vice-president of the Royal Society of London and foreign secretary of the Geological Society of London, 1865. FRS 1845. (*DNB*, *DSB*.)

3 January [1863], [and 6] January [1863], *8 January [1863]*, *9 January 1863*, *18 January [1863]*, 20 [January 1863], *20 April [1863]*, 22 April [1863], *24 April [1863]*, *24 August [1863]*, 25–6 August 1863], *29 August 1863*, 4 [September 1863], *10 September 1863*, *31 December [1863]*

Fall, Phillip Carteret (*fl.* 1840s–60s). Partner at the Southampton and Hampshire Bank. Justice of the peace. (*Banking almanac* 1845–61; *Correspondence* vol. 9, letter to W. E. Darwin, 17 [October 1861], and letter from W. E. Darwin, [17 November 1861].)

Faraday, Michael (1791–1867). Natural philosopher. Apprentice to a bookbinder, 1804. Chemical assistant at the Royal Institution of Great Britain, 1813–24; director of the laboratory, 1825; professor of chemistry, 1833–65. Noted for his popular lectures and for his extensive researches in electrochemistry, magnetism, and electricity. FRS 1824. (*DNB*, *DSB*, *Modern English biography*.)

Field, Edwin Wilkins (1804–71). Lawyer, law reformer, and artist. Instrumental in the reform of laws that disadvantaged dissenters. Worked to establish artistic copyright; one of the founders of the Slade School of Art, London. Member of the committee of the fine art section of the International Exhibition, 1862. Following the outbreak of the Civil War in the United States, his correspondence with Charles Greely Loring concerning that country's relations with Britain was published in Loring 1862. (*DNB*.)

Fischer, Friedrich Ernst Ludwig von (1782–1854). Russian botanist of German origin. Director, St Petersburg botanical garden, 1823–50. (*Taxonomic literature*.)

Fitch, Walter Hood (1817–92). Botanical artist at the Royal Botanic Gardens, Kew, from 1841. Produced over 10,000 published drawings. Illustrated Joseph Dalton Hooker's *Rhododendrons of Sikkim-Himalaya* (1849–51) and *Illustrations of Himalayan Plants* (1855). (R. Desmond 1994, *Proceedings of the Linnean Society of London* (1891–2): 68.)

Flower, William Henry (1831–99). Anatomist and zoologist. Curator of the Hunterian Museum, Royal College of Surgeons of England, 1861–84; Hunterian Professor of comparative anatomy, Royal College of Surgeons, 1870–84. Director of the Natural History Museum, London, 1884–98. President of the Zoological Society of London, 1879–99. President of the British Association for the Advancement of Science, 1889. Knighted, 1892. FRS 1864. (*DNB*.)

13 April [1863], *15 April 1863*, *9 May 1863*, 12 May [1863], 11 July [1863]

Folgate, William (b. 1810/11). Gamekeeper, Down, Kent. (Census returns 1861 (Public Record Office RG9/462: 77).)

Fonvielle, Wilfred de (1824–1914). French journalist. Deported to Algeria, 1851, as a result of his activities following the February 1848 Revolution; returned to Paris in 1859 under amnesty. Left Paris again in 1871 after being condemned by a popular tribunal. Renounced politics after 1876 and devoted himself to literature and popular science. (*DBF*, *Dictionnaire universel des contemporains*.)

Forbes, Edward (1815–54). Zoologist, botanist, and palaeontologist. Naturalist on board HMS *Beacon*, 1841–2. Appointed professor of botany, King's College, London, and curator of the museum of the Geological Society of London in 1842. Palaeontologist with the Geological Survey of Great Britain, 1844–54. Professor of natural history, Edinburgh University, 1854. FRS 1845. (*DNB*, *DSB*.)

Forbes, James David (1809–68). Natural philosopher. Professor of natural philosophy, Edinburgh University, 1833–60. Secretary of the Royal Society of Edinburgh, 1840–51. Principal of the United College, University of St Andrews, 1860–8. FRS 1832. (*DNB*, *DSB*.)

Forgues, Paul Emile Daurand (1813–83). French writer, critic, and translator. Specialised in criticism of English literature under the pseudonym Old Nick. (*Dictionnaire universel des contemporains*.)

Fouché, Joseph (1759–1820). French statesman. Tutor at a number of Catholic colleges before the French Revolution. Leading member of the Jacobin club in Nantes. Deputy to the National Convention for the Lower Loire, 1792. As a commissioner of the republic, led the suppression of the counter-revolution in Brittany, la Vandée, Lyons, and Nièvre. Ambassador to the Cisalpine republic, 1798; to the Hague, 1799. Minister of police, 1799–1802, 1804–10, and 1815. Duke of Otranto, 1808. Governor of Rome, 1810. President of the provisional government after the final defeat of Napoleon Bonaparte. (*EB*.)

Fox, Charles Woodd (1847–1908). Barrister. Son of William Darwin Fox. (*Alum. Oxon.*, *Repton School register*.)

Fox, Ellen Sophia (1820–87). Married William Darwin Fox in 1846. (*Darwin pedigree*.)

Fox, Robert Gerard (b. 1849). Justice of the peace, Hampshire. Son of William Darwin Fox. (*Darwin pedigree*.)

Fox, Samuel William Darwin (b. 1841). Clergyman. Eldest son of William Darwin Fox. Curate of St Paul's, Manningham, Yorkshire, 1865; of St Augustine's, Halifax, Yorkshire, 1867. Rector of St Peter Lymm, Oughtrington, Cheshire, 1874. Vicar of St Paul's, Maidstone, Kent, 1887. (*Alum. Oxon.*; *Crockford's clerical directory*; *Registers of Wadham College, Oxford*.)

Fox, William Darwin (1805–80). Clergyman. CD's second cousin. A friend of CD's at Cambridge; shared his enthusiasm for entomology. Maintained an active interest in natural history throughout his life and provided CD with much information. Rector of Delamere, Cheshire, 1838–73. Spent the last years of his

Fox, William Darwin, cont.

life at Sandown, Isle of Wight. (*Alum. Cantab.*)

6 February [1863], [10 February 1863], *[11 February 1863]*, 9 March [1863], 16 [March 1863], 23 March [1863], *[16–22 May 1863]*, 23 May [1863], *29 May [1863]*, 4 [September 1863], *7 September [1863]*

Francis, William (*fl.* 1852–97). Printer. Partner of Richard Taylor in the printing firm Taylor & Francis from 1852; head of the firm, 1858. Editor of *Annals and Magazine of Natural History*, 1859–97. (*Annals and Magazine of Natural History*, 1859–97; *DNB* s.v. Taylor, Richard.)

Frean, Richard. Surgeon. Trained at the Middlesex Hospital, London; governor's prizeman in clinical surgery and medicine, 1871. Resident obstetrical assistant at the Middlesex Hospital before being appointed surgeon at the Hay Hospital. Emigrated to Australia where he became the resident medical officer at the Sydney Infirmary. (*Medical directory* 1882.)

22 February 1863

Frémont, John Charles (1813–90). American explorer, politician, and soldier. Served with the United States Topographical Corps, and in the early 1840s, led a series of surveying expeditions to the American west on behalf of the War Department. A popular hero for his role in the conquest of California in the war against Mexico. Elected to the United States Senate, 1850; nominated to run for president, 1856 and 1864. Served as a major-general in the Union army during the American Civil War. Governor of Arizona, 1878–83. (*DAB.*)

Froude, James Anthony (1818–94). Historian and man of letters. Disciple of Thomas Carlyle; published *History of England* (1856–70), *The English in Ireland in the eighteenth century* (1872–4), and *The English in the West Indies* (1888). Editor of *Fraser's Magazine* 1860–74. Regius professor of modern history, Oxford University, 1892–4. (*DNB.*)

Fry, James Thomas (*fl.* 1850s–70s). Neighbour of CD. Resident at Baston House, Hayes, near Bromley, Kent, from between 1851 and 1855 until between 1870 and 1874. (*Post Office directory of the six home counties* 1851–74)

Gage, William Hall (1777–1864). Naval officer. Commanded the *Terpsichore* at the blockade of Malta; the *Thetis*, 1805–8; the *Indus*, 1813–14. Rear-admiral, 1821. Commander, East Indies, 1825–30; Plymouth, 1848–51. Admiral, 1846. Admiral of the fleet, 1862. Knighted, 1860. (*DNB.*)

Galton, Douglas Strutt (1822–99). Military engineer and civil servant. Officer in the Royal Engineers. Joined the Ordnance Survey in 1846. Secretary, Railway Commission, 1847; Railway Department, Board of Trade, 1854. Assistant permanent under-secretary of state for war, 1862. Director of public works and buildings, Office of Works, 1869–75. Author of numerous government reports on sanitation, telegraphy, and railways. Associate, Institution of Civil Engineers, 1850. Member, British Association for the Advancement of Science, 1860; general secretary, 1871–95; president, 1895. Knighted, 1887. FRS 1859. (*DNB.*)

Galton, Francis (1822–1911). Traveller, statistician, and scientific writer. CD's cousin. Explored in south-western Africa, 1850–2. Carried out various researches on heredity. Founder of the eugenics movement. FRS 1860. (*DNB*, *DSB*.)

García Moreno, Gabriel (1821–75). President of Ecuador, 1861–5 and 1870–5. Professor of chemistry, University of Quito. (*EB*, *NUC*.)

Gärtner, Karl Friedrich von (1772–1850). German physician and botanist. Practised medicine in Calw, Germany, from 1796. Contracted an eye ailment in the course of his microscopical investigations, which forced his retirement from medical practice. Studied plant hybridisation from *circa* 1824. (*ADB*, *DSB*.)

Geikie, Archibald (1835–1924). Scottish geologist. Appointed member of the Scottish branch of the Geological Survey of Great Britain, 1855. Director of the Geological Survey of Scotland, 1867–82. Director-general of the Geological Survey of Great Britain, 1882–1901. Murchison Professor of geology and mineralogy, Edinburgh University, 1871–81. Knighted, 1891. FRS 1865. (*DNB*, *DSB*.)

Geoffroy Saint-Hilaire, Etienne (1772–1844). French zoologist. Professor of zoology, Muséum d'Histoire Naturelle, 1793. Devoted much attention to embryology and teratology. (*DBF*, *DSB*.)

Geoffroy Saint-Hilaire, Isidore (1805–61). French zoologist. Succeeded his father, Etienne Geoffroy Saint-Hilaire, as professor at the Muséum d'Histoire Naturelle in 1841. Continued his father's work in teratology. Became professor of zoology at the Sorbonne in 1850. (*DBF*, *DSB*.)

George IV (1762–1830). King of Great Britain and Ireland; succeeded to the throne, 1820. (*DNB*.)

Gervais, Paul (1816–79). French zoologist and palaeontologist. Assistant, Muséum d'Histoire Naturelle, Paris, 1835–45. Professor of zoology and comparative anatomy, Faculté des sciences de Montpellier, 1845; head of the faculty, 1856. Professor of comparative anatomy, Muséum d'Histoire Naturelle, 1868. (Tort 1996.)

Giblitt, William. Butcher with premises at 110 Bond Street, London. (*Post Office London directory* 1854.)

Gladstone, William Ewart (1809–98). Statesman and author. Chancellor of the exchequer, 1852–5 and 1859–66. Prime minister, 1868–74. FRS 1881. (*DNB*.)

Glenie, Mary Elizabeth Louisa (d. 1892). Daughter of Mary Selway and Gustavus Adolphus Tranchell (1787–1866). Second wife of Samuel Owen Glenie. Her previous marriages were to Andrew Halliday Hall (1808–38), a surgeon, in 1836, and to George Cochrane (1789/90–1860), an army officer. (J. P. Lewis 1913, pp. 230, 268, 273, 274, and 438.)

Glenie, Samuel Owen (1811–75). Clergyman. Assistant chaplain at St Paul's, Colombo, Ceylon, 1834; chaplain at Trincomalee, 1840; archdeacon of Kandy, 1870; retired 1871. Edited the *Ceylon Chronicle*, 1837–8. Collected plants for George Henry Kendrick Thwaites. (*Alum. Cantab.*; Beven ed. [1946], pp. 53–4; R. Desmond 1994; J. P. Lewis 1913, p. 438.)

Godman, Frederick Du Cane (1834–1919). Ornithologist, entomologist, and traveller. Founder member of the ornithological magazine *Ibis*. Friend of Osbert

Godman, Frederick Du Cane, cont.
Salvin, with whom he undertook an exploration of Guatemala in 1861. Studied the fauna of the Azores, 1865–70. Continued the editorial supervision of the sixty-three-volume *Biologia Centrali-Americana* (Godman and Salvin eds. 1879–1915) after Salvin's death in 1898. Trustee of the British Museum, 1897–1919. FRS 1882. (*Alum. Cantab.*; Godman and Salvin eds. 1879–1915, 1: 4–12.)

Godron, Dominique Alexandre (1807–80). French botanist, zoologist, and ethnologist. In 1854, became head of the science faculty at Nancy, where he established a natural history museum and a botanic garden. (*DBF.*)

Godwin-Austen, Robert Alfred Cloyne (1808–84). Geologist. Noted for his work on the stratigraphy of southern England. Predicted the existence of coal-bearing strata in the south-east. Secretary of the Geological Society of London, 1843–4 and 1853–4. FRS 1849. (*DNB*, *DSB.*)

Golding, George. Surveyor, Bromley Highway District, 1863. Resident of Chislehurst, Kent. (Minutes of the Bromley Highway Board (Bromley Central Library 847/HB/B/1), *Post Office directory of the six home counties* 1862.)

Goodsir, John (1814–67). Scottish surgeon and anatomist. Surgeon in Anstruther, Fifeshire, 1835–40. Conservator, museum of the Royal College of Surgeons, Edinburgh, 1841–3. Conservator, human and comparative anatomy museum, Edinburgh University, 1840; curator, anatomy and pathology museum, 1843; demonstrator in anatomy, 1844; professor of anatomy, 1846. FRS 1846. (*DNB*, *DSB.*)
21 August [1863], *27 August [1863]*, *28 August [1863]*

Gordon, George (1801–93). Scottish botanist, geologist, and clergyman. Minister of Birnie, near Elgin, Morayshire, 1832–89. Specialised in the stratigraphy and fossil vertebrates of Elgin sandstone. Founded the Elgin and Morayshire Literary and Scientific Association. (R. Desmond 1994, Sarjeant 1980–96.)

Gordon, William (*c.* 1796–1836?). Of Haffield House, Ledbury, Herefordshire. (*Cottage Gardener*, 28 November 1854, p. 157; R. Desmond 1994 s.v. Beaton, Donald; *Survey gazetteer of the British Isles.*)

Gosse, Edmund William (1849–1928). Poet and writer. Son of Philip Henry Gosse. Employed in the cataloguing section of the British Museum, 1865–75. Translator, Board of Trade, from 1875. Clark Lecturer in English literature, Trinity College, Cambridge, 1885–90. Librarian, House of Lords, 1904–14. (*DNB.*)

Gosse, Philip Henry. (1810–88). Zoologist, traveller, and writer. Collected zoological specimens in the West Indies for the British Museum, 1844–6. Made his living as a writer from 1847. Lived at St Marychurch, near Torquay, Devonshire, from 1852. Cultivated orchids. Studied marine invertebrates. FRS 1856. (*DNB.*)
30 May 1863, 2 June [1863], *4 June 1863*, 5 June [1863], *13 July 1863*

Gould, John (1804–81). Ornithologist and artist. Taxidermist to the Zoological Society of London, 1826–81. Described the birds collected on the *Beagle* and *Sulphur* expeditions. FRS 1843. (*DNB*, *DSB.*)

Gourlie, William (1815–56). Scottish merchant and botanist. Partner in merchant house of William Gourlie & Son, 8 South Frederick Street, Glasgow. Collected mosses, shells, and fossil plants. An active member of the Glasgow Philosophical Society; local secretary in Glasgow for the Botanical Society of Edinburgh. A local secretary for the British Association for the Advancement of Science meeting in Glasgow in 1855. (*Imperial directory of universal biography*, *Modern English biography*, *Post Office Glasgow directory* 1839–40.)

Gower, William Hugh (1835–94). Nurseryman. Foreman, Royal Botanic Gardens, Kew, until 1865. Specialist on orchids. (R. Desmond 1994.)

Graham, Robert (1786–1845). Physician and botanist. Regius professor of botany, Edinburgh University, 1820–45. Physician to the Edinburgh Infirmary. (*DNB*.)

Grant, Robert Edmond (1793–1874). Scottish physician and zoologist. An early supporter of the theory of the transmutation of species. Befriended CD in Edinburgh. Professor of comparative anatomy and zoology, University College, London, 1827–74. FRS 1836. (*DNB*.)

Gratiolet, Louis Pierre (1815–65). French anatomist and anthropologist. Laboratory assistant, Muséum d'Histoire Naturelle, Paris, 1842–53; lectured on anatomy, 1844–50; director of anatomical studies, 1853–62. Deputy to the professor of zoology, Faculty of Science, Paris, 1862–3; professor, 1863–5. (*DSB*.)

Gray, Asa (1810–88). American botanist. Fisher Professor of natural history, Harvard University, 1842–88. Wrote numerous botanical textbooks and works on North American flora. President of the American Academy of Arts and Sciences, 1863–73; of the American Association for the Advancement of Science, 1872; a regent of the Smithsonian Institution, 1874–88. Foreign member, Royal Society of London, 1873. (*DAB*, *DSB*.)

2 January [1863], 19 January [1863], *27 January 1863*, 23 February [1863], 20 March [1863], *22–30 March 1863*, *11 April 1863*, *13 April 1863*, 20 April [1863], *20 April 1863*, 11 May [1863], *26 May 1863*, 31 May [1863], *[10–16] June [1863]*, 26 June [1863], *6 July 1863* (J. D. Hooker), *7 July 1863*, *21 July 1863*, 4 August [1863], *1 September 1863*, *23 November 1863*

Gray, Jane Loring (1821–1909). Daughter of Charles Greely Loring, Boston lawyer and politician, and Anna Pierce Brace. Married Asa Gray, 4 May 1848. Edited the *Letters of Asa Gray* (1893). (Barnhart 1965; Dupree 1959, pp. 177–84.)

Gray, John Edward (1800–75). Botanist and zoologist. Assistant keeper of the zoological collections at the British Museum, 1824; keeper, 1840–74. President, Botanical Society of London, 1836–56. FRS 1832. (R. Desmond 1994, *DNB*.)

12 May 1863

Green, Joseph Henry (1791–1863). Surgeon. Surgeon, St. Thomas's Hospital, 1820–52. Professor of anatomy, Royal College of Surgeons, 1824; Royal Academy, 1825–52. Professor of surgery, King's College, London, 1832–7. President, Royal College of Surgeons, 1849, 1858. President of the General Medical Council, 1860. Literary executor of Samuel Taylor Coleridge. FRS 1825. (*DNB*, Plarr 1930.)

Gregson, Charles Stuart (1817–99). Shipyard worker, plumber, and entomologist. Worked as a painter in the Liverpool shipyards. Collected British Lepidoptera

Gregson, Charles Stuart, cont.
and was the author of numerous articles on entomology. Active in a number of local scientific societies, including the Northern Entomological Society and the East Lancashire Natural History Society. (*Entomologist's Monthly Magazine* 35 (1899): 96–7; Gregson 1883; *Post Office directory of Lancashire, Liverpool and Manchester* 1864.)

Gresson, John George (b. 1831/2). Clergyman and schoolteacher. Bible clerk, 1849–53. Master at St Andrew's College, Bradfield, Berkshire, 1855–63. Headmaster of West Mansion Preparatory School, Heene, Worthing, Sussex, from 1885 or earlier. (*Alum. Oxon.*, *Clergy list* 1885, *Crockford's clerical directory* 1889.)

Greville, Robert Kaye (1794–1866). Scottish botanist, entomologist, and social reformer. President, Botanical Society of Edinburgh, 1838 and 1866; honorary secretary, from *circa* 1861 until 1866. Edited *The Amethyst, or Christian's Annual*, 1832–4; secretary of the Sabbath Alliance; one of the compilers of the *Church of England hymn-book* (1838). Vice-president of the Anti-slavery Convention, 1840. Earned a living as a landscape painter in his later years. (*DNB*, *Medical directory* 1861–7, *Modern English biography*, *Transactions of the Botanical Society* [*of Edinburgh*] 1 (1844).)

Grisebach, August Heinrich Rudolf (1814–79). German botanist. Travelled through the Balkan Peninsula and north-western Asia Minor, 1839–40, studying the flora of these regions. Professor of botany, Göttingen University, 1847. (*DSB*, *NDB*.)

Grove, William Robert (1811–96). Lawyer, judge, and natural philosopher. Professor of experimental philosophy, London Institution, 1847. An active member of the Royal Society of London; treasurer and chairman of the executive committee of the Philosophical Club, 1847; Royal Medallist, 1847. Member of the royal commission on the law of patents, 1864. Appointed to the bench, Court of Common Pleas, 1871. FRS 1840. (*DNB*, *DSB*.)

Gully, James Manby (1808–83). Physician. Practised medicine in London, 1830–42. Set up a hydropathic establishment in Great Malvern, Worcestershire, in 1842; a successful practitioner of hydropathy until his retirement in 1872. (*DNB*, *Modern English biography*.)

Gunn, Harriet (1806–69). Daughter of Dawson Turner, banker, of Yarmouth, Norfolk. Married John Gunn in 1830. Sister of Maria Hooker, and aunt of Joseph Dalton Hooker. (Allan 1967 s.v. 'Hooker pedigree'; L. Huxley ed. 1918, 1: 18–19.)

Gunn, John (1801–90). Geologist, palaeontologist, and clergyman. Rural dean, vicar of Barton Turf, and rector of Irstead, Norfolk. Leading member of the Norfolk and Norwich Museum; one of the founders of the Norfolk and Norwich Archaeological Society, 1846; of the Norwich Geological Society, 1864; president, Norwich Geological Society, 1864–77 and 1883–8. (L. Huxley ed. 1918, Woodward ed. 1891.)

Haast, John Francis Julius (Julius) von (1824–87). German-born explorer and geologist. Travelled to New Zealand in 1858 to report on the prospect of German

emigration. Explored the western districts of Nelson province at the request of the provincial government, 1859. Appointed provincial geologist, 1861. Conducted the first geological survey of Canterbury province, 1861–8. Became a British national in 1861. Founded the Philosophical Institute of Canterbury in 1862, and the Canterbury Museum in 1863. Professor of geology, Canterbury College, 1876–87. Member of the senate of the University of New Zealand, 1879–87. Knighted, 1886. FRS 1867. (*DNB*, *DSB*, *DNZB*, H. F. von Haast 1948.)
22 January 1863, *5 March 1863*, *13 May 1863*, 18 July [1863], *21 July[–7? August] 1863*, *6 August 1863*

Hacon, William Mackmurdo. CD's solicitor. Offices at 31 Fenchurch Street, London. Admitted to practise, 1854; formed partnerships with David Rowland, James Weston, and Edward Francis Turner at Leadenhall House, Leadenhall Street. Ceased to practise in 1885. Commissioner of oaths and affidavits, and examiner of witnesses in England, and for the High Court of Judicature, Bombay. (*Law list* 1854–85.)

Haeckel, Ernst Philipp August (Ernst) (1834–1919). German zoologist. MD, Berlin, 1857. Lecturer in comparative anatomy, University of Jena, 1861–2; professor extraordinarius of zoology, 1862–5; professor of zoology and director of the Zoological Institute, 1865–1909. Specialist in marine invertebrates. Leading populariser of evolutionary theory. His *Generelle Morphologie der Organismen* (1866) linked morphology to the study of the phylogenetic evolution of organisms. (*DSB*, *NDB*.)
30 December [1863] – 3 January [1864]

Haldeman, Samuel Steman (1812–80). Zoologist and philologist. Assistant to Henry Darwin Rodgers on the geological survey of New Jersey, 1836–42. Professor of natural history, University of Pennsylvania, 1851–5; professor of natural science, Delaware College, 1853–8; professor of comparative philology, University of Pennsylvania, 1868–80. (*DAB*.)

Haliburton, Robert Grant (1831–1901). Canadian lawyer and ethnologist. Active in many Nova Scotian societies; founder (1862) and vice-president of the Nova Scotia Institute of the Natural Sciences. Secretary to the Nova Scotia commissioners for the International Exhibition in London, 1862. Advocated protectionism; a founding member of the Canada First Movement. Published anthropological and ethnological studies. (*DCB*.)

Hanbury, Daniel (1825–75). Pharmacist; partner in the London firm Allen & Hanbury of Lombard Street, London. Member of a number of scientific societies in London, including the Royal Society, the Linnean Society, the Chemical Society, and the Microscopical Society. Member of the board of examiners of the Pharmaceutical Society, 1860–72. FRS 1867. (*DNB*.)

Hance, Henry Fletcher (1827–86). Diplomat and botanist. Stationed in Hong Kong, 1844; vice-consul, Whampoa, 1861–78; consul, Canton, 1878–81 and 1883; acting consul, Amoy, 1886. Published a supplement to George Bentham's *Flora Hongkongensis* (1861) in 1873. (R. Desmond 1994.)
10 May 1863

Harcourt, William George Granville Venables Vernon (William Vernon) (1827–1904). Lawyer and statesman. Acquired a large practice at the parliamentary bar. Regular contributor to publications including the *Morning Chronicle* and the *Saturday Review*; from 1861, published many letters in *The Times*, under the signature Historicus, on matters of international law arising from the American Civil War. Whewell Professor of international law, University of Cambridge, 1869–87. Liberal MP for Oxford, 1868–80; for Derby, 1880–95; for Monmouthshire West, 1895–1904. Home secretary, 1880; chancellor of the exchequer, 1892. Knighted, 1873. (*DNB*, *Dod's parliamentary companion.*)

Hardwicke, Robert (1823–75). Printer and publisher at 192 Piccadilly, London. One of the founders of the Quekett Microscopical Club, 1865. Published *Hardwicke's Science Gossip*, 1865–75. (*Modern English biography.*)

Harley, George (1829–96). Scottish physician and physiologist; a specialist on diseases of the liver. Surgeon and physician at the Edinburgh Infirmary, 1850–1; studied physiology and chemistry in France, 1851–3, and Germany, 1853–5. Curator of the anatomical museum and lecturer in practical physiology and histology, University College, London, 1855; professor of medical jurisprudence, 1859; physician, University College Hospital, 1860. FRS 1865. (*DNB*, *Medical directory*, *Physicians.*)

Harris, William Cornwallis (1807–48). Military engineer and hunter. Commissioned in the Bombay engineering corps, where he achieved the rank of major. Superintending engineer to various Indian provinces. Hunted big game in southern Africa, 1835–7. Led a diplomatic mission to the kingdom of Shoa (now Ethiopia) in 1841. Knighted, 1844. (*DNB.*)

Hart, John Hinchley (1847–1911). Botanist. Landscape gardener, Nova Scotia, 1872. Superintendent of *Cinchona* plantations, Jamaica, 1881. Director of the Botanic Garden, Trinidad, 1887–1908. Collected plants in Jamaica, 1881–7; in Trinidad, 1887–97; in the Panama Canal region, 1885. (R. Desmond 1994.)

Harvey, Reuben (1789–1866). Elder brother of the Irish botanist William Henry Harvey. Died in Limerick, Ireland. (Biographical notes, Religious Society of Friends in Ireland.)

Harvey, William Henry (1811–66). Irish botanist. Colonial treasurer in Cape Town, 1836–42. Keeper of the herbarium, Trinity College, Dublin, from 1844; professor of botany, 1856–66. Specialist in marine algae. FRS 1858. (R. Desmond 1994, *DNB*, *DSB.*)
3 February 1863

Haughton, Samuel (1821–97). Irish clergyman, mathematician, geologist, and palaeontologist. Professor of geology, Dublin University, 1851–81. Became registrar of the medical school after graduating in medicine in 1862. Co-editor of the *Natural History Review*, 1854–60. President of the Royal Irish Academy, 1887. FRS 1858. (R. Desmond 1994, *DNB*, Sarjeant 1980–96.)

Hawkins, Benjamin Waterhouse (1807–89). Artist, sculptor, geologist, and popular natural history lecturer and author. Assistant superintendent of the Great Exhibition, 1851. Commissioned by the Crystal Palace Company to construct life-size replicas of extinct animals for the park in 1852. Illustrated *Fish* (part 4) and *Reptiles* (part 5) for the *Zoology of the voyage of HMS Beagle.* (Freeman 1978, *Men and women of the time* 1865.)

Hawthorn, Miss. Friend of Frances Harriet Hooker (letter from J. D. Hooker, 1 October 1863). Possibly Grace Hawthorn (b. 1824/5), a family friend (see *Correspondence* vol. 8, letter from J. S. Henslow to J. D. Hooker, 10 May 1860, and letter from Leonard Jenyns, 9 May 1868 (*Calendar* no. 6168)), and sister of Sarah Hawthorn (Census returns 1851 (Public Record Office HO107/1759/628a)), who married Leonard Jenyns in 1862 (*DNB* s.v. Blomefield, Leonard).

Hawthorn, Sarah. Eldest daughter of the clergyman Robert Hawthorn, of Stapleford, Cambridgeshire. Married Leonard Jenyns (later Blomefield) in 1862. (*DNB* s.v. Blomefield, Leonard.)

Hector, James (1834–1907). Scottish geologist. Surgeon and geologist on the government expedition to the western parts of British North America, 1857–60. Geologist to the provincial government of Otago, New Zealand, 1861–5. Director of the Geological Survey of New Zealand, 1865. Director of the meteorological and weather department of the New Zealand Institute, of the Colonial Museum, and of the botanical garden, Wellington, 1866–1903. Knighted, 1887. FRS 1866. (*DNZB.*)

Heer, Oswald (1809–83). Swiss biogeographer, palaeontologist, and botanist. An expert on Tertiary flora. Lecturer in botany, University of Zürich, 1834–5; director of the botanic garden, 1834; associate professor, 1835–52; professor of botany and entomology, 1852–83. (*DSB*, *NDB.*)

Henchman, Thomas (d. 1746). Clergyman. BA, Clare College, Cambridge, 1685–6. Vicar of Norton-by-Daventry, 1702. Prebendary of Salisbury Cathedral, 1717–46. (*Alum. Cantab.*)

A. Henderson & Co. London nursery firm at 1 Pineapple Place, Maida Vale, and 1 Garden Road, St John's Wood, London. Founded by Andrew Henderson (1760–1841) and continued by his son John Andrew Henderson (1794–1872). (R. Desmond 1994, *Post Office London directory* 1863.)

Henderson, John Andrew (1794–1872). Nurseryman. Proprietor of A. Henderson & Co., nurseries at 1 Pineapple Place, Maida Vale, and 1 Garden Road, St John's Wood, London. (R. Desmond 1994, *Post Office London directory* 1863.)

Henfrey, Arthur (1819–59). Botanist. Translated German works on physiological botany into English. Lecturer in botany, St George's Hospital, London, 1847. Professor of botany, King's College, London, 1854–9. FRS 1852. (*DNB*, *DSB.*)

Hensel, Reinhold Friedrich (1826–81). German naturalist, zoologist, and palaeontologist. Taught natural history in Berlin, 1850–60. Commissioned by the Berliner Akademie to conduct zoological studies in southern Brazil, 1863–6. Professor of zoology, Forstakademie in Proskau, from 1867; at the Agricultural Institute in Proskau from 1881. (*Leopoldina* 18 (1882): 19–21, Sarjeant 1980–96.)

Henslow, Anne Frances (d. 1863). Eldest daughter of John Prentis Henslow and sister of John Stevens Henslow. (*Gentleman's Magazine* n.s. 15 (1863): 520; Jenyns 1862, pp. 4–5.)

Henslow, John Stevens (1796–1861). Clergyman, botanist, and mineralogist. CD's teacher and friend. Professor of mineralogy, Cambridge University, 1822–7; professor of botany, 1825–61. Extended and remodelled the Cambridge botanic garden. Curate of Little St Mary's Church, Cambridge, 1824–32; vicar of Cholsey-cum-Moulsford, Berkshire, 1832–7; rector of Hitcham, Suffolk, 1837–61. (*DNB*, *DSB*, *Historical register of the University of Cambridge*.)

Henslow, Leonard Ramsay (1831–1915). Eldest son of John Stevens Henslow. Curate of Hitcham, Suffolk, 1854–6; of Bangor-Monachorum, Flintshire, 1856–60; of Great Chart, Kent, 1860–3. Rector of St Mary Magdalene, Pulham, Norfolk, 1863–70; of Zeals, Wiltshire, 1870–1915. (*Alum. Cantab.*)

Henslow, Susan. Daughter of Thomas Barker Wall, army officer. Married Leonard Ramsay Henslow in 1862. (*Alum. Cantab.* s.v. Henslow, Leonard Ramsay.)

Herbert, William (1778–1847). Naturalist, classical scholar, linguist, politician, and clergyman. Noted for his work on plant hybridisation. Rector of Spofforth, Yorkshire, 1814–40. Dean of Manchester, 1840–7. (*DNB*, *DSB*.)

Johann Christian Hermann'sche Verlagsbuchhandlung. German publishers and booksellers, based in Frankfurt, headed by Friedrich Emil Suchsland.

Herschel, John Frederick William, 1st baronet (1792–1871). Astronomer, mathematician, chemist, and philosopher. Member of many learned societies. Carried out astronomical observations at the Cape of Good Hope, 1834–8. Master of the Royal Mint, 1850–5. Created baronet, 1838. FRS 1813. (*DNB*, *DSB*.)

Hibberd, Shirley (1825–90). Horticultural journalist, writer, experimental gardener, and social reformer. Editor, *Floral world*, 1858–75; *Gardener's magazine*, 1861–90. Member of the Royal Horticultural Society. (R. Desmond 1994, *DNB*, *Gardener's magazine* 33 (1890): 743–4.)

Higwood, John (b. 1826/7). Under-gamekeeper, Down, Kent. (Census returns 1861 (Public Record Office RG9/462: 73).)

Hildebrand, Friedrich Hermann Gustav (Friedrich) (1835–1915). German botanist. Lecturer, Bonn University, 1859. Professor of botany, University of Freiburg, 1868–1907. Primarily concerned with ecological aspects of botany. (Correns 1916, Junker 1989, *Taxonomic literature*.)

16 July 1863, [25?] July 1863, *10 November 1863*

Hitchcock, Edward (1793–1864). American geologist and clergyman. Professor of chemistry and natural history, Amherst College, 1825–45; president, 1844–54; professor of geology and natural theology, 1855–64. Served on the Massachusetts, New York, and Vermont geological surveys. (*DAB*, *DSB*.)

Hochstetter, Ferdinand von (1829–84). German geologist. Geologist for the Reichsanstalt in Vienna, 1854–6. Lecturer in geology, University of Vienna, 1856–7. Naturalist on the *Novara* expedition, 1857–9. Professor of geology and mineralogy at the Polytechnic Institute in Vienna, 1860–81. Undertook further travels, acting as geological adviser on the construction of the Turkish (1869) and trans-Siberian (1872) railways. Director of the Hofmuseum, Vienna, from 1876. (*BHGW*, *DBE*, *NDB*.)

Hodgson, Brian Houghton (1800–94). Orientalist and ethnologist. In the service of the East India Company from 1816; assistant resident in Nepal, 1820; resident, 1833–43. Acquired an important collection of Sanskrit and Tibetan manuscripts. Wrote extensively on the geography, ethnography, and natural history of India and the Himalayas. Lived in Darjeeling, 1845–58; thereafter settled in England. FRS 1877. (*DNB.*)

Hofmann, August Wilhelm von (1818–92). German organic chemist. Student of Justus von Liebig at the University of Giessen; assistant to Liebig, 1843. Director, Royal College of Chemistry, London, 1845–65. Professor of chemistry, Berlin University, 1865–92. Founded the Deutsche Chemische Gesellschaft, Berlin, 1867. FRS 1851. (*ADB*, *DBE*, *DSB*, *NDB.*)

Hofmeister, Wilhelm Friedrich Benedikt (Wilhelm) (1824–77). German botanist. Joined the family publishing, music and bookshop business in 1841. Honorary doctor of philosophy, University of Rostock (for his work on the embryology of flowering plants), 1851. Professor of botany and director of the botanic garden, University of Heidelberg, 1863–72; professor of botany, University of Tübingen, 1872–6. (*ADB*, *DSB*, Goebel 1926, *NDB.*)

Hogg, Robert (1818–97). Scottish horticulturist and pomologist. Gained practical horticultural training with Peter Lawson in Edinburgh, and Hugh Ronalds in Brentford, after which he studied in Paris, Belgium, and Germany. Partner in a nursery in Brompton Park, London, 1842–51. Co-editor of the *Cottage Gardener* (from 1861, *Journal of Horticulture*), 1855–95. (R. Desmond 1994.)

Holland, Henry, 1st baronet (1788–1873). Physician. Cousin of the Darwins and Wedgwoods. Physician in ordinary to Prince Albert, 1840; to Queen Victoria, 1852. President of the Royal Institution of Great Britain, 1865–73. Created baronet, 1853. FRS 1815. (Caroe 1985, *DNB*, *Emma Darwin* (1904), *Physicians*, *Record of the Royal Society of London.*)
[10 February 1863]

Hooker, Brian Harvey Hodgson (1860–1932). Fifth child of Frances Harriet and Joseph Dalton Hooker. (Allan 1967 s.v. 'Hooker pedigree'.)

Hooker, Charles Paget (1855–1933). Physician and surgeon. Third child of Frances Harriet and Joseph Dalton Hooker. Trained at St Bartholomew's Hospital, London; made a licentiate of the Royal College of Physicians of London and the Royal College of Surgeons, Edinburgh, 1879, before being appointed to the staff of the Hertfordshire General Infirmary. Cottishall Cottage Hospital, Norfolk, 1880–5; Cirencester Cottage Hospital, Gloucestershire, 1885–1912. (Allan 1967; *Medical directory* 1881–1933; *Medical who's who* 1914.)

Hooker, Frances Harriet (1825–74). Daughter of John Stevens Henslow. Married Joseph Dalton Hooker in 1851. (Allan 1967; *DNB* s.v. Hooker, Joseph Dalton.)

Hooker, Harriet Anne (1854–1945). Second child of Frances Harriet and Joseph Dalton Hooker. (Allan 1967 s.v 'Hooker pedigree'.)

Hooker, Joseph Dalton (1817–1911). Botanist. Worked chiefly on taxonomy and plant geography. Son of William Jackson Hooker. Friend and confidant of CD.

Hooker, Joseph Dalton, cont.

Accompanied James Clark Ross on his Antarctic expedition, 1839–43, and published the botanical results of the voyage. Appointed palaeobotanist to the Geological Survey of Great Britain in 1846. Travelled in the Himalayas, 1847–50. Assistant director, Royal Botanic Gardens, Kew, 1855–65; director, 1865–85. Knighted, 1877. FRS 1847. (*DNB*, *DSB*.)

3 January [1863], *6 January 1863*, *[12 January 1863]*, 13 January [1863], *[15 January 1863]*, *[15 January 1863]*, 30 January [1863], 15 February [1863], *[16 February 1863]*, [21 February 1863], *[23 February 1863]*, 24[–5] February [1863], *[26 February 1863]*, *[1 March 1863]*, 5 March [1863], *[6 March 1863]*, 13 [March 1863], *[15 March 1863]*, 17 March [1863], *[24 March 1863]*, 26 [March 1863], *[28 March 1863]*, [29 March 1863], *[31 March 1863]*, [17 April 1863], *20 April 1863*, 23 April [1863], *[7 May 1863]*, *[8 May 1863]*, [9 May 1863], *[13 May 1863]*, 15 and 22 May [1863], 23 May [1863], *[23–7 May 1863]*, *[24 May 1863]*, 29 May [1863], 8 [June 1863], *10 June 1863*, *19 June 1863*, 23 [June 1863], 25 [June 1863], *[2]9 June 1863*, 1 July [1863], 14 July [1863], *[21 July 1863]*, 22 July [1863], 26 [July 1863], *[31 July 1863]*, 3 August [1863], 12–13 August [1863], 25 [August 1863], *26 August 1863*, *[26 August 1863]*, [28 August 1863], *15 September 1863*, *[28 September 1863]*, *1 October 1863*, [4 October 1863], *23 October 1863*, [30 October 1863], *[1–3 November 1863]*, 10 [November 1863], *11 November 1863* (Emma Darwin), [13 November 1863], 16 [November 1863], [22–23 November 1863], 27 [November 1863], 5 [December 1863]

Hooker, Maria (1797–1872). Eldest daughter of Dawson Turner. Married William Jackson Hooker in 1815; acted as her husband's secretary. Mother of Joseph Dalton Hooker. (Allan 1967 s.v. 'Turner pedigree'; R. Desmond 1994.)

Hooker, Maria Elizabeth (Minnie) (1857–63). Fourth child of Frances Harriet and Joseph Dalton Hooker. (Allan 1967 s.v. 'Hooker pedigree'.)

Hooker, William Henslow (1853–1942). Eldest child of Frances Harriet and Joseph Dalton Hooker. Civil servant, India Office, 1877–1904. Encouraged imperial ties between metropolitan institutions (particularly the Royal Botanic Gardens, Kew) and British East Africa, *circa* 1896–1906. (Allan 1967; *India list* 1904–5; *Zanzibar Gazette*, 5 February 1896, p. 6, and 28 November 1900, p. 5.)

Hooker, William Jackson (1785–1865). Botanist. Father of Joseph Dalton Hooker. Regius professor of botany, Glasgow University, 1820. Appointed first director of the Royal Botanic Gardens, Kew, in 1841. Knighted, 1836. FRS 1812. (*DNB*, *DSB*.)

Hopkins, William (1793–1866). Mathematician and geologist. Tutor in mathematics at Cambridge University. President of the Geological Society of London, 1851–3. Specialised in quantitative studies of geological and geophysical questions. FRS 1837. (*DNB*, *DSB*, *Record of the Royal Society of London*.)

Horner, Ann Susan (Susan) (1815/16–1900). Author and translator. Daughter of Leonard Horner. Author of works on Italian history and other subjects. (*CDEL*,

Freeman 1978, *Modern English biography*, *NUC*.)

Horner, Joanna B. (b. 1822?). Author and translator. Daughter of Leonard Horner. Author, with Susan Horner, of *Walks in Florence and its environs* (1873) and translator, with Leonora Pertz, of German works. (Freeman 1978, *NUC*.)
24 September 1863

Horner, Leonard (1785–1864). Scottish geologist and educationalist. Founded the Edinburgh School of Arts in 1821. Warden of University College, London, 1828–31. Inspector of factories, 1833–56. A promoter of science-based education at all social levels. President of the Geological Society of London, 1846 and 1860–2. Father-in-law of Charles Lyell. FRS 1813. (*DNB*, *DSB*.)

Horsfield, Thomas (1773–1859). American naturalist. Served in the East Indies under the Dutch and British, 1799–1819. Keeper of the East India Company Museum, Leadenhall Street, London, 1820–59. (*DNB*.)

Horwood, John (b. 1823/4). Gardener to George Henry Turnbull of Down, Kent. Supervised the construction of the hothouse at Down House, 1863. (Census returns 1861 (Public Record Office RG9/462: 70); letter to G. H. Turnbull, [16? February 1863]; *Orchids*, p. 158 n.)

Houtte, Louis Benoit van (1810–76). Belgian horticulturalist. (*Taxonomic literature*.)

Howitt, William (1792–1879). Writer, traveller, and pharmacist. Prolific author; published on a wide variety of subjects, including science, country life, foreign travel, history, and the supernatural. Lived in Germany, 1840–3; travelled in Australia, 1852–4; settled in Rome, 1870. (*DNB*, *Modern English biography*.)

Hudson, James (1810–85). Diplomat. Assistant private secretary to King William IV before joining the foreign service and becoming Secretary to the Hague legation, 1843. Envoy to Rio de Janerio, 1850; served on the legation to Turin, 1851–63. Supporter of Italian unification. Knighted, 1863. (*DNB*.)

Huish, Marcus (1815–68). Lawyer. Of Castle Donington, Leicestershire, from 1838. Married Frances Sarah Barton, daughter of Francis Sacheverel Darwin, in 1849. (*Darwin pedigree*, *Law list* 1838–68.)

Humboldt, Friedrich Wilhelm Heinrich Alexander (Alexander) von (1769–1859). Eminent Prussian naturalist, geographer, and traveller. Official in the Prussian mining service, 1792–6. Explored equatorial South America, Cuba, Mexico, and the United States, 1799–1804. Travelled in Siberia in 1829. Foreign member, Royal Society of London, 1815. (*DBE*, *DSB*, *NDB*.)

Hunt, Edward Bissell (1822–63). American engineer. Commissioned in the United States Army Corps of Engineers, 1845; captain, 1859; major, 1863. Assistant professor of engineering, Military Academy at West Point, 1846–9. Coast survey duty, Superintendent's Office, Washington DC, 1851. Corps of Engineers, Key West, Florida, 1857–62. Chief of engineering, Shenandoah Valley, 1862. (Cullum 1891.)

Hunt, James (1833–69). Speech therapist and anthropologist. Honorary secretary of the Ethnological Society of London, 1859–62. Founder and first president of the Anthropological Society of London, 1863. (*DSB*.)

Huxley, George Knight (1810–63). London barrister and businessman; eldest brother of Thomas Henry Huxley. (A. Desmond 1994–7, 1: 5, 321.)

Huxley, Henrietta Anne (1825–1915). Married Thomas Henry Huxley in 1855. (Freeman 1978.)

Huxley, Leonard (1860–1933). Biographer. Son of Thomas Henry Huxley. Editor of the *Cornhill Magazine*, 1916–33. (*DNB*, Clark 1968, pp. 133, 181.)

Huxley, Nettie (1863–1940). Singer and illustrator. Daughter of Henrietta and Thomas Henry Huxley. Married Harold Roller, joint owner of a firm of picture restorers, in 1889. (Bibby 1959, pp. 15, 275, 283; Clark 1968, p. 111 and *passim.*)

Huxley, Thomas Henry (1825–95). Zoologist. Assistant-surgeon on HMS *Rattlesnake*, 1846–50, during which time he investigated Hydrozoa and other marine invertebrates. Lecturer in natural history, Royal School of Mines, 1854; professor, 1857. Appointed naturalist to the Geological Survey of Great Britain in 1855. Hunterian Professor, Royal College of Surgeons of England, 1863–9. Fullerian Professor, Royal Institution of Great Britain, 1863–7. President of the Royal Society of London, 1883–5. FRS 1851. (Clark 1968, *DNB*, *DSB*, *Record of the Royal Society of London.*)

10 [January 1863], [8 February 1863], 16 February [1863], [after 16 February 1863], 18 [February 1863], *20 February 1863*, [before 25 February 1863], *25 February 1863*, 26 [February 1863], 27 June [1863], *2 July 1863*, 3 July [1863]

Innes, Eliza Mary Brodie. Daughter of John Laidlaw. Married John Innes (later John Brodie Innes) in 1847. (*Burke's landed gentry* 1879 s.v. Brodie-Innes, John.)

Innes, John Brodie (1817–94). Perpetual curate of Down, 1846–68; vicar, 1868–9. Left Down at the start of 1862 after inheriting an entailed estate at Milton Brodie, near Forres, Scotland; changed his name to Brodie Innes in 1861 as required by the entail. Priest in charge of Milton Brodie Mission and general licentiate of the diocese of Moray, 1861. Chaplain to the Bishop of Moray, 1861–80 and 1886–94. (*Clergy list*, *County families* 1864, *Crockford's clerical directory*, Freeman 1978, Moore 1985.)

Innes, John George (1815–81). Scottish surgeon and botanist. Practised medicine in, and parochial medical officer for, Forres, Morayshire. Fellow of the Botanical Society of Edinburgh. (R. Desmond 1994, *Medical directory* 1861–82.)

Innes, John William Brodie (1848–1923). Barrister and novelist. Son of John Brodie Innes. Called to the bar, 1876. Advocate at the Scottish bar, 1888. Interested in antiquarian research, romance, demonology, witchcraft, and criminology. (*Alum. Cantab.*, Freeman 1978.)

Jackson, Patrick Tracy III (b. *c.* 1843–6). American cotton broker. Nephew of Jane Loring Gray. First lieutenant in the Fifth Massachusetts Cavalry during the American Civil War. President and director of the family firm, Lowell Weaving Co., Lowell, Massachusetts, 1906. (Dupree 1959, *Lamb's textile industries of the United States* s.v. Jackson, Patrick Tracy, II and IV.)

Jacobson, Eleanor Jane (1811–95). Youngest daughter of Dawson Turner. Sister of Maria Hooker, Joseph Dalton Hooker's mother. Married William Jacobson, regius professor of divinity at Oxford University (1848–65) and bishop of Chester (1865–84), in 1836. (Allan 1967 s.v. 'Hooker pedigree'; *DNB* s.v. Jacobson, William.)

Jäger, Gustav (1832–1917). German physician and zoologist. Co-founder and director of the Vienna zoological garden, 1858–66. Professor of zoology and anthropology, Hohenheim Academy, from 1867; Stuttgart Polytechnic, from 1870. Taught physiology and histology at the Veterinary School, Stuttgart, from 1874. Co-editor of *Kosmos*, 1877–9. Returned to medical practice in 1884. (*DBE*, Freeman 1978, *NDB*, Weinreich 1993.)

Jameson, William (1796–1873). Scottish botanist. Licentiate of the Royal College of Surgeons, Edinburgh. Emigrated to South America in 1820. Practised medicine in Guayaquil, Ecuador, 1822–6. Professor of chemistry and botany, University of Quito, 1827–69. Assayer of the Mint in Quito, Ecuador, 1832; director, 1861. Prepared a synopsis of the flora of Ecuador. (*DNB*, *Transactions and Proceedings of the Botanical Society* [*of Edinburgh*] 12: 19–28.)

Jamieson, Thomas Francis (1829–1913). Scottish agriculturalist and geologist of Ellon, Aberdeenshire. Factor on the Ellon estate for many years; later took the farm of Mains, Waterton. Appointed Fordyce Lecturer on agricultural research, University of Aberdeen, in 1862. Carried out notable researches on Scottish Quaternary geology and geomorphology. Became a fellow of the Geological Society of London in 1862. (*Geological Magazine* 50 (1913): 332–3, Sarjeant 1980–96.)
28 January 1863

Jansen, Frederick Halsey (b. 1813). Solicitor. Partner in the London legal firm Janson, Cobb, Pearson & Co. Member of a number of scientific societies, including the Linnean Society of London, the Royal Statistical Society, and the London Institution. Deputy lieutenant of Tower Hamlets, 1859. (*Law list* 1836–1914, *Leading men of London*.)

Jardine, William, 7th baronet (1800–74). Naturalist. A founder of the *Annals and Magazine of Natural History*, 1841. Commissioner for the salmon fisheries of England and Wales, 1860. FRS 1860. (*DNB*, *Record of the Royal Society of London*.)

Jeffreys, John Gwyn (1809–85). Conchologist, zoologist, and lawyer. Solicitor in Swansea until 1856, when he was called to the bar and moved to London. Justice of the peace for Glamorgan, for Brecon, and for Hertfordshire. Deputy lieutenant, Hertfordshire; high sheriff, 1877. Treasurer for the Linnean Society of London and the Geological Society of London for many years. Conducted extensive deep-sea dredging researches. FRS 1840. (*DNB*, *DSB*.)

Jenner, William, 1st baronet (1815–98). Physician. Professor of pathological anatomy at University College, London, 1849; Holme Professor of clinical medicine, 1860. Assistant, then full, physician to University College Hospital, 1849–76; consulting physician, 1879. Physician extraordinary to Queen Victoria,

Jenner, William, 1st baronet, cont.
1861; physician-in-ordinary, 1862. Established, through clinical studies, definitive evidence that typhus and typhoid were different diseases. FRS 1864. (*DNB.*)

Jenyns, Leonard (1800–93). Naturalist and clergyman. Brother-in-law of John Stevens Henslow. Vicar of Swaffham Bulbeck, Cambridgeshire, 1828–49. Settled near Bath in 1850. Founder and first president of the Bath Natural History and Antiquarian Field Club, 1855. Member of many scientific societies. Described the *Beagle* fish specimens. Adopted the name Blomefield in 1871. (R. Desmond 1994, *DNB.*)

Johnes, Charlotte Anna Maria. Daughter of John and Elizabeth Johnes of Dolaucothy, Llandilo, Carmarthenshire, Wales. Married Charles Cookham of County Wexford, Ireland. (*County families* 1860–76 s.v. Johnes, John.)

Johnes, Elizabeth. Daughter of John Edwardes, clergyman, of Gileston Manor, Glamorgan, Wales; married John Johnes in 1822. Friend of Andrew Crombie Ramsay. (*County families* 1860–76 s.v. Johnes, John; Geikie 1895, p. 76.)

Johnes, Elizabeth. Daughter of John and Elizabeth Johnes of Dolaucothy, Llandilo, Carmarthenshire, Wales. (*County families* 1860–76 s.v. Johnes, John.)

Johnes, John (1800–76). Judge on the Welsh circuit. Called to the bar, 1831; county court judge, 1847; recorder, Carmarthen, 1851; chairman of the Quarter Sessions, Carmarthenshire, 1853. County magistrate, Carmarthen, Cardigan, Glamorgan, and Pembroke. Deputy lieutenant, Carmarthenshire. Friend of Andrew Crombie Ramsay. (*Alum. Oxon.*; *County families* 1860–76; Geikie 1895, p. 76.)

Johnson, Samuel (1709–84). Lexicographer and man of letters. (*DNB.*)

Jussieu, Adrien Henri Laurent de (1797–1853). French botanist. In 1826, succeeded his father, Antoine Laurent de Jussieu, as professor of botany at the Muséum d'Histoire Naturelle, Paris, where he helped to build up a large herbarium. (*DSB*, *NBU.*)

Juvenal (*c.* 60–140). Roman poet and satirist. (*EB.*)

King, Philip Gidley (1817–1904). Australian farmer and mining company manager. Son of Phillip Parker King. Midshipman in HMS *Beagle*, 1831–6. Settled in Australia in 1836. Entered the service of the Australian Agricultural Company in 1842; superintendent of stock, 1851. New South Wales manager for the Peel River Land and Mineral Company, 1852–81. (*Aust. dict. biog.*)
19 April 1863

King, Phillip Parker (1793–1856). Naval officer and hydrographer. Commander of the *Adventure*, accompanying the *Beagle*, on the first surveying expedition to South America, 1826–30. Settled in Australia and became commissioner of the Australian Agricultural Company in 1834. Rear-admiral, 1855. FRS 1824. (*Aust. dict. biog.*, *DNB.*)

Kinglake, Alexander William (1809–91). Barrister, historian and politican. After studying at Cambridge University and Lincoln's Inn, travelled in Algeria. Followed the British expedition to the Crimea and wrote a history of the campaign. Liberal MP for Bridgewater, 1857–68. (*DNB.*)

Kingsley, Charles (1819–75). Author and clergyman. Lecturer on English literature, Queen's College, London, 1848–9. Professor of modern history, Cambridge University, 1860–9. Rector of Eversley, Hampshire, 1844–75. Chaplain to the queen, 1859–75. (*Alum. Cantab.*, *DNB.*)

Kippist, Richard (1812–82). Botanist. Librarian of the Linnean Society of London, 1842–80. Specialist in Australian plants. (R. Desmond 1994, *DNB.*)
13 May [1863 or 1868]

Kirby, William Forsell (1844–1912). Entomologist. Assistant at the Royal Dublin Society, 1867–79. Assistant in the zoological department of the British Museum, 1879. (R. Desmond 1994, Gilbert 1977.)
9 July [1863], *8 September [1863]*

Kitaibel, Pál (Paul) (1757–1817). Hungarian botanist, zoologist, and mineralogist. Qualified in medicine but never practised. Assistant to Jacob Joseph Winterl at the Institute for Chemistry and Botany, University of Pest. Became professor of botany after Winterl's death in 1809. (*DSB*, *Taxonomic literature.*)

Knight, Thomas Andrew (1759–1838). Botanist and horticulturist. Correspondent to the Board of Agriculture from 1795. President of the Horticultural Society of London, 1811–38. Interested in cross-breeding and hybridisation; conducted research on the phenomenon now known as geotropism. FRS 1805. (*DSB*, R. Desmond 1994.)

Koch, Wilhelm Daniel Joseph (1771–1849). German physician and botanist. Professor of botany, University of Erlangen, from 1824. Published a flora of Germany with Franz Carl Mertens. (*ADB*, *DBE*, Koch and Mertens 1823–30.)

Kölreuter, Joseph Gottlieb (1733–1806). German botanist. Assistant keeper of the natural history collections, Imperial Academy of Sciences, St Petersburg, 1756–61. Professor of natural history and director of the gardens of the margrave of Baden, Karlsruhe, 1763–86. Carried out extensive hybridisation experiments on plants. (*ADB*, *DBE*, *DSB*, *NDB*, *Taxonomic literature.*)

Krohn, August David (1803–91). Russian-born zoologist, anatomist and embryologist, working in Bonn; travelled extensively. (Blyakher 1982.)

Kunth, Karl Sigismund (1788–1850). German botanist. Worked in Paris on the plants collected by Alexander von Humboldt on his travels in South America and Mexico 1813–29. Appointed associate professor of botany, Berlin University, and vice-director of the botanical garden, Berlin, 1829. (*ADB*, *DBE.*)

Laird Brothers. Shipbuilders at Birkenhead, Cheshire. (*Post Office directory of Cheshire* 1857.)

Laird, John (1805–74). Shipbuilder. Partner in the family shipbuilding firm of Laird Brothers in Birkenhead. Pioneered the construction of iron vessels. Liberal-conservative MP for Birkenhead, 1861–74. Deputy lieutenant and justice of the peace for the county of Chester. Member of the Mersey docks and harbour board. Chairman of the Birkenhead improvement commission. (*DNB.*)

Lamarck, Jean Baptiste Pierre Antoine de Monet (Jean Baptiste) de (1744–1829). French naturalist. Held various botanical positions at the Jardin

Lamarck, Jean Baptiste de, cont.
du Roi, 1788–93. Appointed professor of zoology, Muséum d'Histoire Naturelle, in 1793. Believed in spontaneous generation and the progressive development of animal types; propounded a theory of transmutation. (*DSB.*)

Lambert, Aylmer Bourke (1761–1842). Botanist. Vice-president of the Linnean Society, 1796–1842. FRS 1791. (R. Desmond 1994, *DNB.*)

Lane, Edward Wickstead (1823–89). Physician. Proprietor of a hydropathic establishment at Moor Park, near Farnham, Surrey, from 1859 or before, and at Sudbrook Park, near Ham, Surrey, 1860–79. Practised in Harley Street, London, 1879–89. Member of the Faculty of Advocates, the Botanical Society, and the Speculative Society, Edinburgh. Author of works on hydropathy. (Freeman 1978, *Medical directory* 1859–89, *Post Office directory of the six home counties* 1859–62.)

Langton, Charles (1801–86). Rector of Onibury, Shropshire, 1832–41. Left the Church of England in 1841. Resided at Maer, Staffordshire, 1841–7, and at Hartfield Grove, Hartfield, Sussex, 1847–63. Married Emma Darwin's sister, Charlotte Wedgwood, in 1832. After her death, married CD's sister, Emily Catherine Darwin, in 1863. (*Alum. Oxon.*, *Emma Darwin* (1915), Freeman 1978.)

Langton, Charlotte (1797–1862). Emma Darwin's sister. Married Charles Langton in 1832. Resided at Maer, Staffordshire, 1840–6, and at Hartfield Grove, Hartfield, Sussex, 1847–62. (*Emma Darwin* (1915), Freeman 1978.)

Langton, Edmund (1841–75). Son of Charles and Charlotte Langton. BA, Trinity College, Cambridge, 1864. Admitted at Lincoln's Inn, 1864. (*Alum. Cantab.*, *Emma Darwin* (1915).)

Langton, Emily Catherine (Catherine) (1810–66). CD's sister. Married Charles Langton in 1863. (*Darwin pedigree.*)

Lankester, Edwin (1814–74). Surgeon, physician, medical reformer, anatomist, botanist, and popular author and lecturer. Founder and editor of the *Journal of Social Science*, 1865–74. Secretary for the biological section of the British Association for the Advancement of Science, 1839–64; of the Ray Society from 1844. President of the Microscopical Society of London, 1859. FRS 1845. (*DNB*, *Medical directory* 1845–75.)

Lartet, Edouard Amant Isidore Hippolyte (Edouard) (1801–71). French palaeontologist and prehistorian. Trained as a lawyer. Began palaeontological research in the 1830s, and became interested in the question of the antiquity of the human species; carried out excavations in 1860 that provided proof of the contemporaneity of man with extinct animal species. Named professor of palaeontology, Muséum d'Histoire Naturelle, 1869, but ill health prevented him from fulfilling his duties. (*DSB*, Sarjeant 1980–96.)

La Touche, James Digues (1824–99). Irish cleric, tutor, geologist, and palaeontologist. Also interested in entomology, botany, meteorology, and astronomy. Vicar of Stokesay, Shropshire, 1855–99. Tutored the sons of Joseph Dalton Hooker, John William Colenso, and Friedrich Max Müller, and the nephews of Charles

Lyell. Published *A hand-book of the geology of Shropshire* (1884). (*Geological Magazine* n.s. 6 (1899): 235–7.)
14 May [1863?]

Laugel, Antoine Auguste (Auguste) (1830–1914). French writer on a wide range of subjects, including science, philosophy, politics, history, and psychology. (*Dictionnaire universel des contemporains* 1893, *NUC*.)

Peter Lawson & Sons. Edinburgh nursery founded in 1770. From 1821 conducted by Charles Lawson, with premises at 1 George IV Bridge, and nurseries at Bangholm, Inverlieth Row, Wardie, and Trinity. (R. Desmond 1994, *Post Office Edinburgh and Leith directory* 1834–64.)

Layard, Edgar Leopold (1824–1900). Served in the Ceylon civil service, 1846–54; in the Cape of Good Hope civil service, 1854–70. Studied the birds and shells of Ceylon. Founded the South African museum; curator, 1855–72. Published on the birds of South Africa (1867). (*DSAB*, *Modern English biography*.)

Le Conte, John Eatton (1784–1860). American entomologist, botanist, and soldier. Major of engineers in the United States Army. Father of John Lawrence Le Conte. (*DAB* s.v. John Lawrence Le Conte; Gilbert 1977.)

Le Conte, John Lawrence (1825–83). American entomologist. Son of John Eatton Le Conte. Trained as a physician but never practised. Published in many fields of natural history, with a particular interest in geographical distribution. Served with the army medical corps during the American Civil War. President, American Association for the Advancement of Science, 1874. Chief clerk to the United States Mint at Philadelphia from 1878. (*DAB*.)

Lecoq, Henri (1802–71). French naturalist and vulcanologist. Professor of natural history, University of Clermont-Ferrand, and director of the town's botanic garden, 1826–54. Taught at the Preparatory School of Medicine and Pharmacy, Clermont-Ferrand, from 1840; professor, science faculty, from 1854. Published widely on botany, agriculture, and meteorology. (*Grande encyclopédie*, Sarjeant 1980–96.)

Ledger, William (*fl.* 1850s–60s). Builder, from Hayes, near Bromley, Kent. (*Post Office directory of the six home counties* 1855–66.)

Lee, James (1754–1824). Nurseryman. Succeeded his father in partnership in Lee & Kennedy's Vineyard Nursery, Hammersmith, London. Sole proprietor from 1818. (R. Desmond 1994.)

Lee, James Prince (1804–69). Bishop of Manchester, 1847–69. Schoolmaster, Rugby, 1830–8; headmaster, King Edward's School, Birmingham, 1838. (*DNB*.)

Leidy, Joseph (1823–91). American anatomist, naturalist, and palaeontologist. Demonstrator of anatomy, Franklin Medical College, 1846. Visited Europe in 1848 and 1850. Professor of anatomy, University of Pennsylvania, 1853–91. President of the Academy of Natural Sciences of Philadelphia, 1881–91. Published extensively on the fossil remains of extinct vertebrates of North America. (*DAB*, *DSB*.)

Leifchild, John R. (b. 1815). Author and reviewer. Son of John Leifchild, a leading evangelical minister (*DNB*). Visiting government commissioner to coal mines in Newcastle and Durham. Author of works on mining and coal. Reviewed *Origin* in the *Athenaeum* in 1859 and *Orchids* in 1862. (*BLC*, *NUC*, *Wellesley index*.)

Leighton, William Allport (1805–89). Botanist, clergyman, and antiquary. Schoolfellow of CD's in Shrewsbury, 1817. Curate of St Giles's, Shrewsbury, 1846–8. Edited the *Transactions of the Shropshire Archaeological Society* for many years. Published *The flora of Shropshire* (1841). (*Clergy list* 1846–8, R. Desmond 1994, *DNB*, Freeman 1978.)

[before 21 June 1863]

Lennard, John Farnaby, 1st baronet (1816–99). Artillery officer and county councillor; of Wickham Court, West Wickham, Kent. Lieutenant-colonel, Kent artillery militia, 1853–76. Justice of the peace; chairman of the Bromley District Highway Board, 1863; chairman of the Kent general sessions, 1874–89; chairman of Kent County Council, 1889–99. Adopted the surname Lennard in 1861. Created baronet, 1880. (Minutes of the Bromley Highway Board (Bromley Central Library 847/HB/B/1), *Modern English biography*, *Post Office directory of the six home counties* 1866–78.)

Lesley, J. Peter (1819–1903). American geologist. Pastor, Congregational church, Milton, Massachusetts, 1847–52. Employed by the Pennsylvania geological survey, 1852. Professor of mining, University of Pennsylvania, 1859; dean of the science department, 1872. Visited Europe in 1863 to study rail manufacture and steel processing on behalf of Pennsylvania Railroad. Secretary, American Iron Association, 1856–64. State geologist of Pennsylvania, 1873–87. Librarian, American Philosophical Society, 1859–85; secretary, 1859–87; vice-president, 1887. (*DAB*.)

Lesquereux, Leo (1806–89). Swiss-born bryologist and palaeontologist. Moved to the United States after the political upheavals of 1848; after brief periods assisting Louis Agassiz and Asa Gray, settled in Columbus, Ohio, where he assisted the bryologist William Starling Sullivant. Engaged by Agassiz to organise the palaeobotanical collections at the Museum of Comparative Zoology, Harvard University, 1867–72. (*DAB*, *DSB*, Rodgers 1940, Sarjeant 1980–96, Sarton 1942.)

Lettington, Henry (b. 1822/3). Gardener of Farthing Street, Down; a gardener at Down House. (CD's Classed account books (Down House MS); Census returns 1861 (Public Record Office RG9/462: 70).)

Liebig, Justus von (1803–73). German organic chemist. Professor of chemistry, University of Giessen, 1825–51; University of Munich, 1852–73. Played a major role in the development of organic chemistry. Developed artificial agricultural fertilisers. President of the Bavarian Academy of Science, 1859–73. Foreign member, Royal Society, 1840. (*ADB*, *DBE*, *DSB*, *NDB*.)

Lightbody, George (d. 1872). Gardener in Falkirk, Scotland. Raised numerous auricula varieties. (R. Desmond 1994.)

Lincoln, Abraham (1809–65). American lawyer and statesman. Republican president of the United States of America during the American Civil War, from 1861 until his assassination in 1865. (*DAB*.)

Lindley, John (1799–1865). Botanist and horticulturist. Assistant in Joseph Banks's library and herbarium, 1818 or 1819. Garden assistant secretary, Horticultural Society of London, 1822–7; general assistant secretary, 1827–41; vice-secretary, 1841–58; honorary secretary, 1858–62. Lecturer on botany, Apothecaries' Company, 1836–53. Professor of botany, London University (later University College, London), 1829–60. Horticultural editor of the *Gardeners' Chronicle* from 1841. FRS 1828. (R. Desmond 1994, *DNB*, *DSB*.)

Lindsay, William Schaw (1816–77). Scottish merchant, shipowner, and politician. Entered the merchant navy in 1831; captain, 1836–40. Fitter to the Castle Eden Coal Company, Hartlepool, before establishing the shipbroking firm W. S. Lindsay & Co, which became one of the largest shipowning concerns in the world. Liberal MP, Tynemouth and North Shields, 1854–9; Sunderland, 1859–65. Leading figure in the Confederate lobby in Britain. (*DNB*; *Dod's parliamentary companion*; Jenkins 1974–80, 2: 308, 336.)

Link, Heinrich Friedrich (1767–1851). German naturalist and philosopher. Professor of zoology, botany, and chemistry, University of Rostock, 1792–1811; professor of chemistry and botany, University of Breslau, 1811–15. Professor of botany and director of the botanic garden at the University of Berlin, 1815. Interested in the philosophical foundations of the natural sciences. Foreign member, Royal Society of London, 1842. (*DSB*, *NDB*, *Record of the Royal Society of London*.)

Linné, Carl von (Carolus Linnaeus) (1707–78). Swedish botanist and zoologist. Professor of practical medicine, University of Uppsala, 1741; professor of botany, diatetics, and materia medica, 1742; court physician, 1747. Proposed a system for the classification of the natural world, and reformed scientific nomenclature. FRS 1753. (*DSB*, *Record of the Royal Society of London*.)

Linné, Carl von (1741–83). Swedish botanist. Son of Carolus Linnaeus (1707–78). Studied at the University of Uppsala; doctor of medicine, 1765; professor of medicine and botany, 1777–83. (*Taxonomic literature*.)

Linnell, Thomas George (Thomas) (1835–1911). Landscape painter. Youngest son of the well-known landscape artist, John Linnell (1792–1882). (*Dictionary of British artists*, Linnell 1994.)

Litchfield, Henrietta Emma. *See* Darwin, Henrietta Emma.

Loder, Robert, 1st baronet (1823–88). Scientific farmer with estates in England, Russia, and Sweden. Sheriff of Northampton, 1877. Conservative MP for Shoreham, Sussex, 1880–5. Created baronet, 1887. (*Alum. Cantab.*, *Dod's parliamentary companion*, *Modern English biography*.)

Logan, William Edmond (1798–1875). Canadian stratigrapher. Director of the geological survey of Canada, 1842–70. Knighted, 1856. FRS 1851. (*DNB*, *DSB*.)

Lombe, Elizabeth (1820–98). Daughter of William Jackson Hooker and Maria Turner. Married Thomas Robert Evans Lombe in 1853. (Allan 1967 s.v. 'Hooker pedigree'; L. Huxley ed. 1918.)

Longman, Green, Longman, Roberts & Green (Longmans). Publishers, founded by Thomas Longman (1699–1755), with premises at 14 Ludgate Hill and 38–41 Paternoster Row, London. The name of the firm changed frequently, with alterations in the partnership; it was colloquially known as Longmans. From the 1840s Thomas and William Longman were the senior partners. (P. A. H. Brown 1982, *DNB*, *Post Office London directory*.)

Lord, John Keast (1818–72). Naturalist and veterinary. Served as veterinary surgeon in the Crimea, 1855–6. Naturalist to the Boundary Commission sent to British Columbia, 1858. Joined the staff of the journal *Land and Water* in 1866 before being engaged by the viceroy in archaeological and scientific research in Egypt. Appointed manager of the newly established Brighton Aquarium four months before his death. (*DNB*.)

Loring, Charles Greely (1794–1867). American lawyer and author. Father-in-law of Asa Gray. Elected to the Massachusetts State Senate, 1862. Member of the American Academy of Arts and Sciences and the Massachusetts Historical Society, and president of the Boston Union Club. Following the outbreak of the Civil War in the United States, his correspondence with Edwin Wilkins Field concerning American relations with Britain was published in Loring 1862. (Adams ed. 1897, *Appleton's cyclopaedia of American biography*, Chamberlain ed. 1899, Dupree 1959.)

Loudon, John Claudius (1783–1843). Landscape gardener and horticultural writer. Travelled in northern Europe, 1813–15; in France and Italy, 1819–20. Urban and rural landscape designer. A prolific author; founded and edited the *Gardener's Magazine*, 1826–43, and the *Magazine of Natural History*, 1828–36. (R. Desmond 1994, *DNB*.)

Louis XIV (1638–1715). King of France; succeeded to the throne in 1643. (*EB*.)

Louis Napoleon (1808–73). President of the French Republic, 1848–52. Proclaimed Emperor of the French, as Napoleon III, in 1852. Defeated in the war with Prussia in 1870 and deposed, after which he lived in Chislehurst, England. (*EB* s.v. Napoleon III.)

Low, Hugh (1793–1863). Nurseryman. Proprietor of nurseries at Clapton, London, 1831–63. (R. Desmond 1994, *Post Office London directory* 1863.)

Hugh Low & Co. Nursery and florist, with premises at Clapton, London. (*Post Office London directory* 1863.)

Low, Stuart Henry (1826–90). Nurseryman. Son of Hugh Low. Proprietor of nurseries at Clapton, London, 1863. Established Bush Hill Park nursery, Enfield, 1882. (R. Desmond 1994.)

Lubbock, Ellen Frances (1834/5–1879). Daughter of Peter Hordern, clergyman, of Chorlton-cum-Hardy, Lancashire. Married John Lubbock in 1856. (*Burke's peerage* 1970; Census returns 1861 (Public Record Office RG9/462: 75).)

Lubbock, John, 4th baronet and 1st Baron Avebury (1834–1913). Banker, politician, and naturalist. Son of John William Lubbock and a neighbour of CD's in Down. Studied entomology and anthropology. An active supporter of

CD's theory of natural selection. A partner in the family bank from 1849. Liberal MP for Maidstone, Kent, 1870 and 1874; for London University, 1880–1900. Succeeded to the baronetcy in 1865. Created Baron Avebury, 1900. FRS 1858. (*DNB*, *DSB*, *Record of the Royal Society of London.*)

4 January [1863], *6 January [1863]*, *20 February 1863*, 23 [February 1863], *28 February 1863*, *6 March 1863*, 5 April [1863], *7 April 1863*, *4 September 1863*, *14 October 1863*, *24 October 1863* (Emma Darwin), *6 November 1863* (Emma Darwin)

Lubbock, John William, 3d baronet (1803–65). Astronomer, mathematician, and banker. A neighbour of CD's in Down. First vice-chancellor of London University, 1837–42. Partner in the family bank, 1825. Treasurer and vice-president of the Royal Society of London, 1830–5 and 1838–45. Succeeded to the baronetcy in 1840. FRS 1829. (*DNB*, *DSB*, *Record of the Royal Society of London.*)

Ludlow, John Malcolm Forbes (1821–1911). Indian-born lawyer and social reformer. Called to the bar, 1843. Founding member of the Christian socialist movement, 1848, and editor of the weekly *Christian Socialist*, 1850. Helped to found the Working Men's College, Great Ormond Street, in 1854 and lectured there. Secretary of the royal commission on friendly and benefit societies, 1870–4. Chief registrar of friendly societies, 1875–91. (*DNB.*)

Lyell, Charles, 1st baronet (1797–1875). Scottish geologist. Uniformitarian geologist whose *Principles of geology* (1830–3), *Elements of geology* (1838), and *Antiquity of man* (1863) appeared in many editions. Professor of geology, King's College, London, 1831. President of the Geological Society of London, 1835–6 and 1849–50; of the British Association for the Advancement of Science, 1864. Travelled widely and published accounts of his trips to the United States. Knighted, 1848; created baronet, 1864. CD's scientific mentor and friend. FRS 1826. (*DNB*, *DSB.*)

4 [February 1863], 17 [February 1863], 6 March [1863], *11 March 1863*, 12–13 March [1863], *15 March 1863*, 17 March [1863], 18 April [1863], [7 May 1863], 14 August [1863]

Lyell, Henry (1804–75). Army officer in India. Married Katharine Murray Horner in 1848. Brother of Charles Lyell. (*Burke's peerage* 1980.)

Lyell, Katharine Murray (1817–1915). Daughter of Leonard Horner. Married Henry Lyell, brother of Charles Lyell, in 1848. Collected plants in India. Edited *Life, letters and journals of Sir Charles Lyell* (1881), and memoirs of Charles James Fox Bunbury and Leonard Horner. (R. Desmond 1994, Freeman 1978.)

Lyell, Leonard, 1st baron (1850–1926). Eldest son of Henry Lyell, Charles Lyell's brother, and Katharine Murray Lyell. MP for Orkney and Shetland, 1885–1900. Created baronet, 1894; peer, 1914. (*Burke's peerage* 1980.)

Lyell, Mary Elizabeth (1808–73). Eldest child of Leonard Horner. Married Charles Lyell in 1832. (Freeman 1978.)

M'Andrew, Robert (1802–73). Liverpool merchant, yachtsman, and naturalist. President, Literary and Philosophical Society of Liverpool, 1856–7. Member of

M'Andrew, Robert, cont.
the dredging committee of the zoological section of the British Association for the Advancement of Science, 1844. FRS 1853. (*Proceedings of the Literary and Philosophical Society of Liverpool*, 1856–7, pp. 14–17; *Record of the Royal Society of London*; Rehbock 1979.)

McCall, Grace Anne (1832–99). Scottish-born writer and illustrator. Eldest daughter of James Milne and Louisa Falconer of Findhorn, Morayshire. Niece of Hugh Falconer, with whom she travelled widely. Married George McCall (died 1856) in 1854, and Joseph Prestwich, later professor of geology at Oxford University, in 1870 (*DNB* s.v. Joseph Prestwich; Milne 1901.)

M'Caul, Joseph Benjamin (d. 1892). Clergyman. Assistant, British Museum, 1846–9 and 1851–65. Divinity lecturer, King's College, London, 1852–4. Curate, St Magnus the Martyr, London, 1851–4; All Saints, Gordon Square, 1854–5; St Edmund's the King, Lombard Street, 1858–65. Rector, St Michael Bassishaw, London, from 1865; chaplain at Amsterdam 1877–9; honorary canon of Rochester, Kent, from 1865. (*Crockford's clerical directory* 1860–91, *Modern English biography.*)

McCosh, James (1811–94). Scottish theologian, philosopher, and botanist. Professor of logic and metaphysics, Queen's College, Belfast, 1851–68. Professor of philosophy, College of New Jersey (later Princeton University), 1868–94; president, 1868–88. (*DNB.*)

Macfadyen, James (1800–50). Scottish botanist and physician in Jamaica. MD, Glasgow University, 1821–2. Island botanist, Jamaica, 1826–8. Established the Jamaica Botanic Garden. (R. Desmond 1994.)

McGibbon, James (*fl.* 1840s–*c.* 1886). Scottish-born gardener. Gardener at the Royal Botanic Gardens, Kew, 1848. Superintendent of the Botanic Garden, Cape Town, from *circa* 1850. (*Cape of Good Hope general directory* 1881, R. Desmond 1994.)

McGilvray, Maria (1819–89). Daughter of William Jackson Hooker and Maria Turner. Married Walter McGilvray in 1846. (Allan 1967 s.v. 'Hooker pedigree'.)

M'Intosh, Charles (1794–1864). Scottish gardener and horticulturalist. Gardener to John Campbell, second marquis of Breadalbane, at Taymouth Castle; to Thomas Baring, Strathan Park, Hampshire; to Prince Leopold of Belgium at Claremont, Esher, Surrey; to Walter Francis Scott, fifth duke of Buccleuch, Dalkeith Palace. Author of *The book of the garden* (1853–5). (*CDEL*, R. Desmond 1994, Hadfield *et al.* 1980)

Mackay, Robert William (1803–82). Philosophical and theological scholar. (*Alum. Oxon.*, *DNB.*)

Mackie, Samuel Joseph. Geologist, palaeontologist, and antiquary. Elected fellow of the Geological Society of London, 1851; of the Society of Antiquaries of London, 1853. Editor of the *Geologist*, 1858–64; of the *Geological and Natural History Repertory*, 1865–7. Published a catalogue of British fossil sponges in 1866. Author of works on the history and topography of Folkestone, Kent. (*BLC*, *Journal of* *the Geological Society of London* 7 (1851): 257, *Proceedings of the Society of Antiquaries of London* 2 (1853): 314.)

Mackintosh, Eva. Daughter of Robert James and Mary Mackintosh. Niece of Frances Emma Wedgwood. (Wedgwood and Wedgwood 1980.)

Maclagan, Andrew Douglas (1812–1900). Scottish surgeon and botanist. MD and fellow of the Royal College of Surgeons of Edinburgh, 1833; president, 1859–60. Fellow of the Royal College of Physicians of Edinburgh, 1864; president, 1884–7. President of the Botanical Society of Edinburgh, 1863. Professor of medical jurisprudence and public health, University of Edinburgh, 1862–96. Surgeon-general of the Royal Company of Archers, 1868–1900. Knighted, 1886. (*Medical directory* 1861–1901, *Modern English biography*.)

MacLeod, William (d. 1875). Scottish physician. MD, University of St Andrews, 1843; appointed physican at the Edinburgh Royal Public Dispensary and lectured at the Argyle Square School of Medicine before moving to Yorkshire where he took up appointments as physician at the Wharfedale hydropathic establishment and the Institute for Diseases of Females in Bradford. Physican at the hydropathic establishment at Ben Rhydding, Otley, Yorkshire, 1849–75. Published widely on hydropathic treatment. (*Modern English biography*, *Medical directory* 1849–76).

McNab, James (1810–78). Scottish botanist. Collected plants in North America, 1834. Superintendent, Caledonian Horticultural Society, 1835. Curator of the Royal Botanic Garden, Edinburgh, from 1849. President of the Edinburgh Botanical Society, 1872. (R. Desmond 1994.)

McNab, William (1780–1848). Scottish gardener. Gardener at the Royal Botanic Gardens, Kew, 1801–10. Superintendent of the Royal Botanic Garden, Edinburgh, 1810–48. (R. Desmond 1994.)

Mann, Gustav (1836–1916). German-born botanist. Gardener at the Royal Botanic Gardens, Kew, 1859. Botanical collector to the Niger expedition, 1859–62. Worked for the Indian Forest Service, 1863–91. (R. Desmond 1994, R. Desmond 1995.)

Mantell, Gideon Algernon (1790–1852). Physician, geologist, and palaeontologist. Scientific lecturer and author. Made noted collections of invertebrate and dinosaur fossils. Secretary, Geological Society of London, 1841–2; vice-president, 1848–9. FRS 1825. (*DNB*, *DSB*.)

Marsh-Caldwell, Anne (1791–1873). Novelist. (*DNB*.)

Marshman, John. Agent at the Canterbury Emigration Office, 16 Charing Cross, London, *circa* 1862–7. (*Post Office London directory* 1863–8.)

Martins, Charles Frédéric (1806–89). French botanist. Qualified as a doctor in Paris in 1834. Professor of botany, Faculté de Montpellier, 1846. Also published on geology and meteorology. (*Dictionnaire universel des contemporains*, *NBU*.)

Martius, Karl Friedrich Philipp von (1794–1868). German botanist and ethnologist. Assistant to the conservator of the Munich botanic garden, 1814. Accompanied the Austrian expedition to Brazil, 1817–20, returning with several thousand specimens for the botanic garden. Professor of botany, Munich, 1826; principal

Martius, Karl Friedrich Philipp von, cont.
conservator of the botanic garden, institute and collections, 1832. Secretary of the physiomathematical section of the Royal Bavarian Academy, Munich, 1840. (*DSB*.)

Mason, Sarah Ann. *See* Bates, Sarah Ann.

Masters, Maxwell Tylden (1833–1907). Botanist, surgeon, and general medical practitioner. Sub-curator, Fielding Herbarium, University of Oxford, *circa* 1853–7. Lecturer on botany at St George's Hospital medical school, 1855–68. Editor of the *Gardeners' Chronicle*, 1865–1907. Active in the Royal Horticultural Society, succeeding Joseph Dalton Hooker to the chair of the scientific committee; secretary of the International Horticultural Congress, 1866. FRS 1870. (Clokie 1964, pp. 106 and 208; R. Desmond 1994; *DNB*; *Medical directory* 1857–1908.)
6 April [1863], [8–13 April 1863], *14 April 1863*

Masters, William (1796–1874). Nurseryman in Canterbury. Founded the Canterbury Museum in 1823; honorary curator, 1823–46. Conducted hybridisation experiments on passion flowers. Designed the formal gardens at Walmer Castle. Father of Maxwell Tylden Masters. (R. Desmond 1994.)

Matthew, Patrick (1790–1874). Scottish gentleman farmer. Author of works on political and agricultural subjects. Advanced a theory of natural selection in the 1830s. (Dempster 1996, R. Desmond 1994.)

Maull & Fox. Photographic artists of Piccadilly and Cheapside, London, from 1879 until after 1908. (Pritchard 1986.)

Maull & Polyblank. Photographic artists of Piccadilly and Gracechurch Street, London, 1856–65; moved to Fulham Road, 1865. A partnership between Henry Maull and George Henry Polyblank. (*Post Office London directory* 1856–65, Pritchard 1986.)

Maw, Arthur (b. 1834). Tile manufacturer. Younger brother of George Maw, with whom he established the encaustic tile company Maw & Co., in Broseley, Shropshire, in 1850; factory premises in Benthall, Shropshire, were opened in 1852; the company's work took first-class prizes at international exhibitions from 1862 onwards. Author of the article 'Encaustic tiles' for the ninth edition of the *Encyclopaedia Britannica* (1875–89). ('Biographical notes', Ironbridge Gorge Museum Trust, Telford; Messenger 1970.)

Maw, George (1832–1912). Tile manufacturer, geologist, botanist, and antiquarian. Partner with his younger brother Arthur in the encaustic tile company, Maw & Co., of Brosley, Shropshire. Established a well-known garden at his residence at Benthall Hall, Shropshire; an expert on crocuses. Wrote on the geology of western England and North Wales. Travelled to Morocco and Algeria with Joseph Dalton Hooker in 1871 and independently in 1873, writing on the geology of these countries. (Benthall 1980; R. Desmond 1994; *Gardeners' Chronicle*, 12 February 1881, pp. 205–6, 208 and 209; Sarjeant 1980–96.)
19 February 1863, *25 February 1863*, 28 February [1863], *25 April 1863*, 12 May [1863]

Max Müller, Friedrich (1823–1900). German-born orientalist and philologist. Published an edition of the 'Rigveda', the most important of the sacred books of the Brahmans, 1849–73. Moved to Paris in 1845; settled in Oxford in 1848 after fleeing the revolution in France. Deputy Taylorian Professor of modern European languages, Oxford University, 1850–4; professor, 1854–68; professor of comparative philology, 1868–75. Curator of the Bodleian Library, Oxford, 1856–63 and 1881–94. (*DNB.*)

Maxwell, James Clerk (1831–79). Scottish physicist. Fellow, Trinity College, Cambridge, 1855. Professor of natural philosophy, King's College, London, 1860–5. Member of the newly formed electrical standards committee of the British Association for the Advancement of Science, 1862. Superintended the building of the Cavendish Laboratory; first professor of experimental physics, University of Cambridge, 1871. Wrote papers on colour vision, the kinetic theory of gases, electricity, and magnetism. FRS 1861. (*DNB, DSB.*)

Mays, J. Aldous Reporter and shorthand writer. His notes on Thomas Henry Huxley's lectures to working men provided the text of T. H. Huxley 1863a. (Freeman 1978, Huxley 1863a.)

Mellersh, Arthur (1812–94). Naval officer. Midshipman and mate on HMS *Beagle*, 1825–36. Served off the coast of Syria, then in command of HMS *Rattler* in the Burma campaign in 1852. Served off the coast of China in the 1850s, suppressing piracy, and in the Caribbean and South America before retiring in 1864. (*Modern English biography*; *The Times*, 28 September 1894, p. 4.)

Ménétriés, Edouard (1802–61). French-born entomologist. Conservator of rarities at the zoological museum, Imperial Academy of Sciences, St Petersburg. (Essig 1965.)

Meryon, Edward (1809–80). Physician. Physician, Hospital for Nervous Diseases; lecturer in comparative anatomy, St Thomas's Hospital, London. Author of books on the nervous system and the history of medicine. Fellow of the Royal Geographical Society, 1857. Fellow of the Royal College of Physicians, 1859. (*List of the Geological Society of London*, *Medical directory* 1863, *Modern English biography*, *Physicians*.)

Meteyard, Eliza (1816–79). Author. Born in Liverpool; lived in Shrewsbury, 1818–29; in Thorpe, near Norwich, 1829–42; thereafter in London. Contributed fiction and social articles to numerous periodicals under the pen-name Silverpen. Her novels include *Struggles for fame* (1845), *Mainstone's housekeeper* (1860), and *Lady Herbert's gentlewoman* (1862). Published a number of works about the Wedgwoods including a two-volume life of Josiah Wedgwood I (1865–6), and *The Wedgwood handbook* (1875). (*DNB.*)

Meyer, Carl Anton von (1795–1855). Russian botanist and explorer. Director, St Petersburg botanical garden, 1850–5. (*Taxonomic literature.*)

Meyer, Christian Erich Hermann (Hermann) von (1801–69). German palaeontologist. Devoted himself to full-time palaeontological research, 1827–37. Entered financial administration of Bundestag, 1837; director of finances, 1863. Refused

Meyer, Hermann von, cont.
appointment as professor in the University of Göttingen in 1861, in order to retain scientific independence. Published chiefly on vertebrate palaeontology. (*DSB.*)

Mill (or Mille), Humphrey (*fl.* 1646). Poet. (*DNB.*)

Mill, John Stuart (1806–73). Philosopher and political economist. Administrator, East India Company, 1823–58. (*DNB*, *DSB.*)

Miller, William Allen (1817–70). Chemist and astronomer. Professor of chemistry, King's College, London, from 1845. Pioneered the use of spectroscopic analysis in chemistry. Developed a telescope to give spectra of celestial objects, providing detailed information on stellar chemistry. FRS 1845. (*DNB*, *DSB.*)

Miller, William Hallowes (1801–80). Mineralogist and crystallographer. Professor of mineralogy, Cambridge University, 1832–80. Foreign secretary, Royal Society of London, 1856–73; Royal Medallist, 1870. FRS 1838. (*DNB*, *DSB.*)

Mitchell, Walter (*fl.* 1834–74). Clergyman and natural philosopher; interested in crystallography and mineralogy. Lecturer in natural philosophy, St Bartholomew's Hospital; vicar of St Bartholomew's-the-less, 1861–9; hospitaller, St Bartholomew's Hospital, 1846–69. Reader at Christ Church, Newgate Street, London, 1846–69. (*Alum. Cantab.*, *Clergy list* 1843–1874, *Crockford's clerical directory* 1860–74.)

Mohl, Hugo von (1805–72). German biologist. Professor of physiology, University of Bern, 1832–5; professor of botany, University of Tübingen, 1835–72. Known for his work on the microscopic anatomy of plants and for his study of the plant cell. Co-founder of the *Botanische Zeitung*, 1843. Foreign member, Royal Society of London, 1868. (*DSB*, *Record of the Royal Society of London.*)

Monro, Alexander (*c.* 1814–83). Scottish Congregational pastor and hydropathic practitioner. Pastor of several Congregational churches in Scotland, *circa* 1840 – *circa* 1844; of Blackhills Evangelical Union Church, Aberdeenshire, *circa* 1844–64. Conducted a hydropathic establishment at Angusfield, 1850–1; at Lochhead, 1851–64. Medical superintendent at Cluny Hill hydropathic establishment, near Forres, Morayshire, 1864–9, 1873–81; at Waverley hydropathic establishment, Melrose, Roxburghshire, 1869–73; at Bishop's Teignton, Devon, 1882–3. Author of works on hydropathy. (McNaughton 1993, Metcalfe 1906.)

Monro, David (1813–77). Scottish-born politician. MD, University of Edinburgh, 1834. Travelled to New Zealand, 1841–42, and became active in provincial and national politics. Speaker of the House of Representatives, 1861–70. Founding member, Nelson Horticultural Society. Contributed specimens to Joseph Dalton Hooker's work on New Zealand flora. Knighted, 1866. (*DNB*; *DNZB*; J. D. Hooker 1864–7, p. 11; *Nature* 16 (1877): 15–16.)

Monteiro, Joachim John (*fl.* 1858–75). Mining engineer and zoologist, resident in Angola from 1858. Involved in trading and mining operations; engaged by the Western African Malachite Copper Mines Company to develop a concession at Bembe, 1858. Author of papers on African ornithology. Associate of the Royal School of Mines; corresponding member of the Zoological Society. (Egerton 1957, Monteiro 1875, *Royal Society catalogue of scientific papers.*)

Moore, John. Secretary to the first lord of the Admiralty *circa* 1860–3. (*Post Office London directory* 1860–3.)

Moreno, Gabriel García. *See* García Moreno, Gabriel.

Morren, Charles François Antoine (1807–58). Belgian botanist and horticultural writer. Taught physics at the industrial school and the university in Ghent, 1831–5; studied medicine and qualified in 1835. Professor extraordinarius of botany, University of Liège, 1835–7; professor, 1837–54. (*BNB*, Gilbert 1977.)

Mortimer, Cromwell (d. 1752). Physician. MD, Leyden, 1724. Editor of the *Philosophical Transactions of the Royal Society* ; second or acting secretary, Royal Society of London, 1730–52. FRS 1728. (*DNB*.)

Morton, Lord. *See* Douglas, George Sholto.

Mota, Vicente Pires da (1799–1882). Brazilian law lecturer, secular priest, and administrator. Professor of civil law, Faculdade de Direito de São Paulo, 1834; director of law faculty, 1865–82. Member of the council of (provincial) government, 1828–34; member of the district (provincial) assembly, 1834–41. President of the province of São Paulo, 1848–51 and 1862–4; of Ceará, 1854–5; of Minas Gerais, 1860–1; of Santa Catarina, 1861–2. (*Dicionario de história de São Paulo*, *Fundação da cidade de São Paulo*, Nogueira 1977.)

Mowbray, William (*c.* 1792–1832). Gardener at Harley Hall, Staffordshire, for eleven years before becoming curator of the Manchester Botanical and Horticultural Garden. (R. Desmond 1994, *Horticultural Register* (1832): 670.)

Mudie's Select Library. Commercial subscription library, founded in 1842 by Charles Edward Mudie (1818–90), with main premises at New Oxford Street, London, from 1852. (*EB* s.v. Mudie, Charles Edward.)

Mulder, Gerrit Jan von (1802–80). Dutch physician and organic chemist. Practised medicine in Amsterdam and Rotterdam, lectured at the Bataafsch Genootschap der Proefondervindelijke Wijsbegeerte, and taught botany to student apothecaries before joining the faculty at the new medical school in Rotterdam in 1828. Professor of chemistry, University of Utrecht, 1840–68. (*DSB*.)

Müller, Friedrich Max. *See* Max Müller, Friedrich.

Müller, Philipp Jakob (Muller, Philippe Jacques) (1832–89). Alsatian botanist at Weissenburg am Rhein. Took French nationality in 1872 after the Franco-Prussian War, and moved to Nyon, Switzerland, abandoning his botanical work. (*Bulletin de la Société Botanique de France* 1890, pp. 203–4; *Taxonomic literature*.)

Murchison, Roderick Impey, 1st baronet (1792–1871). Geologist and army officer. Served in the British army, 1807–14. Noted for his work on the Silurian system. President of the Geological Society of London, 1831–3 and 1841–3; of the British Association for the Advancement of Science, 1846; of the Royal Geographical Society of London, 1843–58. Director-general of the Geological Survey of Great Britain, 1855. Knighted, 1846; created baronet, 1866. FRS 1826. (*DNB*, *DSB*.)

Murray, Andrew (1812–78). Lawyer, entomologist, and botanist. Practised law in Edinburgh until 1860, when he was appointed assistant secretary to the Royal

Murray, Andrew, cont.
Horticultural Society of London. An expert on insects harmful to crops. In entomology, specialised on the Coleoptera; in botany, on the Coniferae. (*Entomologist's Monthly Magazine* 14 (1878): 215–16, Gilbert 1977.)

Murray, George Augustus Frederick John, 6th duke of Atholl (1814–64). Army officer. Admitted as nobleman at Trinity College, Cambridge, 1832. Succeeded his father as Lord Glenlyon, 1837, and his uncle as duke of Atholl, 1846. (*Alum. Cantab.*)

Murray, John (1808–92). Publisher and author of guide-books. CD's publisher from 1845. (*DNB*, Freeman 1978.)
22 January [1863]

Nägeli, Carl Wilhelm von (1817–91). Swiss botanist. Maintained a teleological view of evolution. Originally studied medicine, but transferred to botany under Alphonse de Candolle at Geneva. Spent eighteen months with Matthais Jacob Schleiden in University of Jena, then worked in Zurich, where he collaborated with Carl Cramer, 1845–52. Professor of botany, University of Freiburg, 1852; professor of botany, University of Munich, 1857. (*DSB.*)

Napier, Frances (d. 1872). Married Charles James Napier in 1835. (*DNB* s.v. Napier, Charles James.)

Napier, William Francis Patrick (1785–1860). Irish-born soldier, writer and artist. Served in the army in the Peninsular War, of which he wrote a six-volume history. Lieutenant-governor of Guernsey, 1842–7. Promoted to general, 1859. Knighted, 1848. (*DNB.*)

Napoleon III. *See* Louis Napoleon.

Nasmyth, Alexander (1789–1848). Surgeon-dentist at Hanover Square, London. Surgeon-dentist to Prince Albert and Queen Victoria. Fellow of the Royal College of Surgeons of England; member of the Royal Medical and Chirurgical Society of London, the Linnean Society of London, and the council of the Geological Society of London. (A. Desmond 1989, p. 426; *Medical directory* 1845–9.)

Nash, John. Resident of Widmore Lane, Bromley, Kent. (*Post Office directory of the six home counties* 1862–78.)

Nash & Lukey. Linen-drapers, silk-merchants, and milliners, High Street, Bromley, Kent. (*Post Office directory of the six home counties* 1851–66.)

Naudin, Charles Victor (1815–99). French botanist. Joined the herbarium staff at the Muséum d'Histoire Naturelle and became professor of zoology at the Collège Chaptal, Paris, 1846. Resigned almost immediately owing to a severe nervous disorder. Appointed aide-naturaliste at the Muséum d'Histoire Naturelle, 1854. Established a private experimental garden at Collioure, 1869, earning his living by selling seeds and specimens. First director of the state-run experimental garden at Antibes, 1878. Experimented widely on plants, particularly on acclimatisation and hybridity. Published a theory of transmutation based on hybridisation. (*DSB*, *Taxonomic literature.*)
7 February 1863

Naumann, Karl Friedrich (1797–1873). German geologist and mineralogist. Professor of crystallography, Freiberg Bergakademie, 1826; professor of geognosy, 1835. Professor of mineralogy and geognosy, University of Leipzig, 1842. Corresponding member of the Academies of Berlin, Munich, Paris, and St Petersburg, and of the Royal Society of London. (*ADB*; *DSB*; *EB* s.v. Naumann, G. A. C. F.; *NDB*.)

Neighbour, Alfred (1825–90). Merchant and apiarian specialist in London. Continued his father, George Neighbour's, firm, which sold apiarian appliances. Purchased William Bernhard Tegetmeier's library of bee books. (Fraser 1958.)

Neighbour, George. Merchant and apiarian specialist; established a firm selling apiarian appliances in Holborn, London, in 1814; the firm moved to 149 Regent Street, London, in 1852. (Fraser 1958.)

George Neighbour & Sons. Merchants and apiarian specialists at 149 Regent Street, London, conducted by Alfred Neighbour; established by Alfred's father, George Neighbour, in 1814. (Fraser 1958, *Post Office London directory* 1863.)

Neumann, Louis (*fl.* 1849–93). French gardener. Son of Joseph Henri François Neumann (1800–58). Gardener at the Muséum d'Histoire Naturelle from 1849 (or before) to 1878 (or after). Member of the Société Centrale d'Horticulture de France, 1863; secretary, 1865–6. Editor of *Le Nouveau Jardinier Illustré*, 1865–93. (*Annuaire de la Société [Nationale d'Horticulture de France]* (1878): lxvi, (1884): 60, (1893): 109; *Catalogue de la Bibliothèque Nationale*; *Journal de la Société Impériale et Centrale d'Horticulture de France* 11 (1865): 15 and 44, 12 (1866): 25; Rivière 1866, p. 284.)

Newton, Alfred (1829–1907). Zoologist and ornithologist. Travelled extensively throughout northern Europe and North America on ornithological expeditions, 1854–63. Professor of zoology and comparative anatomy, Cambridge University, 1866–1907. FRS 1870. (*DNB*.)

21 March 1863, 24 March [1863], *31 October 1863*

Newton, Isaac (1642–1727). Natural philosopher. Advanced the theory of gravitation in *Principia mathematica* (1687) and of light in *Opticks* (1704). Lucasian Professor, Cambridge University, 1669–1701. Master of the mint, 1699. President of the Royal Society of London, 1703–27. Knighted, 1705. FRS 1672. (*DNB*, *DSB*.)

Nightingale, Florence (1820–1910). Reformer of hospital nursing. Nursed British troops during the Crimean War, 1854–6. Established a training school for nurses at St Thomas's Hospital, London, in 1860. (*DNB*.)

Nightingale, William Edward (1794–1874). Father of Florence Nightingale. (*DNB* s.v. Florence Nightingale.)

Norgate, Frederick. Formed the science publishing house Williams & Norgate of Covent Garden, London, and South Frederick Street, Edinburgh, in partnership with Edmund Sydney Williams. (*Modern English biography* s.v. Williams, Edmund Sydney.)

Oken, Lorenz (1779–1851). German naturalist and leading exponent of *Naturphilosophie.* Associate professor of medicine, University of Jena, 1807–12; professor of

Oken, Lorenz, cont.

natural history, 1812–19. Professor of physiology, University of Munich, 1827–32; professor of natural history, natural philosophy, and physiology, Zürich University, 1832–51. (*ADB*, *DBE*, *DSB*, *EB*.)

Oldfield, Augustus Frederick (1820–87). Botanist and ethnologist. Collected plants in Australia and Tasmania for Ferdinand von Mueller and others during the 1850s. Placed botanical information and specimens at the disposal of Joseph Dalton Hooker and George Bentham. Acquired extensive knowledge of the Australian aborigines. Returned to England by 1863. Became blind *circa* 1867. (R. Desmond 1994, *Papers and Proceedings of the Royal Society of Tasmania* (1909): 24–6.)

Oldham, Thomas (1816–78). Irish geologist. Chief geological assistant to the Geological Survey of Ireland, 1839–50. Professor of geology, Trinity College, Dublin, 1845–51. In 1849, discovered in the rocks of Bray Head, County Wicklow, fossils or organic marks, which were named *Oldhamia*, after him. Director, Geological Survey of India, 1850–76. (*Modern English biography*, Sarjeant 1980–96.)

Oliver, Daniel (1830–1916). Botanist. Assistant in the herbarium of the Royal Botanic Gardens, Kew, 1858; librarian, 1860–90; keeper, 1864–90. Professor of botany, University College, London, 1861–88. FRS 1863. (R. Desmond 1994, *List of the Linnean Society of London* 1859–91.)

20 [January 1863], *22 January 1863*, *17 February 1863*, 20 [February 1863], *27 February 1863*, 24–5 March [1863], *[26 March 1863]*, *[27 March 1863]*, 28 March [1863], [12 April 1863], *14 April 1863*, [after 14 April 1863], 18 July [1863], *20 July 1863*, *[after 20 July 1863]*, [before 27 November 1863], *27 November 1863*, 28 [November 1863]

Olmsted, Frederick Law (1822–1903). American landscape architect. Farmed on Staten Island, New York, for several years; also pursued literary interests and travelled widely. Wrote extensively on slavery and the economic and social conditions of the American South. Appointed superintendent of Central Park, New York City, 1857; architect-in-chief, 1858. Superintendent, Frémont Mariposa mining estates, California, 1863–5. Reappointed landscape architect, Central Park, 1865, until his removal in 1878. Thereafter worked independently. (*DAB*.)

Overton, Frederick Arnold (1862–1935). Clergyman. Grandson of William Darwin Fox and son of Harriet Emma Overton. Educated at Exeter College, Oxford. (*Alum. Oxon.*, *Darwin pedigree*.)

Overton, Harriet Emma (b. 1837). Daughter of William Darwin Fox. Married Samuel Charlesworth Overton (*Crockford's clerical directory* 1865–92), curate of Hackness, Yorkshire, in 1861. Mother of Frederick Arnold Overton. (*Darwin pedigree*; *Gentleman's Magazine* (1861), pt 2: 558.)

Owen, Richard (1804–92). Comparative anatomist. Assistant conservator of the Hunterian Museum, Royal College of Surgeons of England, 1827; Hunterian Professor of comparative anatomy and physiology, 1836–56. Superintendent of the natural history departments, British Museum, 1856–84; prime mover in establishing the Natural History Museum, South Kensington, in 1881. President of the British Association for the Advancement of Science, 1858. Described the *Beagle* fossil mammal specimens. Knighted, 1884. FRS 1834. (*DNB*, *DSB*.)

Oxenden, George Chichester (1797–1875). Author of satiric verse and parodies, and orchid-fancier. Son of Henry Oxenden, seventh baronet. Lived at the family seat at Broome Park, near Canterbury, Kent. (*Alum. Cantab.*, *BLC*, *Burke's peerage* 1895, *Correspondence* vol. 10.)
13 May 1863

Padmore, Richard (1789–1881). Iron-founder, banker, and politician. Sheriff of Worcester, 1845; mayor, 1848 and 1852. Liberal MP, 1860–68. (*Dod's parliamentary companion*, *Modern English biography*.)

Paget, James, 1st baronet (1814–99). Surgeon. Assistant surgeon at St Bartholomew's Hospital, London, 1847; surgeon, 1861–71. Arris and Gale Professor of anatomy and surgery at the Royal College of Surgeons of England, 1847–52. Lectured on physiology in the medical school, St Bartholomew's, 1859–61; on surgery, 1865–9. Appointed surgeon-extraordinary to Queen Victoria, 1858; serjeant-surgeon, 1877. Created baronet, 1871. FRS 1851. (*DNB*.)
7 February 1863, 11 March [1863], *16 March 1863*

Pallas, Pyotr Simon (1741–1811). German naturalist and geographer. Travelled widely in the Russian empire. Foreign member, Royal Society of London, 1764. (*ADB*, *DSB*, *Record of the Royal Society of London*.)

Palmerston, Viscount. *See* Temple, Henry John.

Pappe, Karl Wilhelm Ludwig (1803–62). German-born physician and botanist in South Africa. Studied medicine and botany at Leipzig before moving to Cape Town in 1831, initially practising as a doctor. First colonial botanist and South Africa's first professor of botany, 1858. International government adviser on botanical issues. (*DSAB*, *Taxonomic literature*.)

Parker, Henry (1827/8–92). Fine art specialist. Scholar, Oriel College, Oxford, 1846–51; fellow, 1851–85. Son of CD's sister, Marianne Parker. (*Alum. Oxon.*, *CDEL*, *Darwin pedigree*.)

Parker, Marianne (1798–1858). CD's eldest sister. Married Henry Parker in 1824. (*Darwin pedigree*.)

Parkes, Edmund Alexander (1819–76). Physician and author of works on tropical diseases. Served as assistant surgeon in India, 1842–5. Special professor of clinical medicine, University College, London, and physician to the University College Hospital, 1849. Professor of hygiene at the Army Medical School at Chatham, 1860. FRS 1861. (*DNB*.)

Parry, Charles Christopher (1823–90). British-born physician and botanist. After a short period in medical practice, joined the geological survey of Wisconsin, Iowa, and Minnesota in 1848. Appointed botanist to the United States and Mexican boundary survey in 1849. The first botanist appointed by the United States Department of Agriculture. Engaged in numerous expeditions of botanical exploration, in particular in the western United States. (*DAB*.)

Parslow, Joseph (1811/12–98). CD's manservant at 12 Upper Gower Street, London, *circa* 1840–2, and butler at Down House until 1875. (Census returns 1861 (Public Record Office RG9/462: 74), Freeman 1978.)

Parsons, Theophilus (1797–1882). American lawyer. Practised in Massachussetts. Specialised in admiralty, patent, and insurance law. Dane Professor of law, Harvard Law School, 1848–69. (*DAB*, *WWWA*.)

Partington, Eliza (*fl.* 1850–64). Lodging-house keeper, Montreal House, Great Malvern, Worcestershire. CD stayed in her lodging-house in April 1851. (*Correspondence* vol. 5, letter to E. A. Darwin, 19 April 1851; *Post Office directory of Birmingham* 1850–64.)

Pasteur, Louis (1822–95). French chemist and microbiologist. Professor of chemistry, Strasbourg University, 1849–54. Professor of chemistry and dean of the science faculty, Lille University, 1854–7. Administrator and director of scientific studies, Ecole Normale, Paris, 1857–67; director of the laboratory of physiological chemistry, 1867–88. Professor of chemistry, the Sorbonne, Paris, 1867–74. Director of the Institut Pasteur, Paris, 1888–95. Renowned for his work on fermentation and for experiments providing evidence against the theory of spontaneous generation. Foreign member, Royal Society of London, 1869. (*Dictionnaire universel des contemporains*, *DSB*, *Record of the Royal Society of London*.)

Patten-Saunders, William Henry. *See* Saunders, William Henry Patten.

Paxton, Joseph (1803–65). Gardener and architect. Gardener at the Horticultural Society gardens in Chiswick, 1823; foreman, 1824–6. Superintendent of the gardens at Chatsworth, Derbyshire, 1826. Designed the lay-out of and constructions for the Great Exhibition in 1851. Superintended the re-erection of the Crystal Palace at Sydenham, 1853–4; director of the gardens there, 1854–65. Liberal MP for Coventry, 1854–65. Knighted, 1851. (R. Desmond 1994, *DNB*, *Modern English biography*.)

Peirce, Benjamin (1809–80). American mathematician and astronomer. Tutor in mathematics, Harvard University, 1831; professor, mathematics and natural philosophy, 1833; Perkins Professor of astronomy and mathematics, 1842–80. Superintendent of the United States coast survey, 1867–74. (*DAB*.)

Pelzel von Pelzeln, August von (1825–91). Austrian naturalist and ornithologist. Assistant curator of birds and mammals at the Kaiserlich-königliche Hofnaturaliencabinet, 1851; curator of the ornithological collection, 1857; of the mammal collection, 1869. Co-editor of *Die Schwalbe* (Ornithological Society magazine). (*DBE*, *OBL*, *Verhandlungen der Zoologisch-botanischer Gesellschaft in Wien* 41 (1891): 791.)

Pengelly, William (1812–94). Mathematical tutor and geologist. An expert on the geology of Devon; explored the plant-bearing deposits at Bovey Tracey, and at Brixham Cave and Kent's Hole, Torquay. Honorary secretary of the Torquay Natural History Society, 1851–90. President of the Devonshire Association for the Advancement of Literature, Science, and Art, 1867–8. FRS 1863. (*DNB*, Sarjeant 1980–96.)

Percy, John (1817–89). Metallurgist. Lecturer and professor of metallurgy at the Royal School of Mines, Jermyn Street, 1851–79. President of the Iron and Steel Institute, 1885 and 1886. FRS 1847. (*DNB*.)

Perthes, Jacques Boucher de. *See* Boucher de Crèvecoeurs de Perthes, Jacques.

Pertz, Annie. Daughter of Georg Heinrich Pertz and his second wife, Leonora. (K. M. Lyell ed. 1890, 2: 359.)

Pertz, Dora. Daughter of Georg Heinrich Pertz and his second wife, Leonora. (K. M. Lyell ed. 1890, 2: 359.)

Pertz, Georg Heinrich (1795–1876). German historian. Historian of medieval Germany; published important source material in *Monumenta Germaniae historica*. Archives secretary, royal library, Hanover, 1823; principal keeper, 1827. Called to Berlin as chief librarian, 1842. Married Leonora Horner in 1853. (*ADB*, *EB*.)

Pertz, Leonora (b. 1818). Daughter of Leonard Horner. Translator, with her sister Joanna B. Horner, of German works. Married Georg Heinrich Pertz in 1853. (*ADB* s.v. Pertz, Georg Heinrich, *NUC*.)

Petschler, Charles (*fl.* 1858–65). Merchant at Vulcan Lane, Auckland, New Zealand. Friend of Julius von Haast and Ferdinand Hochstetter. (*Chapman's New Zealand Almanac* 1860, 1862; *Correspondence* vol. 10, letter from Julius von Haast, 9 December 1862; H. F. von Haast 1948, p. 446.)

Phillips, John (1800–74). Geologist. Keeper of the museum of the Yorkshire Philosophical Society, 1825–40. Assistant secretary, British Association for the Advancement of Science, 1832–62. Professor of geology, King's College, London, 1834–40. Palaeontologist to the Geological Survey of Great Britain, 1840–4. Deputy reader in geology, Oxford University, 1853; professor, 1860–74. FRS 1834. (*DNB*, *DSB*.)

Pictet de la Rive, François Jules (1809–72). Swiss zoologist and palaeontologist. Professor of zoology, University of Geneva, 1835. (Gilbert 1977, Sarjeant 1980–96.)

Planchon, Jules Emile (1823–88). French botanist. Herbarium assistant, Royal Botanic Gardens, Kew, 1844–8. Professor of botany, Institute of Horticulture, Ghent, Belgium, 1849–51. Professor, School of Medicine and Pharmacy, Nancy, 1851–3. Professor of botany, faculty of sciences, Montpellier, 1853–81; director of the school of pharmacy, 1856. Became professor of botany in the faculty of medicine and director of the botanic garden, Montpellier, 1881. (R. Desmond 1994, *Grande encyclopédie*.)

Plato (427–347 B.C.). Athenian philosopher. (*EB*.)

Playfair, Lyon, 1st Baron Playfair of St Andrews (1818–98). Statesman and chemist. Chemist to the Geological Survey of Great Britain and professor of chemistry at the School of Mines, 1845. Secretary at the Department of Science and Art, 1853–8. President of the Chemical Society, 1857–9. Professor of chemistry, Edinburgh University, 1858–69. Liberal MP for the Universities of Edinburgh and St Andrews, 1868–85; for South Leeds, 1885–92. Postmaster-

Playfair, Lyon, 1st Baron Playfair of St Andrews, cont.
general, 1873; chairman and deputy speaker of the House of Commons, 1880–3. Knighted, 1883; created Baron Playfair, 1892. FRS 1848. (*DNB*, *DSB*.)

Poiret, Jean Louis Marie (1755–1834). French clergyman, naturalist, and traveller. Collaborated with Lamarck on the botanical dictionary of the *Encyclopédie*. Taught natural history at the Ecole Centrale de l'Aisne. Lived in Paris after 1806, publishing chiefly botanical works. (*BNB*, Dantès 1875.)

Pollock, George Frederick (1821–1915). Barrister-at-law. Master of the Court of the Exchequer until 1879 and senior master of the Supreme Court of Judicature, 1851–1901. Queen's and king's remembrancer, 1886–1902. (*Burke's peerage* 1970, *Men-at-the-bar*.)

Pouchet, Félix Archimède (1800–72). French biologist and naturalist. Director of the Muséum d'Histoire Naturelle at Rouen. Prolific author and populariser of science. Adversary of Louis Pasteur in the debate over spontaneous generation. (*Dictionnaire universel des contemporains*, *DSB*.)

Poulton, Edward Bagnall (1856–1943). Zoologist. Hope Professor of zoology, University of Oxford, 1893–1933. President of the British Association for the Advancement of Science, 1937. Knighted, 1935. FRS 1889. (*DNB*.)

Powell, Baden (1796–1860). Clergyman and writer on optics and theological topics; active in university reform. Vicar of Plumstead, Kent, 1821–7. Savilian Professor of geometry at Oxford University, 1827–60. Author of an article on the study of the evidences of Christianity in *Essays and reviews* (1860). FRS 1824. (*DNB*, *DSB*.)

Powell, Henrietta Grace. Daughter of William Henry Smyth. Married Baden Powell, becoming his second wife, in 1846. (*DNB* s.v. Powell, Baden.)
11 February 1863

James Powell & Sons. A firm of glaziers in Fleet Street, Whitefriars, London. (*Post Office London directory* 1863.)

Prescott, William Hickling (1796–1859). American historian of Spain, Mexico, and Peru. (*DAB*.)

Prestwich, Joseph (1812–96). Geologist and businessman. Entered the family wine business in London in 1830; became proprietor in 1842. Professor of geology, Oxford University, 1874–88. Fellow of the Geological Society of London, 1833; president, 1870–72. An expert on the Tertiary geology of Europe. Prominent in studies of human prehistory. Knighted, 1896. FRS 1853. (*DNB*, *DSB*.)

Prévost, Louis-Constant (1787–1856). French geologist. Studied the geology of France and Austria. Collaborated with Charles Lyell in comparing the Tertiary and Secondary strata on both sides of the English Channel, 1823–4. A founder of the Société Géologique de France, 1830; president, 1834, 1839, and 1851. Professor of geology, the Sorbonne, 1831. (*DSB*, Sarjeant 1980–96.)

Price, John (1803–87). Scholar, school-teacher, and naturalist. Educated at Shrewsbury School, 1818–22, and St John's College, Cambridge. Assistant master, Shrewsbury, 1826–7. Headmaster of the junior department at Bristol College, then classics principal at the Liverpool High School, before settling in Chester. A founding member of the Chester Natural Science Society. (*Alum. Cantab.*, *Eagle* 15 (1888): 169–72, *Modern English biography.*)

[8 September–13 October 1863]

Pritchard, Charles (1808–93). Clergyman and astronomer. Headmaster of Clapham Grammar School, 1834–62, where he established an observatory. Hulsean Lecturer, Cambridge University, 1867. Savilian Professor of astronomy, Oxford University, 1870–93. President of the Royal Astronomical Society, 1886. FRS 1840. (*Alum. Cantab.*, *DNB*, *DSB.*)

Pycroft, George (1819–94). Surgeon. Practised as a surgeon at Kenton, Exeter, Devon, 1844–90. Member of the Royal College of Surgeons of England. Fellow of the Geological Society of London. One of the originators of the volunteer movement in Devon, 1852. A popular lecturer, and one of the original members of the Devonshire Association for the Advancement of Literature, Science, and Art, founded in 1862. (*List of the Geological Society of London*, *Modern English biography.*)

Quaritch, Bernard (1819–99). German-born book-dealer. Worked for Henry George Bohn (1796–1884), 1842–4 and 1846–7; established an independent business near Leicester Square, London, in 1847; moved to 15 Piccadilly in 1860. (*DNB.*)

Quatrefages de Bréau, Jean Louis Armand de (Armand de Quatrefages) (1810–92). French zoologist and anthropologist. Doctorate in the physical sciences, University of Strasbourg, 1830; doctorate in medicine, 1832. Founded *Journal de médicine et de chirurgie de Toulouse*, 1836. Moved to Paris and took a doctorate in the natural sciences, 1840. Professor of natural history at the Lycée Henri IV, Paris, 1850; professor of the natural history of man, Muséum d'Histoire Naturelle, 1855. Foreign member, Royal Society of London, 1879. (*DSB*, *Record of the Royal Society of London.*)

25 March 1863, 27 March [1863], *[28 March–]11 April 1863*, [14 April 1863], 14 May [1863], *19 May [1863]*

Quenstedt, Friedrich August (1809–89). German palaeontologist and stratigrapher. Associate professor of mineralogy and geology, University of Tübingen, 1837; professor, 1842–89. (*ADB*, *DSB.*)

Ramsay, Andrew Crombie (1814–91). Geologist. Appointed to the Geological Survey of Great Britain, 1841; senior director for England and Wales, 1862; director-general, 1871–81. Professor of geology, University College, London, 1847–52; lecturer on geology at the Royal School of Mines, 1852–71. President of the Geological Society of London, 1862–4. Knighted, 1881. FRS 1862. (*DNB*, *DSB.*)

29 April [1863], *6 May 1863*

Ramsay, Louisa. Daughter of James Williams, clergyman, of Llanfairynghornwy, Anglesey, Wales; married Andrew Crombie Ramsay in 1852. (*DNB* s.v. Ramsay, A. C.)

Ransome, Frederick (1818–93). Inventor of an artificial sandstone. The material was used for building in Britain and in the colonies. (*Modern English biography.*)

Rapin, Daniel (1799–82). Swiss botanist and pharmacist. Worked in pharmacies in Fribourg, Geneva, and Payerne, Switzerland, and in Strasbourg and Paris, France, 1832–8; settled in Rolle, Switzerland, 1838–53. In retirement, lived in Yverdon, 1853–7, and from 1857, in Plainpalais. (*Taxonomic literature.*)

Rawson, Arthur (1818–91). Clergyman. Ordained deacon, 1841; priest, 1842. Curate, Bromsgrove, Worcestershire, 1841–3; perpetual curate, Trinity Church, Bromley, Kent, 1843–82. Cultivated orchids and florists' flowers, particularly pelargoniums. (*Alum. Cantab.*, R. Desmond 1994.)
1 April [1863], [6 April 1863]

Reed, George Varenne (1816–86). Anglican clergyman. Curate of Hayes, Kent, 1837–9; of Tingewick, Buckinghamshire, 1839–54. Rector of Hayes, 1854–86. Tutor to George Howard, Francis, Leonard, and Horace Darwin. (*Alum. Cantab.*, Freeman 1978.)
12 January 1863

Reeve, Henry (1813–95). Man of letters. Began a literary career in London in 1834. Joined the staff of *The Times* in 1840; resigned in 1855. Editor of the *Edinburgh Review*, 1855–95. Member of the Society of Antiquaries, 1852; vice-president, 1879–82. (*DNB.*)

Reeve, Lovell Augustus (1814–65). Conchologist, publisher, and bookseller. Set up a shop in King William Street, Strand, for the sale of natural history specimens and the publication of conchological works; moved to Henrietta Street, Covent Garden, in 1848. Editor and proprietor of the *Literary Gazette*, 1850–6. (*DNB*, Sarjeant 1980–96.)
25 April [1863]

Lovell Reeve & Co. Publishers and booksellers at 5 Henrietta Street, Covent Garden, London. (*Post Office London directory* 1863.)

Reeves (or Reaves), William (b. 1836/7). Blacksmith and farrier. Resident of Down, Kent. (Census returns 1861 (Public Record Office RG9/462: 71), *Post Office directory of the six home counties* 1862.)

Regnault, Henri Victor (1810–78). French chemist and physicist. Chair of chemistry, Ecole Polytechnique, 1840. Professor of physics, Collège de France, 1841. Director of the porcelain factory at Sèvres from 1851. Received the Royal Society of London's Copley Medal in 1869. (*DSB.*)

Renan, Joseph Ernest (1823–92). French philologist. Assistant, department of manuscripts, Bibliothèque Nationale, 1851. Member of the Académie des Inscriptions et Belles Lettres, 1856. Professor of Hebrew and Chaldaic languages, Collège de France, 1862–4. Lost the professorship following publication of his highly controversial *Vie de Jésus* in 1863; reinstated, 1870. Member of the Académie Française, 1878. (*Dictionnaire universel des contemporains*, *EB.*)

Rendall, Charles Edward (1800/1–72). Landowner. Resided at Brigmerston House, near Amesbury, Wiltshire. Associate of Edward Blyth. (Crowley ed. 1995, pp. 138–9, 141; *Field* 21 (1863): 464; *Post Office directory of Hampshire, Dorsetshire and Wiltshire* 1859; *Salisbury and Winchester Journal*, 11 May 1872.)

Rengger, Johann Rudolph (1795–1832). German physician, explorer, and naturalist. Travelled in South America (especially Paraguay), 1818–26. (*ADB*, *Neuer Nekrolog der Deutschen* 10 (1832): 699–700, Schumann 1888.)

Reuss, Jeremias David (1750–1837). German philologist and philosopher. Held various senior posts in the university libraries of Tübingen and Göttingen. Associate professor of philosophy, University of Göttingen, 1783; professor of philosophy and the history of scholarship, 1785–1837; honorary librarian, 1814–37. (*ADB*, Haug 1790, Kordes 1797.)

Ricardo, Osman (1795–1881). Politican. Liberal MP for Worcester, 1847–65; deputy lieutenant of Worcestershire, 1848. (*Alum. Cantab.*, *Dod's parliamentary companion*, *Modern English biography.*)

Riedel, Wilhelm. German naturalist and cleric. Minister and school inspector in Pfuhl, Bavaria. (Riedel 1824.)

Rivers, Thomas (1798–1877). Nurseryman. Succeeded to the family business in Sawbridgeworth, Hertfordshire, in 1827. Specialised in the cultivation of roses and fruit. Author of works on rose and fruit culture; contributed extensively to gardening journals. A founder of the British Pomological Society, 1854. (*DNB.*)

7 January [1863], 11 January [1863], 15 January [1863], 17 [January 1863], *21 January 1863*, 25 January [1863], *26 January 1863*, *30 January 1863*, 1 February [1863], *[3 February 1863]*, [14 February 1863], 5 March [1863], [9 May 1863], 17 August [1863]

Robarts, Lubbock & Co. London bank in which John William Lubbock, his son John Lubbock, and Abraham John Robarts were partners. (*DNB* s.v. Lubbock, Sir John William, and Lubbock, Sir John; *WWW* s.v. Robarts, Abraham John.)

Robinson, Edward W. (1835–77). London-based entomological artist, steel-engraver, and landscape painter. Exhibited, 1859–76. Worked in Britain, France, Switzerland, and northern Italy. Illustrated the *Entomologists Annual*, 1857–74, and many of the papers in the *Transactions of the Entomological Society of London*, the *Journal of the Linnean Society*, the *Journal of Entomology*, and the *Annals and Magazine of Natural History.* Illustrated some of the plates for Henry Walter Bates's *The naturalist on the River Amazons* (1863). (*Dictionary of watercolour artists*, *Entomologist's Monthly Magazine* 14 (1877): 118–19.)

Rogers, John (1807–67). Barrister. Resided at River Hill House, Sevenoaks Weald, Kent. Keen orchid grower; supplied CD with information on orchids. FRS 1839. (*Modern English biography*; *Orchids*, p. 236 n.; *Post Office directory of the six home counties* 1862.)

Roget, Peter Mark (1779–1869). Physician. First Fullerian Professor of physiology at the Royal Institution of Great Britain, 1833–6. Secretary of the Royal Society of London, 1827–48. Compiler of the *Thesaurus of English words and phrases* (1852). FRS 1815. (*DNB*, *Physicians*, *Record of the Royal Society of London.*)

Rolfe, Robert Monsey, 1st Baron Cranworth of Cranworth (1790–1868). Statesman and jurist. Whig MP for Penryn and Falmouth, 1832–9. Solicitor-general, 1834, 1835–9. Created Baron Cranworth of Cranworth, 1850. Lord justice of appeal, 1851–2. Lord chancellor, 1852–8 and 1865–6. Holwood Park, his country residence, was a mile and a half to the north of Down House. (*DNB*, *Dod's parliamentary companion*, Freeman 1978.)
[20 November 1863]

Rolle, Friedrich (1827–87). German geologist, palaeontologist, and natural history dealer. Assistant at the Kaiserlich-königliche Hofmineralien-Cabinett, Vienna, 1857–9; associate, 1859–62. Returned to Bad Homburg in 1862. Author of *Ch. Darwin's Lehre von der Entstehung der Arten und ihre Anwendung auf die Schöpfungsgeschichte* (1863). (*ADB*, *BLKO*, Martin and Uschmann 1969, Sarjeant 1980–96.)
26 January 1863

Rolleston, George (1829–81). Physician and anatomist. Physician to the British Civil Hospital at Smyrna (Izmir), Turkey, 1855–6, during the Crimean War. Physician to Radcliffe Infirmary, Oxford, and Lee's Reader in anatomy at Christ Church, Oxford, 1857. Linacre Professor of anatomy and physiology, Oxford University, 1860–81. Fellow of Merton College, Oxford, 1872. FRS 1862. (*DNB*.)

Rolph, George Frederick (*fl.* 1862–78). Civil servant and anthropologist. War Office clerk, 1862–78. Fellow of the Anthropological Society of London, 1864–9. (*British Imperial Calendar* 1861–79; Anthropological Society council minutes, Royal Anthropological Institute.)

Romer, Robert (1840–1918). Mathematician and judge. Professor of mathematics at Queen's College, Cork, 1865–6. Fellow of Trinity Hall, Cambridge, 1867. Called to the bar, 1867. Judge of the Chancery Division, High Court of Justice, 1890–9. Lord justice of appeal, 1899–1906. Knighted, 1890. Privy councillor, 1899. FRS 1899. (*Alum. Cantab.*, *DNB*.)

Rothrock, Joseph Trimble (1839–1922). American physician, botanist, and forester. In 1860, entered the Lawrence Scientific School, Harvard University, where he was a student of, and an assistant to, Asa Gray. Enlisted in the 131st Pennsylvania Infantry in 1862; commissioned captain in the 20th Pennsylvania Cavalry in 1863; honourably discharged, 1864. Professor of botany, Pennsylvania State Agricultural College, 1867. Botanist and surgeon to the government survey in Colorado, New Mexico, and California, 1873. Professor of botany, University of Pennsylvania, 1877–1904. (*DAB*.)

Rothschild, Jules (b. 1838). Publisher at 14 rue de Buci, Paris. (*NUC*; letter from H. L. F. Saussure, 17 January 1863.)

Roxburgh, William (1751–1815). Scottish botanist and surgeon. Medical officer with the Madras Medical Service, 1776–80. Superintendent, Samalkot Botanic Garden, 1781–93. Chief botanist to the East India Company and superintendent of the Calcutta Botanic Garden, 1793–1813. (*DNB*, R. Desmond 1994.)

Royer, Clémence Auguste (1830–1902). French author and economist. Studied natural science and philosophy in Switzerland. In Lausanne in 1859, founded a course on logic aimed at women. Translated *Origin* into French in 1862. (*Dictionnaire universel des contemporains*, J. Harvey 1997.)

Rucker, Sigismund (1809/10–76). East and West India broker with premises in Great Tower Street, City of London. Cultivated a major orchid collection at his residence at West Hill, Wandsworth, and frequently exhibited at Chiswick. Member of the council of the Royal Horticultural Society. (*Post Office London directory* 1861; *Post Office London suburban directory* 1860; *Gardeners' Chronicle*, 23 October 1875, pp. 532–3.)

Ruprecht, Franz Josef (1814–70). German-born botanist and physician. Practised medicine in Prague. Curator of the botanical collections of the Academy of Sciences at St Petersburg, 1839; assistant director, St Petersburg botanical gardens, 1851–5; director, botanical museum of the academy, 1855; professor of botany at the Paedagogium, St Petersburg, 1855–9. (*ADB*, *Taxonomic literature.*)

Russell, John, 1st Earl Russell (1792–1878). Statesman. Liberal prime minister, 1846–52 and 1865–6. Home secretary, 1835–9; secretary of state for foreign affairs, 1852–3; for the colonies, 1839 and 1855. Foreign secretary under Lord Palmerston, 1859–65. Created Earl Russell of Kingston Russell in 1861. In later life, occupied with literary work. President of the Royal Historical Society, 1872. FRS 1847. (*DNB*, *Record of the Royal Society of London.*)

Rütimeyer, Karl Ludwig (Ludwig) (1825–95). Swiss palaeozoologist and geographer. Professor of zoology and comparative anatomy, University of Basel, 1855; rector, 1865; professor in the medical and philosophical faculties, 1874–93. Made important contributions to the natural history and evolutionary palaeontology of ungulate mammals. (*DSB.*)

Ryan, Mr. Overseer of a private *Cinchona* plantation at Darjeeling, India, 1863. (*Report on Cinchona cultivation at Darjeeling.*)

Sabine, Edward (1788–1883). Astronomer, geophysicist, and army officer. Entered army in 1803; major-general, 1856; general, 1870. Astronomer to expeditions in search of a north-west passage, 1818 and 1819–20. Appointed one of three scientific advisers to the Admiralty, 1828. General secretary of the British Association for the Advancement of Science, 1839–52 and 1853–9. Foreign secretary of the Royal Society of London, 1845; treasurer, 1850; president, 1861–71. Knighted, 1869. FRS 1818. (*DNB*, *DSB.*)

Sageret, Augustin (1763–1851). French naturalist and agronomist. One of the founders of the Society of Horticulture of Paris and member of the Royal Society of Agriculture (Academy of Agriculture). Author of an agronomic survey of Lorris, where he settled in 1819/20 and began research in horticulture, particularly hybridisation. (Roberts 1929.)

Saint-Hilaire, Augustin François César Prouvençal (Auguste) de (1779–1853). French naturalist. Surveyed the flora and fauna of Brazil, 1816–22. Appointed professor at the Faculty of Sciences, Paris, 1830. (*DSB.*)

Saint-Simon, Maximilien Henri, marquis de (1720–99). French–Dutch soldier and horticulturalist. Cultivator of hyacinths. Engaged in the French–Italian wars

Saint-Simon, Maximilien Henri, marquis de, cont.
as aide-de-camp to the Prince de Conti; retired to his estate in the Netherlands, Nieuw Amelisweerd, near Utrecht, in 1758. (*Taxonomic literature.*)

Salimbeni, Leonardo (1830–89). Italian naturalist and political activist. Teacher of natural history and geography, Collegio San Carlo, Modena, 1861–7. Elected parliamentary deputy for the province of Modena, 1864. Translated the first Italian edition of *Origin* (1864) with Giovanni Canestrini. (Barbieri 1973; Commissione Araldica Modenese, busta 30, fascicolo 'Salimbeni', Archivio di Stato di Modena; 'Cronaca del Collegio San Carlo, redatta dal rettore dottor don Luigi Spallanzani dal 1° gennaio fino alla sua morte' and 'Registro II° degli atti dell'amministrazione del Collegio dal 1855 al 1881 inclusive', Archivio Storico del'Collegio San Carlo; Pancaldi 1991, pp. 79–81, 193–4.)

Salter, John (1798–1874). Nurseryman, florist, and cheesemonger. Established a nursery for the sale of English flowers at Versailles, 1838–48. Established a florist's shop specialising in chrysanthemums in Williams Street, Hammersmith, *circa* 1863. (R. Desmond 1994, *Gardener's year-book* 1864, *Post Office directory of the six home counties* 1866.)

Salvin, Osbert (1835–98). Ornithologist and entomologist. Undertook natural history explorations in North Africa, 1857, and Central America, 1857–60, 1861–3, 1873–4. Strickland Curator of ornithology, University of Cambridge, 1874–82. FRS 1873. (*DNB.*)
11 [May 1863], *12 May 1863*

Sanches, Francisco (Franciscus Sanchez) (1551–1623). French Renaissance philosopher, mathematician, and doctor of medicine. Physician, Hôtel Dieu, Toulouse, 1582–1612. Regius professor of philosophy, University of Toulouse, 1585–1612; Regius professor of medicine, 1612–23. (Sanches 1988.)

Sanders, Charles Henry Martyn (1862–90). Son of Eliza Ann Sanders; grandson of William Darwin Fox. (*Darwin pedigree*, *Gentleman's Magazine* (1862) pt 1: 638.)

Sanders, Eliza Ann (1836–74). Daughter of William Darwin Fox. Married Henry Martyn Sanders. Mother of Charles Henry Martyn Sanders. (*Darwin pedigree.*)

Sanders, George Nicholas (1812–73). American political activist. Leader of the Young America movement in the 1850s, advocating American support for republicanism in Europe; editor of *Democratic Review* from 1851. Appointed to a consulship in London, 1853. Confederate agent in Britain and Canada during the American Civil War. (*DAB.*)

Sanson, André (1826–1902). French veterinarian and animal breeder. Author of a number of books and articles on animal physiology, breeding, and veterinary medicine. Teacher at the Ecole Veterinaire de Toulouse, 1856; professor of zoology and zootechny at the Ecole d'Agriculture, Grignon, 1872. Chief editor and contributor to *Culture* from 1859 to 1870. (Glaeser 1878, *Larousse.*)

Saporta, Louis Charles Joseph Gaston (Gaston), comte de (1823–96). French palaeobotanist. Specialist on the Tertiary and Jurassic flora. Wrote extensively on the relationship between climatic change and palaeobotany. (*DSB.*)

Saunders, William Henry Patten (Patten-Saunders) (*fl.* 1861–5). Champion athlete and novelist; published the novel *Black and gold* (1864). (*CDEL*, *Modern English biography.*)

Saunders, William Wilson (1809–79). Insurance broker, entomologist, and botanist. Underwriter at Lloyd's of London. President of the Entomological Society, 1841–2 and 1856–7. Treasurer of the Linnean Society of London, 1861–73. FRS 1853. (R. Desmond 1994, *DNB.*)

Saussure, Henri Louis Frédéric de (1829–1905). Swiss zoologist. Following his doctoral studies at the University of Giessen, embarked on scientific travels (1854–6) in the West Indies, Mexico, and the United States; he returned to Geneva with significant collections. (*Larousse*, *HBLS*, *NUC.*)
17 January 1863

Schacht, Hermann (1814–64). German botanist. Practised pharmacy until 1847. Private lecturer, Berlin, 1853–60. Professor of botany and director of the botanical garden at Bonn, 1860–4. Specialised in plant anatomy and embryology. (*Taxonomic literature.*)

Schleicher, August (1821–68). German linguist and philologist. Professor of philology, University of Prague, 1853–6; University of Jena, 1857–68. (*ADB.*)

Schleiden, Matthais Jacob (1804–81). German botanist, author, and lecturer. Trained and practised as a lawyer but abandoned the profession in 1833 for medical and botanical studies at Göttingen and Berlin. Honorary professor of botany, University of Jena, 1846; professor, 1850–62. Professor of anthropology, University of Dorpat, 1863–4. Thereafter a private lecturer, residing in various German cities. (*ADB*, *DSB.*)

Schmerling, Philippe Charles (1791–1836). Belgian palaeontologist born in the Netherlands. Practised medicine in Liège. Wrote on Quaternary faunas and fossil humans. Taught zoology at the University of Liège. (*BNB*, Sarjeant 1980–96.)

Schneider, Carl Friedrich Adolph (b. 1821). German-born colonial medical officer. Medical officer on the voyage of the *Circe*, 1849–50. Health officer in the Royal Dutch East Indian Army, 1852–66. Practised medicine privately in Soerabaya, Java, 1866–77. (*Natuurkundig Tijdschrift voor Nederlandsch-Indië*; Register of the public servants of the department of colonies (Netherlands) 910: 103; Steenis ed. 1950, p. 473; *Geneeskundig Tijdschrift voor Nederlandsche-Indië* 20 (1880): 1–82; Index biographicus, Institute for the History of Science, University of Utrecht, Netherlands.)

Schomburgk, Robert Hermann (1804–65). German-born explorer, diplomat, and naturalist. Explored British Guiana under the direction of the Royal Geographical Society, 1831–5. Government commissioner for surveying and establishing the boundaries of British Guiana, 1841–3. British consul, San Domingo, 1848; Bangkok, 1857–64. Knighted, 1844. FRS 1859. (*DNB.*)

Schousboe, Peder Kofod Anker (1766–1832). Danish botanist and traveller. Collected in Spain and Morocco for the Copenhagen botanical garden. Danish consul at Tangier, 1801–21; consul-general, 1821–32. (*Taxonomic literature.*)

Schroeder van der Kolk, Jacob Lodewijk Coenraad (Jacob Ludwig Conrad) (1797–1862). Dutch physician, anatomist, physiologist, and zoologist. Resident physician at the Buitengasthuis, Amsterdam, 1821. Professor of anatomy and physiology, University of Utrecht, 1827–62. Governor of the Utrechtsehe Dolhuis, 1827; inspector of lunatic asylums, 1842–62. Involved in reforms in the care of the mentally ill. (*DSB*, *NNBW*.)

Schweizerbart, Christian Friedrich (1805–79). German publisher. Director of E. Schweizerbart'sche Verlagsbuchhandlung of Stuttgart, 1841–67. Publisher of the German translations of *Origin* (1860) and *Orchids* (1862). (*Jubiläums-Katalog*.)

Schweizerbart, Wilhelm Emanuel (1785–1870). German publisher. Founded E. Schweizerbart'sche Verlagsbuchhandlung in Stuttgart in 1830. Publisher of many German scientific works. Retired in 1841. (*Jubiläums-Katalog*.)

E. Schweizerbart'sche Verlagsbuchhandlung. German publishing company in Stuttgart; founded by Wilhelm Emanuel Schweizerbart in 1830; conducted by his nephew Christian Friedrich Schweizerbart from 1841. (*Jubiläums-Katalog*.)

Sclater, Philip Lutley (1829–1913). Lawyer and ornithologist. One of the founders of *Ibis*, 1858; editor, 1858–65 and 1878–1912. Secretary of the Zoological Society of London, 1860–1903. FRS 1861. (*DSB*, Scherren 1905.)

Scott, Alexander (b. 1802/3). Scottish gardener; head gardener, Leigh Park, Hampshire, 1837–61. (Gladwyn 1992, pp. 89, 102, 131.)

Scott, John (1836–80). Scottish botanist. Gardener at Chatsworth House, Derbyshire, before becoming foreman of the propagating department at the Royal Botanic Garden, Edinburgh, in 1859. Through CD's patronage became head of the herbarium department, Calcutta botanic garden, 1864. Fellow of the Linnean Society of London, 1873. Carried out numerous experiments on CD's behalf. (R. Desmond 1994, Freeman 1978, *Transactions of the Botanical Society* [*of Edinburgh*] 14 (1883): 160–1.)

6 January 1863, *16 January 1863*, 21 January [1863], 16 February [1863], *18 February [1863]*, 20 [February 1863], *3 March 1863*, 6 March 1863, *21 March [1863]*, 24 March [1863], *[1–11] April [1863]*, 12 April [1863], *[after 12] April [1863]*, 2 May [1863], *21 May [1863]*, *22 May 1863*, 23 May [1863], 25 and 28 May [1863], *26 May [1863]*, *28 May [1863]*, 31 May [1863], *[3 June 1863]*, 6 June [1863], 11 June [1863], *16 June [1863]*, 20 [June 1863], 2 July [1863], *23 July [1863]*, 25 [July 1863], *[26 July – 2 August 1863]*, 1 and 3 August [1863], *21 September [1863]*, *25 September [1863]* (Emma Darwin), 7 November [1863]

Scott, Walter, 1st baronet (1771–1832). Scottish poet and novelist. President of the Royal Society of Edinburgh, 1820. Created baronet, 1820. (*DNB*.)

Scott, William, 1st Baron Stowell of Stowell Park (1745–1836). Scholar, politician, and maritime judge. Camden Reader in ancient history, University of Oxford, 1773–85. George III's advocate-general, 1788; judge of the High Court of the Admiralty, 1798–1828. MP for Downton, 1784 and 1790; for Oxford

University, 1801–21. Knighted, 1788; created Baron Stowell of Stowell Park, 1821. (*DNB*.)

Scudder, Samuel Hubbard (1837–1911). American entomologist. Graduated from the Lawrence Scientific School in 1862; assistant to Louis Agassiz, 1862–4. Custodian, Boston Society of Natural History, 1864–70. Palaeontologist, United States Geological Survey, 1886–92. (*DAB*.)

Sedgwick, Adam (1785–1873). Geologist and clergyman. Woodwardian Professor of geology, Cambridge University, 1818–73. Prebendary of Norwich Cathedral, 1834–73. President, Geological Society of London, 1829–31; British Association for the Advancement of Science, 1833. FRS 1821. (*DNB*, *DSB*, *Record of the Royal Society of London*.)

Seemann, Berthold Carl (1825–71). German-born traveller and botanist. Studied botany at the Royal Botanic Gardens, Kew, 1844–6. Naturalist to HMS *Herald*, 1847–51. Commissioned in 1860 to report on the Fiji Islands and published a botanical catalogue of the islands. Editor of *Bonplandia*, 1853–62; of the *Journal of Botany, British and Foreign*, 1863–71. Travelled in Venezuela in 1864; in Nicaragua, 1866–7. (R. Desmond 1994, *DNB*.)

Sefström, Nils Gabriel (1787–1845). Swedish chemist, physician, and natural historian. (*BLOS*, *NUC*, *SMK*.)

Selby, Prideaux John (1788–1867). Naturalist. Author and illustrator of works on British ornithology and forestry. In 1837, helped to found the *Magazine of Zoology and Botany*, later known as the *Annals and Magazine of Natural History*, of which he was an editor. (R. Desmond 1994, *DNB*.)

Sella, Quintino (1827–84). Italian mineralogist, crystallographer, and statesman. Professor of geometry (later of mathematics), Instituto Tecnico di Torino (later University of Turin), 1852. An active member of the Accademia dei Lincei; a founding member of the Italian geological society and of the Alpine Club. Italian minister of finance, 1862, 1865, and 1869–73. (*Enciclopedia Italiana*, Sarjeant 1980–96.)

Seringe, Nicolas Charles (1776–1858). French botanist. Professor of botany and director of the botanic garden at Lyon, 1830–58. (*NUC*, *Taxonomic literature*.)

Serres, Antoine Etienne Reynaud Augustin (1786–1868). French comparative anatomist. Professor of comparative anatomy, Jardin des Plantes, Paris, 1839. President of the Académie des Sciences, 1841. (*DSB*.)

Seymour, Edward Adolphus Seymour, 12th duke of Somerset (1804–85). Statesman. Liberal MP for Okehampton, 1830; for Totnes, 1834–55. Lord of the Treasury, 1835; secretary to the Board of Control, 1839; under-secretary, Home Department, 1841; commissioner of works, 1851–2; first lord of the Admiralty, 1859–66. (*DNB*.)

Shakespeare, William (1564–1616). Poet and dramatist. (*DNB*.)

Sharpey, William (1802–80). Scottish physiologist. MD, Edinburgh University, 1823. Joint lecturer on systematic anatomy, Edinburgh University, 1832. Professor

Sharpey, William, cont.
of anatomy and physiology, University College, London, 1836–74. Examiner in anatomy, London University, from 1840. Secretary of the Royal Society of London, 1853–72. Member of the General Medical Council, 1861–76. FRS 1839. (*DNB*, *DSB*.)

Shaw, Henry Norton (d. 1868). Secretary of the Royal Geographical Society of London, 1849–63. Consul at St Croix, West Indies, 1866. (*Modern English biography*.)

Shilleto, Richard (1809–76). Clergyman, schoolmaster, and classics scholar. Assistant master, Harrow, 1843–5. Leading Cambridge coach; fellow of Peterhouse College, Cambridge, 1867; dean, librarian, assistant tutor, and praelector. (*Alum. Cantab.*, *DNB*.)

Sigerson, George (1838–1925). Irish physician and author. Studied medicine at Cork, Dublin, and Paris. Practised as a physician in Dublin, specialising in neurology. Professor of botany and of zoology at the Catholic University of Ireland medical school in Dublin and then at the National University of Ireland. Published on Irish history, culture, and politics as well as on medicine and botany. (R. Desmond 1994, *DIB*.)
8 July 1863

Silfverstråhle, Gustaf (1790–1841). Swedish lawyer, appeal court judge, and botanist. (Krok 1925.)

Silliman, Benjamin (1779–1864). American chemist, geologist, and mineralogist. Professor of chemistry and natural history, Yale University, 1802–53. Founder and first editor of the *American Journal of Science and Arts*, 1818. (*DAB*, *DSB*.)

Silliman, Benjamin Jr (1816–85). American chemist. Teaching assistant to his father Benjamin Silliman, 1837. An editor of the *American Journal of Science and Arts*, 1838–85. Professor of practical chemistry, Yale University, 1846; succeeded his father as professor of chemistry and natural history in 1853. Professor of chemistry in the medical department, University of Louisville, Kentucky, 1849–54. Consultant to petroleum and mining industries. (*DAB*, *DSB*.)

Slack, Henry James (1818–96). Author and journalist. Worked on provincial journals, 1846–52. Proprietor and editor of the *Atlas*, 1852; editor of the *Intellectual Observer*, 1862; continued as the *Student*, 1868–71. Wrote under the name Little John for the *Weekly Times*. Author of liberal tracts and popular science papers. Secretary and president, Royal Microscopical Society. (*DNB*.)

Smith, Albert Richard (1816–60). Surgeon, playwright, novelist, essayist, and travel writer. (*DNB*.)

Smith, Beck & Beck. Instrument-makers at 6 Coleman Street and Pear Tree Cottage, Holloway Road, London; one of the leading manufacturers of microscopes in London. The partners, James Smith (died 1870), Richard Beck (1827–66), and Joseph Beck (1829–91), were members of the Microscopical Society of London. (*Post Office London directory* 1861, Turner 1989.)

Smith, Edmund (1804–64). Surgeon and hydropathic doctor. In practice in Huddersfield, 1847–9; in Sheffield, 1849–58. Proprietor of Ilkley Wells hydropathic establishment, Ilkley, Yorkshire, 1858–64. (*Medical directory* 1848–65; Metcalfe 1906, p. 107.)

Smith, Elder & Company. Partnership of George Smith (1789–1846) and Alexander Elder (1790–1876). Publishers and East India agents. (*DNB*, *Modern English biography*.)

14 January 1863, *3 March 1863*, 5 March [1863], *9 March 1863*, 10 March [1863]

Smith, Frederick (1805–79). Entomologist in the zoology department of the British Museum from 1849. Specialised in the Hymenoptera. President of the Entomological Society of London, 1862–3. (*Entomologist* 12 (1879): 89–92, Gilbert 1977.)

11 February 1863

Smith, Goldwin (1823–1910). Historian and essayist. Fellow at University College, Oxford, 1846–67. Regius professor of modern history, Oxford University, 1858–66; professor of history, Cornell University, 1868. (*DNB*.)

Smith, Henry John Stephen (1826–83). Irish mathematician. Fellow of Balliol College, Oxford, 1849; lecturer in mathematics, 1850–73. Savilian Professor of geometry, University of Oxford, 1860. Fellow of Corpus Christi College, Oxford, 1873. Member of the royal commission on scientific instruction, 1870–5. FRS 1861. (*DNB*.)

Smith, James (1782–1867). Scottish antiquarian, numismatist, and geologist. Known as Smith of Jordanhill. Made major early contributions to Quaternary geology; also wrote on the geology of Madeira, Gibraltar, and Portugal. Partner in the firm of West India merchants Leitch & Smith. FRS 1830. (*DNB*, Sarjeant 1980–96.)

Smith, James Edward (1759–1828). Botanist. Purchased the library and collections of Carolus Linnaeus in 1784. Founded the Linnean Society of London in 1788 and served as president until his death. Knighted, 1814. FRS 1785. (*DNB*, *DSB*.)

Smith, John (b. 1797/8). Farmer of Down Court, Down, Kent, until *circa* 1851. (Census returns 1851 (Public Record Office HO107/482/6: 246), *Post Office directory of the six home counties* 1845–55.)

Smith, John (1798–1888). Scottish gardener and pteridologist. Gardener at the Royal Botanic Garden, Edinburgh, 1818; at the Royal Botanic Gardens, Kew, 1822. Curator of the Royal Botanic Gardens, Kew, 1842–64. (R. Desmond 1994, *Taxonomic literature*.)

Smith, John (b. 1828/9). Farmer of Down Court, Down, Kent, until *circa* 1871. Son of John Smith (b. 1797/8). (Census returns 1871 (Public Record Office RG10/875: 35), *Post Office directory of the six home counties* 1855–74.)

Smith, Josiah (b. 1831/2). Farmer of Down Court, Down, Kent, until *circa* 1862. Son of John Smith (b. 1797/8). (Census returns 1861 (Public Record Office RG9/462: 78), *Post Office directory of the six home counties* 1855–62.)

Smith, Sarah (b. 1806). Wife of John Smith (b. 1797/8). (Census returns 1861 (Public Record Office RG9/462: 78).)

Smith, William (*fl.* 1860s). Resident of York. (*Journal of Horticulture*, 30 December 1862, p. 779.)

Snow, George (1820/1–85). Coal-dealer, Down, Kent. Operated a weekly carrier service between Down and London. (Census returns 1861 (Public Record Office RG9/462: 72); gravestone inscription, Down churchyard; *Post Office directory of the six home counties* 1862.)

Sothern, Edward Askew (1826–81). Actor. Came to public notice in New York in 1858 as Lord Dundreary in Tom Taylor's *Our American cousin*, a part he substantially rewrote and developed for himself. (*DNB*.)

Sowerby, George Brettingham Jr (1812–84). Conchologist and illustrator. Assisted his father, George Brettingham Sowerby, in a business selling natural history specimens; succeeded to the business in 1854. Illustrated numerous works on shells. Fellow of the Linnean Society of London, 1844. (*DNB*.)
22 July 1863

Spearing, William (1782/3–1863). Agricultural labourer in Down, Kent. (Census returns 1861 (Public Record Office RG9/462: 71); letter to J. B. Innes, 1 September [1863].)

Spencer, Herbert (1820–1903). Philosopher. Civil engineer on the railways, 1837–41 and 1844–6. Became sub-editor of the *Pilot*, a newspaper devoted to the suffrage movement, in 1844. Sub-editor of the *Economist*, 1848–53. From 1852, author of papers on evolution and numerous works on philosophy and the social sciences. (*DNB*, *DSB*.)

Sprengel, Christian Konrad (1750–1816). German botanist. Rector of the Great Lutheran Town School, Spandau, where he taught languages and natural science, 1780–94. Moved to Berlin where he worked as a private tutor. Published his major work on the pollination of flowers in 1793. (*ADB*, *DSB*.)

Spruce, Richard (1817–93). Botanist and school-teacher. School-teacher at Haxby and at the Collegiate School, York. Collected plant specimens in the Pyrenees, 1845–6; and in South America, 1849–64. PhD, Berlin University, 1864. (R. Desmond 1994, *DNB*.)

Stanford, Edward. Stationer specialising in geological maps, with offices at 6, 7, and 16 Charing Cross and 5 Trinity Place, London. (*Post Office London directory* 1861.)

Stanley, Arthur Penrhyn (1815–81). Cleric, ecclesiastical historian, and travel writer. Fellow of University College, Oxford, 1838–50; secretary of the Oxford University commission, 1850–2. Canon of Canterbury, 1851. Professor of ecclesiastical history, Oxford University, 1856; canon of Christ Church, 1858. Dean of Westminster, 1864–81. (*DNB*.)

Startin, James (1806–72). Surgeon and skin specialist. Trained at St Bartholomew's, London, and practised at the General Hospital and Town Infirmary, Birmingham, before going to France to study at the Hôpital Saint-Louis and

Montpellier, specialising in skin diseases. Became the senior surgeon at the Hospital for Diseases of the Skin, Blackfriars; also established a private consultancy at 3 Savile Row, Piccadilly. Member, Royal College of Surgeons of England, 1828; fellow, 1852. (*Medical directory* 1863, Plarr 1930.)

Staunton, George Thomas, 2d baronet (1781–1859). Oriental scholar, colonial administrator, diplomat, and politician. MP for St Michael's, 1818–26, Heytesbury, 1830–1, South Hampshire, 1832–5, Portsmouth 1838–51. Founding member of the Royal Asiatic Society. Developed pleasure grounds at Leigh Park, Hampshire. Author of works on China. FRS 1803. (*DNB*, *Dod's parliamentary companion*, Gladwyn 1992.)

Steenstrup, Johannes Japetus Smith (Japetus) (1813–97). Danish zoologist. Professor of zoology and director of the Zoology Museum, University of Copenhagen, 1846–85. Foreign member, Royal Society of London, 1863. (*DBL*, *DSB*.)

Stephens, Thomas Sellwood (b. 1825). Clergyman. Curate at Wanstead, Essex, 1853–9; at Down, Kent, 1859–67. Rector of St Erme, Cornwall, 1867–1904. (*Alum. Oxon.*, *Crockford's clerical directory*.)

Stevens, Edward Cephas John (1837–1915). British-born land-agent and politician in New Zealand. Educated at the Royal Agricultural College, Cirencester. Emigrated to New Zealand in 1858. Land- and commission-agent, Canterbury, New Zealand. Member of the provincial executive, 1863–6; elected representative for Selwyn, 1866–71; for Christchurch, 1875–81. Interested in flower cultivation. Chairman, Canterbury Horticultural Society. Secretary, Philosophical Institute of Canterbury, 1863. (*DNZB*; letter from Philosophical Institute of Canterbury, New Zealand, 14 September 1863.)

Stevens, Samuel (1817–99). Natural-history agent, water-colourist, and entomologist. Founding member of the Entomological Society of London and member of its council for several years. Active member of the Zoological Society of London. Collector of Coleoptera and Lepidoptera. Worked with his brother as an auctioneer and natural-history dealer at 38 King Street, Covent Garden, London, 1840–8. Natural-history agent at 24 Bloomsbury Street, London, 1848–67. Agent for Henry Walter Bates and Alfred Russel Wallace, among others. (*Entomologist* 32 (1899): 264; *List of the Entomological Society of London*; *List of the Linnean Society of London*; *Post Office London directory*; Wallace 1905, 1: 266.)

Stevens, Thomas (1809–88). Clergyman. Assistant poor-law commissioner and perpetual curate, Keele, Staffordshire, 1839–42. Rector of Bradfield, 1843–82, and founder and warden of St Andrew's College, Bradfield, Berkshire, 1849–81. Rector of Newton with Toft, Lincolnshire, from 1883. (*Alum. Oxon.*, Blackie 1976.)

Stevenson, Henry (1833–88). Newspaper proprietor and naturalist. Secretary of the Norfolk and Norwich Museum, 1855–88. Vice-president of the Norfolk and Norwich Naturalists' Society, 1869–71; president, 1871–2. Proprietor of the *Norfolk Chronicle*. (*Modern English biography*.)

Stokes, George Gabriel, 1st baronet (1819–1903). Mathematician and physicist. Lucasian Professor of mathematics, Cambridge University, 1849–1903. Secretary

Stokes, George Gabriel, 1st baronet, cont.
of the Royal Society of London, 1854–85; president, 1885–90. President of the British Association for the Advancement of Science, 1869. Conservative MP for Cambridge University, 1887–91. Master of Pembroke College, Cambridge, 1902–3. Created baronet, 1889. FRS 1851. (*DNB*, *DSB*.)

Stovell, Mary Anne (b. 1841/2). Daughter of the army officer Walter Stovell; married Gustav Mann, 12 November 1863. (Register of marriages for Chichester (General Register Office) 1863, no. 60.)

Stowe, Harriet Elizabeth Beecher (1811–96). American author and novelist. Her first and most famous novel, *Uncle Tom's cabin*, was serialised in the anti-slavery paper *National Era* (1851–2), before being published in Boston by John P. Jewett. (*DAB*.)

Strickland, Sefton West (1839–1910). Conveyancer and equity draftsman. Friend of William Erasmus Darwin. Admitted to Christ's College, Cambridge, 1858. BA, 1862. Admitted to Lincoln's Inn, 1863; called to the bar, 1866. (*Alum. Cantab.*)

Strong, Edward. Printer and stationer with premises on High Street, Bromley, Kent. (*Post Office directory of the six home counties* 1862.)

Strzelecki, Paul Edmund de (1797–1873). Polish-born explorer and geologist. Explored the Americas, the Far East, and Australia, 1834–43. Naturalised as a British subject, 1845. Knighted, 1869. FRS 1853. (*ADB*, *DNB*.)

Suchsland, Friedrich Emil (1808–1903). German publisher and bookseller. Head of the publishing firm Johann Christian Hermann'sche Verlagsbuchhandlung, based in Frankfurt-am-Main. Published Friedrich Rolle's *Ch. Darwin's Lehre von der Entstehung der Arten* (1863) and sought to publish a German translation of *Origin* in 1866. (Letter from Rudolf Suchsland, 16 March 1866 (*Calendar* no. 5035), Martin and Uschmann 1969.)
[after 19 January 1863]

Suchsland, Georg Rudolf Emil (Rudolf) (1839–1921). Son of the German publisher Friedrich Emil Suchsland. Emigrated to Britain and became a mining director in Swansea, 1854. (Family information.)

Suess, Eduard (1831–1914). Austrian geologist, palaeontologist, politician, and educational reformer. Palaeontologist at the Kaiserlich-königliche Hofmineralien-Cabinett, Vienna, 1852. Professor extraordinarius of palaeontology, University of Vienna, 1857–62; professor extraordinarius of geology, 1862–7. Professor of geology, University of Vienna, 1867–1911. President of the Kaiserliche Akademie der Wissenschaften, Vienna, 1898–1911. Foreign member, Royal Society, 1894. (*Almanach der kaiserlichen Akademie der Wissenschaften* 63 (1914); Eisenberg 1893; *Mitteilungen der Geologischen Gesellschaft Wien* (1914): 1–32; *Neue Österreichische Biographie* I.)

Sugden, Edward (*fl.* 1860s). Nurseryman. Established Barr & Sugden, Covent Garden, in partnership with Peter Barr. (*Post Office London directory* 1863.)

Sulivan, Bartholomew James (1810–90). Naval officer and hydrographer. Lieutenant, HMS *Beagle*, 1831–6. Surveyed the Falkland Islands in HMS *Arrow*, 1838–9. Commander of HMS *Philomel*, 1842–6. Resided in the Falkland Islands,

1848–51. Commanded HMS *Lightning* in the Baltic, 1854–5. Naval officer in the marine department of the Board of Trade, 1856–65. Admiral, 1877. Knighted, 1869. (*DNB*.)

4 February [1863], 10 February [1863]

Sumner, John Bird (1780–1862). Archbishop of Canterbury, 1848–62. Held the second prebendal stall in Durham cathedral, 1827–48. Bishop of Chester, 1828–48. Author of works on scriptural geology. (*DNB*.)

Surtees, William Edward (1811–89). Barrister. Resided at Seaton Carew, West Hartlepool, Durham, and Tainfield House, Taunton, Somerset. High sheriff of Durham, 1866. (*Burke's landed gentry* 1872, *Men-at-the-bar*.)

Swan, Mrs. Of Blessington Road, Lee, Kent. Sister of Eliza Mary Brodie Innes. (*Correspondence* vol. 12, letter from J. B. Innes to Emma Darwin, 23 January [1864]; letter from J. B. Innes, 5 June 1871 (*Calendar* no. 7801); *Post Office directory of the six home counties* 1866.)

Swift, Jonathan (1667–1745). Clergyman, essayist, and satirist. Dean of St Patrick's Cathedral, Dublin, from 1713. (*DNB*.)

Swinhoe, Robert (1836–77). Diplomat and ornithologist. Attached to the British consulate in Hong Kong, 1854; in Amoy, China, 1855. British vice-consul, Taiwan, 1860–5; consul, Taiwan, 1865–73. Acting consul, Amoy, 1865–71; Ning-po, 1871–3. Consul, Ning-po, 1873–5. Collected plants and animals in Eastern Asia. FRS 1876. (*Foreign Office list* 1877, P. B. Hall 1987.)

14 April 1863, 29 July 1863, 14 August 1863

Swinhoe, Samuel (1786/7–1866). Army officer, Bengal Army. Attained the rank of lieutenant-general. Retired 1855. (*Modern English biography*, *Army list*.)

Sylvester, James Joseph (1814–97). Mathematician. Professor of natural philosophy, University College, London, 1837–41. Professor of mathematics, University of Virginia, 1841–5; Royal Military Academy, Woolwich, 1855–70; Johns Hopkins University, 1877–83. Savilian Professor of geometry, University of Oxford, 1883–94. FRS 1839. (*DNB*.)

Symonds, William Samuel (1818–87). Geologist, author, and clergyman. Curate of Offenham, Worcestershire, 1843–5. Rector of Pendock, Worcestershire, from 1845. Wrote papers on local archaeology and geology. Visited the Auvergne and the Ardèche in the 1870s to search for traces of ancient glaciers. (*DNB*, Sarjeant 1980–96.)

Tait, William Chester (1844–1928). British-born merchant, landowner, horticulturalist, and sportsman; resident in Portugal. Head of Tait & Co., merchants and agents for the Royal Marine Steam Packet Co. (*Gardeners' Chronicle*, 5 May 1928, p. 327.)

Tausch, Ignaz Friedrich (*c.* 1785–1848). Austrian botanist. Professor of botany and director of the garden of the duke of Canal de Malabaillas in Prague. Published a flora of Bohemia in 1831. (*BLKO*.)

John Edward Taylor. Printers at 10 Little Queen Street, Lincoln's Inn Fields, London. (*Post Office London directory* 1863.)

Taylor, Tom (1817–80). Playwright, classics scholar, lawyer, and public servant. Fellow of Trinity College, Cambridge, 1842–4; professor of English, London University, 1845. Called to the bar, 1846. Assistant secretary to the Board of Health; later, assistant secretary to the sanitary department of the Local Government Board, 1850–54; secretary, 1854–71. Editor of *Punch*, 1874–80. In thirty-five years, more than seventy of his plays appeared in principal London theatres. (*DNB*.)

Tegetmeier, William Bernhard (1816–1912). Editor, journalist, lecturer, and naturalist. Pigeon-fancier and expert on poultry. Secretary of the Apiarian Society of London. (Richardson 1916.)

26 [January 1863], *[after 26 January 1863]*, *18 February 1863*, 19 February [1863], *29 June – 7 July 1863*, 9 July [1863], *[c. 26 September 1863]*

Temminck, Coenraad Jacob (1778–1858). Dutch ornithologist. Director, Imperial Museum of Natural History, Leyden, 1820. Produced systematic studies of birds. (*NNBW*.)

Temple, Frederick (1821–1902). Educator and clergyman. Fellow of Balliol College, Oxford, 1842–8. Headmaster of Rugby School, 1857–69. Bishop of Exeter, 1869–85; of London, 1885–96. Archbishop of Canterbury, 1896–1902. Author of 'The education of the world', published in *Essays and reviews* (1860). (*DNB*.)

Temple, Henry John, 3d Viscount Palmerston (1784–1865). Statesman. Succeeded to the Irish peerage in 1802. Secretary, War Office, 1809–28. Foreign secretary, 1830–41 and 1846–51. Home secretary, 1852–5. Prime minister, 1855–8 and 1859–65. FRS 1853. (*DNB*.)

Tennyson, Alfred, 1st Baron Tennyson (1809–92). Poet. Poet laureate, 1850. Created Baron Tennyson, 1884. (*DNB*.)

Thimm, Franz. Bookseller with premises at 3 Brook Street, Grosvenor Square, London. (*Post Office London directory* 1863.)

Thom, J. P. (b. 1838/9). Journalist. Patient at Edward Wickstead Lane's hydropathic establishment in the 1850s. Sub-editor of the colonial newspaper *Home News*. Emigrated to Queensland, Australia, in 1863. (*Correspondence* vol. 7, letter to W. D. Fox, 24 June [1858]; vol. 10, letter from Mary Butler, [before 25 December 1862]; this volume, letter from J. P. Thom, 14 January 1863.)

14 January 1863

Thomas (d. *c.* AD 72). First-century apostle. Chiefly remembered for his refusal to believe in the Resurrection until he touched the wounds of the crucifixion in the risen Christ's body. (Farmer 1992.)

Thomson, Thomas (1817–78). Naturalist. MD, Glasgow, 1839. Travelled to India as assistant surgeon to the East India Company. Curator of the Asiatic Society's museum, Calcutta, 1840. Accompanied Joseph Dalton Hooker to the Himalayas, 1849–51, and collaborated with him on various botanical publications. Superintendent of the Calcutta botanic garden and professor of botany at the Calcutta Medical College from 1854 until his return to England in 1860 or 1861. FRS 1855. (R. Desmond 1994, *DNB*.)

Thunberg, Carl Peter (1743–1828). Swedish botanist. A protégé of Carolus Linnaeus. Accompanied a Dutch merchant vessel to Japan, spending time in Cape Colony, India, and Ceylon en route, 1772–9. Botanical demonstrator, Uppsala University, 1779; professor of botany, 1784. (*DSB*, *SBL*.)

Thwaites, George Henry Kendrick (1811–82). Botanist and entomologist. Superintendent of the Peradeniya botanic gardens, Ceylon, 1849; director, 1857–80. FRS 1865. (R. Desmond 1994, *DNB*.)

17 February 1863, 30 March [1863], *8 June 1863*, 29 July [1863], *24 September 1863*

Timbs, John (1801–75). Author and editor. Edited the *Mirror of literature*, 1827–38; *Literary world*, 1839–40; and the *Year book of science and art*, which he originated in 1839. Sub-editor of the *Illustrated London News*, 1842–58. Prolific author of popular literature. (*DNB*.)

Tocqueville, Alexis Charles Henri Clerel (Alexis) de (1805–59). French statesman and political writer. Studied law in Paris, 1823–6. Visited the United States, 1831–2, and England in 1833. Wrote his major work, *De la démocratie en Amérique*, between 1832 and 1834. Wrote numerous works on democracy, prison reform, French colonial possessions, and the abolition of slavery. (*NBU*.)

Tomes, Robert Fisher (b. 1823). Farmer and zoologist. Specialist on bats. Farmed in Welford, Gloucestershire. (*NUC*; *Post Office directory of Gloucestershire, with Bath, Bristol, Herefordshire, and Shropshire* 1863; *Royal Society catalogue of scientific papers*.)

Torell, Otto Martin (1828–1900). Swedish zoologist and geologist. Wrote works on the Scandinavian drift deposits and the glacial geomorphology of the Arctic. Professor of geology at the University of Lund, 1862, 1864–6; professor of zoology from 1866. Chief of the geological survey of Sweden, 1871–97. (Sarjeant 1980–96, *SBH*.)

Torrey, John (1796–1873). American botanist and chemist. Collected and described plants from various government expeditions. Professor of chemistry at the College of Physicians and Surgeons, New York City, 1827–54. Assayer of the United States Mint in New York, 1853–73. (*DAB*, *DSB*.)

Travers, William Thomas Locke (1819–1903). Irish-born lawyer, politician, and naturalist. Emigrated to New Zealand in 1849. Established a legal practice in Nelson, 1849, and in Christchurch, 1860. Founding member, New Zealand Institute. Collected plants for the Royal Botanic Gardens, Kew, 1860–80, and wrote papers on agriculture, botany, geology, and anthropology. (R. Desmond 1994, *DNZB*, *Transactions and Proceedings of the New Zealand Institute* 35 (1902): xviii-xix.)

Treub, Melchior (1851–1910). Dutch botanist. Researcher and assistant at the botanical institute, University of Leiden, 1873–80. Director of the botanical gardens at Buitenzorg (now Bogor), West Java, 1880. (*DSB*.)

Treviranus, Ludolph Christian (1779–1864). German botanist. Professor of botany, University of Breslau, 1816–30; University of Bonn, 1830–64. (*ADB*, *DSB*.)

Trimen, Henry (1843–96). Botanist. Assistant in the botanical department of the British Museum, 1869–79. Appointed to succeed George Henry Kendrick

Trimen, Henry, cont.
Thwaites as director of the Peradeniya botanic gardens, Ceylon, in 1879. Editor of the *Journal of Botany*, 1871–96. Brother of Roland Trimen. FRS 1888. (*DNB*.)

Trimen, Roland (1840–1916). Zoologist and entomologist. Emigrated to South Africa, 1858; arranged the Lepidoptera at the South African Museum. Held civil-service positions in the Commission of Land and Public Works, the governor's office, and the colonial secretary's office, before becoming part-time curator of the South African Museum in 1873; full-time curator, 1876. FRS 1883. (*DSAB*.)
31 January [1863], 16 February [1863], *16 March 1863*, 23 May [1863], *16, 17, and 19 July 1863*, 27 August [1863], *10, 13, and 18 October 1863*, 25 November [1863]

Trimmer, Joshua (1795–1857). Geologist. Employed on the Geological Survey of England, 1846–54. Wrote papers on the relationship between geology and agriculture. (*DNB*, Flett 1937, Sarjeant 1980–96.)

Tristram, Henry Baker (1822–1906). Clergyman and ornithologist. Secretary to the governor of Bermuda, 1847–9. Rector of Castle Eden, Durham, 1849–60. Canon residentiary of Durham, 1874. Formed extensive ornithological collections from travels in Algeria, Palestine, Egypt, and Japan. FRS 1868. (*DNB*.)

Trübner, Johann Nicolaus (Nicholas) (1817–84). German-born publisher and oriental scholar. Foreign corresponding clerk for Longmans (publishers), 1843–51. Established a successful London publishing house in partnership with Thomas Delf and David Nutt that specialised in oriental studies and translations of works in philology, philosophy, and religion. (*ADB*, *DNB*.)

Trübner & Co. Booksellers at Paternoster Row, London, specialising in oriental, American, and European publications (*Post Office London directory*). Founded by Nicholas Trübner (*DNB*).

Tulloch, Alexander Murray (1803–64). Army officer. Joined the forty-fifth regiment in Burma in 1826; colonel, 1854; major-general, 1859. Campaigned for army reform. Invited to investigate the commissariat system in the Crimean campaign, 1854. Knighted, 1857. (*DNB*.)

Turnbull, George Henry (b. 1819/20). Building contractor. Resided at The Rookery, Down. (Census returns 1861 (Public Record Office RG9/462: 70); *Post Office directory of the six home counties* 1862.)
[16? February 1863]

Turner, Charles (1818–85). Nurseryman. Acquired the Royal Nurseries, Slough, Buckinghamshire, *circa* 1845. Edited the *Florist*, 1851–60. (R. Desmond 1994.)
[1 April – 16 June 1863?]

Turner, Dawson (1775–1858). Botanist and antiquary. Father-in-law of William Jackson Hooker. Studied cryptogamic plants and algae. After 1820, directed his attention to the study of antiquities and formed an extensive collection of manuscripts, autographs, and drawings. FRS 1802. (*DNB*.)

Turner, Mary (1774–1850). Second daughter of William Palgrave and Elizabeth Thirkettle, of Coltishall, Norfolk; married Dawson Turner in 1796. Joseph Dalton

Hooker's maternal grandmother. (*DNB* s.v. Turner, Dawson; L. Huxley ed. 1918, 1: 18.)

Tyndall, John (1820–93). Irish physicist, lecturer, and populariser of science. Professor of natural philosophy at the Royal Institution of Great Britain, 1853; superintendent, 1867–87. Scientific adviser to Trinity House and the Board of Trade, 1866–83. FRS 1852. (*DNB*, *DSB*.)

Van Houtte, Louis-Benoit (1810–76). Belgian horticulturalist and politician. Assembled extensive plant collections on travels in South and Central America. Associated with the Jardin Botanique des Bruxelles, 1836–8. In 1839, moved to Gentbrugge, where he founded the Ecole d'Horticulture and the horticultural journal *Flora des serres et des Jardins de l'Europe*. (*BNB*.)

Van Voorst, John (1804–98). Publisher. Employed by Longmans (publishers), 1826–33. Had premises in Paternoster Row, London, 1833–86. Published many works on natural history. (*Modern English biography*.)

Vaucher, Jean Pierre Étienne (1763–1841). Swiss botanist and clergyman. Parish priest, Geneva, 1797–1822. Honorary professor of botany, Geneva University, 1798–1807; of ecclesiastical history, 1807–39; rector, 1818–21. Founding member of the Société de Physique et d'Histoire Naturelle de Genève. Wrote extensively on plant physiology. (*DSB*.)

Veitch, James (1792–1863). Nurseryman. Proprietor of a nursery at Mount Radford, near Exeter; in partnership with his son, James Veitch (1815–69), proprietor of a nursery in Chelsea, London, from 1853. (R. Desmond 1994.)

Veitch, James Jr (1815–69). Nurseryman. In partnership with his father, James Veitch (1792–1863), proprieter of a nursery in Chelsea from 1853. (R. Desmond 1994.)

James Veitch & Son. Nurserymen with premises on the King's Road, Chelsea, London. (*Gardener's year-book* 1864, *Post Office London directory* 1863.)

Verlot, Bernard (1836–97). French botanist and horticulturalist. Head of propagation, Muséum d'Histoire Naturelle. (*NUC*, Tort 1996, Verlot 1864.)

Victoria (1819–1901). Queen of the United Kingdom of Great Britain and Ireland, and empress of India. Succeeded to the throne, 1837; designated empress of India, 1876. (*DNB*.)

Vrolik, Willem (1801–63). Dutch anatomist and physiologist. Chair of the medical faculty in Groningen, 1828. Professor of anatomy and surgery, University of Amsterdam, 1831. Secretary, Royal Academy of Sciences. (*NNBW*.)

Wagner, Johann Andreas (Andreas) (1797–1861). German palaeontologist, zoologist, and archaeologist. Lecturer, University of Erlangen, Bavaria, *circa* 1826. Professor of zoology and assistant to the zoological museum, University of Munich, 1832. Wrote in opposition to the materialist views of Karl Hermann Konrad Burmeister and Karl Vogt. (*ADB*, Martius 1862.)

Wagner, Rudolph (1805–64). German anatomist, physiologist, and anthropologist. Professor of physiology, comparative anatomy, and natural history at Göttingen University, 1840. (*ADB*, *DSB*.)

Waldstein, Franz de Paula Adam (1759–1823). Austrian soldier and botanist. Succeeded to the family estate in Bohemia in 1814. (*Taxonomic literature.*)

Walker, Bryan (1840–87). Legal scholar and clergyman. Scholar and tutor, Trinity Hall, Cambridge, 1860–7. Fellow and mathematical tutor, Corpus Christi College, Cambridge, 1866–71. Cambridge University examiner in the law tripos, 1866 and 1877–9. Ordained deacon, 1868; priest, 1870. Rector of Landbeach, Cambridgeshire, 1871–87. (*Alum. Cantab.*, *Modern English biography.*)

Walker, Francis (1809–74). Entomologist. Specialised in chalcids, small parasitic Hymenoptera. Described the Chalcididae CD collected on the *Beagle* voyage. Catalogued a number of the insect collections in the British Museum. (*Entomologist's Monthly Magazine* 11 (1874): 140–1, Gilbert 1977.)
26 February 1863

Wallace, Alfred Russel (1823–1913). Naturalist. Collector in the Amazon, 1848–52; in the Malay Archipelago, 1854–62. Independently formulated a theory of evolution by natural selection in 1858. Lecturer and author of works on protective coloration, mimicry, and zoogeography. President of the Land Nationalisation Society, 1881. Wrote widely on socialism, spiritualism, and vaccination. FRS 1893. (*DNB*, *DSB.*)
14 January [1863], *26 September 1863*, [29? September 1863]

Walsh, John Henry (1810–88). Sports writer under the name Stonehenge. Practised medicine in London and Worcester before becoming editor of the *Coursing calendar* in 1856. Editor of *The Field*, 1857–88. Published many works on dogs, horses, guns, domestic economy, medicine, and cookery. Fellow of the Royal College of Surgeons, 1844. (*DNB.*)

Warner, Thomas. Solicited contributions to a testimonial for James Buckman (see letter from H. E. Darwin to Thomas Warner, 14 October [1863]). Possibly Thomas Warner of Gosditch Street, Cirencester, Gloucestershire, justice of the peace for the county of Gloucestershire. A Thomas Warner was one of the original shareholders in the Royal Agricultural College, Cirencester. (*Post Office directory of Gloucestershire, with Bath, Bristol, Herefordshire, and Shropshire* 1870; Sayce 1992, p. 351.)

Waterhouse, George Robert (1810–88). Naturalist. A founder of the Entomological Society of London, 1833. Curator of the Zoological Society of London, 1836–43. Assistant in the mineralogical and geological branch of the natural history department of the British Museum, 1843–50; keeper, 1851–6; keeper, geology department, 1857–80. Described CD's mammalian and entomological specimens from the *Beagle* voyage. (*DNB*, Gilbert 1977.)

Watson, Hewett Cottrell (1804–81). Botanist, phytogeographer, and phrenologist. Edited the *Phrenological Journal*, 1837–40. Collected plants in the Azores, 1842. Wrote extensively on the geographical distribution of plants. (*DNB*, *DSB.*)

Watts, Thomas (1811–69). Librarian. Library assistant, British Museum, 1838; assistant keeper of printed books, 1856; superintendent of the new reading room, 1857; keeper of printed books, 1866–9. (*Modern English biography.*)

Waugh, George. Chemist. Proprietor of George Waugh & Co., 177 Regent Street, London. (*Post Office London directory* 1863.)

George Waugh & Co. Chemists at 177 Regent Street, London. (*Post Office London directory* 1863.)

Weddell, Hugh Algernon (1819–77). British-born botanist and physician. Assistant naturalist to the Muséum d'Histoire Naturelle, Paris, 1850–7. Specialist on South American flora. (R. Desmond 1994.)
13 May 1863

Wedgwood, Alfred Allen (1842–92). Son of Hensleigh and Frances Emma Elizabeth Wedgwood. (Freeman 1978.)

Wedgwood, Caroline Sarah (1800–88). CD's sister. Married Josiah Wedgwood III, her cousin, in 1837. (*Darwin pedigree.*)

Wedgwood, Charlotte. *See* Langton, Charlotte.

Wedgwood, Frances Emma Elizabeth (1800–89). Second child of James Mackintosh and Catherine Allen. Married Hensleigh Wedgwood in 1832. (Freeman 1978, O'Leary 1989.)

Wedgwood, Francis (Frank) (1800–88). Master-potter and partner in the Wedgwood pottery works at Etruria, Staffordshire, until 1876. Emma Darwin's brother. Married Frances Mosley in 1832. (*Alum. Cantab.*)

Wedgwood, Hensleigh (1803–91). Philologist and barrister. Emma Darwin's brother. Fellow, Christ's College, Cambridge, 1829–30. Metropolitan police magistrate at Lambeth, 1832–7. Registrar of metropolitan carriages, 1838–49. An original member of the Philological Society, 1842. Published the first volume of his *Dictionary of English etymology* in 1859. Married Frances Mackintosh in 1832. (*DNB.*)

Wedgwood, Hope Elizabeth (Dot) (1844–1934). Daughter of Hensleigh and Frances Emma Elizabeth Wedgwood. Second wife of Godfrey Wedgwood. (Freeman 1978.)

Wedgwood, James Mackintosh (1834–64). Son of Hensleigh and Frances Wedgwood. (Wedgwood and Wedgwood 1980.)

Wedgwood, Josiah, I (1730–95). Master-potter. Founded the Wedgwood pottery works at Etruria, Staffordshire. Grandfather of CD and Emma Darwin. Greatly interested in experimental chemistry. Contributed several papers on the measurement of high temperatures to the Royal Society of London's *Philosophical Transactions*. Actively associated with scientists and scientific societies. FRS 1783. (*DNB*, *DSB.*)

Wedgwood, Josiah, II (1769–1843). Master-potter of Etruria. Resident at Maer Hall, Staffordshire. Whig MP for Stoke-on-Trent, 1832–4. Emma Darwin's father. (*Burke's peerage* 1980, *Emma Darwin* (1915).)

Wedgwood, Josiah, III (1795–1880). Master-potter. Partner in the Wedgwood pottery works at Etruria, Staffordshire until 1844, when he moved to Leith Hill Place, Surrey. Emma Darwin's brother. Married CD's sister Caroline, his cousin, in 1837. (Freeman 1978.)

Wedgwood, Katherine Elizabeth Sophy (Sophy) (1842–1911). Daughter of Caroline and Josiah Wedgwood III. CD's niece. (*Darwin pedigree*, Freeman 1978.)

Wedgwood, Lawrence (1844–1913). Son of Emma Darwin's brother, Francis (Frank) Wedgwood, and Frances Mosley. (Wedgwood and Wedgwood 1980.)

Wedgwood, Lucy Caroline (1846–1919). Daughter of Caroline Wedgwood and Josiah Wedgwood III. CD's niece. (*Darwin pedigree*, Wedgwood and Wedgwood 1980.)

Wedgwood, Margaret Susan (1843–1937). Daughter of Caroline Wedgwood and Josiah Wedgwood III. CD's niece. Married Arthur Charles Vaughan Williams (Freeman 1978) in 1869. Mother of Ralph Vaughan Williams (*DNB*). (*Emma Darwin* (1915).

Wedgwood, Sarah Elizabeth (Elizabeth) (1793–1880). Emma Darwin's sister. Resided at Maer Hall, Staffordshire, until 1847, then at The Ridge, Hartfield, Sussex, until 1862. Moved to London before settling in Down in 1868. (*Emma Darwin* (1915), Freeman 1978.)

John Weeks & Co. Ventilation firm located at Sussex Terrace, Chelsea, London. (*Post Office London directory* 1863.)

Weinland, David Friedrich (1829–1915). German theologian and zoologist. Travelled extensively in North America and Haiti. Assistant to Louis Agassiz, 1855–8. Lecturer in zoology at the Senckenberg Museum, Frankfurt, and secretary of the Zoological Society, 1858. Founder and editor of *Zoologische Garten*, 1859. Director of the Senckenberg naturalist society, 1859–63. (*DBA*, *Zoologischer Beobachter* 57 (1916): 160.)

Welwitsch, Friedrich Martin Josef (Friedrich) (1806–72). Austrian-born botanist. Resided in Lisbon, 1839–53, where he became director of the botanic garden at Coimbra. Collected and travelled in Portuguese West Africa (Angola), 1853–61. Resided in London, 1863–72. (*DNB*.)

Westwood, John Obadiah (1805–93). Entomologist and palaeographer. Founding member of the Entomological Society of London. Hope Professor of invertebrate zoology, Oxford University, 1861–93. Entomological referee for the *Gardeners' Chronicle*. Awarded the Royal Medal of the Royal Society of London, 1855. (*DNB*, Gilbert 1977.)

Wheatstone, Charles (1802–75). Experimental physicist and inventor. Professor of experimental physics, King's College, London, 1834. Noted for contributions in optics, electricity, telegraphy, and automatic recording instruments. Knighted, 1868. FRS 1836. (*DNB*.)

Whewell, William (1794–1866). Mathematician, historian, and philosopher of science. Professor of moral philosophy, Cambridge University, 1838–55. Master of Trinity College, Cambridge, 1841–66. FRS 1820. (*DNB*, *DSB*.)

Whitcombe, John Henry (d. 1863). Civil engineer with the Great Western railway until 1854, when he went to India. Moved to New Zealand to become the provincial surveyor for Canterbury in 1857. (*DNZB*.)

White, Charles (1728–1813). Surgeon. One of the founders of the Manchester Infirmary, 1752, and the Manchester Lying-in Hospital, 1790. FRS 1762. (*DNB*.)

Whitley, Charles Thomas (1808–95). Attended Shrewsbury School, 1821–6. BA, St John's College, Cambridge, 1830. Reader in natural philosophy and mathematics, Durham University, 1833–55. Vicar of Bedlington, Northumberland, 1854–95. (*Alum. Cantab.*, *Modern English biography*.)
20 June [1863]

Wickham, John Clements (1798–1864). Naval officer and magistrate. First lieutenant on HMS *Beagle*, 1831–6; commander, 1837–41, surveying the Australian coast. Settled in Australia in 1842. Police magistrate in New South Wales, 1843–57; government resident, 1857. Left Australia and took his family to southern France in 1862. (*Aust. dict. biog.*, R. Desmond 1994.)

Wiegmann, Arend Friedrich (1771–1853). German pharmacist and botanist. Lecturer in natural science at a college in Brunswick from 1821; professor, 1832. His research on plant hybridisation was recognised by the Prussian Academy of the Sciences in 1826. (*BHGW*; Roberts 1929, pp. 160–4; *Taxonomic literature*.)

Wight, Robert (1796–1872). Botanist. Assistant surgeon to the East India Company in Madras, 1819–26; naturalist, 1826–8; garrison surgeon, 1828–31; surgeon, 1834–6; superintendent of cotton cultivation, 1836–53. Published numerous works on the botany of India. FRS 1855. (R. Desmond 1994, *DNB*.)

Wilberforce, Samuel (1805–73). Clergyman. Rector of Brightstone, Isle of Wight, 1830–40; of Alverstoke, Hampshire, 1840–3. Chaplain to Prince Albert, 1841. Dean of Westminster, 1845. Bishop of Oxford, 1845–69. FRS 1845. (*DNB*, Sarjeant 1980–96.)

Williams, Edmund Sydney (1817–91). Formed the publishing house Williams & Norgate of Covent Garden, London, and South Frederick Street, Edinburgh, in partnership with Frederick Norgate. Specialised in scientific works. (*Modern English biography*.)

Williams & Norgate. Booksellers and publishers specialising in foreign scientific literature, with premises at Covent Garden, London, and South Frederick Street, Edinburgh. The business was a partnership between Edmund Sydney Williams (1817–91) and Frederick Norgate. Publishers of the *Natural History Review*. (*Modern English biography* s.v. Williams, Edmund Sidney.)

Williamson, Alexander (1819–70). Scottish gardener. Curator of the pleasure grounds at the Royal Botanic Gardens, Kew, 1848–66. (R. Desmond 1995.)

Willis, Robert (1800–75). Engineer and archaeologist. Fellow of Caius College, Cambridge, 1826. Jacksonian Professor of natural and experimental philosophy, Cambridge University, 1837–75. (*DNB*, *DSB*.)

Wilmot, Edward Woollett (1808–64). CD's first cousin by marriage. Married Emma Elizabeth Darwin in 1842. Son-in-law of Francis Sacheverel Darwin. Resided at Chaddesden, Derbyshire. (*Darwin pedigree*.)

Wilson, Daniel (1816–92). Scottish archaeologist and educational reformer. Journalist in London, 1837–42. Honorary secretary of the Scottish Society of

Wilson, Daniel, cont.
Antiquaries, 1845. Professor of history and English literature, Toronto University, 1853. President of Toronto University, 1881. Knighted, 1888. (*DNB.*)

Wilson, Edward (1814–78). Australian journalist and politician. Left London for Australia in 1842; proprietor of the *Argus*, 1847–64. Founded the Acclimatisation Society of Victoria in 1861. (*DNB.*)

Wolf, Josef (1820–99). German-born painter and illustrator; specialist on animals. Apprenticed as a lithographer in Koblenz before studying art at the Antwerp Academy. Arrived in London in 1848. Established a studio with Johann Baptist Zwecker at 59 Berners Street, London, *circa* 1860. Illustrated many natural history publications including a number of the plates for Henry Walter Bates's *The naturalist on the river Amazons* (1863). (Bates 1863, p. vi; *DNB*; *Post Office London directory* 1859–63.)

Wollaston, Thomas Vernon (1822–78). Entomologist and conchologist. Passed many winters in Madeira where he collected insects and shells. Wrote a series of works on the Coleoptera. (*DNB*, Gilbert 1977.)

Wolstenholme, Joseph (1829–91). Mathematician. Assistant tutor at Christ's College, Cambridge, 1852–69. Professor of mathematics, Royal Indian Engineering College, Cooper's Hill, Berkshire, 1871–89. (*DNB.*)

[27 March 1863?] (William Erasmus Darwin)

Woodbury, Thomas White (1818–70). Journalist and beekeeper. From 1850 devoted himself exclusively to beekeeping. Introduced Ligurian bees to Britain in 1859 and developed the 'Woodbury hive', marketed by the London apiarian specialists George Neighbour & Sons. As 'A Devonshire Beekeeper', a regular contributor to the *Cottage Gardener*, *Journal of Horticulture*, *Gardeners' Chronicle*, and *The Times.* (R. H. Brown 1975, Ron Brown 1994, Dodd 1983, Fraser 1958, Neighbour 1865.)

15 March [1863], *17 March 1863*, [after 17 March 1863]

Woodd, Charles Henry Lardner (1821–93). Resided at Oughtershaw Hall, Yorkshire. Justice of the peace. Fellow of the Geological Society of London, 1846. Brother-in-law of William Darwin Fox. (*Burke's landed gentry* 1952.)

Woodhouse, Alfred J. Dentist at 1 Hanover Square, London. The Darwin family dentist. (CD's Account book (Down House MS), *Post Office London directory.*)

Woodward, Henry (1832–1921). Geologist and palaeontologist. Joined his brother, Samuel Pickworth Woodward, as an assistant in the geological department, British Museum, in 1858; keeper of geology, 1880–1901. Co-founder of the *Geological Magazine*, 1864; editor, 1865–1918. President of many scientific organisations, including the Geological Society of London, Palaeontographical Society, Royal Microscopical Society, Malacological Society, Geologists' Association, and Museums Association. FRS 1873. (*Geological Magazine* (1921) 58: 481–4, *WWW.*)

Woodward, Samuel Pickworth (1821–65). Naturalist. Sub-curator, Geological Society of London, 1839–45. Professor of geology and natural history at the Royal

Agricultural College, Cirencester, 1845. First-class assistant in the department of geology and mineralogy, British Museum, 1848–65. (*DNB*, Sarjeant 1980–96.)
14 February 1863, 5 June 1863

Wooler, William Alexander (1813–91). Landowner, colliery-owner, and agriculturalist, of Halliwell House, Heighington, and Sadberge Hall, Darlington, County Durham. Started the conservative newspaper, the *North-Eastern Independent*, 1879. Founder of Darlington Conservative Association. Member of the Darlington Rural Sanitary Authority. Instrumental in founding the South Durham and North Yorkshire Chamber of Agriculture. (*North Eastern Daily Gazette*, 5 May 1891; *North Star*, 5 May 1891; *Northern Echo*, 5 May 1891; *Post Office directory of the county of Durham* 1873.)
5 May 1863

Woolner, Thomas (1825–92). Sculptor and poet. Established his reputation in the 1850s with medallion portrait sculptures of Robert Browning, Thomas Carlyle, and William Wordsworth. Went on to make acclaimed busts of CD, Charles Dickens, Thomas Henry Huxley, Adam Sedgwick, and Alfred Tennyson, and life-size studies of Francis Bacon, John Stuart Mill, and William Whewell. (*DNB*.)

Wright, Edward Perceval (1834–1910). Irish naturalist. One of the editors of the *Natural History Review*, 1854–66. Director of the university museum in Dublin, 1857. Lecturer in zoology, Trinity College, Dublin, 1858–68; professor of botany and keeper of the herbarium, 1869. Resigned the professorship in 1904 owing to ill health, but continued to superintend the herbarium. (*DNB*.)

Wright, Thomas (1809–84). Scottish surgeon and palaeontologist. Medical officer of health to the urban district, and surgeon, Cheltenham General Hospital. Formed an extensive collection of Jurassic fossils. FRS 1879. (*DNB*, *Medical directory* 1862, Sarjeant 1980–96.)
27 February 1863

Wrigley, Alfred (1817–98). Mathematician and educator. Professor of mathematics at the Royal Military College, Addiscombe, Surrey, 1841–61. Headmaster, Clapham Grammar School, 1862–82. (*Alum. Cantab.*, *Modern English biography*.)

Wyman, Jeffries (1814–74). American anatomist and ethnologist. Curator of the Lowell Institute, Boston, 1840; lecturer, 1840–1 and 1848–9. Travelled in Europe, 1841–3. Professor of anatomy and physiology, Hampden-Sydney Medical College, Virginia, 1843–8. Hersey Professor of anatomy, Harvard University, 1847–74. Professor and curator, Peabody Museum of American Archaeology and Ethnology, Harvard, 1866–74. (*DAB*, *DSB*.)

Ximenès, Augustin-Marie, marquis de (1726–1817). French poet. (*NBU*.)

Young, Sophia. Daughter of vice-admiral James Young. Married Bartholomew James Sulivan in 1837. (*County families* 1871.)

Zanichelli, Nicola (1819–84). Italian bookseller and publisher based in Modena and Bologna. Published the first Italian edition of *Origin*. (*Enciclopedia Italiana*; Freeman 1977, p. 105.)

Zwecker, Johann Baptist (1815–76). German-born painter and illustrator. Studied art in Düsseldorf and Frankfurt. Came to London *circa* 1860 and set up a studio with Josef Wolf at 59 Berners Street, London. Illustrated a number of natural history publications including Henry Walter Bates's *The naturalist on the river Amazons* (1863). (Bates 1863, p. vi; *Bryan's dictionary of painters and engravers*; *Post Office London directory* 1859–63.)

BIBLIOGRAPHY

The following bibliography contains all the books and papers referred to in this volume by author–date reference or by short title. Short titles are used for some standard reference works (e.g., *DNB, OED*), for CD's books, and for editions of his letters and manuscripts (e.g., *Descent, LL, Notebooks*). Works referred to by short titles are listed in alphabetical order according to the title; those given author–date references occur in alphabetical order according to the author's surname. Notes on manuscript sources are given at the end of the bibliography.

Adams, Oscar Fay, ed. 1897. *A dictionary of American authors.* Boston, Mass.: Houghton Mifflin & Co.

ADB: *Allgemeine deutsche Biographie.* Under the auspices of the Historical Commission of the Royal Academy of Sciences. 56 vols. Leipzig: Duncker & Humblot. 1875–1912.

Admissions to Trinity College, Cambridge. Edited by W. W. Rouse Ball and J. A. Venn. 5 vols. London: Macmillan. 1911–16.

Agardh, Jacob Georg. 1858. *Theoria systematis plantarum; accedit familiarum phanerogamarum in seris naturales dispositio, secundum structuræ normas et evolutionis gradus instituta.* Lund, Sweden: C. W. K. Gleerup.

Agassiz, Jean Louis Rodolphe. 1857–62. *Contributions to the natural history of the United States of America.* 4 vols. Boston, Mass.: Little, Brown & Company. London: Trübner.

——. 1863a. *Methods of study in natural history.* Boston, Mass.: Ticknor & Fields.

——. 1863b. The Silurian beach. *Atlantic Monthly* 11: 460–71.

——. 1863c. The formation of glaciers. *Atlantic Monthly* 12: 568–76.

Allan, Mea. 1967. *The Hookers of Kew, 1785–1911.* London: Michael Joseph.

Allen, David Elliston. 1994. *The naturalist in Britain: a social history.* 2d edition. Princeton, N.J.: Princeton University Press.

Allen, H. C. 1954. *Great Britain and the United States: a history of Anglo-American relations (1783–1952).* London: Oldhams Press.

Alum. Cantab.: *Alumni Cantabrigienses. A biographical list of all known students, graduates and holders of office at the University of Cambridge, from the earliest times to 1900.* Compiled by John Venn and J. A. Venn. 10 vols. Cambridge: Cambridge University Press. 1922–54.

Alum. Oxon.: *Alumni Oxonienses: the members of the University of Oxford, 1500–1886: . . . with a record of their degrees. Being the matriculation register of the university.* Alphabetically arranged, revised, and annotated by Joseph Foster. 8 vols. London and Oxford: Parker & Co. 1887–91.

Ames, Mary Lesley, ed. 1909. *Life and letters of Peter and Susan Lesley.* 2 vols. New York: G. P. Putnam's Sons. London: Knickerbocker Press.

An appeal. [By Charles and Emma Darwin.] [Bromley, Kent]: [privately printed.] [1863.]

'Ancient glaciers of Caernarvonshire': Notes on the effects produced by the ancient glaciers of Caernarvonshire, and on the boulders transported by floating ice. By Charles Darwin. *London, Edinburgh, and Dublin Philosophical Magazine and Journal of Science* 21 (1842): 180–8. [*Collected papers* 1: 163–71.]

Anderson[-Henry], Isaac. 1853. [Crossing or hybridising.] In *The book of the garden*, vol. 2, pp. 319–22, by Charles Mackintosh. 2 vols. Edinburgh and London: William Blackwood & Sons. 1853–5. [Reprinted in *Floricultural Cabinet and Florists' Magazine* 26 (1858): 60–7.]

Anderson-Henry, Isaac. 1863. Crossing strawberries. *Journal of Horticulture* n.s. 4: 45–6.

——. 1867a. On the hybridization or crossing of plants. [Read 14 March 1867.] *Transactions of the Botanical Society [of Edinburgh]* 9: 101–15.

——. 1867b. On pure hybridization; or, crossing distinct species of plants. [Read 14 November 1867.] *Transactions of the Botanical Society [of Edinburgh]* 9: 206–31.

Annual register: *The annual register. A review of public events at home and abroad.* New ser. 1863–1946. London: Longmans [& others].

Anon. 1860. [Review of *Origin.*] *Calcutta Review* 35: 64–88.

——. 1863a. *A report of a sad case, recently tried before the lord mayor, Owen* versus *Huxley, in which will be found fully given the merits of the great recent bone case.* London: n.p.

——. 1863b. On the relations of man to the inferior animals. *Anthropological Review* 1: 107–17

——. 1957. *Adam & Charles Black 1807–1957. Some chapters in the history of a publishing house.* London: Adam & Charles Black.

Appleton's cyclopædia of American biography. Edited by James Grant Wilson and John Fiske. 6 vols. New York: D. Appleton. 1887–9.

Armstrong, William George. 1863. [Presidential address.] *Report of the 33d meeting of the British Association for the Advancement of Science, held at Newcastle-upon-Tyne*, pp. li–lxiv.

Army list: *The army list.* London: the compiler of the annual official army list, sold by T. Egerton, Military Library, Whitehall; Her Majesty's Stationery Office. 1815–1900.

Audubon, John James La Forest. 1831–[9]. *Ornithological biography, or an account of the habits of the birds of the United States of America accompanied by descriptions of the objects represented in the work entitled 'The birds of America' and interspersed with delineations of American scenery and manners.* 5 vols. Edinburgh: Adam Black; Adam & Charles Black.

Aust. dict. biog.: *Australian dictionary of biography*. Edited by Douglas Pike *et al.* 14 vols. [Melbourne]: Melbourne University Press. London and New York: Cambridge University Press. 1966–96.

Autobiography: *The autobiography of Charles Darwin 1809–1882. With original omissions restored.* Edited with appendix and notes by Nora Barlow. London: Collins. 1958.

Babington, Charles Cardale. 1862. *Manual of British botany, containing the flowering plants and ferns arranged according to the natural orders.* 5th edition. London: John van Voorst.

Baer, Karl Ernst von. 1828–37. *Über Entwickelungsgeschichte der Thiere. Beobachtung und Reflexion.* 2 vols. in 1. Königsberg: Gebrüder Kornträger.

Bailey, Liberty Hyde and Bailey, Ethel Zoe. 1976. *Hortus third: a concise dictionary of plants cultivated in the United States and Canada.* Revised and expanded by the staff of the Liberty Hyde Bailey Hortorium. New York: Macmillan. London: Collier Macmillan.

Baillon, Ernest Henri, *et al.* 1876–92. *Dictionnaire de botanique.* 4 vols. Paris: Librairie Hachette & Cie.

——. 1861. Sur l'émission des tubes polliniques des *Helianthemum. Adansonia* 2 (1861–2): 56–9.

Baker, Herbert G. 1965. Charles Darwin and the perennial flax——a controversy and its implications. *Huntia* 2: 141–61.

Balfour, John Hutton. 1873. Obituary notice of Thomas Anderson, MD, FLS. *Transactions of the Botanical Society* [*of Edinburgh*] 11: 41–5.

Banking almanac: *The banking almanac, directory, yearbook and diary.* London: Richard Groombridge; Waterlow & Sons. 1845–1919.

Barber, Lynn. 1980. *The heyday of natural history, 1820–1870.* London: Jonathan Cape.

Barbieri, Alberto. 1973. *Modenesi da ricordare.* Modena, Italy: Società Tipografica Editrice Modenese, Mucchi.

Barnett, James Arthur, *et al.* 1990. *Yeasts: characteristics and identification.* 2d edition. Cambridge: Cambridge University Press.

Barnhart, John Hendley. 1965. *Biographical notes upon botanists . . . maintained in the New York Botanical Garden Library.* Compiled by John Hendley Barnhart. 3 vols. Boston, Mass.: G. K. Hall.

Bartholomew, Michael. 1973. Lyell and evolution: an account of Lyell's response to the prospect of an evolutionary ancestry for man. *British Journal for the History of Science* 6 (1972–3): 261–303.

——. 1976. The non-progress of non-progression: two responses to Lyell's doctrine. *British Journal for the History of Science* 9: 166–74.

Bates, Henry Walter. 1860. Contributions to an insect fauna of the Amazon valley. Diurnal Lepidoptera. [Read 5 March and 24 November 1860.] *Transactions of the Entomological Society of London* n.s. 5 (1858–61): 223–8, 335–61.

——. 1861. Contributions to an insect fauna of the Amazon valley. Lepidoptera: Heliconidæ. [Read 21 November 1861.] *Transactions of the Linnean Society of London* 23 (1862): 495–566.

Bates, Henry Walter. 1863. *The naturalist on the river Amazons. A record of adventures, habits of animals, sketches of Brazilian and Indian life, and aspects of nature under the equator, during eleven years of travel.* 2 vols. London: John Murray.

——. 1886–90. *Pectinicornia and Lamellicornia.* Vol. 2, pt 2 of *Biologia Centrali-Americana. Insecta. Coleoptera*, edited by Frederick Ducane Goodman and Osbert Salvin. 7 vols. London: R. H. Porter; Dulau & Co. 1881–99.

——. 1892. *The naturalist on the river Amazons. A record of adventures, habits of animals, sketches of Brazilian and Indian life, and aspects of nature under the equator, during eleven years of travel. With a memoir of the author by Edward Clodd.* Reprint edition. London: John Murray.

Bathurst, Charles Jr and Kinch, Edward, eds. 1898. *Register of the staff and students at the Royal Agricultural College, from 1844 to 1897, with a short historical preface.* Cirencester, Gloucestershire: George H. Harmer.

'Beagle' diary: *Charles Darwin's* Beagle *diary.* Edited by Richard Darwin Keynes. Cambridge: Cambridge University Press. 1988.

Beal, William James. 1868. Agency of insects in fertilizing plants. *American Naturalist* 1: 254–60.

Beale, Lionel S. 1861. *How to work with the microscope: a course of lectures on the practical use of the instrument and microscopical manipulation.* London: John Churchill.

Bean, William Jackson. 1970–88. *Trees and shrubs hardy in the British Isles.* 8th edition, fully revised by D. L. Clarke and George Taylor. 4 vols. and supplement. London: John Murray.

Beaton, Donald. 1860. Crossing flowers. *Cottage Gardener* 24: 253–5

——. 1861. Fertilisation of wheat. *Journal of Horticulture* n.s. 1: 311–13.

——. 1862. The process of fertilisation. *Journal of Horticulture* n.s. 2: 329–30.

Becker, Lydia Ernestine. 1869. On alteration in the structure of *Lychnis diurna*, observed in connexion with the development of a parasitic fungus. [Read August 1869.] *Report of the 39th meeting of the British Association for the Advancement of Science, held at Exeter*, Transactions of the sections, p. 106.

Beer, Joseph Georg. 1863. *Beiträge zur Morphologie und Biologie der Familie der Orchideen.* Vienna: Carl Gerald's Sohn.

Belich, James. 1986. *The New Zealand Wars, and the Victorian interpretation of racial conflict.* Auckland, New Zealand: Auckland University Press.

Bell, Thomas. 1837. *A history of British quadrupeds, including the Cetacea.* London: John van Voorst.

Benthall, Paul. 1980. George Maw: a versatile Victorian. *National Trust Studies* (1980): 11–20.

Bentham, George. 1858. *Handbook of the British flora; a description of the flowering plants and ferns indigenous to, or naturalised in, the British Isles.* London: Lovell Reeve.

——. 1861. *Flora Hongkongensis: a description of the plants and ferns of the Island of Hong Kong.* London: Lovell Reeve.

——. 1862. [Anniversary address, 24 May 1862.] *Journal of the Proceedings of the Linnean*

Society of London (Botany) 6: lxvi–lxxxiii.
Bentham, George. 1863. [Anniversary address, 25 May 1863.] *Journal of the Proceedings of the Linnean Society (Botany)* 7 (1864): xi–xxix.
——. 1863–78. *Flora Australiensis: a description of the plants of the Australian territory.* Assisted by Ferdinand Mueller. 7 vols. London: Lovell Reeve.
——. 1878. *Handbook of the British flora: a description of the flowering plants and ferns indigenous to, or naturalised in, the British Isles.* 4th edition. London: Lovell Reeve.
Bentham, George and Hooker, Joseph Dalton. 1862–83. *Genera plantarum. Ad exemplaria imprimis in herbariis Kewensibus servata definita.* 3 vols. London: A. Black [& others].
Benton, Michael J. 1982. Progessionism in the 1850s: Lyell, Owen, Mantell and the Elgin fossil reptile *Leptopleuron (Telerpeton)*. *Archives of Natural History* 11: 123–36.
Bergey's manual of systematic bacteriology. Edited by Noel R. Krieg *et al.* 4 vols. Baltimore, Md., and London: Williams & Wilkins. 1984–9.
Berkeley, Miles Joseph. 1850–1. Gärtner's observations upon muling among plants. *Journal of the Horticultural Society of London* 5 (1850): 156–72; 6 (1851): 1–13.
[——.] 1862. Fertilization of orchids. *London Review and Weekly Journal of Politics, Arts and Sciences* 4: 553–4.
Beven, Francis Lorenz, ed. [1946.] *A history of the diocese of Colombo. A centenary volume.* [Colombo, Sri Lanka]: n.p.
Bewick, Thomas. 1826. *A history of British birds.* 2 vols. Newcastle: T. Bewick.
BHGW: *Biographisch-literarisches Handwörterbuch zur Geschichte der exacten Wissenschaften enthaltend Nachweisung über Lebensverhältnisse und Leistungen von Mathematikern, Astronomen, Physikern, Chemikern, Mineralogen, Geologen usw.* By Johann Christian Poggendorff. 5 vols. Leipzig: Johann Ambrosius Barth; Verlag Chemie. 1863–1926.
Bibby, Cyril. 1959. *T. H. Huxley. Scientist, humanist and educator.* London: Watts.
Bible: *The Holy Bible: containing the Old and New Testaments.* Cambridge: Cambridge University Press.
Birse, Ronald M. 1994. *Science at the University of Edinburgh, 1583–1993. An illustrated history to mark the centenary of the Faculty of Science and Engineering, 1893–1993.* Edinburgh: Faculty of Science and Engineering, University of Edinburgh.
Blackburn, Helen. 1902. *Women's suffrage: a record of the Women's Suffrage Movement in the British Isles with biographical sketches of Miss Becker.* London: Williams & Norgate.
Blackie, John. 1976. *Bradfield 1850–1975.* Bradfield, Berkshire: Warden and council of St Andrew's College, Bradfield.
Blake, Charles Carter. 1861. On the distribution of Mastodon in South America. *Geologist* 4: 469–72.
——. 1862. On a fossil elephant from Texas (*E. Texianus*). *Geologist* 5: 57–8.
[——.] 1863a. [Review of T. H. Huxley's *Man's place in nature* & other works.] *Edinburgh Review* 117: 541–69.
——. 1863b. On some points in the cranioscopy of South American nations. *Report*

of the 33d meeting of the British Association for the Advancement of Science, held at Newcastle-upon-Tyne, Transactions of the sections, pp. 133–4.

Blake, Charles Carter, ed. 1863. *A manual of zoology.* By Henri Milne-Edwards. Translated by Robert Knox. 2d edition. London: Henry Renshaw.

Blaringhem, Louis. 1913. La notion d'espèce et la disjonction des hybrides, d'après Charles Naudin (1852–1875). *Progressus rei botanicæ* 4: 27–108.

BLC: *The British Library general catalogues of printed books to 1975.* 360 vols. and supplement (6 vols.). London: Clive Bingley; K. G. Saur. 1979–88.

BLKO: *Biographisches Lexikon des Kaiserthums Oesterreich, enthaltend die Lebensskizzen der denkwürdigen Personen, welche seit 1750 in den österreichischen Kronländern geboren wurden oder darin gelebt und gewirkt haben.* By Constant von Wurzbach. 60 vols. Vienna: L. C. Zamarski. 1856–91.

BLOS: *Biographiskt lexicon öfver namnkunnige svenska män.* 23 vols. in 12 and supplement (1836). Uppsala, Sweden: Palmbland, Sebell & Co. Orebro, Sweden: N. M. Lindh. 1835–57.

Blyakher, L. Y. 1982. *History of embryology in Russia from the middle of the eighteenth century to the middle of the nineteenth century.* Translated from the Russian by H. I. Youssef and B. A. Malek, with an introduction by Jane Maienschein. Washington, D.C.: Al Ahram Center for Scientific Translations for the Smithsonian Institution and the National Science Foundation.

Blyth, Edward. 1849. *Catalogue of the birds in the Museum Asiatic Society.* Calcutta: Baptist Mission Press.

——. 1875. Catalogue of mammals and birds of Burma. With a memoir [by A. Grote] and portrait of the author. *Journal of the Asiatic Society of Bengal* pt 2, extra no., 43 (1874): i–xxiv, 1–167.

BNB: *Biographie nationale publiée par l'Académie Royale des Sciences, des Lettres et des Beaux-arts de Belgique.* 44 vols., including 16 supplements. Brussels: H. Thiry-Van Buggenhoudt [& others]. 1866–1986.

Bonham-Carter, Victor. 1960. *In a liberal tradition: a social biography 1700–1950.* London: Constable.

Bonney, T. G. 1919. *Annals of the Philosophical Club of the Royal Society written from its minute books.* London: Macmillan.

Börsenblatt für den Deutschen Buchhandel: *Börsenblatt für den Deutschen Buchhandel und für die mit ihm verwandten Geschäftszweige herausgegeben von den Deputirten des Vereins der Buchhändler zu Leipzig.* Leipzig: Vereins der Buchhändler. 1834–80.

Bosse, J. F. W. 1829. Ueber die Befruchtung der Passions-Blumen. *Verhandlungen des Vereins zur Beförderung des Gartenbaus in den königlich Preussischen Staaten.* 5: 431–2.

Boucher de Crèvecoeur de Perthes, Jacques. 1838–41. *De la création. Essai sur l'origine et la progression des êtres.* 5 vols. Paris: Treuttel & Wurtz.

——. 1847–64. *Antiquités celtiques et antédiluviennes. Mémoire sur l'industrie primitive et les arts a leur origine.* 3 vols. Paris: Treuttel & Wurtz [& others].

——. 1860a. De l'homme antédiluvien et de ses œuvres. [Read 7 June 1860.]

Mémoires de la Société Impériale d'Emulation d'Abbeville (1857–60): 471–571.

Boucher de Crèvecoeur de Perthes, Jacques. 1860b. *De l'homme antédiluvien et de ses oeuvres.* Paris: Jung-Treuttel [& others]. [Reprinted from *Mémoire de la Société Impériale d'Emulation d'Abbeville* (1857–60): 471–571.]

Bowlby, John. 1990. *Charles Darwin: a biography.* London: Hutchinson.

Boylan, Patrick J. 1977. *The Falconer papers, Forres.* Leicester: Leicestershire Museums, Art Galleries and Records Service.

——. 1979. The controversy of the Moulin-Quignon jaw: the role of Hugh Falconer. In *Images of the earth: essays in the history of the environmental sciences*, edited by Ludmilla J. Jordanova and Roy S. Porter. Chalfont St Giles, Buckinghamshire: British Society for the History of Science.

Brace, Charles Loring. 1852. *Hungary in 1851: with an experience of the Austrian police.* London: Richard Bentley.

——. 1863. *The races of the old world. A manual of ethnology.* London: John Murray.

Bradshaw's railway guide: *Bradshaw's railway almanack, directory, shareholders' guide, and manual.* London: W. J. Adams. Manchester: Bradshaw & Blacklock. 1848–62. *Bradshaw's railway manual, shareholders' guide, and official directory.* London: W. J. Adams. Manchester: Bradshaw & Blacklock. 1863–1923.

Braun, Alexander Carl Heinrich. 1851. *Betrachtungen über die Erscheinung der Verjüngung in der Natur, insbesondere in der Lebens- und Bildungsgeschichte der Pflanze.* Leipzig: Wilhelm Engelmann.

Bravais, Auguste and Bravais, Louis. 1838. Essai sur la disposition générale des feuilles rectisériées. *Congrès scientifique de France* 6: 278–330.

Bree, Charles Robert. 1860. *Species not transmutable, nor the result of secondary causes. Being a critical examination of Mr Darwin's work entitled 'Origin and variation of species'.* London: Groombridge & Sons. Edinburgh: Maclachlan & Stewart.

Brewer: *Brewer's dictionary of phrase & fable.* Revised by Adrian Room. 15th edition. London: Cassell. 1995.

Brian, A. D. and Crane, E. E. 1959. Charles Darwin and bees. *Bee world* 40: 297–303.

Bridgman, William Kencely. 1861. On the influence of the venation in the reproduction of monstrosities among ferns. *Annals and Magazine of Natural History* 3d ser. 8: 490–2.

British imperial calendar: *The British imperial calendar or general register of the United Kingdom of Great Britain and Ireland and its colonies.* London: Winchester & Son [& others]. 1812–1919.

British postal guide: *British postal guide; containing the chief public regulations of the Post Office with other information.* London: Post Office [& others]. 1861–99.

Brock, Thomas D., *et al.* 1994. *Biology of microorganisms.* 7th edition. Englewood Cliffs, N.J.: Prentice-Hall International.

Bronn, Heinrich Georg, trans. 1860. *Charles Darwin, über die Entstehung der Arten im Thier- und Pflanzen-Reich durch natürliche Züchtung, oder Erhaltung der vervollkommneten Rassen im Kampfe um's Daseyn.* Stuttgart: E. Schweizerbart.

Bronn, Heinrich Georg, trans. 1862. *Charles Darwin, über die Einrichtungen zur Befruchtung Britischer und ausländischer Orchideen durch Insekten und über die günstigen Erfolge der Wechselbefruchtung.* Stuttgart: E. Schweizerbart.

Brougham, Henry Peter. 1839. *Dissertations on subjects of science connected with natural theology: being the concluding volumes of the new edition of Paley's work.* 2 vols. London: C. Knight.

——. 1863. Opening address. *Transactions of the National Association for the Promotion of Social Science, Edinburgh, 1863*, pp. 1–25.

Brown, Edwin. 1862a. On the mutability of specific or race forms. *Proceedings of the Northern Entomological Society*, 28 July 1862, pp. 7–18. [Reprinted in the *Zoologist* 20 (1862): 8249–53.]

——. 1862b. On the mutability of species. *Proceedings of the Northern Entomological Society*, 22 December 1862, pp. 4–26.

——. 1863. Monograph on the genus *Acentropus*. In *The natural history of Tutbury. Together with the fauna and flora of the district surrounding Tutbury and Burton-on-Trent*, edited by Oswald Mosley. London: John van Voorst.

Brown, Philip A. H. 1982. *London publishers and printers* c. *1800–1870*. London: British Library.

Brown, R. H. 1975. *One thousand years of Devon beekeeping.* Devon: Devon Beekeepers Association.

Brown, Robert. 1831a. *Observations on the organs and mode of fecundation in Orchideæ and Asclepiadeæ.* London: [privately printed].

——. 1831b. On the organs and mode of fecundation in Orchideæ and Asclepiadeæ. [Read 1 and 15 November 1831.] *Transactions of the Linnean Society of London* 16 (1833): 685–745.

Brown, Ron. 1994. *Great masters of beekeeping.* Borrowbridge, Somerset: Bee Books New & Old.

Browne, Janet. 1983. *The secular ark: studies in the history of biogeography.* New Haven, Conn., and London: Yale University Press.

——. 1990. Spas and sensibilities: Darwin at Malvern. In *The medical history of waters and spas*, edited by Roy S. Porter. *Medical History*, supp. 10. London: Wellcome Institute for the History of Medicine.

——. 1995. *Charles Darwin. Voyaging. Volume 1 of a biography.* New York: Alfred A. Knopf.

Bryan's dictionary of painters and engravers. New edition revised and enlarged under the supervision of George C. Williamson. 5 vols. 1903–5. London: George Bell & Sons.

Buckle, Henry Thomas. 1857–61. *History of civilization in England.* 2 vols. London: John W. Parker & Son.

Buckman, James. 1857. Report on the experimental plots in the Botanical Garden of the Royal Agricultural College at Cirencester. *Report of the 27th meeting of the British Association for the Advancement of Science, held at Dublin*, pp. 200–15.

Buffetaut, Eric. 1987. *A short history of vertebrate palaeontology.* London: Croom Helm.

Buffon, George Louis Leclerc, comte de, *et al.* 1749–1804. *Histoire naturelle, générale et particulière, avec la description du cabinet du roy.* 44 vols. Paris: de l'imprimerie royale.

Burke's landed gentry: *A genealogical and heraldic history of the commoners of Great Britain and Ireland enjoying territorial possessions or high official rank but unvisited with heritable honours. Burke's genealogical and heraldic history of the landed gentry.* By John Burke, *et al.* 1st–18th edition. London: Henry Colburn [& others]. 1835–1972.

Burke's peerage: *A general and heraldic dictionary of the peerage and baronetage of the United Kingdom. Burke's peerage and baronetage.* 1st–105th edition. London: Henry Colburn [& others]. 1826–1980.

Burkhardt, Richard W. Jr. 1977. *The spirit of system. Lamarck and evolutionary biology.* Cambridge, Mass., and London: Harvard University Press.

Burnett, William A. S. 1992. Darwin's microscopes. *Microscopy: the Journal of the Quekett Microscopical Club* 36: 604–27.

Busk, George. 1842. On the occurrence of *Sarcina ventriculi* in the human stomach. *Microscopic Journal* 2: 321–3.

——, trans. 1853. On the natural history of *Protococcus pluvialis.* By Ferdinand Cohn. In *Botanical and physiological memoirs*, edited by Arthur Henfrey. London: Ray Society.

——. 1864. On a very ancient human cranium from Gibraltar. *Report of the 34th meeting of the British Association for the Advancement of Science, held at Bath*, Transactions of the sections, pp. 91–2.

Butcher, Barry W. 1988. Darwin's Australian correspondents: deference and collaboration in colonial science. In *Nature in its greatest extent: western science in the Pacific*, edited by Roy MacLeod and Philip F. Rehbock. Honolulu, Hawaii: University of Hawaii Press.

Butler, Samuel. 1863. *A first year in Canterbury Settlement.* London: Longman, Green, Longman, Roberts, & Green.

——. 1923. *A first year in Canterbury Settlement and other early essays.* Vol. 1 of the Shrewsbury edition of the collected works of Samuel Butler, edited by Henry Festing Jones and A. T. Bartholomew. London: Jonathan Cape. New York: E. P. Dutton. 1923–6.

Butler, Stella V. F., *et al.* 1986. *The social history of the microscope.* Cambridge: Whipple Museum of the History of Science.

Butterworths medical dictionary: *Butterworths medical dictionary.* 2d edition. Edited by Macdonald Critchley *et al.* London and Boston, Mass.: Butterworths. 1978.

Bynum, William F. 1984. Charles Lyell's *Antiquity of man* and its critics. *Journal of the History of Biology* 17: 153–87.

Cairnes, John Elliot. 1862. *The slave power: its character, career, & probable designs; being an attempt to explain the real issues involved in the American contest.* London: Parker, Son, & Bourn.

Calendar: *A calendar of the correspondence of Charles Darwin, 1821–1882. With supplement.*

Edited by Frederick Burkhardt *et al.* 2d edition. Cambridge: Cambridge University Press. 1994.

Cambridge University calendar: *The Cambridge University calendar.* Cambridge: W. Page [& others]. 1796–1950.

Cameron, Nigel Malcom de Segur, *et al.* 1993. *Dictionary of Scottish church history & theology.* Edinburgh: T & T Clark.

Campbell, George Douglas. 1860. Opening address, 1860–1 session [3 December 1860]. *Proceedings of the Royal Society of Edinburgh* 4 (1857–62): 350–77.

[——.] 1862. [Review of *Orchids* & other works.] *Edinburgh Review* 116: 378–97.

Candolle, Alphonse de. 1855. *Géographie botanique raisonnée ou exposition des faits principaux et des lois concernant la distribution géographique des plantes de l'époque actuelle.* 2 vols. Paris: Victor Mason. Geneva: J. Kessmann.

——. 1862a. Etude sur l'espèce à l'occasion d'une révision de la famille des Cupulifères. *Bibliothèque Universelle et Revue Suisse (Archives des Sciences Physiques et Naturelles)* n.s. 15: 211–37, 326–65.

——. 1862b. Note sur un nouveau caractère observé dans le fruit des chênes et sur la meilleure division a adopter pour le genre *Quercus. Bibliothèque Universelle et Revue Suisse (Archives des Sciences Physiques et Naturelles)* n.s. 15: 89–101.

——. 1862c. *Mémoires et souvenirs de Augustin-Pyramus de Candolle . . . écrits par lui-même et publiés par son fils.* Geneva and Paris: J. Cherbuliez.

——. 1882. Darwin considéré au point de vue des causes de son succès et de l'importance de ses travaux. *Bibliothèque Universelle (Archives de Sciences Physiques et Naturelles)* ser. 3, 7: 481–95.

Candolle, Augustin Pyramus de and Candolle, Alphonse de. 1824–73. *Prodromus systematis naturalis regni vegetabilis, sive enumeratio contracta ordinum generum specierumque plantarum huc usque cognitarum, juxta methodi naturalis normas digesta.* 17 vols. Paris: Treuttel & Würtz [& others].

Canestrini, Giovanni and Salimbeni, Leonardo, trans. 1864. *Carlo Darwin. Sull'origine delle specie per elezione naturale, ovvero conservazione delle razze perfezionate nella lotta per l'esistenza.* Modena, Italy: Nicola Zanichelli.

Cantor, Geoffrey. 1991. *Michael Faraday: Sandemanian and scientist. A study of science and religion in the nineteenth century.* London: Macmillan.

Cape of Good Hope general directory: *Cape of Good Hope commercial directory and business guide. General directory and guide-book to the Cape of Good Hope and its dependencies.* Cape Town, South Africa: Saul Solomon; Dennis Edwards. 1868–87.

Caroe, Gwendolen Mary. 1985. *The Royal Institution: an informal history.* With a final chapter by Alban Caroe. London: Murray.

Carpenter, William Benjamin. 1850. On the mutual relations of the vital and physical forces. [Read 20 June 1850.] *Philosophical Transactions of the Royal Society* (1850): 727–57.

——. 1854. *Principles of comparative physiology.* 4th edition. London: John Churchill.

[——.] 1860a. Darwin on the origin of species. *National Review* 10: 188–214.

Carpenter, William Benjamin. 1860b. [Review of *Origin.*] *British and Foreign Medico-Chirurgical Review* 25: 367–404.

[——.] 1862. *Introduction to the study of the Foraminifera.* Assisted by W. K. Parker and T. R. Jones. London: Ray Society.

——. 1863. On the fossil human jawbone recently discovered in the gravel near Abbeville. *Proceedings of the Royal Society of London* 12 (1862–3): 524–9.

——. 1865. Researches on the structure, physiology, and development of *Antedon* (*Comatula*, Lamk.) *rosaceus.* [Read 15 June 1865.] *Philosophical Transactions of the Royal Society of London* 156 (1866): 671–756.

——. 1888. *Nature and man: essays scientific and philosophical.* With an introductory memoir by J. Estlin Carpenter. London: Kegan Paul, Trench & Co.

Carr, Raymond. 1981. Country sports. In *The Victorian countryside*, edited by G. E. Mingay. 2 vols. London: Routledge & Kegan Paul.

Carroll, P. Thomas, ed. 1976. *An annotated calendar of the letters of Charles Darwin in the library of the American Philosophical Society.* Wilmington, Del.: Scholarly Resources Inc.

Caspary, Robert. 1861. Vergrünungen der Blüthe des weissen Klees. *Schriften der Königlichen Physikalisch-ökonomische Gesellschaft zu Königsberg* 2 (1863): 51–72.

Castelnau, Francis de. 1850–1. *Histoire du voyage.* 6 vols. Pt 1 of *Expédition dans les parties centrales de l'Amérique du Sud . . . exécutée par ordre du gouvernment Français pendant les années 1843 a 1847, sous la direction de Comte Francis de Castelnau.* 7 pts in 13 vols. Paris: P. Betrand. 1850–7.

Catalogue of graduates, University of Dublin: *A catalogue of graduates who have proceeded to degrees in the University of Dublin, from the earliest recorded commencements to July, 1866: with supplement to December 16, 1868.* Dublin: Hodges, Smith, & Foster. 1869.

Catalogue of the Linnean Society library: *Catalogue of the natural history library of the Linnean Society of London.* 3 pts in 3 vols. London: Linnean Society. 1866–77.

CBD: *A classical dictionary of Greek and Roman biography, mythology, and geography. Based on the larger dictionaries by the late Sir William Smith.* Revised by G. E. Marindin. London: John Murray. 1894.

CDEL. *A critical dictionary of English literature, and British and Americian authors, living and deceased, from the earliest accounts to the middle of the nineteenth century . . . with forty indexes of subjects.* By S. Austin Allibone. 3 vols. London: Trübner. Philadelphia: Childs & Peterson; J. B. Lippincott. 1859–71. *A supplement to Allibone's critical dictionary of English literature and British and Americian authors. Containing over thirty-seven thousand articles (authors), and enumerating over ninety-three thousand titles.* By John Foster Kirk. 2 vols. Philadelphia and London: J. B. Lippincott. 1891.

Cervantes Saavedra, Miguel de. 1605. *El ingenioso hidalgo Don Quixote de la Mancha.* Madrid: I. de la Cuesta.

Chamberlain, Joshua Lawrence, ed. 1898–1900. *Universities and their sons; history, influence and characteristics of American universities, with biographical sketches and portraits of alumni and recipients of honorary degrees.* 5 vols. Boston, Mass.: R. Herndon.

Chambers: *Chambers 20th century dictionary*. New edition. Edited by E. M. Kirkpatrick. Edinburgh: Chambers. 1983.

[Chambers, Robert.] 1844. *Vestiges of the natural history of creation*. London: John Churchill.

Chapman's New Zealand almanac. Auckland, New Zealand: George T. Chapman. 1860–2.

Chittenden, F. J. 1956. *Curtis's Botanical Magazine index; with a revision of the names of the plants depicted from the beginning in 1787 to the end of 1947: volume 1 to volume 164. To which is appended a brief history of the magazine*. London: Royal Horticultural Society.

Claparède, Edouard. 1861. M. Darwin et sa théorie de la formation des espèces. *Revue Germanique Française & étrangère* 16: 523–59; 17: 232–63.

Clapham, A. R., *et al.* 1987. *Flora of the British Isles*. 3d edition. Cambridge: Cambridge University Press.

Clark, Ronald W. 1968. *The Huxleys*. London: William Heinemann.

Clarke, John J. 1955. *A history of local government of the United Kingdom*. London: Herbert Jenkins.

Cleevely, R. J. and Chapman, S. D. 1992. The accumulation and disposal of Gideon Mantell's fossil collections and their role in the history of British palaeontology. *Archives of Natural History* 19: 307–64.

Clergy list: *The clergy list . . . containing an alphabetical list of the clergy*. London: C. Cox [& others]. 1841–89.

'Climbing plants': On the movements and habits of climbing plants. By Charles Darwin. [Read 2 February 1865.] *Journal of the Linnean Society (Botany)* 9 (1867): 1–118.

Climbing plants: *On the movements and habits of climbing plants*. By Charles Darwin. London: Longman, Green, Longman, Roberts & Green; Williams & Norgate. 1865.

Clokie, Hermia Newman. 1964. *An account of the herbaria of the department of botany in the University of Oxford*. Oxford: Oxford University Press.

Colenso, John William. 1861. *St Paul's epistle to the Romans: newly translated, and explained from a missionary point of view*. Cambridge: Macmillan.

——. 1862–79. *The Pentateuch and Book of Joshua critically examined*. 5 vols. London: Longman, Green, Longman, Roberts & Green.

——. 1863. *The Pentateuch and Book of Joshua critically examined*. Pt 1. 5th edition, revised. London: Longman, Green, Longman, Roberts, & Green.

Collected papers: *The collected papers of Charles Darwin*. Edited by Paul H. Barrett. 2 vols. Chicago and London: University of Chicago Press. 1977.

Collingwood, Cuthbert. 1860a. On homomorphism; or, organic representative form. *Proceedings of the Literary and Philosophical Society of Liverpool* 14 (1859–60): 181–216.

Collins dictionary of geology. By Dorothy Farris Lapidus. London and Glasgow: HarperCollins. 1990.

Colonial Office list: *The Colonial Office list . . . or, general register of the colonial dependencies of Great Britain*. London: Edward Stanford; Harrison & Sons. 1862–99.

Colp, Ralph, Jr. 1977. *To be an invalid: the illness of Charles Darwin*. Chicago and London: University of Chicago Press.

Complete peerage: *The complete peerage of England, Scotland, Ireland, Great Britain, and the United Kingdom, extant, extinct, or dormant*. By George Edward Cokayne. Revised edition. Edited by Vicary Gibbs *et al.* 12 vols. London: St Catherine Press. 1910–59.

Concordance: *A concordance to Darwin's* Origin of species, *first edition*. Edited by Paul H. Barrett *et al.* Ithaca, N.Y., and London: Cornell University Press.

Connor, R. D. 1987. *The weights and measures of England*. London: Her Majesty's Stationery Office.

Conry, Yvette. 1972. *Correspondance entre Charles Darwin et Gaston de Saporta*. Paris: Presses Universitaires de France.

——. 1974. *L'introduction du Darwinisme en France au XIXe siècle*. Paris: J. Vrin.

Cooke, Benjamin. 1745. Extract of a letter from Mr Benj. Cooke, FRS, to Mr Peter Collinson, FRS, concerning the effect which the farina of the blossoms of different sorts of apple-trees had on the fruit of a neighbouring tree. *Philosophical Transactions of the Royal Society* 43 (1744–5): 525–6.

Cooper, [Susan Fenimore]. 1855. *Journal of a naturalist in the United States*. 2 vols. London: R. Bentley.

Coral reefs: *The structure and distribution of coral reefs. Being the first part of the geology of the voyage of the* Beagle, *under the command of Capt. FitzRoy RN, during the years 1832 to 1836*. By Charles Darwin. London: Smith, Elder & Co. 1842.

Coral reefs 2d ed.: *The structure and distribution of coral reefs*. By Charles Darwin. Revised edition. London: Smith, Elder & Co. 1874.

Correns, C. 1916. Friedrich Hildebrand. *Berichte der deutschen botanischen Gesellschaft* 34 (pt 2): 28–49.

Correspondence: *The correspondence of Charles Darwin*. Edited by Frederick Burkhardt *et al.* 10 vols. to date. Cambridge: Cambridge University Press. 1985–.

Corsi, Pietro. 1983. 'Lamarckiens' et 'Darwiniens' à Turin (1812–1894). In *De Darwin au Darwinisme: science et idéologie*, edited by Yvette Conry. Paris: J. Vrin.

——. 1988. *Science and religion: Baden Powell and the Anglican debate, 1800–1860*. Cambridge: Cambridge University Press.

Corsi, Pietro and Weindling, Paul J. 1985. Darwinism in Germany, France and Italy. In *The Darwinian heritage*, edited by David Kohn. Princeton, N.J.: Princeton University Press in association with Nova Pacifica.

County families: *The county families of the United Kingdom; or, royal manual of the titled & untitled aristocracy of Great Britain & Ireland*. By Edward Walford. London: Robert Hardwicke; Chatto & Windus. 1860–93. *Walford's county famlies of the United Kingdom or royal manual of the titled and untitled aristrocracy of England, Wales, Scotland, and Ireland*. London: Chatto & Windus; Spottiswoode & Co. 1894–1920.

Craig, Frederick Walter Scott. 1989. *British parliamentary election results 1832–1885*. 2d edition. Aldershot, Hampshire: Parliamentary Research Services, Dartmouth Publishing Company.

Cramer, Pieter, *et al.* 1779–91. *De uitlandsche kapellen voorkomende in de drie waereld-deelen Asia, Africa en America.* 4 vols. and supplement. Amsterdam: S. J. Baalde; N. T. Gravius. Utrecht: Barthelemy Wild.

[Crawfurd, John.] 1859. [Review of *Origin.*] *Examiner*, 3 December 1859, pp. 772–3.

Crawfurd John. 1861. On the Aryan or Indo-Germanic theory. *Transactions of the Ethnological Society of London* n.s. 1: 268–86

——. 1863. On Sir Charles Lyell's 'Antiquity of man' and on Professor Huxley's 'Evidence as to man's place in nature'. [Read 14 April 1863.] *Transactions of the Ethnological Society of London* n.s. 3 (1864–5): 58–70.

Cresy, Edward. 1850. *Public Health Act, 11 & 12 Vict., cap. 63. Report to the General Board of Health on a preliminary inquiry into the sewerage, drainage, and supply of water, and the sanitary condition of the inhabitants of the town of Worthing.* London: General Board of Health.

Crockford's clerical directory: *The clerical directory, a biographical and statistical book of reference for facts relating to the clergy and the church. Crockford's clerical directory.* London: John Crockford [& others]. 1858–1900.

Croll, James. 1875. *Climate and time in their geological relations. A theory of secular changes of the earth's climate.* London: Daldy, Isbister & Co.

Crook, D. P. 1974. *The North, the South and the powers 1861–1865.* London: John Wiley & Sons.

Cross and self fertilisation: *The effects of cross and self fertilisation in the vegetable kingdom.* By Charles Darwin. London: John Murray. 1876.

Crowley, D. A., ed. 1995. *Amesbury hundred; Branch and Dole hundred.* Vol. 15 of *A history of Wiltshire.* Oxford: published for the Institute of Historical Research by Oxford University Press.

Crüger, Hermann. 1851. Ueber Befruchtung bei den Orangen. *Botanische Zeitung* 9: 57–63, 73–80.

——. 1864. A few notes on the fecundation of orchids and their morphology. [Read 3 March 1864.] *Journal of the Linnean Society (Botany)* 8 (1865): 127–35.

Cullum, George Washington. 1891. *Biographical register of the officers and graduates of the US Military Academy at West Point, NY, from its establishment, in 1802, to 1890. With the early history of the United States Military Academy.* 3d edition, revised. 3 vols. Boston, Mass.: Houghton Mifflin.

Cumming, John. 1863. *Moses right, and Bishop Colenso wrong; being popular lectures in reply to the first and second parts of 'Bishop Colenso on the Pentateuch'.* London: John F. Shaw.

Curle, Richard. 1954. *The Ray Society: a bibliographical history.* London: Ray Society.

Curtis's botanical magazine dedications: *Curtis's botanical magazine dedications 1827–1927.* Portraits and biographical notes compiled by Ernest Nelmes and William Cuthbertson. London: Bernard Quaritch for the Royal Horticultural Society. [1931.]

Curwen, E. Cecil, ed. 1940. *The journal of Gideon Mantell, surgeon and geologist. Covering the years 1818–1852.* London: Oxford University Press.

Cuvier, Georges. 1817. *Le règne animal distribué d'après son organisation, pour servir de base a l'histoire naturelle des animaux et d'introduction a l'anatomie comparée.* 4 vols. Paris: Deterville.

DAB: *Dictionary of American biography.* Under the auspices of the American Council of Learned Societies. 20 vols., index, and 10 supplements. New York: Charles Scribner's Sons; Simon & Schuster Macmillan. London: Oxford University Press; Humphrey Milford. 1928–95.

Dana, James Dwight. 1851–2. On coral reefs and islands. *American Journal of Science and Arts* 2d ser. 11 (1851): 357–72; 12 (1851): 25–51, 165–86, 329–38; 13 (1852): 34–41, 185–95, 338–50; 14 (1852): 76–84.

——. 1853. *On coral reefs and islands ... From the author's exploring expedition report on geology, with additions.* New York: G. P. Putnam.

——. 1863a. *Manual of geology: treating of the principles of the science with special reference to American geological history, for the use of colleges, academies, and schools of science.* Philadelphia: Theodore Bliss. London: Trübner.

——. 1863b. On cephalization, and on megasthenes and microsthenes, in classification. *American Journal of Science and Arts* 2d ser. 36: 1–10.

——. 1863c. On the higher subdivisions in the classification of mammals. *American Journal of Science and Arts* 2d ser. 35: 65–71.

Dantès, Alfred Langue. 1875. *Dictionnaire biographique et bibliographique alphabétique et méthodique des hommes les plus remarquables dans les lettres, les sciences et les arts, chez tous les peuples, à toutes les époques.* Paris: A. Boyer.

Dareste, Camille. 1862. Mémoire sur la production artificielle des monstruosités. *Annales des Sciences Naturelles (Zoologie)* 4th ser. 18: 243–76.

——. 1863. Recherches sur les conditions de la vie et de la mort chez les monstres ectroméliens, célosomiens et exencéphaliens, produits artificiellement dans l'espèce de la poule. [Read 23 January 1863.] *Mémoires de la Société Impériale des Sciences de l'Agriculture et des Arts de Lille* 10: 39–82. [Reprinted in *Annales des Sciences Naturelles (Zoologie)* 4th ser. 20: 59–99.]

Darwin, Charles and Wallace, Alfred Russel. 1858. On the tendency of species to form varieties; and on the perpetuation of varieties and species by natural means of selection ... Communicated by Sir Charles Lyell ... and J. D. Hooker. [Read 1 July 1858.] *Journal of the Proceedings of the Linnean Society (Zoology)* 3 (1859): 45–62. [*Collected papers* 2: 3–19.]

[Darwin, Edward Levett.] 1859. *The game-preserver's manual, containing full instructions in all that relates to getting up and maintaining a good head of game.* By 'High Elms'. London: the Field office. Derby: W. Bemrose & Sons.

Darwin, Edward Levett. 1863. *The game-preserver's manual, and keeper's assistant. Containing full directions for getting up and maintaining a good head of game ... and a few words on the 'Night poaching prevention act' of 1862.* 4th edition. Buxton, Derbyshire: the author.

Darwin, Erasmus. 1794–6. *Zoonomia; or, the laws of organic life.* 2 vols. London: J. Johnson.

Darwin, Francis. 1914. William Erasmus Darwin. *Christ's College Magazine* 29: 16–23.

——. 1920. *Springtime and other essays.* London: John Murray.

Darwin pedigree: *Pedigree of the family of Darwin.* Compiled by H. Farnham Burke. N.p.: privately printed. 1888. [Reprinted in facsimile in *Darwin pedigrees*, by Richard Broke Freeman. London: printed for the author. 1984.]

Davies, Gordon Leslie. 1969. *The earth in decay: a history of British geomorphology 1578–1878.* London: MacDonald Technical & Scientific.

Davis, Richard Beale. 1955. The Abbé Correa in America, 1812–1820. The contributions of the diplomat and natural philosopher to the foundations of our national life. *Transactions of the American Philosophical Society* n.s. 45 (pt 2): 85–197.

Dawson, John William. 1860. Review of Darwin on the origin of species by means of natural selection. *Canadian Naturalist* 5: 100–20.

Dawson, Warren R. 1946. *The Huxley papers. A descriptive catalogue of the correspondence, manuscripts and miscellaneous papers of the Rt Hon. Thomas Henry Huxley, PC, DCL, FRS, preserved in the Imperial College of Science and Technology, London.* London: Macmillan for the Imperial College of Science and Technology.

DBE: *Deutsche Biographische Enzyklopädie.* Edited by Walter Killy *et al.* 7 vols. (A–Plessner). Munich: K. G. Saur. 1995–8

DBF: *Dictionnaire de biographie Française.* Under the direction of J. Balteau *et al.* 17 vols. and 7 fascicles of vol. 18. (A–Laglenne). Paris: Librairie Letouzey & Ané. 1933–95.

DBI: *Dizionario biografico degli Italiani.* Edited by Alberto M. Ghisalberti *et al.* 46 vols. (A–Ferrerio). Rome: Istituto della Enciclopedia Italiana. 1960–96.

DBL: *Dansk biografisk leksikon.* Founded by Carl Frederick Bricka, edited by Povl Engelstoft and Svend Dahl. 26 vols. and supplement. Copenhagen: J. H. Schultz. 1933–44.

DCB: *Dictionary of Canadian biography.* Edited by George W. Brown *et al.* 13 vols. and index to first 12 vols. Toronto, Ontario: University of Toronto Press. 1966–94.

De Beer, Gavin. 1954. *Archeopteryx lithographica. A study based upon the British Museum specimen.* London: British Museum (Natural History).

Decaisne, Joseph. 1858–75. *Le jardin fruitier du Muséum ou iconographie de toutes les espèces et variétés d'arbres fruitiers cultivés dans cet établissement avec leur description, leur histoire, leur synonymie, etc.* 9 vols. in 5. Paris: Firmin Didot Frères, fils & Cie.

——. 1863. De la variabilité dans l'espèce du Poirier; résultat d'expériences faites au Muséum d'Histoire Naturelle de 1853 à 1862 inclusivement. *Comptes rendus hebdomadaires des Séances de l'Académie des Sciences* 57: 6–17. [Reprinted in *Annales des Sciences Naturelles (Botanique)* 4th ser. 20: 188–200.]

De Filippi, Fillipo. 1861. Osservazioni zoologiche. *Archivio per la Zoologia, l'Anatomia e le Fisiologia* 1: 200–27.

Dempster, W. J. 1996. *Natural selection and Patrick Matthew: evolutionary concepts in the nineteenth century.* Durham: Pentland Press.

Denney, Robert E. 1992. *The civil war years: a day-by-day chronicle of the life of a nation.* New York: Sterling Publishing.

Descent: *The descent of man, and selection in relation to sex.* By Charles Darwin. 2 vols. London: John Murray. 1870–1.

Desmond, Adrian. 1982. *Archetypes and ancestors: palaeontology in Victorian London, 1850–1875.* London: Blond & Briggs.

——. 1989. *The politics of evolution: morphology, medicine, and reform in radical London.* Chicago and London: University of Chicago Press.

——. 1994–7. *Huxley.* 2 vols. London: Michael Joseph.

Desmond, Ray. 1992. *The European discovery of the Indian flora.* Oxford: Oxford University Press [in association with the] Royal Botanic Gardens.

——. 1994. *Dictionary of British and Irish botanists and horticulturists including plant collectors, flower painters and garden designers.* New edition, revised and completely updated with the assistance of Christine Ellwood. London: Taylor & Francis and the Natural History Museum. Bristol, Pa.: Taylor & Francis.

——. 1995. *Kew: the history of the Royal Botanic Gardens.* London: Harvill Press with the Royal Botanic Gardens, Kew.

Devay, Francis. 1862. *Du danger des mariages consanguins sous le rapport sanitaire.* 2d edition. Paris: Victor Masson & fils

DIB: *A dictionary of Irish biography.* [Compiled by] Henry Boylan. 2d edition. Dublin: Gill & Macmillan. 1988.

Dicionário de história de São Paulo. By Antonio Barreto do Amaral. São Paulo, Brazil: Governo do Estado de São Paulo. 1980.

Dictionary of British artists: *The dictionary of British artists, 1880–1940.* Compiled by J. Johnson and A. Greutzner. Woodbridge, Suffolk: Antique Collectors' Club. 1976.

Dictionary of watercolour artists: *The dictionary of watercolour artists up to 1920.* By H. L. Mallalieu. 2d edition. 3 vols. Woodbridge, Suffolk: Antique Collectors Club. 1986.

Dictionnaire historique & biographique de la Suisse. Under the direction of Marcel Godet *et al.* 7 vols. and supplement (1934). Neuchâtel, Switzerland: Administration du Dictionnaire Historique & Biographique de la Suisse. 1921–34.

Dictionnaire universel des contemporains: *Dictionnaire universel des contemporains contentant toutes les personnes notables de la France et des pays etrangers . . . Ouvrage rédigé et continuellement tenu a jour avec le concours d'écrivains et des savants de tous les pays.* Edited by Louis Gustave Vapereau. Paris: Libraire Hachette. 1858. 3d edition, 1865. 4th edition, 1870. 5th edition, 1880. 6th edition, 1893.

'Dimorphic condition in *Primula*': On the two forms, or dimorphic condition, in the species of *Primula*, and on their remarkable sexual relations. By Charles Darwin. [Read 21 November 1861.] *Journal of the Proceedings of the Linnean Society (Botany)* 6 (1862): 77–96. [*Collected papers* 2: 45–63.]

DNB: *Dictionary of national biography.* Edited by Leslie Stephen and Sidney Lee. 63 vols. and 2 supplements (6 vols.). London: Smith, Elder & Co. 1885–1912. *Dictionary of national biography 1912–90.* Edited by H. W. C. Davis *et al.* 9 vols. London: Oxford University Press. 1927–96.

DNZB: *A dictionary of New Zealand biography.* Edited by G. H. Scholefield. 2 vols. Wellington, New Zealand: Department of Internal Affairs. 1940. *The dictionary of New Zealand biography.* 3 vols. Edited by W. H. Oliver *et al.* Auckland and Wellington, New Zealand: Department of Internal Affairs [& others]. 1990–6.

Dobell, Horace. 1861. *Lectures on the germs and vestiges of disease, and on the prevention of the invasion and fatality of disease by periodical examinations.* London: John Churchill.

——. 1862. A contribution to the natural history of hereditary transmission. [Read 25 November 1862.] *Medico-Chirurgical Transactions* 46: 25–8.

——. 1866. *On winter cough, catarrh, bronchitis, emphysema, asthma, with an appendix on some principles of diet in disease, a course of lectures delivered at the Royal Infirmary for Diseases of the Chest.* London: John Churchill & Sons.

——. 1880. *On loss of weight, blood-spitting and lung disease.* 2d edition. London: J. & A. Churchill.

Dodd, Victor. 1983. *Beemasters of the past.* Hebden Bridge, Yorkshire: Northern Bee Books.

Dod's parliamentary companion: *The parliamentary pocket companion* . . . *compiled from official documents, and from the personal communications of members of both houses. Dod's parliamentary companion.* London: Whittaker, Treacher, & Arnot; Whittaker & Co. 1833–1914.

Don, David. 1834. An attempt at a new arrangement of the *Ericaceæ. Edinburgh New Philosophical Journal* 17: 150–60.

Douglas, Robert. 1934. *Annals of the Royal Burgh of Forres.* Elgin: the author.

Downing, Andrew Jackson. 1845. *The fruits and fruit trees of America: or the culture, propagation, and management, in the garden and orchard, of fruit trees generally; with descriptions of all the finest varieties of fruit, native and foreign, cultivated in this country.* London: Wiley & Putman.

DSAB: *Dictionary of South African biography.* Edited by W. J. de Kock *et al.* 4 vols. Pretoria and Cape Town, South Africa: Nasionale Boekhandel Beperk [& others]. 1968–81.

DSB: *Dictionary of scientific biography.* Edited by Charles Coulston Gillispie and Frederic L. Holmes. 18 vols. including index and supplements. New York: Charles Scribner's Sons. 1970–90.

Duchartre, Pierre Etienne Simon. 1862. Sur un cas de grossissement, sans fécondation, des ovules du *Cycas revoluta.* [Read 26 December 1862.] *Bulletin de la Société Botanique de France* 9: 531–3.

Duchesne, Antoine Nicolas. 1766. *Histoire naturelle des Fraisiers contenant les vues d'économie réunies à la botanique; et suivies de remarques particulières sur plusieurs points qui ont rapport à l'histoire naturelle générale.* Paris: Didot le jeune.

Duncan, James. 1840. *The natural history of bees: comprehending the uses and economical management of the British and foreign honey-bee; together with the known wild species.* Vol. 38 of *The naturalist's library*, edited by William Jardine. 40 vols. 1833–43. Edinburgh: W. H. Lizars.

Duncan, Peter Martin. 1863–7. On the fossil corals of the West Indian Islands. [Read 6 May and 18 November 1863, 11 May 1864, and 4 December 1867.] *Quarterly Journal of the Geological Society of London* 19 (1863): 406–58; 20 (1864): 20–44, 358–74; 24 (1868): 9–33.

Dupree, Anderson Hunter. 1959. *Asa Gray, 1810–1888*. Cambridge, Mass: Belknap Press of Harvard University.

Dutch medical biography: *Dutch medical biography: a biographical dictionary of Dutch physicians and surgeons 1475–1975*. By G. A. Lindeboom. Amsterdam: Rodopi. 1984.

Dutrochet, René Joachim Henri. 1843. Des mouvements révolutifs spontanés qui s'observent chez les végétaux. [Read 6 November 1843.] *Comptes rendus hebdomadaires des séances de l'Académie des Sciences* 17: 989–1008.

——. 1844. Recherches sur la volubilité des tiges de certains végétaux et sur la cause de ce phénomène. *Comptes rendus hebdomadaires des séances de l'Académie des Sciences* 19: 295–303.

Eaton, John Matthews. 1852. *A treatise on the art of breeding and managing tame, domesticated, and fancy pigeons, carefully compiled from the best authors, with observations, containing all that is necessary to be known of tame, domisticated, and fancy pigeons*. London: the author.

EB: *The Encyclopædia Britannica. A dictionary of arts, sciences, literature and general information.* 11th edition. 29 vols. Cambridge: Cambridge University Press. 1910–11.

Edinburgh world atlas: *The Edinburgh world atlas or advanced atlas of modern geography*. By John C. Bartholomew. 7th edition. Edinburgh: John Bartholomew & Son. 1970.

Egerton, F. Clement C. 1957. *Angola in perspective: endeavour and achievement in Portuguese West Africa*. London: Routledge & Kegan Paul.

Ehrlich, Eugene. 1986. *Nil desperandum: a dictionary of Latin tags and useful phrases*. London: Robert Hale.

Eisenberg, Ludwig. 1893. *Das geistige Wien. Künstler- und Schriftsteller-Lexikon*. 2 vols. Vienna: Daberkow.

Ellegård, Alvar. 1990. *Darwin and the general reader: the reception of Darwin's theory of evolution in the British periodical press, 1859–1872*. Reprint edition. Chicago and London: University of Chicago Press.

Ellis, Ieuan. 1980. *Seven against Christ. A study of 'Essays and reviews'*. Leiden, Netherlands: E. J. Brill.

Emerson, Ralph Waldo. 1841. *Essays*. With preface by Thomas Carlyle. London: James Fraser.

Emma Darwin (1904): *Emma Darwin, wife of Charles Darwin. A century of family letters*. Edited by Henrietta Litchfield. 2 vols. Cambridge: privately printed by Cambridge University Press. 1904.

Emma Darwin (1915): *Emma Darwin: a century of family letters, 1792–1896*. Edited by Henrietta Litchfield. 2 vols. London: John Murray. 1915.

Enciclopedia Italiana: *Enciclopedia Italiana di scienze lettere e d'arti*. 36 vols., with 5 appendixes in 13 vols. Rome: Istituto della Enciclopedia Italiana fondata da Giovanni Treccani. 1929–95.

Ernst, Wallace R. 1962. The genera of Papaveraceae and Fumariaceae in the south-eastern United States. *Journal of the Arnold Arboretum* 43: 315–43.

Essays and reviews. London: John W. Parker. 1860.

Essig, E. O. 1965. *A history of entomology*. New York and London: Hafner Publishing Company.

Evans, Joan. 1943. *Time and chance. The story of Arthur Evans and his forebears*. London: Longmans, Green & Co.

Evans, John. 1865. On portions of a cranium and of a jaw, in the slab containing the fossil remains of the Archæopteryx. *Natural History Review* n.s. 5: 415–21.

Ewan, Joseph Andorfer. [1950.] *Rocky Mountain naturalists*. [Denver, Colo.]: University of Denver Press.

Expression: *The expression of the emotions in man and animals*. By Charles Darwin. London: John Murray. 1872.

Falconer, Hugh. 1857a. On the species of Mastodon and elephant occurring in the fossil state in Great Britain. [Read 8 April 1857.] *Quarterly Journal of the Geological Society of London* 13: 307–60.

——. 1857b. Description of two species of the fossil mammalian genus *Plagiaulax* from Purbeck. [Read 11 March 1857.] *Quarterly Journal of the Geological Society of London* 13: 261–82.

——. 1862. On the disputed affinity of the mammalian genus *Plagiaulax*, from the Purbeck beds. [Read 4 June 1862.] *Quarterly Journal of the Geological Society of London* 18: 348–69.

——. 1863a. On the American fossil elephant of the regions bordering the Gulf of Mexico (*E. Columbi*, Falc.); with general observations on the living and extinct species. *Natural History Review* n.s. 3: 43–114.

——. 1863b. The reputed fossil man of Abbeville. *The Times*, 25 April 1863, p. 14.

——. 1868. *Palæontological memoirs and notes of the late Hugh Falconer . . . with a biographical sketch of the author*. 2 vols. Compiled and edited by Charles Murchison. London: Robert Hardwicke.

Falconer, Hugh, *et al.* 1863. An account of the proceedings of the late conference held in France to inquire into the circumstances attending the asserted discovery of a human jaw in the gravel at Moulin-Quignon, near Abbeville; including the *Procès Verbaux* of the sittings of the conference, with notes thereon. *Natural History Review* n.s. 3: 423–62.

Farley, John. 1974. The initial reactions of French biologists to Darwin's *Origin of Species*. *Journal of the history of biology* 7: 275–300.

——. 1977. *The spontaneous generation controversy from Descartes to Oparin*. Baltimore, Md., and London: Johns Hopkins University Press.

Farmer, David Hugh. 1992. *The Oxford dictionary of saints*. 3d edition. Oxford and New York: Oxford University Press.

Ferris, Norman B. 1977. *The* Trent *affair: a diplomatic crisis*. Knoxville, Tenn.: University of Tennessee Press.

'Fertilization of orchids': Notes on the fertilization of orchids. By Charles Darwin. *Annals and Magazine of Natural History* 4th ser. 4 (1869): 141–59. [*Collected papers* 2: 138–56.]

Finlay, Mark R. 1991. The rehabilitation of an agricultural chemist: Justus von Liebig and the seventh edition. *Ambix* 38: 155–67.

Fletcher, Harold R. and Brown, William H. 1970. *The Royal Botanic Garden, Edinburgh, 1670–1970.* Edinburgh: Her Majesty's Stationery Office.

Flett, John Smith. 1937. *The first hundred years of the Geological Survey of Great Britain.* London: His Majesty's Stationery Office.

Flower, William Henry. 1862. On the posterior lobes of the cerebrum of the Quadrumana. [Read 9 January 1862.] *Philosophical Transactions of the Royal Society of London* 152: 185–201.

Fonvielle, Wilfrid de. 1863. La science a la Nouvelle-Zélande. *Revue du Monde Coloniale* 8: 182–8.

Forbes, Edward. 1846. On the connexion between the distribution of the existing fauna and flora of the British Isles, and the geological changes which have affected their area, especially during the epoch of the Northern Drift. *Memoirs of the Geological Survey of Great Britain, and of the Museum of Economic Geology in London* 1: 336–432.

[Forbes, James David.] 1863. [Review of Charles Lyell's *Antiquity of man* and other works.] *Edinburgh Review* 118: 254–302.

Foreign Office list: *The Foreign Office list.* London: Harrison & Sons. 1852–1965.

Forgues, Paul Emile Davrand. 1863. Un naturaliste sous l'équateur. *Revue des Deux Mondes* 2d ser. 46: 703–37.

Forms of flowers: *The different forms of flowers on plants of the same species.* By Charles Darwin. London: John Murray. 1877.

Fossil Cirripedia (1851): *A monograph on the fossil Lepadidæ, or, pedunculated cirripedes of Great Britain.* By Charles Darwin. London: Palaeontographical Society. 1851.

Fossil Cirripedia (1854): *A monograph of the fossil Balanidæ and Verrucidæ of Great Britain.* By Charles Darwin. London: Palaeontographical Society. 1854.

Foster, Michael and Lankester, Edwin Ray, eds. 1898–1903. *The scientific memoirs of Thomas Henry Huxley.* 4 vols. and supplement. London: Macmillan. New York: D. Appleton.

Fox, Grace. 1969. *Britain and Japan 1858–1883.* Oxford: Clarendon Press.

Fraser, H. Malcolm. 1958. *History of beekeeping in Britain.* London: Bee Research Association.

Freeman, Richard Broke. 1977. *The works of Charles Darwin: an annotated bibliographical handlist.* 2d edition. Folkestone, Kent: William Dawson & Sons. Hamden, Conn.: Archon Books, Shoe String Press.

——. 1978. *Charles Darwin: a companion.* Folkestone, Kent: William Dawson & Sons. Hamden, Conn.: Archon Books, Shoe String Press.

Freeman, Richard Broke and Wertheimer, Douglas 1980. *Phillip Henry Gosse: a bibliography.* Folkestone, Kent: Dawson.

Friedl, Hans, *et al.* 1992. *Biographisches Handbuch zur Geschichte des Landes Oldenburg.* Oldenburg, Germany: Isensee Verlag.

Fundação da cidade de São Paulo: *IV centenário da fundação da cidade de Sço Paulo.* Comissão Executiva dos Festejos do IV centenário da cidade de São Paulo. São Paulo, Brazil: Gráfica Municipal. 1954.

Galton, Francis. 1863. *The first steps towards the domestication of animals.* London: privately printed. [Read before the Ethnological Society of London, 22 December 1863, and printed in *Transactions of the Ethnological Society of London* n.s. 3 (1865): 122–38.]

Gardeners' year-book: *The gardeners' year-book, almanack, and directory.* By Robert Hogg. London: Journal of Horticulture and Cottage Gardener office. 1864–86.

Gardner, W. J. 1992. A colonial economy. In *The Oxford history of New Zealand*, 2d ed. Edited by Geoffrey W. Rice. Auckland, New Zealand: Oxford University Press.

Garner, Robert. 1861. On the structure of the *Lepadidæ. Report of the 30th meeting of the British Association for the Advancement of Science, held at Oxford*, Transactions of the sections, p. 130.

Gärtner, Karl Friedrich von. 1826. Nachricht über Versuche, die Befruchtung einiger Gewächse betreffend. *Naturwissenschaftliche Abhandlungen. Herausgegeben von einer gesellschaft in Würtemberg* 1: 35–66.

——. 1827. Correspondenz. *Flora* 10: 74–80.

——. 1838. Verhandeling ter beantwoording der vraag: 'Wat leert de ondervinding aangaande het ontstaan van nieuwe soorten of bijsoorten van planten door kunstige bevruchting van bloemen van de eene met het bloemstof van andere soorten? En welke nieuwe, nuttige of fraaije plantgewassen kunnen op die wijze worden voortgebragt en vermenigvuldigd?'. *Natuurkundige verhandelingen van de Hollandsche maatschappij der wetenschappen te Haarlem* 24: 1–202.

——. 1844. *Versuche und Beobachtungen über die Befruchtungsorgane der vollkommeneren Gewächse und über die natürliche und künstliche Befruchtung durch den eigenen Pollen.* Pt 1 of *Beiträge zur Kenntniss der Befruchtung der vollkommeneren Gewächse.* Stuttgart: E. Schweizerbart.

——. 1849. *Versuche und Beobachtungen über die Bastarderzeugung im Pflanzenreich. Mit Hinweisung auf die ähnlichen Erscheinungen im Thierreiche, ganz umgearbeitete und sehr vermehrte Ausgabe von der Königlich holländischen Akademie der Wissenschaften.* Stuttgart: E. Schweizerbart.

Gascoigne, Robert Mortimer. 1984. *A historical catalogue of scientists and scientific books; from the earliest times to the close of the nineteenth century.* New York and London: Garland Publishing.

Gebhard, Rosa, trans. [1878.] *An English country squire as sketched at Hardwicke Court, Gloucestershire.* By Professor [Franz] von Holtzendorff. Gloucester: John Bellows.

Geikie, Archibald. 1863. *On the phenomena of the glacial drift of Scotland.* Glasgow: John Gray.

——. 1895. *Memoir of Sir Andrew Crombie Ramsay.* London and New York: Macmillan.

General index to the Journal of the Linnean Society: *General index to the first twenty volumes of the Journal (Botany), and the botanical portion of the Proceedings, November 1838 to June 1886, of the Linnean Society.* London: Linnean Society of London. 1888.

Geoffroy Saint-Hilaire, Etienne. 1825a. Sur des déviations organiques provoquées et observées dans un établissment d'incubations artificielles. *Mémoires du Muséum d'Histoire Naturelle* 13: 289–96.

——. 1825b. Considérations zoologiques et physiologiques, relatives à un nouveau genre de monstruosités, nommé Hypognathe, et établi pour trois espèces de veaux bicéphales, à têtes opposées et attachées ensemble par la symphyse de leurs mâchoires inférieures. *Mémoires du Muséum d'Histoire Naturelle* 13: 93–125.

——. 1825c. Sur une monstruosité humaine d'un caractère encore inconnu. *Revue Médicale Française et Etrangère* 1: 372–7.

——. 1825d. Recherches sur l'organization des gavials. *Mémoires du Muséum d'Histoire Naturelle* 12: 97–155.

——. 1826. Description d'un monstre humain né avant l'ère chrétienne, comparé à un pareil monstre de l'époque actuelle; et considérations zootomiques et physiologiques sur le caractère de ces monstruosités, dites Anencéphales; sur l'indépendance de formation de chaque sexe; et sur l'existence de deux noyaux dans l'os basilaire. *Annales des Sciences Naturelles* 7: 357–88.

——. 1827a. Mémoire sur un enfant monstrueux, né dans le département d'Indre et Loire, déterminé et classé sous le nom d'Hétéradelphe de Bénais. *Memoires du Muséum d'Histoire Naturelle* 15: 385–94.

——. 1827b. Des adhérences de l'exterieur du fœtus, considérées comme le principal fait occasionnel de la monstruosité, et observations nouvelles à l'appui de cette théorie. *Archives Générales de Médecine; Journal Publié par une Société de Médecins* 14: 392–406.

——. 1829. *Cours de l'histoire naturelle de mammifères*. Paris: Pichon & Didier.

Geoffroy Saint-Hilaire, Isidore. 1832–7. *Histoire générale et particulière des anomalies de l'organisation chez l'homme et les animaux, ouvrage comprenant des recherches sur les charactères, la classification, l'influence physiologique et pathologique, les rapports généraux, les lois et les causes des monstruosités, des variétés et des vices de conformation, ou traité de tératologie* 3 vols. and atlas. Paris: J. B. Baillière.

Geological observations 2d ed.: *Geological observations on the volcanic islands and parts of South America visited during the voyage of HMS 'Beagle'*. By Charles Darwin. London: Smith, Elder & Co. 1876.

Geology of the 'Beagle': *Geological observations on coral reefs, volcanic islands, and on South America.* By Charles Darwin. London: Smith, Elder & Co. 1851.

Gibbs, Denis. 1997. The demon of dyspepsia: some nineteenth-century perceptions of disordered digestion. In *Gastroenterology in Britain: historical essays*, edited by William F. Bynum. London: Wellcome Institute for the History of Medicine.

Gilbert, Pamela. 1977. *A compendium of the biographical literature on deceased entomologists.* London: British Museum (Natural History).

Gillespie, Neal C. 1979. *Charles Darwin and the problem of creation.* Chicago and London: University of Chicago Press.

Gladstone, William Ewart. 1863. *Wedgwood: an address ... Delivered at the laying of the foundation stone of the memorial building in honour of Josiah Wedgwood at Burslem, Staffordshire, October 26, 1863.* London: John Murray.

Gladwyn, Derek. 1992. *Leigh Park: a 19th century pleasure ground.* Midhurst, West Sussex: Middleton Press.

Glaeser, Ernest. 1878. *Biographie nationale des contemporains, rédigée par une société de gens de lettres, sous la direction de M. Ernest Glaeser.* Paris: Glaeser & Cie.

Godman, Frederick Du Cane and Salvin, Osbert, eds. 1879–1915. *Biologia Centrali-Americana; or contributions to the knowledge of the fauna and flora of Mexico and Central America.* 257 pts in 63 vols. London: R. H. Porter; Dulau & Co.

Godron, Dominique Alexandre. 1859. *De l'espèce et des races dans les êtres organisés et spécialement de l'unité de l'espèce humaine.* 2 vols. Paris: J. B. Baillière.

Goebel, K. von. 1926. *Wilhelm Hofmeister. The work and life of a nineteenth century botanist.* With a biographical supplement by Constanze Ganzenmüller. Translated by H. M. Bower. Botanical editor, F. O. Bower. London: Ray Society.

Goodsir, John. 1842. History of a case in which a fluid periodically ejected from the stomach contained vegetable organisms of an undescribed form, with a chemical analysis of the fluid by George Wilson. *Edinburgh Medical and Surgical Journal* 57: 430–43.

Gorman, W. Gordon. 1910. *Converts to Rome. A biographical list of the more notable converts to the Catholic Church in the United Kingdom during the last sixty years.* New and enlarged edition. London: Sands & Co.

Goss, John. 1822. On the variation in the colour of peas, occasioned by cross impregnation. [Read 15 October 1822.] *Transactions of the Horticultural Society of London* 5 (1824): 234–7.

Gosse, Philip Henry. 1863. Microscopic observation on some seeds of orchids. *Journal of Horticulture* n.s. 4: 287–8.

Gould, John. 1837. Observations on some species of the genus *Motacilla* of Linnæus. *Magazine of Natural History* n.s. 1: 459–61.

Graepel, Peter H. 1978. Carl Friedrich von Gärtner (1772–1850): Familie, Leben, Werk: ein Beitrag zur Geschichte der Sexualtheorie und der Bastarderzeugung im Pflanzenreich. (PhD dissertation, Philipps University, Marburg.)

Graham, Richard. 1968. *Britain and the onset of modernization in Brazil 1850–1914.* Cambridge: Cambridge University Press.

Grande encyclopédie: *La grande encyclopédie inventaire raisonné des sciences, des lettres et des arts.* Edited by F. Camille Dreyfus *et al.* 31 vols. Paris: H. Lamirault; Société Anonyme de la Grande Encyclopédie. [1886–1902.]

Gray, Asa. 1849. On the composition of the plant by phytons, and some applications of phyllotaxis. *Proceedings of the American Association for the Advancement of Science; second meeting, held at Cambridge, Massachusetts, August 1849*, pp. 438–44.

——. 1856. *Manual of the botany of the northern United States: second edition; including*

Virginia, Kentucky, and all east of the Mississippi: arranged according to the natural system. New York: George P. Putnam.

Gray, Asa. 1857. *First lessons in botany and vegetable physiology, illustrated by over 360 wood engravings, from original drawings by Isaac Sprague. To which is added a copious glossary, or dictionary of botanical terms.* New York: G. P. Putnam; Ivison & Phinney.

——. 1858a. *Introduction to structural and systematic botany, and vegetable physiology, being a fifth and revised edition of the botanical text-book.* New York: Ivison & Phinney.

——. 1858b. Note on the coiling of tendrils of plants. [Read 12 August 1858.] *Proceedings of the American Academy of Arts and Sciences* 4 (1857–60): 98–9.

——. 1858–9. Diagnostic characters of new species of phænogamous plants, collected in Japan by Charles Wright, botanist of the US North Pacific Exploring Expedition . . . With observations upon the relations of the Japanese flora to that of North America, and of other parts of the northern temperate zone. [Read 14 December 1858 and 11 January 1859.] *Memoirs of the American Academy of Arts and Sciences* n.s. 6: 377–452.

——. 1859. On the coiling of tendrils. *American Journal of Science and Arts* 2d ser. 27: 277–8.

——. 1860a. Discussion between two readers of Darwin's treatise on the origin of species, upon its natural theology. *American Journal of Science and Arts* 2d ser. 30: 226–39.

[——.] 1860b. Darwin on the origin of species. *Atlantic Monthly* 6: 109–16, 229–39; Darwin and his reviewers. *Atlantic Monthly* 6: 406–25.

——. 1860c. Review of Darwin's theory on the origin of species by means of natural selection. *American Journal of Science and Arts* 2d ser. 29: 153–84.

——. 1861a. *Natural selection not inconsistent with natural theology. A free examination of Darwin's treatise on the origin of species, and of its American reviewers.* London: Trübner. Boston, Mass.: Ticknor & Fields.

——. 1861b. Characters of some Compositæ in the collection of the United States South Pacific Exploring Expedition under Captain Wilkes, with observations, &c. *Proceedings of the American Academy of Arts and Sciences* 5 (1860–2): 114–52.

——. 1862a. Fertilization of orchids through the agency of insects. *American Journal of Science and Arts* 2d ser. 34: 420–9.

——. 1862b. [Review of *Orchids*.] *American Journal of Science and Arts* 2d ser. 34: 138–44.

——. 1862c. Enumeration of the plants of Dr Parry's collection in the Rocky Mountains in 1861. *American Journal of Science and Arts* 2d ser. 33 (1862): 404–11; 34 (1862): 249–61; 330–41; 35 (1863): 137.

——. 1862d. Dimorphism in the genitalia of flowers. *American Journal of Science and Arts* 2d ser. 34: 419–20.

——. 1863a. Variation and mimetic analogy in Lepidoptera. *American Journal of Science and Arts* 2d ser. 36: 285–90.

——. 1863b. Structure and fertilization of certain orchids. *American Journal of Science and Arts* 2d ser. 36: 292–4.

Gray, Asa. 1863c. Dimorphism in the flowers of *Linum*. *American Journal of Science and Arts* 2d ser. 36: 279–84.

——. 1863d. Species, considered as to variation, geographical distribution, and succession. *American Journal of Science and Arts* 2d ser. 35: 431–44.

——. 1863e. A new character in the fruit of oaks. *American Journal of Science and Arts* 2d ser. 35: 430–1.

——. 1863f. [Review of J. D. Hooker's study of *Welwitschia*.] *American Journal of Science and Arts* 2d ser. 36: 434–9.

——. 1863g. [Review of Alphonse de Candolle's *Mémoires et souvenirs de Augustin-Pyramus de Candolle*.] *American Journal of Science and Arts* 2d ser. 35: 1–16.

——. 1863h. A revision and arrangement (mainly by the fruit) of the North American species of *Astragalus* and *Oxytropis*. *Proceedings of the American Academy of Arts and Sciences* 6 (1862–5): 188–236.

——. 1878–84. *Synoptical flora of North America*. 2 vols. New York: Invison, Blakeman, Taylor & Co.

Gray, James. 1952. *History of the Royal Medical Society, 1737–1937*. Edited by Douglas Guthrie and with a foreword by Robert Hutchison. Edinburgh: Edinburgh University Press.

Grayson, Donald K. 1983. *The establishment of human antiquity*. New York: Academic Press.

——. 1985. The first three editions of Charles Lyell's *The geological evidences of the antiquity of man*. *Archives of Natural History* 13: 105–21.

Gregory, Frederick. 1977. *Scientific materialism in nineteenth century Germany*. Dordrecht, Netherlands, and Boston, Mass.: D. Reidel Publishing Company.

Gregson, Charles Stuart. 1883. *Remarks on 'Scientific nomenclature', being a reply to the papers on 'Scientific nomenclature' published in the 'Entomologist' vol. 14*. Liverpool: Meek, Thomas & Co.

Grenand, Pierre, *et al.* 1987. *Pharmacopées traditionnelles en Guyane*. Paris: Institut Francais de Recherche Scientifique pour le Développement en Coopération avec le concours du conseil régional de la Guyane française.

Grisebach, August Heinrich Rudolph. 1845. Gentianaceae. In vol. 9, pp. 39–141, of *Prodromus sytematis naturalis regni vegetabilis*, by Augustin Pyramus de Candolle and Alphonse de Candolle. Paris: Fortin, Masson & Sociorum.

Grzimek, Bernhard, ed. 1975. *Insects*. Vol. 2 of *Grzimek's animal life encyclopedia*. New York: Van Nostrand Reinhold Company.

Gunther, Albert E. 1975. *A century of zoology at the British Museum through the lives of two keepers, 1815–1914*. London: Dawsons of Pall Mall.

Haan, Wilhelm. 1875. *Sächsisches Schriftsteller-Lexicon. Alphabetisch geordnete Zusammenstellung der im Königreich Sachsen gegenwärtig lebenden Gelehrten, Schriftsteller und Künstler, nebst kurzen biographischen Notizen*. Leipzig: R. Schaefer.

Haast, H. F. von. 1948. *The life and times of Sir Julius von Haast, explorer, geologist, museum builder*. Wellington, New Zealand: privately published.

Haast, John Francis Julius von. 1861. *Report of a topographical and geological exploration of the western districts of the Nelson Province, New Zealand.* Nelson, New Zealand: Nelson provincial government.

——. 1862a. [*Address delivered by Mr Haast at the dinner of the Philosophical Institute of Canterbury.*] Canterbury, New Zealand: Philosophical Institute of Canterbury.

——. 1862b. Notes on the geology of the province of Canterbury, NZ. *New Zealand Government Gazette. Province of Canterbury*, 9 (24 October 1862): 121–31.

——. 1863. Bemerkungen über *Strigops habroptilus*, eingesendet aus Canterbury auf Neuseeland. *Verhandlungen der Kaiserlich-Königlichen Zoologisch-Botanischen Gesellschaft in Wien* 13: 1115–20.

——. 1864. Notes on the ground-parrot of New Zealand (*Stigops habroptilus*). *Ibis* 6: 340–6.

——. 1879. *Geology of the provinces of Canterbury and Westland, New Zealand: a report comprising the results of official explorations.* Christchurch, New Zealand: The 'Times' office.

Hadfield, Miles, *et al.* 1980. *British gardeners. A biographical dictionary.* London: A. Zwemmer in association with The Condé Nast Publications.

Haeckel, Ernst. 1862. *Die Radiolarien. (Rhizopoda Radiaria.) Eine Monographie.* 2 vols. Berlin: Georg Reimer.

Haliburton, Robert Grant. 1863. *New materials for the history of man, derived from a comparison of the customs and superstitions of nations.* Halifax, Nova Scotia: n.p.

Hall, Marie Boas. 1984. *All scientists now: the Royal Society in the nineteenth century.* Cambridge: Cambridge University Press.

Hall, Philip B. 1987. Robert Swinhoe (1836–1877), FRS, FZS, FRGS: a Victorian naturalist in Treaty Port China. *Geographical Journal* 153: 37–47.

Hance, Henry Fletcher. 1871. Floræ Hongkongensis supplementum. A compendious supplement to Mr Bentham's description of the plants of the Island of Hongkong. *Journal of the Linnean Society (Botany)* 13 (1873): 95–144.

Harbottle, Thomas Benfield and Dalbiac, Philip Hugh. 1901. *Dictionary of quotations (French and Italian).* London: Swan Sonnenschein.

[Harcourt, William Vernon.] 1863. *Letters by Historicus or some questions of international law. Reprinted from "The 'Times' with considerable additions.* London and Cambridge: Macmillan.

Harkness, Stanley B. 1955. *The career of Samuel Butler (1835–1902): a bibliography.* London: Bodley Head.

Harnack, Adolf. 1900. *Geschichte der Königlich Preussischen Akademie der Wissenschaften zu Berlin im Auftrage der Akademie bearbeitet.* 3 vols. in 4. Berlin: gedruckt in der Reichsdruckerei.

Harvey, Joy. 1997. *'Almost a man of genius': Clémence Royer, feminism, and nineteen-century science.* New Brunswick, N.J., and London: Rutgers University Press.

Harvey, William Henry. 1838. *The genera of South African plants, arranged according to the natural system.* Cape Town, South Africa: A. S. Robertson.

——. 1859–63. *Thesaurus Capensis: or, illustrations of the South African flora, being figures and brief descriptions of South African plants selected from the Dublin University Herbarium.* 2 vols. in 1. Dublin: Hodges, Smith & Co. London: John van Voorst.

Harvey, William Henry and Sonder, Otto Wilhelm. 1859–65. *Flora Capensis: being a scientific description of the plants of the Cape Colony, Caffraria, & Port Natal.* 3 vols. Dublin: Hodges, Smith & Co.

Haug, Balthasar. 1790. *Das gelehrte Wirtemberg.* Stuttgart: Herzoglichen Hoben Carlsschule.

Haughton, Samuel. 1860a. Presidential address. *Journal of the Geological Society of Dublin* 8 (1857–60): 137–56.

[——.] 1860b. [Biogenesis.] *Natural History Review* 7: 23–32.

——. 1862a. On the use of nicotine in tetanus, and cases of poisoning by strychnia. [Read before the King and Queen's College of Physicians in Ireland, 19 March 1862.] *Dublin Quarterly Journal of Medical Science* 34: 172–86.

——. 1862b. On the form of the cells made by various wasps, and by the honey bee; with an appendix on the origin of species. [Read 21 November 1862.] *Proceedings of the Natural History Society of Dublin* 3 (1859–62): 128–40. [Reprinted in the *Annals and Magazine of Natural History* 3d ser. 11 (1863): 415–29.]

——. 1863. *On the form of the cells made by various wasps, and by the honey bee; with an appendix on the origin of species.* Dublin: University Press.

Haughton, W. 1862. On the unicorn of the ancients. *Annals and Magazine of Natural History* 3d ser. 10: 363–70.

HBLS: *Historisch-Biographisches Lexikon der Schweiz.* Under the auspices of the Allgemeine Geschichtsforschende Gesellschaft der Schweiz. Edited by Heinrich Türler *et al.* 7 vols. and supplement. Neuchâtel, Switzerland: Administration des Historisch-Biographischen Lexikons der Schweiz.

Heer, Oswald. 1855. Ueber die fossilen Pflanzen von St. Jorge in Madeira. *Neue Denkschriften der allgemeinen Schweizerischen Gesellschaft für die gesammten Naturwissenschaften* n.s. 5 (1857): paper 2.

——. 1855–9. *Flora tertiaria Helvetiae. Die tertiäre Flora der Schweiz.* 3 vols. Winterthur, Switzerland: J. Wurster.

——. 1860. *Untersuchungen über das Klima und die Vegetationsverhältnisse des Tertiärlandes.* Winterthur, Switzerland: Anstalt von Wurster.

Heilprin, Angelo and Heilprin, Louis, eds. [1931.] *A complete pronouncing gazetteer or geographical dictionary of the world.* 2 vols. Philadelphia and London: J. B. Lippincott.

Hemsley, W. Botting. 1906. *A new and complete index to the Botanical Magazine, from its commencement in 1787 to the end of 1904, including the first, second, and third series. To which is prefixed a history of the magazine.* London: Lovell Reeve.

Henfrey, Arthur, trans. 1852. *Principles of the anatomy and physiology of the vegetable cell,* by Hugo von Mohl. London: John van Voorst.

——, trans. 1853. Reflections on the phenomena of rejuvenescence in nature, especially in the life and development of plants. By Alexander Carl Heinrich Braun. In *Botanical and physiological memoirs*, edited by Arthur Henfrey. London: Ray Society.

Hensel, [Reinhold Friedrich]. 1860. Über Hipparion mediterraneum. [Read 28 June 1860.] *Abhandlungen der Königlichen Akademie der Wissenschaften zu Berlin. Aus dem Jahre 1860 (Physikalische Abhandlungen)* (1861): 27–121.

Henslow, George. 1866. Note on the structure of *Indigofera*, as apparently offering facilities for the intercrossing of distinct flowers. [Read 19 April 1866.] *Journal of the Linnean Society (Botany)* 9 (1867): 355–8.

Henslow, John Stevens. 1837. *Descriptive and physiological botany.* New edition. Part of Dionysius Lardner's *Cabinet cyclopædia.* London: Longman, Orme, Brown, Greene, & Longmans; John Taylor.

Herbert, William. 1818. Instructions for the treatment of the *Amaryllis longifolia*, as a hardy aquatic, with some observations on the production of hybrid plants, and the treatment of the bulbs of the genera *Crinum* and *Amaryllis.* [Read 7 July 1818.] *Transactions of the Horticultural Society of London* 3 (1820): 187–96.

——. 1819. On the production of hybrid vegetables; with the result of many experiments made in the investigation of the subject. [Read 21 December 1819.] *Transactions of the Horticultural Society of London* 4 (1822): 15–50.

——. 1837. *Amaryllidaceæ; preceded by an attempt to arrange the monocotyledonous orders, and followed by a treatise on cross-bred vegetables, and supplement.* London: James Ridgway & Sons.

——. 1846. On hybridization amongst vegetables. [Read 14 October 1846.] *Journal of the Horticultural Society of London* 2: 1–28, 81–107.

Herschel, John Frederick William. 1830. *A preliminary discourse on the study of natural philosophy.* Part of Dionysius Lardner's *Cabinet cyclopædia.* London: Longman, Rees, Orme, Brown & Green; John Taylor.

Highways (District Surveyors) Bill. [British parliamentary papers, Session 1854, 3: 51–8.]

Hildebrand, Friedrich Hermann Gustav. 1861. Die Verbreitung der Coniferen in der Jetztzeit und in den früheren geologischen Perioden. *Verhandlungen des naturhistorischen Vereines der preussischen Rheinlande und Westphalens* 18: 199–384.

——. 1863a. Die Fruchtbildung der Orchideen, ein Beweis für die doppelte Wirkung des Pollen. *Botanische Zeitung* 21: 329–33, 337–45.

——. 1863b. On the impregnation in orchids as a proof of the two different effects of the pollen. *Annals and Magazine of Natural History* 3d ser. 12: 169–74.

——. 1864. Experimente über den Dimorphismus von *Linum perenne* und *Primula sinensis. Botanische Zeitung* 22: 1–5.

Hinchcliff, Peter Bingham. 1964. *John William Colenso, bishop of Natal.* London: Thomas Nelson.

Histoire de l'instruction publique dans le Canton de Neuchâtel: *Histoire de l'instruction publique dans le Canton de Neuchâtel, de l'origine a nos jours … Ouvrage publié à l'occasion de l'Exposition Nationale Suisse, Berne 1914, par le Département de l'Instruction Publique avec la collaboration de membres du corps enseignant.* Neuchâtel, Switzerland: Attinger Frères. [1914.]

Historical register of the University of Cambridge: *The historical register of the University of Cambridge, being a supplement to the* Calendar *with a record of university offices, honours, and distinctions to the year 1910*. Edited by J. R. Tanner. Cambridge: Cambridge University Press. 1917.

Hitchcock, Edward. 1836. Ornithichnology. Description of the foot marks of birds, (Ornithichnites) on New Red Sandstone in Massachusetts. *American Journal of Science and Arts* 29: 307–40.

Hochstetter, Ferdinand Christian von. 1863. *Neu-Seeland.* Stuttgart: Cotta'scher.

Hodge, M. J. S. 1985. Darwin as a lifelong generation theorist. In *The Darwinian heritage*, edited by David Kohn. Princeton, N.J.: Princeton University Press in association with Nova Pacifica.

Hofmeister, Wilhelm. 1858. Neuere Beobachtungen über Embryobildung der Phanerogamen. *Jahrbücher für wissenschaftliche Botanik* 1: 82–188.

Hooker, Joseph Dalton. 1844–7. *Flora Antarctica.* 1 vol. and 1 vol. of plates. Pt 1 of *The botany of the Antarctic voyage of HM discovery ships* Erebus *and* Terror *in the years 1839–1843, under the command of Captain Sir James Clark Ross.* London: Reeve Brothers. 1844–60.

——. 1849. *The rhododendrons of Sikkim-Himalaya; being an account, botanical and geographical, of the rhododendrons recently discovered in the mountains of eastern Himalaya, from drawings and descriptions made on the spot, during a government botanical mission to that country.* Edited by William Jackson Hooker. London: Reeve, Benham & Reeve.

——. 1852. On the climate and vegetation of the temperate and cold regions of East Nepal and the Sikkim Himalaya Mountains. *Journal of the Horticultural Society of London* 7: 69–131.

——. 1853. *Introductory essay to the flora of New Zealand.* London: Lovell Reeve.

——. 1853–5. *Flora Novæ-Zelandiæ.* 2 vols. Pt 2 of *The botany of the Antarctic voyage of HM discovery ships* Erebus *and* Terror, *in the years 1839–1843, under the command of Captain Sir James Clark Ross.* London: Lovell Reeve. 1844–60.

——. 1854a. On the possibility of impregnating ovules after the removal of the stigma. *Gardeners' Chronicle*, 30 September 1854, p. 629.

——. 1854b. *Himalayan journals; or, notes of a naturalist in Bengal, the Sikkim and Nepal Himalayas, the Khasia Mountains, &c.* 2 vols. London: John Murray.

——. 1854c. On the functions and structure of the rostellum of *Listera ovata.* [Read 15 June 1854.] *Philosophical Transactions of the Royal Society of London* 144: 259–63.

——. 1859. *On the flora of Australia, its origin, affinities, and distribution; being an introductory essay to the flora of Tasmania.* London: Lovell Reeve.

——. 1860a. *Flora Tasmaniæ.* 2 vols. Pt 3 of *The botany of the Antarctic voyage of HM discovery ships* Erebus *and* Terror, *in the years 1839–1843, under the command of Captain Sir James Clark Ross.* London: Reeve Brothers; Lovell Reeve. 1844–60.

——. 1860b. Outlines of the distribution of Arctic plants. [Read 21 June 1860.] *Transactions of the Linnean Society of London* 23 (1862): 251–348.

——. 1861. On the vegetation of Clarence Peak, Fernando Po; with descriptions of the plants collected by Mr Gustav Mann on the higher parts of that mountain. [Read 7 March 1861.] *Journal of the Proceedings of the Linnean Society (Botany)* 6: 1–23.

——. 1863a. On *Welwitschia*, a new genus of Gnetaceæ. [Read 18 December 1862 and 16 January 1863.] *Transactions of the Linnean Society of London* 24 (1864): 1–48.

Hooker, Joseph Dalton. 1863b. On the plants of the temperate regions of the Cameroons Mountains and islands in the Bight of Benin; collected by Mr Gustav Mann, government botanist. [Read 5 November 1863.] *Journal of the Proceedings of the Linnean Society (Botany)* 7 (1864): 171–240.

[——.] 1863c. On species. *Natural History Review* n.s. 3: 189–201.

——. 1864. Note on the replacement of species in the colonies and elsewhere. *Natural History Review* n.s. 4: 123–7.

——. 1864–7. *Handbook of the New Zealand flora: a systematic description of the native plants of New Zealand and the Chatham, Kermadec's, Lord Auckland's, Campbell's, and MacQuarrie's Islands.* 2 vols. London: Lovell Reeve.

——. 1875–97. *The flora of British India.* Assisted by various botanists. 7 vols. London: L. Reeve.

Hooker, Joseph Dalton and Thomson, Thomas. 1855. *Flora Indica: being a systematic account of the plants of British India, together with observations on the structure and affinities of their natural orders and genera.* London: W. Pamplin.

Hooker, William Jackson. 1830. *The British flora; comprising the phænogamous, or flowering plants, and the ferns.* London: Longman, Rees, Orme, Brown, & Green.

——. 1840. *Flora Boreali-Americana; or the botany of the northern parts of British America: compiled principally from the plants collected by Dr Richardson & Mr Drummond on the late northern expeditions, under command of Captain Sir John Franklin RN. To which are added . . . those of Mr Douglas, from north-west America, and of other naturalists.* 2 vols. London: Henry G. Bohn.

Hooker, William Jackson and Arnott, George Arnott Walker. 1860. *The British flora comprising the phænogamous or flowering plants and the ferns with numerous figures illustrative of the umbelliferous plants, the composite plants, the grasses, and the ferns.* 8th edition, revised and corrected. London: Longman, Green, Longman, & Roberts.

Hopkins, William. 1860. Physical theories of the phenomena of life. *Fraser's Magazine* 61: 739–52; 62: 74–90.

Horsburgh, E. L. S. 1980. *Bromley, Kent: from the earliest times to the present century, compiled from materials collected from original sources by various hands.* Reprint edition. Chislehurst, Kent: Lodgemark Press.

Huggins, A. K. 1970. *British postal stationery. A priced hand book of the postal stationery of Great Britain.* London: Philatelic Society.

Hull, David L. 1973. *Darwin and his critics: the reception of Darwin's theory of evolution by the scientific community.* Cambridge, Mass.: Harvard University Press.

Humboldt, Alexander von, *et al.* 1815–25. *Nova genera et species plantarum quas in peregrinatione orbis novi collegerunt, descripserunt, partim adumbraverunt.* 7 vols. Pt 6 of *Voyage de Humboldt et Bonpland.* Paris: Libraire Grecque-Latine-Allemande [& others].

Humboldt, Alexander von and Bonpland, Aimé Jacques Alexandre. 1822. *Personal narrative of travels to the equinoctial regions of the new continent, during the years 1799–1804.* Translated by Helen Maria Williams. 3d edition. 2 vols. in 1. London: Longman, Hurst, Rees, Orme, & Brown. Vols. 1 and 2 of *Personal narrative of travels to the equinoctial regions of the new continent, during the years 1799–1804*, by Alexander von Humboldt and Aimé Jacques Alexandre Bonpland. Translated by Helen Maria Williams. 7 vols. in 6. Pt 1 of *Voyage de Humboldt et Bonpland.* London: Longman, Hurst, Rees, Orme, & Brown. 1814–29.

Hunt, Edward Bissell. 1863. On the origin, growth, substructure and chronology of the Florida reef. *American Journal of Science and Arts* 2d ser. 35: 197–210.

Hutchinson, Horace Gordon. 1914. *Life of Sir John Lubbock, Lord Avebury.* 2 vols. London: Macmillan.

Huxley, Leonard, ed. 1900. *Life and letters of Thomas Henry Huxley.* 2 vols. London: Macmillan.

——, ed. 1918. *Life and letters of Sir Joseph Dalton Hooker, OM, GCSI.* Based on materials collected and arranged by Lady Hooker. 2 vols. London: John Murray.

Huxley, Thomas Henry. 1859. On the persistent types of animal life. [Read 3 June 1859.] *Notices of the Proceedings at the Meetings of the Members of the Royal Institution of Great Britain* 3 (1858–62): 151–3.

[——.] 1860. Darwin on the origin of species. *Westminster Review* n.s. 17: 541–70.

——. 1861a. Preliminary essay upon the systematic arrangement of the fishes of the Devonian epoch. *Memoirs of the Geological Survey of the United Kingdom. Figures and Descriptions illustrative of British Organic Remains*, decade 10, pp. 1–40.

——. 1861b. *Glyptolæmus kinnairdi. Memoirs of the Geological Survey of the United Kingdom. Figures and Descriptions illustrative of British Organic Remains*, decade 10, pp. 41–6.

——. 1861c. *Phaneropleuron andersoni. Memoirs of the Geological Survey of the United Kingdom. Figures and descriptions illustrative of British organic remains*, decade 10, pp. 47–9.

——. 1862a. The anniversary address. [Read 21 February 1862.] *Quarterly Journal of the Geological Society of London* 18: xl–liv.

——. 1862b. Description of *Anthracosaurus Russelli*, a new Labyrinthodont from the Lanarkshire coal-field. [Read 3 December 1862.] *Quarterly Journal of the Geological Society of London* 19 (1863): 56–68.

——. 1862c. Description of a new specimen of *Glyptodon* recently acquired by the Royal College of Surgeons of England. [Read 18 December 1862.] *Proceedings of the Royal Society of London* 12 (1862–3): 316–26.

——. 1863a. *On our knowledge of the causes of the phenomena of organic nature. Being six lectures to working men, delivered at the Museum of Practical Geology.* London: Robert Hardwicke.

——. 1863b. *Evidence as to man's place in nature.* London: Williams & Norgate.

——. 1863c. Professor Huxley's lectures at the Royal College of Surgeons. The lectures on the vertebrate skull. *Medical Times and Gazette* n.s. 26: 607–10, 633–5; 27: 1–6, 57–8, 107–10, 189–92, 371–4, 425–9, 475–8, 529–31, 579–80, 607–9, 633–6, 663–8.

——. 1863d. *Evidence as to man's place in nature.* New York: D. Appleton.

——. 1864a. *Lectures on the elements of comparative anatomy. On the classification of animals and on the vertebrate skull.* London: John Churchill & Sons.

Huxley, Thomas Henry. 1864b. On the osteology of the genus *Glyptodon.* [Read 28 January 1864.] *Philosophical Transactions of the Royal Society of London* 155 (1865): 31–70.

——. 1865. On a collection of vertebrate fossils from the Panchet Rocks, Ranigunj, Bengal. *Memoirs of the Geological Survey of India. Palæontologia Indica* ser. 4 (*Indian Pretertiary Vertebrata*), 1 (1865–85), pt 1.

——. 1866a. *Lessons in elementary physiology.* London: Macmillan.

——. 1866b. Illustrations of the structure of the crossopterygian ganoids. *Memoirs of the Geological Survey of the United Kingdom. Figures and Descriptions illustrative of British Organic Remains*, decade 12, pp. 3–4.

——. 1871. *A manual of the anatomy of vertebrated animals.* London: J. & A. Churchill.

——. 1872. *Holophagus gulo. Memoirs of the Geological Survey of the United Kingdom. Figures and Descriptions illustrative of British Organic Remains*, decade 13, plate 10.

——. 1877. *A manual of the anatomy of invertebrated animals.* London: J. & A. Churchill.

——. 1893. *Darwiniana.* Vol. 2 of *Collected essays.* By T. H. Huxley. 9 vols. London: Macmillan. 1893–4.

Huxley, Thomas Henry and Etheridge, Robert. 1865. *A catalogue of the collection of fossils in the Museum of Practical Geology, with an explanatory introduction.* London: Her Majesty's Stationery Office.

IBN: *Index bio-bibliographicus notorum hominum.* Pt C: *Corpus alphabeticum sectio generalis.* Edited by Jean-Pierre Lobies *et al.* 60 vols. (A–Durau, Pierre). Osnabrück, Germany: Biblio Verlag. 1974–92.

'Illegitimate offspring of dimorphic and trimorphic plants': On the character and hybrid-like nature of the offspring from the illegitimate unions of dimorphic and trimorphic plants. By Charles Darwin. [Read 20 February 1868.] *Journal of the Linnean Society (Botany)* 10 (1869): 393–437.

Imperial directory of universal biography: *The imperial directory of universal biography: comprising a series of original memoirs of distinguished men, of all ages and all nations.* Edited by John Francis Waller. 3 vols. London: William Mackenzie. [1857–63.]

Index animalium: *Index animalium sive index nominum quae ab* A.D. MDCCLVIII *generibus et speciebus animalium imposita sunt.* By Charles Davies Sherborn. 10 vols. Cambridge: Cambridge University Press. London: British Museum. 1902–32.

Index biographique de l'Académie des Sciences: *Index biographique de l'Académie des Sciences du 22 Décembre 1666 au 1er Octobre 1978.* Paris: Gauthier-Villars. 1979.

Index juridicus: *The Scottish law list: a legal almanac, directory and remembrancer.* Edinburgh and London: William Blackwood & Sons. 1848–9. *Index juridicus: the annual Scottish law list and legal directory . . . including correct and complete indices or lists of the judges and officers of the several courts of justice; of the advocates, writers to the signet, and certificated practitioners in Edinburgh and Glasgow, and of the certificated county practitioners . . . et cætera.* Edinburgh: Adam & Charles Black [& others]. 1849–1961.

Index Kewensis: *Index Kewensis: plantarum phanerogamarum, nomina et synonyma omnium generum et specierum . . . nomine recepto auctore patria unicuique plantae subjectis.* 4 vols. in

4 pts, and 20 supplements. Compiled by Benjamin Daydon Jackson *et al.* Oxford: Clarendon Press; Royal Botanic Gardens, Kew. 1893–1996.

Index Londinensis: *Iconum botanicarum index Londinensis, sive G. A. Pritzeli iconum botanicarum index locupletissimus emendatus auctus et ad annum* MCMXX *productus*. Prepared under the auspices of the Royal Horticultural Society of London at the Royal Botanic Gardens, Kew, by O. Stapf. 6 vols. and supplement in 2 vols. Oxford: Clarendon Press. 1929–41.

India list: *The East-India register . . . from the official returns received at the East-India House. The India Office list . . . Compiled from official records by direction of the secretary of state for India in council.* London: Wm. H. Allen [& others]. 1806–1937.

Insectivorous plants. By Charles Darwin. London: John Murray. 1875.

Jackson, Benjamin Daydon. 1906. *George Bentham*. London: J. M. Dent. New York: E. P. Dutton.

Jäger, Gustav. 1862. Die Darwin'sche Theorie über die Entstehung der Arten. *Schriften des Vereines zur Verbreitung naturwissenschaftlicher Kenntnisse in Wien 1860–1* 1: 83–110.

——. 1864. *Zoologische Briefe*. Vienna: Wilhelm Braumüller.

Jalland, Pat. 1996. *Death in the Victorian family*. Oxford: Oxford University Press.

Jameson, William. 1858. Excursion made from Quito to the River Napo, January to May 1857. *Journal of the Royal Geographical Society* 28: 337–49.

——. 1865. *Synopsis plantarum æquatoriensium, exhibens plantas praecipue in regione temperata et frigida crescentes, secundum systematam naturalem descriptas viribus medicatis et usibus oeconomicis plurimarum adjectis*. 3 vols. Quito, Ecuador: typis J. P. Sanz.

Jamieson, Thomas Francis. 1863. On the parallel roads of Glen Roy, and their place in the history of the glacial period. [Read 21 January 1863.] *Quarterly Journal of the Geological Society of London* 19: 235–59.

Janal, Martin Joel. [1988.] Classification of the Foraminifera: a case study in taxonomy and its history. (PhD dissertation: University of Cambridge.)

Jenkins, Brian. 1974–80. *Britain & the war for the Union*. 2 vols. Montreal, Quebec, and London: McGill-Queen's University Press.

Jenyns, Leonard. 1862. *Memoirs of the Rev. John Stevens Henslow, late rector of Hitcham, and professor of botany in the University of Cambridge*. London: John van Voorst.

Johnson, George W., ed. 1857. *The cottage gardeners' dictionary*. London: W. Kent & Co.

Jones, Hugh Percy, ed. 1900. *A new dictionary of foreign phrases and classical quotations . . . with English translations or equivalents*. London: Charles William Deacan.

Journal of researches: *Journal of researches into the geology and natural history of the various countries visited by HMS Beagle, under the command of Captain FitzRoy, RN, from 1832 to 1836*. By Charles Darwin. London: Henry Colburn. 1839.

Journal of researches 2d ed.: *Journal of researches into the natural history and geology of the countries visited during the voyage of HMS Beagle round the world, under the command of Capt. FitzRoy RN*. 2d edition, corrected, with additions. By Charles Darwin. London: John Murray. 1845.

Journal of researches (1860): *Journal of researches into the natural history and geology of the countries visited during the voyage of HMS Beagle around the world, under the command of Capt. FitzRoy RN.* By Charles Darwin. Reprint edition. London: John Murray. 1860.

Jubiläums-Katalog: *Jubiläums-Katalog der E. Schweizerbart'schen Verlagsbuchhandlung (Erwin Nägele) G.m.b.H., Stuttgart, 1826–1926.* Stuttgart: E. Schweizerbart. 1926.

Junker, Thomas. 1989. *Darwinismus und Botanik. Rezeption, Kritik und theoretische Alternativen im Deutschland des 19. Jahrhunderts.* Stuttgart: Deutscher Apotheker Verlag.

Jussieu, Antoine de Laurent de. 1789. *Genera plantarum secundum ordines naturales disposita, juxta methodum in horto regio Parisiensi exaratum.* Paris: Herissant & Barrois.

Keynes, Margaret. 1943. *Leonard Darwin, 1850–1943.* Cambridge: privately printed at Cambridge University Press.

Kinglake, Alexander William. 1863–87. *The invasion of the Crimea: its origin, and an account of its progress down to the death of Lord Raglan.* 8 vols. Edinburgh and London: William Blackwood & Sons.

Kirby, William Forsell. 1863. On the geographical distribution of European *Rhopalocera.* [Read 5 January 1863.] *Transactions of the Entomological Society of London* 3d ser. 1 (1862–4): 481–91.

Knight, Thomas Andrew. 1815. On the want of permanence of character in varieties of fruit, when propagated by grafts and buds. [Read 4 April 1815.] *Transactions of the Horticultural Society of London* 2 (1817): 160–1.

——. 1817. An account of a peach tree, produced from the seed of the almond tree, with some observations on the origin of the peach tree. [Read 7 October 1817.] *Transactions of the Horticultural Society of London* 3 (1820): 1–5.

——. 1821. Observations on hybrids. [Read 6 February 1821.] *Transactions of the Horticultural Society of London* 4 (1822): 367–73.

——. 1823. Some remarks on the supposed influence of the pollen in cross breeding, upon the colour of the seed-coats of plants, and the qualities of their fruits. [Read 3 June 1823.] *Transactions of the Horticultural Society of London* 5 (1824): 377–80.

——. 1826. Notice of new or remarkable varieties of fruits, ripened in the summer and autumn of the years 1823 and 1824, which were exhibited at meetings of the Horticultural Society. *Transactions of the Horticultural Society of London* 6: 392–7.

Koch, Wilhelm Daniel Joseph. 1843–4. *Synopsis florae Germanicae et Helveticae, exhibens stirpes phanerogamas rite cognitas, praemissa generum dispositione secundum classes et ordines systematis Linnaeani conscripta.* 2d edition. 2 vols. Frankfurt: Fridericus Wilmans. Leipzig: Gebhardt & Reisland.

Koch, Wilhelm Daniel Joseph and Mertens, Franz Carl. 1823–39. *J. C. Röhlings Deutschlands flora. Nach einem veränderten und erweiterten Plane bearbeitet.* 5 vols. Frankfurt: Friedrich Wilmans.

Kohn, David. 1981. On the origin of the principle of diversity. *Science* 213: 1105–8.

Kölreuter, Joseph Gottlieb. 1761–6. *Vorläufige Nachricht von einigen das Geschlecht der Pflanzen betreffenden Versuchen und Beobachtungen, nebst Fortsetzungen 1, 2 und 3.* 3 pts in 1 vol. Leipzig: Gleditschen Handlung.

Kordes, Berend. 1797. *Lexikon der jetztlebenden schleswig-holsteinischen und eutinischen Schriftsteller.* Schleswig, Germany: Johann Gottlob Röhss.

Krohn, August David. 1859. Beobachtungen über den Cementapparat und die weiblichen Zeugungsorgane einiger Cirripedien. *Archiv für Naturgeschichte* 25 (pt 1): 355–64.

Krok, Thorgny Ossian Bolivar Napoleon. 1925. *Bibliotheca botanica suecana ab antiquissimis temporibus ad finem anni* MCMXVIII . . . *Svensk botanisk litteratur från äldsta tider T.O.M. 1918.* Uppsala and Stockholm: Almquist & Wiksells.

Laird, John. 1864. *Correspondence between Her Majesty's government and Messrs Laird Brothers; . . . respecting the iron-clad vessels building at Birkenhead, 1863–4.* London: Vacher & Sons.

Lamarck, Jean Baptiste Pierre Antoine de Monet de. 1801. *Système des animaux sans vertèbres, ou tableau général des classes, des ordres et des genres de ces animaux; . . . et suivant l'arrangement établi dans les galeries du Muséum d'Hist. Naturelle, . . . Précédé du discours d'ouverture du cours de zoologie, donné dans le Muséum National d'Histoire Naturelle l'an 8 de la République.* Paris: the author; Deterville.

——. 1809. *Philosophie zoologique; ou exposition des considérations relatives à l'histoire naturelle des animaux; à la diversité de leur organisation . . . et les autres l'intelligence de ceux qui en sont doués.* 2 vols. Paris: Dentu; the author.

——. 1815–22. *Histoire naturelle des animaux sans vertèbres, présentant les charactères généraux et particuliers de ces animaux, leur distribution, leurs classes, leurs familles, leurs genres, et la citation des principales espèces qui s'y rapportent; précédée d'une introduction offrant la détermination des caractères essentiels de l'animal, sa distinction du végétal et des autres corps naturels, enfin, l'exposition des principes fondamentaux de la zoologie.* 7 vols. Paris: Verdière [& others].

——. 1830. *Philosophie zoologique, ou exposition des considérations relatives à l'histoire naturelle des animaux; à la diversité leur organisation . . . et les autres l'intelligence de ceux qui en sont doués.* New edition. 2 vols. Paris: J. B. Baillière.

Lamarck, Jean Baptiste Pierre Antoine de Monet de and Poiret, Jean Louis Marie. 1783–1817. *Encyclopédie méthodique. Botanique.* 8 vols. and supplement in 5 vols. Paris: Panckoucke; Agasse.

Lamb's textile industries of the United States: *Lamb's textile industries of the United States, embracing biographical sketches of prominent men and a historical résumé of the progress of textile manufacture from the earliest records to the present time.* 2 vols. Edited by John Howard Brown *et al.* Boston, Mass.: James H. Lamb. 1911–16.

Lankester, Edwin, trans. 1849. *Principles of scientific botany; or, botany as an inductive science.* By M. J. Schleiden. London: Longman, Brown, Green, & Longmans.

Larousse: *Larousse du XXe siècle.* Under the direction of Paul Augé. 6 vols. Paris: Larousse. 1928–33. *Grand dictionnaire encyclopédique Larousse.* 10 vols. Paris: Larousse. 1982–5.

La Touche, James Digues. 1884. *A hand-book of the geology of Shropshire.* London: Edward Stanford. Shrewsbury, Shropshire: Adnitt & Naunton.

Laugel, Antoine Auguste. 1860. Nouvelle théorie d'histoire naturelle. L'origine des espèces. *Revue des Deux Mondes* 2d ser. 26: 644–71.

Law list: *Clarkes' new law list; being a list of the judges and officers of the different courts of justice . . . and a variety of other useful matter.* London: W. Clarke. 1816–40. *The law list; comprising the judges and officers of the different courts of justice: counsel, special pleaders, draftsmen, conveyancers, attorneys, notaries, &c., in England and Wales . . . and a variety of other useful matter.* London: V. & R. Stevens & G. S. Norton [& others]. 1841–1969.

Leading men of London: *Leading men of London: a collection of biographical sketches with portraits.* London: British Biographical Co. 1895.

Lecoq, Henri. 1845. *De la fécondation naturelle et artificielle des végétaux et de l'hybridation, considérée dans ses rapports avec l'horticulture, l'agriculture et la sylviculture . . . Contentant les moyens pratiques d'opérer l'hybridation et de créer facilement des variétés nouvelles.* Paris: Audot.

——. 1854–8. *Etudes sur la géographie botanique de l'Europe et en particulier sur la végétation du plateau central de la France.* 9 vols. Paris: J. B. Baillière.

——. 1862. *De la fécondation naturelle et artificielle des végétaux et de l'hybridation. Considérée dans ses rapports avec l'horticulture, l'agriculture et la sylviculture. Contenant les moyens pratiques d'opérer l'hybridation et de créer facilement des variétés nouvelles.* 2d edition. Paris: Libraire agricole de la Maison rustique.

Leidy, Joseph. 1858. Notice of remains of extinct Vertebrata, from the valley of the Niobrara River, collected during the exploring expedition of 1857, in Nebraska, under the command of Lieut. G. K. Warren, US Top. Eng., by Dr F. V. Hayden, geologist to the expedition. *Proceedings of the Academy of Natural Sciences of Philadelphia* 10: 20–9.

——. 1863. Pliocene fossil fauna of the Niobrara River, in Nebraska. *Annals and Magazine of Natural History* 3d ser. 11: 148–50.

Le Maout, Emmanuel and Decaisne, Joseph. 1873. *A general system of botany, descriptive and analytical.* 2 pts in 1 vol. Translated by Frances Harriet Hooker and edited by Joseph Dalton Hooker. London: Longmans, Green & Co.

Lesquereux, Leo. 1859–63. On some questions concerning the coal formations of North America. *American Journal of Science and Arts* 2d ser. 28 (1859): 21–37; 30 (1860): 63–74, 367–84; 32 (1861): 15–25, 193–205; 33 (1862): 206–16; 35 (1863): 375–86.

Lewis, Charlton T. and Short, Charles. 1969. *A Latin dictionary, founded on Andrews' edition of Freund's Latin dictionary.* Revised, enlarged, and in great part rewritten. Oxford: Clarendon Press.

Lewis, John Penry. 1913. *List of inscriptions on tombstones and monuments in Ceylon, of historical or local interest, with an obituary of persons uncommemorated.* Colombo, Sri Lanka: n.p.

Leyland, John, ed. 1888. *Contemporary medical men and their professional work: biographies of leading physicians and surgeons, with portraits, from the 'Provincial Medical Journal'.* 2 vols. Leicester: Provincial Medical Journal.

Liddon, Henry Parry. 1893–7. *Life of Henry Bouverie Pusey. Doctor of divinity, canon of Christ Church, Regius professor of Hebrew in the University of Oxford.* 4 vols. Edited and prepared for publication by J. O. Johnston *et al.* London and New York: Longmans, Green & Co.

Liebig, Justus von. 1840–4. *Traité de chimie organique.* 3 vols. Paris: Fortin, Masson & Cie.

——. 1862. *Die Chemie in ihrer Anwendung auf Agricultur und Physiologie.* 7th edition. 2 vols. Brunswick, Germany: Friedrich Vieweg & Sohn.

Lindley, John. 1852–5. *Folia orchidacea. An enumeration of the known species of orchids.* London: J. Matthews.

——. 1853. *The vegetable kingdom; or, the structure, classification, and uses of plants, illustrated upon the natural system.* 3d edition with corrections and additional genera. London: Bradbury & Evans.

——. 1855. *The theory and practice of horticulture; or an attempt to explain the chief operations of gardening upon physiological grounds.* 2d edition. London: Longman, Brown, Green, & Longmans.

Linnaeus, Carl. 1753. *Species plantarum, exhibentes plantas rite cognitas, ad genera relatas, cum differentiis specificis, nominibus trivialibus, synonymis selectis, locis natalibus, secundum systema sexuale digestas.* 2 vols. Stockholm: Laurentius Salvius.

Linnell, David. 1994. *Blake, Palmer & Linnell. The life of John Linnell.* Lewes, Sussex: Book Guild.

List of the Geological Society of London. London: [Geological Society of London]. 1864–1934.

List of the Linnean Society of London. London: [Linnean Society of London]. 1805–1939.

List of the Society of Antiquaries of London. London: Society of Antiquaries of London. 1853–86.

Living Cirripedia (1851): *A monograph of the sub-class Cirripedia, with figures of all the species. The Lepadidæ; or, pedunculated cirripedes.* By Charles Darwin. London: Ray Society. 1851.

Living Cirripedia (1854): *A monograph of the sub-class Cirripedia, with figures of all the species. The Balanidæ (or sessile cirripedes); the Verrucidæ, etc.* By Charles Darwin. London: Ray Society. 1854.

LL: *The life and letters of Charles Darwin, including an autobiographical chapter.* Edited by Francis Darwin. 3 vols. London: John Murray. 1887–8.

[Logan, William Edmund.] 1863. *Geological Survey of Canada. Report of progress from its commencement to 1863.* Montreal, Quebec: Dawson Brothers. London: Baillière.

Longmate, Norman. 1978. *The hungry mills. The story of the Lancashire cotton famine, 1861–5.* London: Temple Smith.

Lord, John Keast. 1863. Notes on two new species of mammals. *Proceedings of the scientific meetings of the Zoological Society of London* (1863): 95–8.

——. 1866. *The naturalist in Vancouver Island and British Columbia.* 2 vols. London: Richard Bentley.

Loring, Charles Greely. 1862. *Correspondence on the present relations between Great Britain and the United States of America.* Boston, Mass.: Little, Brown & Co.

Loring, Charles Greely. 1863. *Neutral relations of England and the United States.* Boston, Mass.: W. V. Spencer.

——. 1864. *England's liability for indemnity: remarks on the letter of 'Historicus'* [William Vernon Harcourt] *dated November 4th, 1863; printed in the London 'Times', November 7th; and reprinted in the 'Boston Daily Advertiser', November 25th.* Boston, Mass.: W. V. Spencer.

Loudon, John Claudius. [1835.] *An encyclopædia of gardening; comprising the theory and practice of horticulture, floriculture, arboriculture, and landscape-gardening . . . in the British Isles.* New ed. London: Longmans.

Lubbock, John. 1861. The kjökkenmöddings: recent geologico-archæological researches in Denmark. *Natural History Review* n.s. 1: 489–504.

——. 1862a. Notes on the *Thysanura.* [Read 6 February and 19 June 1862.] *Transactions of the Linnean Society of London* 23: 429–48, 589–601.

[——.] 1862b. Note on the habits of vipers. *Natural History Review* n.s. 2: 118–20.

——. 1862c. On the evidence of the antiquity of man, afforded by the physical structure of the Somme valley. *Natural History Review* n.s. 2: 244–69.

[——.] 1863a. [North American archæology.] *Natural History Review* n.s. 3: 1–26.

[——.] 1863b. [Review of Daniel Wilson's *Prehistoric Man.*] *Natural History Review* n.s. 3: 26–30.

[——.] 1863c. [Review of Charles Lyell's *Antiquity of man.*] *Natural History Review* n.s. 3: 211–19.

——. 1863d. On the ancient lake-habitations of Switzerland. [Read 27 February 1863.] *Notices of the Proceedings at the Meetings of the Members of the Royal Institution of Great Britiain* 4 (1862–6): 29–40.

——. 1863e. A visit to the ancient shell-mounds of Scotland. *Natural History Review* n.s. 3: 415–22.

——. 1863–5. On the development of *Chloëon (Ephemera) dimidiatum.* [Read 15 January 1863 and 21 December 1865.] *Transactions of the Linnean Society of London* 24 (1864): 61–78; 25 (1866): 477–92.

[——.] 1864. Huxley's lectures on the origin of species. *Natural History Review* n.s. 1: 37–43

——. 1866. On the metamorphoses of insects. [Read before the Royal Institution of Great Britain, 9 March 1866.] *Annals and Magazine of Natural History* 3d ser. 17: 375–81.

Lurie, Edward. 1960. *Louis Agassiz: a life in science.* Chicago: University of Chicago Press.

Lyell, Charles. 1830–3. *Principles of geology, being an attempt to explain the former changes of the earth's surface, by reference to causes now in operation.* 3 vols. London: John Murray.

——. 1838. *Elements of geology.* London: John Murray.

——. 1850. *Principles of geology; or the modern changes of the earth and its inhabitants considered as illustrative of geology.* 8th edition. London: John Murray.

——. 1851a. *A manual of elementary geology: or, the ancient changes of the earth and its inhabitants as illustrated by geological monuments.* 3d edition, revised. London: John Murray.

Lyell, Charles. 1851b. Anniversary address of the president. [Read 21 February 1851.] *Quarterly Journal of the Geological Society of London* 7: xxv–lxxvi.

——. 1853. *Principles of geology; or, the modern changes of the earth and its inhabitants considered as illustrative of geology.* 9th edition, entirely revised. London: John Murray.

——. 1857. *Supplement to the fifth edition of a 'Manual of elementary geology'.* London: John Murray.

——. 1863a. *The geological evidences of the antiquity of man with remarks on theories of the origin of species by variation.* London: John Murray.

——. 1863b. *The geological evidences of the antiquity of man with remarks on theories of the origin of species by variation.* 2d edition, revised. London: John Murray.

——. 1863c. *The geological evidences of the antiquity of man with remarks on theories of the origin of species by variation.* 3d edition, revised. London: John Murray.

Lyell, Katherine Murray, ed. 1881. *Life, letters and journals of Sir Charles Lyell, Bart.* 2 vols. London: John Murray.

——, ed. 1890. *Memoir of Leonard Horner . . . consisting of letters to his family and from some of his friends.* 2 vols. London: privately printed.

Lyons, Henry. 1944. *The Royal Society 1660–1940: a history of its administration under its charters.* Cambridge: Cambridge University Press.

Mabberley, David J. 1985. *Jupiter botanicus. Robert Brown of the British Museum.* Brunswick, Germany: J. Cramer. London: British Museum (Natural History).

M'Andrew, Robert and Woodward, Henry. 1864. Species of Mollusca obtained in Corunna Bay in May 1863. *Annals and Magazine of Natural History* 3d ser. 14: 232–4.

McCosh, James. 1851. Some remarks on the plant morphologically considered. *Botanical Gazette* 3: 118–23.

——. 1852. Morphological analogy between the disposition of the branches of exogenous plants and the venation of their leaves. *Report of the 22d meeting of the British Association for the Advancement of Science, held at Belfast*, Transactions of the sections, pp. 66–8.

——. 1854. Some observations on the morphology of pines and firs. *Report of the 24th meeting of the British Association for the Advancement of Science, held at Liverpool*, Transactions of the sections, pp. 99–100.

McCosh, James and Dickie, George. 1856. *Typical forms and special ends in creation.* Edinburgh: Thomas Constable.

Macfadyen, James. 1830. Some remarks on the species of the genus *Citrus*, which are cultivated in Jamaica. *Botanical miscellany* 1: 295–304.

——. 1831. Sketch of a short botanical excursion in Jamaica. *Botanical miscellany* 2: 110–21.

M'Intosh, Charles. 1853–5. *The book of the garden.* 2 vols. Edinburgh and London: William Blackwood.

Mackay, Robert William. 1850. *The progress of the intellect, as exemplified in the religious development of the Greeks and Hebrews.* 2 vols. London: John Chapman.

MacKenzie, John M. 1988. *The empire of nature: hunting, conservation and British imperialism.* Manchester and New York: Manchester University Press.

Mackie, Samuel Joseph. 1863. The aeronauts of the Solenhofen age. *Geologist* 6: 1–8.

MacLeod, Roy M. 1971. Of medals and men: a reward system in Victorian science, 1826–1914. *Notes and Records of the Royal Society of London* 26: 81–105.

Macmillan dictionary of women's biography: *The Macmillan dictionary of women's biography.* 2d edition. Compiled and edited by Jennifer Uglow, assisted (1st edition) by Frances Hinton. London and Basingstoke, Hampshire: Macmillan Press. 1989.

McNab, James. 1872. On the discoveries of Mr John Jeffrey and Mr Robert Brown, collectors to the botanical expeditions to British Columbia between the years 1850 and 1866, with remarks on the cultivation of certain species. [Read 14 March 1872.] *Transactions and Proceedings of the Botanical Society* [*of Edinburgh*] 11: 322–38.

McNaughton, William D. 1993. *The Scottish Congregational ministry 1794–1993*. Glasgow: Congregational Union of Scotland.

McPherson, James M. 1988. *Battle cry of freedom: the Civil War era.* New York and Oxford: Oxford University Press.

Mantell, Gideon Algernon. 1825. Notice on the Iguanodon, a newly discovered fossil reptile, from the sandstone of Tilgate Forest, in Sussex. *Philosophical Transactions of the Royal Society of London* (1825): 179–86.

Marginalia: *Charles Darwin's marginalia.* Edited by Mario A. Di Gregorio with the assistance of Nicholas W. Gill. Vol. 1. New York and London: Garland Publishing. 1990.

Markham, Clements R. 1880. *Peruvian bark. A popular account of the introduction of chinchona cultivation into British India.* London: John Murray.

[Marsh (Marsh-Caldwell), Anne.] 1834. *Two old men's tales: the deformed, and the admiral's daughter.* 2 vols. London: Saunders & Otley.

[——.] 1846. *Emilia Wyndham.* 3 vols. London: Henry Colburn.

Martin, Gerald P. R. and Uschmann, Georg. 1969. *Friedrich Rolle 1827–1887, ein Vorkämpfer neuen biologischen Denkens in Deutschland.* Leipzig: Johann Ambrosius Barth.

Martius, Carl Friedrich Philipp von. 1862. *Denkrede auf Joh. Andreas Wagner. Gehalten in der öffentlichen Sitzung am 28. November 1862.* Munich: [Könighlich-bayerische] Akademie [der Wissenschaften].

Masters, Maxwell Tylden. 1863. On the existence of two forms of peloria. *Natural History Review* n.s. 3: 258–62.

Matthew, Patrick. 1831. *On naval timber and arboriculture; with critical notes on authors who have recently treated the subject of planting.* London: Longman, Rees, Orme, Brown, & Green. Edinburgh: Adam Black.

[Maw, George.] 1861. [Review of *Origin* & other works.] *Zoologist* 19: 7577–611.

Maw, George. 1864. Notes on the drift-deposits of the valley of the Severn, in the neighbourhood of Coalbrook Dale and Bridgnorth. [Read 20 January 1864.] *Quarterly Journal of the Geological Society* 20: 130–44.

Max Müller, Friedrich. 1861. *Lectures on the science of language delivered at the Royal*

Institution of Great Britain in April, May, and June, 1861. London: Longman, Green, Longman & Roberts.

Mayr, Ernst and Cottrell, G. William. 1979. *Check-list of birds of the world.* Vol. 1. 2d ed. Revision of the work of James L. Peters. Cambridge, Mass.: Museum of Comparative Zoology.

Medical directory: *The London medical directory . . . every physician, surgeon, and general practitioner resident in London.* London: C. Mitchell. 1845. *The London and provincial medical directory.* London: John Churchill. 1848–60. *The London & provincial medical directory, inclusive of the medical directory for Scotland, and the medical directory for Ireland, and general medical register.* London: John Churchill. 1861–9. *The medical directory . . . including the London and provincial medical directory, the medical directory for Scotland, the medical directory for Ireland.* London: J. & A. Churchill. 1870–1905.

Medical who's who: *The medical who's who.* London: London & Counties Press Association [& others]. 1913–27.

Men and women of the time: *The men of the time in 1852 or sketches of living notables.* [2d edition, 1853. 3d edition, 1856. 4th edition, 1857. New edition, 1865. 7th edition, 1868. 8th edition, 1872. 9th edition, 1875. 10th edition, 1879. 11th edition, 1884. 12th edition, 1887. 13th edition, 1891. 14th edition, 1895.] *Men and women of the time: a dictionary.* 15th edition. By Victor G. Plarr. 1899. London: David Bogue [& others]. 1852–99.

Men-at-the-bar: *Men-at-the-bar: a biographical hand-list of the members of the various inns of court, including Her Majesty's judges, etc.* By Joseph Foster. London: Reeves & Turner. 1885.

Ménétriès, Edouard. 1859. Lépidoptères de la Sibérie orientale et en particulier des rives de l'Amour. In *Reisen und Forschungen im Amur-Lande in den Jahren 1854–1856 im Auftrage der Kaiserl. Akademie der Wissenschaften zu St Petersburg ausgeführt und in Verbindung mit mehreren Gelehrten*, edited by Leopold von Schrenck. 4 vols. in 5. St Petersburg: Kaiserlich Akademie der Wissenschaften in St Petersburg. 1858–76.

Messenger, Michael. 1970. Maws of Jackfield are no more. *Shropshire Magazine*, June 1970, pp. 24–5.

Metcalfe, Richard. 1906. *The rise and progress of hydropathy in England and Scotland.* London: Simpkin, Marshall, Hamilton, Kent & Co.

Meteyard, Eliza. 1865–6. *The life of Josiah Wedgwood from his private correspondence and family papers . . . with an introductory sketch of the art of pottery in England.* 2 vols. London: Hurst & Blackett.

Meyer, Hermann von. 1861. [Vogel-Federn und *Palpipes priscus* von Solenhofen.] *Neues Jahrbuch für Mineralogie, Geognosie, Geologie, und Petrefakten-Kunde* (1861): 561.

Mill, Humphrey. 1646. *The second part of The Nights Search, discovering the condition of the various fowles of night. Or, the second great mystery of iniquity exactly revealed: with the projects of these times.* In a poem. London: printed for Henry Shepeard and William Ley.

Mill, John Stuart. 1862. *A system of logic, ratiocinative and inductive, being a connected view of the principles of evidence, and the methods of scientific investigation.* 5th edition. 2 vols. London: Parker, Son, & Bourn.

Milne, Louisa E. 1901. Memoir. In *Essays descriptive and biographical*, by Grace Prestwich. Edinburgh and London: William Blackwood.

ML: *More letters of Charles Darwin: a record of his work in a series of hitherto unpublished letters.* Edited by Francis Darwin and Albert Charles Seward. 2 vols. London: John Murray. 1903.

Modern English biography: *Modern English biography containing many thousand concise memoirs of persons who have died since the year 1850.* By Frederick Boase. 3 vols. and supplement (3 vols.). Truro, Cornwall: printed for the author. 1892–1921.

Mohl, Hugo von. 1827. *Ueber den Bau und das Winden der Ranken und Schlingpflanzen.* Tübingen, Germany: Heinrich Laupp.

——. 1851. *Grundzüge der Anatomie und Physiologie der vegetablischen Zelle.* Brunswick, Germany: F. Vieweg.

——. 1853. Die vegetabilische Zelle. In vol. 4 of *Handwörterbuch der Physiologie mit Rücksicht auf physiologische Pathologie*, edited by Rudolph Wagner. 4 vols. Brunswick, Germany: Friedrich Vieweg & Sohn. 1842–53.

——. 1863. Einige Beobachtungen über dimorphe Blüthen. *Botanische Zeitung* 21: 309–15, 321–8.

Monteiro, Joachim John. 1875. *Angola and the River Congo.* 2 vols. London: Macmillan.

Montgomery, William M. 1974. Germany. In *The comparative reception of Darwinism*, edited by Thomas F. Glick. Austin, Tex., and London: University of Texas Press.

Moore, James R. 1977. On the education of Darwin's sons: the correspondence between Charles Darwin and the Reverend G. V. Reed, 1857–1864. *Notes and Records of the Royal Society* 32 (1977–8): 51–70.

——. 1985. Darwin of Down: the evolutionist as squarson-naturalist. In *The Darwinian heritage*, edited by David Kohn. Princeton, N.J.: Princeton University Press in association with Nova Pacifica.

Moss, Arthur W. 1961. *Valiant crusade: the history of the RSPCA.* London: Cassell.

Movement in plants: *The power of movement in plants.* By Charles Darwin. London: John Murray. 1880.

Moyal, Ann. 1986. *'A bright and savage land': scientists in colonial Australia.* Sydney: William Collins.

Mozley, Ann. 1967. Evolution and the climate of opinion in Australia, 1840–76. *Victorian Studies* 10: 411–30.

Mulder, Gerrit Jan van. 1849. *The chemistry of vegetable & animal physiology.* Translated by P. F. H. Fromberg. With introduction and notes by James F. W. Johnston. Edinburgh and London: William Blackwood & Sons.

Müller, Phillipp Jacob. 1859. Versuch einer monographischen Darstellung der gallogermanschen Arten der Gattung *Rubus.* In *Sechszehnter und siebenzehnter Jahresbericht der Pollichia, eines naturwissenschaftlichen Vereins der Rheinpfalz*, pp. 74–298.

Müller, Phillipp Jacob. 1861. Rubologische Ergebnisse einer dreitägigen Excursion in die granitischen Hoch-Vogesen der Umgegend von Gérardmer (Vogesen-Depart.——Frankreich). *Bonplandia. Zeitschrift für die gesammte Botanik* 9: 276–308.

Murchison, Roderick Impey. 1861. Address [of the president of section 'C']. *Report of the 31st meeting of the British Association for the Advancement of Science, held at Manchester*, Transactions of the sections, pp. 95–108.

——. 1863a. Address of the president [of section 'E']. *Report of the 33d meeting of the British Association for the Advancement of Science, held at Newcastle-upon-Tyne*, Transactions of the sections, pp. 126–33.

——. 1863b. Observations upon the Permian group of the north-west of England, in communicating the outline of a memoir thereon by Prof. R. Harkness and himself. *Report of the 33d meeting of the British Association for the Advancement of Science, held at Newcastle-upon-Tyne*, Transactions of the sections, pp. 83–5.

Nägeli, Carl Wilhelm von. 1858–68. *Beiträge zur Wissenschaftlichen Botanik.* 4 pts in 1 vol. Leipzig: Wilhelm Englemann.

Nägeli, Carl Wilhelm von and Cramer, Carl. 1855–8. *Pflanzenphysiologische Untersuchungen.* 4 pts in 1 vol. Zurich: Friedrich Schulthess.

Narrative: *Narrative of the surveying voyages of His Majesty's ships Adventure and Beagle, between the years 1826 and 1836.* [Edited by Robert FitzRoy.] 3 vols. and appendix. London: Henry Colburn. 1839.

Nasmyth, Alexander. 1841. *Three memoirs on the development and structure of the teeth and epithelium, read at the ninth annual meeting of the British Association for the Encouragement of Science, held at Birmingham, in August, 1839; with diagrams in illustration of them.* London: John Churchill.

——. 1842. On the minute structure of the tusks of extinct Mastodontoid animals. *Proceedings of the Geological Society of London* 3 (1838–42): 775–80.

——. 1849. *Researches on the developement, structure and diseases of the teeth.* London: John Churchill.

Natural selection: *Charles Darwin's Natural selection: being the second part of his big species book written from 1856 to 1858.* Edited by R. C. Stauffer. Cambridge: Cambridge University Press. 1975.

Naudin, Charles Victor. 1849–53. Melastomatacearum quæ in Musæo Parisiensi continentur monographicæ descriptionis et secundum affinitates distributionis tentamen. *Annales des Sciences Naturelles (Botanique)* 3d ser. 12 (1849): 196–284; 13 (1850): 25–39, 126–59, 273–303, 347–62; 14 (1850): 53–76, 118–65; 15 (1851): 43–79, 276–345; 16 (1851): 83–246; 17 (1852): 305–82; 18 (1852): 85–154, 257–94.

——. 1853. *Melastomacearum, quæ in Musæo Parisiensi continentur monographicæ descriptionis et secundum affinitates distributionis tentamen.* Paris: V. Masson.

——. 1855. Organographie végétale. Observations relatives à la nature des vrilles et à la structure de la fleur chez les Cucurbitacées. *Annales des Sciences Naturelles (Botanique)* 4th ser. 4: 5–19.

——. 1856. Nouvelles recherches sur les caractères spécifiques et les variétés des plantes du genre *Cucurbita. Annales des Sciences Naturelles (Botanique)* 4th ser. 6: 5–73.

Naudin, Charles Victor. 1859. Essais d'une monographie des espèces et des variétés du genre *Cucumis*. *Annales des Sciences Naturelles (Botanique)* 4th ser. 11: 5–87.

——. 1862. Cucurbitacées cultivées au Muséum d'Histoire Naturelle en 1862. Description d'espèces nouvelles et de quelques formes hybrides obtenues de plantes de cette famille. *Annales des Sciences Naturelles (Botanique)* 4th ser. 18: 159–208.

——. 1863. Nouvelles recherches sur l'hybridité dans les végétaux. *Annales des Sciences Naturelles (Botanique)* 4th ser. 19: 180–203.

——. 1865. Nouvelles recherches sur l'hybridité dans les végétaux. Mémoire présenté a l'Académie des Sciences par M. Ch. Naudin en Décembre 1861 et couronné dans la séance du 29 Décembre 1862. *Nouvelles Archives du Muséum d'Histoire Naturelle* 1: 25–176.

Naumann, Karl Friedrich. 1842. Uber den Quincunx, als Grundgesetz der Blattstellung im Pflanzenreiche. *Annalen der Physik und Chemie* 2d ser. 56: 1–37.

——. 1845. *Ueber den Quincunx als Grundgesetz der Blattstellung vieler Pflanzen.* Dresden and Leipzig: Arnoldische Buchhandlung.

Navy list: *The navy list.* London: John Murray; Her Majesty's Stationery Office. 1815–1900.

NBU: *Nouvelle biographie universelle depuis les temps les plus regulés jusqu'à nos jours, avec les renseignements bibliographiques et l'indication des sources a consulter.* Edited by Jean Chrétien Ferdinand Hoefer. 46 vols. in 23. Paris: Firmin Didot Frères. 1852–66.

NDB: *Neue deutsche Biographie.* Under the auspices of the Historical Commission of the Bavarian Academy of Sciences. 18 vols. (A–Nausea). Berlin: Duncker & Humblot. 1953–96.

Neighbour, Alfred. 1865. *The apiary; or, bees, bee-hives, and bee culture: being a familiar account of the habits of bees, and the most improved methods of management, with full directions, adapted for the cottager, farmer, or scientific apiarian.* London: Kent & Co.; Geo. Neighbour & Sons.

Neue Österreichische Biographie: *Neue Österreichische Biographie 1815–1918.* 22 vols. and bibliography. Vienna: Amalthea-Verlag. 1923–87.

Newspaper press directory: *The newspaper press directory and advertiser's guide. The newspaper press directory … A directory of the class papers and periodicals.* London: C. Michell. 1856–1900.

Newton, Alfred. 1863. On an illustration of the manner in which birds may occasionally aid in the dispersion of seeds. [Read 21 April 1863.] *Proceedings of the Zoological Society of London* (1863): 127–9.

Nicholas, F. W. and Nicholas, J. M. 1989. *Charles Darwin in Australia. With illustrations and additional commentary from other members of the* Beagle's *company including Conrad Martens, Augustus Earle, Captain FitzRoy, Philip Gidley King, and Syms Covington.* Cambridge: Cambridge University Press.

NNBW: *Nieuw Nederlandsch biografisch Woordenboek.* Edited by P. C. Molhuysen *et al.* 10 vols. Leiden: A. W. Sijthoff's Uitgevers-Maatschappij. 1911–37.

Nogueira, José Luís de Almeida. 1977. *A Academia de São Paulo: Tradições e Reminiscências.* São Paulo, Brazil: Saraiva.

Notebooks: *Charles Darwin's notebooks, 1836–1844. Geology, transmutation of species, metaphysical enquiries.* Transcribed and edited by Paul H. Barrett *et al.* Cambridge: Cambridge University Press for the British Museum (Natural History). 1987.

Nott, Josiah Clark and Gliddon, George Robins. 1854. *Types of mankind: or, ethnological researches, based upon the ancient monuments, painting, sculptures, and crania of races and upon their natural, geographical, philological, and biblical history.* Philadelphia: Lippincott, Grambo & Co. London: Trübner.

NSTC: *Nineteenth century short title catalogue.* Ser. 1, 1801–15. 6 vols. Ser. 2, 1816–70. 56 vols. Newcastle-upon-Tyne: Avero Publications. 1984–95.

NUC: *The national union catalog. Pre-1956 imprints.* 685 vols. and supplement (69 vols.). London and Chicago: Mansell. 1968–81.

OBL: *Österreichisches biographisches Lexikon 1815–1950.* Edited by Leo Santifaller *et al.* 10 vols. and 2 fascicles of vol. 11 (A–Schreyvogee). Vienna: Osterreichischen Akademie der Wissenschaften. 1957–97.

ODQ: *The Oxford dictionary of quotations.* Edited by Angela Partington. 4th edition, revised. Oxford: Oxford University Press. 1996.

OED: *The Oxford English dictionary. Being a corrected re-issue with an introduction, supplement and bibliography of a new English dictionary.* Edited by James A. H. Murray *et al.* 12 vols. and supplement. Oxford: Clarendon Press. 1970. *A supplement to the Oxford English dictionary.* 4 vols. Edited by R. W. Burchfield. Oxford: Clarendon Press. 1972–86. *The Oxford English dictionary.* 2d edition. 20 vols. Prepared by J. A. Simpson and E. S. C. Weiner. Oxford: Clarendon Press. 1989. *Oxford English dictionary additional series.* 3 vols. Edited by John Simpson *et al.* Oxford: Clarendon Press. 1993–7.

O grande livro dos Portugueses: *O grande livro dos Portugueses; 4000 personalidades em texto e imagem: nomes, datas, factos, com 980 ilustraçonnes.* [By Manuel Alves de Oliveira *et al.*] [Lisbon]: Círculo de Leitores. 1991.

O'Leary, Patrick. 1989. *Sir James Mackintosh: the Whig Cicero.* Aberdeen: Aberdeen University Press.

[Oliver, Daniel.] 1862a. Botanical bibliography. *Natural History Review* n.s. 2: 416–85.

[——.] 1862b. [Review of 'Dimorphic condition in *Primula*'.] *Natural History Review* n.s. 2: 235–43.

——. 1862c. The Atlantis hypothesis in its botanical aspect. *Natural History Review* n.s. 2: 149–70.

——. 1863a. Notes on the Loranthaceæ, with a synopsis of the genera. [Read 15 January 1863.] *Journal of the Proceedings of the Linnean Society (Botany)* 7 (1864): 90–106.

[——.] 1863b. [Review of *The British flora*, by William Jackson Hooker and George Arnott Walker Arnott, & other works.] *Natural History Review* n.s. 3: 34–40.

[——.] 1863c. [Review of J. D. Hooker's study of *Welwitschia*.] *Natural History Review* n.s. 3: 201–11.

[——.] 1863d. [Review of 'Two forms in species of *Linum*'.] *Natural History Review* n.s. 3: 476–8.

——. 1863e. Note on the structure and mode of dehiscence of the legumes of

Pentaclethra macrophylla, Benth. [Read 19 November 1863.] *Transactions of the Linnean Society of London* 24 (1864): 415–20.

——. 1864a. *Lessons in elementary botany. The part on systematic botany based upon material left in manuscript by the late Professor Henslow.* London and Cambridge: Macmillan.

[——.] 1864b. Dimorphic flowers. *Natural History Review* n.s. 4: 243–8.

Oliver, Daniel and Hanbury, Daniel. 1863. On some new species of *Amomum* from West Africa. [Read 16 April 1863.] *Journal of the Proceedings of the Linnean Society (Botany)* 7 (1864): 109–10.

Olmsted, Frederick Law. 1856. *A journey in the seaboard slave states, with remarks on their economy.* New York: Dix & Edwards. London: S. Low.

——. 1857. *A journey through Texas; or, a saddle-trip on the southwestern frontier: with a statistical appendix.* New York: Dix, Edwards & Co.

——. 1860. *A journey in the back country in the winter of 1853–4.* London: Sampson, Low, Son & Co.

——. 1861. *Journeys and explorations in the cotton kingdom. A traveller's observations on cotton and slavery in the American slave states.* 2 vols. London: Sampson, Low, Son & Co.

Orchids: *On the various contrivances by which British and foreign orchids are fertilised by insects, and on the good effects of intercrossing.* By Charles Darwin. London: John Murray. 1862.

Orchids 2d ed.: *On the various contrivances by which British and foreign orchids are fertilised by insects, and the good effects of intercrossing.* By Charles Darwin. 2d edition, revised. London: John Murray. 1877.

Origin: *On the origin of species by means of natural selection, or the preservation of favoured races in the struggle for life.* By Charles Darwin. London: John Murray. 1859.

Origin 2d ed.: *On the origin of species by means of natural selection, or the preservation of favoured races in the struggle for life.* By Charles Darwin. London: John Murray. 1860.

Origin 3d ed.: *On the origin of species by means of natural selection, or the preservation of favoured races in the struggle for life.* By Charles Darwin. With additions and corrections. London: John Murray. 1861.

Origin 4th ed.: *On the origin of species by means of natural selection, or the preservation of favoured races in the struggle for life.* By Charles Darwin. With additions and corrections. London: John Murray. 1866.

Origin 5th ed.: *On the origin of species by means of natural selection, or the preservation of favoured races in the struggle for life.* By Charles Darwin. With additions and corrections. London: John Murray. 1869.

Origin 6th ed.: *On the origin of species by means of natural selection, or the preservation of favoured races in the struggle for life.* By Charles Darwin. With additions and corrections. London: John Murray. 1872.

Origin US ed.: *On the origin of species by means of natural selection, or the preservation of favoured races in the struggle for life.* By Charles Darwin. A new edition, revised and augmented by the author. New York: D. Appleton. 1860.

Origin 2d US ed.: *On the origin of species by means of natural selection, or the preservation of favoured races in the struggle for life.* By Charles Darwin. A new edition, revised and augmented by the author. New York: D. Appleton. 1869.

Ospovat, Dov. 1981. *The development of Darwin's theory. Natural history, natural theology, and natural selection, 1838–1859.* Cambridge: Cambridge University Press.

Owen, Richard. 1838a. A description of the cranium of the *Toxodon platensis*, a gigantic extinct mammiferous species, referrible by its dentition to the *Rodentia*, but with affinities to the *Pachydermata* and the herbivorous Cetacea. *Proceedings of the Geological Society of London* 2 (1833–8): 541–2.

——. 1838b. Description du crâne du *Toxodon platensis*, grand Mammifère perdu que l'on doit rapporter à l'ordre des Pachydermes, mais qui offre en même temps des affinités avec les Rongeurs, les edentés et les Cétacés herbivores. *Annales de Sciences Naturelles (Zoologie)* 2d ser. 9: 25–45.

——. 1838c. Description d'une mâchoire inférieure et de dents de Toxodon trouvées à Bahia-Blanca, à 39° de latitude sur la côte Est de l'Amèrique méridionale. *Annales des Sciences Naturelles (Zoologie)* 2d ser. 9: 45–54.

——. 1839. Recherches sur la structure et la formation des dents des Squaloïdes, et application des faits observés à une nouvelle théorie du développement des dents. *Comptes rendus hebdomadaires des Séances de l'Académie des Sciences* 9: 784–8.

——. 1840. Description of the Mammalian remains found at Kyson in Suffolk. *Annals of Natural History* 4: 191–4.

——. 1842. *Description of the skeleton of an extinct gigantic sloth,* Mylodon robustus, *Owen, with observations on the osteology, natural affinities, and probable habits of the Megatheroid quadrupeds in general.* London: Royal College of Surgeons.

——. 1844. Description of a fossil molar tooth of a mastodon discovered by Count Strzlecki in Australia. *Annals and Magazine of Natural History* 14: 268–71.

——. 1846a. Report on the archetype and homologies of the vertebrate skeleton. *Report of the 16th meeting of the British Association for the Advancement of Science, held at Southampton*, pp. 169–340.

——. 1846b. *A history of British fossil mammals, and birds.* London: John van Voorst.

——. 1849. *On the nature of limbs. A discourse delivered on Friday, February 9, at an evening meeting of the Royal Institution of Great Britain.* London: John van Voorst.

——. 1850a. On *Dinornis.* (Part 4): containing the restoration of the feet of that genus and of *Palapteryx*, with a description of the sternum in *Palapteryx* and *Aptornis.* [Read 26 February 1850.] *Transactions of the Zoological Society of London* 4 (1862): 1–20.

[——.] 1850b. [Review of Alexander Nasmyth's *Researches on the development, structure, and diseases of the teeth.*] *British and Foreign Medico-Chirurgical Review* n.s. 5: 356–61.

[——.] 1851. [Review of Charles Lyell's anniversary address to the Geological Society of London (1851), *Principles of geology* (8th edition), & other works.] *Quarterly Review* 89: 412–51.

[——.] 1853. *Descriptive catalogue of the osteological series contained in the museum of the Royal College of Surgeons of England.* 2 vols. London: [Royal College of Surgeons of England].

——. 1854. On some fossil reptilian and mammalian remains from the Purbecks. [Read 7 June 1854.] *Quarterly Journal of the Geological Society of London* 10: 420–33.

Owen, Richard. 1857. On the characters, principles of division, and primary groups of the class Mammalia. [Read 17 February and 21 April 1857.] *Journal of the Proceedings of the Linnean Society (Zoology)* 2 (1858): 1–37.

——. 1858. Address of the president. *Report of the 28th meeting of the British Association for the Advancement of Science, held at Leeds*, pp. xlix–cx.

——. 1859a. Palæontology. In *Encyclopædia Britannica, or a dictionary of arts, sciences, and general literature.* 8th edition. [Edited by T. S. Traill.] 21 vols. and index. Edinburgh: Adam & Charles Black. 1853–60.

——. 1859b. *On the classification and geographical distribution of the Mammalia, being the lecture on Sir Robert Reade's Foundation, delivered before the University of Cambridge, in the Senate House, May 10, 1859. To which is added an appendix 'On the gorilla', and 'On the extinction and transmutation of species'.* London: John W. Parker.

[——.] 1860a. [Review of *Origin* & other works.] *Edinburgh Review* 111: 487–532.

——. 1860b. *Palæontology or a systematic summary of extinct animals and their geological relations.* Edinburgh: Adam & Charles Black.

——. 1861a. On the cerebral characters of man and the ape. *Annals and Magazine of Natural History* 3d ser. 7: 456–8.

——. 1861b. *Palæontology or a systematic summary of extinct animals and their geological relations.* 2d edition. Edinburgh: Adam & Charles Black.

——. 1862a. On the *Archaeopteryx* of von Meyer, with a description of the fossil remains of a long-tailed species, from the lithographic stone of Solenhofen. [Read 20 November 1862.] *Philosophical Transactions of the Royal Society of London* 153 (1863): 33–47.

——. 1862b. On the hyracotherian character of the lower molars of the supposed *Macacus* from the Eocene sand of Kyson, Suffolk. *Annals and Magazine of Natural History* 3d ser. 10: 240.

——. 1862c. On the Aye-aye (*Chiromys*, Cuvier; *Ciromys madagascariensis*, Desm.; *Sciurus madagascariensis* Gmel., Sonnerat; *Lemur psilodactylus*, Schreber, Shaw). [Read 14 and 28 January 1862.] *Transactions of the Zoological Society of London* 5 (1866): 33–101.

——. 1866–8. *On the anatomy of vertebrates.* 3 vols. London: Longmans, Green & Co.

——. 1871. *Monograph of the fossil mammalia of the Mesozoic formations.* London: Palaeontographical Society.

Paget, James. 1857. On the hereditary transmission of tendencies to cancerous and other tumours. *Medical Times and Gazette* n.s. 15: 191–3.

Pallas, Pyotr Simon. 1811–31. *Zoographia Rosso-Asiatica, sistens omnium animalium in extenso imperio Rossico et adjacentibus maribus observatorum recensionem, domicilia, mores et descriptiones, anatomen atque icones plurimorum.* 3 vols. St Petersburg: Caes. Academiae Scientiarum Impress.

Palm, Ludwig Heinrich. [1827?] *Ueber das Winden der Pflanzen: eine botanisch-physiologische Abhandlung, welche von der medicinischen Facultät der Universität Tübingen im Jahr 1826 als Preisschrift gekrönt wurde.* Tübingen, Germany: Christian Richter.

Pancaldi, Giuliano. 1991. *Darwin in Italy. Science across cultural frontiers.* Translated by Ruey Brodine Morelli. Updated and expanded edition. Bloomington and Indianapolis, Ind.: Indiana University Press.

'Parallel roads of Glen Roy': Observations on the parallel roads of Glen Roy, and of other parts of Lochaber in Scotland, with an attempt to prove that they are of marine origin. By Charles Darwin. [Read 7 February 1839.] *Philosophical Transactions of the Royal Society of London* (1839), pt 1: 39–81. [*Collected papers* 1: 89–137.]

Paris, John Ayrton. 1820. *Pharmacologia; or the history of medicinal substances, with a view to establish the art of prescribing and of composing extemporaneous formulæ upon fixed and scientific principles; illustrated by formulæ, in which the intention of each element is designated by key letters.* 3d edition, enlarged. London: W. Phillips.

Pasteur, Louis. 1861. Mémoire sur les corpuscules organisés qui existent dans l'atmosphère, examen de la doctrine des générations spontanées. *Annales des Sciences Naturelles (Zoologie)* 4th ser. 16: 5–98.

——. 1862. Mémoire sur les corpuscules organisés qui existent dans l'atmosphère, examen de la doctrine des générations spontanées. *Annales de Chimie et de Physique* 3d ser. 64: 5–110.

Patterson, Alfred Temple. 1962. *The University of Southampton. A centenary history of the evolution and development of the University of Southampton, 1862–1962.* Southampton: University of Southampton.

Peckham, Morse, ed. 1959. *The Origin of Species by Charles Darwin: a variorum text.* Philadelphia: University of Pennsylvania Press.

Peirce, Benjamin. 1849. Mathematical investigation of the fractions which occur in phyllotaxis. *Proceedings of the American Association for the Advancement of Science; second meeting, held at Cambridge, Massachusetts, August 1849*, pp. 444–7.

Pelzeln, August von. 1861. *Bemerkungen gegen Darwin's Theorie vom Ursprung der Spezies.* Vienna: A. Pichler Witwe & Sohn.

Penguin dictionary of botany: *The Penguin dictionary of botany.* Edited by Elizabeth Tootill *et al.* Harmondsworth, Middlesex: Penguin Books. 1984.

Phillips, John. 1873. Sedgwick. *Nature* 7: 257–9.

Phisterer, Frederick. 1912. *New York in the war of the rebellion, 1861 to 1865.* 3d edition. 5 vols. and index. Albany, N.Y.: J. B. Lyon.

Physicians: *The roll of the Royal College of Physicians of London.* By William Munk. 2d edition, revised and enlarged. 3 vols. London: Royal College of Physicians. 1878. *Lives of the fellows of the Royal College of Physicians of London.* Compiled by G. H. Brown *et al.* 5 vols. London: Royal College of Physicians. Oxford and Washington, DC: IRL Press. 1955–89.

Pictet de la Rive, Francois. 1860. Sur l'origine de l'espèce par Charles Darwin. *Bibliothèque Universelle. Revue Suisse et étrangère. (Archive des Sciences, Physiques et Naturelles)* n.s. 7: 233–55.

Pingree, Jeanne. 1968. *Thomas Henry Huxley: a list of his scientific notebooks, drawings and other papers, preserved in the college archives.* London: Imperial College of Science and Technology, University of London.

Planchon, Jules Emile. 1847–8. Sur la famille des Linnes. *London Journal of Botany* 6 (1847): 588–603; 7 (1848): 165–86, 473–501, 507–28.

Plarr, Victor Gustave. 1930. *Plarr's lives of the fellows of the Royal College of Surgeons of England.* Revised by Sir D'Arcy Power. 2 vols. London: Simpkin Marshall.

Plato. 1965. *Timaeus.* Translation with introduction by H. D. P. Lee. London: Penguin Books.

——. 1977. *Plato's Phaedo.* Translated by Georges Maxmilien Antoine Grube. Indianapolis, Ind.: Hackett Publishing.

Playfair, Lyon, trans. 1840. *Organic chemistry in its applications to agriculture and physiology.* By Justis von Liebig. London: Taylor & Walton.

Pole, William, ed. 1877. *The life of Sir William Fairbairn, Bart.* London: Longmans, Green & Co.

Porter, Duncan M. 1985. The *Beagle* collector and his collections. In *The Darwinian heritage*, edited by David Kohn. Princeton, N.J.: Princeton University Press in association with Nova Pacifica.

——. 1986. Charles Darwin's vascular plant specimens from the voyage of HMS *Beagle. Botanical Journal of the Linnean Society* 93: 1–172.

Porter, Roy. 1977. Biliousness. In *Gastroenterology in Britain: historical essays*, edited by William F. Bynum. London: Wellcome Institute for the History of Medicine.

Post Office directory of Birmingham: *Post Office directory of Birmingham, Warwickshire, and part of Staffordshire. Kelly's directory of Birmingham, Staffordshire, Warwickshire, and Worcestershire.* London: W. Kelly; Kelly & Co. 1845–1928.

Post Office directory of Cambridge, Norfolk, and Suffolk: *Post Office directory of Cambridge, Norfolk, and Suffolk. Kelly's directory of Cambridgeshire, Norfolk and Suffolk.* London: W. Kelly; Kelly & Co. 1853–1900.

Post Office directory of Cheshire: *Post Office directory of Cheshire. Kelly's directory of Cheshire.* London: Kelly & Co. 1857–1902.

Post Office directory of Gloucestershire, with Bath, Bristol, Herefordshire, and Shropshire. London: Kelly & Co. 1856–79.

Post Office directory of Hampshire, Dorsetshire, and Wiltshire: *Post Office directory of Hampshire, Dorsetshire, and Wiltshire. Post Office directory of Dorsetshire, Wiltshire, and Hampshire, including the Isle of Wight.* London: Kelly & Co. 1848–75.

Post Office directory of Lancashire, Liverpool and Manchester. London: Kelly & Co. 1858–92.

Post Office directory of the six home counties: *Post Office directory of the six home counties, viz., Essex, Herts, Kent, Middlesex, Surrey and Sussex.* London: W. Kelly & Co. 1845–78.

Post Office directory of Somersetshire, with the city of Bristol: *Kelly's Post Office directory of Somersetshire, with the city of Bristol.* London: Kelly & Co. 1861–97.

Post Office Edinburgh directory: *Post-office annual directory and calendar. Post-office Edinburgh and Leith directory.* Edinburgh: Ballantyne & Hughes [& others]. 1845–1908.

Post Office Glasgow directory: *The Glasgow Post-office directory . . . a street directory and an appendix.* Glasgow: Post Office. 1839. *Post Office Glasgow directory . . . with an appendix containing general and local information.* Glasgow: Post Office; William Mackenzie. 1863–1900.

Post Office London directory: *Post-office annual directory. . . . A list of the principal merchants, traders of eminence, &c. in the cities of London and Westminster, the borough of Southwark, and parts adjacent . . . general and special information relating to the Post Office. Post Office London directory.* London: His Majesty's Postmaster-General [& others]. 1802–1967.

Post Office London suburban directory: *The Post Office London suburban directory. Kelly's London suburban directory.* London: Kelly & Co. 1860–1903.

Pouchet, Félix Archimède. 1858. Note sur des proto-organismes végétaux et animaux, nés spontanément dans l'air artificiel et dans le gaz oxygène. *Comptes rendus hebdomadaires des Séances de l'Académie des Sciences* 47: 979–84.

——. 1859. *Hétérogénie ou traité de la génération spontanée.* Paris: J. B. Baillière.

——. 1862. Etudes expérimentales sur la genèse spontanée. *Annales des Sciences Naturelles (Zoologie)* 4th ser. 18: 277–303.

Poulson, Barry W. 1981. *Economic history of the United States.* New York: Macmillan. London: Collier Macmillan.

Poulton, Edward Bagnall. 1909. *Charles Darwin and the origin of species. Addresses, etc., in America and England in the year of the two anniversaries.* London: Longmans, Green & Co.

Praeger, Robert Lloyd. 1949. *Some Irish naturalists: a biographical note-book.* Dundalk, Co. Louth: W. Tempest, Dundalgan Press.

Prescott, William Hickling. 1847. *History of the conquest of Peru, with a preliminary view of the civilization of the Incas.* 2 vols. London: Richard Bently.

Prestwich, Grace Anne, ed. 1899. *Life and letters of Sir Joseph Prestwich.* Edinburgh and London: William Blackwood.

Prestwich, Joseph. 1834–6. On the geology of Coalbrook Dale. [Read 5 February 1834, 13 and 27 April 1836.] *Transactions of the Geological Society of London* 2d ser. 5 (1840): 413–95.

——. 1862a. Theoretical considerations on the conditions under which the drift deposits containing the remains of extinct mammalia and flint implements were accumulated; and on their geological age. [Read 27 March 1862.] *Proceedings of the Royal Society of London* 12: 38–52.

——. 1862b. Theoretical considerations on the conditions under which the (drift) deposits containing the remains of extinct mammalia and flint implements were accumulated, and on their geological age. On the Loess of the valleys of the south of England, and of the Somme and the Seine. [Read 27 March and 19 June 1862.] *Philosophical Transactions of the Royal Society of London* 154 (1864): 247–309.

——. 1863. On the section at Moulin Quignon, Abbeville, and on the peculiar character of some of the flint implements recently discovered there. [Read 3 June 1863.] *Quarterly Journal of the Geological Society of London* 19: 497–505.

Prete, Frederick R. 1990. The conundrum of the honey bees: one impediment to the publication of Darwin's theory. *Journal of the History of Biology* 23: 271–90.

Price, John. 1863–4. *Old Price's remains; præhumous, or during life.* 12 pts. London: Virtue, Brothers & Co.

Pritchard, Michael. 1986. *A directory of London photographers 1841–1908.* Watford, Hertfordshire: ALLM Books.

Publishers' circular: *The publishers' circular ... A complete alphabetical catalogue of the new works and new editions, their sizes, prices, date of publication, and publishers' names.* London: Sampson Low. 1837–90.

Quatrefages de Bréau, Jean Louis Armand de. 1859. *Etudes sur les maladies actuelles du ver à soie.* Paris: Victor Masson.

——. 1861. *Unité de l'espèce humaine.* Paris: J. Claye.

——. 1862. *Métamorphoses de l'homme et des animaux.* Paris: J. B. Baillière.

——. 1863. Notes sur la mâchoire humaine découverte par M. Boucher de Perthes dans le diluvium d'Abbeville. *Comptes rendus hebdomadaires des Séances de l'Académie des Sciences* 56: 782–8, 809–16, 857–60.

——. 1879. *The human species.* London: C. Kegan Paul.

Quenstedt, Friedrich August. 1856. *Sonst und jetzt. Populäre Vorträge über Geologie.* Tübingen, Germany: H. Laupp'schen Buchhandlung.

Rainger, Ronald. 1978. Race, politics, and science: the Anthropological Society of London in the 1860s. *Victorian Studies* 22: 51–70.

Ramsay, Andrew Crombie. 1855. On the occurrence of angular, subangular, polished, and striated fragments and boulders in the Permian Breccia of Shropshire, Worcestershire, &c.; and on the probable existence of glaciers and icebergs in the Permian epoch. [Read 21 February 1855.] *Quarterly Journal of the Geological Society of London* 11: 185–205.

——. 1862. On the glacial origin of certain lakes in Switzerland, the Black Forest, Great Britain, Sweden, North America, and elsewhere. [Read 5 March 1862.] *Quarterly Journal of the Geological Society of London* 18: 185–204.

——. 1863. The anniversary address of the president. [Read 20 February 1863.] *Quarterly Journal of the Geological Society of London* 19: xxix–lii.

——. 1864. The anniversary address of the president. [Read 19 February 1864.] *Quarterly Journal of the Geological Society of London* 20: xxxiii–lx.

Raphael, Sandra. 1970. The publication dates of the *Transactions of the Linnean Society of London*, series 1, 1791–1875. *Biological Journal of the Linnean Society* 2: 61–76.

Record of the Royal Society of London: *The record of the Royal Society of London for the promotion of natural knowledge.* 4th edition. London: Royal Society. 1940.

Reeve, Henry, trans. 1862. *Democracy in America.* By Alexis de Tocqueville. New edition with an introductory notice by the translator. 2 vols. London: Longman, Green, Longman, & Roberts.

Reeve, Lovell Augustus. 1863. *The land and freshwater mollusks indigenous to, or naturalised in, the British Isles.* London: Reeve & Co.

——, ed. 1863–6. *Portraits of men of eminence in literature, science, and art, with biographical memoirs.* 4 vols. London.

Registers of Wadham College, Oxford: *The registers of Wadham College, Oxford. (Part II) From 1719 to 1871.* Edited by Robert Barlow Gardiner. London: George Bell. 1895.

Registrum Orielense: *Registrum Orielense. An account of the members of Oriel College, Oxford.* Collected and arranged by Charles Lancelot Shadwell. 2 vols. London: Henry Frowde. 1893–1902.

Rehbock, Philip F. 1979. The early dredgers: 'naturalizing' in British seas, 1830–1850. *Journal of the History of Biology* 12: 293–368.

Renan, Ernest. 1863. *Vie de Jésus.* Paris: Michel Lévy Frères.

Rengger, Johann Rudolph. 1830. *Naturgeschichte der Saeugethiere von Paraguay.* Basel, Switzerland: Schweighauserschen Buchhandlung.

Report on Cinchona cultivation in British Sikkim: *First annual report on Cinchona cultivation in British Sikkim, from 1st April 1862 to 30th April 1863.* By Thomas Anderson. [British parliamentary papers, Session 1866, 53: 638–41.]

Report on Cinchona cultivation at Darjeeling: *Report on Cinchona cultivation at Darjeeling from 1st April 1863 to 15th July 1864.* By Thomas Anderson. [British parliamentary papers, Session 1866, 53: 643–7.]

Reptiles: *Reptiles.* Pt 5 of *The zoology of the voyage of HMS Beagle.* By Thomas Bell. Edited and superintended by Charles Darwin. London: Smith, Elder & Co. 1843.

Repton School register: *Repton School register 1557–1910.* Edited for the O. R. Society by the widow of G. S. Messiter. Repton, Derbyshire: A. J. Lawrence. 1910.

Rérolle, Louis, trans. 1870. *De la fécondation des orchidées par les insectes et des bons résultats du croisement.* By Charles Darwin. Paris: C. Reinwald.

Return of highway districts: *Return showing the several highway districts in each county in England constituted under the Highway Acts 1862 and 1864 (25 & 26 Vict. c. 61 and 27 & 28 Vict. c. 101), and the names of the parishes comprised in each district.* [British parliamentary papers, Session 1873, 20: 193–252.]

Reuss, Jeremias David. 1801–21. *Repertorium commentationum a societatibus litterariis editarum.* 16 vols. in 15. Göttingen, Germany: Henricum Dietrich.

'Review of Bates on mimetic butterflies': [Review of H. W. Bates's paper on mimetic butterflies. By Charles Darwin.] *Natural History Review* n.s. 3 (1863): 219–24. [*Collected papers* 2: 87–92.]

Richardson, Edmund William. 1916. *A veteran naturalist; being the life and work of W. B. Tegetmeier.* London: Witherby & Co.

Riedel, Wilhelm. 1824. *Die Taubenzucht in ihrem ganzen Umfange, oder, vollständige Anweisung zur Kenntniss des Taubenschlags.* Ulm, Germany: J. Ebnerschen Buchhandlung.

Ritvo, Harriet. 1990. *The animal estate: the English and other creatures in the Victorian age.* London: Penguin Books.

Rivers, Thomas. 1837. *The rose amateur's guide; containing ample descriptions of all the fine leading varieties of roses . . . The whole arranged so as to form a companion to the descriptive catalogue of the Sawbridgeworth collection of roses, published annually.* London: the proprietor.

Rivière, Auguste. 1866. Observations sur la fécondation et la germination des orchidées, a l'occasion d'un *Lælia* hybride présenté en fleurs, le 25 août 1865. *Journal de la Société Impériale et Centrale d'Horticulture* 12: 268–91.

Roberts, H. F. 1929. *Plant hybridisation before Mendel.* Princeton, N.J.: Princeton University Press.

Rodgers, Andrew Denny III. 1968. *'Noble fellow': William Starling Sullivant.* London and New York: Hafner Publishing.

Rogers, G. Albert. 1865. *A winter in Algeria. 1863–4.* London: Sampson Low, Son, & Marston.

Roget, Peter Mark. 1834. *Animal and vegetable physiology considered with reference to natural theology.* 2 vols. London: William Pickering. [Treatise 5 of the 'Bridgewater Treatises on the power wisdom and goodness of God as manifested in the creation'.]

Rolle, Friedrich. 1863. *Ch. Darwin's Lehre von der Entstehung der Arten im Pflanzen- und Thierreich in ihrer Anwendung auf die Schöpfungsgeschichte.* Frankfurt: J. C. Hermann.

——. 1866. *Der Mensch, seine Abstammung und Gesittung, im Lichte der Darwin'schen Lehre von der Art-entstehung und auf Grundlage der neuern geologischen Entdeckungen dargestellt.* Frankfurt: J. C. Hermann.

Rose, R. N. 1953. *The Field, 1853–1953.* London: Michael Joseph.

Rothrock, Joseph Trimble. 1868. The fertilization of flowering plants. *American Naturalist* 1: 64–72.

Royal Society catalogue of scientific papers: *Catalogue of scientific papers (1800–1900). Compiled and published by the Royal Society of London.* 19 vols. and index (3 vols.). London: Royal Society of London. Cambridge: Cambridge University Press. 1867–1925.

Royer, Clémence Auguste, trans. 1862. *De l'origine des espèces ou des lois du progrès chez les êtres organisés. Par Ch. Darwin. Avec une préface et des notes du traducteur.* [Translated from the 3d English edition.] Paris: Guillaumim & Cie; Victor Masson.

Rupke, Nicolaas A. 1994. *Richard Owen, Victorian naturalist.* New Haven, Conn., and London: Yale University Press.

Ruse, Michael. 1991. William Whewell: omniscientist. In *William Whewell. A composite portrait*, edited by Menachem Fisch and Simon Schaffer. Oxford: Clarendon Press.

Rütimeyer, Karl Ludwig. 1861. *Die Fauna der Pfahlbauten in der Schweiz. Untersuchungen über die Geschichte der wilden und der Haus-Säugethiere von Mittel-Europa.* Basel, Switzerland: Bahnmaier's Buchhandlung (C. Detloff).

Sabine, Edward. 1863. [Aniversary address, 30 November 1863.] *Proceedings of the Royal Society of London* 13: 22–39.

Sachs, Julius von. 1890. *History of botany (1530–1860).* Translated by Henry E. F. Garnsey. Revised by Isaac Bayley Balfour. Oxford: Clarendon Press.

Sageret, Augustin. 1826. Considérations sur la production des hybrides, des variantes et des variétés en général, et sur celles de la famille des Cucurbitacées en particulier. *Annales des Sciences Naturelles* 8: 294–314.

Saint-Hilaire, Auguste de. 1841. *Leçons de botanique: comprenant principalement la morphologie végétale, la terminologie, la botanique comparée, l'examen de la valeur des charactères dans les diverses familles naturelles, etc.* Paris: P. J. Loss.

Saint-Simon, Henri Maximilien, marquis de. 1768. *Des jacintes, de leur anatomie, reproduction et culture.* Amsterdam: C. Eel.

Sanches, Francisco. 1988. *That nothing is known (Quod nihil scitur)*. Introduction, notes, and bibliography by Elaine Limbrick. Latin text established, annotated, and translated by Douglas F. S. Thomson. Cambridge: Cambridge University Press.

Sanchez, Franciscus. [Sanches, Francisco.] 1581. *Quod nihil scitur.* Lyons: Antonius Gryphius.

——. [Sanches, Francisco.] 1618. *De multum nobili et prima universali scientia Quod nihil scitur.* Frankfurt: Ioannis Bernerus.

Saporta, Louis Charles Joseph Gaston de. 1862a. Etudes sur la végétation du sud-est de la France à l'époque Tertiaire. *Annales des Sciences Naturelles (Botanique)* 16: 309–45; 17: 191–311.

——. 1862b. *Notice sur les plantes fossiles de Coumi et d'Oropo.* Paris: [privately printed]. [Reprinted in *Animaux fossiles et géologie de l'Attique d'après les recherches faites en 1855–56 et en 1860 sous les auspices de l'Académie des Sciences.* With atlas. By Albert Gaudry. Paris: F. Savy. 1862–7.]

——. 1869. L'Ecole transformiste et ses dernièrs travaux. *Revue des deux mondes* 2d ser. 83: 635–74.

Sarjeant, William A. S. 1980–96. *Geologists and the history of geology: an international bibliography.* 10 vols. including supplements. London: Macmillan. Malabar, Fla.: Robert E. Krieger Publishing Company.

Sarton, George. 1942. Lesquereux (1806–89). *Isis* 34: 97–108.

Saussure, Henri Louis Frédéric de. 1852–8. *Etudes sur la famille des Vespides.* 3 vols. Paris and Geneva: Victor Masson [& others].

——. 1858. *Observations sur les mœurs de divers oiseaux du Mexique.* Geneva: Ramboz & Schuchardt.

——. 1858–71. *Mémoires pour servir a l'histoire naturelle du Mexique des Antilles et des Etats-unis.* 2 vols. Geneva: H. Georg.

——. 1862. *Carte du Mexique représentant le plateau de l'Anahuac et son versant oriental.* Geneva: Jules-Gme Fick.

Sayce, Roger. 1992. *The history of the Royal Agricultural College, Cirencester.* Stroud, Gloucestershire: Alan Sutton.

SBH: *Svenskt biografiskt handlexikon.* Edited by Herman Hofberg *et al.* 2 vols. Stockholm: Albert Bonniers. 1906.

SBL: *Svenskt biografiskt lexikon.* Edited by Bertil Boëthius *et al.* 28 vols. and 4 fascicles of vol. 29 (A–Ramsay). Stockholm: Albert Bonnier & P. A. Norstedt. 1918–96.

Schacht, Hermann. 1856–9. *Lehrbuch der Anatomie und Physiologie der Gewächse.* 2 vols. Berlin: G. W. F. Müller.

Scherren, Henry. 1905. *The Zoological Society of London. A sketch of its foundation and development and the story of its farm, museum, gardens, menagerie and library.* London: Cassell.

S[chlechtenda]l, [D. F. L.]. 1864. Personal-Nachrichten [Herman Crüger]. *Botanische Zeitung* 22: 119–20.

Schleicher, August. 1863. *Die Darwinsche Theorie und die Sprachwissenschaft. Offenes Sendschreiben an Herrn Dr Ernst Häkel.* Weimar, Germany: Hermann Böhlau.

Schleiden, Matthias Jacob. 1842–3. *Grundzüge der wissenschaftlichen Botanik, nebst einer methodologischen Einleitung als Anleitung zum Studium der Pflanze.* 2 vols. in 1. Leipzig: Wilhelm Engelmann.

Schneider, Carl Friedrich Adolph. 1863. Bijdrage tot de geologische kennis van Timor. *Natuurkundig Tijdschrift voor Nederlandsch-Indië* 25: 87–107

Scholefield, Guy Hardy, ed. 1950. *New Zealand parliamentary record, 1840–1949.* Wellington, New Zealand: R. E. Owen, government printer.

Schomburgk, Robert Hermann. 1836. On the identity of three supposed genera of orchideous epiphytes. [Read 15 November 1836.] *Transactions of the Linnean Society of London* 17 (1837): 551–2.

——. 1857. Description of a remarkable spike or bunch of fruits of the fig banana (*Musa sapientum*), var. [Read 2 June 1857.] *Journal of the Proceedings of the Linnean Society (Botany)* 2 (1858): 130–2.

Schroeder van der Kolk, Jacob Ludwig Conrad and Vrolik, Willem. 1849. Ontleedkundige nasporingen over de hersenen van den chimpansé. *Verhandelingen der eerste klasse van het Koninklijk-Nederlandsche Instituut van wetenschappen, letterkunde en schoone kunsten te Amsterdam* 3d ser. 1 (1848–50): 263–78.

Schumann, Albert. 1888. *Aargauische Schriftsteller.* Aarau, Switzerland: H. R. Sauerlaender.

Sclater, Philip Lutley. 1860. Remarks on the geographical distribution of recent terrestrial vertebrata. *Report of the 30th meeting of the British Association for the Advancement of Science, held at Oxford*, Transactions of the sections, pp. 121–2.

Scott, John. 1862a. Remarks on the nature and peculiarities of the fern-spore. [Read 12 June 1862.] *Transactions of the Botanical Society* [*of Edinburgh*] 7: 352–70. [Reprinted in the *Edinburgh New Philosophical Journal* n.s. 16: 209–27.]

——. 1862b. On the propagation and irritability of *Drosera* and *Dionaea*. [Read 11 December 1862.] *Transactions of the Botanical Society* [*of Edinburgh*] 7 (1863): 429–30. [Abstract of paper. Abstracts also appeared in the notice of the proceedings of the Botanical Society of Edinburgh in the *Edinburgh New Philosophical Journal* n.s. 17 (1863): 317–18 and in the *Gardeners' Chronicle*, 10 January 1863, p. 30.]

——. 1862c. On the homologies of the sporiferous appendices of Anemia, &c. [Read 13 February 1862.] *Transactions of the Botanical Society* [*of Edinburgh*] 7 (1863): 300–5.

——. 1862d. Remarks on *Marsilea macropus* (the Nardoo plant) which had germinated in the Edinburgh Botanic Garden. [Read 13 March 1862.] *Transactions of the Botanical Society* [*of Edinburgh*] 7 (1863): 335–6. [Abstract of paper.]

——. 1863a. Experiments on the fertilisation of orchids in the Royal Botanic Garden of Edinburgh. [Read 14 May 1863.] *Transactions of the Botanical Society* [*of Edinburgh*] 7: 543–50. [An abstract appeared in *Edinburgh New Philosophical Journal* n.s. 18: 322–4.]

——. 1863b. Remarks on the sexual changes in the inflorescence of *Zea Mays.* [Read before the Botanical Society of Edinburgh, 10 December 1863.] *Edinburgh New Philosophical Journal* n.s. 19 (1864): 213–20.

Scott, John. 1864a. Observations on the functions and structure of the reproductive organs in the Primulaceæ. [Read 4 February 1864.] *Journal of the Linnean Society (Botany)* 8: 78–126.

——. 1864b. On the individual sterility and cross-impregnation of certain species of *Oncidium*. [Read 2 June 1864.] *Journal of the Linnean Society (Botany)* 8: 162–7.

——. 1864c. Remarks on the sexuality of the higher Cryptogams, with a notice of a hybrid *Selaginella*. [Read before the Botanical Society of Edinburgh, 10 March 1864.] *Edinburgh New Philosophical Journal* n.s. 19: 192–200.

——. 1864d. Notes on the sterility and hybridization of certain species of *Passiflora*, *Disemma*, and *Tacsonia*. [Read 16 June 1864.] *Journal of the Linnean Society (Botany)* 8: 197–206.

——. 1867. On the reproductive functional relations of several species and varieties of *Verbasca*. *Journal of the Asiatic Society of Bengal* 36 (pt 2): 145–74.

Scudder, Samuel Hubbard. 1862a. On the genus *Colias* in North America. *Proceedings of the Boston Society of Natural History* 9 (1862–3): 103–11.

——. 1862b. *Materials for a monograph of North American Orthoptera*. Cambridge, Mass.: [privately printed]. [Reprinted from the *Boston Journal of Natural History* 7 (1859–63): 409–80.]

——. 1862c. An account of the structure of *Pogonia ophioglossoides* Nutt., and of the probable manner in which its fertilization is effected. [Read 3 December 1862.] *Proceedings of the Boston Society of Natural History* 9 (1862–3): 182–5.

——. 1889. *The butterflies of the Eastern United States and Canada with special reference to New England*. 3 vols. Cambridge, Mass.: the author.

Secord, James A. 1986. *Controversy in Victorian geology: the Cambrian-Silurian dispute*. Princeton, N.J.: Princeton University Press.

Sedgwick, Adam. 1835. Remarks on the structure of large mineral masses, and especially on the chemical changes produced in the aggregation of stratified rocks during different periods after their deposition. [Read 11 March 1835.] *Transactions of the Geological Society of London* 2d ser. 3: 461–86.

Seemann, Berthold Carl. 1852–7. *The botany of the voyage of HMS Herald . . . during the years 1845–51*. London: Lovell Reeve.

Seringe, Nicolas Charles. 1823. *Musée Helvétique d'Histoire Naturelle. (Partie botanique.) Ou collection de mémoires, monographies, notices botaniques . . . par N. C. Seringe*. Berne, Switzerland: J. J. Burgdorfer.

Serres, Antoine Etienne Reynaud Augustin. 1857. Note sur une collection d'ossements fossiles recueillis par M. Séguin dans l'Amérique du Sud. *Comptes rendus hebdomadaires des Séances de l'Académie des Sciences* 44: 954–62.

——. 1867. De l'ostéographie du *Mesotherium* et des ses affinités zoologiques. *Comptes rendus hebdomadaires des Séances de l'Académie des Sciences* 65: 6–17, 140–8, 273–8, 429–37, 593–9, 740–8, 841–8.

Shaw, John. 1877. *A synopsis of the proceedings of the Agri-horticultural Society of Madras, from July 1835 to December 1870, with an index to the synopsis and the proceedings of 1871, 1872, 1873, and 1874*. Madras, India: Gantz Brothers.

Sheail, John. 1976. *Nature in trust: the history of nature conservation in Britain.* Glasgow and London: Blackie & Son.

Sibley, Charles G. and Monroe, Burt L., Jr. 1990–3. *Distribution and taxonomy of birds of the world.* With a supplement. New Haven, Conn., and London: Yale University Press.

Sifakis, Stewart. 1988. *Who was who in the Civil War.* New York and Oxford: Facts on File Publications.

Sigerson, George. 1863. *Some remarks on a proto-morphic phyllotype.* Dublin: n.p. [Reprinted from *Atlantis* 4 (1863): 450–8.]

Silfverstråhle, Gustaf. 1839. Om Blads och Knoppars ställning: å Växters yta. *Königlich Vetenskapsacademiens Handlingar* (1839): 212–21.

Simpson, George Gaylord. 1980. *Splendid isolation: the curious history of South American mammals.* New Haven, Conn., and London: Yale University Press.

Sixth report of the medical officer of the Privy Council: *Sixth report of the medical officer of the Privy Council. With appendix. 1863.* [British parliamentary papers, Session 1864, 28: 1–799.]

Slack, Henry J. 1862. Life changes on the globe. *Intellectual Observer* 1: 325–33.

Smith, Albert Richard. 1843. The physiology of evening parties. In *The wassail-bowl*, by Albert Richard Smith. 2 vols. in 1. London: Richard Bently.

——. 1847. *The natural history of the gent.* London: David Bogue.

——. 1848. *The natural history of the idler upon town.* London: David Bogue.

Smith, Brian S. 1978. *A history of Malvern.* N.p.: Alan Sutton & The Malvern Bookshop.

Smith, Frederick. 1863. Notes on the geographical distribution of the aculeate *Hymenoptera* collected by Mr A. R. Wallace in the Eastern Archipelago. [Read 5 March 1863.] *Journal of the Proceedings of the Linnean Society (Zoology)* 7: 109–45.

Smith, Goldwin. 1863. *The Empire. A series of letters published in 'The Daily News,' 1862, 1863.* Oxford and London: John Henry & James Parker.

Smith, James. 1845. On the scratched boulders and rocks of the coal-field of Scotland. [Read 4 June 1845.] *Quarterly Journal of the Geological Society of London* 2 (1846): 33–7.

——. 1848. On scratched boulders. [Read 19 April 1848 and 17 May 1848.] *Quarterly Journal of the Geological Society of London* 5 (1849): 17–20

——. 1862. *Researches in newer Pliocene and post-Tertiary geology.* Glasgow: John Gray.

Smith, Pleasance. 1832. *Memoir and correspondence of the late Sir James Edward Smith, MD.* 2 vols. London: Longman, Rees, Orme, Brown, Green, & Longman.

Smith, William. 1862. Cross-breeding strawberries. *Journal of Horticulture* n.s. 3: 779.

SMK: *Svenska män och kvinnor biografisk uppslagsbok.* Edited by Nils Bohman *et al.* 8 vols. Stockholm: Albert Bonniers. 1942–55.

Soulsby, Hugh G. 1933. *The right of search and the slave trade in Anglo-American relations 1814–1862.* Ser. 51, no. 2, of *The Johns Hopkins University Studies in Historical and Political Science.* Baltimore, Md.: Johns Hopkins Press.

South America: *Geological observations on South America. Being the third part of the geology of the voyage of the Beagle, under the command of Capt. FitzRoy RN, during the years 1832 to 1836.* By Charles Darwin. London: Smith, Elder & Co. 1846.

Spater, George. 1982. *William Cobbett. The poor man's friend.* 2 vols. Cambridge: Cambridge University Press.

'Specific difference in *Primula*': On the specific difference between *Primula veris*, Brit. Fl. (var. *officinalis* of Linn.), *P. vulgaris*, Brit. Fl. (var.*acaulis*, Linn.), and *P elatior*, Jacq.; and on the hybrid nature of the common oxlip. With supplementary remarks on naturally produced hybrids in the genus *Verbascum*. By Charles Darwin. [Read 19 March 1868.] *Journal of the Linnean Society (Botany)* 10 (1869): 437–54.

Spencer, Herbert. 1860–2. *First principles.* London: George Manwaring; Williams & Norgate.

——. 1864–7. *The principles of biology.* 2 vols. London: Williams & Norgate.

Spokes, Sidney. 1927. *Gideon Algernon Mantell, LLD, FRCS, FRS, surgeon and geologist.* London: John Bale, Sons & Danielsson.

Sprengel, Christian Konrad. 1793. *Das entdeckte Geheimniss der Natur im Bau und in der Befruchtung der Blumen.* Berlin: Freidrich Vieweg.

Stace, Clive Anthony. 1991. *New flora of the British Isles.* Cambridge: Cambridge University Press.

Stafford, Robert A. 1989. *Scientist of empire: Sir Roderick Murchison, scientific exploration and Victorian imperialism.* Cambridge: Cambridge University Press.

[Stanley, Penrhyn Arthur.] 1861. [Review of *Essays and reviews* & other works.] *Edinburgh Review* 113: 461–500.

Stebbins, Robert E. 1974. France. In *The comparative reception of Darwinism*, edited by Thomas F. Glick. Austin, Tex., and London: University of Texas Press.

Steenis, Cornelis Gijsbert Gerrit Jan van, ed. 1950. *Flora Malesiana: being an illustrated systematic account of the Malaysian flora, including ... notes on the ecology of its wild and commonly cultivated plants.* Volume 1 of *Malaysian plant collectors and collections, being a cyclopaedia of botanical explorations in Malaysia and a guide to the concerned literature up to the year 1950.* Ser. 1, *Spermatophyta.* Published under the auspices of the Kebun Raya Indonesia, Bogor, Java, Botanic Gardens of Indonesia, Bogor (Buitenzorg) and the Rijksherbarium, Leyden, Netherlands. Jakarta, Indonesia: Noordhoff-Kolff.

Stepan, Nancy. 1982. *The idea of race in science: Great Britain 1800–1960.* London: Macmillian in association with St Anthony's College, Oxford.

Stocking, George W. Jr. 1968. *Race, culture, and evolution: essays in the history of anthropology.* New York: The Free Press. London: Collier Macmillan.

——. 1987. *Victorian anthropology.* New York: The Free Press. London: Collier Macmillan.

Stowe, Harriet Elizabeth Beecher. 1852. *Uncle Tom's cabin, or life among the lowly.* 2 vols. Boston, Mass.: John P. Jewett.

Stresemann, Erwin. 1975. *Ornithology: from Aristotle to the present.* Translated by Hans J. and Cathleen Epstein. Edited by G. William Cottrell. With foreword and epilogue by Ernst Mayr. Cambridge, Mass., and London: Harvard University Press.

Suess, Eduard. 1863. Über die Verschiedenheit und die Aufeinanderfolge der tertiären Landfaunen in der Niederung von Wien. *Sitzungsberichte der Kaiserlichen Akademie der Wissenschaften. Mathematisch-Naturwissenschaftliche Classe* 47 (pt 1): 306–31.

Sulivan, Henry Norton, ed. 1896. *Life and letters of the late Admiral Sir Bartholomew James Sulivan, KCB, 1810–1890.* London: John Murray.

Sullivan, Alvin, ed. 1984. *British literary magazines: the Victorian and Edwardian age, 1837–1913.* London and Westport, Conn.: Greenwood Press.

Sulloway, Frank J. 1979. Geographic isolation in Darwin's thinking: the vicissitudes of a crucial idea. In vol. 3 of *Studies in history of biology*, edited by William Coleman and Camille Limoges. Baltimore, Md., and London: Johns Hopkins University Press.

——. 1984. Darwin and the Galapagos. *Biological Journal of the Linnean Society* 21: 29–59.

Sumner, John Bird. 1816. *A treatise on the records of the creation, and on the moral attributes of the creator; with particular reference to the Jewish history, and to the consistency of the principle of population with the wisdom and goodness of the deity.* 2 vols. London: J. Hatchard.

Survey gazetteer of the British Isles: *The survey gazetteer of the British Isles including summary of 1951 census.* By John Bartholomew. 9th edition. Edinburgh: John Bartholomew & Son at the Geographical Institute.

Sutton, R. J. 1966. *The stamp collector's encyclopaedia.* 7th edition. 6th edition revised by K. W. Anthony. London: Stanley Paul.

[Swift, Jonathan.] 1726. *Travels into several remote nations of the world. In four parts. By Lemuel Gulliver, first a surgeon, and then a captain of several ships.* 2 vols. London: Benjamin Motte.

Swinhoe, Robert. 1861. *Narrative of the north China campaign of 1860; containing personal experiences of Chinese character, and of the moral and social condition of the country; together with a description of the interior of Pekin.* London: Smith, Elder & Co.

——. 1863a. The ornithology of Formosa, or Taiwan. *Ibis* 5: 198–219, 250–311, 377–135.

——. [1863b.] *Notes on the ethnology of Formosa.* [London]: n.p. [Extracted from a paper read before the Ethnological Society of London, with additional remarks, and read before the British Association for the Advancement of Science in August 1863.]

——. [1863c.] *Notes on the island of Formosa.* [London]: n.p. [Read before the British Association for the Advancement of Science in August 1863 and before the Royal Geographical Society of London.]

——. [1863d.] *The ornithology of Formosa, or Taiwan.* N.p.: n.p. [Reprinted from *Ibis* 5 (1863): 198–219, 250–311, 377–435.]

——. [1863e.] *On the mammals of the island of Formosa (China).* N.p.: n.p. [Reprinted from *Proceedings of the Scientific Meetings of the Zoological Society of London* (1862): 347–65.]

Swinhoe, Robert. [1863f.] *A list of the Formosan reptiles; with notes on a few of the species, and some remarks on a fish (*Orthagoriscus, *sp.).* N.p.: n.p. [Reprinted from *Annals and Magazine of Natural History* 3d ser. 12 (1863): 219–26.]

——. [1863g.] *List of plants from the island of Formosa, or Taiwan.* [London]: n.p.

——. [1863h.] *Catalogue of the birds of China, with remarks principally on their geographical distribution.* N.p.: n.p. [Reprinted from *Proceedings of the Scientific Meetings of the Zoological Society of London* (1863): 259–339.]

——. [1863i.] *Notes on the ornithology of Hongkong, Macao, and Canton, made during the latter end of February, March, April, and the beginning of May, 1860.* N.p.: n.p. [Reprinted from *Ibis* 3 (1861): 23–57.]

——. [1863j.] *Ornithological ramble in Foochow, in December 1861.* N.p.: n.p. [Reprinted from *Ibis* 4 (1862): 253–65.]

——. [1863k.] *On new and little-known birds from China.* N.p.: n.p. [Reprinted from *Proceedings of the Scientific Meetings of the Zoological Society of London* (1863): 87–94.]

——. [1863l.] *Additions and corrections to the ornithology of Northern China.* N.p.: n.p. [Reprinted from *Ibis* 5: 87–97.]

——. [1863m.] *On some Tientsin birds, collected by Mr Fleming, RA, in the possession of Mr Whiteley.* N.p.: n.p. [Reprinted from *Proceedings of the Scientific Meetings of the Zoological Society of London* (1862): 315–20.]

——. [1863n.] *On a bird supposed to be the female of* Crossoptilon auritum, *Pallas, from Northern China.* N.p.: n.p. [Reprinted from *Proceedings of the Scientific Meetings of the Zoological Society of London* (1862): 286–7.]

——. [1863o.] *Notes on the ornithology of Northern Japan.* N.p.: n.p. [Reprinted from *Ibis* 5 (1863): 442–5.]

Tausch, Ignaz Friedrich. 1821. Beobachtungen über das Längenverhältniss der Befruchtungsorgane bei der Gattung Primula. *Flora* (1821): 353–68.

Taxonomic literature: *Taxonomic literature. A selective guide to botanical publications and collections with dates, commentaries and types.* By Frans A. Stafleu and Richard S. Cowan. 2d edition. 7 vols. Utrecht: Bohn, Scheltema & Holkema. The Hague: W. Junk. 1976–88.

Taylor, Tom. 1869. *Our American cousin. A drama in three acts.* N.p.: privately printed.

Tebbel, John. 1972. *A history of book publishing in the United States. Volume 1. The creation of an industry, 1630–1865.* New York and London: R. R. Bowker.

[Tegetmeier, William Bernhard.] 1863a. Philoperiston Society's annual show. *Journal of Horticulture* n.s. 4: 58–9.

Tegetmeier, William Bernard. 1863b. Variations in plumage. *Intellectual Observer* 3: 171–3.

——. 1867. *The poultry book: comprising the breeding and management of profitable and ornamental poultry, their qualities and characteristics; to which is added 'The standard of excellence in exhibition birds', authorized by the Poultry Club.* London and New York: George Routledge & Sons.

Temminck, Coenraad Jacob. 1813–15. *Histoire naturelle générale des pigeons et des gallinacés.* 3 vols. Amsterdam: J. C. Sepp. Paris: G. Dufour.

Tennyson, Alfred. 1859. *Idylls of the king*. London: E. Moxon.

Thomas, Keith. 1984. *Man and the natural world: changing attitudes in England 1500–1800*. Harmondsworth, Middlesex: Penguin Books.

Thompson, F. M. L. 1981. Landowners and the rural community. In *The Victorian countryside*, edited by G. E. Mingay. 2 vols. London: Routledge & Kegan Paul.

Thomson, Charles Wyvill. 1863–4. On the embryology of the Echinodermata. *Natural History Review* n.s. 3 (1863): 395–415; 4 (1864): 581–611.

'Three forms of *Lythrum salicaria*': On the sexual relations of the three forms of *Lythrum salicaria*. By Charles Darwin. [Read 16 June 1864.] *Journal of the Linnean Society (Botany)* 8 (1865): 169–96. [*Collected papers* 2: 106–31.]

'Three sexual forms of *Catasetum tridentatum*': On the three remarkable sexual forms of *Catasetum tridentatum*, an orchid in the possession of the Linnean Society. By Charles Darwin. [Read 3 April 1862.] *Journal of the Proceedings of the Linnean Society (Botany)* 6 (1862): 151–7. [*Collected papers* 2: 63–70.]

Thwaites, George Henry Kendrick. 1848. Further observations on the Diatomaceæ; with descriptions of new genera and species. *Annals and Magazine of Natural History* 2d ser. 1: 161–72.

——. 1858–64. *Enumeratio plantarum Zeylaniæ: an enumeration of Ceylon plants, with descriptions of the new and little-known genera and species, observations on their habits, uses, native names, etc.* Assisted in the identification of the species and synonymy by J. D. Hooker. 5 pts. London: William Paplin; Dulau & Co.

Times atlas: *'The Times' atlas of the world. Comprehensive edition.* 9th edition. London: Times Books. 1992.

Tocqueville, Charles Alexis Henri Maurice Clérel de. 1836. *De la démocratie en Amérique*. 4th edition. 2 vols. in 1. Paris: Charles Gosselin.

Todhunter, Isaac. 1876. *William Whewell, DD, master of Trinity College, Cambridge: an account of his writings with selections from his literary and scientific correspondence.* 2 vols. London: Macmillan.

Tolles, Winton. 1940. *Tom Taylor and the Victorian drama*. New York: Columbia University Press.

Tonks, Edmund. 1883. *General index to the Latin names and synonyms of the plants depicted in the first hundred and seven volumes of Curtis's Botanical Magazine, to which is added a short list of popular names.* London: B. Quaritch.

Topham, Jonathan Richard. 1993. 'An infinite variety of arguments': the *Bridgewater treatises* and British natural theology in the 1830s. (PhD dissertation: University of Lancaster.)

Torell, Otto Martin. 1859. *Bidrag till Spitsbergens molluskfauna. Jemte en allmän öfversigt af Arktiska regionens naturförhållanden och forntida utbredning.* Stockholm: Typografiska föreningens boktryckeri.

Torrey, John and Gray, Asa. 1838–43. *A flora of North America: containing abridged descriptions of all the known indigenous and naturalized plants growing north of Mexico; arranged according to the natural system.* 2 vols. New York and London: Wiley & Putnam.

Tort, Patrick. 1996. *Dictionnaire du Darwinisme et de l'évolution.* 3 vols. Paris: Presses Universitaires de France.

Treub, Melchior. 1891. Sur les Casuarinées et leur place dans le Système naturel. *Annales du Jardin Botanique de Buitenzorg* 10: 145–231.

Treviranus, Ludolf Christian. 1861. [Review of J. D. Hooker's *Flora Tasmaniæ.*] *Botanische Zeitung* 19: 133–5, 142–4.

——. 1863a. Ueber Dichogamie nach C. C. Sprengel und Ch. Darwin. *Botanische Zeitung* 21: 1–7, 9–16.

——. 1863b. [Review of 'Two forms in species of *Linum*'.] *Botanische Zeitung* 21: 188–90.

——. 1863c. Nachträgliche Bemerkungen über die Befruchtung einiger Orchideen. *Botanische Zeitung* 21: 241–3.

Trimen, Roland. 1862–6. *Rhopalocera Africæ Australis; a catalogue of South African butterflies, comprising descriptions of all the known species with notices of their larvæ, pupæ, localities, habits, seasons of appearance, and geographical distribution.* London: Trübner. Cape Town, South Africa: W. F. Mathew.

——. 1863. On the fertilization of *Disa grandiflora*, Linn. ... drawn up from notes and drawings sent to C. Darwin, Esq., FLS, &c. [Read 4 June 1863.] *Journal of the Proceedings of the Linnean Society (Botany)* 7: 144–7.

——. 1864. On the structure of *Bonatea speciosa*, Linn. sp., with reference to its fertilisation. [Read 1 December 1864.] *Journal of the Linnean Society (Botany)* 9 (1865): 156–60.

——. 1887–9. *South-African butterflies: a monograph of the extra-tropical species.* With the assistance of James Henry Bowker. 3 vols. London: Trübner.

Trimmer, Joshua. 1831. On the diluvial deposits of Caernarvonshire, between the Snowdon chain of hills and the Menai strait, and on the discovery of marine shells in diluvial sand and gravel on the summit of Moel Tryfane, near Caernarvon, 1000 ft above the level of the sea. [Read 8 June 1831.] *Proceedings of the Geological Society of London* 1 (1826–33): 331–2.

——. 1850. Generalizations respecting the erratic tertiaries or northern drift, founded on the mapping of the superficial deposits of a large portion of Norfolk. With a description of the freshwater deposits of the Gayton-Thorpe Valley; and a note on the contorted strata of Cromer cliffs. [Read 20 November 1850.] *Quarterly Journal of the Geological Society of London* 7 (1851): 19–31.

Turnbull, Constance Mary. 1989. *A history of Singapore 1819–1988.* 2d edition. Oxford: Oxford University Press.

Turner, Gerard L'E. 1989. *The great age of the microscope: the collection of the Royal Microscopical Society though 150 years.* Bristol and New York: Adam Hilger.

Turner, James Crewdson. 1980. *Reckoning with the beast: animals, pain, and humanity in the Victorian mind.* Baltimore, Md., and London: Johns Hopkins University Press.

'Two forms in species of *Linum*': On the existence of two forms, and on their reciprocal sexual relation, in several species of the genus *Linum.* By Charles Darwin. [Read 5 February 1863.] *Journal of the Proceedings of the Linnean Society*

(Botany) 7 (1864): 69–83. [*Collected papers* 2: 93–105.]

Van Riper, A. Bowdoin. 1993. *Men among the mammoths: Victorian science and the discovery of human prehistory.* Chicago and London: University of Chicago Press.

Variation: *The variation of animals and plants under domestication.* By Charles Darwin. 2 vols. London: John Murray. 1868.

Vaucher, Jean Pierre Etienne. 1841. *Histoire physiologique des plantes d'Europe ou exposition des phénomènes qu'elles présentent dans les diverses périodes de leur développement.* 4 vols. Paris: Marc Aurel Frères.

Vegetable mould and worms: *The formation of vegetable mould, through the action of worms, with observations on their habits.* By Charles Darwin. London: John Murray. 1881.

Verlot, Bernard. 1864. Mémoire sur la production et la fixation des variétés dans les plantes d'ornement. *Journal de la Société Impériale et Centrale d'Horticulture* 10: 243–56, 305–20, 375–84, 420–32, 468–80, 518–28, 571–6, 624–40.

——. 1865. *Sur la production et la fixation des variétés dans les plantes d'ornement.* Paris: J. B. Baillière.

Volcanic islands: *Geological observations on the volcanic islands, visited during the voyage of HMS Beagle, together with some brief notices on the geology of Australia and the Cape of Good Hope. Being the second part of the geology of the voyage of the Beagle, under the command of Capt. FitzRoy RN, during the years 1832 to 1836.* By Charles Darwin. London: Smith, Elder & Co. 1844.

Vrolik, Willem. 1840–2. *Handboek der ziektekundige ontleedkunde.* [*De menschelijke vrucht beschouwd in hare regelmatige en onregelmatige ontwikkeling.*] 2 vols. Amsterdam: Johannes Müller.

Wade, Nicholas J., ed. 1983. *Brewster and Wheatstone on vision.* London: Academic Press.

Wagner, Johann Andreas. 1861a. Zur feststellung des Artbegriffes, mit besonderer Bezugnahme auf die Ansichten von Nathusius, Darwin, Is. Geoffroy und Agassiz. *Sitzungsberichte der königl. bayerischen Akademie der Wissenschaften zu München* 1: 308–58.

——. 1861b. Ein neues, angeblich mit Vogelfedern versehenes Reptil. *Sitzungsberichte der königl. bayerischen Akademie der Wissenschaften zu München* 2: 146–54.

——. 1862. On a new fossil reptile supposed to be furnished with feathers. *Annals and Magazine of Natural History* 3d ser. 9: 261–7. Translated by W. S. Dallas. [Originally published as 'Ein neues, angeblich mit Vogelfedern versehenes Reptil', *Sitzungsberichte der königl. bayerischen Akademie der Wissenschaften zu München* 2 (1861): 146–54.]

Wagner, Rudolph, ed. 1842–53. *Handwörterbuch der Physiologie mit Rücksicht auf physiologische Pathologie.* 4 vols. Brunswick, Germany: Friedr. Vieweg & Sohn.

Wallace, Alfred Russel. 1853. *A narrative of travels on the Amazon and Rio Negro, with an account of the native tribes, and observations on the climate, geology, and natural history of the Amazon valley.* London: Reeve.

——. 1863a. List of birds collected in the island of Bouru (one of the Moluccas), with descriptions of the new species. *Proceedings of the Zoological Society of London* (1863): 18–36.

Wallace, Alfred Russel. 1863b. Remarks on the Rev. S. Haughton's paper on the bee's cell, and on the origin of species. *Annals and Magazine of Natural History* 3d ser. 12: 303–9.

——. 1863c. On the geographical distribution of animal life. *Report of the 33d meeting of the British Association for the Advancement of Science, held at Newcastle-upon-Tyne,* Transactions of the sections, pp. 108–9.

——. 1905. *My life: a record of events and opinions.* 2 vols. London: Chapman & Hall.

Walton, John K. 1983. *The English seaside resort: a social history 1750–1914.* Leicester: Leicester University Press.

War of the rebellion: *The war of the rebellion: a compilation of the official records of the Union and Confederate armies.* 128 vols. in 4 ser. Washington, DC: Government Printing Office. 1880–1901.

Watt, Robert. 1824. *Bibliotheca britannica; or a general index to British and foreign literature.* 4 vols. Edinburgh: Archibald Constable; Longmans.

Webb, Sidney and Webb, Beatrice. 1913. *English local government: the story of the king's highway.* Vol. 5 of *English local government.* By Sidney and Beatrice Webb. 9 vols. London: Longmans, Green & Co. 1906–29.

Weddell, Hugh Algernon. 1849. *Histoire naturelle des quinquinas ou monographie du genre* Cinchona *suivie d'une description du genre* Cascarilla *et de quelques autres plantes de la même tribu.* Paris: V. Masson.

——. 1855–7. *Chloris Andina. Essai d'une flore de la région alpine des Cordillères de l'Amérique du Sud.* 2 vols. Pt 6 of *Expédition dans les parties centrales de l'Amérique du Sud . . . exécutée par ordre du gouvernment français pendant les années 1843 a 1847, sous la direction de Comte Francis de Castelnau.* 7 pts in 13 vols. Paris: P. Betrand. 1850–7.

——. 1860. Mémoire sur le *Cynomorium coccineum* parasite de l'ordre des Balanophorées. *Archives de Muséum d'Histoire Naturelle* 10 (1858–61): 269–308.

Wedgwood, Barbara and Wedgwood, Hensleigh. 1980. *The Wedgwood circle, 1730–1897: four generations of a family and their friends.* London: Studio Vista.

Weinland, David Friedrich. 1861. [Review of *Origin* & other works.] *Der zoologische Garten* 2: 82–5.

Weinreich, Heinrich. 1993. *Duftstofftheorie: Gustav Jaeger (1832–1917); vom Biologen zum 'Seelenriecher'.* Heidelberger Schriften zur Pharmazie- und Naturwissenschaftsgeschichte, Bd. 11. Stuttgart: Wissenschaftliche Verlagsgesellschaft.

Wellesley index: *The Wellesley index to Victorian periodicals 1824–1900.* Edited by Walter E. Houghton *et al.* 5 vols. Toronto, Ontario: University of Toronto Press. London: Routledge & Kegan Paul. 1966–89.

Wells, Stanley and Taylor, Gary, eds. 1988. *William Shakespeare: the complete works.* Compact edition. Oxford: Clarendon Press.

Welwitsch, Friedrich. 1861. Extract from a letter, addressed to Sir William J. Hooker, on the botany of Benguela, Mossamedes, &c., in Western Africa. [Read 17 January 1861.] *Journal of the Proceedings of the Linnean Society (Botany)* 5: 182–7.

Westwood, John Obadiah. 1838. *The entomologist's text book; an introduction to the natural history, structure, physiology, and classification of insects.* London: Wm. S. Orr.

Westwood, John Obadiah. 1860. Mr Darwin's theory of development. *Annals and Magazine of Natural History* 3d ser. 5: 347–8. [Reprinted from *Gardeners' Chronicle*, 11 February 1860, p. 122.]

Wheeler, Alwyne. 1985. *The world encyclopedia of fishes*. New edition. London and Sydney: Macdonald.

White, Charles. 1782. On the regeneration of animal substances. [Read 18 December 1782.] *Memoirs of the Literary and Philosophical Society of Manchester* 1 (1785): 325–41.

——. 1799. *An account of the regular gradation in man, and in different animals and vegetables; and from the former to the latter*. London: C. Dilly.

White, Henry Patrick. 1982. *A regional history of the railways of Great Britain. Vol. 2: Southern England*. 4th edition. Newton Abbot, Devon: David & Charles.

——. 1987. *A regional history of the railways of Great Britain. Vol. 3: Greater London*. 3d edition, revised. Newton Abbot, Devon: David St John Thomas.

Wiegmann, Arend Friedrich. 1828. *Über die Bastarderzeugung im Pflanzenreiche. Eine von der königl. Akademie der Wissenschaften zu Berlin gekrönte Preisschrift*. Brunswick, Germany: Friedrich Vieweg.

Wight, Robert and Arnott, George Arnott Walker. 1834. *Prodromus floræ peninsulæ Indiæ orientalis: containing abridged descriptions of the plants found in the peninsula of British India, arranged according to the natural system*. London: Parbury, Allen & Co.

[Wilberforce, Samuel.] 1860. [Review of *Origin*.] *Quarterly Review* 108: 225–64.

Willey, Basil. 1960. *The Hibbert Lectures 1959. Darwin and Butler: two versions of evolution*. London: Chatto & Windus.

Willis, Iacobus, ed. 1997. *D. Iunii Iuvenalis Saturae sedecim*. Stuttgart and Leipzig: B. G. Teubner.

Willis, John Christopher. 1973. *A dictionary of the flowering plants and ferns*. 8th edition. Revised by H. K. Airy Shaw. Cambridge: Cambridge University Press.

Wilson, Daniel. 1862. *Prehistoric man: researches into the origin of civilisation in the old and the new world*. 2 vols. Cambridge and London: Macmillan.

Wilson, Leonard G., ed. 1970. *Sir Charles Lyell's scientific journals on the species question*. New Haven, Conn., and London: Yale University Press.

——. 1996a. Brixham Cave and Sir Charles Lyell's . . . *the Antiquity of man*: the roots of Hugh Falconer's attack on Lyell. *Archives of Natural History* 23: 79–97.

——. 1996b. The gorilla and the question of human origins: the brain controversy. *Journal of the History of Medicine and Allied Sciences* 51: 184–207

Winkler, Tiberius Cornelis, trans. 1860. *Het ontstaan der soorten door middel van de natuurkeus, of het bewaard blijven van bevoorregte rassen in den strijd des levens*. [*Origin*], by Charles Darwin. Haarlem, Netherlands: A. C. Kruseman.

Wollaston, Alexander Frederick Richmond. 1921. *Life of Alfred Newton, professor of comparative anatomy, Cambridge University, 1866–1907. With a preface by Sir Archibald Geikie*. London: John Murray.

Wood, Casey Albert. 1931. *An introduction to the literature of vertebrate zoology based chiefly on the titles in the Blacker Library of Zoology . . . and other libraries of McGill University, Montreal*. London: Humphrey Milford. Oxford: Oxford University Press.

Wood, Christopher. 1995. *Dictionary of British art. Volume IV. Victorian painters.* 2 vols. Research by Christopher Newall and Margaret Richardson. Woodbridge, Suffolk: Antique Collectors' Club.

Wood, Neville. 1912. *Health resorts of the British Islands.* London: Hodder & Stoughton for University of London Press.

Woodcock, George. 1969. *Henry Walter Bates, naturalist of the Amazons.* London: Faber & Faber.

Woodward, Horace B. 1881. A memoir of Dr S. P. Woodward ALS, FGS, &c., with a list of his published papers. [Read 26 April 1881.] *Transactions of the Norfolk and Norwich Naturalists' Society* 3 (1879–84): 279–312.

——, ed. 1891. *Memorials of John Gunn, MA, FGS . . . Being some account of the Cromer Forest bed and its fossil mammalia . . . with a memoir of the author.* With the assistance of E. T. Newton. Norwich: W. A. Nudd.

Worster, Donald. 1985. *Nature's economy: a history of ecological ideas.* Cambridge: Cambridge University Press.

Wright, Edmund Percival. 1860. Notes on *Tomopteris onisciformis. Report of the 30th meeting of the British Association for the Advancement of Science, held at Oxford*, Transactions of the sections, p. 124.

Wright, Thomas. 1857–80. *Monograph on the British fossil Echinodermata of the Oolitic formations.* 2 vols. London: Palaeontographical Society.

——. 1878–86. *Monograph on the Lias ammonites of the British islands.* With a volume of plates. London: Palaeontographical Society.

WWW: *Who was who: a companion to Who's who, containing the biographies of those who died during the period [1897–1995].* 9 vols. and cumulated index (1897–1990). London: Adam & Charles Black. 1920–96.

WWWA: *Who was who in America. A companion to Who's Who in America.* 9 vols., historical vol., and index. Chicago: A. N. Marquis. 1943–89.

Wyman, Jeffries. 1854. [Cause of contractility in some vegetable tissues.] [Read 8 November 1854.] *Proceedings of the American Academy of Arts and Sciences* 3 (1852–7): 167–8.

——. 1859. On some unusual modes of gestation. *American Journal of Science and Arts* 2d ser. 27: 5–13.

——. 1862. Experiments on the formation of Infusoria in boiled solutions of organic matter, enclosed in hermetically sealed vessels, and supplied with pure air. *American Journal of Science and Arts* 2d ser. 34: 79–87.

——. 1867. Observations and experiments on living organisms in heated water. *American Journal of Science and Arts* 2d ser. 44: 152–69.

Zoologische Buchillustration: *Die zoologische Buchillustration, ihre Bibliographie und Geschichte.* By Claus Nissen. 2 vols. Stuttgart: Anton Hiersemann. 1969–78.

Zoology: *The zoology of the voyage of HMS Beagle, under the command of Captain FitzRoy RN, during the years 1832 to 1836.* Edited and superintended by Charles Darwin. 5 pts. London: Smith, Elder & Co. 1838–43.

NOTES ON MANUSCRIPT SOURCES

The majority of the manuscript sources cited in the footnotes to the letters are located either in the Darwin Archive, Cambridge University Library, or at Down House Museum, Kent. Further details about the Darwin Archive are available from the *Handlist of Darwin papers at the University Library Cambridge* (Cambridge: Cambridge University Press, 1960) and the unpublished supplementary handlist available at the library; a new catalogue of the papers is currently being prepared. Further details about the manuscripts at Down House are available from Philip Titheradge ed., *The Charles Darwin Memorial at Down House, Downe, Kent*, revised ed. ([Downe: Down House Museum], 1981) and from the curator (The Curator, Down House Museum, Downe, Kent, BR6 7JT). In addition, there are a number of named sources that are commonly used in the footnotes: for each of these, the editors have provided brief descriptive notes.

CD's Account books (Down House MS). This series of seventeen account books begins on 12 February 1839, a fortnight after CD and Emma's marriage, and ends with CD's death. The books contain two sets of accounts. From the start, CD recorded his *cash account* according to a system of double-entry book-keeping. On each left-hand page he recorded credits (i.e., withdrawals from the bank, in the form either of cash paid to himself or of cheques drawn for others), and on each right-hand page he recorded debits (i.e., cash or cheques paid to others). CD also recorded details of his *banking account* from the start, but only noted them down in a single column at the bottom of the left-hand page of his cash account. In August 1848, however, he began a system of detailing his banking account according to double-entry book-keeping, in a separate chronological section at the back of each account book. On the left, he recorded credits to the account in the form of income (i.e., investments, rent, book sales, etc.). On the right, he recorded debits to the account (i.e., cash or cheque withdrawals).

CD's Classed account books (Down House MS). This series of four account books, covering the years 1839–81, runs parallel to CD's Account books. For each year, September – August (after 1867, January – December), CD divided his expenditure into different classified headings; in addition, he made a tally for the year of his income, expenditure, cash in hand, and money in the bank. From 1843 CD also compiled at the back of each book a separate account of the total expenditure under the various headings in each year, and from 1844 he added a full account of his income in each year, and of capital invested and 'paid' up. The classification of expenditure varied over time, but for

1862–3 the classes were: 'Miscellaneous', 'Personal', 'Governess & girls School', 'House', 'William & Boys', 'Coals & Firewood', 'Beer', 'Wine', 'Man Servants', 'Repairs, Furniture', 'Medical attendance', 'Stationery, Newspapers Stamps', 'Books', 'Science', 'Garden', 'Cows, Pigs, Field', 'Stable', 'Travelling Expenses', 'Taxes Rates', 'Charities', and 'Gifts, Annual Subscriptions'.

CD's Experimental notebook (DAR 157a). This notebook contains notes on some, though by no means all, of the experiments carried out between 13 November 1855 (with some back references) and 20 May 1868, although the majority of the notes date from before 1863. In many cases only the details of the experiment attempted are given, often with cross-references to results recorded in CD's portfolios of notes. The notebook also contains a number of letters to CD.

CD's Investment book (Down House MS). This book records under each of CD's investments the income received during the period 1846–81.

CD's 'Journal'. *See* Appendix II.

CD's Library catalogue (DAR 240). This manuscript catalogue of CD's scientific library was compiled by Thomas W. Newton, assistant librarian of the Museum of Practical Geology, London, in August 1875. Additions to the catalogue were subsequently made by Francis Darwin (who inherited most of his father's scientific library) and by H. W. Rutherford, who apparently used this catalogue as a basis for compiling his *Catalogue of the library of Charles Darwin now in the Botany School, Cambridge* (Cambridge: Cambridge University Press, 1908). However, there are items listed in this manuscript catalogue that do not appear in Rutherford's published catalogue, and that must have been dispersed after being listed.

Darwin's copy of the *Gardeners' Chronicle*: account of numbers and index to marks & memoranda (DAR 222). This is a bound manuscript index to CD's collection of the *Gardeners' Chronicle and Agricultural Gazette*, 1841–71, which is at the Cory Library, Cambridge Botanic Garden. It comprises a list showing which numbers of the journal are missing from CD's collection, together with a 'List of the numbers of special interest to Darwin and kept by him in separate parcels'. The latter details the page-numbers and subjects of the articles that were of particular interest to (and annotated by) CD in the numbers that he kept separately. The handwriting of the list is unknown, but probably dates from the late nineteenth century.

Down Coal and Clothing Club account book (Down House MS). CD was for some years treasurer of this charitable organisation. The account book records subscriptions made by honorary subscribers between 1841 and 1876; between 1848 and 1869 the entries are in CD's handwriting. For the years 1841–8 and 1868–76, there is also a statement of expenditures, though not in CD's handwriting.

Emma Darwin's diary (DAR 242). This comprises a series of small pocket diaries, in which Emma recorded details of the health of family members, trips made by herself, CD, and the children, school holidays, and visits to Down by others. The collection at CUL comprises diaries for the years 1824, 1833–4, 1839–45, and 1848–96.

H. E. Litchfield's autobiography (DAR 246). This unfinished autobiography, written in 1926 on forty-two loose leaves, and chiefly concerning Henrietta Emma Darwin's childhood, has never been published.

List of pamphlets (DAR 252.4). This is a catalogue of CD's pamphlet collection prepared by CD and Francis Darwin in 1878 (see the letter from Emma Darwin to Henrietta Emma Litchfield, [June 1878] (DAR 219.9: 175)). From about 1850 CD began to arrange the articles, papers, and reprints he received into a numbered collection. CD maintained this reprint collection until his death, when it was taken over by Francis Darwin. Francis continued the collection, adding new items, the numbers running consecutively from those of his father. Evidently, until this catalogue was prepared, CD used a working index similar to that of his 'List of reviews'. The catalogue has two sections, a list of the quarto collection and a list of the general collection. Both sections are arranged alphabetically with the entries pasted on sheets in a loose-leaf folder.

List of reviews (DAR 262 (DH/MS* 8: 9–18)). This manuscript, headed 'List Reviews of Origin of Sp & of C. Darwins Books', was CD's working index to his collection of the reviews of his own books. It corresponds approximately to the review collection in the Darwin Pamphlet Collection–CUL, but includes some items that were dispersed after being listed.

Reading notebooks. *See Correspondence* vol. 4, Appendix IV. These notebooks comprise sections entitled 'Books Read' and 'Books to be Read'. CD's entries in 'Books Read' often include a brief opinion of the work.

Scrapbook of reviews (DAR 226.1 and 226.2). Many of the reviews contained in these two volumes bear CD's annotations and thus were evidently collected by CD. However, the scrapbook seems to have been assembled by Francis Darwin: the tables of contents are in the handwriting of H. W. Rutherford, an assistant at CUL who acted as a copyist for Francis on several occasions (see *ML*, 1: x, and Francis Darwin ed., *The foundations of the Origin of Species. Two essays written in 1842 and 1844 by Charles Darwin* (Cambridge: Cambridge University Press, 1909)). In addition, the scrapbook is identified as Francis's in a note (DAR 226.1:132a) made in 1935 by Arthur Keith, whose appeal led to the purchase of Down House as a Darwin memorial (see Arthur Keith, *An autobiography* (London: Watts & Co., 1950)). DAR 226.1 bears the inscription 'Reviews of C. Darwin's works' on the spine, and contains, among others, reviews of *Origin* and *Orchids*; DAR 226.2 is inscribed: 'Reviews. Descent. Expression. Insect. Pl. Eras. D.'

W. E. Darwin's botanical notebook (DAR 117). This notebook contains observational and experimental notes on plants made by William, often in consultation with CD. The first observation bears the date 13 July 1862, and, although the date of the last observation is 26 June 1870, most of the notes were made between 1862 and 1864. The notebook originally contained letters from CD, but these were subsequently removed. William entered notes made from botanical textbooks in a separate notebook (DAR 234).

W. E. Darwin's botanical sketchbook (DAR 186: 43). This sketchbook, which contains entries dated 1862–72, was evidently begun in parallel to William's botanical notebook. It contains ink drawings of various parts of plants, and of sections, together with descriptions, some of which are extensive.

INDEX

The dates of letters to and from Darwin's correspondents are listed in the Biographical register and index to correspondents and are not repeated here. Darwin's works are indexed under the short titles used throughout this volume and listed in the bibliography.

Table of

Robert Darwin 1682–1754 = Elizabeth Hill 1702–97

- William Alvey Darwin 1726–83 = Jane Brown 1746–1835
  - Samuel Fox 1765–1851 = Ann 1771–1859
    - Mary Ann 1800–29 = Samuel Ellis Bristowe 1800–55
    - Eliza 1801–86
    - Emma 1803–85
    - Ellen Sophia Woodd 1820–87 ≟2 William Darwin 1805–80 ≟1 Harriet Fletcher 1799–1842
    - Frances Jane b. 1806 = John Hughes 1794–1873
    - Julia b. 1809
- Elizabeth Collier Pole 1747–1832 ≟2 Erasmus Darwin 1731–1802 ≟1 Mary Howa[rd] 1740–70
  - (by 1) Charles 1758–78
  - (by 1) Erasmus 1759–99
  - (by 1) Robert Waring 1766–184[8]
  - (by 2) Edward 1782–1829
  - (by 2) Samuel Tertius Galton 1783–1844 = Frances Anne Violetta 1783–1874
    - Elizabeth Ann (Bessy) 1808–1906
    - Lucy Harriot 1809–48
    - Milicent Adèle 1810–83
    - Emma Sophia 1811–1904
    - Darwin 1814–1903
    - Erasmus 1815–1909
    - Francis 1822–1911 = Louisa Jane Butler d. 1897
  - (by 2) Emma 1784–1818
  - (by 2) Francis Sacheverel 1786–1859 = Jane Harriett Ryle 1794–1866
  - (by 2) John 1787–1818
  - (by 2) Harriot 1790–1825 = Thomas James Maling 1778–1849

Henry Parker 1788–1856 = Mariann[e] 1798–185[8]

Susan Elizabeth 1803–66

Erasmus Alvey 1804–81

Emily Catherin[e] 1810–66